The Statesman's Year-Book
World Gazetteer

Fourth Edition

The Statesman's Year-Book World Gazetteer

Fourth Edition

John Paxton

First edition 1975, second edition 1979,
third edition 1986

Fourth edition first published 1991 by
THE MACMILLAN PRESS LTD
London and Basingstoke

Associated companies in Auckland, Delhi,
Dublin, Gaborone, Hamburg, Harare, Hong Kong,
Johannesburg, Kuala Lumpur, Lagos, Manzini,
Melbourne, Mexico City, Nairobi, New York,
Singapore, Tokyo.

British Library Cataloguing in Publication Data

The 'statesman's year-book' world gazetteer.–
    4th ed.
    1. Gazetteers
    I. Paxton, John
    910'.3'21   G103.5

    ISBN 0–333–52527–2
    ISSN 0268–5205

Printed in Hong Kong

JOAN
who put up with a cottage full of paper
(yet again)

Contents

Contents

Introduction

The Statesman's Year-Book was first published in 1864. Letters in the editor's files show that for many years there have been constant requests for a *World Gazetteer* capable of standing as a publication in its own right and also suitable to act as a supplement to *The Statesman's Year-Book*. Although the annual has over fifty pages of index containing over eight thousand entries, readers obviously do need, from time to time, to know more about individual towns or regions in the world; it is not possible to provide this in *The Statesman's Year-Book* both because that is not its prime aim and also because of size and cost.

The editor has not set out to mention *every* place in the world. He has, however, attempted to give details of places of importance or size. Each entry attempts to give information on location, on recent history, where considered particularly relevant, and on industries and population. The essential details about the countries of the world are given, but *no attempt* has been made to duplicate the detailed descriptions which appear in the annual volumes of *The Statesman's Year-Book*.

The Statesman's Year-Book World Gazetteer will continue to be revised from time to time and the editor will be pleased to receive constructive and informed criticism for future editions.

John Paxton

Acknowledgements

Many people and organizations throughout the world have helped in the compilation of *The Statesman's Year-Book World Gazetteer* and I am extremely grateful for the time and trouble taken to answer my many requests. The typing of reference works is an exhausting business and I am particularly grateful to Penny White for her superb work and eagle eye for error. Evelyn Beadle helped in the initial planning of the first edition and undertook a vast amount of sub-editing and some proof-reading. A. J. Bruce was responsible for planning the geographical terms and for writing a very large proportion of them and I have used his geographical knowledge on many occasions. Others who have given substantial help to at least one of the editions, and for which I am very grateful, are: John C. Bartholomew, Sheila Fairfield, Len Jones, S. Mukherjee, Jessica Orebi-Gann, Marian Paxton, Teresa Plenty, Deborah Smith, Rif Winfield. Errors are my own.

J.P.
Bruton, Somerset
August 1990

Abbreviations used within the World Gazetteer

Afr.	Africa
agric.	agriculture/agriculturally
anc.	ancient
approx.	approximately
arch.	archipelago
a.s.l.	above sea level
Aust.	Australia
B.	bight
bbls.	barrels
boro.	borough
Braz.	Brazil
Brit.	Britain/British
b.s.l.	below sea level
c.	circa
C	after a date (1980C)=census
C.	centigrade
Can.	Canada
cap.	capital
cent.	century
Co.	company
cu.	cubic
Czech.	Czechoslovakia
Den.	Denmark
depend.	dependency
dept.	department
E	after a date (1980E)=pop. estimate
E	east/eastern
Eng.	England/English
esp.	especially
est.	estimate
estab.	established
F.	Fahrenheit
FRG	Federal Republic of Germany
Fr.	French
ft.	feet/foot
govt.	government
GDR	German Democratic Republic
in.	inch/inches
inc.	include/including
indust.	industry/industries
intern.	international

Abbreviations

km.	kilometre
L.	lake
Ls.	lakes
lat.	latitude
Lond.	London
long.	longitude
m.	mile(s)
manu.	manufacture/manufacturing
Mex.	Mexico
N	north/northern
Neth.	Netherlands
p.	pass
p.a.	per annum
pop.	population
Pres.	President
R.	river
Rep.	Republic/Republics
S	south/southern
Scot.	Scotland
sit.	situated
S. Afr.	South Africa
sq.	square
Switz.	Switzerland
temp.	temperature
trad.	traditional/traditionally
trib.	tributary/tributaries
U.N.	United Nations
U.S.A.	United States of America
U.S.S.R.	Union of Soviet Socialist Republics
W	west/western

Weights and Measures

On 1 Jan. 1960 following an agreement between the standards laboratories of Britain, Canada, Australia, New Zealand, South Africa and the U.S.A. an international yard and an international pound (avoirdupois) came into existence.
1 yard = 91·44 centimetres; 1 lb. = 453·59237 grammes.

Length
Centimetre	0·394 inch
Metre	1·094 yards
Kilometre	0·621 mile

Foot	0·305 metre
Yard	0·914 metre
Mile (=1,760 yds)	1·609 kilometres

Surface Measure
Square metre	10·76 sq. feet
Hectare	2·47 acres
Square kilometre	0·386 sq. mile

Square foot	9·290 sq. decimetres
Square yard	0·836 sq. metre
Acre	0·405 hectare
Square mile	2·590 sq. kilometres

Liquid Measure
Litre	1·76 pints
Hectolitre	22·00 gallons

Pint	0·568 litre
Gallon	4·546 litres
Quarter	2·909 hectolitres

Dry Measure
Litre	0·91 quart
Hectolitre	2·75 bushels

Weight
Avoirdupois
Gramme	15·42 grains
Kilogramme	2·205 pounds
Quintal (= 100 kg)	220·46 pounds
Tonne (= 1,000 kg)	0·984 long ton / 1·102 short tons

Ounce (= 437·2 grains)	28·350 grammes
Pound (= 7,000 grains)	453·6 grammes
Hundredweight (= 112 lb.)	50·802 kilogrammes
Long ton (= 2,240 lb.)	1·016 tonnes
Short ton (= 2,000 lb.)	0·907 tonne

Troy
Gramme	15·43 grains
Kilogramme	32·15 ounces / 2·68 pounds

A

Aabenraa Sønderjyllands, Denmark. 55 02N 9 26E. Port in SE Jutland also holiday resort. Manu. industries associated with fishing, brewing and agric.

Aachen North Rhine-Westphalia, Federal Republic of Germany. 50 47N 6 04E. Charlemagne's cap., Holy Roman Emperors crowned in cathedral begun A.D. 796. (*Fr., Aix-la-Chapelle*.) Pop. (1984E) 241,000.

Aalborg Nordjyllands, Denmark. 57 02N 9 54E. Port on Lim fjord. Manu. town, shipbuilding important. Pop. (1983E) 154,755.

Aalst *see* **Alost**

Aarau Switzerland. 47 24N 8 04E. Cap. of canton of Aargau. Manu. electrical equipment, footwear, chemicals and precision instruments. Pop. (1983E) 17,000.

Aare River Switzerland. 47 37N 8 13E. Largest Swiss trib. of Rhine. Approx. 180 m. (288 km.) long. Navigable from point of confluence with R. Rhine to Thun.

Aargau Switzerland. 47 30N 8 10E. German-speaking canton estab. 1803. Cap. Aarau. Pop. (1983E) 459,100.

Aarhus Denmark. 56 09N 10 13E. Second largest city in Denmark. Railway junction and manu. centre. Chief manu. iron, electronics and science, machinery, tobacco products and confectionery. Seat of Lutheran bishop, with 13th-cent. cathedral. Pop. (1989E) 259,155.

Aba Imo, Nigeria. 5 06N 7 21E. Town sit. N of Port Harcourt trading in agric. products including palm oil and palm products. Pop. (1981E) 242,000.

Abaco Bahamas. Island in Bahamian group. Pop. (1980C) 7,324.

Abadán Iran. 30 20N 48 16E. Island and port on delta of Shatt-al-Arab. Owes its importance to oil refining and exporting. Pop. (1985E) 294,068.

Abakan U.S.S.R. 53 43N 91 26E. Cap. of Khakass Autonomous Region, Krasnoyarsk Territory, R.S.F.S.R. Sit. SSW of Krasnoyarsk. Industrial centre. Sawmilling, metal working. Pop. (1983E) 143,000.

Abbeville Somme, France. 29 58N 92 08W. Town on R. Somme. Manu. sugar and carpets. Pop. (1982C) 25,998.

Abbotsford Borders Region, Scotland. 55 36N 2 47W. Village on R. Tweed, sit. W of Melrose. Home of the writer, Sir Walter Scott 1812–32.

Abeokuta Ogun, Nigeria. 7 10N 3 36E. Town on Ogun R. N of Lagos, on the railway to Kano. Agric. products include yams, palm oil and kernels, cacao, timber. Main industries cotton weaving and dyeing. Pop. (1981E) 345,000.

Aberavon West Glamorgan, Wales. 51 35N 3 47W. Town sit. ESE of Swansea and now part of Port Talbot.

Abercarn Gwent, Wales. 51 38N 3 08W. Town sit. NW of Newport. Agriculture and light indust. Pop. (1985E) 4,000.

Aberdare Mid-Glamorgan, Wales. 51 43N 3 27W. Town on Cynon R. Light industry and printing. Manu. electrical cables. Pop. (1982E) 38,000.

Aberdeen Grampian Region, Scotland. 57 10N 2 04W. A royal burgh sit. near mouths of R. Don and

R. Dee. Chief seaport of the NE; Scotland's largest resort. Manu. knitwear, engineering, food processing, paper. Shipbuilding. Off-shore oil centre. King's College, part of Aberdeen University, founded 1494. Because many of its buildings are constructed of granite known as the 'Silver City'. Pop. (1988E) 205,180.

Aberdeenshire Scotland. Now part of Grampian Region.

Aberfeldy Tayside Region. Scotland. 56 37N 3 52W. Burgh on R. Tay sit. NW of Perth. Whisky distilling. Pop. (1984E) 1,600.

Abergavenny Gwent, Wales. 51 50N 3 00W. Market town on R. Usk and R. Gavenny, sit. N of Newport. Pop. (1983E) 9,500.

Aberystwyth Dyfed, Wales. 52 25N 4 05W. University town and seaside resort at mouth of R. Ystwyth and R. Rheidol on Cardigan Bay. Pop. (1989E) 15,000.

Abidjan Côte d'Ivoire. 5 19N 4 01W. Cap. of the Republic, sit. on N of Ebrié Lagoon. It became cap. of the French Ivory Coast in 1935. Rail head for exports of timber, cotton, cacao, palm oil and kernels. Industries inc. car assembly, agric. machinery, electrical appliances, food processing. Pop. (1982E) 1,850,000.

Abilene Texas, U.S.A. 38 55N 97 13W. City sit. WSW of Fort Worth. Commercial centre for agric. products and oil. Manu. clothing, food products, electronics, cottonseed oil, building materials, aircraft parts and musical instruments. Pop. (1980C) 98,315.

Abingdon Oxfordshire, England. 51 41N 1 17W. Town on R. Thames sit. S of Oxford. Manu. leather goods. Brewing important. Pop. (1989E) 28,622.

Abitibi, Lake Canada. 51 03N 80 55W. Forms boundary between Ontario and Quebec covering 350 sq. m. (906 sq. km.); drained by R. Abitibi, a trib. of R. Moose.

Abkhazian Autonomous Soviet

Socialist Republic within Georgia, U.S.S.R. 43 00N 41 30E. Area 3,320 sq. m. (8,600 sq. km.). Cap. Sukhumi, sit. NNW of Batumi. The region was annexed from Turkey in 1810 and became an Autonomous Soviet Socialist Rep. in 1921. The Abkhazian coast, along the Black Sea, possesses a famous chain of health resorts, Gagra, Sukhumi, Akhali-Antoni, Gulripsha and Gudauta, sheltered by thickly forested mountains. The Rep. has coal, electric power, building materials and light industries. In 1985 there were 89 collective farms and 56 state farms; main crops are tobacco, tea, grapes, oranges, tangerines and lemons. Pop. (1989E) 537,000.

Abomey Zou, Benin. 7 14N 2 00E. Cap. of dept. and anc. cap. of Kingdom of Dahomey. Centre for groundnut-growing area. Pop. (1982E) 54,418.

Abquaiq Hasa, Saudi Arabia. 25 26N 49 40E. Town sit. SW of Dhahran on the Riyadh railway. Oil town and important oilfield and pipeline terminus. The field was discovered in 1940 and came into production after the 2nd World War. The Buqqa extension field to the N was discovered in 1947.

Abruzzi e Molise Italy. 42 20N 13 45E. Administrative Region in S central Italy. Area 5,880 sq. m. (15,229 sq. km.). Comprises provinces of Aquila, Campobasso, Chieti, Pescara and Teramo. Cap. of Region, Aquila. Main occupations stock rearing, cultivation of cereals, grapes and sugarbeet. Pop. (1981C) 1,546,162.

Abu Dhabi United Arab Emirates. 24 00N 54 00E. (i) Westernmost and largest of the United Arab Emirates on the SW coast of the Gulf. Area 26,000 sq. m. (67,350 sq. km.). Pop. (1982E) 516,000. (ii) Cap. and Federal Cap. Abu Dhabi is sit. SW of Sharja. Petroleum is main industry, with some fishing and pearling. Pop. (1980C) 242,975.

Abuja Nigeria. 9 05N 7 30E. It was confirmed in 1982 that Abuja would be the new capital, replacing Lagos. Building will be completed in 1990-91.

Abu, Mount   Rajasthan. India. Mountain (5,650 ft./1,722 metres a.s.l.) 115 m. (184 km.) N of Ahmedabad also popular hot-weather resort sit. on the mountain. Pop. (1985E) 8,000.

Abu Simbel   Egypt. Site of anc. temples on left bank of R. Nile, 145 m. (232 km.) SW of Aswân. Two temples (13th cent.) were cut out of solid rock with 4 figures of Rameses, 65 ft. (20 metres) high. Because of the possible rise of the water level of the reservoirs behind the Aswân High Dam, the Egyptian Govt. with help from the U.N. had them removed to a site 200 ft. (61 metres) above the original and 12 ft. (3.7 metres) above the highest water level.

Abymes, Les   Guadaloupe, French West Indies. Largest town of dept., on Grand-Terre island. Pop. (1982C) 56,165.

Abyssinia   *see* **Ethiopia**

Acadia   Name first used for all French possessions S of St. Lawrence R.; later applied to New Brunswick, Nova Scotia and part of Maine.

Acapulco de Juàrez   Guerrero, Mexico. 16 51N 99 55W. Pacific seaside resort sit. SSW of Mexico city. Commercial centre. Pop. (1980C) 409,335.

Accra   Ghana. 5 33N 0 15W. City and cap. sit. WSW of Lagos, Nigeria, on the Gulf of Guinea. It is a seaport, railway terminus and commercial centre. It has grown round three trading posts built by the British, the French and the Danes in the 17th cent. but grew most rapidly after the opening of the railway to Kumasi through the cacao-producing hinterland. Industries inc. food processing, fishing, engineering, brewing, and it exports cacao, timber, gold and fruit. Pop. (1984C) 964,879.

Accrington   Lancashire, England. 53 46N 2 21W. Sit. N of Manchester. Indust. inc. textile manu., textile and general engineering, brickmaking. Pop. (1981C) 33,681.

Achill Island (Eagle Island)   Ireland. 54 00N 10 00W. Sit. off coast of County Mayo. Area 57 sq. m. (148 sq. km.). Main occupation fishing and growing potatoes and oats. Tourism has been developed. Pop. (1980E) 3,200, with Dooega and Slievemore.

Acklin's Island   Bahamas. 22 25N 74 00W. Island SE of Crooked Island and of Nassau. Area 120 sq. m. (311 sq. km.). The main settlement is Snug Corner. Exports are cascarilla and timber. Pop. (1980C) 618.

Aconcagua, Mount   Mendoza, Argentina. 32 39S 70 00W. Sit. WNW of Mendoza city. Highest peak in the Andes (22,835 ft. a.s.l.) near the border with Chile.

Aconcagua River   Chile. 32 55S 71 32W. Rises on Mount Aconcagua and flows for 120 m. (192 km.) and is used for irrigation purposes. Enters Pacific 12 m. (19 km.) N of Valparaiso.

Acre   Brazil. 10 00S 70 00W. Federal territory of W Brazil bounded W and S by Peru, SE by Bolivia and N by Amazonas. Area 58,915 sq. m. (152,589 sq. km.) Cap. Rio Branco. Sit. in the Amazon basin rain forest and drained by the rivers Juruá and Purus. The main products are wild rubber, tin and gold, mainly exploited through air and river transport. The territory was gained from Bolivia in 1903 in return for an indemnity and a promise to build the Madeira–Mamoré railway. Pop. (1989E) 406,787.

Acre   Israel. 32 55N 35 04E. Town and former important port on the Bay of Acre near Mount Carmel. Pop. (1982E) 39,100.

A.C.T.   *see* **Australian Capital Territory**

Acton   Greater London, England. Sit. W of central London. Engineering and other manu. industries.

Adam's Bridge   Northern Province, Sri Lanka/India. 9 04N 79 37E. Reef of sand and gravel extending 30 m. (48 km.) off NW Sri Lanka.

Adam's Peak   Sabaragamuwa, Sri

Lanka. 6 49N 80 30E. Granite mountain (7,352 ft./2,240 metres a.s.l.). Foot-like depression on summit ascribed by Moslems to Adam, by Buddhists to Buddha and by Hindus to Siva.

Adana   Turkey. 37 00N 35 19E. Cap. of Adana province on Seyhan R. Industrial and commercial centre. Pop. (1980C) 574,515.

Adapazari   Kocaeli, Turkey. 40 46N 30 23E. Town sit. WNW of Ankara. Manu. tobacco, textiles and leather goods. Pop. (1980C) 130,177.

Addis Ababa   Ethiopia. 9 00N 38 42E. Cap. of Ethiopia and terminus of railway from Djibouti. The city was founded by Menelik II in 1887 and was occupied by the Italians 1936–41. It is the headquarters of the Organization of African Unity. It contains over 50% of the industry of the country which includes the manu. of footwear, textiles and food processing. Administrative region of 84 sq. m. (218 sq. km.). Pop. (1984E) 1,412,574.

Adelaide   Australia. 34 56S 138 36E. Cap. of South Australia, city divided by R. Torrens. Industries inc. manu. of agric. implements, cars, electrical equipment, paint, textiles. Ships dock 16 m. (26 km.) from the city centre, and important exports are wool, fruit, wine and wheat. Pop. (1988E) 1,023,517.

Adelboden   Bern, Switzerland. Holiday resort in Bernese Oberland, sit. SSE of Bern. Altitude 4,450 ft. (1,356 metres) a.s.l. Pop. (1989E) 3,400.

Adélie Land   Antarctica. 67 00S 139 00E. Sit. between Wilkes Land and King George V Land (between long. 136° E and 142° E and S of lat. 60° S). Area, 166,000 sq. m. (432,010 sq. km.). Discovered by Dumont d'Urville 1840. Placed under French sovereignty 1938 and now part of French Southern and Antarctic Territories. A research station is sit. at Base Dumont d'Urville. Pop. (1983C) 26.

Aden   Yemen. 12 45N 45 12E. Town sit. E of the Babel Mandeb

Strait on the Gulf of Aden. Seaport and fishing port and important commercial centre. Former cap. of the People's Democratic Republic of Yemen. Pop. (1981E) 264,326.

Adige River   Italy. 45 10N 12 20E. Second most important R. in Italy (220 m./352 km. long). Rises in the Rhaetian Alps and enters Adriatic just N of the mouth of R. Po. Used for hydroelectricity and irrigation.

Adirondack Mountains   U.S.A. 44 00N 74 00E. Sit. NE of New York. Many isolated peaks, highest being Mount Marcy 5,344 ft. (1,629 metres) a.s.l. Lakes inc. L. Champlain and L. George. Heavily forested. Holiday resorts.

Adiyaman   Malatya, Turkey. Town sit. S of Malatya formerly called Hisnumansur or Husnumansur. Centre of a farming area producing wheat, barley, chick-peas, lentils, tobacco and cotton. Pop. (1980E) 367,595.

Adjarian Autonomous Soviet Socialist Republic   within Georgia, U.S.S.R. 42 0N 42 0E. Area, 1,160 sq. m. (3,004 sq. km.). Annexed to Russia, having been under Turkish rule, after Berlin Treaty 1878, and constituted as an Autonomous Rep. within Georgia 1921. Cap. Batumi. Subtropical products inc. tea, mandarines, lemons, grapes, bamboo, eucalyptus; also livestock breeding. Important shipyards, oil-refining, food-processing and canning, clothing, building materials, drug factories. Pop. (1989E) 393,000.

Admiralty Range   Antarctica. Part of the great mountain range W of Ross Sea. First seen and named by Sir James Ross in 1841.

Adour River   France. 43 32N 1 32E. Rises in Pic du Midi de Bigorre in the Pyrenees and flows for 207 m. (312 km.) into Bay of Biscay 4 m. (6½ km.) below Bayonne.

Adria   Veneto, Italy. 45 03N 12 03E. Town sit. E of Rovigo between R. Adige and R. Po. Formerly on coast but because of silting now 15 m. (24 km.) inland. Was a flourishing port in 12th cent. and gave its name to Adriatic Sea. Manu. bricks and cement.

Adriatic Sea An arm of Mediterranean Sea, between Italy and the coasts of Yugoslavia and Albania. 43 0N 16 0E. Extreme length, 450 m. (720 km.) and average breadth 90 m. (144 km.). Chief industries fishing and tourism. Chief ports are Ancona, Bari, Brindisi, Dubrovnik, Durrës, Rijeka, Split, Trieste and Venice.

Adygei Autonomous Region within Krasnodar Territory, U.S.S.R. Area, 2,934 sq. m. (7,599 sq. km.). Estab. 1922. Cap. Maikop, sit. approx. 220 m. (352 km.) s of Rostov. Chief industries are timber, woodworking, food processing, engineering; also cattle breeding. Pop. (1989E) 432,000.

Adzharian Autonomous S.S.R. *see* **Adjarian Autonomous S.S.R.**

Aegean Sea Part of Mediterranean Sea between Greece and Asia Minor, connected with Sea of Marmara and Black Sea by Dardanelles. 37 0N 25 0E. Studded with islands, inc. the Cyclades, the Dodecanese, Euboea and the Sporades. Chief ports are Izmir, Piraeus and Thessaloníki.

Aetolia Acarnania, Greece. *Nome* of w central Greece, on the N shore of the Gulf of Patras, bounded w by Achelous R. Mainly agric. Chief town, Missolonghi. Pop. (1981C) 219,764, with Acarnania.

Afghánistán 33 0N 65 0E. Afghánistán is bordered by U.S.S.R. in N, Pakistan in E and S, Iran in W and China in NE. Length approx. 600 m. (960 km.) and greatest breadth 700 m. (1,120 km.). Area, 251,773 sq. m. (652,090 sq. km.). Cap Kábul.

Although the greater part of Afghánistán is more or less mountainous and a good deal of the country is too dry and rocky for successful cultivation, there are many fertile plains and valleys, which, with the assistance of irrigation from small rivers or wells, yield very satisfactory crops of fruit, vegetables and cereals. It is est. that there are 14m. hectares of cultivable land in the country, of which 6% is being cultivated (5.34m. hectares of this being irrigated land). Afghánistán is virtually self-supporting in foodstuffs, apart from wheat and sugar. The castor-oil plant, madder and the asafœtida plant abound. Fruit forms a staple food (with bread) of many people throughout the year, both in the fresh and preserved state, and in the latter condition is exported in great quantities. The fat-tailed sheep furnish the principal meat diet, and the grease of the tail is substitute for butter. Wool and skins provide material for warm apparel and one of the more important articles of export. Persian lambskins (Karakuls) are one of the chief exports.

Mineral resources are scattered and little developed. Coal is mined at Karkar in Pul-i-Khumri, Ishpushta near Doshi, N of Kábul, and Dara-i-Suf s of Mazar. Natural gas is found in N Afghánistán around Shiberghan and Sar-i-Pol; this is now being piped to the U.S.S.R. Rich, but as yet unexploited, deposits of iron ore exist in the Hajigak hills about 100 m. (160 km.) w of Kábul; beryllium has been found in the Kunar valley and barite in Bamian province. Other deposits inc. gold; silver (now unexploited, in the Panjshir valley); lapis lazuli (in Badakhshán); asbestos; mica; sulphur (near Maimana); chrome (in the Logar valley and near Herát); and copper (in the N).

At Kábul there are factories for the manu. of cotton and woollen textiles, leather, boots, marble-ware, furniture, glass, bicycles, prefabricated houses and plastics. A large machine shop has been constructed and equipped by the Russians, with a capability of manu. motor spares. There is a wool factory and a cotton ginning plant at Kandahár; a small cotton factory at Jabal-us-Seráj and a larger one at Pul-i-Khumri. A cotton-seed oil extraction plant has been built in Lashkargah and there is a large modern cotton textile factory at Gulbahar, and another at Bagrami. A large cotton plant has been completed in the N at Balkh. Pop. (1988E) 10m.–12m. Over 3m. are refugees in Pakistan and over 1m. in Iran. Several hundred thousand have been killed in the wars.

Africa Second largest continent.

Extends S from Mediterranean Sea, and is almost bisected by equator; greatest length, *c.* 5,000 m. (8,000 km.); greatest breadth, *c.* 4,500 m. (7,200 km.) area, *c.* 11.5m. sq. m. (*c.* 29.8m. sq. km.). Africa approaches Europe at Strait of Gibraltar in NW, and touches Asia at Isthmus of Suez in NE. Off SE coast is the large island of Madagascar, and the Canary Islands lie off NW coast; coast-line singularly unbroken, length about 16,000 m. (25,600 km.); principal inlets, Gulfs of Gabès and Sirte on N, Gulf of Guinea with Bights of Benin and Bonny on W. Land rises rapidly from coast; from fairly continuous outer rim of mountains greater part of surface spreads inwards in two table-lands, of which the S has mean elevation of over 3,500 ft. (1,067 metres) a.s.l. and slopes down to N, which has mean height of *c.* 1,300 ft. (396 metres) a.s.l. The higher plateau has a number of ridges, and in equatorial regions are many craters of extinct volcanoes, such as Kilimanjaro (20,000 ft./6,096 metres) and Mount Kenya (18,500 ft./5,639 metres). Chief mountain ranges near coast are Atlas on N and Drakensberg on SE. Pop. (1985E) 555m.

Afyonkarahisar   Turkey. 38 20N 30 15E. (i) Province of W central Turkey. Area, 5,234 sq. m. (13,556 sq. km.). Rises to Mount Emir in the N. Drained by the Akar R. with L. Aksehir in the E and L. Aci in the S. It is the main opium-producing area in Turkey; also produces sugar-beet, wheat, barley, mohair, wool and fruit. Pop. (1980C) 597,516. (ii) Town sit. SW of Ankara. Provincial cap., market town and rail junction. Pop. (1980E) 60,000.

Agadez   Niger. 16 58N 7 59E. Cap. of dept. in N on the Aïr Massif, sit. NE of Niamey. Trading centre. Pop. (1983E) 27,000.

Agadir   Morocco. 30 30N 9 40W. Port on Atlantic coast sit. S of Essaouira. Industries inc. fishing and fish processing, tourism. An earthquake in 1960, killing 12,000, destroyed the town but it has since been rebuilt. Pop. (1982C) 110,479.

Agalega   Mauritius, Indian Ocean. Island 600 m. (960 km.) N of Mauritius, of which it is a dependency. Area, 27 sq. m. (70 sq. km.). Comprises a N and S island linked by a sandbank, the main settlement being S island. Occupations are horse rearing and coconut growing. There is a meteorological station. Pop. (1983E) 350.

Agaña   Guam. 13 28N 144 45E. Commercial town and seat of govt. of this Pacific island and U.S. Overseas Territory. Sit. 8 m. (13 km.) from the anchorage in Apra Harbour. Pop. (1980E) 3,000.

Agartala   Tripura, India. 23 50N 91 16E. State cap. sit. NE of Dacca, Bangladesh on a trib. of the Meghra R. Trading centre for jute, rice, cotton, tea and mustard. Pop. (1981C) 132,186.

Agen   Lot-et-Garonne, France. 44 12N 0 37E. Cap. of Dept. sit. in fruit-growing and market gardening area. Pop. (1982C) 32,893.

Agincourt   *now* **Azincourt**, Pas-de-Calais, France. Village sit. SE of Boulogne-sur-Mer. Famed for victory of King Henry V of Eng. over French army in 1415.

Agra   Uttar Pradesh, India. 27 12N 77 59E. Sit. on R. Jumna. Famous for Taj Mahal. Industrial and commercial centre. Manu. cotton goods, carpets, footwear. At one time cap. of India but declined in importance after cap. moved to Delhi, 1658. Also cap. North West Provinces 1835–62. Pop. (1981C) 694,191.

Agri   Turkey. Province of E Turkey bounded N by the Aras Mountains, E by Iran. Area, 4,888 sq. m. (12,660 sq. km.). Cap. Karakose. Mountainous and unproductive, drained by the Murat R. Pop. (1980C) 368,009.

Agrigento   Sicily, Italy. 37 18N 13 35E. Cap. of province of same name sit. SSE of Palermo. Sulphur mining. Pop. (1981E) 51,525.

Aguascalientes   Mexico. 21 53N 102 18W. (i) State in central Mex. Area, 2,969 sq. m. (5,589 sq. km.).

Mainly agric. Cereals and stock raising. Pop. (1989E) 702,615. (ii) Cap. of state of same name. Commercial and industrial centre. Pop. (1980C) 359,454.

Agulhas, Cape   Cape Province, Republic of South Africa. 34 50S 20 00E. Also known as 'The Needles'. Most s point of Africa sit *c*. 100 m. (162 km.) ESE of the Cape of Good Hope. Dangerous to shipping because of fogs, rocks and uncertain currents.

Ahaggar Mountains   Algeria. 23 00N 6 30E. Range of mountains of central Sahara in s Algeria, rising to heights of 9–10,000 ft. (2,743–3,048 metres) a.s.l. Highest peak, Tahat 9,840 ft. (2,999 metres).

Ahlen   North Rhine-Westphalia, Federal Rep. of Germany. 51 46N 7 53E. Town sit. SSE of Münster. Coal-mining important occupation. Manu. metal goods and footwear. Pop. (1984E) 52,600.

Ahmadi   Kuwait. 29 05N 48 10E. Town sit. SSE of Kuwait and residential centre for Burgan oilfield. Linked by pipelines with its port and terminal Mena al Ahmadi, 5 m. (8 km.) ESE. There is a tank farm and water-evaporation plant. Pop. (1980E) 24,000.

Ahmadnagar   Maharashtra, India. 19 05N 74 45E. Town sit. E of Bombay. Commercial and manu. centre. Pop. (1981C) 143,937.

Ahmedabad   Gujarat, India. 23 0N 72 40 E. City, centre of the textile industry, sit. on R. Sabarmati. Important commercial, manu. and cultural centre. Many fine mosques and modern Jain temple. Pop. (1981C) 2,159,127.

Ahuachapán   El Salvador. 13 57N 89 49W. (i) Dept. of El Salvador. Area, 472 sq. m. (1,222 sq. km.). Coffee growing important. Pop. (1981E) 183,682. (ii) Cap. of dept. of same name sit. WNW of San Salvador. Pop. (1984E) 53,260.

Ahvenanmaa Islands   Finland. 60 15N 20 0E. Arch. of some 80 inhabited islands and a considerable number of rocks and islets at entrance to the Gulf of Bothnia, coextensive with province of same name. Area, 590 sq. m. (1,527 sq. km.). Ahvenanmaa is largest island with Maarianhamina as cap. Fishing is main occupation but there is some cattle rearing, and rye and flax are grown. Pop. (1983E) 23,435.

Ahwáz   Iran. 31 19N 48 42E. Commercial town, railway junction and port on R. Karun sit. NNE of Abadán. Main occupations connected with the oil industry. Pop. (1983E) 471,000.

Aigues-Mortes   Gard, France. 43 41N 4 03E. Town formerly a port, now 3 m. (5 km.) from the sea, sit. ESE of Montpellier on the Rhône delta. Mainly vineyards. Pop. (1982C) 4,475.

Ailsa Craig   Ayrshire, Scotland. 55 16N 5 07E. Rocky island at mouth of Firth of Clyde 10 m. (16 km.) W of Girvan. Area, 1 sq. m. (2.56 sq. km.) and rises to 1,100 ft. (335 metres) a.s.l.

Ain   France. 46 5N 5 20E. Dept. of Rhône–Alps region, bisected by R. Ain. E mountainous and W flat, bordered on NE by Switz. Area, 2,222 sq. m. (5,756 sq. km.). Important towns Bourg-en-Bresse (the cap.) and Belley. Agric. area producing cheese, cereals, vines. Stock raising and forestry also important. Pop. (1982C) 418,516.

Ain Dar   Hasa, Saudi Arabia. 26 55N 49 10E. Oilfield 28 m. (45 km.) W of Abqaiq and linked to it by pipeline. Established in 1948.

Ain River   France. 45 48N 5 10E. Rises in the Jura mountains and flows 120 m. (194 km.) SSW through depts. of Jura and Ain to join Rhône R. 18 m. (29 km.) above Lyon. It provides hydroelectric power.

Aïr Highlands   Niger. 18 0N 8 0E. Mountainous area rising to 5,000 ft. (1,524 metres) a.s.l. in s Saharan region. Chief centre Agadès.

Airdrie   Strathclyde Region, Scotland. 55 52N 3 59W. Town sit. E of Glasgow. Indust. inc. engineering, electronics, pharmaceuticals and distilling. Pop. (1981C) 45,747.

Aire River England. 53 44N 0 54W. Rises in the Pennines and flows 70 m. (123 km.) S and SE to the R. Ouse above Goole. Its course above Leeds is known as Airedale.

Aisne France. 49 42N 3 40E. Dept. in Picardy region. Area, 2,849 sq. m. (7,378 sq. km.). Chief towns Laon (the cap.), St. Quentin and Soissons. Considerable industry inc. chemicals, glass, paper, pottery, textiles and leather. Agric. products inc. sugarbeet, cereals and stock raising. Pop. (1982C) 533,970.

Aisne, River France. 49 26N 2 50E. Rises in the Argonne and flows for 150 m. (240 km.) into the R. Oise above Compiègne, and is connected by canals to other river systems.

Aix-en-Provence Bouches-du-Rhône, France. 43 32N 5 27E. Town, important spa, and industrial and commercial centre. Considerable trade in olive oil, almonds and fruit. Pop. (1982C) 124,550.

Aix-les-Bains Savoie, France. 45 51N 5 53E. Spa sit. N of Chambéry near L. Bourget. Pop. (1982C) 22,534.

Ajaccio Corse-du-Sud, France. 41 55N 8 44E. Town on Gulf of Ajaccio. Holiday resort and commercial town, and cap. of dept. Fishing and the timber trade main occupations. Birthplace of Napoleon. Pop. (1982C) 55,279.

Ajman United Arab Emirates. 25 25N 55 30E. Sheikdom forming an enclave in Sharjah on SW coast of (Persian) Gulf. Area, 25 sq. m. (65 sq. km.) Cap. Ajman, sit. NE of Sharjah. The main occupations are pearling, fishing and date-growing. Pop. (1980E) 36,000.

Ajmer Rajasthan, India. 26 27N 74 40E. City, railway junction and industrial and commercial centre. Manu. footwear, textiles and soap. Pop. (1981C) 375,593.

Akershus Norway. 60 10N 11 15E. County of SE Norway extending N from Oslo fiord to the S end of L. Mjosa, and inc. suburbs of Oslo. Area, 1,898 sq. m. (4,916 sq. km.). Cap. Oslo, although the city centre is not in the county. Indust., mainly in the suburbs of Oslo, inc. shipping and light industry. Tourism important. Pop. (1989E) 410,881.

Akhaia Greece. 38 05N 21 45E. *Nome* of the N Peloponnese bounded N by Gulf of Corinth and Gulf of Patras. Area, 1,141 sq. m. (3,209 sq. km.). Cap. Patras. Hilly but fertile region, mainly agric. The main products are currants, wine, fruit, olives, sheep and goats. Pop. (1981C) 275,193.

Akita Honshu, Japan. 39 44N 140 05E. Port on W coast sit. NNW of Sendai on R. Omono. Exports petroleum products from local oilfield. Trades in rice and manu. textiles. Pop. (1988E) 295,000.

Aklavik Northwest Territories, Canada. 68 12N 135 00W. Village in Mackenzie District of Arctic Can. Built by the Hudson's Bay Co. in 1912, but importance diminished when Inuvik, 33 m. (53 km.) NE, was built.

Akola Maharashtra, India. Town sit. WSW of Nagpur. Manu. cotton goods and soap. Pop. (1981C) 225,412.

Akron Ohio, U.S.A. 41 05N 81 31W. City sometimes called 'the rubber cap. of the world' because four major rubber companies have estab. intern. headquarters there. Also important chemical and plastics industries and polymer research centre. Although only vestiges of the Ohio Canal remain it is still an important transportation centre. Pop. (1980C) 237,000.

Aktyubinsk Kazakhstan, U.S.S.R. 50 16N 57 13E. Cap of Aktyubinsk Region. Sit. SW of Orsk. Manu. chemicals, metals, electrical equipment. Pop. (1985E) 231,000.

Akyab (Sittwe) Burma. 20 09N 92

55E. Port sit. on Bay of Bengal SSE of Chittagong. Rice is main export. Pop. (1983E) 143,215.

Alabama   U.S.A. 31 0N 87 0E. S state bordered on N by Tennessee, E by Georgia, W by Mississippi, S by Florida and Gulf of Mex. Area, 51,609 sq. m. (133,667 sq. km.). The soil of the coastal plain is sandy but further N there is rich black soil (Black Prairie). Surface flat except Appalachian Mountains in NE. Alabama, settled in 1702 as part of the French Province of Louisiana, and ceded to the Brit. in 1763, was organized as a Territory in 1817, and admitted into the Union in 1819. Chief towns, Birmingham, Mobile, Huntsville, Montgomery (the cap.) Tuscaloosa, Gadsden. Principal crops, soybeans, wheat, cotton and peanuts. Poultry and stock raising also important. Manu. inc. iron, steel, textiles and fertilizers. Pop. (1989E) 4,150,000.

Alabama River   U.S.A. 31 08N 87 57W. Formed by the confluence of the R. Coosa and R. Tallapoosa, flows 312 m. (499 km.) SW to join the Tombigbee R. and eventually forms the Mobile R. 44 m. (70 km.) above Mobile.

Alagoas   Brazil. 9 08S 36 0W. State in the NE. Area, 10,704 sq. m. (27,731 sq. km.). Cap. Maceió. Sugar and cotton main agric. products. Pop. (1980E) 2,303,000.

Alai Mountains   U.S.S.R. Extension of Tian-Shan Mountains. Highest peak in U.S.S.R. in this range, Mount Lenin, 23,382 ft. (7,127 metres) a.s.l.

Alajuela   Costa Rica. 10 00N 84 00W. (i) Province of N Costa Rica bounded N by Nicaragua. Area, 3,700 sq. m. (9,583 sq. km.). The chief towns are Alajuela San Ramón, Grecia, Orotina and Villa Quesada. Mainly tropical lowland extending S across the Central Cordillera and plateau to the Tárcoles valley. Largely a farming area, main products being coffee, sugar-cane, maize, beans, rice, pineapples and livestock. There is some lumbering and gold mining. Pop. (1984C) 430,634. (ii) Town sit.

NW of San José. Provincial cap. and commercial centre; route centre on the railway and Inter-American Highway. Trading centre for grain and livestock. Indust. inc. the manu. of vegetable oils, textiles, food products, soaps and coffee. Sawmilling is important. Pop. (1984C) 33,929.

Alameda   California, U.S.A. 34 46N 122 16W. Residential city and port sit. SSE of Oakland. Pop. (1980C) 63,852.

Alamein, El   Egypt. 30 48N 28 58E. Village sit. WSW of Alexandria. Important battle took place in 2nd World War.

Åland Islands   *see* **Ahvenanmaa Islands**

Alaska   U.S.A. 65 00N 150 0W. State in the NW of N. America, bounded by Arctic Ocean on N, Arctic Ocean and Bering Strait on W, Pacific on S, Canada on E. Area, 586,400 sq. m. (1,518,776 sq. km.). Cap. Juneau. Alaska was sold by Russia to U.S.A. in 1867 for US$7.2m. Traversed by many mountain ranges. Principal peaks, McKinley (20,320 ft./6,194 metres) and Foraker, and there are many active volcanoes and hot springs. The principal R. is the Yukon. In some parts of the country the climate during the brief spring and summer (about 100 days in major areas and 152 days in the SE coastal area) is not unsuitable for agric. operations, because of the long hours of sunlight, but Alaska is a food-importing area. There were 25,000 reindeer in W Alaska in 1980. In SE Alaska timber fringes the shore of the mainland and all the islands, extending inland to a depth of 5 m. (8 km.). Fishing is an important occupation and salmon the largest catch. Commercial production of crude petroleum began in 1959 and is now the most important economic activity. Other minerals inc. natural gas, coal and gold. Pop. (1987E) 537,800.

Alaska Highway   Canada/U.S.A. The Alaska Highway was opened in 1942. It runs for 1,523 m. (2,437 km.) from Dawson Creek, British Columbia (sit. 170 m. (272 km.) NE of Prince

George) *via* Fort Nelson, Watson L., Whitehorse, Boundary and Big Delta to Fairbanks. The Highway forms part of the Inter-American Highway which extends 16,800 m. (26,880 km.) from Alaska to Argentina.

Alava   Spain. 42 48N 2 28W. Basque province in N Spain. Area, 1,176 sq. m. (3,047 sq. km.). Cap. Vitoria. Main occupations agric. also fruits and vines. Pop. (1988E) 275,703.

Albacete   Spain. 38 59N 1 51W. (i) Province in SE Spain. Area, 5,737 sq. m. (14,862 sq. km.). Pop. (1988E) 342,278. (ii) cap. of (i). Mainly agric. and sheep rearing. Pop. (1988E) 127,126.

Alba-Iulia   Hunedoara, Romania. 46 04N 23 35E. Town sit. NW of Bucharest on Mures R. Historic city, trading and communications centre for a wine-producing area. Indust. inc. the manu. of soap, furniture and footwear.

Albania   41 0N 20 0E. Rep. sit. on E shore of Adriatic Sea. Extremely mountainous with fertile coastal plains. Cap. Tirana. Chief towns, Shkodër, Durrës, Vlonë. Area, 11,101 sq. m. (28,748 sq. km.). Agric. is important but cultivation methods are primitive. Main crops, grain, cotton, tobacco, potatoes, sugar-beet. Forest land represents 47% of the Rep. The mineral wealth of Albania is considerable but has only recently been developed. Minerals inc. coal, chromium, copper, nickel and salt. Industry totally nationalized and output is small. The principal industries are agric. product processing, textiles, oil products and cement. Oil is produced mainly at Qytet Stalin and a pipeline extends from there to Vlonë. Pop. (1989E) 3.2m.

Albany   Western Australia, Australia. 34 57S 117 54E. Port with one of the finest natural harbours in Australia and holiday resort on S coast. Industries inc. food processing, woollen mills, fishing, fertilizer manu., wild flower processing. Pop. (1989E) 23,000.

Albany   Georgia, U.S.A. 31 35N 84

10W. City on Flint R. Indust. inc. agric., manu. of aircraft and agric. equipment, paper products, food processing, textiles and telephone equip. Pop. (1988E) 86,900.

Albany   New York, U.S.A. 42 39N 75 45W. City sit. on W bank of Hudson R. Formerly a fur trading centre now state cap. Industries inc. retailing, regional banking centre and manu. of electrical goods, textiles and chemicals. Pop. (1980C) 101,727.

Albany   Oregon, U.S.A. 44 38N 123 06W. City at confluence of Willamette and Calapooia Rs., sit. S of Portland. Industries inc. manu. timber products, extracting and processing rare metals, processing foods and grass seed, high technology. Pop. (1980C) 26,546.

Albert, Lake   *see* **Mobutu Sese Seko, Lake**

Alberta   Canada. 54 40N 115 0W. Prairie province of Can. sit. between British Columbia and Saskatchewan with Montana (U.S.A.) to S. Area, 255,285 sq. m. (661,188 sq. km.). Cap. Edmonton, and chief towns Calgary, Lethbridge, Red Deer and Medicine Hat. Consists of a high sloping plateau descending E and NE from the Rocky Mountains. Important rivers are the Peace and Athabaska. Agric. is extremely important particularly wheat. The most recent est. of the coal resources of Alberta indicates that the province contains about 40% of the coal resources in W Can. Alberta produces approx. 75% of Canada's oil output. Oil was discovered at Leduc in 1947 and there are vast reserves of natural gas. Pop. (1989E) 2,432,400.

Albert Canal   Belgium. Canal in the NE linking the Scheldt estuary to the R. Meuse at Liège. Completed in 1939 and 80 m. (128 km.) long.

Albertville   *see* **Kalemie**

Albi   Tarn, France. 43 56N 2 09E. Industrial town and cap. of dept. Textile and glass manu., aniseed and wine trade. Pop. (1982C) 48,341.

Alborz   Iran. 36 00N 52 00E.

Mountain range separating Caspian Sea from central plateau of Iran, and extending in parallel ranges along S border of Caspian coastal provinces. Rises to Mount Demavend, 18,600 ft. (5,669 metres) a.s.l. The N slopes are heavily forested and receive heavy rainfall; the S slopes are dry.

Albula Pass　Graubünden, Switzerland. 46 35N 9 50E. Pass in the Rhaetian Alps, SE Switz. reaching 7,595 ft. (2,315 metres) a.s.l.

Albuquerque　New Mexico, U.S.A. 35 05N 106 40W. City founded 1706 by Philip V of Spain on Rio Grande. Manu. and trading centre particularly agric. products, and for technology and scientific research. Pop. (1988E) 378,480.

Albury　New South Wales, Australia. 36 05S 146 55E. Town sit. on R. Murray. Commercial centre trading in wool, wheat and fruit. Noted for its wine. Pop. (1981E) 53,214, with Wodonga.

Alcalá de Guadaira　Sevilla, Spain. 37 20N 5 50W. Resort sit. on R. Guadaira, ESE of Seville. Industries inc. food processing and metal products. Pop. (1982E) 46,000.

Alcalá de Henares　Madrid, Spain. 40 28N 3 22W. Town sit. on R. Henares, ENE of Madrid. Industries inc. plastics, pharmaceuticals, electrical goods. Pop. (1980C) 142,862.

Alcázar de San Juan　Ciudad Real, Spain. 39 24N 3 12W. Market town sit. NE of Ciudad Real. Industries inc. wine, woollen garments, olives. Railway junction.

Alcoy　Alicante, Spain. 38 43N 0 30W. Town sit. N of Alicante. Manu. inc. textiles and paper. Pop. (1982C) 62,000.

Aldabra Island　Seychelles. 9 25S 46 20E. Formerly part of British Indian Ocean Territory, a colony created in 1965 consisting of the Chagos Arch. (formerly a depend. of Mauritius), Aldabra, Farquhar and Des Roches but returned to Seychelles in 1976.

Aldan River　U.S.S.R. 63 28N 129 35E. Rises in the Aldan Mountains flowing NE then N and W for 1,700 m. (2,720 km.) to join the R. Lena.

Aldeburgh　Suffolk, England. 52 9N 1 35E. Town holding annual musical festival. Pop. (1989E) 2,911.

Aldermaston　Berkshire, England. Village sit. SW of Reading. Headquarters of Atomic Weapons Research Establishment. Pop. (1985E) 2,200.

Alderney　Channel Islands, United Kingdom. 49 43N 2 12W. Only town is St. Anne's sit. in centre of island. Area, 3 sq. m. (8 sq. km.). Exports early potatoes. Pop. (1985E) 2,000.

Aldershot　Hampshire, England. 51 15N 0 47W. Town containing large military depot. Pop. (1985E) 38,000.

Alegranza　Canary Islands. Northernmost islet of the Canaries, 145 m. (232 km.) NE of Las Palmas. Area, 4.5 sq. m. (12 sq. km.). Rises to 800 ft. (244 metres) a.s.l. with extinct volcanoes. Uninhabited.

Alençon　Orne, France. 48 27N 0 04E. Cap. of dept. sit. at confluence of R. Sarthe and R. Briante. Agric. main occupation. Pop. (1982C) 32,526.

Aleppo　Syria. 36 12N 37 10E. Cap. of Aleppo province. Sit. in NW of Syria. Manu. textiles, carpets, cement and soap. Agric. important. Originally on important caravan route between Europe and Asia, but the route by sea to India caused decline in its economic importance although there was some revival when railways were built. Pop. (1981C) 985,413.

Alès　Gard, France. 44 08N 4 05E. Town sit. at foot of Cévennes Mountains. Coal, iron and zinc mined. Metal, glass and chemical industries. Pop. (1982C) 44,343.

Alessandria　Piemonte, Italy. 44 54N 8 37E. Cap. of Alessandria province. Sit on R. Tanaro. Engineering is an important industry. There is also a large railway workshop. Centre of rich agric. and wine-producing district. Pop. (1981C) 100,523.

Aletsch Glacier   Bernese Oberland, Switzerland. 46 28N 8 00E. Largest glacier in Europe (16 m. (26 km.) long), sit. in S central Switz. To the NW lies the Aletschhorn, the second highest peak in the Bernese Oberland.

Aleutian Islands   Alaska, U.S.A. 52 00N 176 00W. Continuation of the Alaskan Peninsula consisting of over 150 mountainous islands, some rising to 8,000 ft. (2,438 metres) a.s.l. and many volcanic. Agric. development is nearly impossible. Fishing, particularly salmon, is important but the main value of the islands is strategic.

Alexandria   Egypt. 31 13N 29 55E. Important port and second city of Egypt. Built on peninsula between Mediterranean and L. Mareotis. Founded by Alexander the Great in 331 B.C. Large proportion of Egyptian trade goes through Alexandria and the main export is raw cotton. Industries tend to be connected with the cotton industry (ginning and seed-oil pressing). Pop. (1986C) 2,893,000.

Alexandria   Louisiana, U.S.A. 31 18N 92 27W. City sit. on Red. R. NW of Baton Rouge. Service and trading centre for agric. area. Pop. (1980C) 51,565.

Alexandria   Virginia. U.S.A. 45 53N 95 22W. City sit. just S of Washington, D.C. on R. Potomac. Commercial and residential area. Once the home of George Washington and Robert E. Lee. Pop. (1980C) 103,217.

Alexandroupolis   Thrace, Greece. 40 50N 25 52E. Port and cap. of Evros prefecture. Sit. on an inlet of the Aegean Sea, SW of Edirne, Turkey. Railway centre. Pop. (1981C) 34,535.

Alfreton   Derbyshire, England. 53 06N 1 23W. Midlands town, manu. engineering goods and clothing. Pop. (1981C) 21,338.

Algarve   Portugal. 36 58N 8 20W. Smallest province of Portugal, and coextensive with Faro district. Separated in the N from the rest of Portugal by a range of barren mountains ending at Cape St. Vincent. Held by the Moors until 1249 A.D. Cap. Faro. Area, 1,937 sq. m. (5,017 sq. km.). Main occupations fishing, horticulture and tourism.

Algeciras   Cádiz, Spain. 36 08N 5 30W. Port and holiday resort sit. on Algeciras Bay opposite Gibraltar (5 m., 8 km.) Citrus fruit and cork are exported. Pop. (1981C) 86,042.

Alger   *see* **Algiers**

Algeria   35 10N 3 00W. Independent state in N. Africa, former French colony, bordered in N by the Mediterranean Sea, in the E by Tunisia, in the S by Niger, Mali and Mauritania and in the W by Morocco. Cap. Algiers. Chief towns, Oran, Constantine, Annaba. Area, 919,595 sq. m. (2,381,741 sq. km.). Algeria is divided by elevation and climatic conditions into three unequal parallel zones: The Tell, undulating N region (containing the Little Atlas) of forests and arable land; The Steppe region of herbaceous vegetation and of pasture land diversified by ranges of mountains; The Sahara (S of Sahara Atlas), where agric. is possible only by irrigation in oases. Coast, 690 m. (1,104 km.), High and rocky. Rivers nothing but torrents (*wadis*), frequently dried up; some dammed for irrigation; in steppes they form the *Shotts* of the plains, shallow lakes where snow and rain-water gather in winter. Chief crops, wheat, barley, wine, olive oil, citrus fruits and dates. Fisheries are important particularly sardines, anchovy and tunny. Algeria possesses deposits of iron, zinc, lead, mercury, copper and antimony. Kaolin, marble, onyx, salt and coal are also found. Oil and natural gas are important. Pop. (1988E) 23,850,000.

Alghero   Sardinia, Italy. 40 34N 08 19E. Port in NW Sardinia, sit. 17 m. (27 km.) SW of Sassari. Fishing for lobster and coral main occupation.

Algiers   Algeria, North Africa. 36 50N 3 00E. Cap., port and largest city in Algeria. Founded in the 10th cent. and from 1530 when the Spaniards were expelled until its conquest by France in 1830 was a base for pirates. Wine making and associated industries are important occupation. Pop. (1984E) 2,442,303.

Alhambra California, U.S.A. 34 06N 118 08W. City, part of the Los Angeles conurbation. Mainly residential. Pop. (1980C) 64,615.

Alicante Spain. 38 21N 0 29W. City and important port sit. on E coast. Exports wine, olive oil and fruit. There is some manu. industry and an oil refinery. Pop. (1986E) 265,543.

Alice Springs Northern Territory, Australia. 23 42S 133 53E. Town sit. in the MacDonnell Ranges. Major tourist destination. Terminus of the S section of the transcontinental railway from Port Adelaide to Darwin. An important airway centre, and serves cattlemen, tourists and aboriginal communities. Pop. (1988E) 23,600.

Aligarh Uttar Pradesh, India. 27 54N 78 05E. Town sit. SE of Delhi. Trading centre for cereals and cotton. Manu. cotton and metal goods, inc. locks. Pop. (1981C) 320,861.

Alkmaar Noord-Holland, Netherlands. 52 37N 4 44E. Town sit. near N. Sea coast. Centre of the cheese trade. Manu. furniture, clothing and cigars. Pop. (1984E) 83,892.

Allahabad Uttar Pradesh, India. 25 27N 81 50E. City sit. at confluence of R. Jumna and R. Ganges. Administrative and religious centre. An important point of pilgrimage because of Allahabad's position at the confluence of two holy rivers. Trades in cereals, cotton and sugar. Industries inc. textiles and flour milling. Noted for the fair *Kumbh Mela* held every 12 years. Pop. (1981C) 619,628.

Allegheny Mountains U.S.A. 38 30N 80 00W. Part of the Appalachian range extending from Pennsylvania to North Carolina, rising to 4,000 ft. (1,219 metres) a.s.l. The mountains contain coal, iron, limestone.

Allen, Bog of Ireland. Group of morasses in central Ireland. Area, 370 sq. m. (958 sq. km.). Some parts are cultivated and the peat is cut, some being used as fuel for power stations.

Allentown Pennsylvania, U.S.A. 40 36N 75 29W. Town sit NNW of Philadelphia on R. Lehigh. Manu. textile products, vehicles, steel, industrial gases and high technology products. Pop. (1980C) 106,500.

Alleppey Kerala, India. 9 30N 76 20E. Port and city sit. NNW of Trivandrum. Manu. coir ropes and matting, and coconut oil. Pop. (1981C) 169,940.

Alliance Ohio, U.S.A. 40 55N 81 06W. City sit. SSE of Cleveland. Manu. iron and steel. Pop. (1980C) 24,315.

Allier France. 46 25N 3 00E. Dept. in Auvergne region. Area, 2,829 sq. m. (7,327 sq. km.). Chief towns, Moulins (the cap.) and Montluçon. Agric. extremely important. Pop. (1982C) 369,580.

Allier River France. 46 58N 3 04E. Rises in the Cévennes and flows N for 270 m. (432 km.) to join R. Loire.

Alloa Central Region. Scotland. 56 07N 3 49W. Market town sit. E of Stirling at the head of the Firth of Forth. Indust. inc. the manu. of woollens and yarns, glass bottles, whisky distilling, brewing and engineering. Pop. (1981C) 26,390.

Alma-Ata Kazakhstan, U.S.S.R. 43 15N 76 57E. Cap. of Rep. and region. Trades in agric. and horticultural products and manu. machinery, textiles, food and tobacco products. Pop. (1985E) 1,068,000.

Almada Setúbal, Portugal. Town on S bank of Tagus estuary and linked to Lisbon by Salazar bridge. Manu. centre.

Almadén Ciudad Real, Spain. 38 46N 4 50W. Town sit. SW of Ciudad Real in the Sierra Morena Mountains. Mercury mines.

Almelo Overijssel, Netherlands. 52 22N 6 42E. Town sit. NW of Enschede. Manu. textiles and furniture. Pop. (1989E) 62,000.

Almería Spain. 36 52N 2 27W. (i)

Mountainous province of Spain. Area, 3,388 sq. m. (8,775 sq. km.). Important for iron and lead. Pop. (1981c) 405,513. (ii) Cap. of province of same name sit. on s coast. Trade important but little manu. Pop. (1981c) 140,946.

Alnwick Northumberland, England. 55 25N 1 42W. Town sit. N of Newcastle upon Tyne on the Aln R. Market town for an agric. area and site of Alnwick Castle, seat of the Dukes of Northumberland. Indust. inc. agric. engineering, small electricals, research and development of pharmaceuticals, opencast mining and the manu. of fishing rods. Pop. (1989E) 7,200.

Alofi Hoorn Islands, part of Wallis and Futuna Islands. Island 5 m. (8 km.) SE of Futuna. Area, 20 sq. m. (51 sq. km.). Volcanic, rising to 1,200 ft. (366 metres) a.s.l. Main products are coconuts and timber. Uninhabited.

Alost East Flanders, Belgium. 50 56N 4 02E. Town sit. WNW of Brussels. Manu. textiles, clothing. Brewing and trade in hops are important. Pop. (1988E) 76,714.

Alpes-de-Haute-Provence France. 44 08N 6 10E. Formerly Basses-Alpes. Dept. in Provence-Côte d'Azur region, bounded NE by Italy. Area, 2,681 sq. m. (6,944 sq. km.). Cap. Digne. Drained by Durance R. and its tribs. Mountainous and infertile N and E, valleys produce vines, olives and fruit. Pop. (1982c) 119,068.

Alpes-Maritimes France. 43 55N 7 10E. Dept. in Provence-Côte d'Azur region, bordered on the E by Italy, on S by the Mediterranean. It also surrounds the principality of Monaco. Area, 1,658 sq. m. (4,294 sq. km.). Chief towns, Nice (the cap.), Cannes, Antibes and Menton. The range of the Maritime Alps forms the N boundary. The most important R. is the Var. Sheep rearing is important and olives and fruit are grown. Perfume is distilled and wine manu. Fisheries inc. sardines, anchovy and tunny. The coast, the Côte d'Azur, is part of the French Riviera and the tourist industry gives considerable employment. Pop. (1982c) 881,198.

Alphen aan den Rijn Zuid-Holland, Netherlands. (Also, Alfen aan den Rijn.) Town sit. E of Leiden on the Old Rhine R. Railway junction. Indust. inc. shipbuilding, market gardening, and the manu. of asphalt, bricks, roofing tiles, ships' machinery, motors, electrical goods, paper, leather products and fruit products. Pop. (1984E) 54,560.

Alps Crescent-shaped mountain system, extending from Gulf of Genoa to Vienna. Total length, *c.* 700 m. (1,120 km.), and breadth, 30–160 m. (48–256 km.), area, *c.* 80,000 sq. m. (*c.* 207,200 sq. km.). Divided into w Alps (Tenda P. to Simplon P.), inc. Maritime Alps, Cottian Alps, Dauphiné Alps, Graian Alps, chain of Mont Blanc, Pennine Alps; Central Alps (Simplon P. to Reschen P., Scheidegg and Stelvio P.), inc. Bernese Alps, N Swiss Alps, Lepontine and Adula Alps, Tödi group, Rhaetian Alps; E Alps (Reschen, Scheidegg and Stelvio to the Semmering P.), inc. Ortler, Adamello and Brenta groups, Limestone Alps of Bavaria, N Tyrol and Salzburg, Central Tyrolese Alps, S Tyrolese Alps (Dolomites); SE Alps (Carnic, Karawanken and Julian Alps). Snow-line varies from 8,900 ft. (2,712 metres) on S side to 9,200 ft. (2,804 metres) on N; glaciers of upper valleys descend to 4,000–5,000 ft. (1,219–1,524 metres) a.s.l.—Aletsch, 4,400 ft. Best-known peaks of w Alps, Gran Paradiso (13,320 ft.), Mont Blanc (15,780), Dent du Midi (10,690), Matterhorn (14,775), Monte Rosa (15,215), Aletschhorn (13,770), Jungfrau (13,670), Mönch (13,460), and Finsteraarhorn (14,020). Drained by R. Rhine and trib., Aar, Reuss, Limmat; R. Danube and trib., Iller, Lech, Isar, Inn; R. Po and trib., Dora Riparia, Ticino, Mincio, Adige; R. Rhône with Arve, Isère, Durance; and smaller streams flowing to Adriatic and Ligurian Sea. Chief passes: Great St. Bernard, E of Mont Blanc to Rhône; Mont Cenis, from R. Dora Riparia NW to Arc and Isère; Simplon, from Maggiore to Brig; St. Gotthard, from Ticino to Upper Reuss valley; Brenner, from Adige to Innsbrück.

Alpujarras, Las Almería/Grenada, Spain. A mountainous region sit. between the Sierra Nevada and the S coast.

Als   Denmark. 54 59N 9 55E. Island sit. in the Little Belt between mainland and island of Fyn. Chief town, Sönderborg sit. 31 m. (50 km.) SW of Odense. It is connected to the mainland by a bridge across Alsen sound. The island is 19 m. (30 km.) long and 3–12 m. (5–20 km.) broad.

Alsace   France. 48 15N 7 25E. Region in the NE of France sit. on the W side of the R. Rhine valley and the E slopes of Vosges, comprising the depts. of Haut-Rhin and Bas-Rhin. Chief towns are Strasbourg, Mulhouse, Colmar and Belfort. Area, 3,208 sq. m. (8,310 sq. km.). Cap. Strasbourg. Alsace has long been a disputed territory: once part of Germany, in 1648 much of it was ceded to France and further territory later. After the Franco-Prussian War, all Alsace except the Territory of Belfort was incorporated into Germany to form, with part of Lorraine, the territory of Alsace-Lorraine. The Treaty of Versailles returned Lorraine to France, and Alsace became the depts. of Haut-Rhin and Bas-Rhin. The fertile lowlands produce cereals, hops, fruit, tobacco. Wine is also produced. Pop. (1987E) 1,605,300.

Alsace-Lorraine   Borderland between Federal Republic of Germany and France *see* **Alsace**.

Altai Mountains   U.S.S.R. 48 00N 90 00E. The Altai Mountains proper, and the Greater Altai stretch from the Gobi desert NW in two parallel ranges across the Siberian frontier. The upper streams of the R. Ob and R. Irtysh lie within the range. The highest peak in the U.S.S.R. Altai is the Belukha Mountain at 14,783 ft. (4,506 metres) a.s.l. Lead, zinc and silver are mined.

Altai Territory   U.S.S.R. Territory within the Russian Soviet Federal Socialist Republic sit. in SW of Siberia. Cap. Barnaul. Area, 101,000 sq. m. (261,590 sq. km.). Although the Altai mountains form part of the Territory there is considerable fertile agric. land.

Altamura   Puglia, Italy. 40 50N 16 33E. Town sit. SW of Bari. Agric. district; cultivating cereals, grapes and olives. Manu. pasta esp. macaroni.

Alta River   Norway. Rises near the Finnish border and flows N for 120 m. (192 km.) and enters the Arctic Ocean at the Alta Fiord. Noted for salmon.

Alta Verapaz   Guatemala. Dept. in central Guatemala in the Sierra de Chamá bounded W by the Chixoy R. and SE by the Sierra de los Minas. Area, 3,353 sq. m. (8,584 sq. km.). Chief towns are Cobán (cap.), San Cristóbal and Carchá. Mainly agric. with some lumbering and craft indust. The main products are coffee, maize, beans, sugar-cane, cacao, vanilla, fruit and livestock. Pop. (1984E) 390,059.

Altenburg   Leipzig, German Democratic Republic. 50 59N 12 26E. Town sit. S of Leipzig in the Gera district. Commercial and manu. centre. Lignite found in the area.

Alton   Hampshire, England. 51 08N 0 59W. Market town sit. SSE of Basingstoke. Pop. (1980E) 14,000.

Alton   Illinois, U.S.A. 38 55N 90 05W. City sit. NNE of St. Louis on R. Mississippi. Flour milling and manu. sports amenities, paper and fabricated steel products. Oil refining is also important. Pop. (1980C) 34,171.

Altona   Hamburg, Federal Republic of Germany. 53 32N 9 56E. Port sit. on R. Elbe. Boundaries altered in 1938 and Altona became part of Hamburg. Manu. machine tools, chemicals and textiles. Fish processing also important.

Altoona   Pennsylvania, U.S.A. 40 31N 78 24W. City sit. E of Pittsburgh. Important railway and railway engineering centre. Manu. clothing and electrical equipment. Pop. (1980C) 57,078.

Alto Paraná   Paraguay. Dept of SE Paraguay bounded E by Brazil and Argentina across the Paraná R. Area, 7,817 sq. m. (20,248 sq. km.). Cap. Hernandarias. Hilly and heavily forested with sub-tropical vegetation and heavy rainfall. The main products are timber and maté. Pop. (1982E) 188,351.

Altrincham Trafford, Greater Manchester, England. 53 24N 2 21W. Town sit. sw of Manchester. Market town, with service industries and light engineering important occupations. Pop. (1981C) 39,641.

Älvsborg Sweden. 58 20N 12 20E. County of sw Sweden. Area, 4,400 sq. m. (11,395 sq. km.). The chief towns are Vänersborg (cap.), Borås, Trollhattan, Alingsas, Ulricehamn and Åmål. Low-lying, drained by the rivers Göta, Viska and Atra. Mainly agric., the main products being wheat, rye, oats, with cattle raising and dairy farming. Heavy indust. at Trollhattan inc. textiles, paper and pulp milling. Pop. (1980C) 425,189.

Alwar Rajasthan, India. 27 38N 76 34E. Town and former cap. of state of same name now part of Rajasthan, sit. NE of Jaipur. Trades in millet and cotton goods. Industries inc. flour-milling, car components, razor blades, synthetic fibres and electronics. Pop. (1981C) 145,795.

Amagasaki Honshu, Japan. 34 33N 135 25E. Town sit. on Osaka Bay just NNW of Osaka. Manu. textiles, chemicals, glass and metal goods. Pop. (1988E) 497,000.

Amager Zealand, Denmark. Island in the Oresund between Koge Bay and Drogden Strait. Part of Copenhagen is on the N part; other main towns are Taarnby, Kastrup and Dragor. The main occupation is market gardening.

Amalfi Campania, Italy. 40 38N 14 36E. Town on the Gulf of Salerno sit. SE of Naples. Tourist resort.

Amapá Brazil. 1 40N 50 00W. Territory in the NE lying N of the R. Amazon. Cap. Macapá. Area, 54,161 sq. m. (140,276 sq. km.). Largely undeveloped. Pop. (1984E) 207,000.

Amarillo Texas, U.S.A. 35 13N 101 49W. City sit. NW of Fort Worth. Agric. activity wheat and cattle raising. Indust. centre producing oil and

natural gas, meat processing, refining metals. Manu. clothing, fibreglass and aircraft. Pop. (1980C) 149,230.

Amasya Turkey. 40 39N 35 51E. (i) Province of N central Turkey, bounded N by the Canik Mountains. Area, 1,886 sq. m. (4,828 sq. km.). Drained by Rs. Yesil and Cekerek. Mainly agric., the main products being opium, wool, apples, tobacco, hemp and sugar-beet. There are some metal ores including lead, with gold and silver at Gumushacikoy in the NW. Pop. (1985C) 358,289. (ii) Town sit. ssw of Samsun on Yesil R. Provincial cap. Trading centre for apples, tobacco, wheat and onions; trad. centre of tile manu. Pop. (1980C) 60,000.

Amazon River Brazil. 2 00S, 53 30W. Flows from the Andes along equator to the Atlantic Ocean with total length of 4,000 m. (6,400 km.). Largest R. in the world in volume. Area of basin 2.7m. sq. m. (7m. sq. km.). The head stream is the Alto Marañón in Peru which issues from L. Lauricocha at 11,980 ft. (3,652 metres) a.s.l. Numerous affluents form gigantic network of streams which feed the Atlantic at the rate of 2.54m cu. ft. (72,000 cu. metres) per second. Alto Marañón becomes navigable for small craft 2,700 m. (4,320 km.) from Atlantic. Ocean-going steamers penetrate to Iquitos, Peru (1,935 m., 3,096 km.) from mouth. On entering Braz. Amazon is known as Solimões—tribs. Javari (550 m., 880 km.), Jutaí (400 m., 640 km.), Juruá, and Purus, right bank; Putumayo (650 m., 1,040 km.), Japurá (930 m., 1,488 km.), left bank. At confluence of Negro (480 m., 768 km.), on the left bank, is Manaus. Below this main stream known as Amazon. Joined (right bank) by largest of tribs. Madeira. Other notable tribs.: Tapajós, Xingu and Tocantins (500 m., 800 km.). Negro is connected with Orinoco R. system. Madeira formed by Beni (Bolivia) and Mamoré; Madeira–Mamoré railway facilitates transport of produce from Bolivia and Brazilian state of Mato Grosso. Tidal influence felt 400 m. (640 km.) upstream. In flood time enormous areas of the surrounding country are under water.

Breadth, 200 m. (320 km.) at mouth and from 3–4 m. (5–6 km.) or more at 1,000 m. (1,600 km.) from ocean. Numerous marshy islands at mouth; largest Marajó (2,060 sq. m., 5,335 sq. km.). Total length of waterways in system 30,000 m. (48,000 km.). Produce of region: rubber, quinine, caffeine, cotton, indigo, nuts, sugar, cacao, coffee, tobacco. The river is said to contain a fresh-water replica of every salt-water fish, and turtles abound.

Amazonas Brazil. 4 20S 64 0W. Largest state in the upper Amazon basin. Area, 616,148 sq. m. (1,564,445 sq. km.). Cap. Manaus. Forested, and produces timber and rubber. Pop. (1989E) 1,948,508.

Amazonas Peru. Dept. of N Peru, bounded N by Ecuador. Area, 13,948 sq. m. (41,298 sq. km.). Cap. Chachapoyas. Sit. at W edge of the Amazon basin and crossed by ridges of the Andes N to S. Drained by Rs. Marañón, Santiago and Utcubamba. The area is hot and humid. It is undeveloped, densely forested but fertile in the upper valleys. The main products are rubber, timber, resin, vanilla, coffee, sugar-cane, cacao, potatoes, tobacco, fruit and grain. There are some salt, sulphur and coal deposits. The main indust. are alcohol distilling, tanning and the manu. of straw hats. Pop. (1988E) 319,500.

Amazonas Venezuela. Territory of S Venezuela, bounded W by Colombia, S and SE by Brazil. Area, 67,857 sq. m. (175,750 sq. km.). Cap. Puerto Ayacucho. The territory is mainly jungle, rising in the SE to the Guiana highlands. It is the source of the Orinoco R. The climate is tropical with heavy rains. The main products are rubber, balata gum, vanilla and sugar-cane. Pop. (1981C) 45,667.

Ambala Haryana, India. 30 23N 76 46E. Town sit. NNW of Delhi. Railway junction. Trades in grain and cotton. Cotton ginning, flour milling and food processing important occupations. Pop. (1981C) 104,565.

Ambato Tungurahua, Ecuador. 1 15S 78 44W. Resort and cap. of

Tungurahua province. Sit. S of Quito at 8,500 ft. (2,591 metres) a.s.l. Fruit canning and tanning important. Manu. textiles. Pop. (1984E) 110,000.

Amberg Bavaria, Federal Republic of Germany. 49 27N 11 52E. Old garrison town sit. between Nuremberg and border with Czech. Manu. textiles, glass, steel, electronics, enamels and cement. There are 10 breweries and iron mines in vicinity. Pop. (1988E) 42,300.

Ambleside Cumbria, England. 54 26N 2 58W. Town and tourist centre in Lake District, sit. N of L. Windermere. Associated with William Wordsworth (1770–1850) poet. Pop. (1985E) 2,500.

Amboina Indonesia. (i) Mountainous island in the Moluccas group in the Banda Sea 7 m. (11 km.) S of Ceram, rising to 3,405 ft. (1,038 metres) a.s.l., with fertile plain. Area, 386 sq. m. (1,000 sq. km.). (ii) Chief town and port of island of same name. Exports, copra and spices.

Ambrym Vanuatu, South West Pacific. 16 15S 168 10E. Island 65 m. (104 km.) SE of Espiritu Santo. Area, 230 sq. m. (589 sq. km.). Volcanic, rising to Mount Benbow at 3,720 ft. (1,134 metres) a.s.l. The chief products are coffee and cocoa.

American Samoa *see* **Samoa**

Amersfoort Utrecht, Netherlands. 59 09N 5 24E. Town sit. ENE of Utrecht. Indust. chemicals, engineering goods, brewing, canning, tobacco processing, market gardening, and carpets. Pop. (1988E) 90,072.

Amersham Buckinghamshire, England. 51 40N 0 38W. Town sit. SE of Aylesbury. Manu. furniture and radio isotopes. Pop. (1985E) 20,000.

Ames Iowa, U.S.A. 42 02N 93 37W. City sit. N of Des Moines. Shopping centre of agric. area. Pop. (1980C) 45,775.

Amiens Somme, France. 49 54N 2 18E. Cap. of dept. sit. on R. Somme.

The Cathedral of Notre Dame, built 1220–88, is one of finest Gothic buildings in Europe and largest church in France. Manu. textiles and machinery. Trading centre for the many market gardens in the vicinity. Pop. (1989E) 131,332.

Amindivi Islands *see* **Lakshadweep**

Amirante Islands Seychelles, Indian Ocean. 6 00S 53 10E. Arch. 500 m. (800 km.) NNE of Madagascar, forming an out-lying dependency of the Seychelles. Length 100 m. (160 km.) N to S. The main islands are African Islands, Daros, Desneuf, St. Joseph, Poivre, and Marie Louise. The main product is copra.

Amman Amman, Jordan. 31 57N 35 52E. Cap. sit. between Dead Sea and border with Syria and just N of Jerusalem. Built on the site of biblical Rabbat 'Ammon, the Hellenistic Philadelphia. Its first settlers in modern times were the Circassians and it became cap. of Transjordan in 1921. It has increased in size because of influx of Palestinian refugees since 1948 following the Arab–Israeli war. Administrative and tourist centre and noted for marble quarried locally. Pop. (1986E) 1,160,000.

Amoy *see* **Xiamen**

Amravati Maharashtra, India. 20 56N 77 47E. Town sit. WSW of Nagpur. Commercial centre, particularly for cotton. Pop. (1981C) 261,404.

Amritsar Punjab, India. 31 38N 74 53E. Town sit. E of Lahore, Pakistan. Indust. and commercial centre trading in textiles and hides. Manu. carpets, chemicals and silk and woollen goods. The religious centre of the Sikhs, noted for the Golden Temple which is sit. on a small island in the Amrita Saras, or Sacred Tank. Pop. (1981C) 594,844.

Amstelveen Noord-Holland, Netherlands. 52 18N 4 51E. Village sit. S of Amsterdam near the Amstel R. Watersports centre. Pop. (1988E) 69,505.

Amsterdam Noord-Holland, Netherlands. 52 23N 4 54E. Official cap. (the seat of government being at The Hague) and port, and largest city sit. at mouth of Amstel R. in the SW corner of the Ijsselmeer. It is dissected by a radical system of canals, crossed by about 400 bridges and is connected to the North Sea by two canals. Its commercial success began with the establishment of the East India Company in 1602. Chief industries are diamond cutting, sugar refining, engineering, shipbuilding and repairing and chemicals. Pop. (1989E) 691,837.

Amsterdam New York, U.S.A. 42 57N 74 11W. City sit. NW of Albany on Mohawk R. Manu. toys, hi-tech. electronics, textiles and plastics. Pop. (1980C) 21,872.

Amur River U.S.S.R./China. 52 26N 141 10E. Formed by junction of R. Shilka and R. Argun at the U.S.S.R./China border. It flows for 1,800 m. (2,800 km.) SE and then NE and forms part of the border between U.S.S.R. and China, and enters the Sea of Okhotsk at Nikolayevsk opposite the N end of Sakhalin. The Amur R. system has 8,400 m. (13,440 km.) of navigable waterways open from May to Oct. each year, and drains nearly 800,000 sq. m. (2,072,000 sq. km.).

Anambra Nigeria. State comprising the Abakaliki, Enugu and Onitsha provinces of the former E region. Area, 6,824 sq. m. (17,675 sq. km.). Mainly populated by the Ibo people, this area formed the core of the Biafran secession of 1967–70. Main cities Enugu (cap.), Onitsha, Awka, Abakaliki and Nsukka. Pop. (1983E) 5.5m.

Anadyr Russian Soviet Federal Socialist Republic, U.S.S.R. 64 45N 177 29E. Town sit. on Gulf of Anadyr, in the extreme NE. Industries lignite mining and fishing.

Anadyr Range Russian Soviet Federal Socialist Republic, U.S.S.R. Mountain range in NE Siberia extending SE from the E. Siberian Sea.

Anadyr River Rises in the Gydan range of NE Siberia and flows for 500 m. (800 km.) SW and then E and enters the Bering Sea at the Gulf of Anadyr.

Anaheim   California, U.S.A. 33 51N 117 57W. City sit. S of Los Angeles. Manu. and tourist centre. Pop. (1980C) 219,311.

Anáhuac   Mexico. Anc. name of the Aztec kingdom now generally used of the Central Mexican Plateau, but more correctly meaning the area in which Mexico City is sit.

Anápolis   Goiás, Brazil. 16 20S 48 58W. Town sit. SW of Brasília. Centre of an agric. region trading in livestock, cereals and coffee.

Ancash   Peru. Dept of W central Peru, bounded W by the Pacific and E by the Marañón R. Area, 14,705 sq. m. (36,308 sq. km.). Cap. Haurás. Crossed NW to SE by the Cordillera-Occidental. Drained by the Santa R. The main occupations are farming and mining; the main products are grain, vegetables, cattle, sugar, cotton, rice, lead, copper, silver and gold. There are important coal and iron deposits along the Santa R. A railway connects the mountain districts with the port of Chimbote. Pop. (1981C) 818,289.

Anchorage   Alaska, U.S.A. 61 13N 149 53W. City and port in Alaska sit. at head of Cook Inlet and surrounded by the Chugach Mountains. Centre for communications, transport, finance and trade. Pop. (1989E) 218,979.

Ancona   Marche, Italy. 43 38N 13 30W. (i) Province in Marche Region. NE coastal plain rises in SW to central Apennines. Principal R. are the Esino, Misa and Musone. (ii) Cap. of province of same name and of Marche, and port on Adriatic Sea. Industries ship-building, engineering and sugar refining. Pop. (1981C) 106,498.

Andalusia   Spain. 37 35N 5 0W. Region of SW and S Spain comprising provinces of Cádiz, Malaga, Granada, Almeria, Huelva, Sevilla, Cordoba and Jaén. Separated from the central plateau by the Sierra Morena and crossed in the S by the Cordillera Penibetica rising to the Sierra Nevada, 11,411 ft. (3,478 metres) a.s.l. at Mulhacen. Between the 2 ranges the central plain is watered by the Guadalquivir R. and its tribs. and is very

fertile. The Mediterranean coast is a resort area. Mining and fishing are important, as well as agric. The main products are copper, iron, galena, tin, zinc, nickel, sulphur, antimony, tungsten, coal, pyrite, manganese, jasper and gypsum; olives, cereals, fruit, vegetables, vines and cotton; cattle, fighting bulls, and horses for breeding. Indust. are mainly associated with agric. and viticulture and inc. flour milling, wine making, distilling, and the processing of olive oil, fish, sugar, fruit and meat.

Andaman Islands   India. 12 30N 92 30E. Group of islands forming N part of the territory of the Andaman and Nicobar Islands, in the Bay of Bengal. Area, 2,474 sq. m. (6,408 sq. km.). Chief town, Port Blair. The Great Andamans consist of N Andaman, Midde Andaman, Baratang, S Andaman and Rutland, with their offshore islets; Ritchie's Arch. is E across Diligent Strait and Little Andaman is S across Duncan Passage. Mainly hilly, rising to Saddle Peak, 2,400 ft. (732 metres) a.s.l. on N Andaman. Tropical monsoon climate, mean annual rainfall 90–130 ins. (230–330 cm.). Densely forested. The main products are timber, coconuts, rice, pulses, coffee, fruit and rubber. Industries inc. timber and related manu. The indigenous tribes are Negrito aborigines. Pop. (1981C) 158,287.

Andermatt   Uri, Switzerland. 46 38N 8 36E. Village and resort sit. SE of Luzern at junction of St. Gotthard road and Furka Pass, at 4,700 ft. (1,444 metres) a.s.l.

Anderson   South Carolina, U.S.A. 34 31N 82 39W. City sit. SW of Greenville. Manu. textiles, fibres, tools, machinery, chemicals and metal products. Pop. (1980C) 27,313.

Andes, Los   Aconcagua, Chile. Town sit. N of Santiago on Aconcagua R. Centre of an agric. region producing cereals, fruit, wine and tobacco. Pop. (1984E) 106,000.

Andes Mountains   South America. 20 0S 68 0W. Mountain system extending along W side of S. America from Isthmus of Panama and mouth of Orinoco R. to Tierra del Fuego. In the middle section there are 2 main chains and elsewhere 3. Total length, 4,000 m. (6,400 km.). In Ecuador 18 peaks are over 15,000 ft. (4,572 metres) a.s.l. (Chimborazo 20,498 ft.). The highest peak is Aconcagua (22,868 ft.). To the N of 20° S., the breadth exceeds 500 m. (800 km.). L. Titicaca, alt. 12,000 ft. (3,658 metres) a.s.l., has area of 3,300 sq. m. (8,547 sq. km.). There are many volcanoes still active. Minerals: gold, silver, copper, iron, manganese, tin, sodium nitrate, borax and quick-silver. Richest metal-bearing states are Peru and Bolivia; latter contains famous silver mountain of Potosí. In 1910 the Andes were tunnelled and Argentinian and Chilean railways linked up by Uspallata tunnel (*c*. 2 m. (3 km.) long; alt. 10,500 ft. (3,200 metres) a.s.l.).

Andhra Pradesh India. 15 0N 80 0E. State of S India bounded E by the Bay of Bengal and W by Maharashtra and Karnataka. Area, 106,204 sq. m. (275,068 sq. km.). Chief towns, Hyderabad (cap.), Eluru, Guntur, Kakinada, Kurnool, Machilipatnam, Nellore, Nizamabad, Rajahmundry and Vishakapatnam. Deccan plateau on the W; the centre is crossed SW to NE by the E Ghats. About 23% of the state is forest, with *c*. 14m. hectares of cultivatable land. The main products are tobacco, castor, timber, rice, maize and wheat. There are deposits of coal, copper, mica, chrysolite, asbestos, limestone quarries and supplies of manganese, iron ore and barytes. Oil has also been found. The main indust. is at Vishakapatnam which has India's major shipbuilding yards, an oil refinery and iron and steel plants. Industries inc. machine tools, pharmaceuticals, heavy electrical machinery, fertilizers, electronics, aeronautical parts, cement, chemicals, glass, tobacco products, textiles, paper-making and sugar-milling as well as handicraft indust. Pop. (1981C) 53·50m.

Andizhan Uzbekistan, U.S.S.R. 40 48N 72 23E. Cap. of the Andizhan region, sit. ESE of Tashkent. Indust. cotton and food production. Pop. (1985E) 275,000.

Andorra 42 30N 1 30E. Small E Pyrenean semi-independent state sit. on the French/Spanish frontier, in a high valley at 3,000 ft. (914 metres) a.s.l., among mountain peaks rising to 10,000 ft. (3,048 metres). The political status of Andorra was regulated by the *Paréage* of 1278 which placed Andorra under joint suzerainty of the Comte de Foix and the Bishop of Urgel. Area, 180 sq. m. (465 sq. km.). There are 6 villages and tourism is the most important industry. Catalan is the language spoken. Cap. Andorra-la-Vieille. Pop. (1988E) 51,400.

Andover Hampshire, England. 51 13N 1 29W. Town on the R. Anton. Fast-growing commercial centre of an agric. district and London overspill town. Indust. light engineering, financial services and computing. Pop. (1985E) 32,000.

Andria Puglia, Italy. 41 13N 16 17E. Town sit. WNW of Bari. Produces wine, almonds and olive oil.

Andropov *see* **Rybinsk**

Andros Cyclades, Greece. 37 50N 24 50E. Island in Aegean Sea, the most N of the Cyclades. Area, 117 sq. m. (303 sq. km.). Cap. Andros. Mountainous with fertile valleys. Noted for wine.

Andújar Jaén, Spain. 38 3N 4 5W. Town sit. NW of Jaén on the Guadalquivir R. Manu. soap and textiles. Noted for producing *alcarrazas*, porous jars to keep water cool.

Aneityum Vanuatu, South West Pacific. Southernmost of the New Hebrides group. Area, 25 sq. m. (65 sq. km.). The island is volcanic. A source of Kauri pine.

Angara River South East Siberia, U.S.S.R. 58 06N 93 00E. Leaves L. Baikal at the SW end and flows for 1,300 m. (2,080 km.), at first NNW through a deep valley to Irkutsk, and then E to become a trib. of Yenisei R. 35 m. (56 km.) SSE of Yeniseysk.

Angarsk Russian Soviet Federal Socialist Republic, U.S.S.R. 52 34N 103 54E. Town sit. NW of Irkutsk on

Angara R. Manu. machinery, petro-chemicals and building materials. Pop. (1985E) 256,000.

Ångerman River Sweden. 62 48N 17 56E. Rises near Norwegian frontier and flows SSE for 280 m. (448 km.) to enter the Gulf of Bothnia near Härnösand.

Angers Maine-et-Loire, France. 47 28N 0 33E. Cap. of dept. on Maine R. Anc. cap. of Anjou with many fine medieval buildings. Manu. agric. machinery, textiles, footwear and wine. Pop. (1982C) 141,143 (agglomeration, 195,859).

Angkor Cambodia. 13 26N 103 50E. Anc. site sit. N of Siemreap near the N shore of L. Tonte Sap. Area, 40 sq. m. Consists of Khmer cap. of Angkor Thom, temple of Angkor Wat and extensive ruins of other Khmer buildings. Discovered in 1860 in thick jungle which was cleared after 1908.

Anglesey, Isle of　*see* **Ynys Môn**

Anglet Pyrénées-Atlantiques, France. Resort between Bayonne and Biarritz. Pop. (1982C) 30,364.

Angmagssalik Greenland. 65 40N 37 20W. Trading post and settlement just S of Arctic Circle on the E side of Angmagssalik Island.

Angola 12 0S 18 0E. Former Portuguese possession; became independent in 1975. Bounded by Zaïre on N and NE, Zambia on E, Botswana and Namibia on S and the Atlantic Ocean on the W. It has a coastline of over 1,000 m. (1,600 km.). There is an extensive mountain range in N forming watershed between Atlantic-flowing and Zaïre-flowing rivers. The S is a plateau ranging from 5,000–6,000 ft. (1,524–1,829 metres) a.s.l. Area, 481,351 sq. m. (1,246,700 sq. km.). Cap. Luanda. Chief towns, Benguela and Lobito. The principal crops are coffee, maize, sugar, palm-oil and palm kernels. Other products are cotton, wheat, tobacco, cocoa, sisal and wax. The country possesses valuable diamond deposits. Other minerals iron ore, petroleum and salt. Pop. (1988E) 9,387,000.

Angoulême Charente, France. 45 39N 0 09E. Cap. of dept. Ancient cap. of Angoumois. Sit. on a promontory above the Charente R. Manu. paper, pottery and brandy. Pop. (1982C) 50,151 (agglomeration, 103,632).

Angra do Heroísmo Azores, Atlantic. 38 39N 27 13W. (i) District of the central Azores, inc. Terceira, Graciosa and São Jorge Islands. Area, 271 sq. m. (703 sq. km.). (ii) Town sit. NW of Ponta Delgado, São Miguel Island. District cap. Episcopal seat. Indust. inc. fish canning, embroidering, tanning and tobacco processing. Seaport exporting wine, pineapples and dairy products. Pop. (1980E) 75,000.

Anguilla West Indies. 18 14N 63 5W. A dependent territory of U.K. Island sit. NNW of St. Kitts, part of the Leeward Islands group. Cap. The Valley. Area, 35 sq. m. (91 sq. km.). Produces sea-island cotton and salt. Pop. (1989E) 7,019

Angus Scotland. Now part of Tayside Region.

Anhult German Democratic Republic. Former state, now mainly in Halle, sit. on Elbe R. Cap. Dessau.

Anhui (Anhwei) China. 33 15N 116 15E. E province on both sides of the Yangtse R. Area, 54,015,005 sq. m. (139,900,000 sq. km.). Cap. Hefei, sit. WNW of Shanghai. Tea, rice and cotton are grown in the S. The N produces wheat, beans and millet. Industries coalmining and iron and steel works. Pop. (1982C) 49,665,724.

Anhwei　*see* **Anhui**

Aniene Italy. R. rising in the Apennines 10 m. (16 km.) SW of Avezzano and flowing 61 m. (98 km.) NW past Subiaco, then SW past Tivoli and W to the R. Tiber above Rome. The river is used for water supply and power. Below Tivoli it is also called the Teverone R.

Anjou France. Former province sit. in the Paris Basin. Now part of the Maine-et-Loire dept.

Anjouan Island   *see* **Nzani**

Ankara   Turkey. 39 55N 35 50E.
Cap. and second largest city of
Turkey. Ankara is sit. in the centre of
the Anatolian plateau. Formerly
known as Ancyra and Angora. The
city was important on the caravan
route from Istanbul to the E in the
days of the Ottoman Empire. The old
city was replanned by Kemal Atatürk.
It is an industrial and commercial
centre and is important for its breed of
goats which have long silky hair
known as Mohair. Industries inc.
engineering, chemicals, food processing and the manu. of cement, leather
goods and textiles. Pop. (1985E)
2,251,533.

Annaba   Annaba, Algeria. 36 54N
7 46E. Town and port formerly called
Bône. Manu. chemicals. Exports iron
ore, phosphates, wine and cork. Pop.
(1983E) 453,951.

Annan   Dumfries and Galloway
Region. Scotland. 54 59N 3 16W.
Town sit. ESE of Dumfries on the
Annan R. Indust. inc. engineering,
pharmaceutical, nuclear, textile and
food processing. Pop. (1984E) 8,500.

Annan River   Scotland. 54 59N
3 16W. Rises in Moffat Hills and
flows S for 49 m. (78 km.) to enter the
Solway Firth 2 m. (3 km.) S of Annan.

Annapolis   Maryland, U.S.A.
38 59N 76 30W. Cap. of Maryland
state sit. S of Baltimore on Severn R.
Industries inc. government services,
tourism and fish processing. Home of
the U.S. Naval Academy, founded in
1845. Pop. (1986E) 33,360.

Annapolis Royal   Nova Scotia,
Canada. 44 45N 65 31W. Port sit. on
the Annapolis Basin at mouth of Annapolis R., sit. W of Halifax. Tourism
is important. Pop. (1985E) 600.

Ann Arbor   Michigan, U.S.A.
42 18N 83 45W. City sit. W of Detroit.
Manu. car accessories, computers,
scientific instruments and cameras.
Pop. (1980C) 107,966.

Annecy   Haut-Savòie, France.
45 54N 6 07E. Cap. of dept. S of

Geneva, Switz., on N shore of L. Annecy. Indust. centre and tourist resort.
Manu. textiles, watches and paper.
Pop. (1982C) 51,593 (agglomeration,
112,632).

Annecy, Lake   Haute-Savoie,
France. 45 52N 4 40E. L. lying S of
Geneva, Switz. Area, 10 sq. m. (26
sq. km.). Popular tourist area.

Annobón   Equatorial Guinea. 1 25S
5 36E. Island (called Pagalu 1973–79)
sit. SW of Bata. Area, 7 sq. m. (17 sq.
km.). Portuguese creole spoken. Pop.
(1984E) 3,000.

Ansbach   Bavaria, Federal Republic
of Germany. 49 17N 10 34E. Town
sit. WSW of Nuremberg (Nürnberg).
Manu. machinery, buttons and textiles. Pop. (1984E) 37,800.

Anshan   Liaoning, China. 41 3N
122 58E. Town sit. SW of Shenyang
(Mukden). China's leading producer
of iron and steel. Other manu. chemicals and cement. Pop. (1987C)
1,270,000.

Anstruther   Fife Region. Scotland.
Town on the Firth of Forth. Manu. oil-
skins and golf-clubs. Principal industry fishing. Pop. (1985E) 3,000.

Antalya   Antalya, Turkey. 36 53N
30 42E. Port and cap. of Antalya province sit. on the Gulf of Antalya. Industries canning and flour milling.
Trades in grain and timber. Pop.
(1980C) 173,501.

Antananarivo   Madagascar. 18 52S
47 30E. Cap. of rep. and province of
the same name, sit. in the central E
and on the interior plateau at 4,800 ft.
(1,463 metres) a.s.l. Industries inc.
manu. textiles, footwear, food processing. Pop. (1986E) 703,000.

Antarctica   Continent surrounding
the S Pole. Area, between 5m. and 6m.
sq. m. (13m. and 15m. sq. km.).
Roughly circular, with the Ross Sea
indenting the Pacific side and the
Weddell Sea the Atlantic side. The
central area is an icecap with ice sheet
extending over most of the land area;
the movement of ice barriers, glaciers
and shelves makes for continual alter-

ation of the coastline. Very severe climate, winter temps. dropping to –80°F. (–62°C.) and summer temps. only rising to 15°F. (–9°C.). Few forms of life survive the cold; birds and animals live mainly on coastline, inc. penguins, whales and seals. Divided between Norway, Australia, France, New Zealand, Great Britain. Chile and Argentina have overlapping claims. Uninhabited, except for visiting survey teams.

Antibes Alpes-Maritimes, France. 43 35N 7 07E. Port and resort sit. SW of Nice on French Riviera, inc. Juan-les-Pins. Manu. chocolates and perfumes. Trades in flowers, oranges and olives. Pop. (1982C) 63,248.

Anticosti Island Quebec, Canada. 49 20N 62 40W. Sit. in Gulf of St. Lawrence. Area, 135 m. (216 km.) long and a maximum of 30 m. (48 km.) broad at the centre. Mainly forested. Chief occupation, lumbering.

Antifer Seine-Maritime, France. 49 41N 0 10E. Oil terminal sit. N of Le Havre. It is equipped to take 500,000-ton tankers.

Antigua Sacatepéquez, Guatemala. 14 34N 90 41W. Anc. cap. sit. W of Guatemala City. Now cap. of Sacatepéquez dept. and centre of a coffee-growing area. Pop. (1984E) 30,000.

Antigua and Barbuda West Indies. 17 0N 61 50W. Independent state. Island 45 m. (72 km.) N of Guadeloupe, part of the Leeward Islands group. Area, 108 sq. m. (280 sq. km.). Cap. and chief port St. John's Mainly dry and fertile, producing sea-island cotton and sugar-cane. The chief exports are sugar, molasses, cotton and rum. Pop. (1986E) 81,500.

Antilles West Indies. 19 00N 70 00W. Main chain of Caribbean islands extending *c.* 2,500 m. (4,000 km.) from Florida to the N coast of Venezuela between the Atlantic and the Caribbean and Gulf of Mexico. The Greater Antilles consist of Cuba, Jamaica, Hispaniola (Haiti and the Dominican Rep.) and Puerto Rico. The Lesser Antilles consist of the Virgin Islands, Leeward Islands, Wind-

ward Islands, Barbados, Trinidad and Tobago, Curaçao and Margarita.

Antioch Hatay, Turkey. 36 14N 36 07E. City sit. W of Aleppo, Syria, on the Orontes R. 20 m. (32 km.) from the Mediterranean coast. An anc. and important city of which the present city occupies a fraction of the site. Provincial cap. and commercial centre of an agric. area producing cotton, olives and grain. Indust. inc. food processing. Pop. (1980E) 94,942.

Antioquia Colombia. 7 00N 75 30W. Dept. of NW central Colombia. Area, 25,409 sq. m. (66,809 sq. km.). Cap. Medellín. Mountain valley extending along Cauca R. between the Cordillera central and the Cordillera occidental. Fertile, and a commercially developed area. Gold-mining, coffee-growing, textile milling, petroleum and timber are important. Other products inc. sugarcane, cotton, cereals, tobacco, cacao, fruit, livestock, rubber, vanilla, medicinal plants, iron, coal, silver, platinum and copper. Medellín is an indust. city with textile, metal-working and food processing indust. It is linked by rail with Puerto Berrío on the Magdalena R. Pop. (1985C) 4,055,064.

Antipodes Islands New Zealand. 49 30S 177 30E. A small group of uninhabited rocky islands in the S. Pacific Ocean about 450 m. (720 km.) SE of S. Island.

Antisana Ecuador. Volcano sit. in the Andes 30 m. (48 km.) ESE of Quito. 18,885 ft. (5,756 metres) a.s.l.

Antofagasta Antofagasta, Chile. 23 39S 70 24W. Cap. of Antofagasta province and port sit. N of Valparaiso in N Chile. Indust. ore refining, canning and brewing. Exports inc. copper and nitrates. Pop. (1987E) 204,577.

Antrim Northern Ireland. 54 55N 6 10W. (i) Maritime county in the NE. Area, 1,175 sq. m. (3,043 sq. km.). County town Belfast. Principal R., the Bann, the Main and the Bush. Mainly a basalt plateau, hilly in the N and E, and sloping gradually inland to Lough Neagh. The famous Giant's Causeway is on the N coast. There are peat bogs in the interior and coalfields near Ballycastle. Lignite is also found. Antrim is the centre of the Irish linen industry. Other industries are textiles, cattle and sheep rearing, and ship-building at Belfast. Principal agric. products are potatoes, flax and oats. Pop. (1988E) 47,700. (ii) Town in County Antrim sit. NW of Belfast on the NE shore of Lough Neagh. 54 43N 6 13W. Industry linen mills. Pop. (1981C) 22,342.

Antwerp Belgium. 51 13N 4 25E. (i) Province of N Belgium bounded N by the Netherlands. Area, 1,104 sq. m. (2,859 sq. km.). Fertile plain drained by Scheldt, Dyle, Nèthe and Rupel Rs. and the Albert Canal. Chief towns Antwerp (cap.), Mechlin, Lierre, Turnhout, Mol. Mainly industrial E, agric. W. Pop. (1988E) 1,587,500. (ii) City sit. N of Brussels on Scheldt R. and the Scheldt-Meuse Junction Canal. Provincial cap. Major port, commercial and indust. centre, important trading centre for diamonds. Indust. inc. oil refining, vehicle assembly, chemicals, telecommunications. Port handles 96m. tonnes cargo each year. Exports, steel chemicals, metal products, foodstuffs; imports, oil, fertilizer, timber, fruit, ores and coal. There is considerable trade with the Federal Republic of Germany, esp. the Ruhr. The seaport consists mainly of basins off the Scheldt R. It is of exceptional artistic interest, being the birthplace of Van Dyck and the seat of the academy founded by Philip the Good in 1454 which was the basis of the Flemish school; also the home and work place of Rubens. Pop. (1989E) 474,000.

Anuradhapura North-Central Province, Sri Lanka. 8 20N 80 25E. Anc. cap. from 4th cent. B.C. to 8th cent. A.D. and now cap. of the North-Central province sit. NNE of Colombo. Centre for Buddhist pilgrims and containing the famous Bo-tree of Gautama. Major road junction and sit. on railway line. Pop. (1981C) 36,248.

Anvers *see* **Antwerp**

Anzhero Sudzhensk Russian Soviet Federal Socialist Republic,

U.S.S.R. 56 10N 83 40E. Town sit. NNW of Kemerovo in the Kuznetsk basin. An important coalmining centre. Manu. mining equipment and by-products of coal.

Anzio Lazio, Italy. 41 28N 12 37E. Port sit. S of Rome on Tyrrhenian Sea. Main occupation fishing and tourism. Birthplace of Nero. Roman Emperor, A.D. 54–68. Pop. (1984E) 25,000.

Anzoategui Venezuela. State of NE Venezuela on the Caribbean coast. Area, 16,720 sq. m. (43,305 sq. km.) Chief towns Barcelona (cap.), Guanta, Puerto La Cruz. Almost all of it is 'Ilanos' watered by the Unare, Neverí and Orinoco Rs. and has a tropical climate. There are rich mineral resources; petroleum at El Tigre, El Roble, San Joaquín, Santa Ana and coal at Naricaul. The main occupation apart from oil-extraction and mining is cattle grazing; other farm products are cotton, cacao, coffee, sugar-cane, maize, coconuts, bananas, yucca, rice, tobacco. Industries are associated with farming and are centred on Barcelona. The main ports are Guanta and Puerto la Cruz. Pop. (1980E) 634,515.

Aoba Vanuatu, South West Pacific. 17 30S 168 15E. Volcanic island sit. E of Espiritu Santo. Area, 95 sq. m. (246 sq. km.). The main product is copra.

Aomori Honshu, Japan. 40 50N 143 43E. Port and cap. of Aomori prefecture sit. N of Tokyo on Mutsu Bay. Trades in fish, rice and timber. There is a distribution complex for biochemistry and high technology. Pop. (1985E) 295,000.

Aosta Valle d'Aosta, Italy. 45 43N 7 19E. Cap. of Valle d'Aosta Region on Dora Baltea R., and S of the Great St. Bernard P. Industries metal-working and tourism. Pop. (1989E) 36,651.

Apeldoorn Gelderland, Netherlands. 52 13N 5 57E. Town sit. N of Arnhem. Industries inc. electronics and tourism. Manu. paper and blankets. Pop. (1989E) 146,300.

Apennines Italy. 41 0N 15 0E. Mountain range extending S from the Alps through Italy and Sicily for *c.*

800 m. (1,280 km.). The average height is *c.* 4,000 ft. (1,219 metres) a.s.l. Gran Sasso d'Italia is highest peak at 9,560 ft. (2,914 metres). Vesuvius, near Naples, is an active volcano. The range is crossed by several railways. R. rising in the range inc. Arno and Tiber. Marble is found at Carrara.

Apia Upolu, Western Samoa. 13 48S 171 45W. Cap. and port sit. on the N coast of Upolu. Exports copra, cocoa beans and bananas. Robert Louis Stevenson (Scot. novelist, 1850–94) was buried on Mount Vaea. Pop. (1981E) 33,170.

Apolima Western Samoa, South Pacific. 13 49S 172 07W. Island sit. in a strait between Upolu and Savaii.

Appalachian Mountains U.S.A. 41 00N 77 00W. Mountain system extending over 1,500 m. (2,400 km.) from N to S along the Atlantic coast. The range is divided in two groups by the Hudson R., L. Champlain and the Richelieu R. These are the Green Mountains and White Mountains to the N and Allegheny Mountains and Blue Mountains to the S. The highest peak is Mount Mitchell at 6,684 ft. (2,037 metres). The mountains are rich in coal, iron, petroleum and natural gas.

Appenzell Switzerland. 47 20N 9 20E. Canton in NE. Divided in 1597 into two half-cantons—Inner Rhoden (cap. Appenzell) which is pastoral; and Outer Rhoden (cap. Trogen) which is industrial. Embroidery, cotton and silk manu. Area of Inner Rhoden 72 sq. m. (186 sq. km.); of Outer Rhoden 93 sq. m. (241 sq. km.). Pop. (1980C) 60,455.

Appleby in Westmorland Cumbria, England. 54 36N 2 29W. Town sit. SE of Carlisle on R. Eden. Industries inc. creamery, toiletries and furniture manu. Pop. (1989E) 2,700.

Appleton Wisconsin, U.S.A. 44 16N 88 25W. City sit. NNW of Milwaukee on Fox R. Manu. paper, knitted goods, and food products. Insurance, banking and education important activities. Pop. (1989E) 64,500.

Apra Guam, South Pacific. 13 27N 144 38E. Town sit. N of Agana, the only good harbour on Guam. Port of entry and U.S. naval base.

Apulia *see* **Puglia**

Apuré River Venezuela. 7 37N 66 25W. Rises in the E. Cordillera of Colombia and flows E for *c.* 350 m. (560 km.) to join the Orinoco R.

Apurímac River Peru. 12 17S 73 56W. Rises in L. Villafra in the Andes and flows generally NNW for 550 m. (880 km.) and joins the Urubamba R. to form the Ucayali R., which in turn joins the R. Amazon.

Aqaba Jordan. 29 31N 35 0E. Port sit. at N end of the Gulf of Aqaba. Jordan's only seaport. After the 1948 Arab-Israeli war Aqaba was developed because Jordan could no longer use ports in Palestine. In 1960 it was linked by modern road to Amman.

Aquila Abruzzi e Molise, Italy. 42 22N 13 24E. Cap. of Abruzzi e Molise Region and of Aquila province, sit. NE of Rome on the Pescara R. Industries inc. agric. and livestock rearing, chemicals, oil refining. Manu. textiles and macaroni. Pop. (1981E) 64,000.

Aquitaine France. Region in SW France comprising the departments of Dordogne, Gironde, Lot-et-Garonne and Pyrénées-Atlantiques it is bounded by the Pyrenees in the S and the Bay of Biscay in the W and is drained by the Garonne R. and its trib. Area, 15,988 sq. m. (41,408 sq. km.). Cap. Bordeaux. It is an extremely fertile area producing fruit, maize, vegetables, wheat and wine. Pop. (1982C) 2,656,544.

Arabia South West Asia. 25 00N 45 00E. Large peninsula bounded N by the Syrian Desert, E by the (Persian) Gulf and Gulf of Oman, S by the Arabian Sea and Gulf of Aden, W by the Red Sea. Area, 1m. sq. m. (2,590,000 sq. km.). divided politically into Saudi Arabia, occupying two-thirds, Yemen, Southern Yemen, Oman, the United Arab Emirates, Qatar and Kuwait. Rocky plateau rising to mountain ridges in the s and w and sloping gradually E to the Persian Gulf. The interior is alternately steppe and desert, inc. the Natud (N), Dahana (E) and Rub' Al Khali (S) deserts. Dissected by deep wadis, but there are now no permanent Rs. The s highlands receive enough rain (20–40 ins. or 50–100 cm. annually) for agric. mainly coffee and cereals; otherwise cultivation is confined to oases. But by far the most important natural resource is petroleum.

Arabian Desert   Egypt. 28 00N 32 30E. Mountainous rocky desert bounded by R. Nile in the w and the Gulf of Suez and the Red Sea in the E. There are peaks of over 6,000 ft. (1,829 metres) a.s.l. near the coast.

Arabian Sea   15 00N 65 00E. The NW part of the Indian Ocean between India and the 'horn' of Africa. It has two branches, the Gulf of Aden leading to the Red Sea, and the Gulf of Oman leading to the (Persian) Gulf.

Aracajú   Sergipe, Brazil. 10 55S 37 4W. Port and cap. of Sergipe state sit. on E coast 6 m. (10 km.) above mouth of Cotinguiba R. Indust. sugar-refining, cotton mills and tanning. Exports sugar, cotton, hides, rice and coffee. Pop. (1980C) 287,934.

Arad   Arad, Romania. 46 11N 21 20E. Cap. of province of same name sit. NW of Bucharest on Mures R. Railway centre. Indust. railway and textile engineering, tanning and distilling. Pop. (1985E) 185,892.

Arafura Sea   A part of the SW Pacific Ocean between Australia and New Guinea. The Timor Sea and the Torres Strait are adjoining.

Aragón   Spain. 41 00N 1 00W. Anc. kingdom of Spain, now forms provinces of Huesca, Zaragoza and Teruel. Area, 18,382 sq. m. (47,609 sq. km.). Chief town, Zaragoza.

Aragón River   Spain. 42 13N 1 44W. Rises in the Central Pyrenees and flows SW for 80 m. (128 km.) to join the Ebro R.

Aragua   Venezuela. State of N

Venezuela on the Caribbean. Area, 2,160 sq. m. (5,594 sq. km.). Cap. Maracay. Mountainous region in the coastal range with fertile valleys and a tropical climate. Part of an important agric. region, producing cattle, coffee, cacao, sugar-cane, maize, tobacco, rice, bananas, coconuts. Pop. (1980E) 854,121.

Araguaia River   5 21S 48 41W. Brazil. Trib. of R. Tocantins.

Arak   Iran. 34 05N 49 41E. Cap. of Markazi province sit. SW of Tehrán on the Trans-Iranian Railway. Manu. carpets, rugs and matches. Pop. (1976C) 114,507.

Arakan   Burma. 19 00N 94 00E. A division of Lower Burma, sit. on the Bay of Bengal between Pegu and Chittagong. A narrow strip approx. 400 m. (640 km.) long and 15–90 m. (24–144 km.) wide. Chief town Akyab, sit. SSE of Chittagong. A coastal strip growing rice, cotton, fruit and tobacco is bordered by the Arakan Yoma range rising to over 10,000 ft. (3,048 metres) a.s.l.

Arakan Yoma   Burma. 19 00N 94 40E. A mountain range 400 m. (640 km.) long extending from Chin Hills in N to Irrawaddy R. delta in S. A climatic barrier with tropical rain forests on the w slopes and teak forests on the E side.

Aral Sea   U.S.S.R. 44 30N 66 00E. 'Sea of Islands'. Large inland sea separated in w from Caspian Sea by the Ust Urt plateau. Area, 24,000 sq. m. (62,160 sq. km.). It has no outlet but is fed by R. Syr Darya and R. Amu Darya. Fishing, particularly sturgeon, carp and herring, is important. The water is brackish and generally shallow.

Aran Islands   Galway, Ireland. 53 05N 9 42W. Chief islands, Inishmore, Inishman and Inisheer. Area, 18 sq. m. (47 sq. km.). Fishing main occupation. The islands are of antiquarian and archaeological interest.

Aranca   Venezuela/Colombia. R. rising E of Bucaramanga in N central Colombia and flowing 500 m. (800 km.) E along Colombia/Venezuela border to enter Venezuela below Arauca and join the Orinoco R. 70 m. (112 km.) ESE of San Fernando de Apure. It is navigable for small boats.

Araq   *see* **Arak**

Ararat, Mount   Turkey. 39 42N 44 18E. A volcanic mountain mass rising from the Armenian plateau in the NE. There are two main peaks, Great Ararat, 16,916 ft. (5,156 metres) a.s.l., the supposed resting place of Noah's Ark, and Little Ararat, 12,843 ft. (3,915 metres) a.s.l.

Arauca   Colombia. 7 00N 70 40W. (i) District of E Colombia, bounded N and E by Venezuela. Area, 9,973 sq. m. (25,830 sq. km.). Mainly 'Ilano' grassland, sparsely populated. Main occupation farming, with some lumbering. Pop. (1985E) 17,000. (ii) Town sit. NE of Bogotá on the Aranca R. opposite El Amparo, Venezuela. Trading centre for an agric. area, handling corn, cacao, sugar-cane, rice, cattle, furs, hides, rubber and resins. Airport and custom house.

Aravalli Range   Rajasthan, India. A range of hills in NW India running SW to NE for 350 m. (560 km.) at 1,500–3,000 ft. (457–914 metres) a.s.l.

Arbil   Arbil, Iraq. 30 15N 44 05E. Town sit. E of Mosul at the head of the Kirkuk railway. Provincial cap. Trading centre for a farming and oil producing area, linked by road with Iran, Turkey and Syria. An anc. city continuously inhabited since Assyrian times.

Arbroath   Tayside Region. Scotland. 56 34N 2 35W. Port and E coast resort at mouth of Brothock Water. Indust. engineering, sail-making and knitwear manu., fishing and farming. Arbroath is noted for its smoked haddock. Pop. (1981C) 24,119.

Arcachon   Gironde, France. 44 40N 1 10W. Port and resort on shore of Arcachon Basin, near its channel into Bay of Biscay. Noted for oysters. Pop. (1982C) 13,664.

Arcadia   Greece. Prefecture sit. in the central Peloponnessos. Area, 1,706 sq. m. (4,419 sq. km.). Cap. Tripolis, sit. 44 m. (70 km.) SW of Corinth. Mainly mountainous but some crops grown inc. wheat, grapes and tobacco. Pop. (1981C) 107,932.

Archangel   *see* **Arkhangelsk**

Arctic Regions   Area around N. Pole and within Arctic circle (66 30N lat.), but better defined as polar area beyond limit of tree growth; inc. the Arctic Ocean, practically landlocked by Eurasia, N. America and Greenland. The Arctic Ocean, area 5.4m. sq. m. (14m. sq. km.), which surrounds N. Pole, is deepest in centre (over 2,000 fathoms), but shallow over a wide continental shelf where many islands occur. It communicates with the Atlantic Ocean between Greenland and Norway, where it is known as the Greenland Sea (2,000 fathoms), and between Greenland and Baffin Island by shallow Davis Strait; and with the Pacific Ocean through Bering Strait between Asia and N. America. Principal islands are Spitsbergen, Franz Josef Land and Novaya Zemlya, which with Norway bound the shallow Barents Sea; to E of Novaya Zemlya is Kara Sea (100 fathoms); farther E are New Siberian Islands and Wrangel Island; off N. America is a vast arctic arch., forming N part of Canada. Arctic Ocean is largely ice-covered with much of it unexplored and salinity low. Whale, seal, walrus and other fisheries, esp. in shallow parts, are found. The arctic lands are practically uninhabited except by nomadic Lapps, Samoyedes and Chukchees of Eurasia and Eskimos of N. America, who live on reindeer, musk-ox, walrus, and other arctic animals and fish. Within these regions lay the historic routes of search for NE and NW Passages to China, Japan and India: NE Passage found by Nordenskiöld, 1879; NW passage discovered by Sir John Franklin, 1845–6, but first traversed by Amundsen, 1906; during search for NW Passage Ross discovered N. Magnetic Pole (70 30N lat., 97W long.), 1831.

Ardebil   Azerbaijan, Iran. 38 15N 48 18E. Town sit. E of Tabriz. Manu. carpets and rugs. Pop. (1983E) 221,970.

Ardèche France. 44 42N 4 16E. Dept. in Rhône-Alpes region. Bounded on E by Rhône R. and having Vivarais Mountains in w. Area, 2,132 sq. m. (5,523 sq. km.). Principal R. is the Ardèche. Chief towns are Annonay, which manu. silk and paper, and Privas (the cap.) sit. 50 m. (80 km.) s of St. Etienne, and noted for *marrons glacés*. Agric. products inc. cereals, chestnuts, mushrooms, olives and wine. Coal and iron ore are mined. Pop. (1982c) 267,970.

Ardèche River France. 44 16N 4 39E. Rises in the Cévennes and flows SE for 70 m. (112 km.) to enter the Rhône R.

Arden, Forest of England. District which was formerly part of a forest covering an area of the Midlands between the R. Avon and Birmingham. Possibly the setting of Shakespeare's *As You Like It.*

Ardennes Belgium/France. A forested region comprising SE Belgium, part of Luxembourg and the Ardennes dept. of France. The highest peaks are about 2,000 ft. (610 metres) a.s.l. Industries mining and pasturage.

Ardennes France. 49 35N 4 40E. Dept in Champagne-Ardennes region, bordering on Belgium. Area, 2,015 sq. m. (5,219 sq. km.). Cap. Charleville-Mégières. Principal R. are the Aisne and the Meuse. Industries lumbering, stock rearing, slate quarrying and iron mining. Pop. (1982c) 302,338.

Ardnamurchan Point Headland in Highland Region, Scotland. 56 44N 6 14E. The most w point on the British mainland. Lighthouse built in 1849 which is visible for 29 km.

Ardrossan Strathclyde Region, Scotland. 55 39N 4 49W. Port and resort sit. NNW of Ayr on the Firth of Clyde. Indust. shipbuilding and oil-refining. Pop. (1981c) 11,421.

Arequipa Arequipa, Peru. 16 25S 71 32W. Cap. of dept. of same name, sit. between L. Titicaca and the coast, at the base of the extinct volcano El Misti, 19,200 ft. (5,852 metres) a.s.l. The second city of Peru and an important commercial centre. Manu. textiles and soap. Built on the site of an anc. Inca city. Pop. (1981c) 447,431.

Arezzo Toscana, Italy. 43 28N 11 53E. Cap. of Arezzo province sit. near confluence of Arno R. and Chiana R. in central Italy. Built on the site of an anc. Etruscan settlement, with many fine medieval buildings. Now a commercial centre trading in agric. produce, wine and olive oil. Manu. textiles, leather, furniture and pottery. Pop. (1981E) 92,000.

Argenteuil Val-d'Oise, France. 48 57N 2 15E. A NW suburb of Paris on the Seine R. Indust. market gardening and the manu. of aircraft engines, car parts, textiles and electrical equipment. Pop. (1982c) 96,045.

Argentina 35 00S 66 00W. Federal Rep. and second largest country of S. America. It is bounded on N by Bolivia, NE by Paraguay, E by Brazil, Uruguay and the Atlantic Ocean, w by Chile. Length *c.* 2,000 m. (3,200 km.) and breadth *c.* 800 m. (1,280 km.). The surface slopes gradually from the Andes on w towards E and consists mainly of great plains. The N part of the plains, the Gran Chaco, is densely wooded, the central pampas portion has vast areas of treeless pasture and s portion, Patagonia, contains expanses of stony desert with patches of stunted thorn bush. The Paraná R. and Paraguay R. drain the central area. Area, 1,072,515 sq. m. (2,777,815 sq. km.). Cap. Buenos Aires, and other large towns, Rosario, Córdoba, La Plata and Mar del Plata. Argentina's wealth is based on agric. and livestock. Argentina has over 50m. head of cattle and is the world's chief meat exporter. Wool production is important and crops inc. wheat, linseed, maize, oats, barley, rye, sunflower-seed and sugarcane. Cotton, potatoes, vine, tobacco, citrus fruit, olives, rice, soya and yerba maté (Paraguayan tea) are also cultivated. Mining is mainly of local importance. Among minerals exploited are coal, gold, silver, copper, iron-ore, tungsten, beryllium, mica, lead, barites, zinc, manganese and limestone. Manu. inc. cotton yarn, cement, pig iron and steel. Pop. (1989E) 31·93m.

Arges Romania. 41 10N 26 45E. Province of s central Romania, NW of Bucharest. Area, 2,626 sq. m. (6,801 sq. km.). Cap. Pitesti. Drained by Arges R. Mainly agric. esp. fruit and vine growing. Pop. (1982E) 659,289.

Argolis Greece. 37 38N 22 50E. *Nome* in the NE Peloponnessos inc. the islands in the Gulf of Argolos. Area, 855 sq. m. (2,214 sq. km.). Cap. Nauplion, sit. WSW of Athens. Produces wine, sultanas and currants. Pop. (1981C) 93,020.

Argolis, Gulf of Greece. 37 33N 22 45E. An inlet of the Aegean Sea in the E Peloponnessos. Spetsai Island is sit. at the mouth.

Argonne France. 49 00N 5 20E. Forested area in NE France inc. parts of the depts. of Meuse, Marne and Ardennes.

Argos Greece. 37 39N 22 44E. Town sit. SSW of Corinth in the NE Peloponnessos. One of the oldest towns in Greece.

Argostolion Ionian Islands, Greece. 38 10N 20 30E. Cap. of Kefallenia prefecture sit. on the SW coast. Trades in wine, olive oil and currants. It has a stream which flows *from* the sea. Pop. (1981C) 6,788.

Argun River China/U.S.S.R. 53 20N 121 28E. Trib. of the Amur R. Rises in the Great Khingan Mountains in NE China, and flows W past Hilar to the U.S.S.R. frontier. Total length, 900 m. (1,440 km.). Forms part of the frontier after linking with L. Hulun Nor and joins R. Shilka to form Amur R.

Argyllshire Scotland. Now part of Highland Region.

Arica Tarapacá, Chile. 18 29S 70 20W. Port just S of border with Peru in a completely rainless area. Railway and oil terminal. About 50% of Bolivia's foreign trade passes through Arica. Pop. (1982E) 120,046.

Ariège France. 42 56N 1 30E. Dept. in Midi-Pyrénées region, comprising parts of the anc. provinces of Languedoc and Gascony and the county of Foix. Area, 1,888 sq. m. (4,890 sq. km.). Cap. Foix. Principal R. are Ariège and Salat. The N Pyrenees cover much of the S while the fertile N produces cereals, vegetables, fruits and wines. Iron, manganese, bauxite and zinc are mined. Indust. inc. chemicals, textiles, paper and metals. Pop. (1982C) 135,725.

Ariège River France. 43 31N 1 32E. Rises in E Pyrenees near the Andorra border, flows NNW for 100 m. (160 km.) to join the Garonne near Toulouse.

Arizona U.S.A. State of the SW U.S.A. 34 20N 111 30W. Bounded S by Mexico, W by Nevada and California. Area, 113,580 sq. m. (290,765 sq. km.). Chief cities Phoenix (cap.), Mesa, Glendale, Temple, Tucson, Scotdale. Settled from 1752, a Territory in 1863 and a State in 1912. High plateau in drainage basin of Colorado R., with chasms, notably the Grand Canyon of the Colorado R., mountain ridges and desert plains. About 73% of the land area is owned by the Federal Govt., inc. Indian land and forest land; the main cultivated land is irrigated by the Salt, Gila, Santa Cruz, Verde and Agua Fria Rs. Main products are cotton, cereals, fruit, vegetables, timber. There are considerable mineral deposits, inc. copper (about one-third of U.S. output), gold, silver, lead, zinc. The main industries are smelting and refining ores, lumbering, meat packing, food processing, cotton ginning, flour milling, tanning. Tourism is important, esp. for winter resorts. Pop. (1986E) 3,469,000.

Arkansas U.S.A. 35 0N 92 30W. State bordered by Missouri in N, by R. Mississippi in E, by Louisiana in S and by Texas and Oklahoma in W. Area, 53,104 sq. m. (137,539 sq. km.). Cap. Little Rock, and chief towns Fort Smith, North Little Rock and Pine Bluff. Settled in 1686, a Territory in 1819 and a State in 1836. Arkansas is an agric. state. Land erosion is serious. Largest sources of income are soybeans, poultry, cotton, cattle, rice and eggs. Mineral wealth inc. petroleum, natural gas and coal. Arkansas produces about 90% of the U.S.A.'s bauxite. Manu. inc. food products, electronic equipment and wood products. Pop. (1988E) 2,395,000.

Arkansas River U.S.A. 33 48N 91 04W. Rises in the Rocky mountains of central Colorado and flows for 1,500 m. (2,400 km.) ESE through Kansas, Oklahoma and Arkansas to the R. Mississippi. Its chief trib. is the Canadian R.

Arkhangelsk Russian Soviet Federal Socialist Republic, U.S.S.R. 64 34N 40 32E. (i) Region in the N adjacent to the Arctic Ocean. Area, 229,000 sq. m. (593,110 sq. km.). Mainly forested with lumbering and wood-processing indust. (ii) Cap. of region of same name sit. on the N. Dvina R. near the White Sea. Principal saw-milling and timber exporting centre of the U.S.S.R. Other indust. shipbuilding, fish canning and rope manu. The port is kept open from May to Nov. with the aid of icebreakers. Pop. (1985E) 408,000.

Arklow Wicklow, Ireland. 52 48N 6 09W. Port and resort on E coast at mouth of the Avoca R. Manu. pharmaceuticals, fertilizers and pottery. Indust., include boatbuilding and quarry products. Pop. (1985E) 8,750.

Arlberg Austria. 47 09N 10 12E. Mountain and pass sit. in the Alps in W Austria, between Federal States of Tirol and Vorarlberg, which are connected by road and rail.

Arles Bouches-du-Rhône, France. 43 41N 43 8E. Town sit. NE of Marseilles on Grand Rhône R. and Arles-Port-de-Bouc Canal. Commercial centre, near head of Rhône delta, trading in olives, wine, salt and livestock, esp. horses bred in the Camargue and Merino sheep. Indust. inc. boatbuilding, metal working, tourism, publishing, music records and refining sulphur; manu. newsprint, hats, foodstuffs. Noted for important Roman remains inc. arena, theatre and necropolis. Van Gogh (1853–90), painted many of his best-known pictures here. Pop. (1982C) 50,772.

Arlington Texas, U.S.A. 32 44N 97 07W. City sit. E of Fort Worth and 15 m. (19 km.) W of Dallas. Manu. aircraft, automobiles, semi-conductors and health service supplies. Pop. (1985E) 225,000.

Arlington Virginia, U.S.A. 42 25N 71 09W. County sit. on Potomac R. opposite Washington, D.C. The Arlington National Cemetery is here, also the National Airport and the Pentagon. Pop. (1980C) 152,599.

Arlon Luxembourg, Belgium. 49 41N 5 49E. Town sit. SE of Brussels. Provincial cap. Market town for a farming area. Indust. inc. manu. pipes and tobacco products, but many of the pop. work in neighbouring steel plants, inc. those in France and Luxembourg.

Armagh Northern Ireland. 54 21N 6 39W. (i) County in the S. Area, 512 sq. m. (1,326 sq. km.). Chief towns, Armagh, Lurgan and Portadown. Principal Rs. are Blackwater, Callan and Upper Bann. Potatoes and flax are grown and apples are an important crop. Indust. inc. linen manu. (ii) County town of county of same name. Chief indust. textiles manu. Pop. (1971C) 133,969.

Armagnac France. 43 44N 0 10E. Anc. district in the SW, now mainly in the Gers dept. Auch was formerly its cap. Noted for brandy.

Armavir Russian Soviet Federal Socialist Republic, U.S.S.R. 45 00N 41 08E. Town sit. E of Krasnodar on Kuban R. Manu. agric. machinery, food products and vegetable oils. Pop. (1985E) 168,000.

Armenia Quindío, Colombia. 4 31N 75 41W. Cap. of dept. sit. in the Central Cordillera. Centre of a coffee-growing region. Pop. (1980E) 184,000.

Armenian Soviet Socialist Republic U.S.S.R. 40 0N 41 10E. In 1920 Armenia was proclaimed a Soviet Socialist Rep. The Armenian Soviet Govt., with the Russian Soviet Govt., was a party to the Treaty of Kars (1921) which confirmed the Turkish possession of the former Govt. of Kars and of the Surmali District of the

Govt. of Yerevan. From 1922 to 1936 it formed part of the Transcaucasian Soviet Federal Socialist Republic. In 1936 Armenia was proclaimed a constituent Rep. of the U.S.S.R. Cap. Yerevan. Area 11,490 sq. m. (29,800 sq. km.). Mainly mountainous country but valley of the Araks R. yields cotton, orchards and vineries as well as sub-tropical plants. Mineral deposits inc. copper, zinc, aluminium and marble, therefore mining very important. Pop. (1989E) 3·3m.

Armentières   Nord, France. 50 41N 2 53E. Town sit. WNW of Lille on the Lys R. Manu. brewing, textiles, hosiery and copper goods. Pop. (1982C) 25,992.

Armidale   New South Wales, Australia. 30 31S 151 39E. Univ. and cathedral city in NE of State. Centre of a sheep rearing area and for tourism. Pop. (1985E) 21,500.

Arnhem   Gelderland, Netherlands. 51 59N 5 55E. Cap. of province on Rhine R. Manu. chemicals, textiles and food products. Pop. (1985E) 128,140.

Arnhem Land   Northern Territory, Australia. Reservation in N bordering on the Gulf of Carpentaria. Area, *c.* 31,200 sq. m. (*c.* 80,808 sq. km.). Mainly inhabited by aborigines. Pop. (1980E) 96,000.

Arno River   Italy. 43 31N 10 17E. Rises on Mount Falterona in the Apennines and flows W past Florence and Pisa to the Ligurian Sea.

Arnsberg   North Rhine-Westphalia, Federal Republic of Germany. 51 24N 8 02E. Town sit. SSE of Münster. Industries inc. metal products and electrical equipment. Pop. (1987E) 74,091.

Arnstadt   Erfurt, German Democratic Republic. 50 50N 10 57E. Town sit. SSW of Erfurt on Gera R. Manu. machinery, radio equipment, gloves and footwear. Pop. (1989E) 30,400.

Arran   Strathclyde Region, Scotland. 55 35N 5 15W. Island in the Firth of Clyde sit. between Ayrshire and the Mull of Kintyre. Area, 166 sq. m. (430 sq. km.). Mountainous with many anc. relics and abundant wild life. Tourist centre. Pop. (1988E) 4,007.

Arras   Pas-de-Calais, France. 50 17N 2 47E. Cap. of dept. on Scarpe R. formerly famous for tapestry, hence the English word. Manu. agric. implements, hosiery, vegetable oils, beet sugar and metal goods. Pop. (1982C) 45,364.

Arsi   Ethiopia. Region in central S of country. Area, 9,073 sq. m. (23,500 sq. km.). Cap Asela. Pop. (1984C) 1,662,233.

Arta   Greece. 39 09N 20 59E. (i) *Nome* in Epirus, bounded S by Gulf of Ara, N by Tzoumerka mountains. Area, 672 sq. m. (1,740 sq. km.). Mainly agric., producing cereals, cotton and fruit. There is a fishing industry on the Gulf of Arta. Pop. (1981C) 80,044. (ii) Town sit. SSE of Ioannina on Arachthos R. near its mouth. *Nome* cap. Trading centre for an agric. area, handling cotton, cereals, fruit, olives, tobacco and fish. Indust. inc. textiles; manu. leather products. Seat of Greek Orthodox metropolitan bishop. There are important Byzantine remains. Pop. (1981C) 18,283.

Artaxata   Armenian Soviet Socialist Republic, U.S.S.R. (Otherwise Artashat). Town sit. S of Erivan in the Aras R. valley, in a cotton, wheat and wine producing area. Indust. inc. distilling and wine making. Formerly called Kamarlu; renamed in 1945 after the anc. Armenian cap. whose ruins are nearby.

Artemovsk   Ukraine, U.S.S.R. 48 35N 37 55E. Town sit. NNE of Donetsk in the Donbas. The largest centre of salt-mining in the U.S.S.R. Manu. iron and glass.

Artibonite   Haiti. 19 15N 72 46W. Dept. of W Haiti on NE shore of Gulf of Gonaïves. Area, 2,600 sq. m. (6,734 sq. km.). Cap. Gonaïves. Fertile plain producing coffee, bananas, cotton, sugar-cane, rice. Drained by Artibonite R. Pop. (1977E) 748,247.

Artigas   Uruguay. 30 24S 56 28W.

(i) Dept. of NW Uruguay bounded N and NE by Brazil, W by Argentina. Area, 4,393 sq. m. (11,378 sq. km.). Main towns Artigas (cap.), Bella Unión, Tomás Gomensoro. Crossed SE to NW by the Cuchilla de Belén range. Drained by Tres Cruces Grande and Cuaró Grande Rs. Mainly a stock raising and agric. area. Some quartz and amethyst deposits near the Catalán Grande R. Pop. (1975E) 57,528. (ii) Town sit. NE of Salto on the Cuareim R. opposite Quaraí, Brazil. Dept. cap. Rail terminus and airport, trading centre for cattle, wool, cereals and fruit. Pop. (1984E) 14,000.

Artvin Coruh, Turkey. 41 11N 41 49E. Town sit. NNE of Erzurum and S of the U.S.S.R. border. Provincial cap. on Coruh R. Trading centre for grain and tobacco. Pop. (1980E) 229,000.

Aruba 12 30N 70 00W. Island sit. W of Curaçao and near N coast of Venezuela. Area, 74 sq. m. (193 sq. km.). Independence was achieved in 1986. Chief town, Oranjestad. Indust. inc. petrochemicals, rum distilling and tourism. Pop. (1988E) 62,500.

Aru Islands South Moluccas, Indonesia. 6 00S 134 30E. A group of islands in the Arafura Sea sit. between New Guinea and the Tanimbar Islands. Area, 3,306 sq. m. (8,562 sq. km.). Chief town, Dobo. Indust. fishing for pearls and sea-slugs.

Arunachal Pradesh India. 28 00N 95 00E. In 1972 the former NE Frontier Agency of Assam was created a Union Territory and attained statehood in 1987. The state inc. the Kameng, Tirap, Subansiri, Siang and Lohit frontier divisions and has an area of 32,333 sq. m. (83,743 sq. km.). Capital, Itanagar sit. NE of Calcutta. About 60% of the land area is forest. In 1987 there were 196,131 hectares under cultivation, 50,455 hectares of it irrigated. Rice is the principal crop. There are reserves of coal, oil, dolomite and limestone. Pop. (1981C) 631,839.

Arundel West Sussex, England. 51 51N 0 34W. Town sit. WNW of Worthing on Arun R. and sit. below Arundel Castle. Pop. (1980E) 2,500.

Arusha Arusha, Tanzania. 3 21S 36 40E. Cap. of region of same name in the N sit. to SW of Mount Meru. Centre of a coffee-growing region. Pop. (1978C) 55,281.

Arvada Colorado, U.S.A. 39 48N 105 05W. City immediately NW of Denver. Industries inc. food processing and chemicals. Pop. (1980C) 84,576.

Arve River France/Switzerland. 46 12N 6 08E. Rises in Savoy Alps, flows first SW and then NW for 62 m. (99 km.) to join Rhône R. below L. Geneva.

Arvida Quebec, Canada. 48 26N 71 11W. Town sit. N of Quebec on Saguenay R. Centre of one of the world's largest aluminium smelting industries.

Asahikawa Hokkaido, Japan. 43 46N 142 22E. Town sit. NE of Hakodate. Manu. textiles, *saké* and wood products. Pop. (1989E) 364,000.

Asansol West Bengal, India. 23 40N 86 59E. Town sit. NW of Calcutta on Raniganj coalfield. Indust. railway engineering, iron and steel, chemicals, manu. bicycles and glass. Pop. (1981C) 187,039.

Ascension Island South Atlantic. 7 56S 14 25W. Volcanic island *c.* 700 m. (1,120 km.) NW of St. Helena. A Brit. possession now used mainly as a relay and cable station. Area, 34 sq. m. (88 sq. km.). Principal settlement, Georgetown. 10 acres have been cultivated for the production of fruit and vegetables. Limited grazing for sheep and cattle. An important breeding ground for turtles. Pop. (1988E) 1,007 excluding military personnel.

Aschaffenburg Bavaria, Federal Republic of Germany. 49 59N 9 09E. Town sit. ESE of Frankfurt on Main R. Also a R. port trading in coal, timber, paper and wine. Manu. scientific and optical instruments and textiles. Pop. (1985E) 59,643.

Aschersleben Halle, German Democratic Republic. 51 45N 11 27E. Town sit. NW of Halle. Indust. textiles, chemicals and steel. Potash and lignite mines are nearby. Pop. (1984E) 40,000.

Ascoli Piceno   Marche, Italy. 42 51N 13 34E. Cap. of province of same name, sit. 16 m. (26 km.) from Adriatic coast, on Tronto R. Manu. glass, pottery, textiles and macaroni. Pop. (1981E) 353,000.

Ascot   Berkshire, England. 51 25N 0 41W. Village sit. SW of Windsor. Residential with some light industry. Site of famous racecourse. Pop. (1983E) 13,234, with Sunningdale.

Ashanti   Ghana.   Administrative region in S. Area, 9,417 sq. m. (24,390 sq. km.). Cap. Kumasi. Formerly an African kingdom. It is extremely hilly with large areas of tropical rain forest. Extensive clearing has taken place and cassava, cereals, cocoa, groundnuts, sugar and yams are produced. The forests contain much hardwood, particularly mahogany. Pop. (1984C) 2,089,683.

Ashbourne   Derbyshire, England. 53 02N 1 44W. Town sit. NW of Derby on R. Dove. Noted 13th-cent. church. Various light indust. Pop. (1985E) 5,960.

Ashburton River   Western Australia, Australia. 21 40S 114 56E. Rises 150 m. (240 km.) S of Nullagine and flows WNW for 300 m. (480 km.) to enter Indian Ocean near Onslow.

Ashby de la Zouch   Leicestershire, England. 56 46N 1 28W. Town sit. WNW of Leicester. Centre of an agric. and coalmining area. Manu. soap and snackfoods. Pop. (1985E) 11,906.

Ashdod   South Coastal Plain, Israel. 31 45N 34 40E. Village sit. SW of Rehovot on the site of the Philistine fort and temple of Dagus. Pop. (1980E) 40,000.

Ashdown Forest   England. Region of heathland sit. between Horsham and Tunbridge Wells.

Asheville   North Carolina, U.S.A. 35 34N 82 33W. City sit. WNW of Charlotte in the S Appalachian Moun-

tains. Health and tourist resort. Manu. machinery, textiles, furniture and paper. Pop. (1980C) 53,583.

Ashford   Kent, England. 51 08N 0 53E. Market town serving agric. area WNW of Dover. Indust. inc. light engineering and perfume manu. Site of the Channel tunnel's passenger terminal. Pop. (1989E) 50,000.

Ashington   Northumberland, England. 55 12N 1 35W. Town sit. N of Newcastle upon Tyne. Centre of a declining coalmining area. Pop. (1985E) 27,786.

Ashkhabad   Turkmenistan, U.S.S.R. 37 57N 58 23E. Cap. of Turkmenistan S.S.R. sit. at foot of the Kopet Dagh 25 m. (40 km.) from the Iran frontier. Rebuilt following an earthquake in 1948. Indust. inc. textiles, glass, food processing and meat packing. Pop. (1985E) 356,000.

Ashland   Kentucky, U.S.A. 38 28N 82 38W. City and port sit. ENE of Lexington on Ohio R. Manu. nickel, steel, textiles, leather goods and chemicals. Pop. (1980C) 27,064.

Ashmore and Cartier Islands   Indian Ocean. 12 14S 123 05E. Islands 300 m. (480 km.) N of Broome, Western Australia, forming an overseas territory of Australia. The Ashmore Islands consist of 3 coral islets about $1/_2$m. (1 km.) wide and enclosed by a reef. Cartier Island, of a similar size, is sandy and also enclosed by a reef.

Ashton-in-Makerfield   Greater Manchester, England. 53 29N 2 39W. Town sit. S of Wigan. Industries inc. general engineering and distribution. Pop. (1985E) 24,000.

Ashton-under-Lyne   Greater Manchester, England. 53 29N 2 06W. Town sit. E of Manchester. Industries inc. mechanical, electrical and chemical engineering, clothing, footwear, plastics, textiles and tobacco. Cotton milling. Pop. (1987E) 44,406.

Asia   40 00N 90 00E. The largest continent. Bounded on N by Arctic Ocean, on E by Pacific Ocean, on S by Indian Ocean and on W by Europe.

Greatest length, 5,350 m. (8,560 km.) and breadth *c.* 6,000 m. (9,600 km.). Area, *c.* 17m. sq. m. (27.5m. sq. km.). In SW Asia is a plateau at the point where India, Afghánistán, U.S.S.R. and China meet, and from there great mountain chains go in all directions. The principal mountain system is the Himalayas with Mount Everest at 29,002 ft. (8,840 metres) a.s.l. as the highest peak. Inhabitants of Asia form over half pop. of world with a density of 197 persons per sq. m. (76 per sq. km.), and belong to five different groups; Mongolian, Caucasian, Malayan, Dravidian, Negroid. Mongolians, numerically greatest, inhabit China, Japan, Tibet, N Asia; Caucasians predominate in W from Afghánistán to Asia Minor, and in India; Malayans in E peninsula and E Arch.; Dravidians in S India and Sri Lanka; Negroid peoples in SE Asia and Philippine Islands. Other inhabitants inc. Russians, Britons, Jews, Arabs. Pop. (1985E) 2,813m. (excluding U.S.S.R.).

Asia Minor (Anatolia)   Turkey. Peninsula in W Asia. Area, 220,000 sq. m. (569,800 sq. km.). Bounded on N by Black Sea, on E by Iran, on S by Mediterranean Sea and Syria and on W by Bosporus, Sea of Marmara, Dardanelles and the Aegean Sea. There is a central tableland at 3,000–6,000 ft. (914–1,829 metres) a.s.l. bounded by mountain ranges to N and S.

Askja   Iceland. Largest volcano in the Odadahravn lava field 56 m. (90 km.) SE of Akureyri at 4,574 ft. (1,394 metres) a.s.l.

Asmara   Eritrea, Ethiopia. 15 20N 38 58E. Cap. of Eritrea 40 m. (64 km.) from Red Sea coast SW of Massawa, sit. on the Hamasen plateau. Manu. meat canning, tanning, flour milling, textiles, soap and matches. Pop. (1984C) 275,385.

Asnières-sur-Seine   Hauts-de-Seine, France. Town and NW suburb of Paris on the Seine R. Indust. aircraft, cars, machine tools, chemicals and perfume. Pop. (1982C) 71,220.

Aspromonte   Calabria, Italy. Mountains sit. in W Sila Mountains behind the town of Reggio di Calabria. They

rise to nearly 7,000 ft. (2,134 metres) a.s.l.

Aspropotamus River   Greece. Rises in Pindus Mountains and flows S for 130 m. (210 km.) to enter the Ionian Sea opposite Ithaca.

Assam   India. 25 45N 93 30E. Assam first became a Brit. Protectorate at the close of the first Burmese War in 1826. On the partition of India almost the whole of the predominantly Moslem district of Sylhet was merged with E. Bengal (Pakistan). Dewangiri in N Kamrup was ceded to Bhután in 1951. The Naga Hill district, administered by the Union govt. since 1957, became part of Nagaland in 1962. The autonomous state of Meghalaya within Assam, comprising the districts of Garo Hills and Khasi and Jaintia Hills, came into existence in 1970, and achieved full independent statehood in 1972. Area, 30,285 sq. m. (78,438 sq. km.). Total forest area, 22% of Assam. Chief towns, Dispur (cap.), Guwahati, Pandu, Cachar, Doom Dooma, Dibrugarh. The cultivation of tea is the principal industry in Assam and produces 50% of India's output. Agric. employs about 77% of the pop. Other crops inc. rice, oilseeds and jute. Assam also has important oilfields. Industries inc. oil refineries, fertilizers, timber. Sericulture and hand-loom weaving, both silk and cotton, are the most important home industries. There is an important coal industry. Pop. (1981C) 19,896,843.

Assen   Drenthe, Netherlands. 53 00N 6 35E. Cap. of province sit. S of Groningen. Canal and railway junction. Manu. food processing and canning, and clothing. Pop. (1984E) 46,745.

Assiniboine River   Canada. 49 53N 97 08W. Rises in Saskatchewan and flows for 590 m. (944 km.) SE and then E through Manitoba and joins Red R. at Winnipeg. Named after the *Assiniboin* Indians.

Assisi   Umbria, Italy. 43 04N 12 37E. Town in Apennine foothills in central Italy. St. Francis, the founder of the Franciscan order, was born

here. Many superb medieval churches
and other buildings much visited by
pilgrims and tourists. Pop. (1989E)
24,567.

Asti   Piemonte, Italy. 44 54N 8 12E.
Cap. of Asti province ESE of Turin sit.
at confluence of Rs. Tanaro and Bor-
bore. Manu. glass, chemicals, textiles
and food products. Famous for wine.
Pop. (1983E) 76,439.

Astrakhan   Russian Soviet Federal
Socialist Republic, U.S.S.R. 42 22N
48 04E. (i) Region on Lower Volga R.
adjacent to Caspian Sea. Area, 17,200
sq. m. (44,558 sq. km.). Mainly agric.
near the Volga producing cotton and
fruits. Some cattle and sheep reared
esp. lambs for fur. Salt deposits in L.
Baskunchak. (ii) Cap. of region of
same name near Caspian Sea on delta
of Volga R. Principal port for the
Caspian, trading in timber, grain,
cereals, cotton, fruit and rice. Indust.
inc. shipbuilding, sawmilling, textiles
and fish processing esp. caviare. Pop.
(1985E) 493,000.

Asturias   Spain. 43 25N 5 50W.
Anc. kingdom in the NW now the province of Oviedo.

Asunción   Paraguay. 25 15S 57
40W. Cap. of Paraguay sit. on E bank
of Paraguay R. near its junction with
Pilcomayo R. Important commercial
and transportation centre. Much of the
industry of Asunción is involved in
processing maize, cotton, sugar, fruit,
tobacco and cattle in the rich adjacent
agric. area. Pop. (1984E) 729,307.

Aswân   Egypt. 24 05N 32 53E.
Town in Upper Egypt on the right
bank of R. Nile. Commercial centre,
winter health resort and tourist centre.
Quarrying important occupation. The
Aswân dam is 3.5 m. (5.5 km.) above
the town and was completed in 1902.
The dam was increased in height in
1912 and 1933. The new Aswân High
Dam was started in 1960 and inaugu-
rated in Jan. 1971 and as a result none
of the waters of the R. Nile's annual
flood will be lost and the agric. area
will be increased by 20%. The power
produced by the dam should raise
national income by 50% as the result
of greatly increased agric. and indus-

trial production. *See also* Abu Simbel.
Pop. (1986E) 195,700.

Asyût   Asyût, Egypt. 27 11N 31
11E. Cap. of governorate sit. on W
bank of Nile R. Trading centre. Manu.
pottery, cotton goods and wood and
ivory carvings. Pop. (1986E) 291,300.

Atacama Desert   Chile. 22 30S 69
15W. Barren and extremely dry
region stretching from S of border
with Peru, and comprising the provinces of Atacama, Antofagasta and
Tarapacá. Area, 31,000 sq. m. (80,290
sq. km.). Important deposits of
nitrates, iodine, copper and iron-ore.

Atafu   Tokelau, South Pacific.
8 33S 173 30W. Atoll sit. N of Western Samoa consisting of *c.* 60 islets on
a reef 3 m. (5 km.) long and 2¹⁄₂m. (4
km.) wide. Main product coconuts.
Pop. (1981E) 554.

Atakpamé   Plateaux, Togo. 7 32N
1 08E. Cap. of region. Trading centre
for   coffee-producing   area.   Pop.
(1977E) 21,800.

Atbara   Sudan. 17 42N 34 00E.
Town sit. at confluence of R. Atbara
and Nile. Railway junction. Indust.
inc. railway engineering and cement
manu. Pop. (1983C) 73,009.

Atbara River   Or Black Nile.
Sudan. 17 42N 33 56E. Rises in the
Ethiopian Highlands and flows for
700 m. (1,120 km.) NNW joining R.
Nile at Atbara.

Atchafalaya River   U.S.A. 29 53N
91 28W. An outlet of the Red R. and
in time of flood of the R. Mississippi;
flows for 220 m. (352 km.) to
Atchafalaya Bay on the Gulf of Mex.,
70 m. (112 km.) SW of New Orleans.

Athabasca, Lake   Alberta/Saskat-
chewan, Canada. 59 07N 110 00W.
Sit. in NW Saskatchewan and NE Alberta. 200 m. (320 km.) in length. Fed
by Athabasca and Peace R. in SW and
drained by Slave R. in NW. Uranium
ores were discovered on the N shore in
the 1950s.

Athabasca River   Canada. 58 40N
110 50W. Rises in Alberta and flows

for 740 m. (1,184 km.) NE to L. Athabasca.

Athelney, Isle of   Somerset, England. 51 03N 2 56W. A small area near the junction of Rs. Tone and Parrett, 5 m. (8 km.) SSE of Bridgwater, formerly an island. Now part of the village of Stoke St. Gregory.

Athens   Greece. 38 00N 23 44E. Cap. of Greece, sit. on small stretch of flat ground NW of Gulf of Aegina, between R. Ilissus and R. Cephissus. The port is Piraeus. Anc. Athens was built on several low hills rising from Attic plain. The original site of the city was Acropolis on the summit of which have been found traces of a Mycenean fortress and palace dated before 1100 B.C. Near the centre is Parthenon, chief temple of goddess Athena. The foundations of Athens' greatness were laid by two men, Pisistratus who made Athens a commercial power in the second half of the 6th cent. B.C., and Thermisticles who made it a naval power. Industries inc. tourism, metallurgy, chemicals, textiles, tobacco, engineering, oil refining, footwear and food processing. Pop. (1981E) 885,737, Greater Athens 3,037,331.

Athens   Georgia, U.S.A. 33 57N 83 23W. City sit. ENE of Atlanta on Oconce R. Industrial centre of an important cotton-growing area. Manu. textiles, fertilizers and cottonseed oil. Pop. (1980C) 42,549.

Atherstone   Warwickshire, England. 52 34N 1 32W. Town sit. NW of Nuneaton. Manu. hats, footwear and knitwear. Pop. (1981C) 7,429.

Atherton   Greater Manchester, England. 53 31N 2 31W. Town sit. WNW of Manchester. Industries inc. light engineering, textiles, soft drinks. Pop. (1985E) 22,000.

Athlone   Westmeath, Ireland. 53 25N 7 56W. Town in central Ireland on Shannon R. Manu. textiles, plastics, telecommunications equipment. Pop. (1981C) 14,700.

Athos, Mount   Macedonia, Greece. 40 09N 24 22E. Autonomous monas-

tic district co-extensive with the Akte 'prong' of the Chalcidice peninsula. Inhabited entirely by monks under direct rule of Greek patriarch of Constantinople, who inhabit *c.* 20 monasteries, many of them difficult of access on precipitous crags. It was a major centre of Byzantine art and theology.

Atitlán   Sololá, Guatemala. 14 41N 91 12W. Volcano sit. S of L. Atitlán. 11,633 ft. (3,546 metres) a.s.l.

Atitlán Lake   Sololá, Guatemala. 14 41N 91 12W. L. 40 m. (64 km.) W of Guatemala City. Approx. 24 m. (39 km.) long and 8–10 m. (13–16 km.) broad. Possibly the crater of an extinct volcano.

Atiu   Cook Islands, South Pacific. 20 00S 158 07W. Coral island 115 m. (184 km.) NE of Rarotonga. Area, 10·4 sq. m. (27 sq. km.), rising to 394 ft. (120 metres). The main products are fruit and copra. Pop. (1981C) 1,225.

Atlanta   Georgia, U.S.A. 33 45N 84 23W. City and State cap. Estab. in 1837 at the end of a railway line and originally called Terminus. Transport, communications, finance, food processing, retail/wholesale centre. Pop. (1980C) 425,022.

Atlantic City   New Jersey, U.S.A. 39 21N 74 27W. Resort sit. SE of Philadelphia sit. on a sandbar at Absecon Beach. Indust. casino gambling, conventions, tourism. Pop. (1980C) 40,199.

Atlantic Ocean   10 00N 30 00W. Ocean separating the Old from the New World, Europe and Africa lying to E, N. and S. America to W. It opens northward into Arctic Ocean and spreads out S into great Southern Ocean. The extreme length is *c.* 8,500 m. (13,600 km.) along part enclosed by land, and the greatest breadth, *c.* 5,000 m. (8,000 km.). The narrowest part is between Braz. and Afr. coasts: 1,800 m. (2,880 km.). Area, over 31m. sq. m. (80m. sq. km.). Deepest point, 90 m. (144 km.) N of Puerto Rico, has depth of 4,662 fathoms. It is relatively saltier than other oceans.

Atlantico Colombia. Dept. of N Colombia, bounded N by Caribbean Sea. Area, 1,340 sq. m. (3,270 sq. km.). Cap. Barranquilla. Humid, tropical region of thickly wooded lowland drained by the Magdalena R. The main crops are cotton, sugar-cane, cereals; other products timber, cattle, gold, petroleum and coal. Industries are concentrated on Barranquilla, which is a major port. Pop. (1983E) 1,379,100.

Atlántida, Honduras Dept. of N Honduras on Caribbean. Chief towns La Ceiba (cap.), Tela. Rises from coastal lowlands to the Sierra de Nombre de Dios in the S. Mainly agric., producing rice, coconuts, abaca, citronella, plantains, sugarcane, beans, fruit. Pop. (1983E) 242,235.

Atlas Mountains 33 00N 2 00W. An irregular series of mountain groups running along N of Afr. from Cape Ghir in Morocco to Cape Bon in Tunisia, distance of 1,500 m. (2,400 km.). The greatest height is between 14,000 and 15,000 ft. (4,310 and 4,610 metres) a.s.l. The valleys are fertile and the hills are well forested.

Atrato River Colombia. 8 17N 76 58W. Rises in the W. Cordillera and flows N for 400 m. (640 km.) to enter Gulf of Darien. Gold and platinum found in the upper course.

Attica Greece. 38 10N 23 40E. A prefecture in central Greece and inc. the SE peninsula and Saronic Gulf islands. Area, 964 sq. m. (2,496 sq. km.). Cap. Athens. Produces olive oil, wine and cereals. Pop., excluding Athens (1981E) 342,093.

Attleboro Massachusetts, U.S.A. 41 56N 71 17W. City sit. NE of Providence. Manu. jewellery, silverware, chemicals, metals, electrical and electronic equipment. Pop. (1980C) 34,196.

Aube France. 48 15N 4 05E. Dept. in Champagne-Ardennes region, sit. on SE edge of Paris Basin. Area, 2,317 sq. m. (6,002 sq. km.). Cap. Troyes. Other chief towns, Bar-sur-Aube, Nogent-sur-Seine and Arcis-sur-

Aube. Mainly agric., the fertile SW producing cereals, vines and vegetables, and the NE being largely pastoral. Indust. inc. spinning and weaving. Pop. (1982C) 289,300.

Aube River France. 48 34N 3 43E. Rises in plateau of Langres and flows NW for 150 m. (240 km.) to Seine R.

Aubervilliers Seine-St. Denis, France. 48 55N 2 23E. Town and NE suburb of Paris. Manu. paints, varnishes, fertilizers and perfumes. Pop. (1982C) 67,775.

Auburn Maine, U.S.A. 44 06N 70 14W. City sit. N of Portland on Androscoggin R. Manu. plastics, electrical goods and footwear. Pop. (1980C) 23,128.

Auburn New York, U.S.A. 42 56N 76 34N. City sit. WSW of Syracuse at N end of L. Owasco. Manu. agric. machinery, car accessories, plastics, textiles, chemicals, surgical instruments and footwear. Pop. (1980C) 32,548.

Aubusson Creuse, France. 45 57N 2 11E. Town sit. ENE of Limoges on Creuse R. Noted for carpets and tapestries. Pop. (1982C) 6,153.

Auch Gers, France. 43 39N 0 35E. Cap. of dept. on Gers R. in SW France. Manu. furniture and hosiery, and esp. noted for *pâté de foie gras*. Trades in brandy, wine, horses and poultry. Pop. (1982C) 25,543.

Auckland North Island, New Zealand. 36 53S 174 45E. Largest city and important seaport sit. between Manukau Harbour and Waitemata Harbour in the N of North Island. A large proportion of New Zealand's trade goes through the port. Manu. in the region inc. textiles, chemicals, clothing, beverages, motor vehicle assembly, steel products, electronics, carpets, footwear, furniture, household appliances, plastics and food processing. Pop. (1985E) 143,800 (region, 880,500).

Auckland Islands New Zealand. 51 00S 166 30E. Six volcanic islands 300 m. (480 km.) S of Invercargill in S. Island. Area, 234 sq. m. (606 sq. km.).

Aude France. 44 13N 3 15E. Dept. in Languedoc-Roussillon region, sit. on the Gulf of Lions. Area, 2,406 sq. m. (6,232 sq. km.). Chief towns, Carcassonne (the cap.), Narbonne and Castelnaudary. Principal R. is the Aude. Mainly mountainous, but the agric. N produces olives, vines, fruit and cereals. Salt is produced on the coast. Pop. (1982C) 280,686.

Aude River France. Rises in the Pyrenees near the Pic de Carlitte and flows N and E for 140 m. (224 km.) to enter the Gulf of Lions.

Aue Chemnitz, German Democratic Republic. 50 35N 12 42E. Town sit. SE of Zwickau in the Erzgebirge. Centre of a uranium-mining district. Manu. textiles, chemicals and metal goods.

Augsburg Bavaria, Federal Republic of Germany. 48 23N 10 53E. Town in S central F.R.G. on Lech R. near its union with Wertach R. Important industrial and railway centre. Also the chief textile centre of S. Germany. Other manu. heavy and light machinery, chemicals and precision instruments. Many fine medieval churches and other buildings. Pop. (1984E) 246,600.

Augusta Sicily, Italy. 37 13N 15 13E. Port sit. NNW of Syracuse sit. on a small island which is joined to the mainland by two bridges. Indust. salt and fishing. Trades in agric. produce.

Augusta Georgia, U.S.A. 33 29N 81 57W. City and port near the border with South Carolina, on Savannah R. Sit. at the head of navigation. Manu. inc. food, paper board, newsprint, clay refractories, golf cars and textiles. Pop. (1980C) 47,532.

Augusta Maine, U.S.A. 44 19N 69 49W. City on the Kennebec R. near its mouth. State cap. and commercial town. Pop. (1980C) 21,819.

Aulnay-sous-Bois Seine-St. Denis, France. NE suburb of Paris. Pop. (1982C) 76,032.

Aunu'u Island American Samoa, South Pacific. 14 17S 170 33W.

Island off S coast of E Tutuila with a circular volcanic crater at *c.* 275 ft. (84 metres).

Aurangabad Maharashtra, India. Town sit. ENE of Bombay. Manu. textiles and electronic goods. Pop. (1981C) 298,937.

Aurès Algeria. 35 08N 6 30E. Mountain massif in Constantine dept., NE Algeria, forming part of the Saharan Atlas. Bounded N by the High Plateaux, W by the Hodna depression, E by the Nemencha Mountains and S by the Sahara. Highest peaks, Djebel Chélia, 7,641 ft. (2,329 metres), Kef Mahmel, 7,615 ft. (2,321 metres). The higher slopes have sufficient rainfall to support pine and oak forests. Difficult of access, but crossed by one major route through the El-Kantara defile, which carries a road and railway.

Aurillac Cantal, France. 44 56N 2 26E. Cap. of dept. in S central France on Jordanne R. Manu. leather goods and umbrellas. An important livestock market. Pop. (1982C) 33,197.

Aurora Colorado, U.S.A. 39 44N 104 52W. City immediately E of Denver. Employment is in trade, services and some manu. Pop. (1980C) 158,588.

Aurora Illinois, U.S.A. 41 46N 88 19W. City sit. SW of the S end of L. Michigan. Manu. iron, steel and aluminium goods inc. office and school equipment. Pop. (1980C) 81,293.

Aust-Agder Norway. County of S Norway on the Skagerrak coast. Area, 3,609 sq. m. (9,212 sq. km.). Chief towns Arendal (cap.), Risor, Tvedestrand, Grimstad, Lillesand. Coastal lowland strip rising to highlands and the Bykle Mountains. Mainly forested with some mining, nickel, low-grade iron, and very little agric. Some cattle farming. Lumbering and fishing are important. Pop. (1980C) 90,629.

Austin Minnesota, U.S.A. 43 40N 92 59W. Town sit. SW of Rochester. Manu. inc. food products, packaging and building materials. Pop. (1980C) 23,020.

Austin Texas, U.S.A. 30 16N 97 45W. State cap. WNW of Houston on Colorado R. Indust. inc. government, services and tourism, with some manu. and scientific research. Pop. (1980C) 345,496.

Australia The continent is bounded on W by Indian Ocean and on E by the Pacific Ocean and is the largest island in the world. The length of the coastline is 12,210 m. (19,536 km.) and the continent is bisected by the Tropic of Capricorn. Area, 2,967,900 sq. m. (7,686,900 sq. km.). In 1901 New South Wales, Victoria, Queensland, South Australia, Western Australia and Tasmania were federated under the name of the 'Commonwealth of Australia', the designation of 'colonies' being at the same time changed into that of 'states'—except in the case of Northern Territory, which was transferred from South Australia to the Commonwealth as a 'territory' in 1911. Northern Territory acquired responsible self-government in 1979. In 1911 the Commonwealth acquired from the State of New South Wales the Canberra site for the Australian capital. Building operations were begun in 1923 and Parliament was opened at Canberra in 1927.

Territories under the administration of the Commonwealth, but not included in it, comprise Norfolk Island, the territory of Ashmore and Cartier Islands, and the Australian Antarctic Territory (1936), comprising all the islands and territory other than Adélie Land, situated south of 60° S lat. and between 160° and 45° E long. The British Government transferred sovereignty in the Heard Island and McDonald Islands to the Australian Government in 1947. Cocos (Keeling) Islands in 1955 and Christmas Island in 1958 were also transferred to Australian jurisdiction.

Mineral wealth includes copper, gold, iron, lead, rutile, tungsten, zinc, coal and petroleum. Pop. (1989E) 16,806,730.

Australian Alps Australia. 37 00S 148 00E. A mountain range in the SE stretching for about 300 m. (480 km.) through Victoria and New South Wales, to the S of the Great Dividing Range. Mount Kosciusko, 7,328 ft.

(2,234 metres) a.s.l., is the highest peak in Australia. Mainly well forested and a popular area for winter sports.

Australian Antarctic Territory An Order in Council of 1933 placed under Australian authority all the islands and territories other than Adélie Land sit. S of 60° S lat. and lying between 160° E long. and 45° E long.

In 1959 Australia signed the Antarctic Treaty with Argentina, Belgium, Chile, France, Japan, New Zealand, Norway, Rep. of S. Afr., U.S.S.R., U.K. and U.S.A. Poland, Czechoslovakia and Denmark have subsequently acceded to the Treaty. The Treaty reserves the Antarctic area S of 60° S lat. for peaceful purposes, provides for intern. co-operation in scientific investigation and research, and preserves, for the duration of the Treaty, the *status quo* with regard to territorial sovereignty, rights and claims. The Treaty came into force in 1961.

Australian Capital Territory *see* **Canberra**

Austral Islands French Polynesia, South Pacific Ocean. 23 20S 151 00W. A group of islands S of the Society Islands. Total area, 67 sq. m. (174 sq. km.). There are five inhabited islands, Rurutu, Tubuai, Rapa, Rimatara and Raivavae. Pop. (1983C) 6,283.

Austria 47 00N 14 00E. A Federal Republic, bounded on N by the Federal Republic of Germany and Czechoslovakia, on E by Hungary and in the S by Yugoslavia and Italy. The Federal States are Vienna, Lower Austria, Burgenland, Upper Austria, Salzburg, Styria, Carinthia, Tirol and Vorarlberg. Area, 32,366 sq. m. (83,850 sq. km.). Chief towns, Vienna (the cap.), Graz, Linz, Salzburg and Innsbruck. The surface is mountainous but Lower Austria is flat near Vienna. The area is drained by R. Danube. Agric. products inc. wheat, rye, barley, oats, potatoes, sugar-beet. Mineral production inc. lignite, iron, lead, zinc, copper, magnesite. Austria is also one of the world's largest sources of graphite. Petroleum is also found and commercial production began in early 1930s. There is a wide range of manu. industry. Tourism is an important industry; there are over 20,000 hotels and boarding houses and over 16m. tourists visit Austria each year. Pop. (1988E) 7,596,100.

Auvergne   France. 45 00N 8 36W. Region (former province) covering part of the Massif Central and spilling out into the plains that border it in the N. Comprises the departments of Allier, Cantal, Haute-Loire and Puy-de-Dôme. Area, 10,034 sq. m. (25,988 sq. km.). The physical and economic axis of the region is the long, low-altitude corridor of the R. Allier, of which the S part is known as Les Limagnes, and the N part as Sologne Bourbonnaise, but now tends to be called the Val d'Allier. The anc. *Arverni* were among the most enterprising peoples in Gaul in Roman times. Clermont-Ferrand is the regional cap. Pop. (1982C) 1,332,678.

Auxerre   Yonne, France. 47 48N 3 34W. Cap. of dept. in N central France on Yonne R. Manu. wine, metal goods and paints from local ochre. Pop. (1982C) 40,698.

Avebury   Wiltshire, England. 51 27N 1 51W. Sit. SSW of Swindon on R. Kennet. Sit. within a prehistoric earthwork with stone circle and avenue.

Aveiro   Portugal. 40 38N 8 39W. (i) District in N central Portugal bounded W by the Atlantic and consisting of part of Beira Litoral and S Douro Litoral provinces. Drained by the Vouga R. Main occupations inc. farming, viticulture, lumbering, fishing. Area (excluding lagoon), 7,014 sq. m. (2,708 sq. km.). Pop. (1981C) 622,988. (ii) Town sit. S of Oporto at mouth of Vouga R. Sardine fishing port and salt extracting centre. District cap. Connected with the Atlantic by canal across a lagoon and sand bar. Trading centre for an agric. area, handling wine, fruit, olive oil and grain. Indust. inc. fish canning, pottery and porcelain.

Avellaneda   Buenos Aires, Argentina. Industrial suburb of Buenos Aires sit. at union of Riachuelo R. and La Plata estuary. Meat-packing, tanning, oil refining and textile manu. Pop. (1980E) 650,000.

Avellino   Campania, Italy. 40 49N 14 47E. Cap of Avellino province sit. ENE of Naples in the Apennines. Manu. hats. Indust. inc. sulphur mining and refining. Pop. (1981E) 57,000.

Averno   Campania, Italy. 40 50N 14 04E. A lake sit. in a volcanic crater W of Naples. Very dark and 213 ft. (65 metres) deep, it was believed in anc. times to be the gateway to hell.

Aveyron   France. 44 22N 2 45E. Dept. in Midi-Pyrénées region, sit. in the SW of the Massif Central. Area, 3,373 sq. m. (8,735 sq. km.). Mainly mountainous with the Cévennes in the SE and the Monts d'Aubrac in the NE. Agric. is possible in the valleys and products inc. cereals, vines, fruits and vegetables. Cattle and sheep are reared and Roquefort is noted for cheese. Coal and iron are mined. Chief towns, Rodez (the cap.), and Millau. Pop. (1982C) 278,654.

Aveyron River   France. 44 05N 1 16E. Rises on the border between Aveyron and Lozère depts. and flows WSW for 150 m. (240 km.) to enter Tarn R. near Montauban.

Avignon   Vaucluse, France. 43 57N 4 49E. 'Windy city'. Cap. of dept. sit. on R. Rhône in S France. Residence of the Popes (1309–79). Manu. chemicals, man-made fibres, flour, soap and paper. Pop. (1982C) 91,474 (agglomeration, 174,264.

Ávila   Spain. 40 39N 4 42W. (i) Province in Old Castile. Area, 3,107 sq. m. (8,048 sq. km.). Mountainous with fertile valleys producing cereals, olives and chestnuts. Merino sheep raised. Pop. (1981C) 178,997. (ii) Cap. of province of same name sit. WNW of Madrid on the Adaja R., at 3,800 ft. (1,158 metres) a.s.l. Tourist centre with fine medieval buildings. Pop. (1981C) 86,584.

Avoca River   Wicklow, Ireland. Formed by union of R. Avonbeg and Avonmore, flows SE to enter the Irish Sea at Arklow on the E coast.

Avon   England. 51 59N 2 10W. County of SW England bounded on N by Gloucestershire, E by Wiltshire, S by Somerset and W by Severn Estuary. The County is centred on Bristol, the largest town and administrative centre of the SW of England, and has varied countryside including parts of the Cotswold and Mendip Hill areas of outstanding natural beauty. Bath and Weston-super-Mare are important tourist centres. Since 1974 reorganization the county consists of 6 districts, Bath, Bristol, Kingswood, Northavon, Wansdyke, Woodspring. Principal industries are financial services, transport and communications, aerospace, engineering, paper, packaging and printing. Pop. (1988E) 954,300.

Avonmouth   Avon, England. 51 30N 2 40W. Port sit. NW of Bristol at mouth of Lower Avon R. Indust. flour milling, chemical works, engineering and zinc smelting. It is also an important distribution centre.

Avon, River   51 30N 2 43W. The Bristol or Lower Avon rises in Gloucestershire and flows for 75 m. (120 km.) in a wide curve and enters the R. Severn at Avonmouth, near Bristol.

Avon, River   51 59N 2 10W. The Warwickshire or Upper Avon rises in Northamptonshire and flows for 96 m. (154 km.) in a SW direction through Stratford-on-Avon to join the R. Severn at Tewkesbury.

Avon, River   Other R. with name of Avon in Brit. are: (i) the 'East' Avon which rises in Wiltshire and enters the Eng. Channel at Christchurch, Dorset; (ii) A trib. of the R. Clyde in Scot.; (iii) A trib. of the R. Spey in Scot.; (iv) A trib. of R. Forth in Scot.

Avranches   Manche, France. 48 41N 1 22W. Town and port in SW Normandy sit. on hill above estuary of Sée R. Indust. inc. tanning, fishing and horsebreeding. Trades in dairy produce, fruit, cider and grain. Pop. (1982C) 10,419.

Awe, Loch   Scotland. 56 15N 5 15W. Lake 23 m. (37 km.) long, between the W coast and Loch Lomond. Drained by Awe R. which flows into Loch Etive. Ben Cruachan, 3,689 ft. (1,124 metres) a.s.l., is sit. near the N end.

Axminster   Devon, England. 50 47N 3 00W. Town sit. E of Exeter on R. Axe. Noted in the 18th cent. for carpets, now once again a thriving industry and in recent years there has been a growth in the number of small manufacturing companies. Pop. (1988E) 5,250.

Ayacucho   Ayacucho, Peru. 13 07S 74 13W. Cap. of dept. sit. WNW of Cuzco. Tourist centre with many historically interesting buildings. Pop. (1984E) 30,800.

Aycliffe   *see* **Newton Aycliffe**

Aydin   Turkey. 37 51N 27 51E. (i) Province of W Turkey on Aegean Sea. Drained by Buyuk Menderes and Akcay Rs. It has deposits of iron, copper, lignite, mercury, emery, arsenic, antimony and magnesite. Agric. areas produce cotton, tobacco, valonia, olives, figs, raisins, millet. Pop. (1980C) 652,488. (ii) Town sit. SE of Smyrna. Provincial cap. and market for mineral and food producing area. Pop. (1980E) 60,000.

Ayeyarwady   *see* **Irrawaddy**

Aylesbury   Buckinghamshire, England. 51 50N 0 50W. County town sit. just N of the Chiltern Hills, Indust. inc. engineering, printing, flour milling, food processing, recording company, tobacco and hairdressing preparations. Pop. (1988E) 50,000.

Aylesford   Kent, England. 51 18N 0 29E. Town sit. NW of Maidstone on the R. Medway. Indust. inc. paper mills and cement works. Pop. (1981C) 7,960.

Ayr   Strathclyde Region. Scotland. 55 28N 4 38W. Town on SW coast, sit. at confluence of Ayr R. and Firth of Clyde. Industries inc. fishing and manu. of machinery, metal products, woollen, silicon chip and aircraft. Exports coal, iron and agric. produce. Pop. (1985E) 49,481.

Ayrshire   Scotland. Now part of Strathclyde Region.

Azerbáiján Iran. 38 00N 46 00E. A region in the NW comprising the two provinces of E. Azerbáiján (cap. Tabriz) and W. Azerbáiján (cap. Orumiyeh). Chief R., Araks. Mainly mountainous with the extinct volcano Savalan rising to 15,784 ft. (4,819 metres) a.s.l. W of Ardebil. Combined area, 40,908 sq. m. (105,952 sq. km.). Pop. (1976C) 4,612,731.

Azerbaijan U.S.S.R. 40 20N 48 0E. The independence of Azerbaijan was declared in 1918, with cap. first at Ganja, and later at Baku. In 1920 Azerbaijan was proclaimed a Soviet Socialist Rep. With Georgia and Armenia it formed the Transcaucasian Soviet Federal Socialist Republic. In 1936 it became one of the reps. of the U.S.S.R. Area, 33,430 sq. m. (86,600 sq. km.). Sub-tropical agric. possible, therefore cotton-growing, orchards, vineyards and silk culture, as well as tea plantations. Rich in natural resources, the most important industry being oil. Pop. (1989E) 7m.

Azogues Cañar, Ecuador. 2 46S 78 56W. Town sit. NE of Cuenca on the Pan-American Highway in the Andes. Provincial cap. The main industry is manu. panama hats; also tanning and flour milling. Pop. (1982E) 15,000.

Azores 38 44N 29 0W. Three groups of hilly islands belonging to Portugal, sit. about 900 m. (1,440 km.) W of Portugal, in the Atlantic Ocean. They are of volcanic origin and earthquakes have occurred. The largest island is São Miguel, 41 m. by 9.5 m. (65.6 km. by 15 km.) and the smallest Corvo, 4.5 m. by 3 m. (7 km. by 5 km.). The E group consists of Santa Maria, São Miguel and Formigos Rocks; the central group, Terceira, Graciosa, São Jorge, Pico and Faial; the W group, Flores and Corvo. The Azores produce wine, pineapples, oranges, bananas and tobacco. Manu. inc. agric. produce. Pop. (1981C) 243,410.

Azov Russian Soviet Federal Socialist Republic, U.S.S.R. 47 03N 39 25E. Town sit. WSW of Rostov-on-Don sit. near mouth of Don R. Indust. include fishing and fish-canning.

Azov, Sea of U.S.S.R. 46 00N 36 30E. A shallow sea connected to Black Sea, S of European Russia. Area, 14,500 sq. m. (37,555 sq. km.). It has valuable fisheries. Principal ports, Rostov-on-Don, Taganrog and Kerch.

Azua Dominican Republic. 18 27N 70 44W. (i) Province in S on the Caribbean coast. Area, 936 sq. m. (2,424 sq. km.). Mountainous, semi-arid area between the Cordillera Central and the coast. Crops are grown mainly by irrigation and inc. coffee, tobacco, sugar-cane, cereals and vegetables. Some lumbering. Pop. (1987E) 142,770. (ii) City officially called Azua de Compostela at the S end of the Sierra de Ocoa. Provincial cap. and commercial town trading in agric. produce. Intern. airport.

Azuay Ecuador. Province of S Ecuador in the Andes. Area, 3,443 sq. m. (8,917 sq. km.). Cap. Cuenca. Mountainous with fertile valleys and humid, semi-tropical climate. Rich mineral deposits, esp. gold, silver, mercury, platinum. Mainly agric. producing sugar-cane, coffee, cereals, fruit, cotton, livestock and dairy products. Indust. inc. manu. panama hats. Pop. (1982C) 440,571.

Azul Buenos Aires, Argentina. 36 42S 59 43W. Town sit. SSW of Buenos Aires. Indust. are all connected with cattle. Pop. (1984E) 45,000.

B

Baalbek Lebanon. Town sit. ENE of Beirut at the foot of the Anti-Lebanon mountains. Noted for its splendid ruins.

Bab-el-Mandeb 12 40N 43 20E. Strait sit. between Arabia and Afr. and joining the Red Sea to the Gulf of Aden.

Babylon Iraq. 32 33N 44 24E. Site of anc. town *c*. 55 m. (88 km.) S of Baghdad on R. Euphrates.

Bacău Bacău, Romania. 46 34N 26 55E. Cap. of district sit. NNW of Galati on Bistrita R. Manu. oilfield equipment, textiles, paper and leather goods. Pop. (1983E) 156,891.

Bacolod Negros, Philippines. 10 40N 122 57E. Port sit. on NW coast, SSE of Manila. Main occupations are fishing and sugar refining. Trades in sugar-cane and rice. Pop. (1980C) 262,415.

Bacup Lancashire, England. 53 43N 2 12W. Town sit. N of Manchester. Industries inc. quarrying, engineering, printing and the manu. of footwear. Pop. (1981C) 15,259.

Badajoz Spain. 38 53N 6 58W. (i) Province in the W bordering on Portugal. It is the largest province with an area of 8,361 sq. m. (21,657 sq. km.). The principal R. is the Guadiana. Stock-rearing is the major occupation and cereals, olives and vines are also grown. Pop. (1986E) 664,516. (ii) Cap. of province of same name sit. on the Guadiana R. near the Portuguese frontier. Industries inc. flour milling, food processing, brewing and distilling. Pop. (1986E) 126,340.

Badalona Barcelona, Spain. 41 27N 2 15E. Port and suburb of NE Barcelona. Industries inc. shipbuilding, and the manu. of textiles, chemicals and glass. Pop. (1986E) 223,444.

Bad Ems Rhineland-Palatinate, Federal Republic of Germany. 50 20N 7 43E. Town and spa sit. ESE of Koblenz on Lahn R.

Baden Lower Austria, Austria. 48 00N 16 14E. Resort in the E sit. at the foot of the Wiener Wald. Noted for its warm sulphur springs. Pop. (1981C) 123,140.

Baden Aargau, Switzerland. 47 29N 8 18E. Resort sit. NW of Zürich on Limmat R. Noted for its thermal springs. Manu. electrical equipment and textiles.

Baden-Baden Baden-Württemberg, Federal Republic of Germany. 48 46N 8 14E. Resort in the SW sit. in the Oos valley. Noted for its hot mineral springs. Pop. (1989E) 50,800.

Baden-Württemberg Federal Republic of Germany. 48 40N 9 00E. *Land* of the Federal Rep. in the SW bounded N by Hesse, E by Bavaria. Area, 13,965 sq. m. (35,750 sq. km.). Chief cities Stuttgart (cap.), Mannheim, Karlsruhe, Freiburg, Heidelberg. Drained by Rhine and Neckar Rs., both of which water fertile land, the Rhine plain and the Neckar valley, producing fruit, tobacco, cereals, vines, hops. Otherwise mountainous, inc. S, the Black Forest. the W part, formerly Baden, is mainly rural with a concentration of industrial towns in the N, along the Rhine and near L. Constance, S. The main industries are textiles, engineering, chemicals, papermaking. Pop. (1987E) 9,286,000.

Bad Kreuznach Rhineland-Palatinate, Federal Republic of Germany. 49 52N 7 51E. Resort sit. on Nahe R. Noted for its saline springs. Manu. inc. machinery, tyres and optical instruments. Centre for wine-growing and the wine trade. Pop. (1988E) 40,000.

Badlands South Dakota, U.S.A. Arid plateau E of Black Hills, extending *c.* 120 m. (192 km.), bounded S by White R., crossed by Bad R. Deeply and dramatically eroded and with many fossil remains. Part has been reserved as the Badlands National Monument.

Badrinath Uttar Pradesh, India. 30 44N 79 30E. Village and centre of pilgrimage sit. NE of Delhi in the Garwhal Himalayas. The peak of Badrinath is sit. nearby.

Baffin Bay North Atlantic. 73 00N 66 00W. Arm of the Atlantic Ocean extending *c.* 700 m. (1,120 km.) N between Northwest Territories, Canada (W), and Greenland (E). Connected with the Atlantic (S) by Davis Strait and with the Arctic by Smith Sound (N) and Lancaster and Jones Sounds (W). Ellesmere, Devon, Bylot and Baffin Islands are on its Canadian shoreline; trading posts at Thule, Upernavik and Umanak on the Greenland side. Ice-covered for most of the year; the Labrador current carries icebergs through it. There are navigable passages in summer.

Baffin Island Northwest Territories, Canada. 68 00N 70 00W. Sit. in the entrance to Hudson Bay, it is the largest island in the Canadian Arctic. Area, 184,000 sq. m. (476,560 sq. km.). Very mountainous with many glaciers and snow-fields. The principal occupations are whaling, hunting and fur trapping.

Bafoussam Cameroon. 5 29N 10 24E. Cap. of W province and main trading centre for Bamileke people. Area produces bauxite and coffee. Pop. (1981E) 75,832.

Bagé Rio Grande do Sul, Brazil. 31 20S 54 06W. Town in extreme S sit. near the Uruguay frontier. It has an experimental agric. estab. and trades in livestock.

Baghdad Baghdad, Iraq. 33 20N 44 26E. Cap. of Iraq and of Baghdad governorate, sit. NW of Basra on Tigris R. Communications centre at the head of shallow-draft navigation on Tigris R. Commercial centre. Industries inc. distilling, tanning; manu. clothing, cement, tobacco products. Pop. (1985E) 4,648,609.

Bagnères-de-Bigorre Hautes-Pyrénées, France. Resort in SW sit. on Adour R. Noted for its mineral springs. Industries inc. marble and slate quarrying, engineering and the manu. of textiles. Pop. (1982C) 9,850.

Bagnolet Seine-St. Denis, France. A suburb of E Paris. Manu. plaster of Paris, textiles and furniture. Pop. (1982C) 32,557.

Baguio Luzon, Philippines. 16 25N 120 36E. Resort sit. N of Manila. Centre of a gold mining district.

Bahamas West Indies. 24 00N 74 00W. The Commonwealth of the Bahamas is a sovereign state which gained independence from Britain in 1973. It consists of an arch. of 700 islands extending from within 50 m. (80 km.) of SE Florida to within 70 m. (112 km.) of Haiti forming an independent Commonwealth state. Area, 4,404 sq. m. (11,406 sq. km.). Cap. Nassau. The family islands (formerly out islands) are Grand Bahama, Great Abaco, Bimini Islands, Berry Islands, New Providence, Andros, Eleuthera, Cat, Great Exuma, San Salvador, or Watling, Long Island, Crooked Island, Acklin's Island, Mayaguana, Little Inagua, Great Inagua. Tourism is the most important industry, others inc. fishing, salt panning, lumbering, with some agric. Main products are fish, timber, shells, vegetables and fruit. Pop. (1986E) 236,171.

Bahawalpur Bahawalpur, Pakistan. 29 37N 71 40E. Cap. of Bahawalpur district sit. SSE of Multan near Sutlej R. Manu. inc. cotton goods and soap. Trades in cotton and rice.

Bahía Brazil. 12 59S 38 31W. A state on the Atlantic coast, in E central Braz. Area, 216,612 sq. m. (561,026 sq. km.). A low coastal plain rises to the W to the central plateau, which is crossed by the São Francisco R. Cacao, sugar-cane, tobacco and cotton are grown on the fertile coastal plain. Industrial diamonds are mined and there are oil wells on the coast. Cap. Salvador. Pop. (1984E) 10,504,000.

Bahía Blanca Buenos Aires, Argentina. 38 44S 62 16W. Port sit. on Blanca Bay. Industries inc. oil refining, meat packing, tanning and flour milling. Meat, hides, grain and wool are exported. Pop. (1980C) 233,126.

Bahrain 26 00N 50 3E. State lying between Qatar peninsula and Hasa coast of Saudi Arabia, together with neighbouring islands of Muharraq, Sitra and Umm Nasan. Area, 231 sq. m. (598 sq. km.). Cap. Manama. Limestone plateau, mainly arid and rising to Jabal Dukhan, 445 ft. (137 metres) from a central depression. Irrigation from wells supports agric. in the N, esp. dates, citrus fruit, alfalfa, wheat and millet. The main industry is extracting and refining oil. The refinery processes local oil and oil coming by pipeline from Dhahran, Saudi Arabia. Pop. (1988E) 421,040.

Bahr el Ghazal Sudan. (i) Region in S bounded E by Sudd swampland, W by Central African Empire. Area, 77,820 sq. m. (199,219 sq. km.). Cap. Wau. Clay plateau drained by Tonj and Jur Rs. The main occupation is farming, esp. livestock, cotton, peanuts sesame, corn, durra. Pop. (1983C) 2,265,510. (ii) R. rising in L. Ambadi in S Sudan and flowing 150 m. (240 km.) NE past Wankai and joining the Bahr el Jebel in L. No. Together they form the White Nile R. Navigable July–March to Meshra er Req and July–October to Wau, on the Jur R., a headstream. It adds little water volume to the Nile R., owing to the evaporation and dispersal of its waters in the Sudd swamps.

Bahr el Jebel, River Sudan. R. rising in Uganda, as the Albert Nile, and entering Sudan at Nimule to flow 594 m. (950 km.) N through the Sudd swamps and join the Bahr el Gazal in L. No, to form the White Nile. Navigable throughout the year except through the rapids between Juba and Nimule.

Baia Mare Maramures, Romania. 47 40N 23 35E. Cap. of district sit. WNW of Iaşi. Industries inc. lead and zinc smelting and the manu. of chemicals. Pop. (1983E) 123,675.

Baikal Lake, U.S.S.R. 53 00N 107 40E. L. in SE Siberia on the boundary between the Buryat-Mongol Autonomous Soviet Socialist Rep. and Irkutsk oblast. Area, 12,150 sq. m. (31,104 sq. km.), and the world's deepest fresh-water L. Fed by over 300 Rs., drained only by the Angara R. Frozen from January–April.

Baja California, Mexico. Peninsula of NW Mexico extending 760 m. (1,216 km.) between the Gulf of California and the Pacific, and forming a state. Chief towns Tijuana (cap.), Mexicali, Ensenada, La Paz. Plains on the Pacific coast, otherwise mountainous and generally desolate. Only small areas are inhabited and farming is sustained by irrigation. The main occupations are fishing and mining. Chief products silver, lead, gold, copper, magnesite, iron ore, kaolin, nitrates, fish, cotton, wheat, vines, vegetables. Pop. (1980C) 1,177,886.

Bakersfield California, U.S.A. 35 23N 119 01W. City in S centre of state sit. on Kern R. Industries inc. oil refining, railway engineering, food processing and the manu. of oilfield equipment, agric. machinery and chemicals. Pop. (1980C) 105,611.

Bakewell Derbyshire, England. Town sit. W of Chesterfield. Bakewell puddings were first made here. Pop. (1987E) 3,820.

Bakhchisarai Russian Soviet Federal Socialist Republic, U.S.S.R. 44 45N 33 52E. Town sit. SW of Simferopol in the Crimea. Manu. copper and leather products. Pop. 15,000.

Bakhtarán Iran. 34 19N 47 04E. (i) Province in SW, formerly Kermánsháhán. Area, 9,137 sq. m. (23,667 sq. km.). Pop. (1982C) 1,030,714. (ii) Cap. of (i) formerly Kermánsháh. Standing in the centre of a fertile grain-producing area, it has caravan routes from Tehrán, Baghdad and Esfáhán passing through. An oil refinery is joined by pipeline to the Naft-i-Shah oilfield. Pop. (1983E) 531,350.

Bakony Forest Hungary. 46 55N 17 40E. Sit. to the N of L. Balaton, in the NW. There are deposits of coal, bauxite and manganese. The chief town is Veszprém.

Baku Azerbaijan, U.S.S.R. 44 22N 49 53E. Cap. of Azerbaijan and port sit. on S coast of the Apsheron peninsula on the Caspian Sea. Industries inc. shipbuilding, oil refining, and the manu. of oilfield equipment, chemicals, textiles and cement. A pipeline runs to Batumi on the Black Sea and oil is exported. Pop. (1985E) 1,693,000.

Bala Gwynedd, Wales. 52 54N 3 35W. Town sit. at N end of L. Bala on R. Dee, NE of Barmouth. Pop. (1980E) 1,700.

Balaklava Russian Soviet Federal Socialist Republic, U.S.S.R. 44 30N 33 30E. Port sit. S of Sevastopol in the Crimea.

Balaton, Lake Hungary. 46 50N 17 45E. Sit. to SW of Budapest and S of the Bakony Forest. It is the largest L. in central Europe with an area of 231 sq. m. (598 sq. km.). Vines and fruit are grown in the vicinity and it is a tourist centre.

Balboa Panama. 8 57N 79 34W. Port sit. at Pacific end of Panama Canal to the W of Panama City. It has many shipyards, docks and wharves. Pop. (1980E) 3,000.

Baldock Hertfordshire, England. 51 59N 0 11W. Town immediately ENE of Letchworth on Ivel R., mainly residential. Industries inc. electrical goods and light manufacturing. Pop. (1984E) 7,743.

Bale Ethiopia. Region in S Area, 48,108 sq. m. (124,600 sq. km.). Cap. Goba. Pop. (1984C) 1,006,491.

Bâle *see* **Basel**

Balearic Islands Spain. 39 30N 3 00E. Arch. in W Mediterranean, 50–190 m. (80–300 km.) off the E coast of Spain, forming a province. Area, 1,936 sq. m. (5,014 sq. km.). Cap. Palma. There are four islands, Majorca, Minorca, Ibiza, Formentera, which are inhabited, and many uninhabited islets, extending 180 m.

(290 km.) NE to SW. The main occupations are tourism, fishing, farming, with some mining and lumbering. Industries are centred on Majorca and inc. pottery, manu. metalware and leather goods. Pop. (1986E) 754,777.

Bali Lesser Sundas, Indonesia. 8 25S 115 15E. Island 20 m. (32 km.) W of Lombok across Lombok Strait. Area, 2,243 sq. m. (5,742 sq. km.). Chief towns Singaraja, Buleleng, Denpasar. Mountainous, rising to Mount Agung. Moderate climate with low rainfall in N, monsoon rains in S. Main occupation farming, with some handicraft industries. Main products are copra, rice, coffee, tobacco, livestock. Pop. (1981E) 2,469,930.

Balikesir Balikesir, Turkey. 39 39N 27 53E. Cap. of province in central W. Industries inc. flour milling and the manu. of rugs and textiles. Trades in cereals and opium. Pop. (1980C) 124,051.

Balikpapan South Kalimantan, Indonesia. 1 17S 116 50E. Port sit. on SE coast on the Strait of Makassar, and 210 m. (336 km.) NE of Bandjarmasin. Industries inc. oil refining.

Balkan Mountains Bulgaria. 43 15N 25 00E. A range stretching from Black Sea to Yugoslav border. The highest peak is Yamrukchal. The chief pass is the Shipka.

Balkan Peninsula South East Europe. 42 00N 23 00E. Large peninsula extending S from lower Danube and Sava Rs. to Mediterranean, S, and bounded by Adriatic and Ionian Seas, W, and by Aegean and Black Seas (linked by the Dardanelles Strait, the Sea of Marmara and the Bosporus) in the E. Divided politically between Yugoslavia, Bulgaria, most of Romania, Albania, Greece and European Turkey. Crossed E to W by the Balkan Mountains, with also the Pindus Mountains, S, the Dinaric and N Albanian Alps, W, and the Rhodope Mountains which rise to Musala Peak, SE. There are fertile plains separated by the mountain ranges and producing cereals, tobacco, cotton, olives, flax, vegetables, flowers for essential oils, and grapes.

Balkh Mazár-i-Sharif, Afghánistán. Town sit. WNW of Mazár-i-Sharif. Sit. in an oasis it is an important road junction. Pop. 12,000.

Balkhash Kazakhstan, U.S.S.R. 46 49N 74 59E. Town sit. on N shore of L. Balkhash. It is an important centre of the copper industry.

Balkhash, Lake Kazakhstan, U.S.S.R. 46 00N 74 00E. L. 100 m. (160 km.) W of Chinese frontier between Kazakh Hills (N) and Sary-Ishik-Otrau desert. Area, 6,680 sq. m. (17,301 sq. km.). Average depth 20 ft. (6 metres). Fed chiefly by the Ili R. and with no outlet. Fishing and salt extracting are important. Its main ports are Balkhash, Burlyu-Tobe, Burlyu-Baital.

Ballachulish Highland Region, Scotland. 56 40N 5 10W. Village sit. NE of Oban on S shore of Loch Leven, with large slate quarries.

Ballarat Victoria, Australia. 37 34S 143 52E. (Officially, Ballaarat). Town sit. WNW of Melbourne. Railway junction and industrial centre, industries inc. textiles, metal working, brewing. Founded in 1851 as a gold-rush town. Pop. of Urban Ballarat (1981E) 62,641.

Ballater Grampian Region, Scotland. Town sit. WSW of Aberdeen on Dee R. Resort with mineral springs. Pop. (1984E) 1,180.

Ballina Mayo, Ireland. 54 07N 9 09W. Port sit. on R. Moy near its mouth on Killala Bay. Industries inc. flour milling, plastics, manu. of medical appliances, and tools, salmon fishing, sawmilling and printing. Pop. (1980E) 8,000.

Ballinasloe Galway, Ireland. 53 20N 8 13W. Town in W centre on R. Suck. Industries inc. limestone quarrying, electrical equipment and the manu. of footwear and bonemeal. It is noted for its livestock fairs. Pop. (1981C) 6,371.

Ballinrobe Mayo, Ireland. 53 36N 9 13W. Town sit. S of Castlebar, on R. Robe near its mouth on Lough Mask.

Noted for trout and coarse fishing. Tourist centre. There is some light industry. Pop. (1985E) 1,600.

Ballybunion Kerry, Ireland. Resort sit. WNW of Listowel at the mouth of the R. Shannon. The main occupation is tourism. Pop. (1989E) 1,550.

Ballycastle Antrim, Northern Ireland. 55 12N 6 15W. Resort sit. NNW of Belfast on Ballycastle Bay. Pop. (1981C) 3,650.

Ballyclare Antrim, Northern Ireland. 54 45N 6 00W. Market town sit. SW of Larne. Industries include manu. building materials, flax fibres, acrylic yarns and protective clothing. Quarrying is also important. Pop. (1980E) 7,000.

Ballymena Antrim, Northern Ireland. 54 52N 6 17W. Town in NW sit. on R. Braid. Main occupations are tobacco industry, tyre manu. and engineering. Pop. (1981C) 55,000 (district).

Ballymoney Antrim, Northern Ireland. 55 05N 6 30W. Borough sit. SE of Coleraine. Manu. of medical supplies, mineral waters, knitwear and potato products. There is some cotton spinning. Pop. (1989E) 24,000.

Ballyshannon Donegal, Ireland. 54 30N 8 11W. Resort sit. SSW of Donegal at mouth of R. Erne on Donegal Bay. The main occupation is salmon fishing. Pop. (1980E) 2,500.

Baltic Sea North Europe. 57 00N 19 00E. Arm of the Atlantic Ocean linked to the North Sea by the Danish straits and by the Kiel Canal, Federal Republic of Germany, otherwise enclosed by Denmark, Sweden, Finland, U.S.S.R., Poland and Germany. Area, 163,000 sq. m. (422,170 sq. km.). Mainly shallow with an average depth of 180 ft. (55 metres), extending N to the Gulf of Bothnia and E to the Gulf of Finland. It has an exceptional intake of fresh water, inc. the Dal, Angerman, Ume and Torne Rs. from Scandinavia and the Neva, Narva, Western Dvina, Neman, Vistula and Oder Rs. on its S shore.

Baltimore   Maryland, U.S.A. 39 17N 76 37W. City sit. NE of Washington on Patapsco R. estuary. Commercial, industrial and communications centre of Maryland and major sea port with 2 outlets to Atlantic, one through Chesapeake Bay and one through the Chesapeake and Delaware canal. The harbour frontage is over 40 m. (64 km.). Pop. (1980C) 786,775.

Baltistan   Kashmir. A mountainous region in W Ladakh district sit. on S side of Karakoram Mountains. The highest peak is K2. The chief town is Skardu.

Baluchistan   Pakistan. 27 30N 65 00E. Region bounded N by Afghánistán, W by Iran, S by the Arabian Sea. Area, 134,139 sq. m. (347,420 sq. km.). Chief towns Quetta (cap.), Sibi, Fort Sandeman, Chaman, Kalat, Mastung, Nushki, Gawadur, Pasni. Mainly mountainous and barren with long mountain ridges crossing SW to NE. Annual rainfall is 3–10 ins. (7.5–25 centimetres), and Rs. are seasonal. The main occupation is livestock farming, with some agric. in the N valleys. There are important mineral resources inc. coal at Khost, Shahrig and Harnai, gypsum, chromite, limestone, marble, sulphur, salt and lead.

Bamako   Mali. 12 40N 7 59E. Cap. of Mali sit. on R. Niger. It is an important trading centre and became cap. in 1908 after the opening of the railway link in 1904, from Kayes, near the Senegal border. Goods traded inc. cattle and kolanuts. Pop. (1980E) 405,000.

Bamberg   Bavaria, Federal Republic of Germany. 49 53N 10 53E. Town in E centre sit. on canalized Regnitz R. near its confluence with the Main R. Industries inc. printing, precision engineering, metal works, brewing and the manu. of textiles, footwear, electrical components and leather goods. Trades in fruit and vegetables. Pop. (1985E) 70,400.

Bamburgh   Northumberland, England. 55 36N 1 42W. Resort sit. SE of Berwick-upon-Tweed, on the North Sea. Bamburgh Castle is believed to have been the seat of the Kings of

Northumbria. Industries are farming and tourism. Pop. (1981C) 423.

Banaba   *see* **Ocean Island**

Bamenda   Cameroon. 5 56N 10 10E. Cap. of NW province. Trading centre for agric. products inc. coffee and tobacco. Noted for Tikar handicrafts, esp. woodcarving. Pop. (1981E) 58,697.

Banbridge   Down, Northern Ireland. 54 21N 6 16W. Town sit. SW of Belfast on R. Bann. Manu. linen, footwear, rope and fishing nets. Pop. (1985E) 10,000.

Banbury   Oxfordshire, England. 52 04N 1 20W. Town in S centre sit. on R. Cherwell. Industries inc. high technology, printing, food processing and the manu. of aluminium, car components and electrical goods. Pop. (1981C) 35,796.

Banda Islands   Muluku, Indonesia. 4 34S 129 55E. A group of volcanic islands 50 m. (80 km.) S of Ceram. Area, c. 40 sq. m. (c. 104 sq. km.). The largest island is Bandalontar. Agric. crops inc. nutmeg, mace, sago, coconuts and copra.

Bandar Seri Begawan   Brunei. 4 56N 114 58E. Cap. sit. near the mouth of Brunei R. and once known as Brunei Town. Trading centre for agric. products and rubber. Pop. (1983E) 57,558.

Bandjarmasin   South Kalimantan, Indonesia. 3 20S 114 35E. Cap. of province and port sit. on Barito R. 24 m. (38 km.) from its mouth. Timber, rubber and rattan are exported. Pop. (1980C) 381,286.

Bandundu   Zaïre. 3 18S 17 20E. (i) Region in SW producing palm oil, cocoa and timber. Area, 114,154 sq. m. (295,658 sq. km.). Pop. (1981E) 4,119,524. (ii) Cap. of (i) on Kasai R., NE of Kinshasa. Commercial centre, river port and airport. Pop. (1976C) 96,841.

Bandung   West Java, Indonesia. 6 54S 107 36E. Cap. of province sit. SE of Djakarta. Manu. machinery, chemicals, quinine, textiles and rubber goods. Pop. (1980C) 1,462,637.

Banff Alberta, Canada. 51 10N 115 34W. Resort sit. WNW of Calgary on Bow R. in Banff National Park. The main industry is tourism. Pop. (1989E) 5,500.

Banff Grampian Region, Scotland. 57 40N 2 33W. Resort in NE sit. at mouth of R. Deveron on the Moray Firth. The main occupations are fishing and food processing. Pop. (1988E) 4,090.

Banffshire Scotland. Former county now part of Grampian Region.

Bangalore Karnataka, India. 12 58N 77 36E. State cap. in central S. Manu. inc. printing, textiles, machine tools, machinery, aeronautics, telecommunications, pharmaceuticals, electronics, watches, tractors, electrical apparatus and footwear. Pop. (1981C) 2,628,593.

Bangka Indonesia. 1 48N 125 09E. An island in the Java Sea off SE coast of Sumatra from which it is separated by Bangka Strait. Area, 4,611 sq. m. (11,942 sq. km.). One of the richest sources of tin in the world. There are also deposits of iron, lead, manganese, copper and gold.

Bangkok Phra Nakhon, Thailand. 13 44N 100 30E. City opposite Thonburi on Chao Phraya R. near the Gulf of Thailand, known as Krung Thep (city of angels by the Thais). Cap. and provincial cap., commercial, communications and cultural centre. Linked by rail with other Thai centres, Cambodia and Malaysia; accessible to smaller ocean-going vessels; has an intern. airport. Sea port handling most of Thailand's foreign trade, exporting rice, teak, rubber, salt-fish, hides, gold, silver. Industries inc. paper making, sawmilling, rice milling; manu. aircraft, rolling stock, matches, oil refining. The 18th cent. Royal Town is in the centre, and the city has grown in rings around it. Established on the site of a fort in 1782 it became cap. in place of the temporary cap. at Thonburi on the opposite bank. Pop. conurbation (1987E) 5,609,352.

Bangladesh South Asia. Commonwealth state consisting of the former E province of Pakistan, Bounded S by the Bay of Bengal, SE by Burma, E, N and W by India. Following a civil war Bangladesh was declared an independent state in Dec. 1971. Area, 55,126 sq. m. (142,776 sq. km.). Chief towns Dacca (cap.), Chittagong, Chalna, Narayanganj. Divided N to S by Brahmaputra R., which joins Ganges R. W of Dacca; most of the S is sit. in the Ganges delta. Flood plains constitute 80% of the total area. Mainly agric., with *c*. 80% of the pop. employed and 65% of the total area used in cultivation. The main products are rice, jute, sugar, wheat, gram and tea. The main export is jute, *c*. 50% of world production. Forests cover 8,560 sq. m. (22,170 sq. km.) and produce hardwoods, bamboos and honey. Fisheries are important on the coast and on the numerous Rs. Industry is concentrated on Dacca and Chittagong and inc. jute milling, textiles, steel and aluminium processing, hosiery manu. and paper making. There are deposits of natural gas, esp. at Titas. Pop. (1988E) 107m.

Bangor Down, Northern Ireland. 55 40N 5 40W. Resort in E sit. on S side of Belfast Lough. Pop. (1987E) 50,000.

Bangor Maine, U.S.A. 44 49N 68 47W. City in SE of state sit. on Penobscot R. Manu. inc. footwear, paper, medical equip., turbines, wires, resistors and semi-conductors. Pop. (1980C) 31,643.

Bangor Gwynedd, Wales. 53 13N 4 08W. Port in NW sit. on Menai Strait. Exports slate from the Penrhyn quarries. Pop. (1981C) 46,585.

Bangui Central African Rep. 4 23N 18 37E. Cap. of rep. sit. on Ubangi R. and bordering on Zaïre. Manu. inc. timber products, palm oil, office machinery and soap. Pop. (1988E) 596,776.

Bangweulu, Lake Zambia. 11 05S 29 45E. Sit. in the N on a high plateau. It is fed by the Chambezi R. and drained by the Luapula R.

Banias Syria. Port sit. SW of

Aleppo. It is the terminal of an oil pipeline from Kirkuk, Iraq.

Banja Luka   Bosnia-Hercegovina, Yugoslavia. 44 46N 17 11E. Town in NW sit. on Vrbas R. Industries inc. brewing, flour milling and the manu. of textiles. Pop. (1981C) 183,618.

Banjul   The Gambia. 13 28N 16 35W. Formerly Bathurst. Cap. and port at mouth of R. Gambia on St. Mary's Isle. Exports agric. products, chiefly groundnuts, groundnut oil and palm kernels. Pop. (1983C) 44,183.

Banks Island   British Columbia, Canada. Sit. in Hecate Strait off the Pacific coast. It is 43 m. (69 km.) long.

Banks Island   Northwest Territories, Canada. 73 15N 121 30W. Sit. in Arctic Arch. in the SW Franklin District. Area, 26,000 sq. m. (67,350 sq. km.). It is mainly plateau.

Banningville   *see* **Bandundu**

Bannockburn   Central Region, Scotland. 56 06N 3 55W. Town sit. SSE of Stirling on the Bannock Burn. Here in 1314 the battle of Bannockburn was fought in which Robert the Bruce defeated the English under Edward II. Pop. (1981C) 6,068.

Bann River   Northern Ireland. Rises 6 m. (9 km.) ENE of Warrenpoint in the Mourne Mountains and flows 80 m. (128 km.) NW past Banbridge, Gilford and Portadown, to enter L. Neagh at the S end, re-emerging at the N and flowing N along the border between County Antrim and County Londonderry to enter the Atlantic below Coleraine. Above the lough it is known as the Upper Bann R., below it is known as the Lower Bann. Salmon fisheries are important on the lower course.

Banstead   Surrey, England. 51 19N 0 12W. Town sit. E of Epsom, mainly residential with pharmaceutical industry. Pop. (1980E) 45,000.

Bantry   Cork, Ireland. 51 41N 9 27W. Port and resort in SW sit. at head of Bantry Bay. The main occupations are fishing, tourism and the manu. of machine belting and animal food. Pop. (1980E) 3,000.

Bantry Bay   Cork, Ireland. 51 38N 9 48W. Inlet of Atlantic Ocean on SW coast of County Cork, an exceptional natural anchorage. Bantry town is at its head.

Baotau   Inner Mongolia, China. 40 40N 109 59E. Town sit. WNW of Beijing. Route and commercial centre, trading in wool, hides and cereals. Industries inc. steel; manu. rugs and soap. Pop. (1982C) 1,042,000.

Bapaume   Pas-de-Calais, France. Town sit. SE of Calais. Industries inc. brick making and flour milling. Pop. (1982C) 4,085.

Baracaldo   Vizcaya, Spain. 43 18N 2 59W. Town sit. NW of Bilbao. Industries inc. iron and steel manu., engineering. Pop. (1981C) 117,422.

Baranagar   West Bengal, India. 22 38N 88 22E. Town just N of Calcutta sit. on Hooghly R. Industries inc. jute and cotton milling and the manu. of machinery and chemicals. Pop. (1981C) 170,343.

Barbados   West Indies. 13 10N 59 32W. Sovereign state sit. in the Caribbean E of St. Vincent and most E of the West Indies. Area, 166 sq. m. (430 sq. km.). Cap. Bridgetown. Coral limestone overlaid with volcanic ash and encircled by reefs. Temperate climate, but very little water. Manu. sugar, molasses, rum, food products. There is some light industry. Bridgetown is main port for ocean traffic and a cable station. Pop. (1987E) 253,881.

Barberton   Transvaal, Republic of South Africa. Town sit. in De Kaap valley near Swaziland frontier. Founded as a gold mining town which is still important but also the centre of a rich agric. area growing cotton, citrus fruit and tobacco. Forestry is also important. Pop. (1987E) 25,888.

Barberton   Ohio, U.S.A. 41 01N 81 36W. City just SW of Akron. Manu. inc. boilers, chemicals, tools, machin-

ery, auto parts, electrical equipment
and foodstuffs. Pop. (1980c) 29,751.

Barbizon   Seine-et-Marne, France.
Village sit. SSW of Melun near the
forest of Fontainebleau, which gave
its name to the Barbizon school of
artists. Pop. (1982C) 1,273.

Barbuda   Leeward Islands, West
Indies. 17 03N 61 48W. An island,
part of Antigua and Barbuda, 28 m.
(45 km.) N of Antigua. Area, 62 sq. m.
(160 sq. km.). Sea-island cotton and
sugar-cane are grown. Pop. (1984E)
79,000 with Antigua.

Barcellona Pozzo di Gotto   Sicilia,
Italy. 38 09N 15 13E. Town sit. W of
Messina. Trades in fruit, wine, olive
oil and cereals.

Barcelona   Spain. 41 23N 2 11E. (i)
Province of NE Spain bounded E by
the Mediterranean. Area, 2,975 sq. m.
(7,733 sq. km.). Chief cities Barcelona
(cap.), Sabadell, Tarrasa, Manresa,
Mataró. Pyrenean slopes occupy the N
and there is a range of hills along the
coast; inland is a fertile plain. Drained
by Llobregat and Vich Rs. The main
occupations outside the cities are
farming and mining, with tourism on
the coast. Highly industrialized and
heavily populated. Pop. (1986C)
4,598,249. (ii) City sit. ENE of Madrid
on Mediterranean coast. Provincial
cap. and second largest city in Spain.
Major sea port and commercial and
industrial centre, sit. on a fertile plain
between Ter and Llobregat R. mouths.
Industries inc. chiefly textiles, also
manu. textile and farm machinery, rolling stock, vehicles, aircraft and
marine engines; printing and publishing, chemicals, brewing, oil refining,
distilling. The port exports textiles,
machinery, wine, cork, glass, fruit,
potash and pyrites, and imports cotton, grain, coal, metals, mineral oils,
rubber. It is a bishopric and seat of a
university. Pop. (1986C) 1,694,064.

Barcelona   Anzoátegui, Venezuela.
10 08N 64 42W. State cap. sit. on
Neveri R. near its mouth on the Caribbean Sea. Trades in coffee, sugar,
cacao and tobacco. Guanta, its port,
exports cattle, hides and skins. Pop.
(1984E) 65,000.

Bardsey   Gwynedd, Wales. 52 45N
4 48W. An island just off the mainland and separated from the Lleyn
peninsula in the NW by Bardsey
Sound. It is sometimes called the
Island of 20,000 graves, which is the
number of saints believed to be buried
there.

Bareilly   Uttar Pradesh, India. 28
20N 79 25E. Town sit. ESE of Delhi.
Manu. inc. carpets and furniture.
Trades in grain and sugar-cane. Pop.
(1981C) 394,938.

Barents Sea   73 0N 39 0E. Arctic
Ocean. Sea extending 800 m. (1,280
km.) across the N of Norway and the
U.S.S.R. from the Norwegian Sea (W)
to Novaya Zemlya (E). Bounded N by
Spitzbergen and Franz Josef Land.
Shallow, and warmed by the Gulf
Stream in the SW.

Barfleur   Manche, France.. 49 40N
1 17W. Port and resort in the N sit. E
of Cherbourg. Pop. (1982C) 630.

Bari   Puglia, Italy. 41 07N 16 52E.
Cap. of Bari province and port in SE
sit. on Adriatic Sea. Industries inc. oil
refining, iron founding, food canning,
flour milling and the manu. of chemicals, textiles and soap. Wine, fruit and
olive oil are exported. Pop. (1981C)
371,022.

Barinas   Barinas, Venezuela. 8 38N
70 12W. State cap. sit. SW of Barquisimeto. Centre of an oil producing and
cattle rearing area. Pop. (1977C)
56,239.

Barisal   Bangladesh. 22 42N 90
22E. Cap. of district sit. S of Dacca. It
is a trading centre for rice, oilseeds,
jute, cane and betel-nuts. Occupations
inc. the milling of rice, oilseeds and
flour. Manu. bricks, soap, ice and general engineering products. Pop.
(1981E) 1,830,000 (district).

Barking and Dagenham   Greater
London, England. Town sit. ENE of
Lond. on Roding R. Mainly industrial;
manu. chemicals, electrical and radio
equipment. Pop. (1988E) 147,600.

Barkly East   Cape Province, Republic of South Africa. 19 50S 138

40E. Town sit. SE of Springfontein near the S end of the Drakensberg. It is the centre of a sheep farming area.

Bar-le-Duc Meuse, France. 48 47N 5 10E. Cap. of dept. in NE sit. on Ornain R. and Marne–Rhine Canal. Industries inc. metal working, brewing and the manu. of textiles, hosiery and jam. Pop. (1982C) 20,029.

Barletta Puglia, Italy. 41 20N 16 17E. Port in SE sit. on Adriatic Sea. Manu. inc. chemicals, soap and cement. Trades in fruit and wines.

Barmouth Gwynedd, Wales. 52 44N 4 03W. Resort sit. at mouth of R. Mawddach on Cardigan Bay. Pop. (1985E) 2,100.

Barnard Castle County Durham, England. 53 33N 1 55W. Town sit. W of Darlington on R. Tees. Manu. penicillin. Pop. (1989E) 6,000.

Barnaul Russian Soviet Federal Socialist Republic, U.S.S.R. 53 22N 83 45E. Cap. of Altai Territory sit. on Ob R. and Turksib Railway. Industries inc. engineering, steel works, sawmilling, food processing and the manu. of textiles and footwear. Pop. (1985E) 578,000.

Barnes Greater London, England. 51 28N 0 14W. District of SW Lond. on S bank of Thames R. Mainly residential.

Barnet Greater London, England. 51 40N 0 13W. District of N Lond. and site of a battle between the royal houses of York and Lancaster, 1471. Industries inc. manu. electrical equipment, leather. Pop. (1988E) 301,400.

Barnoldswick Lancashire, England. Town sit. WSW of Skipton. Industry includes aero engineering, plastics, textiles and bed manu. Pop. (1981C) 10,126.

Barnsley South Yorkshire, England. 53 35N 1 28W. Metropolitan borough of South Yorkshire. Sit. on R. Dearne N of Sheffield. Industries inc. coalmining, glass manu. and light and heavy engineering. Clothing and carpet manu. important. Pop. (1985E) 225,890.

Barnstaple Devon, England. 51 05N 4 04W. Port, market town and resort in SW sit. on estuary of R. Taw. There is light industry inc. the manu. of pharmaceuticals, plastics, textiles, electronics, packaging and footwear. Pop. (1981C) 19,178.

Baroda *see* **Vadodara**

Barquisimeto Lara, Venezuela. 10 04N 69 19W. State cap. in NW sit. on Barquisimeto R. It is the centre of a coffee growing area and trades in coffee, cacao, sugar and rum. Industries inc. flour milling, tanning, and the manu. of textiles, leather goods, biscuits and cement. Pop. (1984E) 459,000.

Barra Island of Western Isles, Scotland. 57 00N 7 30W. A mountainous island in the Outer Hebrides separated from S. Uist by the Sound of Barra. Area, 35 sq. m. (91 sq. km.). The principal occupations are fishing and tourism. The chief town is Castlebay. Pop. (1981C) 1,364.

Barrackpore West Bengal, India. 22 46N 88 22E. Town sit. N of Calcutta on Hooghly R. Industries inc. engineering, aeronautics, jute and rice milling. Pop. (1981C) 115,516.

Barrancabermeja Santander, Colombia. 7 03N 73 52W. Port sit. on E bank of middle Magdalena R. and W of Bucaramanga. It is an important oil centre with a pipeline to Cartagena. Pop. (1984E) 64,400.

Barranquilla Atlántico, Colombia. 10 59N 74 48W. Cap. of dept. and port sit. on Magdalena R. near its mouth. It is an industrial centre and international airport. Manu. inc. chemicals, textiles, pharmaceuticals, food products, vegetable oils, rubber goods and footwear. Pop. (1980E) 924,000.

Barreiro Setúbal, Portugal. 38 40N 9 06W. Town sit. on estuary of Tagus R. opposite Lisbon. Manu. inc. fertilizers and cork products. Pop. (1981C) 38,546.

Barrhead Strathclyde Region, Scotland. 55 48N 4 24W. Town sit.

sw of Glasgow. Industries inc. the manu. of porcelain and sanitary appliances, animal foods. There is some light engineering. Pop. (1981C) 18,418.

Barrow-in-Furness Cumbria, England. 54 07N 3 14W. Port sit. on sw of Furness peninsula coast. An important centre of engineering and shipbuilding and repairs. Manu. inc. paper and chemicals. There are also the industries associated with the offshore gasfields. Pop. (1986E) 73,200.

Barrow, Point Alaska, U.S.A. The most northerly point of Alaska, on the Arctic coast. Meteorological station, with a neighbouring U.S. naval base.

Barrow River Ireland. 52 13N 6 59W. Rises in the Slieve Bloom Mountains and flows 119 m. (190 km.) E past Portarlington to the border with County Kildare, then along the borders of Kildare, Carlow, Kilkenny, Wexford and Waterford to Waterford Harbour. Navigable up to Athy.

Barrow Strait Canada. Arm of Arctic Ocean extending 150 m. (240 km.) E to w between Bathurst, Cornwallis and Devon Islands (N) and Prince of Wales and Somerset Islands (S). Viscount Melville Sound is at its w end and Lancaster Sound at its E; Peel Sound extends S, McDougall Sound and Wellington Channel extend N.

Barry South Glamorgan, Wales. 51 24N 3 18W. Port and tourist centre sit. sw of Cardiff on Bristol Channel. Service centre with some industries, inc. chemicals. Pop. (1981C) 42,900.

Bartlesville Oklahoma, U.S.A. 34 10N 97 08W. City sit. N of Tulsa on Caney R. It is the centre of an agric. and oil producing district. Industries inc. oil refining, zinc smelting, printing, and the manu. of electrical control gear, metal goods and sulphuric acid. Pop. (1980C) 34,568.

Basel Baselstadt, Switzerland. 47 33N 7 35E. City on Rhine R. at French and German frontiers, and at the head of Rhine navigation. Commercial, communications and manu.

centre. Industries inc. manu. metal goods, chemicals, pharmaceuticals, foodstuffs, textiles. Episcopal see. Co-extensive with Basel-Stadt, a half-canton. Pop. (1989E) 200,000.

Basel Canton Switzerland. 47 35N 7 35E. Canton of N Switzerland consisting of Basel-Stadt and Basel-Land. Area, 180 sq. m. (466 sq. km.). Cap. Basel. Mainly occupied by the city and suburbs of Basel, with farming in the rural areas, esp. cattle and cereals. Pop. (1983E) Basel-Stadt (203,915), Basel-Land (219,822).

Basel-Land Switzerland. Demi-canton in the N. Area, 165 sq. m. (428 sq. km.). Cap. Liestal.

Basel-Stadt Switzerland. Demi-canton in the N. Area, 14 sq. m. (37 sq. km.). Cap. Basel.

Bashkir Russian Soviet Federal Socialist Republic, U.S.S.R. 54 00N 57 00E. Autonomous Rep. sit. in w foothills of Ural Mountains. Area, 55,430 sq. m. (143,564 sq. km.). Annexed to Russia 1557; constituted as Autonomous Soviet Rep. 1919. Cap. Ufa. Chief industries are oil, chemicals, coal, steel, electrical engineering, timber and paper. Cereals, potatoes and sugar-beet are grown. The longest pipeline in the U.S.S.R. connects the oilfield at Tuimazy with the Omsk refineries. Pop. (1981E) 3,860,000.

Basildon Essex, England. 51 35N 0 25E. Town sit. SSW of Chelmsford in a mainly residential area. A 'new town' developed since 1949 originally to accommodate London overspill population. Pop. (1985E) 160,000 (district).

Basilicata Italy. 40 30N 16 00E. Region in S sit. between Tyrrhenian Sea and Gulf of Taranto, and consisting of the provinces of Matera and Potenza. Area, 3,858 sq. m. (9,992 sq. km.). It is mountainous in the W. The chief crops are cereals, vines and olives. The principal towns are Matera and Potenza. Pop. (1981C) 610,186.

Basingstoke Hampshire, England. 51 15N 1 05W. Town in central S. Manu. inc. electronics, light engineering, printing and publishing. Headquarters of the Civil Service Commission and the Automobile Association. Pop. (1984E) 80,000.

Basle *see* **Basel**

Basque Provinces Spain. The three provinces of Álava, Guipúzcoa and Vizcaya on the Bay of Biscay and bounded NE by France. Combined area 2,803 sq. m. (7,260 sq. km.). Chief cities Bilbao Vizcaya, Guernica, San Sebastián and Victoria. Densely populated and highly industrialized; mining (iron, lead, copper and zinc), metal working, shipbuilding and fishing are important. Only Álava is mainly agric., producing maize, sugarbeet, wine and cider. Pop. (1986C) Álava, 275,703; Guipúzcoa, 688,894; Vizcaya, 1,168,405; total, 2,133,002.

Basra Basra, Iraq. 30 30N 47 47E. Cap. of governorate and port sit. 70 m. (112 km.) from Persian Gulf on Shatt-al-Arab R. It is linked to Baghdad by rail and R. steamer. Exports inc. dates, cereals, wool and oil. Pop. (1970E) 333,684.

Bas-Rhin France. 48 40N 7 30E. Department Alsace region. Bounded E and N by Federal Republic of Germany and W by Vosges mountains. Area, 1,848 sq. m. (4,787 sq. km.). Chief towns Strasbourg (prefecture), Haguenau. Mainly lowland along the Rhine R. and rising to the lower slopes of the Vosges in the W. Highly industrialized, the main industries being manu. machinery, textiles, locomotives, leather. The Vosges foothills produce vines, the lowlands are fertile and produce cereals, vegetables, fruit and hops. Pop. (1982C) 915,676.

Bassano del Grappa Veneto, Italy. 45 46N 11 44E. Town in the NE sit. NNE of Vicenza. Industries inc. printing, tanning and the manu. of grappa, metal goods, paper, pottery, porcelain, silk and wax.

Bassein Lower Burma, Burma. 15 56N 94 18E. Port 70 m. (112 km.) from coast on Bassein R., WSW of Rangoon. Industries inc. rice milling and the manu. of pottery. Rice is exported.

Basses-Alpes *see* **Alpes-de-Haute-Provence**

Basse-Normandie France. 48 45N 0 10E. Region in N comprising depts of Calvados, Manche and Orne. Area, 6,788 sq. m. (17,583 sq. km.). Cap. Caen. Extends W to the W coast of the Cherbourg peninsula and E to mouth of Seine R. Drained by Orne, Vire and Touques Rs. Fertile agric. area. Chief cities Caen and Cherbourg. Pop. (1982C) 1,350,979.

Basses-Pyrénées *see* **Pyrénées-Atlantiques**

Basse-Terre Guadeloupe, French West Indies. 16 00N 61 42W. (i) The W half of Guadeloupe. Area, 327 sq. m. (848 sq. km.). Pop. (1982C) 141,300. (ii) Cap. of dept. and chief town of (i) and port for coffee and cacao exports. Pop. (1982C) 13,656.

Bass Rock Lothian Region, Scotland. 56 05N 2 38W. Sit. at mouth of Firth of Forth. Once used as a prison, it is now a seabird sanctuary.

Bass Strait Australia. 39 20S 145 30E. A channel between the Indian Ocean and the Tasman Sea and separating Victoria and Tasmania.

Bastia Haute-Corse, France. 42 42N 9 27E. Cap. of dept. and port sit. on NE coast. Industries inc. fishing and the manu. of tobacco products. Fish, wine and fruit are exported. Pop. (1982C) 45,081.

Basutoland *see* **Lesotho**

Bas-Zaïre Zaïre. Region in W containing Zaïre R. estuary. Area, 20,819 sq. m. (52,920 sq. km.). Mineral-producing area, with oil, iron and gold, besides agric. products. Chief towns Matadi (cap.) and Boma (both sea and river ports). Pop. (1981E) 1,921,524.

Bata Equatorial Guinea. 1 51N 9 45E. Chief town of mainland part (Rio Muni) of rep. Port for timber exports. Pop. (1984E) 24,000.

Bataan Luzon, Philippines. Peninsula extending 30 m. (48 km.) along the W side of Manila Bay. Width 15–20 m. (24–32 km.).

Bath   Avon, England. 51 23N
2 22W. City sit. ESE of Bristol on
Avon R. There are minor industries
including engineering, footwear and
printing, but it is mainly known as a
spa of exceptional architectural interest. The hot springs were known in
Roman times when Bath was called
Aquae Sulis. Pop. (1981C) 79,965.

Bathgate   Lothian Region, Scotland. 55 55N 3 39W. Town sit. WSW
of Edinburgh. Industries inc. engineering, iron and steel foundry. Pop.
(1988E) 14,480.

Bathurst   New South Wales, Australia. 33 25S 149 35E. City in central
E sit. on Macquarie R. Industries inc.
agric., food-processing, brickmaking,
plastics, mapping and education. Pop.
(1989E) 26,500.

Bathurst   New Brunswick, Canada.
47 36N 65 39W. Port sit. on estuary
of Nipisiguit R. in the Gulf of St.
Lawrence, NNW of Moncton. Industries inc. salmon fishing, paper milling
and lumbering. There are important
deposits of lead, zinc and copper in
the vicinity. Pop. (1980E) 16,301.

Bathurst, Gambia   *see* **Banjul**

Bathurst, Cape   Northwest Territories, Canada. A promontory on N
coast extending into the Beaufort Sea.

Bathurst Inlet   Northwest Territories, Canada. An arm of the Arctic
Ocean S of Victoria Island and
stretching S from Coronation Gulf.

Bathurst Island   Canada. 76 00N
100 30W. Island in Parry group N of
Barrow Strait in the Arctic Ocean.
Area, 7,300 sq. m. (18,907 sq. km.).

Batley   West Yorkshire, England.
53 44N 1 37W. Town sit. SSW of
Leeds. Light engineering and the
manu. of bedding, biscuits and clothing have largely replaced the old
woollen cloth and shoddy industries.
Pop. (1989E) 60,000.

Baton Rouge   Louisiana, U.S.A. 30
23N 91 11W. State cap. and port sit.
on E bank of Mississippi R. Industries
inc. oil refining, food processing and

the manu. of chemicals. Pop. (1980C)
219,419.

Battersea   Greater London, England. District of SW London on S bank
of Thames R., opposite Chelsea.
Noted for its extensive park.

Battle   East Sussex, England. 50
55N 0 29E. Town sit. NW of Hastings.
Scene of Battle of Hastings (1066)
resulting in William I becoming King
of England. Pop. (1989E) 5,800.

Battle Creek   Michigan, U.S.A. 42
19N 85 11W. City sit. W of Detroit at
confluence of Battle Creek and Kalamazoo R. Manu. inc. cereals, health
foods, chewing gum, automotive parts,
packaging. The Kellogg bird sanctuary is sit. in the vicinity. Pop. (1989E)
56,339.

Batumi   Georgian Soviet Socialist
Republic, U.S.S.R. 41 38N 41 38E.
City on E coast of Black Sea. Cap. of
the Adzhar Autonomous Soviet Socialist Rep. in SW Georgia. The main industries are refining oil from Baku
and Alyaty, marine and railway engineering, fruit and vegetable canning;
manu. cans and clothing. Chief sea
port, exporting petroleum and manganese. There is also a naval base. Pop.
(1984E) 111,000.

Bauchi   Nigeria. 10 19N 9 50E.
State divided into the emirates of
Bauchi, Gomba, Jama'are, Katagum
and Misau, formerly the Bauchi pro-
vince of the N region. Area, 24,944 sq.
m. (64,605 sq. km.). Chief towns
Bauchi (cap.), Kumo, and Gombe.
Pop. (1983E) 3·5m.

Bauchi Plateau   Nigeria. Highlands
in central Nigeria rising to over 5,000
ft. (1,524 metres). The main product is
tin, chiefly mined at Jos.

Bautzen   Dresden, German Democratic Republic. 51 11N 14 26E.
Town in SE sit. on Spree R. Manu.
railway rolling stock, machinery, textiles and chemicals. Pop. (1989E)
52,700.

Bavaria   Federal Republic of Germany. 48 30N 11 30E. *Land* in S,
bounded NE by German Democratic

55

Republic, E by Czechoslovakia, SE and S by Austria. Area, 27,232 sq. m. (70,551 sq. km.). Chief cities Munich (cap.), Nuremberg, Augsburg, Regensburg, Würzburg. Mountainous and heavily forested with fertile valleys and plains between the ranges. The Alps extend E to W along its S boundary, the Bohemian Forest is on its E and the Fichtelgebirge and Frankenwald Mountains are NE. It is also crossed by the Franconian Jura. Drained by Danube and Main Rs. and their tribs. Brewing is a major industry. The agric. areas produce cereals and potatoes, the pastures support large numbers of cattle and pigs. Most of the forest is commercially valuable. Pop. (1983E) 10,964,220.

Bay City   Michigan, U.S.A. Port sit. NNW of Detroit on Saginaw R. on Saginaw Bay. Industries inc. oil refining, sugar refining, fishing, and the manu. of cranes, boats, prefabricated buildings, car parts, chemicals and cement. Pop. (1980C) 41,593.

Bayern   *see* **Bavaria**

Bayeux   Calvados, France. 49 16N 0 42W. Town near English Channel coast of Normandy. Manu. inc. lace and pottery. The Bayeux Tapestry, a roll of linen 20 inches (50 cm.) wide and 231 ft. (70 metres) long upon which are worked, in thread, the events concerning the conquest of England, is in the museum. Pop. (1982E) 15,237.

Bay Islands   *Islas de La Bahía,* Honduras. 35 14S 174 08E. Dept. comprising a group of islands sit. in Gulf of Honduras. Area, 114 sq. m. (373 sq. km.). The largest island is Roatán. The principal crops are coconuts, bananas and pineapples. Pop. (1984E) 8,863.

Bayonne   Pyrénées-Atlantiques, France. 43 29N 1 29W. Port in SW sit. at confluence of Rs. Nive and Adour near Bay of Biscay. Industries inc. oil refining, distilling, bacon-curing, flour milling and the manu. of fertilizers and leather goods. Exports inc. timber, brandy, crude oil and steel products from the steel works at nearby Boucau. Pop. (1982C) 42,920 (agglomeration, 127,477).

Bayonne   New Jersey, U.S.A. 40 41N 74 07W. City sit. just S of Jersey City on a peninsula between Newark Bay and Upper New York Bay. Industries inc. oil refining, and the manu. of chemicals, machinery, textiles, boilers, radiators, cables and food products. It is the terminal of an oil pipeline from Texas. Pop. (1980C) 65,047.

Bayreuth   Bavaria, Federal Republic of Germany. 49 57N 11 35E. Town in central E. Noted for its annual musical festival of Wagner's works. Wagner, Liszt and Richter are all buried here. Manu. inc. textiles and pottery. Pop. (1984E) 71,800.

Bayswater   Greater London, England. District of W central Lond., W of Marble Arch and N of Hyde Park. Mainly a residential and hotel area.

Bayt Lahm   *see* **Bethlehem**

Baytown   Texas, U.S.A. 29 43N 94 59W. City sit. E of Houston on Galveston Bay and is a water recreation area. Industries inc. oil refining and the manu. of chemicals and synthetic rubber. Pop. (1985E) 59,231.

Beachy Head   England. 50 44N 0 16E. A chalk headland on the S coast sit. between Eastbourne and Seaford.

Beaconsfield   Buckinghamshire, England. 51 37N 0 39W. Town sit. WNW of central London. Mainly residential. Pop. (1981C) 10,909.

Beardmore Glacier   Antarctica. One of the largest glaciers in the world, moving from the Queen Alexandra Range to the Ross Ice Shelf.

Bearsden   Strathclyde Region, Scotland. Town sit. NW of Glasgow developed since 1958 as a residential town.

Beas River   India. 31 10N 75 00E. Rises in the Himalayas and flows 290 m. (464 km.) W and SW to join the Sutlej R. 25 m. (40 km.) ENE of Ferozepur.

Beauce   France. Area in Paris Basin

extending through parts of the Eure-et-Loir, Loir-et-Cher, Loiret, Essonne and Yvelines depts. A fertile lime-stone plain producing mainly wheat. The chief town is Chartres.

Beaufort Sea   Arctic Ocean. 70 30N 146 00W. Sit. between N Alaska, U.S.A., and Banks Island, Canada. It is mainly shallow and covered with drifting ice but is more than 12,000 ft. (3,658 metres) deep in the NW.

Beaujolais   France. Area on NE edge of Massif Central extending through the N of the Rhône dept. and part of the Loire dept. Noted as a major producer of Burgundy wines. The chief town is Villefranche sur Saône.

Beaulieu   Hampshire, England. 43 45N 7 20E. Village sit. SSW of Southampton. The National Motor Museum is sit. in the grounds of Palace House and there are ruins of a Cistercian abbey. Pop. (1985E) 1,200.

Beauly   Highland Region, Scotland. 57 29N 4 29W. Village in NW sit. on Beauly R., at its mouth on Beauly Firth, a continuation of Moray Firth. It has ruins of the Cistercian Priory of St. John. Pop. (1981C) 1,496.

Beaumaris   Isle of Anglesey, Gwynedd, Wales. 53 16N 4 05W. Resort sit. at N end of Menai Strait, on Beaumaris Bay, and 4 m. (6.5 km.) NNE of Telford's suspension bridge. Pop. (1981C) 2,046.

Beaumont   Texas, U.S.A. 30 05N 94 06W. City in extreme SE of state sit. on Neches R., N of the Gulf of Mexico. Industries inc. oil refining, petrochemicals, meat packing, and the manu. of paper and forestry products, laboratory and optical equipment. Trades in cattle, timber, grain, rice and cotton. Pop. (1980C) 118,201.

Beaune   Côte d'Or, France. 47 02N 4 50E. Town in central E. Manu. inc. agric. implements, mustard and casks. It is noted for Burgundy wines. Pop. (1982C) 21,127.

Beauvais   Oise, France. 49 26N 2 05E. Cap. of dept. in N sit. on

Thérain R. Manu. inc. brushes, rayon, tiles, perfumes, brakes, tractors, carpets and foodstuffs. Pop. (1982C) 54,147.

Bebington   Merseyside, England. 53 23N 3 01W. Town immediately S of Birkenhead on Wirral peninsula. The Bromborough district is industrial, with chemical works; Port Sunlight is a model town built in 1888 to house workers at the Lever Bros., now Unilever, soap and margarine factory. Pop. (1981C) 64,174.

Bec Abbey   Eure, France. Ruins of a Benedictine house in the village of Le Bec-Hallouin, sit. SW of Rouen. An important centre of the medieval church from which Anselm and Lanfranc went as Archbishop of Canterbury, England.

Beccles   Suffolk, England. 52 28N 1 34E. Town in East Anglia sit. on R. Waveney. Industries inc. printing, boat building, abbatoir and meat processing plant. Manu. plastic containers, packaging and process machinery equipment. Pop. (1988E) 10,815.

Béchar   Saoura, Algeria. 31 37N 2 13W. Oasis town sit. near frontier with Morocco. Linked by rail with Abadla, on the edge of the Sahara, and with the Kenadsa coalfield.

Bechuanaland   *see* **Botswana**

Beckenham   Greater London, England. District of SE London forming a dormitory suburb. Noted as the site, since 1930, of the Bethlem (or 'Bedlam') mental hospital, founded in London in 1247.

Bedford   Bedfordshire, England. 52 08N 0 29W. County town in SE central Eng. sit. on R. Ouse. Manu. inc. diesel engines, pumps, agric. and electrical and electronic equipment, confectionery and bricks. Aircraft research is also undertaken. Pop. (1987E) 90,450.

Bedfordshire   England. 52 05N 0 30W. Under 1974 reorganization the county consists of the districts of North Bedfordshire, Mid Bedfordshire, South Bedfordshire and Luton. Manufacturing industry is a major employer, and the county's prime location and accessibility has encouraged significant growth in the service sector. Minerals include clay, chalk, Fullers Earth and sand and gravel. Major car and truck factories are based in the Luton-Dunstable area and in mid-Bedfordshire is one of the world's largest brickworks. Agriculture is very important, about 70% of land being in arable use, mainly wheat and barley. Pop. (1988E) 539,700.

Bedlington Northumberland, England. 55 08N 1 35W. Town sit. N of Newcastle upon Tyne. A former coal-mining town it is now developing light industry and is a residential area. It gave its name to a breed of terriers. Pop. (1985E) 14,832.

Bedworth Warwickshire, England. 52 28N 1 29W. Town sit. S of Nuneaton. Industries inc. mechanical engineering, automobile parts and textiles. Pop. (1984E) 42,500.

Beersheba Israel. 31 15N 34 47E. Town sit. SW of Jerusalem. Manu. pottery and glass. Pop. (1982E) 112,600.

Beeston and Stapleford Nottinghamshire, England. 52 56N 1 12W. Town sit. SW of Nottingham. Industries inc. engineering; manu. pharmaceutical products, hosiery, lace. Pop. (1988E) 65,855.

Beira Sofala, Mozambique. 19 49S 34 52E. Cap. of province and port sit. at mouth of R. Pungwe and Busi. It is linked by rail with Zaïre, Malawi, Zambia and Zimbabwe, and handles a large amount of their imports and exports. Pop. (1980E) 214,613.

Beijing (Peking) Hebei, China. 39 55N 116 15E. Cap. of the rep. sit. in NE, and its commercial and cultural centre. The Outer City to the S and Inner City to the N are adjacent, and otherwise known as the Chinese and Tartar cities. The Inner City contains the Imperial City, which in turn contains the Forbidden City. Tiananmen Square is one of the largest public squares in the world, and on one side is the Gate of Heavenly Peace built in 1417 used by Chinese leaders as a rostrum on great political occasions. The Great Wall of China lies 50 m. (80 km.) to the N. Industries inc. engineering, food processing, printing and publishing, tanning and textiles. Pop. (1987E) 5,970,000, metropolitan area, 9,750,000.

Beirut Lebanon. 33 52N 35 30E. Cap. of rep. and port. It controls the main road and railway links along the coast. Beirut was founded by the Phoenicians in the 14th cent. It was occupied by the French in 1918 and became the cap. of a new state in 1920. Before the civil war started in 1975 it was the most important trade and financial centre in the Middle East. Industries inc. engineering, food processing and the manu. of textiles. There are many small workshops making furniture and clothing. Tourism was important before the political troubles. Pop. (1980E) 702,000.

Beisan Israel. 32 30N 35 30E. Town sit. SE of Haifa on the site of an anc. fortress. Many early relics have been found.

Béja Béja, Tunisia. 36 44N 9 11E. Cap. of governorate sit. W of Tunis in Medjerda valley. Trades in wheat. Pop. (1984C) 46,708.

Bejaia Constantine, Algeria. 36 45N 5 05E. Town sit. WNW of Constantine, on the Mediterranean. Sea port and oil pipeline terminal, handling oil from Hassi Messaoud, fruits, cereals, olive oil, iron ore and phosphates. Pop. (1983E) 124,122.

Békés Békés, Hungary. 46 47N 21 09E. Town sit. NNE of Szeged on White Körös R. Manu. textiles, furniture and bricks. Trades in cereals and tobacco. Pop. (1984E) 24,000.

Békéscsaba Békés, Hungary. 46 41N 21 06E. Cap. of county sit. SE of Budapest. Manu. inc. machinery, textiles and bricks. Trades in poultry and agric. produce. Pop. (1984E) 69,000.

Belaya River U.S.S.R. Rises in the S Urals and flows 700 m. (1,120 km.) SW past Beloretsk, then N and NW past Sterlitamak and Ufa to join Kama R. Used for transport and irrigation.

Belaya Tserkov Ukraine, U.S.S.R. 49 49N 30 07E. Town sit. SSW of Kiev. Industries inc. food processing, flour milling and the manu. of leather goods and clothing. Pop. (1985E) 181,000.

Belém Pará, Brazil. 1 27S 48 29W. State cap. and port in N sit. on Pará R. It is the chief port of the R. Amazon basin and exports timber, nuts, rubber, rice, jute and carnauba wax. Pop. (1980C) 755,984.

Bélep Islands South West Pacific. 19 45S 163 40E. French group of islands sit. N of New Caledonia, of which they form depend. Main islands are Art, Pott and Nienane. Area, 27 sq. m. (70 sq. km.). Pop. (1983C) 686.

Belfast Antrim and Down. Northern Ireland. 54 35N 5 55W. Cap. of N. Ireland and County town sit. at mouth of Lagan R. on Belfast Lough. Industries inc. shipbuilding, aircraft and missile construction, light engineering, and the manu. of tobacco and textiles. Seat of Queen's University and of the Northern Ireland administration at Stormont. Received large numbers of English and Scottish settlers during 16th–17th cent. 'plantation' settlement, also many Huguenots after 1685 who stimulated the cloth industry. The city still has marked Nationalist (Catholic) and Unionist (Protestant) areas and since 1968 has been torn by civil unrest. Pop. (1987E) 303,800.

Belfort Territoire de Belfort, France. 47 38N 6 52E. Cap. of dept. in central E sit. in Belfort Gap between Vosges and Jura mountains. Manu. inc. locomotives, turbines, computers and electrical equipment. Pop. (1982C) 52,739.

Belfort, Territoire de France. 47 38N 6 52E. Dept. in Franche-Comté region, part of the former province of Alsace. Area, 236 sq. m. (610 sq. km.). Cap. Belfort. Pop. (1982E) 131,999.

Belgaum Karnataka, India. 15 52N 74 31E. Town sit. SSE of Bombay. Manu. inc. textiles, furniture and leather. Trades in rice. Pop. (1981C) 274,430.

Belgium 51 30N 5 00E. Kingdom bounded N by North Sea and Netherlands, E by Federal Republic of Germany and Luxembourg, S and SW by France. Area, 11,778 sq. m. (30,152 sq. km.). Since 1830 an independent Kingdom, it was part of *Galla Belgica*, so called from the tribe of *Belgae*. Chief cities: Brussels (cap.), Antwerp, Liège, Ghent, Bruges, Mons, Hasselt, Arlon, Namur, Ostend. Divided into 9 provinces: Antwerp, Brabant, West Flanders, East Flanders, Hainaut, Liège, Limbourg, Luxembourg, Namur. Divided physically into 3 regions: the N is sandy and lowlying; the central plain is drained by the Scheldt and Meuse Rs. and their tribs. and canals and is very fertile; the SE is plateau rising to the Ardennes, heavily forested, with a continental climate. There are 1,529,470 hectares of land under cultivation (about half the total), producing cereals, potatoes, sugar-beet, flax and hops; pastures support cattle, pigs and sheep. Natural resources inc. coal in the Hainaut province, esp. at Mons and Charleroi. Iron ore for the metallurgical industry at Liège is imported from France. The other main industries are textiles (centred on Ghent), and engineering. Belgium is mainly bilingual. The French language prevails in the Walloon areas in the S and the E, and Flemish (or Dutch) in the provinces of Antwerp, and the N of Brabant, Flanders and Limbourg. The Belgian Govt. has co-operated with the Dutch Govt. in agreeing to simplified rules for writing the Dutch language. Since 1963, there have been official language zones in Belgium: Flemish is the official language in the N, French in the S, and German in the E border districts (Liège province). Brussels, which is in the Flemish-speaking area, is officially bilingual. Pop. (1987E) 9,875,716.

Belgorod Russian Soviet Federal Socialist Republic, U.S.S.R. 50 36N 36 35E. Town sit. NNE of Kharkov on N. Donets R. Industries inc. chalk quarrying, meat packing, flour milling and tanning. Pop. (1985E) 280,000.

Belgorod-Dnestrovsky Ukraine, U.S.S.R. 46 12N 30 20E. Port sit. at mouth of Dniester R. on the Black

Sea, sw of Odessa. Trades in fish, salt and wine.

Belgrade Serbia, Yugoslavia. 44 50N 20 30E. City at confluence of Danube and Sava Rs. Cap. of Yugoslavia and of the Serbian constituent rep. since 1945 Novi Beograd has been established on the left bank of the Sava R. to the s. Little remains of the old city which was destroyed in the second world war. Commercial centre, R. port and route centre. Industries inc. textiles, chemicals, food processing; manu. electrical goods and farm equipment. Pop. (1981C) 1,470,073.

Belgravia Greater London, England. 51 30N 0 09W. District of w central London bounded N by Knightsbridge and planned mainly in residential squares by Thomas Cubitt. Built 1825–30.

Belize 17 00N 88 00W. Formerly British Honduras. Bounded N by Mexico, w by Guatemala, and s and E by the Caribbean Sea. Area, 8,867 sq. m. (22,963 sq. km.). Cap. Belmopan. The early settlement of the territory was probably effected by Brit. woodcutters about 1638; from that date to 1798, in spite of armed opposition from the Spaniards, settlers held their own and prospered. In 1780 the Home Govt. appointed a superintendent and in 1862 the settlement was declared a colony, subordinate to Jamaica. It became an independent colony in 1884. The main agric. export is sugar, followed by citrus fruit, chiefly grapefruit and oranges. The forests produce mahogany, cedar, Santa Manà, pine and rosewood. There are many secondary hardwoods of known or probable market value as well as wood suitable for pulp production. Fishing is an important industry and fish products represent the second most important export. Pop. (1988E) 179,800.

Belize City Belize. 17 30N 88 12W. Chief city and port and former cap. sit. at mouth of Belize R. Belize City has a modern deep water port. Exports inc. timber, esp. mahogany, citrus fruits, bananas, fish, including lobster, clothing, sugar, honey. Pop. (1988E) 49,700.

Bellagio Lombardia, Italy. 45 59N 9 15E. Resort sit. on L. Como on a promontory dividing the two s arms of the L., and sit. NNE of Como.

Bellary Karnataka, India. 15 09N 76 56E. Town sit. wsw of Madras. Industries inc. sugar milling and the manu. of textiles. Trades in cotton. Pop. (1981C) 201,579.

Bellegarde-sur-Valserine Ain, France. Town sit. sw of Geneva at confluence of R. Rhône and Valserine. Manu. inc. textiles and metal products. Pop. (1982C) 11,787.

Belle-Île-en-Mer Morbihan, France. 47 20N 3 10W. An island sit. in Bay of Biscay s of the Quiberon peninsula. Area, 32 sq. m. (83 sq. km.). The main occupation is fishing, esp. pilchards and sardines. The chief town is Le Palais. Pop. (1980C) 5,500.

Belle Isle Canada. 51 35N 56 30W. Rocky island sit. in the Strait of Belle Isle between Newfoundland and Labrador.

Belleville Ontario, Canada. 44 10N 77 23W. City sit. w of Kingston on L. Ontario. Industries inc. engineering, meat packing, and the manu. of food processing, telecommunications, electronics, metals, plastics and paper products. Pop. (1985E) 40,000.

Belleville Illinois, U.S.A. 38 31N 90 00W. City sit. SE of St. Louis. Manu. inc. stoves, chemicals and clothing. There is coalmining in the vicinity. Pop. (1980C) 41,580.

Belleville New Jersey, U.S.A. 40 48N 74 09W. Town just NE of Newark. Manu. inc. machinery, electrical equipment, chemicals and food products. Pop. (1988C) 38,005.

Bellevue Washington, U.S.A. 47 37N 112 12W. City facing Seattle across L. Washington, from the E shore and joined to it by floating-bridge highways. Residential and office centre; industries inc. food distribution, advanced medical and aerospace equipment, computer services and software. Pop. (1980C) 73,903.

Bellingham Washington, U.S.A. 48
46N 122 29W. City and port sit. N of
Seattle. Industries inc. pulp and paper,
chemicals, aluminium, oil, fishing and
food processing. Pop. (1989E) 46,380.

Bellingshausen Sea Southern
Ocean. 71 00S 85 00W. Sit. to W of
Graham Land and to SW of Drake
Strait.

Bellinzona Ticino, Switzerland. 46
11N 9 02E. Cap. of canton in central S
Switz. sit. ENE of the head of L. Maggiore. A transportation and tourist
centre with industries inc. railway
engineering, printing and woodworking. Pop. (1980C) 16,743.

Bell Island Canada. 47 36N 52
58W. Sit. in Conception Bay off SE
Newfoundland, WNW of St. John's.
Area, 11 sq. m. (28 sq. km.). There are
iron mines.

Bell (Inchcape) Rock Scotland.
Reef 12 m. (19 km.) SE of Arbroath in
the North Sea.

Belluno Veneto, Italy. 46 09N 12
13E. Cap. of Belluno province in NE
sit. on Piave R. Manu. inc. electrical
equipment, silk, furniture and wax.

Belmopan Belize. 17 18N 88 30W.
Cap. of Belize. Following severe hurricane 'Hattie' in 1961 Belmopan was
moved 50 m. (80 km.) inland to be
away from the hurricane zone. Construction began in 1967 and it became
the seat of govt. in 1970. Many of the
public buildings are the Maya style of
architecture. Pop. (1988E) 3,700.

Belo Horizonte Minas Gerais, Brazil. 19 55S 43 56W. State in E. It is the
centre of an agric. and mining area,
producing iron, manganese, cotton
and cattle. Industries inc. iron and
steel works, diamond cutting, food
processing, and the manu. of textiles
and footwear. It was Brazil's first
planned city. Pop. (1980C) 1,441,567.

Beloretsk Russian Soviet Federal
Socialist Republic, U.S.S.R. 53 58N
58 24E. Town sit. ESE of Ufa. It is a
centre of the iron and steel industry,
using local iron and manganese ores.

Belorussia U.S.S.R. The Belorussian Soviet Socialist Rep. was established in 1919. Area, 80,134 sq. m.
(207,600 sq. km.). Valuable forest
land and rich deposits of rock salt. Industries almost completely destroyed
during Second World War, but now
very wide-ranging, with particular
attention to production of peat. Cap.
Minsk. Pop. (1989E) 10·2m.

Belovo Russian Soviet Federal
Socialist Republic, U.S.S.R. 54 25N
86 18E. Town in S Siberia sit. NW of
Novokuznetsk, in the Kuznetsk Basin.
Industries inc. coalmining, zinc smelting, and the manu. of radio equipment
and metal products. Pop. (1977E)
112,000.

Belper Derbyshire, England. 53
01N 1 29W. Town sit. N of Derby on
R. Derwent. Industries inc. cotton and
hosiery mills, oil refinery and chemical works. Manu. domestic boilers.
Pop. (1981C) 16,414.

Belsen Lower Saxony, Federal
Republic of Germany. Village in the
NE sit. NW of Celle. Site of a former
Nazi concentration camp.

Beltsy Moldavia, U.S.S.R. 47 46N
27 56E. Town in the SW sit. NW of
Kishinev. Industries inc. meat packing, sugar refining and flour milling.
Pop. (1981E) 128,000.

Benares *see* **Varanasi**

Benbecula Western Isles area,
Scotland. 57 26N 7 20W. An island
sit. between N. Uist and S. Uist in
Outer Hebrides. Area, 36 sq. m. (93
sq. km.). The main occupations are
farming and fishing.

Bendel Nigeria. State comprising
the Benin and Delta provinces of the
former W region. Area, 13,707 sq. m.
(35,500 sq. km.). The inhabitants are
chiefly Edo people, with minorities of
Ibos in the E and Ijaw in the S. Main
towns Benin (cap.), Sapele and Warri.
Pop. (1988E) 5,391,700.

Bendigo Victoria, Australia. 36
46S 144 17E. City in N centre of state.
It is a service centre sit. in an agric.
district. Other industries inc. tourism

and the manu. of textiles and rubber. There are railway workshops. Pop. (1985E) 62,000.

Benevento   Campania, Italy. 41 08N 14 45E. Cap. of Benevento province sit. NE of Naples sit. on Calore R. Manu. confectionery, leather goods, matches and Strega liqueur. Pop. (1982E) 63,000.

Benfleet   Essex, England. 51 33N 0 35E. Town sit. W of Southend on N shore of Thames R. estuary. S Benfleet is on Benfleet Creek which separates the mainland and Canvey Island. A residential town. Pop. (1982E) 51,000.

Bengal, Bay of   Indian Ocean. 18 00N 89 00E. Sit. between India and Burma, and containing the islands of Andaman, Nicobar and Mergui. The Rs. Ganges, Brahmaputra, Irrawaddy, Mahanadi, Godavari, Krishna and Cauvery discharge into the bay.

Bengal, East   *see* **Bangladesh**

Benghazi   Benghazi, Libya. 32 11N 20 03E. Port and joint cap. (with Tripoli) of division, sit. at E end of the Gulf of Sidra. Pop. (1984E) 650,000.

Benguela   Angola. 12 35S 13 25E. Port sit. SW of Lobito. Linked by railway to Zaïre and Zimbabwe. Industries inc. sugar milling and the manu. of soap, pottery and tools. Pop. (1975E) 41,000.

Benin   10 00N 2 00E. Rep. on Gulf of Guinea bounded E by Nigeria, N by Niger and Burkina Faso and W by Togo. Area, 43,483 sq. m. (112,622 sq. km.). Chief towns, Porto Novo (cap.), Cotonou, Abomey and Parakou. The climate on the coastal plain is equatorial. Chief occupation farming, esp. cassava, millet, maize, yams, cotton and coffee. Chief exports are cotton, cocoa, palm oil and kernels. Pop. (1988E) 4,444,000.

Benin, Bight of   Gulf of Guinea. 5 00N 3 00E. Extends W to E from Achowa Point, Ghana, to the Niger delta, *c.* 520 m. (832 km.) across.

Benin City   Bendel, Nigeria. 6 19N 5 41E. Cap. of State sit. E of Lagos. The forests contain rubber. Trading centre for palm oil and kernels, timber. Handicraft industries inc. wood carving, brass working. It was formerly cap. of the anc. Benin kingdom. Pop. (1988E) 375,430.

Beni Suef   Beni Suef, Egypt. 29 03N 31 02E. Cap. of governorate in the N sit. on the Nile R. Industries inc. cotton ginning. Trades in local produce, esp. of the oasis of El Faiyum. Pop. (1983E) 146,000.

Ben Lomond   Scotland. 56 12N 4 39W. A mountain 27 m. (43 km.) WNW of Stirling sit. on E side of Loch Lomond.

Ben Nevis   Scotland. 56 48N 5 00W. A mountain 4 m. (6.5 km.) ESE of Fort William overlooking Glen Nevis. It is the highest peak in the Brit. Isles.

Benoni   Transvaal, Republic of South Africa. 26 19S 28 27E. Town sit. E of Johannesburg on Witwatersrand. Industries inc. engineering and gold mining. Pop. (1980C) 206,810.

Bensberg   North Rhine-Westphalia, Federal Republic of Germany. 50 58N 7 09E. Town sit. ENE of Cologne. Manu. inc. leather goods.

Benue   Nigeria. State mainly created from the Benue province of the former N region. Area, 17,442 sq. m. (45,174 sq. km.). The main ethnic group are the Tiv people, and the cap. and chief town is Makurdi. Pop. (1988E) 5,317,500.

Benue River   Cameroon/Nigeria. 7 48N 6 46E. R. rising in the Adamawa Mountains N of Ngaoundéré, Cameroon, flowing 900 m. (1,440 km.) N to the Nigerian border then WSW past Yola and Makurdi, Nigeria, to join the Niger R. at Lokoja. Navigable as far as Garona, Cameroon, from July–October.

Beograd   *see* **Belgrade**

Berar   *see* **Madhya Pradesh**

Berbera   Somali Republic. 10 30N 45 02E. Port sit. on Gulf of Aden. Exports inc. goats, sheep, skins and gums. Pop. (1980E) 65,000.

Berbice River   Guyana. 6 20N 57 32W. Rises in Guiana Highlands in SE Guyana and flows 300 m. (480 km.) N to enter Atlantic near New Amsterdam. Some diamond mining on its middle course.

Berchtesgaden   Bavaria, Federal Republic of Germany. 47 38N 13 01E. Town and resort sit. SSW of Salzburg, Austria. The main occupation is tourism. Pop. (1989E) 23,650.

Berdichev   Ukraine, U.S.S.R. 49 54N 28 36E. Town sit. WSW of Kiev. Industries inc. engineering, sugar refining, tanning and food processing. Pop. (1980C) 60,000.

Berezina River   U.S.S.R. 52 33N 30 14E. Rises *c*. 37 m. (59 km.) W of Lepel in N Belorussia and flows 350 m. (560 km.) S to join the Dnieper R. above Rechitsa. Linked by canal with the W Dvina R. and the Baltic, forming a waterway from the Baltic to the Black Sea.

Berezniki   Russian Soviet Federal Socialist Republic, U.S.S.R. 59 24N 56 46E. Town sit. N of Perm. It has deposits of sodium, potassium and magnesium salts and is an important centre of the chemicals industry. Pop. (1985E) 195,000.

Bergamo   Lombardia, Italy. 45 41N 9 43E. Cap. of Bergamo province in central N sit. at foot of Bergamasque Alps. Manu. inc. machinery and textiles. Pop. (1981C) 122,142.

Bergen   Norway. 60 23N 5 20E. County and city at head of the fjords in SW Norway. Commercial and cultural centre and important sea port, trading centre for fish and fish products and offshore operations. Industries inc. shipbuilding, engineering, paper making; manu. furniture, pottery and rope. Birthplace of Grieg the composer (1843–1907). Pop. (1989E) 209,000.

Bergen op Zoom   Noord-Brabant, Netherlands. 51 30N 4 17E. Town sit. ENE of Vlissingen. Industries inc.

cigarettes, food processing, chemicals, metal, engineering, precision goods, distilling. Pop. (1989E) 46,800.

Bergerac   Dordogne, France. 44 51N 0 29E. Town in SW sit. on Dordogne R. Industries inc. distilling, tanning and the manu. of footwear. Trades in truffles, chestnuts and wine. Pop. (1982C) 27,704.

Bering Sea   Pacific Ocean. 66 00N 170 00E. Arm of the Pacific between NE Siberia and Alaska, U.S.A., bounded S by the Aleutian Islands; through the Bering Strait connecting the Pacific and Arctic Oceans.

Berkeley   Gloucestershire, England. 51 42N 2 27W. Town sit. SW of Gloucester in the Vale of Berkeley. Noted for Double Gloucester cheese. Pop. (1980E) 1,600.

Berkeley   California, U.S.A. 37 57N 122 18W. City sit. opposite San Francisco on San Francisco Bay. Mainly residential with some industry. Berkeley University is here and associated professional employment. Pop. (1980C) 103,326.

Berkhamsted   Hertfordshire, England. 51 46N 0 35W. Town sit. W of St. Albans, largely residential but with some light industry. Pop. with Northchurch (1988E) 21,000.

Berkshire   England. Under 1974 reorganization the county consists of the districts of Newbury, Reading, Wokingham, Windsor and Maidenhead, Slough and Bracknell. Bordered on N by R. Thames. County town, Reading. Extremely fertile agric. area and there is considerable industry in Reading, Bracknell and Slough. Pop. (1989E) 731,633.

Berlin   Germany. 52 32N 13 25E. City and cap. on Spree R. formerly divided between the Federal Rep. of Germany (West Berlin) and the German Democratic Rep. (East Berlin). Area of West Berlin, 187 sq. m. (480 sq. km.). Berlin became the cap. of a united Germany in 1871 and again in 1990. Avus Autobahn, the world's first motorway was opened in 1921 and is still in use. From 1949–89 it was separated from *East Berlin* by a wall along the boundary of the former U.S.S.R. occupation zone. This effectively hampered all natural contacts between the two parts of the city. Industries inc. manu. of machine tools, pharmaceuticals, chemicals, electrical cables, clothing and brewing. Pop. (1987E) 2,016,100. The former *East Berlin* had a total area of 157 sq. m. (403 sq. km.) (city and district). Industries inc. food processing, engineering, chemicals and the manu. of motor vehicles. Pop. (1988E) 1,284,536.

Berlin New Hampshire, U.S.A. 44 29N 71 10W. City in N part of state on the Androscoggin R., sit. midway between Boston, Massachusetts and Montreal, Canada. Industries inc. wood, metal and clothing and the production of wood pulp and related paper products. Pop. (1988E) 11,918.

Bermejo River Argentina/Paraguay. 26 51S 58 23W. Rises near Hurbe in extreme NW Argentina and flows 650 m. (1,040 km.) SE to the Paraguayan frontier, there to join the Paraguay R. at Pilar, Paraguay. Divides into two streams in its middle course, of which the N is the Tenco R.

Bermondsey Greater London, England. District of SE London on S bank of R. Thames. Noted for tanneries and R. warehouses.

Bermuda Atlantic Ocean. 32 20N 65 45W. Group of islands sit *c.* 700 m. (1,120 km.) SE of New York, U.S.A., and forming part of British Commonwealth. About 150 islands, 20 inhabited, total area, 21 sq. m. (54 sq. km.). British Colony with representative government. The Spaniards visited the islands in 1515, but, according to a 17th-century French cartographer, they were discovered in 1503 by Juan Bermudez, after whom they were named. No settlement was made, and they were uninhabited until a party of colonists under Sir George Somers was wrecked there in 1609. A company was formed for the 'Plantation of the Somers' Islands', as they were called at first, and in 1684 the Crown took over the government.

Cap. Hamilton. Low-lying and rocky, mainly coral formation which has grown on a submarine volcanic cone; linked by causeways and bridges in a 20 m. (32 km.) arc. The main industry is tourism; there is little cultivable land, *c.* 770 acres, and this produces fruit and vegetables mainly from smallholdings. Chief exports are pharmaceutical oils. Pop. (1988E) 58,616.

Bern Switzerland. 46 57N 7 26E. Canton between Bernese Alps and French frontier. Area, 2,336 sq. m. (6,050 sq. km.). Chief towns Bern, Interlaken, Mürren, Grindelwald, Kandersteg, Meiringen. Mountainous in the S with the Alpine Bernese Oberland rising to Jungfrau and Finsteraarhorn. The central region inc. Aar and Emme valleys, with dairy farming and cheese making, the N or Seeland inc. L. Biel, L. Neuchâtel and the Bernese Jura. Industries inc. watches, textiles and food products. Pop. (1984E) 920,445.

Bern Switzerland. 46 57N 7 26E. City sit. SW of Zürich on Aar R. Cap. of Switzerland and of Bern canton. It was founded in 1191 and became a free imperial city in 1218, joining the Swiss Confederation in 1353. Since 1513 bears have been kept there at the public expense and the bear pit is one of the sights of the city. There is a university and it is the headquarters of the Universal Postal Union. Industries inc. printing and publishing; manu. furniture, confectionery, knitwear, musical instruments. Pop. (1988E) 298,700 (conurbation).

Bernburg Halle, German Democratic Republic. 51 48N 11 44E. Town sit. NNW of Halle on Saale R. Industries inc. potash and rock salt mining, engineering and the manu. of chemicals. Pop. (1989E) 41,200.

Bernkastel-Kues Rhineland-Palatinate, Federal Republic of Germany. 49 55N 7 04E. Town sit. NE of Trier on Mosel R. Noted for wine.

Berre, Étang de Bouches-du-Rhône, France. 43 27N 5 05E. A salt water lagoon containing eel fisheries with area of 60 sq. m. (155 sq. km.).

Salt works and oil refineries are in the vicinity.

Bertoua   Cameroon. 4 30N 13 45E. Cap. of E province and centre for forestry products. Pop. (1981E) 18,254.

Berwickshire   Scotland. former county now part of Border Region.

Berwick-upon-Tweed   Northumberland, England. 55 46N 2 00W. Town at mouth of Tweed R., on the N shore with the suburbs of Tweedmouth and Spittal on the S shore. Industries inc. agric. engineering, economic and commercial services, salmon fishing and food processing. The town was frequently disputed between England and Scotland and changed hands 14 times until the Union of the Crowns in 1603 brought final peace. Pop. (1981C) 12,142.

Berwyn   Illinois, U.S.A. 41 50N 87 47W. Residential city sit. WNW of Chicago. A suburb of Chicago with some manu. Pop. (1980C) 46,849.

Besançon   Doubs, France. 47 15N 6 02E. Cap. of dept. in central E sit. at foot of Jura Mountains and on the Doubs R. It is the chief French centre of watch and clock making. Other industries inc. cars, engineering, brewing, and the manu. of rayon, hosiery, paper and chocolate. Pop. (1982C) 119,687.

Beskids   Czechoslovakia/Poland. 49 15N 22 30E. Two ranges of the Carpathian Mountains. The highest peak is Babia Góra on the frontier.

Bessarabia   U.S.S.R. Region in W bounded N and E by R. Dniester, S by R. Danube and W by Prut R. The N and S parts are in the Ukrainian Soviet Socialist Rep., the main central area is the Moldavian Soviet Socialist Rep. Area, 17,100 sq. m. (44,289 sq. km.). An agric. area producing maize, wheat, sugar-beet, vines, sheep, cattle and pigs.

Bessemer   Alabama, U.S.A. 33 25N 86 57W. City sit. SW of Birmingham. Traditionally an important iron and steel centre but now diversifying. Pop. (1980C) 31,729.

Bethesda   Maryland, U.S.A. 38 59N 77 06W. Town immediately NW of Washington D.C. Pop. (1980C) 62,736.

Bethesda   Gwynedd, Wales. 53 11N 4 03W. Town sit. SE of Bangor. The Penrhyn slate quarries are in the vicinity. Pop. (1980E) 4,000.

Bethlehem (Bayt Lahm)   Israel. 31 43N 35 12E. Town sit. SSW of Jerusalem. It is believed to be the birthplace of Jesus Christ. Pop. (1980C) 14,000.

Bethlehem   Pennsylvania, U.S.A. 40 37N 75 25W. City sit. NNW of Philadelphia on Lehigh R. It is an important centre of the iron and steel industry. Pop. (1980C) 70,419.

Bethnal Green   Greater London, England. District of E central London. *See* Tower Hamlets.

Béthune   Pas-de-Calais, France. 50 32N 2 38E. Town sit. WSW of Lille. It is the centre of an agric. and coalmining district. Manu. inc. beet sugar, textiles and footwear. Pop. (1982C) 26,105 (agglomeration, 258,383).

Betwys-y-Coed   Gwynedd, Wales. 53 05N 3 48W. Resort sit. S of Llandudno on R. Llugwy near its confluence with the R. Conway. Pop. (1981C) 750.

Beverley   Humberside, England. 53 52N 0 26W. Market town in NE, sit. N of R. Humber. Industries inc. tourism, aerospace, metal refining, agric. and horticulture, caravans and plastics. Pop. (1988E) 114,000 (borough).

Beverly   Massachusetts, U.S.A. 42 33N 70 53W. Port sit. SE of Lawrence. Industries inc. fishing and the manu. of shoes, shoe machinery, canvas and clothing. Pop. (1980C) 37,655.

Beverly Hills   California, U.S.A. 34 03N 118 26W. A residential suburb of Los Angeles. Pop. (1980C) 32,367.

Beverwijk   Noord-Holland, Netherlands. 52 28N 4 38E. Town sit. N of Haarlem. Industries inc. engineering, fruit growing and canning and jam making. Pop. (1984E) 34,947.

Bewdley Hereford and Worcester, England. Town sit. SW of Dudley on R. Severn. Manu. inc. leather, combs, rope, brass and iron ware. Pop. (1980E) 9,000.

Bexhill-on-Sea East Sussex, England. 50 50N 0 29E. Resort sit. WSW of Hastings on the English Channel. Pop. (1989E) 38,000.

Bex-les-Bains Vaud, Switzerland. Spa sit. on Avançon R. near its confluence with Rhône R., SE of Lausanne. Noted for brine and sulphur baths. Pop. 6,000.

Bexley Greater London, England. 51 27N 0 10E. District of SE Lond. comprising the suburbs of Bexley, Erith, Crayford and parts of Chislehurst and Sidcup. Industries inc. electrical and mechanical engineering, food processing, furniture and chemicals but mainly residential. Pop. (1988E) 220,400.

Bezhitsa Russian Soviet Federal Socialist Republic, U.S.S.R. Town sit. SE of Smolensk on Desna R. Manu. inc. railway rolling stock, locomotives and agric. machinery.

Béziers Hérault, France. 43 21N 3 15E. Town in central S sit. on Orb R. and Canal du Midi. It is an important centre of the wine trade. Manu. inc. barrels, corks, textiles, chemicals and confectionery. Pop. (1982C) 78,744.

Bhádgáon Nepál. 27 41N 85 26E. Town sit. NNE of Patna, India. It is a Hindu religious centre with many temples. Pop. (1971E) 40,112.

Bhagalpur Bihar, India. Town sit. ESE of Patna on Ganges R. Manu. inc. textiles, esp, silk. Trades in rice and oilseeds. Pop. (1981C) 225,062.

Bhatpara West Bengal, India. 22 52N 88 24E. Town sit. N of Calcutta on Hooghly R. Industries inc. jute and cotton milling and engineering. Pop. (1981C) 265,419.

Bhavnagar Gujarat, India. Port sit. SSW of Ahmedabad on Gulf of Cambay. Manu. inc. textiles, brassware, bricks and tiles. Trades in cotton. Pop. (1981C) 308,642.

Bhopal Madhya Pradesh, India. 23 20N 77 53E. State cap. sit. ENE of Indore. Manu. inc. textiles, chemicals,ectronics, pharmaceuticals, electrical equipment, matches and ghee. Scene of the world's worst industrial accident in the Union Carbide plant, killing over 3,000 and injuring over 100,000. The Taj-ul-Masjid is reputed to be the largest mosque in India. Pop. (1981C) 871,018.

Bhubaneswar Orissa, India. 20 15N 85 50E. State cap. sit. SW of Calcutta. A noted pilgrimage centre with many temples. Pop. (1981C) 219,211.

Bhután South Asia. 27 25N 89 50E. Kingdom in E Himalayas bounded N by Tibet, E and S by India and W by Sikkim. Area, 18,000 sq. m. (46,500 sq. km.). Cap. Thimphu. Mountainous, rising to peaks over 20,000 ft. (6,096 metres). Drained by tribs. of the Brahmaputra R. There is permanent snow on the mountain summits and heavy forestation in the valleys; the forests are commercially valuable. The main occupations are farming and horse breeding; main products rice, millet, wheat, barley, maize, cardomom and fruit. Pop. (1988E) 1·4m.

Biafra, Bight of *Now* B. of Bonny, 1975. Atlantic Ocean. Sit. between R. Niger delta and Cape Lopez in the Gulf of Guinea. The island of Fernando Póo is sit. in the bay.

Bialystok Bialystok, Poland. 53 09N 23 09E. Cap. of province sit. NE of Warsaw. Manu. inc. textiles, esp. woollens, agric. machinery, chemicals and metal goods. Pop. (1982E) 230,000.

Biarritz Pyrénées-Atlantiques, France. 43 29N 1 34W. Resort sit. on Bay of Biscay near Spanish frontier. Pop. (1982C) 26,647.

Bicester Oxfordshire, England. 51 53N 1 09W. Town sit. NNE of Oxford. Trades in cattle and agric. produce. There is some light industry and engineering. Pop. (1988E) 19,000.

Bideford Devon, England. 51 01N 4 13W. Port and resort on Bristol Channel coast, and sit. on estuary of R. Torridge. Industries inc. boatbuilding and the manu. of electrical components, pottery, toys and games. Pop. (1981c) 12,296.

Biel, Lake Bern, Neuchâtel, Switzerland. 47 10N 7 12E. Town in w Switz. sit. on Schüss R. near L. Biel. It is an important industrial centre with manu. of special steels, machine tools, watches, and paper. Pop. (1988e) 52,056.

Biel, Lake Bern/Neuchâtel, Switzerland. 47 05N 7 10E. Sit. in w Switz. at foot of Jura Mountains. Area, 16 sq. m. (41 sq. km.).

Bielefeld North Rhine-Westphalia, Federal Republic of Germany. 52 01N 8 31E. Town in N centre sit. at foot of Teutoburger Wald. Industries inc. textiles, processed foods, publishing and sewing machines. Pop. (1988e) 310,000.

Biella Piemonte, Italy. 45 34N 8 03E. Town sit. NE of Turino. Manu. woollens, cotton goods and hats.

Bielsko-Biala Katowice, Poland. 49 49N 19 02E. Town sit. s of Katowice. Noted for textiles since the 16th cent. Also manu. machinery, chemicals and paper. Pop. (1982e) 170,000.

Bienne *see* **Biel**

Bienne, Lake *see* **Biel, Lake**

Biggleswade Bedfordshire, England. 52 05N 0 17W. Town sit. ESE of Bedford on R. Ivel. Manu. inc. clothing, machine tools and general light industry. Market gardening area. Pop. (1984e) 11,600.

Bihar India. 25 12N 85 33E. State of NE India bounded N by Nepál. Area, 69,920 sq. m. (173,877 sq. km.). Chief towns Patna (cap.), Gaya, Bhagalpur, Darbhanga, Dhanbad, Ranchi, Muzaffarpur, Jamshedpur. Flat in the N, where a fertile plain is watered by the R. Ganges and its tribs. Hilly in the s. Extensive and important mineral deposits, producing over a third of India's mineral revenue. The most important is coal, then copper, of which Bihar is the only Indian source, also iron ore, mica, kyanite, pyrites. Coal and iron are the base of the iron and steel works at Jamshedpur and Bokaro. Other industries inc. fertilizers, oil refining, copper, cable insulators and watch manu., heavy engineering, explosives, aluminium and paper. Pop. (1981c) 69,914,734.

Biisk Russian Soviet Federal Socialist Republic, U.S.S.R. Town in s Siberia sit. SE of Barnaul near confluence of R. Biya and Katun. Industries inc. meat packing, sugar refining and the manu. of textiles. Pop. (1987e) 231,000.

Bijagós Islands Guinea-Bissau. An arch. adjacent to the coast containing four large and many small islands. Area, 1,013 sq. m. (2,624 sq. km.). The main products are rice and coconuts. The port of Bolama is sit. on Bolama island. Pop. (1979c) 26,473.

Bijapur Karnataka, India. Town sit. WSW of Hyderabad. Industries inc. oilseed milling and cotton ginning. Trades in cotton and grain. The mausoleum of Mohammed Adil Shah is believed to be the largest dome in the world. Pop. (1981c) 147,313.

Bikaner Rajasthan, India. Town sit. WNW of Jaipur sit. on edge of the Thar Desert. Manu. carpets and blankets. Trades in wool and hides. Pop. (1981c) 256,057.

Bikini Atoll Marshall Islands, w Pacific. 11 35N 165 23E. An uninhabited atoll where the atom bomb was first tested in 1946.

Bilbao Vizcaya, Spain. 43 15N 2 58W. Cap. of province and port sit. near mouth of Nervión R. on the Bay of Biscay. It has the most important iron and steel works in Spain. Other industries are shipbuilding, fishing, engineering and the manu. of railway rolling stock, chemicals, paper, tyres and cement. Exports inc. iron ore, lead and wine. Pop. (1986c) 378,221.

Billingham Cleveland, England. 54 36N 1 17W. Town sit. NNE of Stockton-on-Tees. It has one of the largest chemical works in the world. Pop. (1982E) 37,000.

Billings Montana, U.S.A. 45 47N 108 29W. City in S centre of state sit. on Yellowstone R. Industries inc. tourism, oil refining, food processing and the manu. of machinery and electrical equipment. Transport, service and distribution centre for an agric. and mining area. Trades in livestock and agric. produce. Pop. (1988E) 80,310 (metropolitan area, 118,100).

Billiton Indonesia. 2 50S 107 55E. An island sit. between the island of Bangka and SW Borneo. Area, 1,860 sq. m. (4,817 sq. km.). It is noted for its tin mines. Exports inc. rice, sago, nuts and tortoiseshell. The chief town and port is Tanjungpandan.

Biloxi Mississippi, U.S.A. 30 24N 88 53W. Port and tourist resort in SE of state sit. on a peninsula on Mississippi Sound with a 26 m. (42 km.) public beach. Industries inc. boatbuilding, oyster fishing and shrimping. Pop. (1980C) 49,311.

Bilston West Midlands, England. 52 34N 2 04W. Town sit. SE of Wolverhampton. Manu. inc. footwear, clothing, automobile components, paper and timber, machine engineering. Pop. (1981C) 26,800.

Bingen Rhineland-Palatinate, Federal Republic of Germany. 49 58N 7 54E. Town sit. W of Mainz on Rhine R. at mouth of Nahe R. The main occupation is tourism. Trades in wine. Pop. (1984E) 22,700.

Binghamton New York, U.S.A. 42 08N 75 54W. City sit. S of Syracuse at confluence of Rs. Chenango and Susquehanna. Manu. inc. machinery, aircraft components, photographic supplies, electrical and metal goods, canvas and footwear. Pop. (1980C) 55,860.

Bingley West Yorkshire, England. Town sit. NW of Bradford on R. Aire. Manu. inc. foundry work, plastics, woollens, worsteds, paper and textile machinery. Pop. (1981C) 18,942.

Bío-Bío River Chile. 36 49S 71 10W. Rises in Andes near Chile–Argentina frontier and flows 220 m. (352 km.) NW to enter the Pacific near Concepción. The upper tribs. are used for hydroelectric power.

Bioko Equatorial Guinea. Formerly Maciás Nguema, before that Fernando Póo. Island in the Bight of Bonny (formerly Bight of Biafra) sit. 20 m. (32 km.) from Cameroon. Cap. Malabo. Area, 779 sq. m. (2,017 sq. km.). Produces and exports cacao. Pop. (1984E) 75,000.

Birdum Northern Territory, Australia. 15 39S 133 13E. Town in central N of state. It is the centre of a large cattle rearing area.

Birganj Nepál. 27 01N 84 54E. Town sit. near Indian border, on the India–Káthmándu road. Trading centre for rice, jute, wheat, barley and oilseeds.

Birkenhead Merseyside, England. 53 24N 3 02W. Town opposite Liverpool on S bank of Mersey R. and connected to it by rail and road tunnels. Industries inc. shipbuilding, flour milling, engineering, food processing. Sea port exporting flour and machinery, importing grain and livestock. Pop. (1981C) 123,884.

Birmingham West Midlands, England. 52 30N 1 50W. City sit. in central England. Important manu. and commercial centre providing 40% of UK exports. The main industry is engineering but other industries include cars, cycles, plastic goods, chocolate, chemicals, electro-plate, guns, machine tools, glass, paint, wire and electrical equipment. Tourism is growing. Pop. (1988E) 993,700.

Birmingham Alabama, U.S.A. 33 31N 86 49W. Town sit. at S end of Appalachian Mountains. Industries inc. iron and steel, based on local deposits of coal, iron ore and limestone, health care, medical research, engineering, computers, construction. Manu. steel, cast and wrought iron, other metal goods, paints, foods, clothing and paper products. Pop. (1980C) 284,413.

Birnie Island Phoenix Islands, South Pacific. The smallest and most central island of the Phoenix group. Area, 44 acres. Uninhabited.

Birobidjhan Khabarovsk, U.S.S.R. Town sit. on the Trans-Siberian railway. Cap. of the Jewish Autonomous Region, to which the name is also given. Industries inc. sawmilling, woodworking; manu. clothing. Pop. (1980E) 67,000.

Birr County Offaly, Ireland. 53 05N 7 54W. Town in w County Offaly on Camcor R. Market town in an agric. area. Industries inc. peat products, and manu. of cables, upholstery fabric, boilers, typewriter ribbons, general engineering and iron foundry products. Pop. (1989E) 4,500.

Biscay, Bay of 45 00N 2 00W. An inlet sit. between the w coast of France and the N coast of Spain.

Bisceglie Bari, Italy. 41 14N 16 31E. Town sit. on Apulian coast. Industries inc. sawmilling, engineering; manu. furniture. Sea port, trading in wine and olives.

Bischoff, Mount Tasmania, Australia. Tin-mining centre near Waratah, 90 m. (144 km.) w of Launceston.

Bishop Auckland County Durham, England. 54 40N 1 40W. Town sit. on Wear R. Industries inc. airconditioning, specialist glass. Pop. (1989E) 18,940.

Bishop's Stortford Hertfordshire, England. 51 53N 0 09W. Town sit. on Stort R. Industries inc. manu. matches, electrical goods. Pop. (1982E) 23,000.

Biskra Constantine, Algeria. 34 51N 5 44E. Town and oasis. An important centre for growing and selling dates, and a resort. Pop. 60,000.

Bisley Surrey, England. 51 45N 2 08W. Village sit. WNW of Woking where the National Rifle Association has met annually since 1890.

Bismarck North Dakota, U.S.A. 46 48N 100 47W. Town in central N Dakota on Missouri R. and Northern Pacific railway. State cap. Main industries are government, agriculture, retailing and services. Pop. (1989E) 48,500.

Bismarck Archipelago Papua New Guinea. 5 00S 150 00E. Island group off E coast of New Guinea mainland. Area, 20,000 sq. m. (51,200 sq. km.). The main islands are New Britain, New Ireland, Lavongai and the Admiralty Islands. Mountainous and volcanic, densely forested. Chief products copra, cacao, copper, gold.

Bissagos Islands *see* **Bijagós Islands**

Bissau Guinea-Bissau. 11 52N 15 39W. Cap. of rep. on an island in Geba R. estuary. It was estab. in 1687 as a slave trading centre. Industries inc. food processing and timber. Sea port exporting copra, palm oil and rice. Pop. (1979C) 109,214.

Bistrita Rodna, Romania. 47 08N 24 30W. Town sit. NNW of Bucharest. Commercial centre trading in timber and agric. produce. Industries inc. wine and vinegar making, cabinet making, brewing, tanning, food processing; manu. shoes, brushes, stoves, earthenware, soap, alcohol. Pop. (1983E) 62,862.

Bitlis Turkey. 38 22N 42 06E. (i) Province of SE Turkey bounded E by L. Van. The area is extremely mountainous. Chief products grain, tobacco, potatoes. Pop. (1985C) 300,843. (ii) Town sit. ENE of Diyarbakir. Provincial cap. The main industry is processing local tobacco. Religious centre of the 'whirling dervishes'. Pop. (1970C) 20,842.

Bitolj Macedonia, Yugoslavia. 41 01N 21 21E. Town sit. S of Skopje. Centre of an agric. area. Industries inc. tanning, carpet making. Pop. (1981C) 137,835.

Bitonto Bari, Italy. 41 06N 16 42E. Town sit. W of Bari. Market town. Industries inc. manu. wine, olive oil.

Biysk *see* **Biisk**

Bizerta   Tunisia. 37 18N 9 52E.
City on Mediterranean coast on channel from L. Bizerta. Sea port and
naval base, with steelworks in the
vicinity. Oil refining is the main
industry. Pop. (1984c) 94,505.

Bjorneboorg   *see* **Pori**

Bjørnøya   Svalbard, Norway. 74
25N 19 00E. Island in Arctic Ocean
between W Spitzbergen and Norway.
Area, 69 sq. m. (177 sq. km.). Whaling station and meteorological station.

Blackburn   Lancashire, England.
53 45N 2 29W. Town sit. E of Preston
on Leeds–Liverpool canal. The main
industry is engineering; also textiles,
electronics, paper and paint. Also
high-tech industries, compact discs
and manu. machinery for tufted carpets. Pop. (1989E) 140,000.

Black Country   England. Region of
W Midlands, named from the smoke
and soot of the original heavy industries with furnaces and foundries.
Main industrial centres; Wolverhampton,   Walsall,   West   Bromwich,
Wednesbury.

Black Forest   Baden-Württemberg,
Federal Republic of Germany. Mountainous area thickly forested and
rising to 4,898 ft. (1,493 metres) in
the Feldberg. Area, 1,800 sq. m.
(4,662 sq. km.). Source of Danube
and Neckar Rs. and bounded W and S
by Rhine R.

Blackfriars   Greater London, England. District of City of London on N
bank of Thames R.

Blackheath   Greater London, England. District of SE Lond. mainly residential,   between   Lewisham   and
Greenwich.

Black Hills   U.S.A. 44 00N 103
50W. Mountains on border between
SW South Dakota and NE Wyoming,
an isolated group rising to Harney
Peak 7,242 ft. (2,207 metres). The
area is mainly national park and inc.
the   Mount   Rushmore   National
Memorial sculptures.

Black Mountain   Wales. 51 52N 3
50W. Ridge extending *c.* 12 m. (19
km.) NE from Amman R. 12 m. (19
km.) N of Swansea.

Black Mountains   Wales. 51 57N 3
08W. Mountain group on English/
Welsh border N of Usk R. rising to
Waun Fach, 2,660 ft. (811 metres).

Blackpool   Lancashire, England. 53
50N 3 03W. Town on Irish Sea coast.
Leading holiday resort for the N and
the Midlands, attracting about 16m.
tourists in summer. Pop. (1987E)
144,100.

Blacksburg   Virginia, U.S.A. 37
14N 80 25W. Town sit. SW of
Roanoke in SW Virginia. Industries
inc. electronics, automotive parts,
ceramics. Pop. (1988E) 35,000.

Black Sea   Europe/Asia. 43 30N 35
00E. Inland sea bounded N and E by
the U.S.S.R., S by Turkey, W by Bulgaria and Romania. Area, 164,000 sq.
m. (424,760 sq. km.), excluding the
NE arm called the Sea of Azov. Connected SW to the Aegean and Mediterranean by the Bosporus strait, leading
to the Sea of Marmara and the Dardanelles strait. It receives the Danube,
Dnieper, Dniester and S Bug Rs., with
the Don and Kuban Rs. emptying into
the Sea of Azov. Almost tideless with
stagnant waters below 80 fathoms.
Anchovy, sardines, mackerel and
tunny fish are caught in the N and W.

Blackwater River   Ireland. 51 51N
7 50W. Rises near Castleisland,
County Kerry and flows 104 m. (166
km.) E through County Cork into
County Waterford where it turns S and
enters the Atlantic at Youghal.

Blackwater River   Republic of Ire-
land/Northern Ireland. 54 31N 6 34W.
Rises near Fivemiletown, Tyrone, N
Ireland, and flows 50 m. (80 km.) E
and N along the border between
Tyrone (N) and Monaghan and
Armagh (S) to enter the SW end of
Lough Neagh.

Blaenavon   Gwent, Wales. Town
sit. NNW of Pontypool. The main
industries are coalmining, iron and
steel. Pop. (1980E) 7,000.

Blagoevgrad   Blagoevgrad, Bulgaria. 42 01N 23 06E. Town sit. SSW of Sofia on Struma R. Industries inc. tourism, tobacco processing.

Blagoveshchensk   Russian Soviet Federal Socialist Republic, U.S.S.R. 50 17N 127 32E. Town sit. on Manchurian border on Zeya R. near its confluence with Amur R., and on a branch of the Trans-Siberian Railway. Industries inc. flour milling, sawmilling; manu. machinery, footwear, furniture. Pop. (1985E) 195,000.

Blair Atholl   Tayside Region, Scotland. 56 46N 3 51W. Tourist village NW of Pitlochry at confluence of Tilt and Garry Rs. Pop. (1981C) 1,070.

Blairgowrie and Rattray   Tayside Region, Scotland. 56 36N 3 21W. Town comprising 2 towns on opposite banks of the R. Ericht. Centre for soft fruit growing, mainly raspberry. Tourism is important. Pop. (1981C) 7,027.

Blanc Cape   Mauritania. Headland at the S end of the Cape Blanc peninsula which contains the port of Nouadhibou.

Blanc Cape   Tunisia. Headland NNW of Bizerta on the N Tunisian coast.

Blandford Forum   Dorset, England. 50 54N 2 11W. Town sit. SW of Salisbury on Stour R. Market town in a farming area with some light industry including agric. engineering, brewing and soft drink production, service industries and tourism. Pop. (1987E) 7,685.

Blantyre   Malawi. 15 47S 35 00E. Town in the Shire highlands on the railway from Beira to L. Malawi. Commercial centre which was founded as a Scottish missionary centre. Industries inc. cement, soft drinks, textiles, footwear, brewing, metal products, milling, food processing and tobacco. Tourism is being developed. Pop. (1985E) 355,200.

Blantyre   Strathclyde Region, Scotland. 55 46N 4 06W. Town sit. SE of Glasgow. Industries inc. engineering. Pop. (1981C) 19,875.

Blarney   County Cork, Ireland. 51 56N 8 34W. Village sit. WNW of Cork noted for the ruins of the 15th cent. Blarney Castle with the Blarney Stone built into its walls. Pop. (1985E) 1,500.

Blasket Islands   County Kerry, Ireland. Group of islands off Slea Head; rocky, uninhabited since 1953.

Blaydon   Tyne and Wear, England. 54 58N 1 42W. Town sit. W of Newcastle upon Tyne on the Tyne R. The main industries coal by-products and engineering. Pop. (1978E) 7,200.

Blekinge   Sweden. 56 15N 15 15E. County of S Sweden, known as the 'Garden of Sweden', on Baltic coast. Area, 1,136 sq. m. (2,941 sq. km.). Cap. Karlskrona. Low-lying, drained by Mörrum and Ronneby Rs. Intensively cultivated. Industries inc. fishing, shipyard, electronics, machinery, pulp and paper. Karlskrona is a major naval base. Pop. (1989E) 149,907.

Blenheim   Bavaria, Federal Republic of Germany. Village sit. NE of Ulm, noted as the site of the Battle of Blenheim in 1704, when English and Austrian forces defeated the French and Bavarians.

Blenheim   Marlborough, New Zealand. 41 31S 173 57E. Town sit. NNE of Christchurch. District cap. and centre of a farming area. Industries inc. vineyards and wineries, clothing manu., seafood processing. Pop. (1989E) 18,500.

Bletchley   Buckinghamshire, England. 52 00N 0 46W. Town sit. E of Buckingham, comprising Bletchley and Fenny Stratford. Market town. Industries inc. engineering, brickmaking. Pop. (1981C) 38,273.

Blida   Algeria. 36 30N 2 49E. Town sit. SW of Algiers on Mitidja plain. Trading centre on the Algiers–Oran railway, for oranges and wine. Industries inc. flour milling; manu. olive oil, soap. Pop. (1983E) 191,314.

Block Island   Rhode Island, U.S.A. 41 11N 71 35W. Island off S coast of Rhode Island across Block Island Sound. Length, 7 m. (11 km.) Summer resort for yachting and fishing.

Bloemfontein Orange Free State, Republic of South Africa. 29 12S 26 07E. City sit. WNW of Durban. Provincial cap. and judicial cap. of the Rep. Trading centre for the Orange Free State and Lesotho. Industries inc. railway engineering and a variety of light and heavy industries. Pop. (1989E) 240,000.

Blois Loir-et-Cher, France. 47 35N 1 20E. Cap. of Dept. sit. SW of Orléans. Market town trading in wine, grain and brandy. Industries inc. manu. furniture, vinegar. Noted for the Château. Pop. (1982C) 49,422.

Bloomfield New Jersey, U.S.A. 40 48N 74 12W. Formerly Wardsesson. Town immediately NNW of Newark. Industries inc. manu. pharmaceutical, metal and electrical goods. Pop. (1980C) 47,792.

Bloomington/Normal Illinois, U.S.A. 40 29N 88 60W. City sit. SW of Chicago forming commercial and service centre in agric. area. Pop. (1980C) Bloomington, 44,189; Normal, 35,672.

Bloomington Indiana, U.S.A. 39 10N 86 32W. City sit. SSW of Indianapolis. Industries inc. furniture, refrigerators, television sets, elevators, electrical equipment, quarrying. Pop. (1988E) 54,850.

Bloomington Minnesota, U.S.A. 44 50N 93 17W. City immediately S of Minneapolis. Industries inc. computers, electronics, hydraulic filters and refrigeration units. Pop. (1988E) 85,300.

Bloomsbury Greater London, England. District of central London, N of Thames R. and seat of the British Museum and London University, with many literary associations.

Bluefield West Virginia, U.S.A. 37 16N 81 13W. City sit. W of Roanoke in the Allegheny Mountains. Mining and distribution centre of the Pocahontas coalfield; also manu. flour, textiles, timber products, mining equipment, beverages. Pop. (1980C) 16,060.

Bluefields Nicaragua. 12 00N 83 45W. Town on Caribbean coast near mouth of Bluefields R. Exports hardwoods and bananas through its outport, El Bluff. Pop. (1984E) 18,000.

Blue Mountains New South Wales, Australia. 33 33S 150 17E. Spur of the Great Dividing Range *c.* 40 m. (64 km.) W of Sydney, formed by a sandstone plateau.

Blue Mountains Jamaica. 18 06N 76 40W. Range in E Jamaica rising to Blue Mountain Peak. Thickly wooded. Main occupations tourism, coffee growing.

Blue Ridge U.S.A. 37 00N 82 00W. Range of Appalachians extending 650 m. (1,040 km.) SW to NE from NE Georgia to NE Virginia, rising to Mount Mitchell.

Blyth Northumberland, England. 55 07N 1 30W. Town sit. NNE of Newcastle upon Tyne at mouth of Blyth R. Industries inc. electronics, clothing and engineering. Sea port exporting coal, importing timber products and alumina. Pop. (1981C) 34,466.

Bnei Brak Israel. Town sit. NE of Tel Aviv on the Yarkon in the Plain of Sharon. Industrial centre; manu. woollens, aluminium tubes, textile machinery, wire, glass, tobacco and food products, soaps, dyes.

Bo Sierra Leone. 7 58N 11 45W. Cap. of province sit. ESE of Freetown. Trading centre for palm oil and kernels, cacao, coffee. Pop. (1974C) 39,371.

Boaco Nicaragua. 12 27N 85 43W. (i) Dept. of S central Nicaragua, bounded N by Río Grande. Area, 2,085 sq. m. (5,400 sq. km.). Cap. Boaco. Mainly a stock-raising area, with some agric. and lumbering. Pop. (1980C) 88,862. (ii) City sit. NE of Managua. Industries inc. processing of farming products, sawmilling; manu. Panama hats, soap, bricks, mineral water. Pop. (1980E) 9,000.

Boa Vista Roraima, Brazil. 2 49N 60 40W. City sit. N of Manaus on Rio Branco near border with Guyana.

Trading centre of a cattle raising area, handling tobacco, rubber and Brazil nuts. Pop. (1980E) 27,000.

Boa Vista   Cape Verde, Atlantic. 16 05N 22 50W. Island between Sal and Maio, 300 m. (480 km.) WNW of Cape Vert. The most E of the Cape Verde Group. Area, 239 sq. m. (620 sq. km.). Chief town, Sal Rei. Main products sand, archil. Pop. (1980C) 3,397.

Bobigny   Seine-St Denis, France. 48 54N 2 27E. Cap. of dept. and NE suburb of Paris. Pop. (1982C) 42,727.

Bobo-Dioulasso   Houet, Burkina Faso. 11 12N 4 18W. City sit. WSW of Ouagadougou on Abidjan railway. Commercial centre trading in groundnuts, shea nuts. Pop. (1982E) 165,171.

Bobruisk   Belorussia, U.S.S.R. 53 09N 29 14E. Town sit. SE of Minsk on Berezina R. Commercial centre trading in timber and grain. Industries inc. engineering; manu. paper, cellulose, clothing, footwear. Pop. (1985E) 223,000.

Boca Raton   Florida, U.S.A. 26 21N 80 05W. City sit. S of West Palm Beach. Resort on the Atlantic coast. Headquarters of companies in data systems, electrical and electronics. Pop. (1980C) 49,505.

Bocas del Toro   Panama. 9 22N 82 14W. (i) Province of W Panama on the Caribbean, bounded W by Costa Rica. Mainly agric., producing fruit, cacao, coconuts, tobacco, livestock, timber. Pop. (1980C) 53,579. (ii) Town sit. WNW of Panama city at S tip of Colón Island. Provincial cap. and commercial centre. Sea port on the Caribbean coast. Pop. (1980C) 2,520.

Bochum   North Rhine-Westphalia, Federal Republic of Germany. 51 28N 7 13E. Univ. town sit. NE of Essen. Industries inc. manu. steel, automobiles, electronics, measurement and control instruments and textiles. Pop. (1988E) 389,100.

Bodh Gaya   Bihar, India. 24 42N 84 59E. Village sit. S of Gaya noted as the site of a temple and the sacred tree associated with Buddha. Pop. (1981C) 15,724.

Bodmin   Cornwall, England. 50 28N 4 44W. Town sit. WNW of Plymouth on SW edge of Bodmin moor. County town. Centre for a farming area with some light industry. Pop. (1989E) 15,000.

Bodø   Nordland, Norway. 67 17N 14 23E. City sit. SW of Narvik at mouth of Salt Fjord on North Sea. Port handling metals, marble and slate. Industries inc. fisheries, shipyards, brewing. Pop. (1989E) 36,000.

Boeotia   Greece. *Nome* in Attica, bounded S by Gulf of Corinth. Area, 1,240 sq. m. (3,211 sq. km.). Cap. Levadia. Mainly agric. Main products wheat, wine, olives, livestock. Pop. (1981C) 117,175.

Bognor Regis   West Sussex, England. 50 47N 0 41W. Town sit. on English Channel. Seaside resort. Industries inc. electronics, electrical and instrument engineering. Pop. (1988E) 20,025.

Bogotá   Cundinamarca, Colombia. 4 38N 74 05W. City in Cundinamarca basin of the E Cordillera sit. on a fertile plateau at 9,200 ft. (2,800 metres). Cap. and dept. cap. Commercial and cultural centre. The city was founded in 1538 and has 15 universities. There is in the Gold Museum a unique collection of pre-Colombian art and the world's largest emeralds. The home of Simón Bolivar (1783–1830) is sit. near the city centre. Industries inc. textiles, chemicals, food processing; manu. tobacco products. Pop. (1985C) 3,967,988.

Bogra   Rajshahi, Bangladesh. 24 51N 89 22E. Town sit. S of Rangpur. Market town for a farming area. Also district. Pop. (1981E) 69,000.

Bohemia   Czechoslovakia. The W part of Czech. bounded W by Germany, S by Austria and E by Moravia. Fertile plateau drained by the Elbe and Ultava Rs. and with important mineral resources, inc. uranium, coal, lignite, iron ore, graphite, silver. The main cities are Prague and Plzeň.

Böhmerwald   Czechoslovakia/Federal Republic of Germany. 49 30N 12 40E. Forested mountain range extending 150 m. (240 km.) along the Czech./Bavarian frontier, rising to Mount Arbor. The main product is timber.

Bohol Philippines. 9 50N 124 10E. Island N of Mindanao. Area, 1,588 sq. m. (4,113 sq. km.). Chief town Tagbilaran. Main occupations growing rice, coconuts and Manila hemp.

Bois de Boulogne Paris, France. Park and residential district of w Paris on Seine R.

Boise Idaho, U.S.A. 43 37N 116 13W. City in sw Idaho on Boise R. State cap. Manu. electronics, data systems, locomotive and aircraft engineering. Pop. (1980C) 102,451.

Boké Guinea. Town sit. NNW of Conakry on Rio Nunez. Produces palm kernels, sesame, honey, beeswax, gum, rice and cattle.

Boksburg Transvaal, Republic of South Africa. 26 12S 28 14E. Town sit. E of Johannesburg, the chief centre of the E Rand gold mining area. Industries inc. manu. railway equipment, electrical goods, soap, pottery. Pop. (1980E) 150,287.

Bolan Pass Pakistan. Pass extending 60 m. (96 km.) through Brahui Range between Sibi and Quetta, rising to *c.* 5,900 ft. (1,798 metres).

Bolgatanga Mamprusi, Ghana. 10 47N 0 51W. Town sit. N of Tamale in N Ghana. District headquarters. Main products nuts, durra, yams, cattle, skins. Pop. (1982E) 48,000.

Boliden Västerbotten, Sweden. Village sit. WNW of Skellefteå near the Skellefteå R. Mining centre, inc. the mining settlements of Strömfors and Bjurliden. Gold, arsenic, sulphur and copper ores are sent to smelters at Rönnskär.

Bolivar Colombia. 2 00N 77 00W. Dept. of N Colombia on Caribbean coast and bounded E by Magdalena R. Area, 10,190 sq. m. (26,392 sq. km.). Chief towns Cartagena (cap.), Car-

men, Sincelejo. Drained into Magdalena R. by San Jorge, Cauca and Sinú Rs., it is mainly lowland and savannah with alluvial forests. There are some hill ranges, foothills of the Cordillera Occidental. It is humid with a tropical climate. Cattle are raised on extensive pasture land; main agric. crops are sugar, tobacco, cotton, maize, rice, beans, bananas, cacao, coconuts. The forests are commercially valuable; mineral resources inc. petroleum, coal, lime, gold. Cartagena is linked by canal with the Magdalena R. Pop. (1983E) 1,076,800.

Bolivar Ecuador. Province of central Ecuador in Andes, bounded E by Chimborazo and w by Cordillera de Guaranda Mountains. Area, 1,252 sq. m. (3,243 sq. km.). Cap. Guaranda. Mountainous, with forests producing valuable timber and cinchona. Main occupations are lumbering, farming, esp. cattle and sheep, cereals, sugarcane, fruit, vegetables, coffee and tobacco. Pop. (1982C) 148,161.

Bolívar Venezuela. 7 28N 63 25W. State in s Venezuela bounded E by Guyana and SE by Brazil. Area, 91,890 sq. m. (237,995 sq. km.). Cap. Ciudad Bolívar. Watered by Orinoco R. and its tribs. inc. Caroní and Caura Rs. Very hot climate with a rainy season May–October. Mainly sit. in the Guiana Highlands, some of it still little explored. Heavily forested. There is farming on the N Orinoco lowlands, esp. cattle, horses, cereals, tobacco, sugar and fruit. Mineral resources inc. gold, mica, diamonds, iron. Pop. (1980E) 503,194.

Bolivia South America. 17 06S 64 00W. Rep. in central s America bounded N and E by Brazil, s by Paraguay and Argentina, w by Chile and Peru. Area, 424,160 sq. m. (1,098,574 sq. km.). Chief cities La Paz (administrative cap.), Sucre (cap.), Chochabamba, Santa Cruz, Potosi. Bolivia is a land of violent contrasts. Structurally it is divided into three main regions. There is the country of the mountains and the high plateau, the country of the high valleys and gorges and the country of the broad plains. Bolivia is entirely in the tropical zone, but the extreme spread in altitude, from about 300 ft. (91 metres) to over 21,000 ft. (6,401 metres) a.s.l., produces a wide variation of climate. The country is endowed with many natural resources. Tin, silver, gold, zinc, antimony, tungsten, lead, copper and a variety of other minerals occur in large quantities in the highlands. Petroleum, gas and hydroelectric resources are plentiful. The fertile valleys and the wide plains provide good soil and climate for rice, cotton, sugar, coffee, tea, cacao and all kinds of fruits. Even the high-lying Altiplano produces potatoes, barley, wheat, quinoa and other basic necessities. Two-fifths of the country is covered with forests rich in valuable timber, fibres and a good supply of wild rubber. There are extensive pastures for cattle in the valleys and on the plains and for sheep and llama on the highlands. Lakes and rivers are rich in fish. Mining has been the leading industry since early colonial days. Metallic minerals have accounted for all or nearly all exports throughout the centuries. Tin replaced silver as the leading export mineral early in this century. As tin ores become uneconomically low in grade, efforts are being made to diversify mining and replace tin by a more intensive mining of copper, lead, zinc, antimony, tungsten and gold, the mining of which has already reached significant proportions. Prospects for increased copper and gold mining are reportedly good. Lack of transportation facilities makes the mining of non-metallic minerals uneconomical at present; nevertheless, deposits of salts, soda and sulphur have been worked to a limited extent. The most important ore deposits actually in process of exploitation, are to be found in Camiri, Sandandita, Bermejo, Monteagudo, Caranda, Colpa, Rio Grande and other places. The total crude oil production in 1988 was 900,000 tonnes. The highlands are virtually treeless, except for a few eucalyptus groves near the principal urban centres. The valleys are much better endowed with trees. Some eucalyptus trees are grown for fuel and mining supports near Cochabamba and Sucre, but the best forests are found on the eastern slopes of the Andes and on the lowlands along the rivers of the Ama-

zon basin. Here, large areas of tropical forests of evergreens and hardwood exist. Rubber of high quality grows wild in the forests of Pando and Beni and constitutes the primary forest export commodity. Brazil nuts constitute the second most important forest export commodity. Dry forests of abundant hardwoods, such as quebracho, walnut and mahogany, exist in the south-eastern part of the country between Yucuiba, Tarija and the slopes of the Andes. Most of the vast timber resources lie beyond the reach of modern means of transportation. Pop. (1988E) 7m.

Bollington   Cheshire, England. 53 18N 2 08W. Town sit. NNE of Macclesfield. Industries inc. coated papers, polyurethane foam, jersey fabrics, bias bindings and dyeing. Pop. (1985E) 7,000.

Bologna   Bologna, Italy. 44 29N 11 20E. Anc. univ. city sit. N of Florence on the Aemilian Way. Provincial cap. The main industry is engineering, also manu. pasta and processed meats. Pop. (1980E) 466,593.

Bolsover   Derbyshire, England. 53 14N 1 18W. Town sit. E of Chesterfield. Industries inc. coalmining, limestone quarrying, fuel and chemical refining. Pop. (1981C) 11,080.

Bolton   Greater Manchester, England. 53 35N 2 26W. Town sit. NW of Manchester. The main industry engineering of all types. Manu. textiles, chemicals and paper. Pop. (1989E) 263,600.

Bolu   Turkey. 40 44N 31 37E. (i) Province of NW Turkey on Black Sea, bounded N by Bolu Mountains. Area, 4,433 sq. m. (11,481 sq. km.). Heavily forested area, drained by Devrek, Soganli and Aladag Rs. Agric. products inc. cereals, flax, tobacco, opium. Pop. (1985C) 504,778. (ii) Town and provincial cap. NW of Ankara on Devrek R. Manu. leather goods. Pop. (1970C) 26,994.

Bolzano   Bolzano, Italy (i) 46 31N 11 22E. Province of N Italy bounded N by Austria and Switzerland and sit. in, and co-extensive with, the Trentino–

Alto Adige area. Area, 3,090 sq. m. (7,910 sq. km.). Chief towns Bolzano, Merano, Bressanone, Brunico, Dobbiaco. Alpine area inc. Dolomite and Ortler ranges, drained by Isarco and Adige Rs. The main occupations are tourism and farming, with some lumbering. (ii) City and provincial cap. SSW of Brenner Pass on Isarco R. Railway junction, tourist centre. Industries inc. steel, fruit canning, distilling, textiles; manu. wine, food products, pianos. Pop. (1981C) 105,180.

Boma   Bas-Zaïre, Zaïre. 5 51S 13 03E. Town sit. WSW of Kinshasa on Zaïre R. R. port exporting palm oil, bananas, timber and coffee. Pop. (1976C) 93,965.

Bombay   Maharashtra, India. 18 58N 72 50E. City on Arabian Sea at S end of Bombay Island. State cap. and a major commercial and industrial centre. The main industries are cotton, oil refining, shipbuilding, film-making, chemicals; manu. machinery, textiles, pharmaceuticals, cosmetics, electronic and electrical equipment, furniture, carpets. It is India's main sea port. Pop. (1981C) 8,243,405 (Greater Bombay).

Bomi Hills (Vaitown)   Western Province, Liberia. 6 52N 10 45W. Area, 40 m. (64 km.) N of Monrovia, and important iron mining district. Pop. (1980E) 7,000.

Bonaire Island   Netherlands Antilles. 12 10N 68 15W. Island 30 m. (48 km.) E of Curaçao island. Area, 99 sq. m. (256 sq. km.). Chief town Kralendjik. Main products sisal, goat dung, salt. Pop. (1983E) 9,704.

Bonanza   Zelaya, Nicaragua. Village sit. W of Puerto Cabezas in Pis Pis Mountains. Centre for the Neptune goldfields.

Bonavista   Newfoundland, Canada. 48 39N 53 07W. Town sit. NNW of St. John's on E shore of Bonavista Bay. The main industries are fishing and fish processing. Pop. (1975E) 4,605.

Bondy   Seine-St. Denis, France. 48 54N 2 28E. Suburb of NE Paris on Ourcq Canal. Industries inc. manu.

glass, chemicals, foodstuffs. Pop. (1975C) 51,555.

Bo'ness (Borrowstounness)   Central Region, Scotland. 56 01N 3 37W. Commuter town sit. WNW of Edinburgh on Firth of Forth. Heritage trust area for tourism. Pop. (1979E) 17,000.

Bonin Islands   Pacific Ocean. 27 00N 142 10E. In 1968 the U.S.A. retroceded the islands to Japan. Group of islands *c.* 600 m. (960 km.) S of Tokyo. Area, 40 sq. m. (102 sq. km.), the main groups being Bailey Island, Beechey Island, Parry Island. The largest island, Chichi-jima, has an important harbour. Main occupation farming. Pop. (1970E) 300.

Bonn   North Rhine-Westphalia, Federal Republic of Germany. 50 43N 7 06E. City sit. SSE of Cologne on Rhine R. The destruction of Berlin and the dividing of Germany caused Bonn to be selected as the provisional cap. in 1949. Industries inc. nutritional, electrotechnical synthetic-material processing, chemicals and mechanical engineering. Pop. (1989E) 282,000.

Bonneville, Lake   U.S.A. Flat plain of salt flats surrounding the Great Salt L., Utah, which is all that remains of an extensive prehistoric L. occupying much of what is now Utah.

Bonny, Bight of   formerly **Biafra, B. of.**

Bonnyrigg and Lasswade   Midlothian, Lothian Region, Scotland. 55 52N 3 08W. Town sit. SE of Edinburgh comprising 2 small towns united in 1929. Industries inc. manu. of kitchen equipment, knitwear and meat products. Pop. (1981C) 14,328.

Boothia   Northwest Territory, Canada. 70 30N 95 00W. Peninsula extending to most N point of N American mainland.

Boothia, Gulf of   Canada. 70 30N 95 00W. Inlet of the Arctic Ocean *c.* 200 m. (320 km.) long, extending NW to SE between Boothia and Baffin Island.

Bootle   Merseyside, England. 53

28N 3 00W. Town adjoining Liverpool on N bank of Mersey R. Sea port, with much of the Mersey docks, exporting timber. Industries inc. tanning, engineering, tin-smelting. Pop. (1981c) 62,463.

Bophuthatswana   26 00S 25 00E. An independent Republic surrounded by the Republic of South Africa and Botswana. Independence was granted in 1977, but by 1990 this was not recognized by the UN or any other country other than the Republic of South Africa. Cap. Mmabatho. Area, 14,773 sq. m. (38,261 sq. km.). There is mining of platinum and chrome. The semi-arid area of grass veld is suitable for stockfarming. Pop. (1980E) *de jure* 2·6m. and *de facto* 1,328,637.

Boppard   Rhineland-Palatinate, Federal Republic of Germany. Town sit. S of Coblenz on Rhine R. Trading centre for a wine producing area. Industries inc. cosmetic manu., road construction machinery and tourism. Pop. (1989E) 16,000.

Boquerón   Paraguay. Dept. of N Paraguay in the Chaco, bounded N and W by Bolivia, SE by Brazil across the Paraguay R. and SW by Argentina across the Pilcomayo R. Area, 64,876 sq. m. (166,083 sq. km.). Consists of low grasslands, marshes, jungle and forest. There is farming along the Paraguay R., mainly cattle raising with some cotton. The other main occupation is lumbering, with some mining. The area was awarded to Paraguay in 1938, after the Chaco War with Bolivia, 1932–35. The dept. was estab. 1944. Pop. (1982E) 14,685.

Bor   Serbia, Yugoslavia. 44 05N 22 07E. Town sit. NNW of Zajecar. Copper mining centre with smelters and refinery, the largest in Yugoslavia. Pop. (1981E) 101,000.

Borås   Älvsborg, Sweden. 57 43N 12 55E. Town sit. E of Göteborg on Viska R. The main industry is textiles; also manu. hosiery, clothing. Pop. (1983E) 100,184.

Bora-Bora   Society Islands, French Polynesia. 16 30S 151 45W. Island in Leeward group, 25 m. (40 km.) NW of Raiatea. Mountainous, rising to Mount Taimanu. Area, 15 sq. m. (39 sq. km.). Chief town, Pahia. It has a large lagoon with a good harbour. Chief products copra, oranges, mother-of-pearl, vanilla and tobacco. Pop. (1977C) 2,572.

Borama   Somalia. Town sit. WNW of Hargeisa, on a plateau, near the Ethiopian frontier. Centre of a stock raising area, also producing sorghum, maize, beans.

Bordeaux   Gironde, France. 44 50N 0 34W. City on Garonne R. Cap. of Dept. Sea port with an important export trade in wines from the Graves, Sauternes, Médoc and St. Émilion districts. Industries inc. shipbuilding, sugar refining, chemicals, food processing; manu. wine bottles, casks, corks and crates. Pop. (1982C) 211,197 (agglomeration, 640,012).

Borders   Region of Scotland. 55 30N 3 00W. Area, 1,803 sq. m. (4,672 sq. km.) comprising districts of Berwickshire, Roxburgh, Tweeddale, Ettrick and Lauderdale. Pop. (1988E) 102,592.

Bordighera   Imperia, Italy. 43 46N 7 39E. Town next to French border on Gulf of Genoa. Resort, with an export trade in flowers and palm branches.

Borgerhout   Antwerp, Belgium. 51 13N 4 26E. Suburb of E Antwerp. Industries inc. manu. food products, chemicals, clothing, leather, tobacco, diamond polishing and cutting.

Borislav   Ukrainian Soviet Socialist Republic, U.S.S.R. Town sit. SW of Lvov in an area producing oil and natural gas. Industries inc. oil refining; manu. oil drilling equipment.

Borisoglebsk   Russian Soviet Federal Socialist Republic, U.S.S.R. 51 23N 42 06E. Town sit. ESE of Voronezh on Khoper R. Industries inc. meat packing, flour milling, tanning.

Borisov   Belorussian Soviet Socialist Republic, U.S.S.R. 54 15N 28 30E. Town sit. NE of Minsk on Berezina R. Industries inc. manu.

matches, enamelware, glass products, food products, musical instruments.

Borku   Chad. 18 15N 18 50E. Region s of Tibesti highlands in L. Chad depression. Chief town Faya. Mainly desert, with intermittent streams and oases, of which the main products are dates, barley, livestock.

Borkum   Borkum Island, Federal Republic of Germany. 51 03N 9 21E. Town at w end of island, a holiday resort. Pop. (1971E) 30,000.

Borkum Island   Federal Republic of Germany. 51 35N 6 41E. Island off entrance to Ems R. estuary, the most w of German Frisian islands. Area, 14 sq. m. (36 sq. km.).

Borlänge   Kopparberg, Sweden. 60 29N 15 25E. Town sit. NW of Stockholm. The main industry is steel. Pop. (1983E) 46,407.

Borneo   South East Asia. 00 30N 114 00E. Largest island of Malay arch. to E of Sumatra. Area, 187,000 sq. m. (484,330 sq. km.). Politically divided into four sections: Kalimantan forms part of Indonesia and covers most of the island except the N; Sarawak and Sabah are parts of the Federation of Malaysia and comprise Eastern Malaysia, they extend the length of the N coast and the British protectorate of Brunei forms an enclave in NE Sarawak. Mountainous, rising to Mount Kinabalu. Forested in the interior with fertile coastal lowlands. The main products are petroleum, rubber, copra, rice, sago and fish.

Bornholm   Denmark. 55 10N 15 00E. Island in Baltic Sea off SE Sweden. Area, 227 sq. m. (588 sq. km.). Chief town Rönne. Main occupations farming, fishing, tourism. Pop. (1983E) 47,000.

Borno   Nigeria. State, formerly the Borno province of the N region, Nigeria's largest state. Area, 44,942 sq. m. (116,400 sq. km.). The state comprises the emirates of Bornu, Dikwa, Biu, Fika and Bedde, chief towns Maiduguri (cap.) and Nguru, mainly Kanuri people. Pop. (1988E) 6,567,200.

Borodino   Russian Soviet Federal Socialist Republic, U.S.S.R. Village sit. W of Moscow, site of a battle in 1812 between Napoleon's army and Russian forces.

Boroughbridge   North Yorkshire, England. 54 05N 1 24W. Town sit. ESE of Ripon on Ure R. Market town, noted for prehistoric monoliths called the Devil's Arrows.

Borroloola   Northern Territory, Australia. Village ESE of Darwin on McArthur R. Cattle raising station. Airfield.

Borromean Islands   Italy. 46 00N 8 40E. Group of 4 islands in L. Maggiore famous for their gardens.

Borrowdale   Cumbria, England. Valley in Lake District extending S and W from S end of L. Derwentwater.

Boscastle   Cornwall, England. 50 41N 4 42W. Town N of Bodmin on the Atlantic. Fishing port and resort.

Bosnia-Hercegovina   Yugoslavia. 44 00N 18 00E. Constituent rep. of the Yugoslav Federation, sit. mainly in the Dinaric Alps. Area, 19,736 sq. m. (51,116 sq. km.). Chief towns Sarajevo (cap.), Banja Luka, Tuzla, Mostar. Mainly agric. in the valleys; main products cereals, vegetables, fruit and tobacco. Pop. (1981C) 4,124,008.

Bosporus   Asia/Europe. 41 00N 29 00E. Strait extending 18 m. (29 km.) SSW to NNE between European and Asiatic Turkey, with Istanbul on its W shore. It links the Black Sea with the Sea of Marmara. The width is $1/2$–$2 1/2$ m. (1–4 km.).

Bossier City   Louisiana, U.S.A. 32 28N 93 38W. Town opposite Shreveport on Red R. Industries inc. oilfield equipment, wire cable, mobile homes, tubing, concrete, plastics. Has the ruins of the Confederate Forces' Fort Smith, now a memorial park. Pop. (1980C) 50,817.

Boston   Lincolnshire, England. 52 59N 0 01W. Town on Witham R. near mouth. Market town. Industries inc.

bedding manu., agric., label manu. Sea port. Pop. (1985E) 27,000.

Boston Massachusetts, U.S.A. 42 21N 71 04W. City and state cap. at head of Massachusetts Bay. Boston was settled by a chartered company of English Puritans in 1630. Indian inhabitants called it Shawmut; it was renamed Boston from a town in Lincolnshire, England, from which many of the Puritans had come. In 1632 it became the capital of the Massachusetts Bay Colony. It is the commercial and cultural centre of New England. It is an industrial city and a major sea port with an ice-free and almost land-locked harbour. Pop. (1980C) 562,994.

Botany Bay New South Wales, Australia. 33 59S 151 12E. Inlet 5 m. (8 km.) S of Sydney on Pacific coast, the site of Captain Cook's original landing in Australia; now a suburb of Sydney with oil, textile, plastics and chemical industries.

Bothnia, Gulf of Scandinavia. 63 00N 21 00E. Arm of Baltic Sea extending N of Ahvenanmaa between Sweden and Finland. Ice-free only in summer.

Botosani Moldavia, Romania. 47 45N 26 40E. Town sit. NW of Iaşi. Industries inc. flour milling, textiles; manu. clothing. Pop. (1983E) 89,606.

Botswana Africa. 23 00S 24 00E. Rep. in Southern Africa bounded NE by Zimbabwe, N and W by Namibia, S and SE by Republic of South Africa. Area, 222,000 sq. m. (575,000 sq. km.). The main business centres are Lobatse, Gaborone and Francistown. Largest towns are Serowe, Kanye and Molepolole. Mainly plateau at over 3,000 ft. (914 metres) with depressions in the N which become extensive Ls. in summer. The mean annual rainfall is 20 ins. (51 cm.). Cattle-rearing and dairying are the chief industries, but the country is more a pastoral than an agricultural one, crops depending entirely upon the rainfall. However, increasing numbers of boreholes are being established where underground supply is adequate. Main products beef, cattle, hides, skins. Pop. (1989E) 1,255,749.

Bottrop North Rhine-Westphalia, Federal Republic of Germany. 51 31N 6 55E. Town immediately N of Essen on Rhine–Hern Canal in the Ruhr coalfield. The main industry is coal-mining and processing coal products. Pop. (1984E) 112,600.

Bouaké Côte d'Ivoire. 7 41N 5 02W. Town sit. NNW of Abidjan on railway to Burkina Faso. Trading centre for coffee and cacao. Industries inc. cotton ginning, textiles; manu. sisal products. Pop. (1975C) 175,264.

Bou Arfa Oujda, Morocco. 32 32N 1 58E. Village sit. S of Oujda on Trans-Saharan railway. The main industry is manganese mining and processing.

Bouches-du-Rhône France. Dept. in Provence-Côte d'Azur region, on Mediterranean coast. Area, 1,974 sq. m. (5,112 sq. km.). Chief towns Marseilles (prefecture), Arles, Aix-en-Provence. Marshy plain in the W (the Camargue), with dry plain to the E and the Maritime Alps in the E and N. Main occupation farming; main products livestock, olive oil, wine, fruit, lignite and salt. Pop. (1982C) 1,724,199.

Bougainville North Solomons, Papua New Guinea. 6 00S 155 00E. Largest island of the province. Area, 3,880 sq. m. (10,049 sq. km.). Cap. Kicta.

Bougie *see* **Bejaia**

Boulder Western Australia, Australia. Town immediately next to Kalgoorlie and ENE of Perth. As the municipality of Kalgoorlie-Boulder they form the principal mining centre of Western Australia, and the centre of the E Coolgardie goldfield. It is a junction in the Trans-Australian railway. Pop. of Kalgoorlie-Boulder (1983E) 21,800.

Boulder Colorado, U.S.A. 40 01N 105 17W. City sit. NW of Denver. Centre of a mining and ranching area, and a health resort. Industries inc. electronics, storage technology, engineering. Pop. (1980C) 76,855.

Boulogne-Billancourt Hauts-de-Seine, France. Suburb of SW Paris on Seine R. Industries inc. manu. vehicles, aircraft, rubber, cosmetics and soap. Pop. (1982C) 102,595.

Boulogne-sur-mer Pas-de-Calais, France. 50 43N 1 37E. Town at mouth of Liane R. on English Channel. Industries inc. fishing, fish processing; manu. iron and steel, cement, tiles. Sea port with heavy passenger traffic for Folkestone and Dover. Pop. (1982C) 48,349 (agglomeration, 98,566).

Bountiful Utah, U.S.A. 40 53N 111 53W. City sit. N of Salt Lake City in an irrigated agric. area. Farming centre producing fruit and poultry. Pop. (1980C) 32,877.

Bounty Islands New Zealand. 48 00S 178 30E. Group of islands 400 m. (640 km.) ESE of Dunedin in the S Pacific. Named after, and discovered by, Captain Bligh's *Bounty*. Uninhabited.

Bourg-en-Bresse Ain, France. 46 12N 5 13E. Town sit. NNE of Lyon on Reyssouze R. Prefecture. Market town trading in grain, wine, poultry. Industries inc. flour milling, cheese making, pottery. Pop. (1982C) 43,675.

Bourges Cher, France. 47 05N 2 24E. Town on Canal du Berry at confluence of Auron and Yèvre Rs. Prefecture. Military town with state ordnance factories; also manu. woollens, linoleum, hardware and agric. implements. Pop. (1982C) 79,408.

Bourget, Lac du France. 45 44N 5 52E. L. in Savoie dept. near Aix-les-Bains, 11 m. (17.5 km.) long and up to 2 m. (3 km.) wide.

Bourget, Le Seine-St. Denis, France. Town sit. NNE of Paris with one of the 3 main metropolitan airports. Pop. (1982C) 11,021.

Bourgogne *see* **Burgundy**

Bourne Lincolnshire, England. 52 46N 0 23W. Town sit. SE of Grantham. Market town in an agric. area. Pop. (1977E) 7,594.

Bournemouth Dorset, England. 50 43N 1 54W. Town on Poole Bay, on English Channel. The main industry is tourism but there has been an increase in financial and banking services in the 1980s. Pop. (1989E) 151,465.

Bournville West Midlands, England. 52 25N 1 56W. Area of S Birmingham founded in 1879 by George Cadbury as a residential area for workers of his cocoa and chocolate factories.

Bouvet Island South Atlantic Ocean. 54 26S 3 24E. Island 1,800 m. (2,880 km.) SSW of Capetown. Area, 19 sq. m. (48 sq. km.). It forms a depend. of Norway and is uninhabited.

Bovey Tracey Devon, England. 50 35N 3 40W. Town sit. SW of Exeter. Market town in a farming area. Industries inc. pottery. Pop. (1981C) 4,226.

Bow (Stratford-le-Bow) Greater London, England. 51 31N 0 02W. District of E London N of Thames R. forming part of the boro. of Tower Hamlets.

Bowie Maryland, U.S.A. 39 00N 76 47W. City sit. E of Washington, D.C. Mainly residential. Pop. (1980C) 32,400.

Bowling Green Kentucky, U.S.A. 37 00N 86 27W. City sit. NNE of Nashville Tennessee on Barren R. at the head of navigation. Market centre for an agric. area, trading in fruit, tobacco, maize, livestock. Industries inc. manu. food products, electrical goods, clothing, building materials, dry cleaning equipment, automobiles, cranes, space heaters and water coolers; processing flour and cattle food, tobacco, timber, oil. Pop. (1986E) 56,321.

Bowling Green Ohio, U.S.A. 41 22N 83 39W. City sit. SSW of Toledo in an agric. area. Industries inc. manu. motor vehicles, machinery, newspaper publishing, metal goods, food products, glass products. Seat of Bowling Green State University. Pop. (1980C) 25,728.

Bowness-on-Windermere Cumbria, England. 54 31N 3 23W. Town on E shore of L. Windermere. Tourist centre. Pop. (1981C) 3,754.

Bow River Canada. Rises in Rocky Mountains in SW Alberta and flows 315 m. (504 km.) SE through Banff National Park, past Banff and Calgary to turn SE near Bassano and joins Saskatchewan R.

Box Hill Surrey, England. 51 15N 0 17W. Spur of N Downs immediately NNE of Dorking.

Boyacá Boyacá, Colombia. Town in the E Cordillera. Scene of the Spanish defeat by Bolívar, 1819, which gained independence for Colombia and Venezuela.

Boyne River Ireland. 53 43N 6 15W. Rises in Bog of Allen and flows 80 m. (128 km.) NE through County Meath to enter Irish Sea 4 m. (6.5 km.) below Drogheda. The Battle of the Boyne, the defeat of James II by William III, was fought 3 m. (5 km.) W of Drogheda in 1690.

Boy's Town Nebraska, U.S.A. 41 16N 96 08W. Village sit. W of Omaha founded in 1917 by Father E. J. Flanagan as a settlement for homeless boys.

Bozcaada Canakkale, Turkey. Island off NW Turkish coast in Aegean Sea, S of the entrance to the Dardanelles Strait. Area, 15 sq. m. (39 sq. km.). Chief town, Bozcaada.

Bozeman Montana, U.S.A. 45 41N 111 02W. City sit. ESE of Butte and S of Bridger mountains. Distribution centre for an agric. area. Pop. (1989E) 27,800.

Bozen *see* **Bolzano**

Brabant Belgium. 49 15N 5 20E. Province of central Belgium. Area, 1,267 sq. m. (3,282 sq. km.). Cap. Brussels. Drained by Dyle, Demer and Senne Rs. Mainly agric. with engineering and textile industries in the towns. Pop. (1987E) 2,221,818.

Brac Yugoslavia. 43 20N 16 40E. The main Dalmatian island sit. SE of

Split in the Adriatic. Area, 152 sq. m. (394 sq. km.). Chief town, Supetar. Industries inc. tourism and fishing.

Brackley Northamptonshire, England. 52 02N 1 09W. Town sit. ESE of Banbury on Ouse R. Industries inc. soap-making, meat processing, light engineering, packaging. Pop. (1989E) 8,500.

Bracknell Berkshire, England. 51 26N 0 45W. 'New town' sit. ESE of Reading. Industries inc. engineering; manu. clothing, furniture. Headquarters of the Meteorological Office. Pop. (1985E) 51,552.

Bradford West Yorkshire, England. 53 48N 1 45W. City sit. W of Leeds and Metropolitan district of West Yorkshire, and centre of the woollen industry; also manu. other textiles, textile machinery. Pop. (1989E) 466,000.

Bradford-on-Avon Wiltshire, England. 51 20N 2 15W. Town sit. ESE of Bath on Avon R. There is some industry inc. the manu. of industrial polymers. Historic town with Saxon church in a farming area. Pop. (1985E) 9,150.

Braemar Grampian Region, Scotland. 57 01N 3 23W. Village sit. WSW of Ballater on Clunie Water. Scene of the annual Highland Games.

Braga Braga, Portugal. 41 33N 8 26W. (i) District, area 1,057 sq. m. (2,730 sq. km.). Pop. (1981C) 708,924. (ii) Town sit. NNE of Oporto. District cap. Industries inc. manu. hats, cutlery. Pop. (1981C) 63,771.

Bragança Bragança, Portugal. 41 49N 6 45W. (i) District in NE, borders N and E with Spain. Agric. area. Area, 2,551 sq. m. (6,608 sq. km.). Pop. (1981C) 184,252. (ii) City 110 m. (176 km.) NE of Oporto. District cap. Railway terminus and trading centre for an agric. area, handling grain, wine, olive oil and livestock. Industries inc. manu. textiles, hats, confectionery. Episcopal see. Pop. (1980E) 10,000.

Brahmaputra River South Asia. 24 02N 90 59E. Rises in SW Tibet in Himalayas and flows 1,800 m. (2,880

km.) E, as Tsangpo R., across S Tibet through Himalayas. At the E end of the range near Namcha Barwa it loops N and then S through deep gorges, as the Dighang R. and enters NE Assam, India, as the Brahmaputra. It flows WSW through Assam and turns S into Bangladesh to join Ganges R. 35 m. (56 km.) W of Dacca. Navigable for steamers up to Ganhati and for small craft up to Dibrugarh.

Brăila   Brăila, Romania. 45 16N 27 58E. Town sit. SSW of of Galati on Danube R. Industries inc. flour milling, railway engineering, timber. R. port trading in grain. Pop. (1983E) 214,516.

Braintree and Bocking   Essex, England. 51 53N 0 32E. Town sit. W of Colchester comprising 2 towns joined in 1934. Industries inc. food processing, textiles; manu. metal windows. Pop. (1981C) 31,139.

Brakpan   Transvaal, Republic of South Africa. 26 13S 28 20E. Town sit. E of Johannesburg on the Witwatersrand, an important gold mining centre. Pop. (1984E) 88,000.

Brampton   Cumbria, England. 54 36N 2 30W. Town sit. ENE of Carlisle. Market town for a farming area. Pop. (1985E) 3,774.

Branco Island   Cape Verde Islands, Atlantic. Island sit. SSE of Santa Luzia. Area, 1 sq. m. (2.5 sq. km.). Desolate and uninhabited.

Brandenburg   Potsdam, German Democratic Republic. 52 24N 12 32E. Town sit. WSW of Berlin on Havel R. Industries inc. textiles, engineering. Pop. (1989E) 95,000.

Brandon   Manitoba, Canada. 49 50N 99 57W. Town sit. W of Winnipeg on Assiniboine R. Trading centre for a wheat-growing area. Industries inc. agric. engineering, oil refining. Pop. (1983C) 36,242.

Brandywine Creek   U.S.A. 39 44N 75 32W. R. rises in SE Pennsylvania and flows 20 m. (32 km.) SE through Delaware to join Christina R. Scene of a battle between American and English forces, 1777.

Brantford   Ontario, Canada. 43 08N 80 16W. Town sit. WSW of Hamilton on Grand R. Industries inc. manu. refrigerators, agric. implements and pottery. Pop. (1981C) 74,315.

Bras d'Or, Lake   Canada. 45 52N 60 50W. Tidal L. in Nova Scotia extending SW to NE through Cape Breton island and nearly dividing it in two. Area, 360 sq. m. (932 sq. km.).

Brasilia   Federal District, Brazil. 15 45S 47 57W. City sit. NE of Rio de Janeiro. Construction began in 1956 and in 1960 it became the federal cap. The main employment is in govt. service. Pop. (1980C) 410,999.

Braşov   Braşov, Romania. 45 39N 25 37E. Town sit. NNW of Bucharest in Transylvanian Alps. District cap. Industries inc. textiles, iron and copper smelting; manu. aircraft, tractors, machinery, oil-drilling equipment. Pop. (1983E) 290,772.

Bratislava   Slovakia, Czechoslovakia. 48 09N 17 07E. Town sit. SE of Brno on Danube R. near Austrian and Hungarian borders. Slovakian cap. Railway centre with important trade in agric. produce. Industries inc. refining oil from Brody (Ukraine), engineering, brewing, printing, chemicals, textiles, paper making, flour milling, sugar refining. Pop. (1989E) 425,000.

Bratsk   Russian Soviet Federal Socialist Republic, U.S.S.R. 56 05N 101 48E. Town sit. NNW of Irkutsk on Angara R. Industries are based on an important hydroelectric power station and inc. sawmilling; manu. wood pulp, cellulose, furniture. Pop. (1987E) 249,000.

Braunschweig   *see* **Brunswick**

Brava   Cape Verde. 14 52N 24 43W. Most S island of arch. Area, 26 sq. m. (67 sq. km.). Chief town Nova Sintra. Pop. (1980C) 6,984.

Bray   Wicklow, Ireland. 53 12N 6 06W. Town sit. SSE of Dublin on Irish Sea coast. Seaside resort. Pop. (1981E) 23,000.

Brazil   South America. 11 00S 54 00W. Rep. bounded E by the Atlantic, and on its NW and S borders by all the S American countries except Ecuador and Chile. Area, 3,287,195 sq. m. (8,513,835 sq. km.), the 4th largest country in the world, comprising 21 states, 4 territories and the Federal District. Chief cities Brasilia (cap.), São Paulo, Rio de Janeiro, Porto Alegre, Salvador, Recife. About nine-tenths of the people and most industry and agric. is found on the coastal belt. The interior is little developed. The Amazon basin rain forests cover the N; the NE is dry scrub, the centre is table-land rising to 8,000 ft. (2,438 metres). 44% of the pop. is rural and 75% of the foreign exchange derived from agric. exports. Coffee is grown in the states of São Paulo, Paraná, Espirito Santo and Minas Gerais and is an important crop. Rubber is a natural product of Brazil and is found chiefly in the states of Acre, Amazonas and Pará. Sugar, tobacco, citrus fruits, bananas and rice are also important crops. Brazil is a livestock producer and ranks ahead of Argentina. Brazil is the only source of high-grade quartz crystal in commercial quantities. Other minerals include, industrial diamonds, chrome, salt, mica, beryllium, graphite, titanium, magnesite, iron ore, manganese, coal, tungsten. Gold is found in every state but large scale mining is at Minas Gerais. The most important manufacturing industry in Brazil is the weaving industry, which employs about 16% of all industrial workers; nearly 50% of the factories are in São Paulo and another 28% in Guanabara and Minas Gerais. Pop. (1989E) 147,404,375.

Brazzaville   Congo. 4 14S 15 14E. City and cap. on Zaïre R. opposite Kinshasa (Zaïre). It was founded in 1880 by the French explorer Pierre de Brazza. Industries inc. railway engineering, ship yards, textiles and the manu. of footwear. It is a major river port and railway centre. Pop. (1984E) 585,812.

Brechin   Tayside Region, Scotland. 56 44N 2 04W. Town sit. NE of Forfar on S Esk R. Industries inc. manu. whisky, linen, paper. Pop. (1980E) 7,500.

Brechou   Channel Islands. Small

island off W coast of Sark. Area, 74 acres.

Brecon   Powys, Wales. 51 57N 3 24W. Town at confluence of Usk and Honddu Rs. Industries inc. electronics and light engineering, manu. textiles. Pop. (1981E) 7,400.

Breda   North Brabant, Netherlands. 51 35N 4 46E. Town sit. N of Antwerp at confluence of Merk and Aa Rs. Industries inc. textiles, engineering, food canning; manu. textile machinery, shoes, paint and matches. Pop. (1989E) 120,200.

Bregenz   Vorarlberg, Austria. 47 30N 9 46E. Town at E end of L. Constance. Provincial cap. Industries inc. tourism and textiles. Pop. (1989E) 26,730.

Breisach   Baden-Württemberg, Federal Republic of Germany. 48 01N 7 40E. Town sit. WNW of Freiburg on Rhine R. The main industry is wine-making.

Breisgau   Baden-Württemberg, Federal Republic of Germany. District extending along E bank of Rhine N and S of Freiburg and inc. the mountains of the S Black Forest.

Breitenfeld   Leipzig,   German Democratic Republic. Village sit. NW of Leipzig, scene of two Swedish victories, 1631 and 1642, during the Thirty Years War.

Bremen   Federal Republic of Germany. 53 04N 8 49E. (i) *Freie Hansestadt Bremen*, a *Land* of the Federal Rep. consisting of 2 areas on the Weser R. in Lower Saxony, one centred on Bremen, the other on Bremerhaven. Area, 156 sq. m. (399 sq. km.). Pop. (1987E) 660,000. (ii) City on Lower Weser R. 50 m. (80 km.) from the sea, a Free Hansa City and major sea port. Industries inc. oil refining, shipbuilding, flour milling, brewing, sugar refining; manu. chocolate and tobacco products. Exports iron, steel and other manu. goods. Pop. (1987E) 533,400.

Bremerhaven   Bremen, Federal Republic of Germany. 53 33N 8 34E.

Town sit. NNW of Bremen on Weser R. estuary. Outport for Bremen. Industries inc. fishing, fish processing, ship repairing. Pop. (1989E) 130,000.

Bremerton Washington, U.S.A. 47 34N 122 38W. City and port on an inlet of Paget-Sound W of Seattle. Industries inc. shipbuilding and naval dockyards. Pop. (1980C) 36,208.

Brenner Pass Austria/Italy. 47 00N 11 30E. Alpine pass between Innsbruck, Austria and Bolzano, Italy. The lowest of the main alpine passes.

Brent Greater London, England. Borough of NW London. Pop. (1988E) 257,200.

Brentford and Chiswick Greater London, England. Suburb of W Lond. on N bank of Thames R. Industries inc. manu. pharmaceutical goods, tyres, soap.

Brentwood Essex, England. 51 38N 0 18E. Town sit. NE of London, mainly residential. Pop. (1981C) 51,643.

Brescia Brescia, Italy. 45 33N 10 15E. Town sit. E of Milan at foot of Alps. Provincial cap. Industries inc. manu. metal goods, firearms, hosiery, textiles. Pop. (1980E) 210,027.

Breslau *see* **Wroclaw**

Bresse France. 46 20N 5 10E. District between Saône R. and Jura mountains, extending through Ain, Jura and Saône-et-Loire depts. Chief town, Bourg; a fertile plain, main products pigs and poultry.

Brest Finistère, France. 48 24N 4 29W. City sit. W of Rennes on W coast of Brittany and N shore of Brest Roads. Sea port exporting fruit and vegetables and importing wheat, wine, coal, timber. Industries inc. fishing, flour milling, engineering, chemicals. Naval station with extensive dockyards, arsenal and naval training school. Pop. (1982C) 160,355 (agglomeration, 201,145).

Brest Belorussia, U.S.S.R. 52 06N 23 42E. City on the Bug R. where it forms the frontier with Poland. Railway centre and R. port trading in timber, grain and cattle. Industries inc. sawmilling, cotton spinning, food processing, engineering. Pop. (1985E) 222,000.

Briançon Hautes-Alpes, France. 44 54N 6 39E. Town sit. ESE of Grenoble in the upper Durance valley. Trading centre for cheese and silks. The main industry is tourism, esp. for winter sports. Pop. (1982C) 11,544.

Briansk Russian Soviet Federal Socialist Republic, U.S.S.R. 53 13N 34 25E. City sit. SW of Moscow on the Desna R. Oblast cap. Railway centre and industrial centre. Industries inc. iron and steel, sawmilling, rope-making, brick making; manu. roadmaking machinery, cement. There is a lumbering and forestry school. It forms the centre of the Briansk-Bezhitsa industrial area. Pop. (1985E) 430,000.

Bridgend Mid Glamorgan, Wales. 51 31N 3 35W. Market town sit. W of Cardiff on Ogwr R. Industries inc. heavy engineering, rubber, motor manu., engineering, plastics, electrical and light and high technology industry. Pop. (1984E) 29,500.

Bridge of Allan Central Region, Scotland. 56 09N 3 57W. Town sit. N of Stirling on the Allan Water. Resort with mineral springs. Pop. (1973E) 3,318.

Bridgeport Connecticut, U.S.A. 41 11N 73 11W. City sit. SW of New-Haven on Long Island Sound. Industries inc. engineering; manu. machinery, sewing machines, electrical equipment, hardware. Sea port. Pop. (1989E) 143,000.

Bridgetown Barbados. 13 06N 59 36W. Cap. and principal town of Barbados, on Carlisle Bay on SW coast. It was founded in 1628 and is a sea port with deepwater harbour, exporting sugar, molasses and rum, and has an airport with heavy tourist traffic. Pop. (1987E) 7,500.

Bridgnorth Shropshire, England. 52 33N 2 25W. Market town sit. WSW of Wolverhampton on Severn R. Industries inc. aluminium and some light industry. Pop. (1987E) 11,160.

Bridgwater Somerset, England. 51 08N 3 00W. Town sit. NNE of Taunton on Parrett R. Industries inc. footwear, plastics and electrical goods. Pop. (1980E) 27,000.

Bridlington Humberside, England. 54 05N 0 12W. Town sit. WSW of Flamborough Head on Bridlington Bay. Fishing port and resort. Pop. (1989E) 31,850.

Bridport Dorset, England. 50 44N 2 46W. Town sit. W of Dorchester on Brit R. Market town for a farming and fishing area. Industries inc. boat building, manu. ropes, fishing nets, sailcloth, cordage. Pop. (1989E) 11,750.

Brie France. 48 35N 3 10E. Area between Seine and Marne Rs., E of Paris. Agric. area producing wheat, sugar-beet and dairy produce esp. cheese. The chief town is Meaux.

Brienz Bern, Switzerland. 46 44N 7 53E. Town on NE side of L. Brienz. The main industry is tourism, with some wood working. Pop. (1980C) 3,000.

Brierley Hill West Midlands, England. 52 29N 2 07W. Town in the Dudley Enterprise Zone, sit. W of Birmingham. Industries inc. iron and steel; manu. pottery, bricks and glassware.

Brig Valais, Switzerland. 46 19N 8 00E. Town at N end of Simplon Pass and tunnel, on Rhône R. An important route centre. Pop. (1980E) 5,000.

Brighouse West Yorkshire, England. 53 42N 1 47W. Town sit. N of Huddersfield on the Calder R. Industries inc. manu. machinery, machine tools, wire goods, textiles, carpets. Pop. (1986E) 30,524.

Brightlingsea Essex, England. 51 49N 1 02E. Town sit. SE of Colchester on Colne R. estuary. The main industry is tourism, esp. for yachting, and some oyster fishing. Pop. (1989E) 8,000.

Brighton East Sussex, England. 50 50N 0 8W. Town sit. S of London on English Channel. Seaside resort with some industries inc. manu. furniture, food products. Pop. (1981C) 146,134.

Brindisi Brindisi, Italy. 40 38N 17 56E. Town sit. SE of Bari on Adriatic coast of Apulia. Provincial cap. Sea port exporting wine and olive oil and handling passenger traffic for Greece. Pop. (1980C) 90,000.

Brisbane Queensland, Australia. 27 28S 153 09E. City near mouth of Brisbane R. on a dredged channel, founded in 1824. State cap. and commercial centre for an extensive farming area. Sea port, exporting frozen meat, dairy products, sugar, wool, fruit, hides. Industries inc. oil refining, chemical plants, engineering, shipbuilding, food processing; manu. cement, clothing, footwear, furniture. Pop. (1981C) 942,836.

Bristol Avon, England. 51 27N 2 35W. City 7 m. (11 km.) above mouth of Avon R. on Bristol Channel. Industries inc. manu. aerospace, chocolate, mechanical engineering, packaging and financial services. Sea port importing grain, timber, petroleum. Pop. (1988E) 377,700.

Bristol Connecticut, U.S.A. 41 41N 72 57W. City sit. SW of Hartford. The main industry is spring making; also manu. machine tools. Pop. (1989E) 61,100.

Bristol Pennsylvania, U.S.A. 40 06N 74 52W. Town sit. NE of Philadelphia on Delaware R. Residential with industrial waterfront. Industries inc. printing, engineering and distribution. Pop. (1980C) 10,867.

Bristol Channel England/Wales. 51 20N 4 00W. Inlet of Atlantic Ocean, extending 80 m. (128 km.) E between SW Eng. and S Wales. The width varies from 50 m. (80 km.) at the mouth to 5 m. (8 km.). On the Welsh (N) coast the chief towns are Cardiff and Swansea, on the English coast Ilfracombe and Weston-super-Mare. It receives the Taff and Towy Rs. from the N, the Severn R. from the NE and the Parrett, Taw and Torridge Rs. from the S.

British Antarctic Territory   South
Atlantic. 66 00S 45 00W. Colony
comprising Antarctic mainland and
islands s of 60°s lat. and between
20°w and 80°w long. Area, 700,000
sq. m. (1,813,000 sq. km.) which inc.
South Orkney and South Shetland
Islands, the Antarctic peninsula
(Palmer Land and Graham Land) and
Coats Land. There is no permanent
pop. although some 300 scientists
man research stations.

British Columbia   Canada. 55 00N
125 15W. Province on w coast bounded s by U.S.A. and E by Alberta.
Area, 366,255 sq. m. (948,600 sq.
km.). Chief towns Victoria (cap.),
Vancouver, New Westminster. Mountainous, with parallel ranges running N
to s and cut by Fraser, Kootenay,
Thompson and Columbia Rs. Heavily
forested with valuable timber. About
3% of the land is suitable for cultivation, with dairying, fruit growing and
mixed farming in the s. Main occupations lumbering, fishing, mining, with
associated industries powered by
hydroelectric power. Pop. (1989E)
3,055,600.

British Honduras   *see* **Belize**

British Indian Ocean Territory
Colony, founded 1965, consisting of
the Chagos Arch., the Aldabra, Farquhar and Desroches islands but in 1976
Aldabra, Farquhar and Desroches islands were returned to the Seychelles,
sit. between Sri Lanka and Mauritius.
There is no permanent population.

British Isles   55 00N 4 00W. An
archipelago on the continental shelf of
Europe but separated from the conti-
nent by the North Sea, the Strait of
Dover and the English Channel. It
consists of two large islands and some
5,000 small islands and islets. The
largest island is Great Britain (Eng-
land, Wales and Scotland); the other
large island is Ireland (Northern
Ireland and the Irish Republic).
Among single islands are the Isle of
Man, Anglesey and Isle of Wight. The
groups include Orkney and Shetland
Islands to the N of Scotland, the
Hebrides to the w and the Isles of
Scilly to the sw of England. The
Channel Islands are geographically

French but are considered part of the
British Isles.

British Virgin Islands   West Indies.
18 40N 64 30W. Group of islands E of
Virgin Islands arch., consisting of 36
islands. Area, 59 sq. m. (151 sq. km.).
Main islands Tortola, Virgin Gorda,
Anegada, Jost Van Dyke. Cap. Road
Town on Tortola. Main occupations
fishing, farming, fruit growing. Pop.
(1987E) 13,300.

Brittany   France. 48 00N 3 00W.
Region occupying peninsula between
English Channel and Bay of Biscay
and consisting of the Côtes-du-Nord,
Morbihan, Ile-et-Vilaine and Finistère
depts. Area, 10,496 sq. m. (27,184 sq.
km.). Chief cities, Rennes, Brest, Lorient, St. Brieuc and Quimper. Mainly
moorland with important prehistoric
sites. Main occupations farming, market gardening and fishing; main products cider, potatoes, onions and other
vegetables. Pop. (1987E) 2,763,100.

Brive-la-Gaillarde   Corrèze, France.
45 10N 1 32E. Town sit. SSE of
Limoges on Corrèze R. Market town
trading in fruit and vegetables, wool,
cattle. Industries inc. manu. food products, furniture, clothing and paper.
Pop. (1989E) 54,440.

Brixen   Bolzano, Italy (also Bressanone). 46 43N 11 39E. Town sit. NE
of Bolzano on Brenner rail route and
Isarco R. Health and tourist resort at
1,834 ft. (559 metres). Industries
powered by an important hydroelectric plant inc. manu. wax, foodstuffs,
alcohol. Pop. (1971E) 19,000.

Brixham   Devon, England. 50 24N
3 30W. Town sit. s of Torquay on
English Channel. Fishing port and
holiday resort. Pop. (1981C) 15,171.

Brixton   Greater London, England.
District of sw Lond. containing a
major prison.

Brno   Jihomoravský, Czechoslovakia. 49 12N 16 37E. City sit. SE of
Prague above confluence of Svitava
and Svratka Rs. The country's 2nd
city, an industrial, commercial and
cultural centre. Industries inc. manu.
textiles, vehicles, machinery, clothing,

furniture and soap. Pop. (1980C) 371,000.

Broads, The   Norfolk, England. 52 30N 1 15E. Region of Ls. and marshland surrounding Bure, Ant and Thurne Rs., the Ls. partly formed from the 'broadening' of the Rs. Tourist area for sailing, bird watching and fishing.

Broadstairs   Kent, England. 51 22N 1 27E. Town in Isle of Thanet in E Kent. Holiday resort with some light industry. Pop. (1984E) 25,326.

Brocken   German Democratic Republic. 51 48N 10 37E. Mountain peak in Magdeburg district, the highest in the Harz mountains. Noted as a trad. meeting place of witches.

Brockton   Massachusetts, U.S.A. 42 05N 71 01W. City sit. S of Boston. Industries inc. manu. shoes, shoe making machinery and leather. Pop. (1980C) 95,172.

Broken Hill   New South Wales, Australia. 31 57S 141 27E. Town near the South Australian border. Mining centre with a 3 m. (5 km.) lode of metallic sulphides. Pop. (1988E) 24,500.

Broken Hill   Zambia *see* **Kabwe**

Bromley   Greater London, England. 51 24N 0 02E. Boro. of SE Lond. extending into NW Kent. Mainly residential. Pop. (1988E) 298,000.

Brompton   Greater London, England. District in Kensington in W Lond. containing the Roman Catholic Brompton Oratory and several major museums.

Bromsgrove   Hereford and Worcester, England. 52 20N 2 03W. Town sit. NNE of Worcester. Many small firms in light industry. Pop. (1988E) 90,300.

Bronx   New York, U.S.A. Boro. of New York City forming NE district. Pop. (1980C) 1,168,972.

Brookings   South Dakota, U.S.A. 44 19W 96 48N. City sit. N of Sioux

Falls. Cultural, educational and small industrial centre for NE of State. Pop. (1980C) 14,971.

Brookline   Massachusetts, U.S.A. 42 21N 71 07W. Town forming a suburb of W Boston, mainly residential with some industries. Pop. (1980C) 55,062.

Brooklyn   New York, U.S.A. District of New York City on extreme W of Long Island and SE from Manhattan across the East R; co-extensive with Kings County. Bounded N and NE by Newtown Creek and SE by Jamaica Bay. Mainly residential but with an important water front and neighbouring industrial area.

Brownsville   Texas, U.S.A. 25 54N 97 30W. City opposite Matamoros, Mex., on Rio Grande, Commercial centre trading in petroleum, citrus fruits, cotton. Industries inc. chemicals, food canning. Pop. (1989E) 104,000.

Bruay-en-Artois   Pas-de-Calais, France. 50 29N 2 33E. Town sit. SW of Béthune. Industries inc. engineering and plastics. Pop. (1989E) 27,200.

Bruck an der Mur   Styria, Austria. 47 25N 15 16E. Town sit. NNW of Graz at confluence of Mur and Mürz Rs. The main industry is iron and steel, also manu. paper. Pop. (1981C) 71,330.

Bruges   West Flanders, Belgium. 51 13N 3 14E. City sit. WNW of Ghent. Provincial cap. Market town, railway and canal centre. Industries inc. tourism, engineering, brewing, lace making, flour milling. Pop. (1989E) 120,000.

Brühl   North Rhine-Westphalia, Federal Republic of Germany. 50 48N 6 54E. Town sit. S of Cologne in a lignite mining area. Industries inc. car assembly, making briquettes and machinery. Pop. (1988E) 40,586.

Brunei   4 52N 115 00E. A sovereign state forming an enclave in Sarawak, Malaysia, on NW coast of Borneo. Area, 2,226 sq. m. (5,765 sq. km.). Cap. Bandar Seri Begawan. Brunei depends primarily on its oil industry, which employs more than 7% of the working population. Other important products are rubber, padi, jelutong, firewood and sago. Local industries include boat-building, cloth weaving and the manufacture of brass- and silverware. Most of the interior is forested, containing large potential supplies of serviceable timber. Pop. (1988E) 241,400.

Brunei Town   *see* **Bandar Seri Begawan**

Brunssum   Limburg, Netherlands. Town sit. N of Heerlen. Pop. (1989E) 29,857.

Brunswick   Lower Saxony, Federal Republic of Germany. 52 19N 10 31E. City sit. ESE of Hanover on Oker R. Industries inc. microelectronics, biotechnology, computers, fruit and vegetable canning, flour milling, sugar refining; manu. car components, pianos, cameras and railway signalling equipment. Pop. (1989E) 254,851.

Bruny Island   Tasmania, Australia. 43 17S 147 18E. Island off SE coast of Tasmania across the D'Entrecasteaux Channel; consists of S and N parts connected by an isthmus 3 m. (5 km.) long. Area, 142 sq. m. (363 sq. km.). Chief town Allonah, on S Bruny although in general N Bruny is the more populous. Main occupation farming.

Brussels   Brabant, Belgium. 50 50N 4 21E. City which developed after A.D. 580 on Senne R. Cap., cultural and commercial centre. Most of the earliest buildings are of the 15th cent. and the Grand' Place is generally accepted as the finest of Europe's medieval squares. Industries inc. textiles, chemicals; manu. paper, rubber, furniture, clothing, lace. Administrative centre of the European Economic Community since 1958 and of NATO. Pop. (1988E) 970,346 (inc. suburbs).

Bryansk   U.S.S.R. *see* **Briansk**

Bucaramanga   Santander, Colombia. 7 08N 73 09W. City in the E Cordillera. Dept. Cap. Commercial centre for a coffee, tobacco and cotton growing area. Manu. tobacco products, tex-

tiles and straw hats. Pop. (1980E) 441,000.

Bucharest   Ilfov, Romania. 44 25N 26 07E. City on Dambovita R. in Walachia. Cap. and district cap., commercial and cultural centre. Since 1945 the city has expanded and high-rise blocks dominate the outskirts but other parts of the centre have been demolished, including buildings of historic note, to allow the building of govt. offices. Industries inc. oil refining, chemicals, textiles, flour milling, tanning; manu. aircraft and machinery. Trading centre for oil, timber and agric. produce. Pop. (1985E) 1,975,808.

Buckfastleigh   Devon, England. 50 29N 3 46W. Town sit. W of Torquay on Dart R. There is a tannery and some light industry. Tourism is important. Noted for the Benedictine Buckfast Abbey, built by the monks between 1907–37. Pop. (1985E) 3,000.

Buckhaven and Methil   Fife Region, Scotland. 56 11N 3 03W. Town, formed by union of 2 towns in 1891, on Firth of Forth in a former coalmining area. Pop. (1987E) 15,900.

Buckie   Grampian Region, Scotland. 57 40N 2 58W. Port sit. on Moray Firth coast. Industries inc. fishing, fish processing, boat building, marine engineering, structural engineering. Pop. (1988E) 8,320.

Buckingham   Buckinghamshire, England. 52 00N 1 00W. Town sit. NW of Aylesbury on Ouse R. Market town for a farming area. Industries inc. manu. dairy products, engineering. Pop. (1973E) 5,290.

Buckinghamshire   England. 51 50N 0 55W. County in S Midlands, bounded on N by Northamptonshire, E by Bedfordshire, Hertfordshire and Greater London, S by Berkshire, Surrey and W by Oxfordshire. Area, 749 sq. m. (1,940 sq. km.). County town Aylesbury; largest town Milton Keynes. Hilly in the S where it is crossed by the Chiltern Hills, where local woods at one time provided the material for the still important furniture industry. Other industries inc. brickmaking, printing and high technology. Mainly agric. in the N in the fertile vale of Aylesbury. Pop. (1988E) 627,300.

Bucureşti   *see* **Bucharest**

Budapest   Pest, Hungary. 47 29N 19 05E. City on Danube R. at the edge of the Hungarian plain. Cap. consisting of Buda on the right bank and Pest on the left. Commercial centre trading in grain, wine, cattle, hides and wool. Industries inc. flour milling, brewing, iron and steel, textiles, chemicals, engineering. Pop. (1989E) 2,115,000.

Budaun   Uttar Pradesh, India. 28 03N 79 07E. Town sit. SW of Bareilly, trading in grain, cotton. Pop. (1981C) 93,004.

Bude   Cornwall, England. 50 49N 4 33W. Town on Atlantic coast sit. near Devon border. There is some light industry but it is primarily a holiday resort. Pop. (1988E) 7,418.

Budleigh Salterton   Devon, England. 50 38N 3 19W. Town sit. ENE of Exmouth on Lyme Bay. Holiday resort. Pop. (1985E) 4,500.

Buea   Cameroon. 4 09N 9 14E. Cap. of South-West Province sit. NNE of Victoria on S slopes of Cameroon Mountain. Trading centre for an agric. area, producing cacao, bananas, palm oil, kernels, hardwoods, rubber. Pop. (1981E) 29,953.

Buenaventura   Valle del Cauca, Colombia. 3 53N 77 04W. Town sit. NW of Cali on the Pacific coast. Industries inc. fishing and fish canning. Port exporting coffee, sugar, hides, gold and platinum from the Cauca valley farming area and the Chocó mines. Pop. (1984E) 122,500.

Buenos Aires   Buenos Aires, Argentina. 34 40S 58 30W. City on W bank of Rio de la Plata estuary. Cap. and chief sea port. It was founded by Spanish settlers in 1536. Industries are largely based on local natural gas and petroleum, and inc. food processing, textiles, chemicals; manu. paper, vehicles, paint and metal goods. The

main exports are grain, beef and wool. Pop. (1989E) 2,900,794.

Buffalo   New York, U.S.A. 42 54N 78 53W. City at E end of L. Erie and W end of main route across the Appalachians. Port and important route and distribution centre. Industries are mainly based on power from Niagara Falls and inc. transport equipment, primary metals, flour milling, printing, chemicals. Service industries are expanding. Pop. (1980C) 357,870.

Buganda   Uganda. One of the 4 regions of Uganda, in S, comprising districts of Mengo, Masaka and Mubende and bounded (S) by Tanzania. Area, 25,600 sq. m. (66,304 sq. km.). Chief towns Kampala (cap.) and Entebbe. Mainly savannah with stretches of dense forest and an extensive L. area. Main products cotton, coffee and bananas.

Bug River   Poland/U.S.S.R. 52 31N 21 05E. Also called the Western Bug, rises in W Ukraine ENE of Lvov and flows 480 m. (768 km.). NW forming Poland/Ukraine frontier from NW of Sokal to NW of Brest and turning W to join Vistula R. below Warsaw. Navigable below Brest. Linked by the Mukhanets R. to the Dnieper-Bug canal.

Bug River   U.S.S.R. Also called the Southern Bug, rises in S Ukraine 37 m. (59 km.) NE of Ternopol and flows 530 m. (848 km.) SE past Vinnitsa and Nikolayev to enter Dnieper estuary on Black Sea. Navigable for *c*. 60 m. (96 km.).

Builth Wells   Powys, Wales. 52 09N 3 24W. Town sit. N of Brecon. Market town for a farming area and tourist centre. There is some light industry. Pop. (1989E) 1,800.

Bujumbura   Burundi. 3 22S 29 21E. Town founded by the Germans in 1899, at NE end of L. Tanganyika. Cap. and chief port, exporting tea, cotton, coffee and hides. Industries inc. manu. canoes, nets, pharmaceutical goods. Pop. (1986C) 272,600.

Buka   North Solomons, Papua New Guinea. 5 15S 154 35E. Island sit. NW of Bougainville across Buka Passage.

Area, 190 sq. m. (492 sq. km.). Chief settlement Queen Carola Harbour. Main product copra.

Bukavu Kivu, Zaïre. 2 30S 28 52E. Town at S end of L. Kivu near the Rwanda frontier. Regional cap. and commercial centre. Industries inc. processing cinchona; manu. cement, insecticides and pharmaceutical goods. Pop. (1976C) 209,051.

Bukhara Uzbek Soviet Socialist Republic, U.S.S.R. 39 50N 64 10E. City sit. SW of Tashkent in an oasis on Zeravshan R. on a spur of the Trans Caspian railway. Industries inc. processing karakul, silk spinning, carpet making. Noted as a medieval centre of Islamic culture and was dominated in turn by Arabs, Persians, Turks and Uzbeks. Pop. (1987E) 220,000.

Bukit Mertajam Penang, Malaysia. 5 22N 100 28E. Town sit. ESE of Georgetown at W foot of Bukit Mertajam hill. Railway junction on the coastal railway.

Bukittinggi Western Sumatra, Indonesia. 0 19S 100 22E. Town sit. N of Padang. Centre of an agric. area producing rice, copra, tobacco, coffee.

Bukoba Tanzania. 1 20S 31 49E. Town sit. NW of Mwanza on W shore of L. Victoria, S of Uganda frontier. Coffee producing centre; also produces tobacco, sisal, maize and fish. Some deposits of glass-sand, tin and tungsten.

Bulawayo Matabeleland, Zimbabwe. 20 09S 28 36E. Town sit. SW of Harare. Commercial and route centre. Industries inc. textiles, agric. engineering, railway engineering, iron founding, sugar refining. Pop. (1977E) 386,000.

Buldana Maharashtra, India. 20 32N 76 11E. Town and district cap. sit. WSW of Akola in the Ajanta Hills. Industries inc. oilseed milling. Pop. (1981C) 35,914.

Bulgaria South East Europe. 42 35N 25 30E. Rep. in Balkan peninsula bounded N by Romania, E by Black Sea, S by Turkey and Greece and W by

Yugoslavia. Area, 42,818 sq. m. (110,912 sq. km.). Chief towns Sofia (cap.), Plovdiv, Varna. Plains in the N, mountainous in the S, the Balkan Mountains rise to 7,000 ft. (2,134 metres) and descend SE to the Maritsa basin; the Rhodope Mountains rise to Peak Musala, 9,596 ft. (2,925 metres), in the SW. The N is drained by tribs. of Danube which forms N frontier, the S by Maritsa, Mesta and Struma Rs. to the Aegean. About half the land is cultivated, main products being cereals, sugar-beet, tobacco, sunflowers and roses. Mineral resources inc. oil and coal. Pop. (1988E) 8·97m.

Bull Run Virginia, U.S.A. Stream noted as the scene of two Confederate victories in the American Civil War, 1861 and 1862.

Bunbury Western Australia, Australia. 33 20S 115 38E. Town sit. S of Perth on Koombana Bay. Industries inc. tourism, forestry and agric. Port exporting timber, wheat and mineral sands. Pop. (1988E) 27,000.

Bundaberg Queensland, Australia. 24 51S 152 21E. Town sit. NNW of Brisbane. Industries inc. heavy and light engineering, brewing, distilling, sugar refining. Port, exporting sugar. Queensland leading tomato grower. Pop. (1988E) 44,500.

Bundelkhand India. Region of central India extending S from Jumna R. across S Uttar Pradesh and consisting of former districts of Jalaun, Jhansi, Hamirpur, Banda and S Allahabad. Mainly hilly with jungles in the S, alluvial plain in the N. Crossed by the Vindhya Range. Drained by Betwa, Dhasan and Ken Rs.

Bungay Suffolk, England. 52 27N 1 26E. Town sit. W of Lowestoft on Waveney R. Market town for a farming area. Industries inc. printing, agric. engineering. Pop. (1981C) 4,106.

Bunker Hill Massachusetts, U.S.A. Small hill forming part of Boston with a monument to mark the site of the first major battle in the American War of Independence.

Buraida   Nejd, Saudi Arabia. Town
sit. NW of Riyadh. Oasis, producing
and trading in grain and dates.

Burao   Somalia. Town sit. E of
Hargeisa in Ogo highlands. Farming
centre, esp. livestock, beans and
cereals. District headquarters.

Burbank   California, U.S.A. 34
12N 118 18W. City on outskirts of
Los Angeles. Industries inc. film and
television studios, aircraft manu. Pop.
(1980C) 84,625.

Burdekin River   Australia. 19 39S
147 30E. Rises 85 m. (136 km.) NW of
Townsville in E Queensland and flows
SE past Sellheim to within 15 m. (24
km.) of Mount McConnel, then turns
N to enter Pacific Ocean by an estuary
below Home Hill. Floods widely in
summer.

Burdur   Turkey. (i) Province of SW
Turkey bounded SE by Elmali Mountains, SW by Mentese Mountains.
Area, 2,876 sq. m. (7,449 sq. km.).
Farming area with oak forests. Main
products, valonia, opium, hemp, anise,
wheat, potatoes. Pop. (1985C)
248,002. (ii) Town sit. W of Konya on
a railway near L. Burdur. Provincial
cap. Manu. carpets, attar of roses.
Pop. (1970C) 37,746.

Burgas   Burgas, Bulgaria. 42 30N
27 28E. Town on Gulf of Burgas on
the Black Sea. Provincial cap. Industries inc. oil refining, agric.
engineering, textiles; manu. soap,
flour milling, fishing. Sea port exporting tobacco, wool, leather. Pop.
(1987E) 197,555.

Burgan   Kuwait. Major oilfield sit.
WSW of Mena al Ahmadi, its terminal,
to which it is connected by pipeline.

Burgdorf   Bern, Switzerland. 47
04N 7 37E. Town sit. NE of Bern on
Emme R. Trading centre for Emmen-
thal cheese. Industries inc. textiles.
Pop. (1977E) 17,800.

Burgenland   Austria. 47 20N 16
20E. Province of E Austria bounded E
by Hungary. Area, 1,530 sq. m. (3,963
sq. km.). Cap. Eisenstadt. Low-lying
in the N with L. Neusiedler, hilly else-

where. The main occupations are
agric. and stock rearing. Pop. (1988E)
266,800.

Burghead   Grampian Region, Scotland. 57 42N 3 29W. Town sit. NW of
Elgin at E end of Burghead Bay. Fishing port and resort. Pop. (1980E)
1,500.

Burgos   Spain. 42 21N 3 42W. (i)
Province in N Spain forming part of
the 'Meseta'. Area, 5,531 sq. m.
(14,269 sq. km.). Mainly forest with
some cultivated land where cereals •
are grown and sheep raised. Drained
by Ebro and Douro Rs. Pop. (1986C)
363,530. (ii) Town sit. SW of Bilbao
on Arlanzón R. Provincial cap. Industries inc. tourism, engineering,
flour milling, chemicals, paper
making; manu. tyres, leather and
woollen goods. Of particular architectural interest. Pop. (1986C) 163,900.

Burgundy   France. Region in E
comprising depts of Côte-d'Or,
Nièvre, Saône-et-Loire and Yonne.
Area, 12,198 sq. m. (31,592 sq. km.).
Chief cities, Dijon, Chalon-sur-Saône,
Mâcon, Nevers and Auxerre. Former
kingdom and famous wine producing
area. Pop. (1982C) 1,596,054.

Burhanpur   Madhya Pradesh, India.
21 19N 76 14E. Town on Tapti R. NE
of Bombay. Industries inc. manu.
shellac and textiles, esp. fine fabrics
decorated with metallic thread. Pop.
(1981C) 140,986.

**Buriat Autonomous Soviet Socialist
Republic**   U.S.S.R. 53 00N 110
00E. Autonomous Rep. bounded S by
Mongolia and W by L. Baikal. Area,
135,650 sq. m. (351,334 sq. km.).
Cap. Ulan-Ude. Mainly plateau rising
to the forested Barguzin Mountains.
The main occupation is rearing cattle
and sheep. Pop. (1989E) 1,042,000.

Burkina Faso   (formerly Upper
Volta), West Africa. 12 00N 0 30W.
The State is totally surrounded by
Mali, Niger, Benin, Togo, Ghana and
Côte d'Ivoire. The country is drained
by R. Volta and its tribs. Area,
105,839 sq. m. (274,122 sq. km.).
Cap. Ouagadougou. Burkina Faso was
formerly a province of French West

Africa and became independent in 1960. Agric. is the dominant occupation including nomads raising cattle. Irrigation is being extended but most of the agric. activity is in R. valleys or at oases. Chief crops are: beans, maize, millet, sorghum and yams. Manganese and phosphates are found. Pop. (1988E) 8·53m.

Burkina River Burkina Faso/Ghana. Rises in 2 head streams, Black Volta (Mouhoun) and White Volta (Nakambe) Rs. The Black Volta rises in W Burkina Faso and flows in a N curve then S along the border between Burkina Faso and the Côte d'Ivoire (W) and Ghana (E), then turns E to join the White Volta above Yeji. The White Volta flows S from N Burkina Faso. From their confluence they form a L. which extends S to the Akosambo Dam and is also fed by the Oti R. (NE). Beyond the dam the Volta flows SE to enter the Bight of Benin at Ada. The course of the Volta R. proper is 400 m. (640 km.).

Burlington Ontario, Canada. 43 19N 79 47W. City sit. W of Toronto and 9 m. (14.5 km.) E of Hamilton. Industries inc. market gardening, high technology, furniture, foods, chemicals, steel related products. Pop. (1985E) 117,500.

Burlington Iowa, U.S.A. 40 49N 91 14W. City sit. ESE of Des Moines on Mississippi R. Industries inc. manu. furniture, electrical switchgear, steam turbines, excavators, spark plugs, electronic equip. explosives and ammunition. Pop. (1980C) 29,529.

Burlington Vermont, U.S.A. 44 29N 73 13W. City and port sit. on L. Champlain. Industries inc. service and high technology. Pop. (1989E) 40,447.

Burma *see* **Myanmar**

Burma Road South Asia. Highway, constructed between 1936 and 1938, extending 800 m. (1,280 km.) from the rail-head at Lashio, E Burma (Myanmar) to Kunming and Chungking in China.

Burnaby British Columbia, Canada.

Suburb of E Vancouver. The main industry is manu. asphalt. Pop. (1981C) 136,494.

Burnham Buckinghamshire, England. 51 33N 0 38W. Town immediately NW of Slough noted for the nearby anc. woodland of Burnham Beeches. Pop. (1981C) 11,185.

Burnham-on-Crouch Essex, England. 51 38N 0 49E. Town on Crouch R. estuary. Industries inc. oyster culture. Yachting resort. Pop. (1989E) 7,500.

Burnham-on-Sea Somerset, England. 51 14N 2 59W. Town on Bridgwater Bay on Bristol Channel. Holiday resort. Pop. (1980E) 12,000.

Burnie Tasmania, Australia. 41 04S 145 54E. Town sit. WNW of Launceston on Bass Strait. Rail junction and centre of an agric. area. Industries inc. manu. dairy products, paper making.

Burnley Lancashire, England. 53 47N 2 15W. Town at confluence of Brun and Calder Rs. Industries inc. engineering, chemicals; manu. aerospace equipment, vehicle components and paper. Pop. (1989E) 85,400.

Burntisland Fife Region, Scotland. 56 03N 3 15W. Town opposite Edinburgh on Firth of Forth. Industries inc. manu. aluminium. Tourist centre. Pop. (1987E) 6,150.

Burry Port Dyfed, Wales. 51 41N 4 17W. Town sit. W of Llanelli. Town serves as residential and service centre. Industries inc. engineering, yacht services, chemicals and tape assembly. Former seaport, now harbour for pleasure craft. Pop. (1981C) 6,035.

Bursa Bursa, Turkey. 40 15N 29 05E. Town sit. S of Istanbul. Provincial cap. Trading centre for tobacco, fruit and grain. Industries inc. manu. textiles, carpets. Pop. (1980C) 445,113.

Burslem Staffordshire, England. 53 02N 2 11W. Town forming part of Stoke-on-Trent in the 'Potteries'; oldest of the pottery towns, the industry there dating from the 17th cent.

Burton-upon-Trent Staffordshire, England. 52 49N 1 36W. Town sit. SW of Derby on Trent R. The main industry is the manu. of tyres and industrial rubber products together with brewing; also manu. food products. Pop. (1988E) 57,740.

Burujird Luristan, Iran. 33 54N 48 46E. Town sit. S of Hamadan. Commercial centre trading in grain and fruit. Manu. textiles, carpets, rugs. Pop. (1976C) 100,103.

Burundi Central Africa. 3 15S 30 00E. Rep. bounded N by Rwanda, E and S by Tanzania, W by L. Tanganyika and Zaïre. Area, 10,759 sq. m. (27,866 sq. km.). Cap. Bujumbura. Mainly high plateau at 4,000–6,000 ft. (1,219–1,829 metres) rising to 8,000 ft. (2,438 metres) in the S. The Luvironza R. which rises in the S is the most S headstream of the Nile. The main occupation is farming; the main products livestock, maize, cassava, plantains, coffee (the main cash crop) and cotton. Pop. (1989E) 5·54m.

Burutu Bendel, Nigeria. 5 21N 5 31E. Town sit. E of Forcados on Forcados R. in Niger R. delta. Industries inc. processing palm-oil, hardwoods, rubber. Port exports palm products, nuts, cotton, hides. Accessible for ocean-going craft.

Bury Greater Manchester, England. 53 36N 2 19W. Town on Irwell R. The main industry is paper; also manu. textile machinery, felts, chemicals, paints, paper making machinery. Pop. (1989E) 173,000.

Buryat U.S.S.R. *see* **Buriat**

Buryat-Aginsky National Area, U.S.S.R. Area sit. in S Siberia in the Chita Region of the Russian Soviet Federal Socialist Rep. Area, 9,000 sq. m. (23,310 sq. km.). Cap. Aginskoye. The main occupations are stock rearing and lumbering. Pop. (1981E) 929,000.

Bury St. Edmunds Suffolk, England. 52 15N 0 43E. Town sit. ENE of Cambridge. Market town in a farming area. Industries inc. brewing, beet sugar refining; manu. electronic equipment, precision engineering, confectionery, cameras and lamps. Pop. (1989E) 32,000.

Bushey Hertfordshire, England. 51 38N 0 21W. Town sit. SE of Watford, mainly residential.

Buskerud Norway. 60 20N 9 0E. County of SE Norway extending NW from Oslo Fjord to Hardangervidda and Hallingskarv mountains. Area, 5,766 sq. m. (14,933 sq. km.). Chief towns Drammen (cap.), Kongsberg, Honefoss. Forested in the SE. Agric. land produces cereals, fruit, vegetables, livestock. Main industries lumbering, paper and pulp making, textiles. Pop. (1989E) 224,578.

Busselton Western Australia, Australia. 33 39S 115 20E. Town sit. SSW of Perth on S shore of Geographe Bay. Holiday resort, also producing potatoes, maize, dairy products. Pop. (1981E) 6,500.

Bussum Noord-Holland, Netherlands. 52 16N 5 10E. Town sit. ESE of Amsterdam. A residential 'dormitory' town with some industries inc. manu. chocolate. Pop. (1984E) 33,401.

Busto Arsizio Varese, Italy. 45 37N 8 51E. Town sit. NW of Milan with an important textile industry, esp. cotton and rayon; also manu. iron and steel, textile machinery, footwear, dyes.

Butaritari Kiribati W. Atoll in N Gilbert Islands. Area, 4.5 sq. m. (12 sq. km.). Chief town Butaritari. Chief product copra. Formerly called Pitt Island and sometimes called Makin Island. Pop. (1985C) 3,622.

Bute Strathclyde Region, Scotland. 55 50N 5 06W. Island in Firth of Clyde. Area, 47 sq. m. (122 sq. km.). Chief town, Rothesay. Occupations farming and tourism.

Buteshire Scotland. Former county, now part of Strathclyde Region.

Butte Montana, U.S.A. 46 00N 112 32W. City sit. SSW of Helena. The main industry is mining, esp. copper also associated manu. Pop. (1989E) 34,500 (Butte-Silver Bow).

Buttermere, Lake   Cumbria, England. 54 31N 3 16W. L. s of Buttermere village, length 1.5 m. (2 km.) and width 0.33 m. (0.5 km.). Tourist area.

Buxton   Derbyshire, England. 53 15N 1 55W. Town sit. w of Chesterfield on Wye R. Tourist centre, former spa town with mineral springs. Pop. (1981C) 20,797.

Buzău   Romania. 45 09N 26 49E. Town sit. NE of Bucharest on Buzău R. Trading centre for grain, timber, petroleum. Industries inc. textiles, oil refining, flour milling. Pop. (1983E) 120,419.

Buzovny   Azerbaijan Soviet Socialist Republic, U.S.S.R. Town sit. NE of Baku on NE shore of Apsheron penin-sula. Coastal resort with industries inc. oilwells.

Bydgoszcz   Bydgoszcz, Poland. 53 08N 18 00E. Cap. of province in the NW sit. on Brda R. Manu. inc. machinery, textiles, paper, footwear and clothing. Trades in timber. Pop. (1982E) 356,000.

Byelgorod   Russian Soviet Federal Socialist Republic, U.S.S.R. City sit. NNW of Kharkov on Northern Donets R. Industries inc. fruit canning, flour milling, meat packing, brewing, metalworking, chalk quarrying.

Byelorussia   *see* **Belorussia**

Bytom   Katowice, Poland. 50 22N 18 54E. Town sit. NNW of Katowice. Industries inc. coal, zinc and lead mining and the manu. of machinery. Pop. (1982E) 238,000.

C

Caacupé La Cordillera, Paraguay. 25 23S 57 09W. Town sit. ESE of Asunción. Dept. cap. Industries inc. sugar refining, tile making. Resort in an area growing oranges, sugar-cane, tobacco. Noted for the Shrine of the Blue Virgin on the central plaza, to which annual pilgrimages are made. Pop. (1985E) 9,200.

Caazapá Paraguay. 26 09S 56 24W. (i) Dept. of E Paraguay 8,345 sq. m. (21,613 sq. km.). Pop. (1982E) 109,500. (ii) Town and cap. of dept. sit. S of Villarica on the Asunción railway.

Cabinda Angola. 5 40S 12 11E. District on Atlantic coast separated from the rest of Angola by Zaïre where it extends to Atlantic along Zaïre (Congo) R. Area, 3,000 sq. m. (7,800 sq. km.). Chief town, Cabinda. The main occupation is farming; main products cocoa, coffee, timber and palm oil. Pop. (1986E) 138,400.

Cabo Delgado Mozambique. Province NE of Mozambique. Area, 30,260 sq. m. (78,374 sq. km.) Cap. Pemba. Pop. (1982E) 977,600.

Cabo Gracias a Dios Nicaragua. (i) Territory in Zelaya dept. bounded N by Honduras and E by Caribbean. Area, 5,520 sq. m. (14,297 sq. km.). Drained by Coco R. Main products cacao, bananas, sugar-cane, livestock. (ii) Town sit. NNE of Puerto Cabezas on an island at mouth of Coco R. on the Caribbean. Territorial cap. Industries inc. manu. beverages, coconut products.

Cáceres Spain. 39 29N 6 22W. (i) Province in central W, bounded W by Portugal. Area, 7,700 sq. m. (19,945 sq. km.). The chief occupation is growing cereals and olives, and raising pigs and sheep. Pop. (1986C) 424,027. (ii) Cap. of (i) sit. on Cáceres R. Market town trading in grain, olive oil, wool, ham and sausages. Manu. textiles, fertilizers, cork and leather products. Pop. (1986C) 79,342.

Cacheu Guinea-Bissau. 12 16N 16 10W. Village sit. NW of Bissau on S bank of Cacheu R. estuary, in a stock raising area; the other main occupation is growing coconuts, almonds. Formerly a slave-trading centre.

Cadenabbia Lombardia, Italy. Village sit. on W side of L. Como. The main industry is tourism. esp. to the nearby Villa Carlotta.

Cader Idris Wales. 52 42N 3 54W. Mountain ridge in NW Wales, sit. NW of Tywyn, rising to Pen-y-Gader.

Cádiz Spain. 36 32N 6 18W. (i) Province bounded SE by Mediterranean, S by Strait of Gibraltar and SW by Atlantic. Area, 2,851 sq. m. (7,384 sq. km.). The main occupations are viticulture, esp. for sherry, and farming. Pop. (1986C) 1,054,503. (ii) Cap. of (i) sit. at the tip of a peninsula extending SE to NW across Bay of Cádiz. Sea port exporting sherry, salt, olive oil and cork. Pop. (1986C) 154,051.

Caen Calvados, France. 49 11N 0 21W. Cap. of dept. in N sit. on Orne R. Commercial centre and R. port trading in iron ore, dairy produce, cement, building stone, coal, timber and grain. Industries inc. steel, textiles, pottery and cement manu. Pop. (1982C) 117,119 (agglomeration, 183,526).

Caerleon Gwent, Wales. 51 37N 2 57W. Town sit. NE of Newport on Usk R., with important Roman sites. Pop. (1981C) 6,711.

Caernarfon Gwynedd, Wales. 53 08N 4 16W. Town in NW sit. on S side of Menai Strait facing Anglesey. Market town for an agric. area. Industries inc. plastics and metal-working. Pop. (1981C) 9,506.

Caerphilly   Mid Glamorgan, Wales. 51 35N 3 14W. Town sit. N of Cardiff. Market town where Caerphilly cheese was originally made.

Cagliari   Sardegna, Italy. 39 13N 9 06E. Cap. of island and region sit. on S coast between salt lagoons. Industries inc. flour milling, tanning and fishing. Port exporting salt, lead and zinc. Pop. (1988E) 220,192.

Caguas   Puerto Rico. 18 14N 66 02W. Town sit. S of San Juan. Commercial centre of an agric. area. Manu. tobacco products and leather goods. Pop. (1980E) 117,959.

Cahir   Tipperary, Ireland. 52 21N 7 56W. Market town sit. on Suir R. NW of Waterford. The main industry is tourism. There is a meat packing plant and some light engineering. Pop. (1981C) 2,120.

Cahors   Lot, France. 44 27N 1 26E. Cap. of dept. in SW sit. in a loop of Lot R. Market town for an agric. area trading in wine and truffles. Industries inc. food processing. Pop. (1982C) 20,774.

Cairngorm Mountains   Scotland. 57 06N 3 30W. Range of the Grampian Mountains extending across the union of Invernessshire, Banffshire and Aberdeenshire, rising to Ben Macdhui, Braeriach, Cairn Toul and Cairn Gorm.

Cairns   Queensland, Australia. 16 55S 145 46E. Town in NE of state sit. on Pacific coast. Major transport and servicing centre for an agric., mainly sugar, lumbering, fishing and mining area. Industries inc. tourism. Port exporting sugar and timber. Pop. (1988E) 76,500.

Cairo   Cairo, Egypt. 30 03N 31 15E. Cap. of Egypt and of Cairo governorate, and the largest city in Africa, sit. on E bank of Nile R. at the head of the delta. It was founded in 642 A.D. and just S of the city on the W bank of the Nile are the pyramids and the sphinx. Al-Azhur university founded in 972 A.D. It is an important communications and commercial centre. Industries inc. textiles and brewing; manu.

cement and vegetable oils. There is a R. port at Bulag. Pop. (1986E) 6,325,000.

Caistor   Lincolnshire, England. Town sit. SW of Grimsby. Market town for an agric. area (particularly duck production). Manu. inc. plastics, polythene, textiles and timber products.

Caistor St. Edmunds   Norfolk, England. Suburb of Norwich S of city centre with an extensive Roman site.

Caithness   Scotland. Former county now part of Highland Region.

Cajamarca   Cajamarca, Peru. 7 10S 78 31W. Cap. of dept. in the W. Cordillera in NW Peru sit. at over 9,000 ft. (2,743 metres) a.s.l. Market town for an agric. and mining area trading in cereals, gold, silver, copper and zinc. Manu. textiles and leather goods. Scene of the ambush and execution of the Inca King Atahualpa, in 1533.

Calabar   Cross River, Nigeria. 4 57N 8 29E. Town on Cross R. estuary. Provincial cap. and sea port, trading in palm oil and kernels. New Calabar is a trading depot 100 m. (160 km.) SW on the New Calabar R. Pop. (1975E) 103,000.

Calabria   Italy. Region of S Italy inc. provinces of Cosenza, Catanzaro and Reggio di Calabria, bounded W by Tyrrhenian Sea and E by Ionian Sea. Area, 5,828 sq. m. (15,095 sq. km.). Cap., Reggio. Mainly mountainous. The main occupations are growing vines, olives and citrus fruit; rearing sheep and goats. Pop. (1980E) 2,078,264.

Calais   Pas-de-Calais, France. 50 57N 1 50E. Town in NE sit. on Strait of Dover. Industries inc. manu. textiles and clothing. Important sea port on the shortest route to Eng., handling raw materials for industries in NE France. Pop. (1982C) 76,935 (agglomeration, 100,823).

Calamata   Messenia, Greece. Cap. of Messenia prefecture sit. on Gulf of Messenia SSE of Patrai. Industries inc. manu. cigarettes and flour milling.

Port exporting olives and citrus fruits. Pop. (1981C) 41,119.

Calcutta   West Bengal, India. 22 30N 88 30E. State cap. sit. on E bank of Hooghly R. Capital of India until 1912. Important commercial centre. Only city in India with trams and an underground railway system. Industries inc. jute milling, railway engineering, pharmaceuticals, tanning, chemicals, textiles and paper-making. Port exporting manganese ore, pig iron, raw jute and jute products, tea and oil seeds. Pop. (1981C) 3,305,006 (agglomeration, 9,194,018).

Caldas   Colombia. Dept. of central Colombia in Cauca valley. Area, 5,160 sq. m. (13,364 sq. km.). Cap. Manizales. Pop. (1983E) 870,600.

Caledonian Canal   Scotland. Waterway extending 60 m. (96 km.) SW to NE across Scot. from Loch Linnhe in W to Moray Firth in NE, consisting of Loch Lochy, Loch Oich and Loch Ness, and linking canals.

Calgary   Alberta, Canada. 51 03N 114 05W. Town in SW sit. at confluence of Rs. Bow and Elbow. Communications and commercial centre of a ranching area, with oil nearby. Industries inc. meat packing, flour milling and oil refining; manu. explosives. Pop. (1989E) 671,138.

Cali   Valle del Cauca, Colombia. 3 27N 76 31W. Cap. of dept. in SW, sit. in Cauca valley. Commercial centre of a rich farming area, handling livestock, rice, coffee, tobacco and sugar-cane. Industries inc. textiles; manu. clothing, footwear and soap. Pop. (1980E) 1,450,000.

Calicut   *see* **Kozhikode**

California   U.S.A. 37 43N 120 00W. State in SW bounded W by Pacific Ocean. Area, 158,693 sq. m. (411,015 sq. km.). Chief cities Sacramento (the cap.), San Francisco, Los Angeles, San Diego and San José. Separated from Mexico, 1846-48, and became a State in 1850. The Coast Range extends N to S between the coast and the central valley which is drained and fertilized by the Sacramento R. in the N, and San Joaquin R. in the S. Behind the central valley is the Sierra Nevada rising to Mount Whitney, 14,495 ft. (4,418 metres) a.s.l. The SE has areas of desert, and depressions b.s.l. The chief products are fruit, cotton, livestock, grapes, lumber, fish, petroleum and minerals, esp. gold and mercury. Industries inc. electronics, aerospace, engineering, chemicals, film and television production and food processing. Pop. (1988E) 23,314,500.

California, Gulf of   27 00N 111 00W. Arm of Pacific Ocean extending *c.* 760 m. (1,216 km.) NW to SE between the mainland of NW Mex. and the peninsula and state of Baja California.

Callan   Kilkenny, Ireland. Town sit. SW of Kilkenny on King's R. Market town for an agric. area. Industries inc. radiators, food processing. Pop. (1985E) 1,300.

Callander   Central Region, Scotland. 56 15N 4 14W. Town sit. near Stirling on Teith R. The main industry is tourism. Pop. (1985E) 2,000.

Callao   Peru. 12 02S 77 05W. Town sit. W of Lima. Port handling most of Peru's imports and *c.* 25% of exports. Industries inc. fishing, fish processing, meat packing, flour milling and brewing. Pop. (1984E) 411,200.

Calne   Wiltshire, England. 51 27N 2 00W. Town sit. SW of Swindon. Town serving an agric. area. Industries inc. electronics and engineering. Pop. (1985E) 11,000.

Calshot   Hampshire, England. 50 48N 1 19W. Promontory W of Southampton Water in S Hampshire.

Caltagirone   Sicilia, Italy. 37 14N 14 31E. Town sit. SW of Catania. Industries inc. pottery, esp. terracotta and majolica. Pop. 50,000.

Caltanisetta   Sicilia, Italy. 37 29N 14 04E. Cap. of Caltanisetta province in S central Italy. The main industry is sulphur. Pop. (1980C) 70,000.

Calvados   France. 49 05N 0 15W. Dept. in Bas-Normandie region, bounded N by English Channel. Area, 2,137 sq. m. (5,536 sq. km.). Chief towns Caen (the cap.), Bayeux, Falaise and Lisieux. Low-lying. The main occupations are stock farming and dairying with some cider making. The main industry is textiles. Pop. (1982C) 589,559.

Camagüey   Cuba. 21 23N 77 55W. (i) Province in E Cuba. (ii) Cap. of (i) and largest inland city in Cuba. Communication and trading centre for an agric. area, handling sugar-cane, fruit, livestock and timber. Pop. (1987E) 711,200. (iii) Arch. off N coast of Camagüey province. The main islands are Romano, Sabinal and Coco.

Camargue, La   Bouches-du-Rhône, France. 43 20N 4 38E. Area, in delta of Rhône R. extending E to W from Grand Rhône R. to Petit Rhône R. Marshy, with salt lagoons. The main occupations are fishing and salt extraction; sheep, cattle and fighting bulls are reared in the N. There is some cultivation on reclaimed land.

Cambay   Gujarat, India. 22 19N 72 37E. Town sit. S of Ahmadabad on Gulf of Cambay. Trading centre for cotton and grain. Industries inc. textiles and match manu. Pop. (1985E) 60,000.

Cambay, Gulf of   India. 21 30N 72 30E. Inlet of Arabian Sea extending N into S Gujarat. Most of its harbours are now silted up.

Camberley   Surrey, England. Town sit. N of Aldershot. Industries inc. light engineering. Seat of the Royal Staff College. Pop. (1981C) 52,271 with Frimley.

Camberwell   Greater London, England. District in S London, mainly residential.

Cambodia   Asia. 12 15N 105 00E. Bounded in N by Laos and Thailand, in W by Thailand, E by Vietnam, S by Kerala and the Gulf of Thailand. Cap. Phnom Penh.
Area, 69,898 sq. m. (181,035 sq. km.). Mainly an alluvial plain of the Mekong R. and is remarkably fertile. The overwhelming majority of the population is normally engaged in agriculture, fishing and forestry. Of the country's total area of 44 m. acres, about 20 m. are cultivable and over 20 m. are forest land. Some 4 m. acres are cultivated, well over half being devoted to rice production. A relatively small proportion of the food production enters the cash economy. The war (1970–74) led to a disastrous reduction in production to a stage in which the country became a net importer of rice. Other products are maize, and, in usual order of value, rubber, paddy, livestock, timber, pepper, haricot beans, soybeans and fish.
Much of the surface is covered by potentially valuable forests. There are substantial reserves of pitch pine. Minerals include phosphate, iron-ore, gold. Some development of industry had taken place before the spread of open warfare in 1970. Pop. (1985E) 6,232,000.

Camborne   Cornwall, England. 50 12N 5 17W. Town sit. WSW of Truro, once a tin-mining centre. Industries inc. engineering and milk production. Pop. (1985E) 18,500.

Cambrai   Nord, France. 50 10N 3 14E. Town in NE sit. at junction of Escaut (Scheldt) R. and St. Quentin canal. Industries inc. textiles, esp. cambric, sugar refining and flour milling. Pop. (1982C) 36,618.

Cambridge   Ontario, Canada. 43 20N 80 20W. City created by the amalgamation of the cities of Galt, Preston and Hespeler in 1973, sit. SW of Toronto. Major indust. inc. fabricated metal products, electrical machinery, transportation equipment, primary metal, food, rubber and plastic products, computer software, furniture and fixtures, scientific instruments, printing, publishing, footwear and textiles. Pop. (1989E) 84,000.

Cambridge   Cambridgeshire, England. 52 13N 0 08E. Town in E sit. on Cam (or Granta) R. Industries inc. high technology, scientific instruments, communications, electronics and printing. Chiefly noted for Cambridge University, whose colleges make up most of the town centre. Pop. (1988E) 103,000.

Cambridge Maryland, U.S.A. 38 34N 76 04W. City on banks of Choptank R., SE of Baltimore. Industries inc. seafood, electronics, wire cloth and publishing. Pop. (1986E) 11,070.

Cambridge Massachusetts, U.S.A. 42 22N 71 06W. City opposite Boston on the Charles R. Industries inc. research and development, business and health services. Manu. inc. electronic and photographic equipment, and instruments. Seat of Harvard University and Massachusetts Institute of Technology. Pop. (1980C) 95,322.

Cambridgeshire England. 52 21N 0 05E. County in E Eng. Bounded by Lincolnshire in N, Norfolk and Suffolk in E, Essex and Hertfordshire in S and Bedfordshire and Northamptonshire in W. Chief towns, Cambridge (county town), Peterborough, Ely and Huntingdon. The districts are Peterborough, Fenland, Huntingdonshire, East Cambridgeshire, South Cambridgeshire and Cambridge. Flat, with much fenland soil, and extremely fertile. Drained by R. Ouse and Nene through partly artificial courses. Chief products cereals and vegetables. Industries inc. food processing, electronics, mechanical engineering. Pop. (1989E) 652,740.

Camden Greater London, England. Boro. of N London.

Camden New Jersey, U.S.A. 39 57N 75 07W. City sit. opposite Philadelphia on Delaware R. Industries inc. oil refining, shipbuilding and textiles; manu. food products, radio and television equipment. Port. Pop. (1988E) 182,500.

Camelford Cornwall, England. Town sit. NNE of Bodmin on Camel R. Town in an agric. area with some light industry.

Camembert Orne, France. Village sit. SE of Caen. Camembert cheese originated here.

Cameroon 3 30N 12 30E. Rep. bounded W by Bight of Bonny, NW by Nigeria and NE by Chad, with L. Chad at its N tip. Area, 179,558 sq. m. (465,054 sq. km.). Chief towns

Yaoundé (the cap.), and Douala. Mainly plateau. Tropical climate with heavy rainfall: 412 in. (1,046 cm.) annually at Debundscha on the coast, 62 in. (157 cm.) at Yaoundé in the interior, and less than 20 in. (51 cm.) in the extreme N. The S has rain forest with ebony and mahogany, the N is savannah. The chief products are coffee, cocoa, crude oil, timber, cassava, plantains, millets, durra and aluminium from a plant at Edéa. Pop. (1988E) 11,082,000.

Cameroon Mountains Cameroon. 4 45N 8 55E. Volcanic group near coast NW of Douala, rising to Great Cameroon, 13,350 ft. (4,069 metres) a.s.l. The last eruptions were in 1959.

Camiri Santa Cruz, Bolivia. Town sit. SSE of Lagunillas. The main industry is oil refining. It is linked by road and oil pipeline with Sucre. Pop. (1984E) 20,000.

Camopi River French Guiana. 3 45N 52 50W. Rises in Tumuc-Humac Mountains on Brazilian frontier and flows 150 m. (240 km.) NE through tropical forest to join Oyapock R. Its course is obstructed by rapids.

Camorta Island Nicobar Islands, Indian Ocean. Small island in Nicobar Group in Bay of Bengal, 60 m. (96 km.) NNW of Great Nicobar. Length 16 m. (26 km.) N to S, width 2–4 m. (3–6 km.).

Campagna di Roma Lazio, Italy. Plain in Roma province bounded SW by Tyrrhenian Sea. Drained by Tiber R. Neglected land much restored by drainage.

Campania Italy. Region in S Italy bordered W by Tyrrhenian Sea, inc. the provinces of Avellino, Benevento, Caserta, Napoli and Salerno, and Capri, Ischia, Procida and Pontine Islands. Area, 5,248 sq. m. (13,592 sq. km.). Chief towns Naples (the cap.), Salerno, Benevento and Caserta. A fertile region producing fruit, vegetables, vines, olives, cereals and tobacco. Pop. (1981C) 5,463,134.

Campbell Island New Zealand. 52

30S 169 00E. Volcanic island sit. S of Invercargill in the S Pacific. Semicircular in shape, circumference *c.* 30 m. (48 km.). The main product is fur seal.

Campbellton New Brunswick, Canada. 48 00N 66 40W. Town on estuary of Restigouche R. Port and centre for a hunting, fishing and lumbering area.

Campbeltown Strathclyde Region, Scotland. 55 26N 5 36W. Town on SE coast of Kintyre peninsula. Industries inc. fishing, distilling, tourism, clothing manu., milk processing and boatbuilding. Pop. (1987E) 6,077.

Campeche Mexico. 19 51N 90 32W. (i) State in SE Mex. on Gulf of Campeche bounded E by Quintana Roo, NE by Yucatán and S by Guatemala. Area, 21,660 sq. m. (56,099 sq. km.). Chief towns Campeche (the cap.) and Ciudad del Carmen. Lowland with forests. Industries inc. timber. Pop. (1980C) 420,553. (ii) Cap. of (i) sit. on Gulf of Campeche 100 m. (160 km.) SW of Mérida. Industries inc. manu. leather and tobacco products. Pop. (1984E) 120,000.

Campinas São Paulo, Brazil. 22 54S 47 05W. Town in SE. Trading centre for coffee. Industries inc. sugar refining and textiles; manu. sewing machines, tyres, wine and cotton-seed oil. Pop. (1980C) 566,627.

Campobasso Abruzzi e Molise, Italy. 41 34N 14 39E. Cap. of Campobasso province sit. NNE of Naples. Market town. Industries inc. manu cutlery and soap. Pop. (1980C) 40,000.

Campo Grande Mato Grosso, Brazil. 20 27S 54 37W. Town on São Paulo–Corumbá railway, SE of Corumbá. Market town for the uplands between Rs. Paraguay and Paraná. Industries inc. packing and distributing livestock, meat, hides and skins, and agric. produce. Pop. (1980C) 282,844.

Campos Rio de Janeiro, Brazil. 21 45S 41 18W. Town sit. on Paraíba R. Industries inc. textiles and sugar refin-

ing; manu. leather goods and soap. Pop. (1984E) 174,000.

Camrose Alberta, Canada. 53 00N 112 50W. Town sit. SE of Edmonton near Bitten L. Railway centre. Industries inc. agric. steel pipe manu. metal fabrication. Pop. (1989E) 12,968.

Canada 60 00N 100 00W. Independent Commonwealth state bounded S by U.S.A. Area, 3,851,809 sq. m. (9,976,185 sq. km.), occupying the whole continent N of the U.S.A. except for Alaska, U.S.A. in the NW and the French Islands of St. Pierre et Miquelon off Newfoundland in the E. There are 10 provinces and 2 territories. The territories which now constitute Canada came under British power at various times by settlement, conquest or cession. Nova Scotia was occupied in 1628 by settlement at Port Royal, was ceded back to France in 1632 and was finally ceded by France in 1713, by the Treaty of Utrecht; the Hudson's Bay Company's charter, conferring rights over all the territory draining into Hudson Bay, was granted in 1670; Canada, with all its dependencies, including New Brunswick and Prince Edward Island, was formally ceded to Great Britain by France in 1763; Vancouver Island was acknowledged to be British by the Oregon Boundary Treaty of 1846, and British Columbia was established as a separate colony in 1858. As originally constituted, Canada was composed of the provinces of Upper and Lower Canada (now Ontario and Quebec), Nova Scotia and New Brunswick. They were united under the provisions of an Act of the Imperial Parliament known as 'The British North America Act, 1867', which came into operation on 1 July 1867 by royal proclamation. The Act provides that the constitution of Canada shall be 'similar in principle to that of the United Kingdom'; that the executive authority shall be vested in the Sovereign, and carried on in his name by a Governor-General and Privy Council; and that the legislative power shall be exercised by a Parliament of two Houses, called the 'Senate' and the 'House of Commons'. The present position of Canada in the British Commonwealth of Nations was defined at the Imperial

Conference of 1926. In 1931 the House of Commons approved the enactment of the Statute of Westminster emancipating the Provinces as well as the Dominion from the operation of the Colonial Laws Validity Act, and thus removing what legal limitations existed as regards Canada's legislative autonomy. The statute received the royal assent in 1931. Provision was made in the British North America Act for the admission of British Columbia, Prince Edward Island, the Northwest Territories and Newfoundland into the Union. In 1869 Rupert's Land, or the Northwest Territories, was purchased from the Hudson's Bay Company; the province of Manitoba was constituted from this territory and admitted into the confederation in 1870. In 1871 the province of British Columbia was admitted, and Prince Edward Island in 1873. The provinces of Alberta and Saskatchewan were formed from the provisional districts of Alberta, Athabasca, Assiniboia and Saskatchewan, and admitted in 1905. Newfoundland formally joined Canada as its tenth province in 1949. In 1931 Norway formally recognized the Canadian title to the Sverdrup group of Arctic islands. Canada thus holds sovereignty in the whole Arctic sector north of the Canadian mainland. In Nov. 1981 the Canadian government agreed on the provisions of an amended constitution, to the end that it should replace the British North America Act and that its future amendment should be the prerogative of Canada. These proposals were adopted by the Parliament of Canada and were enacted by the UK Parliament as the Canada Act of 1982.

The enactment of the Canada Act was the final act of the UK Parliament in Canadian constitutional development. The Act gave to Canada the power to amend the Constitution according to procedures determined by the Constitutional Act 1982, which was proclaimed in force by the Queen on 17 April 1982. The Constitution Act 1982 added to the Canadian Constitution a charter of Rights and Freedoms, and provisions which recognize the nation's multi-cultural heritage, affirm the existing rights of native peoples, confirm the principle of equalization of benefits among the provinces, and strengthen provincial ownership of natural resources.

Chief towns, Ottawa (cap.), Montreal, Toronto, Vancouver, Edmonton, Calgary. The E coast is broken up by Hudson Strait and Bay, James Bay, Gulf of St. Lawrence, Bay of Fundy; N coast commercially useless on account of latitude; SE is cut up by chain of great lakes—greatest freshwater area in world; largest are Lakes Superior and Huron.

Along E coast, in Labrador, Cape Breton I., and Nova Scotia, are low hills, while SE of St. Lawrence basin is bounded by northern extension of Appalachians. Along NW of St. Lawrence basin, of Great Lakes, and of their northern feeders, stretch Laurentian Plateau and Laurentian Range. From NNW to SSE towards Pacific coast, run Rocky Mountains and parallel Selkirk and Cascade Ranges, with great intervening valleys and with average height of *c.* 8,000 ft. (2,438 metres) a.s.l. Highest peaks, Mts. St. Elias, Logan, Robson. Between Laurentian Range, N of Lake Superior, and foothills of Rockies, is enormous stretch of fertile prairie land, important for wheat growing and stock rearing; districts E and W of this are in many parts densely wooded. The N is partly plateau, partly plain, valuable for minerals and fur-bearing animals. The great northern and north-eastern stretch drained by Upper Yukon, Mackenzie, Athabasca, Peace, Coppermine, Great Fish or Back's Rs., flowing to Arctic; Churchill, Nelson, Albany, entering Hudson Bay; E, S, and centre drained by St. Lawrence and its tributaries, Red R., Assiniboine, Saskatchewan; SW by Fraser and upper waters of the Columbia R. Though the manufacturing industries now predominate, agriculture is still very important to the Canadian economy.

Grain growing, dairy farming, fruit farming, ranching and fur farming are all carried on successfully. Fishing is important including, salmon, lobster, cod scallops, halibut, herring, flounders and sole. Mining is a dominant industry and includes copper, nickel, zinc, iron ore, gold, lead, silver, molybdenum, asbestos, petroleum, salt, sulphur and gypsum.

The manu. industry includes motor vehicles, pulp and paper mills, slaughtering and meat processing, petroleum refining, iron and steel mills, dairy products, sawmills and clothing. Pop. (1989E) 26·2m.

Canakkale   Turkey. (i) Province of NW Turkey on Dardanelles strait and mainly on Asian shore. Area, 3,814 sq. m. (9,878 sq. km.). Pop. (1985C) 417,121. (ii) Town on Dardanelles strait. Provincial cap., commercial centre and port. Also known as Dardanelli.

Cañar   Ecuador. Province in the Andes in S Central Ecuador. Area, 1,277 sq. m. (3,307 sq. km.). Cap. Azogues. The climate varies from temperate to semi-tropical; the main occupations are farming and mining. Main products cereals, potatoes, fruit, livestock; gold, silver, copper and mercury. Coal is mined at Biblián. The main industry is manu. panama hats. Pop. (1984E) 180,500.

Canary Islands   North Atlantic. 28 00N 15 30W. Groups of islands off NW coast of Spanish Sahara, forming two overseas provinces of Spain; Las Palmas contains Lanzarote, Fuerteventura and Grand Canary; Santa Cruz de Tenerife contains Tenerife, Palma, Gomera and Hierro. Area, 2,807 sq. m. (7,270 sq. km.). Volcanic, mountainous and arid. The chief crops are bananas, tomatoes and potatoes, grown by irrigation. Industries inc. tourism, fishing and fish canning. Pop. (1986E) 1,614,882.

Canaveral, Cape   Florida, U.S.A. 28 24N 80 37W. From 1963 known as Cape Kennedy but reverted to Canaveral in 1973. NASA still use name Cape Kennedy. Site of rocket launches to the moon.

Canberra   Australian Capital Territory, Australia. 35 18S 149 08E. Cap. of Australia sit. in SE on Molonglo R., planned in 1911–13 as the federal cap., but the Commonwealth Parliament was not brought from Melbourne until 1927. The main employment is in administration. Pop. (1986E) 288,900.

Canchis   Cuzco, Peru. Province of S Peru. Area, 1,244 sq. m. (3,222 sq. km.). Cap. Sicuani. Mainly agric.

Candia   Iràklion, Greece. City and seaport sit. on N coast of Crete. Manu. soap, leather and leather goods and wire.

Canea   Crete, Greece. (i) Prefecture in NW Crete, mainly an agric. area. Pop. (1981C) 125,856. (ii) Cap. of prefecture of same name on NW coast. Sea port trading in citrus fruits, carob beans, wine and olive oil. Pop. (1981C) 47,338.

Canelones   Uruguay. 34 32S 56 17W. Dept. of S Uruguay bounded by Rio de la Plata estuary. Area, 1,835 sq. m. (2,936 sq. km.). Cap. Canelones (or Guadeloupe). Fertile and mainly agric. with flour milling in the cap. Pop. (1975E) 313,858.

Canillo   Andorra. Village sit. NE of Andorra la Vella on a headstream of Valira R. Main occupation farming.

Cankiri   Turkey. (i) Province of N (Asiatic) Turkey. Area, 3,346 sq. m. (8,666 sq. km.). Pop. (1985C) 263,964. (ii) Town sit. NE of Ankara. Provincial cap.

Cannes   Alpes-Maritimes, France. 43 33N 7 01E. Town on the Riviera in SE France. Resort in a fruit and flower growing area. Industries inc. tourism; manu. candied fruits and essential oils. Pop. (1980C) 72,787 (agglomeration, 295,525).

Cannock   Staffordshire, England. 52 42N 2 09W. Town sit. NE of Wolverhampton. Industries inc. light engineering. Pop. (1988E) 62,300.

Canso, Strait (or Gut) of   Canada. 45 37N 61 25W. Strait between Cape Breton Island and mainland Nova Scotia, 1 m. (2 km.) across at its narrowest and crossed by a causeway.

Cantabrian Mountains   Spain. 43 00N 5 00W. Range in N Spain extending 300 m. (480 km.) E to W from the Pyrenees and parallel with the N coast, and rising to Peña Cerredo and Peña Vieja. Rich in coal and iron deposits.

Cantal   France. 45 04N 2 45E. Dept. in Auvergne region. Area, 2,217 sq. m. (5,741 sq. km.). Cap. Aurillac.

Moun'ainous, and volcanic in the centre, rising to Plomb du Cantal. The main occupations are cattle rearing and dairy farming. Pop. (1982C) 162,838.

Canterbury   Kent, England. 51 17N 1 05E. Cathedral city in the SE sit. on Stour R. Leading tourist, educational and shopping centre. The cathedral is the seat of the Archbishop and Primate of the Anglican Church. Pop. (1989E) 127,000.

Canterbury   South Island, New Zealand. 43 45S 117 19E. Provincial district occupying E half of central South Island. Area, 16,769 sq. m. (43,432 sq. km.). Mountainous in the W and S, descending to the Canterbury Plains in the E. Chief towns, Christchurch and Timaru. The main occupations are mixed farming and sheep farming.

Can Tho   Can Tho, Vietnam. 10 02N 105 47E. Town sit. SW of Saigon on Bassac R. in Mekong Delta. Provincial cap. Important rice growing centre with an experimental station. Industries inc. extracting coconut oil, making soap, cigarettes, milling rice and timber.

Canton   China *see* **Gangzhou**

Canton   Ohio, U.S.A. 40 48N 81 22W. City sit. SSE of Akron on the edge of the Pittsburgh iron and steel district. Industries inc. steel; manu. roller bearings, vacuum cleaners and enamelware. Pop. (1980C) 94,730.

Canton   *see* **Phoenix Islands**

Canvey Island   Essex, England. 51 33N 0 34E. Island E of London in Thames Estuary off Benfleet from which it is separated by a narrow channel. Mainly residential but some light industry and 2 oil refineries.

Cape Breton Island   Nova Scotia, Canada. 46 00N 60 30W. Island in NE of province separated from mainland Nova Scotia by the Canso Strait. Area, 3,975 sq. m. (10,295 sq. km.). Cold and foggy; almost bisected SW to NE by L. Bras d'Or. The main occupations are dairy farming, lumbering and fishing. Coal deposits in the Sydney–

Glace Bay area are important and gypsum is mined in the N.

Cape Girardeau   Missouri, U.S.A. 37 19N 89 32W. City sit. SSE of St. Louis on Mississippi R. Industries inc. manu. footwear and clothing. Pop. (1980C) 34,361.

Capelle aan den Yssel   Zuid-Holland, Netherlands. Town sit. ENE of Rotterdam on Hollandsche Ijssel R. Industries inc. shipbuilding, electronics, computer software and packaging material. Pop. (1985E) 54,000.

Cape Mount   Liberia. (Grand Cape Mount). County in W Liberia on the Atlantic. Cap. Robertsport. The main occupations are farming and fishing, the main products palm oil, kernels, kola nuts, coffee, cassava, rice.

Cape Province   Republic of South Africa. 32 00S 23 00E. Southernmost and largest of the four provinces occupying all of the Cape of Good Hope. Area, 249,549 sq. m. (646,332 sq. km.). Chief towns Cape Town (the cap.), Port Elizabeth and East London. Mainly plateau bounded N by Kalahari Desert. Drained by Orange R. to the Atlantic and by Great Kei R. to the Indian Ocean. Mediterranean climate in the SW, heavier rainfall in the E. The chief occupations are farming and mining; the chief products maize, wheat, citrus fruits, sheep, diamonds and copper. Pop. (1986C) 4,901,261.

Capernaum   Galilee, Israel. Town sit. on N shore of Sea of Galilee and a place of Christian pilgrimage.

Cape Town   Cape Province, Republic of South Africa. 33 55S 18 22E. City sit. on Table Bay on SW coast. Legislative cap. of the Rep., and cap. of province. Important sea port handling oil, grain and food products. Pop. (1980C) 213,830.

Cape Verde   Atlantic Ocean. 16 00N 24 00W. Republic sit. WNW of Cape Verde, Senegal. Cap. Praia. Divided into the Barlavento and Sotavento groups. The former inc. São Vicente, Santo Antão, São Nicolau, Santa Luzia, Sal and Boa Vista; the latter, São Tiago, Maio,

Fogo and Brava. Area, 1,557 sq. m. (4,033 sq. km.). Mountainous and volcanic, rising to Pico do Cano, and hot and humid in climate. The chief products are coffee, cattle and goats. Pop. (1980C) 296,093.

Cape York Peninsula Queensland, Australia. 12 00S 142 30E. The N tip of Queensland projecting between the Coral Sea to E and the Gulf of Carpentaria to W.

Cap Haïtien Haiti. 19 45N 72 15W. Town sit. on N coast. Market centre for a fertile hinterland. Sea port handling coffee, sugar-cane and bananas. Pop. (1988E) 133,233.

Caporetto *see* **Kobarid**

Capri Campania, Italy. 40 33N 14 13E. Island at S end of Bay of Naples. Area, 4 sq. m. (10 sq. km.). Industries inc. viticulture and tourism.

Caprivi Strip Namibia. 18 00S 23 00E. Panhandle in NE extending for 300 m. (480 km.) and *c.* 30 m. (48 km.) wide to give access to Zambezi R. Bounded N by Angola and Zambia, E by Zimbabwe and S by Botswana.

Caquetá Colombia. In S Colombia. Area, 34,820 sq. m. (90,185 sq. km.). Cap. Florencia. Undeveloped tropical forest area inhabited by Indians. The forests yield rubber, gums and resins, medicinal plants and cacao. Drained by Caguán and Yarí Rs. Pop. (1984E) 89,000.

Caquetá River Colombia. 3 08S 64 46W. The upper course of the Japurá R., rising in the Andes to the S of Páramo del Buey, Cauca, and flowing 750 m. (1,200 km.) ESE through tropical forests to cross the equator and join Apaporis R. on the Brazilian frontier where together they form the Japurá R. Main tribs. Orteguaza, Caguán, Yarí and Miriti Paraná Rs. Not navigable.

Carácas Federal District, Venezuela. 10 33N 66 55W. Cap. of rep. sit. near N coast in a basin. It was founded in 1567 and sacked by Drake in 1595 and is the birthplace of Simon Bolivar the nationalist leader. Com-

mercial centre dealing mainly in, and supported by, oil. Industries inc. food processing and textiles; manu. inc. vehicles, cement, plastics, chemicals, paper and textiles. Pop. (1981C) 1,044,851.

Carabobo Venezuela. State of N Venezuela on Caribbean coast. Area, 1,800 sq. m. (4,662 sq. km.). Chief towns Valencia (cap.), Puerto Cabello. Mainly agric. Noted for Simon Bolivar's victory over the Spaniards, 1821. Pop. (1984E) 846,000.

Caras River Romania/Yugoslavia. Rises 12 m. (19 km.) NE of Oravita in Transylvanian Alps and flows 50 m. (80 km.) N into Yugoslavia, then SW to enter Danube 7 m. (11 km.) SW of Bela Crkva.

Carazo Nicaragua. Dept. in SW Nicaragua on the Pacific coast. Area, 370 sq. m. (958 sq. km.). Cap. Jinotepe. Mainly hilly with a dry climate. The main occupations are farming (coffee, sugar, rice, sesame, livestock), quarrying, esp. limestone, salt extraction, lumbering. Pop. (1981E) 109,450.

Carbondale Illinois, U.S.A. 37 44N 89 13W. Town sit. SE of St. Louis, Missouri in S Illinois. Pop. (1980C) 26,287.

Carbonia Sardegna, Italy. 39 11N 08 32E. Town sit. W of Cagliari. The main industry is lignite-mining. Pop. (1989E) 33,501.

Carcassonne Aude, France. 43 13N 2 21E. Cap. of dept. in SW, sit. on Aude R. and Canal du Midi. Medieval fortified town and trading centre for wine. Industries inc. tourism and tanning. Pop. (1982C) 42,450.

Carchi Ecuador. Province of N Ecuador bounded N by Colombia. Area, 1,495 sq. m. (3,582 sq. km.). Cap. Tulcán. Mainly agric. Pop. (1982C) 128,113.

Cárdenas Matanzas, Cuba. 23 05N 81 10W. Town sit. E of Matanzas on N coast. Industries inc. sugar refining, rope-making, rice milling and distilling rum. Port exporting sugar.

Cardiff   South Glamorgan, Wales. 51 29N 3 13W. Cap. of Wales, university city and regional centre for government, sit. on Taff R. near its mouth on Severn R. estuary. Commercial, financial and administrative centre. Industries inc. high technology, electrical goods, steel production, ship repairing, engineering, flour milling, paper making and chemicals. Sea port handling timber, fruit, grain, oil and chemical products, steel, coal and coke. Pop. (1988E) 283,900.

Cardigan   Dyfed, Wales. 52 06N 4 40W. Town sit. on Teifi R. above its mouth in Cardigan Bay. Market town for an agric. area. Pop. (1980C) 3,592.

Caribbean Sea   15 00N 75 00W. Sea forming extreme W reach of the Atlantic Ocean, bounded S by mainland of S. America, W by Central American isthmus and N by Antilles. Area, 1m. sq. m. (2.6m. sq. km.).

Cariboo   British Columbia, Canada. 53 00N 121 00W. Mining district in the foothills of the Cariboo Mountains.

Cariboo Mountains   British Columbia, Canada. 53 00N 121 00W. Mountain range extending 200 m. (320 km.) NW to SE within bend of Fraser R. and parallel to Rocky Mountains.

Carinthia   Austria. 46 52N 13 30E. Federal State bordered S by Italy and Yugoslavia. Area, 3,680 sq. m. (9,531 sq. km.). Main towns Klagenfurt (cap.) and Villach. Mainly mountainous, rising to Gross Glockner. Drained by Drau R. with many L. The main occupations are stock farming, tourism, lumbering and mining. Pop. (1982E) 541,800.

Carlisle   Cumbria, England. 54 54N 2 55W. City in NW sit. on Eden R. 10 m. (16 km.) from its mouth on Solway Firth. Industries inc. food processing, distribution, textiles, metal goods, services and engineering. Pop. (1988E) 102,050.

Carlow   Ireland. 52 50N 6 55W. (i) County in SE separated from Irish Sea by Wexford and Wicklow. Area, 346 sq. m. (896 sq. km.). Rises to mountains in the SE with Mount Leinster. Drained by Rs. Barrow and Slaney. Main occupation farming. Pop. (1988E) 40,988. (ii) County town of (i) sit. on Barrow R. Market town for an agric. area. Industries inc. general manu. inc. agric. machinery, timber, grain and seed. Pop. (1985C) 10,000.

Carlsbad   New Mexico, U.S.A. 32 25N 104 14W. City sit. on Pecos R. in SE of state. Commercial centre trading in cotton, wool, cattle and alfalfa. Industries inc. potash mining and processing, oil and gas and tourism, esp. to the Carlsbad Caverns National Park and Guadalupe Mountains National Park. Pop. (1980C) 31,351.

Carlton   Nottinghamshire, England. 53 22N 1 06W. Town immediately E of Nottingham. Industries inc. mining, manu. bricks, furniture and hosiery. Pop. (1983E) 44,800.

Carluke   Strathclyde Region, Scotland. 55 45N 3 51W. Town sit. SE of Glasgow. Industries inc. food processing, general service activities and light engineering. Pop. (1986E) 14,058.

Carmarthen   Dyfed, Wales. 51 52N 4 19W. Market town for an agric. area sit. on the Towy R. near its mouth on the Bristol Channel. Administrative centre for Dyfed county. Industries inc. flour milling; manu. dairy products. Pop. (1981C) 12,302.

Carmel, Mount   Israel. 32 43N 35 03E. Mountain ridge extending 14 m. (22 km.) SE to NW from the Samarian Hills to Haifa on the Mediterranean coast, rising to 1,732 ft. (528 metres) a.s.l.

Carmona   Sevilla, Spain. 37 28N 5 38W. Town sit. ENE of Seville in a vine and olive growing area. Industries inc. flour milling and tanning; manu. olive oil.

Carnac   Morbihan, France. 47 35N 3 05W. Village in NW France, sit. on S coast of Brittany. Noted for the standing stones which extend in parallel rows *c.* 3 m. (5 km.) long. Pop. (1982C) 3,964.

Carnarvon   *see* **Caernarfon**

Carnforth   Lancashire, England.
Town sit. N of Lancaster. Market town
for a farming area. Industries inc.
quarrying, warehousing and distribution Pop. (1981C) 4,800.

Car Nicobar Island   Nicobar
Islands, Indian Ocean. Northernmost
of Nicobar Group sit. S of Little
Andaman Island. Area, 49 sq. m. (127
sq. km.). Chief village Mus. Fairly flat
with numerous coconut groves.

Carnoustie   Tayside Region, Scotland. 56 30N 2 44W. Town sit. ENE of
Dundee on North Sea. Noted as a
golfing centre. Pop. (1978E) 8,250.

Carolina   Puerto Rico. 18 23N 65
57W. Town sit. ESE of San Juan.
Industries inc. processing sugar and
tobacco; manu. blankets. Pop. (1980E)
165,954.

Caroline Islands   West Pacific
Ocean. 8 00N 147 00E. Arch. within
U.N. Trust Territory comprising Rep.
of Belau and Federated States of Micronesia. The main occupation is
farming. The main products are copra,
sugar, arrowroot and tapioca. Mineral
deposits inc. bauxite, iron ore and
phosphates. Pop. (1980E) 86,000.

Caroni River   Venezuela. 8 21N 62
43W. Rises in the Guiana Highlands
in SE Venezuela near Brazil/Guyana
border, and flows 500 m. (800 km.) W
and then N to join Orinoco R. near
Ciudad Guayana. Used for hydroelectric power.

Carpathian Mountains   East
Europe. 48 00N 24 00E. Mountain
system extending 900 m. (1,440 km.)
NW to SE across S Czechoslovakia, the
Ukraine, U.S.S.R. and Romania. The
main ranges are the Little Carpathians, the White Mountains, the W.
and E. Beskids, the High and Low
Tatra and the Transylvanian Alps. The
highest point is Gerlachovka in the
High Tatra. Rich mineral deposits.

Carpentaria, Gulf of   Northern
Territory/Queensland, Australia. 14
00S 139 00E. Inlet in N coast between
Cape York Peninsula to the E and

Arnhemland to the W; depth *c.* 370 m.
(592 km.) and width *c.* 300 m. (480
km.). It contains the Groote Eylandt
and Wellesley Islands group.

Carpentras   Vaucluse, France. 44
03N 5 02E. Town NE of Avignon sit.
on E edge of Rhône R. valley. Market
town. Industries inc. tourism, manu.
confectionery. Pop. (1989E) 30,000.

Carpi   Emilia-Romagna, Italy. 44
47N 10 53E. Town in N central Italy.
Industries inc. wine making; manu.
food products.

Carrara   Toscana, Italy. 44 05N 10
06E. Town sit. near coast of Gulf of
Genoa. The main industry is the quarrying and exporting of marble.

Carriacou   Grenadines, West Indies.
Island sit. NE of St. Georges and
largest of the group. A depend. of
Grenada. Area, 13 sq. m. (34 sq. km.).
Main settlement Hillsborough. Hilly.
Main products cotton, limes. Pop.
(1984E) 6,000.

Carrickfergus   Antrim, Northern
Ireland. 54 43N 5 49W. Town sit. on
N side of Belfast Lough, NE of Belfast.
There is some light industry. Pop.
(1989E) 34,000.

Carrickmacross   Monaghan, Ireland. 53 58N 6 43W. Town in NE, sit.
WSW of Dundalk. Market town for an
agric. area. Industries inc. lace-
making, electronics, food processing
(inc. meat packaging), footwear. Pop.
(1985E) 4,000.

Carrick-on-Shannon   Leitrim, Ireland. 53 57N 8 05W. County town in
N centre, sit. on Shannon R. Market
town for an agric. area. Pop. (1980E)
2,000.

Carrick-on-Suir   Tipperary, Ireland.
52 21N 7 25W. Town in SE sit. on
Suir R. and Kilkenny border. Industries inc. furniture, textiles and
crystal glass, basket weaving and
tourism esp. for salmon fishing. Pop.
(1989E) 5,353.

Carsamba   Samsun, Turkey. Town
sit. ESE of Samsun on Yesil R. Rail
terminus and trading centre for agric.

produce, esp. tobacco, corn. Pop. (1970c) 20,463.

Carshalton   Greater London, England. District of sw London.

Carson City   Nevada, U.S.A. 39 10N 119 46W. State cap. in w of state and sit. in a silver mining area. Commercial and manu. centre and resort. Pop. (1980c) 30,807.

Cartagena   Bolívar, Colombia. 10 25N 75 32W. Cap. of dept. sit. on the Caribbean coast. Industries inc. textiles and chemicals; manu. tobacco products, footwear and soap. Sea port exporting agric. produce, gold and platinum. Oil pipeline terminal. Pop. (1980E) 470,000.

Cartagena   Murcia, Spain. 37 36N 00 59W. Town in SE sit. on Mediterranean coast. Industries inc. lead and iron smelting, boatbuilding and chemicals; manu. glass and bicycles. Sea port exporting lead and iron. Naval base. Pop. (1981c) 172,751.

Cartago   Costa Rica. 9 52N 83 55W. (i) Province of central Costa Rica. Area, 1,207 sq. m. (3,125 sq. km.). Main occupations growing coffee and vegetables, dairy farming. Pop. (1984E) 269,860. (ii) Town at foot of Mount Irazú volcano. Provincial cap. and market town, trading in agric. produce. It has twice been destroyed by earthquakes in 1823 and 1910. Pop. (1984E) 23,884.

Cartwright   Newfoundland, Canada. Village on s shore of Sandwich Bay, s Labrador. The main occupation is lumbering.

Carúpano   Sucre, Venezuela. 10 40N 63 14W. Town sit. on Caribbean coast. Industries inc. manu. pottery, soap and straw hats. Sea port exporting coffee and cacao. Pop. (1984E) 32,000.

Casablanca   Morocco. 33 39N 7 35W. City sit. on Atlantic coast. Industries inc. textiles and fishing; manu. glass, soap and super phosphates and manganese ore and handling over 75% of foreign trade. Pop. (1982c) 2,139,204.

Casale Monferrato   Piemonte, Italy. 45 08N 8 27E. Town sit. E of Turin on Po R. Commercial centre trading in fruit and rice. Industries inc. agric. engineering and textiles; manu. footwear. Pop. 45,000.

Cascade Range   U.S.A. 45 00N 121 30W. Range extending N to S from Canadian border through Washington and Oregon, near the Pacific coast. A general height of 4,000–5,000 ft. (1,219–1,524 metres) a.s.l. rises to Mount Rainier, 14,408 ft. (4,392 metres) a.s.l. and Mount Adams, 12,307 ft. (3,751 metres) a.s.l. Heavily forested.

Cascais   Lisboa, Portugal. 38 42N 9 25W. Town sit. W of Lisbon on Atlantic coast. Industries inc. tourism, fishing, fish canning, esp. sardines, lobster. Pop. (1980E) 16,000.

Caserta   Campania, Italy. 41 04N 14 20E. Cap. of Caserta province sit. NNE of Naples. Market town trading in citrus fruits, wine, olive oil and cereals. Manu. chemicals and soap. Pop. (1981E) 66,300.

Casper   Wyoming, U.S.A. 42 51N 106 19W. City sit. on North Platte R. in E central Wyoming. Industries inc. oil, gas, uranium, tourism. Pop. (1989E) 50,000.

Caspian Sea   Iran/U.S.S.R. 42 00N 50 30E. Sea between Asia and extreme SE Europe; bounded N, W and E by U.S.S.R. and S by Iran. Extends 750 m. (1,200 km.) N to S and is c. 220 m. (352 km.) wide. It is tideless with no outlet and is the largest inland sea in the world. Of lower salinity than the Black Sea, it receives fresh water from Rs. Volga and Ural. Frozen in the N for 2–3 months a year.

Castellammare di Stabia   Campania, Italy. 40 42N 14 29E. Town SE of Naples and sit. on Bay of Naples. Industries inc. marine engineering and textiles; manu. aircraft parts and pasta. Naval dockyard and resort.

Castellón   Spain. Province on E coast bounded SE by Gulf of Valencia. Mountainous in N and W descending to a fertile coastal plain. The chief occupation is farming. The chief products are cereals, wine, olives and fruit. Pop. (1986E) 437,320.

Castellón de la Plana Castellón, Spain. 39 59N 00 02W. Cap. of province sit. near E coast. Industries inc. tourism, textiles and paper-making; manu. tiles and cement. Exports oranges, almonds and other fruit through its port of El Gráo de Castellón 3 m. (5 km.) to the E. Pop. (1989E) 134,021.

Castelo Branco Portugal. 39 49N 7 30W. (i) District of central Portugal bounded E and SE by Spain, S by Tagus R. and NW by the Serra da Estrêla. Area, 2,577 sq. m. (6,674 sq. km.). Agric., with extensive forests. Main products cereals, beans, livestock, olives, wine and vinegar. The main industry is woollen milling. Pop. (1987E) 223,700. (ii) City sit. NE of Lisbon near Spanish border. District and provincial cap. Trading centre for agric. produce. Industries inc. manu. cork, olive oil, cereals, cheese, candles, pottery, concrete, furniture and woollen goods, esp. blankets. There is an episcopal palace, and the ruins of a castle of the Templars who founded the city, 1209. Pop. (1980E) 21,300.

Castelvetrano Sicilia, Italy. 37 41N 12 47E. Town in Trapani province, in extreme W. Industries inc. wine-making (marsala). Pop. 33,000.

Castile Spain. Region of N and central Spain bounded N by Bay of Biscay and S by the Sierra Morena. Plateau at 2,500–3,000 ft. (762–914 metres) a.s.l., bare and arid. Old Castile lies to N of the Sierra de Guadarrama and Sierra de Gredos, New Castile to S. Old Castile, area, 19,390 sq. m. (50,220 sq. km.) comprises six provinces; Ávita, Burgos, Logroño, Santander, Segovia and Soria. New Castile comprises five; Ciudad Real, Cuenca, Guadalajara, Madrid and Toledo.

Castlebar Mayo, Ireland. 53 52N 9 17W. County town sit. at E end of Castlebar Lough, in central W. Market town for an agric. area. Industries inc. manu. of laboratory equipment and medical appliances. Pop. (1981C) 6,049.

Castle Douglas Dumfries and Galloway Region, Scotland. 54 57N 3 56W. Town sit. NE of Kirkcudbright on Carlingwark Loch. Market town trading in livestock. Industries inc. tourism. Pop. (1980E) 3,500.

Castleford West Yorkshire, England. 53 44N 1 21W. Town sit. 10 m. (16 km.) SE of Leeds at confluence of Rs. Aire and Calder. Industries inc. coalmining and chemicals; manu. glass and earthenware. Pop. (1981C) 39,401.

Castlegar British Columbia, Canada. City N of Trail near mouth of Kootenay R. Main occupation lumbering, inc. pulp manu. Pop. (1989E) 6,500.

Castlemaine Victoria, Australia. 37 00S 144 19E. Town NNW of Melbourne. The main industry is iron founding. Market trading in apples, sheep and dairy produce. Once a goldtown, the diggings being among the earliest discovered. Pop. (1984E) 7,000.

Castleton Derbyshire, England. 53 21N 1 46W. Village NE of Buxton, noted for Blue John (fluorspar) mines, an ornamental stone used in jewellery. Industries inc. electronics and tourism. Pop. (1989E) 700.

Castletown Isle of Man, British Isles. 54 04N 4 40W. Town 9 m. (14 km.) SW of Douglas sit. on S coast. Industries inc. fishing and tourism. Pop. (1981C) 3,148.

Castoria *see* **Kastoria**

Castres Tarn, France. 43 40N 3 59E. Town in S sit. on Agout R. Industries inc. textiles—cotton and woollen cloth—engineering, metal working and tanning. Pop. (1982C) 46,877.

Castries Saint Lucia, Windward Islands. 14 01N 60 59W. Cap. sit. S of Fort de France, Martinique, on NW coast of Saint Lucia. It was first settled by the British in 1605. It has many times been destroyed by hurricanes and fire. The port has a landlocked harbour and coaling station with modern docks. Exports sugar-cane, rum, molasses, cacao, coconuts, copra, lime, limejuice, essential oils, bay rum, fruits and vegetables. Industries inc. processing limes, sugar, bay oil and rum. There is a botanic research station. Pop. (1988E) 52,868.

Castrop-Rauxel   North Rhine-Westphalia, Federal Republic of Germany. 51 33N 11 36E. Industrial city sit. immediately N of Dortmund. Pop. (1989E) 80,000.

Catalonia   Spain. 41 40N 1 15E. Autonomous community of NE Spain comprising four provinces: Barcelona, Gerona, Lérida and Tarragona. Area, 12,329 sq. m. (31,932 sq. km.). Mainly hilly, drained by lower Ebro R. and its tribs., and Rs. Llobregat and Ter. Chief products of the rural areas are wine, olive oil, almonds and fruit. Highly industrialized areas served by hydroelectric power produce textiles, chemicals and metal goods. Pop. (1986E) 5,977,008.

Catamarca   Catamarca, Argentina. 28 30S 65 45W. Cap. of province in NW, sit. in a sub-Andean valley. Market centre of an agric. area. Manu. clothing. Pop. (1984E) 29,000.

Catanduanes   Philippines. Island 5 m. (8 km.) off Rungus Point, SE Luzon, in Philippine Sea. Area, 552 sq. m. (1,430 sq. km.). Together with small offshore islands, inc. Panay, it forms the province of Catanduanes. Cap. Virac. Mountainous. The chief occupations are farming and copper mining.

Catania   Sicilia, Italy. 37 30N 15 06E. Town sit. on Gulf of Catania immediately S of Mount Etna, and SE of Palermo. Industries inc. shipbuilding, textiles, sulphur processing, food processing, sugar refining and papermaking. Sea port. Pop. (1983E) 380,370.

Catanzaro   Calabria, Italy. 38 54N 16 36E. Town sit. near Ionian Sea coast and NE of Reggio di Calabria. Residential town in a citrus-fruit growing area. Pop. (1981E) 101,000.

Caterham and Warlingham   Surrey, England. 51 17N 0 04W. Town

sit. S of Croydon on N Downs. Industries inc. engineering and manu. cosmetics. Pop. (1981C) 30,344.

Cat Island   Bahamas. 24 30N 75 30W. Very fertile and hilly island rising (N) to 400 ft. (122 metres) which is the highest point in the Bahamas. Main products coconuts, sisal, sweet potatoes, pineapples, bananas, pine, cedar and mahogany. Pop. (1984E) 2,143.

Catskill Mountains   U.S.A. 42 10N 74 30W. Range in E, on Allegheny Plateau, at N end of Appalachian Mountains at an average height of 3,000 ft. (914 metres) a.s.l. but deeply dissected by R. gorges. Thickly wooded.

Catterick   North Yorkshire, England. Village sit. SSW of Darlington on Swale R.; noted for its military camp.

Cauca   Colombia. Dept. of SW Colombia. Area, 11,660 sq. m. (30,199 sq. km.). Cap. Popayán. Mainly agric. with important mineral resources. Main products rubber, sugar, tobacco and livestock. Pop. (1983E) 821,700.

Cauca River   Colombia. 8 54N 74 28W. Rises at S end of Central Cordillera and flows 840 m. (1,344 km.) N between Central and Western Cordillera ranges to join Magdalena R. Its valley is very fertile.

Caucasus   U.S.S.R. 42 50N 44 00E. Region between Black Sea and Caspian Sea. The N Caucasus is mainly plain, inc. Stavropol Plateau and Kuban Steppe. Drained in the N by Don R. into Sea of Azov and in the S by Kuban R. into the same sea, and by Terek R. into Caspian Sea. Chief crops are cereals and cotton; industry is concentrated on Armavir, Astrakhan, Krasnodar, Rostov and Stavropol. The Great Caucasus is mountainous with ranges extending 750 m. (1,200 km.) WNW to ESE from Taman peninsula in W to Apsheron peninsula in E, and rising to Mount Elbruz. Rainfall is up to 100 in. (254 cm.) annually on the S slopes and 10 in. (25 cm.) in the E. There are important deposits of petroleum and manganese. Transcaucasia is mountainous with the Surami

Range extending N to S between the Great Caucasus and Little Caucasus.

Causses   France. Limestone plateau of S and SW France in the Massif Central dissected by deep R. gorges. Arid and sparsely populated. The main occupation is sheep farming.

Cauterets   Hautes-Pyrénées, France. Town near Spanish frontier in SW France. Spa and resort. Pop. (1986E) 1,183.

Cauvery River   India. 11 10N 79 51E. Rises in the Western Ghats and flows ESE through Karnataka and Tamil Nadu for 470 m. (752 km.) to enter the Bay of Bengal by a delta, of which the N channel is the Coleroon. Used for hydroelectric power.

Caux   Seine-Maritime, France. Region of chalk plateau in central N, extending from Le Havre to Dieppe. Fertile, producing oats, flax and sugar-beet.

Cava de' Tirreni   Campania, Italy. 40 42N 14 42E. Town sit. NW of Salerno. Industries inc. textiles and tourism.

Cavan   Ireland. 53 58N 7 10W. (i) County in N, bounded N by N. Ireland. Area, 730 sq. m. (1,890 sq. km.). Hilly, rising to Mount Cuilcagh in the NW. Drained by R. Erne and Annalee with many L. Pop. (1988E) 53,965. (ii) County town sit. near the E shore of L. Oughter. Market town. Pop. (1981E) 5,035.

Cawdor   Highland Region, Scotland. 57 31N 3 56W. Village in N sit. SW of Nairn, noted for its castle, trad. the scene of Macbeth's murder of Duncan in A.D. 1040.

Cawnpore   *see* **Kanpur**

Cayenne   French Guiana. 5 00N 52 18W. Cap. of dept. sit. on an island at mouth of the Cayenne R. Sea port for shrimping and importing food. Pop. (1982C) 38,135.

Cayman Brac   West Indies. 19 45N 79 50W. One of the smaller Cayman Islands sit. WNW of Jamaica, of which

it is a depend. The main product is turtles.

Cayman Islands   West Indies. 19 30N 80 30W. Group of islands in Caribbean sea NW of Jamaica and forming a Brit. Crown Colony. Area, 100 sq. m. (259 sq. km.). The main island is Grand Cayman, cap. Georgetown: the others are Little Cayman and Cayman Brac. The chief products are turtles and shark skin. Pop. (1983E) 18,750.

Ceanannus Mór (Kells)   Meath, Ireland. 53 44N 6 53W. Town in E, near Blackwater R., which developed around a monastery founded by St. Columba, and noted for the medieval *Book of Kells* which was written and illuminated there. Pop. (1980E) 2,500.

Ceará   Brazil. 5 00S 40 00W. State in NE bounded N by Atlantic Ocean. Area, 58,158 sq. m. (150,630 sq. km.). Cap. Fortaleza. Irrigated by R. to produce cotton, sugar, hides, skins and carnauba wax. Pop. (1980C) 5,288,253.

Cebu   Philippines. 10 18N 123 54E. (i) Island between Negros to W and Bohal and Leyte to E. Together with smaller islands it forms Cebu province. Area, 1,703 sq. m. (4,411 sq. km.). Main occupations farming, mining and quarrying. The chief products are maize, sugar-cane, abacá, tobacco, coal, copper and limestone. (ii) Cap. of (i) sit. on E coast of Cebu Island. Industries inc. manu. cement and pottery. Port exporting abacá, copra and cement. Pop. (1980C) 489,208.

Cedar City   Utah, U.S.A. 37 41N 113 04W. City and trading centre for agric. produce. Industries inc. manu. dairy products, woollen goods, plaster, timber products. There is coal and iron mining in the neighbourhood. Pop. (1986E) 12,380.

Cedar Falls   Iowa, U.S.A. 42 32N 92 27W. City sit. WNW of Waterloo on Cedar R. Industries inc. manu. agric. equipment, pumps, blankets, food products and animal foods. Pop. (1980C) 36,322.

Cedar Rapids   Iowa, U.S.A. 41

59N 91 40W. City in E of state sit. on Cedar R. Industries inc. meat packing and agric. engineering; manu. cereal products and road-building machinery. Pop. (1980C) 110,243.

Ceiba, La   Atlántida, Honduras. 15 45N 86 45W. Port sit. E of Puerto Cortés. Industries inc. flour milling, tanning, and soap and footwear manu. Pop. (1984E) 69,000.

Celaya   Guanajuato, Mexico. 20 31N 100 37W. Town in central Mex. sit. on Laja R. and on the central plateau. Market town trading in grain, livestock and cotton. Industries inc. manu. sweetmeats and textiles. Pop. (1980E) 65,000.

Celle   Lower Saxony, Federal Republic of Germany. 52 37N 10 05E. Town in NE sit. on Aller R. Industries inc. textiles and chemicals; manu. machinery and foodstuffs. Pop. (1984E) 71,200.

Central   Region of Scotland. 56 12N 4 25W. Area, 1,000 sq. m. (2,590 sq. km.) comprising the districts of Clackmannan, Falkirk and Stirling. Bordered by Tayside N, Fife E, Lothian SE and Strathclyde W. The main R. is the Forth. Agric. is important and industry inc. petro-chemicals, aluminium, textiles and brewing. Pop. (1988E) 271,526.

Central African Republic   7 00N 20 00E. Country bounded N by Chad, E by Sudan, S by Zaïre and W by Cameroon. Area, 240,324 sq. m. (622,436 sq. km.). Cap. Bangui. Mainly savannah drained by tribs. of Ubangi R. and other Rs. draining into L. Chad. The main occupations are farming and mining. Chief products are cotton and diamonds. Pop. (1988E) 2,899,376.

Central America   Region between Isthmus of Tehuantepec, Mexico, and Isthmus of Panama, comprising the five S states and territories of Mexico; Belize; Guatemala; Honduras; El Salvador; Nicaragua; Costa Rica and Panama. Area, 230,000 sq. m. (596,000 sq. km.). Tropical, mountainous region with volcanic mountains and earthquakes. The chief crops

are bananas in the lowlands and coffee on the mountain slopes.

Central Asia   U.S.S.R. The name given, since 1917, to the Soviet Reps. of Russian Turkestan; the Kirghiz, Tadzhik, Turkmen and Uzbek Soviet Socialist Reps. The Kazakh Soviet Socialist Rep. is sometimes inc.

Central Provinces   *see* **Madhya Pradesh**

Centre   France. Region in centre of France comprising depts. of Cher, Eure-et-Loire, Indre, Indre-et-Loire, Loiret and Loir-et-Cher. Area, 15,082 sq. m. (39,062 sq. km.) Chief towns, Tours, Orléans. Pop. (1982C) 2,264,164.

Cephalonia   *see* **Kefallenia**

Ceram   Indonesia. 3 00S 129 00E. Island in S Maluku group between Ceram Sea to N, and Banda Sea to S. Area, 6,622 sq. m. (17,151 sq. km.). Chief town, Wahai. Mountainous and forested; the chief products are copra, sago, timber and dried fish. Pop. 100,000.

Cerdaña (Cerdagne)   Spain/France. 42 22N 1 35E. Valley in E Pyrénées extending through the Pyrénées-Orientales dept., France, and Gerona and Lérida provinces, Spain.

Cerignola   Puglia, Italy. 41 16N 15 54E. Town in SE, sit. SE of Foggia. Market town trading in wine, olive oil and wool. Industries inc. manu. footwear, olive oil and wine. Pop. (1985E) 60,000.

Cerne Abbas   Dorset, England. Village 7 m. (11 km.) N of Dorchester noted for the anc. Cerne Giant figure cut in a chalk hill.

Cerro de Pasco   Pasco, Peru. 10 41S 76 16W. Cap. of dept. sit. in Central Andes, in W central Peru. The main industry is mining and smelting copper, zinc, lead and bismuth. Pop. (1980E) 30,000.

Cerro-Lárgo   Uruguay. Dept. of NE Uruguay bounded NW by Río Negro. Area, 5,764 sq. m. (14,929 sq. km.).

Cap. Melo, which is an important market for the whole of NE Uruguay. Cattle ranching area. Pop. (1985E) 75,000.

César River Colombia. Rises on E side of Sierra Nevada de Santa Marta and flows 200 m. (320 km.) SW through Ciénaga de Zapatose to join Magdalena R. at El Banco. Its course is meandering and tortuous through fertile pasture land.

Cesena Emilia-Romagna, Italy. 44 08N 12 15E. Town sit. SE of Forli on Savio R. Industries inc. manu. beet sugar, pasta and wine. Pop. 85,000.

České Budejovice Jihočeský, Czechoslovakia. 48 59N 14 28E. Town sit. S of Prague on Vltava R. Commercial centre for S Bohemia, trading in timber and cereals. Industries inc. brewing; manu. furniture, enamel goods and pencils. Pop. (1980E) 89,000.

Cessnock New South Wales, Australia. 32 50S 151 21E. Town sit. WNW of Newcastle sit. in a fruit-growing, wine-making and dairy farming area. Market town with industries inc. light industry, coal mining and those related to farming. Tourism and aluminium manu. are growth industries. Pop. (1989E) 44,000.

Cetinje Montenegro, Yugoslavia. 42 23N 18 55E. Town sit. ESE of Dubrovnik. Former cap. of Kingdom of Montenegro.

Ceuta North Africa. 32 52N 5 20W. Town and Spanish enclave on a peninsula on NW coast of Africa, and administered as part of Cádiz province, Spain. Bounded on two sides by the Mediterranean, otherwise surrounded by Morocco. Sea port. Pop. (1986C) 71,403.

Cévennes France. 44 00N 3 30E. Mountain range on SE edge of Massif Central extending 150 m. (240 km.) SW to NE at an average height of 3,000–4,000 ft. (914–1,219 metres) a.s.l. and rising to Mont Mézenc. Barren limestone forming a watershed between Rs. Loire and Garonne in W, and Rs. Rhône and Saône in E.

Ceylon *see* **Sri Lanka**

Chachapoyas Amazonas, Peru. 6 10S 77 50W. Town sit. NE of Cajamarca in E Andes. Dept. and provincial cap. Commercial centre of an extensive agric. area. Industries inc. distilling; manu. straw hats. Considered the oldest Peruvian town E of the Andes. Pop. (1984E) 5,200.

Chaco Argentina. 25 00S 60 00W. Province of N Argentina bounded E by Paraguay. Area, 38,041 sq. m. (98,526 sq. km.). Cap. Resistencia. Wooded plain, becoming desert in the NW, with sub-tropical climate. There are sand quarries on the Paraná R. and salt deposits along the Bermejo R. Main occupations farming and lumbering, with associated industries inc. cotton ginning, oil extraction, sugar refining, sawmilling, tanning. Cotton is the most important crop. Pop. (1980C) 701,392.

Chad 12 30N 17 15E. Rep. bounded N by Libya, E by Sudan, S by Central African Rep. and W by Cameroon, Niger and Nigeria. Area, 495,753 sq. m. (1,284,000 sq. km.). Cap. N'djamena, formerly Fort Lamy. Desert in the N rising to the Tibesti Highlands; savannah in the S, mainly dry but with seasonal flooding. The chief occupation is farming; the chief product cotton. Pop. (1988E) 5·4m.

Chad, Lake Africa. 13 20N 14 00E. L. mainly in W Chad with Nigeria bounding it SW, Cameroon on the S, and Niger on the NW. Area varies according to the seasons. Fed mainly by Shari R. and has no visible outlet.

Chadderton Greater Manchester, England. 53 33N 2 08W. Town immediately W of Oldham. Industries inc. textiles, esp. cotton; manu. electrical equipment. Pop. (1981C) 33,518.

Chadron Nebraska, U.S.A. 42 50N 103 00W. Town sit. N of Alliance near the South Dakota border. Trading centre for agric. produce. Pop. (1980C) 5,933.

Chagos Archipelago Indian Ocean. 6 00S 72 00E. Island group 1,200 m. (1,920 km.) NE of Mauritius forming a

Brit. depend. and consisting of five main atolls, the largest being Diego Garcia. The chief product is copra.

Chalatenango El Salvador. 14 03N 88 56W. (i) Dept. of N Salvador. Area, 968 sq. m. (2,507 sq. km.). The main products are cereals and indigo. Pop. (1981E) 190,000. (ii) Town on Tamulasca and Colco Rs. Dept. cap. Pop. (1980E) 15,000.

Chalcidice *see* **Khalkidiki**

Chalcis Euboea, Greece. 38 28N 23 36E. Cap. of prefecture sit. NNW of Athens at narrowest point of Euripus Strait. Sea port trading in wine, citrus fruits, olives, olive oil, cereals and livestock. Pop. (1981C) 44,867.

Chalna (Mangla) Bagerhat, Bangladesh. 22 38N 89 30E. Town sit. S of Khulna on Pussur R. 60 m. (96 km.) from the sea. Port exporting jute. Pop. (1981E) 69,000.

Châlons-sur-Marne Marne, France. 48 57N 4 22E. Cap. of dept. in NE sit. on Marne R. Trading centre for the Champagne region. Industries inc. manu. wooden and leather goods. Pop. (1982C) 54,359.

Chalon-sur-Saône Saône-et-Loire, France. 46 48N 4 50E. Town in central E, at junction of Saône R. and Canal du Centre. Commercial centre for the Saône valley. Industries inc. engineering, boatbuilding and sugar refining. Pop. (1989E) 63,000.

Chaman Baluchistan, Pakistan. Town sit. NW of Quetta on Afghan border. Rail terminus and border control post.

Chambal River India. 26 30N 79 15E. Rises in the Vinghya Range and flows 550 m. (880 km.) N and NE to join Jumna R. below Etaurah.

Chambéry Savoie, France. 45 34N 5 56E. Cap. of dept. in SE. Market town and tourist centre. Industries inc. manu. aluminium, silk and vermouth. Pop. (1982C) 54,896 (agglomeration, 96,163).

Chambord Loir-et-Cher, France.

47 37N 1 31E. Village 28 m. (45 km.) SW of Orléans noted for the Renaissance château of Chambord. Pop. (1982C) 206.

Chamdo Tibet, China. Town in E Chamdo Area, sit. ENE of Lhasa on Mekon R. Trading centre on the Lhasa–S China route.

Chamonix-Mont Blanc Haute-Savoie, France. Town sit. E of Annecy near the border with Switzerland in the Arve R. valley. Mountain resort. The main industry is tourism. Pop. (1982C) 9,255.

Champagne-Ardennes France. 49 00N 4 00E. Region in NE France comprising the Ardennes, Marne, Aube and Haute-Marne depts. Area, 9,884 sq. m. (25,600 sq. km.). Mainly plains, with vineyards in the W, sheep farming on dry chalk in the centre, and dairy farming on clay in the E. Drained by Rs. Aisne, Marne, Aube and Seine. Chief towns Reims, Troyes, Charleville-Mezieres and Chalons-sur-Marne. Pop. (1982C) 1,345,935.

Champaign Illinois, U.S.A. 40 07N 88 14W. City sit. SSW of Chicago. Industries inc. manu. food products and clothing. Pop. (1990E) 68,172.

Champerico Retalhulen, Guatemala. 14 18N 91 55W. Town sit. SW of Retalhulen on Pacific coast. Rail terminus and port handling grain, livestock, coffee, lumber and sugar. Industries inc. fishing.

Champigny-sur-Marne Val-de-Marne, France. 48 49N 2 31E. Town forming a suburb of E Paris sit. on Marne R. Industries inc. flour milling; manu. furniture and pharmaceutical products. Pop. (1982C) 76,260.

Champlain, Lake U.S.A. 44 45N 73 15W. Narrow L. 1–14 m. (2–22 km.) wide extending 107 m. (171 km.) N to S along N border between Vermont and New York State, with many fishing resorts.

Chandannagar West Bengal, India. 22 52N 88 22E. Town, formerly a French settlement, sit. N of Calcutta on Hooghly R. Industries inc. jute milling; manu. cotton goods. Pop. (1981C) 101,925.

Chandigarh   India. 30 44N 76 47E. Union territory and city sit. N of Delhi (Area, 44 sq. m. (114 sq. km.). The city is the cap. of both Punjab and Haryana states. Industries inc. manu. of hosiery and knitting machines, needles, meters, cycles, antibiotics, soft drinks and electronics. Pop. (1971C) 233,004 (city); (1981C) 451,610 (territory).

Chang chia kou   *see* **Kalgan**

Changchun   Jilin, China. 43 53N 125 19E. Cap. of Jilin province in NE, and sit. SW of Harbin. Industries inc. railway engineering, sawmilling and food processing. Important railway junction. Pop. (1982C) 1,740,000.

Chang-hua   Taiwan. 24 05N 120 32E. Town sit. SW of Taichung. Processing and trading centre for agric. produce of the area. Industries inc. flour and rice milling, weaving, sawmilling, hatmaking. Hot springs. One of the oldest towns in Taiwan. Pop. (1985E) 198,929.

Chang Jiang (Yangtse-kiang)   China. Largest R. and chief commercial highway of China. It rises in mountains of Tibet on the N side of the Tangla Mountains and traverses central provinces of China, and after tortuous course of *c.* 3,400 m. (5,440 km.) through densely populated region, enters Yellow Sea near Shanghai. It receives numerous tribs. and drains 750,000 sq. m. (1,942,500 sq. km.) and falls 15,000 ft. (4,572 metres) in the first 1,500 m. (2,400 km.) but after that it falls 1 ft. (0.3 metres) every 12 m. (19 km.). Flooding very often reaches disaster proportions.

Changsha   Hunan, China. 28 12N 113 01E. Cap. of Hunan province in E central China, sit. SSW of Wuhan, on Siang R. Port trading in rice, tea and timber. Industries inc. metal smelting and textiles; manu. glass and fertilizers; handicraft industries. Pop. (1982C) 1,072,000.

Channel Islands   United Kingdom.

49 20N 2 40W. Group of islands at SW end of English Channel. Area, 75 sq. m. (194 sq. km.). The main islands are Jersey, Guernsey, Alderney and Sark. Fertile, with a mild climate. Main towns St. Helier, on Jersey and St. Peter Port on Guernsey. The main occupations are farming, tourism and fishing. Main products tomatoes, potatoes, flowers and dairy cattle. Pop. (1981C) 128,878.

Channel–Port aux Basques   Newfoundland, Canada. 47 35N 59 11W. Town sit. W of St. John's on Cabot Strait, comprising the 2 towns of Channel and Port aux Basques, united in 1945. The main industry is cod and halibut fishing and processing. Pop. (1986C) 5,901.

Chantilly   Oise, France. 49 12N 2 28E. Town sit. N of Paris near the Forest of Chantilly. Resort. Pop. (1982C) 10,208.

Chapada Diamantina   Bahia, Brazil. Plateau extending *c.* 75 m. (120 km.) N from Lençóis and rising to 3,000 ft. (914 metres) a.s.l. The main product is black diamonds, for which it is a major source.

Chapalé, Lake   Mexico. 20 15N 103 00W. L. 30 m. (48 km.) SE of Guadalajara in W central Mexico, at 6,000 ft. (1,828 metres) a.s.l. on the central plateau, extending 70 m. (112 km.) E to W and 20 m. (32 km.) N to S at its widest. Resort area.

Chaparé River   Bolivia. Rises in the Cordillera de Cochabamba and flows 180 m. (288 km.) NE through tropical lowlands past San Antonio, to join Ichilo R. 80 m. (128 km.) S of Trinidad and form Mamoré R. Navigable for 90 m. (144 km.) above its mouth.

Chapel-en-le-Frith   Derbyshire, England. Market town sit. NE of Macclesfield. Industries inc. manu. brake and clutch linings, and industrial cranes.

Chapel Hill   North Carolina, U.S.A. 35 55N 79 04W. Town sit. SW of Durham. Site of the University of North Carolina which was chartered in 1789. Pop. (1989E) 34,612.

Chapra　Bihar, India. 25 46N 84
45E. Town sit. WNW of Patna. Market
centre trading in cereals. Pop. (1981C)
111,564.

Chard　Somerset, England. 50 52N
2 59W. Town sit. SSE of Taunton.
Market town for an agric. area. Industries inc. processing, light and general
engineering, food processing and shirt
manu. Pop. (1989E) 11,500.

Chardzhou　Turkmenistan, U.S.S.R.
39 06N 63 34E. Cap. of Chardzhou
region sit. ENE of Ashkhabad on Amu
Darya R. and Trans-Caspian Railway.
Railway junction and R. port. Industries inc. textiles, esp. cotton. Pop.
(1980E) 140,000.

Charente　France. 45 50N 0 36W.
Dept. in Poitou-Charentes region.
Area, 2,298 sq. m. (5,953 sq. km.).
Chief towns Angoulême (the cap.),
and Cognac. Hilly in the NE and used
for cattle farming; fertile elsewhere,
drained by Rs. Charente and Vienne.
The chief products are wine, cereals
and potatoes. Pop. (1982C) 340,770.

Charente-Maritime　France. 45
50N 0 35W. Dept. in Poitou-Charentes region, bounded W by Bay of
Biscay. Area, 2,644 sq. m. (6,848 sq.
km.). Chief towns La Rochelle (the
cap.), Rochefort and Saintes. Low-
lying and drained by Rs. Charente,
Seudre and Sèvre Niortaise. Mainly
agric.; the chief products are cereals,
fodder crops, wine, dairy products,
mussels and oysters. Pop. (1982C)
513,220.

Charente River　France. 45 57N
1 05W. Rises in the Haut-Vienne
dept. and flows 220 m. (352 km.) W
past Angoulême, Cognac and Saintes
to enter Bay of Biscay below
Rochefort. Navigable for small boats
below Angoulême.

Charikár　Kábul, Afghánistán. 35
02N 69 11E. Town sit. N of Kábul in a
vine and cotton growing area. Industries inc. iron founding and other
metal work.

Charleroi　Hainaut, Belgium. 50
25N 4 26E. Town in central Belgium
sit. on Sambre R. in a coalmining

area. Industries inc. treated steel,
brewing, aeronautics, aerospace, nuclear, electronics and electrical engineering; manu. pharmaceuticals, glass
and cement. Pop. (1988E) 209,000.

Charleston　Illinois, U.S.A. 39 30N
88 10W. Town sit. ESE of Decatur.
Trading centre for an extensive farming area. Industries inc. manu. of concrete pipes, iron castings, truck
trailers, steel industry ceramics, business forms. Pop. (1980C) 19,355.

Charleston　South Carolina, U.S.A.
City sit. on a peninsula at confluence
of Rs. Ashley and Cooper. Important
container port. Commercial centre
trading in cotton, fruit and timber.
Industries inc. tourism and steel;
manu. pulp and paper, fertilizers and
timber products. Pop. (1988E) 80,900.

Charleston　West Virginia, U.S.A.
38 21N 81 38W. State cap. sit. on
Kanawha R. and ENE of Lexington,
Kentucky. Industries inc. oil refining
and chemicals; manu. glass and
paints. Pop. (1980C) 63,968.

Charlestown　Nevis, Windward
Islands. 17 08N 62 37W. Town on W
coast sit. SE of Basseterre on St. Kitts.
Island cap. Industries inc. tourism and
the port. Noted for hot springs. Pop.
(1983E) 1,200.

Charleville-Mézières　Ardennes,
France. 49 46N 4 43E. Cap. of dept.
on Meuse R. Manu. inc. mechanical
equipment. Pop. (1982C) 61,588.

Charlotte　North Carolina, U.S.A.
35 16N 80 46W. City in S of state sit.
near S. Carolina border. Industries
inc. furniture, textiles and tobacco,
and chemicals, manu. vehicles. Pop.
(1980C) 314,447.

Charlotte-Amalie　St. Thomas, U.S.
Virgin Islands. 18 21N 64 56W. Town
on S coast of St. Thomas. Cap. and sea
port with a natural harbour. Named
after the consort of Christian V of
Denmark. Pop. (1980E) 11,756.

Charlottenburg　West Berlin, Federal Republic of Germany. 52 32N 13
24W. Suburb, mainly residential. Industries inc. manu. glass and paper.

Charlottesville Virginia, U.S.A. 38 02N 78 29W. City sit. WNW of Richmond. Industries inc. communications equipment, navigational systems and electronic goods. Pop. (1980C) 39,916.

Charlottetown Prince Edward Island, Canada. 46 14N 63 08W. Cap. of province sit. on S coast. Industries inc. services, manu. and tourism. Sea port exporting potatoes and dairy produce. Pop. (1986E) 15,776.

Charnwood Forest Leicestershire, England. Upland area to E of Leicestershire coalfield rising to 912 ft. (278 metres) a.s.l. at Bardon hill. Barren.

Charters Towers Queensland, Australia. 20 05S 146 16E. Town in NE of state. Market centre of a cattle-rearing area.

Chartres Eure-et-Loir, France. 48 27N 1 30E. Town in central N sit. on Eure R. Market town for a farming area, trading in cereals and livestock. Industries inc. cosmetics, pharmaceuticals, brewing, tanning and agric. electrical and electronic engineering. Noted for the 12th–13th cent. cathedral of Notre Dame. Pop. (1982C) 39,243.

Châteauroux Indre, France. 46 49N 1 42E. Cap. of dept. in central France sit. on Indre R. Commercial centre trading in grain and livestock. Industries inc. brewing; manu. woollen goods, tobacco products and furniture. Pop. (1982C) 53,967.

Château-Thierry Aisne, France. 49 03N 3 24E. Town in NE sit. on Marne R. Industries inc. manu. of musical and precision instruments and food products. Pop. (1982C) 14,920.

Chatham Ontario, Canada. 42 24N 82 11W. Town sit. E of Detroit, U.S.A. Commercial centre of an agric. area. Industries inc. sugar refining, canning and meat packing. Pop. (1989E) 42,200.

Chatham Kent, England. 51 23N 0 32E. Town sit. on Medway R. estuary. Industries inc. navigation

equipment, cement, electronics and insurance. Pop. (1983E) 146,200 (with Rochester and Strood).

Chatham Islands New Zealand. 44 00S 176 35W. Group of islands sit. ESE of Wellington, North Island, in the Pacific Ocean. The main islands are Chatham, 348 sq. m. (901 sq. km.), and Pitt, 24 sq. m. (62 sq. km.). Volcanic. The chief occupations are sheep farming and fishing.

Chatsworth Derbyshire, England. Village sit. WSW of Chesterfield noted for Chatsworth House, the seat of the Dukes of Devonshire.

Chattanooga Tennessee, U.S.A. 35 03N 85 19W. City in SE Tennessee sit. on Tennessee R. Industries developed after the provision of hydroelectric power by the Tennessee Valley Authority, and inc. diverse manu. Pop. (1988E) 162,670.

Chaumont Haut-Marne, France. 48 07N 5 08E. Cap. of dept. sit. ESE of Troyes at confluence of Rs. Marne and Suize. Industries inc. tourism. Pop. (1982C) 28,429.

Chautauqua, Lake New York, U.S.A. L. 17 m. (27 km.) long and up to 3 m. (5 km.) wide in W of state.

Chaux-de-Fonds, La Neuchâtel, Switzerland. 47 07N 6 51E. Town in NW Switzerland. Centre of the Jura clock and watch-making industry.

Cheadle and Gatley Greater Manchester, England. 53 24N 2 13W. Town WSW of Stockport. Industries inc. engineering and chemicals; manu. pharmaceutical products and bricks.

Cheb West Bohemia, Czechoslovakia. 50 01N 12 25E. Town sit. W of Prague on Ohre R. Important railway junction. Industries inc. brewing, textiles and agric. engineering. Pop. (1984E) 21,300.

Cheboksary Russian Soviet Federal Socialist Republic, U.S.S.R. 56 09N 47 15E. Cap. of Chuvash Autonomous Rep. of the Russian Soviet Federal Socialist Rep. sit. on S bank of Volga R. Industries inc. textiles and

electrical engineering; manu. matches.
Pop. (1985E) 389,000.

Chechaouen   *see* **Xauen**

**Checheno-Ingush   Autonomous
Soviet Socialist Republic**   U.S.S.R.
Autonomous Rep. of the Russian
Soviet Federal Socialist Rep. on the N
slopes of the Great Caucasus; bounded E by Dagestan and S by Georgia.
Area, 7,451 sq. m. (19,300 sq. km.).
Cap. Grozny. Industries are based on
the Grozny oilfield and inc. engineering, chemicals and food canning;
manu. building materials, timber products and furniture. Pop. (1984E)
1·2m.

Cheddar   Somerset, England. 51
17N 2 47W. Town sit. ESE of Weston-
super-Mare. Market town for an agric.
area. Noted for Cheddar Gorge with
its limestone cliffs and caves. Cheddar
cheese was originally made here. Pop.
(1983E) 3,994.

Cheektowaga   New York, U.S.A.
42 55N 78 46W. Town sit. E of Buffalo. Industries inc. manu. machinery,
steel and wood products, canned
foods; quarrying stone. Pop. (1980C)
92,145.

Cheju Island   South Korea. 33 31N
126 32E. Island sit. S of South Korean
coast. Area, 718 sq. m. (1,860 sq.
km.). Cap. Cheju. Mountainous and
well wooded. The main occupations
are fishing and cattle rearing. Main
products are cereals and soya beans.

Chekiang   *see* **Zhejiang**

Chelm   Lublin, Poland. 51 10N 23
28E. Town in E sit. ESE of Lublin.
Market town. Industries inc. flour
milling   and   brickmaking.   Pop.
(1981E) 89,000.

Chelmsford   Essex, England. 51
44N 0 28E. Town in SE on Chelmer R.
Cathedral and market town for a residential and agric. area. Industries inc.
packaging, flour milling, brewing and
malting; manu. electrical and electronic equipment, bearings. Pop.
(1985E) 144,000.

Chelsea   Greater London, England.

Area of W London on N bank of R.
Thames, with a long history of literary
and artistic associations.

Chelsea   Massachusetts, U.S.A. 42
24N 71 02W. City sit. NE of Boston
on opposite bank of Mystic R. Industries inc. manu. shoes, paints and
chemicals. Pop. (1980C) 25,431.

Cheltenham   Gloucestershire, England. 51 54N 2 04W. Town in W sit.
on Chelt R. Residential town and spa
with light industries. Pop. (1988E)
86,400.

Chelyabinsk   Russian Soviet Federal Socialist Republic, U.S.S.R. 55
10N 61 24E. Cap. of Chelyabinsk
Region sit. in S Ural Mountains and
on Trans-Siberian railway. Industries
inc. iron and steel, zinc, chemicals,
agric. engineering and flour milling.
Trade in grain and coal. Pop. (1985E)
1,096,000.

Chelyuskin, Cape   Russian Soviet
Federal Socialist Republic, U.S.S.R.
77 45N 104 20E. The most N point of
continental U.S.S.R., extending into
Boris Vilkitski Strait in the Arctic
Ocean at N tip of Taimyr peninsula.

Chemnitz   (Known as Karl-Marx-
Stadt, 1953–90) German Democratic
Republic. 50 50N 12 55E. (i) District,
area, 2,320 sq. m. (6,009 sq. km.).
Pop (1990E) 1,800,000. (ii) Cap. of
same. Sit. WSW of Dresden. Important
industries are textiles and engineering.
Pop. (1990E) 300,000.

Chenab River   India/Pakistan. 29
23N 71 02E. Rises in the Himalayas
in N Himachal Pradesh, India, and
flows 700 m. (1,120 km.) NW into
Kashmir and then SW through Pakistan to join Sutlej R. near Alipur,
receiving Rs. Jhelum and Ravi. Used
for irrigation.

Chengchow   *see* **Zhengzhou**

Chengdu   Sichuan, China. 30 39N
104 04E. Cap. of province in S central
China sit. NW of Changqing. Trading
centre in an irrigated agric. area.
Industries inc. railway engineering
and textiles; manu. bricks and tiles.
Pop. (1982C) 2,470,000.

Chengtu *see* **Chengdu**

Chepstow Gwent, Wales. 51 39N 2 41W. Town sit. on R. Wye near its mouth on the Severn estuary, and 14 m. (22 km.) ENE of Newport. Tourism, light industry and farming are the main occupations. Pop. (1980E) 10,500.

Chequers Buckinghamshire, England. 47 10N 2 30E. Estate sit. SSE of Aylesbury; the official country residence of the Prime Minister of the U.K.

Cher France. Dept. in Centre region. Area, 2,791 sq. m. (7,228 sq. km.). Chief towns Bourges (the cap.), and Vierzon. Fertile, drained by Rs. Cher and Loire. Extensive pasture. The chief crops are wine, vegetables, fodder crops and cereals. Pop. (1982C) 320,174.

Cherbourg Manche, France. 49 39N 1 39W. Town on N coast of Cotentin peninsula on the English Channel coast. Industries inc. shipbuilding and repairing. Sea port equipped for transatlantic traffic, and a naval base. Pop. (1989E) 802,000 (agglomeration).

Cheremkhovo Russian Soviet Federal Socialist Republic, U.S.S.R. 53 09N 103 05E. Town in S centre sit. NW of Irkutsk on Trans-Siberian railway. Industries inc. coalmining, engineering and chemicals. Pop. (1980E) 120,000.

Cherepovetz Russian Soviet Federal Socialist Republic, U.S.S.R. 59 08N 37 54E. Town sit. NW of Yaroslavl on N side of the Rybinsk Reservoir. Industries inc. shipbuilding, sawmilling, iron and steel, and agric. engineering; manu. footwear and matches. Pop. (1985E) 299,000.

Cherkassy Ukraine, U.S.S.R. 49 26N 32 04E. Town sit. SE of Kiev on S bank of Dnieper R. Industries inc. sawmilling, engineering, metal working and food processing. Pop. (1985E) 273,000.

Cherkessk Russian Soviet Federal Socialist Republic, U.S.S.R. 44 14N 42 04E. Town sit. ESE of Krasnodar on Kuban R. Cap. of Cherkessk region. Railway terminus for the N Caucasus. Industries inc. metal working, flour milling and chemicals. Pop. (1980E) 75,000.

Chernigov Ukraine, U.S.S.R. 51 30N 31 18E. Town sit. NNE of Kiev on Desna R. Industries inc. textiles; manu. knitwear, footwear and chemicals. R. port handling grain, flax and potatoes. Pop. (1985E) 278,000.

Chernobyl Ukraine, U.S.S.R. 51 16N 30 15E. Town N of Kiev and the scene of nuclear accident at a power station in April 1986. The fallout covered many thousands of miles and reached parts of Scandinavia, Poland and Britain.

Chernovtsy Ukraine, U.S.S.R. 48 18N 25 56E. City sit. SE of Lvov on Prut R. Cap. of Chernovtsy region. Industries inc. sawmilling, engineering, food processing, textiles and rubber products. Pop. (1985E) 244,000.

Chernyakhovsk Russian Soviet Federal Socialist Republic, U.S.S.R. 54 38N 21 49E. Town sit. E of Kaliningrad on Angerapp R. Industries inc. food processing and chemicals.

Cherrapunji Meghalaya, India. 25 17N 91 44E. Town sit. NE of Calcutta in the Khasi Hills. Noted for exceptionally high rainfall, 428 in. (1,087 cm.) annual average. Pop. (1981C) 6,097.

Cher River France. 47 21N 1 29E. Rises in the Massif Central, and flows 199 m. (318 km.) N and then W to join Loire R. 10 m. (16 km.) WSW of Tours. Its lower course is navigable.

Cherry Hill New Jersey, U.S.A. Township in S New Jersey. Residential, convention and retailing centre; industries inc. telecommunications, engineering, coffee. Pop. (1980C) 68,785.

Cherski Range Russian Soviet Federal Socialist Republic, U.S.S.R. 52 00N 114 00E. Mountain system extending 600 m. (960 km.) NW to SE in NE Siberia, rising to Pobeda, 10,325 ft. (3,147 metres) a.s.l.

Chertsey   Surrey, England. 51 24N
0 30W. Town sit. WSW of London on
Thames R. Market gardening centre.
Pop. (1977E) 12,468.

Cherwell River   England. 51 44N
1 15W. Rises in SW Northamptonshire
and flows S through Oxfordshire to
join Thames R. at Oxford.

Chesapeake   Virginia, U.S.A. 39
32N 75 49W. City S of Norfolk. Industries inc. fertilizers, cement and
timber products. Pop. (1980C)
114,486.

Chesapeake Bay   U.S.A. 38 40N
76 25W. Inlet of the Atlantic Ocean
bounded W by mainland areas of
Maryland and Virginia and E by Delmarva peninsula. Length 200 m. (320
km.) and up to 30 m. (48 km.) wide.
Rs. Susquehanna, York, James, Potomac and Rappahanock flow into it.

Chesham   Buckinghamshire, England. 51 43N 0 38W. Town sit. NW of
London. Market town. Manu. timber
products, footwear and brushes. Pop.
(1983E) 20,853.

Cheshire   England. 53 14N 2 30W.
County in NW England bounded W by
Wales, N by Merseyside and Greater
Manchester. Under 1974 reorganiza-
tion its districts are Warrington,
Halton, Ellesmere Port, Vale Royal,
Macclesfield, Chester, Crewe and
Congleton. Low-lying. Dairy farming
is important with industry, inc. petro-
chemicals and vehicle manu., concen-
trated near R. Mersey estuary and at
Crewe. Chief towns Chester, the
county town, Warrington and Crewe.
Pop. (1988E) 955,800.

Cheshunt   Hertfordshire, England.
51 43N 0 02W. Town sit. N of London. Residential with market gardening centre and horticultural research
station. Some light industry. Pop.
(1981C) 49,718 with Waltham Cross.

Chesil Bank   Dorset, England.
Shingle beach extending SE to NW for
9 m. (14 km.) as a spit of land from
Portland to Abbotsbury on the Dorset
coast. Separated from mainland by the
Fleet inlet, it then continues a further
7 m. (11 km.) as part of the mainland

coast. Towards the seaward end, larger pebbles are found. The width of
the spit is up to 200 yards (183
metres).

Chester   Cheshire, England. 52 12N
2 54W. County town sit. on Dee R.
near its estuary. Commercial and
tourist centre. Industries inc. high
technology. Town centre medieval in
appearance. Pop. (1984E) 116,657.

Chester   Pennsylvania, U.S.A. 39
51N 75 21W. Town sit. SW of Philadelphia on Delaware R. on 3 m. (5
km.) waterfront. Industries inc. shipbuilding, haulage. Manu. paper, chemicals and steel. Pop. (1980C) 45,794.

Chesterfield   Derbyshire, England.
53 15N 1 25W. Town in N midlands
sit. on Rother R. and near the Peak
District National Park. Industries inc.
engineering. Manu. include steel
tubes, tunnelling and mining equipment, packaging, glass and pottery.
Pop. (1988E) 72,000.

Chesterfield Islands   South West
Pacific. 19 52S 158 15E. Group of 11
coral islands 340 m. (544 km.) NW of
New Caledonia, of which they form a
depend. Area, 4 sq. m. (10 sq. km.).
The main product is guano. Uninhabited.

Chester-le-Street   County Durham,
England. 54 52N 1 34W. Town sit. S
of Newcastle. Industries inc. general
services, manu. of electrical goods
and chemicals. Pop. (1981C) 34,975.

Chetumal   Quintana Roo, Mexico.
18 30N 88 18W. Town on Belize border on Chetumal Bay at mouth of Río
Hondo. Territorial cap. Trading centre
for an agric. area, handling pigs,
cattle, chicle, henequen, fruit and timber. Known as Payo Obispo until
1935.

Cheviot Hills   England/Scotland. 55
24N 2 20W. Range extending for 35
m. (56 km.) SW to NE along border between England and Scotland, rising to
The Cheviot, 2,676 ft. (816 metres)
a.s.l. Sheep farming is the main occupation.

Cheyenne   Wyoming, U.S.A. 41
08N 104 49W. State cap. sit. in SE of

state. Employment is in government, the air force base, retailing, professional services and transport, with some manu. Pop. (1980c) 47,283.

Chiana River Italy. 42 52N 12 14E. Rises in Apennines and flows for *c.* 60 m. (96 km.) through the Val di Chiana and discharges into R. Tiber and *via* canals to R. Arno.

Chiangmai Chiangmai, Thailand. 18 46N 98 58E. Cap. of province in NW sit. on Ping R. Railway terminus and commercial centre trading in teak. Pop. (1983E) 105,000.

Chianti Monti Toscana, Italy. Range in the Apennines extending 15 m. (24 km.) through Toscana and rising to 2,900 ft. (884 metres) a.s.l. Grapes for Chianti wine are grown on the slopes.

Chiapas Mexico. 17 00N 92 45W. State in S Mex. bounded W by the Pacific and S and E by Guatemala. Area, 28,520 sq. m. (73,866 sq. km.). Cap. Tuxtla Gutiérrez, sit. 450 m. (720 km.) ESE of Mexico City. Mountainous, descending to a narrow coastal plain. The main occupation is farming; the main products are fruit, coffee, cocoa, cotton and hardwoods. Pop. (1980c) 2,084,717.

Chiatura Georgia, U.S.S.R. 42 19N 43 18E. Town sit. NW of Tbilisi on Kvirila R. The main industry is manganese mining.

Chiayi Taiwan. Town sit. NNE of Tainan linked by rail to the Ali Shan lumbering area. The main industries are sawmilling, sugar refining, distilling; manu. tiles and cement. Pop. (1985E) 252,637.

Chiba Honshu, Japan. 35 30N 140 07E. Cap. of Chiba prefecture sit. ESE of Tokyo on Tokyo Bay. Industries inc. steel, paper making and textiles. Pop. (1979E) 718,000.

Chicago Illinois, U.S.A. 41 53N 87 38W. City on W shore of L. Michigan, the second city of the U.S.A. Area, 228 sq. m. (591 sq. km.). Commercial and commodity-dealing centre. Industries inc. steel engineering, printing,

publishing, food processing, paper products. L. port handling steel, grain and other bulk cargo. Pop. (1980c) 3,005,072.

Chicago Heights Illinois, U.S.A. 41 30N 87 38W. City sit. E of Joliet. Main industries are iron and steel, chemicals, textiles; manu. soap, boxes, furniture, railway equipment, tiles, asphalt, linseed. Pop. (1980c) 37,026.

Chichester West Sussex, England. 50 50N 0 48W. Cathedral city sit. E of Portsmouth. Market town for an agric. area. Pop. (1989E) 25,000.

Chickamauga Georgia, U.S.A. City sit. S of Chattanooga, Tennessee, on Chickamauga R. and noted as the site of a Civil War battle.

Chickamauga Dam Tennessee, U.S.A. Dam on Tennessee R. near its confluence with Chickamauga R. NE of Chattanooga. Part of the Tennessee Valley Authority Scheme for power and flood control.

Chiclayo Lambayeque, Peru. 6 46S 79 50W. Cap. of dept. near NW coast in Lambayeque valley. Commercial centre of a farming area. Industries inc. brewing, tanning, rice milling and cotton ginning.

Chicopee Massachusetts, U.S.A. 42 10N 72 36W. City sit. WSW of Boston at confluence of Rs. Chicopee and Connecticut. Industries inc. manu. of paper goods and juices. Pop. (1989E) 57,100.

Chicoutimi Quebec, Canada. 48 26N 71 04W. Town sit. N of Quebec at confluence of Rs. Chicoutimi and Saguenay. Industries, based on hydroelectric power and a lumbering neighbourhood, inc. pulp milling, paper and furniture making. Pop. (1981c) 60,064.

Chicoutimi River Quebec, Canada. Rises in SE Quebec and flows 100 m. (160 km.) E to join Saguenay R. at Chicoutimi. Used for hydroelectric power.

Chiemsee Bavaria, Federal Republic of Germany. 47 54N 12 29E. L. in SE Bavaria, ESE of Munich. Area, 33 sq. m. (85 sq. km.). Fed by Ache R. and drained by Alz R. into Inn R.

Chieti   Abruzzi e Molise, Italy. 42 21N 14 10E. Town sit. SSW of Pescara. Cap. of Chieti province. Industries inc. textiles; manu. bricks and pasta. Pop. (1982E) 55,000.

Chignecto Bay   Canada. 45 35N 64 45W. Inlet of Bay of Fundy extending 35 m. (56 km.) SW to NE between New Brunswick and Nova Scotia. Width 10 m. (16 km.).

Chignecto Isthmus   Canada. 45 55N 64 10W. Strip of land across the head of Chignecto Bay separating it from Northumberland Strait. Width 15 m. (24 km.) at its narrowest point on the New Brunswick/Nova Scotia border.

Chigwell   Essex, England. 51 38N 0 05E. Town sit. NE of central London and forming a residential suburb. Pop. (1985E) 9,866.

Chihuahua   Mexico. 28 38N 106 05W. (i) State in N Mexico bounded N by the U.S.A. Area, 95,376 sq. m. (247,024 sq. km.). Mountainous in the W, with the Sierra Madre Occidental; plateau in the centre, and desert in the E. The main occupations are mining and farming. Pop. (1980C) 2,005,477. (ii) Cap. of state of same name in central Chihuahua in a valley of the Sierra Madre Occidental, sit. in a mining and cattle rearing area. Industries inc. textiles and smelting. Pop. (1984E) 375,000.

Chile   36 00S 72 00W. Rep. on the Pacific bounded N by Peru, and E by Bolivia and Argentina with the Andes forming E frontier. Area, 292,135 sq. m. (756,626 sq. km.). The Republic of Chile threw off allegiance to the crown of Spain, constituting a national government in 1810, finally freeing itself from Spanish rule in 1818. Main cities Santiago (the cap.), Valparaiso and Concepción. Rainless desert in N with mineral deposits, more fertile in the centre with rich farmland in the central valley, and in S mountainous and forested with heavy rains. Agric. and forestry contribute one-ninth of the national product and one-third of the pop. is employed in these industries. Chile imports two-thirds of its food requirements. The wealth of the country consists mainly of its minerals and these include: copper (the most important source of foreign exchange), nitrate of soda, iron ore, coal, petroleum, gold, silver, molybdenum, zinc, manganese, salt, sulphur and lead. Pop. (1988E) 12,683,000.

Chilkoot Pass   Canada/U.S.A. Pass on the Alaska/British Columbia border N of Skagway, Alaska, at 3,500 ft. (1,067 metres) a.s.l.

Chillán   Nuble, Chile. 36 36S 72 07W. Town in S central Chile sit. ENE of Concepción. Market town trading in fruit and wine. Industries inc. flour milling; manu. leather goods. Pop. (1982E) 120,941.

Chillicothe   Ohio, U.S.A. 39 20N 82 59W. City sit. S of Columbus on Scioto R. Industries inc. manu. paper, footwear and furniture. Pop. (1980C) 23,420.

Chilliwack   British Columbia, Canada. 49 10N 121 57W. District municipality on Fraser R., E of Vancouver. Trading centre for agric. area. Pop. (1989E) 45,643.

Chiloé Island   Chile. 42 30S 73 55W. Island off S central Chile and N of the Chonos Arch. Area, 3,241 sq. m. (8,394 sq. km.). Chief town Ancud. Forested, with cultivation in clearings. The chief products are timber, barley, potatoes and fish. Pop. (1984E) 116,000.

Chiltern Hills   England. 51 42N 0 48W. Range extending NE from R. Thames valley to continue as the East Anglian heights, and rising to Coombe Hill, 852 ft. (260 metres) a.s.l., in Buckinghamshire.

Chilung   Taiwan. 25 08N 121 44E. Port on N coast, sit. ENE of Taipei. It exports sugar and tea. Shipbuilding and flour milling are carried on and manu. inc. cement, chemicals and fertilizers. Pop. (1985E) 351,904.

Chimborazo   Chimborazo, Ecuador. 7 28S 78 48W. Inactive volcano in the W centre, at 20,577 ft. (6,272 metres) a.s.l. the highest peak in the Ecuadorian Andes.

Chimbote   Ancash, Peru. 9 05S 78 36W. Town in w centre sit. on the Pacific coast. Industries inc. iron and steel and fishing. Port exporting sugar and fish products. Pop. (1984E) 185,000.

Chimkent   Kazakhstan, U.S.S.R. 42 18N 69 36E. Town sit. N of Tashkent on Turksib Railway. Industries inc. refining lead and zinc; chemicals and textiles; manu. cement. Pop. (1985E) 369,000.

China   35 00N 100 00E. Rep. bounded E by South China Sea, East China Sea and Yellow Sea, and by North Korea and the U.S.S.R.; N by Mongolia, NW by the U.S.S.R., w by India, Nepál, Sikkim and Bhután, sw by Burma, and s by Laos and Vietnam. Area, 3,695,000 sq. m. (9,571,300 sq. km.). China is traversed by 2 folded mountain systems which stretch from Pamir plateau. In products and climate the Kunlun Mountains divide N from s China. In the E there is a great alluvial plain which covers over 200,000 sq. m. (518,000 sq. km.), and supports the larger part of the pop. Principal Rs. are Yangtze Kiang, Hwang-Ho and Si-Kiang. There are numerous canals, inc. the Grand Canal, which cross the plain from Hangzhou to Tianjin 850 m. (1,360 km.). The chief Rs. are Tungting, Po-yang and Tai-hu. As a defence against N tribes the Great Wall of China was completed in 214 B.C. Its length is *c.* 1,500 m. (2,400 km.). China remains essentially an agricultural country. Intensive agriculture and horticulture have been practised for millennia. Present-day policy aims to avert the traditional threats from floods and droughts by soil conservancy, afforestation, irrigation and drainage projects, and to increase the 'high stable yields' areas by introducing fertilizers, pesticides and improved crops. Crop priorities: food grains; raw materials for industry (especially cotton); crops for export (especially oil seeds). Among livestock, priority is given to pig production.

In 1950 the land belonging to the feudal nobility and to monasteries and other institutions was confiscated by

the State. By the end of 1958 the socialization of agriculture was declared to be complete and the peasant population of some 500m. had been organized into roughly 24,000 'communes', each consisting of a number of villages and 5,000–10,000 families. Since 1958 some modifications have been made in the commune system, and the number of communes raised to 74,000 by reducing their size. Small private plots are permitted. Crops include, rice, wheat, potatoes, soy beans, groundnuts, tobacco, cotton-seed. The chief forested areas are in Heilongjiang, Sichuan and Yunnan. The most important tree is the tung (*Jatropha Curcas* L.), from which oil is produced: it grows chiefly in Szechwan. The most important timber product is teak. Most provinces contain coal, and there are 70 major production centres, the larger ones in the north. Coal reserves are estimated at 769,180m. tonnes. Iron ores are abundant in the anthracite field of Shanxi, in Hebei, in Shandong and other provinces, and iron (found in conjunction with coal) is worked in Manchuria. China has made rapid progress in oil extraction and refining. Crude oil production (1987) 928 tonnes. Natural gas is available from fields near Canton and Shanghai. Tin ore is plentiful in Yunnan, where the tin-mining industry has long existed. China is the world's principal producer of wolfram (tungsten ore). Mining of wolfram is carried on in Hunan, Guangdong and Yunnan. Other minerals include phosphate rock, salt, sulphur, asbestos, bauxite, aluminium, antimony, copper, lead, manganese, zinc, barite, bismuth, gold, graphite, gypsum, mercury, molybdenum, silver. 'Cottage' industry is very old in the economy and persists into the 20th century. Modern industrial development began with the manufacture of cotton textiles, and the establishment of some silk filatures, steel plants, flourmills and match factories. Expanding sectors of manufacture are: steel, chemicals, cement, agricultural implements, plastics and lorries. Pop. (1990E) 1,110m.

Chinandega   Chinandega, Nicaragua. 12 35N 87 10W. Cap. of dept. WNW of Managua. Market town. Industries inc. sugar refining, flour milling and textiles; manu. furniture. Pop. (1980c) 114,291.

China Sea   10 00N 113 00E. A section of the Pacific Ocean bounded w by China, and known as the East China Sea to the N of Taiwan, and as the South China Sea to the s of it.

Chincha Alta   Ica, Peru. Town sit. SE of Lima. Industries inc. cotton ginning and brandy distilling.

Chinchow   *see* **Jinzhou**

Chindwin River   Burma. 21 26N 95 15E. Formed in the extreme N from various headstreams and flows 650 m. (1,040 km.) s to join Irrawaddy R. 13 m. (21 km.) NE of Pakokku, sit. 79 m. (126 km.) SW of Mandalay. Navigable for small vessels for about 300 m. (480 km.).

Chingford   Greater London, England. 51 37N 0 01E. Town sit. NE of central London forming a residential suburb on the edge of Epping Forest.

Chinghai   *see* **Qinghai**

Chingola   Western Province, Zambia. 12 32S 27 52E. Small town sit. NW of Ndola on a railway spur. Commercial centre for nearby copper mining centre at Nchanga. Pop. (1980E) 145,869.

Chinon   Indre-et-Loire, France. 47 10N 0 15E. Town in NW central France sit. on Vienne R. Manu. wine, ropes, baskets and leather goods. Pop. (1982c) 8,873.

Chioggia   Veneto, Italy. 45 13N 12 17E. Town and commercial port sit. on an island at s end of Venetian lagoon. Industries inc. fishing, tourism and market gardening. Pop. (1990E) 57,000.

Chios   *see* **Khios**

Chipata   Eastern District, Zambia. 19 39S 32 40E. Town in extreme SE Zambia near Malawi border. Formerly Fort Jameson. Industries associated with a thermal power station. Pop. (1980c) 32,291.

Chipinge   Mashonaland, Zimbabwe. Village sit. s of Umtali in the Chimanimani Mountains. Centre of a farming area producing tobacco, maize, wheat, citrus fruit, livestock and dairy produce.

Chippenham   Wiltshire, England. 51 28N 2 07W. Town sit. ENE of Bath on Avon R. Market town for a dairy farming area. Industries inc. light and heavy engineering, food processing, distribution and warehousing. Pop. (1990E) 26,000.

Chipping Campden   Gloucestershire, England. 52 03N 1 46W. Town sit. w of Banbury in the Cotswold Hills. Industries inc. agric., horticulture and tourism. Pop. (1983E) 2,000.

Chipping Norton   Oxfordshire, England. 51 56N 1 32W. Town sit. SW of Banbury. Market town. Industries inc. light industry and manu. of furniture. Pop. (1989E) 5,200.

Chiquinquirá   Boyacá, Colombia. Town in N centre sit. in E. Cordillera. Market town trading in coffee, sugarcane, cotton and cereals. Emerald mines lie to the SW. Pop. (1980E) 50,000.

Chirchik   Uzbekistan, U.S.S.R. 41 29N 69 35E. Town sit. NE of Tashkent on Chirchik R. Industries inc. agric. engineering; manu. fertilizers and footwear. Pop. (1970E) 108,000.

Chiriqui, Gulf of   Panama. 8 00N 82 10W. Gulf extending *c.* 90 m. (144 km.) SE along Pacific coast.

Chiriqui, Mount   Panama. 8 55N 82 35W. Inactive volcano near Costa Rica frontier, at 11,397 ft. (3,507 metres) a.s.l., the highest mountain in Panama.

Chiriqui Lagoon   Panama. 9 10N 82 00W. Lagoon 35 m. (56 km.) NW to SE along w end of Caribbean coast.

Chislehurst   Greater London, England. 51 24N 0 05E. Town sit. SE of central London and forming a residential suburb.

Chiswick   Greater London, England.

51 29N 0 15W. District of W London on N bank of R. Thames.

Chita Russian Soviet Federal Socialist Republic, U.S.S.R. 52 03N 113 30E. Town in S Siberia sit. on Chita R. near its confluence with Ingoda R. and on Trans-Siberian railway. Industries inc. railway engineering, tanning and flour milling. Pop. (1985E) 336,000.

Chitral Peshawar, Pakistan. District in N. Area, 5,700 sq. m. (14,750 sq. km.). Cap. Chitral, 180 m. (288 km.) NNW of Rawalpindi. Mountainous, rising to 25,260 ft. (7,699 metres) a.s.l. in Tirich Mir. There is small farming in the valleys. Pop. 115,000.

Chitré Herrera, Panama. 7 58N 80 26W. Town sit. SW of Panama City on a branch of the Inter-American Highway. Provincial cap. and commercial centre for an agric. area producing onions, tomatoes, lettuce, corn, rice, beans, livestock. Industries inc. distilling brandy; manu. soap, ice and beverages. Pop. (1980E) 14,635.

Chittagong Chittagong, Bangladesh. 22 20N 91 50E. City sit. SE of Dacca on Karnaphuli R. near the mouth on the Bay of Bengal. Industries inc. cotton ginning, flour, rice, oil and timber milling, chemicals, textiles and engineering; manu. soap, candles and bricks. Port trading in rice, oilseeds, jute, tea, sugar and tobacco. Pop. (1987E) 1·84m.

Chittaranjan West Bengal, India. Town sit. NW of Asansol. The main industry is building railway locomotives. Pop. (1981C) 52,443.

Chocó Colombia. (i) Bay on the Pacific coast between Point Chirambira and Point Guascama. (ii) Dept. on the Pacific coast. Cap. Quibdó. Pop. (1983C) 253,500.

Choiseul Solomon Islands, South West Pacific. 7 00S 157 00E. Island sit. SE of Bougainville. Area, 980 sq. m. (2,538 sq. km.). The main product is copra.

Cholet Maine-et-Loire, France. 47 04N 0 53W. Town in NW, sit. SW of

Angers. Industries inc. textiles; manu. footwear and clothing. Pop. (1982C) 56,528.

Cholo Southern Province, Malawi. Town sit. SSE of Blantyre. Provincial administration centre and centre of a tobacco growing area; also produces tea, soybeans, tung, sisal, maize, rice.

Cholon Saigon, Vietnam. 10 46N 106 40E. Town sit. SW of Saigon and administered as part of it. Industries inc. mainly rice milling, also sawmilling, tanning and pottery. R. port trading in fish and rice.

Choluteca Choluteca, Honduras. 13 18N 87 12W. Cap. of dept. of same name sit. S of Tegucigalpa on Choluteca R. and Inter-American Highway. Market town for a farming area. Pop. (1980E) 49,000.

Chomutov Bohemia, Czechoslovakia. 50 28N 13 25E. Town sit. NW of Prague in a coalmining area. Industries inc. steel and paper making; manu. glass. Pop. (1984E) 57,398.

Chongjin Hamgyong, North Korea. 41 47N 129 50E. City sit. SW of Vladivostok, U.S.S.R., on E coast. Provincial cap., principal port and industrial centre. Industries inc. manu. matches, staple-fibre; steel, pig-iron, fish processing. Port exports timber, iron, and cotton textiles. There is a marine experimental station. Since 1943 Chongjin has inc. the town of Naman. Pop. (1981E) 265,000.

Chongqing (Chungking) Sichuan, China. 31 08N 104 23E. City on Yangtse R. at its confluence with Kialing R. Trading centre for tung oil, hog bristles and tea. Industries inc. iron and steel, textiles and chemicals; manu. paper and matches. Pop. (1987E) 2·83m.

Chonju Cholla North, South Korea. 35 49N 127 08E. Town sit. S of Seoul. Industries inc. rice milling, textiles and paper making.

Chontales Nicaragua. Dept. of S Nicaragua along E shore of L. Nicaragua. Chief cities Juigalpa (cap.), La Libertad, Santo Domingo. Hilly in N with the Cordillera Amerrique and Huapi Mountains. Main occupations agric., esp. sugar-cane, beans, maize, coffee, bananas; livestock farming; lumbering; mining, esp. gold and silver at La Libertad and Santo Domingo. Chief· shipping ports on L. Nicaragua, Puerto Diaz, San Ubaldo. Pop. (1981c) 98,462.

Chorzów   Katowice, Poland. 50 19N 18 57E. Town sit. NW of Katowice in Upper Silesia coal and iron district. Industries inc. iron and steel, railway engineering, chemicals and glass making. Pop. (1972E) 153,000.

Choumen   *see* **Shumen**

Chowkowtien   Hebei, China. Village sit. SW of Beijing where bones of *Sinanthropus pekinensis* (Peking Man) were found in 1929.

Christchurch   Dorset, England. 50 44N 1 45W. Resort immediately ENE of Bournemouth sit. on Christchurch Harbour at confluence of R. Avon and Stour. Pop. (1983E) 40,300.

Christchurch   South Island, New Zealand. 43 32S 172 38E. City in E Canterbury Plains. Industries inc. meat processing, woollen mills, tanning, canning and chemicals; manu. fertilizers and footwear; railway workshops. Pop. (1986c) 299,373.

Christianshaab   Greenland. Settlement on Disko Bay, W Greenland, and base for hunting and fishing. The only industry is refining seal-oil.

Christiansted   U.S. Virgin Islands, West Indies. 17 45N 64 42W. Town on N shore of St. Croix Island. Chief town and port, and once cap. of Danish West Indies. Pop. (1980c) 2,856.

Christmas Island   Indian Ocean. 10 25S 105 39E. An Australian depend. 200 m. (320 km.) S of W Java. Area, 52 sq. m. (135 sq. km.). Plateau rising to 1,000 ft. (305 metres) a.s.l. The chief occupations are fishing and phosphate mining. Pop. (1980E) 2,700.

Christmas Island   (Kiritimati) Kiri-

bati, Pacific Ocean. 1 58N 157 27W. Island in the Line Islands group in the Central Pacific. Area, 223 sq. m. (578 sq. km.). Chief product is copra. Pop. (1980E) 1,265.

Chu River   U.S.S.R. R. rising in the SE Kazakh desert and flowing 600 m. (960 km.) E to enter L. Issyk Kul in the Kirghiz Soviet Socialist Rep. Used for power and irrigation, esp. for cotton crops.

Chubut   Patagonia, Argentina. 43 15S 65 00W. Territory of Patagonia. Area, 87,152 sq. m. (225,724 sq. km.). Chief towns Rawson (cap.), Port Madrya. Cool, dry climate. Mountainous in the W, otherwise irrigated steppe where sheep and grain farming are the main occupations. There is an important oilfield at Comodoro Rivadavia. The territory was originally a Welsh settlement. Pop. (1980c) 283,116.

Chudskoye, Lake   *see* **Peipus, Lake**

Chukot   Russian Soviet Federal Socialist Republic, U.S.S.R. 68 00N 175 00W. National Area in extreme NE Siberia. Area, 275,000 sq. m. (712,000 sq. km.). Cap. Anadyr. Mainly tundra. The chief occupations are reindeer farming, hunting for furs and fishing.

Chungking   *see* **Chongqing**

Chuquisaca   Bolivia. Dept. of S Bolivia. Area, 33,148 sq. m. (85,853 sq. km.). Cap. Sucre. Pop. (1982c) 435,406.

Chur   Graubünden, Switzerland. 46 51N 9 32E. Town in E sit. on Plessur R. Commercial centre trading in wine. Industries inc. tourism. Pop. (1985E) 33,000.

Churchill   Manitoba, Canada. 58 46N 94 10W. Town sit. at mouth of Churchill R. on W shore of Hudson Bay. Port and railway terminus trading in grain. Pop. (1986E) 1,003.

Churchill River   Canada. 58 47N 94 12W. (i) Rises in L. Methy in NW Saskatchewan and flows 925 m.

(1,480 km.) E through various L. to enter Hudson Bay at Churchill. Used for hydroelectric power. (ii) Rises in L. Ashuanipi in Labrador, near the Quebec border, and flows 560 m. (896 km.) N then SE to enter L. Melville at Goose Bay and thence to the Atlantic. Hydroelectric power is developed at Churchill Falls below Lobstick L., where R. descends over 800 ft. (244 metres) in 10 m. (16 km.) inc. one 245 ft. (75 metres) drop. Formerly named the Hamilton R., and Hamilton Falls.

Church Stretton   Shropshire, England. Town sit. S of Shrewsbury. Market town for a farming area. There is some light industry. Pop. (1981C) 3,781.

Chusan Archipelago   China. Islands off Hangchow Bay in E. China Sea. The largest island is Chusan, 230 sq. m. (596 sq. km.). Rich fishing grounds, but dangerous to navigation.

Chuvash   Russian Soviet Federal Socialist Republic, U.S.S.R. 55 30N 48 00E. Autonomous Rep. in Volga valley bounded N by R. Volga and S by Sura R. Area, 7,064 sq. m. (18,296 sq. km.). Cap. Cheboksary, sit. 360 m. (576 km.) E of Moscow. The main occupations are farming and lumbering, with related industries; also engineering, chemicals and textiles. Pop. (1984E) 1,314,000.

Cicero   Illinois, U.S.A. 41 51N 87 45W. Town immediately W of Chicago. Industries inc. manu. radio, telephone and electrical equipment, metal products and building materials. Pop. (1980C) 61,232.

Ciénaga   Magdalena, Colombia. 11 01N 74 15W. Town sit. on Caribbean coast. Sea port exporting bananas and cotton. Pop. (1980E) 75,000.

Cienfuegos   Cuba. 22 09N 80 27W. Town sit. ESE of Havana on S coast, on Cienfuegos Bay. Industries inc. sugar refining and coffee processing; manu. cigars and soap. Sea port handling sugar, tobacco, coffee and molasses. Pop. (1981E) 235,293.

Cieszyn   Katowice, Poland. 49 45N 18 38E. Town sit. SW of Katowice on

the Czech. border in a coal and iron district. Pop. (1981E) 30,000.

Cimarron River   U.S.A. 36 10N 96 17W. Rises in NE New Mexico and flows 650 m. (1,040 km.) E through Oklahoma and SE Colorado, briefly into SW Kansas, and re-enters Oklahoma to join Arkansas R. above Tulsa.

Cincinnati   Ohio, U.S.A. 39 06N 84 31W. City in extreme SW Ohio sit. on Kentucky border and Ohio R. Route and important inland port. Industries inc. food processing and chemicals; manu. printing, jet engines, machine tools, fabricated metal products, transport equipment and plastics. Pop. (1987E) 1,714,000.

Cinderford   Gloucestershire, England. Town sit. WSW of Gloucester. Industries inc. light engineering, forestry and farming. There are some free coal mines but the main collieries closed in the 1960s. Pop. (1988E) 7,674.

Cirencester   Gloucestershire, England. 51 44N 1 59W. Town sit. on Churn R. Market town for a rural area. Industries inc. engineering. Pop. (1989E) 17,000.

Ciskei   33 00S 27 00E. An independent homeland granted independence by the Rep. of South Africa in 1981. Cap. Bisho. Area, 3,205 sq. m. (8,300 sq. km.). Industries inc. agric. and manu. textiles, wood and leather goods, metal products. Pop. (1987E) *de jure* 2m., *de facto* 1m.

Città di Castello   Umbria, Italy. 43 27N 12 14E. Town sit. E of Arezzo on Tiber R. Industries inc. agric. engineering; manu. cement.

Città Vecchia   Malta. Town sit. W of Valletta. Industries inc. tourism.

Ciudad Bolívar   Bolívar, Venezuela. 8 08N 63 33W. State cap. sit. on Orinoco R. Industries inc. food processing. R. port, accessible to ocean-going vessels, exporting cattle, hides, balata gum, chicle and gold. Pop. (1980E) 115,000.

Ciudad Real   Spain. 38 59N 3 56W.

(i) Province in s centre. Area, 7,622 sq. m. (19,741 sq. km.). The main occupations are farming and mining. Pop. (1986E) 477,967. (ii) Cap. of (i) and market town trading in cereals, olive oil and wine. Industries inc. flour milling, brandy distilling and textiles. Pop. (1981E) 51,000.

Ciudad Trujillo *see* **Santo Domingo**

Ciudad Victoria Tamaulipas, Mexico. 23 44N 99 08W. State cap. sit. NW of Tampico at E end of the Sierra Madre Oriental at 1,100 ft. (335 metres) a.s.l. Centre of an agric. and mining area. Industries inc. textiles and tanning. Pop. (1980C) 153,206.

Civitavecchia Lazio, Italy. 42 06N 11 48E. Town in central Italy sit. on w coast. Industries inc. fishing; manu. cement and calcium carbide.

Clackmannan Central Region, Scotland. 56 06N 3 46W. Town sit. ESE of Alloa near R. Forth and sit. in a coalmining area. Pop. (1981C) 3,597.

Clackmannanshire Scotland. Former county now in Central Region.

Clacton-on-Sea Essex, England. 51 48N 1 09E. Town in SE Eng. sit. on North Sea coast. The main industry is tourism. Pop. (1981C) 43,000.

Clamart Hauts-de-Seine, France. 48 48N 2 16E. Residential suburb of sw Paris. Pop. (1982C) 48,678.

Clapham Greater London, England. Residential district of sw London, and an important railway junction.

Clare Suffolk, England. Town sit. ssw of Bury St. Edmunds on Stour R. Market town for an agric. area. Pop. (1990E) 2,070.

Clare Ireland. County in w between Galway Bay and R. Shannon Estuary. Area, 1,231 sq. m. (3,188 sq. km.). Chief towns Ennis (county town), Shannon, and Kilrush. Lowlying, drained by R. Shannon and Fergus. The main occupation is farming; other industries inc. tourism, esp. for salmon fishing, industrial diamonds and chemicals. Chief products are oats, potatoes and dairy products. Pop. (1988E) 91,344.

Claremont New Hampshire, U.S.A. 42 23N 72 20W. City sit. s of Lebanon on Sugar R. near its confluence with Connecticut R. Industries inc. manu. mining and milling machinery, textiles, shoes, paper. Resort, esp. for winter sports. Pop. (1988E) 13,901.

Clarence Tasmania, Australia. City sit. on the E shore of the R. Derwent adjacent to Hobart City. Primarily a dormitory residential area of Hobart with commercial and industrial areas. Pop. (1989E) 48,500.

Clarendon Jamaica. Parish of s central Jamaica. Area, 468 sq. m. (1,212 sq. km.). Chief town May Pen. Mountainous in the N, irrigated plain in the s. Main products sugar-cane, bananas, citrus fruit, ginger, sisal, yams, breadfruit, cacao, coffee, honey, livestock, copper. There are thermal springs at Milk R. Spa. Pop. (1980E) 200,000.

Clarksburg West Virginia, U.S.A. 39 17N 80 21W. City in N West Virginia on West Fork R. in a coal and natural gas area, and ssw of Pittsburgh, Pennsylvania. Industries inc. aircraft engines, carbon products and support services for gas, oil and mining industries. Pop. (1980C) 22,371.

Clausthall-Zellerfeld Lower Saxony, Federal Republic of Germany. Town in upper Harz Mountains. The main industry is mining, esp. iron, lead, copper, silver, zinc. There is also a 200-year old technical university.

Clay Cross Derbyshire, England. 53 11N 1 26W. Town sit. s of Chesterfield. Industries inc. iron founding, engineering and brick-making.

Clear, Cape Cork, Ireland. 51 26N 9 30W. Headland, with lighthouse, at sw tip of Clear Island, off sw coast of County Cork. The most s point in Ireland.

Clearwater Florida, U.S.A. 27 58N 82 48W. City sit. w of Tampa. Resort for holidays and retirement, on Gulf of Mexico. Manu. electronic and medical products. Pop. (1988E) 99,866.

Cleethorpes Humberside, England. 53 34N 0 02W. Resort sit. on Humber R. estuary. Pop. (1987E) 35,540.

Clent Hills England. Range sw of Birmingham.

Clerkenwell Greater London, England. District of E London sit. N of Thames R.

Clermont-Ferrand Puy-de Dôme, France. 45 47N 3 05E. Cap. of dept. in s centre and in Massif Central. Industries inc. rubber and chemicals; manu. clothing and footwear. Pop. (1982C) 151,092 (agglomeration, 256,189).

Clevedon Avon, England. 51 26N 2 52W. Resort sit. w of Bristol on Bristol Channel. Pop. (1989E) 22,020.

Cleveland England. 54 35N 1 08E. County in NE Eng. bounded E by North Sea. Area, 225 sq. m. (583 sq. km.). Under 1974 reorganization its districts are: Hartlepool, Stockton-on-Tees, Middlesbrough and Langbaurgh. (Langbaurgh is an old Danish name). Rural and industrial, esp. petro-chemicals, engineering, iron and steel. North Sea oil is a dominant industry. Pop. (1983E) 565,000.

Cleveland Ohio, U.S.A. 41 30N 81 41W. City on s shore of L. Erie sit. at mouth of Cuyahoga R. The main industry is steel, based on local coal and iron ore, also oil refining, meat packing and chemicals; manu. cement, electrical appliances and vehicles. L. port, handling oil and steel. Pop. (1980C) 573,822 (metropolitan area, 1,898,325).

Cleveland Heights Ohio, U.S.A. 41 30N 81 34W. City immediately NE of Cleveland, of which it forms a residential satellite. Pop. (1980C) 56,438.

Cleves North Rhine-Westphalia, Federal Republic of Germany. Town sit. ESE of Nijmegen, Netherlands. Industries inc. manu. foodstuffs and footwear.

Clichy Hauts-de-Seine, France. 48 54N 2 18E. Suburb of NW Paris. Industries inc. oil refining, chemicals and plastics; manu. aircraft and vehicle parts, and electrical equipment. Pop. (1982C) 47,000.

Clifton New Jersey, U.S.A. 40 53N 74 08W. City sit. 6 m. (10 km.) NNW of Newark. Industries inc. textiles and chemicals; manu. aircraft parts, clothing, machine tools. Pop. (1980C) 74,788.

Climax Colorado, U.S.A. Village NNE of Leadville in Rocky Mountains at 11,300 ft. (3,444 metres) a.s.l. with road and railway running through Fremont Pass. It has a mine which is a major source of molybdenum. High-altitude observatory.

Clinton Iowa, U.S.A. 41 51N 90 12W. Town sit. ESE of Cedar Rapids. Industries inc. manu. machine tools, packaging, pet food and wire products. Pop. (1980C) 32,828.

Clipperton French Polynesia, Pacific Ocean. 10 17N 109 13W. Atoll sit. SW of Mexico, a coral ring 2.5 m. (4 km.) in diameter enclosing volcanic rock. Once a valuable source of phosphates. Uninhabited.

Clitheroe Lancashire, England. 53 53N 2 23W. Town sit. NW of Burnley on R. Ribble. Industries inc. cement, chemicals and animal feed manu. Pop. (1981C) 13,671.

Clonmacnoise Offaly, Ireland. Village sit. E of Ballinasloe, Galway, on Shannon R.

Clonmel Tipperary, Ireland. 52 21N 7 42W. Town sit. on Suir R. Industries inc. tourism, food processing, fibre board, computers, clothing, pharmaceuticals and furniture. Pop. (1981C) 16,000.

Clovelly Devon, England. 51 00N 4 24W. Fishing village and resort on N coast of Devon, W of Bideford, noted for its steep stepped streets. Pop. (1989E) 500.

Clovis New Mexico, U.S.A. 34 24N 103 12W. City sit. NE of Roswell on

Llano Estacado near Texas border. Important trading centre for a farming area. Railway junction. Industries inc. distributing livestock feed, railway repair shops, flour milling; manu. dairy products and beverages. The E New Mexico State Park is in the neighbourhood. Pop. (1989E) 32,767.

Cluj   Cluj, Romania. 46 47N 23 36E. Cap. of district NW of Bucharest sit. on Little Somes R. Industries inc. textiles and chemicals; manu. metal goods and hardware. Pop. (1983E) 270,820.

Cluny   Saône-et-Loire, France. 46 26N 4 39E. Town sit. in the E, NW of Mâcon sit. on Grosne R. and noted as the site of the Benedictine Abbey. Pop. (1982E) 4,734.

Clutha River   South Island, New Zealand. 46 20S 169 49E. Rises in L. Wanaka and L. Hawea and flows 150 m. (240 km.) SE through Otago, to enter S. Pacific by a delta near Kaitangata, 45 m. (72 km.) SW of Dunedin. Navigable up to Roxburgh, where it is dammed for hydroelectric power.

Clwyd   Wales. 53 00N 3 15W. County in N Wales bounded N by Irish Sea and E by Cheshire, England. Created under reorganization, 1974. Its districts are Colwyn, Rhuddlan, Delyn, Alyn-Deeside, Glyndwr and Wrexham Maelor. Hilly and fertile, drained by Rs. Dee and Clwyd. Along the coast tourism is the main industry and the industrial area is centred on Deeside and the Wrexham area. There is 1 coal pit in production and a reduced steel industry at Shotton and Brymbo. Other industries inc. clothing, plastics, microprocessors, chemicals, light engineering, paper and aircraft. The rest of the county depends on the agricultural industry. Pop. (1989E) 402,800.

Clydebank   Strathclyde Region, Scotland. 55 54N 4 24W. Town sit. WNW of Glasgow on Clyde R. The main industries are light and heavy engineering. Pop. (1987E) 30,060.

Clyde River   Scotland. 55 56N 4 29W. Rises in S Lanarkshire as the Daer Water and flows 106 m. (170 km.) N and NW past Lanark, Hamilton, Glasgow and Clydebank, widening into the Firth of Clyde at Dumbarton. At the Falls of Clyde, Lanark, it descends 230 ft. (70 metres) in under 4 m. (6 km.). Passing through the main industrial region of Scot. it serves 20 m. (32 km.) of shipyards below Glasgow.

Coahuila   Mexico. 27 00N 112 30W. State in NE Mexico bounded N by Texas, U.S.A. Cap. Saltillo. Mainly plateau crossed by the Sierra Madre Oriental, sloping to the valley of the Río Grande R. to the N. Cotton is grown in the W. The main occupations are farming, and mining for coal, copper, silver and lead. Pop. (1980C) 1,557,265.

Coalville   Leicestershire, England. 52 44N 1 20W. Town sit. NW of Leicester. Industries inc. some coal-mining, quarrying, engineering and brickmaking. Pop. (1987E) 30,592 inc. Turingstone and Whitwick.

Coatbridge   Strathclyde Region, Scotland. 55 52N 4 01W. Town sit. E of Glasgow. Industries inc. steel, heavy and precision engineering, food processing and pharmaceuticals. Pop. (1986E) 46,136.

Coatzacoalcos   Veracruz, Mexico. 18 09N 94 25W. Town sit. near mouth of Coatzacoalcos R. on S shore of the Gulf of Campeche. Port exporting petroleum products. Pop. (1980E) 186,129.

Cobán   Alta Verapaz, Guatemala. 15 29N 90 19W. Town sit. N of Guatemala. Market town for a coffee-growing area. Pop. (1989E) 120,000.

Cóbh   Cork, Ireland. 51 51N 8 17W. Town on S shore of Great Island, Cork Harbour. Resort and sea port. Pop. (1986E) 8,282.

Cobija   Pando, Bolivia. 11 02S 68 44W. Town sit. NNW of La Paz opposite Brasiléia, Brazil, on Acre R. Major trading centre for rubber. R. port, airport and border customs point. Pop. (1984E) 4,523.

Coblenz Rhineland-Palatinate, Federal Republic of Germany. 50 21N 7 36E. City in central w sit. at confluence of R. Rhine and Moselle. Trading centre for wine. Industries inc. manu. paper, pianos, furniture and footwear. Pop. (1987E) 108,200.

Coburg Bavaria, Federal Republic of Germany. 50 15N 10 58E. Town near the s border of German Democratic Republic sit. on Itz R. Industries inc. porcelain, toys and machinery. Pop. (1989E) 44,000.

Cochabamba Cochabamba, Bolivia. 17 24S 66 09W. Cap. of dept. in w centre sit. in the E Andes. Centre of an extensive agric. area. Industries inc. oil refining; manu. furniture and footwear. Pop. (1982E) 281,962.

Cochin Kerala, India. 9 58N 76 15E. Town in sw sit. on the Malabar coast. Industries inc. shipbuilding and engineering and manu. coir products and plywood. Port handling copra, coir, tea, rice and petroleum products. Pop. (1981C) 551,567.

Cockenzie and Port Seton Lothian Region, Scotland. Town on s shore of Firth of Forth comprising neighbouring fishing ports. Pop. (1984E) 3,600.

Cockermouth Cumbria, England. 54 40N 3 21W. Town sit. sw of Carlisle, at confluence of Cocker and Derwent Rs. Industries inc. manu. footwear and gland packing. Pop. (1985E) 7,000.

Coclé Panama. Province of central Panama. 8 30N 80 15W. Area, 1,470 sq. m. (3,807 sq. km.). Cap. Penonomé. A cattle-rearing area. Pop. (1980C) 140,320.

Coco River Honduras/Nicaragua. 15 00N 83 08W. Rises in sw Honduras and flows 400 m. (640 km.) ENE into Nicaragua and then along the Honduras/Nicaragua border, to enter the Caribbean by a delta at Cabo Gracias á Dios. It is used for logging.

Cocos Island Pacific Ocean. 5 32N 87 04W. Island sit. sw of the Osa Peninsula, Costa Rica. Area, 9 sq. m. (23 sq. km.). Noted as the legendary site of buried treasures. Uninhabited.

Cocos (Keeling) Islands Indian Ocean. 12 10S 96 55E. Group of 27 islands sit. NW of Perth, Australia, and forming a Territory of Australia. Area, 5 sq. m. (13 sq. km.). The main islands are West, Home and Direction. West Island has an aircraft refuelling station. Low-lying, with many coconut groves. The chief products are coconuts and coconut oil, and copra. Pop. (1984E) 584.

Cod, Cape Massachusetts, U.S.A. 41 41N 70 15W. Peninsula extending 65 m. (104 km.) NE along seaward side of Cape Cod Bay on the Atlantic coast. Low and sandy, with many fishing villages and resorts; it was the landfall of the Pilgrim Fathers in 1620.

Coëtivy Seychelles, Indian Ocean. 7 08S 56 16E. Island sit. NE of Madagascar forming an outlying depend. of the Seychelles. Area, 7 m. (11 km.) by 2 m. (3 km.) with some settlement on the w coast. The main products are copra, fish. Transferred from Mauritius to the Seychelles in 1907.

Coëtquidan Morbihan, France. Military camp near Guer village, sit. E of Ploërmel. Site of the noted military academy formerly at St. Cyr.

Coffin Bay South Australia, Australia. Inlet of the Great Australian Bight between mainland and westward projection of sw Eyre Peninsula.

Cognac Charente, France. 45 42N 0 20W. Town in central w sit. on Charente R. The main industry is brandy distilling, together with manu. bottles, corks, crates, barrels, etc. Pop. (1982C) 20,995.

Coimbatore Tamil Nadu, India. 11 02N 76 59E. Town sit. sw of Madras on Noyil R. in the Nilgiri Hills. Industries inc. rice and flour milling, and textiles; manu. fertilizers. Pop. (1981C) 704,514.

Coimbra Coimbra, Portugal. 40 15N 8 27W. (i) District, drained by the Mondego R. inc. part of Beira Litoral province. Pop. (1987E) 446,500. (ii) Cap. of district of the same name in central w sit. on right bank of Modego R. Market and university town trading in wine, grain and olives. Industries inc. manu. pottery. Pop. (1987E) 74,616.

Cojedes   Venezuela. 9 20N 68 20W. State in N Venezuela in the 'Ilanos' of the Orinoco. Area, 5,700 sq. m. (14,763 sq. km.). Cap. San Carlos. Pop. (1980E) 116,784.

Colac   Victoria, Australia. 38 20S 143 35E. City sit. W of Melbourne. Railway junction. Industries inc. food processing and timber. Pop. (1984E) 10,110.

Colchester   Essex, England. 51 54N 0 54E. Town in SE sit. on Colne R. Market town for an agric. and horticultural area. Industries inc. computers, engineering and printing. Pop. (1989E) 87,500.

Coldstream   Borders Region, Scotland. 55 39N 2 15W. Town sit. SW of Berwick-on-Tweed in an agric. area. Pop. (1984E) 1,677.

Coleraine   Londonderry, Northern Ireland. 55 08N 6 40W. University town sit. near N coast and on Bann R. Industries inc. tourism, food processing, electronic components, engineering and manu. shirts and linen. Pop. (1989E) 19,000.

Colima   Mexico. 19 14N 103 43W. (i) State in SW Mex. bounded SW by the Pacific. Area, 1,966 sq. m. (5,191 sq. km.). Mainly agric. Pop. (1980C) 346,293. (ii) Cap. of (i) sit. 312 m. (499 km.) W of Mexico City on Colima R. Market town. Industries inc. processing maize, rice and sugarcane; manu. cigars and shoes. Pop. (1984E) 58,000. (iii) Volcano in Jalisco state in the Nevado de Colima, 14,000 ft. (4,267 metres) a.s.l.

College Station   Texas, U.S.A. 30 37N 96 21W. City sit. S of Bryan near Brazos R. The Agric. and Mechanical University of Texas is sit. here. Pop. (1980C) 37,272.

Collie   Western Australia, Australia. 33 21S 116 09E. Town sit. SE of Perth. Industries inc. coalmining, timber, agric., electricity and alumina

refinery. There is a satellite town, Collie Cardiff. Pop. (1985E) 10,000.

Colmar   Haut-Rhin, France. 48 05N 7 22E. Town in extreme central E, and sit. on the plain E of the Vosges Mountains. Trading centre for Alsatian wine. Industries inc. brewing, pottery, porcelain, electronics, flour milling and textiles; manu. starch. Pop. (1982C) 63,764 (agglomeration, 82,468).

Colne   Lancashire, England. 53 52N 2 09W. Town sit. NNE of Burnley. Industries inc. tanning and textiles, esp. cotton, wallcoverings, precision engineering and textile finishing. Pop. (1981C) 18,150.

Colne River   England. (i) Rises in NW Essex and flows 35 m. (56 km.) SE past Colchester to enter North Sea at Mersea Island. (ii) Rises near Hatfield, Hertfordshire, and flows 35 m. (56 km.) SW past Uxbridge to join Thames R. at Staines.

Colôane Island   Macao, South China Sea. 22 08N 113 33E. Island sit. SSE of the Macao peninsula. Area, 2 sq. m. (5 sq. km.). Chief town, Colôane. The main occupation is fishing.

Cologne   North Rhine-Westphalia, Federal Republic of Germany. 50 56N 6 59E. City in W sit. on Rhine R. Commercial and industrial centre. Industries inc. manu. motor vehicles, diesel engines, cables, petro-chemicals, mechanical and electrical engineering, pharmaceuticals, handicrafts and insurance. Pop. (1987E) 927,500.

Colomb-Béchar   *see* **Béchar**

Colombes   Hauts-de-Seine, France. 48 55N 2 15E. Suburb of NW Paris. Industries inc. engineering and chemicals; manu. hosiery. Pop. (1982C) 78,783.

Colombia   3 45N 73 00W. Rep. bounded N by Caribbean Sea, NW by Panama, W by Pacific Ocean, SW by Ecuador and Peru, NE by Venezuela and SE by Brazil. Area, 440,829 sq. m. (1,141,748 sq. km.). Chief towns Bogotá (the cap.), Barranquilla, Cartagena, Buenaventura, Medellin and

Cali. About half the country is sparsely inhabited tropical lowland. Mountainous in the W with parallel Andean ranges, the Cordillera Occidental and Cordillera Central, extending SW to NE and rising to Mount Huila, in the Cordillera Central, at 18,865 ft. (5,750 metres) a.s.l. Inland is the Cordillera Oriental, also parallel, and in the extreme N are isolated ranges, with Cristóbal Colón near the coast. There is great variation in altitude, temp. and vegetation. The *tierra caliente* lies below 3,000 ft. (914 metres) a.s.l. and produces tropical crops; the *tierra templada* lies at 3,000–6,500 ft. (914–1,981 metres) a.s.l. and produces maize, sugar and coffee; the *tierra fria* lies at 6,500–10,000 ft. (1,981–3,048 metres) a.s.l. and produces cereals, potatoes and fruit; the *paramos* lie above that up to the snowline and are mainly pasture. The main occupation is farming, esp. food crops and coffee. Main products are petroleum from the NW, bananas, timber and platinum. Pop. (1985C) 29,481,852.

Colombo Western Province, Sri Lanka. 6 55N 79 52E. Cap. of rep. sit. on W coast S of Kelani R. Founded by Arab traders and captured by the Portuguese in 1517. The Dutch occupied it in 1658 and the British in 1796. It is a commercial centre and a major sea port exporting tea, rubber, coconut oil and copra. Pop. (1981C) 587,647.

Colón Honduras. Dept. in NE. Area, 17,104 sq. m. (44,299 sq. km.). Cap. Trujillo. Hilly in the W, fertile plain in the E. Main occupation farming. Pop. (1983E) 128,370.

Cólon Cólon, Panama. 9 22N 79 54W. Cap. of province sit. on Manzanillo Island at the Caribbean end of the Panama Canal. Industries associated with the canal. Pop. (1980C) 59,043.

Colón Archipelago Ecuador *see* **Galapagos Islands**

Colonia Uruguay. 34 29S 57 48W. (i) Dept. in SW Uruguay between Uruguay and Plata Rs. Area, 2,194 sq. m. (5,682 sq. km.). Mainly agric., with important production of cereals, flax, grapes and dairy produce. Pop.

(1985C) 112,100. (ii) Town opposite Buenos Aires, Argentina, on Rio de la Plata. Dept. cap., port and trading centre for agric. produce. Pop. (1985C) 19,077.

Colorado U.S.A. 34 40N 106 00W. State in W bounded E by Kansas and W by Utah, N by Wyoming and S by New Mexico. Area, 104,247 sq. m. (270,000 sq. km.). Chief towns Denver (the cap.), Lakewood, and Colorado Springs. Settled from 1858 and became a State in 1876. Mountainous at an average height of 6,800 ft. (2,073 metres) a.s.l. in the Rocky Mountains and rising to Mount Elbert, 14,430 ft. (4,398 metres) a.s.l. The main occupations are trade and services, stock rearing, farming, esp. by irrigation, and manufacturing. Main products are alfalfa, wheat, sugarbeet; coal, petroleum, molybdenum, and precious minerals, esp. uranium. Manu. inc. machinery, food products, printed goods. Pop. (1988E) 3,301,458.

Colorado Desert California, U.S.A. 33 15N 115 15W. Area in SE California bounded E by Colorado R. Contains the Salton Sea in a depression 249 ft. (76 metres) b.s.l.

Colorado River Argentina. 39 50S 62 08W. Formed on borders of Mendoza and Neuquén provinces by confluence of R. Grande and R. Barrancas, and flows 530 m. (848 km.) SE to the Atlantic below Pedro Luro, directly S of Bahía Blanca.

Colorado River U.S.A. 31 45N 114 40W. Rises in Rocky Mountains National Park, N Colorado, and flows 1,440 m. (2,304 km.) SW through Utah and Arizona where it bends W through the Grand Canyon, then S along Arizona's border with Nevada and California, into Mexico above Mexicali and thence into the Gulf of California. The Grand Canyon gorge is 218 m. (349 km.) long and up to 15 m. (24 km.) wide. Used extensively for irrigation and power.

Colorado River Texas, U.S.A. 28 36N 95 58W. Rises on Llano Estacado in W Texas and flows 970 m. (1,552 km.) SE past Austin to enter the

Gulf of Mexico in Matagorda Bay. Used for irrigation and power.

Colorado Springs Colorado, U.S.A. Resort in centre of state and located at the foot of Pikes Peak. The main industry is tourism. Pop. (1989E) 300,000.

Columbia Missouri, U.S.A. 38 57N 92 20W. City sit. ESE of Kansas City. Industries inc. flour milling; manu. clothing and furniture. Pop. (1980C) 62,061.

Columbia South Carolina, U.S.A. 34 00N 81 03W. State cap. sit. on Congaree R. at the head of navigation. Industries inc. manu. nuclear fuels, chemicals, fibres and electronic goods. Pop. (1988E) 94,810.

Columbia, District of *see* **District of Columbia**

Columbia River Canada/U.S.A. 46 15N 124 05W. Rises in L. Columbia in the Rocky Mountains, British Columbia, and flows 1,150 m. (1,840 km.) first N round the Selkirk Range, then S to the U.S.A. border to Franklin D. Roosevelt L., Washington, further S to the Oregon border below Kennewick, and W along the border to enter the Pacific Ocean SW of Tacoma. Extensively used for power.

Columbus Georgia, U.S.A. 32 29N 84 59W. Town in extreme W of state sit. at the head of navigation on Chattahoochee R. Commercial centre. Industries inc. textiles; manu. bricks and tiles, fertilizers, agric. implements and food products. Pop. (1980C) 169,441.

Columbus Mississippi, U.S.A. 33 15N 89 56W. City near E state border on Tombigbee R. Pop. (1980C) 27,383.

Columbus Ohio, U.S.A. 39 57N 83 00W. State cap. sit. on Scioto R. Centre of government, services and distribution with some manu. industry. Pop. (1980C) 564,871.

Colwyn Bay Clwyd, Wales. 53 18N 3 43W. Town on N coast, ESE of Llandudno. Seaside resort. Pop. (1985E) 26,000.

Comayagua Honduras. 14 25N 87 37W. (i) Dept. of central Honduras. Area, 1,919 sq. m. (4,970 sq. km.). Fertile, watered by Humuya R. Main product cereals. Pop. (1983C) 211,465. (ii) Town on Humuya R. Dept. cap. noted for its cathedral and law school. Pop. (1986E) 28,800.

Comilla Chittagong, Bangladesh. 23 28N 91 10E. Cap. of district sit. ESE of Dacca just S of Gumti R. Trading centre for rice, jute, oilseeds, cane and tobacco. Industries inc. engineering; manu. metal goods and soap. Pop. (1984E) 150,000.

Comino Malta. 36 00N 14 20E. Islet between Malta (SE) and Gozo (NW) and NW of Valletta. Area, 1 sq. m. (2.6 sq. km.). The main occupations are farming and fishing, the main products beeswax, honey, grapes, goats, sheep and fish.

Commenijne River Suriname, South America. Rises 20 m. (32 km.) N of Dam and flows 100 m. (160 km.) N and W through tropical forest and alluvial lowland to join the Suriname R. at Nieuw Amsterdam. Used in its lower course by ocean-going vessels to transport bauxite. There are gold deposits on the upper course.

Communism, Mount *see* **Kommunizma Peak**

Como Lombardia, Italy. 45 47N 9 05E. Town in extreme central N sit. at the SW end of L. Como. Industries inc. manu. textile machinery, motor cycles, silk, and furniture. L. port and tourist resort. Pop. (1989E) 94,634.

Como, Lake Lombardia, Italy. 46 00N 9 20E. L. in the N at 650 ft. (198 metres) a.s.l. Area, 55 sq. m. (142 sq. km.). *c.* 30 m. (48 km.) long and 2.5 m. (4 km.) wide; on Adda R. which feeds it from the N and drains it to the SE. The promontory of Bellagio in the S divides it into L. Como, the W arm, and the L. of Lecco, the E arm.

Comodoro Rivadavia Argentina. 45 50S 67 40W. Town sit. on SW coast and on Gulf of San Jorge. The main industry is oil extraction and refining. There is a natural gas pipeline to Buenos Aires. Pop. (1984E) 120,000.

Comoros Indian Ocean. 12 10S 44 15E. Republic, comprising 3 islands, gaining independence from France in 1976 although Mayotte, the fourth of the islands, opted to become an overseas department of France. Between the coast of Mozambique and the island of Madagascar. The 3 islands of Njazídja, Nzwani and Mwali have an area, 719 sq. m. (1,862 sq. km.). They are volcanic, mountainous, forested and very fertile. The chief occupation is farming; chief products are vanilla, copra and essential oils. Pop. (1987E) 422,500.

Compiègne Oise, France. 49 25N 2 49E. Town sit. NNE of Paris on Oise R. Resort, with some industries inc. engineering and sawmilling. Pop. (1982C) 43,311.

Compton California, U.S.A. 33 54N 118 13W. City sit. S of Los Angeles. Industries inc. oil refining; manu. oil-drilling equipment, and steel and glass products. Pop. (1980C) 81,286.

Conakry Guinea. 9 30N 13 43W. Cap. of rep. sit. partly on the island of Tumbo off the Atlantic coast and partly on the adjacent mainland; connected to the mainland by a causeway. It was laid out by the French in 1889. Sea port handling iron ore, alumina and fruit. Pop. (1983C) 705,280.

Concarneau Finistère, France. 47 52N 3 56W. Town in NW, sit. on S coast of Brittany. Fishing port and resort. Pop. (1982C) 18,225.

Concepción Concepción, Chile. 36 50S 73 03W. Cap. of province in S central Chile sit. on Bio-Bio R. Commercial centre with important coalmines in the area. Industries inc. textiles and paper making; manu. cement, leather goods and glass. Pop. (1982E) 206,107.

Concepción Concepción, Paraguay. 23 25S 57 17W. Cap. of dept. in central Paraguay sit. on Paraguay R. Commercial centre trading in *yerba maté*, timber, hides and livestock. Industries inc. sawmilling, cotton ginning and flour milling. Pop. (1983C) 24,000.

Concepción del Uruguay Entre Rios, Argentina. 32 30S 58 14W. Town in central W sit. on Uruguay R. R. port handling grain and beef. Pop. (1980C) 40,000.

Concord California, U.S.A. 37 59N 122 02W. Residential town sit. NE of Oakland. Pop. (1980C) 103,255.

Concord Massachusetts, U.S.A. 42 28N 71 21W. Residential town sit. WNW of Boston on Concord R. Noted as the scene of the first battle of the American War of Independence, 1775. Pop. (1985E) 15,636.

Concord New Hampshire, U.S.A. 43 05N 71 30W. State cap. sit. on Merrimack R. Centre of government and services with some manu. Pop. (1988E) 37,024.

Concordia Entre Rios, Argentina. 31 20S 58 02W. Town sit. opposite Salto, Uruguay, on Uruguay R. Industries inc. manu. vegetable oils and leather. R. port exporting cereals and citrus fruits. Pop. (1980C) 93,618.

Coney Island New York, U.S.A. Seaside resort in SW of Long Island.

Congleton Cheshire, England. Town sit. ENE of Crewe, on Dane R. Industries inc. manu. textiles, ribbons, clothing, electrical goods, transport bodywork, shelving, billiard balls, pottery, soft drinks. Pop. (1989E) 26,000.

Congo 1 00S 16 00E. The Republic of the Congo became independent in 1960, after having been one of the 4 territories of French Equatorial Africa. Bounded N by Cameroon and the Central African Rep., E and S by Zaïre, SW by the Atlantic and W by Gabon. Area, 132,000 sq. m. (342,000 sq. km.). Chief towns Brazzaville (the cap.), and Pointe Noire. The climate is humid; mainly equatorial forest with some savannah. The main occupations are lumbering, farming and mining. Main products are crude oil, tropical hardwoods, palm oil and kernels, and lead. Pop. (1988E) 2,266,000.

Congo (Kinshasa) *see* **Zaïre**

Coniston Water   Cumbria, England. Lake W of L. Windermere at foot of the Old Man of Coniston, a mountain 2,633 ft. (810 metres) a.s.l. Length 5 m. (8 km.), width ½m. (0.8 km.). Used for water-speed trials.

Connaught   Ireland. 53 23N 8 40W. Province in the W bounded E by Shannon R. and W by Atlantic and comprising the counties of Galway, Mayo, Leitrim, Roscommon and Sligo. Area, 6,611 sq. m. (17,122 sq. km.). Pop. (1981C) 424,410.

Connecticut   U.S.A. 41 40N 72 40W. State in New England bounded S by Atlantic, E by Rhode Island, N by Massachusetts and W by New York. Area, 5,009 sq. m. (12,973 sq. km.). Chief towns Hartford (the cap.), Bridgeport, New Haven and Stamford. Settled from 1634 and became a Commonwealth in 1637 and a State in 1818. Mountainous in the NW rising to Bear Mountain, 2,322 ft. (708 metres) a.s.l. Well wooded. Agric. is mainly poultry and dairy farming, market gardening, and tobacco growing. Industries inc. transport equipment, machinery and metals. Pop. (1980C) 3,107,576.

Connemara   Ireland. 53 25N 9 40W. Boggy area in W Galway with mountains and L. The chief occupation is tourism, esp. for fishing.

Consett   County Durham, England. 54 51N 1 49W. Town sit. SW of Newcastle upon Tyne. Industries inc. engineering. Pop. (1981E) 31,000.

Constance   Baden-Württemberg, Federal Republic of Germany. 47 39N 9 10E. Town sit. on Rhine R. at its exit from L. Constance. Industries inc. computers, electronics, tourism, textiles, machinery and chemicals. Pop. (1989E) 70,000.

Constance, Lake   (Bodensee), Federal Republic of Germany/Austria/Switzerland. 47 35N 9 25E. L. extending 40 m. (64 km.) NW to SE along the frontier between Switzerland to the S and F.R.G. to the N, with Austria surrounding its E end. Area, 205 sq. m. (531 sq. km.). The Rhine R. enters in the SE and leaves in the W past Con-

stance. The greater part of it is called the Obersee; in the W it is divided by a peninsula into the Untersee and Überlingersee.

Constanta   Constanta, Romania. 44 10N 28 40E. Cap. of district sit. on W coast of Black Sea. Industries inc. textiles; manu. bricks, furniture and soap. Terminal of the oil pipeline from Ploești. Sea port handling grain, timber and oil. Resort. Pop. (1985E) 323,236.

Constantine   Constantine, Algeria. 36 22N 6 40E. Cap. of dept. in NE sit. on a plateau nearly enclosed by a gorge of the Rummel R. Trading centre for grain, wool and leather for the High Plateau. Industries inc. flour milling; manu. woollen and leather goods. Pop. (1983E) 448,578.

Constantinople   *see* **Istanbul**

Conwy   Gwynedd, Wales. 53 17N 3 50W. Market town and seaside resort on R. Conwy. Pop. (1981C) 12,969.

Conwy River   Wales. 53 17N 3 50W. Rises near Blaenau Ffestiniog and flows 30 m. (48 km.) N to enter Irish Sea at Aberconwy.

Cook Inlet   Alaska, U.S.A. 60 30N 152 00W. Inlet of the Gulf of Alaska, running *c.* 200 m. (320 km.) SW-NE beside the Kenai Peninsula. Anchorage (port) lies at the head.

Cook Islands   South Pacific Ocean. 20 00S 158 00W. Island groups forming a self governing territory in free association with New Zealand in Polynesia, sit. SE of Western Samoa. Total area, 113 m. (293 sq. km.). The Northern group inc. Manihiki, Pukapuka, Penrhyn and Rakahanga; the Southern group, Rarotonga, Mangaia, Aitutaki and Atiu. Cap. is Rarotonga. The chief products are copra, pearlshell and fruits, esp. bananas. Pop. (1986C) 17,185.

Cook, Mount   South Island, New Zealand. 43 46S 170 09E. Mountain peak in the Southern Alps, at 12,349 ft. (3,764 metres) a.s.l. the highest peak in New Zealand. Permanently snow-capped with extensive glaciers.

Cookstown Tyrone, Northern Ireland. 54 39N 6 45W. Town sit. W of Belfast on Ballinderry R. Industries inc. manu. cement, clothing, dairy and bacon products. Pop. (1985E) 8,400.

Cook Strait New Zealand. 41 14S 174 30E. Strait separating N. and S. Island. Width 16 m. (26 km.) at its narrowest.

Copán Honduras. (i) Dept. of W Honduras. Area, 1,430 sq. m. (3,704 sq. km.). Mainly mountainous. Pop. (1983E) 217,258. (ii) Town sit. near the ruins of an important Maya city. Pop. (1983E) 20,000.

Copenhagen Zealand, Denmark. 55 43N 12 34E. Cap. of Denmark sit. on E coast of Zealand and at S end of the Öresund, the main entrance to the Baltic. Founded in the 13th cent. it is a commercial centre. Industries inc. shipbuilding, engineering and brewing; manu. porcelain, textiles, paper and confectionery. Sea port exporting dairy products and bacon. Pop. (1989E) 467,850.

Copiapó Atacama, Chile. 27 22S 70 20W. Cap. of province in central Chile sit. on Copiapó R. Industries inc. copper and iron mining. Pop. (1982C) 70,241.

Coppermine Northwest Territories, Canada. 68 00N 116 00W. Settlement above mouth of Coppermine R. and NE of Great Bear L. The main occupations are hunting and whaling. Eskimo trading post.

Coppermine River Northwest Territories, Canada. 67 49N 115 04W. Rises in NE Mackenzie District, and flows 525 m. (840 km.) S into L. de Gras, then NW through the L. and on past Coppermine to enter Coronation Gulf, between Victoria Island and the mainland.

Coquilhatville *see* **Mbandaka**

Coquimbo Chile. 38 00S 71 20W. Market town trading in grain, fruit and copper. Winter station of the Chilean navy. Pop. (1980E) 78,000.

Coral Gables Florida, U.S.A. 25

45N 80 16W. City immediately SW of Miami, and a residential resort. Pop. (1980C) 43,241.

Coral Sea 15 00S 150 00E. An arm of the Pacific Ocean between Australia to the W and the Vanuatu to the E, containing the Great Barrier Reef.

Corby Northamptonshire, England. 52 29N 0 40W. Town sit. N of Kettering. Successful diverse industrial regeneration since closure of steelworks in 1980. Pop. (1988E) 51,500.

Cordillera Paraguay. Dept. in E Paraguay. Cap. Caacupé. Pop. (1982E) 194,826.

Córdoba Córdoba, Argentina. 31 25S 64 10W. Cap. of province in central Arg. sit. on Primero R. Commercial centre. Industries inc. motor engineering and textiles; manu. glass and cement which are based on hydroelectric power from the San Roque Dam. Pop. (1980C) 1m.

Córdoba Veracruz, Mexico. 18 53N 96 56W. Town sit. WSW of Veracruz in the Sierra Madre Oriental. Market town for a coffee growing area. Industries inc. coffee and sugar processing. Pop. (1980E) 72,000.

Córdoba Spain. 37 53N 4 46W. (i) Province in central S. Area, 5,296 sq. m. (13,717 sq. km.). Mountainous in the N descending to Guadalquivir R. and a fertile plain in the S. Mainly agric. The chief products are olives, fruit, cereals, silver, lead, coal and copper. Pop. (1986C) 745,175. (ii) Cap. of (i) sit. on Guadalquivir R. Industries inc. engineering, brewing, distilling and tourism; manu. textiles, pottery and leather. Noted for the Moorish influence of its architecture. Pop. (1986C) 304,826.

Corfe Castle Dorset, England. 50 38N 2 04W. Village sit. SSW of Poole. Tourist resort with castle ruins. Industries inc. oil, clay, electronics and stone polishing. Pop. (1985E) 1,500.

Corfu *see* **Kerkyra**

Corinth *see* **Korinthos**

Corinth   Canal   *see*   **Korinthos Canal**

Corinto   Chinandega, Nicaragua. 12 29N 87 12W. Town on Pacific coast. Railway terminus and port exporting coffee, timber, sugar, hides. Pop. (1980E) 12,000.

Corisco   Equatorial Guinea. Island in Gulf of Guinea 18 m. (29 km.) off mouth of Rio Muni. Area, 6 m. (15 km.) by 2 m. (3 km.). The main occupations are fishing and quarrying silica-sand.

Cork   Ireland. 51 54N 8 28W. (i) County in s bounded s by Atlantic. Area, 2,880 sq. m. (7,460 sq. km.). Chief towns Cork (county town), Cóbh, Mallow and Youghal. Mountain ranges cross w to E, rising to the Boggeragh Mountains, 2,000 ft. (610 metres) a.s.l. The valleys are fertile. The chief occupation is dairy farming. Pop. (1978E) 279,464. (ii) County town sit. at head of Cork Harbour. Industries inc. brewing, distilling, assembling motor vehicles and bacon curing; manu. tyres, fertilizers and woollen goods. Sea port exporting dairy produce and cattle. Pop. (1986E) 173,694.

Cornella   Barcelona, Spain. 41 21N 2 04E. Suburb of sw Barcelona. Industries inc. chemicals, cotton, aluminium; manu. electrical goods, motor car accessories, watches, pottery.

Corner Brook   Newfoundland, Canada. 48 57N 57 57W. City sit. w of Gander near mouth of Humber R., on w coast of Newfoundland. The main industry is paper making, esp. newsprint; also manu. iron, cement and gypsum products. Pop. (1981C) 24,339.

Corniche   France/Monaco. Three roads running s of the Maritime Alps between Nice and Menton. The Grande Corniche was built as a military road under Napoleon I and climbs to over 1,700 ft. (518 metres) a.s.l.; it is augmented by the Moyenne and Petite Corniches.

Corn Islands   Caribbean Sea. 12 15N 83 00W. Two small islands sit. off Bluefields, Nicaragua, called Great Corn Island and Little Corn Island. Leased to the U.S.A. The main occupation is growing and processing coconuts.

Cornwall   Ontario, Canada. 45 02N 74 44W. Town ESE of Ottawa sit. on St. Lawrence R. Industries inc. manu. textiles, paper and chemicals. Pop. (1981E) 46,000.

Cornwall   England. 50 26N 4 40W. County in extreme sw forming a peninsula between Atlantic Ocean to the N and W and English Channel to the s and E and bounded E by Devon. Area, 1,370 sq. m. (3,547 sq. km.). Chief towns Bodmin, Truro, Newquay, Falmouth, Penzance, Saltash, Camborne-Redruth and St. Austell. Mainly pasture and moors, with a rocky coastline. Drained by R. Camel, Fal, Fowey and Lynher. Under 1974 reorganization its districts are North Cornwall, Restormel, Caradon, Penwith, Kerrier and Carrick. Main economic activities are agric. and fishing, mining for tin, quarrying (clay), light industry and tourism. Pop. (1988E) 460,000.

Coro   Falcón, Venezuela. 11 25N 69 41W. State cap. in NW, sit. on an isthmus between Gulf of Coro and Caribbean Sea. Industries inc. manu. cigars and soap. Has a sea port 7 m. (11 km.) ENE, La Vela, through which it trades in coffee and maize. Pop. (1981C) 96,339.

Coromandel Coast   India. 36 35S 175 31E. The SE coast, on the Indian Ocean, extending *c*. 430 m. (688 km.) from Kistma R. delta in the N to Point Calimere in the s. There are no natural harbours.

Coronation Gulf   Northwest Territories, Canada. 68 00N 114 00W. An arm of the Arctic Ocean between the mainland and Victoria Island.

Coronel   Concepción, Chile. 37 01S 73 08W. Town sit. ssw of Concepción in a coalmining area. Industries inc. flour milling and soap manu., and it is the chief coaling station on the Chilean coast. Pop. (1980C) 65,000.

Coronel Oviedo Casguazú, Paraguay. 25 25S 56 27W. Cap. of province sit. E of Asunción. Commercial centre. Industries inc. sawmilling and sugar refining. Pop. (1984E) 22,000.

Corovodë Skrapari, Albania. 40 30N 20 12E. Village sit. SE of Berat on Osum R. A lignite mining centre.

Corozal Belize. 18 24N 88 24W. Town near the Mexican border, the second most important in the colony. Trading centre for cereals, sugar, citrus fruit, coconuts. Pop. (1980E) 23,000.

Corpus Christi Texas, U.S.A. 27 48N 97 24W. City sit. on Corpus Christi Bay, Gulf of Mexico. Industries inc. oil refining, chemicals and cotton ginning; manu. cottonseed oil and cement. Resort and sea port connected with the sea by a deepwater channel, exporting cotton, petroleum, sulphur and fish. Oil and natural gas in the immediate area. Pop. (1980C) 231,999.

Corregidor Luzon, Philippines. 14 23N 120 35E. Fortified island at entrance to Manila Bay.

Corrèze France. 45 20N 1 50E. Dept. in Limousin region, crossed in N by the Monts du Limousin. Area, 2,263 sq. m. (5,860 sq. km.). Chief towns Tulle (the cap.) and Brive. Drained by Rs. Corrèze, Vézère and Dordogne. Fertile only in R. valleys. The main products are cereals, fruit, vegetables, wine, sheep and pigs. Pop. (1982C) 241,448.

Corrèze River France. 45 20N 1 50E. Rises in Corrèze dept. and flows 60 m. (96 km.) SW past Tulle and Brive to enter Vézère R.

Corrib, Lough Ireland. 53 05N 9 10W. L. mainly in W Galway, with its NE end in Mayo. Area, 68 sq. m. (176 sq. km.), extending S to Galway and draining into Galway Bay by Corrib R.

Corrientes Corrientes, Argentina. 27 30S 58 50W. Cap. of province in the NE sit. on Paraná R. 25 m. (40 km.) below confluence of Alto Paraná

R. with Paraguay R. Trading centre for agric. produce. Industries inc. tanning and sawmilling; manu. textiles and vegetable oils. Pop. (1980C) 62,000.

Corse France. 42 00N 9 00E. Island in the Mediterranean, forming region divided into two departments, Corse-du-Sud and Haute-Corse. Area, 3,352 sq. m. (8,680 sq. km.). Chief towns Ajaccio (the cap.) and Bastia. Mountainous, rising to Monte Cinto, 8,891 ft. (2,710 metres) a.s.l., and descending to plains on the E coast. The main occupations are rearing sheep and goats, and cultivating olives, vines and citrus fruits. Pop. (1982C) 240,178.

Corse-du-Sud France. 42 00N 9 00E. Dept. in Corse region. Area, 1,550 sq. m. (4,014 sq. km.). Cap. Ajaccio. Pop. (1982C) 108,604.

Corsica *see* **Corse**

Cortés Honduras. Dept. of NW Honduras. Area, 2,787 sq. m. (7,218 sq. km.). Cap. San Pedro Sula. Pop. (1983E) 624,090.

Corum Turkey. 40 33N 34 58E. (i) Province of N Asiatic Turkey. Area, 4,340 sq. m. (11,241 sq. km.). Pop. (1985C) 599,204. (ii) Town 115 m. (184 km.) NE of Ankara. Provincial cap. Industries inc. carpet manu. and fruit growing.

Corumbá Mato Grosso, Brazil. 19 01S 57 39W. Town on W bank of Paraguay R. on Bolivian border, in SW Mato Grosso. Market town and R. port handling the products of an extensive beef producing area. Pop. (1980E) 180,000.

Coruña, La Spain. 43 20N 8 25W. (i) Province in NW Spain bounded N and W by Atlantic. Area, 3,041 sq. m. (7,876 sq. km.). Very heavy rainfall. The main occupations are farming and fishing. Pop. (1986C) 1,102,376. (ii) Cap. of (i) sit. on NW coast. Industries inc. sardine fishing and canning; also manu. cigars, glassware and cotton goods. Pop. (1986C) 241,808.

Corvallis Oregon, U.S.A. 44 34N

123 16W. City in NW Oregon on Willamette R. Commercial centre trading in fruit, farm produce. Industries inc. high technology and education. Pop. (1989E) 42,000.

Corvo Island   Azores, Atlantic. 39 42N 31 06W. Smallest and northernmost of the Azores group. Area, 6.8 sq. m. (17.6 sq. km.), consisting of a single extinct volcano, Monte Gordo. Only settlement, Vila Nova de Corvo. The main occupations are dairying, fishing and fish salting.

Corwen   Clwyd, Wales. Town sit. on Dee R. at the foot of Berwyn Mountains, and WSW of Wrexham. Market town for a farming area, and a fishing and tourist resort.

Cos   *see* **Kos**

Coseley   West Midlands, England. 52 33N 2 06W. Town sit. SSE of Wolverhampton. Industries inc. metal working; manu. machinery.

Cosenza   Calabria, Italy. 39 17N 16 15E. Cap of Cosenza province sit. on Crati R. Market town for a farming area. Industries inc. textiles; manu. furniture. Pop. (1981E) 106,801.

Cosmoledo Islands   Seychelles, Indian Ocean. 9 43S 47 35E. Atoll sit. NW of Madagascar in the Aldabra group, forming a Seychelles depend. Area, 11 m. (18 km.) by 8 m. (13 km.). The main products are copra and fish.

Costa Mesa   California, U.S.A. 33 38N 117 55W. Town sit. SE of Los Angeles. Industries inc. manu. computer and electronic equipment and research and development. Pop. (1980C) 86,145.

Costa Rica   10 00N 84 00W. Rep. bounded N by Nicaragua, E by Caribbean Sea, SE by Panama, and S and W by Pacific. Area, 19,730 sq. m. (51,100 sq. km.). Cap. San José. Mountainous and volcanic, rising to peaks over 10,000 ft. (3,048 metres) a.s.l. About three-quarters is forest; the central plain is fertile with a temperate climate. The main occupations are farming and lumbering.

Main products are coffee, bananas, cacao, sugar-cane and timber. Pop. (1989E) 2,886,990.

Côte d'Azur   Alpes-Maritimes/Var, France. 45 25N 6 50E. Coastline of French Riviera. The main resorts are Antibes, Cannes, Juan-les-Pins, Menton, Monte Carlo (Monaco), Nice and Ville Franche.

Côte d'Ivoire (Ivory Coast)   West Africa. 7 30N 5 00E. Rep. bounded N by Mali, NE by Burkino Faso, E by Ghana, W by Liberia, S by Gulf of Guinea, NW by Guinea. Area 124,503 sq. m. (322,463 sq. km.). Cap. Abidjan, to be moved to Yamoussoukro. Low-lying coastal belt rises to interior plateau, heavily forested. Equatorial climate. Drained by Sassandra, Bandama and Camoé Rs. flowing N to S into the Gulf of Guinea. Fertile soil producing cacao and mahogany, also fruit, coffee, cotton, rubber, palm products, cereals, yams and livestock. Industries are concerned with processing these products. Pop. (1988E) 11,634,000.

Côte d'Or   France. 47 30N 4 50E. Dept. in Burgundy region. Area, 3,384 sq. m. (8,765 sq. km.). Cap. Dijon. Crossed SW to NE by the Côte d'Or range, on the E slopes of which are famous Burgundy vineyards. The main occupations are viticulture and farming. Pop. (1982C) 473,548.

Cotentin   France. 49 30N 1 30W. Peninsula forming the N of Manche dept. and extending into the English Channel. Northernmost points are Cap. de la Hague in the NW and Pte. de Barfleur in the NE. Cherbourg is sit. on the N coast.

Côtes-du-Nord   France. 48 28N 2 50W. Dept. in Brittany region, bounded N by English Channel. Area, 2,656 sq. m. (6,878 sq. km.). Chief towns St. Brieuc (the cap.), Dinan, Guingamp and Lannion. Low-lying, mainly agric. The main products are cider apples, cereals and hemp. Pop. (1982C) 538,869.

Cotonou   Benin. 6 21N 2 26E. Largest city of Benin and cap. of dept. sit. WSW of Porto Novo on a sandspit be-

tween L. Nokoué and the Bight of
Benin. Industries inc. manu. vegetable
oils and soap. Sea port exporting
agric. produce. Pop. (1982E) 487,020.

Cotopaxi   Ecuador. 0 40S 78 26W.
Volcano in the Andes in N central
Ecuador. At 19,344 ft. (5,896 metres)
a.s.l., it is the world's highest active
volcano.

Cotswold Hills   England. 51 45N
2 10W. Range extending SSW to NNE
from Bath to Chipping Campden at an
average of 500–600 ft. (152–183
metres) a.s.l., rising to Cleeve Cloud,
1,083 ft. (333 metres) a.s.l.

Cottbus   Cottbus, German Demo-
cratic Republic. 51 45N 14 19E.
Town in E central G.D.R. sit. on Spree
R. Industries inc. manu. woollen
goods, carpets, machinery and soap.
Pop. (1989E) 128,500.

Council Bluffs   Iowa, U.S.A. 41
16N 95 52W. City in SW of state sit.
on Missouri R. Industries inc. railway
workshops and agric. engineering.
Pop. (1980C) 56,449.

Coupar-Angus   Tayside Region,
Scotland. Town sit. NNE of Perth on
Isla R. Industries inc. printing, agricultural engineering, chicken processing. Pop. (1980E) 2,100.

Courbevoie   Hauts-de-Seine,
France. 48 54N 2 15E. Suburb of NW
Paris, on Seine R. Industries inc.
manu. vehicles and pharmaceutical
goods. Pop. (1982C) 59,931.

Courland   Latvia, U.S.S.R. Region
between Gulf of Riga and Lithuanian
border. Mainly agric. The main products are cereals, flax and potatoes.

Courtrai   *see* **Kortrijk**

Coutances   Manche, France. 49
03N 1 26W. Town in NW sit. on
Cotentin peninsula. Market town for a
farming area. Industries inc. textiles
and leather manu. Pop. (1982C)
13,439.

Coventry   West Midlands, England.
52 25N 1 30W. Midland city. The
main industries are the manu. of cars,

telephone exchanges, turbines, tractors and man-made fibres. Pop.
(1984E) 314,100.

Covilhã   Beira Baixa, Portugal. 40
17N 7 30W. Town sit. E of Coimbra.
The most important wool milling
town in Portugal, with dye works and
textile training school. Tourist centre
esp. for winter sports. Pop. (1981E)
22,000.

Covington   Kentucky, U.S.A. 39
05N 84 30W. Town in extreme N of
state, sit. immediately opposite Cincinnati on Ohio R. Industries inc.
manu. machine tools, tiles, bricks and
paper. Pop. (1980C) 49,563.

Cowdenbeath   Fife Region, Scotland. 56 07N 3 21W. Town sit. NE of
Dunfermline. Industries inc. petrochemical and oil related concerns,
open-cast coalmining. Pop. (1987E)
10,350.

Cowes   Isle of Wight, England. 50
45N 1 18W. Town and famous yachting centre sit. on N coast, and extending to both sides of Medina R.
estuary. Industries inc. boat-building,
building hovercraft, radar and tourism. Pop. (1981C) 19,591.

Cowley   Oxfordshire, England. 51
44N 1 11W. Suburb of E Oxford. Industries inc. mainly vehicle manu.

Craçow   *see* **Kraków**

Cradock   Cape Province, Republic
of South Africa. 32 08S 25 36E. Town
sit. on Great Fish R. Market town for
a sheep-farming area.

Craigavon   Armagh, Northern Ireland. New town (1965) consisting of
Lurgan and **Portadown**

Crail   Fife Region, Scotland. 56
16N 2 38W. Town sit. near mouth of
Firth of Forth. Industries inc. fishing
and tourism. Pop. (1990E) 1,350.

Craiova   Dolj, Romania. 44 19N 23
48E. Cap. of district sit. W of Bucharest on Jiu R. Industries inc. textiles
and food processing; manu. machine
tools and leather goods. Pop. (1983E)
243,117.

Cramlington   Northumberland, England. 55 04N 1 35W. 'New town' under construction with projected pop. of 62,000. Industries inc. pharmaceuticals and toiletries, textiles, industrial rubber hosing, adhesive stickers and razor blades. Pop. (1988E) 30,000.

Cranbrook   British Columbia, Canada. 49 31N 115 46W. City sit. in the SE of the province on the W edge of the Rocky Mountain trench. Major service and supply centre for SE British Columbia, an area rich with enormous coal, lead and zinc deposits. Important indust. inc. mining, forestry, service, supply, ranching, transportation and tourism. Pop. (1989E) 15,024.

Cranbrook   Kent, England. 51 06N 0 32E. Town S of Maidstone. Pop. (1981C) 4,209.

Cranston   Rhode Island, U.S.A. 41 47N 71 26W. City immediately S of Providence sit. on Pawtucket R. Industries inc. textiles; manu. medical products and machinery. Pop. (1980C) 71,992.

Cranwell   Lincolnshire, England. 53 01N 0 27W. Village sit. NE of Grantham. The Royal Air Force College is sit. here.

Crawley   West Sussex, England. 51 07N 0 12W. Town sit. NE of Horsham created as a 'new town' in 1947. Industries inc. pharmaceuticals, printing, light engineering; manu. electronic equipment and furniture. Gatwick international airport here. Pop. (1981C) 81,250.

Crayford   Greater London, England. 51 26N 0 10E. Town sit. on Cray R. and is part of the London borough of Bexley in SE. Industries inc. engineering, flour milling and furniture making.

Crediton   Devon, England. 50 47N 3 39W. Town NW of Exeter sit. on Creedy R. Market town for an agric. area. Industries inc. engineering. Pop. (1988E) 6,396.

Crema   Lombardia, Italy. 45 22N 9 41E. Town sit. NW of Cremona. Industries inc. textiles.

Cremona   Lombardia, Italy. 45 07N 10 02E. Town in central N sit. on Po R. Cap. of Cremona province, and the commercial centre of the Po valley. Industries inc. manu. dairy products, pasta, confectionery, silk and musical instruments, famous in the 16th–18th cent. for making violins (Amati, Guarneri and Stradivari). Pop. (1981E) 81,000.

Cres   Croatia. Yugoslavia. 44 50N 14 25E. Island sit. in Gulf of Kvarner, an inlet of the N Adriatic Sea. Area, 130 sq. m. (337 sq. km.). Chief village, Cres. The main occupations are farming and fishing.

Crete   Greece. 35 10N 25 00E. Island in Mediterranean Sea, extending W to E across S end of Aegean Sea. Area, 3,217 sq. m. (8,332 sq. km.). Chief towns, Canea (the cap.) and Candia. Mountainous, rising to Mount Ida, 8,058 ft. (2,456 metres) a.s.l. The main occupation is farming. Main products are olive oil, citrus fruits, wine, goats and sheep. Pop. (1981C) 502,165.

Créteil   Val-de-Marne, France. 48 48N 2 28E. Cap, of dept. on Marne R. S suburb of Paris. Industries inc. copper smelting. Pop. (1982C) 71,705.

Creuse   France. 46 00N 2 00E. Dept. in Limousin region. Area, 2,146 sq. m. (5,559 sq. km.). Chief towns. Guéret (the cap.) and Aubusson. Infertile land, rising to the Plateau de Millevaches in the S. Agric. is possible only in the valleys. The chief products are cereals and potatoes; and livestock from upland areas. Pop. (1982C) 139,968.

Creuse River   France. 47 00N 0 34E. Rises in Plateau de Millevaches and flows 150 m. (240 km.) NNW to join Vienne R. Its upper course flows through a deep gorge.

Crewe   Cheshire, England. 53 05N 2 27W. Town in NW Midlands. An important railway centre. Industries inc. railway engineering and chemicals; manu. cars and clothing. Pop. (1981C) 47,759.

Crewkerne   Somerset, England.

Town sit. SE of Taunton. Market town for an agric. area. Industries inc. textiles and light engineering. Pop. (1980E) 5,100.

Criccieth   Gwynedd, Wales. 52 55N 4 14W. Town sit. on S coast of Lleyn peninsula on Cardigan Bay, and ENE of Pwllheli. Seaside resort. Pop. (1980E) 1,600.

Crieff   Tayside Region, Scotland. 56 23N 3 52W. Town and tourist resort in central Scotland, sit. on Earn R. Industries inc. tourism, glass-making, pottery and distilling. Pop. (1989E) 5,569.

Crimea   Ukraine, U.S.S.R. 45 00N 34 00E. Peninsula extending 120 m. (192 km.) into Black Sea and *c.* 210 m. (336 km.) W to E, joined to mainland by Perekop Isthmus. Area, 9,880 sq. m. (25,590 sq. km.). Cap. Simferopol. Dry but fertile steppe with mountains parallel to the S coast. The climate on the steppes is arid; on the coast Mediterranean. The main occupations are farming and fishing, with metallurgical industries based on the iron fields of Kerch, and tourism on the coast.

Crimmitschau   Chemnitz, German Democratic Republic. 50 49N 12 23E. Town sit. W of Chemnitz. Industries inc. textiles; manu. textile machinery. Pop. (1989E) 24,000.

Cristóbal   Panama. 9 20N 79 55W. Town at Atlantic end of Panama canal opposite Colón, which is on the outside of the zone boundary.

Croagh Patrick   Mayo, Ireland. 53 15N 10 18W. Mountain in W, on S shore of Clew Bay; a place of pilgrimage trad. held to be the site where St. Patrick first preached.

Croatia   Yugoslavia. 45 20N 16 00E. Constituent rep. of the Yugoslav Federal Rep. Bordered N by Hungary and SW by Adriatic Sea, where it extends S along a narrow coastal strip as far as the Gulf of Kotor, and inc. many off-shore islands. Area, 21,824 sq. m. (56,524 sq. km.). Chief towns Zagreb (the cap.), Rijeka, Split and Osijek. Barren mountains along the S

coastal strip with holiday resorts and fishing ports; fertile in the NE. The main occupations are farming, fishing and tourism. Main products are cereals, vegetables and fruits. Pop. (1988E) 4·68m.

Cromer   Norfolk, England. 52 56N 1 18E. Seaside resort sit. on N coast of Norfolk, and on North Sea. Pop. (1981C) 6,192.

Crompton   Greater Manchester, England. 53 54N 2 05W. Town forming a suburb of Oldham. The main industry is cotton spinning.

Crook and Willington   County Durham, England. 54 43N 1 44W. Town sit. SW of Durham. Manu. firebricks, carpets, refrigeration equipment, clothing and furniture.

Crosby   Merseyside, England. 53 30N 3 02W. Residential suburb of NW Sefton on Mersey R. estuary. Pop. (1981E) 53,557.

Cross Fell   Cumbria, England. 54 44N 2 29W. Mountain sit. SE of Carlisle, the highest in the Pennine Range.

Cross River   Nigeria. 4 42N 8 21E. State comprising the Annang, Calabar, Ogoja and Uyo provinces of the former E region. Area, 10,516 sq. m. (27,237 sq. km.). The main ethnic groups are the Ibibio, Efik and Ekoi; chief towns, Calabar (cap.), Opobo, Uyo, Ikot Ekpene, Oron and Ogoja. Pop. (1983E) 5·7m.

Crotone   Calabria, Italy. 39 05N 17 08E. Town sit. ESE of Cosenza on Gulf of Taranto. Industries inc. mainly chemicals and also zinc smelting. Pop. 45,000.

Croton River   U.S.A. Rises in NE New York and flows 60 m. (96 km.) SW to join Hudson R. at Croton Point. Used for water storage, esp. for New York City.

Crouch River   England. Rises near Brentwood, Essex, and flows 24 m. (38 km.) E past Burnham-on-Crouch to enter North Sea at Foulness Point.

Crowborough East Sussex, England. 51 03N 0 09E. Dormitory town sit. SW of Tunbridge Wells. Pop. (1981C) 15,100.

Crowland (Croyland) Lincolnshire, England. 52 41N 0 11W. Town sit. between Peterborough and Spalding. Market town for an agric. area. Pop. (1981C) 3,070.

Crow's Nest Pass Canada. 49 40N 114 40W. Mountain pass at 4,450 ft. (1,356 metres) a.s.l. in the Rocky Mountains between Alberta and British Columbia and *c.* 45 m. (72 km.) N of the U.S.A. frontier. A railway line passes through it between Coleman, Alberta, and Natal, British Columbia.

Croydon Greater London, England. 51 23N 0 06W. District of S London forming a residential suburb. Industries inc. engineering and electronics; manu. pharmaceutical goods and foodstuffs. Large shopping and office complex. Pop. (1988E) 317,200.

Crozet Islands Indian Ocean. 46 00S 52 00E. Arch. sit. ESE of Port Elizabeth, S. Africa. Area, 195 sq. m. (505 sq. km.) in two groups. The W group inc. Apostles, Pigs and Penguins Islands; the E Possession and Eastern Islands, also 15 smaller islands, all form part of French Southern and Antarctic Territories. Mountainous, with a meteorological station on Possession Island, but no other inhabitants. Pop. (1983C) 33.

Cuba 22 00N 79 00W. Island rep. forming the largest and most W of the Greater Antilles group, sit. S of the S tip of Florida, U.S.A. Area, 44,206 sq. m. (114,493 sq. km.). Divided into 14 political divisions: Pinar del Rio, Ciudad de la Habana, Matanzas, Villa Clara, Sancti Spiritus, Ciego de Avila, Las Tunas, Holguin, La Habana, Cienfuegos, Camagüey, Granma, Santiago de Cuba, Guantánamo. Chief towns Havana (the cap.), Marianao, Santa Clara, Sancti Spiritus, Camagüey and Santiago de Cuba. Undulating country rising to the Sierra Maestra in the SE with Pico Turquino at 6,496 ft. (1,980 metres) a.s.l. the N coast is steep and rocky, the S flat and swampy. Sub-tropical climate with

hurricanes. The main occupation is farming esp. sugar, tobacco and fruit. Industries are based on processing the chief crops. The U.S.A. maintains a naval base at Guantánamo. Pop. (1986E) 10·24m.

Cúcuta Norte de Santander, Colombia. 7 54N 72 31W. Cap. of dept. formally known as San José Cúcuta. Sit. in NE, near the Venezuelan frontier, and on Pan-American Highway. A commercial centre trading in coffee. Industries inc. coffee roasting and textiles; manu. soap. Pop. (1980E) 420,000.

Cuddalore Tamil Nadu, India. 11 45N 79 46E. Town sit. SSW of Madras on Coromandel coast. Industries inc. weaving and dyeing. Sea port exporting groundnuts and cotton goods. Pop. (1981C) 127,625.

Cuenca Azuay, Ecuador. 2 52S 78 54W. Cap. of province in S centre. Commercial centre trading in cinchona bark, sugar-cane and cereals. Industries inc. textiles and flour milling; manu. tyres and panama hats. Pop. (1982C) 272,397.

Cuenca Spain. 40 04N 2 08W. (i) Province in E central Spain. Area, 6,586 sq. m. (17,058 sq. km.). Mountainous. The main occupation is rearing sheep and goats, with agric. in the valleys and some lumbering. Pop. (1986C) 210,932. (ii) Cap. of province of same name sit. above Júca R. Market town for a farming area. Industries inc. tanning and sawmilling.

Cuernavaca Morelos, Mexico. 18 55N 99 15W. Town sit. S of Mexico City. State cap. Resort noted for mountain scenery, the ruins of Xochicalco and the nearby Cacahuamilpa caverns. Pop. (1980C) 232,355.

Cuiabá Mato Grosso, Brazil. 15 35S 56 05W. State cap. in W centre sit. at head of navigation on Cuiabá R. Distribution centre for cattle, hides and dried meats. Pop. (1980E) 168,000.

Cuiabá River Brazil. 17 05S 56 36W. Rises on Plateau of Mato

Grosso and flows 300 m. (480 km.) SSW past Cuiabá to join Paraguay R. 80 m. (128 km.) N of Corumbá.

Cuillin Hills   Highland Region, Scotland. 57 14N 6 15W. Mountains in S of the island of Skye, in the Inner Hebrides, rising to Sgurr Alasdair, 3,309 ft. (1,009 metres) a.s.l.

Culebra Island   Puerto Rico, U.S.A. 18 19N 65 18W. Small island off E coast of Puerto Rico with a U.S. naval base. The chief town is Dewey. Pop. (1984E) 1,250.

Culiacán   Sinaloa, Mexico. 24 48N 107 24W. State cap. sit. NNW of Mazatlan in NW and sit. at foot of Sierra Madre Occidental. Industries inc. textiles. Pop. (1980C) 560,011.

Cullinan   Transvaal, Republic of South Africa. Town sit. ENE of Pretoria. The main industry is the Premier Diamond Mine where the 'Cullinan' diamond was found in 1905 and the 599 carat 'Centenary' diamond in 1986. Pop. (1989E) 7,000.

Culross   Fife Region, Scotland. 56 03N 3 37W. Town sit. WSW of Dunfermline. Holiday resort. Pop. (1990E) 560.

Cumaná   Sucre, Venezuela. 10 28N 64 10W. State cap. in NE sit. near the mouth of Manzanares R. Formerly known as Nuera Toledo. Industries inc. fishing, fish processing and cotton milling. Exports, through its sea port at Puerto Sucre, are coffee, cacao and tobacco. It was founded in 1521 and is the oldest European settlement in S. America. Pop. (1981C) 179,814.

Cumberland   Maryland, U.S.A. 39 39N 78 46W. City sit. on Potomac R., in the Appalachian Mountains. Trading centre for locally-mined coal. Industries inc. railway engineering; manu. tyres and fine paper. Pop. (1980C) 25,933.

Cumberland Peninsula   Northwest Territories, Canada. 67 00N 65 00W. Mountainous peninsula in E Baffin Island, extending 200 m. (320 km.) E into the Davis Strait.

Cumberland Plateau   U.S.A. 36 00N 84 30W. Plateau in SW Appalachian Mountains, mainly in NE Tennessee but extending N into SW Virginia.

Cumberland River   U.S.A. 37 09N 88 25W. Rises on Cumberland Plateau and flows 690 m. (1,104 km.) W across S Kentucky and N Tennessee, then N to join Ohio R. above Paducah on Kentucky/Illinois border.

Cumbernauld   Strathclyde Region, Scotland. 55 58N 3 59W. Town sit. ENE of Glasgow created as a 'new town' in 1955. Industries inc. light engineering and textiles. Pop. (1981C) 47,901.

Cumbraes, The   Strathclyde Region, Scotland. 55 46N 4 57W. Two islands in Firth of Clyde separated by Tan Strait. Area: Great Cumbrae, 5 sq. m. (13 sq. km.); Little Cumbrae, 1 sq. m. (3 sq. km.). Chief town, Millport, on Great Cumbrae.

Cumbria   England. 54 44N 2 55W. County in NW Eng. containing Lake District, bounded NW by Solway Firth, and W by the Irish Sea. Area, 2,629 sq. m. (6,809 sq. km.). Almost 40% of the land area lies within the Lake District and the Yorkshire Dales National Parks. Chief towns Carlisle, Workington, Whitehaven and Barrow-in-Furness. Created under 1974 re-organization, its districts are Carlisle, Allerdale, Eden, Copeland, South Lakeland and Barrow-in-Furness. Industries inc. agric., marine engineering, offshore gas, nuclear reprocessing, electronics, chemicals, foodstuffs and footwear. Pop. (1988E) 489,200.

Cumnock and Holmhead   Strathclyde Region, Scotland. 55 27N 4 16W. Town sit. E of Ayr. Industries inc. open cast coalmining; manu. clothing and carpets. Pop. (1989E) 7,400.

Cundinamarca   Colombia. Dept. of central Colombia on an extensive plateau in the Andes. Area, 9,106 sq. m. (23,585 sq. km.). Cap. Bogotá. Important mineral deposits. Pop. (1983E) 1,288,000.

Cunene River   *see* **Kunene River**

Cuneo   Piemonte, Italy. 44 23N
7 32E. Cap. of Cuneo province, sit. in
NW. Trading centre for raw silk. Industries inc. textiles, esp. silk, metal
working and food processing. Pop.
(1981E) 56,000.

Cupar   Fife Region, Scotland. 56
19N 3 01W. Market town in E central.
Industries inc. textiles and engineering. Pop. (1989E) 7,510.

Curaçao   Netherlands Antilles. 12
11N 69 00W. Island in Caribbean Sea
between Aruba and Bonaire and sit. N
of coast of Venezuela. Area, 171 sq.
m. (444 sq. km.). Cap. Willemstad.
The main occupation is farming; main
industry refining Venezuelan oil.
Main products are sisal and citrus
fruits. Pop. (1983E) 165,011.

Curicó   Curicó, Chile. 34 59S 71
14W. Cap. of province in S central
Chile. Market town trading in cattle.
Industries inc. flour milling, wine
making and distilling. Pop. (1980E)
74,000.

Curieuse Island   Seychelles, Indian
Ocean. Island sit. N of Praslin Island.
The main products are copra, essential
oils and fish.

Curitiba   Paraná, Brazil. 25 25S 49
15W. State cap. in SE sit. in the Sierra
do Mar. Commercial centre trading in
coffee, timber and *maté*. Industries
inc. textiles and paper making. Pop.
(1980C) 842,818.

Curragh, The   Kildare, Ireland. Plain
E of Kildare noted for horse breeding
and training. Area, *c.* 5,000 acres. Site
of a well-known racecourse.

Cuscatlán   El Salvador. Dept. of
central Salvador. Area, 296 sq. m.
(766 sq. km.). Cap. Cojutepeque. Pop.
(1981E) 160,000.

Cutch, Rann of   *see* **Kutch, Rann
of**

Cuttack   Orissa, India. 20 25N 85
57E. Town in E Orissa sit. SW of Calcutta at head of Mahanadi R. delta.
Commercial centre trading in rice and

oilseeds. Industries inc. engineering,
tanning and work in precious metals.
R. port. Pop. (1981C) 295,268.

Cuxhaven   Lower Saxony, Federal
Republic of Germany. 53 53N 8 42E.
Town in central N sit. at mouth of
Elbe R. estuary. Industries inc. fish-
ing, fish canning, shipbuilding and
tourism. Pop. (1984E) 57,200.

Cuyahoga Falls   Ohio, U.S.A. 41
08N 81 29W. City sit. SSE of Cleveland on Cuyahoga R. In 1986 the city
merged with Northampton township.
Residential, with some manu. industries. Pop. (1980C) 52,526.

Cuzco   Peru. 13 31S 71 59W. (i)
Dept. on E slopes of the Andes and S
Montaña, bounded E by Bolivia. Area,
32,487 sq. m. (84,141 sq. km.).
Mainly tropical forest. Pop. (1981C)
832,504. (ii) City in a valley of the
Andes at 11,400 ft. (3,475 metres).
Dept. cap. and anc. cap. of the Incas;
extensive ruined Inca sites remain.
Pop. (1981E) 181,604.

Cwmbran   Gwent, Wales. 51 39N 3
00W. Town sit. N of Newport. Created
as a 'new town' in 1949. Industries
inc. engineering. Pop. (1978E) 43,500.

Cyclades   Greece. 37 30N 25 00E.
Group of islands in Aegean Sea forming a prefecture. Area, 995 sq. m.
(2,578 sq. km.). There are over 200
islands. Cap. Hermoupolis, on Syros
island, and 69 m. (110 km.) SE of
Athens. The main occupation is farming; main products are wine, tobacco,
olives, emery and iron ore. Pop.
(1981C) 88,458.

Cyprus   35 00N 33 00E. Rep. within
the Commonwealth forming an island
lying 50 m. (80 km.) S of Turkey and
60 m. (96 km.) W of Syria. In 1914 the
island was annexed by Great Britain
and in 1925 given the status of a
Crown Colony. Following guerilla
warfare by Greek Cypriots (1955–60)
against the British, Cyprus became an
independent republic in 1960. In 1974
Turkey invaded N Cyprus. Area, 3,572
sq. m. (9,251 sq. km.). Chief towns
Nicosia (the cap.) and Famagusta
(completely evacuated since Turkish
invasion). Mountainous, with the

Kyrenia Mountains stretching along the N coast, and in the S the Olympus Mountains rising in the SW to Mount Olympus (Troodos), 6,403 ft. (1,952 metres) a.s.l. The central plain is drained by Rs. which flood in winter and run dry in summer. Agric. contributes about a quarter of the national income and over one third of the island's exports. Fresh meat comes mainly from the free ranging fat-tailed sheep and small hardy goats. These animals are able to maintain themselves, with little supplementary feed, on the short-lived grazing of the common and waste lands and on the crop residues and the weed growth of fallow lands. Minerals include iron pyrites, cupreous concentrates, copper, cement, asbestos, chromium ores and concentrates. These minerals provide approx. 20% of total exports. Industries inc. wine-making and tourism. Pop. (1988E) 687,500.

Cyrenaica Libya. 27 00N 20 00E. Province occupying E Libya, bounded E by Egypt, SE by the Sudan and S by Chad. Area, 330,000 sq. m. (855,000 sq. km.). Chief towns Benghazi (the cap.) and Derna. Arid depression in the S, rises to the Libyan Desert in the centre with fertile land on the coastal plain in the N. The main occupations are stock rearing, farming and fishing. Main products are barley, wheat, grapes, olives, dates and tunny fish.

Cyrene (Shahat) Cyrenaica, Libya. Town in the NE. The original cap. of Cyrenaica, with many anc. Greek and Roman sites.

Cythera *see* **Kithira**

Czechoslovakia 49 00N 17 00E. Rep. bounded E by U.S.S.R., N by Poland and the German Democratic Republic, W by Federal Republic of Germany, S by Austria and Hungary. Area, 49,359 sq. m. (127,840 sq. km.). The Czechoslovak State came into existence in 1918, upon the dissolution of Austria–Hungary. Chief cities Prague (the cap.), Pilsen, Brno, Ostrava and Bratislava. There are 10 administrative regions. Plateau in the W surrounded by the Sudeten Mountains, Moravian Heights, Böhmerwald and Erz Gebirge and drained by R. Elbe and Vltava. Moravia in the centre is lowland, drained by Morava R. to Danube R. and Slovakia in the E is highland, rising to the W Carpathian Mountains and the High Tatra with Gerlachovka, 8,737 ft. (2,663 metres) a.s.l. Main crops are, wheat, rye, barley, oats, maize, potatoes and sugar-beet. Czechoslovakia is a richly forested country and the timber industry is important. Some minerals are found inc. coal (hard and soft), iron ore, graphite, copper, lead, uranium, glass, sand and salt. Industrialization is well developed and antedates the Communist régime. Pop. (1989E) 15,624,021.

Czestochowa Katowice, Poland. 50 49N 19 06E. Town in S sit. on Warta R. Industries inc. iron and steel, textiles, chemicals and paper. There is a monastery at Jasna Góra which is a place of pilgrimage. Pop. (1982E) 238,000.

D

Dabrowa Górnicza Katowice, Poland. 53 40N 23 20E. Town sit. ENE of Katowice; an important coalmining centre. Industries are based on coal, inc. iron-founding. Pop. (1983E) 141,600.

Dacca *see* **Dhaka**

Dachau Bavaria, Federal Republic of Germany. 48 16N 11 26E. Town sit. NNW of Munich. Industries inc. textiles; manu. electrical equipment and paper. Site of a former concentration camp. Pop. (1984E) 33,200.

Dadra and Nagar Haveli India. Union Territory in central W India forming an enclave in Maharashtra State, just SE of Daman and on the border with Gujarat. Area, 192 sq. m. (491 sq. km.). Chief occupation farming. The major crop is paddy. Forests cover 70% of the area. Cap. Silvassa. Pop. (1981C) 103,676.

Dagenham Greater London, England. 51 32N 0 10E. Town sit. E of city centre on N bank of Thames R. The main industry is manu. of motor vehicles, paints and veneers. *See* Barking.

Dagestan Russian Soviet Federal Socialist Republic, U.S.S.R. Autonomous rep. lying between the E ranges of Great Caucasus Mountains and Caspian Sea. Annexed from Persia in 1723 and constituted an Autonomous Republic in 1921. Area, 19,416 sq. m. (50,287 sq. km.). Cap. Makhachkala. Mountainous with a narrow coastal plain. Chief occupation is farming, esp. cattle and sheep, wheat, vines and cotton. Some deposits of oil and natural gas. Industries inc. engineering, oil, chemicals, textiles, wood working and food processing. Pop. (1984E) 1·7m.

Dahlak Islands Eritrea, Ethiopia. Arch. in Red Sea off Mesawa formed by 2 large and many smaller islands. The 2 large islands are Nora and Dahlak Kebir. Area, 340 sq. m. (881 sq. km.). Chief occupation fishing, esp. pearl fishing.

Dahomey *see* **Benin**

Dajabón Libertador, Dominican Republic. 19 34N 71 34W. Town opposite Ouanaminthe on Massacre R. S of Monte Christe and on Haiti border. Trading centre and provincial cap., trading in timber, hides, bananas, honey, coffee.

Dakar Senegal. 14 40N 17 26W. Cap. of Senegal sit. at S end of Cape Verde peninsula on the Atlantic coast and one of the most important towns in West Africa. It was first occupied permanently in 1857. Commercial centre for an agric. hinterland. Important airport, railway centre and cable station. Industries inc. groundnut oil extraction, food processing, titanium refining and cement making. Sea port exporting groundnuts, oil and oilcake. Pop. (1985E) 1,382,000.

Dalat Haut-Donnai, Vietnam. 11 56N 108 25E. Town sit. NE of Saigon at 4,940 ft. (1,506 metres) a.s.l. Important hill station and once summer cap. of Vietnam. Administration and commercial centre of a plantation area producing tea, coffee and rubber.

Dalian (Lüta) Liaoning, China. 38 53N 121 35E. A port sit. at S end of the Liaotung peninsula, in the NE, comprising the former ports of Port Arthur and Dairen. An important industrial centre with shipbuilding, oil refining, railway engineering and food processing. Exports inc. coal, grain and soya bean products. Pop. (1982E) 1,480,000.

Dalkeith Lothian Region, Scotland. 55 54N 3 04W. Town sit. SE of Edinburgh between N. and S. Esk R.

Admin. centre for the area. Industries inc. printing, woollens, binding, electronics and light engineering. Pop. (1981C) 11,844.

Dallas Texas, U.S.A. 32 47N 96 48W. City in NE of state. Communications and commercial centre trading in oil and cotton. Industries inc. petroleum products and chemicals; manu. aircraft, cotton goods, clothing machinery, transport equipment, food products and electrical and electronic equipment. Pop. (1985E) 983,851.

Dalmatia Yugoslavia. 43 00N 17 00E. Coastal strip extending NW to SE from Velebit Channel to Gulf of Kotor: a mountainous and largely barren area rising to the Dinaric Alps in the NW. The chief industries are fishing and tourism: chief towns Dubrovnik, Split, Sibenik, Zadar and Hercegnovi.

Daloa Côte d'Ivoire. 6 53N 6 27W. Town sit. NNW of Sassandra. Trading centre of an agric. area producing rice, coffee, cacao, palm kernels, castor beans, cotton, kola nuts, rubber. Gold deposits are nearby. Pop. (1975C) 60,837.

Dalton-in-Furness Cumbria, England. Town sit. NE of Barrow-in-Furness, near the heart of the lake district. There is some light industry. Pop. (1990E) 14,000.

Daman Daman and Diu, India. 20 25N 72 57E. Town and surrounding district on Gulf of Cambay, an inlet of the Arabian Sea, bordering NW Maharashtra, and forming part of the Union Territory of Daman and Diu. Area, 44 sq. m. (112 sq. km.). Chief occupations are fishing and farming, trading and fish processing. Capital, Daman. Pop. (1981C) town, 21,003; district, 48,560.

Damanhûr Beheira, Egypt. 31 02N 30 28E. Cap of Beheira governorate, sit. ESE of Alexandria in Nile delta. Trading centre for an agric. area, esp. in cotton. Industries inc. cotton ginning. Pop. (1986E) 225,900.

Damascus Syria. 33 30N 36 18E. Cap. of Syria, sit. at E end of Anti-Lebanon Highlands on Barada R. Important commercial centre and craft centre; industries inc. silver-smithing, other metal working and silk weaving. Pop. (1981C) 1,112,214.

Damietta Egypt. 31 26N 31 48E. Town sit. near mouth of Damietta Nile. Industries inc. manu. cotton goods.

Dammam Al Hasa, Saudi Arabia. 26 20N 50 05E. Town on (Persian) Gulf. Developed as an oil port after 1940. Pop. (1974E) 127,844.

Damodar River India. 23 14N 87 39E. Rises in Chota Nagpur plateau and flows 370 m. (592 km.) ESE through Bihar entering W. Bengal S of Dhanbad and joining Hooghly R. *c.* 20 m. (32 km.) above Bhatpara. Used extensively for power and irrigation. Its valley inc. India's most important coalfield.

Dampier, Mount South Island, New Zealand. Peak of Southern Alps in the Tasman National Park, 11,287 ft. (3,440 metres) a.s.l.

Dampier Archipelago Indian Ocean. Arch. of rocky islands off NW coast of Western Australia. Enderby Island is the largest; others inc. Lewis, Legendre, Dolphin, Rosemary and Delambre. Area, 21 sq. m. (54 sq. km.). Chief occupation sheep farming.

Danakil Land Ethiopia/Territory of Afars and Issas. Desert area bounded NE by Red Sea and SW by an escarpment of the Great Rift Valley. Mainly highland, but it inc. the Danakil Depression, lowest point 400 ft. (122 metres) b.s.l. Chief occupation nomadic stock farming.

Da Nang Vietnam. 16 04N 108 13E. Port and second largest town sit. 50 m. (80 km.) SE of Hué. Pop. (1972E) 500,000.

Danbury Connecticut, U.S.A. 41 23N 73 27W. City sit. NW of Bridgeport on Still R. Industries inc. hat-making; manu. machinery, metal goods, silverware, clothing, chemicals, aircraft, rubber and paper products. Noted for its annual fair. Pop. (1980C) 60,470.

Danger Islands (Pukapuka) Cook Islands, Pacific. 10 53S 165 49W. Atoll *c.* 715 m. (1,144 km.) NW of Rarotonga consisting of 3 islets joined by reefs; Pukapuka, Motu Koe, Motu Kavata. Area, 2 sq. m. (5 sq. km.). Chief products copra, pearl shell. Pop. (1986C) 761.

Dangs, The Gujarat, India. District in N Western Ghats bounded N and W by Surat. Area, 681 sq. m. (1,764 sq. km.). Headquarters, Ahwa. Forested hill country, producing millet, rice and teak. Pop. (1981C) 113,664.

Danube River Central/South East Europe. 45 20N 29 40E. Rises as two head streams, Brigach R. and Brege R. in the Black Forest, Federal Republic of Germany, becomes one R. near Donaueschingen, Bavaria, flows 1,750 m. (2,800 km.) NE then SE past Regensburg to join Inn R. near the Austrian frontier; flows past Linz and Vienna, Austria, to enter Czech. at Bratislava and forms part of the Czech./Hungarian frontier, until it turns S to Budapest and across the Hungarian plain to enter Yugoslavia. It flows past Novi Sad and Belgrade and on to form the Yugoslav/Romanian and Romanian/Bulgarian borders, then N below Bucharest, Romania, to Galati and then E to enter the Black Sea by a delta. Navigable from Ulm, Bavaria; navigation controlled by the intern. Danube Commission. Main tribs. Rs. Altmühl, Inn, Drava, Tisza, Sava, Morava and Prut. It is linked by the Altmühl R. with canals to Rs. Main and Rhine.

Danville Illinois, U.S.A. 40 08N 87 37W. City sit. S of Chicago on Vermilion R. Commercial centre for an agric. and mining area. Industries inc. automotive, machine and agric. engineering. Pop. (1980C) 38,985.

Danville Kentucky, U.S.A. 37 39N 84 46W. City sit. SSW of Lexington. Railway junction and industrial centre of a farming area. Industries inc. flour, cattlefood and timber milling, distribution of livestock and tobacco; manu. clothing, furniture, concrete blocks and food products. Noted for precivil-war sites, esp. the Weisiger Memorial State Shrine commemorat-

ing the former seat of Govt. Pop. (1980C) 12,942.

Danville Virginia, U.S.A. 36 35N 79 24W. City sit. SW of Richmond on Dan R. Trading centre for tobacco. Industries inc. textiles and furniture-making. Pop. (1980C) 45,642.

Danzig *see* **Gdańsk**

Daqahliya Egypt. Province of NE Lower Egypt bounded W by Damietta branch of Nile R. Area, 1,337 sq. m. (3,462 sq. km.). Chief towns Mansura (cap.), Mit Ghamr, Manzala. Irrigated agric. areas produce cotton and cereals. Industries inc. cotton ginning, textiles, fisheries. Pop (1985E) 3,469,000.

Darbhanga Bihar, India. 26 10N 85 56E. Town sit. NE of Patna on Little Baghmati R. Trading centre for wheat, fruits, oilseeds and rice. Industries inc. food processing. Pop. (1981C) 176,301.

Dardanelles Turkey. 40 15N 26 25E. Strait separating European and Asiatic Turkey, and connecting Aegean Sea with Sea of Marmara. About 45 m. (72 km.) long and 1–5 m. (1.6–8 km.) wide.

Dar-es-Salaam Tanzania. 6 48S 39 17E. City, chief port and former cap. of Tanzania, sit. on Indian Ocean. Industries inc. oil refining, tanning and rice milling; manu. soap and paint. Port and railway terminus handling about half Tanzania's exports esp. sisal, cotton, coffee, hides, skins, gold and diamonds. Pop. (1978C) 870,020.

Darfur Sudan. Province of W Sudan bounded W by Central African Empire. Area, 75,890 sq. m. (196,555 sq. km.). Main towns El Fasher (cap.), Geneina. Extensive plain rising to Marra Mountains W and drained only by intermittent streams. Bush forest (S) peters out to scrub grassland (N). Main products livestock, gum arabic. Pop. (1983C) 3,093,699.

Darién Panama. 9 00N 77 00W. E part of Rep. between Gulf of Darién, an inlet of the Caribbean Sea, and

Gulf of San Miguel, an inlet of the Pacific. Inc., but not co-extensive with, Darién province of which the area is 6,000 sq. m. (15,500 sq. km.). Bordered s by Colombia. Chief occupation farming, esp. maize, rice and beans. Pop. (1984E) 26,487.

Darjeeling   West Bengal, India. 20 02N 88 20E. Town and hill station sit. in the NE at 6,000–8,000 ft. (1,829–2,438 metres) a.s.l. on s slopes of the Himalayas. Centre of a tea-growing area. Pop (1981C) 57,603.

Darkhan   Mongolia. Mountain sit. SE of Ulan Bator with a monument to Jenghiz Khan which is a place of pilgrimage.

Darlaston   West Midlands, England. Town sit. ESE of Wolverhampton. Industries inc. manu. hardware, metal castings and equipment.

Darling Downs   Queensland, Australia. 27 30S 150 30E. Plateau in SE Queensland w of Great Dividing Range, at 2,000–3,000 ft. (610–914 metres) a.s.l. Chief occupation farming, esp. sheep and dairy cattle, wheat, barley and fodder crops. Chief town Toowoomba.

Darling Range   Western Australia, Australia. 31 25S 116 00E. Ridge extending 250 m. (400 km.) N to s parallel to coast behind Perth and Fremantle, at 800–1,500 ft. (244–457 metres) a.s.l. rising to Mount Cooke, 1,910 ft. (582 metres) a.s.l.

Darling River   New South Wales, Australia. 34 07S 141 55E. Rises in headstreams in SE Queensland joining inside the New South Wales border, and flows 1,900 m. (3,040 km.) SW across N and w of the state to join Murray R. at Wentworth on the border with Victoria. The upper course is called Barwon R. Used for storage and irrigation, but its flow fluctuates violently from very low to extensive flooding.

Darlington   Durham, England. 54 31N 1 34W. Town in NE sit. on Skerne R. Industries inc. engineering and construction, newspapers; manu. woollen goods. Noted as a terminus of

the Stockton and Darlington Railway (first passenger line) in 1825. Pop. (1989E) 99,200.

Darmstadt   Hessen, Federal Republic of Germany. 49 53N 8 40E. Town sit. near the right bank of R. Rhine. Industries inc. railway engineering; manu. machinery, computer software, chemicals and pharmaceutical products. Pop. (1984E) 135,600.

Dartford   Kent, England. 51 27N 0 14E. Town sit. WNW of Chatham on Darent R. Market town. International ferry terminal. Industries inc. papermaking and engineering. Manu. pharmaceutical products and cement. Linked with Essex by Dartford tunnel under Thames R. Pop. (1989E) 81,000.

Dartmoor   Devon, England. 50 38N 4 00W. Moorland region with granite masses in s Devon. Area, *c.* 300 sq. m. (800 sq. km.). Source area for most Devon rivers; land rises to 2,039 ft. (621 metres) a.s.l. at High Willhays and 2,028 ft. (618 metres) a.s.l. at Yes Tor. Chief occupation livestock farming.

Dartmouth   Nova Scotia, Canada. 44 40N 63 34W. City just NE of Halifax and connected with it by two bridges over Halifax harbour. Industries inc. electronics, brewing, packaging, foodstuffs and oil refining. Pop. (1985E) 62,277.

Dartmouth   Devon, England. 50 21N 3 35W. Town on s coast of Devon and sit. on w side of R. Dart estuary. The main industry is tourism, particularly for yachting. Seat of the Royal Naval College. Pop. (1981C) 6,211.

Daru   Papua New Guinea, Australia. Town sit. SW of Pendembu on Moa R. Commercial centre trading in palm products, cacao, coffee.

Darwen   Lancashire, England. 53 42N 2 28W. Town sit. s of Blackburn. Industries inc. manu. cotton goods, paper, plastics, paints and chemicals. Pop. (1981C) 30,048.

Darwin   Falkland Islands. Small

settlement at head of Choiseul Sound, E Falkland Island.

Darwin Northern Territory, Australia. 12 28S 130 50E. Cap. of Northern Territory, sit. in extreme NW of the territory on Port Darwin harbour. Communications and commercial centre for the interior. Important as a sea port, airport, telegraph and cable terminus. Pop. (1984E) 63,900.

Dasht-i-Kavir West Khurasan, Iran. 34 30N 55 00E. Salt desert in N central Iran with a salt crust and marshy ground beneath.

Daugavpils Latvia, U.S.S.R. 55 53N 26 32E. Town sit. on W. Dvina R. Industries inc. railway engineering; manu. textiles and food products. Trading centre for grain, flax and timber. Pop. (1980E) 117,000.

Davao Mindanao, Philippines. 7 04N 125 36E. Town on E coast, on Davao Gulf. Sea port exporting abacá, copra and timber. Industries inc. fishing. Pop. (1980C) 610,375.

Davenport Iowa, U.S.A. 41 32N 90 41W. City in extreme E of state opposite Rock Island on Missouri R. Industries inc. construction equipment, food processing and agric. implements. Pop. (1980C) 103,264.

Daventry Northamptonshire, England. 52 16N 1 09W. Town sit. W of Northampton with an important radio transmitter. Pop. (1989E) 17,000.

David Chiriquí, Panama. 8 25N 82 27W. Town sit. WSW of Panama City on David R. and Inter-American Highway. Commercial centre. Industries inc. tanning, sugar milling; manu. shoes, clothing, soap, furniture, ceramics, beverages. Its sea port is Pedregal 4 m. (6 km.) to the S. Pop. (1984E) 45,000.

Davis Strait North Atlantic Ocean. Arm of Atlantic between SW Greenland and Baffin Island, Canada, 400 m. (640 km.) long and up to 400 m. (640 km.) wide. Navigable from midsummer to late autumn.

Davos Graubünden, Switzerland.

Mountain valley in E watered by Landwasser R. with the Davosersee at the N end. There are two resorts, both at *c.* 5,118 ft. (1,560 metres) a.s.l.; Davos Platz and Davos Dörfli. Pop. (1988E) 13,000.

Dawley *see* **Telford**

Dawlish Devon, England. 50 35N 3 28W. Town sit. SW of Exmouth on English Channel coast. Industries inc. tourism and horticulture. Pop. (1981C) 10,846.

Dawson Yukon Territory, Canada. 64 04N 139 25W. City sit. near border with Alaska, U.S.A. and on Yukon R. Trading and communications centre of the Klondyke mining area, and founded during the 1896 gold rush when its pop. was *c.* 25,000. Industries inc. mining and tourism. Pop. (1989E) 1,800.

Dawson Creek British Columbia, Canada. 55 46N 120 14W. City sit. on Dawson Creek near the Alberta border. Centre of a lumbering and farming area. Industries inc. waferboard and chopstick manu., coalmining, electricity, oil, natural gas, tourism and transport services. Pop. (1990E) 10,500.

Dax Landes, France. 43 43N 1 03W. Spa resort and tourist centre on Adour R. Pop. (1982C) 19,636.

Dayton Ohio, U.S.A. 39 45N 84 15W. City in W of state sit. on Great Miami R. Industries inc. manu. aircraft parts, machine tools and electrical equipment. Important as a centre of aeronautical research and site of the Wright Brothers aircraft research factory. Pop. (1980C) 203,371.

Daytona Beach Florida, U.S.A. 29 12N 80 60W. City sit. SE of Jacksonville on Atlantic coast, on Halifax R. lagoon. The main industry is tourism. The beach is 30 m. (48 km.) long and extends 500 ft. (152 metres) at low tide and is used for motor speed trials. Pop. (1988E) 68,694.

De Aar Cape Province, Republic of South Africa. Town in N of province; important railway junction. The main industry is railway engineering.

Dead Sea   Israel/Jordan. 31 30N 35 30E. Salt L. on the Jordan/Israel border, its N tip 15 m. (24 km.) from Jerusalem, at 1,286 ft. (392 metres) b.s.l. Depth ranges from 3 ft. (1 metre) at the S end to 1,300 ft. (396 metres) at the N. Receives R. Jordan in N but has no outlet. Water is only lost by evaporation and the salt content is *c.* 25%. Exploited for common salt, potassium, magnesium and calcium salts. It is 46 m. (74 km.) long and up to 9 m. (14 km.) wide.

Deal   Kent, England. 51 14N 1 24E. Town sit. NNE of Dover on Strait of Dover. Industries are mainly tourism and fishing. Pop. (1989E) 27,000.

Dean, Forest of   Gloucestershire, England. 51 47N 2 33W. District and anc. royal hunting ground between Rs. Severn and Wye in W Gloucestershire. A small extent of forest remains, and small deposits of coal. Industries inc. farming, tourism, light manu. and mining.

Dearborn   Michigan, U.S.A. 42 18N 83 10W. City sit. W of Detroit on Rouge R., formed in 1929 by union of 2 cities, Dearborn and Fordson. Headquarters and original factory of Ford Motor Co. The main industry is still manu. motor vehicles. Pop. (1980C) 90,660.

Dearborn Heights   Michigan, U.S.A. 42 20N 83 18W. City immediately NW of Dearborn. Pop. (1980C) 67,706.

Death Valley   California/Nevada, U.S.A. 36 00N 116 40W. Arid rift valley to N of the Mojave Desert dropping to 282 ft. (86 metres) b.s.l. at Badwater, and rising to Telescope Peak, 11,045 ft. (3,367 metres) a.s.l. in W. Mainly sand dunes, salt flats and desert vegetation.

Deauville   Calvados, France. 49 22N 0 04E. Town sit. at mouth of Touques R. on English Channel coast. The main industry is tourism. Pop. (1989E) 4,800.

Debrecen   Hajdú-Bihar, Hungary. 47 32N 21 38E. Cap. of county sit. E of Budapest. Market town trading in livestock and tobacco. Industries inc.

flour milling, agric. engineering, pharmaceutical products and foodstuffs. Pop. (1989E) 220,000.

Decatur   Alabama, U.S.A. City sit. N of Birmingham on Tennessee R. Industries developed after the estab. of the Tennessee Valley Authority, inc. manu. metal goods, food products, automobile parts, chemicals, pet food and synthetic fibres. Pop. (1989E) 50,078.

Decatur   Illinois, U.S.A. 39 51N 88 59W. City sit. between Chicago and St Louis on Sangamon R. Industries inc. railway engineering and foodstuffs processing; manu. agric. machinery, motorcar accessories, plastics and metal products. Pop. (1980C) 93,897.

Deccan   India. Plateau of S central India bounded N by Satpura Range and W and E by W. and E. Ghats. Rises to *c.* 3,000 ft. (914 metres) a.s.l. in W and slopes to 1,000 ft. (305 metres) a.s.l. in E; drained by Rs. Godavari, Krishna and Cauvery flowing into the Bay of Bengal.

Děčín   Severočeský, Czechoslovakia. 50 48N 14 13E. Town sit. near German Democratic Republic frontier at confluence of Elbe and Polsen Rs. Industries inc. manu. paper, textiles and chemicals. R. port. Pop. (1985E) 55,000.

Dedza   Central Province, Malawi. 14 22S 34 20E. Town sit. SE of Lilongwe on Mozambique border. Route centre in a farming area, producing tobacco, cotton, cereals, livestock.

Dee River   Scotland. (i) Rises on Ben Braeriach in SW Aberdeenshire and flows 87 m. (139 km.) E through Aberdeenshire past Braemar, Balmoral and Ballater to form E end of Aberdeenshire/Kincardineshire border and enter the North Sea at Aberdeen, by an artificial channel. Noted for salmon. (ii) Rises in Loch Dee, central Kirkcudbrightshire, and flows 38 m. (61 km.) S to enter the Irish Sea at Kirkcudbright.

Dee River   Wales/England. Rises in

Bala L. in Gwynedd and flows 70 m. (112 km.) NE to Corwen, then E past Llangollen to form the English/Welsh border; turns N and flows for c. 16 m. (26 km.) before entering Cheshire c. 7 m. (11 km.) above Chester. It turns W around Chester and enters the Irish Sea by an estuary 13 m. (21 km.) long and up to 5 m (8 km.) wide, which becomes a stretch of sand at low tide.

Dehra Dun Uttar Pradesh, India. Town sit. NNE of Delhi. Commercial centre trading in rice, wheat and oilseeds. Noted as the HQ of the Indian Military Academy and the Oil and Natural Gas Commission. Pop. (1981C) 220,530.

Deir ez Zor Syria. 35 20N 40 09E. Town sit. ESE of Aleppo on Euphrates R. Communications centre for W Syria, S Turkey and Iraq. Industries inc. flour milling and tanning.

Delagoa Bay Mozambique. 25 50S 32 45E. Inlet of the Indian Ocean in extreme S, otherwise called Lourenço Marques Bay. Width c. 25 m. (40 km.). The port of Lourenço Marques is on the NW shore.

Delaware U.S.A. 39 05N 75 15W. State on E seaboard bounded N by Pennsylvania and S and W by Maryland. Area, 2,057 sq. m. (5,328 sq. km.). Cap. Dover. One of the original 13 states. It forms the E half of the Delmarva Peninsula and is generally lowlying. Farms produce poultry, wheat, maize, fruit and vegetables, and fishing, with industries concentrated on Wilmington inc. important chemical plants. Manu. transport equipment and food products. Chief towns are Dover, Wilmington, Newark and Milford. Pop. (1988E) 660,000.

Delaware River U.S.A. 39 20N 75 25W. Rises in SE New York State and flows 280 m. (448 km.) generally SE forming the NE border between Pennsylvania and New York, and the Pennsylvania/New Jersey border. It enters the Atlantic by an estuary below Chester, Pennsylvania.

Delémont Jura, Switzerland. 47 22N 7 21E. Cap. of canton of Jura (created 1979). Pop. (1981E) 12,000.

Delft Zuid-Holland, Netherlands. 52 00N 4 21E. Town sit. SSE of The Hague on Schie canal. Industries inc. important pottery and porcelain works, and chemicals; manu. electronic equipment and cables. Pop. (1984E) 86,988.

Delfzijl Groningen, Netherlands. Town sit. NE of Groningen on Eems estuary and Eems Canal. Industries inc. shipbuilding, manu. salt, soda, alumina, catalysts, polymers, cement, sawmilling. Port exports dairy and agric. produce. Pop. (1990E) 32,000.

Delhi India. 28 37N 77 12E. (i) Union Territory of N India bounded E by Uttar Pradesh and on all other sides by Haryana. Area, 573 sq. m. (1,483 sq. km.). Cap. Delhi. Consists of the city of Delhi and the surrounding agric. area. Pop. (1981C) 6,220,406. (ii) Cap. of India and territorial cap. sit. on W bank of Jumna R. Old Delhi was rebuilt by Shah Jahan in the 17th cent. New Delhi was chosen as the cap. of India in 1912 and designed mainly by Lutyens. Commercial and communications centre, esp. by air and railway. Industries inc. electronics, railway wagons, radio and television, cycles, pharmaceuticals, textiles, chemicals and craftwork. Pop. (1981C) 273,036 (New Delhi).

Delmarva Peninsula U.S.A. 38 45N 75 42W. Peninsula lying N to S between Delaware Bay to E and Chesapeake Bay to W on the Atlantic coast; contains all of Delaware and part of Maryland and Virginia. Main occupation agric., forestry and fishing.

Delmenhorst Lower Saxony, Federal Republic of Germany. 53 03N 8 38E. Town sit. WSW of Bremen. Industries inc. textiles; manu. machinery, linoleum, furniture, clothing and soap. Pop. (1985E) 74,440.

Delos Cyclades, Greece. 37 26N 25 16E. One of the smallest islands of the Cyclades group in the Aegean Sea. Sometimes called Mikra Delos to distinguish it from Megali Delos to W Area, 1.2 sq. m. (3.1 sq. km.). Chief occupations fishing, farming.

Delphi Greece. 38 29N 22 30E. Classical site at S foot of Parnassus, now excavated, and seat of the Delphic Oracle.

Delta British Columbia, Canada. 49 06N 123 00W. Suburb of Vancouver at mouth of Fraser R. Industries inc. fish and food processing. Pop. (1981C) 74,692.

Delta Amacuro Venezuela. Territory of NE Venezuela bounded E by Atlantic, SE by Guyana and sit. mainly in the Orinoco delta. Area, 15,520 sq. m. (40,197 sq. km.). Cap. Tacupita. Tropical climate with frequent rains. Swampy terrain with numerous Rs. and streams; some agric. produces sugar, rice, maize, cacao, bananas, tobacco. Tropical rain-forests produce gum, balata, divi-divi and mangrove bark. Mineral resources inc. petroleum, asphalt, iron, bauxite. Pop. (1980E) 58,063.

Demerara River Guyana. 6 50N 58 10W. Rises in the central forest and flows 215 m. (344 km.) N to enter the Atlantic at Georgetown. Used for shipping bauxite from Mackenzie, 60 m. (96 km.) upstream, to the coast.

Denain Nord, France. 50 20N 3 23E. Town sit. WSW of Valenciennes on Escaut R. Industries inc. mechanical and light engineering, service. Pop. (1984E) 20,500.

Denbigh Clwyd, Wales. 53 11N 3 25W. Town sit. S of Rhyl in the vale of Clwyd. Market town for an agric. area. Industries inc. manu. food products. Pop. (1981C) 9,000.

Denby Dale West Yorkshire, England. 53 35N 1 40W. Town sit. SE of Huddersfield. Industry mainly woollen industry.

Den Helder Noord-Holland, Netherlands. 52 54N 4 45E. Town sit. N of Amsterdam at North Sea end of North Holland Canal. Port with important off-shore activities, naval base and fishing fleet. Industries inc. fish processing, agric. especially market gardening, livestock and flower bulbs. Pop. (1989E) 62,000.

Denizli Denizli, Turkey. Cap. of province sit. ESE of Izmir. Market

town trading in agric. produce. Industries inc. flour milling. Pop. (1980C) 135,373.

Denmark 56 00N 10 00E. Kingdom occupying N part of the Jutland peninsula and adjacent islands, bounded S by Federal Republic of Germany, W by North Sea, N by Skagerrak strait and E by straits leading to the Baltic Sea. Consists of the mainland, Jutland, and the islands of Fünen, Zealand, Laaland, Falster and Bornholm with some smaller islands. Area, 16,631 sq. m. (43,075 sq. km.). Chief towns: Copenhagen (cap. comprising Copenhagen, Frederiksberg and Gentofte municipalities), Aarhus, Odense, and Aalborg. Low-lying and mainly agric. The chief crops are barley, wheat, rye, oats, potatoes and root crops but dairy farming and bacon production particularly important. Although only very few industrial raw materials are produced within the country, a wide range of industries have been developed. Pop. (1989E) 5,129,778.

Dent Blanche Valais, Switzerland. Alpine peak reaching 14,318 ft. (4,405 metres) a.s.l. lying to W of Zermatt, and opposite to and N of the Matterhorn.

Dent du Midi Valais, Switzerland. Mountain massif in SW Switz. near French border, inc. Haute Cime, 10,696 ft. (3,260 metres), Cime de l'Est, 10,433 ft. (3,180 metres) a.s.l., Dent Jaune, 10,456 ft. (3,187 metres) a.s.l., Cathédrale, 10,387 ft. (3,166 metres) a.s.l.

Denton Texas, U.S.A. 33 13N 97 08W. City sit. NNE of Fort Worth. Industries inc. flour milling; manu. dairy products and clothing. Pop (1980C) 48,063.

D'Entrecasteaux Islands Pacific Ocean. 9 30S 150 40E. Group off SE coast of New Guinea forming part of Papua New Guinea. The main islands are Goodenough, Fergusson and Normanby. Area, 1,200 sq. m. (3,108 sq. km.). Chief occupation farming; chief product copra.

Denver Colorado, U.S.A. 39 43N 105 01W. State cap. sit. on the South

Platte R. Centre of government service industries and education. Trading centre for a ranching area, with some manu. There are important federal agencies inc. a branch of the U.S. Mint. Pop. (1980c) 492,365.

Deptford Greater London, England. 51 28N 0 02W. District in SE, mainly residential. Industries inc. engineering and chemicals. Noted as a naval dockyard 1513–1869.

Der'a Hauran, Syria. 32 37N 36 06E. Town sit. ssw of Damascus near Jordan border. Centre of a farming area producing cereals.

Dera Ghazi Khan Punjab, Pakistan. 30 03N 70 38E. Town sit. wsw of Multan. Trading centre for an agric. area. Manu. tractors. Destroyed by Indus R. floods 1908–9 and rebuilt to W of R. Pop (1981E) 102,000.

Dera Ismail Khan   North West Frontier Province, Pakistan. 31 51N 70 56E. Town sit. NNW of Multan on Indus R. Trading centre for grain, oilseeds and other agric. produce. Pop. (1981E) 68,000.

Derbent   Russian Soviet Federal Socialist Republic, U.S.S.R. 42 03N 48 18E. Town in Dagestan, sit. on Caspian Sea. Industries inc. fishing and fish processing, textiles and wine making, with glassworks nearby. Sea port.

Derby Derbyshire, England. 52 55N 1 29W. County town sit. on Derwent R. Railway junction. Industries inc. aero engines and railway engineering; manu. electrical equipment, porcelain, synthetic textiles, plastics and chemicals, printing and paints. Pop. (1985E) 215,000.

Derbyshire   England. 52 55N 1 29W. County in central England. County town, Derby. Under 1974 reorganization consists of districts of High Peak, Chesterfield, West Derbyshire, North East Derbyshire, Bolsover, Amber Valley, Derby, Erewash and South Derbyshire. Hilly in the N and NW rising to The Peak, 2,088 ft. (636 metres) a.s.l. Drained by Rs. Derwent, Dove, Wye and Trent. Chief

occupations stock-farming and coal-mining in E and S, quarrying in the NE. Industries inc. tourism with several spas, iron smelting, textiles and engineering. Chief towns Derby, Chesterfield, Ilkeston, Heanor and Alfreton. Pop. (1988E) 924,200.

Dereham   *see* **East Dereham**

Derg, Lough   Ireland. (i) L. in the NW of Tipperary, forming most of its border with Clare and Galway; 25 m. (40 km.) long and up to 3 m. (5 km.) wide on Shannon R. It contains Holy Island (Inishcaltra) with the ruins of 4 churches. (ii) L. in SW Donegal near the Ulster border, containing Station Island, a place of pilgrimage to the scene of St. Patrick's purgatory.

Derna   Cyrenaica, Libya. 32 46N 22 39E. Port on Mediterranean in NE Libya. Oasis. Chief occupation is farming, esp. dates. Pop. (1970E) 97,000.

Derry   *see* **Londonderry**

Derwent River Tasmania, Australia. 43 03S 147 22E. Rises in L. St. Clair and flows 130 m. (208 km.) SE to enter the Tasman Sea by an estuary into Storm Bay. Used for hydroelectric power.

Derwent River   England. (i) Rises in Borrowdale Fells, Cumbria and flows 35 m. (56 km.) N through Derwentwater and Bassenthwaite L., then W past Cockermouth to the Irish Sea at Workington, 7 m. (11 km.) NNE of Whitehaven. (ii) Rises near The Peak, Derbyshire, and flows 60 m. (96 km.) S past Matlock and Derby to join Trent R. on the Leicestershire border. (iii) Rises in the Pennine hills in extreme SW Northumberland and flows 30 m. (48 km.) ENE forming the Northumberland/Durham border for *c.* 16 m. (26 km.), and crossing Tyne and Wear to join Tyne R. above Newcastle. (iv) rises in the N. York Moors *c.* 7 m. (11 km.) from the North Sea coast, flows 70 m. (112 km.) SSW through Vale of Pickering and past Stamford Bridge to join Ouse R. 5m. (8 km.) SE of Selby.

Derwentwater   Cumbria, England. 54 34N 3 08W. L. immediately S of

Keswick in the Lake District, 3 m. (5 km.) long and up to 1 m. (1.6 km.) wide, on Derwent R.

Désirade Guadeloupe, West Indies. Island sit. ENE of Grande-Terre and forming a depend. of Guadeloupe. Area, 8 sq. m. (22 sq. km.). Main settlement, Grande-Anse. Main occupations farming and fishing; main products sugar, cotton, sisal. Pop. (1982C) 1,602.

Des Moines Iowa, U.S.A. 45 35N 93 37W. State cap. sit. in S centre of state on Des Moines R. Important commercial and route centre for agric. area. Manu. tyres, and agric equipment. Pop. (1980C) 371,800.

Des Moines River Minnesota/Iowa, U.S.A. 40 22N 91 26W. Rises in SW Minnesota and flows 540 m. (864 km.) SSE entering Iowa near Estherville, and flowing through Fort Dodge and Des Moines to form extreme E end of the Iowa/Missouri border and joins Mississippi R. below Keokuk. Used for hydroelectric power.

Desna River Russian Soviet Federal Socialist Republic, U.S.S.R. 50 33N 30 32E. Rises *c.* 50 m. (80 km.) ESE of Smolensk and flows 700m. (1,120 km.) SSW past Bryansk to join Dnieper R. above Kiev. Navigable below Bryansk; used for carrying timber, grain and other agric. produce.

Des Plaines Illinois, U.S.A. 42 02N 87 54W. City forming a suburb of NW Chicago on Des Plaines R. Industries inc. general manu., retail and wholesale trade. Pop. (1980C) 55,374.

Desroches Seychelles. Island sit. NNE of the N tip of Madagascar. Area, 4 m. (6 km.) by $^1/_2$ m. (0.8 km.). Coral, forming the S side of an atoll reef. Main products copra, fish.

Dessau Halle, German Democratic Republic. 51 50N 12 14E. Town sit. in W central G.D.R. on Mulde R. near its confluence with Elbe R. Industries inc. railway engineering and chemicals; manu. machinery. Pop. (1989E) 104,000.

Dessie Wollo, Ethiopia. 11 08N 39 37E. Town sit. 160 m. (256 km.) NNE of Addis Ababa at 8,000 ft. (2,438 metres) a.s.l. Route and market centre, trading in coffee, cereals and hides. Pop. (1980E) 75,616.

Detmold North Rhine-Westphalia, Federal Republic of Germany. 51 56N 8 52E. Town sit. ESE of Bielefeld. Industries inc. manu. furniture and foodstuffs. Pop. (1984E) 66,600.

Detroit Michigan, U.S.A. 42 20N 83 03W. City sit. on Detroit R. opposite Windsor, Ontario, Canada. The area's main industry is the motor industry, accounting for three-quarters of its manu. and producing *c.* 25% of all U.S. cars and trucks. Other important industries are chemicals, machine tool manu., aeronautical engineering and manu. electrical goods. A major port, with oil refineries, shipyards and salt works. Pop. (1980C) 1,203,339.

Deurne Antwerp, Belgium. Suburb of E Antwerp and site of Antwerp airport.

Deux-Sèvres France. 46 30N 0 20W. Dept. in Poitou-Charentes region. Area, 2,318 sq. m. (6,004 sq. km.). Cap. Niort. Mainly agric. esp. for dairy cattle, wheat, oats, potatoes and other vegetables. Drained by R. Sèvre Nantaise and Sèvre Niortaise. Chief towns Niort and Parthenay. Pop. (1982C) 342,812.

Deva Hunedoara, Romania. 45 53N 22 55E. Town sit. NW of Bucharest on Mures R. Provincial cap. and tourist centre, trading in livestock, timber and fruit. Industries inc. meat and vegetable canning, flour milling. There is a 17th-cent. Franciscan Monastery, baroque castle and 13th-cent. fortress. Pop. (1982E) 73,420.

Deventer Overijssel, Netherlands. 52 15N 6 10E. Town sit. NE of Arnhem on Ijssel R. and at W end of Overijssel Canal. Industries inc. engineering; manu. textiles, chemicals and confectionery, esp. 'Deventer Koek'. Pop. (1989E) 66,500.

Devizes Wiltshire, England. 51 22N 1 59W. Town sit. 16 m. (26 km.) ESE of Bath on Kennet-Avon Canal.

Market town for an agric. area. Industries inc. building, light engineering, brewing. Pop. (1982E) 12,750.

Devon England. 50 45N 3 50W. County of SW Eng., bounded N by the Atlantic and SW by English Channel. Under 1974 reorganization its districts are Torridge, North Devon, West Devon, Mid-Devon, Teignbridge, Exeter, East Devon, Plymouth, South Hams and Torbay. Area, 2,591 sq. m. (6,709 sq. km.). County town, Exeter. Hilly in the E and NE and again on Dartmoor, in the SW, rising to 2,039 ft. (621 metres) a.s.l. Drained by Rs. Tamar, Exe, Dart, Teign, Taw, Torridge, Tavy, Axe and Otter. The chief occupations are farming, esp. cattle and sheep, and tourism; chief products are meat and dairy products. Industries inc. tourism, and agric. and marine engineering. Chief towns Exeter, Plymouth and Torquay. Pop. (1988E) 1,021,100.

Devonport Tasmania, Australia. 41 11S 146 21E. City sit. WNW of Launceston at mouth of Mersey R. Regional centre for an agric. area. Industries inc. tourism, food processing, textiles, carpets, furniture, paper products. Port exporting dairy produce, vegetables, paper, cement, tallow, sheep and timber. Pop. (1989E) 26,500.

Devonport Devon, England. 52 22N 4 10W. Town forming a district of Plymouth on Tamar R. estuary. Industries associated with important naval dockyards.

Devonport North Island, New Zealand. 36 49S 174 48E. Residential suburb of Auckland on NE shore of Waitemata Harbour. Site of Royal New Zealand naval base. Pop. (1981C) 10,410.

Dewsbury West Yorkshire, England. 53 42N 1 37W. Town sit on Calder R. Industries inc. textiles (wool and shoddy); manu. carpets and leather goods. Pop. (1981C) 48,987.

Dhahran Hasa, Saudi Arabia. 23 42N 90 22E. Town sit. on (Persian) Gulf, developed as centre and pipeline station for Saudi Arabian oifields; air-

port and railway. Industries are centred on the oilfield.

Dhaka Bangladesh. 23 43N 90 25E. Cap. of Bangladesh and Dhaka division, sit. on Burhi Ganga R. Before the partition of India in 1947 it was a small university town and is known as the city of mosques; there are over 1,000. Commercial centre trading in rice, oilseeds, cane and hide. Industries inc. tanning, jute processing and chemicals; manu. glass and electrical goods; a major airport. Pop. (1987E) 4,777,000.

Dhamar San'a, Yemen Arab Republic. 14 46N 44 23E. Town sit. S of San'a on the Yarim road. Centre of a grain growing area; industries inc. tanning, rug making.

Dharwad Karnataka, India. Town sit. NW of Bangalore. Trading centre for cotton and grain; industries inc. oilseed milling, textiles esp. cotton, and soap manu. Pop (1981C) 527,105 (Hubli-Dharwad).

Dhaulagiri Nepál. 28 42N 83 31E. Mountain peak in the Himalaya range 140 m. (224 km.) NW of Kathmándu; 26,810 ft. (8,172 metres) a.s.l.

Dhenkanal Orissa, India. 20 45N 85 35E. Town sit. NW of Cuttack. District cap., formerly cap. of the princely state of Dhenkanal. Trade in rice, oilseeds, timber. Industries inc. handloom weaving. Pop. (1981E) 35,653.

Dhotar Oman. District of S Oman extending 135 m. (216 km.) between Yemen border and cape Ras Nus. Chief towns Salala (cap.), Murbat. Coastal plain rising to the Jabal Samhan hills.

Dhule Maharashtra, India. Town sit. NE of Bombay on Panjhra R. Trading centre for oilseeds, cotton and millet; industries inc. oilseed milling and textiles; manu. soap. Pop. (1981C) 210,750.

Dhursing Nepál. Town sit. SW of Kathmándu; station on an aerial ropeway from Kathmándu to Kisipidi.

Diamantina Minas Gerais, Brazil.

18 15S 43 36W. Town NNE of Belo Horizonte in the Serra do Espinhaço. Industries inc. some diamond mining and processing, textiles and tanning. Pop. (1980E) 25,000.

Dibra *see* **Dibrë**

Dibrë Albania. District with an area, 606 sq. m. (1,569 sq. km.). Cap. of district, Peshkopi. Chief industry cattle raising. Pop. (1982E) 134,800.

Dickson Island Russian Soviet Federal Socialist Republic, U.S.S.R. Island at mouth of Yenisei gulf in Kara Sea, Arctic Ocean, off Krasnoyarsk Territory. Area, 12 sq m. (31 sq. km.). Chief settlement Dickson Harbour, a govt. polar station and coaling station.

Didcot Oxfordshire, England. 51 27N 1 15W. Town sit. S of Oxford. Mainly residential with some light industry. Pop. (1985E) 15,500.

Diégo-Suarez Madagascar. 12 16S 49 17E. Town at extreme N of island on Cap d'Ambre. Industries inc. fishing. Sea port exporting coffee, sisal and meats.

Dienbienphu Laichau, Vietnam. Town sit. SSW of Laichau near the Laos border. Noted as the site of a French defeat in the French Indo-China war.

Dieppe Seine-Maritime, France. 49 56N 1 05E. Town on English Channel coast at mouth of Arques R. Industries inc. shipbuilding, fishing, rope making and tourism; manu. cars, coffee processing. Sea port handling wine, fish and textiles. Pop. (1982C) 35,360.

Differdange Luxembourg. 49 32N 5 53E. Town sit. WNW of Esch-sur-Alzette near French border. Industries inc. iron mining, steel and fertilizer manu. Pop. (1989E) 16,500.

Digne Alpes-de-Haute-Provence, France. 44 06N 6 14E. Cap. of dept. Tourist centre. Pop. (1982C) 16,391.

Dijon Côte-d'Or, France. 47 19N 5 01E. Cap. of dept. in E central France sit. on Ouche R. and Burgundy canal. Commercial and route centre trading in Burgundy wine. Industries inc. chemicals; manu. foodstuffs, bicycles and motor-cycles. Pop. (1982C) 145,569 (agglomeration 215,865).

Dillon Montana, U.S.A. City sit. S of Butte on Beaverhead R. Industries inc. talc mining and processing, cattle and sheep ranching in the area. Pop. (1990E) 4,000.

Dimitrovgrad Bulgaria. 42 03N 25 36E. Town sit. ESE of Plovdiv on Maritsa R. Industries are mainly the chemical industry, esp. fertilizers, and also manu. earthenware and cement.

Dinajpur Dinajpur, Bangladesh. 25 38N 88 44E. District cap. sit. NW of Dhaka. Market centre for an agric. area, trading in rice, jute, cane, rape and mustard and barley. Industries inc. rice and oilseed milling, and soap manu. Pop. (1981E) 1,804,000.

Dinan Côtes-du-Nord, France. 48 27N 2 02W. Town in NW sit. at head of Rance R. estuary, on the English Channel. Industries inc. brewing, cider-making and hosiery manu. Pop. (1989E) 14,157.

Dinant Namur, Belgium. 50 16N 4 55E. Town sit. S of Namur on Meuse R. Industries inc. tourism, esp. for the Belgian Ardennes.

Dinard Ille-et-Vilaine, France. 48 38N 2 04W. Town opposite St Malo on Rance R. estuary, and on the English Channel. The main industry is tourism. Pop. (1982C) 10,016.

Dinaric Alps Yugoslavia. 43 50N 16 35E. Mountain range extending *c.* 180 m. (288 km.) NW to SE behind the Dalmatian coast, rising to Troglav, 6,277 ft. (1,913 metres) a.s.l.

Dindigul Tamil Nadu, India. 10 21N 77 58E. Town sit. NNW of Madurai. Industries inc. manu. of tobacco products. Pop. (1981C) 164,103.

Dingle Kerry, Ireland. 52 08N 10 15W. Town in SW on Dingle Bay. Main industries are fishing and tourism. Pop. (1980E) 1,400.

Dingwall   Highland Region, Scotland. 57 35N 4 36W. County town in N central Scot. at head of Cromarty Firth. Market town for an agric. area. Industries inc. tourism. Pop. (1982E) 4,950.

Dinslaken   North Rhine-Westphalia, Federal Republic of Germany. 51 34N 6 44E. Town sit. N of Duisburg in a coalmining area. Industries inc. mining; manu. iron and steel products. Pop. (1984E) 60,200.

Diourbel   Senegal. 14 40N 16 15W. Town sit. E of Dakar on Dakar–Niger railway. Centre of a peanut-growing area; the main industry is peanut oil manu. Pop. (1979E) 55,307.

Direction Island   Cocos Islands, Australia. Island in Cocos group *c.* 800 m. (1,280 km.) WSW of Singapore, one of the three largest with a cable station. Length, 1.5 m. (2 km.). Main product copra.

Dire Dawa   Harar, Ethiopia. 9 37N 41 52E. Town in E central Ethiopia on railway line to Djibouti. Commercial centre for an agric. area, trading in hides, skins and coffee. Industries inc. textiles; manu. cement. Pop. (1984E) 98,104.

Diriamba   Carazo, Nicaragua. 11 53N 86 15W. Town sit. SSE of Managua on Inter-American Highway. It is a railway terminus. The main industry is coffee milling. There is some trade in timber, limestone and salt. Pop. (1984E) 26,500.

Dismal Swamp   Virginia/North Carolina, U.S.A. Coastal marshes extending N to S across Virginia/N. Carolina border along the Atlantic coast. Much of the land has been reclaimed. Dismal Swamp canal crosses it to link James R. estuary with Albemarle Sound.

Dispur   Assam, India. New city near Gauhati on Brahmaputra R. built as cap. of Assam to replace Shillong, now cap. of Meghalaya.

Diss   Norfolk, England. 52 23N 1 06E. Town sit. SSW of Norwich on Waveney R. Market town for an agric.

area. Industries inc. pottery; manu. brushes and mats. Pop. (1985E) 5,000.

District of Columbia   U.S.A. 38 55N 77 00W. Federal district co-extensive with City of Washington on Potomac R., bounded S by Virginia and N by Maryland. Area, 69 sq. m. (179 sq. km.). The District of Columbia, organized in 1790, is the seat of the Government of the U.S.A., for which the land was ceded by the state of Maryland to the U.S. as a site for the national capital. It was established under Acts of Congress in 1790 and 1791. Congress first met in it in 1800 and federal authority over it became vested in 1801. The District is a service and trade centre. Pop. (1989E) 628,600.

Diu   Daman and Diu, India. District consisting of the island and town of Diu, and some of the adjacent mainland, on the coast of Gujarat state, but forming part of Daman and Diu Union Territory. Area, 15 sq. m. (40 sq. km.). Chief occupation fishing. Pop. (1981C) town 8,020; district, 30,421.

Diwaniya   Diwaniya, Iraq. Cap. of governorate sit. SSE of Baghdad on Hilla branch of Euphrates R. Market centre for a farming area, with some craft industries.

Diyala   Iraq. Province of E Iraq bounded E by Iran. Cap. Ba'quba. The main product is oil with a refinery at Khanaqin. Also some agric. land, producing dates, oranges, apples, plums, along Diyala R. Pop. (1977C) 587,754.

Diyarbakir   Diyarbakir, Turkey. 37 55N 40 14E. Cap. of province sit. ESE of Malatya, on Tigris R. in Kurdistan. Market centre for a farming area, trading in grain, wool and mohair. Industries inc. textiles and leather-working. Pop. (1980C) 235,617.

Dizful   Khuzistan, Iran. 32 23N 48 24E. Town in W central Iran sit. on Diz R. Industries inc. flour milling and dyeing. Pop. (1986C) 151,420.

Djakarta   *see* **Jakarta**

Djerba   Tunisia. 33 48N 10 54E.

Island at s end of Gulf of Gabès connected to the mainland by causeway. Area, 197 sq. m. (510 sq. km.). Chief occupations farming, esp. olives, dates and fishing. Manu. textiles and pottery. Pop. (1984E) 92,269.

Djibouti   11 36N 43 09E. Cap. and free port of Republic of Djibouti, sit. on Gulf of Aden. NE railway terminus from Addis Ababa and sea port. Industries inc. fishing and tourism. Pop. (1988E) 290,000.

Djibouti   East Africa. 11 36N 43 09E. Independence from France was achieved in 1977. Sit. on Gulf of Aden bounded w and s by Ethiopia and E by the Somali Rep. Area, 8,960 sq. m. (23,200 sq. km.). Cap. Djibouti. Plain rising to plateau (s). The main occupation is nomadic farming. Pop. (1988E) 484,000.

Dneprodzerzhinsk   Ukraine, U.S.S.R. 48 30N 34 37E. Town sit. WNW of Dnepropetrovsk on Dnieper R. with important metallurgical industries, esp. iron and steel; also manu. fertilizers and cement. Pop. (1985E) 271,000.

Dneproges   Ukraine, U.S.S.R. Suburb of Zaporozhye sit. on Dnieper R. with the largest dam and power station in Europe. The level of the Dnieper was raised more than 100 ft. (30 metres).

Dnepropetrovsk   Ukraine, U.S.S.R. 48 27N 34 59E. City sit. on Dnieper R. Cap. of region of same name; railway centre. Industries are based on nearby coal, iron and manganese, and power from Dneproges and inc. iron and steel, agric. engineering and chemicals; manu. machine tools. Pop. (1985E) 1,153,000.

Dnieper River   U.S.S.R. 46 30N 32 18E. Rises s of Valdai Hills, w of Moscow in the Smolensk region, flows 1,400 m. (2,240 km.) s and then w through Smolensk, then s through Belorussia to form part of the border with the Ukraine. It passes Kiev, and turning SE widens to a L. *c.* 77 m. (123 km.) long; passes Kremenchug, forms a further L.; passes Dnepropetrovsk and turns s with a further L. before it

enters the Black Sea by an estuary below Kherson. Navigable in the upper course for 8 months of the year; in the lower for 9. Forms the Dneproges reservoir for hydroelectric power. Chief tribs. are the Rs. Berezina, Pripet, Sozh and Desna.

Dniester River   U.S.S.R. 46 18N 30 17E. Rises in the Carpathian Mountains in w Ukraine and flows 870 m. (1,392 km.) on a meandering course, SE into the Moldavian Soviet Socialist Republic and through it to enter the Black Sea by an estuary w of Odessa. Ice-free for 10 months.

Döbeln   Leipzig, German Democratic Republic. 51 07N 13 07E. Town sit. ESE of Leipzig on the Freiberger Mulde R. Industries inc. agric. engineering; manu. steel and soap.

Dobrich   *see* **Tolbukhin**

Dobruja   Bulgaria/Romania. 44 30N 28 30E. Region between the Danube R. and the Black Sea, crossed by the Bulgarian/Romanian frontier and the Danube-Black Sea canal. Mainly agric. esp. cereal and vegetable farming. Chief town Constanta.

Dodecanese   Greece. 36 10N 27 00E. Group of islands in the Aegean Sea off the coast of Turkey. Area, 1,028 sq. m. (2,663 sq. km.). Cap. Rhodes. The main islands are Rhodes and Kos. Chief occupations farming and fishing. Chief products olives, fruit and sponges. Pop. (1981C) 145,071.

Dodoma   Central Province, Tanzania. 6 10S 35 40E. Cap. of Tanzania sit. WNW of Kilosa at 3,713 ft. (1,217 metres) a.s.l. It became cap. in 1974 replacing Dar-es-Salaam. Route centre on the N to s main road and Dar-es-Salaam railway. Trading centre for livestock and agric. produce. Pop. (1984E) 180,000.

Doetinchem   Gelderland, Netherlands. Town sit. E of Arnhem on the Old Ijssel R. Industries inc. meat canning, metal working, printing; manu. bricks, asphalt, motors and bicycle tyres. Pop. (1989E) 41,255.

Dogger Bank   North Sea. 54 40N 1 90E. Sandbank sit. E of the Northumberland coast with water at 10–20 fathoms. Important cod-fishing ground.

Dogs, Isle of   Greater London, England. 51 29N 0 01W. District of E London on the N bank of the Thames R., enclosed on 3 sides by the R. in a horseshoe bend.

Doha   Qatar. 25 17N 51 32E. Cap. of Qatar sit. on the E coast. It is Qatar's only main town. Its modern transformation has been funded by oil revenues. Commercial centre and sea port. Industries inc. oil refining, engineering, shipping, food processing (prawns), construction materials. Pop. (1986E) 217,294.

Dole   Jura, France. 47 06N 5 30E. Town sit. on the Doubs R. and birthplace of Louis Pasteur. Commercial centre for trade in wine and cheese. Industries inc. manu. agric. implements and chemicals. Pop. (1982C) 27,959.

Dolgellau   Gwynedd, Wales. 52 44N 3 53W. Town to the N of Cader Idris mountain on the Wnion R. Market town for a stock-farming area. Industries inc. fellmongery and tourism. Pop. (1989E) 2,261.

Dolgelley   *see* **Dolgellau**

Dolina   Ukrainian Soviet Socialist Republic, U.S.S.R. Town sit. SSE of Stry in a mining area. Industries inc. iron founding, oil refining; manu. potash fertilizer, flour milling, sawmilling and salt extraction.

Dolisie   Congo. 4 12S 12 41E. Town sit. W of Brazzaville. Commercial centre of a gold and lead mining and farming area. Industries inc. tanning and fibre processing (sisal).

Dolj   Romania. Province of S Romania. Cap. Craiova. Mainly agric. Pop. (1982E) 767,127.

Dollar   Central Region, Scotland. 56 09N 3 40W. Residential town sit. WSW of Kinross and S of the Ochil Hills. Pop. (1980E) 2,800.

Dolomites   Italy. 46 25N 11 50E. Alpine range in NE Italy rising to Marmolada, 10,965 ft. (3,595 metres) a.s.l. Jagged limestone peaks and sheer faces. A tourist area; chief resort Cortina d'Ampezzo.

Dom   Valais, Switzerland. 46 06N 7 50E. Highest peak in the Mischabelhörner range, 14,923 ft. (4,549 metres) a.s.l.

Dominica   West Indies. 15 30N 61 20W. Island in Windward group between Martinique and Guadeloupe. It became independent within the Commonwealth in 1978. Area 290 sq. m. (751 sq. km.). Cap. Roseau. Mountainous, with a range crossing the island N to S rising to Morne Diablotin, 4,747 ft. (1,447 metres) a.s.l. Some volcanic activity. Chief occupations farming and fishing. Chief products rum, lime juice, copra and citrates. Pop. (1988E) 81,200.

Dominican Republic   19 00N 70 40W. Rep. forming E half of Hispaniola island and bounded W by Haiti. Area, 18,699 sq. m. (48,430 sq. km.). Cap. Santo Domingo. Divided into the National District around the cap., and 25 provinces. Mountainous, rising to over 10,000 ft. (3,048 metres) a.s.l. in the Cordillera Central. Fertile, chief occupation agric., esp. sugar plantations; other crops grown and exported are coffee, cacao, bananas and tobacco. Industries inc. mining bauxite and rock salt, refining sugar; manu. tobacco products and foodstuffs. Chief towns are Santo Domingo, Santiago and San Francisco de Macorís. Pop. (1987E) 6,708,000.

Donaghadee   Down, Northern Ireland. 54 39N 5 33W. Town sit. ENE of Belfast on Irish Sea. Industries inc. tourism and the manu. of carpets. Pop. (1989E) 6,461.

Doncaster   South Yorkshire, England. 53 32N 1 07W. Metropolitan borough sit. NE of Sheffield on Don R., in a coalmining area. Industries inc. mining, railway engineering; manu. nylon, tractors, chemicals glass, plastics, electric motors, clothing, optical fibres, office equipment, toys and musical instruments. Pop. (1989E) 290,000.

Donegal Ireland. 54 39N 8 07W. (i) County in extreme NW bounded E by Northern Ireland and N and W by the Atlantic. Area, 1,865 sq. m. (4,830 sq. km.). County town, Lifford. Mountainous, with many Rs. and rocky coastal inlets. The Derryveach Mountains, in the NW, rise to 2,466 ft. (752 metres) a.s.l. Drained mainly by Rs. Foyle, Finn and Erne. Infertile, but farming is the main occupation, on a small scale. Chief towns Lifford, Ballyshannon, Donegal and Letterkenny. Pop. (1981c) 125,112. (ii) Town of same name in SW of county sit. at head of Donegal Bay. Market town for a farming area. Industries inc. manu. foodstuffs. Pop. (1971c) 1,725.

Donetsk Ukraine, U.S.S.R. 48 00N 37 48E. Cap. of region sit. N of Sea of Azov, and in a large industrial region in the Donets Basin coalfield, with an important metallurgical industry. Manu. iron and steel machinery, chemicals and cement. Pop. (1985E) 1,073,000.

Donets River U.S.S.R. 47 35N 40 54E. Rises in SW of the Russian Soviet Federal Socialist Republic NNE of Kharkov, and flows S and SE through the Ukraine, which it enters at Volchansk and leaves again in the SE near Kadiyevka, to join Don R. below Konstantinovski. It flows through an extensive coalfield and industrial area.

Dong Hoi Quangbinh, Vietnam. 17 18N 106 36E. Town sit. NW of Hué on Gulf of Tonkin. Industries inc. fishing, fish curing, wood carving, with mining and quarrying in the neighbourhood. Port, trading in cereals.

Dongola Northern Province, Sudan. 19 10N 30 29E. Town sit. above third cataract on Nile and 105 m. (168 km.) NW of Merowe. Trading centre on a caravan route.

Dongting Hu (Tung Ting Lake) Hunan, China. 29 18N 112 45E. L. in SE sit. SW of Wuhan. It is fed by Rs. Siang and Yuan, and drains to Yangtse R. in N.

Don River England. 53 39N 0 59W. Rises in Pennines W of Penistone and flows 70 m. (112 km.) SE to Sheffield, then NE past Rotherham, Conisborough and Doncaster to join Ouse R. at Goole, 11 m. (18 km.) NW of Scunthorpe.

Don River Scotland. 57 10N 2 04W. Rises near Banffshire border in SW Aberdeenshire and flows 82 m. (131 km.) E to enter North Sea at Aberdeen. Used for salmon fishing.

Don River U.S.S.R. 47 04N 39 18E. Rises near Tula in the Russian Soviet Federal Socialist Republic, and flows 1,200 m. (1,920 km) S to Voronezh, then SE to a canal-link with Volga R. near Volgograd, then SW by an extensive L. to Tsimlyanski and on to Rostov. It enters the Sea of Azov by a delta. Navigable to Voronezh, but closed by ice for 3–4 months annually. Used for transporting coal, grain and timber, and for fisheries. Chief tribs. the R. Voronezh, Donets and Medveditsa.

Doorn Utrecht, Netherlands. Town 11 m. (18 km.) SE of Utrecht; a health resort with tourism the main industry. Pop. (1989E) 10,500.

Doornik *see* **Tournai**

Dorchester Dorset, England. 50 42N 2 26W. County town sit. on Frome R. Market centre for an agric. area. Pop. (1989E) 15,000.

Dorchester Oxfordshire, England. 51 39N 1 10W. Village sit. SSE of Oxford on Thame R. near its confluence with Thames. Noted for the Abbey church.

Dordogne France. 45 10N 0 45E. Dept. in Aquitaine region, bounded E by Massif Central. Area, 3,546 sq. m. (9,184 sq. km.). Cap. Périgueux. Dry in the E, fertile in the W. Chief occupations farming and viticulture. Chief products grapes, wheat, tobacco and nuts. Pop. (1982c) 377,356.

Dordogne River France. 45 02N 0 35W. Rises on Mont Dore in the Auvergne Mountains and flows 290 m. (464 km.) SW and W to enter Gironde estuary NNE of Bordeaux. Used for hydroelectric power.

Dordrecht   Zuid-Holland, Netherlands. 51 49N 4 39E. Town sit. on S bank of Old Maas R. Industries inc. shipbuilding, engineering and chemicals; manu. machinery and glass. Pop. (1989E) 108,000.

Dorking   Surrey, England. 51 14N 0 20W. Town sit. 25 m. (40 km.) SSW of London on W bank of Mole R. and 1 m. (1.6 km.) SE of Box Hill, 596 ft. (182 metres) a.s.l. Market town for a residential and agric. area. Pop. (1985E) 19,900 (inc surrounding villages).

Dornakal   Andhra Pradesh, India. village sit. SW of Yellandlapad. Railway junction for the Singareni coalfield serving many mining centres of Yellandlapad and Kottagudem. Pop. (1981C) 14,558.

Dornbirn   Vorarlberg, Austria. 47 25N 9 44E. Town 40 m. (64 km.) NW of Landeck. Industries inc. textile manu. Pop. (1981C) 38,641.

Dornoch   Highland Region, Scotland. 57 52N 04 02W. County town in N, sit. on NW shore of Dornoch Firth. Industries inc. tourism and fishing. Pop (1985E) 1,100.

Dorset   England. 50 47N 2 20W. County of SW Eng. bounded S by English Channel and N by Somerset and Wiltshire. Under 1974 reorganization there are 4 borough councils (Bournemouth, Christchurch, Poole, and Weymouth and Portland); 4 district councils (North Dorset, West Dorset, Purbeck and East Dorset). County town Dorchester. The eastern part is low-lying, with extensive heathlands providing a rare wildlife habitat. The centre is made up of chalk down intersected by deep valleys cut by clear rivers and streams. The western part has an intricate, rich landscape of hills and valleys, small fields and woodlands. A main feature of North Dorset is the Blackmoor Vale. Drained by Rs. Frome and Stour. Mainly agric., esp. stock-farming, with quarrying in S. Chief towns are Dorchester, Bournemouth, Sherborne, Poole and Weymouth. Pop. (1988E) 655,700.

Dortmund   North Rhine-Westphalia, Federal Republic of Germany. 51 31N 7 28E. City in E of Ruhr industrial area, sit. on Emscher R. at head of Dortmund-Ems canal. Important centre for coalmining, steel and engineering industries. Pop. (1984E) 584,800.

Dorval   Quebec, Canada. 45 26N 73 44W. Town 10 m. (16 km.) SW of Montreal on S shore of Montreal Island, L. St. Louis. Resort and site of one of Montreal's airports.

Douai   Nord, France. 50 22N 3 04E. Town in NE sit. on the Scarpe R. in the N coalfield. Manu. iron and steel, glass, machinery and coke. It has an important engineering school. Pop. (1982C) 44,515 (agglomeration, 202,366).

Douala   Cameroon. 4 03N 9 42E. Largest city and main port of the Rep. Cap. of Coastal province. Sit. on Gulf of Guinea, opposite the island of Bioko. Industries inc. brewing, flour milling and textiles; manu. footwear. Sea port, exporting timber, cacao and bananas. Pop. (1986E) 1,029,731.

Douarnenez   Finistère, France. 48 06N 4 20W. Town in extreme NW sit. on S shore of Douarnenez Bay. Industries inc. sardine and tunny fishing, boatbuilding, net making and fish processing. Pop. (1982C) 17,813.

Doubs   France. 47 10N 6 25E. Dept. in Franche-Comté region bounded E by Switzerland. Area, 2,019 sq. m. (5,228 sq. km.). Cap. Besançon. Mountainous, crossed NE to SW by four parallel chains of the Jura Mountains. Lower and more fertile in the NW. Chief products cereals and vegetables. Industries inc. manu. clocks and watches, bicycles and motor-bicycles. Pop. (1982C) 477,163.

Doubs River   France/Switzerland. 46 54N 5 02E. Rises in E Jura near Frasne, France, flows 267 m. (427 km.) NE through limestone gorges to Franco/Swiss frontier near La Chaux-de-Fonds, Switz., and forms part of the frontier for *c.* 25 m. (40 km.) NE. It makes a brief loop into Switzerland, and returns W into France, past St. Hippolyte, then turns N past Pont-de-Roide, and finally SW passing Besançon and Dôle to join Saône R. at Chalon-sur-Saône. Its source is only 56 m. (90 km.) from its confluence with the Saône, in a direct line E to W.

Douglas Isle of Man, England. 54 09N 4 28W. Cap. sit. on E coast. Holiday resort. Industries inc. fishing, tourism, banking and financial services. Pop. (1986C) 20,368.

Douro River Portugal/Spain. 41 08N 8 40W. Rises at W end of the Sierra de Cebollera, flows 480 m. (768 km.) S then W to cross the Meseta, passes Zamora and joins the frontier with Portugal NE of Miranda. It turns SW to form the frontier down to N of La Fregeneda, then W across Portugal to enter the Atlantic 3 m. (5 km.) W of Oporto. Used for irrigation, esp. of vineyards, and hydroelectric power.

Dover Kent, England. 51 08N 1 19E. Town in extreme SE, sit. on Strait of Dover; terminus of shortest English Channel crossing: 22 m. (35 km.). Pop. (1989E) 36,000.

Dover Delaware, U.S.A. 39 10N 75 32W. City and State cap. sit. S of Wilmington on Dover R. Market and shipping centre for a rich agric. area. Industries inc. canning; manu. dairy products, hosiery, paint, rubber products. There is an air force base. Pop. (1988E) 28,000.

Dover New Hampshire, U.S.A. 43 12N 70 56W. Town sit. NW of Portsmouth on Cocheco R. near its confluence with Piscataqua R. Industries inc. manu. electronics, automobile equipment, plastics, athletic equipment, machinery. Pop. (1980C) 22,265.

Dover, Strait of English Channel. 51 00N 1 30E. Strait connecting English Channel with North Sea, extending *c.* 60 m. (96 km.) between SE Eng. and NE France 22 m. (35 km.) at its narrowest.

Dove River England. 54 20N 0 55W. Rises SW of Buxton in W Derbyshire and flows 40 m. (64 km.) S and SE down the Derbyshire/Staffordshire border to join Trent R. above Burton-on-Trent.

Dovrefjell Norway. 62 06N 9 25E. Mountain plateau in S Norway separated from Jotunheim Mountains to S by Gudbransdal valley, and rising to Snöhetta, 7,498 ft. (2,285 metres) a.s.l. The Oslo–Trondheim railway crosses it *c.* 100 m. (160 km.) S of Trondheim.

Down Northern Ireland. 54 24N 5 55W. County in E on Irish Sea coast, bounded N by Belfast Lough. Area, 952 sq. m. (2,466 sq. km.). Chief towns Downpatrick (county town), Bangor, Newry and Newtownards. Hilly, with a much-indented coastline, and rising to Slieve Donard in the SE. Chief occupations farming, esp. oats, potatoes and vegetables, and stockrearing. The main industry is linen manu. Pop. (1981C) 339,229.

Downey California, U.S.A. 33 56N 118 08W. City and suburb of SE Los Angeles 10 m. (16 km.) from city centre. Industries inc. manu. aircraft and soft drinks. Pop. (1984E) 83,582.

Downpatrick Down, Northern Ireland. 54 20N 5 43W. County town in SE. Market town for a farming area. Pop. (1981C) 8,245.

Downs England. Chalk hills in S Eng., particularly the parallel ranges across Surrey/Kent and Sussex respectively, called the North Downs and South Downs. The North Downs end in chalk cliffs at Dover, the South at Beachy Head. Noted as sheep pasture. The North Downs rise to Leith Hill, 965 ft. (294 metres) a.s.l., the South Downs to Butser Hill, 865 ft. (264 metres) a.s.l. Farther W are less extensive downs in Hampshire, Wiltshire and S Oxfordshire.

Downs, The Kent, England. Roadstead off Deal on E Kent coast, protected by Goodwin Sands in most weathers.

Drachenfels Federal Republic of Germany. Mountain in the Siebengebirge on W bank of Rhine R. 8 m. (13 km.) SE of Bonn, height 1,053 ft. (321 metres) a.s.l.

Draguignan Var, France. 43 32N 6 28E. Former cap. of dept. on Nartuby

R. Industries inc. grapes, olive oil and leather. Pop (1982C) 28,194.

Drakensberg Republic of South Africa. 27 00S 30 00E. Mountain range extending NE to SW through the Transvaal, Natal and Cape Province, forming the W frontier of Swaziland with Transvaal, and the E frontier of Lesotho with Natal and Cape Province. The highest peaks are Thaban Ntlenyana, 11,245 ft. (3,427 metres) a.s.l. and Mont-aux-Sources, 10,822 ft. (3,299 metres) a.s.l.

Drama Greece. 41 09N 24 08E. (i) Prefecture in Macedonia bounded N by Bulgaria. Area, 1,339 sq. m. (3,468 sq. km.). Drained by Mesta R. Mainly agric., esp. tobacco growing. Pop. (1981C) 94,772. (ii) Cap. of (i) sit. ENE of Thessaloniki. Market town trading in tobacco. Pop. (1981C) 36,109.

Drammen Buskerud, Norway. 59 44N 10 15E. Town and seaport in SE at the mouth of Drammen R. where it enters Drammen Fjord. Industries inc. food processing, brewing, cables, high-tech water power equip., engineering; manu. paper, wood pulp and cellulose. Pop. (1990E) 52,000.

Drammen River Norway. 59 44N 10 14E. Rises as Hallingdal R. on the Hallingskarv Mountains and flows 190 m. (304 km.) E and then SSE to enter the Skagerrak through Drammen Fjord. Used for hydroelectric power.

Drancy Seine St-Denis, France. 48 55N 2 26E. Residential suburb of NE Paris. Pop. (1982C) 60,224.

Drava River Italy/Austria/Yugoslavia. 45 33N 18 55E. Rises in the Carnic Alps, on the Italian/Austrian border, and flows E through S Austria to enter Yugoslavia E of Bleiburg, then flows ESE to form the Yugoslav/Hungarian frontier, from its confluence with Mur R., until it turns SE past Osijek to join Danube R. Navigable for steamers up to Barcs.

Drenthe Netherlands. 52 45N 6 30E. Province in NE bounded E by Federal Republic of Germany. Area, 1,021 sq. m. (2,644 sq. km.). Cap. Assen. Generally infertile with much heathland, some of it being reclaimed for farming. Chief occupations growing rye and potatoes; rearing cattle; extracting peat. There are some oil and natural gas deposits. Pop. (1984E) 427,336.

Dresden Dresden, German Democratic Republic. 51 03N 13 44E. Cap. of district in SE, sit. on Elbe R. Industries inc. manu. machine tools, musical instruments, electrical equipment, chemical and pharmaceutical goods, optical instruments and tobacco products. The city was reduced to ruins by bombing in 1945 when more than 50,000 people may have died. Since rebuilt. Pop. (1989E) 519,000.

Driffield Humberside, England. 54 00N 0 27W. Town in NE, near N. Sea coast. Market town for an agric. area. Industries inc. agric. engineering, flour milling and manu. clothing, nursery products, spectacle frames and lenses. Pop. (1987E) 9,600.

Drina River Yugoslavia. 44 53N 19 21E. Formed by confluence of Tara and Piva R., on N border of Montenegro, and flows 290 m. (464 km.) NNE to join Sava R. *c.* 65 m. (104 km.) W of Belgrade.

Drin River Albania. Formed by confluence of White Drin and Black Drin R. at Kukës and flows 100 m. (160 km.) W on to the plain of Shkodër where it divides. The older course flows S to enter the Adriatic at Lezhë; the W arm joins Bojana R.

Drogheda Louth, Ireland. 53 43N 6 21W. Town sit. near mouth of Boyne R. on NE coast. Industries inc. engineering; manu. clothing, footwear, laboratory products and cement. Sea port trading in cattle. Pop (1981C) 23,247.

Drogobych Ukraine, U.S.S.R. 49 21N 23 30E. Town sit. SW of Lvov. Centre of production for petroleum and natural gas. Industries inc. oil refining, metal working, chemicals.

Droitwich Spa Hereford and Worcester, England. 52 16N 2 09W. Town sit. NNE of Worcester on Salwarpe R. Conference town with new brine spa

(1985). Industries inc. cold storage warehousing, distribution, packaging, metal castings. Manu. heating equipment and vacuum cleaners. Spa with salt springs. Pop. (1990E) 24,500.

Drôme   France. 44 35N 5 10E. Dept. in Rhône-Alpes region, bounded w by Rhône R. Area, 2,519 sq. m. (6,525 sq. km.). Chief towns Valence (the cap.), and Montélimar. Mountainous in the E, where the Cottian Alps rise to 7,890 ft. (2,405 metres) a.s.l. Chief occupations farming and viticulture; chief products wine, maize and olives. Industries inc. textiles. Pop. (1982C) 389,781.

Drôme River   France. 44 46N 4 46E. Rises on w side of Cottian Alps and flows 65 m. (104 km.) WNW to join Rhône R. near Montélimar.

Dromore   Down, Northern Ireland. 54 25N 6 09W. Town sit. SW of Belfast on Lagan R. Market town for an agric. area. Industries inc. linen manu.

Droylsden   Greater Manchester, England. 53 29N 2 09W. Town immediately E of Manchester. Industries inc. textiles, chemicals and engineering. Pop. (1987E) 22,484.

Drumheller   Alberta, Canada. 51 28N 112 42W. Town sit. NE of Calgary on Red Deer R. Centre of a farming area. There is oil and gas. Pop. (1989E) 6,366.

Drummondville   Quebec, Canada. 45 53N 72 30W. Town sit. ENE of Montreal on St. Francis R. Industries, powered by a hydroelectric plant, inc. manu. paper, synthetic fibres, electrical machinery, electronics, printing, furniture, publishing and hosiery. Pop (1985E) 39,000.

Drygalski Island   Antarctica. Small island *c.* 550 m. (880 km.) off Mirnyy on coast of Queen Mary Land, in Australian Antarctic Territory.

Dschang   Cameroon. 5 27N 10 04E. Town in w province. Agric. centre and tourist resort. Pop. (1981E) 21,705.

Duarte   Dominican Republic. Province of central Dominican Rep.

bounded S by the Cordillera Central. Area, 1,090 sq. m. (2,823 sq. km.). Cap. San Francisco de Macorís. Densely populated farming area, producing cacao, coffee, rice, fruit. Some iron deposits at Cotuí. Pop. (1987E) 235,544.

Dubai   United Arab Emirates. 25 18N 55 18E. (i) Sheikhdom on SE shore of (Persian) Gulf. Cap. Dubai. Chief occupations trading and fishing. Chief industry petroleum. Pop. (1985C) 419,104. (ii) Cap. of (i) on Gulf, chief sea port of Dubai Sheikhdom. Commercial and communications centre. Pop. (1980C) 265,702.

Dubawnt Lake   North West Territories, Canada. 63 00N 102 00W. L. in w Keewatin District on the border with Mackenzie District. Area, 1,600 sq. m. (4,144 sq. km.). Fed by Dubawnt R. and draining through Baker L. and Chesterfield Inlet into Hudson Bay.

Dubbo   New South Wales, Australia. 32 15S 148 36E. Town sit. NW of Sydney on Macquarie R. Market centre for a sheep and arable farming area. Industries inc. flour milling; manu. clothing. Pop. (1981C) 23,986.

Dublin   Ireland. 53 20N 61 15W. Cap. of Ireland and county town sit. at mouth of Liffey R. on Dublin Bay, and entrance to the great central plain of Ireland. Dublin Centre, which was to become a synonym of British rule, was completed in 1220. Commercial centre and sea port trading in livestock and farm produce. Industries inc. brewing, chemicals, distilling, engineering, flour milling and textiles. Pop. (county boro. 1986E) 502,749.

Dublin   Ireland. County on E seaboard, bounded S by Wicklow, w and N by Kildare and Meath. Area, 356 sq. m. (922 sq. km.). Chief towns Dublin (cap. and county town) and Dun Laoghaire. Mountainous in the S rising to Kippure, 2,473 ft. (754 metres) a.s.l. in the Wicklow Mountains. Chief occupation farming, esp. oats, potatoes and cattle. Pop. (1981C) 1,003,164.

Dubrovnik   Croatia, Yugoslavia. 42 38N 18 07E. Town on Dalmatian coast; a tourist resort; industries inc. manu. of liqueurs, silk, leathergoods and cheeses. Pop. (1985E) 33,000.

Dubuque   Iowa, U.S.A. 42 30N 90 41W. Town sit. NE of Cedar Rapids on Mississippi R. Industries inc. meat packing; manu. agric. machinery, furniture and clothing. Pop. (1980c) 62,321.

Ducie   Pitcairn Island, South Pacific. 24 47S 124 50W. Small uninhabited islet off Pitcairn, annexed 1902 as part of Pitcairn district.

Dudelange   Luxembourg. 49 28N 6 05E. Town sit. ESE of Esch-sur-Alzette near French border. Industries inc. iron and steel; manu. synthetic fertilizer, tobacco products, beer. Pop. (1984E) 14,074.

Dudley   West `Midlands, England. 52 30N 2 05W. Metropolitan district of West Midlands county, sit. in an industrial area, with one of the biggest shopping complexes in Europe. Manu. wrought iron products, boilers, chains, plastics, textiles, chemicals, electronics, leather goods, bricks and tiles. Pop. (1981c) 187,228.

Dudweiler   Saarland, Federal Republic of Germany. 49 17N 7 02E. Town sit. NE of Saarbrücken in a coal-mining area. Industries inc. metal working; manu. electrical equipment. Pop. (1985E) 34,000.

Duff Islands   South West Pacific. Group of islands in the Solomon group. Volcanic.

Dugi Otok   Bosnia and Hercegovina, Yugoslavia. Island extending 28 m. (45 km.) NW to SE off Dalmatian coast opposite Zadar. Width 3 m. (5 km.) Chief occupations fishing and tourism. Chief town Sali.

Duisburg   North Rhine-Westphalia, Federal Republic of Germany. 51 27N 6 42E. Town in W sit. at confluence of Rs. Rhine and Ruhr. Industries inc. coal, iron and steel, shipbuilding, engineering, tourism, textiles and chemicals. An important inland port. Pop. (1989E) 528,062.

Dukeries, The   Nottinghamshire, England. District of NW Nottinghamshire in Sherwood Forest, consisting of the parks of former ducal seats at Welbeck, Worksop, Clumber and Thoresby.

Dukhan   Qatar. 25 25N 50 50E. Oilfield on W coast of Qatar peninsula, linked by pipeline to its terminal at Umm Said. Oil discovered in 1939 but only developed after 1945.

Dukinfield   Greater Manchester, England. Town sit. E of Manchester. Industries inc. engineering, textiles, rubber and plastics. Pop. (1987E) 22,165.

Dulain   *see* **Ramadi**

Duluth   Minnesota, U.S.A. 46 47N 92 06W. City in NE of state at mouth of St. Louis R. at W end of L. Superior. Industries inc. publishing, medical, tourism, and maritime related goods. Manu. inc. machinery, food products, paper and wood products, and cement. Important port handling grain, coal, taconite and iron ore from extensive regional deposits. Pop. (1980c) 92,811.

Dulwich   Greater London, England. Residential suburb of S London noted for Dulwich College, and art gallery.

Dumbarton   Strathclyde Region, Scotland. 55 57N 4 35W. Town at confluence of Rs. Leven and Clyde. Industries inc. heavy and light engineering and whisky distilling, blending and bottling. Pop. (1981c) 23,204.

Dum-Dum   West Bengal, India. 23 27N 88 25E. Town immediately NE of Calcutta. Industries inc. engineering, pharmaceuticals, railway and tram carriages. Site of Calcutta airport. Pop (1981c) 360,288.

Dumfries   Dumfries and Galloway Region, Scotland. 55 04N 3 37W. Town in S sit. on Nith R. Market town for a farming area. Tourist centre. Industries inc. manu. plastics, rubber, engineering, tweed and hosiery. Pop. (1981c) 31,600.

Dumfries and Galloway   Scotland.

55 12N 3 30W. Region of sw Scot. bounded s by Solway Firth. Area, 2,500 sq. m. (6,475 sq. km.) comprising the counties of Dumfries, Kirkcudbright and Wigtown. Industries inc. stained glass, woollen, engineering, food processing, rubber and aluminium. Pop. (1990E) 147,036.

Dumfriesshire Scotland. Former county now part of Dumfries and Galloway Region.

Dunajec River Poland. 50 15N 20 44E. Rises in High Tatra and flows 130 m. (208 km.) E and N, in the upper courses as Rs. Czarny (Black) Dunajec and Bialy (White) Dunajec. It passes Nowy Sacz and Tarnow and joins Vistula R. *c.* 40 m. (64 km.) below Kraków.

Dunbar Lothian Region, Scotland. 56 00N 2 31W. Town sit. E of Edinburgh on North Sea. Industries inc. tourism, brewing, cement and operation of nuclear power station and fishing. Pop. (1981C) 5,797.

Dunbartonshire Scotland. Former county now part of Strathclyde Region.

Dunblane Central Region, Scotland. 56 12N 3 59W. Town sit. N of Stirling on Allan Water. Market town for a farming area. Industries inc. light engineering and woollen manu. Pop. (1980E) 3,000.

Dundalk Louth, Ireland. 54 01N 6 25W. County town near mouth of Castletown R. on NE coast midway between Dublin and Belfast. Industries inc. engineering, breweries, footwear, electronics, tobacco, foodstuffs and textiles. Sea port exporting grain, cattle and meat. Pop. (1981C) 25,610.

Dundee Tayside Region, Scotland. 56 28N 3 00W. City in E sit. on N bank of R. Tay. Industries inc. computers, textiles, light engineering, high technology and North Sea oil supplies and servicing. Pop. (1988E) 174,255.

Dunedin South Island, New Zealand. 45 53S 170 30E. Town on E coast at Head of Otago Harbour. Industries inc. food processing and the

manu. woollen goods, agric. machinery, footwear and clothing. Sea port, through Port Chalmers, 8 m. (13 km.) NE exporting meat, timber, wool, fruit and dairy produce. Pop. (1989E) 116,000.

Dunfermline Fife Region, Scotland. 56 04N 3 29W. Town sit. on N shore of Firth of Forth. Manu. and electronics. Industries inc. service, coalmining, engineering, petrochemicals and oil-related. Pop. (1988E) 42,720.

Dungannon Tyrone, Northern Ireland. 54 31N 6 46W. Town sit. WSW of Belfast. Market town for an agric. area. Manu. inc. linen and brick-making. Pop. (1980C) 8,000.

Dungarvan Waterford, Ireland. 52 05N 7 37W. Town on Dungarvan Bay, an inlet on the s coast. Sea port and market town. Industries inc. glass, iron, pharmaceuticals, leather and dairy products. Pop. (1986E) 6,849.

Dungeness Kent, England. 50 55N 0 58E. Shingle headland at extreme s tip of Kent, still extending as shingle accumulates, with a lighthouse serving w end of Strait of Dover.

Dungun Trengganu, Malaysia. Town sit. SSE of Kuala Trengganu at mouth of Besut R. on South China Sea. The port ships iron ore from Bukit Besi. Industries inc. boatbuilding, sawmilling.

Dunkirk *see* **Dunkerque**

Dunkerque Nord, France. 51 03 2 22E. Town in extreme NE on North Sea. Industries inc. shipbuilding, oil refining, nuclear energy; manu. fishing equipment. Scene of the evacuation of Allied troops during the Second World War, 1940. Pop. (1982C) 73,618 (agglomeration, 211,475).

Dunkwa Western Province, Ghana. 5 59N 1 45W. Town sit. N of Takoradi on Ofin R. Rail junction serving Awaso bauxite mines. Industries inc. gold mining, sawmilling.

Dún Laoghaire Dublin, Ireland. 53

17N 6 08W. Town sit. in E on S shore of Dublin Bay. Passenger sea port for Dublin. Industries inc. tourism. Pop. (1986E) 54,715.

Dunmow *see* **Great Dunmow**

Dunoon Strathclyde Region, Scotland. 55 57N 4 56W. Town sit W of Greenock on Firth of Clyde. Industries inc. tourism. Pop. (1981E) 9,372.

Dunstable Bedfordshire, England. 51 53N 0 32W. Town sit. W of Luton at N end of Chiltern Hills. Centre for the distributive industries. Industries inc. engineering and printing, manu. commercial vehicles, hospital equipment, photographic materials, paper and cement. Pop. (1984E) 35,716.

Dunwich Suffolk, England. 52 16N 1 38E. Village sit. N of Aldeburgh on North Sea. An important port in Roman, Saxon and early medieval times, destroyed by sea erosion.

Duque de Caxias Rio de Janeiro, Brazil. 22 47S 43 18W. Town sit. NNW of centre of Rio city near Guanabara Bay. Industries inc. manu. motors, flax and jute milling. Pop (1980C) 306,057.

Durance River France. 43 55N 4 44E. Rises near Briançon in the High Alps near the border with N Italy, and flows 218 m. (349 km.) SSW through deep gorges, turning W to join Rhône R. below Avignon. Used for hydroelectric power in the upper course and irrigation in the lower.

Durango Mexico. 24 02N 104 40W. (i) State in N Mex. separated from Gulf of California by Sinaloa. Area, 47,560 sq. m. (123,181 sq. km.). High plateau rising to the Sierra Madre Occidental in the W. Chief occupations mining and farming. Chief products silver, gold, iron and copper in the mountains; cotton and wheat on lower ground esp. in the Nazas R. valley. Pop. (1980C) 1,182,320 (ii) Cap. of (i) sit. SW of Torreón at 6,314 ft. (1,925 metres) a.s.l. in the Sierra Madre Occidental. Mining town. Industries inc. iron founding, sugar refining and textiles; manu. glass. Pop. (1980E) 137,000.

Durango Colorado, U.S.A. 37 16N 107 53W. City at the base of San Juan mountains in SW Colorado. Originally a transport centre, now a resort town. Tourism is the main industry. Pop. (1989E) 12,700.

Durazno Durazno, Uruguay. 32 22S 56 31W. Cap. of dept. sit. NNW of Montevideo on Yí R. Communications centre and market town trading in livestock, hides and grain. Industries inc. flour milling and meat packing. Pop. (1986E) 27,602.

Durazzo *see* **Durrës**

Durban Natal, Republic of South Africa. 29 55S 30 56E. City sit. in E on Indian Ocean. Industries inc. railway engineering, vehicle assembly, oil refining; manu. machinery, metal goods and furniture. Sea port with greater overseas trade than Cape Town, handling coal, ores, maize, wool, sugar, oranges and pineapples. Important holiday resort. Pop. (1985E) 932,075.

Düren North Rhine-Westphalia, Federal Republic of Germany. 50 48N 6 28E. Town sit. WSW of Cologne on Roer R. Industries inc. manu. textiles, paper, glass and metal goods. Pop. (1984E) 85,000.

Durgapur West Bengal, India. 23 29N 87 20E. Town sit. SE of Raniganj on Damodar R. Site of major thermalelectric plant. Other industries inc. heavy engineering, mining machinery, fertilizers, glass, cement, carbon block, graphite, alloy steel and steel. Pop. (1981C) 313,798.

Durham England. 54 47N 1 34W. County in NE Eng. bounded E by North Sea. Under 1974 re-organization consists of the districts of Derwentside, Chester-le-Street, Wear Valley, Durham, Easington, Teesdale, Sedgefield and Darlington. Hilly in W rising to the Pennine Hills, sloping to a coastal plain; drained by Rs. Wear, Derwent and Tees. Engineering and services are the main occupations although coalmining remains important in E and farming in rural areas. Durham and Darlington are the two main towns. Pop. (1988E) 596,800.

Durham Durham, England. 54 47N 1 34W. County town sit. in loop of R. Wear. Episcopal seat and university town, noted for the cathedral begun 1093, and castle. Industries inc. electrical engineering and carpet making. Pop (1983E) 88,600.

Durham New Hampshire, U.S.A. 43 08N 70 56W. Town immediately s of Dover on Oyster R. in an agric. area. Seat of the University of New Hampshire. Pop. (1980C) 10,652.

Durham North Carolina, U.S.A. 35 59N 78 54W. City in N centre of state. Commercial centre for education, medicine and high technology industries. Pop. (1990E) 140,000.

Durrës (Enver Hoxha) Albania. 41 18N 19 28E. Town sit. on Adriatic coast. Sea port handling olive oil and tobacco. Small industries inc. milling. Pop. (1980E) 65,900.

Durrsi *see* **Durrës**

D'Urville Island Antarctica. Island off NE tip of Palmer Peninsula and N of Joinville Island, s Atlantic.

D'Urville Island New Zealand. 40 50S 173 55E. Island near N coast of S Island, New Zealand, at E entrance to Tasman Bay. Area, 90 sq. m. (233 sq. km.). Main occupation lumbering.

Dushanbe Tadzhikistan, U.S.S.R. 38 35N 68 48E. Cap. of the Tadzhik Soviet Socialist Rep. sit. N of border with Afghánistán. Industries inc. textiles and meat packing; manu. cement and leather. Connected with Trans-Caspian Railway. Formerly known as Stalinabad. Pop. (1985E) 552,000.

Düsseldorf North Rhine-Westphalia, Federal Republic of Germany. 51 12N 6 47E. City sit. on Rhine R. Industries inc. manu. iron and steel, cars, chemicals, machinery, glass, textiles and clothing. Important R. port and commercial centre. Pop. (1985E) 569,361.

Dvina River U.S.S.R. 57 04N 24 03E. (i) Northern Dvina, formed by confluence of R. Sukhona and Yug, flows 470 m. (752 km.) NW through N Russian Soviet Federal Socialist Republic to enter White Sea above Arkhangelsk. Chief tribs. R. Vychegda, Pinega and Vaga; linked with the Mariinsk canal system. Navigable May–Nov. (ii) Western Dvina rises in the Valdai Hills in NW Russian Soviet Federal Socialist Republic and flows 640 m. (1,024 km.) sw into Belorussia, past Vitebsk, then NW to Riga and the Gulf of Riga. Partly navigable May–Nov.

Dyfed Wales. 52 00N 4 30W. County of sw Wales bordered w by Irish Sea. Comprises former counties of Cardiganshire, Carmarthenshire and Pembroke and includes the 6 districts of Carmarthen, Ceredigion, Direfwr, Llanelli, Preseli and South Pembrokeshire. There are 4 refineries and 5 tanker terminals. Many new industries have been developed in recent years and new investment in the coal and tinplate industry has taken place. Tourism is important. Pop. (1988E) 348,400.

Dzaoudzi Mayotte, Indian Ocean. Cap. of Fr. island depend. sit. on small offshore islet. Pop. (1978C) 4,256.

Dzerzhinsk Russian Soviet Federal Socialist Republic, U.S.S.R. 56 15N 43 24E. Town sit. w of Gorki on Oka R. Industries inc. sawmilling and engineering; manu. chemicals esp. fertilizers. Pop. (1985E) 274,000.

Dzhambul Kazakhstan, U.S.S.R. 42 54N 71 22E. Town sit. w of Alma-Ata on Turkish Railway. Cap. of Dzhambul region. Industries inc. fruit canning and sugar refining, manu. super-phosphates and prefabricated buildings. Pop. (1980E) 270,000.

Dzierzonlow Wroclaw, Poland. 50 44N 16 39E. Town sit. ssw of Wroclaw. Industries inc. textiles; manu. electrical machinery.

E

Ealing Greater London, England. 51 31N 0 18W. District of w Lond. Ealing proper is residential, with industries at Greenford and Hanwell. Pop. (1988E) 297,300.

Earby Lancashire, England. 55 55N 2 08W. Town sit. NNE of Burnley. Industries inc. plastics, textiles, esp. cotton and precision engineering. Pop. (1981C) 4,995.

Earl's Court Greater London, England. District of w Central Lond. bounded w by Kensington and E by Fulham. Mainly residential with hotels and apartment houses. Noted as the site of the Earl's Court exhibition hall.

Earlston Borders Region, Scotland. 55 39N 2 40W. Town sit. ENE of Galashiels on Leader Water. Industries inc. timber processing and engineering. Pop. (1981C) 1,677.

Earn River Scotland. 56 23N 4 14W. Emerges from Loch Earn, Perthshire, and flows E for 46 m. (74 km.) to enter Firth of Tay 6 m. (10 km.) SE of Perth.

Easington Durham, England. Town sit. E of Durham. Pop. (1985E) 9,100.

East Anglia England. Area of E England forming promontory into North Sea between Thames Estuary (S) and Wash (N) and inc. counties of Norfolk and Suffolk. Flat and fertile land producing cereals and vegetables, with an important fishing industry based on Yarmouth and Lowestoft, mainly herring.

East Barnet Greater London, England. 51 38N 0 10W. District of N London. Mainly residential.

East Bengal *see* **Bangladesh**

Eastbourne. East Sussex, England. 50 46N 0 17E. Resort sit. on English Channel at foot of S. Downs. Beachy Head is 3 m. (5 km.) SW. Pop. (1989E) 82,000.

East Chicago Indiana, U.S.A. 41 38N 87 27E. City and port sit. SE of Chicago on L. Michigan. Industries inc. steel, railways, railway engineering, oil refining and storage. Manu. chemicals. Pop. (1988E) 35,990.

East China Sea Pacific Ocean. 30 05N 126 00E. Arm of Pacific Ocean extending 600 m. (960 km.) SSW to NNE between E China coast and Kyushu and Ryukyu Islands. Area, 480,000 sq. m. (1,243,200 sq. km.). Width 300–500 m. (480–800 km), average depth 615 ft. (187 metres) but reaching 8,920 ft. (2,719 metres). Connected SW with South China Sea through Formosa Strait and NE with Sea of Japan through Korea Strait.

East Cleveland Ohio, U.S.A. 41 32N 81 35W. City next to Cleveland; mainly residential with some industries inc. manu. metal goods. Pop. (1980C) 36,957.

East Dereham Norfolk, England. 52 41N 0 56E. Market town and industries inc. trailer manu., light engineering, toys and footwear. Pop. (1982E) 12,364.

Easter Island South Pacific. 27 08S 109 23W. Island forming a possession of Chile and *c.* 2,350 m. (3,760 km.) w of the Chilean coast. Area, 46 sq. m. (118 sq. km.). Volcanic, rising to 1,765 ft. (538 metres) a.s.l. Fertile, producing tobacco, sugar-cane, vegetables, fruit. Main occupations farming and fishing. Noted for the carved stone heads 30–40 ft. (9–12 metres) tall which are of unknown origin. Pop. (1982E) 1,867.

Eastern Ghats *see* **Ghats**

East Germany *see* **German Democratic Republic**

171

East Grinstead   West Sussex, England. 51 07N 0 01W. Town sit. near Ashdown Forest. Pop. (1984E) 22,000.

Eastham   Merseyside, England. Town sit. SSE of Birkenhead on Mersey R. at W end of Manchester Ship Canal.

East Ham   Greater London, England. District of ENE London sit. at mouth of Roding R. *See* Newham.

East Hartford   Connecticut. U.S.A. 41 46N 72 39W. Town sit. on Connecticut R. opposite Hartford, and NNE of New Haven. Manu. inc. aircraft engines, machinery, furniture and paper. Pop. (1980C) 52,563.

East Indies   South East Asia. 5 00N 120 00E. Term describing the Malay Arch. and often used to refer to the former Dutch East Indies, now Indonesia. Many of the islands have volcanoes. Their climate is hot and moist and they fall within the influence of the monsoons. Area, *c.* 1m. sq. m. (2.59m. sq. km.).

East Kilbride   Strathclyde Region, Scotland. 55 46N 4 10W. Town sit. SSE of Glasgow. Industries inc. engineering, printing, electronics and the manu. of aircraft engines, clothing and electrical equipment. Designated a New town in 1947 with proposed pop. of 82,500. Pop. (1985E) 72,000.

East Lansing   Michigan, U.S.A. 42 44N 84 29W. City sit. E of Lansing and a residential suburb of that town. Pop. (1980C) 51,392.

Eastleigh   Hampshire, England. 50 58N 1 22W. Town sit. NNE of Southampton. Market and shopping centre. Industries inc. high technology, railway, and communications engineering. Pop. (1985E) 99,000.

East London   Cape Province, Republic of South Africa. 33 02S 27 55E. Port and resort on Indian Ocean, sit. at mouth of Buffalo R. Industries inc. railway engineering, fishing, and the manu. of soap, leather goods, furniture and clothing. Exports inc. maize, meat, hides, wool, fruit and dairy products. Pop. (1980E) 160,582.

East Lothian   Scotland. Former county now part of Lothian Region.

East Meadow   New York, U.S.A. Village sit. E of Hempstead on W Long Island. Residential. Pop. (1980C) 39,317.

Easton   Pennsylvania, U.S.A. 40 42N 75 12W. City sit. N of Philadelphia at confluence of Rs. Delaware and Lehigh. Part of Allentown–Bethlehem–Easton industrial area. Manu. inc. machinery, chemicals, steel, paper and textiles. Pop. (1980C) 26,027.

East Orange   New Jersey, U.S.A. 40 46N 74 13W. City adjoining Newark and near New York. Mainly residential but also some diverse manu. Pop. (1980C) 77,690.

East Pakistan   *see* **Bangladesh**

East Point   Georgia, U.S.A. 33 40N 84 27W. City sit. just SSW of Atlanta. Manu. inc. fertilizers, textiles and furniture. Pop. (1980C) 37,486

East Providence   Rhode Island, U.S.A. 41 49N 71 37W. Town sit. on Seekonk R. and Providence R. opposite Providence. Industries inc. shipbuilding, textile dyeing, engineering, and manu. of petroleum products, wire and steel goods, chemicals and paper. Pop. (1980C) 50,980.

East Retford   *see* **Retford**

East River   China. 21 53N 108 35E. Rises in the Kiulen Mountains in S Kiangsi province and flows 250 m. (400 km.) SW then W to join Canton R. through a delta below Shekling. Main trib. Tseng R. Navigable below Laolung for 200 m. (320 km.).

East River   New York, U.S.A. 40 48N 73 48W. Tidal strait connecting Upper New York Bay and Long Island Sound between Manhattan and the Bronx and Long Island. Connected through Harlem R. with Hudson R. Length 16 m. (26 km.), width varies from 600–4,000 ft. (183–1,219 metres). Crossed by Brooklyn, Manhattan, Williamsburg, Queensboro, Triborough and Bronx-Whitestone road bridges and Hell Gate railway bridge. There are commercial and naval docks and waterfront with heavy traffic on the Brooklyn shore.

East St. Louis   Illinois, U.S.A. 38 38N 90 08W. City and port sit. on Mississippi R. opposite St. Louis. An important railway centre with industries inc. oil refineries, aluminium and steel works, meat packing plants, grain mills and manu. of chemicals, fertilizers, glass and paint. Pop. (1980c) 55,200

East Sussex   England. 50 55N 0 20E. Under 1974 reorganization the county consists of the boroughs of Brighton, Eastbourne, Hastings and Hove and the districts of Lewes, Rother and Wealden. The predominant economic activity is in the financial service sector but industries include electronics, pharmaceuticals, furniture manu. and light industries. An important tourist area. Pop. (1988E) 698,000.

Eau Claire   Wisconsin, U.S.A. 44 49N 91 31W. City in w centre of state, sit. at confluence of Rs. Chippewa and Eau Claire. Trade and service centre of an agric. district with manu. of tyres, paper, computers, electronics, plastics and food. Pop. (1985E) 54,731.

Ebbw Vale   Gwent, Wales. 51 47N 3 12W. Town sit. N of Cardiff on Ebbw R. Industries inc. the manu. of tinplate, batteries and vehicle brake systems. Pop. (1988E) 24,100.

Ebebiyín   Equatorial Guinea. Town sit. ENE of Bata on border with Cameroon and Gabon. Main products cacao, coffee.

Eberswalde   Frankfurt, German Democratic Republic. 52 50N 13 49E. Town sit. in NE, near border with Poland, on Finow Canal. Manu. iron and steel goods, esp. cranes, building materials and chemicals. Pop. (1989E) 54,400.

Ebolowa   Cameroon. 2 54N 11 09E. Town. South province sit. ssw of Yaoundé. Trading centre for cacao. There is an agric. school. Pop. (1981E) 22,222.

Ebro River   Spain. 40 43N 0 54E. Rises in Cantabrian Mountains in N and flows *c.* 550 m. (880 km.) on an irregular and meandering course first SE between Álava and Navarre provinces and Logroño province, then across the plains of Aragon where it feeds an extensive canal system and continues to Mequinenza, Catalonia, where it turns SSE to enter the Mediterranean by a wide delta below Tortosa and 80 m. (128 km.) SW of Barcelona. The main cities on its course are Saragossa, Tortosa, Miranda de Ebro, Logroño, Calahorra, Tudela, Caspe, San Carlos de la Rápita. Ocean-going ships can go up to Tortosa. Main tribs. Aragon, Gállego, Segre, Jalón, Huerva, Guadalope Rs., the system drains *c.* one-sixth of Spain and supplies about half its hydroelectric power.

Eccles   Greater Manchester, England. 53 29N 2 21W. Town sit. w of Manchester on Irwell R. and Manchester Ship Canal. Industries inc. cotton milling, textile printing; manu. chemicals, pharmaceutical goods, machinery. Site of Manchester airport; noted as the original source of 'Eccles cakes'. Pop. (1981c) 37,166.

Ecuador   2 00S 78 00W. South America. Rep. in NW South America bounded w by Pacific, N by Colombia, E and S by Peru, and inc. the Galápagos Islands *c.* 650 m. (1,040 km.) offshore. Colony founded 1532 by Spaniards; became republic in 1830. The frontier with Peru has long been a source of dispute between the two countries. The latest delimitation of it was in the Treaty of Rio, in 1942, when, after being invaded by Peru, Ecuador ceded the latter over half her Amazonian territories. Ecuador unilaterally denounced this Treaty in 1961. No definite figure of the area of the country can yet be given, as a portion of the frontier has not been delimited. One estimate shows 177,912 sq. m. (455,454 sq. km.), including the Archipelago of Colón (the Galápagos Islands) with 2,902 sq. m. (7,430 sq. km.). The United Nations excludes the 'Región Oriental' and the Galápagos Islands and gives the settled portion of Ecuador as 104,505 sq. m. (270,670 sq. km.). Chief cities Quito (cap.), Guayaquil, Cuenca, Ibarra, Latacunga, Riobamba, Loja.

There are 3 distinct zones: the *Sierra* or uplands of the Andes, consisting of high mountain ridges with valleys, with 3·76m. of the population and high-priced farming land; the *Costa*, the coastal plain between the Andes and the Pacific, with 4·03m. whose permanent plantations furnish bananas, cacao, coffee, sugar-cane and many other crops; the *Oriente*, the upper Amazon basin on the E consisting of tropical jungles threaded by large rivers.

The population is predominantly of Amerindians, with small proportions of people of European or African descent. The official language is Spanish. The Amerindians of the highlands speak mainly the Quechua language; in the Oriental Region various tribes have languages of their own. Pop. (1986E) 9·64m.

Edam Noord-Holland, Netherlands. 52 31N 5 03E. Town sit. NNE of Amsterdam. Noted for cheese. It also manu. earthenware. Pop. (1989E) 24,572 (with Volendam).

Eddystone Rocks Cornwall, England. 50 11N 4 16W. Sit. SSW of Plymouth. Lighthouse sit. in dangerous reef.

Ede Gelderland, Netherlands. 52 03N 5 40E. Town sit. WNW of Arnhem. Industries inc. food processing and manu. of rayon and metal products. Pop. (1990E) 93,500.

Edea Cameroon. 3 48N 10 08E. Town sit. W of Yaoundé on Sanaga R. Communications centre at the head of navigation. Trade in rubber and cacao. Industries inc. aluminium, lumbering, quarrying, palm-oil processing. Hydro-electric dam. Pop. (1981E) 31,016.

Eden River England. 54 57N 3 01W. (i) Rises in Pennines and flows NW for 65 m. (104 km.) to enter Solway Firth just below Rockcliffe, and 5 m. (8 km.) NW of Carlisle. (ii) Rises in SE Surrey and flows E for 12 m. (19 km.) to join Medway R. at Penshurst, 5 m. (8 km.) NW of Tunbridge Wells.

Eden River Scotland. R. rising in Fife and flows for 30 m. (48 km.) ENE into St. Andrew's Bay.

Eder River Federal Republic of Germany. 51 13N 9 27E. Rises on Ederkopf and flows 110 m. (176 km.) E to join Fulda near Grifte. Used for water storage behind Eder dam at Hemfurth which supplies Ems-Weser canal.

Edessa Macedonia, Greece. 40 48N 20 03E. Cap. of Pella prefecture sit. WNW of Thessaloniki. Manu. textiles, carpets and rugs. Trades in wine, fruit and tobacco. Pop. (1981C) 16,054.

Edfu Egypt. 24 58N 32 52E. Town sit. N of Aswân on R. Nile. Manu. earthenware products. Trades in cereals, cotton and dates.

Edgware Greater London, England. 51 37N 0 17W. District of NW Lond. Mainly residential.

Edinburgh Lothian Region, Scotland. 55 57N 3 13W. Cap. and university city on Firth of Forth E of Glasgow. Cultural, administrative and educational centre. Industries inc. tanning, distilling, chemicals; manu. machinery, machine tools, paper, printing, electronics, publishing, food products. Tourism is stimulated by the annual Edinburgh Festival of the arts and by the city's exceptional architectural interest. Edinburgh Castle is on Castle Rock. Pop. (1988E) 433,480.

Edirne Edirne, Turkey. Cap. of province in extreme NW, sit. at confluence of Rs. Maritsa and Tundzha. Manu. inc. textiles, carpets, leather, soap and perfumes.

Edison New Jersey, U.S.A. 40 31N 74 24W. Township W of Woodbridge in the conurbation W of the Hudson R. Pop. (1989E) 85,000.

Edjélé Algeria. Town sit. S of Ghadames. Centre of an important oil-producing district.

Edmond Oklahoma, U.S.A. 35 39N 97 29W. City sit. N of Oklahoma city. Trading centre for an agric. and oil-producing area. Industries inc. manu.

petroleum products, flour, animal feed, wood products, leather goods, furniture, glass and concrete blocks. Pop. (1989E) 50,000.

Edmonds　Washington, U.S.A. 47 49N 122 23W. City sit. N of Seattle on Puget Sound. Residential city and resort. Pop. (1980C) 27,679.

Edmonton　Alberta, Canada. 53 33N 113 28W. Cap. of province sit. on N. Saskatchewan R. N of Calgary. The industrial centre of a district concerned with coal, oil, agric. and furs. Industries inc. printing, paper, oil refining, tanning, meat packing, and the manu. of chemicals and clothing. Pop. (1989E) 583,872.

Edmonton　Greater London, England. District of N Lond.

Edward, Lake　Uganda/Zaïre. 0 25S 29 30E. One of the great sources of the R. Nile. Area, *c.* 820 sq. m. (*c.* 2,124 sq. km.) and it is 44 m. (70 km.) long, and 32 m. (57 km.) broad with a mean depth of 365 ft. (111 metres), and is sit. at 3,000 ft. (914 metres) a.s.l. The L. was discovered by Stanley in 1888–9.

Efate　Vanuatu, South West Pacific. 17 40S 168 25E. Island sit. SE of Espiritu Santo. Area, 50 m. (80 km.) by 20 m. (32 km.). Chief towns Vila (cap. of the Anglo-French condominium), Havannah, an important harbour. Main products copra, coffee, cacao, sandalwood.

Egadi Islands　Sicilia, Italy. 37 56N 12 16E. Island group sit. off W Sicily. Area, 15 sq. m. (39 sq. km.). Main islands, Favignana, Marettimo, Levanzo. The main industry is tunny fishing.

Eger　Heves, Hungary. 47 54N 20 23E. Cap. of county sit. NE of Budapest on Eger R. Noted for wine, and trades in agric. produce. Pop. (1984E) 205,000.

Egham　Surrey, England. 51 26N 0 34W. Town sit. WSW of London on R. Thames. Mainly residential but some light industry. Pop. (1981C) 11,695.

Egmont, Mount　North Island, New Zealand. 39 18S 174 03E. An extinct volcanic peak in SW, in Taranaki. 8,260 ft. (2,542 metres) a.s.l.

Egypt　Arab Republic of, North Africa. 28 00N 31 00E. Rep. in NE Africa and the Sinai peninsula, bounded W by Libya, S by Sudan, E by Red Sea, Gulf of Suez and Israel, N by Mediterranean. Area, 386,200 sq. m. (1m. sq. km.). Chief cities Cairo (cap.), Alexandria, Suez, Port Said, Ismailia. About $3^1/_2\%$ of the land area is inhabited and cultivated, the rest being desert; the pop., agric. and industries being concentrated on Nile R. and Suez Canal ports. The R. and its marshes and Ls. cover 2,850 sq. m. (7,381 sq. km.). The country is divided into 25 governorates and two provinces, Upper and Lower Egypt. Upper Egypt consists of the land above the Nile delta and is mainly arid with infrequent rain. Rainfall is light even in the delta, and the economy depends heavily on irrigation, fed by the Nile. The main water-storage dams are at Aswan, Esna, Gebel Aulia, Nag'Hammadi, Asyut, Zifta, Mohammed Ali (below Cairo). The main crops raised are cotton, cereals, vegetables, sugar-cane. Raw cotton accounts for about 75% of agric. exports. Mineral resources inc. phosphates, sea salt, iron ore and petroleum. Industries are concentrated in the Delta and at Suez, and inc. oil refining, iron and steel, cotton ginning and manu. cotton textiles. Pop. (1989E) 50·74m.

Eider River　Federal Republic of Germany. 54 19N 8 58E. Rises S of Kiel and flows 117 m. (187 km.) N to Kiel Canal, which flows through it as far as Rendsburg, then E on a meandering course to enter North Sea 3 m. (5 km.) below Tönning. Navigable for most of its course.

Eifel　Rhineland-Palatinate, Federal Republic of Germany. A plateau of volcanic origin sit. between Rs. Rhine, Mosel, Ahr and Our.

Eiger　Switzerland. 46 34N 8 01E. Mountain peak at 12,697 ft. (3,970 metres) in the Bernese Oberland. The first ascent was made in 1858 and the N face is one of the most difficult climbs in the alps.

Eigg (Egg) Island   Highland Region, Scotland. 56 54N 6 10W. Sit. near mainland. Area, 6¹/₂ m. (10 km.) long and 4 m. (6 km.) broad. The highest point is the Scuir of Eigg. 1,289 ft. (397 metres) a.s.l. Pop. (1980E) 200.

Eilat   Israel. 29 33N 34 56E. Port sit. S of Jaffa–Tel Aviv at head of Gulf of Aqaba, the E arm of the Red Sea.

Eildon Hills   Scotland. 55 35N 2 43W. Three high hills sit. S of Melrose. 1,385 ft. (422 metres) a.s.l.

Eindhoven   Noord-Brabant, Netherlands. 51 26N 5 28E. Town in central S sit. on R. Dommel. Its main industry is radio, electronics and electrical equipment. Headquarters of Philips international organization. Other manu. inc. vehicles, plastics, textiles, cigars and trucks. Pop. (1988E) 190,736.

Einsiedeln   Schwyz, Switzerland. 47 08N 8 45E. Town sit. SE of Zürich. Noted for its 10th cent. Benedictine Abbey. Pop. (1980E) 10,000.

Eire   *see* **Ireland**

Eisenach   Erfurt, German Democratic Republic. 50 59N 10 19E. Town and resort in extreme W sit. at W end of the Thürinerwald. Manu. motor vehicles, machine tools and chemicals. J. S. Bach was born here in 1685. Pop. (1989E) 48,800.

Eisenhower, Mount   Alberta, Canada. A peak sit. in Banff National Park, and in Rocky Mountains. 9,390 ft. (2,889 metres) a.s.l.

Eisenhüttenstadt   Frankfurt, German Democratic Republic. 52 10N 14 39E. Town and port sit. SSE of Frankfurt-an-der-Oder on Oder R. It is the amalgamation of Stalinstadt and Fürstenberg together with numerous blast furnaces and steel works, and is terminus of Oder-Spree Canal.

Eisleben   Halle, German Democratic Republic. 51 31N 11 32E. Town sit. WNW of Halle. Industries inc. copper smelting, engineering, and the manu. of textiles, cigars and clothing. Martin Luther, 1483–1546, was born, and died, here. Pop. (1989E) 26,900.

Elâziğ   Elâziğ. Turkey. 38 41N 39 14E. (i) Province of E Turkey bounded N by Euphrates R., Murat R. and Peri R., also SW by Euphrates R. and SE by Taurus Mountains. Area, 3,554 sq. m. (9,205 sq. km.). Mountainous, producing lignite, chromium, cereals, vegetables, fruit and cotton. Pop. (1980C) 440,808. (ii) Town sit. ENE of Malatya. Provincial cap. Rail terminus and centre of an agric. and fruit-growing area. Pop. (1980C) 142,983.

Elba   Toscana, Italy. 42 46N 10 17E. An island sit. in the Tuscan Arch. near the mainland, and separated from it by Strait of Piombino. Area, 86 sq. m. (223 sq. km.). Mainly mountainous rising to Monta Capanne, 3,340 ft. (1,018 metres) a.s.l. Iron ore is mined and exported together with marble, lead, sulphur and alabaster. Agric. products inc. vines, olives, cereals and fruit and there are anchovy, sardine and tunny fisheries. Napoleon I was exiled here 1814–15. Pop. 35,000.

Elbasan   Elbasan, Albania. 41 06N 20 04E. Cap. of district SE of Tiranë sit. on Shkumbi R. Manu. textiles, olive oil, soap and leather. Trades in maize, olives and tobacco. Pop. (1980E) 56,500.

Elbe River   Czechoslavakia/Germany. 53 50N 9 00E. Rises in Riesengebirge NW of Spindleruv Mlyn and flows 706 m. (1,130 km.) NW through NW Czechoslovakia, turns N into German Democratic Republic, forming the Czech.–G.D.R. border for *c.* 2 m. (3 km.); it flows N through the limestone mountains of Saxony to the N Germ. lowlands and meanders past Dresden, Torgau, Dessau and Magdeburg to Hamburg in Federal Republic of Germany; below the city it enters the North Sea at Cuxhaven at the mouth of a 60 m. (96 km.) estuary. Main tribs. Eger, Mulde, Ilmenan, Saale, Vlatva, Black Elster, Elde,

Havel and Jizera Rs. Navigable for 500 m. (800 km.) and linked by Mitelland canal system with Weser and Rhine Rs., and by Havel-Hohenzollern and Plaue-Havel Spree systems with Oder R. Connected also with Lübeck and Kiel by Elbe-Trave and Kiel canals.

Elblag Gdańsk, Poland. 54 10N 19 25E. Town and port sit. on Elblag R. and near coast of Gulf of Gdynia. Industries inc. shipbuilding, brewing and engineering. Pop. (1983E) 113,000.

El Bluff Zelaya, Nicaragua. Town sit. E of Bluefields on Caribbean, and forming its out-port.

Elbruz (El'brus) Mount Kabardino-Balkar/Georgia, U.S.S.R. 43 21N 42 26E. The highest peak in the Caucasus Mountains, comprising two extinct volcanoes, the W 18,482 ft. (5,633 metres) a.s.l. and the E 18,356 ft. (5,595 metres) a.s.l.

El Callao Bolívar, Venezuela. Town sit. ESE of Ciudad Bolívar on Yuruari R. Gold mining centre.

Elche Alicante, Spain. 38 15N 0 42W. Town sit. WSW of Alicante. Noted for large groves of date-palms which supply the whole of Spain with leaves for Palm Sunday. Manu. textiles, fertilizers, footwear and soap. Pop. (1981C) 162,873.

El Dorado Arkansas, U.S.A. 33 13N 92 40W. City sit. SSW of Little Rock. An important centre of the oil industry with associated trades. Also deals in timber and cotton. Pop. (1980C) 25,270.

Eldoret Rift Valley, Kenya. 0 31N 35 17E. Town sit. NW of Nakuru on Uasin Gishu Plateau. Trading centre for a farming area, with an airport. Pop. (1984E) 70,000.

Elektrostal Russian Soviet Federal Socialist Republic, U.S.S.R. 55 47N 38 28E. Town sit. E of Moscow. Industries inc. heavy engineering, steel works and the manu. of stainless steel. Pop. (1977E) 135,000.

Elephanta Island Maharashtra, India. 18 57N 72 56E. Sit. in Bombay Harbour about 5 m. (8 km.) from Bombay. Noted for its cave temples which are dedicated to Siva.

Eleuthera Island Bahamas. 25 15N 76 20W. Island between Great Abaco and Cat Islands, its N point and NE of Nassau. Area, 80 m. (128 km.) by 6 m. (10 km.). Main settlements Governor's Harbour, Palmetto Point, Tarpum Bay. Main products tomatoes, pineappples, dairy products. Pop. (1980C) 10,600.

Elgin Grampian Region, Scotland. 57 39N 3 20W. Town near N coast, sit. on R. Lossie. Industries inc. textiles, structural engineering and whisky distilling. Pop. (1988E) 20,020.

Elgin Illinois, U.S.A. 42 02N 88 17W. City sit. WNW of Chicago on Fox R. Manu. watches, electrical equipment, motorcar accessories and pianos. Pop. (1980C) 63,798.

Elgon, Mount Kenya/Uganda. 1 08N 34 33E. An extinct volcano 60 m. (96 km.) NE of L. Victoria. 14,178 ft. (4,321 metres) a.s.l. Coffee and bananas are grown on the lower slopes.

Elia Greece. *Nome* of W Peloponnesus bounded E by Alpheus R. and W by Ionian Sea. Area, 1,035 sq. m. (2,681 sq. km.). Cap. Pyrgos. Mainly mountainous with W coastal lowland producing currants, fruits, olives, vines. Cereals and stock farming are the main occupations on the mountain slopes. Industries are concentrated on Pyrgos. Pop. (1980C) 160,305.

Elisabethville *see* **Lubumbashi**

Elista Kalmyk Autonomous Soviet Socialist Republic, U.S.S.R. 46 16N 44 14E. Town sit. W of Astrakhan. Communications and trading centre. Cap. of the Autonomous Soviet Socialist Rep. Industries inc. processing of farming products, esp. sheepskins, wool, meat, grain; manu. bricks and tanning extract.

Elizabeth New Jersey, U.S.A. 40 40N 74 11W. City and seaport sit.

SSW of Newark. Manu. diverse inc. machinery, electronic equipment and chemicals. Pop. (1980C) 106,201.

Elkhart Indiana, U.S.A. 41 41N 85 58W. City sit. N of Indianapolis at confluence of R. Elkhart and St. Joseph. Manu. musical instruments, esp. wind and percussion, producing nearly 75% of the country's total. Also manu. machinery, metal, rubber and paper products. Pop. (1988E) 45,250.

Elk Island National Park Alberta, Canada. Sit. E of Edmonton, it is the largest reserve in the country, 75 sq. m. (194 sq. km.) in area. Mainly forested providing a home for buffalo, moose, elk and deer.

Elko Nevada, U.S.A. 40 50N 115 46W. City sit. W of Salt Lake City, Utah on Humboldt R. at 5,063 ft. (1,543 metres) a.s.l. Goldmining is important achieving 64% of N America's production. Trading and distribution centre for cattle, sheep and wool. Industries inc. railway repair shops. Pop. (1980C) 8,758.

Ellesmere Shropshire, England. 53 17N 2 55W. Town sit. NNW of Shrewsbury on Ellesmere Canal which joins R. Severn with R. Mersey. Industries inc. waste disposal and milking machine manu. Pop. (1987E) 2,940.

Ellesmere Island North West Territories, Canada. 81 00N 80 00W. Arctic island sit. off NW Greenland. There is a meteorological station at Eureka Sound in the NW.

Ellesmere Port and Neston Cheshire, England. 53 17N 2 54W. Port sit. SSE of Liverpool on Manchester Ship Canal. An important oil and petroleum centre with industries inc. engineering and the manu. of motorcars, chemicals and paper. Pop. (1988E) 79,700.

Ellice Islands *see* **Tuvalu**

Ellicott City Maryland, U.S.A. City sit. WSW of Baltimore on Patapsco R. Tourist centre of historical interest. Pop. (1980C) 21,784.

Ellis Island New York, U.S.A. 40 42N 74 02W. Sit. in Upper New York Bay, and formerly used as an immigrant station.

Ellsworth Mountains Antarctica. 79 00S 85 00W. Range extending SSW from base of the Palmer Peninsula to Rockefeller Plateau.

Elmhurst Illinois, U.S.A. 41 53N 87 56W. City sit. W of Chicago. There are limestone quarries in the vicinity. Pop. (1988E) 47,439.

Elmira New York, U.S.A. 42 06N 76 49W. City sit. SSE of Rochester on Chemung R. Manu. inc. machine tools, television tubes, aircraft, glass containers and packaging products. Pop. (1985E) 34,000.

El Mochito Santa Bárbara, Honduras. Town sit. SE of Santa Bárbara. Mining centre, esp. silver, lead, zinc and copper.

Elmont New York, U.S.A. Village immediately E of Jamaica on W Long Island. Mainly residential; manu. clothing. Pop. (1980C) 27,592.

Elmshorn Schleswig-Holstein, Federal Republic of Germany. 53 45N 9 39E. Town sit. NW of Hamburg. Industries inc. food processing and the manu. of metal and plastic goods, textiles and leather goods. Pop. (1988E) 43,000.

El Muerto Chile/Argentina. Mountain peak in the Andes, 21,450 ft. (6,537 metres) a.s.l., 30 m. (48 km.) WSW of Cerro Incahuasi on the Chile/Argentina frontier.

El Obeid Kordofan, Sudan. 13 08N 30 18E. City sit. SW of Khartoum. Railway terminus and communications centre for the oil-seed and gum-arabic region of W Sudan. Airport. Pop. (1984E) 140,024.

Elobey Chico Equatorial Guinea. Islet of Elobey group in Corisco Bay off mouth of Río Muni. Area, 50 acres (20 hectares). Uninhabited.

Elobey Grande Equatorial Guinea. The larger islet of the Elobey group. Area, 1 sq. m. (2 sq. km.).

El Oro   Ecuador, Province of S
Ecuador on an inlet of the Pacific.
Area, 3,134 sq. m. (8,117 sq. km.).
Main cities Machala (cap.), Santa
Rosa, Pasaje, Puerto Bolivar. Coastal
plains rising to Andean highlands.
The main occupation is mining; chief
minerals gold (at Zaruma), silver, copper, lead. Agric. is also important, esp.
cacao, coffee, sugar-cane, rice, fibres,
rubber, fruit and nuts. Pop. (1982C)
335,630.

El Paraíso   Honduras. Dept. of S
Honduras on Nicaragua border. Area,
3,310 sq. m. (8,573 sq. km.). Cap.
Yuscarán. Mountainous, drained by
the Jalán and Choluteca Rs. The main
occupations are farming and mining,
esp. gold at Agua Fría and silver at
Yuscarán. Pop. (1980E) 105,372.

El Paso   Texas, U.S.A. 31 45N 106
29W. Port sit. on Río Grande opposite
Ciudad Juárez in Mexico. Industries
inc. tourism, electronics, copper
smelting, processing and distributing
oil and natural gas, and the manu. of
clothing and building materials.
Trades in silver, copper, lead, meat,
hides and wool. Pop. (1988E)
537,000.

Elphinstone Island   Andaman
Islands, Indian Ocean. Island sit. W of
Mergui in lower Burma. Area, 13 m.
(21 km.) long and 2–5 m. (3–8 km.)
wide.

El Progreso   Guatemala. 14 51N 90
04W. (i) Dept. of E Central Guatemala
sit. in Motagua R. valley and bounded
N by the Chuacús and Sierra de las
Minas. Area, 742 sq. m. (1,922 sq.
km.). Mainly agric. producing maize,
beans, sugar-cane and fodder crops,
with some livestock farming. Pop.
(1982E) 101,203. (ii) Town sit. ENE of
Guatemala city on Guastatoya R.
Dept. cap. Commercial centre trading
in agric. produce.

El Progreso   Yoro, Honduras. 15
21N 87 49W. Town sit. SW of Tela on
Ulúa R. Commercial centre for an
agric. and fruit growing area, trading
in bananas. Pop. (1980E) 105,372.

El Rahad   Kordofan, Sudan. Town
sit. SE of El Obeid, producing cotton,
gum arabic, nuts, cereals.

El Saïd   Egypt. Otherwise called
'Upper Egypt', one of the two main
districts and comprising most of the
country above the Nile delta, which
comprises Lower Egypt.

El Salvador   13 50N 89 00W. Rep.
bounded to N and W by Guatemala, N
and E by Honduras and E by Gulf of
Fonseca. Divided into 14 depts. Cap.
San Salvador. Area, 8,236 sq. m.
(21,331 sq. km.). In 1839 the Central
American Federation, which had comprised the states of Guatemala, El Salvador, Honduras, Nicaragua and Costa
Rica, was dissolved, and El Salvador
declared itself formally an independent republic in 1841. There are volcanic mountains, fertile uplands and a
hot, humid coastal belt. Chief crops
are cotton, coffee, balsam (the world's
main source), rice and maize. Agric. is
the mainstay of the economy. Industries: coffee processing and sugar
refining; manu. textiles and leather
goods, mainly in San Miguel and
Santa Ana. Pop. (1987E) 5,009,000.

El Seibo   Dominican Republic. 18
46N 69 02W. (i) Province in E bounded N by Samaná Bay and crossed W
to E by the Cordillera Central. Area,
1,287 sq. m. (3,333 sq. km.). Fertile
agric. area; main products being coffee, cacao, sugar-cane, rice, corn,
fruit, livestock and dairy products.
Pop. (1980E) 135,000. (ii) Town sit.
ENE of Santo Domingo on Soco R.
Provincial cap. and centre of an agric.
area. (iii) Plain extending *c.* 100 m.
(160 km.) E along Caribbean from
Ozama R. to Mona Passage, and *c.* 20
m. (32 km.) wide. Fertile lowland
producing sugar, cacao, rice, maize,
livestock, coffee, tobacco and fruit.

Elsmere   Delaware, U.S.A. 39 44N
75 36W. Town immediately W of Wilmington in N Delaware. Mainly residential. Pop. (1980C) 6,493.

Eltham   Greater London, England.
Residential district of SE London.

Eltham   North Island, New Zealand.
36 26S 174 19E. Town sit. SSE of New
Plymouth; centre of a dairy farming
area. Pop. (1983E) 2,500.

Eluru   Andhra Pradesh, India. 16
42N 81 07E. Town sit. ESE of Hyder-

abad at junction of Godavari and Krishna Rs. canal systems. Manu. carpets, textiles and leather goods. Trades in rice, oilseeds and tobacco. Pop. (1981C) 168,154.

Elvas Portalegre, Portugal. 38 53N 7 10W. Town sit. W of Badajoz, Spain, near Spanish border. Trades in grain, wine, olive oil and plums. Pop. (1981E) 13,000.

El Wejh Hejaz, Saudi Arabia. Town sit. NW of Jidda on Red Sea. Trading centre and port with a good natural harbour. Sometimes called Al Wajh or Al Wijh.

Ely Cambridgeshire, England. 52 24N 0 16E. City sit. NNE of Cambridge on Ouse R. on a hillock in surrounding drained fenland, called the Isle of Ely. Waterway cruising centre. Industries inc. engineering, agricultural machinery, clothing, chemicals and plastic products. Pop. (1983E) 11,030.

Elyria Ohio, U.S.A. 41 22N 82 06W. City sit. WSW of Cleveland on Black R. Manu. inc. aircraft, motorcars, pumps, electrical goods, chemicals and plastics. Pop. (1980C) 57,538.

Emba Kazakh Soviet Socialist Republic, U.S.S.R. 48 50N 58 08E. Town sit. SSE of Aktyubinsk on transCaspian railway. The main industry is metal-working.

Emba River U.S.S.R. Rises in S Mugodzhar Hills, Kazakh Soviet Socialist Republic and flows 384 m. (614 km.) SW past Zharkamys to enter Caspian Sea near Zhilaya Kosa. Its lower course flows through the Emba oil field. Main trib. Temir R.

Emden Lower Saxony, Federal Republic of Germany. 53 22N 7 12E. Port in NW sit. on R. Ems estuary and Dortmund-Ems Canal. Industries inc. shipbuilding and car manu. Port exports oil, cars, ores and coal. Pop. (1984E) 50,500.

Emek Hefer Israel. District in valley of Jezreel on coastal plain; the main agric. centre of Israel.

Emilia Romagna Italy. 44 33N 10 40E. Region of N Italy. Area, 8,540 sq. m. (22,119 sq. km.). The Apennines are sit. along the S border with the fertile plains of R. Po lying to the N of them. Agric. products inc. cereals, sugar-beet, vines and fruit. The main occupation is food processing. Cap. Bologna. Pop. (1981C) 3,957,513.

Emirau Papua New Guinea. Island in the Mussau group of the Bismarck arch. sit. NW of New Ireland. Area, 20 sq. m. (52 sq. km.). Volcanic. Contains coconut plantations.

Emmen Drenthe, Netherlands. 52 47N 7 00E. Town sit. SSE of Groningen. Manu. inc. agric. machinery, nylon, chemicals and cement. Pop. (1989E) 92,303.

Emmental Bern, Switzerland. A valley of upper Emme R. in W central Switzerland noted for cheese.

Empangeni Natal, Republic of South Africa. 28 50S 31 52E. Town sit. NE of Durban. Railway junction and commercial centre in a fruitgrowing area. Pop. (1989E) 12,200.

Empoli Toscana, Italy. 43 43N 10 57E. Town sit. on Arno R. between Florence and W coast. Manu. textiles, pottery, glassware, hosiery, macaroni and straw goods. Pop. (1980E) 38,000.

Emporia Kansas, U.S.A. 38 24N 96 11W. Town sit. SSW of Topeka. Railway and commercial centre for an extensive farming area, handling grain and livestock. Industries inc. meat packing, flour milling, education, dairying, railway repair shops. Pop. (1980C) 25,287.

Emsland Federal Republic of Germany. Region bounded E by Ems R. and W by Dutch border. Swampy area, much drained and with considerable oil resources.

Ems River Federal Republic of Germany. 51 09N 9 26E. Rises N of Paderborn and flows *c.* 250 m. (400 km.) E and N on a meandering course past Rheine to enter North Sea by a 20 m. (32 km.) estuary. Below Münster it is joined briefly by Dortmund-Ems Canal.

Encamp   Andorra. 42 32N 1 35E. Village sit. NE of Andorra la Vella on a headstream of Valira R. Main occupation livestock farming. Pop. (1982E) 4,500.

Encarnación   Itapúa, Paraguay. 27 20S 55 54W. Cap. of province and port on Alto Paraná R. in extreme SE. Industries inc. sawmilling and tanning. Exports inc. timber, cattle, hides, rice, cotton, tobacco and *maté*. Pop. (1983C) 28,800.

Endeavour Strait   Queensland, Australia. 10 45S 142 00E. Sit. between Cape York Peninsula and Prince of Wales Island, forming S part of Torres Strait.

Enderbury Island   *see* **Phoenix Islands**

Enderby Land   Antarctica. Antarctic mainland extending from Ice Bay to Edward VIII Bay at 66° 50' S lat.

Enfield   Greater London, England. 51 40N 0 05W. Town forming a suburb of NE Lond., noted as the site of a royal small-arms factory manu. Enfield rifles. Also manu. metal products and plastics. Pop. (1988E) 260,900.

Enfield   Connecticut, U.S.A. 41 58N 72 36W. Town sit. NNE of New Haven on Connecticut R. Manu. plastics, machinery, castings, stationery, clothing, tools, pharmaceuticals, food processing. Pop. (1988E) 47,350.

Engadine   Graubünden, Switzerland. 46 45N 10 25E. The upper valley of Inn R. in E Switzerland extending from Maloja Pass to Austrian frontier. It is divided into Upper Engadine and Lower Engadine, the latter containing the Swiss National Park.

Engelberg   Obwalden, Switzerland. 46 49N 8 25E. Resort sit. SSE of Luzern, surrounded by mountains.

Engels   Russian Soviet Federal Socialist Republic, U.S.S.R. 51 30N 46 07E. Town sit. on Volga R. opposite Saratov. Industries inc. railway

engineering, meat packing, flour milling and the manu. of textiles and leather goods. Pop. (1985E) 177,000.

England   53 00N 2 00W. United Kingdom. SE part and largest political division of the island of Great Britain bounded E by North Sea, S by English Channel, N by Scotland and W by Atlantic Ocean, Wales and Irish Sea. Area, 50,331 sq. m. (130,357 sq. km.). Chief cities London (cap.), Birmingham, Liverpool, Manchester, Newcastle upon Tyne, Bristol, Plymouth, Sheffield, Leeds, Nottingham, Coventry, Wolverhampton, Southampton. Generally undulating lowland with downland and low hills in the S and E, rising to the Pennine range which extends centrally N to S, the high moorland of the SW and NE and the Cumbrian mountains, NW, which rise to Scafell Pike, 3,210 ft. (978 metres) a.s.l. Drained into North Sea by Tyne, Ouse, Trent, Norfolk Ouse and Thames Rs. and their tribs.; into Bristol Channel by Severn and Avon Rs. and into Irish Sea by Eden, Ribble, and Mersey Rs. The land surface is 32m. acres, of which 3m. acres are rough grazing land, 8m. permanent pasture and 13m. arable. The main products are cereals, vegetables, beef and dairy cattle, sheep. Mineral resources inc. coal (about half coming from the Yorkshire–Leicester–Nottingham field), off-shore petroleum, mainly in the North Sea, iron, building stone and clays. Heavy industries based on coal supplies inc. iron and steel, chemicals, textiles from natural and man-made fibres, paper making, engineering. Other industries inc. manu. vehicles, aircraft, electrical goods, agric. machinery. The conurbations of the Midlands form the main industrial area. Fishing is important on the E coast, esp. cod and herring. Pop. (1981C) 46,229,955.

Englewood   Colorado, U.S.A. 39 39N 104 59W. City forming a S suburb of Denver on South Platte R. Mainly a residential and retail industry area. Pop. (1990E) 31,500.

Englewood   New Jersey, U.S.A. 40 54N 73 59W. City sit. NNE of Jersey City. Residential with some general manu. Pop. (1980C) 23,701.

English Channel   England/France. 50 20N 1 00W. A narrow sea between England and France; connected with North Sea by Strait of Dover which is the narrowest part, 21 m. (34 km.), and also the shallowest, 12–13 fathoms. It stretches w and joins Atlantic near the Isles of Scilly. Length, about 350 m. (560 km.). The Isle of Wight and the Channel Islands are in the Channel and the only important river flowing into it is the Seine. Mathieu first suggested the idea of a channel tunnel to Napoleon I.

Enid   Oklahoma, U.S.A. 36 19N 9 48W. City sit. NNW of Oklahoma City. A leading grain and cattle centre in an oil-producing region. Industries inc. oil refining and the manu. of oil-field equipment, and flour milling. Air force base main employer. Pop. (1980c) 50,363.

Enna   Sicilia, Italy. 37 34N 14 17E. Cap. of Enna province sit. SE of Palermo. Trades in sulphur and rock-salt. Pop. (1981E) 28,000.

Ennerdale Water   England. Lake sit. ESE of Whitehaven, on NW coast, which it supplies with water.

Ennis   Clare, Ireland. 52 50N 8 59W. Town sit. in w centre, on R. Fergus. Industries inc. distilling, brewing and flour milling. Pop. (1981E) 15,000.

Enniscorthy   Wexford, Ireland. 52 30N 6 34W. Town sit. in SE on R. Slaney. Industries inc. engineering, aluminium extrusion, maltsters, garment and pottery manu., mineral waters and agriculture. Pop. (1986E) 4,483.

Enniskillen   Fermanagh, Northern Ireland. 54 21N 7 38W. Town sit. on an island in R. Erne between Upper and Lower Lough Erne, in SW. Trades in cattle and agric. produce. Pop. (1989E) 10,500.

Enschede   Overijssel, Netherlands. 52 12N 6 53E. Town sit. near Federal Republic of Germany border and joined to Rs. Ijssel and Rhine by Twente Canal. Industry consists of high technology. Pop. (1990E) 145,223.

Ensenada   Baja California, Mexico. 31 52N 116 37W. Port sit. SSE of San Diego, U.S.A. Industries inc. fishing, fish-canning and wine making. Cotton is exported. Pop. (1980c) 175,425.

Entebbe   Buganda, Uganda. 00 04S 32 28E. Town sit. on NW shore of L. Victoria. An administrative centre with an intern. airport. Pop. (1983E) 21,000.

Entre Minho e Douro, Portugal *see* **Minho**

Entre Ríos   Argentina. 14 57S 37 20E. Province of E Argentina bounded w by Paraná R. and E by Uruguay across Uruguay R. Area, 28,487 sq. m. (73,781 sq. km.). Main cities Paraná (cap.), Bajada Grande, Concordia, Concepción del Uruguay, Colón, Gualeguaychú. Low-lying and swampy with fertile alluvial soil and a sub-tropical climate. Mainly a stock raising and farming area with associated industries and some lumbering and quarrying. Pop. (1980c) 908,313.

Entreves   Val d'Aosta, Italy. Village at Italian end of Mont Blanc road tunnel, which links it with Les Pelerins, France (opened in 1965).

Enugu   Anambra, Nigeria. 6 30N 7 30E. Town in SE. Centre of the coalmining industry. Industries inc. railway engineering and sawmilling. Pop. (1981E) 256,000.

Eolie Islands   *see* **Lipari Islands**

Epe   Gelderland, Netherlands. 52 21N 5 59E. Town sit. E of Apeldoorn. Market town trading in eggs and wooden shoes. Pop. (1988E) 33,000.

Épernay   Marne, France. 49 03N 3 57E. Town sit. in NE on Marne R. An important centre of the Champagne wine industry together with Reims. Vast wine cellars are hollowed out of the chalk hills. Manu. inc. corking equipment, corks, casks and bottles. Other industries are sugar refining and the manu. of textiles. Pop. (1990E) 28,000.

Epi   Vanuatu, South West Pacific. 16 43S 168 15E. Island sit. SE of

Espiritu Santo. Area, 27 m. (43 km.) by 11 m. (18 km.). Main products copra, cacao.

Epinal Vosges, France. 48 11N 6 27E. Cap. of dept. in E sit. on Moselle R. Manu. cotton, cheap engravings and lithographs, images, d'Epinal, and Kirsch liqueur. Pop. (1982C) 40,954.

Epirus Greece. 39 30N 20 30E. *Nome* of NW Greece bounded NW by Albania, W by Ionian Sea, S by Gulf of Arta and E by Pindus R. system. Area, 3,495 sq. m. (9,052 sq. km.). Cap. Ioannina. Mainly a limestone area noted for stock farming, mainly goats and sheep, and producing also cereals, and fruit. It is divided administratively into 4 'nomes': Arta, Joannina, Preveza, Thesprotia. Pop. (1981C) 324,541.

Epping Greater London, England. 51 43N 0 07E. Town forming a suburb of NE Lond. at the edge of the anc. Epping Forest, now reduced to about 5,600 acres and a public park.

Epsom and Ewell Surrey, England. 51 20N 0 16W. Borough comprising the two towns sit. SW of Lond. Noted for horse racing on Epsom Downs, esp. the 'Derby' and the 'Oaks'. Pop. (1988E) 67,500.

Equateur Zaïre. Region in NW. Primarily agric. economy, producing rice, cotton and palm products. Cap. Mbandaka. Area, 155,712 sq. m. (403,293 sq. km.). Pop. (1981E) 3,418,296.

Equatoria Sudan. Region of S Sudan bounded S by Zaïre and Uganda, SE by Kenya and E by Ethiopia. Area, 29,535 sq. m. (76,495 sq. km.). Cap. Juba. Mainly savannah and tropical woodlands rising to Imatong Mountains, s. Drained by upper White Nile, here called Bahr el Jebel. The main occupation is farming, the main products being cotton, peanuts, sesame, durra and livestock. Pop. (1983C) 1,406,181.

Equatorial Guinea West Africa. 2 00N 8 00E. Rep. on Gulf of Guinea bounded S and E by Gabon, N and E by

Cameroon. Area, 10,831 sq. m. (28,051 sq. km.). Chief towns Santa Isabel (cap.), Bata. Rio Muni is heavily forested; the main occupations are lumbering and farming. On Bioko the main occupations are farming and fishing, with some fish processing. Pop. (1984E) 321,000.

Erciyas Dağil Turkey. An extinct volcano 13 m. (21 km.) S of Kayseri. 12,850 ft. (3,917 metres) a.s.l.

Ercolano Campania, Italy. 40 46N 14 20E. Town sit. SE of Naples. Indust. inc. tourism, leather goods, glass and wine.

Erebus, Mount Ross Island, Antarctica. An active volcano. 13,202 ft. (4,024 metres) a.s.l.

Eregli Konya, Turkey. 37 31N 34 04E. Town sit. ESE of Konya. The main industry is cotton textile manu.

Erfurt German Democratic Republic. 50 58N 11 01E. Cap. of district of same name. Manu. machinery, typewriters, electrical and electronic equipment, textiles and footwear. Trades in flowers, vegetables and seeds. It is one of the oldest G.D.R. towns. Pop. (1990E) 220,000.

Ericht, Loch Scotland. L. in S Inverness and NW Perthshire extending 14 m. (22 km.) SW to NE along foot of Ben Alder. Its outlets are Ericht and Truim Rs. It forms part of the Lochaber hydroelectric system.

Erie Pennsylvania, U.S.A. 42 08N 80 04W. Port and city in extreme NW of state sit. on L. Erie. Industrial centre. Coal, petroleum, timber, pig iron, locomotives and heavy machinery are exported. Pop. (1980C) 119,123.

Erie Canal New York, U.S.A. Waterway extending 350 m. (560 km.) from Albany on Hudson R. to Buffalo on L. Erie, now improved and widened as the New York State Barge Canal.

Erie, Lake U.S.A./Canada. 42 30N 82 00W. Fourth in size of the Great Lakes, 241 m. (386 km.) long and

30–57 m. (48–91 km.) wide. Area, 9,940 sq. m. (25,745 sq. km.). Bounded N and half sit. in Ontario, Can., bounded S by New York, Pennsylvania and Ohio, U.S.A., and Michigan, U.S.A. to the W. Fed by Detroit R. from L. Huron, drained by Niagara R. to L. Ontario, where the navigable Welland Canal by-passes Niagara Falls. It is connected with New York and Atlantic by Hudson R. and New York State Barge Canal. Ice-free April–mid December, but subject to violent winds. Chief ports on the U.S.A. side, Buffalo, Dunkirk, Erie, Conneaut, Ashtabula, Cleveland, Lorain, Sandusky, Toledo, Detroit. Chief Can. ports, Port Colborne, Port Dover, Port Stanley, Leamington. The main islands are Bass, Kelleys Islands, Pelee Island.

Erigavo Somalia. Town sit. ENE of Hargeisa in Ogo highlands. Livestock farming centre.

Erith Greater London, England. 51 28N 0 10E. Town sit. W of Gravesend on Thames R. Industries inc. chemicals; cables, food products, engineering, furniture, radio.

Eritrea Ethiopia. Region bordering Red Sea. Area, 45,405 sq. m. (117,600 sq. km.). Cap. Asmara. Became part of Ethiopia in 1962. Pop. (1984C) 2.7m.

Erlangen Bavaria, Federal Republic of Germany. 49 36N 11 01E. Town sit. N of Nuremberg at confluence of Rs. Regnitz and Schwabach. Manu. textiles, electrical equipment, gloves, glass and beer, the latter being esp. well-known. Pop. (1981E) 101,400.

Ermelo Gelderland, Netherlands. 52 17N 5 37E. Town sit. SSE of Amsterdam. The centre of a dairy-farming district.

Erne, Lough Fermanagh, Northern Ireland. 54 10N 7 30W. L. consisting of two smaller Ls. joined by a 10 m. (16 km.) stretch of Erne R. Upper Lough Erne, SE is 30 sq. m. (78 sq. km.); Lower Lough Erne, NW is 90 sq. m. (233 sq. km.). There are numerous islands.

Erne River Republic of Ireland/Northern Ireland. 54 30N 8 16W. Rises in Lough Gowna on border of Longford and Cavan, and flows 72 m. (115 km.) NW through Cavan, Fermanagh, Donegal, through Lough Oughter and Lough Erne to enter the Atlantic in Donegal Bay. Used for hydroelectric power.

Erromanga Vanuatu, South West Pacific. 18 45S 169 05E. Island sit. SE of Espiritu Santo. Area, 345 sq. m. (894 sq. km.). Mainly savannah with extinct volcanoes. Main products copra, wool, sandalwood.

Ersekë *see* **Kolonje**

Erzgebirge Czechoslovakia/German Democratic Republic. 50 25N 13 00E. A mountain range stretching WSW to ENE for about 90 m. (144 km.) along the frontier. The highest peak is Klinovec in Czech., 4,081 ft. (1,244 metres) a.s.l.

Erzincan Erzincan, Turkey. Cap. of province sit. near Euphrates R. in E central Turkey. Manu. textiles, copper goods and clothing.

Erzurum Erzurum, Turkey. 39 57N 41 15E. Cap. of province sit. in NE and on Armenian plateau. Industries inc. iron and copper working, sugar refining and tanning. Trades in cattle, furs, grain and vegetables. Pop. (1980C) 190,241.

Esbjerg Denmark. 55 28N 8 27E. Port in SW on North Sea opposite Fanö Island. There is a ferry service to Britain. Industries inc. fishing and offshore activities. Agric. products are exported. Pop. (1989E) 81,000.

Esbo *see* **Espoo**

Esch-Alzette Luxembourg. 49 30N 5 59E. Town sit. SW of Luxembourg on Alzette R. The second largest town and centre of the steel industry. Manu. inc. steel products and fertilizers. Pop. (1989E) 25,000.

Eschwege Hessen, Federal Republic of Germany. Town sit. ESE of Kassel on Werra R. Manu. soap, footwear and textiles. Pop. (1984E) 23,200.

Eschweiler North Rhine-
Westphalia, Federal Republic of
Germany. 50 49N 6 16E. Town sit
ENE of Aachen. Industries inc. open
cast lignite, steel, cable, iron casting,
paint. Pop. (1988E) 53,516.

Escolín Mexico. Oilfied near the
rich field at Poza Rica.

Escorial Madrid, Spain. Town sit.
WNW of Madrid at foot of Sierra de
Guadarrama. Nearby is the famous El
Escorial.

Escuintla Guatemala. 14 18N 90
47W. (i) Dept. of S Guatemala on Pacific coast. Area, 1,693 sq. m. (4,385
sq. km.). Chief towns Escuintla, Santa
Lucia, Palin, San José. Mainly a farming area, producing sugar-cane, fruit,
coffee, maize, cattle. Pop. (1982E)
496,522. (ii) Town SW of Guatemala.
Dept. cap., commercial centre of an
agric. district. Main industries sugar
milling, cotton ginning.

Esdraelon, Plain of Israel. 32 36N
35 14E. Sit. between Mount Carmel
and the Jordan valley. It is traversed
by Kishon R. and is very fertile.

Esfáhán Esfáhán, Iran. 32 40N 51
38E. Cap. of province in W central
Iran sit. on Zayindeh R. It is the second largest town with many fine
buildings and mosques. Manu. carpets, rugs, textiles and brocades. Pop.
(1983E) 926,000

Esher Surrey, England. 51 23N
0 22W. Residential town sit. SW of
London. Pop. (1986E) 5,565.

Eskilstuna Södermanland, Sweden.
59 22N 16 30E. Town sit. just S of L.
Malar. Centre of an important iron
and steel industry noted for cutlery
and swords. Other manu. inc. machinery, electrical equipment and precision instruments. Pop. (1988E)
88,850.

Eskişehir Eskişehir, Turkey. 39
47N 30 30E. Cap. of province in NW.
Industries inc. sugar refining and the
manu. of agric. implements, cement
and textiles. Trades in agric. produce.
It is noted for hot sulphur springs.
Pop. (1980C) 309,431.

Esk River England. Rises on Scafell and flows SW for 20m. (32 km.) to
enter Irish Sea near Ravenglass, 14 m.
(22 km.) SSE of Whitehaven.

Esk River England. Rises in Cleveland Hills and flows E for 24 m. (38
km.) to enter North Sea at Whitby.

Esk River England/Scotland.
Formed by union of Black Esk and
White Esk, flows SSE and SSW for
36m. (58 km.) to enter Solway Firth 8
m. (13 km.) NNW of Carlisle.

Esk River Scotland. 54 58N 3 02W.
Rises in S foothills of Pentland Hills
and flows NE for 21 m. (34 km.) to
enter Firth of Forth at Musselburgh.

Esk River, North Scotland. 56 54N
2 38W. Rises in Grampians and flows
SE for 29 m. (46 km.) to enter North
Sea 4 m. (6 km.) NNE of Montrose.

Esk River, South Scotland. 56 40N
2 40W. Rises in Grampians and flows
SE and E for 49m. (78 km.) to enter
North Sea at Montrose.

Esmeraldas Esmeraldas, Ecuador.
0 59N 79 37W. Cap. of province and
port in extreme NW sit. at mouth of
Esmeraldas R. Timber and bananas
are exported. Pop. (1982C) 141,030.

Esna, Egypt Town sit. SSW of Qena
on Nile R. Manu. pottery, cotton and
shawls. There is an important R. barrage here.

Espaillat Dominican Republic.
Province on N coast, crossed by the
Cordillera Setentrional. Area, 433 sq.
m. (1,121 sq. km.). Main towns Moca
(cap.), Salcedo, Tenares, Veragua.
Very fertile and populous area,
producing coffee, cacao, tobacco, rice,
corn. Pop. (1980E) 140,508

Esperance Western Australia. Australia. 33 51S 121 53E. Town sit. on
Esperance Bay serving a wheat and
wool producing area. Railway terminus, resort and port. Pop. (1989E)
11,000.

Espírito Santo Brazil. A maritime
state in SE. Area, 17,605 sq. m.
(45,597 sq. km.). A low-lying coastal

area rising to heavily forested mountains in the w. The chief R. is the Doce. Coffee, sugar-cane, timber and fruit are produced. Cap. Vitória. Pop. (1980c) 2,239,000.

Espiritu Santo Vanuatu, South West Pacific. 15 50S 166 50E. Island with area 1,485 sq. m. (3,846 sq. km.). British administration centre at Hog Harbour; French at Segond Canal. Volcanic, rising to 6,195 ft. (1,888 metres) a.s.l. Produces copra, cocoa, coffee.

Espoo Uusimaa, Finland. 60 13N 24 40E. Village sit. w of Helsinki. Industries inc. granite quarrying. Pop. (1982E) 145,000.

Esquimalt British Colombia, Canada. 48 26N 123 24W. Port and naval base sit. at s end of Vancouver Island, just w of Victoria. Industries inc. shipbuilding and salmon fishing.

Essaouira Morocco. 31 31N 9 46W. Port sit. on Atlantic coast. Industries inc. sugar refining, fish canning and tanning. Wool and olive oil are exported. Pop. (1982E) 42,000.

Essen North Rhine-Westphalia, Federal Republic of Germany. 51 28N 7 01E. Town in w, sit. between Ruhr R. and Rhine-Herne Canal. An important industrial centre of the Ruhr coalfield with large steel works. Other industries inc. railway engineering, and the manu. of textiles, chemicals, glass and furniture. Pop. (1984E) 628,800.

Essendon Victoria, Australia. Town sit. NW of Melbourne. Retail and commercial centre for a farming area, trading in oats, barley. Pop. (1990E) 55,000.

Essequibo River Guyana, South America. 6 50N 58 30W. Rises in Guiana Highlands near the Brazil border and flows 600 m. (960 km.) N to enter Atlantic by a wide estuary 13 m. (21 km.) WNW of Georgetown. Navigable for ocean-going craft up to Bartica *c.* 50 m. (80 km.), for small craft to Monkey Jump. Used extensively to carry the minerals and timber from the interior to the coast, although

its course is much impeded by rapids which must be avoided by railway. Main tribs. Rupununi, Potaro and Mazaruni-Cuyuni Rs.

Essex England. 51 48N 0 40E. Under 1974 reorganization the county consists of the districts of Uttlesford, Braintree, Colchester, Tendring, Harlow, Epping Forest, Chelmsford, Maldon, Brentwood, Basildon, Rochford, Thurrock, Castle Point, and Southend-on-Sea. Area, 1,530 sq. m. (3,963 sq. km.). It has long, flat coastline with shallow bays, occasionally fringed with marshland with important oyster beds. Inland it is undulated and wooded, much agricultural land devoted to grain production. The economy of the county is extremely varied from the petrochemical industries along Thameside to the electronic engineering activities that exist in Chelmsford, Harlow, Basildon and Colchester. The automotive industry is well represented in the Brentwood and Basildon districts. Considerable sand and gravel extraction takes place within Essex. The county is particularly well connected with Europe and worldwide through its airports at Stansted and Southend and the ports of Tilbury and Harwich. The largest towns are Southend, Chelmsford, Colchester, Harlow, Basildon and Brentwood. Pop. (1988E) 1,529,500.

Esslingen Baden-Württemberg, Federal Republic of Germany. 48 45N 9 16E. Town sit. ESE of Stuttgart on Neckar R. Manu. machinery, textiles, electrical equipment, leather goods and gloves. Pop. (1984E) 87,300.

Essonne France. 48 30N 2 20E. Dept. in Île-de-France region, sit. to s of Paris. Area, 699 sq. m. (1,811 sq. km.). Cap. Ivry. Pop. (1982C) 988,000.

Estelí Estelí, Nicaragua. 13 05N 86 23W. Cap. of dept. 60m. (96 km.) NNW of Managua. Industries are sawmilling and tanning. Pop. (1980E) 30,000.

Estevan Saskatchewan, Canada. 49 07N 103 05W. City sit. SE of Regina on Souris R. at mouth of Long Creek. The main industries are crude oil, natural gas and mining lignite-coal, also flour milling; manu. mobile homes, coal briquettes. Pop. (1989E) 10,500.

Esthwaite Water England. L. sit. between Coniston Water and L. Windermere, in the Lake District.

Eston Cleveland, England. 54 34N 1 07W. Town sit. ESE of Middlesbrough on estuary of R. Tees. Industries inc. shipbuilding, steelworks and the manu. of chemicals. Pop. (1985E) 8,500.

Estonian Soviet Socialist Republic U.S.S.R. 48 30N 25 30E. Constituent rep. of U.S.S.R. bounded N by Gulf of Finland, S by Latvia, W by Baltic Sea and E by the Russian Soviet Federal Socialist Rep. Area, 17,400 sq. m. (45,100 sq. km.). After a brief period of independence following the 1917 revolution, the secret protocol of the Soviet-German agreement in 1939 assigned Estonia to the Soviet sphere of interest. An ultimatum in 1940 led to the formation of a government acceptable to the U.S.S.R., which applied for Estonia's admission to the Soviet Union; this was effected by decree of the Supreme Soviet later that year. The incorporation has been accorded *de facto* recognition by the British Government, but not by the U.S. Government, which continues to recognize an Estonian consul-general in New York. In 1989 a strong movement for independence from the U.S.S.R. emerged. Chief towns Tallinn (cap.), Paldiski. Mainly lowland rising to glacial moraine ridges in the S with poor, sandy soil and about 20% forest. Agric. is based on intensive dairy farming and cereals. Industries inc. distilling gas, petrol and asphalt from local oil shale of which deposits are considerable; textiles; shipbuilding; manu. matches, paper and furniture—all centred on Tallinn. The chief exports are dairy and oil-shale products. Pop. (1989E) 1·6m.

Estoril Lisboa, Portugal. 38 42N 9 23W. Seaside resort sit. W of Lisbon. Pop. 50,000.

Estremadura Portugal. Province of central Portugal. Area, 2,064 sq. m. (5,346 sq. km.). Chief cities Lisbon (cap.), Setúbal, Caldas da Rainha. Mainly agric., esp. grain, vines, fruit and market gardening produce, rice. Drained by Tagus R. Natural resources inc. salt and cork; commerce and industry are concentrated in the Lisbon area. Tourism is important on the coast.

Esztergom Komárom, Hungary. 47 48N 18 45E. Town sit. NNE of Budapest on Danube R. almost opposite its confluence with Hron R. Manu. inc. machinery and textiles. Noted for thermal springs.

Etawah Uttar Pradesh, India. 26 47N 79 01E. Town sit. ESE of Agra. Manu. textiles. Trades in cotton and agric. produce. Pop. (1981C) 112,174.

Ethiopia East Africa. 8 00N 40 00E. Republic bounded N by Red Sea, NE by Djibouti, E by Somali Rep., S by Kenya and W by Sudan. Modern Ethiopia dates from the reign of the Emperor Theodore (1855–68). Eritrea was incorporated into N Ethiopia in 1962. The Emperor was deposed in 1974. Area, 471,800 sq. m. (1,221,900 sq. km.). Chief cities Addis Ababa (cap.), Asmara, Harar, Dire Dawa. It is divided into 15 regions. Consists mainly of mountain plateau rising from the lowlands of the White Nile, W and divided into two blocks by the Great Rift Valley which extends NE to SW. The highlands descend gradually E across the semi-arid Ogaden plateau to the Indian Ocean. The Danakil desert occupies the NE. The N is drained into the Nile basin by Rs. inc. the Blue Nile and headstreams of the White Nile. The SE is drained into the Somali Rep. The climate and vegetation are divided into t.iree zones according to altitude; below 5,500 ft. (1,676 metres) is hot and arid; from 5,500–8,000 ft. (1,676–2,438 metres) has an average temp. of 62°F and rainfall between 25 and 80 ins., supporting most of the country's pop. and agric.; above 8,000 ft. (2,438 metres) is cool, with extensive grazing lands and forests. The main crops are cereals, tobacco, cotton, plantains, coffee; main exports are coffee, beeswax, hides and skins, grain. There are deposits of gold and potash. Industry is little developed, and mainly centred on Addis Ababa and Asmara. It inc. processing cotton, cement, salt, food and tobacco products and oil, from the refinery at Assab. Pop. (1987E) 46m.

Etna, Mount Sicilia, Italy. 37 46N 15 00E. Volcano NNW of Catania, the highest active volcano in Europe at 10,705 ft. (3,263 metres) a.s.l. It has four types of vegetation according to altitude; sub-tropical inc. citrus fruit, olives and figs; temperate, producing vines, almonds, hazelnuts, apples, pears; forest, producing chestnut, pine, birch and some cereals; desert, with lava and ash, snow-covered 9 months of the year. The area occupied by the mountain is *c.* 460 sq. m. (1,191 sq. km.).

Etobicoke Ontario, Canada. 43 42N 79 44W. City within Metropolitan Toronto. Industries inc. electronics and telecommunications. Manu. tyres and rubber. Pop. (1990E) 302,973.

Eton Berkshire, England. 51 31N 0 37W. Town sit. on R. Thames opposite Windsor. Eton College was founded here in 1440. Pop. (1983E) 3,473.

'Eua Tongatabu, Tonga. 21 22S 174 56W. Island of the Tongatabu group. Area, 33 sq. m. (85 sq. km.). Formerly called Middleburg.

Euboea Greece. 38 34N 23 50E. The largest island and prefecture sit. off E coast in Aegean Sea. Area, 1,509 sq. m. (3,908 sq. km.). Mainly mountainous rising to Mount Delphi, 5,718 ft. (1,743 metres) a.s.l. Agric. products inc. cereals, vines and olives, and sheep and goats are raised. Cap. Chalcis, sit. 35 m. (56 km.) N of Athens. Pop. (1981C) 188,410.

Euclid Ohio, U.S.A. 41 34N 81 32W. A suburb of Cleveland sit. on L. Erie. Manu. inc. tractors, metal goods and electrical equipment. Pop. (1980C) 59,999.

Euganean Hills Veneto, Italy. A range of volcanic hills sit. between Padua and Este. The highest point is Mount Venda, 1,980 ft. (604 metres) a.s.l.

Eugene Oregon, U.S.A. 44 02N

123 05W. City in W centre of state, sit. on Willamette R. Industries inc. wood products, food processing, tourism. Pop. (1980C) 105,624.

Eupen and Malmédy Liège, Belgium. 50 37N 6 03E. Towns sit. SE of Liège near the Federal Republic of Germany border and administratively united. Industries inc. wool spinning, weaving; manu. paper, tanning fluids, cables. Pop. (1981E) 17,000.

Euphrates River South West Asia. 33 30N 43 00E. Formed from headstreams rising in Armenian highlands between Black Sea and L. Van: the W Euphrates rises 14 m. (22 km.) N of Erzurum; the Murat rises 40 m. (64 km.) SW of Mount Ararat. Flows *c.* 1,700 m. (2,720 km.) SW past Keban and Birecik to the Syrian border. Total course in Turkey, 685 m. (1,096 km.). It then swings S and SE across the Syrian Desert into Iraq and flows through central Iraq to join the Tigris R. 40m. (64 km.) NW of Basra. The joint stream is called the Shatt al Arab and enters the head of the (Persian) Gulf. Not navigable except for shallow draught vessels in Iraq. Used in Syria and Iraq for irrigation.

Eure France. 49 06N 1 00E. Dept. in Haute-Normandie region. Area, 2,318 sq. m. (6,004 sq. km.). Mainly flat and wooded. The chief Rs. are Seine and Eure. Agric. products inc. cereals, flax, fruit, cheese and cider. Cattle and horses are bred, the latter esp. famous. Industries inc. metalworking and textile manu. Cap. Évreux. Pop. (1982C) 462,323.

Eure-et-Loir France. Dept. in Centre region, formed from parts of Orléanais and Normandy. Area, 2,269 sq. m. (5,876 sq. km.). In the E is an undulating plain called the Beauce, and in the W are the Perche Hills, noted for Percheron horses. The chief Rs. are Eure and Loir. Agric. products inc. wheat, oats and apples. Cap. Chartres. Pop. (1982C) 362,813.

Eureka California, U.S.A. 40 47N 124 09W. Port sit. NNW of San Francisco on Humboldt Bay. Industries inc. timber and pulp, newspapers, food processing. Pop. (1989E) 25,150.

Eure River France. 49 18N 1 12E. Rises in Perche hills SW of La Ferté-Vidame and flows 140 m. (224 km.) ESE to Chartres, then N past Maintenon and Louviers to join Seine R. above Pont-de-L'Arche. Main tribs. Avre and Iton Rs.

Europe 50 00N 15 00E. The most W and smallest continent of the old world forming an extensive peninsula of the Eurasian land-mass, occupying *c.* 8% of the earth's surface and supporting *c.* 25% of its pop. Bounded N and NE by Arctic Ocean, NW and W by Atlantic Ocean, S by Mediterranean. There is no well-defined E boundary, which has been variously placed between 65°E lat. and 45°E lat. through the U.S.S.R. in the Ural mountains. Area, *c.* 3.7m. sq. m. (9.58m. sq. km.). Politically divided into Albania, Andorra, Austria, Belgium, Bulgaria, Czechoslovakia, Denmark, Finland, France, German Democratic Republic, Federal Republic of Germany, Greece, Hungary, Iceland, Ireland, Italy, Liechtenstein, Luxembourg, Monaco, the Netherlands, Norway, Poland, Portugal, Romania, San Marino, Spain, Sweden, Switzerland, Turkey, and parts of the U.S.S.R. (inc. the Latvian, Lithuanian and Estonian Soviet Socialist Reps., the Belorussian Soviet Socialist Rep., the Ukrainian and Moldavian Soviet Socialist Reps. and the Transcaucasian Soviet Socialist Reps.), the United Kingdom, the Vatican and Yugoslavia. Because of the Atlantic drift current along the NW and the large amount of inland water Europe has a more temperate but more varied climate than any other continent. Pop. (1985E) 497m.

Evanston Illinois, U.S.A. 42 03N 87 42W. City sit. on L. Michigan just N of Chicago. Educational centre and headquarters of Rotary International. Pop. (1980C) 76,706.

Evansville Indiana, U.S.A. 37 58N 87 35W. City sit. on Ohio R. near Illinois border, and SSW of Indianapolis. Manu. inc. pharmaceuticals, tents, flags, food processing, agric. machinery, refrigerators and furniture. Pop. (1988E) 137,719.

Evenki National Area Russian Soviet Federal Socialist Republic, U.S.S.R. Sit. in Krasnoyarsk Territory, it consists mainly of tundra and coniferous forest. Area, 285,900 sq. m. (740,481 sq. km.). The chief Rs. are Lower Tunguska and Stony Tunguska. The principal occupations are reindeer breeding and fishing. Cap. Tura. Pop. (1980E) 15,000.

Everest, Mount Nepál/China. 27 59N 86 56E. Part of the Great Himalayan Range and the highest peak in the world 29,028 ft. (8,864 metres) a.s.l., sit. on the Nepál/ Tibet border. The summit was first reached by a British expedition in 1953.

Everett Massachusetts, U.S.A. 42 24N 71 03W. City sit. N of Boston. Manu. high technology products, steel, machinery, chemicals, electronic equipment and printing. Involved in energy production, produce distribution, petrol storage and distribution. Pop. (1985E) 35,773.

Everett Washington, U.S.A. 47 59N 122 13W. Port sit. on Puget Sound at entrance to Snohomish R. Manu. inc. wood-pulp, paper products, aerospace and high technology equipment. Timber, pulp and apples are exported and alumina imported. Pop. (1989E) 64,170.

Everglades Florida, U.S.A. 25 50N 80 40W. A swampy sub-tropical region in the S. A large area has been reclaimed and planted with sugarcane. There is an Everglades National Park with abundant wild life.

Evesham Hereford and Worcester, England. 52 05N 1 56W. Town on R. Avon. Noted for fruit and vegetable growing. Pop. (1990E) 18,000.

Évora Évora, Portugal. 38 34N 7 54W. (i) District covering part of Alentejo province. Agric. inc. wheat and corn production, pigs and sheep raising. Area 2,854 sq. m. (7,393 sq. km.). Pop. (1981C) 180,277. (ii) Town, cap. of district of same name.

Évreux Eure, France. 49 01N 1 09E. Cap. of dept. in central N France. Manu. inc. printing, electronics, pharmaceuticals, metal and rubber goods, chemicals and textiles. Pop. (1989E) 50,000.

Evros　Thrace, Greece. 41 10N 26 00E. A prefecture in NE bordering on Bulgaria and Turkey. It inc. the island of Samothrace. Area, 1,638 sq. m. (4,242 sq. km.). Wheat, cotton and tobacco are grown. Cap. Alexandroupolis. Pop. (1981C) 148,486.

Evrytania　Greece. 39 05N 21 30E. A prefecture sit. in central Greece. Area, 789 sq. m. (2,045 sq. km.). Mainly mountainous, and sheep and goats are reared. Cap. Karpenissi, sit. NW of Athens. Pop. (1981C) 26,182.

Exe River　England. 50 37N 3 25W. Rises on Exmoor and flows SSE for 60 m. (96 km.) to enter English Channel at Exmouth.

Exeter　Devon, England. 50 43N 3 31W. City sit. NE of Plymouth on Exe R. Cathedral city and episcopal seat. Commercial, financial and educational centre of farming and tourist area. Industries inc. manu. metal products, leather goods, wood products, agric. machinery, textiles, paper, food products and pharmaceutical goods. Seat of university. Pop. (1983E) 101,800.

Exmoor　England. 51 10N 3 45W. A high moorland rising to Dunkery Beacon, 1,706 ft. (520 metres) a.s.l. There are many Rs., inc. Exe and Barle, noted for trout fishing. Red deer and Exmoor ponies are numerous. It was estab. as a National Park in 1954.

Exmouth　Devon, England. 50 37N 3 25W. Port and resort sit. on estuary of R. Exe, in Lyme Bay. There are docks and harbour facilities and some light industry. Pop. (1982E) 28,787.

Exploits River　Canada. Rises, as Lloyds R., in SW Newfoundland and flows 200 m. (320 km.) ENE through King George IV L., Lloyds L. and Red Indian L. to enter Bay of Exploits S of Botwood. Used for hydroelectric power. Main trib. Victoria R.

Exuma　Bahamas. Islands between Andros (W) and Eleuthera, Cat and Long Islands (E); the N end and ESE of Nassau. Area, 100 sq. m. (259 sq. km.). The group forms Exuma district and extends 140 m. (224 km.) NW to SE across the Tropic of Cancer. Pop. (1980C) 3,672.

Eyre, Lake　South Australia, Australia. 29 30S 137 20E. A shallow salt L. in NW quarter of state, 35 ft. (11 metres) b.s.l. It is dry except during the rainy season.

Eyre Peninsula　South Australia, Australia. 34 00S 135 45E. Sit. between Spender Gulf and Great Australian Bight. Iron ore is mined in the Middleback Range in the NE. Wheat and barley are grown.

F

Faenza   Emilia-Romagna, Italy. 44 17N 11 53E. Town in N central Italy. Noted for majolica ware since the Middle Ages and possesses a museum of ceramics. Other manu. inc. textiles, chemicals and ironware. Fruit and wine are exported. Pop. (1989E) 54,115.

Faeroe Islands   Denmark. 62 00N 7 00W. Group of islands in N Atlantic between Iceland (NW) and the Shetlands (SE) and forming an overseas part of Denmark. Area, 540 sq. m. (1,399 sq. km.). Cap. Thorshavn, on Stromo Island. Consists of 21 rugged, volcanic islands, 17 of them inhabited. Subject to a harsh climate and with sparse vegetation, they support sheep farming, fishing and whaling. The main islands are Stromo, Ostero. Pop. (1988E) 47,663.

Faial   Azores, Atlantic Ocean. 38 34N 28 42W. Westernmost island of Azores group 4 m. (6 km.) w of Pico Island across Faial Channel. Area, 66 sq. m. (171 sq. km.). Main town Horta. Volcanic, rising to Pico Gordo, 3,350 ft. (1,021 metres) a.s.l. Main products cereals, vines, fruit, livestock, dairy products, whale products. Noted for its flowers. Administered as a county. Pop. (1980E) 17,000.

Fairbanks   Alaska, U.S.A. 64 51N 147 43W. City in centre of state. It is the terminus of the Alaska Railroad and the Alaska Highway. Industries are mining and oilfield services. Pop (1980C) 22,645.

Fairfield   Connecticut, U.S.A. 41 09N 73 15W. Town sit. SW of New Haven on Long Island Sound. Manu. inc. motorcar accessories and chemicals. Pop. (1980C) 54,849.

Fair Isle   Scotland. 59 30N 1 40W. A rocky island sit. between Shetland Islands and Orkney Islands. Area, 6 sq. m. (16.5 sq. km.). The principal occupations are fishing, sheep-rearing and the production of handmade multicoloured knitted goods.

Fairmont   West Virginia, U.S.A. 39 29N 80 09W. City sit. S of Pittsburgh, Pennsylvania, on Monongahela R. Industries inc. coalmining, and the manu. of lighting, motors and generators, steel and aluminium. Pop. (1980C) 23,818.

Fairweather, Mount   Canada/U.S.A. 58 54N 137 32W. Mountain sit. NE of Cape Fairweather at SW end of Glacier Bay. National Park, on the Alaska/British Columbia border at 15,300 ft. (4,663 metres) a.s.l.

Faisalabad   Pakistan. 31 25N 73 09E. Town sit. WSW of Lahore. Industries inc. engineering, cotton ginning, flour milling, textiles, fertilizers and oilseed milling. Pop. (1981E) 1,104,000.

Faiyûm   Faiyûm, Egypt. 29 10N 30 50E. Cap. of governorate sit. in a fertile oasis, in the NE. The principal occupations are cotton spinning and weaving, dyeing and tanning. Nearby is the anc. site of Crocodilopolis. Pop. (1986E) 227,300.

Faizabad   Uttar Pradesh, India. 26 47N 82 08E. Town sit. E of Lucknow on Gogra R. Trades in sugar-cane, grain and oilseeds, and refines sugar. Pop. (1981C) 101,873.

Fakaofo   Tokelau, South Pacific. 9 22S 171 14W. Atoll sit. N of W Samoa consisting of 55 islets. Area, 700 acres. Main product coconuts. Pop. (1970E) 700.

Falaise   Calvados, France. 48 54N 0 12W. Town sit. SSE of Caen on Ante R. Trades in livestock and cheese. Pop. (1982C) 8,820.

Falcon   Venezuela. State of NW

Venezuela bounded N by Gulf of Venezuela. Area, 9,575 sq. m. (24,799 sq. km.). Chief towns Coro (cap), La Vela, Puerto Cumarebo, Tucacas. Coastal lowlands rise to spurs of the Andes. It has a hot, dry climate with rains in the E, November–February. Its main product is oil, esp. in the W oil fields which are extensions of the Maracaibo field and linked by pipeline with Altagracia, also at Pueblo Cumarebo and on the Paraguaná Peninsula. There is coal at Coro and Pueblo Cumarebo, salt at Mitare. The agric. areas produce coffee, cacao, sugar-cane, maize, livestock, tobacco. Pop. (1980E) 499,676.

Falkirk   Central Region, Scotland. 56 00N 3 48W. Town sit. on Forth–Clyde Canal. Manu. inc. iron-founding, coach building, aluminium, bookbinding and whisky distilling. Pop. (1981C) 37,734.

Falkland Islands   South Atlantic. 51 30S 59 00W. Island group sit. off SE coast of Argentina forming a British colony. Area, 4,618 sq. m. (11,961 sq. km.). Cap. Stanley. Consists of East Falkland and West Falkland separated by Falkland Sound which varies in width 1½–20 m. (2–32 km.). There are about 200 small islets. Bleak and desolate with a cold climate, the islands support sheep farming and fishing. Main products are wool, skins, tallow and peat. An international airport at Mount Pleasant was completed in 1986. Pop. (1986E) 1,916.

Fall River   Massachusetts, U.S.A. 41 43N 71 08W. City and port sit. at mouth of Taunton R. and SE of Providence, Rhode Island. It was once an important centre of the cotton industry. Manu. inc. rubber, paper and brass goods, clothing and hats. Pop. (1980C) 92,574.

Falmouth   Cornwall, England. 50 08N 5 04W. Port and resort on E coast of county, and sit. on Carrick Roads. Industries are ship repairing and fishing, esp. oysters. Pop. (1981C) 18,525.

Falster   Maribo, Denmark. 54 48N 11 58E. An island sit. in Baltic Sea and linked to Zealand and Lolland by

bridges. Area, 198 sq. m. (513 sq. km.). Mainly agric. producing sugar-beet and fruit. The chief town is Nyköbing.

Famagusta   Cyprus, (Gazi Magusha). 35 07N 33 57E. Port and resort sit. on E coast on Famagusta Bay, and E of Nicosia. Manu. inc. footwear and clothing. Fruit and vegetables are exported. Considerable construction of hotels has taken place to cater for tourism but the resort has been evacuated since the Turkish invasion in 1974 and remained derelict in 1990. Pop. (1973E) 38,960 but (1985E) 8,000 in Turkish quarter of the town.

Fanning Island   (Tabuaeran) Line Islands, Pacific Ocean. 3 52N 159 20W. Atoll sit. NW of Kiritimani. Area, 13 sq. m. (34 sq. km.). It was inc. in the British Colony of the Gilbert and Ellice Islands in 1916. Main product copra. Pop (1980E) 434.

Fano   Marche, Italy. 43 50N 13 01E. Port and resort sit. on Adriatic Sea. Industries inc. fishing, sugar refining, engineering, packaging and manu. of bricks. Pop. (1989E) 34,000.

Fanö   North Frisian Islands, Denmark. 55 25N 8 25E. An island sit. opposite Esbjerg. Area, 22 sq. m. (57 sq. km.). Fishing is the main occupation.

Farah   Afghánistán. Province of SW Afghánistán bounded W by Iran, S by Baluchistan, Pakistan. Area, 30,000 sq. m. (77,700 sq. km.). Cap. Farah. Mainly desert, drained into Seistan L. depression by Harut Rud, Farah Rud, Khash Rud and Helmand Rs. Pop. and agric. are confined to oases in the depression and in the NE foothills of the Hindu Kush. There is some nomadic livestock rearing in the S.

Far East   Term used to describe the countries on the Asian shores of the Pacific *i.e.* Soviet Far East, Manchuria, Korea (N and S), China, Japan. Often used to describe the countries of SE Asia generally; India, Pakistan, Bangladesh, Sri Lanka, Burma, Thailand, Cambodia, Malaysia, Singapore, Indonesia, Philippines, Vietnam and Laos.

Fareham Hampshire, England. 50
51N 1 10W. Town sit. on Portsmouth
harbour. Industries inc. boatbuilding,
pharmaceuticals, marine engineering
and strawberry growing. Pop. (1988E)
100,000.

Farewell, Cape Egger Island,
Greenland. Sit. at S tip and rising to
over 2,000 ft. (610 metres) a.s.l.

Farewell, Cape South Island, New
Zealand. 39 36S 143 55E. Sit. at N tip,
with a lighthouse.

Fargo North Dakota, U.S.A. 46 52N
96 48W. City sit. on W bank of Red R.
of the North. It is the largest town in
the state and manu. agric. machinery,
electrical equipment, glass and food
products. Pop. (1980C) 66,040.

Faridpur Bangladesh. 23 36N 89
50E. Cap. of district sit. WSW of
Dacca. Industries inc. oilseed milling.
Trades in rice, cane, jute and oilseeds.
A noted Moslem shrine. Pop. (1984E)
130,000.

Farmington New Mexico, U.S.A.
36 44N 108 12W. City in NW corner
of state, SW of San Juan mountains.
Pop. (1988E) 38,470.

Farmington Hills Michigan, U.S.A.
42 28N 83 22W. City forming part of
Detroit industrial area, NW of Detroit
city centre. Pop. (1980C) 58,056.

Farnborough Hampshire, England.
51 17N 0 46W. Town sit. WNW of
Guildford. Site of aeronautical research centre, the Royal Aerospace
Establishment. Also site of biennial
air displays. Pop. (1987E) 42,800.

Farne Islands (The Staples) England. 55 38N 1 38W. A group of
basaltic islets sit. in North Sea just NE
of the mainland. It is a bird sanctuary
and a breeding station for Atlantic
seals.

Farnham Surrey, England. 51 13N
0 49W. Town sit. WSW of Guildford
on R. Wey. Pop. (1981C) 34,993.

Farnworth Greater Manchester,
England. 53 33N 2 24W. Town sit. SE
of Bolton. Industries inc. textile

manu., electrical and electronic
engineering.

Faro Faro, Portugal. 37 01N
7 56W. Cap. of district and port sit. on
S coast near Cape Santa Maria. Industries inc. cork processing, basket
making and fishing, esp. sardines and
tunny. Wine, cork, figs and almonds
are exported. Pop. (1981C) 27,974.

Farquhar Islands British Indian
Ocean Territory. Atoll sit. NE of N tip
of Madagascar in Indian Ocean.
Length *c.* 3 m. (5 km.). Main products
copra, fish.

Farrukhabad Uttar Pradesh, India.
27 24N 79 35E. Town sit. NW of Kanpur near Ganges R. Industries inc.
sugar refining and the manu. of cotton
and metal goods. Trades in sugar-
cane, grain and oilseeds. Pop. (1981C)
145,793.

Fárs Iran. 29 30N 55 00E. Province
in SW bordering on the (Persian) Gulf.
Area, 51,466 sq. m. (133,298 sq. km.).
Mountain chains running parallel to
the coast cross the province, inter-
spersed with fertile valleys. Agric.
crops inc. cereals, cotton, rice, olives,
dates, tobacco and opium. Cap.
Shiráz. Pop. (1986C) 3,294,916.

Fatehpur Uttar Pradesh, India. 25
56N 80 49E. Town sit. SE of Kanpur.
Trades in cereals and oilseeds. Pop.
(1981C) 84,831.

Faversham Kent, England. 51 20N
0 53E. Town and port sit. on Swale R.
Industries inc. food processing, packing and distribution, iron-founding,
brewing and engineering. Pop.
(1989E) 16,750.

Fayetteville Arkansas, U.S.A. 36
00N 94 05W. City sit. NW of Little
Rock in the Boston Mountains. Main
employers are the University of
Arkansas and the manu. of soup and
wheels for cars. Pop. (1985E) 39,641.

Fayetteville North Carolina, U.S.A.
35 03N 78 54W. City sit. SSW of
Raleigh on Cape Fear R. Industries
inc. sawmilling, food processing and
the manu. of textiles and furniture.
Trades in cotton and tobacco. Pop.
(1984E) 68,157.

Fear, Cape   North Carolina, U.S.A. 33 50N 77 58W. Sit. at the S tip of Smith Island in the Atlantic, and S of Wilmington.

Fécamp   Seine-Maritime, France. 49 45N 0 22E. Resort and port in N sit. on English Channel. Noted for Bénédictine liqueur. Industries inc. boatbuilding, fish-curing and the manu. of fishing nets and cod-liver oil. Pop. (1982C) 21,696.

Fejer   Hungary. 47 09N 18 30E. County of W central Hungary bounded E by Danube R. Area, 1,689 sq. m. (4,374 sq. km.). Cap. Szekesfehervar. Rises to the Vertes Mountains in the NW. Mainly agric. producing grain, wine, potatoes, vegetables, livestock and dairy produce. There are bauxite deposits at Gant and Iszkaszent-györgy. Manu. is concentrated in Szekesfehervar, Velence and Ercsi. Pop. (1984E) 423,000.

Felicité   Seychelles, Indian Ocean. Island sit. NE of La Digne Island in the Indian Ocean. Area, 689 acres. Main products copra, fish.

Felixstowe   Suffolk, England. 51 58N 1 20E. Resort and port sit. on North Sea. Industries are flour milling and tourism. Pop. (1985E) 20,858.

Feltham   Greater London, England. 57 26N 0 24W. Town sit. WSW of London adjoining Heathrow airport. Industries inc. manu. vehicles, aircraft, cables, electrical goods.

Feni Islands   Papua New Guinea, South Pacific. Group in Bismarck Arch. Area, *c.* 45 sq. m. (117 sq. km.). Main islands are Ambitle and Babase. The main products are fish and coconuts.

Fens, The   England. 52 45N 0 02E. Flat marshy tract of land around the Wash of E England covering parts of Lincolnshire, Norfolk, Cambridgeshire, Suffolk, and extending *c.* 70 m. (112 km.) N to S and 35 m. (56 km.) E to W. Crossed by Witham, Welland, Nene and Ouse Rs. Formed from the gradual silting up of a former bay of the North Sea. The first attempted drainage was under the Romans; after

A.D. 1621 major reclamation was carried out with Dutch help under Cornelius Vermuyden. The most successful drainage was the Earl of Bedford's in the 17th cent. The area is now under cultivation and produces vegetables, fruit, flowers and good grazing. Some areas are retained as fens forming reserves for wildfowl etc.

Feodosiya   Russian Soviet Federal Socialist Republic, U.S.S.R. 45 02N 35 23E. Port and resort sit. on Black Sea in SE Crimea, and sit. ENE of Sevastopol. Industries inc. fishing, fish-canning, flour milling, engineering and brewing. Wheat is exported. Pop. 60,000.

Ferghana   Uzbekistan, U.S.S.R. (i) A region sit. in the Ferghana Valley and almost completely surrounded by mountains. The principal R. is Syr Darya. Crops inc. cotton, alfalfa and grapes. The chief towns are Andizhan and Namangan. (ii) Town in region of same name 145 m. (232 km.) ESE of Tashkent. Manu. inc. textiles and food products. Pop. (1985E) 195,000.

Fermanagh   Northern Ireland. County in SW. Area, 715 sq. m. (1,852 sq. km.). Mainly hilly in NE and SW rising to 2,188 ft. (667 metres) a.s.l. in Cuilcagh. The principal R. is Erne, and Upper and Lower Lough Erne are sit. in the centre. Industry mainly agric. County town, Enniskillen. Pop. (1989E) 52,000.

Fernando de Noronha   Brazil. 3 51S 32 25W. An island in the S. Atlantic sit. NE of Cape São Roque, on NE coast. Area, 10 sq. m. (26 sq. km.). Pop. (1980E) 1,300.

Fernando Póo   *see* **Bioko**

Ferrara   Emilia-Romagna, Italy. 44 50N 11 35E. Cap of Ferrara province in NE. Manu. inc. fertilizers, plastics, beet-sugar, pottery and clothing. Trades in wine, fruit and agricultural products. There are many fine renaissance buildings. Pop. (1989E) 141,381.

Ferrol, El   Coruña, Spain. 43 29N 3 14W. Port and naval base sit. NE of

Corunna (Coruña.). Industries inc. boat-building, fishing and fish processing. Pop. (1980C) 91,764.

Fez Morocco. 34 02N 4 59W. Town in central N. It is an important trading and religious centre with two noted mosques, and gave its name to the trad. red felt cap. Manu. inc. carpets, textiles, musical instruments, soap and leather goods. Pop. (1982C) 448,823.

Fezzan Libya. 27 00N 15 00E. Division S of Tripolitania in Sahara desert. Bounded NW by Tunisia, W by Algeria and S by Niger. Area, 280,000 sq. m. (725,200 sq. km.). Cap. Sebha. Desert,with pop. concentrated in oases and along the Tripolitania boundary. Mainly rock, gravel and sand dunes, with practically no rain. Some dates are produced and there are some handicraft industries.

Fianarantsoa Fianarantsoa, Madagascar. 21 26S 47 05E. Cap. of province in central E. Industries inc. rice processing and meat canning. Pop. (1982E) 72,901.

Fier Albania. Town sit. W of Berat at S edge of Myzeqe plain. Centre of an agric. area. The main industry is cotton ginning.

Fiesole Toscana, Italy. 43 48N 11 17E. Town sit. NE of Florence on a hill overlooking Arno R. Mainly a residential suburb making straw hats. There are Etruscan and Roman remains.

Fife Scotland. 53 13N 3 02W. Region and former county of E Scot. forming a peninsula between the Firths of Tay and Forth. Under the reorganization of local government boundaries the region is conterminous with the former county. Area, 505 sq. m. (1,307 sq. km.). Chief towns Cupar, Leven, Dunfermline, Kirkcaldy, Cowdenbeath, St. Andrews, Glenrothes, Rosyth. Hilly, fertile in the centre and levelling towards the coast. Industries inc. agric., coalmining, oil and gas related activities, tourism, electrical engineering, textiles, blending and bottling whisky, paper-making, plastics, electronics, shipbuilding; manu. machinery, bricks and tiles. Pop. (1988E) 344,717.

Fiji South West Pacific. 17 20S 179 00E. Island group sit. NE of Sydney, Australia, consisting of 250 islands. Area, 7,078 sq. m. (18,333 sq. km.). Cap. Suva. The Fiji Islands were discovered by Tasman in 1643 and visited by Capt. Cook in 1774, but first recorded in detail by Capt. Bligh after the mutiny of the *Bounty* (1789). Fiji was ceded to Britain in 1874, after a previous offer of cession had been refused. Fiji became an independent member of the Commonwealth in 1970. It consists of about 332 islands of which 110 are inhabited. The main islands are Viti Levu, Vanua Levu, Taveuni, Kandava, Koro, Ngau, Ovalau. The Koro Sea is in the centre with the Lau group of coral and limestone islets to the E, and the Yasawa and Mamanutha groups to the W. Volcanic, mountainous and with dense forests on their windward side, grassy plain on the leeward. Annual rainfall about 60 ins. Fertile. Main products sugar-cane, taro, rice, cotton, pineapple. Main exports sugar, gold, canned fish, molasses. Pop. (1989E) 727,104.

Filey North Yorkshire, England. 53 13N 0 10W. Resort on NE coast sit. on Filey Bay. Pop. (1983E) 6,000.

Finchley Greater London, England. 51 36N 0 11W. District of N London sit. NNW of city centre. Mainly residential.

Findhorn River Scotland. 57 30N 3 45W. Rises in the Monadhliath Mountains and flows 62 m. (99 km.) NE past Forres to enter Moray Firth at Findhorn.

Findlay Ohio, U.S.A. 41 02N 83 39W. City sit. S of Toledo on Blanchard R. Manu. agric. machinery, rubber goods, beet sugar and pottery. Pop. (1980C) 35,594.

Finger Lakes New York, U.S.A. 53 09N 93 30W. A group of Narrow Ls. sit. to W and SW of Syracuse.

Finistère France. 48 20N 4 00W. Dept. in Brittany region, bounded by English Channel, Bay of Biscay and Atlantic Ocean. Area, 2,620 sq. m. (6,785 sq. km.). Two mountain ranges run from E to W, the Montagnes d'Arrée and the Montagnes Noires. Heathland covers much of the surface but there are fertile valleys producing cereals and fruit, esp. cider apples, and vegetables. Horses and cattle are reared. Cap. Quimper. Pop. (1982C) 838,364.

Finisterre, Cape   Coruña, Spain. 42 50N 9 19W. A headland sit. on the NW coast.

Finland   North Europe. 70 00N 27 00E. Rep. of N Europe bounded E by U.S.S.R., N by Norway, W by Gulf of Bothnia and Sweden, S by Gulf of Finland. Area, 130,091 sq. m. (305,475 sq. km.). Chief cities Helsinki, Turku, Tampere, Lahti, Pori, Oulu, Vaasa, Kuopio, Jyväskylä, Kemi, Kotka, Hämeenlinna. Since the Middle Ages Finland was a part of the realm of Sweden. In the 18th cent. parts of SE Finland were conquered by Russia, and the rest of the country was ceded to Russia by the peace treaty of Hamina in 1809. After the Russian revolution Finland declared itself independent. Mainly low plateau at *c.* 500 ft. (152 metres) with numerous Ls., rising towards Lapland in the N and reaching 4,343 ft. (1,324 metres) in Mount Haltia. Rs. are important and extensively used for logging and hydroelectric power. About two-thirds of the land is forested. Agric. produces rye, oats, barley, potatoes, sugar-beet, from *c.* 9% of the land area. Lapland (N) has nomadic livestock farming, esp. reindeer. Mineral resources inc. copper (at Outokumpu), iron (at Otanmäki). The main industries are lumbering and the manu. of the products; textiles and engineering; wood, wood pulp, paper and paper board are the chief exports. Pop. (1988E) 4,954,359.

Finland, Gulf of   Small arm of the Baltic Sea. 60 00N 27 00E. Ls. Lodoga and Onega drain into it. Generally shallow and of low salinity. It is 260 m. (416 km.) long and up to 80 m. (128 km.) wide.

Finnart   Strathclyde Region, Scotland. 56 07N 4 50W. Oil terminal sit. N of Garelochhead. Linked to the Grangemouth refinery by pipeline.

Finnmark   Norway. 69 30N 25 00E. County of N Norway, forming most N part of Scandinavian peninsula and bounded S by Finland, E by the U.S.S.R., N by Arctic Ocean. Area, 18,799 sq. m. (48,689 sq. km.). Cap. Vadso. Low-lying, with a climate mild enough for crops, but with no sun for 2 winter months and no darkness for 2 summer months. Reindeer rearing and fishing are the main occupations. Pop. (1984E) 77,000.

Finsbury   Greater London, England. 51 31N 0 06W. District of London. Mainly residential and commercial.

Finsteraarhorn   Bernese Oberland, Switzerland. 46 32N 8 08E. Peak sit. SE of Bern. 14,026 ft. (4,316 metres) a.s.l.

Firozabad   Uttar Pradesh, India. 28 38N 77 15E. Town sit. E of Agra. Manu. glass bangles. Trades in grain and oilseeds. Pop. (1981C) 202,338.

Firozpur   Punjab, India. Town sit. SSW of Lahore, Pakistan. The main occupation is cotton ginning. Trades in cotton and grain. Pop. (1981C) 61,162.

Fishguard   Dyfed, Wales. 51 59N 4 59W. Town sit. N of Haverfordwest on Fishguard Bay. Passenger port, terminus of the cross-channel service from Rosslare, Ireland. Pop. (1985E) 5,000.

Fitchburg   Massachusetts, U.S.A. 42 35N 71 48W. City sit. WNW of Boston on Nashua R. Manu. inc. machinery, paper, textiles, leather and electrical goods, footwear and saws. Pop. (1980C) 39,580.

Fitzroy River   Queensland, Australia. 23 32S 150 52E. Formed by union of the Dawson and Mackenzie Rs. and flows 174 m. (278 km.) N then SE past Rockhampton to enter Keppel Bay at Port Alma. Navigable below Rockhampton and used to transport wool, meat, gold, copper and coal.

Fitzroy River   Western Australia, Australia. 17 31S 123 35E. Rises at E end of King Leopold Mountains and flows 325 m. (520 km.) SW then NW through mountainous country to enter King Sound near Derby. Main tribs., Margaret and Christmas Rs.

Flamborough Head   England. 54 08N 0 04W. A chalk headland sit. at N end of Bridlington Bay.

Flanders   France/Belgium. Area extending W of Scheldt R. along North Sea coast. Main cities Ghent, Bruges, Ypres, Courtrai, Lille. Many times disputed between powers, now divided between East and West Flanders (Belgian provinces) and French Flanders. Low-lying area, its main trad. source of prosperity was the cloth industry. Pop. (1989E) 5,720,000 (Belgian provinces).

Fleetwood   Lancashire, England. 53 56N 3 01W. Port and resort on NW coast, sit. at mouth of the R. Wyre. There is a daily container service to Northern Ireland and the Irish Republic and there is a steamer service to the Isle of Man in the summer. Pop. (1981C) 24,467.

Flensburg   Schleswig-Holstein, Federal Republic of Germany. 54 47N 9 26E. Port in extreme N sit. at head of Flensburg Fjord. Industries inc. shipbuilding, metal working, paper making and fishing. Trades in coal, timber, cereals and rum. Pop. (1984C) 86,700.

Flinders Island   Tasmania, Australia. 40 00S 148 00E. Sit. in Bass Strait off NE coast. Area, 802 sq. m. (2,077 sq. km.). The chief settlement is Whitemark.

Flinders Range   South Australia, Australia. 31 25S 138 45E. Sit. between L. Frome and Torrens and rising to St. Mary Peak, 3,822 ft. (1,165 metres) a.s.l. There are deposits of uranium, copper and lead.

Flinders River   Queensland, Australia. 17 36S 140 36E. Rises W of Charters Towers in the Great Dividing Range and flows 520 m. (832 km.) W past Hughenden and Richmond then NW and N to enter Gulf of Carpentaria. Main tribs. the Saxby and Cloncurry Rs.

Flin Flon   Manitoba, Canada. 54

46N 101 53W. Town in E centre of province. A centre of mining and smelting concerned with gold, silver, copper and zinc. Pop. (1983C) 7,894.

Flint   Michigan, U.S.A. 43 21N 84 03W. City sit. NW of Detroit on Flint R. An important centre of the motorcar industry. Pop. (1980C) 159,611.

Flint   Clwyd, Wales. 53 15N 3 07W. Town sit. WNW of Chester on Dee R. estuary. Industries inc. coalmining, smelting, and the manu. of rayon and paper. Pop. (1981C) 16,454.

Flint Island   Kiribati, Central Pacific. 11 26S 151 48W. Island sit. SW of Caroline Island. Area, 2.5 m. (4 km.) by $\frac{1}{2}$ m. (0.8 km.). Once worked for guano, now abandoned and uninhabited.

Flintshire   Wales. Now incorporated in Clwyd.

Florence   Tuscany and Firenze, Italy. 43 46N 11 15E. City sit. NNW of Rome on Arno R. Industries inc. iron and steel; manu. vehicles, machinery, electrical goods, food products, textiles, chemicals, paper, furniture, leather goods, pottery, plastics, and fertilizers. Market centre for wine, olive oil, fruit, flowers, vegetables. The city is of outstanding historic, artistic and architectural interest, having reached its peak during the 14th–16th cents. Famous palaces include Palazzo Vecchio, Uffizi (containing many masterpieces in its galleries) connected to the Pitti Palace by the Ponte Vecchio. Tourism an important industry. Pop. (1988E) 417,487.

Florence   Alabama, U.S.A. 34 49N 87 40W. City sit. NNW of Birmingham on Tennessee R. Industries inc. meat packing, coalmining and the manu. of textiles. Pop. (1980C) 37,029.

Florencia   Caquetá, Colombia. 1 36N 75 36W. Town sit. SSW of Neiva on Orteguaza R. Commissary cap. Trading centre for an agric. and forest area, trading in rice, cattle, rubber, gum, timber, resins, cacao.

Flores   Nusa Tenggara, Indonesia. 8 30S 121 00E. An island sit. between the Flores Sea and the Sawu Sea. Area, 6,662 sq. m. (17,255 sq. km.). Mainly mountainous, with several active volcanoes, and heavily forested. The chief town and port is Ende. Sandalwood and copra are exported.

Flores   Azores, Portugal. 39 26N 31 13W. Island sit. in w. Area, 55 sq. m. (142 sq. km.). Noted for its flowers and hot springs. The chief town is Santa Cruz.

Flores   Uruguay. Dept. of SW central Uruguay bounded N and NW by Yí R. and W by the Arroyo Grande. Area, 1,745 sq. m. (4,520 sq. km.). Cap. Trinidad. Crossed E to W by the Cuchilla Grande Inferior. Mainly a sheep raising area, with some agric. inc. cereals, fruit and vines.

Florianópolis   Santa Catarina, Brazil. 27 35S 48 34W. State cap. and port in the S, sit. on Santa Catarina island which is joined to the mainland by the longest steel suspension bridge in the country. Sugar, fruit and tobacco are exported. Pop. (1984E) 154,000.

Florida   Solomon Islands, South Pacific. Island just N of Guadalcanal. Area, 20 m (32 km.) by 10 m. (16 km.). Volcanic. The main product is copra.

Florida   U.S.A. 28 30N 82 00W. State of extreme SE U.S.A. extending as a peninsula between Atlantic and Gulf of Mexico. Area, 54,262 sq. m. (140,539 sq. km.). Chief cities Tallahassee, Jacksonville, Miami, Tampa, St Petersburg, Fort Lauderdale, Hollywood, Hialeah. Settled from Spain in 1565; ceded to U.S.A. in 1821 and a State in 1845. Mainly lowland with many coastal swamps and Ls., inc. the Everglades, the Big Cypress Swamp and L. Okeechobee in the S and the Okefenokee Swamp in the N. Drained by St. John's and St. Mary's Rs. to the Atlantic, and by numerous Rs. to the Gulf. Subtropical climate, with extensive forest, mainly pine, cypress and mangrove. The principal source of income is tourism. Resorts are mainly on the sandy Atlantic coast, inc. Miami, Palm Beach, Fort Lauderdale, Daytona Beach, St Augustine. Other

industries inc. market gardening and fruit growing esp. citrus fruits, with associated industries, metal-working, lumbering, chemicals and instruments. Pop. (1988E) 12,335,000.

Florida   Florida, Uruguay. 34 06S 56 13W. City sit. N of Montevideo on Santa Lucía Chico R. Dept. cap. Centre of an agric. area. Industries inc. flour milling, sawmilling; manu. mosaics, textiles, hosiery. Pop. (1985E) 28,000.

Florida Keys   Florida, U.S.A. 24 45N 81 00W. Chain of coral limestone islands extending *c.* 150 m. (240 km.) around the tip of the Florida peninsula from Virginia Key, S of Miami Beach, to Key West. They separate Biscayne and Florida Bays from the Straits of Florida. Mainly scrub and mangrove swamps, but some islands are developed as resorts esp. Key Largo and Key West. Linked by a highway from Key West which crosses to the mainland at Key Largo, the largest island, 28 m. (45 km.) long.

Florina   Greece. 40 47N 21 24E. (i) A prefecture in Macedonia bordering on Yugoslavia and Albania. Area, 719 sq. m. (1,863 sq. km.). Mainly agric. Pop. (1981C) 52,430. (ii) Cap. of prefecture of same name sit. WNW of Thessaloníki. Trades in livestock, cereals and fruit. Pop. (1981C) 12,562.

Florissant   Missouri, U.S.A. 38 48N 90 20W. Town sit. NNW of St Louis near Missouri R. Mainly residential. Pop. (1980C) 55,372.

Flushing   *see* Vlissingen

Fly River   Papua New Guinea. 7 45S 141 45E. Rises in central Victor Emmanuel range and flows 650 m. (1,040 km.) SE to Gulf of Papua. Navigable for 500 m. (800 km.) from the mouth. Main tribs. Strickland and Tedi Rs.

Focşani   Vrancea, Romania. 45 41N 27 11E. Cap. of district sit. NE of Bucharest on Milcov R. Manu. inc. leather and soap. Trades in cereals and wine. Pop. (1981E) 74,341.

Foggia Puglia, Italy. 41 27N 15 34E. Cap. of Foggia province sit. in SE. Industries inc. engineering, flour milling, cheese and paper making and the production of olive oil and macaroni. Pop. (1981C) 156,467.

Fogo Cape Verde, Atlantic. 14 55N 24 25W. Island sit. ENE of Brava in Leeward group. Area, 184 sq. m. (476 sq. km.). Chief town São Filipe. Circular in shape with an active volcano, Cano Peak, 9,280 ft. (2,829 metres) a.s.l. Main products beans, coffee, oranges, tobacco. Pop. (1980C) 31,115.

Foix Ariège, France. 42 58N 1 36E. Cap. of dept. On Ariège R. Agric and tourist centre. Pop. (1982C) 10,064.

Foligno Umbria, Italy. 42 57N 12 42E. Town in central Italy. Industries inc. sugar refining, brick and paper making, and the manu. of textiles, leather goods and soap.

Folkestone Kent, England. 51 05N 1 11E. Resort and cross channel passenger port on the English Channel. Pop. (1983E) 44,200.

Fond du Lac Wisconsin, U.S.A. 43 47N 88 27W. City sit. NNW of Milwaukee at S end of L. Winnebago. Industries inc. precision tools, marine engineering, machinery, textiles and clothing. Pop. (1980C) 35,863.

Fontainebleau Seine-et-Marne, France. 48 24N 2 42E. Town sit SSE of Paris near Seine R. in the Forest of Fontainebleau. Noted for its palace where Napoleon signed his abdication in 1814. Pop. (1982C) 18,753.

Fontenoy Hainaut, Belgium. 50 34N 3 28E. Village sit. SE of Tournai. Scene of battle in 1745 when France gained control of Flanders.

Foraker, Mount Alaska, U.S.A. 62 59N 151 29W. Mountain sit. NNW of Anchorage in Mount McKinley National Park, at 17,280 ft. (5,267 metres) a.s.l.

Forbach Moselle, France. 49 11N 6 54E. Town sit. ENE of Metz near Saar border. Commercial centre at the edge of the Saar coalfield. Industries inc. manu. mining equipment, paper. Pop. (1982C) 27,321 (agglomeration, 99,606).

Foreland Point Devon, England. A headland sit. ENE of Ilfracombe.

Forest Hills New York, U.S.A. A residential district of New York City on Long Island. Noted for the West Side Tennis Club.

Forfar Tayside Region, Scotland. 56 38N 2 54W. Town in E central Scot. Manu. jute, rayon, linen, ladders and soft drinks. Pop. (1981C) 12,742.

Forli Emilia-Romagna, Italy. 44 13N 12 03E. Cap. of Forli province in E central Italy. Manu. inc. textiles, felt, footwear and furniture. Pop. (1981C) 110,806.

Formby Merseyside, England. 53 34N 3 05W. Town sit. SW of Southport. Mainly residential. Pop. (1981C) 25,773.

Formosa *see* **Taiwan**

Formosa Formosa, Argentina. 26 10S 58 11W. Cap. of province and R. port in NE sit. on Paraguay R. Industries inc. stock- breeding, meat packing and tanning. Pop. (1980E) 40,000.

Forres Grampian Region, Scotland. 57 37N 3 38W. Town in central N Scot. sit. on R. Findhorn. Industries inc. agric. engineering and related activities. Pop. (1988E) 9,180.

Forst Cottbus, German Democratic Republic. 51 44N 14 39E. Town in extreme E sit. on Neisse R. Manu. machinery, chemicals and textiles.

Fortaleza Ceará, Brazil. 3 43S 38 30W. State cap. and port sit. on NE coast. Industries inc. sugar refining, flour milling, and the manu. of textiles and soap. The principal exports are cotton, sugar, coffee, rice, hides and carnauba wax. Pop. (1980C) 647,917.

Fort Augustus Highland Region, Scotland. 57 09N 4 40W. Village sit. SW of Inverness on Caledonian Canal, at head of Loch Ness. The fort was built in 1730 and is now a religious house.

Fort Collins Colorado, U.S.A. 40 35N 105 05W. City sit N of Denver. It is the centre of an agric. and cattle-rearing district with associated industries. Industries inc. photographic supplies, biotech and engineering research and electronics. Pop. (1990E) 92,000.

Fort-de-France Martinique. 14 36N 61 05W. Cap. and port sit. on W coast on Fort-de-France Bay. Exports inc. sugar, rum and cacao. Pop. (1982E) 99,840.

Fort Dodge Iowa, U.S.A. 42 30N 94 10W. City sit. NNW of Des Moines on Des Moines R. There are local deposits of coal, clay and gypsum and manu. inc. gypsum products, bricks and tiles. Pop. (1980C) 29,423.

Fortescue River Western Australia, Australia. 21 20S 116 05E. Rises SSW of Nullagine and flows 340 m. (544 km.) NE then WNW to enter Indian Ocean 75 m. (120 km.) WSW of Roebourne. Its course is intermittent in dry weather.

Forth River Scotland. 56 04N 3 42W. Formed at Aberfoyle, Perthshire, by confluence of Avondhu R. and Duchray Water, and flows 35 m. (56 km.) E along border between Stirling and Perthshire/Stirling and Clackmannan to Alloa. Here it widens to become the Firth of Forth which extends 51 m. (82 km.) E to enter the North Sea between Fife Ness (N) and Dunbar (S) by a mouth 19 m. (30 km.) wide. Chief towns on the R. are Kincardine, Grangemouth, Bo'ness, Burntisland, Edinburgh, Leith, Kirkcaldy, Leven, North Berwick, Crail. Connected to Clyde R. by Forth-Clyde canal. The Firth receives Leven, Esk, Almond, Avon and Carron Rs.

Fort Jameson *see* **Chipata**

Fort Knox Kentucky, U.S.A. An army depot sit. SSW of Louisville. The gold Bullion Depository is here. Pop. (1989E) 38,277.

Fort Lamy *see* **N'djamena**

Fort Lauderdale Florida, U.S.A. 26 07N 80 08W. Resort on the SE coast near Everglades. Pop. (1980C) 153,279.

Fort McMurray Alberta, Canada. 54 40N 111 07W. City sit. NE of Edmonton. Indust. inc. timber, fur, fish, sands, petroleum products, oil, tourism. There are large tar sands plants. Pop. (1989E) 33,698.

Fort Rosebery *see* **Mansa**

Fort St John British Columbia, Canada. 56 15N 120 51W. City sit. on the Alaska Highway near the Peace R. Service centre for oil and natural gas production and exploration. Indust. inc. agric., logging and sawmilling. Pop. (1989E) 12,660.

Fort Smith Arkansas, U.S.A. 35 23N 94 25W. City sit. WNW of Little Rock on Arkansas R. It is the principal industrial centre in the state and manu. inc. metal products, wood and paper goods, textiles, glass, furniture and bricks. Pop. (1980C) 71,626.

Fort Victoria *see* **Masvingo**

Fort Wayne Indiana, U.S.A. 41 04N 85 09W. City sit. NE of Indianapolis at confluence of R. St. Mary's and St. Joseph. Manu. inc. transport and telecommunications equipment, and electric motors. Pop. (1988E) 179,810.

Fort William Highland Region, Scotland. 56 49N 5 07W. Town in W sit. at NE end of Loch Linnhe near foot of Ben Nevis. Industries inc. fish farming and processing, aluminium works, sawmill and paper mills. Pop. (1989E) 4,400.

Fort Worth Texas, U.S.A. 32 45N 97 20W. City in N central Texas. Industries inc. aircraft, computers, electronics, transportation and distribution. Pop. (1989E) 459,000.

Fougères Ille-et-Vilaine, France. 48 21N 1 12W. Town in NW. Industries inc. tourism and the manu. footwear, electronic equipment, glass and clothing. Pop. (1982C) 25,131.

Foumban   Cameroon. 5 43N 10 55E. Town in w province. Chief centre of Bamoun area for woodcarving and agric. products. Pop. (1981E) 41,358.

Fowey   Cornwall, England. 50 20N 4 38W. Port and resort sit. w of Plymouth on estuary of R. Fowey. The principal occupations are tourism, fishing and the exporting of china clay. Pop. (1988E) 2,600.

Foyle, Lough   Northern Ireland/Irish Republic. 55 07N 7 08W. An inlet of the Atlantic between Londonderry and Donegal.

Framingham   Massachusetts, U.S.A. 42 17N 71 25W. Town forming part of the Boston conurbation, sit. sw of Boston. Pop. (1980C) 65,113.

Framlingham   Suffolk, England. 52 13N 1 21E. Town sit. NNE of Ipswich. There is some light industry. Pop. (1989E) 2,580.

Francavilla Fontana   Puglia, Italy. 40 32N 17 35E. Town sit. ENE of Taranto. Trades in wheat, wine and olive oil.

France   Western Europe. 47 00N 1 00E. Rep. bounded NE by Belgium and Luxembourg, E by Federal Republic of Germany (across Rhine R.), Switzerland and Italy, s by Mediterranean and Spain, w by Atlantic Ocean and N by English Channel. Monaco forms an enclave in the SE and Andorra an enclave in the S. It inc. Corsica (Corse), *c.* 140 m. (224 km.) SE of Nice. Area, 210,036 sq. m. (543,994 sq. km.). inc. Corsica. Chief cities Paris (cap.), Marseilles, Lyons, Lille, Bordeaux, Toulouse, Nantes, Nice, Rouen, Toulon, Strasbourg, Grenoble, St. Etienne. There are 96 depts. divided into 22 regions. The land rises to 7 major mountain systems; the Ardennes (NE) extend into Belgium and Luxembourg; the Vosges (E) bordering the Rhine R. valley; the Armorican Massif (NW) in Brittany and Normandy; the Massif Central; the Alps (SE); the Jura (E) and the Pyrénées (SW). The 3 extensive lowland areas are the Paris Basin, the Aquitaine Basin and the Rhône-Saône corridor. Other narrower valleys be- tween the mountain systems have developed as important communications for strategic and commercial uses, mainly the Saverne Gap (N Vosges), the Col de Naurouze between the Massif Central and the Pyrénées and the Belfort Gap between the Vosges and the Jura. The main Rs. are Seine (main tribs. Oise, Marne and Yonne Rs.), Loire, Allier, Garonne, Rhône, Saône, Doubs and Isère Rs., all watering the lowland areas; Meuse and Moselle Rs, which are part of the Rhine system. There is an extensive canal system. About 18·2m. hectares are under cultivation and 11·9m. are pasture. The main crops are wheat, rye, barley, oats, potatoes, beet and maize. About 1m. hectares are under vines. Mineral resources inc. iron ore, coal, potash, lignite, bauxite, oil (mainly from Parentis), natural gas and sulphur (mainly from Lacq). The main industries are engineering; manu. wine and food products, oil refining, chemicals, textiles and tourism. Pop. (1989E) 56,184,000.

Franche-Comté   France. 46 30N 5 50E. Region of E France, once a province, bounded E by Switzerland, s and w by Burgundy and N by the Vosges, and consisting of the Jura, Belfort (Territory of) Doubs and Haute-Saône depts. Area, 6,252 sq. m. (16,189 sq. km.). Lowland along the upper Saône R., rising to the central Jura (N). Main occupations farming, lumbering; main industries manu. clocks, machinery and plastics, centred on Besançon, Dôle, Lons-le-Saunier, Vesoul, Belfort, Pontarlier. Pop. (1982C) 1,084,049.

Francisco Morazán   Honduras. Dept. of s central Honduras. Area, 2,432 sq. m. (6,298 sq. km.) Cap. Tegucigalpa. Mountainous, spanning the continental divide and rising to the Sierra de Comayagua (NW) and the Sierra de Lepaterique (centre). Mainly agric., producing cereals, coffee, sugar-cane, cotton, plantains, fruit, dairy products and poultry. Industry is concentrated at Tegucigalpa and Comayagüela, and inc. tanning, milling and manu. food products. There is now little gold or silver mining; there is some iron (at Agalteca) and limestone. Pop. (1983E) 203,753.

Frankfort   Kentucky, U.S.A. 38 12N 84 52W. Town sit. E of Louisville on Kentucky R. Commercial centre of a 'Bluegrass' agric. and stone quarrying region, trading in tobacco, livestock, dairy products, grain. Industries inc. distilling, sawmilling; manu. vehicles, concrete products, furniture, clothing, hemp twine. Pop. (1980C) 25,973.

Frankfurt-am-Main   Hessen, Federal Republic of Germany. 50 07N 8 40E. Town and port in central F.R.G. sit. on Main R. It is an important transportation, industrial, financial and commercial centre. Industries inc. printing and publishing, engineering, metalworking, and the manu. of chemicals, electrical and electronic equipment. There are many fine buildings and it is noted for trade fairs. Pop. (1989E) 622,500.

Frankfurt-an-der-Oder   German Democratic Republic. 52 20N 14 33E. (i) A district formed from part of Brandenburg. Area, 2,774 sq. m. (7,185 sq. km.). Pop. (1983E) 709,000. (ii) Cap. of district of same name sit. on Oder R. Industries inc. engineering, sugar refining, vegetable canning, and the manu. of machinery, textiles, furniture, leather goods and sausages. Pop. (1989E) 87,300.

Franklin   North West Territories, Canada. 72 00N 96 00W. A district comprising the Canadian Arctic Arch. and the peninsulas of Melville and Boothia. Area, 549,253 sq. m. (1,422,565 sq. km.). The main occupation is fur trapping.

Franz Josef Land   Russian Soviet Federal Socialist Republic, U.S.S.R. Arch. of *c.* 85 islands in Arctic Ocean off Novaya Zemlya and N of the Barents Sea. Area, 8,000 sq. m. (20,720 sq. km.). Three main sections are separated by the British Channel (W) and Austrian Sound (E); main islands are Aleksandra, George, Wilczek, Graham Bell, Hooker and Rudolf Islands. Formed of basalt and almost covered by glacier ice and lichen, the land rises to 2,410 ft. (735 metres) a.s.l. on Wilczek Land. There are govt. observation stations with permanent settlements.

Frascati   Lazio, Italy. 41 48N 12 41E. Town and resort sit. SE of Rome. Tourist centre and noted for its white wine.

Fraserburgh   Grampian Region, Scotland. 57 42N 2 00W. Port in N Scot. sit. near Kinnaird Head on North Sea. Noted for herring fishing, curing and canning. Light industry and tourism also important. Pop. (1986E) 12,960.

Fraser River   British Columbia, Canada. 49 09N 123 12W. Rises in Rocky Mountains at Yellowhead Pass and flows 850 m. (1,360 km.) NW to end of Cariboo Mountains, then S to Hope, then W to enter the Strait of Georgia by a delta 10 m. (16 km.) S of Vancouver. Main tribs. Nechako, Quesnel, Thompson, Chilcotin, Blackwater and Lillooet Rs. Navigable below Yale; above Yale it enters a canyon between mountains rising to 3,000 ft. (914 metres) a.s.l.

Fray Bentos   Río Negro, Uruguay. 33 08S 58 18W. Cap. of dept. and port in extreme W sit. on Uruguay R. An important meat-packing centre with allied trades. Pop. (1985E) 20,000.

Frederick   Maryland, U.S.A. 39 25N 77 25W. City in N central Maryland. Centre for commerce and retailing; Fort Detrick is here (U.S. Army medical research and defence communications), and associated biomedical and electronics industries. Manu. dairy products, leather goods, cement, plastics, aluminium. Pop. (1980C) 28,086.

Fredericia   Jutland, Denmark. 55 36N 9 45E. Port sit. at N end of the Little Belt, SE of Vejle. Industries inc. iron and silver works, and the manu. of fertilizers and cotton goods. Pop. (1977E) 45,226.

Fredericton   New Brunswick, Canada. 45 58N 66 39W. Cap. of province sit. on Saint John R. Industries inc. tourism, food processing, electrical products, metal working, footwear, timber, printing and publishing. Pop. (1989E) 44,722.

Frederiksberg   Copenhagen, Denmark. 55 41N 12 32E. Suburb and separate commune in Copenhagen city, noted for the 17th cent. Frederiksberg Palace.

Fredrikstad Östfold, Norway. 59 13N 10 57E. Port in SE sit. at mouth of Glomma R. on Oslo Fjord. Industries inc. canning, food products, paint, porcelain, plastics. Pop. (1989E) 26,500.

Freeport Bahamas. 10 27N 61 25W. Chief town on Grand Bahama. Pop. (1980E) 24,423.

Freeport Illinois, U.S.A. 42 17N 89 36W. City sit. WNW of Chicago. Centre for insurance and financial services. Manu. inc. electrical equipment, domestic products and tyres. Pop. (1980C) 26,266.

Freeport New York, U.S.A. 40 30N 73 35W. Resort sit. on SW coast of Long Island. A fishing centre and industries inc. electronics and printing. Pop. (1980C) 39,252.

Freetown Sierra Leone. 8 30N 13 17W. Cap. of rep. and also cap. of Freetown district sit. SSE of Dakar on Atlantic coast and N shore of the Sierra Leone Peninsula. It was founded in 1787 by the British as a settlement for freed slaves. Commercial centre and port trading in palm kernels, piassava, ginger, kola nuts, diamonds, platinum, gold, chromite. Industries inc. oil refining, sugar, cement, plastics, footwear, fish processing, soap man., ship repairs. Pop. (1988E) 469,776.

Freiberg Chemnitz, German Democratic Republic. 50 54N 13 20E. Town sit. WSW of Dresden at foot of the Erzgebirge. Manu. inc. smelter for tin, lead and zinc ore, machinery for paper making, linen, precision instruments, leather goods and porcelain. Pop. (1989E) 51,505.

Freiburg-im-Breisgau Baden-Württemberg, Federal Republic of Germany. 47 59N 7 51E. Town in SW on W edge of the Black Forest. Manu. inc. pharmaceuticals, chemicals, metal goods, electronics and wine. Pop. (1989E) 176,500.

Freising Bavaria, Federal Republic of Germany. 48 23N 11 44E. Town sit. NNE of Munich on Isar R. Manu. inc. tractors, gloves and hosiery. It has the oldest brewery in the world. Pop. (1989E) 36,500.

Freital Dresden, German Democratic Republic. 51 00N 13 39E. Town sit. SSW of Dresden. Manu. glass, musical instruments and cameras.

Fréjus Var, France. 43 26N 6 44E. Town sit. SW of Nice. Produces wine, olive oil and cork. There are some fine Roman remains. Pop. (1982C) 32,698.

Fremantle Western Australia, Australia. 32 03S 115 45E. Port and railway terminus on W coast. Manu. inc. shipbuilding, clothing, leather processing and products, paint and furniture. Exports inc. grain, iron, aluminium, non-metal sands, fruit and vegetables, meat, livestock and malt. Pop. (1986E) 22,709.

Fremont California, U.S.A. 37 34N 122 01W. City sit. S of San Francisco in the San Francisco Bay Area. Industries inc. micro-technology and data systems, diverse small industry and commerce. Pop. (1980C) 131,945.

Fremont Ohio, U.S.A. 41 21N 83 07W. City sit. SE of Toledo on Sandusky R. Commercial centre of an agric. area, trading in cereals, sugar beets and market gardening produce. Industries inc. manu. vehicle parts, clothing, cutlery, electrical goods, rubber goods; sugar refining. Pop. (1980C) 17,834.

French Guiana South America. 4 00N 53 00W. Overseas dept. of France on the Atlantic coast of NE South America, bounded W by Suriname, E and S by Brazil. Area, 32,252 sq. m. (83,533 sq. km.). Cap. Cayenne. Mainly forest with rich mineral resources which are little exploited. There are deposits of iron, copper, silver, lead, mercury, platinum, diamonds and alluvial gold. The coastal lowlands are cultivated and produce sugar-cane, coffee, cacao, rice, cassava, maize, spices and fruit from about 15 sq. m. (39 sq. km.). Chief exports shrimps and timber. Devil's Island is among the islets off the coast (N). Pop. (1989E) 93,540.

French Island   Victoria, Australia. 38 21S 145 21E. Island N of Phillip Island in the Bass Strait off S Victoria. Area, 84 sq. m. (218 sq. km.). Low-lying, marshy and uninhabited.

French Polynesia   South Pacific. Overseas territory of France, comprising 5 main arch. Area, 1,522 sq. m. (3,941 sq. km.). Main groups the Society Islands, Marquesas Islands, Austral (Tubuai) Islands, Tuamotu Islands, Gambier Islands. Cap. Papeete on Tahiti, Society Islands. An important product is copra (coconut trees cover the coastal plains of the mountainous island and the greater part of the low-lying islands). Fruit is grown for local consumption. Tourism is extremely important. Pop. (1988E) 188,814.

French Southern and Antarctic Territories   An overseas territory of France formed in 1955 and consisting of Saint Paul and Nouvelle Amsterdam islands, the Kerguelen and Crozet archs. and Terre Adélie. Pop. (1989E) 180.

French Territory of the Afars and Issas   *see* **Djibouti**

Fresnillo Zacatecas, Mexico. 23 10N 102 53W. Town in central Mexico. An important mining and agric. centre, with deposits of gold, silver, lead, copper, tin and zinc.

Fresno   California, U.S.A. 36 45N 119 45W. City in centre of state sit. in a rich farming district producing cotton, grapes, seed crops, cattle and poultry. Industries inc. food processing and the manu. of agric, machinery and transport equipment. Pop. (1980C) 218,202.

Fribourg   Switzerland. 46 48N 7 09E. (i) Canton in the W sit. to the W of the Bernese Oberland. Area, 645 sq. m. (1,670 sq. km.). Mainly hilly but principally agric., producing cereals, sugar-beet and cheese, esp Gruyère. It joined the Confederation

in 1481. Pop. (1989E) 200,005. (ii) Cap. of canton of same name sit. SW of Bern. Manu. inc. agric. and electrical equipment, sewing machines, fertilizers and chocolate. Pop. (1989E) 35,000.

Friedrichshafen   Baden-Württemberg, Federal Republic of Germany. 47 39N 9 28E. Town and port sit. SE of Stuttgart on N shore of L. Constance. Industries inc. cars, aerospace, high technology. Pop. (1985E) 52,272.

Friendly Islands   *see* **Tonga**

Friesland   Netherlands. 53 05N 5 50E. Province in N bounded by Wadden Zee and Ijsselmeer. Area, 1,278 sq. m. (3,311 sq. km.). Mainly flat with many canals and noted for Friesian cattle. Cap. Leeuwarden, sit. NNE of Amsterdam. Pop. (1984E) 597,236.

Frinton-on-Sea   Essex, England. 51 54N 1 17E. Town consisting of the adjacent resorts of Clacton-on-Sea and Walton-on-Naze, S of Harwich. The main industry is tourism. Pop. (1981C) 12,689.

Frisian Islands   North Sea. 53 35N 6 40E. Chain of islands extending parallel to the N coasts of the Netherlands, Federal Republic of Germany and Denmark, at 3–20 m. (5–32 km.) off-shore. Low-lying, their area being affected by erosion as well as by land reclamation, and long subject to serious flooding. Consist of 3 main groups: the West Frisian Islands are Dutch, and inc. Texel, Vlieland, Terschelling, Ameland and Schiermonnikoog. Main occupation sheep and cattle farming; the East Frisian Islands are in F.R.G. and inc. Baltrum, Borkum, Juist, Langeoog, Norderney, Spikeroog, Wangerooge. Main occupation tourism: the North Frisian Islands are divided between Federal Republic of Germany and Denmark and inc. Amrum, Föhr, Sylt, Romo, Fano, Mano, Trischen and Hallig Islands. Main occupation cattle grazing.

Friuli   Italy/Yugoslavia. 46 00N 13 00E. Region between Carnic Alps and Adriatic Sea extending through Friuli-Venezia-Giulia region of NE Italy and Slovenia in NW Yugoslavia. Mainly fertile plain rising to Alpine foothills. Chief cities Udine, Gorizia.

Friuli-Venezia-Giulia Italy. Region of NE Italy bounded N by Austria, E by Yugoslavia, W by Veneto and S by the Adriatic. Area, 3,029 sq. m. (7,846 sq. km.). Cap. Udine. Mainly fertile plain rising to the Carnic Alps (N). Pop. (1981c) 1,233,984.

Frome Somerset, England. 51 14N 2 20W. Market town in the SW sit. on R. Frome. Industries inc. light engineering and printing. Pop. (1989E) 25,000.

Frosinone Lazio, Italy. 41 38N 13 19E. Cap. of Frosinone province sit. ESE of Rome. Trades in agric. produce.

Frunze Kirghizia, U.S.S.R. 42 54N 74 36E. Cap of Kirghizia sit. WSW of Alma Ata in Chu R. valley. Industries inc. meat packing, flour milling, tanning and the manu. of agric. machinery and textiles. Pop. (1985E) 604,000.

Fuerteventura Canary Islands, Spain. 28 20N 14 00W. Island of the Canaries group nearest to the African mainland, W of Cape Juby, Morocco. Area, 666 sq. m. (1,725 sq. km.). Chief town, Puerto de Cabras. Mainly arid plain with volcanic peaks dividing more fertile areas. Main products cereals, vegetables, fruit, livestock, fish (tuna), alfalfa, lime, gypsum, cochineal.

Fujairah United Arab Emirates. 25 10N 56 20E. Sheikhdom and part of the federation of UAE. Area 450 sq. m. (1,170 sq. km.). Oil has been produced since 1966 but agric is the chief economic activity. Pop. (1980c) 32,200.

Fujian (Fukien) China. 26 00N 118 00E. A maritime province in the E on Formosa Strait. Area, 46,718 sq. m. (121,000 sq. km.). Mainly mountainous and forested with an important timber trade, esp. fir, pine and camphor. The principal R. is Minkiang. Crops inc. rice, tea, sugar-cane and fruit. Cap. Fuzhou. Pop. (1982E) 26m.

Fujisawa Honshu, Japan. 35 21N 139 29E. Town sit. SW of Yokohama on N shore of Sagami Bay. Centre of an agric. area, trading in potatoes and wheat. Pop. (1983E) 314,000.

Fujiyama Honshu, Japan. 35 22N 138 44E. An extinct volcano sit. WSW of Tokyo in Shizuoka prefecture. 12,395 ft. (3,778 metres) a.s.l.

Fukien *see* **Fujian**

Fukui Honshu, Japan. 36 04N 136 12E. Cap. of Fukui prefecture sit. NNE of Kyoto. Noted for silk and rayon. Also manu. leather goods, food products and paper. Pop. (1980E) 241,000.

Fukuoka Kyushu, Japan. 33 39N 130 21E. Cap. of Fukuoka prefecture and port sit. NE of Nagasaki. Industries inc. shipbuilding, and the manu. of textiles, chemicals, paper, pottery and metal goods. Machinery and porcelain are exported. Pop. (1983c) 1,082,000.

Fukushima Honshu, Japan. 37 44N 140 28E. Cap. of Fukushima prefecture sit. NNE of Tokyo. It is an important centre of the silk industry. Pop. (1983c) 265,000.

Fukuyama Honshu, Japan. 34 29N 133 22E. Town sit. ENE of Hiroshima at mouth of Ashida R. on Hiuchi Sea. Manu. dyes, rubber products, cotton textiles, sake. Port exporting textiles, mats, fertilizers and sugar. Pop. (1983c) 355,000.

Fulda Hessen, Federal Republic of Germany. 50 33N 9 41E. Town in the central E sit. on Fulda R. Manu. inc. textiles, chemicals, ball bearings and tyres. Trades in agric. produce. Pop. (1984E) 55,800.

Fulda River Federal Republic of Germany. 51 25N 9 39E. Rises on the Wasserkuppe and flows 95 m. (152 km.) N past Fulda, Hersfeld and Kassel to join Werra R. at Münden; together they form Weser R. Navigable for small boats below Hersfeld. Main trib, the Eder.

Fulham Greater London, England. Suburb of SW Lond. on N shore of Thames R. Mainly residential.

Fullerton California, U.S.A. 33 52N 117 55W. City sit. SE of Los Angeles. Industries inc. petroleum production, and fruit packing, canning and processing. Pop (1980C) 102,034.

Funabashi Honshu, Japan. 35 42N 139 59E. City sit. E of Tokyo on N shore of Tokyo Bay. Industries inc. fish hatcheries, woodworking, flour milling; manu. fountain pens. Pop. (1983E) 489,000.

Funafuti Tuvalu, Pacific. 8 31S 179 13E. Atoll forming the cap. of Tuvalu and comprising 30 islets on a 13 m. (21 km.) reef. The main settlement is on the largest island. Fogafale, on the E side of the atoll. Infertile, but produces palms and coconuts. A grass runway and deep sea wharf are sit. on Fogafale. Pop. (1983E) 2,620.

Funchal Madeira. 32 38N 16 54W. Cap. of the autonomous region of Madeira sit. on SE coast. The chief port and commercial centre of the island. Tourism is the main economic activity to which embroidery manu. is linked and Madeira wines are exported. Pop. (1988E) 120,000.

Fundy, Bay of Canada. 45 00N 66 00W. An inlet of the Atlantic Ocean separating New Brunswick and Nova Scotia. Noted for the extreme tides which are used to generate electricity.

Furka Pass Valais/Uri, Switzerland. 46 34N 8 25E. One of the highest passes in Europe, connecting Andermatt and Gletsch, in central E Switzerland. 7,971 ft. (2,453 metres) a.s.l.

Furneaux Islands Tasmania, Australia. Sit. in Bass Strait, the principal islands are Flinders, Clarke and Cape Barren. The main occupation is sheep-farming.

Fürstenwalde Frankfurt, German Democratic Republic. 52 21N 14 04E. Town sit. W of Frankfurt on Spree R. Manu. inc. electrical equipment, tyres and footwear.

Fürth Bavaria, Federal Republic of Germany. 49 28N 10 59E. Town in S central F.R.G. Manu. inc. textiles, chemicals, electrical equipment, toys and mirrors. Pop. (1984E) 98,500.

Fuse *see* **Higashiosaka**

Fushun *see* **Fuxin**

Futuna Wallis and Futuna Islands, Pacific. 14 15S 178 09W. Island in Hoorn group sit. SW of Wallis Islands. Area, 25 sq. m. (64 sq. km.). Volcanic, rising to 2,500 ft (762 metres) a.s.l. Main products coconuts, breadfruit, timber. Pop. (1988E) 4,100.

Fuxin Liaoning, China. 42 08N 121 29E. Town in NE. Industries are coalmining and the manu. of mining equipment and cement. Pop. (1984E) 653,000.

Fylde Lancashire, England. Area of W Lancashire on Irish Sea coast, between Wyre and Ribble estuaries. Low-lying, mainly agric. with dairy and poultry farming. Tourism important. Chief towns Blackpool, Fleetwood, Lytham St. Anne's.

Fyn Denmark. 55 20N 10 30E. The second largest island sit. between the Kattegat, the Little Belt and the Great Belt. Area, 1,150 sq. m. (2,978 sq. km.). Mainly flat and fertile producing cereals, sugar-beet and dairy produce. Cap. Odense.

Fyne, Loch Scotland. 56 00N 5 20W. A sea loch in the W, extending N and NE for about 40 m. (64 km.) from the Sound of Bute. Noted for herring fishing.

G

Gabès Tunisia. 33 53N 10 07E. Town sit. s of Tunis on Gulf of Gabès. Industries inc. tourism, chemicals, engineering, fishing and fruit growing in the oasis. Sea port exporting dates. Pop. (1984E) 92,000.

Gabès, Gulf of Tunisia. 34 00N 10 25E. Inlet of Mediterranean Sea between Sfax to N and Djerba Island to s.

Gabon 0 10S 10 00E. Rep. bounded w by Atlantic Ocean, N by Equatorial Guinea and Cameroon, E and s by Congo. Area, 103,346 sq. m. (267,667 sq. km.). Chief towns Libreville (cap.), Port-Gentil. Mainly rain forest. The main products are manganese ore, petroleum and tropical hardwoods. Pop. (1988E) 1,226,000

Gaborone Botswana. 24 45S 25 55E. Capital sit. in SE. The main occupation is govt. service. It is a commercial centre and industries inc. diamonds, textiles, breweries, soap, printing, publishing and construction. Pop. (1990E) 341,149.

Gabrovo Gorna Oryakhovitsa, Bulgaria. 42 52N 25 19E. Town sit. SE of Pleven on Yantra R. Railway terminus. Industries inc. flour and sawmilling, wood-working, woollen, cotton and linen textiles; manu. knitwear, metalware, gunpowder, furniture, leather goods. Pop. (1982E) 80,901.

Gadag Karnataka, India. 15 30N 75 45E. Town sit. SW of Hyderabad. Market town trading in textiles and grain. Industries inc. manu. cotton goods and leather. Pop. (1981C) 117,368.

Gadsden Alabama, U.S.A. 34 02N 86 02W. Town in NE of state sit. on Coosa R., in a coalmining area. Industries inc. iron and steel, and textiles; manu. tyres. Pop. (1980C) 47,565.

Gaeta Lazio, Italy. 41 12N 13 35E. Town on w coast of s central Italy sit. on Bay of Gaeta. Industries inc. fishing, glass making and tourism.

Gafsa Tunisia. 34 25N 8 48E. Town sit. WSW of Sfax on the Sfax–Tozeur railway, in a phosphate mining region, and with an oasis producing dates, olives and fruit. Pop. (1985E) 61,000.

Gagry Abkhaz Autonomous Soviet Socialist Republic, U.S.S.R. Town sit. NW of Sukhumi on Black Sea. Health resort and port. Industries inc. distilling, sawmilling, metal working. Coast guard station.

Gainesville Florida, U.S.A. 28 40N 82 20W. City sit. SW of Jacksonville. Industries inc. education, government administration, medicine. Pop. (1989E) 87,793.

Gainsborough Lincolnshire, England. 53 24N 0 46W. Town sit. NW of Lincoln on Trent R. Market town for an agric. area. Industries inc. flour milling; manu. machinery, maltsters, poultry packing and clothing. Pop. (1985E) 18,715.

Gairdner, Lake South Australia, Australia. 31 35S 136 00E. Salt L. lying N of Eyre Peninsula, extending 100 m. (160 km.) N to s and 30 m. (48 km.) E to w at its widest, and sometimes dry in summer.

Gairloch Highland Region, Scotland. 57 42N 5 40W. Sea loch on NW coast, N of L. Torridon, extending 5 m. (8 km.) inland, and 3 m. (5 km.) wide at its entrance.

Gaithersburg Maryland, U.S.A. 39 08N 77 12W. City sit. N of Washington, D.C. Manu. electrical and electronic machinery and systems, photographic supplies. Pop. (1980C) 26,424.

Galápagos Islands Ecuador. 0 30S

90 30W. Island group sit. W of
Ecuador and lying across the Equator.
Area, 2,868 sq. m. (7,428 sq. km.).
There are 12 large islands and many
islets; the largest are Isabela, 75 m.
(120 km.) long, and San Cristóbal.
There are unique wild life forms, for
which the group has been made a
sanctuary. Pop. (1980E) 6,000.

Galashiels   Borders Region, Scotland. 55 37N 2 49W. Town in SE. The
main industry is manu. woollen cloth
and electronics. Pop. (1981C) 13,314.

Galati   Galati, Romania. 45 26N 28
03E. Cap. of district in E sit. on
Danube R. Industries inc. textiles and
chemicals; manu. hardware. Naval
base and R. port exporting grain and
timber. Pop. (1983E) 254,636.

Galesburg   Illinois, U.S.A. 40 57N
90 22W. City in NW, sit. SW of Chicago, in a coalmining area. Industries
inc. manu. of refrigerators, dehumidi-
fiers, metal buildings, rubber hose,
clothing, railway and mining equipment. Pop. (1980C) 35,305.

Galicia   Poland/U.S.S.R.   Region
extending N from the Carpathian
Mountains across SE Poland and NW
Ukraine, U.S.S.R. Rich in minerals,
esp. oil and natural gas.

Galicia   Spain. 42 43N 8 00W. Region in NW Spain bounded N by Bay
of Biscay, W by Atlantic Ocean and S
by Portugal. Comprises the provinces
of La Coruña, Lugo, Orense and Pontevedra. Mountainous, rising to 6,000
ft. (1,829 metres) a.s.l. in the E.
Drained by Miño R. The main occupations are farming and sardine fishing. Main products are maize,
potatoes, cattle and pigs.

Galilee   Israel. Region in N Israel
bounded N by Lebanon, and E by
Syria and Jordan. Hilly in the N, lower
and more fertile in the S. Chief towns,
Tiberias and Nazareth.

Galilee, Sea of, (Lake Tiberias)
Israel. 32 49N 35 36E. L. in E Galilee
extending 13 m. (21 km.) N to S along
the Syrian border at 696 ft. (212
metres) b.s.l. Fed and drained by
Jordan R.

Gallarate   Lombardia, Italy. 45
40N 8 47E. Town sit. WNW of Milan.
Industries inc. textiles.

Galle   Southern   Province,   Sri
Lanka. 06 02N 80 13E. Town sit. SSE
of Colombo on SW coast. Commercial
centre and port exporting coconut products. Pop. (1981E) 77,183.

Gällipoli   Turkey. 40 24N 26 40E.
(i) Peninsula extending 55 m. (88
km.) SW between Dardanelles Strait
and Gulf of Saros. (ii) Town of same
name sit. on neck of Gallipoli Peninsula. Trading centre for grain and livestock. Industries inc. fishing.

Galloway   Scotland. 55 00N 4 25W.
District of SW Scot. comprising the
counties of Wigtown and Kirkcudbright. The main occupation is dairy
farming. Noted for Galloway cattle.

Galt   *see* **Cambridge**

Galveston   Texas, U.S.A. 29 18N
94 48W. City in SE of state sit. at
entrance to Galveston Bay, an inlet of
the Gulf of Mexico. Industries inc.
tourism and chemicals; manu. metal
goods and hardware. Port handling
sulphur, cotton and wheat. Pop.
(1980C) 61,902.

Galway   Ireland. 53 16N 9 03W. (i)
County in W bounded W by Atlantic
Ocean, and S by Galway Bay and
County Clare. Area, 2,293 sq. m.
(5,939 sq. km.). Chief towns Galway,
Ballinasloe and Tuam. Mountainous
in the W, with bogs and L. and almost
divided N to S by Lough Corrib; to the
E of the L. it is low-lying, and fertile
in parts. The main occupation is farming. Pop. (1981C) 172,018. (ii) County
town of (i) sit. at the mouth of Corrib
R. on Galway Bay. Industries inc.
tourism, electronics and high technology. Pop. (1981C) 37,835.

Gambia   13 20N 15 45W. Rep.
within the Commonwealth, bounded
W by Atlantic Ocean and on all other
sides by Senegal, and extending 200
m. (320 km.) inland up both banks of
Gambia R. Area, 3,451 sq. m. (8,938
sq. km.). Cap. Banjul, formerly named
Bathurst. The main products are
groundnuts and groundnut derivatives.
Pop. (1988E) 788,163.

Gambia River West Africa. 13 28N 16 34W. Rises in the Fouta Djallon plateau, Guinea, and flows 700 m. (1,120 km.) NW into Senegal, and W through Gambia and past Georgetown to enter Atlantic by an estuary below Banjul. Navigable to ocean-going vessels for *c.* 200 m. (320 km.).

Gambier Islands Pacific Ocean. 23 09S 134 58W. Group of Islands SE of the Tuamotu Islands. Area, 14 sq. m. (36 sq. km.). There are four main islands and several islets, forming part of French Polynesia. The main island is Mangareva; the chief town, Rikitea. Chief products are copra, coffee and mother of pearl. Pop. (1988E) 582.

Gamlakarleby Vaasa, Finland. 63 50N 23 08E. Town sit. NE of Vaasa on Gulf of Bothnia. Railway junction and sea port. Industries inc. woodworking; manu. machinery, leather products. Port exports timber and timber products. About half the pop. speaks Swedish; Gamlakarleby is the Swedish name; Finnish is Kokkola. Pop. (1982E) 34,000.

Gamu-Gofa Ethiopia. Region of S Ethiopia bounded S by Kenya. Area, 15,000 sq. m. (38,850 sq. km.). Cap. Chancha. Sit. in Great Rift Valley between Galana Sagan (E) and Omo (W) Rs. Mountainous and forested, rising to 13,780 ft. (4,200 metres) a.s.l. in Mount Gughe. Main occupation farming; main products cereals, plantain, cotton, tobacco, livestock. Pop. (1980E) 1,003,400.

Gander Newfoundland, Canada. 48 57N 54 34W. Town in centre of country with a major airport. Industries inc. service and communications. Pop. (1989E) 12,000.

Ganges Canals Uttar Pradesh, India. Upper Ganges Canal leaves Ganges R. at Hardwar and travels in two branches to Kanpur and Jumna R. Lower Ganges Canal leaves Ganges R. near Dibai and splits into several branches. Both canals are used for irrigation.

Ganges River India/Bangladesh. 23 22N 90 32E. Formed in Himalayas, in N Uttar Pradesh, India, by union of R. Bhagirathi and Alaknanda. Flows 1,557 m. (2,491 km.) SW past Hardwar, then SE through Uttar Pradesh to enter Bihar above Patna; E through Bihar past Bhagalpur to cross the narrow neck of W. Bengal. It forms the frontier with Bangladesh for *c.* 70 m. (112 km.) and then continues SE, receiving Brahmaputra R., and enters Bay of Bengal by a broad delta. It begins to divide into the streams that make up this delta *c.* 220 m. (352 km.) from its many mouths. The main stream, Meghna R. formed by Ganges R. main stream and Brahmaputra R., is to the E of the delta. Used extensively for irrigation. Regarded as the most sacred R. by the Hindus.

Gangtok Sikkim, India. 27 20N 88 37E. Cap. of Sikkim sit. NE of Darjeeling. Commercial trading centre in rice, maize and fruit. Industries inc. brewing and watch making. Pop. (1981E) 36,747.

Gansu (Kansu) China. 36 00N 104 00E. NW province, sit. between Tibet and Inner Mongolia. Area, 175,290 sq. m. (454,000 sq. km.). Cap. Lanzhou. Mountainous area with fertile valleys growing millet, beans and tobacco, despite variable rainfall. There are abundant minerals inc. coal. The main products are dyes, silk, tobacco, mercury and wool. Pop. (1982C) 19,569,261.

Ganta Nimba, Liberia. 7 15N 8 59W. Town sit. NE of Monrovia on border with French Guinea. Trading centre for cacao, palm oil, kernels, kola nuts.

Gao Mali. 16 19N 0 09W. Cap. of region sit. NW of Niamey (Niger) on Niger R. and trans-Saharan road. Trading centre for butter, hides, wool, livestock, in a farming area. R. and airport. Pop. (1976C) 30,714.

Gap Hautes-Alpes, France. 44 34N 6 05E. Cap. of dept. on Luye R. Tourist centre. Pop. (1982C) 31,272.

Gard France. 44 02N 4 10E. Dept. in Languedoc-Roussillon region, bounded E by R. Rhône delta and S by Gulf of Lions. Area, 2,258 sq. m. (5,848 sq. km.). Chief towns Nîmes and Alès. Flat land in s, marshy and with lagoons, rising to hills in the NW over 5,000 ft. (1,524 metres) a.s.l. The main occupation is growing olives, mulberries, vines and fruit; industries are concentrated in Nîmes and Alès and inc. metal working, textiles and paper. Pop. (1982C) 530,478.

Garda, Lake Lombardia/Veneto, Italy. 45 40N 10 41E. L. in central N. Area, 143 sq. m. (370 sq. km.). Fed from N by Sarca R., and drained to SE by Mincio R. There are several resorts sit. on its banks.

Garden Grove California, U.S.A. 33 46N 117 57W. City immediately NW of Santa Ana in s California. Residential, commercial and retailing centre. Pop. (1980C) 123,307.

Garden of the Gods Colorado, U.S.A. Tourist area immediately NW of Colorado Springs, with eroded sandstone in bizarre shapes.

Garden Reach West Bengal, India. Town sit. WSW of Calcutta on Hooghly R. Industries inc. jute and cotton processing, tanning, docks. Pop. (1981C) 191,107.

Gardez Afghánistán. 33 36N 69 06E. Cap. of province sit. s of Kabul on Kandahar road. Commercial centre.

Gardner Island Phoenix Islands, South Pacific. Atoll 230 m. (368 km.) s of Canton. Area, 1.5 sq. m. (4 sq. km.). The only product is copra.

Garfield New Jersey, U.S.A. 40 53N 74 07W. City sit. on Passaic R. Industries inc. textiles, chemicals, recycling of paper products and rubber products. Pop. (1980C) 26,803.

Garfield Heights Ohio, U.S.A. 41 26N 81 37W. City immediately SW of Cleveland. Pop. (1980C) 34,938.

Garian Tripolitania, Libya. Town sit. s of Tripoli on Gebel Nefusa plateau. Railway terminus. Industries inc. flour milling, olive oil processing, weaving.

Garibaldi Park British Columbia, Canada. Provincial park in SW Brit.

Columbia in Coast Mountains, NE of Vancouver. Area, 973 sq. m. (2,520 sq. km.). Tourist area.

Garland Texas, U.S.A. 32 54N 96 39W. City forming a suburb of Dallas NE of city centre. Industries inc. manu. aircraft, seismic instruments, vehicles. Pop. (1980C) 138,857.

Garmisch-Partenkirchen Bavaria, Federal Republic of Germany. 47 29N 11 05E. Town sit. SW of Munich near the Austrian border. Mountain and health resort. Industries inc. tourism, plastics and woodworking. Pop. (1989E) 26,000.

Garmsar Second Province, Iran. 35 20N 52 25E. Town sit. ESE of Tehrán. Railway junction for Bándár Sheh on the Caspian Sea and Meshed.

Garonne River France. 45 02N 0 36W. Rises in the Val d'Aran in the central Pyrénées, and flows 360 m. (576 km.) first NE to Toulouse, then NW to Bordeaux. It joins the Dordogne R. c. 20 m. (32 km.) below Bordeaux, and enters the Atlantic by the Gironde estuary. Main tribs. are Rs. Ariège, Tarn and Lot. Linked to the Mediterranean at Toulouse by the Canal du Midi.

Garoua Cameroon. 9 18N 13 24E. Cap. of North province sit. NNE of Yaoundé on Benoué R. River port and market handling peanuts, cotton. Pop. (1981E) 77,856.

Gartok Tibet, China. Town in W Tibet, at foot of Kailas Range, at 15,000 ft. (4,572 metres) a.s.l. Trading centre for wool, barley and salt.

Garut Java, Indonesia. 7 14S 107 53E. Town sit. SE of Djakarta. Commercial centre for an agric. area trading in tea, rubber and cinchona. Industries inc. tourism.

Gary Indiana, U.S.A. 41 36N 87 20W. Town sit. at s end of L. Michigan in NW Indiana. The main industry is steel; also manu. tin plate, chemicals and cement. Pop. (1988E) 132,460.

Garzan River Turkey. Rises in the

Bitlis Mountains 12 m. (19 km.) SE of
Mus and flows 65 m. (104 km.) S past
Besiri to join Tigris R. SW of Siirt.

Gazi Magusha   *see* **Famagusta**

Gaspé Peninsula   Quebec, Canada.
48 30N 65 00W. Peninsula between
St. Lawrence R. to N and Chaleur Bay
to S. Area, 11,000 sq. m. (28,500 sq.
km.). Mountainous, rising to 4,160 ft.
(1,268 metres) a.s.l. in Mount Jacques
Cartier, and wooded. The main occupations are fishing and lumbering; the
main industry is pulp milling.

Gastein   Salzburg, Austria. Valley
in W central Austria at E end of Höhe
Tauern at 3,000–3,500 ft. (914–1,067
metres) a.s.l., with health resorts and
spas, inc. Bad Gastein and Hof
Gastein.

Gastonia   North Carolina, U.S.A.
35 16N 81 11W. City sit. W of Charlotte. Industries inc. textiles, plastics
and non-electrical machinery. Pop.
(1988E) 54,606.

Gateshead   Tyne and Wear, England. 54 58N 1 37W. Town sit. on S
bank of Tyne R. opposite Newcastle
upon Tyne. Industries inc. engineering, glass, paints, plastics, rope, printing, textiles, rubber, electrical
engineering, metal working and chemicals. Pop. (1981C) 210,934.

Gatineau   Quebec, Canada. 45 29N
75 39W. City on Ottawa R. opposite
Ottawa, Ontario. Industries inc. paper
and pulp. Pop. (1981C) 74,988.

Gatineau River   Canada. 45 27N
75 40W. Rises on Laurentian Plateau
in SW Quebec, and flows 230 m. (368
km.) SSW, through Baskatong Reservoir, to join Ottawa R. below Ottawa.
Used for logging and hydroelectric
power.

Gatooma   *see* **Kadoma**

Gatwick   West Sussex, England. 51
08N 0 10W. Village sit. SSE of
Reigate. Important intern. airport.

Gävle   Gävleborg, Sweden. 60 41N
17 13E. Cap. of county in W sit. at
mouth of the Gävle R. on the Gulf of

Bothnia, N of Stockholm. Industries
inc. papermaking, electronics and
food processing. Sea port exporting
iron, timber, wood pulp and paper.
The port is ice-free for 9 months a
year. Pop. (1989E) 88,000.

Gaya   Bihar, India. 24 48N 85 03E.
Town sit. S of Patna. Commercial
centre trading in rice, oilseeds and
sugar-cane. Important place of pilgrimage for Hindus. Pop. (1981C)
344,941.

Gaza   Israel. 31 30N 34 28E. Town
sit. SSW of Tel Aviv at NW extremity
of Negev desert. Noted as the anc.
Philistine city whose temple was
destroyed by Samson. The Gaza Strip
was established in 1948. Pop. (1971E)
118,300.

Gaza   Mozambique. Province in S,
area, 32,022 sq. m. (82,937 sq. km.).
Cap. Xaixai. Pop. (1982E) 1,030,536.

Gaziantep   Gaziantep, Turkey. 37
05N 37 22E. Cap. of province in S.
Market town trading in agric. produce.
Industries inc. textiles. Pop. (1980C)
374,290.

Gdańsk   Gdańsk, Poland. 54 23N
18 40E. Cap. of province sit. near
mouth of Vistula R. on Gulf of
Gdańsk. Industries inc. shipbuilding,
food processing, distilling, paper
making and pulp milling; manu. machinery. Sea port exporting coal, grain
and timber. Pop. (1989E) 1,300,000.

Gdynia   Gdańsk, Poland. 53 31N 18
30E. City sit. on Gulf of Gdańsk. Sea
port exporting coal and timber. Pop.
(1983E) 240,200.

Gedaref   Kassala, Sudan. 14 01N
35 24E. Town sit. SW of Kassala.
Trading centre for cotton, wheat, barley, fruit and livestock.

Geelong   Victoria, Australia. 38
08S 144 21E. City in central S of state,
sit. on W shore of Corio Bay, an inlet
of Port Phillip Bay. Industries inc.
tourism, oil refining; manu. cars,
chemicals and cement. Pop. (1988E)
191,360.

Geislingen   Baden-Württemberg,

Federal Republic of Germany. Town sit. in Jura Mountains. Industries inc. manu. cutlery and other metal goods and glassware. Pop. (1985E) 26,000.

Gela Sicilia, Italy. 37 03N 14 15E. Town sit. on s coast. Industries inc. fishing.

Gelderland Netherlands. 52 05N 6 10E. Province in central Neth. bounded N by province of Overijssel, and E by F.R.G. Area, 1,937 sq. m. (5,017 sq. km.). Chief towns Arnhem (the cap.), Nijmegen and Apeldoorn. Pop. (1989E) 1,794,678.

Geldrop North Brabant, Netherlands. Town sit. E of Eindhoven. Industries inc. manu. wool, cotton, plastics, steel, furniture, agric. machinery and linen textiles. Pop. (1989E) 26,000.

Geleen Limburg, Netherlands. Town sit. ESE of Antwerp, Belgium, in a coalmining area. Industries inc. textiles, chemicals and fertilizers. Pop. (1984E) 34,828.

Gelligaer Mid Glamorgan, Wales. 51 40N 3 15W. Town sit. N of Cardiff. Industries inc. mining, furniture, window manu. and engineering. Pop. (1981C) 5,592.

Gelsenkirchen North Rhine-Westphalia, Federal Republic of Germany. 51 31N 7 07E. Town immediately NE of Essen. An important coalmining centre; also manu. steel, machinery, chemicals and glass. Pop. (1985E) 295,000.

Gembloux Namur, Belgium. 50 34N 4 41E. Town sit. NW of Namur. Railway and roadway junction. The main industry is manu. cutlery; also sugar refining. Agric. college. Pop. (1989E) 18,987.

Gemsah Egypt. Town on Red Sea sit. NNW of Safaga.

Gemu Gofa Ethiopia. Region in SW. Area, 15,251 sq. m. (39,500 sq. km.). Cap. Arba Minch. Pop. (1984C) 1,248,034.

Geneva Switzerland. 46 12N

6 09E. (i) Canton in SW Switz. almost encircled by France. Area, 109 sq. m. (282 sq. km.). The main occupations in the rural districts are viticulture and farming. Pop. (1988E) 380,000. (ii) Cap. of (i) sit. at SW end of L. Geneva on Rhône R. Commercial centre and centre for intern. agencies. Manu. watches, clocks, jewellery and precision instruments. Pop. (1988E) 170,000.

Geneva, Lake France/Switzerland. 46 25N 6 30E. Lake in SW Switz. with most of its S shore in Haute-Savoie, France. It is crescent-shaped, area 223 sq. m. (577 sq. km.), about 45 m. (72 km.) long and 8 m. (13 km.) at its widest point. It is fed and drained by Rhône R.

Genk Limbourg, Belgium. 50 58N 5 30E. Town sit. N of Liège. Manu. mining machinery. There are reserves of natural gas in the area. Pop. (1988E) 61,499.

Gennevilliers Hauts-de-Seine, France. 48 56N 2 18E. Suburb of NW Paris sit. in loop of Seine R. Manu. aircraft and electrical equipment. R. port. Pop. (1982C) 45,445.

Genoa Liguria, Italy. 44 25N 8 57E. Cap. of Genoa province sit. on Gulf of Genoa. Industries inc. shipbuilding, oil refining, engineering, food canning, brewing, sugar refining and iron and steel; manu. textiles, aircraft and paper. Important sea port exporting olive oil, wine and textiles. Pop. (1988C) 714,641.

Gentofte Copenhagen, Denmark. Suburb of N Copenhagen, mainly residential.

George, Lake New York, U.S.A. 43 26N 73 43W. L. in NE of state and in Adirondack Mountains. It is a narrow L., 32 m. (51 km.) long and up to 3 m. (5 km.) wide. There is an outlet in the N to L. Champlain. Summer resort.

Georgetown Guyana. 6 46N 58 10W. Cap. of rep. sit. at mouth of Demerara R. on the Caribbean Sea. Founded by the French in 1781 and then came under Dutch influence until taken by the British in 1812. Sea port exporting sugar and rice. Bauxite loaded at Linden 65 m. (105 km.) up R. Pop. (1983E) 188,000.

George Town   Cayman Islands, West Indies. Cap. of the Cayman Islands on the W end of Grand Cayman Island, WNW of Montego Bay, Jamaica. Main industries inc. off-shore finance with over 500 banks, and tourism. There is a meteorological station. Pop. (1989E) 12,972.

George Town   Tasmania, Australia. Town sit. NW Launceston on Tamar R. estuary. Industries inc. tourism, clothing, plastics, engineering, food processing, vineyards. Pop. (1986E) 6,938.

Georgia   U.S.A. 32 00N 79 15W. State in SE bounded E by Atlantic, N by Tennessee and N. Carolina, W by Alabama and S by Florida. Area, 58,876 sq. m. (152,489 sq. km.). Chief towns Atlanta (the cap.), Savannah, Columbus, Augusta, Albany and Macon. Founded as a British colony, 1732 and became the fourth original state of U.S.A. Coastal plain with swamp and forest, rising on the Appalachian Mountains in the NW with Brasstown Bald in the NE, 4,768 ft. (1,453 metres) a.s.l. Humid, subtropical climate. Mainly agric. The chief products are maize, cotton, tobacco, groundnuts and kaolin. Manu. vehicles, textiles, food and timber products, and chemicals. Pop. (1987E) 6,222,000.

Georgian Bay   Ontario, Canada. 45 15N 80 50W. The NE arm of L. Huron, in central S of province. Area, 5,800 sq. m. (15,022 sq. km.). A summer resort area.

Georgian Soviet Socialist Republic   U.S.S.R. 41 00N 45 00E. Constituent rep. in SW U.S.S.R., bounded W by Black Sea and SW by Turkey. Area, 26,900 sq. m. (69,670 sq. km.). Chief towns Tbilisi (the cap.), Kutaisi, Batumi, Sukhumi, Poti, Rustavi and Gori. Georgia inc. the Abkhaz Autonomous Soviet Socialist Rep., Adzhar Autonomous Soviet Socialist Rep. and S. Ossetian Autonomous Rep. Mountainous in the N and S, sloping to central valleys, drained by Rs. Kura

and Rion. The main products are manganese, coal, oil, baryta, tobacco, tea, fruit and wines. Pop. (1989E) 5,449,000.

Gera   German Democratic Republic. 50 52N 12 04E. (i) District created in 1952 from part of the *Land* of Thuringia. Area, 1,546 sq. m. (4,004 sq. km.). Pop. (1988E) 742,000. (ii) Cap. of (i) sit. on White Elster R. Industries inc. textiles and engineering; manu. carpets and furniture. Pop. (1988E) 134,834.

Geraldton   Western Australia, Australia. 28 46S 114 36E. Town in central W of state sit. on Indian Ocean. Industries inc. rock lobster fishing, building, tourism, superphosphates, boat building. A commercial centre for agric. and mining hinterland. Sea port exporting mineral sands, ores and grain. Pop. (1988E) 20,660.

Germantown   Pennsylvania, U.S.A. Suburb of NW Philadelphia. Mainly residential.

Germany   Europe. 51 30N 10 00E. Bounded on N by Denmark and the Baltic, in E by Poland and Czechoslovakia, in S by Austria and Switzerland, in W by France, Luxembourg, Belgium and the Netherlands. Area, 138,172 sq. m. (357,866 sq. km.). In October 1990 the re-unification of the Federal Republic of Germany and the German Democratic Republic came into force.

The German Democratic Republic (East Germany) came into being following the estab. of the Federal Republic of Germany. In 1952 the 5 *Länder* of Mecklenburg, Saxony-Anhalt, Brandenburg, Saxony and Thuringia were replaced by 14 *Bezirke*, but in 1990 the 5 *Länder* were reinstated. Chief towns, Berlin (cap.), Leipzig, Dresden and Chemnitz. Two-thirds of N of the former G.D.R. is low-lying and forms part of the great N plain of Europe. The S is hilly. Agric. is extensive and is organized in collective or state farms. Important crops are wheat, rye, barley, oats, potatoes and sugar-beet. There are rich mineral resources and coal and lignite predominate but uranium, cobalt, bismuth, arsenic and antimony are being exploited. Industry produces over 60% of the national income. Iron and steel production, chemicals, fertilizers, cement, textiles and footwear are growing industries. Pop. (1988E) 16,674,632.

The Federal Republic of Germany became a sovereign independent country in 1955. The former F.R.G. is divided into the *Länder* of Schleswig-Holstein, Hamburg, Lower Saxony, Bremen, North Rhine-Westphalia, Hessen, Rhineland-Palatinate, Baden-Württemberg, Bavaria, Saarland, and West Berlin. Chief towns, Bonn (provisional cap.), West Berlin, Hamburg, Munich, Cologne, Essen, Frankfurt, Dortmund, Düsseldorf, Stuttgart and Bremen. Physically it is divided into 3 areas: the lowlands of the N, the table-land of the S and the basin of the middle R. Rhine. The chief drainage basins are R. Danube to S and R. Rhine to W and R. Elbe to N. Most Rs. are navigable for the greater part of their length and together with the canal system make an important communication system. Agric. is important and the chief crops are wheat, rye, barley, oats, potatoes, sugar-beet. There is considerable cattle rearing and wine manu. Forestry is also of great importance. Coalmining is important and other minerals include lignite, iron ore, potash and oil. The manu. industries employ the majority of the pop. and the industries inc. textiles, steel, precision instruments and optical goods. printing and publishing, cars, locomotives, beer, paper and sugar. Pop. (1988E) 61,241,700. Pop. of unified Germany (1988E) 77,916,332.

Germiston   Transvaal, Republic of South Africa. 26 15S 28 05E. Town in SW of province. The main industry is refining gold from the Witwatersrand mines. Also manu. textiles, chemicals and steel products. Pop. (1988E) 173,000.

Gerona   Spain. 41 59N 2 49E. (i) Province in NE Spain bounded N by Pyrénées and E by Mediterranean. Area, 2,272 sq. m. (5,884 sq. km.). From the Pyrénées it descends to fertile plain and the Costa Brava. Pop. (1986C) 490,667. (ii) Cap. of (i) sit. NE of Barcelona on Ter R. at its con-

fluence with Oñar R. Industries inc. flour milling, textiles and chemicals. Pop. (1986C) 67,578.

Gers   France. 43 45N 0 38E. Dept. in Midi-Pyrénées region, bounded N by Garonne R. Area, 2,415 sq. m. (6,254 sq. km.). Chief towns Auch, the cap. and Condom. Drained by tribs. of Garonne R. and by Adour R. Mainly agric. The chief products are Armagnac brandy, cereals and fruit. Pop. (1982C) 174,154.

Gersoppa Falls   Kharnataka, India. Falls on Sharavati R. NW of Bangalore, inc. 4 cascades, the highest 829 ft. (253 metres). Used for hydroelectric power.

Getafe   Madrid, Spain. 40 18N 3 43W. Town sit. S of Madrid on a wide plain near Los Angeles hill, the geographical centre of Spain. Industries include manu. electrical equipment, pharmaceutical goods, aircraft, radio and telephones, boxes, brushes, alcohol and flour. Pop. (1986C) 130,971.

Gettysburg   Pennsylvania, U.S.A. 39 83N 79 27W. Town sit. NNW of Baltimore, Maryland, noted for the battle of July 1863 when Confederate forces were defeated by Federal forces during the American Civil War. Pop. (1980C) 7,194.

Gevelsberg   North Rhine-Westphalia, Federal Republic of Germany. 51 19N 7 20E. Town sit. ENE of Wuppertal. Industries inc. manu. stoves, locks and bicycles. Pop. (1984E) 30,700.

Gezira, El   Sudan. Plain between Rs. White and Blue Nile above their confluence at Khartoum. Irrigated to grow cotton.

Ghadames   Libya. 30 08N 9 30E. Town in extreme W sit. in an oasis. Route centre near the frontiers of Libya, Tunisia and Algeria.

Ghana   6 00N 1 00W. Rep. within the Commonwealth. Ghana came into existence in 1957 when the former colony of the Gold Coast and the Trusteeship Territory of Togoland attained dominion status. Bounded N by

Burkina Faso, E by Togo, W by the
Côte d'Ivoire and S by the Gulf of
Guinea. Area, 92,010 sq. m. (238,305
sq. km.). Chief towns Accra (the cap.),
Takoradi and Kumasi. Drained by
Volta R. flowing through the extensive L. Volta in E. Ghana. Savannah
in the N, otherwise mainly tropical
rain forest. Cocoa is by far the most
important crop and covers about 5m.
acres. Coffee, improved types of oil
palm and coconut are being planted
on an increased scale and production
from these crops is increasing. Progress has been made in the planting
of Clonal rubber in south-west Ghana.
In the south-east coastal belt irrigation
works have been constructed and
black-clay farming is being successfully undertaken in the Accra plains.
Of the main foodstuffs in south and
central Ghana, maize, rice, cassava,
plantain, groundnuts, yam and coco-
yam predominate. Tobacco is proving
an attractive and very important cash
crop in food-crop-producing areas. In
northern Ghana the chief food crops
are groundnuts, rice, maize, guinea
corn, millet and yams, with tobacco as
an important cash crop. The fishing
industry employs 150,000. The chief
minerals are gold, diamonds, manganese and bauxite. Pop. (1988E)
13,812,000.

Ghardaka   Red Sea Frontier, Egypt.
Town sit. NE of Qena on Red Sea at
entrance to Gulf of Suez. The main industry is crude oil production.

Ghat   Fezzan, Libya. Town sit. WSW
of Murzuk in a Saharan oasis. Caravan centre with some handicraft industries.

Ghats   India. Two mountain ranges
in S India forming W and E boundaries
of central plateau. The Western Ghats
(14 00N 75 00E) extend NNW to SSE
from the Tapti R. to the S coast, at an
average 3,000 ft. (914 metres) a.s.l.
rising to 8,840 ft. (2,694 metres) a.s.l.
at Anai Mudi. The Eastern Ghats (16
00N 79 00E) extend NNE to SSW from
S Orissa to join the Western Ghats in
the Nilgiri Hills, in W Tamil Nadu.
Average height 1,500 ft. (457 metres)
a.s.l. rising to Doda Betta, 8,640 ft.
(2,633 metres) a.s.l.

Ghazipur   Uttar Pradesh, India. 25
38N 83 35E. Town sit. ENE of Varanasi on Ganges R. Industries inc. flour
and oilseed milling. Pop. (1981C)
60,725.

Ghazni   Afghánistán. 33 34N 68
26E. Town in central Afghánistán sit.
on Kandahar road. Trading centre for
wool and fruit. Pop. (1984E) 33,351.

Gheg   *see* **Vlonë**

Ghent   East Flanders, Belgium. 51
03N 3 43E. Cap. of province in NW,
sit. on Scheldt R. at its confluence
with Lys R. Industries inc. brewing,
sugar refining, flour milling and tanning; manu. fertilizers, glassware,
paper, textiles and chemicals. Port
linked by canal to Terneuzen. Pop.
(1988E) 232,620.

Ghor   Herat, Afghánistán. Town sit.
E of Sabzawar on trib. of Farah Rud in
the Hindu Kush foothills. Route and
trading centre of the Gorat district.
The nearby ruined town of Ghor was
the seat of a ruling dynasty in the 12th
cent.

Ghor, El   Jordan/Israel. Depression
2–15 m. (3–24 km.) wide extending
across Jordan and Israel between
Dead Sea and Sea of Galilee, and
bounded S by Gulf of Aqaba. A part
of the Great Rift Valley.

Gialam   Bacninh, Vietnam. Town
opposite Hanoi on Red R. Industries
inc. paint manu. Site of Hanoi airport.

Giant's Causeway   Antrim, Northern Ireland. 55 15N 6 30W. Promontory on N coast of Antrim made of
columnar basalt, and inc. several thousand pillars.

Gibraltar   35 55N 5 40W. British
colony ceded to Britain by the Treaty
of Utrecht in 1713. It is bounded by
the extreme S tip of Spain and at the
entrance to the Mediterranean. Area,
2¹/₂ sq. m. (6¹/₂ sq. km.). Joined to the
Spanish mainland by a sandy isthmus,
and consisting of the Rock, 1,396 ft.
(426 metres) a.s.l., the town to the W,
and the harbour. Free port and naval
base, also a resort. Pop. (1988E)
30,077.

Gibraltar, Strait of   Africa/Europe. 35 57N 5 36W. Channel extending for 36 m. (58 km.) W to E between Atlantic Ocean and Mediterranean, and between Spain and Gibraltar in N to Morocco in S. 8 m. (13 km.) at its narrowest point.

Giessen   Hessen, Federal Republic of Germany. 50 35N 8 40E. Town in central F.R.G. sit. on Lahn R. Manu. machine tools, rubber, leather and tobacco products. Pop. (1984E) 71,800.

Gifu   Honshu, Japan. Cap. of Gifu prefecture sit. ENE of Kyoto. Manu. textiles, cutlery and paper goods; some tourism for cormorant fishing. Pop. (1988E) 408,000.

Gijón   Oviedo, Spain. 43 32N 5 40W. Town sit. on Bay of Biscay. Industries inc. iron and steel, chemicals and oil refining. Port exporting coal and iron ore. Pop. (1989E) 263,154.

Gilan   Iran. 33 00N 49 00E. Former province now inc. in the First Province which is frequently referred to by this name. Sit. on SW Caspian shore and bounded S by the Alburz Mountains. Sub-tropical climate favouring agric. but unhealthy. The main commercial crops are silk, rice, tea, fruit and olives. Chief towns, Resht, Pahlevi, Fumen, Lahijan.

Gila River   U.S.A. 32 43N 114 33W. Rises in SW New Mexico and flows 650 m. (1,040 km.) SW across Arizona to join Colorado R. near Yuma. Used for irrigation, esp. at the Coolidge Dam.

Gilbert Islands   *see* **Kiribati**

Gilboa   Israel. Mountainous district of N. Israel on border with Jordan, between Plain of Esdraelon and Jordan R. Noted in biblical history as the scene of King Saul's defeat.

Gilead   Jordan. District of NW Jordan sit. E of Jordan R. Mountainous, rising to Mount Gilead, 3,652 ft. (1,113 metres) a.s.l.

Gilgil   Rift Valley Province, Kenya. Town 18 m. (29 km.) NNW of Naivasha

in Great Rift Valley. Centre of a farming area. Railway junction serving Thomson's Falls.

Gilgit   Kashmir. 35 55N 74 20E. Mountainous district of NW Kashmir in area claimed by Pakistan. Area, 14,680 sq. m. (38,021 sq. km.). Rises to Nanga Parbat, 26,660 ft. (8,126 metres) a.s.l., and Rakaposhi, 25,550 ft. (7,788 metres) a.s.l. Drained by Rs. Gilgit and Indus.

Gillingham   Dorset, England. Town sit. NNE of Dorchester on Stour R. Town serving an agric area. Industries inc. engineering, chemicals, glove making and food processing. Pop. (1989E) 8,000.

Gillingham   Kent, England. 51 24N 0 33E. Town sit. SE of London on Medway R. estuary. The former Royal Naval Base and Dockyard has been redeveloped for light industrial and commercial development. Pop. (1985E) 94,000.

Gippsland   Victoria, Australia. District of SE Victoria with fertile plains in S rising to wooded mountains in N. There are extensive lignite deposits in the Latrobe valley. Other important products are dairy produce, cereals and hops.

Girardot   Colombia. 4 18N 74 48W. Town in W centre sit. on Magdalena R. in a coffee growing area. Industries inc. manu. textiles, leather and tobacco products. R. port exporting coffee, cattle and hides. Pop. (1980E) 70,000.

Giresun   Turkey. 40 55N 38 24E. (i) Province of N Turkey on Black Sea coast. Area, 2,630 sq. m. (6,812 sq. km.). Drained by Kelkit and Harsit Rs. Heavily forested with deposits of copper and zinc; some agric. produces maize. Pop. (1985C) 502,151. (ii) Town sit. NE of Sivas on Black Sea. Provincial cap., port and trading centre for agric. and forest products.

Girga   Egypt. 26 17N 31 55E. (i) Province of central Upper Egypt, in Nile Valley. Area, 609 sq. m. (1,577 sq. km.). Chief towns Sohag (cap.), Akhmim, El Balyana, Giheina, Tahta,

Tema, Girga, El Manshah. Mainly agric., chief products cotton, cereals, dates, sugar. The main industries are cotton ginning, silk and cotton weaving, sugar milling. (ii) Town sit. WNW of Qena on W bank of Nile R. Industries inc. cotton ginning, pottery, dairying.

Girne   *see* **Kyrenia**

Gironde   France. 45 20N 0 45W. Dept. in Aquitaine region bounded W by Bay of Biscay. Area, 3,861 sq. m. (10,000 sq. km.). Chief towns Bordeaux (the cap.), Archachon and Libourne. Drained by Rs. Garonne and Dordogne uniting to form the Gironde estuary. The main occupations are viticulture and farming. Industries are concentrated in Bordeaux. Pop. (1982C) 1,127,546.

Girvan   Strathclyde Region, Scotland. 55 15N 4 51W. Town in SW sit. on Firth of Clyde. Industries inc. tourism and fishing. Pop. (1989E) 7,839.

Gisborne   North Island, New Zealand. 38 40S 178 02E. City on E coast sit. at N end of Poverty Bay. Sea port exporting agric. and forestry produce. Main wine-producing area of New Zealand. Popular summer tourist industry. Pop. (1986C) 28,485.

Giuba   Somali Republic/Ethiopia. R. formed in Ethiopia at Dolo by the confluence of Ganale Dorya and Dawa and flowing 545 m. (872 km.) S into the Somali Rep. to enter the Indian Ocean 10 m. (16 km.) NE of Kismayu. Navigable for small craft up to Bardera. Its course supports the chief agric. area of the Somali Rep.

Giurgiu   Ilfov, Romania. 43 53N 25 57E. Town sit. SSW of Bucharest on Danube R. Industries inc. boat building, sugar refining, brewing and flour milling. Exports are petroleum from the Ploeşti oilfield, grain and timber.

Giza, El   Giza, Egypt. 30 36N 32 15E. Cap. of governorate sit. opposite Cairo on Nile R. Manu. textiles and footwear. Pop. (1974E) 853,700.

Gizan   Asir, Saudi Arabia. Town

sit. S of Abha on Red Sea. Industries inc. boat building, fishing, pearling, salt extraction. Port shipping grain, sesame, dried fish, dates. Sometimes called Qizan, Jizan.

Gizo   Solomon Islands, South West Pacific. Volcanic island sit. W of New Georgia. Area, 10 sq. m. (26 sq. km.).

Gjirokastër   Albania. 40 05N 20 10E. Town sit. SE of Valona on Drin R. Market centre trading in dairy products, tobacco and wine. Industries inc. tanning, silver smithing.

Gjøvik   Oppland, Norway. Town sit. N of Oslo on NW shore of L. Mjøsa. Industries inc. manu. motors, snowploughs, wallboard, flour, dairy products. Pop. (1989E) 26,126.

Glace Bay   Nova Scotia, Canada. 46 12N 59 57W. Town on E coast of Cape Breton Island, sit. ENE of Sydney, in a coalmining area. Industries inc. mining and fishing. Pop. (1976C) 21,484.

Glacier National Park   Montana, U.S.A. 40 40N 114 00W. Park in Rocky Mountains in NW Montana, bounded N by Canada. Area, 1,560 sq. m. (4,040 sq. km.). Contains glaciers, L. and forests, and rises to Mount Cleveland. Together with the Waterton L. National Park in Alberta, Canada, it forms the Waterton-Glacier Intern. Peace Park.

Gladstone   Queensland, Australia. 23 51S 151 15E. Town sit. SE of Rockhampton on Port Curtis inlet. Port, exporting coal, alumina, aluminium, wheat. The main industries are aluminium and chemicals. Pop. (1989E) 23,800.

Glamorgan, Mid   *see* **Mid Glamorgan**

Glamorgan, South   *see* **South Glamorgan**

Glamorgan, West   *see* **West Glamorgan**

Glarus   Switzerland. 47 02N 9 04E. (i) Canton in E central Switz. Area, 264 sq. m. (684 sq. km.). The main occupation is dairy farming. Pop. (1987E) 37,158. (ii) Cap. of (i) sit. SE of Zürich on Linth R. Industries inc. textiles and woodworking. Pop. (1987E) 5,600.

Glasgow Strathclyde Region, Scotland. 55 53N 4 15W. City in SW sit. on Clyde R. The main industries are mechanical and marine engineering, printing and publishing. Tourism is expanding. Pop. (1983E) 740,000.

Glastonbury Somerset, England. 51 06N 2 43W. Town sit. SW of Bath on Brue R. Former importance as a market town for an agric area. Industries inc. tanning; manu. footwear and sheepskin goods now superseded by light industry and tourism. Pop. (1981C) 6,773.

Glauchau Chemnitz, German Democratic Republic. 50 49N 12 32E. Town sit. NNE of Zwickau. Industries inc. textiles; manu. machinery. Pop. (1989E) 27,800.

Glen Canyon Arizona, U.S.A. Dam across SW end of L. Powell, which lies mainly in Utah, built for hydroelectric power and irrigation.

Glencoe Highland Region, Scotland. 56 40N 5 00W. Valley of Coe R. in central Scot. between Rannoch Moor and Loch Leven; scene of the massacre of the Clan Macdonald in 1692.

Glendale Arizona, U.S.A. 33 32N 112 11W. City immediately NW of Phoenix. Industries inc. aerospace components, high-tech electronics, health care and education. Pop. (1990E) 144,000.

Glendale California, U.S.A. 34 10N 118 17W. Suburb of NE Los Angeles. Professional, financial and car manu. centre. Pop. (1984E) 147,100.

Glendalough, Vale of Wicklow, Ireland. Valley in E sit. SSW of Bray, noted for its ecclesiastical ruins and great natural beauty.

Glenelg South Australia, Australia. 34 59S 138 31E. Resort sit. immediately SW of Adelaide on Spencer Gulf.

Glenelg River Australia. 38 03S 141 00E. Rises in the Grampians in SW Victoria and flows 280 m. (448 km.) W and then SW past Casterton to enter Discovery Bay at the border with South Australia.

Glenorchy Tasmania, Australia. City sit. N of Hobart on Derwent R. Tasmania's chief industrial city. Pop. (1989E) 42,000.

Glenrothes Fife Region, Scotland. 56 12N 3 10W. Town sit. N of Kirkcaldy. Industries inc. electronics, engineering and paper making. Pop. (1989E) 38,070.

Gliwice Katowice, Poland. 50 17N 18 40E. Town in SW sit. on Klodnica R. Industries inc. coalmining, steel, chemicals and food processing; manu. machinery. Pop. (1982E) 202,000.

Glomma River Norway. 59 12N 10 57E. Rises in SE, in a L. sit. SE of Trondheim and flows 350 m. (560 km.) S past Röros, Elverum and Kongsvinger, into L. Öyeren and thence into the Skagerrak at Frederikstad. Used for logging and hydroelectric power.

Glossop Derbyshire, England. 53 27N 1 57W. Market town sit. ESE of Manchester. There is some light industry. Pop. (1981C) 25,539.

Gloucester Gloucestershire, England. 51 53N 2 14W. County town, cathedral city and river port in W sit. on Severn R. Market town trading in timber and grain. Principally a commercial, tourist and shopping centre. Pop. (1987E) 90,500.

Gloucester Massachusetts, U.S.A. 42 41N 70 39W. City sit. NE of Boston on Atlantic coast. Industries inc. boat building, fishing and fish processing; manu. fishing equipment. Pop. (1980C) 27,768.

Gloucestershire England. 51 44N 2 10W. County in SW Midlands bounded W by Gwent, N by Hereford and Worcester, and Warwickshire, E by Oxfordshire and S by Wiltshire and Avon and drained SW by Severn R. estuary into the Bristol Channel.

Under 1974 re-organization its districts are Tewkesbury, Forest of Dean, Gloucester, Cheltenham, Cotswold and Stroud. Chief towns are Gloucester (the county town), Cheltenham and Stroud. Considerable industrial and commercial growth has taken place in and near these towns in recent years. Elsewhere the economy is mainly agricultural especially dairy and sheep farming and fruit growing with extensive arable farming on the Cotswold Hills. Pop. (1988E) 527,500.

Gloversville New York, U.S.A. 43 03N 74 20W. City sit. NW of Albany. The main industry is tanning and glovemaking; also manu. sports goods, textiles, aluminium doors and windows. Pop. (1990E) 17,400.

Gmünd Baden-Württemberg, Federal Republic of Germany. Town sit. E of Stuttgart on Rems R. The main industry is working in precious metals; also manu. precision instruments and clocks.

Gmunden Upper Austria, Austria. 47 55N 13 48E. Town and spa in central N sit. in the Salzkammergut on the Traunsee. Industries inc. tourism, brewing and wood carving. Pop. (1990E) 14,000.

Gniezno Posnań, Poland. 52 31N 17 37E. Town in central W. Industries inc. brewing, sugar refining and flour milling. Pop. (1983E) 66,000.

Goa 15 33N 73 59E. State of W India bounded on the N by Maharashtra and on the E and S by Karnataka. Area, 1,429 sq. m. (3,702 sq. km.). Goa was a prosperous Portuguese port in the 16th cent. In 1961 India annexed Goa (together with Daman and Diu) and it achieved statehood in 1987. Panaji is the largest town. Agric. and fishing are the main occupations. Industries inc. the manu. of fishing nets, clothing, pesticides, footwear and pharmaceuticals. Pop. (1981C) 1,007,749.

Gobi Central Asia. 44 00N 111 00E. Desert extending from the Pamirs to Manchuria, esp. the E end, extending 1,000 m. (1,600 km.) W to E at 3,000–5,000 ft. (914–1,524 metres)

a.s.l., and up to 600 m. (960 km.) S to N. Part sand, part stone, part steppe with some livestock rearing.

Godalming Surrey, England. 51 11N 0 37W. Town sit. SSW of Guildford. Residential with some industries inc. light engineering. Pop. (1981C) 18,204.

Godavari River India. 16 37N 82 18E. Rises at N end of the Western Ghats NE of Bombay, flows 900 m. (1,440 km.) ESE across Deccan plateau, through the Eastern Ghats by a gorge, and into Bay of Bengal by a delta below Rajahmundry.

Godesberg North Rhine-Westphalia, Federal Republic of Germany. Town sit. immediately S of Bonn on Rhine R. Spa with some industries, inc. manu. pharmaceutical products.

Godthaab Greenland. 64 11N 51 44W. Settlement in SW Greenland at entrance to Godthaab Fjord. Sea port. Pop. (1983E) 10,000.

Godwin Austen (K2) India. 36 00N 77 00E. Mountain peak in the Karakoram Range, Kashmir, at 28,250 ft. (8,611 metres) a.s.l. the second highest in the world.

Goes Zeeland, Netherlands. Town sit. ENE of Flushing on South Beveland Island. Market centre for an agric. area; industries inc. manu. paints, organ parts. Pop. (1984E) 31,155.

Gogra River China/Nepál/India. 25 45N 84 40E. Rises in Himalayas, SW of Rakas L., Tibet, China, flows 600 m. (960 km.) SE into Nepál, then S into Uttar Pradesh, India and ESE to join Ganges R. near Chapra.

Goiânia Goiás, Brazil. 16 40S 49 16W. State cap. in S central Braz. sit. on central plateau. Market town trading in coffee and livestock. Pop. (1980C) 702,858.

Goiás Brazil. 15 56S 50 08W. State in central Braz., bounded W by Araguaia R. and enclosing, in SE, the Federal District of Brasília. Area, 247,912 sq. m. (642,092 sq. km.).

Cap. Goiânia. Forest in N, savannah in S. The main products are coffee, tobacco and rice. Pop. (1980C) 3,859,602.

Gojjam Ethiopia. Region of NW Ethiopia bounded N by Sudan. Area, 23,784 sq. m. (61,600 sq. km.). Cap. Debra Markos. Forested plateau, rising to 13,625 ft. (4,153 metres) a.s.l. in the Choke Mountains (E), dropping to grasslands (W). Mainly agric., producing cereals, honey, coffee, with livestock farming. Pop. (1984E) 3,244,882.

Golcük, Lake Turkey. Lake sit. SE of Elazig in E central Turkey. Area, 27 sq. m. (70 sq. km.). Source of Tigris R.

Gold Coast Africa *see* **Ghana**

Golden Gate California, U.S.A. Strait, 5 m. (8 km.) long and up to 2 m. (3 km.) wide between San Francisco Bay and the Pacific, forming the entrance to the Bay, and crossed by the Golden Gate Bridge.

Goldsboro North Carolina, U.S.A. 35 23N 77 59W. City sit. SE of Raleigh on Neuse R. Industries inc. textiles, vehicle parts, electrical products, footwear and furniture. Pop. (1987E) 34,722.

Golfito Puntarenas, Costa Rica. 8 38N 83 11W. Town sit. SE of Puerto Cortés on an inlet of the Golfo Dulce on the Pacific coast. Major banana port.

Gomel Belorussia, U.S.S.R. 52 25N 31 00E. Cap. of Gomel region sit. on Sozh R. Industries inc. railway engineering; manu. electrical goods, agric. implements, footwear, textiles and furniture. Pop. (1985E) 465,000

Gomera Canary Islands, Atlantic Ocean. 28 06N 17 08W. Island of the Canaries group sit. W of Tenerife. Area, 146 sq. m. (378 sq. km.). Chief town San Sebastián. Volcanic and mountainous, with a sub-tropical climate but little water. Main occupations are farming, fishing, lumbering. Main products are vegetables, fruit, cereals, tobacco and sugar-cane.

Gonaïves Artibonite, Haiti. 19 30N 72 40W. Town sit. NNW of Port-au-Prince on NE shore of Gulf of Gonaïves. Port exporting agric. produce of the Artibonite plain, esp. coffee, bananas, cotton, sugar-cane, rice; also timber, beeswax, hides and skins. Pop. (1980E) 18,000.

Gonda Uttar Pradesh, India. 27 09N 81 58E. Town sit. ENE of Lucknow. Industries include rice and sugar milling. Trading centre for grain and oilseeds. Pop. (1981C) 70,847.

Gondar Ethiopia. 12 40N 37 30E. (i) Region in NW. Area, 28,649 sq. m. (74,200 sq. km.). Pop. (1984C) 2,905,362. (ii) Cap. of region of same name. Pop. (1980E) 76,932.

Gongola Nigeria. 9 39N 12 04E. State formed from the Adamawa and Sardauna provinces of the former N region. Area, 35,286 sq. m. (91,390 sq. km.). The largest ethnic group is of Fulani people. The main towns are Yola (cap.), Jimeta and Mubi. Pop. (1983E) 3.7m.

Good Hope, Cape of Cape Province, Republic of South Africa. 34 24S 18 30E. Headland at SW tip of the province along W side of False Bay.

Goodwin Sands England. Sandbanks off E coast of Kent and extending 10 m. (16 km.) N to S. They protect The Downs roadstead but form a hazard to shipping. They are partly exposed at low tide.

Goole Humberside, England. 53 42N 0 52W. Town in E sit. at confluence of Rs. Ouse and Don. Industries inc. transport and distribution, port-related activities, food processing, clothing and chemicals. Port handles fertilizers, timber, steel, grain and vehicles. Pop. (1988E) 18,450.

Goose Bay *see* **Happy Valley–Goose Bay.**

Gopalpur Orissa, India. Port sit. ESE of Berhampur on Bay of Bengal. Pop. (1981C) 4,503.

Göpping Baden-Württemberg, Federal Republic of Germany. 48 42N

9 40E. Town. sit. ESE of Stuttgart on Fils R. Industries inc. textiles; manu. machinery, precision instruments and toys. Pop. (1984E) 51,800.

Gorakhpur Uttar Pradesh, India. Town sit. NNE of Varanasi. Industries inc. fertilizers, railway engineering, textiles and paper making. Pop. (1981C) 307,501.

Gordo, Cerro Mexico. Mountain pass in the Sierra Madre Oriental, WNW of Veracruz.

Gorey Jersey, Channel Islands. 52 41N 6 18W. Village sit. ENE of St. Helier. Main occupations fishing, tourism. Noted for 12th cent. Mont Orgueil Castle.

Gorgol Mauritania. R. formed by union of Gorgal Blanc and Gorgol Noir Rs. and flowing 40 m. (64 km.) W to join Senegal R. at Kaédi. Navigable for small craft between August and October.

Gorgonzola Lombardia, Italy. 45 32N 9 24E. Town sit. ENE of Milan, noted for cheese. Pop. (1971E) 13,000.

Gori Georgia, U.S.S.R. 41 58N 44 07E. Town sit. WNW of Tbilis in a horticultural area. Industries inc. fruit and vegetable canning and sawmilling. Pop. (1973E) 51,000.

Gorinchem Zuid-Holland, Netherlands. Town sit. E of Dordrecht on Upper Merwede R. Railway junction. Industries inc. manu. steel and copper wire, machinery, cement, footwear; shipbuilding, sugar refining. fish processing. Pop. (1984E) 27,538.

Gorizia Friuli-Venezia Giulia, Italy. 45 57N 13 38E. Cap. of Gorizia province in NE, on Yugoslav border. Industries inc. textiles, chemicals, machinery and tourism.

Gorj Romania. Province of SW Romania. Cap. Targu-Jiu. Mainly a farming and lumbering area. Industry is concentrated on the cap. and inc. tanning, brick making, wood-working; manu. cloth, furniture. Pop. (1982E) 363,707.

Gorky (Nizhni Novgorod). Russian Soviet Federal Socialist Republic, U.S.S.R. 56 20N 44 00E. (i) Region drained by Volga R. Area, 28,900 sq. m. (74,851 sq. km.). Coniferous forests in the N and wooded steppe in the S. Industrial area with power from peat, wood, oil and coal. (ii) Cap. of (i) sit. at confluence of R. Volga and Oka. Industries inc. shipbuilding, sawmilling, oil refining, chemicals and textiles; manu. locomotives, aircraft, cars, machine tools and electrical goods. Pop. (1985E) 1,399,000.

Görlitz Dresden, German Democratic Republic. 51 09N 14 59E. Town in SE sit. on Neisse R. where it forms the border with Poland. Industries inc. railway engineering, chemicals and textiles; manu. glass and machinery. Pop. (1989E) 78,000.

Gorlovka Ukraine, U.S.S.R. 48 18N 38 03E. Town sit. NNE of Donetsk in the Donbas coalfield. Industries inc. mining; manu. mining machinery, chemicals and fertilizers. Pop. (1985E) 342,000.

Gorna Dzhumaya   *see* **Blagoevgrad**

Gorno-Altai Autonomous Region Russian Soviet Federal Socialist Republic, U.S.S.R. Region in the Altai Mountains on the Mongolian frontier. Area, 35,740 sq. m. (92,567 sq. km.). Cap. Gorno-Altaisk. Forested. Main occupations lumbering, cattle rearing and gold-mining; some 133,000 hectares are under crops. The main industries are mining (gold, mercury and brown coal) and chemicals. Pop. (1984E) 178,000.

Gorno Altaisk Gorno Altgai Autonomous Region, U.S.S.R. Cap. of region sit. SE of Barnaul. Industries inc. meat packing and textiles; manu. furniture. Pop. (1970E) 98,000.

Gorno-Badakhshan Autonomous Region Tadzhikistan, U.S.S.R. Autonomous region in the Pamirs between the borders of China and Afghánistán. Area, 24,590 sq. m. (63,688 sq. km.). Cap. Khorog, sit. SE of Dunsanbe. Mountainous. The chief occupations are farming esp. cattle and sheep, cereals and fodder crops, and mining esp. gold, coal, salt, mica and rock crystal. Pop. (1984E) 143,000.

Gorzó Wielkopolski Zielona Góra, Poland. 52 44N 15 15E. Town sit. SE of Szczecin on Warta R. Industries inc. textiles and chemicals; manu. wood products.

Gosforth Tyne and Wear, England. 55 01N 1 36W. Suburb of Newcastle upon Tyne. Pop. (1981E) 25,000.

Goslar Lower Saxony, Federal Republic of Germany. 51 54N 10 25E. Town sit. S of Brunswick on NW slope of the Harz Mountains. Industries inc. tourism and the manu. chemicals, furniture and clothing. Pop. (1989E) 46,000.

Gosport Hampshire, England. 50 48N 1 08W. Town sit. on Portsmouth Harbour. Industries inc. engineering, electronics and yacht building. Port and naval base. Pop. (1985E) 82,000.

Göta Canal Sweden. Canal system extending 54 m. (86 km.) between Göteborg on the Kattegat, in the SW, and Söderköping in the SE, on the Baltic Sea coast. The W reach makes use of Göta R. and L. Väner, then a canal passes through small L. to L. Vätter. A second canal leads from L. Vätter through L. Rox to the Baltic.

Göteborg Göteborg and Bohus, Sweden. 57 43N 11 58E. Cap. of county sit. at mouth of Göta R. on the Kattegat, in the SE. Industries inc. shipbuilding, engineering and textiles; manu. cars and ball-bearings. Important ice-free sea port. Pop. (1988E) 430,763.

Gotha Erfurt, German Democratic Republic. 50 57N 10 43E. Town in SW. Industries inc. railway engineering and textiles; manu. food products and machinery. Pop. (1989E) 57,200.

Gotland Sweden. 57 30N 18 30E. Island and county in Baltic Sea. Area, 1,225 sq. m. (3,173 sq. km.). Cap. Visby. The main occupations are farming and tourism. Pop. (1988E) 56,383.

Göttingen Lower Saxony, Federal Republic of Germany. 51 32N 9 55E. Town in central NE sit. on Leine R. Industries inc. manu. machinery and precision instruments. Noted for its university, founded 1734. Pop. (1990E) 134,269.

Gottwaldov Moravia, Czechoslovakia. 49 13N 17 41E. Town sit. E of Brno. The main industry is shoe-making. Pop. (1983C) 84,000.

Gouda Zuid-Holland, Netherlands. 52 01N 4 43E. Town sit. NE of Rotterdam. Industries inc. manu. cheese, candles and pottery. Noted for its cheese-market. Pop. (1989E) 62,000.

Gough Island Tristan da Cunha, Atlantic. 40 10S 9 45W. Islet sit. SSE of Tristan da Cunha. Area, 30 sq. m. (78 sq. km.). Volcanic and uninhabited. Valuable for guano.

Goulburn New South Wales, Australia. 34 45S 149 43E. Town in SE of state sit. on Hawkesbury R. Market centre for wool and livestock. Industries inc. granite quarrying. Pop. (1981E) 22,000.

Goulburn River Victoria, Australia. Rises in the Great Dividing Range and flows 345 m. (552 km.) NW to join Murray R. near Echuca, at the New South Wales border. Used for irrigation.

Goulette, La Tunisia. Town sit. ENE of Tunis on a sand bar, between Lake of Tunis and Gulf of Tunis. Industries inc. fishing and tourism. Port exporting iron ore.

Gourock Strathclyde Region, Scotland. 55 58N 4 49W. Town sit. WNW of Greenock on S shore of Firth of Clyde. Industries inc. marine engineering, yacht building and tourism.

Gower West Glamorgan, Wales. 51 35N 5 10W. Peninsula between Swansea Bay and Burry Inlet, with holiday resorts.

Gozo Malta. 36 03N 14 15E. Island off NW coast of Malta, in the Mediterranean. Area, 26 sq. m. (67 sq. km.). Chief town Victoria. Moderately fertile. The main occupations are growing fruit and vegetables; fishing and tourism. Pop. (1980E) 23,117 with Comino.

Graaff-Reinet Cape Province, Republic of South Africa. 32 14S 24 32E. Town sit. NNW of Port Elizabeth on Sundays R. and in an irrigated horticultural area in a dry veld.

Gracias Lempira, Honduras. 14 35N 88 35W. Town sit. SE of Santa Rosa de Copán. Dept. cap. Trading centre for an agric area.

Graciosa Azores, Atlantic Ocean. 39 04N 28 00W. Island sit. NW of Terceira and N most of the Azores. Area, 23 sq. m. (91 sq. km.). Chief port, Santa Cruz. Volcanic and fertile. The main occupation is farming. Pop. (1980E) 8,000.

Graciosa Canary Islands, Atlantic Ocean. Islet immediately N of Lanzarote across the El Río channel. Area, 10½ sq. m. (27 sq. km.). Volcanic; mainly covered in Sahara sand and uninhabited.

Grafton New South Wales, Australia. 29 41S 152 56E. Town in NE of state sit. on Clarence R. Trading centre for a farming area. Industries inc. forestry and brewing. Pop. (1985E) 17,600.

Graham Land Antarctica. 65 00S 64 00W. Peninsula on W side of the Weddell Sea, sit. S of Cape Horn, Chile, forming part of British Antarctic Territory. It is barren and mountainous.

Grahamstown Cape Province, Republic of South Africa. 33 19S 26 13E. City sit. near the SE coast. Centre of education (Rhodes University). Pop. (1985E) 75,000.

Grain, Isle of Kent, England. Island between R. Thames and Medway in N Kent. Area, 5 sq. m. (13 sq. km.). Connected to mainland across Yantlet Creek, and is the site of an oil refinery.

Grampian Scotland. 57 20N 2 45W. Region of NE Scotland. Area,

3,360 sq. m. (8,700 sq. km.). It consists of the former county of the City of Aberdeen, the counties of Aberdeen, Banff, Kincardine and Moray. Main industries are agric., fishing, food processing, paper, whisky and oil. Pop. (1988E) 497,450.

Grampian Mountains Scotland. 56 50N 4 00W. Mountain system extending from S central W to NE across Scotland. The highest points are Ben Nevis, the highest peak in Great Britain, 4,406 ft. (1,343 metres) a.s.l., Ben Macdhui, 4,296 ft. (1,309 metres) a.s.l., Braeriach, 4,248 ft. (1,295 metres) a.s.l., and Cairn Gorm, 4,084 ft. (1,245 metres) a.s.l. Rs. flowing N inc. Spey, Don, Dee and Findhorn; those flowing S inc. Esk. Tay and Forth.

Granada Granada, Nicaragua. 11 56N 85 57W. Cap. of dept. sit. SE of Managua on NW shore of L. Nicaragua. Trading centre for coffee and sugar-cane. Industries inc. distilling; manu. furniture, clothes and soap. Pop. (1980E) 42,500.

Granada Spain. 37 05N 4 30W. (i) Province in S Spain bounded S by Mediterranean. Area, 4,838 sq. m. (12,530 sq. km.). Mountainous in the S, with the Sierra Nevada rising to 11,421 ft. (3,481 metres) a.s.l. in Mulhacén. Fertile on the coastal plain and in the valleys. The main crops are cereals, tobacco and sugar-cane. Pop. (1986C) 796,857. (ii) Cap. of (i) sit. on Genil R. Industries inc. textiles, distilling and sugar refining; manu. leather and soap. Noted for the Alhambra palace and other Moorish features. Pop. (1986C) 280,592.

Granby Quebec, Canada. 45 24N 72 43W. Town sit. E of Montreal. Industries inc. manu. textiles, plastic and rubber products, and furniture. Pop. (1981E) 38,000.

Gran Chaco South America. Extensive plain bounded W by Andes, E by R. Paraguay and Paraná and extending through N Argentina, Paraguay and E Bolivia. Area, *c.* 250,000 sq. m. (*c.*650,000 sq. km.). Drained by R. Pilcomayo and Berkejo into Paraná R. Hot wet summers with extensive floods and mild, dry winters. Mainly scrub forest and grassland, with some useful timber, and some ranching.

Grand Bahama   Bahamas, West Indies. 26 40N 78 30W. Island immediately w of Great Abaco Island and NNW of Nassau. Area, 530 sq. m. (1,373 sq. km.). Chief town, West End. Main industries inc. tourism, chemicals, oil refining. Pop. (1982E) 45,938.

Grand Banks   47 06N 55 48W. North Atlantic. Section of the N. American continental shelf extending *c.* 500 m. (800 km.) SE from Newfoundland at between 50 and 100 fathoms. Important fishing ground, esp. for cod.

Grand Caicos   Turks and Caicos Islands, West Indies. Island immediately w of East Caicos and the largest of the group. Area, 73 sq. m. (189 sq. km.). The main occupation is growing cotton.

Grand Canal   China. Canal in the E, 1,000 m. (1,600 km.) long, linking Tientsin in N with Hangchow in S. The central section between Rs. Yangtse and Hwangho dates from the 6th cent. A.D.; the S section from the 7th cent., and the N from the 13th cent. The S section is still in use, but much of the rest is silted up.

Grand Canary   Canary Islands, Spain. 28 00N 15 36W. Main island, between Tenerife and Fuerteventura. Area, 592 sq. m. (1,533 sq. km.). Cap. Las Palmas. Volcanic and fertile. The main occupations are farming and tourism.

Grand Canyon   Arizona, U.S.A. 36 10N 112 45W. Gorge of the Colorado R. in N Arizona extending 218 m. (349 km.) E to W from Little Colorado R. to L. Mead. It is up to 15 m. (24 km.) wide and in places over 1 m. (2 km.) deep. Part of it constitutes the Grand Canyon National Park.

Grand Cayman   Cayman Islands, West Indies. 19 20N 81 20W. Westernmost and largest Cayman Island sit. WNW of Jamaica. Area, 71 sq. m. (184 sq. km.). Chief town, George

Town, cap. of the Cayman Islands. Main occupations are lumbering, fishing, fruit growing. Pop. (1980E) 15,000.

Grand Combin   Valais, Switzerland. Mountain peak of the Pennine Alps sit. SE of Orsières and 4 m. (6 km.) E of Bourg St. Pierre, near the Italian border. Height 14,164 ft. (4,317 metres) a.s.l.

Grande, Río   Mexico/U.S.A. Known in Mexico as Río Bravo del Norte, and in U.S.A. as Río Grande. Rises in the San Juan Mountains in SW Colorado, U.S.A., and flows 1,880 m. (3,008 km.) SE across Colorado, then S through New Mexico, U.S.A., and SE to El Paso to form the Mexico/U.S.A. frontier, emptying into Gulf of Mexico between Brownsville, U.S.A., and Matamoros, Mexico. Used for storage, irrigation and power, particularly at Elephant Butte Reservoir, New Mexico, U.S.A. The main trib. is Pecos R.

Grande Chartreuse   France. Massif in the SE between Grenoble and Chambéry rising to 6,847 ft. (2,087 metres) a.s.l. in Chamechaude. Limestone cliffs and gorges. The monastery of La Grande Chartreuse is NNE of Grenoble.

Grande Comore   *see* **Njazidja**

Grande Prairie   Alberta, Canada. 55 10N 118 48W. Town sit. NW of Edmonton on Bear R. near its confluence with Wapiti R. Distribution centre for a farming and lumbering area. Pop. (1981E) 25,000.

Grande-Terre   Guadeloupe, West Indies. The E half of Guadeloupe, separated from Basse-Terre (w) by the Rivière Salée channel. Area, 226 sq. m. (585 sq. km.) Chief towns, Les Abymes and Point-à-Pitre. Arid limestone surface. The main product is sugar-cane. Pop. (1982C) 157,696.

Grand Falls   Newfoundland, Canada. 48 56N 55 40W. Town NW of St. John's sit. on Exploits R. Industries, powered by the Falls, inc. manu. newsprint, mining fishing. Pop. (1986E) 9,121.

Grand Forks North Dakota, U.S.A. 47 55N 97 03W. City in NE of state sit. at confluence of Red R. of the North and Red Lake R. on the Minnesota border. Trading centre for wheat, potatoes and livestock. Industries inc. wood products, meat packing and flour milling. Pop. (1989E) 47,700.

Grand Island Nebraska, U.S.A. 40 55N 98 21W. City in centre of state. Industries inc. food processing. Manu. agric. implements and buildings. Pop. (1980C) 33,781.

Grand Junction Colorado, U.S.A. 39 05N 108 33W. Town in W Colorado, railway junction near confluence of Gunnison and Colorado Rs. Pop. (1980C) 28,144.

Grand Prairie Texas, U.S.A. 32 45N 97 00W. City sit. WSW of Dallas. The main industry is manu. aircraft, and also rubber products. Pop. (1985E) 91,494.

Grand Rapids Michigan, U.S.A. 42 58N 85 40W. City in the SW of state sit. at the rapids on the Grand R. The main industries are furniture making and printing, also manu. motor car parts, electronic and electrical goods, hardware and chemicals. Pop. (1986E) 186,500 (urban area, 500,000).

Grand Turk Turks and Caicos Islands, West Indies. 21 28N 71 08W. Cap. of the crown colony of Turks and Caicos sit. NNW of Cap Haitien, Haiti. The main occupations are fishing, salt panning. Pop. (1985E) 3,098.

Grangemouth Central Region, Scotland. 56 02N 3 45W. Town sit. ENE of Falkirk on Firth of Forth. It is the E terminus of the Forth-Clyde Canal. The main industry is oil-refining, also chemicals, ship-building and engineering. Pop. (1973E) 24,430.

Grange-over-Sands Cumbria, England. 54 12N 2 55W. Resort in NW sit. on N shore of Morecambe Bay. Pop. (1989E) 3,460.

Granite City Illinois, U.S.A. 38 42N 90 09W. Sit. NNE of St. Louis. The main industries inc. iron and

steel, chemicals and railway engineering; manu. tinplate. Pop. (1980C) 36,815.

Gran Paradiso Italy. 45 32N 7 16E. Mountain peak of the Graian Alps sit. SSE of Aosta, at 13,323 ft. (4,061 metres) a.s.l. Centre of a national park.

Grantham Lincolnshire, England. 52 55N 0 39W. Town in E central Eng. Industries inc. engineering; manu. agric. machinery, clothing, food processing and road-making machinery. Pop. (1988E) 35,235.

Grasmere Cumbria, England. 54 28N 3 02W. Village and lake sit. N of Kendal. Tourist resort. Pop. (1985E) 1,100.

Grasse Alpes-Maritimes, France. 43 40N 6 55E. Town in SE. Centre of a horticultural area, producing perfumes from the flowers grown there, esp. roses and jasmin. Pop. (1990E) 50,000.

Graubünden Switzerland. Canton in E Switz. bounded NE by Austria, and SE and S by Italy. Area, 2,744 sq. m. (7,107 sq. km.). Chief towns Chur (the cap.), Davos, St. Moritz, and Arosa. Very mountainous. The main occupations are farming and tourism. Pop. (1980C) 164,641.

Graves France. Area in Gironde dept. in SW, along S bank of Garonne R. from Bordeaux in NW to Langon in SE. Noted for wines made from the grapes grown on its gravel soil.

Gravesend Kent, England. 51 27N 0 24E. Town sit. on Thames R. opposite Tilbury. Industries inc. printing, engineering and paper-making. Port and customs and pilot station for the Port of London. Pop. (1981C) 52,963.

Graz Styria, Austria. 47 05N 15 27E. City in SE sit. on Mur R. Commercial centre. Industries inc. iron and steel, and textiles; manu. paper, glass, leather and precision instruments. Resort and spa. Pop. (1981C) 339,401.

Great Australian Bight Australia. 35 00S 135 00E. Wide bay of the

Southern Ocean on the s coast of Australia, extending 700 m. (1,120 km.) w to E between Cape Pasley, Western Australia, and Cape Carnot, South Australia.

Great Barrier Reef   Australia. 37 12S 175 25E. Coral reef off the coast of Queensland extending about 1,200 m. (1,920 km.) SSE through the Coral Sea from Torres Strait. The largest coral reef in the world.

Great Basin   U.S.A. 40 00N 116 30W. Area bounded w by the Sierra Nevada and the Cascade Range, California, E by the Wasatch Range and the Colorado Plateau. Area, *c.* 200,000 sq. m. (520,000 sq. km.). Extends N to S from S Oregon and S Idaho through most of Nevada, w Utah and SE California; consists of about 100 drainage basins divided by mountains, deserts and salt L. Arid, and only cultivated through irrigation.

Great Bear Lake   Northwest Territories, Canada. 66 00N 120 00W. L. in NW Mackenzie District. Area, 12,275 sq. m. (31,792 sq. km.). Drained by Great Bear R. into Mackenzie R. Navigable for 4 months a year.

Great Belt   Denmark. 55 30N 11 00E. Strait extending 40 m. (64 km.) NNW to SSE between the islands of Fyn and Zealand, linking the Kattegat with the Baltic Sea.

Great Britain   54 00N 2 15W. Largest of the British Isles, comprising England, Scotland and Wales and inc. the Isle of Wight, Scillies, Hebrides, Orkneys and Shetlands. Total area, 88,752 sq. m. (229,868 sq. km.). Bounded N, NW and SW by Atlantic, E by North Sea, s by English Channel and w by St. George's Channel, Irish Sea and North Channel. Greatest length N to S is 605 m. (968 km.), greatest width 325 m. (520 km.). The Isle of Man and the Channel Islands are not included administratively.

Great Dividing Range   Australia. 32 00S 151 00E. Range running parallel to E coast, extending almost the length, N to S, of the country, and comprising the Australian Alps,

Gourock Range, Cullarin Range, Blue Mountains, Hunter Range, Liverpool Range, Hastings Range, New England Range, Atherton Plateau and the Great Divide ranges of Queensland. Watershed between Rs. flowing to the Coral and Tasman Seas, and those flowing w.

Great Dunmow   Essex, England. 51 52N 0 21E. Town NNW of Chelmsford, sit. on Chelmer R. Market town for an agric. area. Pop. (1988E) 5,705.

Greater London   England. Local government for London was reorganized in 1965. The former London County Council was extended and with a number of surrounding authorities formed the Greater London Council. Greater London consists of London, the City of London, parts of Essex, Hertfordshire, Kent, Middlesex and Surrey. The London boroughs are Barking, Barnet, Bexley, Brent, Bromley, Camden, Croydon, Ealing, Enfield, Greenwich, Hackney, Hammersmith, Haringey, Harrow, Havering, Hillingdon, Hounslow, Islington, Kensington and Chelsea. Kingston-upon-Thames, Lambeth, Lewisham, Merton, Newham, Redbridge, Richmond-upon-Thames, Southwark, Sutton, Tower Hamlets, Waltham Forest, Wandsworth and the City of Westminster. Area, 610 sq. m. (1,580 sq. km.). Pop. (1988E) 6,735,000.

Greater Manchester   England. 53 30N 2 15W. Metropolitan county in NW Eng. bounded N by Lancashire, w by Merseyside, s by Cheshire, and E by Derbyshire and W. Yorkshire. Created by 1974 reorganization as a metropolitan county inc. the districts of Wigan, Bolton, Bury, Rochdale, Salford, Oldham, Manchester, Trafford, Stockport and Tameside. Pop. (1988E) 2,577,800.

Great Falls   Montana, U.S.A. 47 30N 111 17W. City in w centre of state, sit. on Missouri R. in an agric. area. Industries inc. refining oil; manu. electrical equipment and milling flour. Pop. (1980C) 59,725.

Great Lake   Australia. 41 52S 146 45E. Lake in Tasmania on the central plateau. Area, 44 sq. m. (114 sq. km.). It has been dammed on the S shore by the Miena Dam.

Great Lakes   Canada/U.S.A. Group of 5 L. on the Can./U.S.A. frontier. L. Michigan is entirely in the U.S.A., but the frontier divides the other 4 L. The St. Lawrence Seaway passes through them, making them accessible to ocean-going ships for 8 months a year. The furthest w is L. Superior, then L. Michigan, L. Huron, L. Erie and L. Ontario. Total water area, 95,000 sq. m. (246,000 sq. km.).

Great Plains   Canada/U.S.A. 42 00N 100 00W. Region E of the Rocky Mountains extending s from Alberta and Saskatchewan, Canada, to New Mexico and Texas, U.S.A. Width *c.* 380 m. (608 km.).

Great Rift Valley   Middle East/East Africa. Depression extending s from Jordan R. valley in Syria through the Dead Sea and the Gulf of Aqaba to the Red Sea and into Ethiopia, Kenya, Tanzania, Malawi and Mozambique. In E Afr. it contains a chain of L. The altitude varies from 6,000 ft. (1,829 metres) a.s.l. in s Kenya to 1,286 ft. (392 metres) b.s.l. on the bed of the Dead Sea.

Great Salt Lake   Utah, U.S.A. 41 10N 112 30W. Salt L. between the Wasatch Range and the Great Salt Lake Desert in NW Utah. Salinity is about 25%, the area alters with evaporation and there is no outlet.

Great Slave Lake   Northwest Territories, Canada. 61 30N 114 00W. L. in s Mackenzie District. Area, 11,000 sq. m. (28,000 sq. km.). Fed by Yellowknife, Slave and Hay R. and drained by Mackenzie R. Navigable for 4 months a year.

Great Torrington   Devon, England. Town sit. SSW of Barnstaple on the Torridge R. Market town for an agric. area. Manu. dairy products, Dartington crystal glass, meat products and gloves. Pop. (1989E) 4,107.

Great Wall   China. 38 30N 109 30E. Defensive wall system extending *c.* 1,500 m. (2,400 km.) from the E coast, on the Gulf of Liaotung, across N China to N Kansu province. Built by Shih Huang Ti in the 3rd cent. B.C. and restored by Hsientung in the 15th cent. A.D. The E half is well preserved.

Great Yarmouth   Norfolk, England. 52 40N 1 45E. Town in East Anglia sit. on a small peninsula between North Sea and Yare R., on the edge of the Norfolk Broads. Industries inc. Ro-Ro ferry route with the Netherlands and Denmark, oil and gas supply services, engineering, tourism, food processing and packaging. Pop. (1988E) 89,000.

Greece   40 00N 23 00E. Rep. occupying the s part of the Balkan Peninsula, bounded N by Albania, Yugoslavia and Bulgaria, E by Turkey and Aegean Sea, s by Mediterranean Sea and w by Ionian Sea. Area, 50,534 sq. m. (130,833 sq. km.). Chief cities Athens (the cap.), Thessaloníki, Larisa, Volos and Patrai. The coastline is high, rocky and much indented. N consists of Macedonia and Thrace, farther s is Thessaly. Greece is traversed on w by the Pindus Mountains and along the E coast a ridge runs through Olympus, Pelion and Euboea, enclosing the plain of Thessaly. The lower courses of Rs. Stuma and Vardar are in N Greece. Of the total area 33% is cultivable. Among crops cultivated are cereals, fodder, cotton, tobacco, citrus fruits, grapes, olives, vegetables, fruits. Fishing is important, also for sponges. A variety of minerals are mined including iron, manganese and chromite. There is little coal but oil was found in 1963. Industries include fruit and vegetable canning, wire, beer, textiles, footwear, chemicals etc. Pop. (1987E) 9·99m.

Greeley   Colorado, U.S.A. 40 25N 104 42W. City sit. NNE of Denver. Commercial and educational centre of an agric. area. Industries inc. photographic, beef processing, computer manu. and publishing. Pop. (1989E) 61,702.

Green Bay   Wisconsin, U.S.A. 44 30N 88 01W. City sit. at head of Green Bay, L. Michigan. Industries inc. engineering and paper making; manu. clothing and furniture. Port handling dairy products. Pop. (1980C) 87,899.

Greencastle   Indiana, U.S.A. 39 38N 86 52W. City sit. wsw of Indianapolis. Trading and communications centre for a farming area. Industries inc. manu. zinc, crushed stone, cement, sawmilling. Pop. (1980C) 8,403.

Greenland 66 00N 45 00W. Large island off NE Canada, sit. mainly in the Arctic Circle, and being part of the Kingdom of Denmark. Area, 840,000 sq. m. (2,176,000 sq. km.). Cap. Godthaab. At the NW extremity it is separated from Ellesmere Island, Can., by the narrow Nares Strait. Mainly icecap in the form of a depression, with a coastal mountain ring rising to Mount Gunnbjorn, 12,139 ft. (3,700 metres) a.s.l. in the SE. Inhabited mainly on the W coast; the main occupations are cod and halibut fishing, with some sheep farming. There are deposits of cryolite at Ivigtut. Pop. (1989E) 55,171.

Greenland Sea Arctic Ocean. 73 00N 10 00W. The S part of the Arctic Ocean between Greenland to the W and Spitsbergen to the E. Mainly ice-covered.

Greenock Strathclyde Region, Scotland. 55 57N 4 45W. Town sit. on S bank of Clyde R. estuary. Industries inc. electronics, engineering, sugar refining and chemicals. Claimed to be the wettest town in Scot., annual rainfall 64 in. (163 cm.). Pop. (1981C) 56,194.

Greensboro North Carolina, U.S.A. 36 04N 79 47W. City sit. NE of Charlotte. Industries inc. manu. textiles, chemicals, furniture and clothing. Pop. (1980C) 155,642.

Greenville Mississippi, U.S.A. 33 25N 91 05W. City sit. NW of Jackson on Mississippi R. in a cotton, rice and soyabean growing area. Industries inc. carpets, aircraft, tools, food processing, metal goods and chemicals. Pop. (1980C) 40,613.

Greenville South Carolina, U.S.A. 34 51N 82 23W. City in NW of state sit. on Reedy R. Commercial centre. Industries inc. chemicals and textiles; manu. metal goods, electronics, pharmaceuticals and plastics. Pop. (1990E) 65,719.

Greenwich Greater London, England. 51 28N 0 02E. Town sit. on S bank of Thames R. forming a district of SE London. Site of the original Royal Observatory; the source of the Greenwich Mean Time, and the point from which longitude is reckoned. Pop. (1981C) 209,871.

Greenwich Connecticut, U.S.A. 41 01N 73 38W. Town sit. NE of New York. Industries inc. boatbuilding; manu. marine engines and pumps. Pop. (1980C) 59,578.

Greenwich Village New York, U.S.A. Part of Manhattan, New York City, bounded W by Hudson R. Residential area with artistic and literary associations.

Grenada West Indies. 12 07N 61 40W. Independent country since 1974. The most southerly island in the Windward Islands Group. Area, 120 sq. m. (311 sq. km.). Cap. St. George's. Mountainous, rising to 2,756 ft. (840 metres) a.s.l. in Mount St. Catherine. The main occupations are farming and cotton ginning. Industries also inc. distilling rum. Tourism is of growing importance. Pop. (1988E) 99,205.

Grenadines St. Vincent and the Grenadines, West Indies. 12 40N 61 15W. Chain of small islands extending for 60 m. (96 km.). The largest island is Carriacou, area, 11 sq. m. (28 sq. km.).

Grenoble Isère, France. 45 10N 5 43E. Cap. of dept. in SE sit. on Isère R. Manu. kid gloves, paper, cement, ferro-alloys, turbines, electrical equipment and machine tools. Pop. (1982C) 159,503 (agglomeration, 392,021).

Gretna Green Dumfries and Galloway Region, Scotland. 54 59N 03 04W. Village sit. NW of Carlisle on the English border. Once famous as the place where runaway couples from Eng. could be married, usually by the village blacksmith, until it became illegal in 1940. Pop. (1981E) 3,000.

Grevena Kozane, Greece. Town sit. SW of Kozane on a trib. of Aliakmon R. Route centre in farming area, trading in wheat, tobacco, dairy products, timber products.

Greymouth   South Island, New Zealand. 42 27S 171 12E. Town sit. at mouth of Grey R. Fishing and fish-processing port. Pop (1989E) 9,000.

Grimsby   Humberside, England. 53 35N 0 05W. Town sit. on S shore of R. Humber estuary, near its mouth in the North Sea. Industries inc. fishing, also manu. nets, ice, fertilizers, chemicals and food processing. Port handling steel, foodstuffs, grain and timber. Pop. (1987E) 89,900.

Grindelwald   Bern, Switzerland. 46 37N 8 02E. Village in central Switzerland. Mountain resort and climbing centre at the foot of the Eiger. Pop. (1989E) 3,600.

Grinnell   Iowa, U.S.A. 41 45N 92 43W. City sit. ENE of Des Moines. Industries inc. processing maize, dairy products, poultry; manu. cosmetics, wood products, metal products, concrete blocks, leather products. Pop. (1980C) 8,868.

Gris-Nez, Cap   Pas-de-Calais, France. 50 50N 1 35E. Limestone headland on the N coast of France, SW Calais, at the narrowest part of the Strait of Dover. Height 180 ft. (55 metres) a.s.l.

Grisons   *see* **Graubünden**

Grodno   Belorussia, U.S.S.R. 53 41N 23 50E. Town sit. SW of Vilnius near Polish border, on Neman R. Industries inc. sugar refining and textiles; manu. fertilizers, leather goods and electrical equipment. Pop. (1985E) 247,000.

Groningen   Netherlands. 53 13N 6 33E. (i) Province in the NE. Area, 899 sq. m. (2,328 sq. km.). Fertile region; the main occupations are dairy farming and growing vegetables. Pop. (1989E) 655,200. (ii) Cap. of (i). Industries inc. sugar refining and chemicals; manu. clothing and furniture. Pop.(1988E) 167,779.

Groote Eylandt   Gulf of Carpentaria, Australia. 14 00S 136 40E.

Largest island in the Gulf, just off NE Australia. Area, 950 sq. m. (2,461 sq. km.). Barren and rocky. Aboriginal settlement.

Grootfontein   Namibia. 19 32S 18 05E. Town sit. NNE of Windhoek in a mining area, esp. copper, lead, zinc, vanadium. Pop. (1988E) 9,000.

Grosseto   Toscana, Italy. 42 46N 11 08E. Cap. of Grosseto province in central Italy. Market town in a reclaimed swamp area, trading in cereals, olive oil and wine. Industries inc. agric. engineering. Pop. (1981E) 70,000.

Gross Glockner   Austria. 47 04N 12 42E. Mountain peak of the Hohe Tauern at 12,460 ft. (3,798 metres) a.s.l. the highest in Austria. The source of the Pasterze glacier.

Groton   Connecticut, U.S.A. 41 19N 72 12W. City opposite New London on Long Island sound. Submarine base. Industries inc. building submarines; manu. chemicals and pharmaceuticals. Pop. (1980C) 10,086.

Grozny   Russian Soviet Federal Socialist Republic, U.S.S.R. 43 20N 45 42E. Cap. of Checkeno-Ingush Autonomous S.S.R. sit. W of Caspian Sea. Oilfield centre connected by pipelines with the oil fields and oil ports of Makhachkala on the Caspian Sea, and Tuapse on the Black Sea. Manu. oil-drilling machinery and chemicals. Pop. (1985E) 393,000.

Grudziaz   Bydgoszcz, Poland. 53 29N 18 45E. Town in NW sit. on Vistula R. Industries inc. chemicals and agric. engineering; manu. glass and footwear. Pop. (1983E) 93,000.

Gruyère   Fribourg, Switzerland. 46 36N 7 04E. District in W Switz. in the R. Saane valley, noted for the cheese which originated there.

Guadalajara   Jalisco, Mexico. 20 40N 103 20W. State cap. in W central Mex. Sit. on the central plateau. Commercial and communications centre. Industries inc. flour milling, tanning and textiles; manu. soap, vegetable oils and clothing, pottery and glass.

Pop. (1980c) 2,244,715 (metropolitan area).

Guadalajara   Spain. 40 38N 3 10W. (i) Province in central Spain. Area, 4,705 sq. m. (12,186 sq. km.). The main products are cereals, olives and vines from the valleys, and salt. Pop. (1986c) 146,008. (ii) Cap. of (i) sit. on Henares R. Commercial centre of an agric. area. Industries inc. tanning and flour milling.

Guadalcanal   Solomon Islands, Pacific Ocean. 9 32S 160 12E. Largest island in the state. Area, 2,060 sq. m. (5,336 sq. km.). Mountainous and forested, rising to 8,000 ft. (2,438 metres) a.s.l. The main product is copra. Pop. (1984e) 64,335.

Guadalquivir River   Spain. 36 47N 6 22W. Rises in the Sierra de Segura in Jaén province, and flows 360 m. (576 km.) wsw through Córdoba to Seville, then divides into branches across the swamps of Las Marismas, and unites again to enter the Gulf of Cádiz by an estuary above Sanlúcar de Barrameda, NNE of Cádiz. Used for irrigation between Córdoba and Seville; navigable below Seville.

Guadalupe, Sierra de   Spain. 39 26N 5 25W. Range in w Spain extending 30 m. (48 km.) sw to NE through Cáceres province between R. Tagus and Guadiana. Rises to 4,734 ft. (1,443 metres) a.s.l.

Guadarrama, Sierra de   Spain. 41 00N 4 00W. Range in central Spain extending *c*. 110 m. (176 km.) sw to NE along the border between Madrid and Segovia provinces, rising to 8,100 ft. (2,469 metres) a.s.l. Developed for timber, and as a winter resort.

Guadeloupe   West Indies. 15 40N 90 30W. Two islands in the Leeward Islands group, forming part of an overseas dept. of France. Area of dept., 657 sq. m. (1,702 sq. km.). The main towns are Basse-Terre (the cap.), sit. on Guadeloupe, and Point-à-Pitre, sit. on Grande Terre. The two main islands, Guadeloupe or Basse-Terre, and Grande Terre, are separated by the Rivière Salée strait. Guadeloupe in the w rises to 4,869 ft. (1,484 metres)

a.s.l. at La Soufrière which is a dormant volcano. The main occupation is farming. Main products are bananas, sugar, rum and coffee. The dept. also includes Désirade, Les Saintes, St. Martin and St. Barthélemy. Pop. (1988e) 336,300.

Guadiana River   Spain/Portugal. 37 14N 7 22W. Rises in headstreams which join near Ciudad Real, Spain, and flows 510 m. (816 km.) w to Badajoz, then s along the Spanish–Portuguese boundary, ssw into Portugal above Mourão, then sse back to the frontier above Alcautum, and s to enter the Gulf of Cádiz below Ayamonte, Spain, which is sit. close to the frontier.

Guahan   *see* **Guam**

Guairá   Paraguay. 24 05S 54 10W. Dept. of s Paraguay. Area, 1,236 sq. m. (3,201 sq. km.). Cap. Villarrica. Fertile lowland drained by Tebicuary-mi R. Mainly a lumbering and agric. area. Main products sugar-cane, maté, cotton, tobacco and fruit. Pop. (1982e) 143,374.

Guaira, La   Venezuela. Federal District. Chief port N of Caracas, sit. on Caribbean Sea. Exports coffee, tobacco and cacao.

Gualeguay   Entre Rios, Argentina. Town sit. ESE of Rosario with a port, Puerto Ruiz, s on Gualeguay R. Trading centre for livestock and cereals. Pop. (1980c) 26,000.

Gualeguaychú   Entre Rios, Argentina. 33 00S 58 30W. Town sit. on Gualeguaychú R. near its confluence with Uruguay R., and opposite Fray Bentos, Uruguay. Industries inc. tanning and meat packing. Pop. (1980c) 51,000.

Guam   Marianas Islands, Pacific Ocean. 13 28N 144 47E. An outlying territory of the U.S.A. Largest of the Marianas group, lying in the w Pacific, and the most southerly. Area, 209 sq. m. (541 sq. km.). Cap. Agaña. Mountainous and volcanic in the s, rising to 1,334 ft. (407 metres) a.s.l., low-lying in the N. Major U.S. naval and air base. Main agric. products are maize, sweet potatoes, taro, bananas, citrus and sugar-cane. Guam is the only U.S. territory with complete 'free trade'. Pop. (1987E) 130,400.

Guanacaste   Costa Rica. Province of NW Costa Rica on the Pacific coast, bounded N by Nicaragua and E by the Cordillera de Guanacaste. Area, 3,915 sq. m. (10,140 sq. km.). Cap., Liberia. Mainly a stock raising area. Other products inc. cereals, sugar-cane, timber, coffee. There is some gold on the Abangares R. Pop. (1984C) 193,024.

Guanajuato   Mexico. 21 01N 101 15W. (i) State in central Mex. Area, 11,773 sq. m. (30,491 sq. km.). Plateau at an average height of 6,000 ft. (1,829 metres) a.s.l. Rich in mineral deposits inc. silver, gold, zinc, copper, mercury and lead. Pop. (1980C) 3,006,110. (ii) Cap. of (i) sit. in central Mex. on the Sierra Madre Occidental. Mining centre, also manu. textiles and pottery. Pop. (1983E) 45,000.

Guanare   Portuguesa, Venezuela. 9 03N 69 45W. Town sit. WSW of Caracas with a port, Guerrilandia, on the Guanare R. nearby. State cap. and centre of an agric. area producing cotton, coffee, cacao, sugar-cane, cereals, livestock. Noted for the shrine of Our Lady of Coromoto, patron of Venezuela. Pop. (1971C) 37,715.

Guangdong (Kwangtung)   China. 23 00N 113 00E. Province of S China bounded SE by the South China Sea and bordering the enclaves of Macao and Hong Kong; inc. the Chinese islands of the South China Sea. Area, 81,853 sq. m. (212,000 sq. km.). Cap. Guangzhou (Canton). Mountainous, except for alluvial lowlands of Canton R. delta where most agric. and production are concentrated. The main crops are sugar-cane, wheat, tobacco, fruit, silk, vegetables, timber. Minerals inc. tungsten, manganese, coal, iron. The silk industry is important. Pop. (1982C) 59,299,220.

Guangxi Zhuang (Kwangi Chuang) Autonomous Republic   China. 24 00N 109 00E. An autonomous region in the S bordering on N. Vietnam. Area, 91,120 sq. m. (236,000 sq. km.).

Cap. Nanning, sit. 197 m. (315 km.) SW of Kweilin. Almost entirely in the Sikiang basin with important crops of rice and sugar-cane. Minerals, inc. manganese and tin, are mined. Pop. (1982C) 36,420,960.

Guangzhou (Kwangchow)   Guangdong, China. 23 08N 113 20E. Otherwise called Canton; city sit. SW of Nanking on Canton R. at N edge of delta. Major commercial, communication and industrial centre. Industries inc. manu. steel, chemicals, machinery, cement, textiles, sugar, matches, paper, furniture and handicraft products. The port has docks and shipyards on both banks of the Canton R. accessible to sea-going ships. Pop. (1984E) 3,222,000.

Guantánamo   Oriente, Cuba. 20 08N 75 12W. Town sit. ENE of Santiago de Cuba. Commercial centre of an agric. area. Industries inc. sugar milling, coffee roasting, and tanning; manu. confectionery. Pop. (1981E) 467,000.

Guantánamo Bay   Oriente, Cuba. 20 00N 75 07W. Inlet on the S coast sit. S of Guantánamo. Site of a U.S. naval and air base. Since 1959 Cuba has refused to accept the rent and periodically threatens to seize the base.

Guaporé   *see* **Rondônia**

Guaqui   La Paz, Bolivia. Town sit. W of La Paz on the Gulf of Taraco, L. Titicaca. Main Bolivian port on the L., rail terminus (from La Paz).

Guaranda   Bolívar, Ecuador. 1 39S 78 57W. Town sit. SSW of Quito on a headstream of the Chimbo R. beneath Chimborazo. Centre of an agric. area, trading in cinchona, timber, cereals, cattle. Pop. (1984E) 14,000.

Guatemala   15 40N 90 30W. Rep. bounded N and W by Mexico, S by Pacific Ocean and E by El Salvador, Honduras and Belize. Area, 42,042 sq. m. (108,889 sq. km.). Cap. Guatemala City. Mountainous with volcanoes, rising to Tajamulco, at 13,812 ft. (4,210 metres) a.s.l. the highest peak in Central America, and volcanic. The climate is humid on the coastal lowlands, and temperate in the uplands. Rainfall is *c.* 50 in. (127 cm.) per annum, falling May–Oct. The main occupations are farming, esp. coffee and cotton, and lumbering. Industries are concentrated on Guatemala City. Pop. (1989E) 9m.

Guatemala City Guatemala. 14 38N 90 22W. Cap. of rep. and dept. sit. in SE Guatemala on the Inter-American Highway. It was founded in 1776. Commercial centre trading in coffee. Industries inc. manu. textiles, silverware, soap and cement. Pop. (1989E) 2m.

Guayama Puerto Rico. 17 59N 66 07W. Town sit. S of San Juan. Industries inc. textiles and sugar milling.

Guayaquil Guayas, Ecuador. 2 16S 79 53W. Cap. of province in central W, sit. near mouth of Guayas R. Industries inc. sugar refining, tanning, sawmilling, brewing, iron-working and textiles. Port exporting bananas, cacao and coffee. Pop. (1982C) 1,300,868.

Guayas Ecuador. Province of W Ecuador, the largest and most densely inhabited. Area, 8,208 sq. m. (21,259 sq. km.). Chief cities, Guayaquil (cap.), Yaguachi, Daule, Balzar, Santa Elena, Milagro, Salinas. The Santa Elena peninsula is arid, with saltworks and oil refineries; most of the coast is lowland watered by Guayas R. and its many tribs.; further E there is forest rising to the Andes. Mainly an agric. area producing cacao, sugar-cane, coffee, tobacco, cotton, cereals, fruit. Manu. industries are centred on Guayaquil. Pop. (1982C) 2,116,819.

Gubbio Umbria, Italy. 43 21N 12 35E. Town sit. NE of Perugia. The main industry is pottery, esp. majolica ware.

Gudauty Georgian Soviet Socialist Republic, U.S.S.R. Town sit. WNW of Sukhumi on the Black Sea. Industries inc. iron founding, tanning; manu. alabaster. Port, handling wines, tobacco, citrus fruit.

Gudbrandsdal Norway. 62 00N 9 14E. Valley of Lågen R. in S central

Norway, extending NW to SE between the Rondane Mountains to the E and the Jotunheim Mountains to the W. Fertile area. The main occupation is farming and the valley is a long-estab. trade route.

Guelph Ontario, Canada. 43 33N 80 15W. City sit. NW of Hamilton on Speed R. Manu. electrical equipment, hardware and leather goods. Pop. (1981C) 76,658.

Guéret Creuse, France. 46 10N 1 52E. Cap. of dept. Pop. (1982C) 16,621.

Guernica Vizcaya, Spain. 43 19N 2 41W. Town sit. ENE of Bilbao. Industries inc. metal-working and furniture-making. Pop. (1989E) 16,378.

Guernsey Channel Islands. 49 28N 2 35W. Island NW of Jersey and the second largest of the Channel Islands group off the NW coast of Normandy, France. Area, 24 sq. m. (62 sq. km.). Cap. St. Peter Port. The main occupations are farming, esp. dairy cattle, horticulture and tourism. Pop. (1986C) 55,482 (inc. Alderney and small islands).

Guerrero Mexico. 17 30N 100 00W. State on Pacific coast of SW Mex. Area, 24,433 sq. m. (63,281 sq. km.), sit. S of Mexico City, and Acapulco. Mountainous, crossed W to E by the Sierra Madre del Sur. The main occupation is farming; main products are cereals, cotton, sugar-cane and coffee. Pop. (1989E) 2,604,947.

Guiana, French *see* **French Guiana**

Guiana, Netherlands *see* **Suriname**

Guianas South America. 5 00N 60 00W. Region in the NE, bounded N and W by Orinoco R., S by Rs. Negro and Amazon, and E by Atlantic. Comprises the Guiana Highlands which extend through E Venezuela, N Brazil, Guyana, Suriname and French Guiana. Sandstone plateaux in the S with forests and deep ravines and some savannah; rocky hills to the N with deposits of gold, diamonds and bauxite, and bounded N and E by marshy coastlands containing most of the pop.

Guildford   Surrey, England. 51
14N 0 35W. Town in the s sit. on
Wey R. Market, cathedral, university
and residential town. Industries inc.
financial services, engineering and
pharmaceuticals. Pop. (1981E) 56,976.

Guimarães   Minho, Portugal. 41
26N 8 19W. Town sit. SE of Braga. Industries inc. textiles; manu. shoes,
cutlery. Noted for castle, *c.* 1100 A.D.
and other medieval buildings. Pop.
(1981C) 2?,947.

Guinea   10 20N 10 00W. Rep.
bounded NW by Guinea-Bissau, and s
by Liberia and Sierra Leone. Area,
94,926 sq. m. (245,857 sq. km.). Cap.
Conakry. Mainly savannah on the
Fouta Djallon plateau. Tropical
monsoon climate with coastal rain-
forests. The main occupation is farming, main products are bananas and
palm kernels. Iron ore is found on the
Kaloum peninsula, bauxite at Boké,
Kindia and on the Los Islands. Pop.
(1988E) 6,533,000.

Guinea-Bissau   12 00N 15 00W.
Bounded on N by Senegal, E and SE by
Guinea and W by the Atlantic. Area,
13,948 sq. m. (36,125 sq. km.). Cap.
and chief port, Bissau. Mainly rain-
forest becoming savannah further inland, with mangrove swamps on the
coast. Tropical monsoon climate, annual rainfall 50–80 in. from May–
November. The main products are
rice, copra, beeswax, groundnuts and
hides. Pop. (1988E) 932,000.

Guinea, Gulf of   West Africa. 3 00N
2 30E. Atlantic inlet extending *c.*
1,200 m. (1,920 km.) W to E from
Cape Palmas, Liberia, to Cape Lopez,
Gabon. Inc. the Bight of Benin and
the Bight of Bonny, which are separated by the Niger delta, and receives
Rs. Volta, Niger and Ogowé.

Guipúzcoa   Spain. 43 12N 2 15W.
Province in N Spain bounded N by
Bay of Biscay and NE by France. The
smallest province, but densely populated. Area, 771 sq. m. (1,997 sq.
km.). Cap. San Sebastián. Mainly industrial, with some stock rearing and
fishing. Pop. (1986C) 688,894.

Guisborough   Cleveland, England.

54 32N 1 04W. Market town sit. ESE
of Middlesbrough. Industries inc. tanning and brewing. Pop. (1989E)
19,242.

Guizhou (Kweichow)   China. 27
00N 107 00E. Province in the SE between R. Yangtse-kiang and Si-kiang.
Area, 67,954 sq. m. (176,000 sq. km.).
Cap. Guiyang. Mainly high plateau
where maize is grown. Rice, tea and
tobacco are grown in the valleys.
Minerals inc. mercury, coal, iron-ore
and copper. Guiyang is noted for its
horses. Pop. (1987E) 30,080,000.

Gujarat   India. 23 20N 71 00E.
State in NW India bounded W and S by
Arabian Sea, SE by Maharashtra, E by
Madhya Pradesh, NE by Rajasthan and
N by Pakistan. Area, 76,556 sq. m.
(195,024 sq. km.). Chief towns
Ahmedabad, Vadodara, Gandhinagar
(cap.), Surat, Rajkot and Jamnagar.
Mainly fertile plain, but the NW inc.
the Rann of Kutch marshes, and the
Kathiawar peninsula is partly arid.
Highly industrialized, esp. in textiles,
engineering, petrochemicals, pharmaceuticals, cement, fertilizers, machine
tools, oil refining and chemicals. Produces oil, salt, soda ash, azo dyes, sulphur black and natural gas. Cereals
and cotton are produced in the rural
areas. Pop. (1981C) 34,085,799.

Gujranwala   Punjab, Pakistan. 32
06N 74 11E. Town sit. NNW of
Lahore. Trading centre for cereals and
cotton. Industries inc. textiles; manu.
leather, electrical goods and metal
goods. Pop. (1981E) 601,000.

Gujrat   Punjab, Pakistan. 32 35N
74 06E. Town sit. N of Gujranwala.
Trading centre for grain and cotton.
Manu. inc. ceramics, footwear, textiles and furniture. Pop. (1981C)
154,000.

Gulbarga   Karnataka, India. Town
sit. W of Hyderabad. Industries inc.
textiles, flour and oilseed milling.
Noted for ruins of the palaces and
mosques of the 14th and 15th cent.
Bahmani   Kings.   Pop.   (1981C)
221,325.

Gulfport   Mississippi, U.S.A. 30
22N 89 06W. City sit. WSW of Mobile,

Alabama, on the Mississippi Sound. Industries inc. tourism and textiles; manu. fertilizers. Seaport handling wood and cotton seed. Pop. (1990E) 58,800.

Gulf Stream   North Atlantic Ocean. Warm ocean current originating in the Gulf of Mexico and flowing E through the Florida Strait, then NE along the U.S. coast to the Grand Banks off Newfoundland where it is dispersed, its waters reaching W and NW Europe by the North Atlantic Drift. Its warmth has a strong influence on the W European climate.

Gumal Pass   Afghánistán/Pakistan. Pass sit. W of Dera Ismail Khan at 5,000 ft. (1,524 metres) a.s.l. on the Afghan border, traversed by Gumal R. and used as a trade route.

Gumti River   India. 23 32N 90 43E. Rises in N Uttar Pradesh and flows 500 m. (800 km.) SE past Lucknow and Jaunpur to join Ganges R. below Varanasi. Navigable for small boats in its lower course.

Gümüsane   Turkey. Province of NE Turkey bounded N by the Rize Mountains and S by the Erzincan Mountains. Area, 3,884 sq. m. (10,060 sq. km.). Drained by Harsit, Kelkit and Coruh Rs., and forested. Some copper deposits. Agric area produces cereals. Pop. (1985C) 283,753.

Guna   Madhya Pradesh, India. 24 40N 77 19E. Town sit. SSW of Agra. Commercial centre trading in grain and oilseeds. Pop. (1981C) 64,659.

Gunnison   Colorado, U.S.A. Town sit. SE of Grand Junction. Tourism is the main industry. Originally a silver mining town; there are still silver, coal and gold mines nearby. Pop. (1988E) 6,400.

Guntur   Andhra Pradesh, India. Town sit. ESE of Hyderabad. Commercial centre trading in cotton and tobacco. Industries inc. engineering, milling rice, jute and oilseeds. Pop. (1981C) 367,699.

Gurgaon   Haryana, India. Town sit. SW of Delhi. Industries inc. auto-

mobiles and engineering. Trading centre for grain, cotton, salt, oilseeds. Pop. (1981C) 100,877.

Guryev   Kazakhstan, U.S.S.R. 47 07N 51 56E. Town sit. on Ural R. near its mouth on the Caspian Sea. Industries inc. fishing and fish canning, and oil refining. Terminus of the oil pipeline from the Emba field.

Gusau   Sokoto, Nigeria. 12 12N 6 40E. Town sit. SSE of Kaura Namoda on Sokoto R. Trading and distribution centre for millet, cattle, skins, timber and cotton. Some gold and diamond deposits nearby. Pop. (1975E) 94,000.

Güstrow   Schwerin, German Democratic Republic. 53 48N 12 10E. Town sit. S of Rostock. Industries inc. engineering, food processing and woodworking. Pop. (1989E) 40,000.

Gütersloh   North Rhine-Westphalia, Federal Republic of Germany. 51 54N 8 23E. Town sit. SW of Bielefeld. Industries inc. manu. machinery, furniture, textiles and food products. Pop. (1985E) 85,000.

Guwahati   Assam, India. Town sit. ESE of Darjeeling on Brahmaputra R. Commercial centre trading in rice, jute and cotton. Industries inc. oil refining, cotton, plywood and flour milling.

Guyana   5 00N 59 00W. Rep. in the NE 'shoulder' of S. America, bounded W by Venezuela, N by the Caribbean, E by Suriname and S and SW by Brazil. The territory, including the countries of Demerara, Essequibo and Berbice, named from the 3 Rs., was first partially settled by the Dutch East India Company in 1620. It was finally ceded to Britain in 1814 and named British Guiana. In 1966 British Guiana became an independent member of the Commonwealth under the name of Guyana. Area, 83,000 sq. m. (215,000 sq. km.). Cap. Georgetown. Bisected by Essequibo R. Mainly jungle; mountainous in the W, savannah in the S, agric. land on the coastal belt. The main occupations are farming and mining; main products are sugar-cane, bauxite, timber, gold and rice. Pop. (1989E) 990,000.

Guyane   *see* **French Guiana**

Gwadar   Baluchistan, Pakistan. 25 07N 62 19E. Town sit. w of Karachi, on Arabian Sea. Industries inc. fishing.

Gwalior   Madhya Pradesh, India. 26 12N 78 09E. Town sit. SSE of Agra. Industries inc. textiles, flour and oilseed milling; manu. pottery and footwear. Pop. (1981C) 539,015.

Gwelo   *see* **Gweru**

Gwent   Wales. 51 45N 3 00W. County of SE Wales created by 1974 reorganization with districts of Newport, Islwyn, Blaenau Gwent, Torfaen and Monmouth. Eastern Gwent is devoted to dairying and mixed farming with market gardening near Newport, Abergavenny and in the Wye Valley. West Gwent is highly industrialized. Aluminium, steel, car components, chemicals, high technology and foodstuffs are large employers of labour. Pop. (1986E) 441,800.

Gweru   Zimbabwe. 19 25S 29 50E. Town in central Zimbabwe. Commercial centre for a farming area. Industries inc. engineering; manu.

asbestos and metal products, footwear and cement. Pop. (1982C) 79,000.

Gwynedd   Wales. 53 00N 4 00W. County of NW Wales created under 1974 reorganization, with the boroughs/districts of Ynys Môn (Isle of Anglesey), Arfon, Aberconwy, Dwyfor and Meirionnydd. The people of Gwynedd are predominantly Welsh-speaking. Agric. is important particularly cattle and sheep. Tourism and outdoor pursuits, in Snowdonia and on the coast, are important. Slate quarrying and the woollen industry are now less important and newer industries such as electronics, and light engineering are growing. Holyhead is an important ferry port for the Republic of Ireland. Pop. (1984E) 232,000.

Gympie   Queensland, Australia. 26 11S 152 40E. Town in SE of state sit. on Mary R. Market centre for a dairy-farming and banana and pineapple growing area. Industry inc. timber processing, dairying, small crops. Pop. (1989E) 30,000.

Györ   Györ-Sopron, Hungary. 47 42N 17 38E. Cap. of county sit. WNW of Budapest on Raba R. Industries inc. flour milling, distilling and textiles. Pop. (1989E) 132,000.

H

Ha'apai Tonga, South Pacific. Island group of central Tonga comprising 30 small coral islands. Area, 46 sq. m. (119 sq. km.). There are guano deposits but lack of good harbours prevents export. Cap. Pangai, on Lifuka island. Pop. (1986C) 8,919.

Haarlem Noord-Holland, Netherlands. 52 23N 4 38E. Cap. of province sit. near the N. Sea coast. Centre of a bulb growing industry. Other industries are printing, and the manu. of textiles, chemicals and chocolate. Pop. (1988E) 149,198.

Haarlemmermeer Noord-Holland, Netherlands. 52 15N 4 38E. Area of drained fenland between Amsterdam and Leiden. Area, 72 sq. m. (186 sq. km.). Agric. area. Main products wheat, vegetables, flax.

Habana *see* Havana

Habarovsk *see* Khabarovsk

Hachinohe Honshu, Japan. 40 30N 141 29E. Port sit. SE of Aomori. An important centre of the fishing industry with manu. of iron, steel, pulp and paper, chemicals and cement. Pop. (1985E) 243,000.

Hachioji Honshu, Japan. 35 39N 139 20E. Town sit. W of Tokyo. Manu. textiles, esp. silk, and chemicals. Pop. (1988C) 433,000.

Hackensack New Jersey, U.S.A. 40 53N 74 03W. City sit. N of Jersey City on Hackensack R. Manu. metal goods, chemicals, paper, glass, furniture, jewellery and food products. Pop. (1980C) 36,039.

Hackney Greater London, England. 51 33N 0 03W. Boro. of E Lond. Inc. districts of Clapton, Homerton, Shoreditch and Stoke Newington. Pop. (1988E) 189,000.

Haddington Lothian Region, Scotland. 55 58N 2 47W. Town sit. E of Edinburgh on R. Tyne. Industries inc. malting, woollens, electronic, precision and electrical engineering. Pop. (1989E) 8,241.

Hadhramaut Arabia. 15 30N 49 30E. Area on S coast of Arabian peninsula corresponding roughly to the area that was Southern Yemen, Wadi Hadhramaut, an intermittent R. valley beginning at Husn al' Abr and extending 350 m. (560 km.) E past Shibam, Seiyun and Tarim to the Arabian Sea near Seihut. Important agric. area, producing dates, grain, sesame.

Hadleigh Suffolk, England. 52 03N 0 58E. Town sit. W of Ipswich. Industries inc. flour milling, agric. engineering, clothing and sail making. Pop. (1980E) 5,500.

Hagen North Rhine-Westphalia, Federal Republic of Germany. 51 22N 7 28E. Town sit. S of Dortmund. Manu. machinery, iron and steel, paper and food products. Pop. (1989E) 208,000.

Hagerstown Maryland, U.S.A. 39 39N 77 43W. City sit. WNW of Baltimore. The industrial centre of a rich agric. district. Manu. truck and sandblasting equipment, ice-cream, pipe organs, metal products, leather goods and furniture. Pop. (1980C) 34,132.

Hague, The Zuid-Holland, Netherlands. 52 06N 4 18E. City sit. SW of Amsterdam and near the North Sea. Provincial cap. Residential and administration centre, seat of the Royal Court, Parliament and the Intern. Court of Justice. Pop. (1984E) 672,127 (agglomeration).

Haguenau Bas-Rhin, France. 48 49N 7 47E. Town sit. N of Strasbourg on Moder R. Manu. water flow meters, clothing, automats, ball bearings, electric motors and confectionery. Pop. (1989E) 31,993.

Haifa Israel. 32 49N 35 00E. Port in the N sit. at foot of Mount Carmel. Manu. textiles, chemicals, metal goods, glass, soap and electrical and electronic equipment. Pop. (1988E) 222,600.

Hail Nejd, Saudi Arabia. 27 33N 41 42E. Town sit. in N central Saudi Arabia, in Jebel Shammar. Sit. on a caravan and pilgrim route from Iraq to Medina and Mecca.

Hainan Guangdong, China. 19 00N 109 30E. An island sit. between China Sea and Gulf of Tonking and separated from the Luichow peninsula by Kiungchow Strait. Area, 13,500 sq. m. (35,000 sq. km.). Mountainous in the centre and S with an undulating plain in the N. Timber, rice, sugar-cane, cotton and tobacco are produced. The chief town and port is Haikou.

Hainaut Belgium. 50 30N 4 00E. A province sit. in S adjoining France. Area, 1,437 sq. m. (3,722 sq. km.). The principal Rs. are Scheldt, Sambre and Dender. Agric. is carried on in the fertile N. The industrial S contains important coalmines and iron and steel works. Cap. Mons. Pop. (1987E) 1,271,649.

Haines Alaska, U.S.A. 64 45N 137 30W. City sit. N of Juneau on W coast of the Lynn Canal. Industries inc. fishing, timber, saw-milling, seasonal tourism. Pop. (1985E) 1,200 (city); 2,380 (surrounding boro.).

Haiphong Vietnam. 20 52N 106 41E. Port sit. on Red R. delta. The second largest town with manu. of textiles, phosphates and plastics. Minerals and rice are exported. Pop. (1980E) 1,305,163.

Haiti West Indies. 19 00N 73 30W. Rep. occupying W part of Hispaniola Island, and bounded E by Dominican Republic. Haiti was discovered by Christopher Columbus in 1492. The Spanish colony was ceded to France in 1697 and became her most prosperous colony. After the extirpation of the Indians by the Spaniards (by 1533) large numbers of African slaves were imported whose descendants now populate the country. The country declared its independence in 1804, and its successful leader, Gen. Jean-Jacques Dessalines, proclaimed himself Emperor of the newly-named Haiti. From 1822 to 1844 Haiti and the eastern part of the island (later the Dominican Republic) were united. After one more monarchical interlude, under the Emperor Faustin (1847–59), Haiti has been a republic. From 1915 to 1934 Haiti was under United States occupation. Area, 10,700 sq. m. (27,750 sq. km.). Inc. 2 large peninsulas extending W into Windward Passage on either side of the Gulf of Gonaïves. Main centres Port-au-Prince (cap.), Gonaïves, Cap-Haïtien, Port-de-Paix, Les Cayes. Mainly forested mountains, rising to 8,793 ft. (2,680 metres) a.s.l. in the Massif de la Selle to the SE. The plains in between the ranges are fertile and densely populated. Tropical climate and rains April–June, October–November. The economy is mainly agric., with coffee, cotton and sugar as the main crops. Industries are mainly concentrated on Port-au-Prince and inc. textiles (cotton denim), tanning, flour milling, food processing and packing; manu. cement, pharmaceutical products, plastics, paints and soap. Pop. (1987E) 5,658,124.

Hakkâri Turkey. (i) Province of SE Turkey bounded E by Iran and S by Iraq. Area, 3,722 sq. m. (9,640 sq. km.). Drained by Great Zab R.; a mountainous and infertile area, with some naphtha. Pop. (1985C) 182,645. (ii) Provincial cap. and grain growing centre.

Hakodate Hokkaido, Japan. 41 45N 140 43E. Port sit. on Tsugaru Strait. An important centre of the fishing industry with exports of salted and canned fish. Other industries are shipbuilding, woodworking and the manu. of food products, matches and cement. Pop. (1988E) 312,000.

Halberstadt Magdeburg, German Democratic Republic. 51 54N 11 02E. Town in W central G.D.R. sit. in foothills of Harz Mountains. Manu. machinery, textiles, chemicals, rubber and paper. Pop. (1989E) 47,600.

Halesowen   West Midlands, England. 52 27N 2 02W. Town sit. w of Birmingham. Industries inc. manu. machinery, machine tools, iron founding. Pop. (1985E) 55,000.

Halifax   Nova Scotia, Canada. 44 39N 63 36W. Cap. of province and port sit. on w shore of Halifax Harbour. A leading transatlantic port and railway terminus with industries inc. ship-building and oil refining. Fish, fish products and timber are exported. Manu. inc. cars, furniture, clothing and food products. The town was founded as a naval base in 1749. Pop. (1986E) 300,000 (metropolitan area).

Halifax   West Yorkshire, England. 53 44N 1 52W. Town sit. on R. Calder. Its cloth industry dates from the 15th cent. Other manu. inc. carpets, textiles, engineering, clothing and confectionery. Pop. (1981C) 87,500.

Halland   Sweden. County of sw Sweden on the Kattegat. Area, 2,106 sq. m. (5,454 sq. km.). Chief cities Halmstad (cap.), Laholm, Falkenberg, Varberg. Low-lying and drained by Laga, Nissa and Atra Rs. Fishing and farming are important; the main industry is metal-working. Pop. (1988E) 247,417.

Halle   German Democratic Republic. 51 29N 11 58E. (i) A district bounded by Magdeburg, Potsdam, Leipzig, Gera and Erfurt. Area, 3,386 sq. m. (8,771 sq. km.). Pop. (1988E) 1,776,500. (ii) Cap. of district of same name in s central G.D.R. sit. on Saale R. The centre of a lignite and potash mining area with engineering, chemical, paper, brewing and sugar industries. Pop. (1988E) 236,044.

Halmahera   Indonesia. 1 00N 128 00E. An island sit. between Sulawesi and W. Irian in Maluku. Area, 7,000 sq. m. (18,130 sq. km.) Volcanic and mountainous, its chief products are sago, rice, nutmeg and coconuts. The chief town is Djailolo.

Halmstad   Halland, Sweden. 56 39N 12 50E. Cap. of county and port sit. in sw on the Kattegat. Manu. textiles, wood pulp, jute and paper. Exports inc. timber, local granite and dairy produce. Pop. (1988E) 78,607.

Halstead   Essex, England. 51 59N 0 39E. Town sit. on R. Colne. Industries include general, electrical and medical products. Pop. (1985E) 9,600.

Haltemprice   Humberside, England. 53 45N 0 24W. Town sit. NW of Hull forming a suburb of the city. Residential and industrial.

Haltwhistle   Northumberland, England. 54 58N 2 27W. Town sit. w of Hexham on S. Tyne R. Industries inc. coalmining, paint and varnish works, plastics, chemicals and steam cleaning equipment manu. Pop. (1982E) 3,522.

Hama   Hama, Syria. Cap. of district in NW, sit. on Orontes R. Industries inc. tanning and weaving. Trades in cereals and fruit. Noted for the large wooden waterwheels used for irrigation. Pop. (1981C) 177,208.

Hamadán   Hamadán, Iran. 34 48N 48 30E. Town in NW, sit. at 6,000 ft. (1,829 metres) a.s.l. Noted for carpets and rugs. Trades in wool, hides and skins. Pop. (1986C) 272,499.

Hamamatsu   Honshu, Japan. 34 42N 137 44E. City sit. halfway between Tokyo and Ōsaka. Manu. textiles, musical instruments, esp. pianos and organs, motor cycles and optical high-tech industries. Pop. (1989E) 530,000.

Hamburg   Federal Republic of Germany. 53 33N 9 59E. City and *Land* of F.R.G. on Elbe R. *Land* area 292 sq. m. (755 sq. km.). Major port importing industrial raw materials and exporting most of F.R.G. manu. goods. Industries inc. shipbuilding, food processing, engineering. It has been a major sea port since the 13th cent. Pop. (1988E) 1,603,070.

Hämeenlinna   Häme, Finland. 61 00N 24 27E. Cap. of province in sw sit. on L. Vanajavesi. Manu. inc. textiles and plywood. Pop. (1988C) 42,760.

Hameln   Lower Saxony, Federal Republic of Germany. 52 07N 9 24E. Town sit. sw of Hanover on Weser R.

Industries inc. electricals, machinery, carpets, clothing, food processing and chemicals. Pop. (1984E) 56,300.

Hamersley Range   Western Australia, Australia. 21 53S 116 46E. A mountain range sit. between Rs. Fortescue and Ashburton, and stretching for about 170 m. (272 km.). There are rich deposits of iron ore. The highest peaks are Mount Bruce, 4,024 ft. (1,227 metres) a.s.l., and Mount Brockman, 3,654 ft. (1,114 metres) a.s.l.

Hamheung   South Hamkyung, North Korea. 39 54N 127 32E. Cap. of province sit. N of Wonsan. Manu. textiles.

Hamilton   Victoria, Australia. 37 45S 142 02E. City sit. W of Melbourne. Centre of an important sheep farming district with related industries. Pop. (1984E) 10,200.

Hamilton   Bermuda Island, Bermuda. 32 18N 64 48W. Cap. of Bermuda and port sit. on an inlet, founded in 1790. Commercial and tourist centre. It became a free port in 1956. Pop. (1989E) 3,000.

Hamilton   Ontario, Canada. 43 15N 79 51W. Town and port sit. at W end of L. Ontario. An industrial and railway centre with important steel works. Also manu. heavy and agric. machinery, typewriters, electrical equipment and textiles. Pop. (1981C) 306,434 (metropolitan area, 542,095).

Hamilton   North Island, New Zealand. 37 47S 175 17E. Town near W coast sit. on Waikato R. Manu. wood products, clothing and dairy produce. Pop. (1986E) 101,814.

Hamilton   Strathclyde Region, Scotland. 55 47N 4 03W. Town sit. on R. Clyde. Manu. metal goods, instruments, clothing, carpets and electrical equipment. Pop. (1988E) 51,294.

Hamilton   New Jersey, U.S.A. Township in Mercer County, NE of Trenton. Pop. (1980C) 82,801.

Hamilton   Ohio, U.S.A. 39 26N 84 30W. City sit. N of Cincinnati on Great Miami R. Manu. paper, machinery and diesel engines. Pop. (1980C) 63,189.

Hamilton River   Labrador, Canada. Rises in Ashuanipi L. as the Ashuanipi R. and flows 600 m. (960 km.) through Dyke and Lobstick Ls. into L. Melville.

Hamlin   *see* **Hameln**

Hamm   North Rhine-Westphalia, Federal Republic of Germany. 51 41N 7 48E. Town sit. ENE of Dortmund on Lippe R. Industries inc. nuclear power station, coal mining, wire manu. machinery, tube mill, chemicals and textiles. Pop. (1985E) 178,000.

Hammerfest   Finnmark, Norway. 70 40N 23 42E. Town sit. on W coast of Kvalöy Island, being the most N town in the world. Its ice-free port allows fishing to be carried on throughout the year. Pop. (1989E) 7,100.

Hammersmith   Greater London, England. 51 30N 0 14W. District of W central Lond. on N bank of Thames R. Residential and industrial.

Hammond   Indiana, U.S.A. 41 36N 87 30W. City adjoining Chicago, Illinois, on Grand Calumet R. Manu. railway equipment, agric. machinery, steel and petroleum products. Pop. (1988E) 84,630.

Hampshire   England. County of S England bounded on N by Berkshire, W by Wiltshire and Dorset, S by Eng. Channel and Isle of Wight, E by West Sussex and Surrey. On the coast are Portsmouth Harbour and Southampton Water. Chief towns, Basingstoke, Portsmouth, Southampton and Winchester. Inland are fertile valleys, hills and woods. The county is crossed by the North Downs in NW and W. W of Southampton is the New Forest. Rs. include Avon, Lymington, Test and Itchen. Under 1974 re-organization the county consists of the city councils of Portsmouth, Southampton, and Winchester; the borough councils of Basingstoke and Deane, Eastleigh, Fareham, Gosport, Havant, Rushmoor, and Test Valley; and the district councils of East Hampshire, Hart, and

New Forest. Industries inc. agric., sheep, cattle and horse rearing, ship-building, oil refining, chemicals, electrical and electronic equipment, pharmaceuticals, computers and information technology. There are many holiday resorts. Pop. (1989E) 1,542,900.

Hampstead Greater London, England. 51 33N 0 10W. District of N Lond. forming part of the boro. of Camden. Residential and picturesque, inc. Hampstead Heath and Kenwood mansion with many literary and artistic associations.

Hampton Virginia, U.S.A. 37 01N 76 22W. City opposite Norfolk on Hampton Roads, on lower estuary of James R. Industries inc. fishing, fish packing and tourism. Residents mainly of Polish extraction. Pop. (1980C) 122,617.

Hamtramck Michigan, U.S.A. 42 24N 83 03W. City sit. in centre of Detroit. Manu. cars, machinery, electrical equipment, paint and varnish. Pop. (1980C) 21,300.

Hanau Hessen, Federal Republic of Germany. 50 08N 8 56E. Town sit. E of Frankfurt at confluence of Rs. Kinzig and Main. Centre of a jewellery and diamond-cutting industry. The Grimm brothers and Paul Hindemith were born here. Pop. (1984E) 85,000.

Hangchow *see* **Hangzhou**

Hangzhou (Hangchow) Zhejiang, China. 30 15N 120 10E. Cap. of province sit. on E coast, and on Fuchun R. and Grand Canal. A commercial centre trading in silk, rice and tea. Manu. inc. textiles and matches. Pop. (1982C) 4,020,000.

Hangzhou Bay Zhejiang, China. 30 15N 120 10E. The estuary of the Fuchun R. on the E. China Sea.

Hanley Staffordshire, England. 53 01N 2 11W. Town forming part of Stoke-on-Trent on Trent R., one of the 'five towns' of the 'potteries' district. Pottery is still the main industry, also iron, steel, engineering. Noted as the birthplace of Arnold Bennett.

Hannibal Missouri, U.S.A. 39 42N 91 22W. City sit. NW of St. Louis on Mississippi R. A railway and river-traffic centre with manu. of cement, steel, metal goods, foodstuffs, heating elements, agric. products and tools. Mark Twain's home was here. Pop. (1980C) 18,811.

Hanoi Vietnam. 21 01N 105 52E. Cap. of Vietnam sit. on Red R. It became the cap. in 1010 A.D. and was cap. of French Indo-China 1887–1946. Much destroyed by bombing in the Vietnam war, 1965–75. Industrial and transportation centre. Industries inc. engineering, rice milling, chemicals, tanning and brewing and the manu. of footwear, textiles, matches and leather goods. Pop. (1979E) 2,570,905.

Hanover Lower Saxony, Federal Republic of Germany. 52 23N 9 44E. City sit. SE of Bremen on Leine and Ihme Rs. *Land* cap. Commercial and communications centre. Industries inc. textiles, paper making; manu. rubber, machinery, metal goods. Noted as the former cap. of the Kingdom of Hanover. Pop. (1987E) 495,300.

Hao Tuamotu Archipelago, French Polynesia. Atoll sit. E of Tahiti. Main base for Fr. nuclear tests on Mururoa and Fangataufa.

Happy Valley–Goose Bay Labrador, Newfoundland, Canada. 53 15N 60 20W. Settlement at mouth of Churchill R. at SW end of L. Melville, E Labrador. There is an intern. airport and a deep sea harbour, Terrington Basin, accessible mid-June to mid-Nov. Pop. (1986C) 7,248.

Haradh Hasa, Saudi Arabia. 24 15N 49 00E. Oil field sit. SSW of Dhahran on Riyadh railway. Pop. (1986E) 100,000.

Hararge Ethiopia. Region in SE, area, 100,270 sq. m. (259,700 sq. km.). Cap. Hararge. Pop. (1984C) 4,151,706.

Hararge Ethiopia. 9 20N 42 10E. Cap. of province in E, sit. at over 5,000 ft. (1,524 metres) a.s.l. Trades in coffee, cotton and hides. The principal Moslem centre with several mosques. Pop. (1984E) 62,160.

Harare Zimbabwe. 17 43S 31 05E. Cap. and centre of gold-mining and agric. area. Founded in 1890 as Salisbury. Trade in tobacco. Industries tobacco processing and packing, brewing, flour milling, sugar refining; manu. clothing, furniture, agricultural equipment, fertilizers. Pop. (1982C) 656,100.

Harbin Heilongjiang, China. 45 45N 126 41E. Cap. of province in NE sit. on Sungari R. Important industrial and transportation centre. Industries inc. railway engineering, soya-bean processing, sugar refining and flour milling. Manu. machinery, textiles, chemicals and glass. It is often called the 'Moscow of the East'. Pop. (1982C) 2,550,000.

Harbour Island Bahamas, West Indies. Islet sit. NE of Nassau. Area, 1.5 sq. m. (3.9 sq. km.). Chief settlement, Dunmore Town. Densely populated and mainly agric., producing fruit, coconuts.

Hardanger Fjord Hordaland, Norway. 60 10N 6 00E. An inlet of N. Sea *c.* 80 m. (128 km.) long, with many cataracts, and surrounded by magnificent scenery.

Hardenberg Overijssel, Netherlands. Town sit. N of Almelo on Vecht R. Industries inc. synthetic materials, metal, transport. Pop. (1988E) 32,065.

Harderwijk Gelderland, Netherlands. 52 21N 5 36E. Town sit. NW of Apeldoorn on Ijsselmeer. Industries inc. fishing, fish processing; manu. soap, furniture, tobacco products, building stone. Pop. (1989E) 35,000.

Hardoi Uttar Pradesh, India. 27 24N 80 07E. Town sit. NW of Lucknow. Trades in grain and oilseeds. Pop. (1981C) 67,259.

Hardwar Uttar Pradesh, India. 29 58N 78 16E. Town sit. NE of Delhi on Ganges R. Industries inc. pharmaceuticals. A noted Hindu centre of pilgrimage. Pop. (1981C) 115,513.

Harfleur Seine-Maritime, France. 49 30N 0 10E. Town sit. E of Le Havre on Seine estuary. Manu. metal goods, pottery and vinegar.

Hargeisa Somali Republic. Town sit. SSE of Djibouti. Trading centre. Pop. (1985E) 150,000.

Harie Rud River SW Asia. Rises in the Hindu Kush W of Kábul and flows for *c.* 600 m. (960 km.) W through Afghánistán, then N to form Afghánistán/Iranian border and later the border of Iran and the Turkmen Soviet Socialist Republic and finally disappears into the Sands of Kara Kum desert. Used extensively for irrigation.

Haringey Greater London, England. 51 35N 0 07W. Pop. (1988E) 192,300.

Harlech Gwynedd, Wales. 52 52N 4 07W. Resort sit. in Snowdonia National Park NNW of Barmouth on Cardigan Bay. Pop. (1989E) 1,313.

Harlingen Texas, U.S.A. 26 11N 97 42W. Port sit. on a branch of the Gulf Intracoastal Waterway, NNW of Matamoros, Mex. Industries inc. fruit and vegetable canning and the manu. of chemicals. Pop. (1980C) 43,543.

Harlow Essex, England. 51 47N 0 08E. Town sit. W of Chelmsford. Industries inc. electronics, engineering, medical research, petroleum, glass and publishing. Pop. (1981C) 79,150.

Harpenden Hertfordshire, England. 51 49N 00 22W. Residential town sit. WNW of St. Albans. Rothamsted agric. experimental station is sit. here. Pop. (1981C) 28,797.

Harpers Ferry West Virginia, U.S.A. 39 13N 77 45W. Town sit. at confluence of Rs. Potomac and Shenandoah, and NW of Washington, D.C. Pop. (1989E) 435.

Harris *see* **Lewis-with-Harris**

Harrisburg Pennsylvania, U.S.A. 40 16N 76 52W. State cap. sit. WNW of Philadelphia on Susquehanna R. Industries inc. railway engineering, steel and the manu. of machinery, textiles, paper, bricks, shoes and food products. Pop. (1980c) 53,264.

Harrogate North Yorkshire, England. 54 00N 1 33W. Town and spa in N central Eng. Noted for its sulphur, chalybeate and saline springs. Conference and exhibition centre. Pop. (1987E) 69,270.

Harrow Greater London, England. 51 35N 0 20W. Town forming a suburb of NW Lond. Pop. (1989E) 201,140.

Hartford Connecticut, U.S.A. 41 46N 72 41W. State cap. sit. NNE of New Haven. It is a leading world centre of insurance. Manu. aeroplane parts, vehicles, typewriters, firearms and electrical equipment. Pop. (1980c) 136,392.

Hartlepool Cleveland, England. 54 42N 1 11W. Town and port sit. N of Middlesbrough on Hartlepool Bay. Industries inc. light engineering, clothing, food processing and hi-tech. Pop. (1989E) 90,900.

Harvey Illinois, U.S.A. 40 37N 87 39W. City 19 m. (30 km.) S of Chicago. Industries inc. engineering and metal working. Pop. (1980c) 35,810.

Harwell Oxfordshire, England. 51 36N 1 17W. Village S of Oxford. An atomic research station was estab. here in 1947.

Harwich Essex, England. 51 57N 1 17E. Port sit. on estuary of Rs. Stour and Orwell in E. Anglia. Industries inc. engineering, electronics and clothing. It is the second largest passenger port in Britain. There is a ferry service to Denmark, Sweden, Germany and the Netherlands. Pop. (1988E) 15,543.

Haryana India. 29 00N 76 10E. State formed from the Hindi-speaking parts of the Punjab by which it is bounded NW; bounded N by Himachal Pradesh, E by Uttar Pradesh, S and W by Rajasthan. Area, 17,274 sq. m. (44,222 sq. km.). Cap. Chandigarh. Sandy plain in S rising to the foothills of the Himalayas, N. Agric. survives by irrigation of *c*. 40% of the land

area; chief crops are rice, wheat, potatoes, maize, bajra, cereals, sugar, oilseeds, cotton. The main industries are textiles, cement, paper, sugar, tyres, steel tubes, refrigerators, vehicles, tourism, oil refining, tractors, motorcycles, electronic and electrical products, agric. engineering. Pop. (1981c) 12,922,618.

Harz Mountains Federal Republic of Germany/German Democratic Republic. 51 45N 10 30E. A range of mountains sit. between Rs. Elbe and Weser. Rise to 3,747 ft. (1,142 metres) a.s.l. in the Brocken and are heavily wooded with mineral springs, making it a favourite tourist centre.

Hasa, Al Saudi Arabia. An E district sit. on (Persian) Gulf, and a major oil-producing region. Dates, cereals and cotton are grown in the oases. The chief town is Hofuf.

Haslemere Surrey, England. 51 06N 0 43W. Town sit. SSW of Guildford. Mainly residential and site of the annual Dolmetsch Festival of early Eng. music. Pop. (1989E) 14,000.

Haslingden Lancashire, England. 53 43N 2 18W. Town sit. ESE of Blackburn. Manu. textiles, footwear, chemicals and rubber goods. Pop. (1981c) 15,600.

Hassan Karnataka, India. 13 01N 76 06E. Town sit. NW of Bangalore at 3,000 ft. (914 metres) a.s.l. Industries inc. engineering and rice milling. Pop. (1981c) 71,534.

Hasselt Limburg, Belgium. 50 56N 5 20E. Cap. of province ENE of Brussels sit. on Demer R. Administrative and commercial centre. Industries inc. manu. gin, distilleries and laser-optics. Pop. (1989E) 66,000.

Hassi Messaoud Sahara, Algeria. 35 15N 6 35E. Centre of the oil-producing industry, sit. S of Biskra.

Hastings East Sussex, England. 50 51N 0 36E. Resort sit. on Eng. Channel. Its principal occupation is tourism, and there is a small fishing industry. Industries inc. electrical and electronic products, engineering, plastics, clothing and instruments. Pop. (1981C) 74,803.

Hastings North Island, New Zealand. 39 38S 176 51E. Town on the E of N. Island. Centre of a fruit growing district with canneries and 3 meat freezing plants. Manu. cycles and motormowers. Pop. (1986E) 54,909.

Hatay Turkey. Province of S Turkey bounded W by Mediterranean, E and S by Syria. Area, 2,205 sq. m. (5,711 sq. km.). Cap. Antioch. Rises to the Amanos mountains in the W and N and has some iron deposits. Pop. (1985C) 1,002,252.

Hatfield Hertfordshire, England. 51 45N 0 13W. Town sit. ENE of St. Albans. Industries inc. engineering and aircraft production. Pop. (1981C) 29,000.

Hathras Uttar Pradesh, India. 27 36N 78 03E. Town sit. NNE of Agra. Cotton ginning and spinning is carried on. Trades in grain, sugar and cotton. Pop. (1981C) 92,962.

Hatteras, Cape North Carolina, U.S.A. 35 13N 75 32W. A promontory sit. between Pamlico Sound and the Atlantic Ocean. It is often called the 'Graveyard of the Atlantic'.

Hattiesburg Mississippi, U.S.A. 31 19N 89 16W. City sit. SE of Jackson. Manu. machinery, metal products, textiles and fertilizers. Pop. (1980C) 40,829.

Hattingen North Rhine-Westphalia, Federal Republic of Germany. 51 23N 7 10E. Town sit. N of Wuppertal on Ruhr R. Industries inc. steel production and engineering. Pop. (1985E) 85,000.

Haugesund Rogaland, Norway. 59 25N 5 18E. Port in SW, sit. on North Sea. The main occupation is off-shore industries and allied trades. Pop. (1990E) 27,500.

Hauraki Gulf North Island, New Zealand. 36 33S 175 05E. An inlet of Pacific Ocean sit. to E of the Auckland Peninsula. The Great Barrier Island forms a breakwater at the entrance and there are many wooded islands.

Haute-Corse France. 42 30N 9 20E. Dept. in Corse region. Area, 1,802 sq. m. (4,666 sq. km.). Cap. Bastia. Pop. (1982C) 131,574.

Haute-Garonne France. 43 28N 1 30E. Dept. in the Midi-Pyrénées region, bordering on Spain. Area, 2,433 sq. m. (6,301 sq. km.). The principal Rs. are Garonne and Ariège, and in the NE there is the Canal du Midi. It is mainly agric. growing cereals and vines. Cap. Toulouse. Pop. (1982C) 824,501.

Haute-Loire France. Dept. sit. in the Auvergne region. Area, 1,917 sq. m. (4,965 sq. km.). Mainly mountainous, with the Monts du Vivarais in the E, the Monts du Velay in the centre and the Monts de la Margeride in the SW. The principal Rs. are Loire and Allier. Cattle and sheep are raised in the mountains and cereals are grown in the district around Le Puy (the cap.). Pop. (1982C) 205,895.

Haute-Marne France. 48 10N 5 20E. Dept. sit. in the Champagne-Ardenne region. Area, 2,400 sq. m. (6,216 sq. km.). The Plateau de Langres lies in the S. The principal Rs. are Aube, Marne and Meuse, and the Marne-Saône Canal traverses the dept. Cap. Chaumont. Pop. (1982C) 210,670.

Haute-Normandie France. Region in N bounded N by the English Channel and extending from Seine R. estuary in W to the Bray and Vexin areas. Area, 4,733 sq. m. (12,258 sq. km.). Drained by the lower Seine R. Comprises depts. of Eure and Seine-Maritime. Chief cities Rouen, Le Havre, Dieppe and Evreux. Pop. (1987E) 1,699,600.

Hautes-Alpes France. 44 40N 6 30E. Dept. sit. in the Provence-Côte d'Azur region and bordering on Italy in NE. Area, 2,131 sq. m. (5,520 sq. km.). Mainly mountainous with the Cottian Alps in the E and the Massif du Pelvoux in the N. The principal R. is Durance. The main occupation is sheep-rearing. Cap. Gap. Pop. (1982C) 105,070.

Haute-Saône France. Dept. sit. in the Franche-Comté region and bounded by the Vosges Mountains in the NE. Area, 2,063 sq. m. (5,343 sq. km.). The principal Rs. are Saône and Ognon. Mainly agric. growing cereals, vegetables and fruit, esp. cherries which are used to make kirsch. Cap. Vesoul. Pop. (1982C) 222,254.

Haute-Savoie France. 48 00N 6 20E. Dept. in the Rhône-Alpes region bounded by L. Geneva, Switzerland, Italy and Rhône R. Area, 1,695 sq. m. (4,391 sq. km.). Mainly mountainous containing the N French Alps and Mont Blanc rising to 15,781 ft. (4,810 metres) a.s.l. in SE. Mainly agric. growing cereals and vines in the lower districts with livestock rearing in the mountains. Cap. Annecy. Pop. (1982C) 494,505.

Hautes-Pyrénées France. 43 00N 0 10E. Dept. sit. in Midi-Pyrénées region, bordering on Spain. Area, 1,740 sq. m. (4,507 sq. km.). Mountainous in the S with Mont Perdu, 10,995 ft. (3,339 metres) a.s.l. and Vignemale, 10,820 ft. (3,298 metres) a.s.l. The principal R. is Adour. Cap. Tarbes. Pop. (1982C) 227,922.

Haute-Vienne France. 45 50N 1 10E. Dept. sit. in the Limousin region. Area, 2,128 sq. m. (5,512 sq. km.). Mainly hilly containing the Monts du Limousin and the Monts de la Marche. The principal R. is Vienne. Agric. crops are limited to potatoes, rye and buckwheat, and livestock raised. There are kaolin quarries which supply the porcelain industry of Limoges (the cap.). Pop. (1982C) 355,737.

Haut-Rhin France. Dept. in the Alsace region, bordering on Switzerland and Federal Republic of Germany. Area, 1,360 sq. m. (3,523 sq. km.). The principal R. is the Ill, and the dept. is traversed by the Rhône-Rhine Canal. Cereals, vines and fruit are grown, esp. cherries which are used to produce kirsch. There are important textile industries and potash is mined. Cap. Colmar. Pop. (1982C) 650,372.

Hauts-de-Seine France. Dept. in the Île de France region. Area, 68 sq. m. (175 sq. km.). Cap. Nanterre. Pop. (1982C) 1,387,039.

Haut-Zaïre Zaïre. Region in NE producing gold, coffee, cotton and palm products. Area, 194,301 sq. m. (503,239 sq. km.). Cap. Kisangani. Pop. (1981E) 4,541,655.

Havana Havana, Cuba. 23 07N 82 25W. City on N coast. Cap., provincial cap. and major port. Industries inc. manu. tobacco products. The port exports sugar, tobacco and products, coffee, fruit. Pop. (1986E) 2,014,800.

Havant and Waterlooville Hampshire, England. 50 51N 0 59W. District comprising the townships of Havant, Emsworth, Purbrook, Waterlooville, Cowplain and Hayling Island on the Eng. Channel coast. Residential, with some tourist and light industrial development. Pop. (1981C) 116,080.

Havel German Democratic Republic. 52 53N 11 58E. R. rising in a L. in Mecklenburg NW of Neustrelitz and flowing 215 m. (344 km.) S to Berlin then W to enter Elbe R. 6 m. (10 km.) NW of Havelberg. Main trib. Spree, which links it with Oder R. Linked with Elbe R. at Plaue by canal.

Haverford Pennsylvania, U.S.A. 40 01N 75 18W. Town just W of Philadelphia and a residential suburb of that city.

Haverfordwest Dyfed, Wales. 51 49N 4 58W. Market town sit. NNW of Pembroke on W. Cleddau R. The centre of an agric. district. Pop. (1985E) 10,000.

Haverhill Massachusetts, U.S.A. 42 47N 71 05W. City sit. N of Boston on Merrimack R. Centre of a boot and shoe industry and also diverse manu. and commerce. Pop. (1980C) 46,865.

Havering Greater London, England. 51 34N 0 14E. Borough sit. NE of the city centre; residential and industrial. Pop. (1988E) 235,600.

Havre Montana, U.S.A. 48 33N 109 41W. Town in central N Montana on Milk R., SE of Fresno Reservoir. Pop. (1988E) 12,500.

Havre, Le   Seine-Maritime, France. 49 30N 0 06E. Port sit. at mouth of Seine R. on Eng. Channel. It is the second largest port of France and Europe's fifth. Imports inc. oil, cotton, fruit, seafood and coffee. Industries inc. oil refining and shipbuilding. Manu. machinery, wire, textiles, chemicals and food products. Pop. (1982C) 199,156 (agglomeration, 254,595).

Hawaii   U.S.A. 20 00N 155 00W. State comprising a group of islands in mid-Pacific extending NW to SE between 19° and 22° N lat. The Hawaiian Islands, formerly known as the Sandwich Islands, were a group of independent Chiefly states until united as a Kingdom in 1819. European and American interest grew during the 19th century. In 1893 the reigning Queen, Liliuokalan, was deposed and a provisional government formed; in 1894 a Republic was proclaimed. The Republican government requested annexation by the U.S. and this was granted in 1898. In 1900 the islands were constituted as a Territory of Hawaii. Statehood was granted to Hawaii in March 1959. Area, 6,450 sq. m. (16,706 sq. km.). Cap. Honolulu. The 4 main islands are Hawaii, Maui, Oahu, Kauai. Volcanic highlands with fertile valleys and coastal plains built up by alluvial deposits. Some volcanoes are still active, inc. Mauna Loa, 13,675 ft. (4,168 metres) a.s.l. and Kilauea on Hawaii, and Mauna Haleakala on Maui. The main occupations are farming, tourism. Main products sugar-cane, pineapples. Pop. (1980C) 964,691.

Hawes Water   Cumbria, England. Lake sit. ESE of Keswick, used as a reservoir for Manchester.

Hawick   Borders Region, Scotland. 55 25N 2 47W. Town sit. SE of Edinburgh on R. Teviot. Centre of a woollen industry esp. high quality knitwear. It has an important sheep and cattle market. Pop. (1984E) 16,584.

Hawke's Bay   North Island, New Zealand. 39 20S 177 30E. A district sit. between Mahia Peninsula and Cape Kidnappers on the E coast. Area, 4,260 sq. m. (11,033 sq. km.). Agric.

products inc. fruit and fodder crops and sheep and cattle are raised.

Haworth   West Yorkshire, England. 53 49N 1 57W. Village sit. SW of Keighley, and the home of the Brontë family. Pop. (1981C) 3,271.

Hawthorne   California, U.S.A. 33 55N 118 21W. City sit. SW of Los Angeles. Manu. aircraft and electrical equipment. Pop. (1980C) 56,447.

Haydock   Merseyside, England. 53 27N 2 39W. Town ENE of St. Helens. Centre of a coalmining district. Haydock Park racecourse is nearby.

Hayes and Harlington   Greater London, England. Towns forming a district of W Lond., SE of Uxbridge. Industries inc. manu. aircraft, radio and electrical equipment, foodstuffs.

Hayling Island   Hampshire, England. 50 48N 0 58W. Resort sit. off S coast and connected by a road bridge to Havant, NE of Portsmouth.

Hay River   Northwest Territories, Canada. 60 51N 115 44W. Village on SW shore of Great Slave L. at mouth of Hay R. Route centre trading in furs. Airfield, radio and meteorological station.

Hayward   California, U.S.A. 37 40N 112 05W. City sit. SE of Oakland. Industries inc. fruit canning and the manu. of electrical and metal products. It is one of the largest poultry centres in the country. Pop. (1980C) 94,167.

Haywards Heath   West Sussex, England. Town sit. N of Brighton. Mainly residential. Pop. (1981C) 20,079.

Hazaribagh   Bihar, India. Town sit. S of Patna. Trades in oilseeds and rice. Minerals inc. mica. Pop. (1981C) 80,155.

Hazleton   Pennsylvania, U.S.A. 40 58N 75 59W. City sit. NNE of Philadelphia. Manu. textiles, paper, metal goods, electrical equipment, shoes and clothing. Pop. (1980C) 27,318.

Heard Island Australian Antarctic Territory. 53 06S 73 30E. A volcanic island sit. SSE of the Kerguelen Islands, S Indian Ocean. Mainly covered with snow and ice and used only occasionally as a weather station.

Heathrow Greater London, England. 51 28N 0 27W. The site of London Airport sit. 15 m. (24 km.) W of the city centre.

Hebburn Tyne and Wear, England. 54 59N 1 30W. Town sit. E of Newcastle upon Tyne on R. Tyne. Industries inc. shipbuilding, engineering, lead-smelting and coalmining. Also manu. chemicals and electrical equipment.

Hebei (Hopei) China. 39 00N 116 00E. Province in NE. Area, 72,587 sq. m. (188,000 sq. km.). Mountainous in N and W and in S forming part of Great Plain. The principal Rs. are Paiho, Hungho, Lwanho, Hutoho and Shongho. The main agric. products are cereals, cotton, sugar, soy beans, tobacco and fruit. Coal is mined in the Taiang Mountains in the W and in the Yenshan Mountains in the N. Cap. Shijiazhuang. Pop. (1982C) 53,005,875.

Hebrides Scotland. 57 00N 6 30W. Group of *c*. 500 islands off W coast of Scotland. The Outer Hebrides inc. Lewis with Harris, N and S Uist, Benbecula, Barra. The Inner Hebrides inc. Skye, Eigg, Coll, Tyree, Mull, Iona, Staffa, Jura, Islay. Under 1974 reorganization the Hebrides are part of the Island Area of the Western Isles. Mainly infertile and sparsely inhabited. Occupations inc. crofting, fishing, weaving tweed.

Hebron Judaea, Israel. 31 32N 35 06E. Town sit. SSW of Jerusalem in Eshcol Valley, sit. in a rich agric. area. Formerly a West Bank of the R. Jordan possession of Jordan. Industries inc. glass making. Noted as one of the oldest cities, found *c*. 1730 B.C., and home of Abraham who is trad. buried there in the Cave of Machpelah with Sarah, Isaac, Rebecca, Leah and Jacob.

Heckmondwike West Yorkshire, England. 53 42N 1 40W. Market town sit. NE of Huddersfield. Manu. carpets, coach upholstery. Pop. (1988E) 16,775.

Hedmark Norway. 61 45N 11 00E. County of SE Norway bounded E by Sweden. Area, 10,592 sq. m. (27,344 sq. km.). Cities Hamar (cap.), Kongsvinger. Drained by Glomma R. through the Osterdal valley N to S. Undulating in S and rising to mountains N with E end of the Dovrefjell and Rondane mountains. Heavily forested with some pyrites deposits. The main industries are lumbering and associated manu.; stock raising and agric. Pop. (1989E) 186,806.

Heerenveen Friesland, Netherlands. 52 57N 5 55E. Town sit. SSE of Leeuwarden. Manu. coach work for buses, animal foods, bicycles and oscilloscopes. Pop. (1990E) 37,900.

Heerlen Limburg, Netherlands. 50 54N 5 59E. Town sit. ENE of Maastricht, Belgium. Administrative and computer centre. Tourism is important. Pop. (1989E) 94,000.

Hefei (Hofei) Anhui, China. 31 51N 117 17E. Cap. of province sit. WSW of Nanjing. Manu. textiles and chemicals and trades in agric. produce. Pop. (1982C) 815,000.

Hegoumenitsa Thesprotia, Greece. Town sit. WSW of Ioannina on an inlet of the Strait of Corfu. Port for a lumbering and farming area. Industries inc. fishing.

Heidelberg Baden-Württemberg, Federal Republic of Germany. 49 25N 8 43E. Town sit. on Neckar R. Industries inc. printing and publishing, brewing and the manu. of cigars, metal goods and electrical equipment. It has a famous university with a valuable library, as well as many fine medieval buildings. Pop. (1988E) 137,850.

Heidenheim Baden-Württemberg, Federal Republic of Germany. 48 41N 10 09E. Town sit. NNE of Ulm. Manu. machinery, textiles, turbines and furniture. Pop. (1984E) 47,400.

Heilbronn Baden-Württemberg,

Federal Republic of Germany. 49 08N 9 13E. Town and port sit. N of Stuttgart on Neckar R. Industries inc. metal, electrical, paper, printing, salt, chemicals. Agric. inc. wine. Pop. (1985E) 110,094.

Heilongjiang (Heilungkiang) China. 48 00N 126 00E. Province sit. in NE and separated from the U.S.S.R. by Amur R. Area, 181,081 sq. m. (469,000 sq. km.). Mainly mountainous with the Great and Little Khingan Mountains in the N and the Changkwansai Mountains in the S. The principal Rs. are Amur and Sungari. Agric. products inc. timber, furs, sugar-beet, cereals and soy beans. Gold and coal are mined. Cap. Harbin. Pop. (1982C) 32,665,546.

Heilungkiang   *see* **Heilongjiang**

Hejaz   Saudi Arabia. Province sit. between the Gulf of Aqaba and Asir on the Red Sea. Area, 150,000 sq. m. (385,500 sq. km.). Cap. Mecca. A barren area.

Hekla   Iceland. 64 00N 19 39W. An active volcano sit. ESE of Reykjavik. 4,747 ft. (1,447 metres) a.s.l.

Helder, Den   Noord-Holland, Netherlands. 52 54N 4 45E. Port sit. on Marsdiep channel, which separates it from Texel Island. It is an important naval and military station with shipbuilding and fishing industries. Pop. (1984E) 63,826.

Helena   Montana, U.S.A. 46 36N 112 01W. State cap. in the Missouri R. valley of SW Montana. The commercial centre of an agric. and gold mining district. Pop. (1980C) 23,938.

Helensburgh   Strathclyde Region, Scotland. 56 01N 4 44W. Resort sit. NW of Glasgow on Firth of Clyde at mouth of Gare Loch. The home of steam navigation in Europe. Pop. (1981C) 16,478.

Helicon   Boeotia, Greece. 38 18N 22 53E. A mountain range sit. between Gulf of Corinth and L. Copais. Noted as the home of the Muses in Greek mythology.

Heligoland   Schleswig-Holstein, Federal Republic of Germany. 54 12N 7 53E. A small island sit. in North Sea. It consists of a rocky plateau, the Oberland, overlooking a sand bank, the Unterland.

Heligoland Bight   Federal Republic of Germany. A bay sit. between Heligoland and mouths of Rs. Elbe and Weser.

Hellendoorn   Overijssel, Netherlands. Town sit. SE of Zwolle. Manu. textiles. Pop. (1984E) 33,508.

Helmand   Afghánistán. 31 12N 61 34E. R. rising in the Hindu Kush mountains W of Kábul and flowing 680 m. (1,088 km.) SW through Afghánistán then W to the Afghan/Iranian frontier, turning N on the Afghan side to enter the swamps of Hamun-i-Helmand L. The name Helmand is also given to the desert SE of the R.

Helmond   Noord-Brabant, Netherlands. 51 29N 5 40E. Town sit. ENE of Eindhoven. Manu. textiles, cars and food products. Pop. (1988E) 68,000.

Helmstedt   Lower Saxony, Federal Republic of Germany. 52 13N 11 00E. Town sit. E of Brunswick, almost on the frontier between F.R.G. and G.D.R. Manu. machinery, textiles and soap. Lignite is mined. Pop. (1984E) 26,000.

Helsingborg   Malmöhus, Sweden. 56 03N 12 43E. Port in SW sit. on The Sound. Industries inc. copper refining and the manu. of machinery, food processing and fertilizers. Timber, paper and chemicals are exported. Pop. (1989E) 108,000.

Helsingör   Zealand, Denmark. 56 02N 12 37E. Port sit. on The Sound opposite Helsingborg in Sweden. Industries inc. shipbuilding, food processing and brewing. Pop. (1980E) 56,566.

Helsinki   Finland. 60 08N 25 00W. City on Gulf of Finland. Cap., commercial and cultural centre, seat of a university. The city was founded in 1550. Port with a good natural harbour which has to be cleared of ice January–April. Industries inc. food processing, paper making; manu. transport and electrical equipment. Exports timber, iron and steel, chemicals, wood pulp. Pop. (1989E) 491,777.

Helston Cornwall, England. 50 05N 5 16W. Town sit. WSW of Falmouth on the R. Cober. There is some light industry. Noted for its 'Furry' or 'Flora' Dance held annually usually on 8 May. Pop. (1989C) 11,715.

Helvellyn Cumbria, England. 54 32N 3 02W. A mountain sit. between Thirlmere and Ullswater, SE of Keswick. It is one of the highest peaks in the country, 3,118 ft. (950 metres) a.s.l.

Helwan Giza, Egypt. Town and resort, sit. SSE of Cairo with sulphur springs. Manu. iron, steel and cement. Pop. 55,000.

Hemel Hempstead Hertfordshire, England. 51 46N 0 28W. Town sit. W of St. Albans on R. Gade. Manu. scientific equipment, aircraft systems, printing, photographic and electronic equipment, paper, office equipment and computers. Pop. (1986E) 81,000.

Hempstead New York, U.S.A. 40 42N 73 37W. Town and resort on W Long Island. Mainly residential with some manu. Pop. (1980C) 40,404.

Henan (Honan) China. 34 00N 114 00E. Province sit. in E central China. Area, 64,462 sq. m. (167,000 sq. km.). Mountainous in SW with the Funiu Shan mountains stretching E. The principal R. is Yellow. It is fertile in E, the chief products being cereals, cotton, soy beans and fruit. There are deposits of coal, iron, sulphur and salt-petre. A very densely populated province with an expanding industry. Cap. Zhengzhou. Pop. (1982C) 74,422,739.

Hendaye Pyrénées–Atlantiques, France. 43 23N 1 47W. Border town across Bidassoa R. from Spain. Pop. (1982C) 11,112.

Hendon Greater London, England.

51 35N 0 13W. District of NW Lond. Residential and industrial.

Hengelo Overijssel, Netherlands. 52 15N 6 45E. Town sit. NW of Enschede on Twente Canal. Industries inc. railway engineering, brewing, and the manu. of diesel engines, textiles and electrical equipment. Pop. (1984E) 76,855.

Henley-on-Thames Oxfordshire, England. 51 32N 0 56W. Town sit. NNE of Reading on R. Thames. Noted for the Henley Royal Regatta held annually. Pop. (1984E) 10,976.

Herát Herát, Afghánistán. 34 20N 62 07E. Cap. of province in the NW sit. on Hari Rud R. at 3,000 ft. (914 metres) a.s.l. It is built on an artificial mount and enclosed by a wall. Manu. carpets and textiles.

Hérault France. 43 34N 3 15E. Dept. in the Languedoc–Roussillon region bordering on the Gulf of Lions. Area, 2,360 sq. m. (6,113 sq. km.). There are three coastal lagoons separated from the sea by a narrow strip of land, while inland the ground rises to the Garrigues in the NE and the Cévennes in the N and W. The principal Rs. are Hérault, Aude and Orb. It is one of the most important wine-producing areas in France and also grows cereals, fruit and olives. Coal, bauxite, iron and salt are found. Cap. Montpellier. Pop. (1982C) 706,499.

Herculaneum Campania, Italy. Excavated Roman town sit. E of Naples on NW slope of Mount Vesuvius. Originally buried by the eruption of 79 A.D.

Heredia Costa Rica. 10 00N 84 07W. (i) Province of N Costa Rica bounded N by San Juan R. and Nicaragua. Area, 1,100 sq. m. (2,816 sq. km.). Chief cities Heredia, Santo Domingo, San Isidro, Barba. Mainly lowland rising to the volcanic section of the Central Cordillera plateau where live the bulk of the pop. Mainly agric. Crossed by the Inter-American Highway. Pop. (1984E) 195,389. (ii) Town sit. NNW of San José on the Inter-American Highway. Provincial cap. and commercial centre of a coffee growing area. Industries inc. coffee processing; manu. soap, matches. Pop. (1984E) 20,867.

Hereford Hereford and Worcester, England. 52 04N 2 43W. Cathedral city sit. NW of Gloucester on the R. Wye. Service centre of an agric. area. Industries inc. cider and tourism. Pop. (1989E) 51,000.

Hereford and Worcester England. 52 14N 1 42W. The county is bounded on N by Shropshire, NE by West Midlands and Staffordshire, E by Warwickshire, S by Gloucestershire and W by Gwent and Powys. Agric. (particularly cattle breeding) and horticulture (fruit, vegetables and hops) are important industries. The county attracts tourism and has the Valley of the R. Wye, famous for its salmon, the Malvern Hills, the Black Mountains and the Vale of Evesham within the boundaries. Industry, particularly the newer technologies, is also expanding. Under 1974 reorganization the county consists of the districts of Wyre Forest, Bromsgrove, Redditch, Leominster, South Herefordshire, Hereford, Malvern Hills, Worcester and Wychavon. Pop. (1989E) 671,000.

Herefordshire *see* **Hereford and Worcester**

Herford North Rhine-Westphalia, Federal Republic of Germany. 52 06N 8 40E. Town sit. NE of Bielefeld. Manu. machinery, carpets, textiles, furniture and chocolate. Pop. (1989E) 62,500.

Herm Channel Islands, United Kingdom. 49 30N 2 28W. Island just E of Guernsey. Area, 320 acres. Tourist resort. Pop. (1981C) 96.

Hermannstadt *see* **Sibiu**

Hermon, Mount Syria/Lebanon. 33 26N 35 51E. A snow-capped mountain sit. WSW of Damascus. 9,232 ft. (2,814 metres) a.s.l.

Hermosillo Sonora, Mexico. 29 10N 111 00W. State cap. sit. near confluence of Rs. Sonora and Zanjón, in NW Mex. The commercial centre of an agric. area producing cereals, cotton, vegetables and fruit. Gold, silver and copper are mined. Pop. (1980C) 340,779.

Hermoupolis Cyclades, Greece. 37 26N 24 56E. Town sit. ESE of Athens on E coast of Syros island. Major port, industrial and commercial centre. Industries inc. shipbuilding, textiles; manu. glass and leather.

Herne North Rhine-Westphalia, Federal Republic of Germany. 51 32N 7 13E. Town sit. NE of Düsseldorf on Rhine-Herne Canal. A coalmining centre with iron and steel works, textile mills and chemical plants. Pop. (1984E) 174,800.

Herne Bay Kent, England. 51 23N 1 08E. Resort sit. NE of Canterbury on Thames estuary. Pop. (1985E) 27,799.

Herrera Panama. Province of S central Panama on the Pacific coast, bounded N by Santa María R. Main cities Chitré (cap.), Ocú, Parita. The main occupations are farming, esp. livestock and mining, esp. gold. Pop. (1980C) 81,866.

Herstal Liège, Belgium. 50 40N 5 38E. Town sit. NE of Liège on Meuse R. Manu. firearms and electrical equipment. Pop. (1989E) 36,343.

Hertford Hertfordshire, England. 51 48N 0 05W. Town sit. N of Lond. on R. Lea. Industries inc. printing, brewing and the manu. of brushes. Pop. (1989E) 25,000.

Hertfordshire England. 51 51N 0 05W. Inland county bounded on N by Cambridgeshire, E by Essex, S by Greater London, W by Buckinghamshire and Bedfordshire. Area, 632 sq. m. (1,618 sq. km.). Chief towns, Hertford, St. Albans, Hemel Hempstead, Bishop's Stortford, Royston, Ware, Rickmansworth, Potters Bar, Borehamwood, Watford, Welwyn Garden City, Hatfield, Stevenage, Hitchin and Letchworth Garden City. Agric. and horticulture are important including wheat and cattle rearing. Manu. inc. brewing, papermaking, printing, computers, electronic equipment, telecommunications, aerospace, pharmaceuticals, light industry and bricks. Under 1974 reorganization the county consists of the districts of North Hertfordshire, Stevenage, East Hertfordshire, Dacorum, St. Albans, Welwyn Hatfield, Broxbourne, Three Rivers, Watford and Hertsmere. Pop. (1988E) 986,000.

Herzegovina Yugoslavia. 44 00N 18 00E. The s region of the constituent rep. of Bosnia-Herzegovina. Area, *c.* 3,500 sq. m. (5,600 sq. km.). Chief city Mostar. Mountainous and drained by the Neretva R.

Hessen Federal Republic of Germany. 50 57N 9 20E. *Land* of the F.R.G. bounded w by Rhine Rs., s by Baden-Württemberg, N by North Rhine-Westphalia. Area, 8,153 sq. m. (21,111 sq. km.). Chief towns Wiesbaden (cap.), Homberg, Ems, Kassel, Mainz, Frankfurt, Giessen, Darmstadt. Rolling countryside, heavily forested with iron, coal and manganese deposits. Main industries agric., viticulture, lumbering, mining; manu. industries are concentrated in Frankfurt, Darmstadt, Giessen, Kassel, Dillenberg, Fulda, Wetzlar, inc. metallurgy, engineering, chemicals. Pop. (1988E) 5,568,892.

Heston and Isleworth Greater London, England. 51 29N 0 22W. District of w Lond., near Heathrow Airport. It is mainly residential.

Hetton-le-Hole Tyne and Wear, England. 54 59N 1 26W. Town sit. NE of Durham. Food processing, printing, engineering, electronics and furniture. Pop. (1981C) 15,720.

Heves Hungary. 47 50N 20 00E. County of N Hungary. Area, 1,404 sq. m. (3,637 sq. km.). Cap. Eger. Heavily forested in N and rising to the Matra mountains N and Alföld SE. Agric. and livestock rearing area. Pop. (1989E) 336,000.

Hexham Northumberland, England. 54 58N 2 06W. Market town sit. in N on R. Tyne. Industries inc. chipboard manu. Pop. (1989E) 10,000.

Heywood Greater Manchester, England. 53 36N 2 13W. Town sit. N of Manchester. Industries inc. ware-

housing and distribution, textiles, dyeing and finishing, engineering. Pop. (1981C) 29,686.

Hialeah Florida, U.S.A. 25 49N 80 17W. Residential suburb just NW of Miami. Noted for its racecourse. Pop. (1980C) 145,254.

Hidalgo Mexico. 20 30N 99 10W. A mountainous central state with the Sierra Madre Oriental in the N and E. Area, 8,101 sq. m. (20,987 sq. km.). One of the leading mining states producing about two-thirds of Mex. silver. Gold, copper and lead are also mined. Agric. products inc. cereals, sugar-cane, cotton, coffee and tobacco. Cap. Pachuca. Pop. (1980C) 1,547,493.

Hidalgo del Parral Chihuahua, Mexico. 26 56N 105 40W. Town sit. in N Mex., SSE of Chihuahua at 6,400 ft. (1,950 metres) a.s.l. The principal industry is silver mining.

Hierro Canary Islands, Atlantic Ocean. 27 57N 17 56W. Island SW of Gomera, smallest and westernmost of the Canary group. Area, 107 sq. m. (274 sq. km.). Chief town Valverde. Volcanic, rising to 4,300 ft. (1,311 metres). Heavily wooded with little water. Main products are cereals, fruit, vegetables, wine and dairy products.

Higashiosaka (formerly Fuse) Honshu, Japan. 34 39N 135 35E. Town just E of Osaka. Manu. machinery, chemicals, textiles, rubber and leather goods. Pop. (1980C) 499,000.

Higham Ferrers Northamptonshire, England. 52 18N 0 36W. Town sit. E of Northampton on R. Nene. Industries inc. leather dressing. Manu. footwear. Pop. (1989E) 5,500.

Highbury Greater London, England. 51 33N 0 06W. District of N Lond. inc. Highbury Fields, a public green, and adjoining 18th cent. properties.

Highgate Greater London, England. 51 34N 0 08W. Village forming a district of N Lond. next to Hampstead Heath. Picturesque, residential, with many artistic and literary associations. Highgate cemetery is the burial place of Karl Marx.

Highland Region of Scotland. 57 30N 4 50W. Area, 9,805 sq. m. (25,391 sq. km.) comprising the districts of Caithness, Sutherland, Ross and Cromarty, Skye and Lochalsh, Lochaber, Inverness, Badenoch and Strathspey, Nairn. The main sources of employment derive from farming, fishing and forestry, but there is a wide range of manufacturing industries and in particular electronics and light industry. Tourism and service industries are also of increasing importance. Specific industries are the generation of nuclear electricity at Dounreay, Caithness, aluminium smelting at Fort William and Kinlochleven, oil platform construction at Nigg and Ardersier, other oil related industries including repair and maintenance of drilling rigs in the Cromarty Firth, timber processing in Inverness and distilling and winter sports in the Spey Valley. Fish farming is also a recent growth industry. An Enterprise Zone has been established at Invergordon. Pop. (1989E) 205,000.

Highlands Scotland. Mountainous area sit. N and NW of a line between Stonehaven and Dumbarton. Trad. the predominantly Gaelic part of Scotland, in language and customs.

High Point North Carolina, U.S.A. 35 58N 80 01W. City sit. W of Durham. Manu. inc. furniture, textiles, chemicals, paint, hosiery and timber products. Pop. (1980C) 63,380.

High Wycombe Buckinghamshire, England. 51 38N 0 46W. Town sit. WSW of Watford on R. Wye. Mainly residential. Industries inc. the manu. of furniture, paper, precision instruments and clothing. Pop. (1981C) 60,500.

Hildesheim Lower Saxony, Federal Republic of Germany. 52 09N 9 57E. Town in NE, sit. on Innerste R. Manu. textiles, machinery, carpets, glass and cosmetics. Contains an 11th cent. cathedral. Pop. (1989E) 109,076.

Hilla Hilla, Iraq. 32 28N 44 29E. Cap. of governorate sit. S of Baghdad on Euphrates R. Manu. textiles and leather goods and trades in cereals and dates. It was built with materials taken from anc. Babylon nearby. Pop. (1980E) 130,000.

Hillingdon Greater London, England. 51 32N 0 27W. Borough to W of Lond.

Hilo Hawaii, U.S.A. 19 43N 155 05W. Cap. of Hawaii island and port sit. on E coast of Hilo Bay. Industries inc. tourism and fruit processing. Exports inc. flowers, sugar and fruit. Pop. (1980C) 35,269.

Hilversum Noord-Holland, Netherlands. 52 14N 5 10E. Town and resort sit. SE of Amsterdam. It is the chief centre of radio and television. Industries inc. printing and publishing and the manu. of electrical equipment. Pop. (1989E) 84,983.

Himachal Pradesh India. 31 30N 77 00E. State formed from small hill states of the NE Punjab. Area, 21,747 sq. m. (55,673 sq. km.). Cap. Simla. Mountainous, rising to the Himalayas, and heavily forested. Forests occupy 39% of the state. The main industries inc. lumbering, saw-milling, turpentine, breweries, fertilizers, electronics, cement and rock salt extracting. Fruit production is important and minerals inc. rock salt, slate, gypsum, limestone, baryte, dolomite and pyrite. There are some handicraft industries. Pop. (1981C) 4,280,818.

Himalaya Central Asia. 28 30N 84 00E. Mountain system extending W to E along N border of the Indian Plain between Indus and Brahmaputra Rs. and inc. many parallel ranges, the N ranges being higher. Length *c.* 1,500 m. (2,400 km.) from the Pamir range in the U.S.S.R. (NW). For most of this length there are three main ranges (Outer, Middle and Inner), but in Kashmir there are five, Lesser and Great Himalayas, Zaskar, Ladakh and Karakoram ranges. The Outer range in India rises to 12,000 ft. (3,658 metres) a.s.l., with Simla and Darjeeling 'hill stations' at *c.* 8,000 ft. (2,438 metres). There are numerous peaks over 25,000 ft. (7,620 metres). Everest

29,028 ft. (8,848 metres), in Nepál, Godwin-Austen, 28,250 ft. (8,611 metres), in Kashmir (Karakoram range), Kinchinjunga, 28,146 ft. (8,579 metres), Nanga Parbat, 26,629 ft. (8,117 metres), Nanda Devi, 25,643 ft. (7,816 metres), Annapurna, 26,502 ft. (8,078 metres) geologically new mountains; the Himalaya ranges were formed later than the Rs. which flow on either side and which have cut deep gorges through them as the earth was pushed up.

Himeji   Honshu, Japan. 34 49N 134 42E. Town sit. WNW of Osaka. Noted for its 'White Heron' castle. Manu. steel, chemicals, electricity and machinery. Pop. (1989E) 453,963.

Hinckley   Leicestershire, England. 52 33N 1 21W. Town sit. NNE of Coventry. Industries inc. mechanical and electrical engineering. Manu. hosiery, footwear and cardboard boxes. Pop. (1987E) 25,710.

Hindu Kush   Central Asia. 36 00N 71 30E. Extension of Himalayas on W side of Indus R. and extending *c.* 360 m. (576 km.) WSW from the Pamir plateau to the Bamian valley. Source of the Oxus and Helmand Rs. Rises to 25,420 ft. (7,748 metres) in Tirach Mir.

Hirosaki   Honshu, Japan. 40 35N 140 28E. City sit. NNE of Sendai. Industries inc. woodworking, soy bean and rice processing, general food processing and the manu. of textiles. It is also famous for its green lacquer ware. Pop. (1985E) 176,082.

Hiroshima   Honshu, Japan. 34 24N 132 27E. Town and port in SW sit. on Ota R. delta. Industries inc. shipbuilding, engineering, textile mills and the manu. of vehicles, machinery and rubber products. First atomic bomb dropped on the town in 1945. Pop. (1988E) 1,043,000.

Hispaniola   *see* **Dominican Republic** *and* **Haiti**

Hissar   Haryana, India. 29 09N 75 44E. Town sit. WNW of Delhi on a branch of the W. Jumna Canal. It has a large govt. livestock farm and trades

in cattle, cotton and grain. Pop. (1981C) 137,369.

Hitchin   Hertfordshire, England. 51 57N 0 17W. Town sit. N of London. Industries inc. engineering, flour milling and parchment making. Pop. (1985E) 30,000.

Hjälmar, Lake   Sweden. Sit. SW of L. Mälar into which it is drained by Eskilstuna R. Area, 190 sq. m. (492 sq. km.).

Ho   Eastern Province, Ghana. 6 36N 0 28E. Town sit. NE of Akuse. Route junction, producing cacao, palm oil, kernels, cotton. Pop. (1984C) 37,777.

Hobart   Tasmania, Australia. 42 53S 147 19E. City on Derwent R. estuary below Mount Wellington. State cap. and sea port. Industries inc. zinc, paper, chocolate, woollen goods and textiles. Pop. (1986C) 180,250.

Hobbs   New Mexico, U.S.A. 32 42N 103 08W. City on Llano Estacado in SE of state near border with Texas. Industries inc. oil refining. Pop. (1980C) 29,228.

Hoboken   Antwerp, Belgium. 51 11N 4 20E. A residential suburb of Antwerp sit. on Scheldt R. Industries inc. shipbuilding, metal working and wool combing.

Hoboken   New Jersey, U.S.A. 40 45N 74 03W. Port sit. on Hudson R. opposite New York. Industries inc. shipbuilding, and the manu. of textiles, chemicals, electrical equipment, scientific instruments and furniture. Pop. (1980C) 42,460.

Hochheim   Hessen, Federal Republic of Germany. Town sit. SE of Wiesbaden on Main R. Noted for wine production.

Ho Chi Minh City   *see* **Saigon**

Hoddesdon   Hertfordshire, England. 51 45N 0 01W. Town sit. SE of Hertford. Industries inc. market gardening and the manu. of pharmaceutical products and electrical equipment. Pop. (1981C) 38,111.

Hodeida   Yemen Arab Republic. Port sit. on Red Sea, SW of San'a. Exports inc. coffee, skins, senna, myrrh and sesame. Pop. (1981C) 126,386.

Hódmezövásárhely   Csongrád, Hungary. 46 25N 20 20E. Cap. of county sit. NE of Szeged near Tisza R. Centre of a rich agric. district with manu. of agric. implements, pottery, textiles and wood products. Pop. (1980E) 54,444.

Hof   Bavaria, Federal Republic of Germany. 50 18N 11 55E. Town sit. N of Munich on Saale R. Manu. machinery, textiles and chemicals. Pop. (1984E) 51,800.

Hofei   *see* **Hefei**

Hofuf   Nejd, Saudi Arabia. Town sit. SSW of Dhahran. Trades in cereals, dates and fruit. Pop. (1977E) 102,000.

Hohenlimburg   North Rhine-Westphalia, Federal Republic of Germany. 51 21N 7 35E. Town sit. S of Dortmund. Industries inc. steel mills and the manu. of wire.

Hohhot (Huhehot)   Inner Mongolia, China. 40 49N 111 37E. Cap. of Autonomous Region sit. WNW of Beijing. Industries inc. iron and steel works, textile and flour mills, and the manu. of rugs, bricks, tiles and dairy products. Pop. (1982C) 747,000.

Hokkaido   Japan. 44 00N 143 00E. Most N island of Japan and second largest, bounded S by Honshu across the Tsugaru Strait. Area, 34,276 sq. m. (88,775 sq. km.). Chief towns Sapporo (cap.), Hakodate, Muroran. Mountainous, rising to 7,000 ft. (2,134 metres) a.s.l. Agric. areas produce rice in S, oats, beans and potatoes in colder N. Fisheries are important, esp. salmon, cod, sardines. Deposits of coal, sulphur and iron support metallurgical and manu. industries.

Holbeach   Lincolnshire, England. 52 50N 0 03E. Town sit. E of Spalding. Centre of an agric. and bulb-growing district. Pop. (1980E) 7,500.

Holborn   Greater London, England.

51 31N 0 07W. District NW of City of Lond., mainly commercial, inc. Bloomsbury and the Brit. Museum and some 15th and 16th cent. Inns of Court, inc. Gray's Inn, Lincoln's Inn, Staple Inn.

Holderness   Humberside, England. 53 51N 0 22W. A peninsula sit. between North Sea and Humber estuary.

Holguín   Oriente, Cuba. 20 53N 76 15W. Town sit. NNW of Santiago de Cuba. A commercial centre with manu. of furniture and tiles. Tobacco, coffee, sugar and maize are exported from the port of Gibara, to the NNE.

Holland   *see* **Netherlands**

Holland, Hook of   *see* **Hook of Holland**

Hollywood   California, U.S.A. 34 06N 118 20W. A suburb of Los Angeles. Noted principally for its film industry.

Hollywood   Florida, U.S.A. 26 00N 80 09W. Resort sit. N of Miami. Manu. inc. cement and electronics. Pop. (1980C) 121,323.

Holyhead   Isle of Anglesey, Gwynedd, Wales. 53 19N 4 38W. Port and resort sit. on N side of Holy Island. It has a car and lorry roll-on/off ferry service to Ireland and a container terminal. Industries inc. aluminium, fish processing, toys and switchgear. Pop. (1988E) 13,000.

Holy Island   (Lindisfarne), Northumberland, England. 55 41N 1 48W. Island in North Sea SE of Berwick on Tweed and connected to the E coast of Northumberland by a causeway. Area, 3 m. by 1¼ m. (5 km. by 2 km.). Site of Lindisfarne priory which developed from a settlement by Columban monks and where the Lindisfarne Gospels were made.

Holy Island (Inishcaltra)   Clare, Ireland. An island sit. in Lough Derg. It contains the ruins of four 7th cent. churches.

Holy Island   Scotland. A rocky island sit. in Firth of Clyde, with a lighthouse.

Holy (Holyhead) Island   Gwynedd, Wales. 53 17N 4 37W. An island sit. off W coast and joined to Anglesey (island) which itself is joined to the mainland by a road and rail bridge.

Holyoke   Massachusetts, U.S.A. 42 12N 72 37W. City sit. WSW of Boston on Connecticut R. Manu. paper, textiles, machinery, steel products and electrical equipment. Pop. (1980C) 44,678.

Holy See   *see* **Vatican City**

Holywell   Clwyd, Wales. 53 17N 3 13W. Town sit. ESE of Rhyl. Industries inc. general light manufacturing. Pop. (1989E) 9,000.

Homberg   North Rhine-Westphalia, Federal Republic of Germany. Town and port sit. on Rhine R. opposite Duisburg. Industries inc. coalmining, iron foundries and textile mills.

Homburg vor der Höhe   Hessen, Federal Republic of Germany. A spa sit. N of Frankfurt at foot of Taunus Mountains. Manu. machinery, leather goods, textiles and biscuits. Pop. (1984E) 41,800.

Home Island   Cocos Islands, Australia. Main island of Cocos group in Indian Ocean WSW of Singapore. Length 1¼m. (2 km.) and containing the main settlement.

Homs   Tripolitania, Libya. Town sit. ESE of Tripoli on Mediterranean. Market and tourist centre noted for nearby remains of the city of Leptis Magna. Industries inc. olive oil pressing, processing esparto grass.

Homs   Homs, Syria. 34 44N 36 43E. Cap. of district sit. on Orontes R. A transportation centre with industries inc. oil refining, sugar-beet processing, and the manu. of textiles and jewellery. Pop. (1981C) 346,871.

Honan   *see* **Henan**

Honduras   Central America. 14 40N 86 30W. Rep. bounded N by Caribbean, E and NE by Nicaragua, W by Guatemala, SW by Salvador, S by the Pacific. Area, 44,411 sq. m. (113,692

sq. km.). Chief towns Tegucigalpa (cap.), Puerto Cortés, San Pedro Sula. Mountainous, with a narrow plain on the Pacific (40 m./64 km.) and Atlantic (400 m./640 km.) coasts. Highland plateau extends NW to SE across the country, broken by the plain of Comayagua, with R. valleys extending from it N and S—Humuya R. N to the Atlantic, Goascoran R. S to the Pacific. About 31% is forested with some agric. on the coastal plains; main products bananas, coffee, coconuts, maize, tobacco, sugar-cane, timber, rubber and resins. The main industries are mining, esp. silver, lead and zinc; lumbering, esp. mahogany and hardwoods; manu. panama hats, mainly in Copán and Santa Bárbara. Pop. (1986E) 4·3m.

Honduras, British   *see* **Belize**

Hong Kong   East Asia. 25 15N 114 10E. British crown colony, partly held on lease, adjoining Kwangtung province, China, SE of Canton. Hong Kong Island and the Kowloon peninsula are British-owned; the New Territories on the mainland N of the Kowloon peninsula (area, 359 sq. m./919 sq. km.) are held on lease until 1997. Total area, 405 sq. m. (1,049 sq. km.). Cap. Victoria (commonly known as Hong Kong city). Hong Kong Island is rocky and mountainous, rising to 1,823 ft. (556 metres) a.s.l., infertile. There is some granite quarrying, fishing and livestock farming. Victoria is on N coast facing Kowloon across a strait 1 m. (1.5 km.) wide which encloses Hong Kong Harbour, a major deep-water anchorage which, owing to the density of pop., is permanently inhabited by an extensive house-boat settlement. Victoria is an important commercial centre and headquarters of numerous banks and trading corporations. The port handles an extensive transit trade with China, inc. coal, cotton, iron, steel and manu. goods. There is an intern. airport. The mainland area is intensively agric., main products rice, vegetables, nuts, fruit, sugar-cane, pigs and poultry. Pop. (1987C) 5,613,000.

Honiara   Solomon Islands, South Pacific. 9 28S 159 57E. Cap. on NW coast of island of Guadalcanal, developed since 1945. Pop. (1986E) 30,499.

Honiton Devon, England. 50 48N 3 13W. Town in sw sit. on R. Otter. Tourist centre and noted for lace-making and pottery. Pop. (1981C) 6,627.

Honolulu Hawaii, U.S.A. 21 19N 157 59W. State cap. and port sit. on s coast of Oahu. The main industries are agriculture, horticulture and tourism. Fruit and sugar are exported. Pop. (1980C) 365,114.

Honshu Japan. 36 00N 138 00E. The largest of the 4 main islands and containing most of the principal towns. Area, 88,919 sq. m. (230,211 sq. km.).

Hoogeveen Drenthe, Netherlands. 52 44N 6 30E. Town sit. s of Groningen. Industries inc. food canning. Pop. (1989E) 46,000.

Hoogezand Groningen, Netherlands. 53 11N 6 45E. Town sit. ESE of Groningen. Manu. inc. machinery and tyres. Pop. (1984E) 34,980.

Hooghly (Hugli) Chinsura West Bengal, India. Town sit. N of Calcutta on Hooghly R. The chief occupation is rice milling and banana cultivation. Pop. (1981C) 128,918.

Hooghly River India. 21 56N 88 04E. R. forming most w channel of Ganges, flowing 64 m. (102 km.) s from the main stream at Nadia to Calcutta and a further 81 m. (130 km.) to enter the Bay of Bengal by a delta. Its course is kept open by dredging. Width at the mouth, 15 m. (24 km.). There is a strong tidal bore.

Hook of Holland Zuid-Holland, Netherlands. 51 59N 4 09E. Port sit. WNW of Rotterdam at mouth of the New Waterway. It has a ferry service to U.K.

Hoover Dam Arizona/Nevada, U.S.A. 36 00N 114 45W. A dam on Colorado R. in Black Canyon sit. SE of Las Vegas. Completed in 1936 it is 726 ft. (221 metres) high and 1,282 ft. (391 metres) long.

Hopei *see* **Hebei**

Hopen Svalbard, Norway. Island SE of Spitzbergen in the Barents Sea. Area, 23 sq. m. (60 sq. km.). Rises to 1,198 ft. (365 metres) a.s.l. There is a meteorological station.

Hopkinsville Kentucky, U.S.A. 36 52N 87 29W. City in sw Kentucky; route and distribution centre. Pop. (1980C) 27,318.

Hordaland Norway. 60 25N 6 45E. County of sw Norway. Area, 6,126 sq. m. (15,683 sq. km.). Chief centres Bergen (cap.), Odda. Inc. the area around mouth of Hardangerfjord and islands off the coast. The main occupations are fishing and agric. Pop. (1980C) 391,463.

Horn, Cape Tierra del Fuego, Chile. 56 00S 67 16W. A rocky headland sit. at extreme s end of Horn Island, rising to 1,390 ft. (424 metres) a.s.l.

Horncastle Lincolnshire, England. 53 13N 0 07W. Town sit. E of Lincoln on R. Bain. Industries are tanning, brewing and malting. Formerly noted for its horse fair. Pop. (1985E) 4,207.

Hornchurch Greater London, England. 51 33N 0 12E. Town sit. ENE of London. Residential. *Now* Havering.

Hornsea Humberside, England. 53 64N 0 10W. Resort on N. Sea coast. Nearby is Hornsea Mere, a large bird sanctuary. Pop. (1980E) 7,250.

Hornsey Greater London, England. 51 35N 0 07W. District of N London, mainly residential.

Horsens Denmark. 55 52N 9 52E. Port in E central Den. on Horsens Fjord. Industries inc. shipbuilding, iron works, and the manu. of textiles and tobacco products. Dairy produce is exported. Pop. (1989E) 55,130.

Horsham West Sussex, England. 51 04N 0 21W. Town in s, sit. on R. Arun. Industries inc. engineering, brick making, brewing and the manu. of electronic and pharmaceutical products. Pop. (1988E) 40,000.

Horta Faial, Azores. 38 32N 28 38W. Town sit. WSW of Angra do Heroísmo on SE shore of Faial opposite Pico Island. Port and cable station. Industries inc. ship repairing, whaling. Exports cattle, dairy produce, fish, wine, oranges. Pop. (1980E) 6,200.

Horwich Greater Manchester, England. 53 35N 2 33W. Town sit. NE of Wigan. Industries inc. railway engineering, stone quarrying, paper and cotton mills.

Hoshiarpur Punjab, India. Town sit. NW of Simla. Manu. furniture, rosin and turpentine. Trades in grain and sugar. Pop. (1981C) 85,648.

Hot Springs Arkansas, U.S.A. 34 30N 93 03W. Resort and national park in W centre of state in the Ozark Hills. Noted for its thermal springs. Pop. (1980C) 35,781.

Houghton-le-Spring Tyne and Wear, England. 54 51N 1 28W. Town sit. SW of Sunderland. Centre of a coalmining district. Pop. (1984E) 34,000.

Hounslow Greater London, England. 51 29N 0 22W. Area forming a suburb of W London on Great West Road. Now adjacent to Heathrow Airport. Varied manu. industries. Pop. (1988E) 190,700.

Houston Texas, U.S.A. 29 46N 95 22W. City and port sit. at head of Houston Ship Channel *c.* 20 m. (32 km.) from the Gulf of Mexico. Corporate, financial and commercial centre. Johnson Space Center is here. Manu. chemicals, refined petroleum, oilfield equipment, milled rice. Exports inc. petroleum, chemicals, grain. Pop. (1980C) 1,505,138.

Hove East Sussex, England. 50 49N 0 10W. Resort sit. on S coast. Pop. (1987E) 90,431.

Howland and Baker Islands Central Pacific. 0 48N 176 38W. Island with area of 1 sq. m. (2.5 sq. km.). Formerly worked for guano. Now uninhabited.

Howrah (Haora) West Bengal,

India. 22 37N 88 27E. Suburb of Calcutta on opposite bank (W) of Hooghly R. Railway terminus. Industries inc. processing jute, cotton, engineering and iron foundries. Pop. (1981C) 744,429.

Howth Dublin, Ireland. 53 21N 6 00W. Resort sit. ENE of Dublin, and a residential suburb of that city.

Hoxton Greater London, England. 51 32N 0 04W. District of E central London. Industrial and residential.

Hoylake Merseyside, England. 53 23N 3 11W. Resort sit. W of Birkenhead. Pop. (1980E) 32,000.

Hradec Králové Východočeský, Czechoslovakia. 50 12N 15 50E. Cap. of region sit. ENE of Prague at confluence of Rs. Orlice and Elbe. Manu. machinery, chemicals, textiles, photographic equipment and musical instruments. Pop. (1989E) 100,000.

Hsinchu Taiwan. 24 48N 120 58E. Town sit. SW of Taipei on W coast. A science-based industrial park was opened in 1980. Industries inc. manu. synthetic oil, fertilizer, glass, paper, timber products. Pop. (1985E) 299,807.

Huacho Peru. 11 10S 77 35W. Pacific port in central Peru. Commercial centre for fertile Huaura R. valley, which produces cotton and sugar.

Huahine Society Islands, South Pacific. 16 45S 151 00W. Island sit. E of Raiatea in the Leeward group. Chief town, Fare. Area, 28 sq. m. (73 sq. km.), comprising 2 islands joined by an isthmus. Mountainous, rising to 2,230 ft. (679 metres) a.s.l. Fertile. Produces copra.

Hualien Taiwan. 23 58N 121 36E. Town sit. S of Ilan on E coast. Manu. inc. chemicals, non-ferrous metal refining, marble dolomite, Taiwan jade. Pop. (1985E) 1,004,246.

Huallaga Peru. 5 07S 75 30W. R. rising in the Cerro de Pasco and flowing 650 m. (1,040 km.) N along the E flank of the central Cordillera, cutting through the range by Chasuta gorge and joining Marañón R.

Huambo Benguela, Angola. 12 42S
15 54E. Formerly Nova Lisboa. Town
sit. E of Lobito on the Bié Plateau, at
5,560 ft. (1,711 metres). District cap.
Industries inc. railway repair shops,
processing agric. produce. Trade and
distribution centre, handling maize,
wheat, rice, fruit, hides and skins.
Pop. (1980E) 80,000.

Huancavelica Peru. 12 50S 75
00W. (i) Dept. of S central Peru. Area,
8,300 sq. m. (22,871 sq. km.). Chief
cities Huancavelica, Acobamba,
Pampas. Mountainous, rising to the
Cordillera Central and Cordillera
Occidental on either side of Mantaro
R. Main occupations farming and
mining. Main products cereals, sugar-
cane, livestock, silver, lead, mercury.
Pop. (1988E) 372,900. (ii) Town sit.
ESE of Lima on Huancavelica R. Dept.
and provincial cap. Industries inc.
mining and smelting mercury, silver
and lead, flour milling. Trade in wool.
Pop. (1988E) 24,700.

Huancayo Junín, Peru. 12 04S 75
14W. Cap. of dept. in W central Peru,
sit. on Mantaro R. Manu. textiles.
Trades in wheat, maize and alfalfa.
Pop. (1988E) 199,200.

Huánuco Huánuco, Peru. Cap. of
dept. sit. in W central Peru, on upper
Huallago R. Sit. in a rich agric. and
mining district growing sugar, cotton,
coffee and cacao. Pop. (1988E)
85,500.

Huaráz Ancash, Peru. 9 32S 77
32W. Town NNW of Lima on Santa R.
Dept. and provincial cap. Commercial
centre for an agric. area. Industries
inc. mining silver, lead, copper, coal;
weaving, brewing. Bishopric and
tourist resort.

Huascarán Ancash, Peru. 9 07S 77
37W. An extinct volcano in W central
Peru. 22,205 ft. (6,768 metres) a.s.l.

Hubbard, Mount U.S.A./Canada.
Mountain peak, 14,950 ft. (4,557
metres) a.s.l., on Yukon/Alaska border in the St. Elias range, 140 m. (224
km.) W of Whitehorse.

Hubei (Hupei) China. 31 00N 112
00E. Province in E central China,

traversed by Yangtse R. and Han R.
Area, 72,375 sq. m. (187,500 sq. km.).
Mountainous in W, but very fertile
elsewhere with many L. along the
river courses giving rise to an important fishing industry. Agric. products
inc. rice, cotton, wheat, beans and
tobacco. There are deposits of coal,
iron, sulphur and salt, and small
amounts of gold. Cap. Wuhan. Pop.
(1982C) 47,804,150.

Hubli Karnataka, India. 15 20N 75
08E. Town sit. SE of Dharwad. Commercial centre and railway junction,
trading in cotton. Cotton processing is
the main industry. Pop. (1981C)
527,108 (Hubli–Dharwad).

Hucknall Nottinghamshire, England. 53 02N 1 11W. Town sit. NNW
of Nottingham. A coalmining centre
and manu. clothing, electrical equip.
Pop. (1981C) 27,506.

Huddersfield West Yorkshire,
England. 53 39N 1 47W. Town in
central N Eng. sit. on R. Colne. An
important centre of the woollen and
worsted industry, with associated
trades. Manu. inc. textile machinery,
metal products, carpets, clothing and
dyes. Pop. (1988E) 121,830.

Hudson Bay Canada. 60 00N 86
00W. Inland sea connected to Atlantic
Ocean by Hudson Strait and to the
Arctic by Foxe Channel. Length *c.*
1,000 m. (1,600 km.), greatest width
600 m. (960 km.). There are iron
deposits on the rocky E shore; forest
products are important on the W. Fed
mainly by Churchill and Nelson Rs.
There are ports with railheads on
James Bay and at Churchill, handling
furs, mineral and forest products.

Hudson River U.S.A. 40 42N 74
02W. Rises in the Adirondacks and
flows 350 m. (560 km.) S through
Hudson Falls (70 ft./21 metres descent), past Troy and Albany to New
York City, where it enters the Atlantic. Navigable by large craft up to Albany and used extensively for commerce, esp. at New York. Tidal for
150 m. (240 km.).

Hué Vietnam. 16 28N 107 36E.
Town sit. near mouth of Hué R. in NE.

The cap. until superseded by Saigon in 1954. Manu. textiles and cement. Pop. (1973E) 209,043.

Huehuetenango Guatemala. 15 20N 91 28W. (i) Dept. of w Guatemala bounded N by Mexico. Area, 2,857 sq. m. (7,314 sq. km.). Rises w to the Cuchumatanes mountains. The higher slopes produce cereals and livestock; the lower, coffee, sugarcane and fruit. There is lead mining at Chiantla and San Miguel Acatan. Pop. (1985E) 571,292. (ii) Town sit. NW of Guatemala. Dept. cap. and commercial centre of NW Highland Guatemala. Industries inc. flour milling, tanning, wool processing.

Huelva Spain. 37 18N 6 57W. (i) Province in SW bordering Portugal and Gulf of Cádiz. Area, 3,893 sq. m. (10,085 sq. km.). Mountainous in N with the Sierra de Aracena descending to the plains in the S. The principal Rs. are Tinto and Odiel. There are important deposits of iron and copper. Pop. (1986C) 430,918. (ii) Cap. of province of same name and port sit. on the peninsula formed by estuaries of Rs. Tinto and Odiel. Industries inc. fishing, fish canning and flour milling. Copper, manganese, iron ore, cork and wine are exported. Pop. (1986C) 135,427.

Huesca Spain. 42 08N 00 25W. (i) Province in NE separated from France by the Pyrénées. Area, 6,052 sq. m. (15,680 sq. km.). Very mountainous containing highest peak in the Pyrénées, Pico de Aneto, 11,168 ft. (3,404 metres) a.s.l. Agric. is possible with extensive irrigation, and cereals, fruit and wine are produced. Pop. (1986C) 220,824. (ii) Cap. of province of same name sit. NE of Zaragoza. Manu. agric. machinery, pottery, chemicals and wine. Trades in agric. produce.

Huhehot *see* **Hohhot**

Huila Huila, Colombia. 3 00N 76 59W. A volcano sit. NE of Popayán in the Andes. 18,865 ft. (5,750 metres) a.s.l.

Hull Quebec, Canada. 45 26N 75 43W. Town sit. at confluence of Rs. Ottawa and Gatineau opposite Ottawa.

An important centre for the production of timber and timber products, pulp and paper. Also manu. cement, mica, clothing and jewellery. Pop. (1981C) 56,225.

Hull Humberside, England. 53 45N 0 20W. City sit. E of Leeds on N shore of Humber estuary. Major commercial port. Industries inc. fish-trawling, fish processing; manu. flour, oils, paints, chemicals, fertilizers, engineering, boat-building. Exports chemicals and metals, imports timber, oilseeds, foodstuffs. Pop. (1984E) 269,100.

Humber River England. 53 32N 0 08E. Estuary formed by Trent and Ouse Rs. extending *c.* 38 m. (61 km.) w to E and entering the North Sea below Spurn Head. Main ports Hull, Grimsby, Goole.

Humberside England. 53 45N 0 20W. Under 1974 reorganization the county was formed from parts of the East and West Ridings of Yorkshire and Lincolnshire (parts of Lindsey). Area, 1,356 sq. m. (3,512 sq. km.) and consists of the districts of East Yorkshire, Boothferry, Beverley, Kingston upon Hull, Holderness, Scunthorpe, Glanford, Cleethorpes and Grimsby. Agric. is important especially sugarbeet. The fishing industry based at Hull and Grimsby supplies a large proportion of the English market. Ports are Immingham and Goole. Industries inc. chemicals, food processing, steel, aerospace, and engineering. Pop. (1988E) 850,500.

Humphrey Island (Manihiki) Cook Islands, South Pacific. 11 00S 161 00W. Atoll sit. NNW of Rarotonga. Area, 2 sq. m. (5 sq. km.), comprising 12 islets. Main products copra, pearl shells. Pop. (1981C) 405.

Hunan China. 27 30N 112 00E. Province in the S. Area, 81,253 sq. m. (210,500 sq. km.). Mainly hilly with a plain surrounding Tung Ting L. The principal Rs. are Yangtse, on N border, Siang and Yuan. There are deposits of coal, lead, antimony and zinc. The chief agric. products are tea, cotton, rice, wheat and tobacco. Cap. Changsha. Pop. (1987E) 56,960,000.

Hunedoară Hunedoară, Romania. 45 45N 22 54E. Town sit. w of Braşov. It is an important iron and steel centre, producing large quantities of pig iron and chemicals.

Hungary Central Europe. 47 20N 19 20E. Rep. bounded N by Czech., NE by Ukrainian Soviet Socialist Rep., E by Romania, s by Yugoslavia, w by Austria. Area, 35,912 sq. m. (93,030 sq. km.). Chief cities Budapest (cap.), Miskolc, Debrecen, Pécs, Szeged, Györ. Drained by Danube R. flowing N to s and dividing the country into plain (E) and hilly country. w of the Danube the Bakony Forest, Vertes and Pilis ranges of the Alps extend sw to NE. Flooding is common on the plain. There are about 8.3m. hectares of agric. land, of which 5m. is arable. The main crops are maize, wheat, sugar-beet, potatoes, barley. Vines are grown in Tokaj and N of L. Balaton. There are deposits of coal, bauxite and iron ore, some oil and natural gas. The main industries are mining coal and bauxite, manu. steel, iron, cement, fertilizers, super-phosphates, textiles. Pop. (1989E) 10,590,000.

Hungerford Berkshire, England. 51 26N 1 30W. Town sit. WNW of Newbury on R. Kennet. Industry inc. agric. engineering and manu. of electrical components. Pop. (1981C) 5,034.

Hunstanton Norfolk, England. 52 57N 0 30E. Resort sit. on NE shore of The Wash. Pop. (1980E) 4,250.

Hunter River New South Wales, Australia. 32 50S 151 42E. Rises in the Liverpool Range and flows sw and SE for 290 m. (464 km.) to enter the Pacific Ocean at Newcastle. Its basin consists mostly of a large coalfield.

Huntingdon Cambridgeshire, England. 52 25N 0 17W. Town sit. on Great Ouse R. Industries are printing and the manu. of plastics, electrical equipment and other light industry. Pop. (1988E) 15,140.

Huntingdonshire *now* **Cambridgeshire**

Huntington West Virginia, U.S.A. 38 25N 82 26W. City and R. port sit. ENE of Lexington, Kentucky, on Ohio R. Industries inc. railway engineering, steel mills, and the manu. of glass, electrical equipment and chemicals. There are deposits of coal, oil and natural gas in the vicinity. Pop. (1980C) 63,684.

Huntington Beach California, U.S.A. 33 39N 118 00W. City sit. SE of Long Beach on the s California coast. Residential and industrial, inc. aerospace systems, precision instruments, data systems and petroleum processing. Pop. (1985E) 185,000.

Huntington Park California, U.S.A. 33 59N 118 13W. A suburb of Los Angeles. Industries inc. retailing; manu. pumps, metal goods, clothing and oilfield equipment. Pop. (1980C) 46,223.

Huntly Grampian Region, Scotland. 57 27N 2 47W. Town in NE sit. at confluence of Rs. Deveron and Bogie. Manu. agric. implements, woollen goods, prefabricated buildings, stainless steel products and oil-related engineering. Pop. (1985E) 5,000.

Huntsville Alabama, U.S.A. 34 44N 86 35W. City sit. N of Birmingham. It is a rocket and guided missile centre, with associated professional services and high technology industry. Also trade and service centre for agric. area. Pop. (1978C) 142,500.

Huon Islands New Caledonia, South West Pacific. 18 03S 162 58E. Group of coral islands 170 m. (272 km.) NW of New Caledonia. Area, 160 acres. Uninhabited.

Hupei *see* **Hubei**

Huron South Dakota, U.S.A. 44 22N 98 13W. City sit. w of Minneapolis on the James R. Industries inc. processing and packing beef and pork, light manu. and assembly. Pop. (1980C) 13,000.

Huron, Lake U.S.A./Canada. 44 30N 82 15W. Second largest of the Great Lakes, E of L. Superior and connected to it by St. Mary's R. Area, 23,010 sq. m. (58,905 sq. km.). Maximum depth, 750 ft. (229 metres). Drained by St. Clair R. (s). Used extensively for through traffic carrying iron ore, coal, grain and oil. Chief ports, Bay City, Alpena, Cheboygan (Michigan, U.S.A.); Goderich, Collingwood, Midland (Ontario, Canada). Icebound mid-December–April.

Hutchinson　　Kansas, U.S.A. 38 05N 97 56W. City in s central Kansas sit. on Arkansas R. Industries inc. food distribution, hydraulics, vehicles, print graphics, agric. equipment and salt mining. Pop. (1989E) 45,200.

Huyton-with-Roby　　Merseyside, England. 53 25N 2 52W. Suburb of s Liverpool. Residential and industrial.

Hvar　　Yugoslavia. 43 09N 16 45E. (i) An island in the Adriatic Sea, part of the Dalmatian arch. A popular tourist centre, producing fruit, olives, honey and wine. (ii) Cap. of island of same name sit. on w coast.

Hwange　　Zimbabwe. 18 20S 26 25E. Town sit. NW of Bulawayo. Centre of a chief coalmining area of Zimbabwe with allied trades. The game reserve site is one of the finest in Africa. Pop. (1989E) 40,000.

Hwanghai　　North Korea. Province of central Korea bounded w by Korea Bay, s by Yellow Sea. Area, 6,463 sq. m. (16,545 sq. km.). Cap. Haeju. Low-lying, fertile, with gold, iron and coal. The chief occupations are farming, fishing and mining. Main products soy-beans, wheat, fruit, rice, tobacco, fish.

Hwang-Hai (Yellow Sea)　　China. 35 00N 123 00E. Arm of the Pacific Ocean between Korea and China, divided into the Gulfs of Korea, Liaotung and Chihli. The Yellow Sea receives the R. Hwango-ho, R. Sangtu, R. Liao, R. Han and R. Yalu. About 620 m. (992 km.) long, with a maximum width of 400 m. (640 km.), it derives its name from the yellow mud deposited by various rivers.

Hwang-ho　　China. 37 55N 118 50E. R. rising at E end of the Kunlun Shan

on the Tibetan plateau and flows *c.* 2,500 m. (4,000 km.) to enter the Yellow Sea on the Gulf of Po Hai. It passes through Kansu to Lanchow then turns N and E through the Ordos desert to the Shansi mountains where it turns s to Tungkwan and receives the Wei R. It then turns E across the plain to the Gulf of Po Hai. Its main importance is in watering the plain, since the caprices and complexities of its course make it useless for navigation. It is estimated to have changed course 11 times in its recorded history of 2,500 years. It carries large amounts of yellow loess which raise its bed above the surrounding country, necessitating embankment and frequently causing serious flooding as well as changes in course.

Hyde　　Greater Manchester, England. 53 27N 2 04W. Town sit. SE of Manchester on Tame R. Industries inc. textiles, footwear, clothing, engineering, rubber, plastics, fibre optics and telecommunications. Pop. (1987E) 34,799.

Hyderabad　　Andhra Pradesh, India. 17 10N 78 29E. State cap. in s central India, sit. on Musi R. An important transportation centre with industries inc. tobacco products, railway engineering, and the manu. of building materials, paper, glass, pharmaceuticals, electronics, offshore drilling rigs and textiles. Pop. (1981C) 2,187,262.

Hyderabad　　Hyderabad, Pakistan. 25 23N 68 36E. Town sit. NE of Karachi near Indus R. Manu. machinery, textiles, leather goods and cement. Trades in cereals, fruit and cotton. Pop. (1981C) 795,000.

Hydra　　Aegean Islands, Greece. An island sit. off coast of Argolis. Industries inc. boatbuilding, fishing and tanning. Cap. Hydra.

Hyères　　Var, France. 43 07N 6 07E. First resort on French Riviera, E of Toulon. Pop.(1982C) 41,739.

Hymettus　　Attica, Greece. 37 55N 23 47E. A mountain near Athens at 3,370 ft. (1,027 metres) a.s.l. Noted for honey.

Hythe Kent, England. 51 05N 1 05E. Resort sit. W of Folkestone on English Channel. Terminus of the Romney, Hythe and Dymchurch Light Railway. Pop. (1981C) 12,723.

Hyvinge   *see* **Hyvinkää**

Hyvinkää Uusimää, Finland. 60 38N 24 52E. Town sit. N of Helsinki. Railway junction and tourist centre for winter sports. Industries inc. woollen textile manu., granite quarrying. Pop. (1988E) 39,808.

I

Ialomita River Romania. 44 42N 27 51E. Rises in the Bucegi Mountains sit. NW of Sinaia and flows 200m. (320 km.) S to Targoviste then E to join the lower Danube R. W of Harsova. Main trib. Prahova R. The upper valley has noted stalactite caverns.

Iaşi Iaşi, Romania. 47 10N 27 35E. Cap. of district sit. W of Prut R. and frontier with the U.S.S.R. A commercial, railway and cultural centre. Manu. metal goods, textiles and furniture. Pop. (1985E) 314,156.

Ibadan Oyo, Nigeria. 7 17N 3 30E. State cap. in SW and the largest town in West Africa. Centre of an agric. region trading in cacao, cotton, palm oil and kernels. Industries inc. cotton weaving, tobacco and fruit canning. Pop. (1983E) 1,060,000.

Ibagué Tolima, Colombia. 4 27N 75 14W. Cap. of dept. in W central Colombia on E slopes of the Central Cordillera. Industries inc. flour milling, brewing and the manu. of leather products. Trades in coffee. Pop. (1980E) 360,000.

Ibarra Imbabura, Ecuador. 0 21N 78 07W. Cap. of province sit. NE of Quito at foot of the Imbabura volcano. Manu. textiles, furniture, and native silver and wood products. Trades in cotton, coffee and sugar-cane. Pop. (1982E) 59,000.

Iberian Peninsula 40 00N 5 00W. Name applied to Spain and Portugal taken together. The name derived from the Iberian people who lived along the R. Ebro (Latin *Iberus*). It is separated from the rest of Europe by the Pyrénées. The flora and fauna resemble those of N Africa, hence it is frequently stated that 'Africa, begins at the Pyrénées'. Area, 229,054 sq. m. (593,250 sq.km.).

Ibiza Spain. 38 54N 01 26E. Third largest island of the Balearics sit. SW of Majorca. Area, 221 sq. m. (572 sq. km.). Industries inc. tourism and fishing. Pop. (1990E) 71,000.

Ica Ica, Peru. 14 00S 75 48W. Cap. of dept. in SW, sit. on Ica R. Centre of an irrigated agric. region producing cotton, sugar and vines. Manu. wine, brandy, textiles and soap. Pop. (1981C) 111,087.

Ica River Peru. Rises in the W Andes, and flows W and S for 120 m. (192 km.) to enter the Pacific Ocean *c.* 152 m. (243 km.) NW of Atico.

Iceland 65 00N 19 00W. Republic and island in North Atlantic sit. 500 m. (800 km.) NW of the Shetland Islands and SE of Greenland. The first settlers came to Iceland in A.D. 874. Between 930 and 1264 Iceland was an independent rep., but by the 'Old Treaty' of 1263 the country recognized the rule of the King of Norway. In 1381 Iceland, together with Norway, came under the rule of the Danish kings, but when Norway was separated from Denmark in 1814, Iceland remained under the rule of Denmark. Since 1918 it has been acknowledged as a sovereign state. It was united with Denmark only through the common sovereign until it was proclaimed an independent rep. in 1944. Area, 39,738 sq. m. (103,000 sq. km.). Cap. Reykjavik. The N, E and W shores are much indented by fjords, and the surface consists of ice-covered plateaux 1,500–2,000 ft. (457–610 metres) a.s.l., and mountains culminating in Oräfa Jökull, 6,425 ft. (1,958 metres) a.s.l., near the SE coast. There are numerous small L., many being crater basins or moraine L., and glacier fields cover 5,000 sq. m. (13,000 sq. km.). In the interior there are large areas covered by recent lavas. Many volcanoes have been active in modern times; the best known are Hecla, 5,108 ft. (1,557 metres) a.s.l.,

Katla and Askja; hot springs and geysers are common. Of the total area about 85% is unproductive, but only about 0.5% is under cultivation, which is confined to hay, potatoes and turnips. Fishing is the dominating industry and Iceland had (1988) 956 fishing vessels. Pop. (1989E) 253,482.

Ichinomiya   Honshu, Japan. 35 18N 136 48E. Town sit. NW of Tokyo. Manu. textiles, metal goods and saké. Pop. (1983C) 253,000.

Idaho   U.S.A. 44 10N 114 00W. State bounded on N by British Columbia, Canada and Montana, on E by Montana and Wyoming, on S by Utah and Nevada, on W by Oregon and Washington. Idaho was first permanently settled in 1860, although there was a mission for Indians in 1836 and a Mormon settlement in 1855. It was organized as a Territory in 1863 and admitted into the Union as a state in 1890. Area, 83,557 sq. m. (216,413 sq. km.). Chief towns, Boise (the cap.), Pocatello and Idaho Falls. Agric. is the leading industry, although a great part of the state is naturally arid. Extensive irrigation works have been carried out. Crops: potatoes, sugar-beet, wheat, alfalfa, oats, barley, field peas, dry beans, apples, prunes and hops. Mining is important and minerals inc. lead, silver, zinc, phosphate rock, cobalt and antimony, colombium-tantalum, copper, gold, mercury, nickel, rare-earth metals, tungsten, thorium barite and clays. Beryllium ore has been discovered. Pop. (1984E) 1,001,000.

Idaho Falls   Idaho, U.S.A. 43 30N 112 01W. City in the SE on Snake R. at 4,700 ft. (1,433 metres) a.s.l. A shipping centre for an irrigated region producing wheat, sugar-beet, vegetables and alfalfa. Gold, silver and lead mines are in the vicinity. Ski resort. Pop. (1990E) 51,000.

Idlib   Aleppo, Syria. 35 56N 23 28E. Town sit. SSW of Aleppo in an agric. area producing cotton, cereals, tobacco, olives.

Ife   Oyo, Nigeria. 7 29N 4 34E. Town sit. E of Ibadan. Important as the spiritual home of the Yoruba tribe.

Trades in cacao, palm oil and kernels. Pop. (1983E) 214,500.

Igarka   Russian Soviet Federal Socialist Republic, U.S.S.R. 67 28N 86 35E. Port sit. in the Krasnoyarsk Territory. The main industries are timber and graphite.

Iglesias   Sardinia, Italy. 39 19N 8 32E. Town sit. WNW of Cagliari. Centre of an important mining area dealing with lead, zinc, silver and lignite. A school of mining is sit. here.

Iguassú River   Brazil. 25 41N 54 26W. Rises E of Curitiba in the Serra do Mar and flows *c.* 820 m. (1,312 km.) W on a meandering course to the Brazil/Argentina frontier which it forms for *c.* 75 m. (120 km.) before joining the Paraná where the Brazil, Argentina and Paraguay frontiers meet. 14 m. (22 km.) above this point it drops *c.* 210 ft. (64 metres) through hundreds of falls and cataracts at the Iguassú Falls. It is a major tourist attraction. Main trib., Rio Negro. Navigable between Pôrto Amazonas and União da Vitória (Brazil).

Ijsselmeer   Netherlands. 52 45N 5 25E. L. formed in N and central Netherlands by damming of Zuider Zee for 19 m. (30 km.) from Den Over to 1$^{1}/_{2}$ m. (2$^{1}/_{2}$ km.) SW of Zürich. The L. receives Vecht R., Ijssel R. and Zwartewater. Extensive areas have been reclaimed for agric., esp. the Wieringermeer, Beemster and North East Polders.

Ijssel River   Netherlands. 52 30N 6 00E. Leaves the Lower Rhine SE of Arnhem and flows 72 m. (115 km.) N to enter Ijsselmeer 4 m. (7 km.) WNW of Kampen. The Old Ijssel R. joins it at Doesburg.

Ilan   Taiwan. 24 45N 121 44E. Town sit. SE of Taipei. Market centre for agric. area, trading in rice, sugarcane, ramie, citrus fruit. The main industries are food processing, inc. fish canning and preserves, tea. Pop. (1985E) 84,793.

Ilebo   Kasai-Occidental, Zaïre. 4 17S 20 47E. Town sit. NNW of Luebo on Kasai R. R. port and rail terminus, handling copper from Katanga. Pop. (1976c) 142,036.

Île-de-France   Paris Basin, France. 49 00N 2 20E. A region comprising the departments of Essonne, Hauts-de-Seine, Paris, Seine-et-Marne, Seine-Saint-Denis, Val-de-Marne, Val-d'Oise and Yvelines. Area, 4,637 sq. m. (12,008 sq. km.). Cap. Paris. Combines agric. areas with the forest of Compiègne and Fontainebleau, and the industrial region surrounding Paris. Pop. (1987E) 10,259,400.

Ilesha   Oyo, Nigeria. 7 38N 4 45E. Town sit. ENE of Ibadan. Trades in cotton, cacao, palm oil and kernels. Pop. (1981E) 306,200.

Ilford   Greater London, England. 51 33N 0 05E. Town sit. ENE of central London. Industries inc. manu. photographic materials, paper, telecommunications equipment, dyes, tools. *See* Redbridge.

Ilfracombe   Devon, England. 51 13N 4 08W. Resort and fishing port sit. NNW of Barnstaple on Bristol Channel. There is some light industry inc. the manu. of plastics, clothing and electronics. Pop. (1981c) 10,479.

Ilhéus   Bahía, Brazil. 14 49S 39 02W. Port sit. S of Salvador near mouth of Cachoeira R. Exports inc. timber and piassava. Pop. (1984E) 108,000.

Ili River   China/U.S.S.R. 45 24N 74 02E. Formed E of Kuldja, Sinkiang, by union of Kunges and Tekes Rs. and flowing 590 m. (944 km.) W between the Dzungarian Ala-Tan and the Trans Ili Ala-Tan into the Kazakh Soviet Socialist Rep., then NW through desert into L. Balkhash. Main tribs. Chilik and Kaskelen Rs. Used for irrigation.

Ilkeston   Derbyshire, England. 52 59N 1 18W. Town ENE of Derby. Sit. in a former coalmining region. Industries inc. pipe making, light engineering and the manu. of textiles, hosiery, furniture and packaging. Pop. (1989E) 35,000.

Ilkley   West Yorkshire, England. 53 55N 1 50W. Town and resort sit. WSW of Harrogate on R. Wharfe. Mainly residential. Pop. (1981c) 13,058.

Illampu, Mount   Andes, Bolivia. 15 51S 68 30W. The highest mountain in the E. Cordillera with two peaks; Illampu, 21,275 ft. (6,485 metres) a.s.l., and Aneohuma, 21,490 ft (6,550 metres) a.s.l.

Ille-et-Vilaine   France. 48 10N 1 30W. Maritime dept. in Brittany region, bordering English Channel and Gulf of St. Malo, and formed from the province of Brittany. Area, 2,609 sq. m. (6,758 sq. km.). Mainly flat, forested in N and W, with coastal marshes. Agric. products inc. cereals, flax, potatoes, tobacco and apples the latter giving rise to an important cider-making industry. The principal Rs. are Ille and Vilaine. Cap. Rennes. Pop. (1982c) 749,764.

Illimani Mountain   Bolivia. 16 37S 67 48W. Peak 21,185 ft. (6,457 metres) a.s.l. in the Cordillera de la Paz, Bolivian Andes, 25 m. (40 km.) SE of La Paz. Some tungsten deposits.

Illinois   U.S.A. 40 15N 89 30W. State in Middle West bounded N by Wisconsin, W by Iowa and Missouri, S by Kentucky, E by Indiana. Area, 56,345 sq. m. (145,932 sq. km.). Main cities Springfield (cap.), Chicago, Rockford, Peoria, Decatur, Joliet, Evanston, Aurora, East St. Louis. Mainly an extensive prairie sloping from over 1,000 ft. (305 metres) a.s.l. on Wisconsin border to c. 280 ft. (85 metres) a.s.l. at confluence of Mississippi and Ohio Rs. Continental climate with extreme temps. Agric. is important, main products being cereals, soy-beans, pigs, cattle and poultry. About 70% of the crop acreage is maize, mainly in the N, with wheat in the W and SW. There are deposits of coal extending to 50% of the land area, petroleum in the S, fluorspar, natural gas. Industries inc. packing and processing farm products, esp. meat, dairy and cereal products; also manu. machinery, steel, petroleum products, electrical goods, railway equipment, furniture, cement, clothing. Most industry is in Cook county, in or around Chicago. Pop. (1987E) 11,582,000.

Illinois River U.S.A. 38 58N 90 27W. Formed by union of Des Plaines and Kankakee,Rs. SW of Chicago, and flowing 273 m. (437 km.) W to Depue then SW past Peoria to join Mississippi R. at Grafton. It forms part of the Illinois Waterway System, a section of the links between the Great Lakes and the Gulf of Mexico. Used for cargoes of coal, grain and petroleum. Its lower course is through a wide flood plain with many bayou Ls. Main tribs. Fox, Spoone, La Moine, Vermilion, Mackinaw and Sangamon Rs.

Ilmen, Lake Russian Soviet Federal Socialist Republic, U.S.S.R. 58 17N 31 20E. A L. sit SSE of Leningrad. Area, 350 sq. m. (906 sq. km.). Drained by Volkhov R. into L. Ladoga.

Iloilo Panay, Philippines. 10 43N 122 45E. Town and port in SE of Panay island. Exports inc. rice, sugar and copra. Pop. (1980C) 244,827.

Ilorin Kwara, Nigeria. 8 30N 4 32E. Cap. of Kwara state in central W. Manu. handwoven cloth, pottery, matches and sugar. Pop. (1983E) 343,900.

Imandra, Lake Russian Soviet Federal Socialist Republic, U.S.S.R. A L. sit. S of Murmansk; 50 m. (80 km.) long and 15 m. (24 km.) wide. Area, 330 sq. m. (855 sq. km.). Drained by Neva R. into the Kandalaksha Gulf of the White Sea.

Imathía Macedonia, Greece. 61 14N 28 50E. A prefecture sit. in N. Area, 656 sq. m. (1,699 sq. km.). Mainly agric. producing wheat, wine and vegetables. Cap. Verria, sit. W of Thessaloníki. Pop. (1981C) 133,750.

Imatra Kymi, Finland. 61 10N 28 46E. Town in SE, near the U.S.S.R. frontier, and ENE of Lappeenranta. It possesses an important hydroelectric power station and has a metallurgical industry. Pop. (1988C) 34,143.

Imbabura Ecuador. 0 12N 78 02W. Province N of the equator in the Andes. Area, 2,136 sq. m. (4,903 sq. km.). Chief towns Ibarra (cap.), Otavalo, Cotacachi, Atuntaqui. Mountain-

ous, with volcanic peaks and crater Ls. Mainly an agric. and stock raising area; chief products cereals, potatoes, sugar-cane, cotton, coffee, fruit. Industries inc. textiles, food processing; manu. leather. Pop. (1981C) 244,421.

Immingham Humberside, England. 53 37N 0 12W. Port on S shore of R. Humber estuary. Oil is an important import and there are oil refineries. Pop. (1981C) 11,506.

Imo Nigeria. State comprising the Owerri and Umuahia provinces of the former E region. Area, 4,575 sq. m. (11,850 sq. km.), the pop. is almost entirely Ibo. Chief cities Owerri (cap.), Aba and Umuahia. Pop. (1988E) 8,046,500.

Imola Emilia-Romagna, Italy. 44 21N 11 42E. Town SE of Bologna sit. at end of Santerno valley. It has many fine medieval buildings. Manu. pottery, machinery and wine.

Imperia Liguria, Italy. 43 53N 8 03E. Cap. of Imperia province sit. on Gulf of Genoa. Formed in 1923 from the two communities of Porto Maurizio and Oneglia, it is now the centre of an important olive oil industry. Pop. (1981E) 42,000.

Imphal Manipur, India. 24 48N 93 56E. State cap. SSW of Sadiya sit. on Manipur R. Centre of an agric. region trading in rice, tobacco and sugarcane. Pop. (1981C) 156,622.

Inaccessible Islands South Atlantic. Islets of the South Orkney group.

Inagua Bahamas. District comprising 2 islands SE of Nassau. Area, 645 sq. m. (1,671 sq. km.). Inagua Island is inhabited with chief settlement at Matthew Town. The main occupation is salt-panning. Little Inagua Island is uninhabited. Pop. (1980C) 939.

Inari, Lake Lapland, Finland. 69 00N 28 00E. A L. sit. in N containing hundreds of islets. It drains into Varanger Fiord *via* Pasvik R. and is a tourist centre for fishing. Area, 535 sq. m. (1,386 sq. km.).

Ince-in-Makerfield Greater Manchester, England. 53 32N 2 37W.

Town sit. SE of Wigan on Leeds and Liverpool Canal. Industries inc. heavy and light engineering and glass bottle manu. Pop. (1980E) 15,500.

Inchcape Rock   *see* **Bell Rock**

Inchon   South Korea. 37 28N 126 38E. Town and port sit. WSW of Seoul on Yellow Sea. Manu. steel, chemicals, glass and textiles. Exports inc. rice, soybeans and dried fish. Pop. (1983C) 1,220,311.

Independence   Missouri, U.S.A. 39 05N 94 24W. Town sit. E of Kansas City on Missouri R. Manu. agric. equipment, stoves and cement. Pop. (1980C) 111,806.

Independencia   Dominican Republic. Province of SW sit. between L. Enniquilo and the Haiti frontier. Cap. Jimani. Pop. (1980E) 32,700.

India   South Asia. 23 00N 77 30E. Rep. extending S from the Himalayas to occupy a peninsula between the Bay of Bengal (E) and the Arabian Sea (W). Bounded NW by Pakistan, N by China, Nepál, Sikkim, Bhután, E by Burma, with Bangladesh as an E enclave. Area, 1,222,714 sq. m. (3,166,829 sq. km.) (excluding Jammu and Kashmir). Chief cities Delhi (cap.), Calcutta, Bombay, Madras, Hyderabad, Bangalore, Ahmedabad, Kanpur. The Himalayas extend W to E across the N frontier states, descending to the Ganges R. plain, S of which the land rises to central uplands, with the Deccan plateau occupying most of the S peninsula. The peninsula is drained by Godavari, Kistna and Cauvery Rs. into the Bay of Bengal, the N is drained by Ganges, Jumna and Brahmaputra Rs. from the Himalayas into the Bay of Bengal, and Sutlej R. into Indus R. Central India is drained by Narbada and Tapti Rs. into the Arabian Sea. Land area under crops is *c*. 160m. hectares; the main crops being rice, wheat, jowar, bajra, sugar, groundnuts, cotton, jute, tea, coffee, rubber (mainly in Kerala). Irrigation is extensively used, the irrigated area exceeding that of any other country except China. Mineral deposits inc. coal (about 33% in Bihar), iron, oil, manganese, gypsum, bauxite,

salt, gold, silver, copper, mica. Industries inc. manu. cotton textiles, railway coaches, steam, diesel and electric locomotives, computers, military equipment, iron and steel, food processing; manu. chemicals, aircraft, helicopters, ocean going vessels, vehicles, and machinery. Pop. (1988E) 800m. (excluding the Pakistan-occupied area of Jammu and Kashmir).

Indiana   U.S.A. 40 00N. 86 00W. State in Middle West bounded N by Michigan, W by Illinois, S by Kentucky and E by Ohio. Area, 36,205 sq. m. (93,771 sq. km.). Chief cities, Indianapolis (cap.), Fort Wayne, South Bend, Evansville, Gary. Drained by Wabash R. and its tribs. White and Tippecanoe Rs. Mainly level prairie rising to rugged hills in S. Farming is important; main crops are maize (N), winter wheat (S), oats, hay, soybeans and tobacco; vegetables, tomatoes and mints are important in the N. There are extensive coal fields W and SW (bituminous: now used for the steel and iron industries of the NW); limestone is quarried at Bloomington and Bedford and supplies most of the U.S. need. There is petroleum in the E and SW. The main industries are steel, iron and chemicals in the Calumet area (NW); also manu. motor vehicles, agric. machinery, metal goods, railway stock and equipment, electrical goods, soap, furniture, cement, glass products. Pop. (1988E) 5,575,000.

Indianapolis   Indiana, U.S.A. 39 46N 86 09W. State cap. sit. in central Indiana on West Fork of White R. Industrial and transportation centre. Distribution centre. Industries inc. transport equipment, pharmaceuticals, machinery electronics, fabricate metals, printing and publishing. Pop. (1988E) 727,130.

Indian Ocean   10 00S 75 00E. Smallest ocean, bounded N by Asia, E by Malaysia and Australia, S by Antarctica and W by Africa. Connected with the Pacific Ocean through the Malacca and Sunda Straits of the Timor Sea, and (S of Australia) the Bass Strait. Connected with the Atlantic S of the Cape of Good Hope and by the Suez Canal and Mediterranean Sea. Area, *c*. 29.34m. sq. m. (75.99m.

sq. km.) and average depth 13,000 ft. (3,962 metres); greatest depth 24,440 ft. (7,449 metres) in the Java Trench. It receives the water from many important Asian rivers. The major inlets are the Arabian Sea and the Bay of Bengal, W and E of the Indian peninsula. Large ports are Bombay, Colombo, Calcutta, Rangoon, Perth, Lourenço Marques, Durban and Mombasa. N of the equator the direction of the currents varies with the monsoons. Chief islands in the ocean are: Madagascar, Mauritius. Réunion and Sri Lanka.

Indigirka River   U.S.S.R. 70 48N 148 54E. Rises SE of Oimyakon on the Oimyakon Plateau and flows 1,113 m. (1,780 km.) N through the Cherski range into the tundra, past Khonu, Druzhina and Chokurdakh to the E Siberian Sea, entering by a delta. Ice-free June–September, navigable up to Khonu. Main tribs. Selennyakh and Moma Rs.

Indochina   East Asia. Descriptive term for the states of Vietnam, Laos and Cambodia; used historically to denote French Indochina, the 3 states having been members of the French Union.

Indonesia   East Asia. 5 00S 115 00E. Rep. in the Malay Arch. consisting of an island group extending 3,000 m. (4,800 km.) from E of the Malay peninsula to N of Australia. Area, 575,000 sq. m. (1.9m. sq. km.). In the sixteenth century Portuguese traders, in quest of spices, settled in some of the islands, but were ejected by the British, who in turn were ousted by the Dutch (1595). From 1602 the Netherlands East India Company conquered the Netherlands East Indies, and ruled them until the dissolution of the company in 1798. Thereafter the Netherlands Government ruled the colony until 1942 when the Japanese invaded the islands. Indonesia became an independent Rep. in 1945. East Timor, former Portuguese colony, became 27th province in 1976. The main cities are Jakarta (cap.), Tanjungpriok, Surabaya, Semarang, Bandung, Solo, Jogjakarta, Bogor, Cheribon, Palembang, Medan, Padang, Macassar. The main islands are Java, Sumatra, part of Borneo, Celebes, Bali, Flores, part of Timor, the Moluccas and the Riouw arch. Most are mountainous, volcanic and heavily forested with extensive and valuable mineral deposits. Agric. in cleared areas and coastal plains produces rice, tapioca, spices, palm products, coconuts, tobacco, rubber, tea, coffee, sugar and cinchona. Minerals inc. petroleum, tin, coal, manganese, nickel, silver, gold, platinum, phosphate, sulphur, diamonds. The main industries are processing industries based on local raw materials; fishing and craft industries are important on the smaller islands. Pop. (1988E) 175.6m.

Indore   Madhya Pradesh, India. 22 43N 75 50E. Town sit. ESE of Ahmedabad on Saraswati R. Manu. chemicals, cotton, hosiery and furniture. Trades in cotton, grain and oilseeds. Pop. (1981C) 829,327.

Indre   France. 46 45N 1 30E. Dept. in Centre region, comprising three districts; Champagne, Brenne and Boischaut. Of these, the Champagne district is the most fertile, producing cereals, vegetables, fruit and wine. Cattle and sheep are reared. The principal Rs. are Indre and Creuse. Area, 2,617 sq. m. (6,778 sq. km.). Cap. Châteauroux. Pop. (1982C) 243,191.

Indre-et-Loire   France. 47 12N 0 40E. Dept. in Centre region, formed from Touraine and parts of Anjou, Orléanais and Poitou. Area, 2,364 sq. m. (6,124 sq. km.). The principal Rs. are Loire, Cher, Indre and Vienne. Mainly agric. esp. in the fertile R. valleys, growing grain, vegetables, fruit and vines. Cap. Tours. Pop. (1982C) 506,097.

Indre River   France. 47 16N 0 19E. Rises N of Boussac in the Massif Central, and flows NW for 160 m. (256 km.) through Indre and Indre-et-Loire to join Loire R. below Tours.

Indus River   South Asia. 24 20N 67 47E. Rises in SW Tibet near Senge in the Kailas range at 17,000 ft. (5,182 metres) a.s.l. and flows *c.* 1,900 m. (3,040 km.) first N and W as the Senge Khambab R., then entering Kashmir and following a twisting course to

Gilgit, where it turns S along the NW side of the Punjab Himalayas into the North West Frontier Province of Pakistan. It continues S across plains past Sukkur and Hyderabad to enter the Arabian Sea by a delta SE of Karachi. Fed by glaciers of the Karakoram and Hindu Kush ranges. Main tribs. Shyok, Kabul, Kurram and Panjnad Rs., the last bringing the waters of the Sutlej, Beas, Ravi, Chenab and Jhelum Rs. The volume of water varies widely, with frequent flooding; the main course is sometimes swollen to a width of 25 m. (40 km.). Total catchment area, 380,000 sq. m. (972,800 sq. km.). Used extensively for power and irrigation, esp. at the Sukkur Barrage, but little for transport.

Inglewood California, U.S.A. 33 58N 118 21W. An industrial and residential suburb sit. SW of Los Angeles. Manu. aircraft, electronics, toys and metal goods. Pop. (1988E) 103,200.

Ingolstadt Bavaria, Federal Republic of Germany. 48 46N 11 27E. Town in S centre sit. on Danube R. Manu. motor cars, textile machinery and textiles. Pop. (1984E) 90,700.

Inhambane Mozambique. 23 54S 35 30E. (i) Province in SE. Area, 26,436 sq. m. (68,470 sq. km.). Pop. (1987E) 1,167,022. (ii) Cap. of (i) sit. NE of Maputo on Indian Ocean. Industries inc. sugar milling; manu. soap, bricks, ceramics. Port and railhead exporting sugar, cotton, timber, peanuts, rubber, maize, copra, nuts.

Inland Sea Japan. Sit. between Honshu, Shikoku and Kyushu, it is about 240 m. (384 km.) in length. In the centre is sit. the Inland Sea National Park. Area, 257 sq. m. (666 sq. km.).

Innerleithen Borders Region, Scotland. 55 38N 3 05W. Spa sit. near confluence of Leithen Water and R. Tweed, and ESE of Peebles. Manu. tweeds and hosiery. Pop. (1985E) 2,500.

Inner Mongolia (Nei Monggol) China. The Inner Mongolian Autonomous Region forms a division of Manchuria, bounded W by Mongolia and the U.S.S.R. across Argun R. Area, 456,757 sq. m. (1,183,000 sq. km.). Cap. Ulan Hoto. Mainly plateau rising to the Khingan Mountains (centre), extensive steppe supports nomadic stock farming which is the main occupation, with some agric. in the SE producing cereals. Pop. (1982C) 19,274,279.

Innisfree Sligo, Ireland. A small island sit. in Lough Gill, and ESE of Sligo.

Inn River Switzerland/Austria/Federal Republic of Germany. 48 35N 13 28E. Rises in Graubünden canton and flows for 320 m. (512 km.) NE past Innsbruck and joins R. Danube at Passau, in extreme SE F.R.G.

Innsbruck Tirol, Austria. 47 16N 11 24E. Cap. of Federal State sit. at confluence of Rs. Inn and Sill, and at the junction of the roads over the Brenner and Arlberg passes. Commercial and tourist centre with many fine medieval buildings. Manu. textiles, glass and food products. Pop. (1981C) 116,025.

Inowroclaw Bydgoszcz, Poland. 52 48N 18 15E. Town and spa in NW central Poland. Manu. machinery, chemicals, glass and brass goods. Salt and gypsum are mined in the vicinity.

Interlaken Bern, Switzerland. 46 41N 7 51E. Resort sit. on Aar R. between L. Thun and L. Brienz at 1,864 ft. (568 metres) a.s.l. Tourist centre. Pop. (1990E) 15,000.

Intibucá Honduras. Dept. of SW Honduras bounded SW by Salvador. Area, 1,057 sq. m. (2,738 sq. km.). Chief towns La Esperanza (cap), Itibucá. Straddles the continental divide with the valleys of Río Grande do Otoro and Lempa R. draining N and S. Mainly agric. and stock farming area. Main products cereals, beans, tobacco, coffee. Pop. (1981C) 111,412.

Inuvik Northwest Territories, Canada. 68 25N 133 30W. Town which has been built on piles sit. on Mackenzie R. delta, near Beaufort Sea coast. Service centre for oil and gas installations in the Beaufort Sea. Pop. (1982E) 3,389.

Inveraray Strathclyde Region, Scotland. 56 13N 5 05W. Town sit. NNW of Greenock at mouth of R. Aray on Loch Fyne. The main occupation is fishing. Pop. (1989E) 470.

Invercargill South Island, New Zealand. 46 25S 168 27E. City sit. on New River Harbour, an inlet of Foveaux Strait. Industries are connected with meat, butter, cheese and wool, as well as timber and coal. Pop. (1986C) 52,809.

Inverkeithing Fife Region, Scotland. 56 02N 3 24W. Port sit. SE of Dunfermline on the Firth of Forth. Industries inc. shipbreaking and papermaking. Pop. (1988E) 6,200.

Inverness Highland Region, Scotland. 57 27N 4 15W. Town on a narrow stretch between Beauly Firth and Moray Firth on E coast at mouth of Ness R. Commercial and administrative centre, county town and port. Industries inc. tourism—as a centre for highland Scotland, light engineering, communications. Pop. (1989E) 61,000 (district).

Inverness-shire Scotland. Former county and now part of Highland Region.

Inverurie Grampian Region, Scotland. 57 17N 2 23W. Town in NE sit. at confluence of Rs. Ury and Don. Pop. (1981C) 7,680.

Iona Strathclyde Region, Scotland. 56 19N 6 25W. Island just off SW coast of Mull across the Sound of Iona. Area, 3$\frac{1}{2}$ m. by 1$\frac{1}{2}$ m. (6 km. by 2.5 km.). Hilly, rising to 332 ft (101 metres). Noted as an early centre of Christianity; St. Columba founded a monastery in A.D. 563 which became the centre for the conversion of Scotland.

Ionia Michigan, U.S.A. 42 59N 85 07W. City sit. E of Grand Rapids on Grand R. Centre of an agric. area. Industries inc. manu. vehicle parts, food products, ceramics. Pop. (1980C) 5,920.

Ionian Islands Ionian Sea, Greece. 38 30N 20 30E. A group of islands about 20 m. (32 km.) from the mainland. Area, 891 sq. m. (2,307 sq. km.). The four largest islands Kefellenia, Corfu, Lefkas and Zante give their names to the four prefectures of the region. Olive oil, fruit and wine are produced in the fertile valleys.

Iowa U.S.A. 42 18N 93 30W. State in Middle West bounded N by Minnesota, E by Wisconsin and Illinois, S by Missouri and W by Nebraska. Area, 56,280 sq. m. (145,765 sq. km.). Chief cities Des Moines (cap.), Sioux City, Davenport, Cedar Rapids, Waterloo, Dubuque, Council Bluffs. Plain rising to hills in the NE and steep bluffs above Mississippi and Missouri Rs., tribs. of which drain the E and W parts respectively. About 95% of the area is farm land, 65% crops, 30% stock. The main products are maize, oats, pigs, poultry, cattle, hay, soybeans, barley, wheat, rye, flax. The main industry is processing and packing farm products. There are extensive deposits of coal. Pop. (1987E) 2,834,000.

Iowa City Iowa, U.S.A. 41 40N 91 32W. Town SSE of Cedar Rapids sit. on Iowa R. Manu. high technology, pharmaceuticals, health care products, electronics, testing and communication systems. (1980C) 50,500.

Ipoh Perak, Peninsular Malaysia. 4 35N 101 04E. Town sit. NNW of Kuala Lumpur. Industries inc. tin-mining and rubber production. Pop. (1980E) 294,000.

Ipswich Queensland, Australia. 27 38S 152 40E. Town in SE of state sit. on Bremer R. Centre of a coal mining district with railway workshops and woollen mills. Other industries inc. engineering, food processing, building materials, furniture. Pop. (1990E) 75,500.

Ipswich Suffolk, England. 52 04N 1 10E. Town and port sit. at head of R. Orwell estuary. Industries inc.light and heavy engineering, printing, insurance and flour milling. Manu. agric. machinery, electrical equipment, plastics, fertilizers, textiles, hi-fi equipment and tobacco products. There are many fine medieval buildings. Pop. (1981C) 120,447.

Iquique Tarapacá, Chile. 20 13S 70 10W. Cap. of province and port in NW sit. on the edge of the Atacama Desert. Industries inc. fishing and fish-canning, sugar and oil refining. There are important exports of nitrates, iodine and salt. The climate is rainless and water is piped 50 m. (80 km.) from the Pica oasis. Pop. (1982E) 109,033.

Iquitos Loreto, Peru. 3 50S 73 15W. Cap. of dept. and port sit. on Amazon R. The commercial centre of a vast area and head of navigation for large steamers. Exports inc. timber, rubber and nuts. Pop. (1982E) 175,000.

Iraklion *see* **Candia**

Iran South West Asia. 33 00N 53 00E. Islamic Rep. bounded S by Gulf of Oman and (Persian) Gulf, W by Iraq and Turkey, N by Azerbaijan and Armenian Soviet Socialist Reps., Caspian Sea and Turkmen Soviet Socialist Rep., E by Afghánistán and Pakistan. Area, 630,000 sq. m. (1,648,000 sq. km.). Chief cities Tehrán (cap.), Tabriz, Esfahán, Meshed, Shiráz, Ahwáz, Abadan, Qum, Bakhtarán, Rasht, Hamadán, Ardebil, Yezd, Kermán. Central plateau at *c.* 4,000 ft. (1,219 metres) a.s.l., ringed by the Alburz mountain system in the W and S, with narrow lowland strips beyond the mountains on the Caspian and Gulf shores. Much of the country is arid with high winds, with extensive salt deserts. There are few perennial Rs., the main ones being Aras, Sefid Rud, Gurgan and Atrek Rs., draining into the Caspian and Karun R. into the Gulf. The Zaindeh R. drains into the interior. About one-tenth of the land area is under crops; main products wheat, rice, cotton, sugar-beet, tobacco. Livestock farming, mainly sheep and goats, in highland areas provides wool for the carpet making industry. The most important mineral is oil; there are also deposits of iron ore, copper, lead, chromite, coal and salt—much of it undeveloped. The oil industry is the most important, followed by textiles and carpet making, food processing and vehicle assembly. Pop. (1988E) 53.92m.

Irapuato Guanajuato, Mexico. 20 41N 101 28W. Town sit. on the central plateau, SE of León at 5,660 ft. (1,725 metres) a.s.l. An important road and rail junction with iron-founding and tanning industries. Trades in tobacco, cereals and fruit and is particularly noted for strawberries. Pop. (1980C) 246,308.

Iraq South West Asia. 33 00N 44 00E. Republic bounded W by Syria and Jordan, SW by Saudi Arabia, NE and E by Iran, NW by Turkey. Area, 168,040 sq. m. (438,446 sq. km.). Chief cities Baghdad (cap.), Basra, Mosul, Kirkuk, Najaf. Extensive desert in the centre (the Syrian desert) rises to highlands in the NE with some fertile plain; between them there is fertile lowland watered by Tigris and Euphrates Rs. In this last area is found most of the pop. and production. The N end of the R. plain has extensive oil fields, esp. Mosul, Kirkuk, Khanaquin. Agric. in the highlands produces tobacco and fruit; on the plain the main crops are barley, wheat, rice, maize, millet, sorghum, vegetables, dates, cotton. The main industry is oil extracting and refining; also cotton ginning, brick making, distilling, tobacco processing. Petroleum is the chief export, followed by dates. Development following the exploitation of oil has led to important growth for the construction industry. Pop. (1988E) 17,064,000.

Irbid Irbid, Jordan. Cap. of district sit. N of Amman. Trades in grain. Pop. (1970E) 110,000.

Ireland West Europe. 53 00N 8 00W. Rep. occupying the greater (S) part of Ireland, *c.* 50 m. (80 km.) W of Great Britain across St. George's Channel. Bounded N, W and S by the Atlantic and E by the Irish Sea, NE by Northern Ireland. Area, 26,599 sq. m. (68,893 sq. km.). Chief cities Dublin (cap.), Cork, Limerick, Waterford. In 1916 an insurrection against British rule took place and a republic was proclaimed. In 1920 an Act was passed by the British Parliament, under which separate Parliaments were set up for 'Southern Ireland' (26 counties) and 'Northern Ireland' (6 counties). The

Unionists of the 6 counties accepted this scheme, and a Northern Parliament was duly elected in 1921. The rest of Ireland, however, ignored the Act.

In 1921 a Treaty was signed between Great Britain and Ireland by which Ireland accepted dominion status subject to the right of Northern Ireland to opt out. This right was exercised, and the border between *Saorstát Éireann* (26 counties) and Northern Ireland (6 counties) was fixed in 1925 as the outcome of an agreement between Great Britain, the Irish Free State and Northern Ireland. The agreement was ratified by the three parliaments.

Subsequently the constitutional links between *Saorstát Éireann* and the U.K. were gradually removed by the *Dáil* (National Parliament). The remaining formal association with the British Commonwealth by virtue of the External Relations Act, 1936, was severed when the Republic of Ireland Act, 1948, came into operation on 18 April 1949.

Low-lying in the centre with mountain ranges around the coast; a deeply indented coastline with many loughs on the w coast. Land area inc. 11,847,800 acres of crop and pasture land and 5m. acres of rough grazing and undeveloped land. Main crops are hay and silage, turnips, potatoes, sugar-beet, barley, mangels, wheat, oats. Livestock inc. dairy and beef cattle, sheep, pigs and horses. The country is mainly rural. Industries are concentrated on the E coast and inc. processing dairy products; manu. tobacco products, milling grain and animal feed; manu. bacon and other meat products; manu. vehicles, metal goods. Noted exports are bloodstock (horses), linen, stout and glassware. Pop. (1988E) 3,540,643.

Irian Jaya   *see* **West Irian**

Irish Republic   *see* **Ireland**

Irish Sea   54 00N 5 00E. An arm of the Atlantic Ocean separating Ireland from England, Wales and Scotland. It is about 100 m. (160 km.) from N to S and connected to the Atlantic Ocean by St. George's Channel in the S, and the North Channel in the N.

Irkutsk   Russian Soviet Federal Socialist Republic, U.S.S.R. 52 16N 104 20E. (i) a region in SE Siberia sit. to N and W of L. Baykal. Mainly forested with large deposits of coal, iron ore, salt, gold and mica. The chief R. is Angara which has large hydroelectric plants. The principal occupations are fur-trapping, cattle breeding and fishing. (ii) Cap. of region of same name sit. at confluence of Rs. Angara and Irkut. An important centre with industries inc. engineering, mica processing, motor car works, meat packing and sawmilling. Pop. (1985E) 597,000.

Irlam   Greater Manchester, England. 53 28N 2 25W. Town sit. WSW of Manchester on Manchester Ship Canal. Industries inc. tar distilling, soap making, farming and market gardening. Pop. (1981C) 19,799.

Iron Gate   Romania. A narrow gorge of Danube R. sit. on the border between Orsova and Turnu-Severin, and WNW of Bucharest. It is about 2 m. (3 km.) in length and navigable.

Irrawaddy (Aye yar wady) River   Burma. 3 50N 95 06E. Rises in Kachin State in 2 headstreams which join 25 m. (40 km.) N of Myitkina, and flows *c.* 1,000 m. (1,600 km.) S through many rapids past Bhamo, Mandalay and Prome to enter the Indian Ocean by a delta which begins 180 m. (288 km.) from the coast, and empties over 158,000 sq. m. (404,480 sq. km.) of coastline. The principal delta arms are the Irrawaddy proper and the Bassein R. The main tribs. are Chindwin, Mogaung, Mu, Mon, Taping, Shweli and Myitnge Rs. Navigable to Bhamo. Used mainly for carrying petroleum and teak.

Irtysh River   U.S.S.R. 61 04N 68 52E. Rises in Sinkiang in the Mongolian Altai range and flows 1,844 m. (2,950 km.) W, as the Black Irtysh, into L. Zaisan in the Kazakh Soviet Socialist Rep. Leaves as the Irtysh R. and flows N and NW through mountains and the Kulunda Steppe into the Russian Soviet Federal Socialist Rep., past Omsk and Tobolsk, to enter Ob R. at Khanty-Mansisk. Navigable to Semipalatinsk April–November and to Tobolsk May–November. Its middle course drains the agric. areas of w Siberia.

Irvine Strathclyde Region, Scotland. 56 37N 4 40W. New town and port sit. near mouth of R. Irvine on Firth of Clyde. Manu. pharmaceuticals, heavy goods vehicles, fork lift trucks, heavy engineering, explosives, electronics, glass products, papermaking, chipboard, golf clubs, clothing and food processing. Pop. (1989E) 57,000.

Irving Texas, U.S.A. 32 49N 96 56W. City NW of Dallas. Commercial, manu. and service centre; industries inc. electronics, jewellery retail; manu. health-care products and soft drinks. Pop. (1985E) 141,400.

Irvington New Jersey, U.S.A. 40 44N 74 14W. Town sit. SW of Newark. Residential. Manu. machinery, tools, paper, paint and rubber goods. Pop. (1980C) 62,000.

Isar River Austria/Federal Republic of Germany. 48 49N 12 58E. Rises in Austrian Tirol and flows NNE for 163 m. (261 km.) to join Danube R. just below Deggendorf, in SE F.R.G.

Ischia Campania, Italy. 40 44N 13 57E. (i) A volcanic island sit. in Bay of Naples. Area, 18 sq. m. (47 sq. km.). Monte Epomeo, 2,589 ft. (789 metres) a.s.l., is sit. in the centre. Noted for mineral springs, fruit, wine and fish. (ii) Chief town of island of the same name, sit on NE coast.

Iseo, Lake Lombardia, Italy. 45 43N 10 04E. A L. sit. between Bergamo and Brescia, in central N. The Oglio R. traverses it, and in the centre is a large mountainous island, Monte Isola, 1,965 ft. (599 metres) a.s.l.

Isère France. 45 10N 5 50E. Dept. in Rhône–Alpes region. Area, 2,886 sq. m. (7,474 sq. km.). Very mountainous and forested in SE. Mainly infertile but wheat, potatoes, vines and tobacco are grown in the R. valleys. The principal Rs. are Rhône, forming the w border, Isère, Drac and Romanche. Coal, iron and lead are mined and livestock reared. Cap. Grenoble. Pop. (1982C) 936,771.

Isère River France. 44 59N 4 51E. Rises in the Graian Alps and flows w and sw for 179 m. (286 km.) to enter Rhône R. near Valence.

Iserlohn North Rhine-Westphalia, Federal Republic of Germany. 51 22N 7 41E. Town sit. SE of Dortmund. Manu. brass and bronze products, beer and needles. Pop. (1989E) 93,000.

Isfahan *see* **Esfahan**

İskenderun Hatay, Turkey. 36 37N 36 07E. Town and port sit. SE of Adana. A naval base and railway terminus. Industries inc. steelworks.

Islamabad Pakistan. 33 40N 78 08E. Cap. of Pakistan and of Sind province sit. just NE of Rawalpindi. Construction of the modern new cap. began in 1961 and functioned as cap. in 1967. Pop. (1981C) 201,000.

Islands, Bay of Newfoundland, Canada. 49 11N 58 15W. Large, safe harbour sit. on w coast.

Islas de la Bahia Honduras. 16 45N 86 15W. Group in Caribbean Sea off N coast of Honduras. Area, 144 sq. m. (368 sq. km.). Cap. Roatán. The important islands are Roatán, Guanaja, Utila, Barbareta, Santa Elena, Morat, Hog. Main occupations farming and lumbering, fishing. Main products coconuts, sugar-cane, plantains, pineapples. The main industries are boatbuilding and fish processing.

Islay Strathclyde Region, Scotland. 55 46N 6 10W. An island sit. off sw coast of Jura with the Sound of Islay running between. Area, 234 sq. m. (606 sq. km.). The highest point is Ben Bheigeir, 1,609 ft (490 metres) a.s.l. The chief occupations are agric., fishing and whisky distilling.

Isle of Ely England. 52 24N 0 16E. Area of high ground surrounding Ely in fenland district of East Anglia.

Isle of Man 54 15N 4 30W. A dependency of the British crown. Island in Irish Sea sit. 10 m. (16 km.) from Scotland, 28 m. (45 km.) from England and 38 m. (61 km.) from Northern Ireland. Area, 227 sq. m. (572 sq.

km.) The surface is undulating sloping up from a rocky coast to a central ridge which reaches 2,024 ft. (617 metres) a.s.l. at Snaefell. Chief towns, Douglas, Ramsey, Peel and Castletown. Tourism and agric. are main industries but minerals are found. The principal agric. products consist of oats, wheat, barley, potatoes, grasses, fatstock, dairy products. Pop. (1986C) 64,282.

Isle of Pines (Isles of Youth) Cuba. Island sit. W of Cuba across the Gulf of Batabanó. Area, 1,182 sq. m. (3,061 sq. km.). Chief town Nueva Gerona. Low plateau with swamps and forests (S) rising to fertile hill country (N). Mainly agric., chief products fruit, vegetables, tobacco. The main industry is tourism. Pop. (1984E) 60,794.

Isle of Pines New Caledonia, South Pacific. Island sit. SE of New Caledonia. Area, 59 sq. m. (153 sq. km.). Forested and mountainous, supporting deer, wild goats, pigs. The main industries are tourism and fishing. Pop. (1983C) 1,287.

Isle of Wight, England *see* **Wight, Isle of**

Isles of Los *see* **Los Islands**

Isles of Scilly England. Group of about 140 islands and islets WSW of Land's End. Area, 6 sq. m. (15¹/₂ sq. km.). Town, Hugh Town. Five inhabited islands: St. Mary's, Tresco, St. Martin's, St. Agnes, Bryher. Industry horticulture and tourism. Pop. (1980E) 2,000.

Islington Greater London, England. 51 34N 0 06W. Residential and industrial borough of N London. Pop. (1988E) 169,200.

Ismailia Egypt. 30 35N 32 16E. Town S of Port Said sit. on Suez Canal on NW shore of L. Timsah. It is connected by rail with Cairo, Suez and Port Said. Pop. (1986E) 236,300.

Isparta Isparta, Turkey. 37 46N 30 33E. Cap. of province ESE of Izmir. Manu. carpets and noted for roses and attar of roses. Pop. (1985C) 101,784.

Israel West Asia. 32 00N 34 50E. Rep. bounded W by Mediterranean, N by Lebanon, NE by Syria, E by Jordan, S by Gulf of Aqaba. Area, 8,017 sq. m. (20,770 sq. km.) (1949 agreement), 34,493 sq. m. (89,337 sq. km.) (1967 cease-fire agreement). Chief towns Jerusalem (cap.), Tel Aviv, Haifa. Narrow, fertile coastal plain rises to the highlands of Galilee and Judaea (E), rising to 2,500 ft (762 metres) a.s.l. near Jerusalem. There is then a steep drop to the Jordan valley, falling to the Dead Sea, 1,292 ft. (394 metres) b.s.l. The S is flat and formerly arid land, now extensively irrigated. The highlands of the N rise to 3,693 ft. (1,126 metres) a.s.l. in Upper Galilee and are divided between Mount Carmel (NW) and the Jordan valley by the Plain of Jezreel. Agric. on the coastal plains, the Valley of Jezreel and the irrigated Negev (S) produces citrus fruit, olives, tobacco, vines, vegetables, cotton, sugar-beet; livestock inc. cattle, sheep, goats and poultry. Most agric. is carried on in varied kinds of co-operative holding. Minerals inc. mainly the potash, bromine and other salts of the Dead Sea; there is some copper at Timna and oil in the Negev, with natural gas near the Dead Sea. Industry is widely diversified and inc. manu. chemicals, metal products, textiles, tyres, paper, plastics, leather goods, glass, ceramics, building materials, tobacco products, food products and electrical goods, polished diamonds. The main export is citrus fruit but industrial products are becoming more important. Pop. (1988E) 4,476,800.

Issyk-Kul Kirghizia, U.S.S.R. 42 25N 77 15E. Lake sit. between two mountain ranges, at 5,200 ft (1,584 metres) a.s.l. Area, 2,390 sq. m. (6,190 sq. km.). Many streams run into its salty waters which contain quantities of fish.

Issy-les-Moulineaux Hauts-de-Seine, France. 48 49N 2 17E. A SW suburb of Paris sit. on Seine R. Manu. aircraft, motor cars, chemicals, textiles and electrical equipment.

Istanbul Istanbul, Turkey. 41 01N 28 58E. City on Bosporus strait at entrance to Sea of Marmara. Provincial cap. Commercial and industrial centre, and major sea port, handling c. 60% of Turkish trade. An important railway terminus linking the European and Asian systems. Industries inc. shipbuilding, textiles, pottery, food processing; manu. tobacco products, cement, glass, leather products. Formerly Constantinople. Noted for the Byzantine church of Hagia Sophia and the 16th cent. mosques built by Bajazet II, Suleiman I and Ahmed I. Pop. (1985c) 5,494,916.

Istria Yugoslavia. 45 15N 14 00E. A peninsula sit. in N Adriatic Sea between Gulf of Trieste and Bay of Kvarner. Mainly hilly with a rocky coast. Produces cereals, olive oil, wine and livestock and has boat-building and fishing industries. Chief town, Pula.

Itajaí Santa Catarina, Brazil. 26 53S 48 39W. Port in the SE sit. at mouth of Itajaí Açu R. and sit. N of Florianópolis. Centre of an agric. district with exports of sugar, tobacco and timber. Pop. (1984E) 64,000.

Italy South Europe. 42 00N 13 00W. Rep. occupying a peninsula extending SE from Alps between the Tyrrhenian Sea (W) and the Adriatic (E), and inc. the islands of Sicily, Sardinia, Elba, Capri, Ischia, the Lipari, Pantelleria and Lampedusa islands. Area, 116,224 sq. m. (301,020 sq. km.). Chief cities Rome (cap.), Milan, Turin, Naples, Palermo, Bologna, Genoa, Florence. Mountainous in the N with the Alpine and Dolomite ranges, and in the centre where the Apennine range extends from the Ligurian coast (NW) to the extreme S. The Apennines are volcanic and often desolate. The climate is mediterranean, drier on the E than elsewhere. The Po R. valley extends between the Alps and the Apennines and supports c. 40% of the pop. on a fertile alluvial plain. Agric. here produces rice, wheat, maize, oats, rye, barley, potatoes, sugar-beet; also meat, dairy products and silk. Elsewhere the main products are fruit, wine, olives, wheat, livestock. Industries are concentrated in Lombardy and inc. textiles and manu. vehicles, electrical goods and food products; also chemicals,

engineering, printing, publishing. Tourism is important. Pop. (1988E) 57,504,691.

Itapúa Paraguay. Dept. of SE Paraguay, bounded S and E by Upper Paraná R. Area, 6,380 sq. m. (16,524 sq. km.). Cap. Encarnación. Heavily forested lowland, the main occupations being lumbering and farming with associated processing industries. Chief products maté, cotton, maize, rice, tobacco, sugar, timber. Pop. (1982E) 263,021.

Ithaca Ionian Sea, Greece. 42 27N 76 30W. An island sit. off NE Kefallenia. Area, 33 sq. m. (85 sq. km.). Agric. products inc. currants, wine and olive oil. Chief town Ithaca.

Ithaca New York, U.S.A. 42 25N 76 30W. City sit. at S end of L. Cayuga, and SSW of Syracuse. Manu. firearms, electronics, fertilizers, salt and electric clocks. Cornell university is here. Pop. (1980c) 28,732.

Itzehoe Schleswig-Holstein, Federal Republic of Germany. 53 55N 9 31E. Town sit NW of Hamburg on Stör R. Manu. machinery, cement, paper and fishing nets. Pop. (1984E) 32,500.

Ivano-Frankovsk Ukraine, U.S.S.R. 48 55N 24 43E. Town sit. SSE of Lvov. Industries inc. railway engineering, oil refining, woodworking, textiles and food processing. Pop (1987E) 225,000.

Ivanovo Russian Soviet Federal Socialist Republic, U.S.S.R. 57 00N 40 59E. (i) A region NE of Moscow. Area, 9,230 sq. m. (23,965 sq. km.). Mainly forested. Agric. products inc. cereals, vegetables and dairy produce. (ii) Cap. of region of same name, sit. NE of Moscow on Uvod R. Manu. cotton, textile machinery and chemicals. Pop. (1987E) 479,000.

Ivory Coast *see* **Côte d'Ivoire**

Ivry-sur-Seine Val-de-Marne, France. A SE suburb of Paris sit. on Seine R. Manu. chemicals, earthenware and cement. Pop. (1982c) 55,948.

Iwakuni Yamaguchi, Japan. 34 09N 132 11E. City sit. SW of Hiroshima on Nishiki R. Railway junction. Industries inc. manu. petrochemicals, textiles, flour and sauces. Noted for the 750 ft. (229 metres) Kintai Bridge.

Iwo Oyo, Nigeria. 7 38N 4 11E. Town sit. NE of Ibadan. Industries inc. weaving and dyeing. Trades in cacao, palm oil and kernels. Pop. (1983E) 261,600.

Ixtaccihuatl Mexico. 19 11N 98 39W. A volcano sit. SE of Mexico City.

Izabal, Lake Guatemala. A L. sit. in E. Drained by Río Dulce into the Caribbean Sea.

Izhevsk *see* **Ustinov**

Izmail Ukraine, U.S.S.R. 45 21N 28 50E. Cap. and port of the Izmail region sit. SW of Odessa on Dniester R. Industries inc. flour milling and tanning. Trades in cereals and hides.

Izmir Turkey. 38 25N 27 09E. (i) Province of W Turkey on the Aegean. Area, 4,952 sq. m. (12,826 sq. km.). Heavily forested with rich mineral resources, inc. copper, silver, antimony, emery, iron, manganese, arsenic, mercury, zinc and chromium. Main occupations mining, lumbering, farming. Pop. (1986C) 2,317,829. (ii) City sit. SSW of Istanbul at head of Gulf of Smyrna. Provincial cap. Major port and rail centre. Industries inc. manu. soap, dyes, tobacco products, textiles, leather goods, tannin extract, olive oil. Pop. (1986C) 1,489,817.

Izmit Kocaeli, Turkey. 40 46N 29 55E. Cap. of province sit. at head of Gulf of İzmit on the Sea of Marmara, and ESE of Istanbul. Centre of a tobacco-growing district with manu. of chemicals and cement. Pop. (1980C) 190,423.

J

Jabalpur Madhya Pradesh, India. 23 09N 79 58E. City sit. NNE of Nagpur. Rail and road junction and industrial centre of an agric. area. Industries inc. manu. armaments, cement, iron, ceramics, glass, textiles, telephone equipment, furniture; sawmilling, flour milling, jewellery making and stone cutting. Pop. (1981C) 649,085.

Jablonec Severočeský, Czechoslovakia. 50 44N 15 10E. Town sit. on R. Nisa in N Czech. Industries inc. glass, paper and textiles. Pop. (1984E) 45,000.

Jáchymov Západočeský, Czechoslovakia. 50 20N 12 55E. Sit. on slopes of Erzgebirge, N of Karlovy Vary. Mining town producing uranium and radium, also important research laboratories. Earlier noted for silver production. The *Thaler* silver coin first minted here. Pop. 10,000.

Jackson Michigan, U.S.A. 42 15N 84 24W. Sit. w of Detroit. Important manu. city esp. vehicles, aircraft components and metal-working machinery. Pop. (1980C) 39,739.

Jackson Mississippi, U.S.A. 32 18N 90 12W. State cap. sit. in central Mississippi on Pearl R. Manu. city particularly cottonseed products. Named after Andrew Jackson, former pres. of U.S.A. Pop. (1980C) 202,895.

Jackson Tennessee, U.S.A. 35 37N 88 49W. Sit. on Forked Deer R., sw of Nashville. Manu. city and R. port. Manu. hardwood flooring, switchgear, food products, textiles. Pop. (1988E) 53,320.

Jacksonville Alabama, U.S.A. City sit. NNE of Anniston. Industries inc. manu. cotton goods, valves, auto parts and cutlery. Pop. (1980C) 9,735.

Jacksonville Florida, U.S.A. 30 20N 81 40W. City on St. John's R. in NE Florida. Named after Andrew Jackson, former pres. of U.S.A. Commercial, transport and trade centre of NE Florida; banking, insurance and financial services, and diverse manu. Part of the Mayo Clinic is here. Port of entry for motor industry. Pop. (1984E) 804,447.

Jadotville *see* Likasi

Jaén Spain. 37 46N 3 47W. (i) Province of S Spain. Area, 5,209 sq. m. (13,492 sq. km.). Pop. (1986C) 633,612. (ii) Town and cap. of Jaén province. Manu. olive oil, soap and agric. products. Pop. (1986C) 102,826.

Jaffa *see* Tel Aviv

Jaffna Northern Province, Sri Lanka. 9 38N 80 02E. Town and port on peninsula of same name, sit. N of Colombo. Intensive agric. including tobacco, curry stuffs and mangoes. Pop. (1981C) 118,215.

Jagersfontein Orange Free State, Republic of South Africa. 29 44S 25 29E. Town sit. sw of Bloemfontein. Woolfarming is the chief occupation. The old open mine hole is a tourist attraction. Pop. (1990E) 9,561.

Jaipur Rajasthan, India. 25 56N 75 50E. Cap. of State and important commercial centre, sit. sw of Delhi. Industries inc. pharmaceuticals. Pop. (1981C) 977,165.

Jakarta Java, Indonesia. 6 08S 106 45E. Cap. of Indonesia since 1949 and formerly the headquarters of the Dutch East India Company, when it was known as Batavia, sit. on NW coast of Java. Industries inc. shipbuilding; manu. textiles, paper and timber products. Sea port exporting rubber, petroleum, tin, coffee and palm-oil. Pop. (1980C) 6,503,449.

Jalálábád Afghánistán. 34 26N 70

25E. Town near Khyber Pass, E of Kábul. Trades in fruit and timber. Pop. (1984E) 62,000.

Jalandhar Punjab, India. Town sit. NNW of Delhi. Industries inc. textiles, engineering and agric. implements. Trades in wheat, sugar and cotton. Pop. (1981C) 408,196.

Jalapa Guatemala. 14 38N 89 59W. (i) Dept. of E central Guatemala sit. in the highlands of the continental divide. Area, 797 sq. m. (2,063 sq. km.). Agric. and stock farming area; main products maize, wheat, beans, tobacco, rice, dairy products. Some lumbering and mining particularly chromite. Pop. (1985E) 171,542. (ii) Town sit. E of Guatemala. Dept. cap. Commercial centre for an agric. area trading in cereals, cattle and pigs.

Jalapa Veracruz, Mexico. 19 32N 96 55W. State cap. sit. NW of Veracruz city. Agric. and manu. town. Medicinal plant from which jalap is prepared grows here but of declining importance. Pop. (1980C) 212,769.

Jalgaon Maharashtra, India. 21 00N 75 42E. Town sit. NE of Bombay. District cap. and trading centre for cotton, millet, linseed, sesame, wheat. Industries inc. cotton processing, oilseed milling and handicraft industries. Pop. (1981C) 145,335.

Jalisco Mexico. 20 00N 104 00W. State on Pacific. Cap., Guadalajara. Area, 31,211 sq. m. (80,836 sq. km.). Mountainous. Cereals, tobacco, sugarcane and cotton are cultivated and it is also an important mining centre. Pop. (1989E) 5,269,816.

Jamaica 18 10N 77 30W. Island in the Caribbean. Jamaica was discovered by Columbus in 1494, and occupied by the Spaniards between 1509 and 1655, when the island was captured by the British; Jamaica achieved complete independence within the Commonwealth in 1962. Cap. Kingston. Area of Jamaica is 4,411 sq. m. (11,525 sq. km.). The island is crossed from E to W by the Blue Mountains, rising to 7,423 ft. (2,263 metres) a.s.l. From processing only a few agric. products such as sugar, rum con-

densed milk, oils and fats, cigars and cigarettes, Jamaica is now producing a wide range of goods such as clothing, footwear, textiles, paints, building materials including cement, agric. machinery and toilet articles. Bauxite is found and a refinery exists in central Jamaica. An oil refinery is in operation at Kingston. Tourism is important, attracting approx. 1m. visitors p.a. Pop. (1988E) 2.4m.

James Bay Canada. 53 30N 80 30W. Inlet in S end of Hudson Bay, 300 m. (480 km.) long and 150 m. (240 km.) wide. Important trading centre of Hudson's Bay Company is sit. at Moose Factory, at mouth of Moose R.

James River Virginia, U.S.A. 37 35N 77 50W. Rises in the Allegheny Mountains and flows to Chesapeake Bay; 340 m. (544 km.) long and navigable, by smaller vessels, for 100 m. (160 km.).

Jamestown St. Helena. 15 56S 5 44W. Cap. and port of island sit. 1,200 m. (1,920 km.) from the W coast of Africa. Exports flax fibre and lily bulbs. Pop. (1984E) 1,600.

Jamestown New York, U.S.A. 42 06N 79 14W. City sit. SW of Buffalo at E end of Chautauqua L. Industries inc. metal working, engineering and food processing. Pop. (1990E) 35,700.

Jamestown Virginia, U.S.A. First Eng. settlement, (1607) on island in James River, but situation proved to be unhealthy and original pop. died of disease. Ruins incorporated in National Park.

Jammu Jammu and Kashmir, India. 32 46N 75 57E. Winter cap. of state. Pop. (1982C) 214,737. *See* **Kashmir.**

Jamnagar Gujarat, India. City sit. WNW of Rajkot on the Kathiawar peninsula. Commercial and communications centre trading in agric. produce. Industries inc. cotton, oilseed and flour milling, tanning, chemicals, ceramics, metalworking, sawmilling; manu. paint, soap, matches. Its port is at Bedi on the Gulf of Cutch. Pop. (1981C) 317,362.

Jamshedpur Bihar, India. 22 48N 86 11E. City sit. WNW of Calcutta near Subarnarekha and Karkai Rs. Tata iron and steelworks founded here in 1909. Now manu. wide range of metallic goods, esp. tin-plate, commercial vehicles and agric. implements. Pop. (1981c) 669,580.

Jämtland Sweden. 62 40N 13 50E. County of NW Sweden bounded W by Norway. Area, 19,273 sq. m. (49,916 sq. km.). Cap. Östersund. Hilly, rising to mountains on the Norwegian border, with Helagsfjäll at 5,892 ft. (1,796 metres). There are many Ls. The main occupations are dairy farming, lumbering, quarrying, tourism, esp. for winter sports. Pop. (1988E) 134,116.

Janesville Wisconsin, U.S.A. 42 41N 89 1W. City sit. SW of Milwaukee on Rock R. Mainly dairying and automotive production. Pop. (1985E) 51,928.

Jan Mayen A bleak, desolate and mountainous island, sit. NNE of Iceland. 71 00N 8 20E. Area, 147 sq. m. (380 sq. km.). Mount Beerenberg, 7,467 ft. (2,297 metres) a.s.l. Volcanic activity, which had been dormant, reactivated in 1970. Incorporated in Kingdom of Norway 1929/30. Radio and meteorological station.

Japan 36 00N 136 00E. The state consists of four main islands: Honshu (mainland), Kyushu, Hokkaido and Shikoku with a total area of 145,882 sq. m. (377,835 sq. km.). Japan's government is based on the Constitution of 1947 which superseded the Meiji Constitution of 1889. The Emperor is a symbol of the States and the unity of the people. Sovereign power rests with the people. Administratively Japan is divided into 47 prefectures. Cap., Tokyo. A mountainous country with many active and dormant volcanoes, and subject to frequent earthquake shocks. Rice is still Japan's most important crop occupying over 40% of the cultivated area. Forests and grasslands cover nearly 70% of the land surface. In 1986 the Japanese catch of fish represented 13.1% of world's total fishing. (In 1939 it was over 50%.) Mining is important and petroleum and natural gas are produced. In 1986 there were 746,734 industrial plants employing 11.5m. production workers. Manu. include cars, electrical appliances, electronic machinery, television sets, radios, cameras, computers and automation equipment. Chemicals, textiles and iron and steel industries are expanding. Pop. (1988E) 122,783,000.

Jarrow Tyne and Wear, England. 54 59N 1 29W. Industrial town and port on R. Tyne. There is light industry, oil and chemicals. Pop. (1988E) 26,960.

Jarvis Island Line Islands, Pacific. 0 23S 160 02W. U.S.-owned island sit. E of Howland near the equator. Area, 1 sq. m. (2.5 sq. km.). Originally worked for guano, now uninhabited.

Jassi *see* **Iaşi**

Jaunpur Uttar Pradesh, India. 25 46N 82 44E. Town sit. on R. Gumti and NW of Varanasi. Produces agric. goods and perfume. Pop. (1981c) 105,140.

Java Indonesia. 7 30S 110 00E. The island is divided into 3 provinces: West Java, Central Java, and East Java. Volcanic mountains exist throughout the island, some peaks reaching 12,000 ft. (3,658 metres) a.s.l. (tree-line 10,000 ft. (3,048 metres) a.s.l.). In N are fertile alluvial plains. Area, 51,200 sq. m. (132,608 sq. km.). Forests produce teak, coconuts, bamboo and spice trees. Agric. products include rice, maize, cassava, sugar, coffee, tea and tobacco. Oil wells exist in the N. Jakarta, formerly Batavia, the national cap., is sit. on the N coast.

Java Sea 4 35S 107 15E. Sit. between Java and Borneo and stretches from W of Sulawesi to E of Sumatra.

Jedburgh Borders Region, Scotland. 55 29N 2 38W. Royal burgh sit. SE of Edinburgh on Jed Water. Manu. woollen goods, precision tools and plastic products. Pop. (1981c) 4,168.

Jedda *see* **Jidda**

Jefferson City   Missouri, U.S.A. 38
34N 92 10W. State cap. sit. w of St.
Louis on R. Missouri. Commercial
centre for agric. area. Employment is
in government services and diverse
manu. Pop. (1980c) 33,619.

Jelgava   Latvia, U.S.S.R. 56 39N
23 42E. Town sit. sw of Riga on
Leilupe R., formerly Mitau. The main
industries are textiles and sugar refining; also manu. linen goods, oilcloth,
rope, woollen goods, bricks, tiles,
leather goods; sawmilling, food processing.

Jena   Gera, German Democratic
Republic. 50 56N 11 35E. Town sit.
between Weimar and Leipzig, renowned for optical and precision
instrument industries. Pop. (1989E)
105,900.

Jerada   Oujda, Morocco. 34 40N
2 10W. Town sit. ssw of Oujda, an
important coalmining centre, esp. for
anthracite.

Jerez de la Frontera   Cádiz, Spain.
36 41N 6 08W. City sit. in sw Spain
and generally known as Jerez (formerly Xeres), and famous for wine to
which it has given its name: sherry.
Once occupied by the Moors, it has
many fine buildings and bodegas.
Pop. (1990E) 184,595.

Jersey   Channel Islands, United
Kingdom. 49 15N 2 10W. Sit. off NW
coast of France and part of the former
Duchy of Normandy. The largest of
the Channel Islands. Area, 45 sq. m.
(116 sq. km.). St. Helier, on s coast, is
the only town. Land highly cultivated,
esp. potatoes and tomatoes. Important
tourist resort. Pop. (1981E) 77,000.

Jersey City   New Jersey, U.S.A. 40
44N 74 02W. Sit. opposite New York
on R. Hudson, connected by tunnel,
also ferry and underground. Industries
include banking, bank office systems,
medical services, transport. Manu.
clothing, foods and electrical goods.
Pop. (1980c) 223,532.

Jerusalem   Jerusalem Hills, Israel.
31 47N 35 15E. Cap. city. Holy city
of Christians, Jews and Moslems.
From 1922–48 cap. of Palestine. In

1948 E part and Old City occupied by
Jordan except for Mount Scopus.
Since the 6-day war of 1967 Israel
holds the city and environs. Jerusalem
became cap. of Israel in 1950. Pop.
(1988E) 493,500.

Jervis Bay   Australian Capital
Territory, Australia. 35 05S 150 44E.
In 1915 28 sq. m. (73 sq. km.) of New
South Wales known as Jervis Bay
were transferred to Australian Capital
Territory as an outlet port for Can-
berra.

Jesselton   *see* **Kota Kinabalu**

Jessore   Bangladesh. 23 10N 89
12E. District cap. sit. on Ganges R.
delta, sw of Dacca. Trades in rice,
jute, linseed, tobacco, cane and tama-
rind. Manu. celluloid and plastics.
Milling rice and oilseeds. Pop.
(1984E) 160,000.

Jethou   Channel Islands, United
Kingdom. Island sit. just E of
Guernsey. Area, 44 acres.

Jewish Autonomous Region   Habarovsk Territory, U.S.S.R. Bordered
in s by R. Amur. Part of the Territory
was established as a Jewish National
District in 1928; as an autonomous
region in 1934. Cap. Birobidzhan.
Area, 13,895 sq. m. (35,988 sq. km.).
Chief industries are non-ferrous
metallurgy, building materials, timber,
engineering, textiles, paper and food
processing. There are 50 factories and
140,000 hectares under crops. Pop.
(1989E) 216,000.

Jhang-Maghiana   Punjab, Pakistan.
31 19N 72 23E. City sit. wsw of
Lahore. Trades in grain. Pop. (1981E)
196,000.

Jhansi   Uttar Pradesh, India. 25
27N 78 35E. Town sit. sw of Kanpur.
Market town and railway junction.
Pop. (1981c) 246,172.

Jhelum   Punjab, Pakistan. 32 58N
73 45E. Town sit. on R. Jhelum N of
Lahore. Commercial centre for timber
and tobacco. Manu. textiles. Pop.
(1981E) 106,000.

Jhelum River   Pakistan. 31 12N 72

08E. One of the 'five rivers' of the Punjab, and an important irrigation source. 450 m. (720 km.) long, rises in Kashmir and flows SSE to join R. Chenab W of Lahore.

Jiangsu (Kiangsu) China. 33 00N 120 00E. Maritime province to E, bordering on the Yellow Sea. A great plain, inc. Yangtse delta, and traversed by Grand Canal which runs N to S. Area, 39,768 sq. m. (103,000 sq. km.). Cap. Nanjing. Chief port, Shanghai. Agric. products inc. tea, rice, cotton, wheat, sugar and silk. Pop. (1987E) 63,130,000.

Jiangxi (Kiangsi) China. 27 30N 116 00E. Province in SE, encircled by mountains except in the N. Area, 65,261 sq. m. (169,000 sq. km.). Cap. Nanchang. Main agric. products are rice, tea, tangerines, cotton and tobacco. Coal and tungsten are mined. Pop. (1987E) 35,090,000.

Jibuti *see* Djibouti

Jidda Saudi Arabia. 21 30N 39 12E. Chief port, sit. on Red Sea, and is the reception centre for pilgrims arriving by air and sea. Pop. (1986E) 1,400,000.

Jihlava Západočeský, Czechoslovakia. 49 24N 15 36E. Sit. on Jihlava R. Industrial and commercial town. Manu. textiles and timber products.

Jilin (Kirin) Manchuria, China. 43 51N 126 33E. (i) Province in NE bordering on Korea in SE. Area, 72,200 sq. m. (187,000 sq. km.). Cap. Changchun. Fertile region, even in the mountainous SE, with important crops of soya-beans, cereals, sugar-beet, tobacco and potatoes. Ginseng is cultivated. Minerals found inc. gold, iron, coal and copper. Pop. (1987E) 23,150,000. (ii) Town and port in province of same name on Sungari R. Manu. chemicals, cement, paper and matches. Connected by railway to Changchun. Pop. (1987E) 1,170,000.

Jimma Kefa, Ethiopia. Cap. of region sit. SW of Addis Ababa on trib. of Gojab R. at 5,740 ft. (1,750 metres) a.s.l. Commercial centre. Trades in coffee, cereals, hides, wax, salt; industries inc. flour milling, sawmilling, tanning, cotton weaving, metal working. Pop. (1984E) 60,992.

Jinan (Tsinan) Shandong, China. 36 38N 117 01E. Cap. of province sit. S of Peking. A railway and industrial centre with manu. of machinery, cement, textiles and paper. Pop. (1987E) 1,460,000.

Jinja Uganda. 00 26N 33 12E. Port on L. Victoria sit. NE of Kampala. Industries inc. copper smelting, brewing, sugar refining. Manu. textiles. Expansion since 1954 because of installation of Owen Falls hydroelectric system. Pop. (1980C) 65,060.

Jinotega Nicaragua. 13 06N 86 00W. Town sit. NNE of Managua. Dept. cap. and trading centre. Industries inc. processing coffee, flour and hides; manu. straw hats. Pop. (1980E) 20,000.

Jinotepe Carazo, Nicaragua. 11 51N 86 12W. Town sit. SSE of Managua on Inter-American Highway. Dept. cap. Trading centre of an agric. area. Industries inc. processing coffee, rice, sugar-cane, salt, limestone, timber. Pop. (1985E) 23,538.

Jinzhou (Chinchow) Liaoning, China. 41 05N 121 03E. Town in the NE sit. WSW of Shenyang (Mukden) on the Peking railway. Industries inc. chemicals and textiles; manu. food products, paper and synthetic fuels. Pop. (1982C) 712,000.

João Pessoa Paraíba, Brazil. 7 07S 34 52W. Town sit. near mouth of R. Paraíba, in NE Brazil. Exports cotton, sugar and coffee through out-port Cabadelo. Pop. (1980C) 290,424.

Jodhpur Rajasthan, India. 26 17N 73 01E. Commercial town sit. WSW of Jaipur. Trades in cotton, wool and hides, and manu. textiles and metal tools. Pop. (1981C) 506,345.

Joensuu Finland. 62 36N 29 46E. Town sit. ESE of Kuopio on a L. in the Saimaa R. system. Railway junction and L. port, handling most of the trade of N Karelia, esp. timber. Lumbering and sawmilling are the main industries. Pop. (1990E) 47,100.

Johannesburg Transvaal, Republic of South Africa. 26 15S 28 00E. Largest city and the most important commercial and industrial centre in S. Afr. Founded 1886 after discovery of gold. Pop. (1985C) 1,609,408.

Johnson City Tennessee, U.S.A. 36 19N 82 21W. Sit. ENE of Knoxville in the Great Appalachian Valley. Manu. inc. timber and dairy products, bricks, textiles, electronics and high technology. Pop. (1988E) 45,420.

Johnston Atoll Pacific. Two small islands sit. SW of Hawaii. A U.S. Defense Dept. base. Area, 1 sq. m. (2.6 sq. km.). Pop. (1985E) 300.

Johnstone Strathclyde Region, Scotland. 55 50N 4 31W. Burgh sit. SW of Paisley. Iron and brass foundries and manu. machine tools.

Johnstown Pennsylvania, U.S.A. 40 20N 78 55W. Manu. city sit. ESE of Pittsburgh. Manu. machinery, iron and steel products, clothing, food processing and chemicals. Pop. (1980C) 35,496.

Johor Peninsular Malaysia, Malaysia. 1 28N 103 45E. The most S state of Peninsular Malaysia, formerly Federation of Malaya. Area, 7,330 sq. m. (18,958 sq. km.). State cap., Johore Baharu. Chief products are rubber, black pepper, timber and coffee. Rubber is principal export shipped through Singapore. Pop. (1980C) 1,601,504.

Joliet Illinois, U.S.A. 41 32N 88 05W. City sit. SW of Chicago on Des Plaines R. Manu. petroleum products, gasoline, steel, wire, chemicals, paper products, baking and packaging machinery, road building equipment. Pop. (1980C) 77,956.

Jonesboro Arkansas, U.S.A. 35 50N 90 42W. Town sit. NW of Memphis (Tennessee). Distribution centre for a farming area. Industries inc. printing, rice milling; manu. flour, footwear, electric motors, agric. machinery. Pop. (1990E) 41,248.

Jönköping Jönköping, Sweden. 57 47N 14 11E. Town sit. at S end of L.

Vätter. Manu. paper, textiles and machinery. Pop. (1990E) 110,000.

Jonquière Quebec, Canada. 48 25N 71 15W. Town sit. N of Quebec. Industries inc. aluminium and timber. Manu. pulp, food, furniture, metal products and paper. Pop. (1989E) 58,467.

Joplin Missouri, U.S.A. 37 06N 94 31W. City sit. S of Kansas City. Centre for distribution of agric. produce. Industries inc. health services, truck haulage. Manu. power units, chemicals, explosives, food products, pet food, batteries. Pop. (1990E) 41,000.

Jordan 31 00N 36 00E. The Hashemite Kingdom of Jordan is a constitutional monarchy. Bordered on N by Syria, NE by Iraq, E and S by Saudi Arabia and W by Israel. Pre-1967 area 37,000 sq. m. (354,830 sq. km.) but 2,165 sq. m. (5,607 sq. km.) were lost in 6-day war. By a treaty of 1946 Brit. recognized Transjordan as a sovereign independent state. The Arab Federation between the Kingdoms of Iraq and Jordan lapsed after the Iraqi revolution in 1958. Jordan is divided into eight districts (*muhafaza*): Amman, Irbid, Balqa, Karak, Ma'an, Jerusalem, Hebron and Nablus. The last three named districts are known collectively as the West Bank and since the hostilities of 1967 have been occupied by Israel. In 1988 Jordan abandoned its efforts to administer the Israeli-occupied West Bank and surrendered its claims to the Palestine Liberation Organization. Cap. Amman. The country E of the Hejaz railway line is largely desert; NW Jordan is potentially of agric. value but entirely dependent on rainfall. Hillsides are being terraced, fruit trees planted and irrigation works constructed. Most of the farms are owner-operated and are less than 25 acres. Phosphates are mined and potash is found in the Dead Sea. Oil prospecting in the S area is being undertaken. Pop. (1988E) 2·97m.

Jordan River 31 46N 35 33E. The principal and only perennial R. of Jordan rises on slopes of Mount Hermon and flows through the Sea of Galilee and Rift Valley into the Dead Sea at 1,292 ft. (394 metres) b.s.l. 160 m. (256 km.) in length and has a total fall of 3,000 ft. (914 metres) b.s.l. The R. provides electric power and irrigation.

Jotunheim Mountains Oppland, Norway. 61 38N 8 18E. Highest mountain range in N. Europe. Highest peaks Galdhöpiggen (8,100 ft./2,469 metres a.s.l.) and Glittertind (8,040 ft./2,451 metres a.s.l.).

Juan Fernández Islands South Pacific. 33 00S 80 00W. Three volcanic island possessions of Chile sit. W of Valparaiso. Formerly penal settlements. Largest Más a Tierra where Alexander Selkirk, prototype of Robinson Crusoe, lived 1704–9. Pop. (1982E) 516.

Juan-les-Pins Alpes-Maritimes, France. Holiday resort sit. SSW of Nice.

Juba River Somalia. 0 12S 42 40E. Rises in the Ethiopian highlands and flows for 1,000 m. (1,600 km.) into the Indian Ocean near Kismayu. Lower reaches fringed by a narrow fertile belt.

Júcar River Spain. 39 09N 0 14W. Rises in the Montes Universales and flows for 314 m. (630 km.) s through Cuenca province, and E through the provinces of Albacete and Valencia to Mediterranean at Cullera. R. used for irrigation and hydroelectricity.

Judaea Israel. Area of SW Israel bounded E by Jordan R. Hilly, rising to 3,340 ft. (1,018 metres) a.s.l. near Hebron. Chief city Jerusalem. Descending to plain near Dead Sea and W near Mediterranean. The lowland areas are densely populated and support agric. and stock farming.

Jugoslavia *see* **Yugoslavia**

Juiz de Fora Minas Gerais, Brazil. 21 45S 43 20W. Industrial town on Paraibuna R. Manu. textiles, especially knitted goods. Trades in coffee, timber and livestock. Pop. (1980C) 299,728.

Jujuy Argentina. Cap. of Jujuy pro-

vince sit. NW of Buenos Aires at 4,000 ft. (1,219 metres) a.s.l. Industries inc. flour milling and timber products. Pop. (1984E) 120,000.

Julfa Third Province, Iran. Town NNW of Tabriz on Aras R. forming the frontier with U.S.S.R. Railway terminus and route centre.

Jumet Charleroi, Belgium. 50 26N 4 25E. Town sit. N of Charleroi. Glass making is an important industry. Pop. (1983E) 25,716.

Jumna River India. Rises in Uttar Pradesh, in the Himalayas, and flows for 860 m. (1,376 km.) past Delhi, Agra and Etawah joining Ganges R. near Allahabad. It feeds the Eastern and Western Jumna canals which were built in the 19th cent. for irrigation.

Junagadh Gujarat, India. Town sit. NNW of Diu. Manu. pharmaceutical products and metal goods. Pop. (1981E) 120,416.

Jundíaí São Paulo, Brazil. 23 11S 46 52W. Town sit. NW of São Paulo. Manu. cotton goods and steel. Trades in fruit, grain and coffee. Pop. (1984E) 210,000.

Juneau Alaska, U.S.A. 58 20N 134 27W. Cap. of state and ice-free port. Chief occupations are state and federal government employment, fishing and tourism. Pop. (1980C) 22,680.

Jungfrau Bernese Oberland, Switzerland. 46 32N 7 58E. Mountain, 13,669 ft. (4,166 metres) a.s.l., first climbed in 1811. Railway from Scheidegg reaches 11,340 ft. (3,456 metres) a.s.l. Opened 1912.

Junín Argentina. 34 35S 60 58W. Town sit. W of Buenos Aires. Manu. timber products; railway engineering. Trades in grain and cattle. Pop. (1984E) 60,000.

Jura France. Dept. in Franche-Comté region, adjoining border with Switzerland. Area, 1,934 sq. m. (5,008 sq. km.). Mountainous and drained by Rs. Doubs and Ain. Cap. Lons-le-Saunier. Pop. (1982C) 242,925.

Jura   Inner Hebrides, Scotland.
56 00N 5 50W. Mountainous island,
with peak of Paps of Jura 2,571 ft.
(784 metres) a.s.l. Fishing and sheep
farming main occupations. The whirlpool of Corrievrecken is sit. off N
coast.

Jura   Switzerland. 46 47N 5 45E.
Canton created 1979 from part of the
canton of Bern. Area, 323 sq. m. (837
sq. km.). Cap. Delémont. Pop.
(1980C) 69,986.

Jura Mountains   France/Switzerland.
46 45N 6 30E. Limestone range running for 150 m. (240 km.) SW–NE with
a breadth of *c*. 40 m. (64 km.).

Juruá River   Brazil/Peru. 2 37S 65
44W. Rises in E Peru near Brazilian
border and flows to the NE for 1,200
m. (1,920 km.), joining Amazon R. *c*.
100 m. (160 km.) N of Tefé.

Jutland   56 25N 9 30E. Peninsula
containing areas of Denmark and Federal Republic of Germany. Bordered
by the Skagerrak in N, the Kattegat
and Little Belt in the E and the North
Sea in the W. Dairy farming chief
occupation.

Jylland   *see* **Jutland**

Jyväskylä   Vaasa, Finland. 62 14N
25 44E. Town sit. on L. Päijänne.
Railway junction. Manu. timber products and paper machinery. Pop.
(1989E) 66,500.

K

Kabardino-Balkar Autonomous Soviet Socialist Republic Russian Soviet Federal Socialist Republic, U.S.S.R. Area, 4,825 sq. m. (12,500 sq. km.). Annexed to Russia 1557 and constituted an Autonomous Republic 1936. Cap. Nalchik, 320 m. (512 km.) SE of Rostov. Chief industries are ore-mining, engineering, coal, food processing, timber and light industries, building materials and varied agric. Pop. (1989c) 235,000.

Kábul Afghánistán. 34 30N 69 10E. Cap. of Afghánistán. An anc. town commanding passes to N and W inc. the Khyber Pass. Manu. textiles, footwear, vehicle components, matches and timber products. Pop. (1984E) 1,179,341.

Kábul River Afghánistán. Rises in the Hindu Kush and flows for 315 m. (510 km.) SE and joins R. Indus at Attock.

Kabwe Zambia. 14 27S 28 27E. Formerly Broken Hill. Town on railway from Lusaka. Mining for lead, zinc and vanadium is the main occupation. Pop. (1980E) 143,635.

Kadievka Ukraine, U.S.S.R. 48 34N 38 40E. Coalmining town sit. NNW of Rostov, with important metallurgical and chemical industries. Pop. (1977E) 141,000.

Kadiyevka *see* **Kadievka**

Kadoma Zimbabwe. Town sit. SW of Harare in a farming and cotton growing area. Industries inc. nickel refining, cotton milling, processing agric. products; manu. glass, paper, bricks, timber products. There is a cotton research station. Pop. (1990E) 100,000.

Kaduna Nigeria. 10 30N 7 21E. State comprising the Katsina and Zaria provinces of the former N region. Area, 27,122 sq. m. (70,245 sq. km.), inc. the emirates of Zaria, Katsina and Daura. The majority of the inhabitants are Hausa, with smaller groups of Fulani and (in the S) Gbari. Chief cities Kaduna (cap.), Zaria and Katsina. Pop. (1983E) 6.3m.

Kaduna Kaduna, Nigeria. 10 33N 7 27E. Town, railway junction and manu. centre chiefly textiles and timber products. Pop. (1981E) 276,000.

Kaesong North Korea. 37 59N 126 33E. City sit. NW of Seoul. Famous for porcelain. Trades in ginseng for medicinal purposes.

Kaffa (Kefa), Ethiopia. Region in SW bordering Sudan. Cap. Jimma. Mountainous with peaks reaching over 10,000 ft. (3,048 metres) a.s.l. Area, 20,849 sq. m. (54,600 sq. km.). Coffee grown on mountain slopes. Pop. (1980E) 1,615,000.

Kafr el Sheikh Gharbiya, Egypt. 31 07N 30 56E. Town sit. NNW of Tanta. Industries inc. cotton ginning; manu. cigarettes. Pop. (1970E) 64,000.

Kafue River Zambia. 15 56S 28 55E. Rises in Zaïre and flows for 600 m. (971 km.) S and then E joining the Zambezi at Chirundu 125 m. (200 km.) NW of Sinoia.

Kagera River Central Africa. 0 57S 31 47E. Formed by confluence of R. Ruvuvu and R. Nyavarongo, flows 300 m. (480 km.) N and E along the Rwanda–Tanzania and Tanzania–Uganda frontiers, finally entering L. Victoria.

Kagoshima Japan. 31 36N 130 33E. Cap. of Kagoshima prefecture. Sit. SSE of Nagasaki in Kyushu Island. Manu. pottery, textiles, metal goods. Pop. (1980E) 505,000.

Kagul Moldavian Soviet Socialist

Republic, U.S.S.R. Town sit. NNW of
Galati on Prut R. where it forms the
frontier with Romania. Industries inc.
wine making, flour milling.

Kahoolawe   Hawaii, U.S.A. 20 33N
156 37W. Island SW of Maui across
the Alalakeiki Channel. Rocky, rising
to 1,444 ft. (440 metres) a.s.l. Infertile
and uninhabited. Military target-prac-
tice zone.

Kaikoura Ranges   New Zealand.
42 25S 173 43E. Two mountain
ranges in S. Island, separated by
Clarence R. and rising to 9,465 ft.
(2,885 metres) a.s.l.

Kailas Range   Tibet, China. 31 00N
82 00E. Mountain range in SW between N chain of the Himalaya and
the Kangri Mountains. The two highest peaks are Kailas (22,028 ft., 6,714
metres), sacred to Hindus and Buddhists, and Lombo Kangra (23,165 ft.,
7,061 metres). It is the source of R.
Indus, R. Sutlej and R. Brahmaputra.

Kaimanawa Mountains   New Zealand. Sit. in centre of N. Island,
stretching 60 m. (96 km.) E and S of L.
Taupo. Highest peak 5,665 ft. (1,727
metres).

Kairiru   Papua New Guinea. Island
off the NE coast of New Guinea.
About 9 m. (15 km.) long; volcanic,
rising to 3,350 ft. (1,021 metres). The
main product is coconut.

Kairouan   Tunisia. 35 42N 10 01E.
City and a Moslem holy place of pilgrimage with numerous mosques and
a citadel. Manu. carpets, leather goods
and copper wares. Pop. (1984E)
72,000.

Kaiserslautern   Rhineland-Palatinate, Federal Republic of Germany. 49
26N 7 46E. Industrial town on Lauter
R. W of Ludwigshafen. Railway junction. Manu. textiles, furniture and
tobacco. Pop. (1984E) 98,700.

Kakhetia   Georgian Soviet Socialist
Republic, U.S.S.R. Region of E
Georgia drained by upper Alazan R.
Chief towns Telavi, Signakhi, Gurdz-
haani. Important wine producing area.

Kakinada   Andhra Pradesh, India.

16 57N 82 14E. Port sit. ESE of
Hyderabad. Manu. cotton and flour
milling. Cotton and groundnuts are
main exports. Pop. (1981C) 226,409.

Kalahari   Southern Africa. 24 00S
22 00E. Semi-desert area, 3,000–
4,000 ft. (914–1,219 metres) high, between R. Zambezi and R. Orange,
forming a large area of Botswana. All
rivers are periodic except the
Okavango R. which flows into L.
Ngami, now only a swamp. Inhabited
by nomadic Bushmen and Bakalahari.
Gemsbok, springbok, ostriches, etc.,
are found in the Kalahari National
Park in the SW.

Kalámai   *see* **Calamata**

Kalamazoo   Michigan, U.S.A. 42
17N 85 32W. City sit. S of Grand
Rapids on Kalamazoo R. Manu.
paper, aircraft and missile components, medical equipment, light
engineering, pharmaceutical products.
Pop. (1990E) 77,230.

Kalat   Pakistan. 29 08N 66 31E.
Division of SW Pakistan, bounded by
Afghánistán in N, Arabian Sea in S
and Iran in W. Area, 98,975 sq. m.
(256,345 sq. km.). Mainly mountainous but grows fruit, inc. dates.
Pop. (1981E) 11,000.

Kalemie   Shaba, Zaïre. 5 56S 29
12E. Port sit. on W shore of L.
Tanganyika; formerly Albertville.
Railway terminus. Manu. cotton
goods. Pop. (1976C) 172,297.

Kalgoorlie-Boulder   Western Australia, Australia. 30 45S 121 28E.
Formerly 2 towns of Kalgoorlie and
Boulder amalgamated in 1966. Town
and important gold-mining centre,
water being pumped nearly 350 m.
(560 km.) from Mundaning Weir in
the Darling Range. Railway junction,
airport and is also centre for Flying
Doctor. Pop. (1981E) 20,000.

Kalimantan   Indonesia. 0 00 115
00E. Largest part of the island of
Borneo. Area, 208,000 sq. m.
(538,720 sq. km.). Oil is important
industry. Pop. (1980E) 6,723,086.

Kalinin   (Tver) U.S.S.R. 56 52N 35

55E. Cap. of Kalinin region. Sit. NW of Moscow, at confluence of R. Volga and R. Tversta. There are large textile and engineering industries. Pop. (1985E) 438,000.

Kaliningrad U.S.S.R. 54 43N 20 30E. Cap. of Kaliningrad region, and important ice-free Baltic port, at mouth of Pregel R. Exports inc. grain, flax, timber, etc. Industries inc. shipbuilding, engineering, chemicals, food products, also paper and flour milling. Pop. (1987E) 394,000.

Kalisz Poznań, Poland. 51 46N 18 06E. Town sit. on Prosna R. Manu. textiles, leather, metal goods and food products. Pop. (1983E) 103,000.

Kalmar Sweden. 56 40N 16 20E. (i) Country in SE, inc. island of Öland. Area, 4,450 sq. m. (11,171 sq. km.). Pop. (1983E) 240,134. (ii) Cap. of (i) and port on E coast. Manu. matches, paper, shipbuilding. The famous Orrefors and Kosta glassworks are nearby. Pop. (1983E) 53,516.

Kalmyk Autonomous Soviet Socialist Republic Part of Russian Soviet Federal Socialist Republic, U.S.S.R. 46 05N 46 01E. Area, 29,300 sq. m. (75,887 sq. km.). First constituted an Autonomous Republic 1935 but this was dissolved 1943; reconstituted 1958. Cap. Elista, sit. W of Astrakhan. Chief industries are fishing, canning, building materials; also cattle breeding and irrigated farming. Pop. (1984E) 315,000.

Kaluga U.S.S.R. 54 31N 36 16E. Cap. of Kaluga region. Sit. SSW of Moscow, on Oka R. Industries inc. sawmilling, engineering, iron and steel goods, bricks, glass, matches and food products. Pop. (1987E) 307,000.

Kalyan Maharashtra, India. 19 15N 73 08E. Town and port for coastal trade sit. NE of Bombay. Manu. rayon, bricks, tiles. Also rice-milling. Pop. (1981C) 136,052.

Kalymnos Greece. (i) Island in the Dodecanese Islands, in the Aegean Sea. Area, 41 sq. m. (106 sq. km.). (ii) Chief town near SE coast of island.

Agric. products inc. olive oil and citrus fruits.

Kamaran Islands Republic of Yemen. 15 21N 42 34E. A group of islands sit. in the Red Sea, of which Kamaran is the largest. Quarantine base for pilgrims to Mecca. Area, 22 sq. m. (57 sq. km.). Pop. (1980E) 3,000.

Kama River U.S.S.R. 55 45N 52 00E. Rises in N of the Udmurt and flows N, E and finally SW for 1,200 m. (1,920 km.) to join Volga R. below Kazan.

Kamchatka U.S.S.R. 56 00N 160 00E. Mountainous, volcanic region in the Khabarovsk Territory in NE U.S.S.R., inc. Kamchatka peninsula, the Chukot and Koryak National Areas and the Komandorski Islands. Area, 490,425 sq. m. (1,270,200 sq. km.). Industries inc. fishing, fish-processing, fur-trapping and woodworking.

Kamenetz-Podolsk *see* **Khmelnitsky**

Kamensk-Uralski U.S.S.R. 56 28N 61 54E. Town sit. ESE of Sverdlovsk. Manu. machine tools, pipes, etc. Also aluminium refining and bauxite mining. Pop. (1987E) 204,000.

Kamloops British Columbia, Canada. 50 40N 120 20W. City sit. on Thompson trib. of Fraser R. Railway and highway junction. Lumbering, mining and agric. are the chief industries. Pop. (1989E) 62,261.

Kampala Uganda. 0 19N 32 35E. Cap. since 1962, and commercial centre. Chief market for L. Victoria region exporting coffee, tea, cotton, tobacco, hides, beans, maize and vegetables. Manu. metal products, textiles, beverages and furniture. Food processing and tea blending are also important. Connected by railway to Mombasa, and to Dar-es-Salaam by ferry wagon and railway. Pop. (1990E) 800,000.

Kampen Overijssel, Netherlands. 52 33N 5 54E. Town sit. on Ijssel R. Manu. machinery, photographic equipment, cigars, enamelware. Pop. (1990E) 32,700.

Kampot   Kampot, Cambodia. 10 36N 104 10E. Town sit. SSW of Phnom Penh at the foot of the Elephant Mountains on the Gulf of Siam. The main product is pepper, with some local limestone and phosphates. Shallow water port. Pop. (1981E) 337,879.

Kampuchea, Democratic   *see* **Cambodia**

Kamyshin   Volgograd, U.S.S.R. 50 06N 45 24E. Town and port on Volga R. opposite Nikolayevski. Agric. trade inc. melons, grain, timber.

Kananga   Kasai-Occidental, Zaïre. 5 54S 22 25E. Cap. of region, sit. ESE of Kinshasa on Lulua R., a trib. of Kasai R. Commercial centre. Pop. (1976C) 704,211.

Kanazawa   Honshu, Japan. 36 34N 136 39E. Cap. of the Ishikawa prefecture. Sit. WNW of Tokyo. Manu. textiles, porcelain, metal leaf, machinery. Pop. (1983C) 412,000.

Kanchenjunga   Himalayas. 27 42N 88 09E. Third highest mountain in the world (28,168 ft./8,586 metres). Sit. on Nepál/India (Sikkim) border.

Kandahár   Afghánistán. 31 36N 65 47E. Cap. of Kandahár province. Industries inc. textiles, fruit-canning. Pop. (1984E) 203,000.

Kandal   Cambodia. Province around Phnom Penh. Area, 1,500 sq. m. (3,885 sq. km.). Cap. Phnom Penh, which is independently administered. Drained by Tonle Sap and Bassac Rs. into Mekong R. Pop. (1981E) 701,000.

Kandersteg   Bern, Switzerland. 46 30N 7 40E. Tourist resort. Sit. at 3,940 ft. (1,201 metres) a.s.l. at N end of Lötschberg tunnel in SW central Switz. Pop. (1989E) 960.

Kandi   Borgou, Benin. 11 08N 2 56E. Town in N in centre of iron-rich area. Pop. (1982E) 53,000.

Kandla   Gujarat, India. 23 02N 70

13E. Port sit. SE of Anjar. Railway terminus and port on the Gulf of Cutch, trading in grain, textiles, timber. Some salt extraction. Pop. (1981C) 23,978.

Kandy   Central Province, Sri Lanka. 7 18N 80 38E. Town sit. on Mahaweli R. ENE of Colombo. Centre of tea plantations and spices. Famous temple containing supposed tooth of Buddha. Pop. (1985E) 140,000.

Kangaroo Island   South Australia, Australia. 35 50S 137 06E. Third largest offshore island. Sit. in St. Vincent Gulf. Area, 1,710 sq. m. (4,429 sq. km.). Products inc. salt, feldspar, tourmaline and gypsum.

Kangwon   North Korea. Province bounded E by the Sea of Japan. Cap. Chunchon. Mountainous and forested, drained by Han R. Minerals inc. coal, gold, tungsten. Main industries inc. shipbuilding, vehicles, refining and cement manu. Pop. (1977E) 1m.

Kankakee   Illinois, U.S.A. 41 07N 87 52W. City sit. S of Chicago on Kankakee R. Manu. agric. machinery, bricks and furniture. Pop. (1980C) 30,141.

Kankakee River   U.S.A. 41 23N 88 16W. Rises near South Bend, Indiana, flows 230 m. (368 km.) SW and then NW joining Des Plaines R. and eventually becomes Illinois R.

Kankan   Guinea. 10 23N 9 18W. Town sit. ENE of Conakry. Important agric. production inc. rice. Industries inc. food and cotton processing, bicycle manu. Pop. (1972C) 85,310.

Kanniyakumari   Tamil Nadu, India. Southernmost tip of India, meeting point of the Bay of Bengal, the Arabian Sea and the Indian Ocean, a place of pilgrimage for the Hindus. Pop. (1981C) 14,084.

Kano   Nigeria. 12 02N 8 30E. State formed from the former Kano province of the N region. Area, 16,712 sq. m. (43,285 sq. km.). Almost entirely populated by the Hausa people, the state comprises the emirates of Kano, Gumel, Hadejia and Kazaure. Chief cities Kano (cap.), Gumel and Hadejia. Pop. (1983E) 8.8m.

Kano   Kano, Nigeria. 12 00N 8 30E. Chief town of State. Important trading centre for cotton, groundnuts and hides. Manu. cotton goods, leather, furniture and soap. Pop. (1981E) 545,000.

Kanpur   Uttar Pradesh, India. 26 28N 80 20E. City, sit. on Ganges R. Important industrial centre, dealing in textiles, cotton, fertilizers, leather, chemicals, sugar, grain and oilseeds. Engineering, aeronautics and flour milling also important. Pop. (1981C) 1,486,522.

Kansas   U.S.A. 38 40N 98 00W. Central state bounded by Missouri on E, Oklahoma on S, Colorado on W and Nebraska on N. Area, 82,264 sq. m. (213,064 sq. km.). Cap. Topeka. Chief towns, Wichita, Kansas City, Over-land Park. Mainly undulating lowland, rising to 4,135 ft. (1,260 metres) in W. Chief rivers, Missouri, Kansas and Arkansas. Almost wholly agric. with wheat as main crop; other crops inc. maize, soybeans and sorghums. Kan-sas is one of 4 greatest cattle-pro-ducing states, with allied industries of meat packing, dairy products and manu. of agric. machinery. Minerals inc. natural gas, in large quantities, coal, lead, zinc and petroleum. There are oil refineries, flour mills, sugar-beet and glass works. Pop. (1985E) 2,450,000.

Kansas City   Kansas, U.S.A. 39 07N 94 39W. Sit. at confluence of Missouri R. and Kansas R. Commer-cial centre of region with warehous-ing; automobile assembly and food processing important. Other manu. inc. agric. machinery, steel and foundry products, cement, soap and paper. Pop. (1980C) 161,087.

Kansas City   Missouri, U.S.A. 39 05N 94 35W. Sit. on Missouri R. at confluence with Kansas R. opposite Kansas City, Kansas. Important trans-port, telecommunications and indus-trial centre, also distributing agric. produce. Industries inc. agric. machin-ery, greeting cards, refrigerators, cars, chemicals and food products. Also

flour milling and livestock centre. Pop. (1980C) 448,159.

Kansas River   U.S.A. 39 07N 94 36W. Formed by union of Smoky Hill R. and Republican R., it flows E for 170 m. (273 km.) joining Missouri R. between Kansas City, Kansas, and Kansas City, Missouri.

Kansk   Krasnoyarsk, U.S.S.R. 56 13N 95 41E. Industrial town sit. E of Krasnoyarsk, on Kan R. Mining, inc. coal, lignite and mica, is important. Industries inc. textiles, wood-processing and sawmills.

Kansu   *see* **Gansu**

Kanye   Botswana. 24 59S 25 19E. Town sit. NNW of Mafeking, asbestos mining centre. District cap. and centre for the Bangwaketse tribe. Pop. (1984E) 22,000.

Kaohsiung   Taiwan. 22 38N 120 17E. Largest port and second largest city on SW coast S of Taipei. Important industries inc. shipbuilding, oil-refin-ing, steel and cement works, textiles and chemicals. Pop. (1988E) 1·3m.

Kaolack   Senegal. 14 09N 16 04W. Port on Saloum R. sit. ESE of Dakar. Large exports of groundnuts. Pop. (1985E) 132,400.

Kapfenberg   Styria, Austria. 47 26N 15 18E. Town sit. NNW of Graz on Mürz R. Industries inc. iron and steel works and paper mills. Pop. (1981C) 25,716.

Kaposvár   Somogy, Hungary. 46 22N 17 47E. Cap. of county, sit. SW of Budapest on Kapos R. Manu. wine, tobacco, textiles and food products. Pop. (1989E) 74,000.

Kara Bogaz Gol   Turkmenistan, U.S.S.R. 41 00N 53 15E. Gulf on E side of Caspian Sea, with large deposits of chemical salts caused by rapid evaporation. Area, approx., 8,000 sq. m. (20,720 sq. km.).

Karachayevo-Cherkess   Autono-mous   Region   Part of Stavropol Territory, U.S.S.R. Area, 5,442 sq. m. (14,048 sq. km.). Present Autonomous

Region re-estab. 1957, having been originally estab. 1926 and dissolved 1943. Cap. Cherkessk, sit. approx. 225 m. (360 km.) NNW of Tbilisi. Chief industries are ore-mining, engineering, chemical and wood-working; also livestock breeding and grain growing, with 203,000 hectares under crops. Pop. (1984E) 384,000.

Karachi   Pakistan. 24 52N 67 03E. Port sit. on Arabian Sea NW of R. Indus delta. Area, inc. neighbouring islands, 812 sq. m. (2,103 sq. km.). The main exports are cotton, grain, oilseeds, wool and hides. Manu. inc. chemicals, textiles, tyres, cement, steel and metal products. Pop. (1981C) 5,103,000.

Karad   Maharashtra, India. Town sit. SSE of Satara on Krishna R. Route junction and trading centre for cereals and peanuts. Industries inc. cotton and silk manu., manu. agric. implements. Pop. (1981C) 54,364.

Karaganda   U.S.S.R. 49 50N 73 10E. (i) Region of Kazakhstan, sit. in Kazakh hills. Area, 156,700 sq. m. (405,853 sq. km.). Semi-desert and large deposits of coal, copper, iron ore and manganese. (ii) Cap. of region of same name sit. NNW of L. Balkhash. Manu. inc. iron, steel and cement. Pop. (1985E) 617,000.

Karaj   Iran. 35 48N 50 58E. Town sit. WNW of Tehrán on Karaj R. Industrial centre of an agric. area. Industries inc. manu. chemical products, beet sugar. There is an agric. college. Pop. (1986C) 275,100.

Kara-Kalpak Autonomous Soviet Socialist Republic   Part of Uzbekistan, U.S.S.R. 43 00N 59 00E. Area, 63,920 sq. m. (165,553 sq. km.). Constituted as an Autonomous Region within the then Kazakh Autonomous Republic 1925, then became an Autonomous Republic within the Russian Federation 1932 and then part of Uzbek Soviet Socialist Republic 1936. Cap. Nukus, sit. 500 m. (800 km.) WNW of Tashkent. Chief industries are bricks, leather goods, furniture, canning, wine. Agric. inc. cotton growing and rearing of cattle, sheep and goats. Pop. (1984E) 1,044,000.

Karakoram   North Kashmir. 35 30N 77 00E. A mountain range between Sinkiang-Uighur, China and India, extending 300 m. (480 km.) NW to the Pamir. Contains 33 peaks rising above 24,000 ft. (7,315 metres) inc. K2 (28,250 ft./8,611 metres) and many glaciers. The main passes are the Karakoram, the Muztagh and the Hispar.

Kara-Kul, Lake   Tadzhikistan, U.S.S.R. 39 05N 73 25E. The 'Black Lake' sit. on the plateau between the Pamir and the Alai Mountains, near China–U.S.S.R. border. Nearly 13,000 ft. (3,962 metres) a.s.l. Area, 140 sq. m. (363 sq. km.).

Kara-Kum   Turkmenistan, U.S.S.R. The 'Black Sands', a desert between Caspian Sea and Amu Darya R., S and SE of the Ust Urt plateau. Area, 110,000 sq. m. (284,900 sq. km.). The 500 m. (800 km.) Kara-Kum Canal runs from the Amu Darya through two oases to Ashkhabad.

Kara Sea   Arctic Ocean. Sit. between Novaya Zemlya and Severnaya Zemlya, and off N coast of the Russian Soviet Federal Socialist Republic. Outlet for Ob R. and Yenisei R. Chief port is Novy Port.

Karbala   Karbala, Iraq. City sit. SSW of Baghdad at the edge of the Syrian Desert. Provincial cap. and pilgrims' city, with the tomb of Mohammed's grandson Husein, an important Shiite shrine. Trading centre for dates, hides, wool. Pop. (1985E) 184,574.

Karditsa   Thessaly, Greece. 39 21N 21 55E. (i) Prefecture, area, 995 sq. m. (2,576 sq. km.). Chiefly agric. Pop. (1981C) 124,930. (ii) Cap. of Prefecture, trading in agric. products. Pop. (1981C) 27,291.

Karelia   Finland. The E marches of Finland. In 1944 some 18,000 sq. m. (46,620 sq. km.) inc. the city of Vyborg were ceded to U.S.S.R.

Karelian Autonomous Soviet Socialist Republic   Part of Russian Soviet Federal Socialist Republic, U.S.S.R. 65 30N 32 30E. Area, 66,564 sq. m. (172,400 sq. km.).

Formed as Republic 1923, formerly a Labour Commune. Cap. Petrozavodsk, sit. NE of Leningrad. Chief industries are timber, paper-cellulose, mica, chemicals, electrical and furniture, also varied farming. Pop. (1978E) 769,000.

Karelian Isthmus U.S.S.R. Tongue of land, strategically important, between the Gulf of Finland and L. Ladoga. Chief city, Leningrad.

Kariba, Lake Zimbabwe/Zambia. 17 00S 28 00E. Artificial L. created by the building of the Kariba Dam, on the Zambezi R. Length, 175 m. (280 km.) width, 12 m. (19 km.).

Karkar Papua New Guinea. Island sit. NE of New Guinea. Area, 140 sq. m. (363 sq. km.). Volcanic and active, rising to 4,900 ft. (1,494 metres) a.s.l. Chief products coconuts, cacao.

Karl-Marx-Stadt *see* **Chemnitz**

Karlovac Croatia, Yugoslavia. 45 29N 15 34E. Town sit. on railway from Zagreb to Split. Manu. inc. textiles and footwear.

Karlovy Vary West Bohemia, Czechoslovakia. 50 11N 12 52E. Spa sit. WNW of Prague on Ohře R. Manu. glass, porcelain and footwear. Kaolin is found in the area. Pop. (1983E) 59,696.

Karlskoga Sweden. 59 20N 14 31E. Industrial town sit. NE of L. Väner, famous for Nobel munition works. Pop. (1988E) 34,316.

Karlskrona Blekinge, Sweden. 56 10N 15 35E. Cap. of country and port on the Baltic coast. Chief naval station of Sweden, and manu. naval equipment. Other industries inc. granite quarrying, sawmilling and brewing. Exports inc. granite, timber, tobacco. Pop. (1983E) 59,696.

Karlsruhe Baden-Württemberg, Federal Republic of Germany. 49 03N 8 24E. Town sit. just E of R. Rhine. Railway centre. Important river harbour. Manu. machinery, bicycles, cosmetics, pharmaceuticals, electrical equipment, electronics and food products. Pop. (1985E) 268,033.

Karlstad Värmland, Sweden. 59 22N 13 30E. Cap. of county and port standing at N end of L. Väner at outlet of Klar R. Manu. inc. machinery and coffee preparation. It has timber, textile and pulp mills. Pop. (1983E) 73,939.

Karnal Haryana, India. 29 41N 76 59E. Town near the Jumna R. and S of Simla. Manu. inc. cotton goods. Pop. (1981C) 132,107.

Karnataka India. 13 15N 77 00E. State formerly called Mysore on the Deccan plateau of S India, bounded W by Arabian Sea, N by Maharashtra. Area, 74,054 sq. m. (191,791 sq. km.). Chief cities Mysore, Bangalore (cap.), Mangalore, Belgaum. Plateau, bounded SW and SE by W and E Ghats, average height a.s.l. 2,500 ft. (762 metres), rising to 6,310 ft. (1,923 metres) in Mulliangiri (SW). Tropical climate with temps. 60°F–90°F (15°C–32°C) and heavy rain in W Ghats. Mainly agric. with some valuable forest and mineral deposits. Chief products coffee, cardamom, pepper, cereals, oilseeds, tobacco, mangoes, silk (E and S), sandalwood (S) and other timber, bamboo, livestock. The Kolar gold fields produce 95% of Indian gold; there is manganese (Shimoga), iron (Baba Budan range), asbestos, copper, mica, limestone, iron ore, corundum, kaolin. Industries inc. textiles (inc. silk), sugar, cement, chemicals, pharmaceuticals, fertilizers, glass, ceramics, telephonic, electronic and electrical equipment, earth movers, iron and steel, building aircraft, paper making; manu. food products. Pop. (1981C) 37,135,714.

Kärnten *see* **Carinthia**

Kárpathos Greece. 35 40N 27 10E. Island of the Dodecanese group in the Aegean Sea, SW of Rhodes. Area, 111 sq. m. (287 sq. km.). Chief town is Pigádhia. Mainly agric.

Karpenission Eurytania, Greece. Town sit. W of Lamia at 3,280 ft. (1,000 metres) a.s.l. Bishopric. *Nome* cap. Centre of a dairy farming area manu. dairy products. Pop. (1981C) 5,100.

Karroo Cape Province, Republic of

South Africa. High plateau between coastal mountains and Orange R. The area is divided by the Zwarteberg range into the Little Karroo in the S 1,000–2,000 ft. (305–610 metres) a.s.l. and the Great Karroo in the N 2,000–3,000 ft. (610–915 metres) a.s.l. Mainly very dry but some agric. possible with irrigation.

Kars   Turkey. 40 36N 43 05E. (i) Mountainous province in Asiatic Territory. Area, 6,710 sq. m. (17,379 sq. km.). There are salt deposits in the S. Pop. (1985C) 722,431. (ii) Cap. of above, sit. SW of Tbilisi, U.S.S.R. Manu. textiles and carpets.

Karst   Yugoslavia. Limestone area joining the E Alps to the Dinaric Alps, E of Istria. Ridged surface with poor drainage, and many underground streams and caves containing stalactites and stalagmites.

Karun River   Iran. 30 25N 48 12E. Rises W of Esfáhán, flows for 530 m. (848 km.) W and S to enter delta of Shatt-al-Arab at Khorramshahr. Navigable to Ahwaz.

Karviná   Severomoravský, Czechoslovakia. 49 50N 18 30E. Town sit. E of Ostrava, in the Ostrava-Karviná coal basin. Industries connected with gas, coke and iron. Pop. (1989E) 72,000.

Kasai-Occidental   Zaïre. Region in S. Diamonds and gold are mined, but region is mainly agric. producing cassava, yams, cotton and coffee. Area, 60,605 sq. m. (156,967 sq. km.). Chief towns Kananga (cap.) and Ilebo. Pop. (1981E) 2,935,036.

Kasai-Oriental   Zaïre. Region in S. Area, 64,948 sq. m. (168,216 sq. km.). The majority of the world's industrial diamonds are found here in the numerous rivers. Cap. Mbuji-Mayi. Pop. (1981E) 2,336,951.

Kasai River   Angola/Zaïre. 3 02S 16 57E. Rises in Angola and flows 1,300 m. (2,080 km.) N forming frontier with Zaïre, before joining Congo R., of which it is a main trib. Alluvial diamonds found in quantity.

Kasama   Northern Province, Zambia. 10 13S 31 12E. Town sit. NE of Lusaka. Provincial cap. and centre of a farming area.

Káshán   Iran. 33 59N 51 29E. Town sit. S of Tehrán. Manu. inc. carpets, cotton, silk and copperware.

Kashmir   35 00N 78 00E. The state of Jammu and Kashmir, which had earlier been under Hindu rulers and Moslem sultans, became part of the Mogul Empire under Akbar from 1586. After a period of Afghan rule from 1756, it was annexed to the Sikh kingdom of the Punjab in 1819. In 1820 Ranjit Singh made over the territory of Jammu to Gulab Singh. After the decisive battle of Sobraon in 1846 Kashmir also was made over to Gulab Singh under the Treaty of Amritsar. Brit. supremacy was recognized until the Indian Independence Act, 1947, when all states decided on accession to India or Pakistan but later Kashmir asked for standstill agreements with both. Area is 85,805 sq. m. (222,236 sq. km.), of which about 28,159 sq. m. (72,932 sq. km.) is occupied by Pakistan and 16,500 sq. m. (42,735 sq. km.) by China. More than 80% of the pop. is supported by agric. The chief industries are tourism, sericulture, handicrafts, manu. of watches and telephones. Pop. (1981C) 5,987,389, Indian side of the line only.

Kassala   Sudan. 15 28N 36 24E. Town sit. near border with Ethiopia. Stands on Gash R. and is a trading centre for cotton. Pop. (1980E) 98,750.

Kassel   Hessen, Federal Republic of Germany. 51 19N 9 29E. Town sit. on Fulda R. Important industries inc. railway and automobile engineering, electronics, optical and precision instruments. Pop. (1989E) 192,000.

Kassérine   Kassérine, Tunisia. Town sit. SW of Tunis at the foot of the Djebel Chambi Mountain. District cap. Route junction in an irrigated agric. area. Pop. (1984C) 47,606.

Kastamonu   Turkey. 41 22N 33 47E. (i) Province of N Turkey on Black Sea coast, extending S to the

Ilgaz Mountains. Area, 5,491 sq. m. (14,222 sq. km.). Heavily forested, drained by Gok, Devrez, Arac and Koca Rs. Minerals inc. coal, lignite, copper, chromium, arsenic. Agric. areas produce cereals, hemp, apples, goats (mohair). Pop. (1985c) 450,353. (ii) Town sit. NNE of Ankara. Provincial cap. Industries inc. manu. textiles and arsenic.

Kastoria Greece. 40 30N 21 19E. (i) Prefecture of Macedonia, bordering Albania and Yugoslavia. Area, 650 sq. m. (1,685 sq. km.). Products inc. wheat and tobacco. Pop. (1981c) 53,169. (ii) Cap. of Kastoria. Sit. W of Thessaloníki. Has an important fur trade. Pop. (1981c) 17,133.

Katanga *see* **Shaba**

Kateríni Macedonia, Greece. 40 16N 22 30E. Town sit. SW of Thessaloníki, cap. of Pieria prefecture. Trades in agric. produce. Pop. (1981c) 38,016.

Kathiawar India. 22 00N 71 00E. Peninsula on the W coast between Gulf of Cambay and Gulf of Kutch. It is now part of Gujarat State and is noted for its religious sites. Area, 23,000 sq. m. (60,000 sq. km.).

Káthmándu Nepál. 27 42N 85 19E. Cap. of Nepál sit. at 4,270 ft. (1,301 metres) a.s.l. on Baghmati R. and N of Patna, India. Founded in 723 A.D. it has many temples and palaces. Pop. (1981c) 235,160.

Katmai, Mount Alaska, U.S.A. 58 20N 154 59W. Active volcano sit. in Aleutian Range at over 7,000 ft. (2,134 metres) a.s.l. Last violent eruption was in 1912, which created the Valley of Ten Thousand Smokes.

Katmandu *see* **Káthmándu**

Katowice Poland. 50 16N 19 00E. (i) Province in the S, bordering on Czechoslovakia. Area, 2,568 sq. m. (6,650 sq. km.). Important mining inc. coal, iron and zinc. Pop. (1989E) 3,932,000. (ii) Cap. of province of same name. Railway and industrial centre. Manu. inc. iron and steel, machinery, bricks and chemicals. Pop. (1985E) 363,000.

Katrine, Loch Scotland. 56 15N 4 31W. Freshwater L. 8 m. (13 km.) long and less than 1 m. (2 km.) wide, in Central Region. It is drained by R. Teith through Lochs Achray and Vennachar, and is the source of Glasgow's water supply.

Katsina Kaduna, Nigeria. 13 00N 7 32E. Town sit. NW of Kano. Commercial centre trading in cotton and groundnuts. Manu. inc. pottery, metal and leather products. Pop. (1983E) 149,300.

Kattegat Sweden/Denmark. 57 00N 11 20E. Strait of the N. Sea, 150 m. (240 km.) long, running from the Skagerrak in the N to the Baltic Sea in the S.

Katwijk aan Zee Zuid-Holland, Netherlands. 52 13N 4 24E. Holiday resort sit. NE of The Hague, at mouth of Old Rhine R. Pop. (1990E) 39,800.

Kauai Hawaii, U.S.A. 22 00N 159 30W. Island 98 nautical m. NE of Honolulu across the Kauai channel. Area, 551 sq. m. (1,427 sq. km.). Cap. Lihue. Central mountain mass with deep fissures and fertile valleys, rising to Kawaikiui, 5,170 ft. (1,576 metres). Chief products sugar, rice, pineapples.

Kaufbeuren Bavaria, Federal Republic of Germany. 47 53N 10 37E. Town sit. WSW of Munich, on Wertach R. Industries inc. textiles, brewing, computers, electronics, jewellery, glass, brewing and plastics. Pop. (1990E) 41,500.

Káunas Lithuania, U.S.S.R. 54 54N 23 54E. Town sit. at confluence of Nemen R. and Vilia R. Railway junction and river port. It has been an important commercial centre since the Middle Ages. Manu. agric. machinery, textiles, metal goods, chemicals and food products. Pop. (1985E) 405,000.

Kavalla Macedonia, Greece. 40 56N 24 25E. (i) Prefecture sit. on Aegean Sea and inc. the island of Thásos. Area, 814 sq. m. (2,109 sq. km.). The main product is tobacco. Pop. (1981c) 135,218. (ii) Port, sit. on Gulf of Kavalla E of Thessaloníki, and built on the site of anc. Neapolis. Pop. (1981C) 56,375.

Kavaratti   Lakshadweep, India. Island of the Lakshadweep group in the Arabian Sea. Coral formation. The main product is coconuts. Pop. (1981C) 6,604.

Kawaguchi   Honshu, Japan. 35 48N 139 43E. City immediately N of Tokyo. Industries inc. iron founding, textiles. Pop. (1983C) 392,000.

Kawasaki   Kanagawa, Japan. 35 32N 139 43E. Town sit. between Tokyo and Yokohama in the Keihin Industrial Zone. It is an important centre of heavy industries inc. shipbuilding, iron and steel, and machinery and also manu. textiles and chemicals. Pop. (1983C) 1,039,000.

Kayes   Mali. 14 27N 11 26W. Cap. of region in W on trib. of Senegal R. Commercial centre and R. port. Pop. (1976C) 44,736.

Kayseri   Turkey. 38 43N 35 30E. Cap. of Kayseri province sit. in central Turkey. Manu. textiles, carpets, rugs and tiles. Pop. (1970C) 167,696.

Kazakhstan   U.S.S.R. 50 00N 58 00E. In 1920 Uralsk, Turgai, Akmolinsk and Semipalatinsk provinces formed the Kazakh Soviet Socialist Republic within the Russian Soviet Federal Socialist Republic. It was made a constituent Rep. of the U.S.S.R. in 1936. To this Rep. were added the parts of the former Governorship of Turkestan inhabited by a majority of Kazakhs. Area is 1,049,155 sq. m. (2,717,300 sq. km.). Cap. Alma-Ata. Agric. of all kinds, esp. grain and cotton, is important, and sheep produce particularly high quality wool. The country is rich in mineral resources inc. coal, tungsten, oil, copper, lead, zinc and manganese, therefore heavy engineering is important. Pop. (1989E) 16·5m.

Kazan   Tatar, U.S.S.R. 55 45N 49 08E. Cap. of the Rep. sit. at confluence of Kazanka R. and Volga R. It is an important transport and industrial centre with large engineering works and oil refineries. Other manu.

are chemicals, soap. textiles, typewriters and musical instruments. Its fur industry handles half the Russian output. Pop. (1987E) 1,068,000.

Kazan-retto   *see* **Volcano Islands**

Kazbek, Mount   U.S.S.R. 42 42N 44 31E. Mountain peak in the Caucasus, N of Tbilisi. 16,546 ft. (5,043 metres) a.s.l.

Kázerun   Fárs, Iran. Town sit. W of Shiráz, in an agric. area, producing rice, tobacco, opium and cotton.

Kázvin   Gilán, Iran. 36 16N 50 00E. Town sit. WNW of Tehrán, at the S foot of the Elburz Mountains. Manu. carpets, textiles and soap. Subject to frequent earthquakes.

Kearney   Nebraska, U.S.A. 40 42N 99 05W. City sit. W of Omaha on Interstate Highway 80. Industries inc. manu. auto valves and gears, oil-and-air filters, pharmaceutical stoppers and closures, portable generators, canvas bags and cookbooks. Pop. (1989E) 24,350.

Kearny   New Jersey, U.S.A. 40 46N 74 09W. Town sit. on Passaic R. E of Newark, and connected to it by bridges. It has shipyards and docks. Manu. are very varied. Pop. (1980C) 35,735.

Kecskemét   Bács-Kiskun, Hungary. 46 57N 19 42E. Cap. of county. Sit. SE of Budapest. It is the centre of a fertile fruit-growing area with associated industries of distilling, preserving of fruit and vegetables and the manu. of agric. implements. Pop. (1989E) 106,000.

Kedah   Peninsular Malaysia, Malaysia. 5 50N 100 40E. State in NW bordering on Thailand. Area, 3,639 sq. m. (9,425 sq. km.). Cap. Alor Star. The chief agric. products are rice and rubber; other crops of commercial value are coconuts, areca-nuts, tapioca and fruit. Tin and tungsten are mined. Pop. (1980C) 1,116,140.

Kediri   Java, Indonesia. 7 51S 112 01E. Town sit. SW of Surabaya. Commercial centre of an agric. region. Pop. (1980C) 221,830.

Keeling Islands   *see* **Cocos Islands**

Keelung   *see* **Chilung**

Keewatin   Northwest Territories, Canada. 63 20N 94 40W. District of E mainland inc. the islands of Hudson Bay but excluding the Melville and Boothia peninsulas. Area, 228,160 sq. m. (590,934 sq. km.). The scattered pop., mainly Indians and Eskimos, live by fur-trapping.

Kefa   Ethiopia. Region in SW. Area, 21,081 sq. m. (54,600 sq. km.). Cap. Jimma. Pop. (1984C) 2,450,369.

Kefallenia   Ionian Islands, Greece. 36 28N 20 30E. Largest of the Ionian Islands, W of the Gulf of Patras. Area, 361 sq. m. (935 sq. km.). Cap. Argostolion. Rocky with limestone hills rising to 5,315 ft. (1,620 metres) a.s.l. Mainly agric., producing currants, wine, olive oil, cotton, citrus fruit, grain. Pop. (1981C) 31,297.

Keflavik   Suderland, Iceland. 64 01N 22 35W. Fishing port SW of Reykjavik. Important for fish processing, and the second export town in Iceland. Pop. (1988E) 7,305.

Keighley   West Yorkshire, England. 53 52N 1 54W. Town sit. NW of Bradford on R. Aire. The main products are woollen and worsted goods, textile machinery, travel goods and footwear, joinery products and machine tools. Pop. (1990C) 57,451.

Keith   Grampian Region, Scotland. 57 32N 2 57W. Burgh sit. WSW of Banff on R. Isla. It has whisky distilleries and manu. textiles and woollens. Pop. (1988E) 4,850.

Kelantan   Peninsular Malaysia, Malaysia. 6 11N 102 16E. State in NE bordering on Thailand. Area, 5,765 sq. m. (14,931 sq. km.). Cap. Kota Bharu. The chief agric. products are rice, coconuts, palm oil, pepper and tapioca. Rubber and copra are important industries, and silk-weaving and boat-building are also carried on. Pop. (1980C) 893,753.

Kelowna   British Columbia, Canada. 49 53N 119 29W. City on E shore of Okanagan L. midway between Penticton and Vernon. There are large manu. and processing plants, and other indust. inc. forestry, agric., wine manu., fruit processing and tourism. Pop. (1989E) 69,000.

Kelso   Borders Region, Scotland. 55 36N 2 25W. Town sit. SE of Edinburgh at the confluence of R. Tweed and R. Teviot. Manu. agric. implements, oil cake, electrical and electronic equipment, plastics, printed board circuits, and fertilizers. Pop. (1985E) 6,000.

Kemerovo   Siberia, U.S.S.R. 55 20N 86 05E. (i) Region in S, mainly in the Kuznetsk basin. Large mineral deposits with associated industries make it the most densely pop. province in Siberia. Area, 39,000 sq. m. (101,000 sq. km.). (ii) Cap. of region of same name sit. on Tom R. Important industries are chemicals, engineering and coalmining. Pop. (1987E) 520,000.

Kemi   Lappi, Finland. 65 49N 24 32E. Town and port near mouth of Kemi R. Sawmilling. Pop. (1983C) 26,600.

Kempten   Bavaria, Federal Republic of Germany. 47 43N 10 19E. Town sit. SSW of Augsburg on Iller R. An important commercial and agric. centre with manu. of textiles, paper and engineering. Pop. (1989E) 60,000.

Kempton Park   Transvaal, Republic of South Africa. 26 07S 28 14E. Town sit. NNE of Germiston at N edge of the Witwatersrand. Industries inc. iron founding; manu. bricks, aircraft, food processing, pharmaceutical and chemical products, cement. Pop. (1986E) 350,000.

Kendal   Cumbria, England. 54 20N 2 45W. Market town and tourist centre near Lake District, on R. Kent. The woollen industry has been important since the 14th cent. Other manu. are footwear, carpets, paper, water turbines and pumps. Pop. (1981C) 23,550.

Kenema   Sierra Leone. 7 52N 11 12W. Cap. of Eastern Province. Centre for diamond mining, cocoa, wood carvings and timber. Pop. (1974c) 31,458.

Kenilworth Warwickshire, England. 52 21N 1 34W. Town sit. sw of Coventry. Industry mainly associated with motor, aircraft and agric. engineering. Manu. plastics, caravan and boat equipment, tools, industrial floors. Pop. (1986E) 21,440.

Kenitra, Morocco *see* **Mina Hassan Tani**

Kennedy, Cape Florida, U.S.A. 28 28N 80 31W. Cape on E coast. Headquarters of the National Aeronautical and Space Administration, and from here U.S. moon probes departed. Named after the assassination of Pres. Kennedy in 1963 but returned to the name Cape Canaveral in May 1973 as the inhabitants complained that Canaveral was the oldest named place on E seaboard. NASA still use name Cape Kennedy.

Kennington Greater London, England. 51 28N 0 07W. Sit. just s of R. Thames and is famous for the Oval cricket ground.

Kenosha Wisconsin, U.S.A. 42 35N 87 49W. City and port sit. s of Milwaukee on L. Michigan. Manu. cars, metal products, mattresses and furniture. Pop. (1980c) 77,685.

Kensington and Chelsea Greater London, England. 51 29N 0 11W. District of w central London on N bank of Thames R. Residential. Pop. (1988E) 125,600.

Kent England. 51 12N 0 40E. Sit. in SE Eng. Under 1974 re-organization the county consists of districts of Dartford, Gravesham, Rochester upon Medway, Gillingham, Swale, Canterbury, Dover, Thanet, Sevenoaks, Tonbridge and Malling, Maidstone (county town), Tunbridge Wells, Ashford, Shepway. Area, 1,441 sq. m. (3,732 sq. km.). The surface is undulating in the interior, rising to 824 ft. (251 metres) a.s.l. on the North Downs. The county is drained by Rs. Thames, Medway, Stour, Rother and Darent. The s area is known as the Weald. Kent is known as the 'garden of England'. Agricultural products include cereals, fruit, hops, vegetables. Manu. include bricks, cement, paper, electronics and pharmaceuticals. There has been considerable growth in banking and financial services, and in distribution. There is fishing round the county's 140 m. (224 km.) coastline. Dover, Folkestone, Ramsgate, and Sheerness are the principal cross-channel ports for ships and hovercraft. Pop. (1988E) 1,520,400.

Kentish Town Greater London, England. 51 32N 0 08W. Sit. N of R. Thames.

Kentucky U.S.A. 37 20N 85 00W. State of U.S.A. bounded on the N by Illinois, Indiana and Ohio, on the E by W. Virginia and Virginia, on the s by Tennessee and on the w by Missouri. Surface undulating, rising to about 3,000 ft. (914 metres) in Allegheny plateau. In sw are great cypress swamps. Principal R. are Ohio, which forms N boundary; R. Mississippi, which forms part of w boundary. Area, 40,395 sq. m. (104,623 sq. km.) of which 745 sq. m. (1,930 sq. km.) are water. Chief towns, Frankfort (cap.), Louisville and Lexington. Climate is temperate. Kentucky is largely agric., esp. Blue Grass country; produces tobacco, also maize, wheat, soy beans and fruit. Famous for horses; mules, cattle, sheep, pigs also raised. There are large forests containing valuable timber. Extensive coalfields in both E and w, also iron, lead, salt, fluorspar, sandstone, natural gas, petroleum. Among principal industries are lumbering, clothing, cigars. Manu. machinery and electrical equipment. Pop. (1988E) 3,727,000.

Kentucky River Kentucky, U.S.A. 38 41N 85 11W. Rises in the Cumberland Mountains and flows NW for 260 m. (416 km.) to join Ohio R. at Carrollton.

Kenya 2 20N 38 00E. Until Kenya became independent in 1963, it consisted of the colony and the protectorate. In 1906 the protectorate was placed under the control of a governor and Commander-in-Chief and (except the Sultan of Zanzibar's dominions) was annexed to the Crown in 1920 under the name of the Colony of Kenya, thus becoming a Crown Colony. The territories on the coast became the Kenya Protectorate. The N boundary is defined by an agreement with Ethiopia in 1947. Bounded by Ethiopia in N, Uganda in W, Tanzania in S and the Indian Ocean and Somali Rep. in E. Area, 224,960 sq. m. (582,000 sq. km.). Cap. Nairobi. Chief towns, Mombasa, Nakuru and Kisumu. As agric. is possible from sea-level to altitudes of over 9,000 ft. (2,743 metres) tropical, sub-tropical and temperate crops can be grown and mixed farming can be practised. Four-fifths of the country is range-land which produces mainly livestock products and wild game which constitutes the major attraction of the country's tourist industry. The main areas of crop production are the Central, Rift Valley, Western and Nyanza Provinces and parts of Eastern and Coastal Provinces. Coffee, tea, sisal, pyrethrum, maize and wheat are crops of major importance in the Highlands, while coconuts, cashew nuts, cotton, sugar, sisal and maize are the principal crops grown at the lower altitudes. The livestock industry is important, and considerable quantities of corned beef, butter, bacon, ham, and hides and skins are exported. Groundnuts, sim-sim, potatoes, beans, essential oils and other miscellaneous crops are grown according to elevation and rainfall. An export trade is developing in mangoes, fresh fruits, flowers and vegetables sent by air to Europe. The total area of gazetted forest reserves in Kenya amounts to 6,486 sq. m. (16,800 sq. km.), of which the greater part is sit. between 6,000 and 11,000 ft. (1,829–3,353 metres) a.s.l., mostly on Mount Kenya, the Aberdares, Mount Elgon, Tinderet, Londiani, Mau watershed, Elgeyo and Charangani ranges. These forests may be divided into coniferous, broad-leafed or hardwood and bamboo forests. The upper parts of these forests are mainly bamboo, which occurs mostly between altitudes of 8,000 and 10,000 ft. (2,438–3,048 metres) and occupies some 10% of the high-altitude forests. Minerals inc. cement, soda ash, copper, gold, limestone, diatomite, carbon dioxide, salt, kaolin, vermiculite, barytes, magnesite, felspar, beryl, aquamarine fluorite, silver, sapphires, galena, guano, wollastrite and corundum. Tourism is a developing industry. Pop. (1988E) 22·8m.

Kenya, Mount Kenya. 0 10S 37 20E. An extinct volcano sit. on E of the Great Rift Valley just S of the equator. First climbed in 1899 the central cone is 17,040 ft. (5,198 metres) a.s.l. There is equatorial forest up to 11,000 ft. (3,353 metres) and then an alpine region and glaciers above 14,000 ft. (4,267 metres).

Kerala India. The state of Kerala, created under the States Reorganization Act, 1956, consists of the previous state of Travancore-Cochin, except for 4 *taluks* of the Trivandrum district and a part of the Shencottah *taluk* of Quilon district. Area, 15,181 sq. m. (38,863 sq. km.). Cap. Trivandrum. Cochin is a major port. Agric. inc. rice, tapioca, coconut, areca-nut, oilseeds, pepper, sugar-cane, rubber, tea, coffee and cardamom. About 25% of the area is comprised of forests, inc. teak, sandalwood, ebony and black-wood and varieties of softwood. Seafood production is important. Next to Bihar, Kerala possesses the widest variety of economic mineral resources among the Indian States. The beach sands of Kerala contain monazite, ilmenite, rutile, zircon, sillimanite, etc. There are extensive white-clay deposits; other minerals of commercial importance inc. mica, graphite, limestone, quartz-sand and lignite. Among the privately owned factories are the numerous cashew and coir factories. Other important factory industries are tyres, electronics, rubber, tea, tiles, oil, textiles, ceramics, fertilizers and chemicals, sugar, cement, rayon, glass, matches, pencils, monazite, ilmenite, titanium oxide, rare earths, aluminium, electrical goods, paper, shark-liver oil, etc. Among the cottage industries, coir-spinning and hand-loom-weaving are the most important ones, forming the means of livelihood of a large section of the people. Other industries are the village oil industry, ivory carving, furniture-making, bell metal, brass and copper ware, leather goods, mat-making, rattan-work, bee-keeping, pottery, etc. These have been organized on a co-operative basis. Pop. (1981C) 25,453,680.

Kerch   Ukraine, U.S.S.R. 45 20N 36 26E. Port at the E end of the Kerch Peninsula, and chief industrial centre of the Crimea, sit. NW of Novorossiysk. Manu. iron and steel, shipbuilding and fisheries. Kerch was founded by the Greeks in the 6th cent. B.C. and has many anc. monuments. Pop. (1985E) 168,000.

Kerch Peninsula   Ukraine, U.S.S.R. Peninsula between the Sea of Azov in the N, and the Black Sea in the S. 60 m. (96 km.) long and a maximum of 30 m. (48 km) wide, it is separated from the mainland of Crimea by the Kerch Strait.

Kerch Strait   Ukraine, U.S.S.R. 45 22N 36 38E. A shallow strait joining the Sea of Azov to the Black Sea.

Kerguelen Islands   Indian Ocean. 48 15S 69 10E. Volcanic and desolate arch. sit. at 40°–50° S lat. and 68°–70° E long. The islands are noted for their prolific bird life and the extraordinary Kerguelen cabbage. The arch. now forms part of the French Southern and Antarctic Territories. Area, 2,786 sq. m. (7,215 sq. km.).

Kerkrade   Limburg, Netherlands. 50 52N 6 04E. Town sit. ENE of Maastricht and the centre of a coal-mining area. Pop. (1988E) 52,994.

Kerkyra   Greece. (i) Island and prefecture in the Ionian group off the coasts of Epirus, Greece, and Albania. Area, 246 sq. m. (641 sq. km.). Mountainous in the N. Fertile, producing olives, vines and citrus fruits. Main occupation farming. Pop. (1981C) 99,477. (ii) Cap. of prefecture of same name sit. on the E coast. Industries inc. textiles, fishing and tourism. Sea port exporting olive oil and fruit. Pop. (1981C) 33,561.

Kermadec Islands   SW Pacific Ocean. 29 50S 178 15W. A group of volcanic islands sit. NE of Auckland, and owned by New Zealand. Area, 13 sq. m. (34 sq. km.). There is a meteorological station on Raoul Island, area 11 sq. m. (28 sq. km.).

Kermán   Iran. 30 17N 57 05E. (i) Province in SE, mainly barren. Area,

71,997 sq. m. (186,472 sq. km.). Pop. (1986C) 1,622,958. (ii) Cap. of province of same name. Manu. carpets, brassware and shawls. Pop. (1986C) 257,284.

Kermánsháh   *see* **Bakhtarán**

Kerry   Ireland. 52 07N 9 35W. County of province of Munster, in SW of Rep. Area, 1,815 sq. m. (4,700 sq. km.). County town, Tralee. A mountainous region with a deeply indented coastline, with many bays and off-shore islands. The highest range in the Rep., Macgillycuddy's Reeks, is inland with the highest peak Carrantuohill rising to 3,414 ft. (1,041 metres) a.s.l. The beautiful Killarney Ls. are in this county. Tourism is now a major industry. Other industries inc. the manu. of engineering products, footwear, woollen goods and cutlery. Pop. (1988E) 124,159.

Keswick   Cumbria, England. 54 37N 3 08W. Town sit. SSW of Carlisle on Greta R. near N shore of Derwentwater. Market town for a farming and quarrying area. The main industry is tourism but also pencil manu. Pop. (1990E) 5,000.

Ketchikan   Alaska, U.S.A. 52 21N 131 35W. City on SE coast of Revillagigedo Island, SE Alaska. Noted for heavy rainfall. Trade and service centre for a mining, lumbering and fishing area, with associated manu. Pop. (1980C) 7,198.

Kettering   Northamptonshire, England. 52 24N 0 44W. Town sit. NE of Northampton on Ise R. Industries inc. manu. leather goods, esp. footwear; breakfast cereals, computer software and clothing. Pop. (1987E) 46,500.

Kettering   Ohio, U.S.A. 39 41N 84 10W. City sit. S of Dayton, and forming a suburb. Residential, service and professional economy; other industries inc. automobiles, health services and electronics. Pop. (1986E) 61,836.

Kew   Greater London, England. 51 29N 0 16W. District of W Lond. on S bank of Thames R. Residential, noted for the Royal Botanical Gardens (288 acres) and Observatory.

Keweenaw Peninsula Michigan, U.S.A. 46 56N 88 23W. Peninsula which is part of the Upper Peninsula, and curves 60 m. (96 km.) NE into L. Superior. Previously very rich in copper deposits, it is now mainly a resort area.

Key West Florida, U.S.A. 24 33N 81 48W. Port sit. SSW of Miami, on Key West Island, the most westerly in the Florida Keys group of islands. Manu. cigars, turtle processing and sponge fishing. There is a govt. naval and air station here, and it is a popular tourist resort. Pop. (1990E) 30,000.

Khabarovsk U.S.S.R. 48 30N 135 06E. (i) Territory in the Russian Soviet Federal Socialist Rep. sit. along Sea of Okhotsk coastline and on Lower Amur R. Area, 965,400 sq. m. (2,500,400 sq. km.). Mainly forested with large deposits of gold, coal and iron. (ii) Cap. of territory of the same name, sit. at junction of Amur R. and Trans-Siberian Railway. The chief industrial centre of E Siberia occupied with oil refining, engineering, brewing, tanning and flour milling. Pop. (1985E) 576,000.

Khairpur Sind, Pakistan. 23 30N 69 08E. Cap. of state of same name, sit. NNE of Karachi. Irrigation allows cereals and cotton to be grown. Manu. textiles and carpets. Pop. (1981E) 62,000.

Khakass Autonomous Region Part of Krasnoyarsk Territory, U.S.S.R. Area, 23,855 sq. m. (61,784 sq. km.). Estab. 1930. Cap. Abakan, sit. 150 m. (240 km.) SSW of Krasnoyarsk. Chief industries coal and ore-mining, timber and woodworking. Also livestock breeding, dairy and vegetable farming, with 621,000 hectares under crops. Pop. (1984E) 533,000.

Khalkidiki Macedonia, Greece. 40 25N 23 20E. Prefecture formed by a peninsula extending into the Aegean Sea in 3 parallel prongs; Kassandra, Sithonia, and Akte. Area, 1,237 sq. m. (2,945 sq. km.). Cap. Polyghyros, sit. 31 m. (50 km,) SE of Thessaloníki. Main products wheat, olive oil and wine. Mount Athos stands at the tip of the Akte peninsula. Pop. (1981C) 79,036.

Khanaqin Diyala, Iraq. 34 21N 45 22E. Town sit. NE of Baghdad near Iranian border. Important oil refining centre.

Khanka, Lake U.S.S.R./China. 45 00N 132 24E. L. N of Vladivostok on China/U.S.S.R. frontier. Area, 1,700 sq. m. (4,400 sq. km.). Fed by Mo and Lefu Rs., drained by Sungacha R. into Ussuri R.

Khanty-Mansi National Area Tyumen, U.S.S.R. 61 00N 69 00E. A National Area in W Siberia crossed by Ob R. Important deposits of oil and natural gas. Area, 215,000 sq. m. (556,850 sq. km.). Cap. Khanty-Mansiisk, sit. at confluence of Ob R. and Irtysh R. and 300 m. (480 km.) NNE of Tyumen. Pop. (1989E) 1,269,000.

Kharagpur West Bengal, India. 22 20N 87 19E. Town sit. WSW of Calcutta. A railway junction occupied with railway and other engineering. Pop. (1981C) 150,475.

Kharga Oasis Egypt. Otherwise called El Kharga or Al-Kharijah (The Great Oasis). Sit. WSW of Nag Hammadi in a basin *c*. 200 m. (320 km.) N to S and 30 m. (48 km.) E to W. Railway and caravan routes connect it with the Dakhla oasis and Nile valley. Main settlement Kharga and the surrounding agric. area which produces dates, cotton, wheat, rice, barley, bananas, vines.

Kharkov Ukraine, U.S.S.R. 50 00N 36 15E. (i) Region in NE Ukraine. Mainly agric. Area, 290,000 sq. m. (751,000 sq. km.). (ii) Cap. of region of same name. Important industrial and transportation centre. Manu. agric. machinery inc. tractors, equipment for coalmining, oil-drilling and electrical products, locomotives, machine tools and many other goods. Pop. (1987E) 1,587,000.

Khartoum Sudan. 15 33N 32 35E. Cap. of Sudan sit. on left bank of Blue Nile near confluence with White Nile. It was founded in 1823 but destroyed by the Mahdi in 1885 and was rebuilt by the British. A road and rail bridge across the Blue Nile connects the town with Khartoum North, and a bridge across the White Nile connects with Omdurman. Khartoum is an important trading centre, esp. for cotton, with good railway and steamer services. There is an oil pipeline running NE to Port Sudan. Pop. (1983C) 476,218.

Khartoum North Khartoum, Sudan. 15 38N 32 33E. Town on right bank of Blue Nile opposite Khartoum. Trades in cotton and cereals. Industries inc. tanning, textile weaving, food processing. Pop. (1983C) 341,146.

Khasi Hills Assam, India. A range of hills extending some 140 m. (224 km.) E to W and forming a 60 m. (96 km.) wide barrier between the valleys of Brahmaputra R. and Surma R. Nearly 10,000 ft. (3,048 metres) a.s.l. On the S side is Cherrapunji, with an annual rainfall of 428 in. (1,087 cm.).

Khaskovo · Bulgaria. Cap. of Khaskovo province, sit. ESE of Plovdiv. Manu. textiles and tobacco. Pop. (1987E) 91,409.

Khatanga River Krasnoyarsk, U.S.S.R. 73 30N 109 00E. Rises in central Siberia being a union of the Kotui R. and the Moiero R. and flows NE for 412 m. (659 km.) to enter the Laptev Sea at Khatanga Bay and abounds in fish.

Kherson Ukraine, U.S.S.R. 46 38N 32 35E. (i) Region in S, a flat agric. region crossed by Dnieper R. Area, 10,600 sq. m. (27,450 sq. km.). Chief industries are engineering, textiles and food products. (ii) Cap. of region of same name, near the mouth of Dnieper R. A R. port exporting grain and timber. Manu. as for province. Also industries of flour milling, shipbuilding and brewing. Pop. (1987E) 358,000.

Khingan Mountains China. 49 40N 122 00E. Two ranges of volcanic mountains. The Great Khingan runs SW to NE at a height of 3,000–6,000 ft. (914–1,829 metres) a.s.l., separating Manchuria from Mongolia. The Little Khingan extend SE parallel to the Amur R. from the extension of the Great Khingan. Lumbering is important.

Khios Greece. 38 27N 26 09E. (i) Island and prefecture in the Aegean Sea off the W coast of Turkey. Area, 349 sq. m. (904 sq. km.). Main occupation farming, esp. vines, figs, olives, almonds, fruits, sheep and goats. Pop. (1981C) 49,865. (ii) Cap. of prefecture of same name, sit. on E coast. Industries inc. tanning, boatbuilding and winemaking. Sea port exporting wine, mastic and fruit. Pop. (1981C) 24,070.

Khiva Uzbek Soviet Socialist Republic, U.S.S.R. 41 24N 60 22E. Town sit. SW of Urgench on Khiva oasis. Industries inc. cotton milling, metal-working, carpet making. Formerly cap. of Khiva Khanate.

Khmelnitsky Ukraine, U.S.S.R. 49 25N 27 00E. (i) Region SW of Kiev, mainly agric. (ii) Cap. of region of same name, WSW of Kiev, on Southern Bug R. Industries, engineering, food processing and furniture. Pop. (1980E) 120,000.

Khmer Republic *see* **Cambodia**

Khoper River Russian Soviet Federal Socialist Republic, U.S.S.R. 52 00N 43 20E. Rises WSW of Penza and flows SSW for 600 m. (960 km.) to join Don R.

Khorasan Iran. 34 00N 58 00E. Province in E. Bounded N by U.S.S.R., E by Afghánistán. Area, 120,980 sq. m. (313,337 sq. km.). Chief city, Meshed. Mountainous in the N, levelling to the dry uplands (E) and edges of the Dasht-i-Kavir and Dasht-i-Lut deserts (SW). Watered by Atrek and Kashef Rs. and Qain and Birjand oases which support some agric., main products cereals, cotton, tobacco, sugar-beets, fruit, nuts. The main industries are rug-weaving, food processing, tanning. Pop. (1986C) 5,280,605.

Khorramshahr Khuzistán, Iran. 30 25N 48 11E. Port sit. at confluence of Shatt-al-Arab R. and Karun R. and near frontier with Iraq. The most important port on the (Persian) Gulf, exporting cotton, dates, hides and skins. Pop. (1976C) 146,709.

Khouribga Casablanca, Morocco. Town sit. SE of Casablanca. Important phosphate mining centre, with associated industries. Pop. (1982C) 127,181.

Khulna Bangladesh. 22 45N 89 34E. Cap. of division. Sit. on Ganges delta, SW of Dacca. Trade in rice, jute, oilseeds, sugar, betel-nut, coconut, salt. Milling rice, flour, oilseeds. Manu. cotton cloth. Shipbuilding. Pop. (1984C) 420,000.

Khurramabad Urestán, Iran. 33 30N 48 20E. Town sit. N of Ahwáz. Trade in wool and fruit. Pop. (1986C) 208,592.

Khuzistán Iran. 31 00N 50 00E. Province of Iran in SW at the head of the (Persian) Gulf. Area, 24,981 sq. m. (64,702 sq. km.). Cap. Ahwáz. Pop. (1986C) 2,681,978.

Khyber Pass Afghánistán/Pakistan. 34 05N 71 10E. Pass 33 m. (53 km.) long between plains of Peshawar, Pakistan and valley of Kabul R. Afghánistán. Varying in width from 450 ft. (137 metres) to 15 ft. (4 metres) with a railway which runs to the Afghán frontier, it is an important trade route.

Kiangsi *see* **Jingxi**

Kiangsu *see* **Jiangsu**

Kicking Horse Pass British Columbia/Alberta, Canada. 51 27N 116 25W. Pass at 5,339 ft. (1,627 metres) a.s.l. in the Rocky Mountains. The highest place on the Canadian Pacific Railway.

Kidderminster Hereford and Worcester, England. 52 24N 2 13W. Town sit. N of Worcester on R. Stour. The principal centre of the Brit. carpet-making trade, with other industries of light engineering, worsted spinning and tin-plating. Manu. electrical equipment, chemicals and drop forgings. Pop. (1988E) 54,194.

Kidwelly Dyfed, Wales. 51 45N 4 18W. A borough sit. NW of Llanelli on Carmarthen Bay. Manu. bricks, silica and optical glass products. Pop. (1983E) 3,100.

Kiel Schleswig-Holstein, Federal Republic of Germany. 54 20N 10 08E. Cap. of Schleswig-Holstein sit. at E end of Kiel Canal on the Kieler Förde. Main industries shipbuilding, engineering, fishing, flour milling and brewing. Pop. (1984E) 246,900.

Kiel Canal Federal Republic of Germany. 54 15N 9 40E. Canal 61 m. (98 km.) in length connecting Elbe estuary on the N. Sea with Kieler Förde on the Baltic Sea. Opened by Wilhelm II in 1895, and formerly called the Kaiser Wilhelm Canal.

Kielce Poland. 50 52N 20 37E. (i) Province in SE. Area, 3,556 sq. m. (9,211 sq. km.). Chief R., Vistula and Pilica. Cereals and flax are grown. Pop. (1989E) 1,123,000. (ii) Cap. of province of same name, sit. NNE of Kraków. Industries, iron-works, tanning, sawmilling. Manu. munitions, rolling stock, textiles, cement, chemicals and food products. Pop. (1989E) 201,000.

Kiev Ukraine, U.S.S.R. 50 30N 30 28E. (i) Region in central Ukraine, traversed by Dnieper R. The N is mainly forested with large peat deposits, and rye, flax and potatoes are grown. Wheat, sugar-beet and sunflowers are grown in the S. (ii) Cap. of region of same name, the third largest city in the U.S.S.R. sit. on right bank of Dnieper R. Commercial, industrial, transportation and cultural centre. Manu. extremely diverse, inc. light and heavy engineering products, chemicals, textiles, and food products. Many fine anc. buildings inc. the Byzantine Cathedral of St. Sophia and Pechersky monastery. Kiev, dating from the 8th cent. is the historical centre of Russian culture. Pop. (1989E) 2,587,000.

Kigali Rwanda. 1 59S 30 05E. Cap. of Rwanda sit. in central Rwanda. Before independence of Rwanda in 1962 the Belgian govt. administered the country from Bujumbura, the cap. of Burundi. Trades in coffee, cattle and hides. Pop. (1981E) 156,650.

Kigoma Tanzania. 5 30S 30 00E. Town sit. W of Tabora on L. Tanganyika. Rail terminus and L. port, exporting rice, vegetable oils, cotton, fish. Pop. (1978C) 50,044.

Kikinda Vojvodina, Yugoslavia. 45 50N 20 28E. Town sit. N of Belgrade. Trades in wheat, fruit and flour.

Kikwit Bandundu, Zaïre. 5 02S 18 48E. Largest city in region on Kwilu R. Pop. (1976C) 172,450.

Kildare Ireland. 53 10N 6 50W. (i) County in the province of Leinster. Area, 654 sq. m. (1,694 sq. km.). Mainly low-lying and containing the Bog of Allen in the N. Chief towns, Kildare (county town) and Naas. Principal Rs. are Boyne, Liffey, Barrow and Lesser Barrow. In the fertile Curragh district, cattle and horses are raised, and potatoes, oats and barley are grown. Pop. (1988E) 116,247. (ii) Market town in county of same name. Breeding and training centre of the Irish horse industry.

Kilimanjaro, Mount Tanzania. 3 04S 37 22E. Mountain sit. between L. Victoria and the coast. The highest mountain in Afr., it has two peaks, both of which are extinct volcanoes, Kibo 19,340 ft. (5,950 metres) a.s.l., and Mawenzi 17,300 ft. (5,323 metres) a.s.l. Coffee, maize and bananas are grown on the SW slopes from 4,000–6,000 ft. (1,230–1,846 metres) a.s.l. Above 6,000 ft. there are forests, followed by grasslands to about 12,000 ft. (3,690 metres), above which there is permanent snow and ice. The first ascent was made in 1889.

Kilinailau Solomon Islands, South Pacific. Atoll sit. NE of Buka consisting of a 7-island reef. Main product coconuts.

Kilindini Mombasa, Kenya. 4 04S 39 40E. Harbour for Mombasa city, with deep water, modern docks. coal and oil terminals.

Kilkenny Ireland. 52 39N 7 15W. (i) County in province of Leinster, sit. in SE. Area, 796 sq. m. (2,061 sq. km.). Mainly hilly with Mount Brandon in the SE. Principal Rs. are Barrow, Suir and Nore. Mainly agric. with cereals, potatoes, beet and turnips grown. Pop. (1988E) 73,186. (ii) County town of

same name, on R. Nore. Many fine anc. buildings. Industries are brewing and tanning.

Killarney Kerry, Ireland. 52 03N 9 30W. Town sit. SE of Tralee. Tourist centre in area of outstanding beauty, containing the three noted Ls. of Killarney, Upper, Middle and Lower. The latter is also known as Lough Leane. Nearby are the Torc and Purple Mountains, Macgillicuddy's Reeks and the Gap of Dunloe. Industries inc. hosiery, footwear, glass and manu. of cranes and tools. Pop. (1990E) 7,963.

Killicrankie Pass Scotland. 56 43N 3 40W. Pass sit. NW of Perth in valley of R. Garry. It extends 1½ m. (2.5 km.) from Killiecrankie Station to Garry Bridge.

Kilmarnock Strathclyde Region, Scotland. 55 36N 4 30W. Industrial town sit. on Kilmarnock Water near its confluence with R. Irvine. Industries, engineering, carpet-making. Manu. lace, woollen goods, whisky and footwear. Pop. (1990E) 48,000.

Kilrush Clare, Ireland. 52 39N 9 30W. Port sit. on Shannon estuary. Industries, tourism, fishing and stone quarrying. Pop. (1985E) 2,784.

Kilwinning Strathclyde Region, Scotland. 55 40N 4 42W. Town sit. WNW of Kilmarnock on R. Garnock. Part of Irvine New Town. Industries, engineering, electronics and worsted spinning. Pop. (1984E) 16,730.

Kimberley Western Australia, Australia. District in N, mainly fertile grassland with a mountainous core. Principal Rs. are Ord and Fitzroy. Mainly a cattle producing area but gold is mined at Hall's Creek.

Kimberley Cape Province, Republic of South Africa. 28 43S 24 46E. Town sit. WNW of Bloemfontein. Important diamond-mining centre. Manu. cement, metal goods, clothing and furniture. Some agric. and cattle ranching is possible based on irrigation schemes. Pop. (1980E) 144,923.

Kincardineshire Scotland. Former county now part of Grampian Region.

Kindia   Guinea. 10 04N 12 51W.
Cap. of Guinée-Maritime. Trading
centre. Pop. (1983C) 55,904.

Kineshma   Russian Soviet Federal
Socialist Republic, U.S.S.R. 57 26N
42 09E. Industrial town sit. ENE of
Ivanovo on Volga R. Centre for textiles, sawmilling and paper manu.

King Edward VII Point   Canada.
Extreme SE tip of Ellesmere Island in
NE Franklin District, North West
Territories.

King Feisal Port   Saudi Arabia.
Deep-water port at Jidda, between
Suez and Aden was commissioned in
Jan. 1973 and serves as the main port
of entry for pilgrims to Mecca.

Kinghorn   Fife Region, Scotland.
56 04N 3 10W. Town sit. S of Kirkcaldy. Tannery. Pop. (1985E) 4,000.

King Island   Tasmania, Australia.
39 50S 144 00E. Island sit. off NW
coast of Tasmania in the Bass Strait.
Area, 425 sq. m. (1,101 sq. km.). Fertile plain supporting dairy farms.
Minerals inc. tungsten, tin, zircons.

Kingsbridge   Devon, England. 50
14N 3 46W. Market town sit. WSW of
Dartmouth and at the head of Kingsbridge Estuary. Pop. (1981C) 4,236.

Kings Canyon National Park   California, U.S.A. Region on W slopes of
the Sierra Nevada, crossed by Middle
Fork and S. Fork of Kings R. Area,
708 sq. m. (1,834 sq. km.).

King's Lynn   Norfolk, England. 52
45N 0 24E. Market town and port, sit.
S of Wash at mouth of Great Ouse R.
Agric. industries, sugar-beet refining,
vegetable and fruit canning, and the
manu. of agric. implements, glass and
refrigeration equipment. There are
many fine medieval buildings. Pop.
(1987E) 35,500.

King's Peak   Utah, U.S.A. 40 46N
110 22W. Mountain peak sit. E of Salt
Lake City in the Uinta range, at
13,498 ft. (4,114 metres) a.s.l. the
highest point in Utah.

Kingsport   Tennessee, U.S.A. 36

32N 82 33W. Industrial town sit. ENE
of Knoxville, on Holston R. Manu.
chemicals, textiles and paper. Pop.
(1988E) 31,440.

Kingston   Ontario, Canada. 44 14N
76 30W. Port sit. on NE of L. Ontario
at mouth of Cataraqui R. Cap. of
Frontenac county. Manu. locomotives,
textiles, chemicals, aluminium goods
and mining machinery. Pop. (1981C)
60,313.

Kingston   Jamaica. 17 58N 76 48W.
Cap. of the island, sit. on SE coast, it is
the largest and most important town
and port in the Caribbean. It was
founded in 1693 after Port Royal was
destroyed in an earthquake. The land-
locked harbour can accommodate
ocean-going vessels. Centre of the
coffee trade with industries based on
the agric. produce of the island. Manu.
tobacco products, matches, clothing
and jam. Pop. (1983E) 100,637.

Kingston   New York, U.S.A. 41 56N
74 00W. Town sit. N of New York on
Hudson R. Manu. bricks, tiles, cement
and clothing. Pop. (1980C) 24,481.

Kingston   Pennsylvania, U.S.A. 41
16N 75 54W. Town sit. on Susquehanna R. opposite Wilkes-Barre, and
NNW of Philadelphia. Industries inc.
manu. furniture and clothing. Pop.
(1980C) 15,681.

Kingston-upon-Hull   *see* **Hull**

Kingston upon Thames   Greater
London, England. 51 25N 0 19W.
Royal borough sit. SW of London
forming a suburb on Thames R. Residential with some industries inc.
metalworking; manu. aircraft, plastics,
paint. Has the Saxon coronation stone.
Pop. (1988E) 134,546.

Kingstown   St. Vincent and the
Grenadines, West Indies. 13 12N 61
14W. Town S of Fort-de-France, Martinique, cap. and port exporting arrowroot, cotton, sugar-cane, molasses,
cacao, tropical fruit. Industries inc.
processing cotton and arrowroot. Pop.
(1987E) 28,942.

Kingussie   Highland Region, Scotland. 47 05N 4 02W. Town sit. S of

Inverness on R. Spey. Tourist centre. Pop. (1981C) 1,190.

Kinlochleven Highland Region, Scotland. Village sit. SE of Fort William at head of Loch Leven. There is an aluminium smelter and hydroelectric plant. Pop. (1990E) 1,100.

Kinross Tayside Region, Scotland. 56 13N 3 27W. Town sit. N of Dunfermline, on W shore of Loch Leven. Industries inc. woollen mills, cashmere spinning. Pop. (1989E) 3,800.

Kinross-shire Scotland. Former county, now part of Tayside Region.

Kinsale Cork, Ireland. 51 42N 8 32W. Port and holiday resort on estuary of R. Bandon. Pop. (1989E) 2,500.

Kinshasa Zaïre. 4 18S 15 18E. Cap. sit. on left bank of Zaïre R. below Malebo Pool. It was founded by Stanley as Léopoldville in 1881 and renamed Kinshasa in 1966. Commercial and transportation centre. Manu. textiles and footwear. It is now one of the most modern cities in Africa. Pop. (1984C) 2,653,558.

Kintyre Scotland. 55 30N 5 35W. Peninsula sit. between Firth of Clyde and Atlantic, and extending 42 m. (67 km.) SSW from Tarbert. Principal town, Campbeltown. Mainly hilly rising to 1,462 ft. (446 metres) a.s.l. The chief occupations are agric., forestry, coalmining, quarrying and fishing.

Kioga Lake Uganda. 1 30N 33 00E. Shallow L. 80 m. (128 km.) long in headwaters of Victoria Nile. One of a chain of L.

Kipini Coast Province, Kenya. Town sit. NNE of Mombasa at mouth of Tana R. on the Indian Ocean. Industries inc. fishing; manu. sisal. Port, handling cotton, copra, sugar-cane.

Kirghizia U.S.S.R. 42 00N 75 00E. After the revolution, Kirghizia became part of Soviet Turkestan, which itself became an Autonomous Soviet Socialist Republic within the Russian Soviet Federal Socialist Republic in 1921. In 1926 the Govt. of the Rus-

sian Soviet Federal Socialist Republic transformed Kirghizia into an Autonomous Soviet Socialist Republic within the Russian Soviet Federal Socialist Republic, and finally in 1936 Kirghizia was proclaimed one of the constituent Soviet Socialist Republics of the U.S.S.R. Area, 76,640 sq. m. (198,500 sq. km.). Cap. Frunze. Generally mountainous, rising to 24,406 ft. (7,439 metres) a.s.l. in Pobeda. Fertile valleys growing wheat, sugar-beet and cotton. Famous for its livestock breeding. General engineering industries. Pop. (1989E) 4·3m.

Kiribati Pacific Ocean. 1 00N 176 00E. Formerly Gilbert Islands. Gained independence in 1979 after the former Ellice Islands severed its constitutional links and took the name Tuvalu. Three groups of islands with neighbouring Ocean Island, area, 2 sq. m. (5 sq. km.). Cap. Tarawa. Area, Gilbert Islands 102 sq. m. (264 sq. km.); Phoenix Islands, 11 sq. m. (28 sq. km.); Line Islands *c.* 200 sq. m. (*c.* 520 sq. km.). The Gilbert group inc. 16 islands; the Phoenix group 8; and the Line group 8. The main occupations are growing coconuts and fishing. Main products are copra and phosphates. Pop. (1987E) 66,250.

Kirin *see* **Jilin**

Kirkby-in-Ashfield Nottinghamshire, England. 53 06N 1 15W. Town sit. NNW of Nottingham. Industries inc. textiles and engineering. Pop. (1989E) 18,140.

Kirkcaldy Fife Region, Scotland. 56 07N 3 10W. Town and port on Firth of Forth, Centre of linoleum industry. Other industries are engineering, iron-founding and textiles. Pop. (1987E) 49,200.

Kirkcudbright Dumfries and Galloway Region, Scotland. 54 50N 4 03W. Town serving an agric. area sit. SW of Dumfries on R. Dee. Industry, fishing. Pop. (1985E) 3,500.

Kirkcudbrightshire Scotland. Former county, now part of Dumfries and Galloway Region.

Kirkenes Finnmark, Norway. 69

40N 30 05E. Port sit. on Varanger Fiord in NE near the U.S.S.R. border. Exports iron ore from the nearby Syd-Varanger mines.

Kirkintilloch Strathclyde Region, Scotland. 55 57N 4 10W. Town sit. NE of Glasgow. Industries inc. electronics, food processing. Pop. (1988E) 36,960.

Kirklareli Turkey. (i) Province of NW Turkey bounded N by Bulgaria, E by the Black Sea. Area, 2,599 sq. m. (6,731 sq. km.). Heavily forested with extensive agric. area producing sugar-beet and canary grass. Pop. (1985C) 297,098. (ii) Town NW of Istanbul. Provincial cap., railway terminus and trading centre for grain, sugar-beet, canary grass.

Kirkuk Kirkuk, Iraq. 35 30N 44 21E. Cap. of governorate N of Baghdad. Centre of an important oilfield with pipelines to Syria and Lebanon. Trades in naphtha, salt, sheep, grain and fruit.

Kirkwall Orkney, Scotland. 58 59N 2 58W. Town and port sit. on NE coast of Orkney mainland. Centre of agric. region, exporting agric. produce and fish products. Manu. gold and silver jewellery, knitwear. Pop. (1990E) 6,681.

Kirov Russian Soviet Federal Socialist Republic, U.S.S.R. 58 38N 49 42E. (i) Region in the NE. Area, 46,600 sq. m. (120,700 sq. km.). Partially forested basin of the Vyatka R. (ii) Cap. of region of same name NE of Gorki on Vyatka R. Industries railway and agric. engineering. Manu. textiles, footwear and matches. Pop. (1987E) 421,000.

Kirovabad Azerbaijan, U.S.S.R. 40 40N 46 22E. Town sit. SE of Tbilisi. The second largest town of the Azerbaijan Rep. and the industrial and cultural centre of the region. Manu. textiles and wine. Pop. (1987E) 270,000.

Kirovakean Armenian Soviet Socialist Republic, U.S.S.R. 40 48N 44 30N Town sit. E of Leninakan. Industries inc. manu. chemical products, tanning, food processing, rug making.

Kirovograd Ukraine, U.S.S.R. 48 30N 32 18E. (i) Region in S, drained by Dnieper R. Area, 9,340 sq. m. (24,190 sq. km.). Mainly agric. with large deposits of lignite and some coalmining. (ii) Cap. of region of same name SE of Kiev. Manu. agric. machinery, clothing, soap and food products. Pop. (1987E) 269,000.

Kirriemuir Tayside Region, Scotland. 56 41N 3 01W. Town sit. WNW of Forfar. Manu. jute cloth. Grain milling. Pop. (1981C) 5,122.

Kirsehir Turkey. (i) Province of central Turkey extending from Delice R. (NE) to Kizil Irmak (SW). Forested and agric. Main products grain, linseed oil, mohair. Pop. (1985C) 260,156. (ii) Town SE of Ankara. Provincial cap. The main industry is carpet making.

Kiruna (Lapland) Norrbotten, Sweden. 67 53N 20 15E. Town sit. WNW of Luleå between the ore-bearing mountains of Kiirunavaara and Luossavaara. In area it is the largest town in the world, containing vast iron ore minefields. Pop. (1990E) 26,500.

Kisangani Haut-Zaïre, Zaïre. Cap. of Region sit. on upper Congo R. below Stanley Falls. Important transportation centre. Pop. (1976C) 339,210.

Kiselyovsk Russian Soviet Federal Socialist Republic, U.S.S.R. Town sit. SE of Novosibirsk in Novosibirsk region. An important coalmining and mining equipment centre in the Kuznetsk basin.

Kishinev Moldavian Soviet Socialist Republic, U.S.S.R. 47 00N 28 50E. Town sit. WNW of Odessa. Cap., commercial and cultural centre in a rich agric. area with extensive food industry. Manu. leather goods, wine and hosiery. Pop. (1988E) 663,000.

Kiskunfélegyháza Bács-Kiskun, Hungary. Town sit. SE of Budapest. Centre of a rich agric. district trading in fruit, cereals, wine, livestock and tobacco.

Kislovodsk Russian Soviet Federal

Socialist Republic, U.S.S.R. Health resort sit. NW of Ordzhonikidze in North Ossetian A.S.S.R.

Kisumu Nyanza, Kenya. 0 03S 34 47E. R. port sit. at head of Kavirondo gulf on the NE shore of L. Victoria. Pop. (1984E) 210,000.

Kitakyushu Kyushu, Japan. 33 52N 130 49E. Town in SW comprising the former towns of Moji, Kokura, Tobata, Wakamatsu and Yawata. One of Japan's largest centres for heavy industry and the chemical industry. Pop. (1988E) 1,035,000.

Kitchener Ontario, Canada. 43 27N 80 29W. Town sit. WNW of Hamilton in Waterloo county. Industries inc. food processing and brewing. Manu. rubber products, vehicle parts, fabricated metal products. Pop. (1990E) 155,000 (metropolitan area, 372,300).

Kithira Greece. 36 09N 23 00E. Island at the entrance to the Gulf of Laconia off the SE coast of Peloponnessos. Area, 106 sq. m. (274 sq. km.). Chief town, Kithira. Main occupation growing vines and olives.

Kitimat British Columbia, Canada. 53 55N 129 00W. Port sit. ESE of Prince Rupert at head of Douglas Channel. Has the largest aluminium smelter in the world.

Kitwe Western Province, Zambia. 12 49S 28 13E. Town sit. WNW of Ndola. Commercial centre for the copper-mining centre of Nkana which is in the vicinity. Pop. (1980E) 314,794.

Kitzbühel Tirol, Austria. 47 27N 12 23E. Winter sports resort sit. ENE of Innsbruck. A former copper-mining town. Pop. (1985E) 9,000.

Kivu Zaïre. 1 48S 29 00E. Region in E containing the major part of the Albert National Park. Area, 99,097 sq. m. (256,662 sq. km.). Cap. Bukavu, sit. at extreme S end of L. Kivu. Coffee, pyrethrum and cinchona are grown, and cattle raised in the E. The W is mainly tropical rain-forest. Pop. (1981E) 4,713,761.

Kivu, Lake Zaïre/Rwanda. 2 00S 29 10E. Lake, 60 m. (96 km.) long, sit. between L. Edward and L. Tanganyika. In the S it discharges into L. Tanganyika by Rusizi R. The chief ports are Bukavu, Goma and Kisenyi.

Kizil Irmak River Turkey. 41 44N 35 58E. Largest R. of Asia Minor. Rises in Kizil Dagh and flows for *c.* 600 m. (960 km.) to the Black Sea at Cape Bafra NW of Samsun.

Kizyl Tuva. Autonomous Soviet Socialist Republic, U.S.S.R. Town sit. S of Krasnoyarsk on Yenisei R. Cap. of Tuva, centre of a mining, farming and lumbering area. Industries inc. tanning; manu. sheepskin and timber products.

Kladno Bohemia, Czechoslovakia. 50 08N 14 05E. Town sit. WNW of Prague in the Středočeský region. Important industries of coal, iron and steel, and engineering. Pop. (1989E) 73,000.

Klagenfurt Carinthia, Austria. 46 38N 14 18E. Cap. of Federal State of Carinthia sit. SW of Graz on Glan R., near the E end of the Wörthersee. Holiday resort. Manu. chemicals, textiles, leather and metal goods. Pop. (1981C) 87,321.

Klaipéda Lithuania, U.S.S.R. 55 43N 21 07E. Port sit. at entrance to Kurskiy Zaliv, a large lagoon on the Baltic Sea. An important centre since the 16th cent. trading in timber, grain and fish. Manu. textiles and products associated with timber. Pop. (1987E) 201,000.

Klaksvig Faroe Islands. Town sit. NNE of Thorshavn on SW Bordo Island. Port.

Klerksdorp Transvaal, Republic of South Africa. 26 58S 26 39E. Town sit. WSW of Johannesburg on Schoonspruit R. near confluence with Vaal R. Centre of an agric and gold mining region. Pop. (1980E) 239,000.

Klondike River Yukon Territory, Canada. 63 03N 139 26W. Small R. which joins Yukon R. at Dawson.

Klosterneuburg   Vienna, Austria. 48 19N 16 20E. Town sit. on NW outskirts of Vienna on Danube R. Produces wine. Pop. (1981C) 22,975.

Klyuchevskaya Sopka   U.S.S.R. 56 04N 160 38E. Active volcano in Kamchatka peninsula sit. NNE of Petropavlosk-Kamchatski at 15,666 ft. (4,775 metres) a.s.l.

Knaresborough   North Yorkshire, England. 54 00N 1 27W. Town sit. ENE of Harrogate on R. Nidd. Industries inc. tourism and electronics. Limestone quarrying in the vicinity. Pop. (1987E) 14,160.

Knob Lake   *see* **Schefferville**

Knockmealdown Mountains   Ireland. 52 14N 7 57W. Mountain range extending 15 m. (24 km.) E to W between the counties of Tipperary and Waterford. The highest peak is 2,609 ft. (765 metres) a.s.l.

Knoxville   Tennessee, U.S.A. 35 58N 83 56W. City and port on Tennessee R. Industrial centre of a region of coal, iron and zinc mines and marble quarries. Manu. clothing, electrical goods, metal and plastic goods, foods. Also the administrative centre of the Tennessee Valley Authority. Pop. (1988E) 172,080.

Knutsford   Cheshire, England. 53 19N 2 22W. Town sit. SW of Manchester. Manu. photographic materials and textiles. Pop. (1990E) 14,000.

Kobarid   Slovenia, Yugoslavia. Village in NW sit. WNW of Ljubljana, on Isonzo R. near the Italian frontier. Formerly Italian and ceded to Yugoslavia in 1947.

Kobe   Honshu, Japan. 34 45N 135 10E. Port and cap. of Hyogo prefecture sit. W of Osaka on Osaka Bay. Important industrial centre, which has expanded rapidly in recent years. Main occupations are ship-building, engineering, sugar-refining and production of rubber goods, textiles and chemicals. Pop. (1983E) 1,370,000.

København   *see* **Copenhagen**

Kocaeli   Turkey. Province of NW Turkey bounded N by Black Sea, W by Sea of Marmara, S by Samauli Mountains. Area, 3,257 sq. m. (8,436 sq. km.). Cap. Izmet. Heavily forested, with valuable beechwoods. Minerals inc. lead, manganese, iron, copper, zinc. Agric areas produce maize, sugar-beet, cotton, flax, tobacco. Pop. (1985C) 742,245.

Kochi   Shikoku, Japan. 33 30N 133 35E. Port and cap. of Kochi prefecture sit. SW of Osaka on Gulf of Kyuko. Esp. noted for coral goods, paper and dried fish (bonito). Pop. (1988E) 312,000.

Kodiak Island   Alaska, U.S.A. 57 48N 152 23W. Island in the Gulf of Alaska, 100 m. (160 km.) in length. First European settlement in Alaska took place by the Russians in 1784. Chief occupation commercial fishing. Pop. (1984E) 6,469 (city), 13,389 (island).

Koforidua   Eastern Province, Ghana. 6 05N 0 15W. Town sit. N of Accra. Distribution centre for cacao. Pop. (1984C) 58,731.

Koidu   Sierra Leone. 8 38N 10 59W. Town, trading centre in Eastern Province.

Kokand   Uzbekistan, U.S.S.R. 40 30N 70 57E. Town sit. ENE of Bokhara in Fergana Valley. Manu. textiles inc. silk, and fertilizers.

Kokkola   Kaarlela, Finland. 63 50N 28 08E. Town sit. NNE of Vaasa, on the Gulf of Bothnia. There is heavy industry and also textiles and leather. Pop. (1989E) 35,000.

Kokomo   Indiana, U.S.A. 40 29N 86 08W. City sit. N of Indianapolis. Manu. glass, iron, steel and brass goods and parts for cars and tractors. Pop. (1988E) 43,950.

Koko Nor   Chinghai, China. 36 50N 100 20E. Salt L. sit. WNW of Lanchow. Area, 2,300 sq. m. (5,957 sq. km.). Over 10,000 ft. (3,048 metres) a.s.l.

Kokura   *see* **Kitakyushu**

Kola Peninsula Murmansk, U.S.S.R. 68 45N 33 08E. Peninsula between Barents Sea in N and White Sea in S. Area, 50,000 sq. m. (129,500 sq. km.). Low granite plateau yielding important minerals, esp. apatite and nepheline which are exported from Kandalaksha.

Kolar Karnataka, India. 13 12N 78 15E. (i) District sit. NE of the Kolar Gold Fields. Area, 3,188 sq. m. (8,257 sq. km.). The gold fields produce 90% of India's gold. The gold deposits were first worked in 1880. (ii) Town producing woollen blankets, leather goods. Pop. (1981E) 77,679.

Kolarovgrad *see* **Shumen**

Kolberg *see* **Kolobrzeg**

Kolding Vejle, Denmark. 55 29N 9 29E. Port sit. SSW of Horsens on Kolding Fiord. Manu. machinery, textiles, cement and food products. Pop. (1984E) 56,519.

Kolhapur Maharashtra, India. 16 43N 74 15E. Town sit. SSE of Bombay. Manu. pottery, textiles and matches. Pop. (1981C) 340,625.

Köln *see* **Cologne**

Kolobrzeg Koszalin, Poland. 54 12N 15 33E. Port and resort sit. on Baltic Sea at mouth of Prosnica R. Manu. machinery. Tourism is important.

Kolombangara Solomon Islands, Pacific Ocean. Island sit. W of New Georgia across the Kula Gulf. Area, 20 m. (32 km.) by 15 m. (24 km.). Volcanic.

Kolomna Russian Soviet Federal Socialist Republic, U.S.S.R. 55 05N 38 49E. Town sit. SE of Moscow on Moskva R. near its confluence with Oka R. Railway engineering centre with manu. of locomotives, diesel engines, wagons and textile machinery.

Kolomyya Ukraine, U.S.S.R. 48 32N 25 04E. Town sit. SE of Lvov on Prut R. Formerly Polish but ceded to the U.S.S.R. in 1945. Manu. chemicals and textiles. Pop. (1974E) 141,000.

Kolonia Pohnpei, Federated States of Micronesia. 06 55N 158 10E. Cap. of the Federated States. Pop. (1989E) 5,500.

Kolonjë Kolonjë, Albania. Town sit. S of Koritsa at foot of Grammos mountains, a lignite mining centre. Otherwise called Ersekë.

Kolwezi Shaba, Zaïre. Town sit. W of Likasi. Commercial centre of a copper and cobalt mining region. Pop. (1976C) 77,277.

Kolyma Range Russian Soviet Federal Socialist Republic, U.S.S.R. Mountain range in Khabarovsk, extending NE for 500 m. (800 km.) between the Kolyma R. and the Sea of Okhotsk.

Kolyma River Russian Soviet Federal Socialist Republic, U.S.S.R. 69 30N 161 00E. Rises on the Pacific divide N of the Sea of Okhotsk and flows for 1,600 m. (2,560 km.) through the Cherskogo Mountains to the tundra of the Kolyma depression. Principal trib. are Omolon R. 700 m. (1,120 km.) and Anyui R.

Komandorski Islands Russian Soviet Federal Socialist Republic, U.S.S.R. 55 00N 167 00E. A group of islands in Khabarovsk Territory, between the Kamchatka peninsula and the Aleutian Islands. The two largest are Bering and Medny and the main village is Nikolskoye.

Komárom Hungary. 47 46N 18 07E. Town on right bank of Danube R. opposite Komarno, Czech. Railway centre and R. port. Industries inc. textiles, sawmilling. Pop. (1984E) 35,000.

Komati River Republic of South Africa/Swaziland. 25 46S 32 43E. R. in SE Africa rising in the Transvaal and flowing for *c.* 500 m. (800 km.) to Delagoa Bay where it enters the Indian Ocean, NE of Lourenço Marques.

Komi Autonomous Soviet Socialist Republic Part of Russian Soviet Federal Socialist Republic, U.S.S.R. 64 00N 55 00E. Area, 160,540 sq. m. (415,800 sq. km.). Constituted as an

Autonomous Republic 1936. Cap. Syktyvkar, sit. ENE of Kotlas. Mainly forested with tundra in the N. Chief industries are coal, oil, timber, gas, asphalt and building materials; also livestock breeding and dairy farming. Pop. (1989E) 1,263,000.

Komi-Permyak National Area Russian Soviet Federal Socialist Republic, U.S.S.R. A National Area in the Perm Region. Area, 160,540 sq. m. (415,900 sq. km.). Cap Kudymkar, sit. NE of Perm. Mainly forested. Industries associated with timber.

Kommunarsk Russian Soviet Federal Socialist Republic, U.S.S.R. 48 30N 38 47E. Town sit. NW of Abakan in the Kuznetsk Ala-Tau, Krasnoyarsk Territory. Gold mining centre.

Kommunizma Peak Tadzhikistan, U.S.S.R. 39 00N 72 02E. Highest mountain in U.S.S.R. sit. in Akademiya Nauk Range of the Pamirs. 24,590 ft. (7,495 metres) a.s.l.

Komotini Thrace, Greece. 41 08N 25 25E. Cap. of the Rodopi prefecture sit. ENE of Thessaloníki. Trades in cereals, vegetables and tobacco. Pop. (1981C) 34,051.

Kompong Cham Kompong Cham, Cambodia. Town sit. NE of Phnom Penh on Mekong R. Provincial cap. Centre of a rubber growing area. Industries inc. manu. rubber, beverages.

Kompong Chhang Kompong Chhang, Cambodia. Town sit. NNW of Phnom Penh, on Tonle Sap R. at the head of navigation. Trading centre for livestock, industries inc. manu. bricks, pottery.

Kompong Speu Kompong Speu, Cambodia. Town sit. WSW of Phnom Penh. Industries inc. silk spinning. Trade in timber.

Kompong Thom Kompong Thom, Cambodia. Town sit. N of Phnom Penh on Stung Sen R. Centre of a thickly forested area hunted for big game. R. port.

Komsomolsk-on-Amur Russian Soviet Federal Socialist Republic,

U.S.S.R. 50 35N 137 02E. Town sit. NE of Khabarovsk on Amur R. Industries ship-building and sawmilling. Manu. steel, chemicals, wood pulp and paper. Pop. (1987E) 316,000.

Kong Karls Land Svalbard, Norway. Island group E of Spitsbergen in Barents Sea. Area, 128 sq. m. (332 sq. km.). The main island is Kongsoya, rising to 1,050 ft. (320 metres) a.s.l.

Königsberg   *see* **Kaliningrad**

Konstantinovka Ukraine, U.S.S.R. 48 32N 37 43E. Town sit. N of Donetsk in the Donbas. Manu. iron and steel, glass and chemicals.

Konya Konya, Turkey. 37 51N 32 30E. Cap. of Konya province sit. S of Ankara. Noted for breeding of horses and camels. Manu. carpets, textiles and leather goods. Pop. (1980C) 329,129.

Kootenay River Canada/U.S.A. 49 15N 117 39W. R. in Brit. Columbia rising in Rocky Mountains and flowing for 407 m. (651 km.) looping into Montana and Idaho (U.S.A.), flowing through Kootenay L., and joining Columbia R. at Castlegar N of Trail in Brit. Columbia.

Kopeisk Russian Soviet Federal Socialist Republic, U.S.S.R. Town sit. SE of Chelyabinsk. Centre for lignite mining.

Kopparberg Sweden. 61 20N 14 15E. County of central Sweden on the Norwegian border. Area, 10,968 sq. m. (28,194 sq. km.). Main cities Falun (cap.), Ludvika, Borlänge, Avesta, Hedemora, Säter. Surface rises to mountains on the Norwegian border. Agric. and dairy farming in the centre around L. Silja. Grängesberg is the centre of a major iron, copper, lead and zinc mining area, with steelworks and other industries inc. lumbering, textiles, tanning. Pop. (1988E) 287,407.

Korca   *see* **Korçë**

Korçë Korçë, Albania. 40 37N 20 46E. Cap. of Korçë district in SE, and the fifth largest town in Albania, sit.

SE of Tirana. Industries, sugar-refining, brewing and flour milling Pop. (1983E) 57,000.

Korčula Croatia, Yugoslavia. 42 57N 16 50E. Island in Adriatic Sea, part of the Dalmatian arch. Area, 107 sq. m. (277 sq. km.). Chief town Korčula. Chief occupations are vine and olive growing, fishing and marble quarrying.

Korea East Asia. 39 00N 124 00E. Peninsula extending SE into Sea of Japan and divided politically between North Korea and South Korea, the frontier corresponding roughly with the 38th parallel of lat. The area of the Rep. of Korea (South) is 38,232 sq. m. (99,022 sq. km.). Chief cities Seoul (cap.), Pusan. The area of the People's Democratic Rep. of Korea (North) is 47,225 sq. m. (122,313 sq. km.). Chief cities Pyongyang (cap.), Kaesong, Chungjin. Mainly mountainous, with the main range extending down the E coast and fertile lowland on the W. The coastline is deeply indented with many offshore islands; most good harbours are on the E and S coast. The major Rs. flow W to drain the fertile plains and inc. Taedong, Han, Kum and Somjin Rs., and Naktong R. flows S into the Korea Strait. South Korea has *c.* 22.5m. acres of arable land, of which *c.* 5m. are cultivated, main crops being rice, barley, wheat, beans, tobacco. The only important mineral deposit is tungsten. Fisheries are important. Industries are mainly concentrated on producing consumer goods, also textiles, fertilizers, cement, plastic products, newsprint. In North Korea *c.* 20% of the land is cultivable with irrigation and soil conservation, producing mainly rice. There is stock farming on larger state farms. There are valuable forest and mineral resources, esp. coal, iron, lead, zinc, copper, tungsten, nickel, manganese, graphite, oil. There is a deep-sea fishing fleet. Industries inc. manu. textiles, chemical fertilizers, iron and steel. About 60% of workers are employed in industry. Pop. of South Korea (1989C) 42,519,000. Pop. of North Korea (1989E) 22·42m.

Korhogo Côte d'Ivoire. Cap. of dept. in N. Centre for livestock and food processing.

Korinthos Korinthia, Greece. 37 26N 22 55E. Cap. of prefecture sit. on Gulf of Korinthos at W end of Korinthos Canal. Trading centre for wine and currants. Pop. (1981C) 22,658.

Korinthos Canal Greece. 37 48N 23 00E. Ship canal across the isthmus of Korinthia, extending 4 m. (6 km.) NW to SE from Gulf of Korinthos to Saronic Gulf.

Koritza *see* Korçë

Koriyama Honshu, Japan. 37 24N 140 23E, Town sit. S of Sendai. Manu. machinery, textiles and chemicals. Pop. (1983E) 290,000.

Kortrijk (Courtrai) West Flanders, Belgium. Town sit. W of Brussels on Lys R. Industries inc. textiles, engineering, electronics, rubber, plastics and chemicals. Pop. (1989E) 76,273.

Koryak National Area Russian Soviet Federal Socialist Republic, U.S.S.R. A National Area in NE Siberia, inc. part of the Kamchatka peninsula. Area, 152,000 sq. m. (393,680 sq. km.). Cap. Palana. Main occupations are reindeer breeding, fishing and hunting. Pop. (1970E) 31,000.

Kos Dodecanese, Greece. 36 50N 27 15E. (i) Island in Dodecanese group off the Turkish coast in the Aegean Sea. Area, 109 sq. m. (282 sq. km.). Lowlying and fertile. Main products cereals, olive oil, fruits and wine. (ii) Chief town of (i) sit. on the NE coast. Trading centre for agric. produce. Industries inc. fishing.

Kosciusko, Mount New South Wales, Australia. 36 27S 148 16E. Mountain peak sit. in the Australian Alps. Highest summit in Australia, 7,328 ft. (2,234 metres) a.s.l.

Košice Slovakia, Czechoslovakia. 48 43N 21 15E. Cap. of the Východoslovenská region sit. ENE of Bratislava, and on Hornád R. Industries, engineering, food processing and sawmilling. Manu. chemicals, textiles and fertilizers. Pop. (1989E) 232,000.

Koso Gol Mongolia. L. sit. in N

near U.S.S.R. frontier, and NW of Ulan Bator, at 5,300 ft. (1,615 metres) a.s.l. Drained in the S by Selenga R.

Kosovo Serbia, Yugoslavia. An autonomous region bordering on Albania. Area, 3,997 sq. m. (10,352 sq. km.). Cap. Priština, sit. S of Belgrade. Mainly agric. producing cereals and fruit. Pop. (1987E) 1·85m.

Kostroma Russian Soviet Federal Socialist Republic, U.S.S.R. 57 46N 40 55E. Cap. of region of same name sit. ENE of Yaroslavl at confluence of Volga R. and Kostroma R. Noted for linen products since the 16th cent. Manu. footwear, paper and clothing. Pop. (1985E) 269,000.

Kostroma River Russian Soviet Federal Socialist Republic, U.S.S.R. 57 47N 40 55E. A 250 m. (400 km.) trib. of R. Volga.

Koszalin Koszalin, Poland. 54 12N 16 09E. Cap. of province of same name near Baltic Sea coast. Manu. agric. machinery, textiles, food products and bricks.

Kota Rajasthan, India. Town sit. S of Jaipur on Chambal R. Manu. textiles and synthetic yarn. Pop. (1981C) 358,241.

Kota Kinabalu Sabah, Malaysia. 5 58N 116 04E. Cap. of Sabah and port sit. on W coast, W of Sandakan. Industries rice-milling and fishing. Exports rubber and timber. Pop. (1980C) 108,725.

Köthen Halle, German Democratic Republic. 51 45N 11 58E. Town sit. NNW of Leipzig. Industries, textiles, engineering and sugar refining. Pop. (1989E) 34,300.

Kotka Finland. 60 28N 26 55E. Port sit. on an island in the Gulf of Finland. Centre of a timber and woodpulp industry. Pop. (1989E) 57,201.

Kotlas Russian Soviet Federal Socialist Republic, U.S.S.R. 61 16N 46 35N. Town in the Arkhangelsk region, sit. near confluence of N. Dvina R. and Vychegda R. Important transportation centre and R. port. Industries

shipbuilding and repairing, wood-processing and cellulose paper works.

Kotor Gulf Montenegro, Yugoslavia. 42 25N 18 40E. Gulf in the Adriatic Sea, 20 m. (32 km.) in length. Chief port Kotor.

Koudougou Burkina, Burkina Faso. 12 15N 2 22W. Centre for agric area producing cotton, tobacco, groundnuts. Textile manu. Pop. (1985E) 59,644.

Kovno *see* **Káunas**

Kovrov Russian Soviet Federal Socialist Republic, U.S.S.R. 56 25N 41 18E. Town sit. SSE of Ivanovo, on Klyazma R. Important railway centre. Industries railway engineering, textiles and excavator production.

Kowloon China. 22 18N 114 10E. Town sit. on Kowloon peninsula opposite, and forming part of, Hong Kong. Important port and railway centre. Manu. rubber and cotton goods, ropes and cement. Pop. (1989E) 860,000.

Kozáni Greece. 40 18N 21 47E. (i) A prefecture in Macedonia, sit. E of the Pindus Mountains and drained by Aliákmon R. Area, 1,375 sq. m. (3,562 sq. km.). Mainly agric. Pop. (1981C) 147,051. (ii) Cap. of prefecture of same name sit. WSW of Thessaloníki, on Aliákmon R. Pop. (1981C) 30,994.

Kozhikode Kerala, India. 11 16N 75 48E. Port on the Malabar coast. An important calico-producing town in the 17th cent. Its former name was Calicut. Manu. ropes, mats and soap. Exports inc. coffee, tea, spices, copra and coconuts. Pop. (1981C) 394,447.

Kpalimé Plateaux, Togo. Town sit. NW of Lomé. Railway terminus serving an agric. area, handling cacao, palm products, cotton. The main industry is cotton ginning. Pop. (1980E) 25,500.

Kragujevac Serbia, Yugoslavia. 44 01N 20 55E. Town sit. SSE of Belgrade on Lepenica R. Manu. munitions and metallurgical products. Pop. (1981C) 132,972.

Kra Isthmus Thailand. An isthmus in the s connecting the Malay peninsula with the mainland of Asia.

Krakow Poland. 50 00N 19 57E. (i) Province of s Poland extending from upper Vistula R. to the High Tatra on the Czech. border. Area, 1,257 sq. m. (3,255 sq. km.). Chief cities Cracow (cap.), Tarnow, Nowa Huta, Nowy Sacz. Agric. areas produce cereals, potatoes, livestock. Minerals inc. salt (at Wieliczka and Bochnia), iron, lead, zinc and coal. Industries inc. lumbering, mining; manu. chemicals, machinery, railway stock, wood and metal products. Pop. (1989E) 1,220,000. (ii) Cap. of province of same name. Pop. (1985E) 716,000.

Kramatorsk Ukraine, U.S.S.R. 48 43N 37 32E. Town sit. SSE of Kharkov. Centre of heavy engineering and a railway junction. Pop. (1985E) 192,000.

Krasnodar Russian Soviet Federal Socialist Republic, U.S.S.R. 42 02N 39 00E. (i) Territory in N Caucasus adjacent to the Sea of Azov and the Black Sea. Area, 34,200 sq. m. (88,578 sq. km.). Principal R. is Kuban. Mainly agric. producing wheat, rice, fruit, tobacco, sunflowers and livestock. (ii) Cap. of territory of the same name on lower Kuban R. Principal industries, food processing, oil refining and railway engineering. Pop. (1987E) 623,000.

Krasnoyarsk Russian Soviet Federal Socialist Republic, U.S.S.R. 56 08N 93 00E. (i) Territory in central Siberia extending from the Sayan Mountains in the s to the Taimyr peninsula in the Arctic Ocean. Area, 928,000 sq. m. (2,404,000 sq. km.). Principal R. is Yenisei. Mainly forested with some agric. in the s. Large deposits of gold, coal, graphite, iron-ore and uranium. (ii) Cap. of territory of the same name sit. on Yenisei R. and Trans-Siberian railway. Industrial centre of a gold-producing area. Manu. heavy engineering products, textiles, paper and cement. Pop. (1987E) 899,000.

Krefeld North Rhine-Westphalia, Federal Republic of Germany. 51 20N 6 34E. Town adjacent to Rhine R. An important textile centre since the 17th cent. Other manu. machinery, steel and chemicals. Pop. (1989E) 234,000.

Kremenchug Ukraine, U.S.S.R. 49 05N 33 25E. Town sit. ESE of Kiev on Dnieper R. Industries, engineering, textiles, food processing and saw-milling. Pop. (1987E) 230,000.

Krems-an-der-Donau Lower Austria, Austria. 48 25N 15 36E. Town on Danube R. Trades in wine, fruit and preserves. Pop. (1981C) 23,056.

Kribi Cameroon. 2 57N 9 55E. Port in s province for timber exports and fishing. Pop. (1981E) 13,720.

Krishna River India. 15 43N 80 55E. Rises in w Ghats inland from the Arabian Sea near Mahabaleshwar and flows 800 m. (1,280 km.), first SE past Sangli, then ENE into Tamil Nadu (along the state frontier), then ESE to enter the Bay of Bengal sw of Masulipatam in a broad delta. Used extensively for irrigation and power. Linked by navigable canal with Godavari R. delta and Buckingham canal. Main tribs. Bhima and Tungabhadra Rs. The main stream is only navigable by small craft below Bezwada.

Kristiania *see* **Oslo**

Kristiansand Vest-Agder, Norway. 58 10N 8 00E. Cap. of Vest-Agder county and port sit. at mouth of Otra R. on the Skagerrak. Centre for offshore oil and gas industry. Industries textiles, smelting, flour and woollen milling and fishing. Popular tourist resort. Pop. (1989E) 64,395.

Kristianstad Kristianstad, Sweden. 56 .05N 14 07E. Cap. of Kristianstad county sit. on Helge R., in s Sweden. Industries inc. food processing, plastics, engineering and sugar refining. Pop. (1990E) 71,119.

Kristiansund Møre og Romsdal, Norway. 63 10N 7 45E. Cap. of Møre og Romsdal county, sit. on four small islands in the N. Sea. Principal industries fish canning and processing. Pop. (1980E) 18,013.

Krivoi Rog Ukraine, U.S.S.R. 47

55N 33 21E. Town sit. SW of Dnepropetrovsk on Ingulets R. Centre of a
very rich iron-ore area. Manu. chemicals, iron and steel, and machine tools.
Pop. (1987E) 698,000.

Krk Croatia, Yugoslavia. 45 05N
14 35E. Island sit. in the Adriatic Sea
SSE of Rijeka. Area, 166 sq. m. (430
sq. km.). Cap. Krk. Noted for fruit and
wines.

Kronoberg Sweden. 56 45N 14
30E. County of S Sweden. Area, 3,266
sq. m. (8,458 sq. km.). Main cities
Växjö (cap.), Ljungby. Undulating
area with numerous marshes, Ls. and
Rs. Main occupations lumbering, sawmilling, paper making. Important
glass industries at Orrefors and Kosta.
Pop. (1988E) 175,427.

Kronstadt Russian Soviet Federal
Socialist Republic, U.S.S.R. 60 05N
29 35E. City W of Leningrad on SE
Kotlin Island in the E Gulf of Finland.
Naval base with arsenal, docks and
shipyards and naval forts and batteries
commanding the approach to Leningrad. Founded by Peter the Great,
1703. Industries inc. sawmilling;
manu. clothing, shoes.

Kroonstad Orange Free State, Republic of South Africa. 27 46S 27
12E. Town NNE of Bloemfontein on
Vals R. Centre of an agric. district.
Manu. clothing. Industries engineering and motor assembly plants.

Kropotkin Russian Soviet Federal
Socialist Republic, U.S.S.R. 45 26N
40 34E. Town sit. ENE of Krasnodar
on Kuban R. Industry railway engineering. Trades in agric. products.

Kruger National Park Republic of
South Africa. Protected area sit. along
frontier between NE Transvaal and
Mozambique. 200 m. (320 km.) in
length and 30–60 m. (48–96 km.)
wide. Almost every species of native
game is found here.

Krugersdorp Transvaal, Republic
of South Africa. 26 05S 27 35E. Town
sit. WNW of Johannesburg on the Witwatersrand. Centre of a gold, uranium
and manganese mining area. Pop.
(1984E) 141,100.

Krujë Albania. Town sit. N of
Tirana at 2,000 ft. (610 metres) above
Ishm R. valley. Main occupations
olive growing, handicraft industries.

Kuala Lumpur Selangor, Malaysia.
3 08N 101 42E. City sit. NW of Singapore on Klang R. also known as KL.
Cap. and state cap., administration
and communications centre, it developed around the rubber and tin industries from 1873. Tin mining and
rubber growing are industries. Pop.
(1980C) 937,875.

Kuban River U.S.S.R. 45 20N 37
30E. Rises on Mount Elbruz in the
Caucasus and flows for 450 m. (720
km.) entering the Black Sea S of the
Taman peninsula, and sends a branch
N to the Sea of Azov.

Kubena, Lake Russian Soviet Federal Socialist Republic, U.S.S.R. L.
sit. in the Vologda region. Area, 140
sq. m. (363 sq. km.). Drained by Sukhona R. to N. Dvina R., and linked by
canal with Sheksna R.

Kuching Sarawak, Malaysia. 1 32N
110 19E. Cap. and port on the
Sarawak R. Exports inc. rubber and
pepper. Pop. (1988E) 152,000.

Kucovë *see* **Qytet Stalin**

Kufra Oases South Cyrenaica,
Libya. 24 20N 23 15E. A group of
five oases in the Libyan Desert, sit. SE
of Benghazi. Important caravan junction. Dates and barley are grown.

Kuibyshev (Samara) Russian
Soviet Federal Socialist Republic,
U.S.S.R. Cap. of the Kuibyshev region, sit. at confluence of Volga R.
with Samara R. Important industrial,
commercial and transportation centre.
Manu. aircraft, locomotives, tractors,
chemicals and textiles. Pop. (1985E)
1,257,000.

Kukes Albania. 42 05N 20 24E.
Town sit. E of Scutari on Drin R. near
the Yugoslav border. Commercial
centre of Albanian Kosovo.

Kula Himachal Pradesh, India. Village sit. ESE of Dharmsala on Beas R.
Trading centre for grain, wool, fruit,

tea, honey, timber, with handicraft industries. Pop. (1981C) 11,869.

Kumamoto   Kyushu, Japan. 32 50N 130 42E. Cap. of Kumamoto prefecture sit. E of Nagasaki. Commercial centre. Manu. textiles and food products. Pop. (1988E) 555,000.

Kumasi   Ashanti, Ghana. 6 41N 1 37W. Cap. of Ashanti region. The second largest city of Ghana. Important transportation and commercial centre of a main cocoa-growing area. Industries metal working, weaving and local crafts. Pop. (1982E) 439,717.

Kumba   Cameroon. 4 38N 9 25E. Chief town of SW province, centre for agric., esp. cocoa, oil palms and rubber. Pop. (1981E) 53,823.

Kumbakonam   Tamil Nadu, India. 10 57N 79 23E. Town sit. SSW of Madras on Cauvery R. delta. Anc. cap. of the Chola Kings and a religious centre. Manu. metal goods and textiles. Pop. (1981C) 132,832.

Kunaitra   Damascus, Syria. Otherwise called El Quneitra, town sit. SW of Damascus near the Israeli border. Centre of an agric. district producing tobacco, olives, fruit, cereals.

Kunar   Eastern Province, Afghánistán. Town NE of Jalalabad on Kunar R. near the Pakistan frontier. sit. in an irrigated oasis producing rice, maize and fruit. The main industry is furniture making.

Kunduz   Kataghan, Afghánistán. Town WNW of Khanabad sit. in an oasis on Kunduz R. which produces rice and cotton. The main industries are cotton ginning, milling cottonseed oil, rice and flour; manu. soap.

Kunene River   Angola. 17 20S 11 50E. Rises in the Benguela Plateau at 5,800 ft. (1,768 metres) a.s.l. and flows for 720 m. (1,152 km.) SSW entering the S. Atlantic Ocean. For 200 m. (320 km.) from its mouth it forms the boundary between Angola and Namibia.

Kunlun   Asia. 36 00N 82 00E. Extensive mountain range in central Asia sit. E of the Pamirs between the Tarim Basin in the N and the Tibetan plateau in the S. The highest peak is Ulugh Muztagh, 25,338 ft. (7,723 metres) a.s.l.

Kunming   Yunnan, China. 25 01N 102 41E. Cap. of Yunnan province in S central China. Commercial centre and terminus of the Burma Road. Manu. chemicals, textiles, machinery and cement. Pop. (1987E) 1·52m.

Kuopio   Kuopio, Finland. 62 54N 27 41E. Cap. of the Kuopio province NE of Helsinki on L. Kallavesi. Industry wood products. Pop. (1988C) 79,495.

Kurashiki   Okayama, Japan. 34 35N 133 46E. Town sit. WSW of Okayama. Railway junction. Industries inc. manu. textiles, floor mats, wood and metal products. Pop. (1988E) 416,000.

Kurdistan   Central Asia. Area of plateau and mountain extending across parts of SE Turkey, NE Iraq and NW Iran and inhabited by the Kurds, who are a Moslem, tribal people. Most settlements are in the SE Anatolian and NW Zagros mountains, and survive by nomadic stock farming in summer, on the mountain pastures, and some agric. and rug making in the valleys in winter. Their claims for national autonomy since the Paris Peace Conference of 1919 have so far not materialized.

Kurdufan   Sudan. 36 00N 124 00E. Province of central Sudan. Area, 56,731 sq. m. (146,932 sq. km.). Main cities El Obeid (cap.) and Nahud. Plateau at average 2,000 ft. (615 metres) with ranges rising to 4,000 ft. (1,219 metres). Main products livestock, gum arabic, of which it is a major source. Pop. (1983C) 3,093,294.

Kure   Honshu, Japan. 34 14N 132 34E. Port sit. SSE of Hiroshima. Industries shipbuilding, engineering, iron and steel. Pop. (1983E) 233,000.

Kurgan   Russian Soviet Federal Socialist Republic, U.S.S.R. 55 26N 65 18E. Cap. of the Kurgan region E of Chelyabinsk on Tobol R. Centre of an agric. area with associated industries. Pop. (1985E) 343,000.

Kurgan-Tyube Tadzhik Soviet Socialist Republic, U.S.S.R. Town sit. s of Stalinabad. The main industry is cotton processing, also food processing.

Kuria Kiribati, w, West Pacific. Island formerly called Woodle atoll in the N Gilbert group. Area, 6 sq. m. (16 sq. km.). Pop. (1985C) 1,052.

Kuria Muria Islands Oman. 17 30N 56 00E. A group of five islands sit. off s coast of Oman in the Arabian Sea. Area, 28 sq. m. (72 sq. km.). Hallaniya, the largest island, is the only inhabited one. Pop. (1980E) 85.

Kuril Islands Russian Soviet Federal Socialist Republic, U.S.S.R. 46 10N 152 00E. Island chain extending 650 m. (1,040 km.) between Cape Lopatka on the Kamchatka peninsula to Hokkaido, Japan, and separating the Sea of Okhotsk from the Pacific Ocean. Area, 5,700 sq. m. (14,763 sq. km.), consisting of 50 islands and numerous rocks. The main chain is volcanic, and inc. Shumshu, Paramushir, Onekotan, Simushir, Urup, Hurup, Kunashir; there is a parallel non-volcanic chain extending *c*. 65 m. (104 km.) ENE of Hokkaido, inc. Shikotan, Shibotsu, Shuishio. Chief towns Severo-Kurilsk, Kurilsk, Yuzhno-Kurulsk. Humid climate, with hot springs, active volcanoes and sulphur deposits. Other minerals inc. iron, copper, gold. Main occupations, fishing, mining, mainly sulphur, lumbering, fur trapping, market gardening.

Kurnool Andhra Pradesh, India. Town sit. ssw of Hyderabad on Tungabhadra R. Manu. carpets, textiles and leather products. Pop. (1981C) 206,362.

Kursk Russian Soviet Federal Socialist Republic, U.S.S.R. 51 42N 36 12E. Cap. of Kursk region on Seim R. Industrial centre of a rich agric. district. Manu. chemicals, electrical equipment, food products and alcohol. Pop. (1987E) 434,000.

Kurskiy Zaliv Lithuania/Russian Soviet Federal Socialist Republic, U.S.S.R. A large coastal lagoon separated from the Baltic Sea by a narrow sandy bar, the Kurskaya Kosa. Length 56 m. (90 km.).

Kushtia Bangladesh. 23 55N 89 05E. District cap. on Ganges delta. Sit. w of Dacca. Sugar milling, cotton cloth weaving and textile manu. Trade in rice, jute, linseed, cane, wheat. Pop. (1984E) 120,000.

Kustanai Kazakhstan, U.S.S.R. 53 20N 63 45E. Cap. of Kustanai region sit. SE of Chelyabinsk on Tobol R. Centre of an agric. district with associated industries. Pop. (1987E) 212,000.

Kütahya Turkey. 39 25N 29 59E. (i) Province of w Turkey bounded NE by the Ulu range, NW by the Demirci range. Area, 5,889 sq. m. (15,253 sq. km.). Drained by Simav, Porsuk, Gediz, Buyuk Menderes Rs. Minerals inc. chromium, copper, mercury, magnesite, iron, lignite, clay. Farming areas produce mohair goats, cotton, sugar-beet, apples, pears, plums, figs, raisins. Pop. (1985C) 543,384. (ii) Town sit. wsw of Ankara on Porsuk R. Provincial cap. Industries inc. ceramics; manu. carpets, cotton. Meerschaum clay in the neighbourhood. Pop. (1985C) 120,354.

Kutaisi Georgia, U.S.S.R. 42 15N 42 40E. Town sit. wNw of Tbilisi on Rion R. Industries, engineering, inc. vehicle assembly, chemicals, silk and food production. Pop. (1987E) 220,000.

Kutch, Rann of Gujarat, India. Extensive salt desert bounded w by the Arabian Sea and N by the Thar desert. Divided into the larger Great Rann (N), 220 m. (352 km.) by 25–50 m. (40–80 km.) wide; Little Rann, 65 m. (104 km.) by 35 m. (56 km.). Formed by the silting up of a shallow arm of the Arabian Sea which became first a L. then a mud-flat. Extensive salt deposits are exploited.

Kuwait North East Arabia. 29 20N 47 59E. Sheikhdom at head of Gulf on the NE coast of the Arabian peninsula, bounded N and w by Iraq, s by Saudi Arabia. On 2 Aug. 1990 Iraq invaded Kuwait and on 8 Aug. formally annexed the Sheikhdom. Area, 6,880 sq. m. (17,819 sq. km.). Cap. Kuwait.

Pop. (1980c) 60,525 and its suburbs Havalli (152,402) and as-Salmiyah (145,991). Mainly desert, with some agric. in irrigated areas near Jahra. Rainfall *c*. 4–5 ins. (10–12.5 cm.) annually. Industries in Kuwait city inc. building dhows, but the other trad. industries of pearling and fishing have been eclipsed by oil production and distribution. Kuwait is the 7th largest oil producer in the world; main production is from the Burgan field, with outlets to storage at Ahmadi and the port at Mena al-Ahmadi. Pop. (1988e) 1·96m.

Kuznetsk Russian Soviet Federal Socialist Republic, U.S.S.R. 53 12N 46 40E. Town sit. E of Penza. Manu. agric. machinery, bricks and rope.

Kuznetsk Basin Russian Soviet Federal Socialist Republic, U.S.S.R. Important coalmining area in the Kemerovo region. Chief centre Novokuznetsk.

Kvarner Croatia, Yugoslavia. 44 50N 14 10E. A gulf in the N Adriatic Sea between the peninsula of Istria and the main Croatian coast.

Kvitoya Svalbard, Norway. Island between the Spitsbergen group (W) and Franz Josef Land in the Barents Sea. Area, 102 sq. m. (264 sq. km.), rising to 886 ft. (270 metres).

Kwangchow *see* **Gangzhou**

Kwangsi-Chuang Autonomous Republic *see* **Guangxi Zhuang**

Kwangtung *see* **Guangdong**

Kwara Nigeria. State comprising the Ilorin and Kabba provinces of the former N region. Area, 25,818 sq. m. (66,869 sq. km.). The chief ethnic groups are the Fulani, Nupe and Yoruba. The main cities are Ilorin (cap.), Offa, Lafiagi and Lokoja. Pop. (1988e) 3,685,100.

Kweichow *see* **Guizhou**

Kweisui *see* **Huhehot**

Kwekwe Zimbabwe. 18 58S 29 48E. City in central region. Industries include steel and chrome processing and gold roasting. Distribution centre for agric. products inc. tobacco. Pop. (1982c) 48,000.

Kwinana Western Australia, Australia. 32 15S 115 48E. Town sit. S of Fremantle. Industries, oil refining, steel and cement works.

Kyle of Lochalsh Highland Region, Scotland. 57 17N 5 43W. Port sit. at entrance to Loch Alsh, WSW of Inverness.

Kymi Finland. County of SE Finland bounded S by Gulf of Finland, E by U.S.S.R. Area, 4,163 sq. m. (10,783 sq. km.). Chief cities Kotka (cap.), Lappeenranta, Hamina. Inc. part of Saimaa L. system and Kymi, Vuoksi Rs. and Saimaa canal. The Kymi R. valley is industrial, with lumbering, timber processing, iron and steel, metalworking, textiles, glass making. Agric. areas produce rye, oats, barley, livestock, dairy products. Pop. (1988e) 255,893.

Kyoga, Lake *see* **Kioga Lake**

Kyoto Honshu, Japan. 35 00N 135 45E. Cap. of Kyoto prefecture near L. Biwa, to which it is joined by canal. Noted for its diverse crafts inc. silks, brocades, porcelain and lacquer ware. Principal Buddhist centre with many fine temples. Pop. (1988e) 1,419,000.

Kyrenia Cyprus (Girne). 35 20N 33 19E. Port sit. N of Nicosia. Tourism is important. Pop. (1973c) 3,892.

Kyushu Japan. 32 30N 131 00E. Southernmost and most densely populated of the four main islands. Area, 14,990 sq. m. (38,824 sq. km.). Chief cities Nagasaki, Yawata, Wakamatsu, Kokura, Fukuoka, Moji. Mountainous, rising to 5,850 ft. (1,783 metres) at Mount Kuju, with hot springs. Drained by Chikugo R. flowing through an extensive rice growing area. Heavily forested elsewhere. Main crops rice, grain, sweet potatoes, fruit; other important products are raw silk, timber, fish. Minerals inc. a major coal field at Chikuho, supporting an area of heavy industries. Noted for porcelain from Satsuma and Hizen. Pop. (1980e) 13m.

L

Laaland *see* **Lolland**

La Altagracia Dominican Republic.
18 00N 68 00W. Province of SE bounded S and E by Caribbean. Area, 1,124
sq. m. (2,911 sq. km.). Cap. La Romana. Tropical lowlands rising to hills
in the NW. The main products are
cattle, sugar, bananas, coffee, rice,
cacao, corn, hides, timber. Industries
inc. sugar refining at La Romana. Pop.
(1981C) 100,112.

La Asunción Nueva Esparta, Venezuela. 11 02N 63 53W. Town sit. ENE
of Caracas on Margarita Island. State
cap. and centre of an agric. area. Industries inc. cotton ginning, milling
sugar and maize, distilling. Pop.
(1984E) 8,000.

Labasa Vanua Levi, Fiji. Town on
S bank of Lambasa R., Indian settlement producing sugar.

Labé Guinea. 11 19N 12 17W.
Cap. of Moyenne-Guinea sit. NE of
Conakry in Fouta Djallon mountains.
Industries inc. slate quarrying; manu.
essential oil, honey, beeswax, food
products. Pop. (1975E) 80,000.

Labrador Newfoundland, Canada.
53 20N 61 00W. The most N district
of Newfoundland, extending from
L'Anse Éclair at the NE entrance of
the Straits of Belle Isle to Cape Chidley at the E entrance of Hudson's
Strait. In 1927 the Privy Council decided the boundary between Canada
and Newfoundland in Labrador. The
area now under the jurisdiction of
Newfoundland is approx. 110,000 sq.
m. (285,000 sq. km.). Iron deposits
are found in the forests which cover
large areas. Cod fishing is the main
occupation. Large hydroelectric plants
have been developed at Twin Falls
and Churchill Falls.

Labuan Malaysia. 5 21N 115 13E.
An island sit. near the NW coast of

Sabah. Area, 35 sq. m. (75 sq. km.).
Formerly part of Sabah, it became
federal territory in 1984. Chief town
and port, Victoria. Copra, rubber, rice,
coconuts and sago are produced. It is
a free port. Pop. (1980C) 12,219.

Laccadive Islands *see* **Lakshadweep**

La Ceiba Atlántida, Honduras. 15
47N 86 50W. Port sit. E of Tela. Important banana port. Industries flour
milling, tanning, and soap and footwear manu. Pop. (1984E) 55,000.

La Chaux-de-Fonds Neuchâtel,
Switzerland. 47 06N 6 50E. Town sit.
NNW of Neuchâtel, in Jura mountains
near the French border, at 3,270 ft.
(997 metres) a.s.l. The main industry
is watch making; also printing, brewing, flour milling. There is a museum
of clocks and watches. Pop. (1978E)
39,000.

Lachine Quebec, Canada. 45 26N
73 40W. Port and resort sit. on S coast
of Montreal Island at SW end of
Lachine Canal. Manu. iron and steel,
wire, electrical apparatus, chemicals
and tyres. Pop. (1981E) 38,000.

Lachlan River New South Wales,
Australia. 34 21S 143 57E. Rises in
Blue Mountains and flows for 922 m.
(1,475 km.) W to join R. Murrumbidgee. The Wyangala Dam, which
was completed in 1935, irrigates a
vast area.

La Ciotat Bouches-du-Rhône,
France. 43 12N 5 36E. Resort on
Mediterranean coast. Naval shipbuilding. Pop. (1982C) 31,727.

Lackawanna New York, U.S.A. 42
49N 78 50W. City sit. at E end of L.
Erie and just S of Buffalo. Industries
iron and steel, cement and wood products. Pop. (1980C) 22,701.

La Concepción Zulia, Venezuela.

Town sit. ssw of Maracaibo on N shore of L. Maracaibo. Oil-producing centre with refineries.

La Condamine Monaco. Resort and financial and commercial district of Monaco immediately sw of Monte Carlo at the head of Monaco harbour.

Laconia New Hampshire, U.S.A. 43 31N 71 29W. City sit. N of Concord on Winnipesaukee R. Resort for nearby Ls. and market centre for an agric. area. Industries inc. hosiery, textiles, textile machinery, metal products, skis, boats. Pop. (1980c) 15,575.

Lacq Pyrénées-Atlantiques, France. 42 25N 0 35W. Village sit. NW of Pau on the Gave de Pau. Noted for petroleum deposits. Pop. (1982c) 564.

La Crosse Wisconsin, U.S.A. 43 49N 91 15W. City sit. WNW of Madison, at junction of Rs. Mississippi, Black and La Crosse. Centre of an agric. region with associated industries. Pop. (1980c) 48,347.

Ladakh Range Kashmir. 34 00N 78 00E. A mountain range running parallel to, and to the N of, upper Indus R. Some peaks exceed 20,000 ft. (6,096 metres) a.s.l.

Ladário Mato Grosso, Brazil. Town sit. E of Corumbá on Paraguay R. Port handling iron ore and manganese from Morro do Urucum. Naval station with ship repair yards.

La Digue Seychelles, Indian Ocean. 4 21S 55 50E. Island sit. NE of Victoria, Mahé Island. Area, 4 m. by 2 m. (6 km. by 3 km.). Main products vanilla, copra, essential oils, fish.

Ladoga, Lake Russian Soviet Federal Socialist Republic, U.S.S.R. 61 00N 31 00E. Area, 7,100 sq. m. (18,389 sq. km.). It discharges into R. Neva in Gulf of Finland. A railway, built for the winters of 1941–3, across the S part, helped to sustain Leningrad during the siege.

Ladysmith Natal, Republic of South Africa. 28 34S 29 45E. Town sit. on Klip R. Industries railway

engineering, tyres, prefabricated houses, furniture, footwear, textiles and food processing. Pop. (1989E) 56,599.

Lae Papua New Guinea. 6 45S 147 00E. Town sit. N of Port Moresby on Huon Gulf. There is a harbour and airfield. Pop. (1980E) 61,617.

Lafayette Indiana, U.S.A. 40 25N 86 53W. City sit. NW of Indianapolis on Wabash R. Centre of an agric. area. Manu. paper and electrical products, foods, metal goods, plastics, vehicle engines and parts, instruments and chemicals. Pop. (1988E) 44,290.

Lafayette Louisiana, U.S.A. 30 14N 92 01W. City sit. WSW of Baton Rouge on Vermilion R. Centre of an agric. and mining area. Industries oil and sugar refining, cotton and cottonseed processing. Pop. (1990E) 89,754.

Lagan River Down, Northern Ireland. 54 37N 5 53W. Rises near Ballynahinch and flows NW and NE for 45 m. (72 km.) to enter Belfast Lough at Belfast.

Lågen River Buskerud, Norway. 59 03N 10 05E. Rises in the Hardangervidda and flows SSE for 200 m. (320 km.) to enter the Skagerrak at Larvik.

Lågen River Oppland, Norway. 61 08N 10 25E. Rises N of Jotunheim Mountains and flows SSE for 120 m. (192 km.) to L. Mjosa.

Lagos Nigeria. State comprising the City of Lagos (a former Federal Capital Territory) and the province of Ikeja (formerly colony province of the w region). Area, 1,292 sq. m. (3,345 sq. km.). Main towns Lagos (cap.), Mushin, Shomolu and Epe. Pop. (1988E) 4,569,400.

Lagos Lagos, Nigeria. 37 05N 8 41W. Cap. of Nigeria and port sit. on an island in a lagoon in the Bight of Benin. Founded in the 13th cent. by the Yoruba. Joined by bridge to Iddo Island which in turn is joined by bridge to the mainland. Industries sawmilling, brewing and soap manu. Exports groundnuts, palm oil and kernels. Aruba is to be the future cap. of Nigeria. Pop. (1983E) 1,097,000.

La Grande Oregon, U.S.A. 45 20N 118 05W. City sit. SE of Pendleton on Grande Ronde R. at foot of Blue Mountains, at 2,786 ft. (849 metres) a.s.l. Industries inc. railways, sawmilling. Manu. particle board, trailers and chemical products. Pop. (1980C) 11,354.

La Guaira Federal District, Venezuela. 10 36N 66 56W. Chief port N of Caracas on Caribbean Sea. Exports coffee, tobacco and cacao.

La Guajira Colombia. Department of N Colombia on Caribbean Sea, comprising a peninsula to the NE of the Sierra Nevada de Santa Marta, bounded S by Venezuela across the Montes de Oca. Area, 8,049 sq. m. (20,848 sq. km.). Mostly arid plain with rich deposits of pearl and salt on the coast, gold, copper, phosphates, gypsum, petroleum, coal. Main products divi-divi bark, used for tanning, hides, livestock, maize, beans, yucca coconuts. Pop. (1983E) 510,200.

Lahad Datu Sabah, Malaysia. Town sit. S of Sandakan on N shore of Darvel Bay. Main occupations fishing, cultivating hemp and tobacco.

Lahore Punjab, Pakistan. 31 34N 74 22E. Cap. of province and second largest city of Pakistan sit. near Ravi. R. Industrial and commercial centre. Railway engineering. Manu. carpets, electrical goods, footwear, textiles and metal goods. Just to the E are the famous Shalamar Gardens. Pop. (1981E) 2,953,000.

Lahti Häme, Finland. 60 58N 25 40E. Town and winter sports resort sit. at S end of L. Päijänne. Sawmilling, furniture manu. and associated industries. Pop. (1983C) 94,466.

Laillahue Mountain Peru. Mountain peak in extreme SE Peru on Bolivian frontier, 16,995 ft. (5,180 metres) a.s.l.

La Joya Oruro, Bolivia. Town sit. NW of Oruro on Desaguadero R. at 12,542 ft. (3,823 metres) a.s.l. in the Altiplano. Centre of a farming area producing sheep, potatoes.

Lake Charles Louisiana, U.S.A. 30 13N 93 12W. City sit. on Calcasieu R., and near the boundary with Texas. Industries lumber and rice milling, oil refining, soyabeans, petrochemicals. Pop. (1980C) 75,226.

Lake District Cumbria, England. 54 30N 3 10W. Area, c. 700 sq. m. (c. 1,813 sq. km.). Famed for beautiful lake and mountain scenery. It inc. L. Windermere, Ullswater and Derwentwater, and among mountain peaks are Scafell, Skiddaw and Helvellyn. The rainfall is heavy. Closely associated with Wordsworth, Coleridge and Southey, who are called the Lake Poets. Important tourist centre. Chief towns, Ambleside, Bowness and Keswick.

Lake Edward Uganda/Zaïre. 0 25S 29 40E. One of the great sources of the R. Nile. Area, c. 820 sq. m. (c. 2,124 sq. km.) and it is 44 m. (70 km.) long, and 32 m. (57 km.) broad with a mean depth of 365 ft. (111 metres), and is sit. at 3,000 ft. (914 metres) a.s.l. The L. was discovered by Stanley in 1888–9.

Lakeland Florida, U.S.A. 28 03N 81 57W. City sit. ENE of Tampa. Centre of a citrus fruit growing district with associated industries. Phosphate mining is carried on. Pop. (1990E) 70,900.

Lakewood Colorado, U.S.A. 39 44N 105 05W. City sit. between Denver and the Rocky Mountains, at 5,440–6,600 ft. (1,659–2,013 metres) a.s.l. Retailing and commercial centre. Pop. (1990E) 130,000.

Lakewood Ohio, U.S.A. 41 29N 81 48W. City and suburb of Cleveland sit. SW of Cleveland on L. Erie. Manu. plastics, hardware and tools. Pop. (1980C) 61,963.

Lakonia Greece. 36 25N 22 37E. Prefecture sit. in the SE Peloponnessos. Area, 1,404 sq. m. (3,636 sq. km.). Cap. Sparte. Chief occupation agric. with wheat, olives and citrus fruits grown. Some sheep and goats are reared. Pop. (1981C) 93,218.

Lakshadweep　Indian Ocean. 10 00N 73 30E. Island group sit. off coast of Kerala in the Arabian Sea. Consists of the Laccadive, Minicoy and Amindivi Islands comprising a Territory of India. Area, 12 sq. m. (32 sq. km.). Chief settlement Kavaratti. There are about 36 islands, 10 inhabited. Main occupations farming, fishing. Pop. (1981C) 40,249.

La Laguna　Tenerife, Canary Islands. Town sit. WNW of Santa Cruz de Tenerife. Resort and trading centre for an agric. area. Industries inc. manu. foodstuffs, ceramics, tobacco products, film. Formerly the chief town of the Canaries; has a university and cathedral. Pop. (1986E) 114,223.

La Libertad　El Salvador. 13 29N 89 19W. (i) Dept. of W Salvador on Pacific coast. Area, 665 sq. m. (1,662 sq. km.). Chief towns Nueva San Salvador (cap.), La Libertad, Quezaltepeque. Crossed E to W by the coastal mountain range. Mainly an agric. and stock farming area with some lumbering. Pop. (1981E) 388,538. (ii) Town sit. S of Nueva San Salvador on the Pacific coast. Important port exporting coffee, sugar, balsam, timber. Resort with some industries inc. fishing. Pop. (1984E) 14,500.

La Libertad　Peru. Dept. of NW Peru bounded W by Pacific, E by Cordillera Central of the Andes. Area, 10,209 sq. m. (26,441 sq. km.). Chief towns Trujillo (cap.), San Pedro, Santiago de Chuco, Otusco, Pacasmayo, Puerto Chicama, Salaverry. Rises E to the Cordillera Occidental, descending again to the Marañón R. Irrigated coastal areas produce sugar-cane, rice and cotton. Cereals and livestock are farmed in the highlands. Minerals inc. gold, silver, copper. Pop. (1981C) 962,949.

La Línea　Cádiz, Spain. 36 10N 5 19W. Town sit. ESE of Cádiz on the isthmus of Algeciras Bay immediately N of frontier zone with Gibraltar. Trading centre for fruit and vegetables. Industries inc. manu. cement, fish products, wines and foodstuffs.

Lama-Kara　Togo. Town sit. N of Sokode in a farming area producing cotton, nuts, livestock.

Lambayeque　Peru. 6 42S 79 55W. (i) Dept. of NW Peru bounded W by Pacific, E by Cordillera Occidental. Area, 6,404 sq. m. (16,586 sq. km.). Chief towns Chiclayo (cap.), Pimentel, Puerto Eten, Farrañafe and Lambayeque. Mainly coastal plain intensively irrigated and producing most of Peru's rice and sugar-cane. Industries inc. sugar refining, rice milling. Pop. (1988E) 881,000. (ii) Town sit. NW of Chiclayo on Lambayeque R. Provincial cap. Trading centre for a rice growing area. Industries inc. rice milling, cotton ginning, distilling. Pop. (1985E) 20,000. (iii) R. formed by confluence of Chancay and Cumbil Rs. 9 m. (14 km.) ENE of Chongoyape and flowing 70 m. (112 km.) W to enter the Pacific 5 m. (8 km.) SW of Lambayeque. Used to irrigate an extensive rice and sugar growing area. In its upper course it is called Chongoyape R. Main trib. Eten R.

Lambeth　Greater London, England. 51 30N 0 07W. A district S of R. Thames. Pop. (1981E) 246,246.

Lamego　Viseu, Portugal. Town sit. E of Oporto. Trading centre for an area producing port wine and fruit. It has a cathedral and the ruins of a Moorish castle.

Lamia　Phthiotis, Greece. 38 54N 22 26E. Cap. of Prefecture sit. NW of Athens. Trades in agric. produce. Pop. (1981C) 41,667.

Lamington, Mount　Papua New Guinea. 8 55S 148 10E. An active volcano NE of Port Moresby sit. in Owen Stanley Range.

Lampedusa　Italy. 35 31N 12 35E. The largest of the Pelagian Islands sit. between Malta and Tunisia, in the Mediterranean Sea. Area 8 sq. m. (21 sq. km.). Main occupation fishing and there is a penal colony. Pop. (1980E) 4,500.

Lampeter　Dyfed, Wales. 52 07N 4 05W. Town sit. E of Cardigan on R. Teifi. Pop. (1981C) 1,908.

Lanai　Hawaii, U.S.A. 20 50N 156 55W. Island sit. W of Maui. Area, 141 sq. m. (365 sq. km.). Most of it is a pineapple plantation, with exporting port and trading centre at Lanai City. Pop. (1980C) 2,119.

Lanark Strathclyde Region, Scotland. 55 41N 3 46W. Market town and administrative centre sit. SE of Glasgow near R. Clyde. Industries inc. textiles, electronics and light engineering. Pop. (1980E) 9,778.

Lanarkshire Scotland. Former county now in Strathclyde Region.

Lancashire England. 53 40N 2 30W. Under 1974 reorganization the county consists of the districts of Lancaster, Wyre, Ribble Valley, Blackpool, Fylde, Preston (administrative centre), South Ribble, West Lancashire, Chorley, Blackburn, Hyndburn, Burnley, Pendle and Rossendale. The county contains important advanced technology industries notably British Aerospace. North East Lancashire (Blackburn, Hyndburn, Burnley, Pendle, Rossendale, Ribble Valley) is a former textile area, now served by M65, which is successfully attracting new industry and benefits from attractive surrounding countryside. West Lancashire is an extensive area of Grade I agricultural land and contains the new town of Skelmersdale, linked to M65 by M58. The Fylde Coast, linked to M6 by M55, is dominated by the holiday resort of Blackpool, facilities including Blackpool Airport and the Port of Fleetwood. In Lancaster District is the county's university whilst at the Port of Heysham is located the central base of the Morecambe Bay Offshore Gasfield. Pop. (1988E) 1,382,000.

Lancaster Lancashire, England. 54 03N 2 48W. County town and port in NW England at head of R. Lune estuary. Manu. textiles, linoleum, oilcloth, plastics and furniture. Pop. (1983E) 126,400.

Lancaster Ohio, U.S.A. 39 43N 82 36W. City sit. SE of Columbus on Hocking R. Manu. glass, machinery, kitchen cabinets, televisions, gas meters and footwear. Pop. (1989E) 36,000.

Lancaster Pennsylvania, U.S.A. 40 02N 76 19W. City sit. W of Philadel-

phia. Centre of a rich agric. district with cotton and silk mills, breweries, tanneries, tobacco and chocolate works. Manu. linoleum, clocks and watches and electrical goods. Pop. (1980C) 54,725.

Landau Rhineland-Palatinate, Federal Republic of Germany. 49 12N 8 07E. Town NW of Karlsruhe. Centre of a fruit and wine trade. Manu. tobacco products. Pop. (1990E) 37,000.

Lander Wyoming, U.S.A. 42 50N 108 44W. City sit. W of Casper on Popo Agie R. at 5,360 ft. (1,634 metres) a.s.l. Resort and trading centre for farm produce. Pop. (1980C) 7,867.

Landes France. 43 57N 0 48W. Dept. in Aquitaine region, bounded W by Bay of Biscay and S by Pyrénées. Area, 3,566 sq. m. (9,237 sq. km.). Chief towns Mont-de-Marsan (prefecture), Dax. Much of it is pine forests planted to reclaim a marshy waste along the coast. Some agric. S of Adour R., but the main industry is lumbering of which the main products are resin, cork. Pop. (1982C) 297,424.

Land's End Cornwall, England. 50 03N 5 44W. A granite headland 60 ft. (18 metres) high, and the most westerly point of England.

Landshut Bavaria, Federal Republic of Germany. 48 33N 12 09E. Town sit. on Isar R. Manu. chemicals, metal goods, rope and soap. Pop. (1984E) 56,400.

Landskrona Malmöhus, Sweden. 55 52N 12 50E. Port sit. NNW of Malmö on The Sound. Industries flour milling, chemicals, components for vehicles, packaging, sugar refining, tanning and fertilizer manu. Pop. (1989E) 35,847.

Langeland Denmark. 55 00N 10 50E. An island in the Baltic Sea sit. between the islands of Fyn and Laaland. Chief town Rudköbing. Mainly agric. Pop. (1975E) 18,000.

Langholm Dumfries and Galloway Region, Scotland. 55 09N 2 59W. Town sit. near border with England on R. Esk. Manu. tweeds, ceramics and sheepskin products. Pop. (1990E) 2,600.

Langreo Oviedo, Spain. 43 18N 5 41W. Mining region in Nalón valley, sit. SE of Oviedo, producing bituminous coal, iron. Chief towns La Felguera, Sama de Langreo.

Languedoc-Roussillon France. 43 58N 3 22E. Region of S France bounded S by Pyrénées, SE by Mediterranean, E by lower Rhône R. and W by Massif Central. Comprises the departments of Aude, Gard, Hérault, Lozière and Pyrénées-Orientales. Area, 10,597 sq. m. (27,447 sq. km.). Important wine producing area. Principal towns, Toulouse, Montpellier, Nîmes, Béziers, Carcassonne, Narbonne, Albi. Pop. (1982C) 1,926,514.

Lansing Michigan, U.S.A. 42 43N 84 34W. State cap. sit. on Grand R. Important centre of the car industry. Also manu. buses, lorries, tools, metal, cement and wood products. Pop. (1980C) 130,414.

Lanzhou Gansu, China. 36 03N 103 41E. Cap. of province in central China on Huang He R., and sit. at 5,200 ft. (1,585 metres) a.s.l. Important trading and oil centre. Manu. oilfield equipment, cement, chemicals, textiles, matches and soap. Pop. (1982C) 1,430,000.

Lanzarote Canary Islands. 29 00N 13 40W. Island sit. NE of Las Palmas between Fuerteventura (S) and Graciosa. Area, 307 sq. m. (795 sq. km.). Chief town, Arrecife. Volcanic and rocky, rising to 2,215 ft. (675 metres) a.s.l. The Montaña de Fuego is active. Agric. areas produce onions, cereals, peas, potatoes, fruit. Fisheries are important.

Laoag Luzon, Philippines. 12 32N 125 08E. Port sit. on Laoag R. near its mouth, and 240 m. (384 km.) NNW of Manila. Trades in rice, corn, sugar, cotton and tobacco.

Laoighis Leinster, Ireland. 53 00N 7 20W. An inland county, mainly flat but rising to 1,700 ft. (518 metres) a.s.l. in the Slieve Bloom Mountains in the NW. Area, 664 sq. m. (1,720 sq. km.). County town Port Laoighise (Maryborough). Chief R. Barrow and the Nare. Mainly agric. with dairy farming. Pop. (1988E) 53,284.

Laokay Laokay, Vietnam. Town sit. NW of Hanoi and opposite Hokow, China, on Red R. Frontier trading centre with phosphates and graphite deposits.

Laon Aisne, France. 49 34N 3 40E. Cap. of dept. in NE France sit. on an isolated hill rising 330 ft. (101 metres) above the plain. Manu. sugar and metal goods. Has a splendid 12th/13th cent. cathedral. Pop. (1982C) 29,074.

Laos 17 45N 105 00E. Bordered on W by Thailand and Burma, on N by China, on E by Vietnam and on S by Cambodia, and inc. basin of R. Mekong and upper basin of R. Menam, Chao Braya and Salween. The Kingdom of Laos, once called Lanxang (the Land of a Million Elephants), was founded in the 14th cent. In 1893 Laos became a French protectorate and in 1907 acquired its present frontiers. Under a new Constitution of 1947 Laos became a constitutional monarc, y under the Luang Prabang dynasty, and in 1949 became an independent sovereign state within the French Union. Area, 91,000 sq. m. (235,700 sq. km.). Chief towns, Vientiane (cap.), Luang Prabang, Pakse and Savvanakhet. The chief agric. products are rice, maize, tobacco, cotton, citrus fruits, sticklac, benjohn tea and in the Boloven plateau coffee, potatoes, cardamom and cinchara. Opium is produced but is the subject of new legislation designed to control its manu. and trafficking. Cattle, buffalo and pigs are numerous. The forests in the N produce valuable woods, teak in particular; the logs are floated S on the Mekong R. Elephants are trained in forest work. Various minerals are found, but only tin is mined at present. There are extremely rich deposits of high-quality iron ore in Xieng Khouang province. Industry is limited to rubber sandals, cigarettes, matches, soft drinks, plastic bags, sawmills, rice-mills, weaving, pottery, distilleries, ice, bricks, etc. Pop. (1987E) 3·83m.

La Pampa Argentina. Province of

central Argentina, bounded S by Colorado R. Area, 55,367 sq. m. (143,400 sq. km.). Chief towns Santa Rosa (cap.), General Pico, General San Martin. Low-lying and grassy with several salt marshes, and a generally dry climate. Mainly a stock raising area with cereal farming in the N, lumbering in the NE. Industries inc. meat packing, flour milling. Pop. (1980C) 208,260.

La Paz   Bolivia. 16 30S 68 09W. Cap. of La Paz dept. La Paz is the actual cap. of Bolivia and seat of the Govt. but Sucre is the legal cap. and seat of the judiciary. At 12,400 ft. (3,780 metres) a.s.l. it is the highest cap. in the world. An important commercial city and trades in copper, alpaca and wool. Pop. (1982E) 881,404.

La Paz   El Salvador. Dept. of S Salvador bounded S by the Pacific. Area, 909 sq. m. (2,354 sq. km.). Chief towns Zacatecoluca (cap.), San Juan Nonualco, Santiago Nonualco. Mountainous N, descending to the coast and drained by Jiboa R. Main occupations grain and stock farming, salt extraction, lumbering. Pop. (1981C) 194,196.

La Paz   Honduras. 14 16N 87 40W. (i) Dept. of SW Honduras bounded SW by El Salvador. Area, 900 sq. m. (2,331 sq. km.). Chief towns La Paz (cap.), Marcala. Sit. on the continental divide and drained by Goascóran and Comayagua Rs. Mainly a farming area, producing coffee, wheat, henequen, cattle. Pop. (1981E) 86,627. (ii) Town SSW of Comayagua. Dept. cap. and commercial centre of a farming area. Industries inc. ropemaking. Pop. (1984E) 4,000.

Lapland   Norway/Sweden/Finland/ U.S.S.R. 68 07N 24 00E. Extensive area of the European Arctic on the Barents Sea coast, stretching across N Norway, Sweden, Finland and the U.S.S.R. as far as the Kola Peninsula, N of 65°N lat. Mountainous in Norway and Sweden, becoming tundra to the E; heavily forested (S) with vegetation becoming sparser northwards. Some agric. and dairying, but the main occupation is mining (in Sweden) and fishing on the Norwegian coast and in

the Ls. Minerals inc. iron (esp. at Kiruna, Gällivare, Malmberget, Sor-Varanger), copper, pyrites, nickel, apatite, gold.

La Plata   Buenos Aires, Argentina. 34 52N 57 55W. Cap. of province sit. near Río de la Plata, and near the port Ensenada. Industries oil refining, meatpacking, cement and textile works. Exports grain, meat and wool. Pop. (1980C) 560,341.

Lappeenranta   Kymi, Finland. 61 04N 28 11E. Town sit. NE of Kotka on S shore of L. Saimaa. L. port and resort. Industries inc. manu. chemicals, lime, cement, cellulose, machinery. Pop. (1983E) 53,967.

Lara   Venezuela. State of N Venezuela. Area, 7,640 sq. m. (19,788 sq. km.). Cap. Barquisimeto. Mountainous, drained by Tocuyo R. Dry, tropical climate. Mainly agric. producing coffee, cacao, sugar, maize, cotton, tobacco, bananas, sisal. wheat, barley, potatoes, livestock. Pop. (1980E) 833,718.

Larache   Morocco. 35 12N 6 09W. Port sit. SSW of Tangier on Atlantic coast. Exports wool, hides, cork, wax, fruit and vegetables. Pop. (1982E) 64,000.

Laramie   Wyoming, U.S.A. 41 19N 105 35W. Town sit. on Laramie R. at 7,145 ft. (2,178 metres) a.s.l. Centre of a stock-rearing, lumbering and mining area. Manu. cement, bricks and tiles. Pop. (1980C) 24,410.

Laramie Mountains   Wyoming, U.S.A. 42 00N 105 40W. A range of the Rocky Mountains in the SE, rising to 10,272 ft. (3,131 metres) a.s.l. in Laramie Peak.

Laramie River   Colorado/Wyoming, U.S.A. 42 12N 104 32W. Rises in the Front Range, Colorado and flows N and NE for 216 m. (345 km.) to join N. Platte R. at Fort Laramie, Wyoming.

Laredo   Texas, U.S.A. 27 31N 99 30W. City sit. on Río Grande opposite Nuevo Laredo in Mexico. Two intern. bridges join Laredo and Nuevo Laredo. Industries oil refining, meat packing, antimony smelting, fruit growing and cattle rearing. Pop. (1980C) 91,449.

Largo   Fife Region, Scotland. Fishing village and resort sit. on N shore of Firth of Forth midway between Kirkcaldy and Anstruther.

Largo   Florida, U.S.A. 27 55N 82 47W. City N of St. Petersburg on the coast of Gulf of Mexico. Pop. (1980C) 58,977.

Largs   Strathclyde Region, Scotland. 55 48N 4 52W. Port and resort sit. SSW of Greenock on Firth of Clyde. Pop. (1981C) 9,905.

La Rioja   Argentina. 29 25S 66 50W. (i) Province on W Argentina bounded NW by Chile. Area, 35,649 sq. m. (92,331 sq. km.). Mountainous with fertile valleys producing cereals, olives, citrus fruits, cotton, goats. Minerals inc. zinc, lead, copper, silver, nickel, gold, tungsten, barium sulphate, coal. Industries inc. lumbering, mining, textiles, food processing, wine making. Pop. (1980C) 164,217. (ii) Town sit. SW of Caramarca on Rioja R. at the foot of the Sierra de Velasco. Provincial cap. and trading centre for a farming area. Industries inc. sawmilling, textiles. Pop. (1980C) 67,000.

Larissa   Thessaly, Greece. 39 38N 22 25E. (i) Prefecture in E. Area, 2,067 sq. m. (5,354 sq. km.). Mainly agric. producing olives, fruit and wheat. Pop. (1981C) 254,295. (ii) Cap. of prefecture, otherwise Larisa, sit. NW of Athens and on Peneus R. Manu. textiles. Trades in agric. produce. Pop. (1981C) 102,048.

Larkana   Sind, Pakistan. 27 33N 68 16E. Town sit. NNE of Karachi. Manu. textiles and metal goods. Pop. (1981E) 124,000.

Larnaca   Cyprus. 34 47N 33 38E. (i) District of SE Cyprus. Area, 435 sq. m. (1,127 sq. km.). Chief towns Larnaca, Athienou, Lefkara. Hilly in N, descending to lowland. Mainly agric. producing cereals olives, vines, livestock, fruit. Fishing is important. Minerals inc. salt, umber, perlite. (ii) Town sit. SE of Nicosia on Larnaca

Bay. District cap. and port trading in wine, fruit, olive oil, barley, livestock. Industries inc. petroleum refining, textiles, clothing, furniture, plastics and food processing. International airport. Pop. (1985E) 45,000.

Larne   Antrim, Northern Ireland. 54 51N 5 49W. Port on NE coast sit. at entrance to Lough Larne. Ferry terminal. Industries inc. engineering, foundry, paper manu. and clothing. Pop. (1981E) 18,000.

La Romana   La Altagracia, Dominican Republic. 18 25N 68 58W. Port sit. E of Santo Domingo. Industries sugar refining and coffee and tobacco processing. Pop. (1981C) 91,571.

La Salle   Quebec, Canada. 45 26N 73 39W. City, S suburb of Montreal. Primarily residential with a variety of industries. Pop. (1981C) 76,299.

Las Alpujarras   *see* **Alpujarras, Las**

Las Bela   Baluchistan, Pakistan. Town sit. NNW of Karachi, former cap. of the princely state of Las Bela. Trading centre for oilseeds, millet, wool, rice. Industries inc. shipbreaking, carpet weaving. Pop. (1981E) 11,000.

Las Cruces   New Mexico. U.S.A. 32 23N 106 29W. City sit. NNW of El Paso, Texas. Centre of an agric. and cattle raising district. Pop. (1988E) 56,000.

La Serena   Coquimbo, Chile. 29 54S 71 16W. Cap. of province of same name in N Central Chile. Resort and market town trading in fruit and flowers. Pop. (1982C) 87,456.

Lashio   Shan, Burma. Town sit. in E Central Burma. Railway terminus and end of the Burma Road.

Las Palmas   Spain. 28 06N 15 24W. (i) Spanish province in the Canary Isles, Atlantic Ocean, consisting of the islands of Gran Canaria, Lanzarote and Fuerteventura plus several smaller barren islands. Pop. (1986C) 855,494. (ii) Cap. of province of same name sit. in NE of Gran Canaria. Tourist centre. The adjacent port La Luz exports fruit, vegetables and wine. Pop. (1986C) 372,270.

La Spezia   Liguria, Italy. City sit. on Gulf of Spezia, an inlet of the Gulf of Genoa. Important naval base. Industries inc. ship repair and construction. Pop. (1987E) 124,600.

Lassithi   Crete, Greece. Prefecture in the E. Area, 702 sq. m. (1,818 sq. km.). Cap. Aghios Nikolaos, sit. SE of Iraklion. Mainly agric. Mount Lasithi (Dhikti) rises to 7,048 ft. (2,148 metres) a.s.l. in the w. Pop. (1981C) 70,053.

Las Vegas   Nevada, U.S.A. 36 11N 115 08W. City and resort in SE Nevada sit. in a mining and farming region. Noted for its gambling casinos. Pop. (1980C) 164,674.

Las Vegas   New Mexico, U.S.A. 35 36N 105 13W. City sit. ESE of Santa Fé on Galinas R. at 6,470 ft. (2,121 metres) a.s.l. in the Sangre de Cristo mountains. Resort, trading and service centre. Pop. (1986E) 15,620.

Latacunga   Cotopaxi, Ecuador. 0 56S 78 35W. Town sit. s of Quito on Patate R. Provincial cap. sit. at 9,055 ft. (2,760 metres) a.s.l. and surrounded by high peaks inc. Cotopaxi. Trading centre for a stock raising and cereal farming area. Industries inc. processing farm products. Airport. Pop. (1980E) 30,000.

Latin America   Those countries of South America where French, Portuguese or Spanish is the spoken language, including roughly all countries s of the U.S.A. border with Mexico, but excluding parts of the West Indies. Spanish is the main language, although the language of the largest country, Brazil, is Portuguese.

Latium   *see* **Lazio**

La Tortue   Haiti. Otherwise called Tortuga Island, sit. N of Port-de-Paix. Area, 70 sq. m. (181 sq. km.). Main products bananas, food crops. Formerly notorious as the headquarters of pirates.

Lattakia   Lattakia, Syria. 35 30N 35 45E. Cap. of district of same name and port. Chief industry tobacco. Pop. (1981C) 196,791.

Latvia   U.S.S.R. 57 00N 25 00E. Bounded on E and s by Gulf of Riga, on N by Estonia, E by U.S.S.R., on s by Lithuania and w by Baltic Sea. The secret protocol of the Soviet-German agreement of 1939 assigned Latvia to the Soviet sphere of interest. An ultimatum in 1940 led to the formation of a govt. acceptable to the U.S.S.R. which applied for Latvia's admission to the Soviet Union, which was later effected. The incorporation has been accorded *de facto* recognition by the Brit. Govt. but not by the U.S. Govt. Area, 25,590 sq. m. (63,700 sq. km.). Cap. Riga. Agric. not now as important as industrial production which inc. the manu. of railway transport. Pop. (1989E) 2.7m.

Launceston   Tasmania, Australia. 41 26S 147 08E. Second largest city of Tasmania sit. on confluence of Rs. Tamar, N. Esk and S. Esk. A port and industrial centre, it was first settled in 1806. The city area is 10.5 sq. m. (27.2 sq. km.) and varies in height from 10–700 ft. (3–213 metres) a.s.l. Industries sawmilling, car parts, engineering, furniture, transport, brewing and manu. of woollen goods. Small ships reach the wharves of the city, and larger ships are accommodated down R. at Bell Bay, Inspection Head and Beauty Point. Pop., Greater Launceston (1986E) 66,286, and the city (1990E) 64,000.

Launceston   Cornwall, England. 50 38N 4 21W. Town sit. NW of Plymouth on R. Kensey. Industries inc. light engineering. Pop. (1990E) 6,500.

La Unión   El Salvador. 13 20N 87 51W. (i) Dept. of El Salvador bounded N by Honduras. Area, 770 sq. m. (1,995 sq. km.). Mountains sloping to coastal lowland, mainly agric., with some mining and fishing. Main products grain, coffee, sugar, livestock, salt, tortoiseshell, gold, silver. Pop. (1981C) 309,879. (ii) Town sit. ESE of San Salvador at the foot of the volcano Conchagua. Dept. cap. and the main Pacific port of Salvador, exporting coffee, indigo, sugar, cotton, henequen, silver, gold. Pop. (1984E) 27,186.

Laurel Mississippi, U.S.A. 31 42N 89 08W. City SW of Meridian and sit. on Tallahala Creek. Produces timber, oil and poultry. Pop. (1980C) 21,897.

Lausanne Vaud, Switzerland. 46 31N 6 38E. Educational, cultural and tourist centre, and cap. of canton, sit. $^1/_2$ m. (1 km.) from Ouchy, the port, on N shore of L. Geneva. On the Simplon route from Paris to Milan. Industries inc. clothing, printing and confectionery. Pop. (1988E) 124,000.

Lauterbrunnen Bern, Switzerland. 46 36N 7 55E. Resort sit. NNE of Brig on White Lütschine R. Sit. in a steep-sided valley with 72 waterfalls. Pop. (1989E) 2,652.

Lautoka Viti Levu, Fiji. 17 37S 177 27E. Town on NW coast of Viti Levu. Port and centre of a sugar producing area, with refineries. Pop. (1986C) 28,728.

Laval Mayenne, France. 48 04N 0 46W. Cap. of Mayenne dept. sit. on Mayenne R. Noted for linen since the 14th cent. Also produces cheese, hosiery, furniture and leather goods. Pop. (1982C) 53,766.

Lavalleja Uruguay. Dept. of SE Uruguay. Area, 4,820 sq. m. (12,484 sq. km.). Chief towns Minas (cap.), José Batlle y Ordóñez, Solís. Cattle and sheep raising area with some lead mining. Pop. (1980C) 65,240.

La Vega Dominican Republic. Province of the interior, bounded S by the Cordillera Central, N by the Cordillera Setentrional. Area, 1,337 sq. m. (3,463 sq. km.). Cap. Concepción de la Vega. Mountainous, descending to the fertile La Vega Real valley. Main products tobacco, cacao, coffee, rice, maize, cattle, wheat. Pop. (1980C) 293,573.

Lawrence Kansas, U.S.A. 38 58N 95 14W. City sit. WSW of Kansas City on Kansas R. Centre of an agric. area. Manu. pipe organs, agric. chemicals, greeting cards, plastics and paper products. Pop. (1980C) 54,307.

Lawrence Massachusetts, U.S.A. 42 42N 71 09W. City sit. on Merrimack R. Centre of an important textile industry with some of the largest mills in the world. Other manu. machinery, paper, plastics, rubber and leather goods. Pop. (1990E) 67,375.

Lawton Oklahoma, U.S.A. 34 37N 98 25W. City and industrial centre of an agric. area with associated trades. Also manu. tyres, tools, leather goods, tiles, concrete and wood products. Pop. (1980C) 80,054.

Lazio Italy. 42 10N 12 30E. Region of central Italy bounded W by the Tyrrhenian Sea. Area, 6,634 sq. m. (17,182 sq. km.). Chief cities Rome (cap.), Civitavecchia. Comprises 5 provinces; Frosinone, Latina, Rieti, Roma, Viterbo, and inc. the Pontine Islands. Mostly sit. in the Apennine hills which enclose a narrow coastal plain. Drained by Tiber, Aniene, Liri, Marta, Rapide Sacco and Velino Rs. Mainly agric., esp. since the reclamation of the Pontine marshes and Campagna di Roma. Fishing, quarrying, mining and lumbering also important. Main products cereals, grapes, olives, fruit, vegetables, livestock, alum, asphalt, marble, alabaster. Industries inc. tourism, paper making, chemicals, manu. furniture and iron-work. Pop. (1988E) 5,156,053.

Lea, River England. 51 30N 0 01E. Rises NW of Luton and flows for 46 m. (74 km.) SE through Hertfordshire and forms the border of Essex and Hertfordshire and then of Essex and Greater London, and joins R. Thames near Blackwall. A source of much of London's water supply.

Leamington Spa Warwickshire, England. 52 18N 1 31W. Health resort sit. on R. Leam. Noted for its parks and saline springs. Manu. motor car parts and gas cookers. Pop. (1985E) 56,538.

Leatherhead Surrey, England. 51 18N 0 20W. Town sit. S of Kingston-upon-Thames on R. Mole. Manu. light electrical and consumer goods. Pop. (1985E) 40,300 (inc. surrounding villages).

Leavenworth Kansas, U.S.A. 39 19N 94 55W. City sit. NW of Kansas City on Missouri R. where it forms the Missouri border. Trading centre for a farming area. Industries inc. meat packing, metalworking; manu. batteries, paper products, candles. Pop. (1980C) 33,656.

Lebanon South West Asia. 34 00N 36 00E. Rep. on E shore of Mediterranean, bounded S by Israel, E and N by Syria. Area, 4,036 sq. m. (10,452 sq. km.). After 20 years' French mandatory régime, the Lebanon was proclaimed independent at Beirut in 1941. In Dec. 1943 an agreement was signed between representatives of the French National Committee of Liberation and of Lebanon, by which most of the powers and capacities exercised hitherto by France were transferred from Jan. 1944 to the Lebanese Government. Chief cities, Beirut, Tripoli, Saida, Tyre. Mountainous, with the Lebanon range extending N to S parallel to the coast and beyond them the fertile Bekaa valley which supports grain, vegetables, maize, fruit, cotton, tobacco, mulberry (for silk), olives. Forests have decreased to *c.* 80,000 hectares. The main industries, small on the whole, are food and drink, textiles, tobacco processing, metalworking, brick and cement making, processing oil and chemicals. Tourism, banking and other commercial operations were an important source of revenue before 1975. Pop. (1984E) 3·4m.

Lebanon Pennsylvania, U.S.A. 40 20N 76 25W. City sit. WNW of Philadelphia. Manu. iron and steel products, chemicals, textiles and food products. Pop. (1980C) 25,711.

Lecce Puglia, Italy. 40 23N 18 11E. Cap. of Lecce province sit. SSE of Brindisi. Manu. *papier-mâché* goods and pottery. Trades in textiles, olive oil, wine and tobacco. Pop. (1981E) 91,000.

Lecco Lombaria, Italy. 45 51N 9 23E. Town sit. NE of Monza and on SE arm of L. Como in Como province. Manu. textiles and brass goods. Exports cheese.

Lech River Austria/Federal Republic of Germany. Rises W of Landeck in the Vorarlberg Alps and flows NE and N for 177 m. (283 km.) to join Danube R. W of Ingolstadt.

Le Creusot Saône-et-Loire, France. 46 48N 4 26E. Town in E central France. Centre of a coalfield with important iron and steel, locomotives and armaments industries. Pop. (1982C) 32,309.

Ledbury Hereford and Worcester, England. 52 03N 2 25W. Town sit. ESE of Hereford on R. Leadon. Trades in cider, perry and hops. Limestone quarrying nearby. Pop. (1980E) 4,000.

Leduc Alberta, Canada. 53 16N 113 33W. City sit. S of Edmonton. Industries inc. agric., oil, gas, sand and gravel. Pop. (1989E) 13,363.

Leeds West Yorkshire, England. 53 50N 1 35W. Metropolitan district of West Yorkshire. The city has a wide variety of manu. industries and includes cloth, clothing, defence systems, heavy and light engineering, chemical and leather products. It is a major financial, commercial and administrative centre. Pop. (1989E) 706,000, metropolitan district.

Leek Staffordshire, England. 53 06N 2 01W. Town sit. NE of Stoke-on-Trent. Manu. textiles, dairy products and chemicals. Pop. (1990E) 20,000.

Leeuwarden Friesland, Netherlands. 52 12N 5 46E. Cap. of province in N. Railway and canal junction. Industries inc. boatbuilding and manu. of glass. Trades in cattle and agric. produce. Pop. (1989E) 85,175.

Leeuwin, Cape Western Australia, Australia. 34 22S 115 08E. Cape sit. at W end of Flinders Bay.

Leeward Islands French Polynesia, South Pacific. W group of the Society Islands, inc. Raiatéa, Tahaa, Huahine, Bora-Bora and Maupiti. Area, 195 sq. m. (507 sq. km.). Pop. (1983C) 19,000.

Leeward Islands Lesser Antilles, West Indies. 16 30N 63 30W. Group inc. islands of Dominica, Antigua (inc. Barbuda and Redonda), St. Kitts (with Nevis and Anguilla), Montserrat, and the Virgin Islands. St. Thomas, Santa Cruz and St. John in the Virgin Islands belong to the U.S.A. Others of the group are Guadeloupe and dependencies. Islands volcanic with lofty peaks, highest Morne Diablotin 4,746 ft. (1,447 metres) a.s.l. in Dominica. They produce sugar and molasses, cotton, citrate of lime, cacao. Lime juice is exported from Montserrat and Dominica.

Lefkas Ionian Islands, Greece. Island off coast of Acarnania. Area, 125 sq. m. (325 sq. km.). Chief town Lefkas. Pop. (1981C) 6,415. Limestone ridge separated from the mainland NE by a narrow channel. The main occupation is farming. Chief products currants, wine, olive oil. Pop. (1981C) 21,863.

Leghorn (Livorno) Toscana, Italy. 43 33N 10 19E. Cap. of Livorno province and port on Ligurian Sea. Industries shipbuilding, engineering, cement and soap works. Exports olive oil, wine, marble and straw (Leghorn) hats. Pop. (1979E) 176,757.

Legnano Lombardia, Italy. 45 36N 8 54E. Town sit. NW of Milan on Olona R. Manu. textiles, soap and shoes.

Legnica Wroclaw, Poland. 51 13N 16 09E. Town sit. WNW of Wroclaw on Katzbach R. Manu. textiles, paints, chemicals and machinery. Pop. (1980E) 81,000.

Leguan Island Essequibo, Guyana. Island sit. WNW of Georgetown in Essequibo R. estuary E of Wakenaam Island. Area, 17 sq. m. (45 sq. km.). The main occupation is rice growing.

Leh Kashmir, India. 34 10N 77 36E. Cap. of Ladakh district sit. E of Srinagar N of Indus R. at 11,500 ft. (3,505 metres) a.s.l. Trading centre. Pop. (1981C) 8,718.

Le Havre *see* **Havre, Le**

Leicester Leicestershire, England.

52 38N 1 05W. City on Soar R. Industries inc. tanning, chemicals, textiles; manu. hosiery, textile machinery, textiles, footwear, printing and light engineering. Pop. (1989E) 279,791.

Leicestershire England. 52 40N 1 10W. Bounded on N by Derbyshire and Nottinghamshire, on E by Lincolnshire, on S by Northamptonshire and W by Warwickshire. Under 1974 re-organization the county consists of the districts of North West Leicestershire, Charnwood, Melton, Hinckley and Bosworth, Leicester, Blaby, Oadby and Wigston, Harborough and Rutland. Agric., particularly arable, and cheesemaking, is important. Industries inc. coalmining and limestone quarrying, and there are a considerable number of engineering firms. Pop. (1988E) 885,000.

Leiden Zuid-Holland, Netherlands. 52 09N 4 30E. Town sit. on Old Rhine R., NE of The Hague. Bioscience park and there is off-shore engineering. Pop. (1984E) 176,000.

Leigh Greater Manchester, England. 53 30N 2 33W. Town sit. W of Manchester. Industries coalmining and engineering. Manu. agric. machinery, electrical goods and cottons. Pop. (1981C) 45,412.

Leighton-Linslade Bedfordshire, England. Town sit. WNW of Luton. Industries sand quarrying, fork lift manu., clothing, tiles, tea importers and light engineering. Pop. (1990E) 35,000.

Leinster Ireland. 53 00N 7 10W. SE province comprising the counties of Louth, Longford, Meath, Westmeath, Dublin, Offaly, Kildare, Laoighis, Wicklow, Carlow, Kilkenny and Wexford. Area, 7,580 sq. m. (19,632 sq. km.). Pop. (1988E) 1,852,649.

Leipzig German Democratic Republic. 51 19N 12 20E. (i) District, formerly part of Saxony. Area, 1,915 sq. m. (4,960 sq. km.). (ii) City and cap. of district sit. at confluence of Elster, Parther and Pleisse Rs. Industries include heavy engineering, chemicals, electrical goods, leather and fur goods, textiles, printing machines.

Centre for international trade fair and for publishing. Pop. (1990E) 530,000.

Leiria Beira Litoral, Portugal. 36 46N 8 53W. Town sit. ssw of Coimbra on Liz R. Trading centre of an agric. area and district cap. Industries inc. tanning, wood-working; manu. cement. Pop. (1981E) 11,000.

Leith Lothian Region, Scotland. 55 59N 3 10W. Port sit. on Firth of Forth, now an integral part of Edinburgh. Industries shipbuilding, engineering, distilling, brewing, manu. of chemicals and paper.

Leitha River Austria/Hungary. 47 54N 17 17E. Rises in E Austria and flows E, forming part of the border, for 110 m. (176 km.) to join Danube R.

Leitrim Connacht, Ireland. 54 20N 8 20W. Maritime county bounded on NW by Donegal Bay. Area, 589 sq. m. (1,525 sq. km.). Mainly hilly in the N, undulating in the S, with many L., Lough Allen being the largest. County town Carrick-on-Shannon. Principal Rs. are Shannon, Bonnet, Drowes and Duff. Chief occupations agric. and dairy farming. Pop. (1988E) 27,035.

Le Kef Le Kef, Tunisia. Town sit. SW of Tunis near the Algerian border. Rail terminus. Trading centre for agric. produce. Industries inc. processing flour, olive oil.

Lekemti Wallaga, Ethiopia. Cap. of region sit. w of Addis Ababa. Trading centre for coffee, hides, beeswax, honey, gold, grain. Pop. (1978E) 21,694.

Le Mans Sarthe, France. 48 00N 0 12E. Cap. of dept. in the NW sit. at the confluence of the R. Sarthe and Husne. Industries inc. railway engineering, flour milling, tanning and food processing. Manu. inc. agric. machinery, textiles, paper and tobacco products. Trades in livestock and agric. produce. The annual sportscar race is held here. Pop. (1982C) 150,331 (agglomeration, 191,080).

Lempira Honduras. Dept. of w Honduras on the border with El Sal-

vador. Area, 1,295 sq. m. (3,354 sq. km.). Chief towns Gracias (cap.), Erandique, Candelaria, Guarita. Mainly mountainous and drained by Jicatuyo and Mocal Rs. Agric. area producing coffee, tobacco, rice, wheat, indigo, livestock. Minerals inc. opals. Pop. (1980E) 175,000.

Lena River Siberia, U.S.S.R. 72 25N 126 40E. The largest R. in the U.S.S.R. and fifth largest in the world. It rises in Baikal Mountains w of L. Baikal and flows for 1,200 m. (1,920 km.) to reach the Arctic Ocean through a large delta. It drains *c*. 1m. sq. m. (*c*. 2.6m. sq. km.). Its chief tribs. are R. Aldam, Olekma, Vitim and Vilui.

Leninabad Tadzhikistan, U.S.S.R. 40 17N 69 37E. Town sit. NNE of Dushanbe on Syr Darya R. Manu. textiles and footwear. Pop. (1980E) 130,000.

Leninakan Armenia, U.S.S.R. 40 48N 43 50E. Town sit. NW of Yerevan. Manu. textiles, carpets, bicycles and food products. Pop. (1985E) 223,000.

Leningrad Russian Soviet Federal Socialist Republic, U.S.S.R. 59 55N 30 15E. Originally St. Petersburg, and later Petrograd, former cap. of Russia, and second city of the U.S.S.R. Founded by Peter the Great. Sit. on 42 islands, at mouth of R. Neva where it enters Gulf of Finland. There are many magnificent palaces and important educational institutions, particularly the Leningrad State University. Industrial estabs. inc. metal works, electronics, iron foundries, sugar refineries, distilleries, breweries, shipbuilding yards and printing works. Manu. inc. chemicals, TV sets, watches, tobacco, soap, crystal and glass, cotton and cloth, leather, cordage, pottery, porcelain and machinery. The city is connected by river, lake and canal with R. Volga and Dnieper, and so with the Caspian and Black Seas. Pop. (1989E) 5,023,500.

Leninogorsk Kazakh Soviet Socialist Republic, U.S.S.R. 50 22N 83 32E. Town sit. NE of Ust-Kamenogorsk in the NW Altai Mountains. Important lead and zinc mining centre, with smelting industry.

Leninsk-Kuznetski Russian Soviet Federal Socialist Republic, U.S.S.R. 54 38N 86 10E. Town sit. S of Kemerovo in the Kuznetsk basin. Industries coalmining, iron-mining, railway engineering and brickmaking.

Lenkoran Azerbaijan Soviet Socialist Republic, U.S.S.R. 38 45N 48 50E. Town sit. SSW of Baku on Caspian Sea, near the Iranian frontier. Industries inc. sawmilling, food canning, fishing.

Lennoxville Quebec, Canada. Town just SE of Sherbrooke at mouth of Massawippi R. on St. Francis R. Industries inc. paper, bronze and steel, hosiery. Centre of a farming and lumbering area. Pop. (1985E) 4,000.

Lens Pas-de-Calais, France. 50 26N 2 50E. Town sit. in NE France, on an important coalfield. Manu. metal goods and chemicals. Pop. (1982C) 38,307 (agglomeration, 327,383).

Leoben Styria, Austria. 47 23N 15 06E. Town sit. on Mur R. Centre of a lignite and iron mining region with a school of mining. Pop. (1981C) 31,989.

Leominster Hereford and Worcester, England. 52 14N 2 45W. Town near border with Wales, on R. Lugg. Trades in antiques, livestock and agric. produce. Industries inc. agric. machinery, plastics, cider and tourism. Pop. (1988E) 9,988.

Leominster Massachusetts, U.S.A. 42 32N 71 45W. City sit. WNW of Boston. Manu. plastic toys and furniture. Pop. (1980C) 34,508.

León Guanajuato, Mexico. 21 07N 101 40W. Sit in central Mexico on R. Gómez at 6,182 ft. (1,884 metres) a.s.l. Industrial centre of an agric. and mining region. Manu. textiles, cement and footwear. Pop. (1980C) 655,809.

León León, Nicaragua. 12 26N 86 54W. Cap. of dept. of same name and the second largest town in the country. Manu. textiles, leather goods, soap,

food products and footwear. Pop. (1980E) 63,000.

León Spain. 42 36N 5 34W. (i) NW region consisting of the provinces of León, Palencia, Salamanca, Valladolid and Zamora. (ii) NW province of same name. Area, 5,972 sq. m. (15,468 sq. km.). Mountainous in the N and W, the S and E being part of the Castilian plateau. Mainly agric. Pop. (1986C) 528,502. (iii) Cap. of province of same name sit. NW of Valladolid. Manu. linen, pottery and leather. Pop. (1986C) 137,414.

Léopold II, Lake Zaïre. 2 00S 18 20E. L. sit. in the W. Area, 900 sq. m. (2,331 sq. km.). Fed by Rs. Lokoro and Lukenye, and drained by Fimi R.

Léopoldville *see* **Kinshasa**

Lérida Spain. 41 37N 0 37E. (i) NE province bounded by France and Andorra. Area, 4,644 sq. m. (12,028 sq. km.). Produces wine and olive oil. Main R. is Segre. Pop. (1986C) 356,811. (ii) Cap. of province of same name sit. WNW of Barcelona on Segre R. Manu. textiles, silk, glass and paper. Pop. (1986C) 111,507.

Lérins, Îles de Alpes-Maritimes, France. A group of Islands in the Mediterranean just SE of Cannes.

Lerwick Shetland Islands, Scotland. 60 09N 1 09W. Cap. of Shetland and port sit. on the E coast of the mainland. Industries fishing, oil, agriculture, tourism and hand-knitting. Pop. (1987E) 7,505.

Leseru Rift Valley, Kenya. Town sit. NW of Eldoret at 6,489 ft. (1,978 metres) a.s.l. Railway junction serving a farming area.

Leskovac Serbia, Yugoslavia. 42 59N 21 57E. Town sit. S of Niš on S Morava R. Manu. textiles, furniture and soap. Pop. (1981C) 159,001.

Lesotho 29 40S 28 00E. Bounded by Rep. of South Africa; Orange Free State on N and W, Cape Province on S and Natal on E and NE. In 1966 Basutoland became an independent and sovereign member of the Commonwealth under the name of the Kingdom of Lesotho. The altitude ranges from 5,000–11,000 ft. (1,524–3,353 metres) a.s.l. Area, 11,716 sq. m. (30,340 sq. km.). Cap. Maseru. The chief crops are wheat, maize and sorghum; barley, oats, beans, peas and other vegetables are also grown. The land is held in trust for the nation by the King and may not be alienated. Soil conservation and the improvement of crops and pasture are matters of vital importance. Efforts are being made to secure the general introduction of rotational grazing in the mountain area. Pop. (1988E) 1·67m.

Lesser Sundas *see* **Nusa Tenggara**

Lesvos Aegean Island, Greece. 39 10N 26 20E. Island in the Aegean off the Turkish coast across the Mytilene and Muselim Remma channels. Area, 832 sq. m. (2,154 sq. km.). Chief city, Mytilene. Hilly and heavily forested, rising to 3,176 ft. (968 metres) in the N and 3,175 ft. in the S, the peaks divided by a fertile lowland. Main products wheat, olives, vegetables, fruit, vines, livestock. Pop. (1981C) 104,620.

Leszno Poznań, Poland. 51 51N 16 35E. Town in W central Poland. Manu. rolling-stock, machinery and footwear. Pop. (1983E) 53,000.

Letchworth Hertfordshire, England. 51 58N 0 14W. Town sit. NNW of Hertford. Industries include engineering, computers, high technology, furniture and publishing. The first 'garden' city founded in 1903. Pop. (1989E) 32,664.

Lethbridge Alberta, Canada. 49 42N 110 50W. City in S near border with U.S.A. on Oldman R. Centre of an irrigated farming area. Industries inc. food processing, electronic assembly, metal fabrication, agric. equipment manu. Pop. (1989E) 60,614.

Leticia Amazonas, Colombia. 4 09S 69 57W. Town sit. SE of Bogotá on upper Amazon R. where it forms the border with Peru and Brazil. Cap. of the commissary. R. port and airfield in a tropical forest area. Pop. (1980E) 22,000.

Letpadan Tharrawaddy, Burma. 17 45N 96 00E. Town sit. NNW of Rangoon on the Rangoon–Mandalay railway. Railhead for Tharrawaw.

Letterkenny Donegal, Ireland. Town sit. W of Londonderry, Northern Ireland, at the head of L. Swilly. Fishing port and market for agric. produce. Industries inc. manu. of yarn and medical products. Pop. (1985E) 8,000.

Leuna Halle, German Democratic Republic. 51 19N 12 01E. Town sit. S of Merseburg near Saale R. The main industry is manu. chemical products.

Leuven *see* **Louvain**

Levadeia Boeotia, Greece. 38 26N 22 53E. Town sit. NW of Athens. *Nome* cap. Trading centre for agric. produce. Industries inc. cotton and woollen textiles. Pop. (1981C) 16,864.

Levallois-Perret Hauts-de-Seine, France. 48 54N 2 18E. NW suburb of Paris. Manu. automobiles. Pop. (1982C) 53,777.

Levant The E end and shores of the Mediterranean Sea and inc. Egypt, Greece, Israel, Lebanon, Syria and Turkey.

Leven Fife Region, Scotland. 56 12N 3 00W. Resort sit. NE of Kirkcaldy at mouth of R. Leven. Industries sawmilling and engineering. Pop. (1987E) 9,210.

Leven, Loch Scotland. 56 41N 5 07W. (i) Sea inlet extending 9 m. (14 km.) E from Loch Linnhe between Argyll and Inverness. Fed by Coe R. and Leven R. (ii) L. in E Kinross, area, 8 sq. m. (21 sq. km.). Inc. Castle Island with the ruins of Lochleven castle where Mary Stuart, Queen of Scots, was imprisoned 1567–8. Drained by Leven R. into Firth of Forth.

Leven, River Scotland. (i) Flows from S end of Loch Lomond and continues 7 m. (11 km.) S past Jamestown, Alexandria and Bonhill to enter Clyde R. at Dunbarton. (ii) Flows from SE end of Loch Leven, Kinross, and continues 15 m. (24 km.) E to enter Firth of Forth at Leven.

Leverkusen North Rhine-West-phalia, Federal Republic of Germany. 51 03N 6 59E. Town and port sit. N of Cologne on Rhine R. Manu. chemicals, machinery and textiles. Pop. (1984E) 156,500.

Levkás Ionian Islands, Greece. 38 39N 20 27E. (i) Prefecture which inc. the islands of Ithaca and Meganesi. Area, 125 sq. m. (325 sq. km.). Produces olive oil, wine and currants. Pop. (1981C) 21,863. (ii) Cap. of prefecture of same name sit. on NE coast. Pop. (1981C) 6,415.

Lévrier Bay Mauretania. Inlet of the Atlantic between Cap Blanc peninsula and the Mauretanian mainland. Length 28 m. (45 km.), width up to 20 m. (32 km.). Important fishing ground for lobster.

Lewes East Sussex, England. 50 52N 0 01E. Market town in SE on R. Ouse. Industries include brewing and manu. of laboratory supplies. Pop. (1989E) 15,000.

Lewisham Greater London, England. 51 27N 0 01E. SE district. Mainly residential. Pop. (1988E) 228,900.

Lewiston Idaho, U.S.A. 46 25N 117 01W. City and inland port at the confluence of Clearwater and Snake Rs. in N central Idaho. Tourist centre for Hell's Canyon, Snake, Clearwater and Salmon Rs. Industries inc. lumber, pulp, paper and plywood; manu. sporting ammunition. Pop. (1980C) 27,896.

Lewiston Maine, U.S.A. 44 06N 70 13W. City sit. on Androscoggin R. opposite Auburn. There is an important textile industry. Manu. footwear, electronic goods and systems, dyes and bleaches. Pop. (1980C) 40,481.

Lewis-with-Harris Scotland. 58 15N 6 40W. Largest group of islands of the Outer Hebrides in the extreme N, with the Minch separating it from the mainland. Area, 859 sq. m. (2,225 sq. km.).

Lexington Kentucky, U.S.A. 38 03N 84 30W. City in NE of state. Sit. in the centre of the Bluegrass region, it is the principal centre of thoroughbred horse breeding in the U.S.A. Trades in livestock and tobacco. Manu. metal products, electrical equipment, clothing and furniture. Pop. (1980C), with Fayette 204,165.

Leyland Lancashire, England. 53 42N 2 42W. Town sit. S of Preston. Centre of bus and truck industry. Also manu. transport equip., plastics, health care products and rubber products. Pop. (1988E) 100,200.

Leyte Philippines. 10 55N 124 50E. Island lying between Luzon and Mindanao. Chief town, Tacloban. Produces rice and copra.

Leyton Greater London, England. 51 33N 0 01W. District in NE London on R. Lea. Manu. furniture and cables.

Lezhë Albania. Town sit. on old Drin R. SSE of Scutari. Copper mining in the area.

Lhasa Tibet (Xizang), China. 29 40N 91 09E. Cap. of autonomous region of China. Name means 'Abode of the Gods' and is a sacred city of Buddhists. Sit. on fertile plain about 12,000 ft. (3,658 metres) a.s.l. and is encircled by mountains. Long known as the 'Forbidden City'. Principal building is the Potala, former residence of the Dalai Lama. Pop. (1987E) 130,000.

Liaoning China. 42 00N 122 00E. NE province bordering on N. Korea. Area, 56,371 sq. m. (146,000 sq. km.). Cap. Shenyang (Mukden). A highly industrialized region containing important coalfields and steel works. Agric. products inc. soya beans, cereals and apples. Pop. (1982C) 35,721,693.

Liard River Canada. 61 52N 121 18W. Rises E of White Horse in the Yukon and flows 570 m. (912 km.) ESE to British Columbia and N from Nelson Forks past Fort Liard in Mackenzie District, Northwest Territories, to enter Mackenzie R. at Fort Simpson.

Libau *see* **Liepāja**

Liberec   Czechoslovakia. 50 46N 15 03E. Town in NW Czech. sit. on Neisse R. Centre of a textile industry since the 16th cent. Pop. (1990E) 104,000.

Liberia   6 30N 9 30W. Independent rep. extending SE of Sierra Leone for 350 m. (560 km.) along coast to Côte d'Ivoire. The Rep. of Liberia had its origin in the efforts of several American philanthropic societies to estab. freed American slaves in a colony on the W Afr. coast. In 1822 a settlement was formed near the spot where Monrovia now stands. The coast is low and swampy; interior rises, and has excellent timber. It is watered by R. Kavalli and other streams. Area, 38,250 sq. m. (99,067 sq. km.). Cap. Monrovia. The soil is very fertile and produces coffee, palm oil and kernels, rubber, cocoa, hides and kola nuts. Iron is widely distributed and is worked. Interior little developed and in parts unexplored. Pop. (1988E) 2·44m.

Libreville   Estuaire, Gabon. 0 30N 9 25E. Cap. of Gabon and port sit. on estuary of Gabon R. Founded in 1849 as a settlement for freed slaves. Industries sawmilling and manu. of plywood. Exports palm oil, kernels and timber. Pop. (1985E) 350,000.

Libya   28 30N 17 30W. Bounded by Mediterranean Sea in N, Algeria and Tunisia in W, Niger and Chad in S and Egypt in E. There are 3 provinces: Tripolitania, Cyrenaica and Fezzan. Area, 679,358 sq. m. (1,759,540 sq. km.). Chief towns, Tripoli (the cap.) and Benghazi. Tripolitania has 3 zones from the coast inland—the Mediterranean, the sub-desert and the desert. The first, which covers an area of about 17,231 sq. m. (44,628 sq. km.) is the only one properly suited for agric., and may be further subdivided into: (i) the oases along the coast, the richest in N. Africa, in which thrive the date palm, the olive, the orange, the peanut and the potato; (ii) the steppe district suitable for cereals (barley and wheat) and pasture; it has olive, almond, vine, orange and mulberry trees and ricinus plants; (iii) the dunes, which are being gradually afforested with acacia, robinia, poplar and pine; (iv) the Jebel (the mountain district, Tarhuna, Garian, Nalut-Yefren), in which thrive the olive, the fig, the vine and other fruit trees, and which, on the E, slopes down to the sea with the fertile hills of Msellata. Of some 25m. acres of productive land in Tripolitania, nearly 20m. are used for grazing and about 1m. for static farming. The sub-desert zone produces the alfa plant. The desert zone and the Fezzan contain some fertile oases, such as those of Ghadames, Ghat, Socna, Sebha and Brak. Cyrenaica has about 10m. acres of potentially productive land, most of which, however, is suitable only for grazing. Certain areas, chief of which is the plateau known as the Barce Plain (about 1,000 ft. (305 metres) a.s.l.), are suitable for dry farming; in addition, grapes, olives and dates are grown. With improved irrigation, production, particularly of vegetables, could be increased, but stock raising and dry farming will remain of primary importance. About 143,000 acres are used for settled farming; about 272,000 acres are covered by natural forests. In the Fezzan there are about 6,700 acres of irrigated gardens and about 297,000 acres are planted with date palms. Among the most important industries of Tripolitania and Cyrenaica are sponge fishing, tunny fishing, tobacco growing and processing, dyeing and weaving of local wool and imported cotton yarn, and olive oil. Tripolitania also produces bricks, salt, leather and esparto grass for paper-making. Home industries of both territories inc. the making of matting, carpets, leather articles and fabrics embroidered with gold and silver. Industries inc. milk products, electric cables, glassware, steel pipes, footwear and vegetable and fruit processing. Oil production was 53m. tonnes in 1989, with reserves of 23,000m. barrels. Pop. (1986E) 3·96m.

Libyan Desert   Libya/Egypt. 25 00N 25 00E. The E area of the Sahara. Also known as the Western Desert.

Licata   Sicilia, Italy. 37 05N 13 56E. Port sit. at mouth of Salso R. SE of Agrigento. Sulphur industry.

Lichfield   Staffordshire, England.

52 42N 1 48W. Town sit. SW of Burton-on-Trent. Noted cathedral with three spires. Industry mainly light engineering and agriculture. Pop. (1987E) 28,300.

Lida Belorussian Soviet Socialist Republic, U.S.S.R. 53 53N 25 18E. Town sit. ENE of Grodno. Industries inc. manu. agric. machinery, textiles, shoes, cement; sawmilling, processing agric. produce.

Lidice Středočeský, Czechoslovakia. 50 03N 14 08E. Village sit. W of Prague. Completely destroyed during the Second World War, and rebuilt on a nearby site in 1947. Pop. (1980E) 500.

Lidingö Stockholm, Sweden. 59 22N 18 08E. Town on Lidingo Island in Baltic Sea, NE of Stockholm city. Some industries, inc. manu. electrical equipment; shipbuilding. Pop. (1989E) 38,819.

Lidköping Skaraborg, Sweden. 58 30N 13 10E. Town sit. ENE of Troll-hättan at the mouth of Lida R. on L. Väner. Railway junction. Industries inc. sugar refining, sawmilling, metal-working, stone quarrying; manu. matches. Resort. Pop. (1988C) 35,168.

Liechtenstein 47 08N 9 35E. The Principality of Liechtenstein, sit. between the Austrian federal state of Vorarlberg and the Swiss cantons of St. Gallen and Graubünden, S of L. Constance, is a sovereign state whose history dates back to 1342 A.D. when Count Hartmann I became ruler of the county of Vaduz. It consists of the two former counties of Schellenberg and Vaduz (until 1806 immediate fiefs of the Roman Empire) which in 1719 were constituted as the Principality of Liechtenstein. Area, 61.8 sq. m. (160 sq. km.). Cap., Vaduz. There is a great variety of light industry inc. textiles, ceramics, steel screws, precision instruments, canned food, pharmaceutical products, heating appliances, etc. The rearing of cattle, for which the fine alpine pastures are well suited, is highly developed. Tourism is the most important source of income. Pop. (1988E) 28,181.

Liège Belgium. 50 38N 5 34E. (i) E province bounded by Federal Republic of Germany and the Netherlands. Area, 1,525 sq. m. (3,950 sq. km.). The N and S are agric. with dairy farming in the foothills of the Ardennes in the S. Industry is concentrated in the central area with steel in the valleys of the R. Meuse and Vesdre. (ii) Cap. of province of same name and commercial centre, sit. at confluence of R. Meuse and Ourthe. Industries iron and steel works and manu. of armaments, machinery, tools, textiles and chemicals. Pop. (1984E) 250,000 (Greater Liège, 650,000).

Lienz Tirol, Austria. 46 50N 12 47E. Town sit. on Drava and Isel Rs. An all-year tourist centre. There is also some manu. of textiles and leather goods. Pop. (1983E) 12,079.

Liepāja Latvia, U.S.S.R. 56 31N 21 01E. Port on Baltic Sea. Industries steelworks, engineering, woodworking, food processing and fish-canning. Exports timber and grain. Pop. (1981E) 108,000.

Lierre (Lier) Antwerp, Belgium. 51 08N 4 34E. Town SE of Antwerp sit. at confluence of Rs. Grande Nèthe and Petite Nèthe. Manu. cutlery, brass products, lace and shoes. Pop. (1983E) 31,209.

Liffey River Ireland. 53 21N 6 16W. Rises in Wicklow Mountains and flows for 50 m. (80 km.) W through Kildare, and then E to reach Dublin Bay.

Lifou New Caledonia, South Pacific. 20 53S 167 13E. Island of the Loyalty group E of New Caledonia. Area, 466 sq. m. (1,207 sq. km.). Chief town Chépénéhé. Main product copra. Pop. (1983C) 8,128.

Liguria Italy. 44 30N 9 00E. Mountainous NW administrative region sit. between Ligurian Alps and Apennines in N, and Gulf of Genoa in S. Consists of the provinces of Genoa, Imperia, La Spezia and Savona. Area, 2,089 sq. m. (5,413 sq. km.). Cap., Genoa. Industries shipbuilding, engineering, iron and steel works and agric. Pop. (1981C) 1,807,893.

Ligurian Sea 43 20N 9 00E. A

branch of the Mediterranean Sea sit. between Liguria and Tuscany, Italy, and Corsica, France; also inc. the Gulf of Genoa.

Lihir Islands   Papua New Guinea. Group in New Ireland district sit. NE of New Ireland. Area, 70 sq. m. (181 sq. km.). Volcanic, comprising 5 islands of which the largest is Lihir, rising to 1,640 ft. (500 metres) a.s.l.

Lihou   Channel Islands, United Kingdom. Island off W coast of Guernsey. Area, 38 acres, with the ruins of a medieval priory.

Likasi   Shaba, Zaïre. 10 59S 26 44E. Copper-smelting and chemical industry. Pop. (1975E) 185,238.

Lille   Nord, France. 50 38N 3 04E. Town in NE sit. on Deûle R. Important industrial centre manu. textiles, loco-motives, machinery, chemicals and biscuits. Many fine medieval build-ings. Pop. (1982C) 174,039 (agglom-eration, 936,295).

Lillehammer   Oppland, Norway. 61 08N 10 30E. Town sit. N of Oslo on Lagen R. at its mouth on L. Mjoesa. Resort and trading centre for a mountain farming area. Pop. (1989E) 22,500.

Lilongwe   Central Region, Malawi. 13 58N 33 49E. City and cap. since 1975, sit. NW of Zomba on Lilongwe R. The main industries are processing tobacco, milling and clothing. Pop. (1985E) 186,800.

Lima   Lima, Peru. 12 06S 77 03W. City sit. on Rimac R. E of its port, Callao. Cap. and dept. cap., commer-cial and administration centre. It was founded in 1535. Sit. in a dry coastal area with frequent fogs. Industries inc. textiles, food processing, tanning, oil refining, flour and lumber milling, brewing, vehicle assembly; manu. to-bacco products, metal goods, cement, glassware, pharmaceutical goods, clothing, footwear. Noted for many religious and secular buildings surviv-ing from the Spanish colonial period. Pop. (1983E) 5,258,000.

Lima   Ohio, U.S.A. 40 46N 84 06W.

City sit. NW of Columbus on Ottawa R. Manu. diesel engines, motorcars, machine tools, electrical goods and chemicals. Pop. (1980C) 47,381.

Limassol   Cyprus. 34 40N 33 03E. Port sit. on S coast of Akrotiri Bay, and SW of Nicosia. Main port and tou-rist centre. Main industries, tourism, wine, clothing, footwear, plastics, fur-niture, food processing. Pop. (1990E) 121,300.

Limavady   Londonderry, Northern Ireland. 55 03N 6 57W. Town sit. ENE of Londonderry on R. Roe. Pop. (1980E) 5,600.

Limbe   Cameroon. 4 01N 9 12E. Formerly Victoria, main port in SW province at foot of Mount Cameroon. Pop. (1981E) 32,917.

Limbe   Malawi. 15 50S 35 03E. Town sit. SE of Blantyre at 3,800 ft. (1,158 metres) a.s.l. Commercial and communications centre for the high-land area. Industries inc. manu. cigarettes, soap.

Limbourg   Liège, Belgium. Town sit. ENE of Verviers on Vesdre R. In-dustries inc. spinning and weaving wool. Formerly cap. of the duchy of Limburg.

Limburg   Netherlands. SE province bordered by Belgium and Federal Re-public of Germany. Area, 837 sq. m. (2,169 sq. km.). Cap. Maastricht. Mainly agric. producing cereals, fruit and sugar-beet with coalmining in the S. Pop. (1985E) 720,000.

Limehouse   Greater London, Eng-land. 51 30N 0 02W. *See* Tower Hamlets.

Limeira   São Paulo, Brazil. 22 34S 47 24W. Town NW of Campinas, sit. in an orange-growing area. Industries fruit-packing and the manu. of coffee-processing machinery. Pop. (1984E) 138,000.

Limerick   Munster, Ireland. 52 30N 9 00W. (i) SW county with estuary of R. Shannon on N. Area, 1,037 sq. m. (2,685 sq. km.). Hilly in S and SE ris-ing to 3,015 ft. (919 metres) a.s.l. at

Gattymore. Mainly level elsewhere with agric. and pastoral farming. Pop. (1971E) 83,000. (ii) County town of county of same name, and port, sit. at head of R. Shannon estuary. Industries flour milling, bacon curing, butter production, tanning and brewing. Many fine medieval buildings. Pop. (1981C) 60,736.

Limoges Haute-Vienne, France. 45 50N 1 16E. Cap. of dept. sit. on Vienne R. Noted for porcelains. Also manu. textiles, shoes and paper. Pop. (1982C) 144,082 (agglomeration (1989E) 175,646).

Limón Limón, Costa Rica. 10 00N 83 02W. Cap. of province and chief port sit. E of San José on Caribbean Sea. Exports cattle, coffee, coconuts and bananas. Pop. (1984E) 49,600.

Limousin ˙France. 46 00N 1 00E. Region in W centre comprising depts. of Correze, Creuse and Haute-Vienne. Area, 6,537 sq. m. (16,931 sq. km.). Crossed by ranges of the Massif Central; infertile in the uplands with R. valleys producing fruit and vegetables. Minerals inc. kaolin. Chief towns Limoges, Brive-la-Gaillarde, Tulle. Pop. (1982C) 737,153.

Limpopo River Republic of South Africa/Mozambique. 25 15S 33 30E. 'The Crocodile River' rises in the Magaliesberg W of Pretoria in the Transvaal and flows for *c.* 1,000 m. (1,600 km.) in a semicircle, reaching the Indian Ocean NE of Delagoa Bay, and forming the boundary between N Transvaal and Zimbabwe. Called the Sehujwane when it flows through Bophuthatswana. Its principal trib. is R. Olifante.

Linares Linares, Chile. 35 51S 71 36W. Cap. of province of same name in S central Chile. Industries tanning, milling and trade in agric. produce. Pop. (1982E) 56,000.

Linares Jaén, Spain. 38 05N 3 38W. Town in S central Spain, sit. in the Sierra Morena foothills. Important lead and copper mining centre. Manu. sheet lead, pipes, chemicals and explosives. Pop. (1981E) 54,500.

Lincoln Lincolnshire, England. 53 14N 0 33W. City and bishopric, sit. on Witham R. Industries inc. manu. machinery, metal products, vehicle parts, cattle feed. Noted for its Cathedral, begun in 1075 A.D. Pop. (1988E) 80,600.

Lincoln Nebraska, U.S.A. 40 48N 96 42W. Cap. of state in E Nebraska. Centre of an agric. region with associated industries inc. food processing, packaging, agric. machinery. Manu. pharmaceuticals, vehicles and components, railway wagons and machinery. Pop. (1980C) 171,932.

Lincolnshire England. 53 14N 0 32W. Under 1974 reorganization the county consists of the districts of West Lindsey, East Lindsey, Lincoln, North Kesteven, Boston, South Kesteven and South Holland. A flat county where agric. (highly mechanized and computerized) is main industry, together with food processing, but tourism important. Pop. (1990E) 602,155.

Linden New Jersey, U.S.A. 40 38N 74 15W. City sit. SW of Newark. Manu. chemicals and clothing. Pop. (1980C) 37,836.

Lindesnes Norway. Cape at extreme S at the entrance to the Skagerrak and the lighthouse.

Lindi Southern Province, Tanzania. 9 58S 39 38E. Town sit. SSE of Dar es Salaam at mouth of Lukuledi R. on the Indian Ocean. Port handling sisal, copra, cotton, tobacco. Industries inc. fishing.

Line Islands Kiribati, Pacific. 0 05N 157 00W. Group extending across the equator from 5°N lat.–11°S lat. and inc. Flint, Vostok, Caroline, Starbuck, Malden, Jarvis, Palmyra, Kingman Reef, Kiritimati, Tabuaeran and Teraina. Only last 3 are inhabited. Jarvis, Palmyra and Kingman Reef are U.S. owned. Pop. (1985E) 2,500.

Lingeh Iran. Town sit. WSW of Bandar Abbas on the (Persian) Gulf. Trading centre for pearls.

Lingfield Surrey, England. 51 10N 0 01W. Village sit. SE of Reigate on

R. Eden. Pop. (1980E) 7,000 with Dormans Land.

Lingga Archipelago Sumatra, Indonesia. A group of islands, inc. Lingga and Singkep, sit. off E coast. Produces copra, tin and sago.

Linguera Senegal. Town sit. ENE of Dakar. Railway terminus exporting peanuts, gums.

Linköping Östergötland, Sweden. 58 25N 15 37E. Cap. of county in SE. Industries aircraft manu., high-tech engineering and computers. Pop. (1990E) 120,500.

Linlithgow Lothian Region, Scotland. 55 59N 3 37W. Town sit. W of Edinburgh. Manu. computer components. Pop. (1987E) 11,328.

Linnhe, Loch Scotland. 56 36N 5 25W. An inlet on W coast at SW end of Glenmore.

Linz Upper Austria, Austria. 48 18N 14 18E. Cap. of federal state and port sit. on Danube R. Manu. steel, machinery, textiles, chemicals and paper. Pop. (1983E) 199,910.

Lions, Gulf of France. 43 00N 4 00E. Bay of the Mediterranean Sea stretching from the Franco-Spanish border to Toulon. Principal port Marseille.

Lipa Luzon, Philippines. 13 57N 121 10E. Town sit. SSE of Manila in Batangas province. Manu. cutlery and textiles. Trades in sugar, maize, rice and tobacco.

Lipari Islands Italy. 38 30N 14 50E. Also known as the Eolie Islands. Group in the Tyrrhenian Sea off NE Sicily. Area, 44 sq. m. (114 sq. km.). Chief islands Lipari, Salina, Vulcano, Stromboli, Filicudi, Panaria, Alicudi. Formed as a submarine mountain chain, volcanic and active, rising to 3,156 ft. (962 metres) a.s.l. on Salina. Main occupations farming, fishing. Main products wine, lobsters, pumice stone.

Lipetsk Russian Soviet Federal Socialist Republic, U.S.S.R. 52 37N

39 35E. Town sit. NNE of Voronezh on Voronezh R. Industries iron and steel works, manu. of tractors and chemicals. Noted as a health spa with chalybeate springs. Pop. (1985E) 447,000.

Lipova Arad, Romania. Town sit. NE of Timisoara on Mures R. Resort with mineral springs and trading centre for livestock. Industries inc. brewing, flour milling; manu. mineral waters, bricks and tiles.

Lippe North Rhine-Westphalia, Federal Republic of Germany. A former principality sit. between Hanover and Westphalia. Cap. Detmold. Area, 469 sq. m. (1,214 sq. km.). Mainly agric.

Lippe River North Rhine-Westphalia, Federal Republic of Germany. 51 39N 6 38E. Rises in the Teutoburgerwald and flows W for 147 m. (235 km.) to join Rhine R. at Wesel.

Lippstadt North Rhine-Westphalia, Federal Republic of Germany. 51 40N 8 19E. Town sit. E of Hamm on Lippe R. Industries iron founding, manu. of wire, metal goods and textiles. Pop. (1984E) 60,400.

Lisboa Portugal. 38 42N 9 10W. District of W central Portugal in Estremadura and W Ribatejo provinces. Area, 1,066 sq. m. (2,762 sq. km.). Extending from the Atlantic (W) to the Tagus R. estuary. Cap. Lisbon. A wine growing and tourist area. Pop. (1981C) 2,069,467.

Lisbon Estremadura, Portugal. 38 42N 9 10W. City on N shore of Tagus R. estuary. Cap. and district cap., commercial, cultural, administration and industrial centre. Following the earthquake of 1755 when much of the city was destroyed the Marquis of Pombal planned and built the new Lisbon. Industries inc. manu. textiles, chemicals, paper, tobacco products, pottery, soap, armaments, explosives, metal goods, flour and sugar. The port exports wine, olive oil, cork, fruit, resins, salt, metal ores, fish. Pop. (1987E) 830,500.

Lisburn Antrim, Northern Ireland.

54 30N 6 09W. Town in E central N.
Ireland on R. Lagan. Industries inc.
engineering, textiles, clothing and furniture. Pop. (1989E) 95,500.

Lisdoonvarna   Clare, Ireland. 53
02N 9 17W. Town and spa sit. NW of
Ennis, with sulphur and chalybeate
springs. Pop. (1990E) 600.

Lisieux   Calvados, France. 49 09N
0 14E. Town sit. E of Caen on the
Touques R. Trades in dairy produce,
esp. Camembert cheese. Pop. (1989C)
24,985.

Liskeard   Cornwall, England. 50
28N 4 28W. Town sit. ESE of Bodmin,
with nearby stone and slate quarries.
Pop. (1990E) 7,000.

Lismore   New South Wales, Australia. 37 57S 143 20E. City in NE of
state sit. on Wilson R. Trades in exotic
fruits, tropical fruit, nuts, light manu.
and tourism. Pop. (1988E) 38,630.

Lismore   Waterford, Ireland. 52 08N
7 57W. Town sit. WSW of Waterford
on R. Blackwater. Pop. (1980E) 1,100.

Listowel   Kerry, Ireland. 52 27N
9 29W. Market town in the SW, sit. on
R. Feale. Pop. (1980E) 3,000.

Lith   Hejaz, Saudi Arabia. Town sit.
SE of Jidda on Red Sea. Port handling
agric. produce.

Lithgow   New South Wales, Australia. 33 29S 150 09E. Town in E central
N.S.W. sit. in the Blue Mountains. Industries coalmining, electricity generation, tile making and the manu. of
small arms. Pop. (1985E) 14,400.

Lithuania   U.S.S.R. 53 30N 24 00E.
Rep. sit. on Baltic Sea and bounded
on N by Latvia, E and S by Poland and
W by the Baltic Sea. The secret protocol of the Soviet-German frontier
treaty of 1939 assigned the greater part
of Lithuania to the Soviet sphere of
influence. In that year the province
and city of Vilnius were ceded by the
U.S.S.R. An ultimatum in 1940 led to
the formation of a govt. acceptable to
the U.S.S.R. This incorporation has
been accorded *de facto* recognition by
the Brit. Govt., but not by the U.S.

Govt. Area, 25,170 sq. m. (65,200 sq.
km.). Cap. Vilnius. Lithuania before
1940 was a mainly agric. country, but
has since been considerably industrialized. The urban pop. was 23% of the
total in 1937 and 66% in 1986. The
resources of the country consist of
timber and agric. produce. Of the total
area, 49.1% is arable land, 22.2%
meadow and pasture land, 16.3% forests and 12.4% unproductive lands.
Heavy engineering, shipbuilding and
building material industries are developing. Pop. (1989E) 3·7m.

Little Belt   Denmark. 55 20N 9 45E.
A strait between Jutland and Fyn
island, joining the Baltic Sea with the
Kattegat.

Little Cayman   Cayman Islands,
West Indies. Island sit. ENE of Grand
Cayman and E of Cayman Brac. Area,
9.24 sq. m. (24 sq. km.). Main products turtle shell, coconuts. Pop.
1,300.

Littlehampton   West Sussex, England. 50 48N 0 33W. S coast resort sit.
at mouth of R. Arun. Pop. (1981E)
22,000.

Little Rock   Arkansas, U.S.A. 34
44N 92 15W. Cap. of state and largest
city, sit. in central Arkansas on
Arkansas R. Important transportation
and commercial centre of a mining
and agric. region. Manu. computers,
clothing, wood and food products,
chemicals, building materials and
foundry goods. Pop. (1989E) 187,473.

Liverpool   Merseyside, England. 53
25N 2 55W. Metropolitan district of
Merseyside, sit. on right bank of R.
Mersey, near the Irish Sea. Once one
of the greatest trading centres of the
world and the principal port in the
U.K. for the Atlantic trade. Pop.
(1984E) 497,300.

Livingston   Lothian Region, Scotland. 55 53N 3 32W. Former village
now new town, designated in 1962.
Industries inc. industrial research, asbestos, paper, scientific instruments
including the manu. of high technology products and electronics. Pop.
(1985E) 40,000.

Livingstone   *see* **Maramba**

337

Livonia Michigan, U.S.A. 42 23N 83 33W. City W of Detroit in industrial Wayne county. Manu. vehicles and parts. Pop. (1980C) 104,814.

Livorno *see* **Leghorn**

Lizard Point Cornwall, England. 49 57N 5 12W. The most S point of Britain. Noted for its serpentine rock. Lizard Town is adjacent.

Ljubija Bosnia, Yugoslavia. Village sit. WNW of Banja Luka. Iron mining centre.

Ljubljana Slovenia, Yugoslavia. 46 03N 14 31E. Cap. of Slovenia sit. near confluence of R. Ljubljanica and R. Sava. Industries electrical, chemical and mechanical engineering, food processing, textiles, paper manu. Pop. (1989E) 335,798.

Llanberis Gwynedd, Wales. 53 07N 4 09W. Town sit. ESE of Caernarvon at W end of Llanberis Pass. The mountain railway to the summit of Snowdon starts here.

Llandovery Dyfed, Wales. 51 59N 3 48W. Town in S central Wales sit. on R. Towy. Industries inc. pet foods, wire wheels. Pop. (1981C) 1,696.

Llandrindod Wells Powys, Wales. 52 15N 3 23W. Spa sit. on R. Ithon. Noted for its mineral springs. Pop. (1989E) 5,020.

Llandudno Gwynedd, Wales. 53 19N 3 49W. Resort on N coast, sit. at mouth of R. Conwy. Pop. (1981C) 18,911.

Llanelli Dyfed, Wales. 51 41N 4 09W. Seaside town sit. on Carmarthen Bay. Tinplate manu. important, also automotive parts, inflatable craft, lenses, bearings, chemicals, mining and tunnelling machinery. Pop. (1989E) 41,440.

Llanfairfechan Gwynedd, Wales. 53 15N 3 58W. Resort sit. SW of Llandudno on Conwy Bay. Pop. (1980E) 3,800.

Llangollen Clwyd, Wales. 52 58N 3 10W. Resort sit. SW of Wrexham on

R. Dee. Centre of an agric. area. Industries inc. tourism, printing, packaging, hide and skin dressing, and crafts. Pop. (1981E) 3,072.

Llanidloes Powys, Wales. 52 27N 3 32W. Town sit. WSW of Newtown on R. Severn. Market for an agric. area. The main industry is manu. textiles, and engineering. Pop. (1981C) 2,500.

Llanquihue, Lake Llanquihue/ Osorno, Chile. 41 10S 72 50W. L. sit. N of Puerto Montt. Area, *c.* 240 sq. m. (622 sq. km.). Drained by Maullin R. which runs into the Pacific Ocean.

Llanrwst Gwynedd, Wales. 53 08N 3 47W. Town sit. S of Llandudno on R. Conwy. Pop. (1980E) 3,000.

Lleyn Peninsula Wales. 52 54N 4 27W. Peninsula sit. between Cardigan Bay and Caernarvon Bay.

Lloydminster Alberta/Saskatchewan, Canada. 53 17N 110 00W. Town NW of North Battleford on the border between the 2 states and sit. in both. Commercial and distribution centre for a grain growing and lumbering area. Industries inc. agric. products, oil and oil by-products, natural gas, salt, gravel and coal. Deposits of oil and natural gas. Pop. (1989E) 16,254, Alberta 9,457; Saskatchewan 6,797.

Llullaillaco, Mountain Argentina/ Chile. 24 43S 68 33W. Peak sit. WSW of San Antonio de los Cobres on the Argentina/Chile frontier; height 22,015 ft. (6,710 metres) a.s.l., an extinct volcano.

Lobatse Lobatse, Botswana. 25 11S 25 40E. Town sit. N of Mafeking near the border of the Republic of South Africa, in a dairy farming area. Headquarters of the Baralong tribe. Pop. (1981C) 19,034.

Lobito Angola. 12 20S 13 34E. Principal port built on reclaimed land. Railway terminus for the trans-Africa railway. Exports cotton, sugar, maize, coffee, salt, sisal and ores. Pop. (1976E) 70,000.

Lobos Canary Islands. Islet NE of

Fuerteventura in La Bocayna channel. Area, 2.4 sq. m. (6.2 sq. km.). The main occupation is fishing.

Lobos Islands   Peru. Two groups of islands sit. off N coast, with guano deposits.

Locarno   Ticino, Switzerland. 46 10N 8 48E. Resort in S central Switzerland sit. at N end of L. Maggiore.

Lochaber   Highland Region, Scotland. 56 56N 5 00W. District extending between the Great Glen and the borders of Perthshire and Argyll. Mountainous and picturesque, rising to Ben Nevis. Tourist area. Many lochs and Rs. supply hydroelectric power.

Lochgelly   Fife Region, Scotland. 56 08N 3 19W. Town sit. W of Kirkcaldy. The main industry is coalmining. Pop. (1987E) 7,500.

Lockerbie   Dumfries and Galloway Region, Scotland. 55 07N 3 22W. Market town sit. NE of Dumfries. Industries inc. furniture manu., road haulage, abbatoirs, packaging and food processing. Pop. (1981C) 3,250.

Lockport   New York, U.S.A. 43 10N 78 42W. City sit. NNE of Buffalo on Erie Canal. Manu. steel, textiles, chemicals, paper, wood, glass and leather products. Pop. (1980C) 24,844.

Locle, Le   Neuchâtel, Switzerland. Town sit. WNW of Neuchâtel near the French border. Centre of a watchmaking industry since the 18th cent.

Lod   Lod, Israel. 31 57N 34 54E. Town sit. SE of Tel Aviv. Site of Israel's international airport and also a railway junction.

Lodi   Lombardia, Italy. 45 19N 9 30E. Town SE of Milan sit on Adda R. Important cheese trade esp. Parmesan. Manu. silk, linen and majolica. Pop. (1980E) 46,000.

Łódź   Łódź, Poland. 51 46N 19 30E. Town in central Poland. The second largest town in Poland and also a province together with the towns of Leçzyca, Pabianice and Zgierz. Manu.

textiles, machinery, chemicals, electrical and leather goods, food products, paper and tobacco. Pop. (1985E) 849,000.

Lofoten Islands   Norway. 68 15N 14 00E. Group in North Sea off coast of N Norway within the Arctic circle. Area, *c*. 550 sq. m. (*c*. 1,425 sq. km.). Main islands are Austvagoy, Vestvagoy, Moskenesoy, Flakstadoy, Vaeroy, Rost. Mountainous with a mild climate caused by the N Atlantic Drift. The straits are subject to violent currents. Important cod and herring grounds offshore. Fishing is the main occupation.

Logan   Utah, U.S.A. 41 44N 111 57W. City sit. N of Ogden on a branch of Little Bear R. Trading centre for an irrigated agric. area, handling sugarbeet, grain, vegetables. Industries inc. food processing, textiles, oil refining, metal working. Pop. (1990E) 32,000.

Logan, Mount   Yukon, Canada. 60 34N 140 24W. Sit. in the St. Elias Mountains, in the NW, it is the highest known peak in Canada and the second highest in N. America, 19,850 ft. (6,050 metres) a.s.l.

Logar River   Afghánistán. Rises in the SW Hindu Kush mountains and flows 150 m. (240 km.) E past Shaikhabad and Baraki Rajan, then N to join Kábul R. The middle course is known as the Wardak R. Used for logging.

Logroño   Spain. 42 28N 2 27W. Town sit. NW of Zaragoza on Ebro R. Centre of the Rioja wine industry. Pop. (1986C) 118,770.

Loire   France. 45 40N 4 05E. Dept. in Rhône-Alpes region sit. on E of the Massif Central. Area, 1,843 sq. m. (4,774 sq. km.). Mainly mountainous but agric. possible. on the plains, producing cereals and vines. Contains an important coalfield and large deposits of iron and lead. Cap. St. Étienne. Principal R. is Loire. Pop. (1982C) 739,521.

Loire-Atlantique   France. 47 25N 1 40W. A maritime dept. in the Pays-de-la-Loire region. Area, 2,661 sq. m. (6,893 sq. km.). Very flat with coastal salt marshes. Agric. products inc.

cereals, vines, flax, sugar-beet and potatoes, and horse and cattle breeding is important. The Grande-Lieu L., the largest in France, is sit. s of the Loire estuary. Industries inc. shipbuilding and foundries at Nantes and St. Nazaire. Also manu. of hemp, linen, paper and sugar. Cap. Nantes. Principal R. is Loire. Pop. (1982C) 995,498.

Loire River   France. 47 16N 2 11W. The longest R. in France, it rises in the Cévennes and flows for 638 m. (1,020 km.) first N and then W and enters the Bay of Biscay.

Loiret   France. 47 58N 2 10E. Dept. in Centre region. Area, 2,603 sq. m. (6,742 sq. km.). Mainly agric. producing grain, sugar-beet, fruit and vines. There is a large forest round Orléans which is the cap. Principal R. is Loire. Pop. (1982C) 535,669.

Loir-et-Cher   France. 47 40N 1 20E. Dept. in Centre region. Area, 2,438 sq. m. (6,314 sq. km.). Large forests but agric. possible in the R. valleys and in the Beauce plateau, the main products being cereals, vines and fruit. Cap. Blois. Principal Rs., Loire, Loir and Cher. Pop. (1982C) 296,220.

Loir River   France. 47 33N 0 32W. Rises in the Collines du Perche, Eure-et-Loir dept., and flows s and W for 190 m. (304 km.) to join Sarthe R. near Angers.

Loja   Ecuador. 3 59N 79 16W. (i) Province of s Ecuador, bounded s by Peru. Area, 11,158 sq. m. (28,900 sq. km.). Mountainous, drained by Río Grande and Catamayo R., with a semitropical climate. Mainly agric., producing cereals, potatoes, sugar-cane, coffee, fruit, cattle, sheep and mules. Minerals inc. gold, silver, copper, iron, kaolin, marble. Industries inc. manu. woollen goods. Pop. (1982E) 358,558. (ii) Town sit. ssw of Quito on the Pan-American Highway at the foot of Cordillera de Zamora at 7,300 ft. (2,225 metres) a.s.l. Provincial cap. and trading centre for an agric. area, handling sugar, coffee, tobacco, cereals, vegetables, cattle. Industries inc. tanning, textiles. Pop. (1982E) 71,130.

Lokeren   East Flanders, Belgium. 51 06N 4 00E. Town sit. ENE of Ghent on Durme R. Manu. textiles, chemicals and rope. Pop. (1989E) 34,881.

Lolland   Denmark. 54 46N 11 30E. Low-lying island in the Baltic Sea s of Zealand. Area, 480 sq. m. (1,243 sq. km.). Mainly agric. producing cereals and sugar-beet. Chief towns Maribo and Naksov.

Lombardy   Italy. 45 35N 9 45E. Region of N Italy bounded N by Switzerland, E by Trentino-Alto Adige and Veneto, s by Emilia-Romagna, W by Piedmont. Area, 9,211 sq. m. (23,856 sq. km.), comprising 9 provinces, Bergamo, Brescia, Como, Cremona, Mantova, Milano, Pavia, Sondrio, Varese. Mountainous in the N, descending to the plain of Lombardy drained by R. Po. Important industrial area, with textiles, iron and steel, chemicals, publishing, paper making, tanning, manu. vehicles, all concentrated on Milan and cities in R. Po Valley. Pop. (1988E) 8,898,951.

Lombok   Indonesia. 8 45S 116 30E. Island sit. between Bali and Sumbawa. Area, 1,825 sq. m. (4,725 sq. km.). Mountainous on the N and s coasts with a fertile plain between, producing rice, coffee, sugar and cotton. Cap. Mataram.

Lomé   Maritime, Togo. 6 10N 1 21E. Cap. of Togo and port sit. on the Bight of Benin. It became cap. of German Togoland in 1897. Exports coffee, cocoa and palm kernels. Meetings held in the 1970s between the European Communities and Third World countries led to aid and trade agreements. Pop. (1983E) 366,476.

Lomond, Loch   Scotland. 56 08N 4 38W. L. extending N to s 23m. (37 km.) from Ardlui (N) to Balloch (s) in the Strathclyde Region. Drained by R. Leven at the s end. Picturesque tourist area.

London   Ontario, Canada. 42 59N 81 14W. City sit. sw of Toronto on Thames R. Important industrial centre with over 500 varied trades. Manu. inc. steel products, textiles, food products, electrical and electronic equipment and printed materials. Pop. (1982E) 266,319.

London England. 51 30N 0 10W. City on R. Thames. Cap. of the United Kingdom. The administration area of Greater London inc. 32 boros. covering *c.* 462,000 acres. London was founded by the Romans in 43 A.D. The city is divided into the City of London (E) and the City of Westminster (W) both N of Thames and surrounded by mixed residential, industrial and shopping areas which developed later, as did those S of the R. The City of Lond. is the financial sector, inc. the Bank of England and the stock exchange; it covers approx. the area of the medieval city of which little remains except the Tower of Lond. fortress and the churches, and is known as the 'square mile'. The City of Westminster inc. Parliament, Govt. offices, the Abbey of Westminster and the royal palaces. The Port handles decreasing traffic and some dock systems are closing, while others have been redeveloped for residential accommodation and offices. Industries cover diverse manu., printing, publishing, brewing, food processing. Pop. (1988E) of Greater London 6,735,400.

Londonderry (also known as Derry), Northern Ireland. 55 00N 7 20W. (i) County of NW Northern Ireland bounded N by the Atlantic, W by the Irish Rep. Area, 804 sq. m. (2,082 sq. km.). Hilly, drained by Bann, Roe and Foyle Rs., and mainly agric. producing seed potatoes, dairy produce. (ii) City sit. NW of Belfast on a hill above R. Foyle, near its mouth on a sealough. Port exporting seed potatoes, scrap metal and livestock. Industries inc. flour milling, tanning, iron founding; manu. linen, alcohol, food products. Pop. (1981C) 62,697.

Londrina Paraná, Brazil. 23 18S 51 09W. Town sit. NW of Curitiba. Rapidly developing with the coffee trade. Pop. (1980C) 349,200.

Long Beach California, U.S.A. 33 46N 118 11W. Resort in S of state, possessing a 7 m. (11 km.) long bathing beach. Manu. aircraft, tyres, chemicals and soap. Industries inc. car assembly plants, oil refining and fish

and fruit canning. It has a good harbour with 10 sq. m. (26 sq. km.) of anchorage. Pop. (1989E) 378,900.

Long Branch New Jersey, U.S.A. 40 18N 74 00W. Resort sit. ENE of Trenton. Manu. silk and rubber goods, clothing and boats. Pop. (1990E) 35,000.

Longchamp France. A famous racecourse sit. on SW of Bois de Boulogne, Paris.

Long Eaton Derbyshire, England. 52 54N 1 15W. Town sit. E of Derby. Mainly light industry but furniture manu. important. Pop. (1987E) 34,690.

Longford Leinster, Ireland. 50 42N 74 5W. (i) County bounded by Leitrim, Cavan, West Meath and Lough Ree. Area, 403 sq. m. (1,043 sq. km.). Mainly low-lying and boggy but rises to 912 ft. (278 metres) a.s.l. in N. The principal R. is Shannon. Agric. products inc. oats and potatoes, and cattle and sheep are raised. Pop. (1988E) 31,496. (ii) County town of county of same name. Manu. textiles. Pop. (1980E) 4,000.

Long Island Bahamas, West Indies. 23 20N 75 10W. Island between Exuma (NW) and Crooked Island (SE) and SE of Nassau. Area, 173 sq. m. (448 sq. km.). Main occupations salt panning, farming, fishing. Pop. (1980C) 3,358.

Long Island New York, U.S.A. 40 50N 73 00W. Island off coast of New York and Connecticut, stretching ENE from mouth of Hudson R. It is mainly residential with many towns and resorts, being separated from the mainland by Long Island Sound. The W end, containing the boroughs of Brooklyn and Queens, is part of New York City.

Long Island Madang, Papua New Guinea. Island sit. NE of New Guinea. Area, 160 sq. m. (414 sq. km.), rising to 4,278 ft. (1,304 metres) in Reumur Peak. Main product copra.

Longton Staffordshire, England. 53 59N 2 08W. Town now forming part of Stoke-on-Trent in the 'Potteries' district.

Longview Texas, U.S.A. 32 30N 94 45W. City sit. E of Dallas, on the E Texas oilfield. Industries oil refining, chemical works and the manu. of oilfield equipment. Pop. (1980C) 62,762.

Longview Washington, U.S.A. 46 08N 122 57W. City and estuary port sit. S of Seattle, N of Portland, Oregon, on the Columbia R. Industries inc. forest products, fibres, paper and metals. Pop. (1980C) 31,055.

Lons-le-Saunier Jura, France. 46 40N 5 33E. Cap. of dept. Manu. optical goods and wines. Pop. (1982C) 21,886.

Looe Cornwall, England. 50 22N 4 28W. Port sit. W of Plymouth, comprising E and W Looe straddling R. Looe. Pop. (1981E) 4,278.

Lopevi Vanuatu, South West Pacific. Island sit. SSE of Ambrym, length 4 m. (7 km.). Volcanic, rising to 4,755 ft. (1,449 metres) a.s.l.

Lorain Ohio, U.S.A. 41 28N 82 10W. City and port sit. SW of Cleveland on L. Erie. Industries shipbuilding and a coal and iron ore trade. Manu. iron and steel products, machinery and cranes. Pop. (1980C) 75,416.

Lord Howe Island New South Wales, Australia. 31 33S 159 05E. Island sit. NE of Sydney. Area, 5 sq. m. (13 sq. km.). Hilly, rising to 2,840 ft. (866 metres) in Mount Gower, and with dense vegetation. Resort.

Lorengau Admiralty Islands, Papua New Guinea. 2 00S 147 15E. Town on E coast of Manus Island. District cap.

Loreto Peru. Dept. of NE Peru bounded NW by Ecuador, NE and E by Brazil. The largest dept., covering *c.* one-third of the country. Area, 119,300 sq. m. (308,987 sq. km.). Cap. Iquitos. Extends from W limit of the Andes across Amazon basin, drained by Amazon R. and numerous tribs. Heavily forested and little developed, with agric. only in limited clearings. Forest products inc. rubber,

balata, chicle, cascarilla, tagua nuts, tanning bark, medicinal plants and tropical timber. Pop. (1982C) 445,368.

Lorient Morbihan, France. 47 45N 3 22W. Port in Brittany, sit. on an inlet of the Bay of Biscay. Industries shipbuilding, fishing and submarine base. Pop. (1982C) 64,675 (agglomeration, 104,025) .

Lorraine France. 49 00N 6 00E. Region of E France bounded N by Belgium and Luxembourg, NE by Federal Republic of Germany. Comprises the Moselle, Meurthe-et-Moselle, Meuse and Vosges depts. Table-land drained by Moselle and Meuse Rs., rising in the E to the Vosges mountains. Area, 9,089 sq. m. (23,540 sq. km.). Agric areas produce vines and hops. Minerals inc. extensive iron-fields in the Briey, Longwy, Thionville and Nancy basins. The main industry is manu. iron and steel. Chief cities Nancy, Metz, Verdun, Toul. Pop. (1982C) 2,319,905.

Los Alamos New Mexico, U.S.A. 35 53N 106 19W. Town sit. NW of Santa Fe, at 7,300 ft. (2,225 metres) a.s.l. Scientific research institutes provide most of the employment. Pop. (1980C) 11,039.

Los Andes *see* **Andes, Los**

Los Angeles California, U.S.A. 34 03N 118 15W. City sit. in S of the state. Founded by a Franciscan, Father Crespi, in 1781 as Nuestra Señora la Reina de Los Angeles. Important industries inc. fruit canning, petroleum, meat packing, timber. Manu. inc. flour, steel and textiles. Hollywood (annexed to Los Angeles in 1910) was the great centre for the production of cinema films, but this has declined in importance in recent decades. Pop. (1980C) 2,966,850.

Los Islands Guinea. 9 30N 13 48W. A small group of islands just off the coast near Conakry. British until 1904, they are now French. There are large bauxite deposits.

Los Ríos Ecuador. Province of W central Ecuador bounded E by the Andes. Area, 2,295 sq. m. (5,944 sq.

km.). Cap. Babahoyo. Densely forested lowland with a humid climate.
Main products cacao, coffee, sugar-
cane, rice, fruit, balsa wood, rubber,
timber, nuts. Pop. (1980E) 451,064.

Los Santos   Panama. (i) Province of
central Panama on the Pacific coast.
Area, 1,411 sq. m. (3,654 sq. km.).
Chief towns Las Tablas (cap.), Los
Santos. Agric. area with deposits of
nickel, cobalt and sulphur. Pop.
(1990E) 82,311. (ii) Town sit. NW of
Las Tablas on a branch of the Inter-
American Highway. Commercial
centre of an agric. area. Industries inc.
distilling, salt extracting. Pop. (1984E)
13,999.

Lossiemouth   Grampian Region,
Scotland. 57 43N 3 18W. Town sit. N
of Elgin on Moray Firth at mouth of
Lossie R. Fishing port and holiday
resort. Industries inc. fishing, fish processing, marine engineering. Town
serves the RAF Lossiemouth air station. Pop. (1988E) 7,340.

Los Teques   Miranda, Venezuela.
10 21N 67 02W. Town sit. SW of
Caracas in the coastal mountains at
3,864 ft. (1,178 metres) a.s.l. State
cap. and trading centre for an agric.
area, handling coffee, grain, cacao,
sugar. Mountain resort. Pop. (1971C)
62,747.

Lostwithiel   Cornwall, England. 50
25N 4 40W. Town sit. SSE of Bodmin
on R. Fowey. Pop. (1990E) 2,700.

Lot   France. 44 39N 1 40E. Dept.
sit. in Midi-Pyrénées region, formed
from the district of Quercy. Area,
2,019 sq. m. (5,228 sq. km.). Cap.
Cahors. It is highest in the NE, where
large amounts of chestnuts are grown.
The land slopes from the NE to the SW
and is drained by Rs. Lot and Dordogne. Mainly agric. growing cereals,
potatoes and fruit; wine is esp. important. Pop. (1982C) 154,533.

Lota   Concepción, Chile. 37 05S 73
10W. Port sit. SSW of Concepción.
Centre of a coalmining area. Manu.
copper goods and ceramics. Pop.
(1980E) 52,000.

Lot-et-Garonne   France. 44 22N

0 30E. Dept. sit. in Aquitaine region,
formed from parts of Guyenne and
Gascony. Area, 2,069 sq. m. (5,358
sq. km.). It is mainly flat and traversed by Rs. Garonne and Lot. Mainly
agric., the fertile R. valleys producing
cereals, vines, fruit and tobacco. Cap.
Agen. Pop. (1982C) 298,522.

Lothian Region   Region of SE Scotland. 55 50N 3 00W. It includes the
former county of the city of Edinburgh, part of Midlothian, East
Lothian. Industries inc. clothing,
glass, high technology and manu. of
television sets. Pop. (1989E) 741,149.

Lot River   France. 44 32N 1 08E.
Rises in the Lozère Mountains, flows
W for 298 m. (477 km.) through Aveyron, Lot and Lot-et-Garonne depts. to
join Garonne R. NW of Agen.

Lötschberg Tunnel   Switzerland.
Railway tunnel extending 9 m. (14
km.) from Thun to Brig under
Lötschen Pass, at 4,078 ft. (1,243
metres) a.s.l. Built in 1911.

Louga   Senegal. Cap. of region sit.
NE of Dakar. Railway junction serving
an agric. area. Industries inc. extracting vegetable oils. Pop. (1976C)
37,665.

Loughborough   Leicestershire, England. 52 47N 1 11W. Town sit. NNW
of Leicester on R. Soar. Industries inc.
footwear, gas, publishing, hosiery,
pharmaceutical products, electrical
equipment, high technology, printing,
bell-founding; the great bell of St.
Paul's, London, was cast here in 1881.
Pop. (1988E) 52,000.

Loughrea   Galway, Ireland. 53 12N
8 34W. Town sit. on N shore of Lough
Rea. Pop. (1971C) 3,075.

Louis Gentil   *see* **Youssoufia**

Louisbourg   Nova Scotia, Canada.
45 55N 59 58W. Town sit. SE of Sydney on E coast of Cape Breton Island.
Formerly an important French fortress. The main industry is fishing and
tourism. Pop. (1990E) 1,450.

Louisiade Archipelago   Papua New
Guinea. 11 12S 153 00E. A group of

volcanic islands sit. to SE of Papua New Guinea. Alluvial gold has been discovered on some of the islands.

Louisiana   U.S.A. 30 50N 92 00W. One of the S states of U.S.A., and was first settled in 1699. That part lying E of the Mississippi R. was organized in 1804 as the Territory of New Orleans, and admitted into the Union in 1812. The section W of the R. was added shortly afterwards. Bounded on N by Arkansas, on E by Mississippi, on S by Gulf of Mexico and on W by Texas. The surface is flat and marshy particularly near the sea. R. Mississippi delta occupies one-third of total area. Area, 52,453 sq. m. (135,844 sq. km.) inc. inland water. Chief towns, New Orleans, Shreveport and Baton Rouge (the cap.). Agric. is important and crops inc. sugar-cane, rice, grain, sweet potatoes, soybeans, pecans, cotton and strawberries. There are 14.5m. acres of forest representing 47% of the state's area. Main minerals, petroleum, salt and sulphur. The manu. industries are chiefly those associated with petroleum, chemicals, lumber, food and paper. Pop. (1987E) 4,460,578.

Louisville   Kentucky, U.S.A. 38 16N 85 45W. Chief trading city of the state sit. on Ohio R. Important tobacco market and meat packing centre. Manu. inc. electrical appliances, trucks, cigarettes, spirits, chemicals and metals. Pop. (1980C) 298,451.

Lourdes   Hautes-Pyrénées, France. 43 06N 0 00W. Town in SW sit. at the foot of the Pyrénées. A leading centre of pilgrimage since 1858 when Bernadette Soubirous, a peasant girl, had visions of the Virgin Mary. Many miraculous cures are said to have followed. Pop. (1982C) 17,619.

Lourenço Marques, Mozambique *see* **Maputo**

Louth   Lincolnshire, England. 53 23N 0 00. Town sit. S of Humber estuary ENE of Lincoln on R. Lud. Industries malting, plastics and manu. of agric. machinery and cartons. Pop. (1988E) 14,462.

Louth   Leinster, Ireland. 53 47N 6 33W. A maritime county bordered

on E by the Irish Sea. Area, 317 sq. m. (821 sq. km.). Generally low-lying, undulating country but rises to over 1,900 ft. (579 metres) a.s.l. in the N. The principal Rs. are Fane, Lagan, Glyde and Dee. Mainly agric. with cattle rearing and production of cereals and potatoes. Also fishing and linen mills. County town Dundalk. Pop. (1988E) 91,810.

Louvain   Brabant, Belgium. 50 53N 4 42E. City sit. E of Brussels on R. Dyle. Cultural centre noted for its university (1425). Industries inc. brewing, distilling; manu. machinery, chemicals, leather goods. Pop. (1989E) 85,157.

Lovech   Pleven, Bulgaria. 43 08N 24 43E. Town sit. S of Pleven on R. Osam. Leather tanning and leathercraft centre.

Lowell   Massachusetts, U.S.A. 42 39N 71 18W. City sit. NW of Boston on Merrimack R. An important textile centre since the early 19th cent. Manu. machinery, munitions, chemicals, plastics, computers, defence equip., textiles, electrical goods and paper. Pop. (1989E) 104,000.

Lower Austria   Austria. 48 25N 15 40E. Province of NE Austria, bounded N by Czechoslovakia and W by Federal Republic of Germany. Area, 7,402 sq. m. (19,170 sq. km.). Chief towns Krems, Wiener Neustadt and Sankt Pölten. Vienna is a provincial enclave within the province. The province is divided W–E by R. Danube. Agric. includes stock-raising and vine-growing Manu. inc. chemicals, glass, sugar, textiles and tobacco. Pop. (1988E) 1,428,700.

Lower Hutt   North Island, New Zealand. 41 10S 174 55E. Town sit. NE of Wellington. Industries car assembling, engineering, meat freezing, and manu. of textiles and furniture. Pop. (1973E) 62,800.

Lower Merion   Pennsylvania, U.S.A. Town forming a suburb of W Philadelphia. Residential. Haverford and Bryn Mawr colleges sit. here.

Lower Saxony   Federal Republic of

Germany. 52 45N 9 00E. *Land* of the Federal Rep. extending from the Harz range to the North Sea, with Bremen and Hamburg as enclaves, and bounded W by the Neth. Area, 18,311 sq. m. (47,426 sq. km.). Chief cities Hanover (cap.), Brunswick, Celle, Delmenhorst, Goslar, Hameln, Lüneberg, Wilhelmshaven, Cuxhaven, Emden, Göttingen. Heath and moorland with reclaimed 'polders' in the N. Agric. areas produce cattle, grain, sugar-beet, vegetables. Minerals inc. oil, iron ore, salt, potash, coal, lignite. Industries inc. iron and steel, textiles, chemicals. Pop. (1988E) 7,184,943.

Lowestoft   Suffolk, England. 52 29N 1 45E. Port and resort in E. Anglia. Industries shipbuilding, fishing and canning, and the manu. of radar and electrical equipment. Lowestoft Ness is the most E point of Eng. Pop. (1990E) 58,000.

Loyada   Djibouti, East Africa. Village sit. SE of Djibouti on the S shore of the Gulf of Tadjoura. Main occupation fishing. Somali frontier control point.

Loyalty Islands   New Caledonia, South West Pacific. 21 00S 167 00E. A group of coral islands sit. E of New Caledonia. Area, 800 sq. m. (2,071 sq. km.). The principal islands are Lifou, Maré and Uvea. Copra is exported. Pop. (1983C) 15,510.

Lozère   France. 44 35N 3 30E. Dept. in Languedoc-Roussillon region. Area, 1,995 sq. m. (5,168 sq. km.). Mainly mountainous, containing the Cévennes, and rising in the N to over 5,000 ft. (1,524 metres) a.s.l. The principal Rs. are Allier, Lot and Tarn. Mainly agric., cattle and sheep being reared on the mountain slopes. Olives, vines, mulberries, cereals and fruits are grown. Cheese and silk are manu. Cap. Mende. Pop. (1982C) 74,294.

Lualaba River   Zaïre. 0 26N 25 20E. The W headstream of the R. Congo and rises in Katanga near the Zimbabwe border and flows for *c.* 1,100 m. (1,760 km.) and becomes R. Zaïre until reaching the Congo/Zaïre border then called R. Congo.

Luanda   Angola. 8 50S 13 15E. City sit. N of Cuanza R. estuary. Cap., administration centre and sea port, founded by Portuguese settlers in 1576. Industries inc. oil refining, processing cotton, sugar, tobacco, timber; manu. textiles, paper. Port exports coffee, sugar, maize, beeswax, cotton, palm products, diamonds, sisal. Pop. (1988E) 1·2m.

Luang Prabang   Luang Prabang, Laos. 19 45N 102 10E. Royal cap. sit. NNW of Vientiane at confluence of Mekong and Nam Khan Rs. R. port trading in catechu, benzoin, lac, teak, cloth. Industries inc. fishing, salt mining. Royal residence. Pop. (1984E) 44,244.

Luapula River   Zaïre/Zambia. Flows from the S end of L. Bangweulu and continues 350 m. (560 km.) S to join Chambezi R. in the swamps SE of L. Bangweulu and along the Zaïre/Zambia border NW and N to enter L. Mweru. Navigable from L. Mweru to Kasenga. Considered as the upper course of Luvua R. which is a headstream of Zaïre R.

Lubango   Huila, Angola. 14 55S 13 30E. Formerly Sá de Bandeira. Town sit. S of Benguela at the W edge of the central plateau. Railway terminus and trading centre of an agric. area, handling hides, skins, dairy produce, rice and flour; there are associated processing industries. Pop. (1970E) 31,674.

Lubbock   Texas, U.S.A. 33 35N 101 50W. City in NW of state. Centre for agric., medical care, transport and education, grain, cattle and poultry. Pop. (1990E) 194,148.

Lübeck   Schleswig-Holstein, Federal Republic of Germany. 53 52N 10 40E. Port sit. SW of Baltic Sea Coast and on Trave R. It is joined by canal to Elbe R. Industries shipbuilding, iron and steel works, manu. of machinery, textiles, wood and food products. There are many fine medieval buildings. Pop. (1984E) 213,400.

Lublin   Lublin, Poland. 51 15N 22 35E. Cap. of province in SE, sit. on Bystrzyca R. Industrial and transportation centre. Manu. aircraft, agric. machinery, textiles, and electrical and food products. Pop. (1982E) 314,000.

Lubombo   Southern Province, Zambia. Town sit. E of Mazabuka. Centre of an agric. area. Railway.

Lubumbashi   Shaba, Zaïre. 11 40S 27 28E. City sit. near the Zambia border, formerly called Elisabethville. Copper smelting and refining centre. Pop. (1976C) 451,322.

Lucca   Toscana, Italy. 43 50N 10 29E. Cap. of Lucca province near Ligurian Sea, and sit. on Serchio R. Manu. olive oil, silk, wines and macaroni. Pop. (1981E) 91,000.

Lucena   Córdoba, Spain. 37 24N 4 29W. Town sit. SSE of Córdoba. Manu. textiles, earthenware jars, olive oil, wines and brandy. Pop. (1981E) 30,000.

Lucerne   *see* **Luzern**

Luckenwalde   Potsdam, German Democratic Republic. 52 05N 13 10E. Town sit. SSE of Potsdam. Manu. textiles, metal products and paper.

Lucknow   Uttar Pradesh, India. 26 50N 80 52E. State cap. sit. on Gumti R. Industries inc. aeronautics, textiles, carpets, chemicals, paper, copper and brass goods. Pop. (1981C) 916,954.

Lüdescheid   North Rhine-Westphalia, Federal Republic of Germany. 51 13N 7 38E. Town sit. ESE of Wuppertal. Manu. aluminium and plastics. Pop. (1984E) 73,700.

Lüderitz   Namibia. 26 38S 15 10E. Town sit. W of Keetmanshoop on the Atlantic coast. Sea port serving a diamond mining area. The main industry is fishing and fish processing, esp. cray fish. Railway terminus and airfield.

Ludhiana   Punjab, India. Town sit. WSW of Shimla. An important grain market with manu. of engineering products, hosiery, textiles inc. silk, machinery, agric. tools and furniture. Pop. (1981C) 607,052.

Ludlow   Shropshire, England. 52 22N 2 43W. Town sit. near confluence of R. Teme and Corve, and not far from the Welsh border. Industries inc. the manu. of agric. machinery, clothing, and precision engineering. Pop. (1990E) 8,000.

Ludvika   Kopparberg, Sweden. Town sit. SW of Falun on L. Väsman. Railway junction and industrial centre. Industries inc. manu. electrical equipment, copper smelting, manu., metal working, sawmilling; bricks. Pop. (1989E) 29,324.

Ludwigsburg   Baden-Württemberg, Federal Republic of Germany. 48 53N 9 11E. Town sit. N of Stuttgart near Neckar R. Manu. pianos, organs, textiles and toys. Pop (1984E) 77,600.

Ludwigshafen   Rhineland-Palatinate, Federal Republic of Germany. 49 29N 8 26E. Town and R. port sit. on Rhine R. opposite Mannheim. An important centre of the chemicals industry with manu. of pharmaceutical products, fertilizers, plastics, dyes, machinery and textiles. Pop. (1989E) 162,000.

Lugano   Ticino, Switzerland. 46 01N 8 58E. Italian speaking resort in extreme S on N shore of L. Lugano. It is a popular tourist and financial centre. Pop. (1990E) 30,000.

Lugano, Lake   Italy/Switzerland. 46 00N 9 00E. Sit. between L. Maggiore and L. Como. Area, 19 sq. m. (49 sq. km.). It drains into L. Maggiore *via* Tresa R.

Lugansk   Ukraine, U.S.S.R. Cap. of Lugansk region sit. on Lugan R. N of Rostov and in the Donets Basin. An important industrial centre. Manu. diesel locomotives, coalmining equipment, motor car parts, machine tools, steel pipes and ball bearings.

Lugo   Galicia, Spain. 43 02N 7 35W. (i) A maritime province on the Bay of Biscay. Area, 3,784 sq. m. (9,803 sq. km.). The principal Rs. are Miño and Sil. The main occupations are fishing, agric. and lumbering. Pop. (1986C) 399,232. (ii) Cap. of province of same name sit. SE of Corunna on Miño R. Manu. textiles. Pop. (1986C) 77,728.

Lugovoya Kazakh Soviet Socialist Republic, U.S.S.R. Town sit. E of Dzhambul on the Turksib railway. Centre of an agric. wheat growing area. Industries inc. metal working.

Luik *see* **Liège**

Luleå Norrbotten, Sweden. 65 34N 22 10E. Cap. of county and port sit. at mouth of Lule R. on the Gulf of Bothnia, in NE Sweden. Manu. iron, steel and wood pulp. Exports inc. iron ore, timber and wood pulp. Pop. (1989E) 67,903.

Lule River Sweden/Norway. Rises SW of Narvik, Norway, and flows 280 m. (448 km.) SE through Stora Lule L. and past Boden to enter the Gulf of Bothnia at Luleå. Used for logging and power. Main trib. Lilla Lule R.

Luluabourg *see* **Kananga**

Lulworth Dorset, England. 50 36N 2 14W. Village sit. SW of Wareham on Eng. Channel. Resort in an agric. area noted for nearby Lulworth Cove, a deep inlet surrounded by high cliffs.

Lund Malmöhus, Sweden. 55 42N 13 11E. Town in SW. Industries inc. sugar refining, printing, publishing and the manu. of paper, woollen goods and furniture. Pop. (1983E) 80,458.

Lundy England. 51 10N 4 40W. An island sit. in the Bristol Channel NNW of Hartland Point. Area, $3\frac{1}{2}$ m. (6 km.) by 1 m. (1.6 km.). Mainly pasture land with botanically interesting flora providing a sanctuary for wild birds. There are two lighthouses.

Lüneburg Lower Saxony, Federal Republic of Germany. 51 15N 10 23E. Town in N, sit. on Ilmenau R. Industries inc. chemicals, paper, electronics, food processing, iron products, cement and clothing. There are fine medieval buildings. Pop. (1990E) 61,000.

Lüneburg Heath Lower Saxony, Federal Republic of Germany. 53 10N 10 20E. A low sandy district sit. S of Elbe R. mainly scrub land. Sheep are reared and potatoes and honey are produced.

Lünen North Rhine-Westphalia, Federal Republic of Germany. 51 36N 7 32E. Town NE of Düsseldorf sit. on Lippe R. Centre of a coalmining area with iron, glass and aluminium industries. Pop. (1984E) 84,400.

Lunéville Meurthe-et-Moselle, France. 48 36N 6 30E. Town ESE of Nancy sit. on Meurthe R. Manu. railway carriages, textiles and porcelain. Pop. (1982C) 23,231.

Lungi Northern Province, Sierra Leone. Town sit. N of Freetown on the Atlantic coast. Site of intern. airport.

Luqa Malta. Town sit. SSW of Valletta. Site of intern. airport. Industries inc. limestone quarrying.

Lurestan Iran. Province of SW Iran in the Zagros mountains and trad. territory of the Lur and Bakhtiari tribes. Area, 12,117 sq. m. (31,383 sq. km.). Main occupations farming, fruit growing. Chief towns Burujird, Khurramabad. Pop. (1976C) 932,297.

Lurgan Armagh, Northern Ireland. 54 28N 6 20W. Town sit. WSW of Belfast near L. Neagh and is now part of the new town of Craigavon. Industries inc. linen, optical goods, furniture and clothing. Pop. (1990E) 62,000 (Craigavon).

Lurín Lima, Peru. Town sit. SE of Lima. Railway terminus in an agric. area producing cotton, sugar-cane, vegetables.

Lusaka Zambia. 15 26S 28 20E. Cap. of the rep. sit. N of Kafue R. in S central Zambia. Founded in 1905 it is a commercial and transportation centre of an agric. area. Manu. tobacco products, vehicles, textiles and cement. Pop. (1987E) 818,994.

Lushnjë Albania. 40 56N 19 41E. Town sit. NW of Berat at the N edge of the Myzeqe plain. Centre of an agric. area. Pop. (1980E) 22,000.

Lushun-Talien *see* **Lü-ta**

Luso Aveiro, Portugal. Village sit. NNE of Coimbra. Spa with mineral springs.

Luton   Bedfordshire, England. 51 53N 0 25W. Town sit. on R. Lea. Industries inc. cars and vans, ball bearings, pumps, electrical products, instruments, hats. Pop. (1988E) 167,600.

Lutong   Brunei. Village sit. N of Miri on N coast. Oil port and refinery serving the oil fields at Seria and Miri.

Lutsk   Ukraine, U.S.S.R. 50 44N 25 20E. Town sit. NE of Lvov on Styr R. Industries inc. flour milling, tanning and the manu. of agric. machinery. Pop. (1970E) 96,000.

Lutterworth   Leicestershire, England. Town sit. NNE of Rugby. Industries inc. clothing and light engineering. Pop. (1981C) 6,673.

Luxembourg   West Europe. 50 00N 6 00E. Grand duchy bounded W by Belgium, S by France, E by Federal Republic of Germany. Area, 999 sq. m. (2,587 sq. km.). Cap. Luxembourg. Undulating country drained by Alzette and Sauer Rs. and heavily wooded. Mainly agric. with small farms producing potatoes, cereals, beets, grapes, livestock. The SW has important iron deposits, with major iron and steel production at Esch-sur-Alzette. Other industries inc. wine making, textiles, brewing, tanning; manu. metal goods, chemicals, explosives. The industrial centres are Differdange, Dudelange, Pétange and Luxembourg city. Tourism is important. Pop. (1989E) 377,100.

Luxembourg   Belgium. A province in the SE of the Ardennes. Area, 1,706 sq. m. (4,417 sq. km.). Mainly forested but agric. carried on in the fertile valleys. Iron ore is found in the S. Cap. Arlon. Pop. (1983E) 223,813.

Luxembourg   Luxembourg. 49 36N 6 09E. City sit. SE of Brussels, Belgium, on heights above Alzette R. Cap. and administration centre with industries in the suburbs, inc. food processing, canning, brewing; manu. metal goods, chemicals, machinery, tobacco products. Noted for the Cathedral and Grand Ducal palace, both 16th cent. Pop. (1988E) 76,600.

Luxor   Qena, Egypt. 25 41N 32 24E.

Town sit. SSW of Qena on E bank of Nile R. Centre of an agric. area; industries inc. pottery, sugar refining. Noted for the ruins of anc. Thebes inc. the temple built by Amenhotep III.

Luzern   Switzerland. 47 03N 8 18E. (i) Canton of central Switzerland. Area, 576 sq. m. (1,492 sq. km.). Cap. Luzern. Farming area heavily forested, drained by Reuss and Kleine Emme Rs. L. Luzern is a noted tourist area. Pop. (1988E) 312,211. (ii) City on Reuss R. on W shore of L. Luzern. Canton cap. and resort with some industry, inc. printing, brewing, chemicals; manu. metal goods. Pop. (1987E) 60,079.

Luzern, Lake   Switzerland. L. bordering on the cantons of Unterwalden, Uri, Schwyz, Luzern. Area, 44 sq. m. (114 sq. km.) at 1,424 ft. (434 metres) a.s.l. Maximum depth 702 ft. (214 metres). It has 3 arms: L. Küssnacht, L. Alpach, L. Uri. Fed by Muota, Sarner Aa and Engelberger Aa Rs. Neighbouring mountains inc. Rigi (N) and Pilatus (S). Resorts inc. Luzern, Küssnacht, Weggis, Vitznau, Gersau, Brunnen, Morschach, Flüelen, Seelisberg, Beckenried, Bürgenstock, Stansstad-Fürigen, Hergiswil.

Luzon   Philippines. 16 00N 121 00E. The largest, most N island in the arch. Area, 40,420 sq. m. (104,647 sq. km.). Mainly mountainous.

Lvov   Ukraine, U.S.S.R. 49 50N 24 00E. Cap. of the Lvov region. An important industrial and railway centre with industries inc. railway engineering, motor car assembling, oil refining. Also manu. chemicals, textiles, agric. machinery and glass. Pop. (1987E) 767,000.

Lyallpur   *see* **Faisalabad**

Lydd   Kent, England. 50 38N 4 16W. Town sit. E of Rye. The explosive, Lyddite, was first tested here. Industries inc. welding, concrete blocks and other small firms. Pop. (1990E) 5,000.

Lydda   *see* **Lod**

Lyme Regis   Dorset, England. 50

44N 2 57W. Resort sit. NW of Wey-mouth on Lyme Bay, with an anc. stone pier, the Cobb, which dates from the 13th cent. A geologically interesting region. Pop. (1981C) 3,447.

Lymington Hampshire, England. 50 46N 1 33W. Market town and yachting centre SW of Southampton sit. at mouth of the Lymington R. It runs a ferry service across the Solent to Yarmouth in the Isle of Wight. Pop. (1990E) 15,000.

Lynchburg Virginia, U.S.A. 37 24N 79 10W. City sit. W of Richmond on James R. Commercial, industrial and tourist centre. Also manu. shoes, textiles, paper, clothing, foundry and food products. Pop. (1980C) 66,743.

Lynn Massachusetts, U.S.A. 42 28N 70 57W. City and port NE of Boston sit. on Massachusetts Bay. A centre of the footwear industry since the 17th cent. Also manu. machinery, electrical equipment and equipment for the shoe trade. Pop. (1980C) 78,471.

Lynton Devon, England. 51 15N 3 50W. Town sit. E of Ilfracombe on a 400 ft. (122 metres) cliff above Lyn-mouth Harbour on the Bristol Chan-

nel. Market town and resort. Pop. (1989E) 2,000.

Lyons Rhône, France. 45 46N 4 50E. City sit. N of Marseilles at confluence of Rhône and Saône Rs. Prefecture, and 3rd largest French city. Industries inc. textiles, chemicals; manu. electrical and metallurgical products, vehicle assembly, tanning, distilling, printing, food processing. R. port and communications centre trading on Saône R. Cultural and financial centre with the oldest stock exchange in France (1506). Pop. (1982C) 418,476 (agglomeration, 1,220,844).

Lyonnais France. Area of E central France extending across the Rhône and Loire depts. and bounded E by the Saône–Rhône valley. Chief town Lyons.

Lytham St. Anne's Lancashire, England. 53 45N 2 57W. Town sit. W of Preston on Ribble R. estuary. Seaside resort with some light industry. Pop. (1982E) 39,641.

Lyubertsy Russian Soviet Federal Socialist Republic U.S.S.R. 55 41N 37 53E. Town sit. SE of Moscow. Manu. electrical equipment, machinery and plastics. Pop. (1987E) 162,000.

M

Ma'an Hejaz, Jordan. 30 12N 35 44E. Town sit. s of Amman at 3,497 ft. (1,066 metres) a.s.l. Railway terminus, road junction and airfield. Commercial centre trading in grain. Industries inc. manu. tobacco.

Maasluis Zuid-Holland, Netherlands. Railway and canal junction sit. w of Rotterdam on the New Waterway. Tug-boat and pilot service headquarters. There is some light industry. Pop. (1990E) 33,150.

Maastricht Limburg, Netherlands. 50 52N 5 43E. Cap. of province in sw sit. on Maas R. It is an important transportation and industrial centre. Manu. inc. glass, pottery, paper, wine and beer. Pop. (1990E) 115,000.

Mablethorpe and Sutton Lincolnshire, England. 53 21N 0 14E. Resort sit. N of Skegness on E coast. Pop. (1981C) 7,456.

McAllen Texas, U.S.A. 26 12N 98 15W. City and winter resort, sit. ssw of Corpus Christi near the border with Mexico. Culturally Spanish/ American. Main industry processing and shipping local fruit and vegetables. Pop. (1980C) 66,281.

Macao East Asia. 22 16N 113 35E. Chinese territory under Portuguese admin., forming an enclave in Guangdong Province, SE China. Area, 6 sq. m. (16 sq. km.). Cap. Macao city. Port and casino resort. The majority of exports are clothing and knitwear. Over 90% of the population live in the city of Macao. Pop. (1987E) 434,300.

Macapá Amapá, Brazil. 0 02N 51 03W. Cap. of federal territory in NE, sit. on the equator and on the N channel of Amazon R. delta. Manganese mining in vicinity. Trades in cattle and rubber. Pop. (1984E) 115,000.

Macas Santiago-Zamora, Ecuador.

2 19S 78 07W. Town sit. SE of Riobamba on the E slopes of the Andes. Army base and airfield; centre of a stock farming, lumbering and agric. area.

Macaulay Island New Zealand. Island of the Kermadec group sit. NE of Auckland. Mountainous, fertile soil. Area of the group, 13 sq. m. (34 sq. km.).

Macclesfield Cheshire, England. 53 16N 2 07W. Town in NW central Eng. sit. on R. Bollin. It is the principal centre of the silk industry. Other manu. inc. textile machinery, textiles, high technology, pharmaceutical products, paper and plastics. Pop. (1990E) 69,000.

Macdonnell Ranges Northern Territory, Australia. 23 45S 133 20E. Sit. in S and extending for about 400 m. (640 km.) along the tropic of Capricorn. Over 4,000 ft. (1,231 metres) a.s.l.

Macduff Grampian Region, Scotland. 57 40N 2 29W. Port and resort sit. at mouth of R. Deveron on Moray Firth. The main industries are tourism, fishing and boatbuilding. Pop. (1984E) 4,028.

Macedonia Greece. 40 39N 22 00E. Region of N Greece, bounded N by Yugoslavia and Bulgaria. Area, 13,206 sq. m. (34,203 sq. km.). Consists of 13 *nomoi*, Athos, Kavalla, Chalcidice, Drama, Imathia, Florina, Kilkis, Kastonia, Kozani, Pella, Pieria, Serrai, Thessaloniki. Mountainous, with the Balkan range (N) descending to fertile plains towards the Aegean coast. Mainly agric. and stock farming area; crops inc. grain, tobacco, olives, vines. Pop. (1981C) 2,121,953.

Macedonia Yugoslavia. 41 53N 21 40E. Constituent rep. bounded E by

Bulgaria, s by Greece. Area, 9,928 sq. m. (25,713 sq. km.). Cap. Skoplje. Sit. mainly in the Vardar basin with L. Okhrida and L. Prespa (sw). Mainly an agric. and dairy farming area, with some mining, esp. for chrome. Pop. (1981c) 1,912,257.

Maceió   Alagoas, Brazil. 9 40S 35 43W. State cap. in the NE sit. on the Atlantic coast. Industries inc. sugar refining, sawmilling, distilling, tanning and the manu. of textiles, tobacco products and soap. Cotton and sugar are exported through its port Jaraguá sit. to the E. Pop. (1980c) 375,771.

Macgillicuddy's Reeks   Kerry, Ireland. 51 59N 9 47W. A mountain range sit. to the SW of Killarney and extending for about 7 m. (11 km.) W from the Gap of Dunloe. The highest peak is Carrantuo-hill, 3,414 ft. (1,041 metres) a.s.l.

Machala   El Oro, Ecuador. 3 18S 79 54W. Town sit. s of Guayaquil and just NE of its port, Port Bolivar. Trading centre for an agric. area, handling cacao, coffee, hides. Pop. (1982c) 117,243.

Machilipatnam   Andhra Pradesh, India. Port ESE of Hyderabad sit. on N side of Krishna R. delta. Industries inc. carpet weaving, rice and cotton milling. Groundnuts and castor seeds are exported. Pop. (1981E) 138,530.

Machynlleth   Powys, Wales. 52 35N 3 51W. Town sit. E of Towyn. The main occupations are agric., tourism and forestry. Pop. (1989E) 1,969.

Maciás Nguema   *see* **Bioko**

Mackay   Queensland, Australia. 21 09S 149 12E. Port on the NE coast of the state. Industries inc. sugar growing and milling, tourism, coalmining, beef cattle and grain. Raw sugar, beef and grain are exported. Pop. (1989E) 56,000.

McKeesport   Pennsylvania, U.S.A. 40 21N 79 52W. City SE of Pittsburgh sit. on the Monongahela R. Industries inc. coalmining, iron and steel works, and the manu. of tinplate, pipes and tubes, boilers and heating equipment. Pop. (1980c) 31,012.

Mackenzie   Northwest Territories, Canada. 60 00N 144 30W. District N of 60° lat. extending N to the Arctic Ocean. Area, 527,490 sq. m. (1,364,620 sq. km.), of which 34,265 sq. m. (88,746 sq. km.) are fresh water. Forested (s) thinning out to tundra (N) with numerous Ls. and Rs. Main occupations lumbering, hunting, fishing, mining.

Mackenzie   Guyana. 6 00N 58 17W. A mining centre sit. on Demerara R. and s of Georgetown. Bauxite is exported. Pop. (1980E) 10,000.

Mackenzie River   Canada. 69 15N 134 08W. Issues from Great Slave L. in Mackenzie district and flows 1,000 m. (1,600 km.) NW through tundra and forest to enter the Arctic Ocean by a delta in Mackenzie Bay. Ice-free July–September near the mouth, otherwise navigable throughout the year. Its headstreams, above the L., are Finlay, Athabaska, Slave and Peace Rs. Main tribs. Lizard and Great Bear Rs.

McKinley, Mount   Alaska, U.S.A. 63 30N 150 00W. Sit. in the Alaska Range, it is the highest peak in N. America, 20,269 ft. (6,178 metres) a.s.l.

Mâcon   Saône-et-Loire, France. 46 18N 4 50E. Cap. of dept. and port in E central France sit. on Saône R. Manu. inc. agric. machinery, textiles, hardware and casks. Trades in brandy and wine. Pop. (1982c) 38,719.

Macon   Georgia, U.S.A. 32 50N 83 38W. City and port in the centre of state sit. on Ocmulgee R. Manu. inc. clothing, chemicals, paper and metal goods, bricks and tiles. Pop. (1980c) 116,896.

Macquarie Island   Tasmania, Australia. 54 36S 158 55E. An uninhabited volcanic island sit. SE of Tasmania in the s Pacific. Rocky rising to 1,420 ft. (433 metres). Area, 21 m. (34 km.) by 3 m. (5 km.).

Macquarie River   Australia. 30 05S

147 30E. Rises in the Great Dividing Range of New South Wales and flows 590 m. (944 km.) NW to join Darling R. above Brewarrina.

Macroom   Cork, Ireland. 53 54N 8 57W. Market town w of Cork sit. on Sullane R. Tourist centre for w Cork. Pop. (1981C) 2,493.

Madagascar   Indian Ocean. 20 20S 47 00E. Large island, the fourth largest in the world, in the Indian Ocean 250 m. (400 km.) off the SE coast of Africa across the Mozambique Channel. Area, 226,662 sq. m. (587,051 sq. km.). Republic since 1960. Cap. Antananarivo. Plateau at an average 4,000 ft. (1,219 metres) a.s.l. extends NNW to SSE down the centre. The climate is tropical with rains December–February reaching 60 ins. (152 cm.). Main occupation farming, with some associated industry, esp. meat-packing, also mining for mica, graphite. Pop. (1988E) 10,919,000.

Madang   Papua New Guinea. 5 15S 145 50E. Town sit. N of Port Moresby on Astrolabe Bay. Port, with a land-locked harbour, exporting gold, copra. Pop. (1980E) 21,335.

Maddalena Island   Sardegna, Italy. 41 13N 09 24E. The largest island in the arch. sit. off the NE coast. Area, 8 sq. m. (21 sq. km.). The main occupation is fishing. The chief town and port is La Maddalena.

Madeira   Portugal. 32 40N 16 45W. Island off NW Africa in the Atlantic Ocean forming with the island of Porto Santo and some smaller islands an Autonomous Region of Portugal. Area, 307 sq. m. (796 sq. km.). Cap. Funchal. Thickly grown with vines, bananas, sugar-cane and vegetables on the coastal lowlands with temperate zone crops on the higher ground and mountains in the centre covered in heathers and sparse pasture. Intensively cultivated by terracing and irrigation. The main product is madeira wine; the main industry is tourism. Pop. (1988E) 280,000.

Madeira River   South America. 3 22S 58 45W. Formed on the Bolivia/ Brazil frontier by confluence of Beni and Mamoré Rs. and flowing 2,000 m. (3,200 km.) NE to join Amazon R. near Manáos. Navigable.

Madhya Pradesh   India. 21 50N 81 00E. State of central India bounded NE by Uttar Pradesh, SW by Maharashtra. Area, 171,215 sq. m. (443,446 sq. km.). Chief cities Bhopal (cap.), Jabalpur, Gwalior, Indore. Plateau (N) descending to the Nagpur plain (S) and Chatisgarh plain (E) giving on to forest. Hot, dry, climate with 45 ins. (114 cm.) of rain annually, most of it June–September. Forests cover about 32% of the total area. Agric. areas produce cereals, oil-seeds, cotton. The most important minerals are manganese and iron ore, coal, dolomite, diamonds, bauxite, limestone. Industries inc. spinning and weaving cotton, newsprint, iron, steel, aluminium, cement, fertilizers, pharmaceuticals, electronics, vehicles, cable, electrical equipment, manganese mining, lumbering. Pop. (1981C) 52,178,844.

Madison   Wisconsin, U.S.A. 43 05N 89 22W. State cap. in the S of state sit. on an isthmus between L. Monona and Mendota. Centre of a rich agric. district. Government, commercial and industrial city. Pop. (1980C) 170,616.

Madiun   Java, Indonesia. Town sit. WSW of Surabaya. Industries inc. railway engineering. Trades in sugarcane, coffee and rice. Pop. (1980E) 150,502.

Madras State   *see* **Tamil Nadu**

Madras   Tamil Nadu, India. 11 00N 78 15E. City sit. SE of Bombay on the Coromandel coast of the Bay of Bengal. State cap. and third city of India. Communications, commercial and cultural centre. Seat of two Christian cathedrals and trad. burial-place of the apostle Thomas. Industries inc. textiles, esp. cotton, tanning, tyres, electronics, railway carriages, petrochemicals, engineering, chemicals; manu. glass, jewellery, railway coaches, clothing. Port with an artificial harbour exporting leather, cotton, hides, nuts, tobacco, wool, mica, magnesite. Pop. (1981C) 3,276,622.

Madre, Sierra   Mexico. 16 00N 93
00W. Mountain system extending
1,500 m. (2,400 km.) SE from the N
frontier in three main ranges enclosing the central Mexican plateau. The
Sierra Madre Occidental runs parallel
to the Gulf of California and the Pacific coast; the Sierra Madre Oriental
runs parallel to the Gulf of Mex.; the
Sierra Madre del Sur in S Mex. follows the Pacific Coast. The highest
peak is Pico de Orizaba, 18,700 ft.
(5,700 metres) a.s.l.

Madre de Dios   South America. 10
59S 66 08W. R. rising in the Caravaya
range, Peru, and flowing 870 m.
(1,392 km.) NW then NE into Bolivia,
to enter Beni R. at Riberalta.

Madrid   Spain. 40 25N 3 45W. Province in central Spain. Area, 3,089 sq.
m. (8,002 sq. km.). Mainly dry plateau
with the Sierra de Guadarrama on the
NW boundary and Tagus R. on the SE
boundary. The principal R. is Jarama.
Pop. (1986C) 4,854,616.

Madrid   Spain. 40 25N 3 45W. Cap.
of Spain and of province of same
name. Sit. in the centre of Spain it was
founded in the 9th cent. as a Moorish
fortress. It became cap. in 1561 but
because of its sit. it only grew with the
building of railways in the 19th cent.
Industries inc. electrical and electronic manu., aircraft, agric. machinery, food processing, tourism, leather
goods, clothing. Pop. (1986C)
3,123,713.

Madriz   Nicaragua. Dept. of the NW
bounded N by Honduras. Area, 530
sq. m. (1,373 sq. km.). Chief towns
Somoto (cap.), Telpaneca. Agric. and
stock farming area, producing coffee,
grain, tobacco, sugar, rice. Industries
inc. manu. straw hats and matting,
dairy products. Pop. (1981E) 72,408.

Madura   Indonesia. 7 00S 113 20E.
Island separated from NE Java by the
Surabaya Strait. Area, 1,762 sq. m.
(4,563 sq. km.). Mainly hilly with salt
mines. The chief products are cattle,
rice, maize and cassava.

Madurai   Tamil Nadu, India. 9 55N
78 10E. Town in SE sit. on Vaigai R.
Manu. inc. textiles and brassware.
Pop. (1981C) 820,891.

Maebashi   Honshu, Japan. 36 23N
139 04E. Town sit. NW of Tokyo.
Noted for its silk industry. Pop.
(1983E) 271,000.

Maesteg   Mid Glamorgan, Wales.
Town sit. N of Bridgend. The main industry is coalmining with some heavy
and light engineering. Manu. clothing,
cosmetics, car components, high technology. Pop. (1981C) 22,780.

Maewo   Vanuatu, South West Pacific. 15 10S 168 10E. Island sit. E of
Espiritu Santo. Area, 135 sq. m. (350
sq. km.). Volcanic, with central
mountains.

Mafeking   *see* **Mafikeng**

Mafeteng   Mafeteng, Lesotho. Town
sit. SSE of Maseru. Centre of a farming
area on a main road. Pop. (1980E)
3,500.

Mafikeng   Bophuthatswana. 25 53S
25 39E. Town sit. in the central N near
the Botswana frontier. Industries inc.
railway engineering.

Magadan   Russian Soviet Federal
Socialist Republic, U.S.S.R. 59 34N
150 48E. Port in E sit. on the Sea of
Okhotsk in Khabarovsk Territory.
Industries inc. ship repairing, fishing
and the manu. of mining equipment.

Magadi, Lake   Rift Valley, Kenya.
1 52S 36 17E. Sit. SW of Nairobi.
Area, 240 sq. m. (621 sq. km.). Carbonate of soda is found in the lake and
is exploited commercially. Magadi
town is sit. on the E shore.

Magallanes   Chile. Region of S
Chile extending to Cape Horn. Area,
51,364 sq. m. (133,033 sq. km.) inc.
numerous islands of Tierra del Fuego
and N of the Strait of Magellan. Cap.
Punto Arenas. Bleak, mountainous
and cold with frequent fogs and rainfall up to 200 ins. (508 cm.) annually.
Some small-scale farming, lumbering
and fishing. Industries at Punto
Arenas inc. meat packing, tanning,
sawmilling, fish processing. Deposits
of coal, oil and gold. Pop. (1982C)
132,333.

Magdalena   Colombia. Dept. of N

Colombia bounded N by Caribbean, W by Magdalena R. and E by Cordillera Oriental, forming the border with Venezuela. Area, 20,819 sq. m. (53,921 sq. km.). Chief towns Santa Marta (cap.), Ciénaga. Rises abruptly on the N coast to the Sierra Nevada de Santa Marta massif, highest point Cristóbal Colón, 18,950 ft. (5,776 metres). Marshy lowland in the Magdalena R. basin (W), rise towards the E Cordillera. Agric. areas produce bananas, cotton, maize, rice, beans, coffee. Coastal forests produce mahogany, cedar, rubber, balata, kapok, balsam, medicinal plants. Fishing and fish processing are the main industries of Santa Marta. Pop. (1983E) 610,100.

Magdalena River Colombia. 11 06N 74 51W. Rises SW of Garzón, flowing *c.* 1,000 m. (1,600 km.) N along the E side of the central Cordillera past Neiva and Honda through a swampy lower course to enter the Caribbean by a delta below Barranquilla. Principal R. of Colombia used extensively for traffic and navigable for large vessels for 600 m. (960 km.) up to Honda. Main trib. Cauca R.

Magdalen Islands Quebec, Canada. 47 20N 61 50W. A group sit. in Gulf of St. Lawrence. Area, 102 sq. m. (264 sq. km.). The main occupation is fishing.

Magdeburg German Democratic Republic. 52 07N 11 38E. (i) District bounded on W by the Federal Republic of Germany. Area, 4,450 sq. m. (11,525 sq. km.). Pop. (1983E) 1,392,000. (ii) Cap. of district of same name in central W sit. on Elbe R. Industries inc. sugar refining using local sugar-beet, engineering, paper milling and the manu. of textiles, chemicals and glass. Pop. (1988E) 290,579.

Magellan, Strait of Chile. 54 00S 71 00W. Sit. between the mainland and Tierra del Fuego and connecting the S. Pacific Ocean with the S. Atlantic Ocean.

Maggiore, Lake Lombardia, Italy. 46 00N 8 40E. Mainly sit. in Lombardia but with its N end in Ticino, Switz. Area, 82 sq. m. (212 sq. km.). The Ticino R. traverses it and the Borromen Islands are sit. in the SW part.

Maghreb The states of Algeria, Morocco and Tunisia.

Magnessia Greece. 39 24N 22 46E. *Nome* of SE Thessaly bounded SW by the Orthrys mountains and inc. the Northern Sporades Islands except Skyros. Area, 1,018 sq. m. (2,636 sq. km.). Cap. Volos. Farming area producing olives, citrus fruit, cereals, tobacco, almonds, livestock. Fisheries are important on the Gulf of Volos. Pop. (1981C) 188,222.

Magnitnaya, Mount Russian Soviet Federal Socialist Republic, U.S.S.R. Sit. in the S Ural Mountains, 2,000 ft. (610 metres) a.s.l.

Magnitogorsk Russian Soviet Federal Socialist Republic, U.S.S.R. 53 27N 59 04E. Town SW of Chelyabinsk sit. on Ural R. It is an important metallurgical centre using magnetite iron ore from nearby Mount Magnitnaya. Other manu. inc. machinery, cement, chemicals and clothing. Pop. (1987E) 430,000.

Mahajunga Madagascar. 15 40S 46 20E. Formerly Majunga. Cap. of province of same name. Port on NW coast sit. at mouth of Betsiboka R. Industries inc. meat packing, rice milling and the manu. of soap and cement. Exports inc. coffee, sugar, rice and vanilla. Pop. (1980E) 80,881.

Mahalla el Kubra Egypt. 30 59N 31 10E. Town N of Cairo sit. on Nile R. delta. Manu. and trades in cotton. There are also rice and flour mills. Pop. (1986C) 385,300.

Mahanadi River 20 00N 86 25E. Rises S of Raipur and flows 520 m. (832 km.) NE and E through Madhya Pradesh then S and SE through Orissa to enter the Bay of Bengal by a delta below Cuttack.

Maharashtra India. 19 30N 75 30E. State of W India bounded W by the Arabian Sea. Area, 118,799 sq. m. (307,690 sq. km.). Chief cities Bombay (cap.), Nagpur, Sholapur, Kolhapur, Pune, Thane, Aurangabad, Nashik. Coastal lowland rising to the interior plateau. Drained by Ghodavari, Bhima and Kistna Rs. E to the

Bay of Bengal. Agric. areas produce rice, wheat, jowar, bajra, sugar-cane, oilseeds, cotton and livestock, esp. goats, buffalo and sheep. Minerals inc. iron, coal, manganese, bauxite, limestone. There are vast reserves of oil and gas. Fishing and lumbering are important; other industries inc. manganese mining and diverse manu. industries mainly centred on Bombay; textiles, paper, pharmaceuticals, cosmetics, oil refining, fertilizers, automobiles, tyres, petrochemicals, copper wire, film making, pharmaceuticals, chemicals, engineering, food processing, sugar refining. Pop. (1981c) 62,784,171.

Mahé Seychelles, Indian Ocean. 4 40S 55 28E. Island of the Seychelles group sit. NE of Madagascar. Area, 56 sq. m. (145 sq. km.). Cap. Victoria, also Seychelles cap. Granite hills rise to 2,993 ft. (912 metres) a.s.l. and descend again to a coastal lowland strip. Main products coconuts, vanilla, cinnamon, patchouli, palmarosa. Pop. (1980E) 46,000 with dependencies.

Maiana Kiribati, W Pacific. Atoll in the N Gilbert group. Area, 10.3 sq. m. (27 sq. km.). The main product is copra. Pop. (1985E) 2,141.

Maidenhead Berkshire, England. 51 32N 0 44W. Town W of London sit. on R. Thames. Industries inc. high technology, computer software, pharmaceuticals, light engineering and printing. Pop. (1983E) 48,473.

Maidstone Kent, England. 51 17N 0 32E. Town in SE sit. on R. Medway. Industries inc. brewing, paper making, fruit canning, printing, confectionery and cement. Pop. (1986E) 72,000.

Maiduguri Borno, Nigeria. 11 51N 13 10E. Town in the NE. Trades in groundnuts, cotton and gum arabic. Pop. (1975E) 189,000.

Maikop Russian Soviet Federal Socialist Republic, U.S.S.R. 44 ,35N 40 07E. Cap. of the Adygei Autonomous Region sit. SE of Krasnodar on Belaya R. Manu. inc. furniture, leather, food and tobacco products. Pop. (1980E) 128,000.

Maine U.S.A. 45 20N 69 00W. State in New England bounded SE by Atlantic, N, E, W by Canada, SW by New Hampshire. Area, 33,215 sq. m. (86,027 sq. km.). Chief cities Augusta (cap.), Portland, Lewiston, Bangor. Crossed SW to NE by the Appalachian range, descending E to coastal lowlands, W to a hilly region with swampy valleys. Drained by St. John, Penobscot, Kennebec and Androscoggin Rs. Main industries are lumbering, fishing, paper making, tourism, farming. Main crops are potatoes, hay, oats, apples, blueberries. Pop. (1986E) 1,174,000.

Maine-et-Loire France. 47 31N 0 30W. Dept. in Pays-de-la-Loire region. Area, 2,753 sq. m. (7,131 sq. km.). Mainly low-lying with fertile plains. The principal Rs. are Loire, Mayenne, Oudon, Sarthe and Authion. The chief crops are vines, fruit, cereals, beet and hemp. Coal is mined near Chalonnes. Cap. Angers. Pop. (1982c) 675,321.

Main River Germany. 50 00N 8 18E. Rises in the Fichtelgebirge and flows 300 m. (480 km.) NW past Aschaffenburg, Würzburg and Frankfurt to enter Rhine R. at Mainz. Linked with Danube by canal between Bamberg and Kehlheim.

Mainz Rhineland-Palatinate, Federal Republic of Germany. 50 01N 8 16E. Cap. of province and port sit. on Rhine R. opposite influx of Main R. Manu. inc. machinery, chemicals, textiles and furniture. Trades in wine, grain and timber. Pop. (1984E) 187,100.

Maio Cape Verde, Atlantic. 15 15N 23 10W. Island between Boa Vista (NNE) and São Tiago (SW). Area, 104 sq. m. (269 sq. km.). Chief town Pôrto Inglês. Low sandy coastline rises to a hilly interior, with Monte Penoso, 1,430 ft. (432 metres) a.s.l. Main occupations salt panning, farming. Pop. (1980c) 4,103.

Maisons-Alfort Val-de-Marne, France. 48 48N 2 26E. A suburb of Paris. Manu. inc. soap, furniture and cement. Pop. (1982c) 51,591.

Maitland New South Wales, Austra-

lia. 32 44S 151 33E. Town in central E
of state sit. on Hunter R. Industries
inc. coalmining, railway engineering,
and the manu. of textiles, pottery,
bricks and tiles. Some of the world's
richest coal seams are sit. to the S.

Majorca   Spain. 39 30N 3 00E.
Largest island of the Balearic group
sit. N of Algiers. Area, 1,300 sq. m.
(3,367 sq. km.). Cap. Palma. Mountainous in the NW, rising to Puig
Mayor, 4,740 ft. (1,445 metres) a.s.l.
A central plain rises again to hills
(SE). Main industries tourism and
farming, pottery; manu. leather products. Pop. (1990E) 582,000.

Majunga   *see* **Mahajunga**

Makassar   *see* **Ujung Padang**

Makeni   Sierra Leone. 8 53N 12
03W. Cap. of Northern Province. Pop.
(1988E) 12,000.

Makeyevka   Ukraine, U.S.S.R. 48
02N 37 58E. Town sit. E of Donetsk
in the Donets basin. It is an important
iron and steel centre, also concerned
with coalmining and coking plants.
Pop. (1987E) 455,000.

Makhachkala   Russian Soviet Federal Socialist Republic, U.S.S.R. 42
58N 47 30E. Cap. of the Dagestan
Autonomous Soviet Socialist Rep.
and port sit. on the W shore of the
Caspian Sea. Industries inc. oil refining, by means of the pipeline to the
Grozny oilfield, shipbuilding, railway
engineering, fish canning and food
processing. Also manu. textiles and
footwear. Pop. (1987E) 320,000.

Makin   Kiribati, W Pacific. 3 07N
172 48E. Atoll of the N Gilbert group.
Area, 4.5 sq. m. (12 sq. km.). The main
product is copra. Pop. (1980E) 1,419.

Makó   Csongrád, Hungary. 46 13N
20 29E. Town sit. E of Szeged on
Maros R. Trades in cereals and vegetables, esp. onions. Pop. (1984E)
30,000.

Makran   Baluchistan, Pakistan.
District of S Baluchistan bounded S by
Arabian Sea, E by Kalat. Area, 26,000
sq. m. (67,340 sq. km.). Chief town

Gwadar. Mountainous and barren.
Main occupations fishing, fish processing, farming. Exports inc. salt
fish, dates.

Malabar   Kerala, India. 11 00N 75
00E. District on the W coast. Cap.
Kozikhode. Kerala was formerly
known as the Malabar coast and was
the first part of India to make contact
with Europe through early explorers,
esp. Vasco de Gama.

Malabo   Equatorial Guinea. 3 45N
8 50E. Cap. and port of rep. sit on N
edge of the island province of Bioko.
Main occupation, export of cocoa,
coffee and timber. Pop. (1986E)
10,000.

Malacca   *see* **Melaka**

Malacca, Strait of   Indonesia/
Malaysia. 2 30N 101 20E. Sit. between Sumatra and Peninsular Malaysia, and joining the Indian Ocean with
the S. China Sea.

Málaga   Spain. 36 43N 4 23W. (i)
Province in the central S bordering on
the Mediterranean Sea. Area, 2,809
sq. m. (7,276 sq. km.). Mainly mountainous with a fertile coastal plain.
The principal R. is Guadalhorce. The
chief crops are vines, fruit, cotton and
olives. Pop. (1986C) 1,214,479. (ii)
Cap. of province of same name and
port sit. on the central S coast. Industries inc. sugar refining, brewing,
distilling, wine making, and the manu.
of textiles, pottery and tobacco products. Wine and fruit are exported.
Pop. (1986C) 595,264.

Malagasy Republic *see* **Madagascar.**

Malaita   Solomon Islands, South
Pacific. 9 00S 161 00E. Island sit. NE
of Guadalcanal. Area, 1,638 sq. m.
(4,243 sq. km.). Chief town Auki. The
main product is copra. Pop. (1984E)
74,036.

Malakoff   Hauts-de-Seine, France.
48 49N 2 19E. A suburb of Paris.
Manu. inc. textiles, pharmaceutical
goods, electrical equipment and musical instruments. Pop. (1982E) 35,000.

Malang   East Java, Indonesia. 7 59S

112 35E. Town sit. S of Surabaya on a plateau ringed by volcanoes. Manu. textiles. Trades in sugar, rice and coffee. Pop. (1980E) 511,780.

Mälar, Lake   Sweden. 59 30N 17 12E. Sit. in the SE on a narrow strait joined to the Baltic Sea. Area, 440 sq. m. (1,140 sq. km.).

Malatya   Malatya, Turkey. 38 22N 38 18E. Cap. of province in the W centre sit. near Euphrates R. Trades in grain and opium. Pop. (1980C) 179,074.

Malawi   Central Africa. 13 00S 34 00E. Republic bounded NE by L. Malawi, W by Zambia, S and E by Mozambique. Area, 36,325 sq. m. (94,082 sq. km.). Chief cities, Zomba (cap.), Blantyre, Limbe, Lilongwe. Sit. in a rift valley drained by Shiré R. flowing from L. Malawi, with plateaux on either side (W, SE and S) at 2,000–3,000 ft. (610–915 metres) a.s.l. and rising to 10,000 ft. (3,048 metres) in Mount Mlanje. Mainly agric. with about two-thirds under maize. Agric. accounts for 90% of exports. Other products inc. groundnuts, cotton, tobacco. Pop. (1985E) 7,058,800.

Malawi, Lake   Central Africa. 12 30S 34 30E. Formerly L. Nyasa. L. on the E border of Malawi at 1,500 ft. (457 metres) a.s.l. About 350 m. (560 km.) long and 40 m. (64 km.) wide, probably formed as a submerged section of the Great Rift Valley. Drained periodically by Shiré R. when the water level is high.

Malaya, States of   *see* **Peninsular Malaysia**

Malay Peninsula   East Asia. 5 00N 102 00E. Peninsula extending S from Thailand, which occupies the N end, to the island of Singapore, and mainly occupied by Peninsular Malaysia.

Malaysia   South East Asia. 5 00N 110 00E. State consisting of 13 constituent states covering most of the Malay peninsula, Sabah and Sarawak. Area, 127,318 sq. m. (329,752 sq. km.). Chief towns Kuala Lumpur (cap.), Ipoh, Penang, Port Klang,

Melaka, Kuching, Brunei. Mountainous, with thick forest and deep valleys and a hot, humid climate. Agric. areas produce rice, tea, palm products, coconuts. Forest areas produce rubber, timber. Mineral resources inc. tin, petroleum, iron, bauxite, gold, coal. Peninsular Malaysia comprises the states of Johor, Pihang, Negeri Sembilan, Selangor, Perak, Kedah, Perlis, Kelantan, Terengganu, Penang, Melaka. Pop. (1989E) 17,360,000.

Malden   Massachusetts, U.S.A. 42 26N 71 04W. Suburb of N Boston. Mainly residential. Industries inc. financial services and the manu. of food products, radio equipment, paint and chemicals. Pop. (1989E) 50,983.

Maldive, Republic of   Indian Ocean. 3 15N 73 00E. Independent republic since 1965 consisting of an island group sit. SW of Sri Lanka. Area, 115 sq. m. (298 sq. km.), consisting of 2,000 islands of which only 220 are inhabited. Chief settlement Malé. The islands are covered with coconut palms and yield millet and fruit as well as coconut produce. Fishing is important to the economy, especially dried bonito which is exported. Pop. (1988E) 200,000.

Maldon   Essex, England. 51 45N 0 40E. Port and riverside town in SE sit. on estuary of R. Blackwater. Industries inc. flourmilling, sawmilling, boatbuilding, engineering, brewing, sea-salt and oyster fishing. Pop. (1985E) 15,500.

Maldonado   Uruguay. 34 54S 54 57W. (i) Dept. of S Uruguay on N shore of Rio de la Plata estuary. Area, 1,587 sq. m. (4,110 sq. km.). Hilly and wooded area with a pleasant climate and many seaside resorts. Farming is important inland, esp. cattle, sheep, cereals, vines, sugar-beet, vegetables. Industries inc. tourism, flour milling, sugar refining, wine making. Pop. (1985E) 93,000. (ii) City sit. E of Montevideo near mouth of Rio de la Plata. Dept. cap. Trading centre for grain and wool. The main industry is seal hunting. Pop. (1985E) 33,000.

Malegaon   Maharashtra, India. 20 33N 74 32E. Town sit. NE of Bombay on Girnn R. Manu. textiles. Trades in grain. Pop. (1981C) 258,219.

Malé Republic of Maldives, Indian Ocean. 4 00N 73 28E. Atoll forming the central group of the Maldives sit. WSW of Sri Lanka. The main island is Malé Island, and is a trading and administration centre; seat of the Sultan's residence and the Maldive Assembly. Tourism is expanding. Main products inc. copra, palm mats, breadfruit, fish. Pop. (1988E) 46,334.

Mali West Africa. 15 00N 10 00W. Rep. bounded N by Algeria, E by Burkina Faso and Niger, S by Guinea and Côte d'Ivoire. Area, 478,832 sq. m. (1,240,142 sq. km.). Cap. Bamako. Desert and savannah; pop. concentrated along Senegal and Niger Rs. (S). Irrigated areas produce rice, cotton, millet. Industries concentrate mainly on food processing, with some cotton ginning and handicrafts. Chief exports are cotton, groundnuts, hides, gum arabic. Pop. (1988E) 7,784,000.

Malines *see* **Mechelen.**

Malin Head Donegal, Ireland. 55 18N 7 16W. Sit. at the N of Inishowen peninsula, in the NE 230 ft. (75 metres) a.s.l.

Mallaig Highland Region, Scotland. 57 00N 5 50W. Port on the W coast, sit. on the Sound of Sleat. There are steamer services to the Hebrides and Skye.

Mallow Cork, Ireland. 52 08N 8 39W. Resort in SW sit. on R. Blackwater. Industries inc. vegetable and milk processing, sugar refining, engineering, angling. Pop. (1981C) 6,572.

Malmédy *see* **Eupen and Malmédy**

Malmesbury Wiltshire, England. 51 36N 2 06W. Town sit. WNW of Swindon on R. Avon. Manu. electrical and lighting equipment. Pop. (1989E) 3,980.

Malmö Malmöhus, Sweden. 55 36N 13 00E. Cap. of county and port in SW sit. on The Sound. Industries inc. shipbuilding, engineering, food, printing,

publishing, and chemicals. Pop. (1988E) 231,575.

Malta Mediterranean Sea. 35 50N 14 30E. Independent Republic within the Commonwealth. Malta was held in turn by Phoenicians, Greeks, Carthaginians and Romans, and was conquered by Arabs in 870. The islands were finally annexed to the British Crown by the Treaty of Paris in 1814. Island group S of Sicily and consisting of Malta, Gozo and Comino. Area, 122 sq. m. (316 sq. km.). Chief town, Valletta (cap.). Rocky with shallow soil supporting cereals, vegetables. Industries inc. agric., fisheries, textiles and tourism. Manu. inc. footwear, clothing, rubber, chemicals and electrical machinery. Pop. (1988E) 349,014.

Maltby South Yorkshire, England. 53 25N 1 12W. Town sit. E of Rotherham. The main industry is coalmining. Pop. (1990E) 16,690.

Malton North Yorkshire, England. 54 09N 0 48W. Market town sit. NE of York on R. Derwent. Industries inc. the manu. of agric. implements, mineral water and other light industry. There is limestone quarrying in the vicinity. Pop. (1981C) 4,119.

Malvern Hereford and Worcester, England. 52 05N 2 21W. Town sit. SW of Worcester on the E slopes of the Malvern Hills. Spa and resort. Pop. (1980E) 32,000.

Malvern Hills England. 52 05N 2 21W. A range about 10 m. (16 km.) long running N to S and rising to 1,395 ft. (425 metres) a.s.l. at the Worcestershire Beacon.

Malvinas, Islas *see* **Falkland Islands**

Mamoundzou Mayotte, Indian Ocean. 12 47S 45 14E. Chief town (but not cap.) of this Fr. island depend. Pop. (1982E) 7,800.

Man Côte d'Ivoire. 7 24N 7 33W. Cap. of dept. in W, noted for wood and ivory carvings. Pop. (1975C) 48,521.

Man, Isle of *see* **Isle of Man**

Manabi   Ecuador. Province of W
Ecuador on the Pacific coast. Area,
7,602 sq. m. (19,689 sq. km.). Chief
towns, Portoviejo (cap.), Bahia de
Caráques, Manta, Jipijapa, Montecristi, Chone. Forested lowland
drained by the Portoviejo and Chone
Rs. Agric. areas yield coffee, cacao,
rice, sugar-cane, cotton, bananas. The
main industry is manu. Panama hats.
Pop. (1982E) 907,000.

Managua   Managua, Nicaragua. 12
06N 86 18W. Cap. of rep., founded
1858, and largest commercial and industrial centre sit. on SE shore of L.
Managua. Following an earthquake in
1931 and a fire in 1936, the town was
completely rebuilt. Again in 1972 the
town was devastated by a further
earthquake and evacuated. Manu. inc.
textiles, cigarettes, matches and cement. Pop. (1985E) 682,111.

Managua, Lake   Managua, Nicaragua. 12 18N 86 20W. Sit. in the W it is
38 m. (61 km.) long and drains into L.
Nicaragua *via* the Tipitapa R.

Manam   Papua New Guinea. Island
sit. NE of New Guinea. Area, 32 sq. m.
(83 sq. km.). Mountainous with an active volcano at 4,265 ft. (1,300
metres) a.s.l.

Manama   Bahrain. 26 12N 50 38E.
Cap. of the State of Bahrain and port
sit. at the N end of Bahrain island. It is
the Gulf's financial and banking
centre; there were in 1990 60 offshore
banking units and 54 banks had representative offices. It is also a centre for
natural pearls. Industries inc. boatbuilding, fishing and the manu. of
plastics and fibreglass. Just outside
the city boundary is an aluminium
smelter and electric cables are manu.
Pop. (1981C) 121,986.

Manasarowar Lake   Tibet, China.
30 40N 81 25E. Sit. in the W Himalayas at 15,000 ft. (4,572 metres) a.s.l.
and to the NNE of Gurla Mandhata
peak, it is a Hindu place of pilgrimage. Area, 200 sq. m. (518 sq. km.).

Manaus   Amazonas, Brazil. 3 08S
60 01W. State cap. and port sit. on
Negro R. near its confluence with
Amazon R. It is only accessible by

steamer and aeroplane. Industries inc.
electronics, consumer durables,
motorcycles, oil refining, jute and
sawmilling, and rubber processing.
Pop. (1980C) 613,068.

Mancha, La   Spain. Region of S
central Spain consisting mainly of
Ciudad Real province. Arid, with
steppe and salt wastes, on a plateau at
2,000 ft. (610 metres) a.s.l. Drained
by Rs. which disappear underground;
water is pumped by windmills. Main
products esparto grass, cereals, wine.

Manche   France. 49 10N 1 20W.
Dept. in Basse-Normandie region,
bordering on the English Channel.
Area, 2,296 sq. m. (5,947 sq. km.).
Mainly agric. concerned with dairy
farming. Crops inc. cereals, fruit and
vegetables. Cap. St. Lô. Pop. (1982C)
465,948.

Manchester   Greater Manchester,
England. 53 30N 2 15W. Metropolitan district sit. on Irwell R., centre of a
densely populated conurbation. Financial, commercial and industrial centre.
Industries inc. chemicals, engineering,
textiles, printing, publishing, paper
making; manu. rubber, food products,
electrical goods. Railway centre and
international airport. Seat of several
noted libraries, schools and art galleries and 2 universities. Pop. (1987E)
450,100.

Manchester   Connecticut, U.S.A.
41 47N 72 31W. Town sit. NNE of
New Haven on Hockanum R. Manu.
inc. electrical equipment, fabricated
metals, aircraft parts, machinery,
paper goods and textiles. Pop. (1989E)
51,100.

Manchester   New Hampshire,
U.S.A. 42 59N 71 28W. City in SE of
state sit. on Merrimack R. Manu. inc.
textile machinery, textiles, electrical
equipment, leather, food products and
footwear. Pop. (1980C) 90,936.

Manchester, Greater   *see* **Greater
Manchester**

Manchester Ship Canal   England.
53 19N 2 57W. Artificial channel dug
along S side of Mersey R. estuary
from Eastham *via* Ellesmere Port and

Runcorn to upper Mersey and Irwell Rs., linking Manchester with Irish Sea through Mersey R. estuary. Length 35 m. (56 km.).

Manchuria China. 42 00N 125 00E. Division of NE China bounded W by Mongolian Peoples' Republic, N and E by U.S.S.R. across the Argun, Amur and Ussuri Rs., SE by Korea, across the Tumen and Yalu Rs., S by the Yellow Sea. The term Manchuria is not used by the Chinese but is roughly identical with the provinces of Liaoning, Jilin and Heilonjiang.

Mandalay Upper Burma, Burma. 22 00N 96 05E. Town and R. port in central Burma sit. on Irrawaddy R. Manu. inc. silk, gold and silver ware, and matches. There are many temples and pagodas. Pop. (1983E) 500,000.

Mandsaur Madhya Pradesh, India. Town sit. NNW of Indore. Trades in grain, textiles and cotton. Pop. (1981E) 77,603.

Manfredonia Puglia, Italy. 41 38N 15 55E. Port in the SE sit. on the Gulf of Manfredonia. Industries inc. fishing and the manu. of leather goods and cement. Tourism is increasing and a hydraulics works has been developed.

Mangaia Cook Islands, S. Pacific. 21 55S 157 55W. Volcanic island sit. SE of Rarotonga. Area, 20 sq. m. (52 sq. km.), with a coral limestone cliff surrounding central hills. Main products fruit, copra. Pop. (1981C) 1,364.

Mangalore Karnataka, India. 12 52N 74 52E. Port sit. SSE of Bombay on the Malabar coast. Manu. inc. textiles, hosiery and tiles. Exports inc. coffee, cashew nuts and spices. Pop. (1981C) 193,699.

Mangla *see* **Chalna**

Mango Togo. 10 20N 0 30E. Town sit. NNW of Sokodé on Oti R. Trading centre for nuts, kapok, butter.

Manhattan New York, U.S.A. 40 46N 73 55W. District of New York city sit. mainly on Manhattan Island but extending to neighbouring islets, all linked by a complex network of road and rail bridges. Commercial, financial and cultural centre of New York; communications centre with important passenger and freight traffic by sea and rail. The best known streets and buildings of New York are mainly in Manhattan; Times Square, Broadway, Wall Street, Greenwich Village, Fifth Avenue, the Rockefeller Center and most of the tallest 'skyscrapers'. Pop. (1980C) 1,428,285 (boro.).

Manica Mozambique. Province of central Mozambique. Area, 19,792 sq. m. (51,261 sq. km.). Cap. Chimoio. Pop. (1982E) 666,848.

Manihiki *see* **Humphrey Island**

Manila Luzon, Philippines. 14 36N 120 59E. Cap. and port founded 1574, sit. in the SW on Manila Bay. Industries inc. sugar refining, coconut-oil processing, and the manu. of textiles and tobacco products. Exports inc. sugar and Manila hemp. (abacá). Pop. (1980C) 1,630,485; 5,925,884 (metropolitan area).

Manipur India. 24 30N 94 00E. State of NE India bounded E by Burma. Area, 8,620 sq. m. (22,327 sq. km.). Cap. Imphal. Plateau in the centre, at an average 2,500 ft. (762 metres); surrounded by the Manipur Hills and drained by Barak and Manipur Rs. The land descends (S) to a marshy L. district. Agric. areas produce bamboo, rice, mustard, sugarcane, tobacco, fruit. Some lumbering and silk production. Industries inc. sugar refining, spinning mills, television and cycle assembly. Pop. (1981C) 1,420,953.

Manisa Manisa, Turkey. 38 36N 27 26E. Cap. of province sit. NE of Izmir. Trades in grain, olives, raisins and tobacco.

Manitoba Canada. 53 30N 97 00W. Province of central Canada on the E edge of the central prairie, bounded NE by Hudson Bay, S by the U.S.A. Area, 246,512 sq. m. (638,466 sq. km.). Chief cities Winnipeg (cap.), Brandon, Portage la Prairie, St. Boniface. Undulating country rising towards hills in the W and with many Ls. and Rs., inc. Ls. Winnipeg, Manitoba, Winnipegosis, Moose, Southern Indian, Cedar, Island and Gods. The s prairie supports most of the pop., main products grain, hay, clover, potatoes. Mining, fishing and lumbering (50% of land area is wooded) are important in the N. Grain is a major export. Pop. (1988E) 1,084,000.

Manitoba, Lake Manitoba, Canada. 51 00N 98 45W. Sit. SW of L. Winnipeg into which it drains *via* Dauphin R. Area, 1,817 sq. m. (4,706 sq. km.).

Manitoulin Islands Canada/U.S.A. 45 50N 82 20W. Sit. in L. Huron in Ontario, Canada, and Michigan, U.S.A. The largest island is Great Manitoulin, 80 m. (128 km.) in length, which is the largest freshwater L. island in the world. The main occupations are farming, fishing and lumbering.

Manitowoc Wisconsin, U.S.A. 44 06N 87 40W. Port sit. at confluence of Manitowoc R. with L. Michigan. Industries inc. yacht building and the manu. of ice machines, textile needles, machinery, aluminium ware, furniture and food products. Pop. (1980C) 32,547.

Manizales Caldas, Colombia. 5 05N 75 32W. Cap. of dept. in the central W, and in the Central Cordillera, sit. at 7,000 ft. (2,134 metres) a.s.l. It is the centre of a rich coffee producing district. Manu. inc. textiles, chemicals, leather goods, shoes and beer. Pop. (1980E) 252,000.

Mannar, Gulf of Indian Ocean. 8 30N 79 00E. Sit. between India and Sri Lanka. Noted for pearl fishing.

Mannheim Baden-Württemberg, Federal Republic of Germany. 49 29N 8 29E. R. Port in SW sit. at confluence of Rs. Rhine and Neckar opposite Ludwigshafen. Industries inc. railway engineering, and the manu. of motorcars, agric. machinery, chemicals, textiles, electrical equipment and tobacco products. Trades in coal and grain. Pop. (1984E) 297,200.

Manono Island Western Samoa, West Pacific. 13 50S 172 05W. Island in a strait extending 10 m. (16 km.) between Upolu and Savaii. Area, 1 sq. m. (2.59 sq. km.), rising to 230 ft. (71 metres) a.s.l.

Manora Karachi, Pakistan. Headland protecting the s entrance to Karachi harbour.

Manresa Barcelona, Spain. 41 44N 1 50E. Town sit. NNW of Barcelona on Cardoner R. Manu. inc. textiles, paper and leather goods.

Mansfield Nottinghamshire, England. 53 09N 1 11W. Town sit. N of Nottingham on Maun R. Industries inc. textiles, coalmining, engineering, electronics, brewing and printing. Pop. (1988E) 67,880

Mansfield Ohio, U.S.A. 40 46N 82 31W. City sit. SW of Cleveland. Manu. inc. steel goods, electrical equipment and tyres. Pop. (1980C) 53,927.

Mansûra Daqahlîya, Egypt. 30 08N 31 04E. Cap. of governorate sit. on Damietta branch of Nile R. Manu. cottons and linens Pop. (1986E) 357,800.

Mantua Lombardia, Italy. 45 09N 10 48E. Cap. of Mantua province in central N sit. on Mincio R. which encloses it on three sides. Industries inc. sugar refining, printing, brewing and tanning. There are many fine medieval and renaissance buildings. Pop. (1981E) 61,000.

Manu'a Islands American Samoa, South Pacific. 14 13S 169 35W. Group consisting of Tau, Ofu and Olosega Islands, trad. the original home of the Samoan people.

Manuae Cook Islands, s Pacific. 19 21S 158 56W. Atoll sit. NE of Rarotonga, consisting of Manuae (W) and Te-Au-o-Tu (E) islets, joined by a coral reef. Area, 2.4 sq. m. (6.2 sq. km.). The main product is copra. Pop. (1981C) 12.

Manus Papua New Guinea. Islands in the Bismarck Arch. sit. NW of New Britain. Area, 633 sq. m. (1,639 sq. km.). Volcanic, rising to 3,000 ft. (914 metres) a.s.l. The main product is copra.

Manych Depression U.S.S.R. Broad valley in the s Russian Soviet Federal Socialist Rep. extending *c.* 350 m. (560 km.). SE from lower Don R. to the Caspian Sea and drained by Western and Eastern Manych Rs.

Manzala, Lake Lower Egypt, Egypt. A coastal lagoon extending from Damietta branch of Nile R. to Suez Canal. Area, 660 sq. m. (1,709 sq. km.). A narrow spit separates it from the Mediterranean Sea.

Manzanillo Cuba. 20 21N 77 07W. Port sit. at head of Gulf of Guacanayabo on SE coast, and WNW of Santiago de Cuba. Industries inc. sugar refining, fish canning and sawmilling. Sugar, molasses and coffee are exported. Pop. (1984E) 26,000.

Maputo Mozambique. Province of s Mozambique. Area, 6,480 sq. m. (16,783 sq. km.). Pop. (1982E) 1,297,000.

Maputo Mozambique. 25 58S 32 32E. Cap. of rep. and province, port and resort, sit. NW of Maputo Bay. Its development began in 1895 and it became cap. in 1907. Joined by road and rail with Rep. of S. Africa and offering a good harbour. Manu. footwear, furniture and cement. Pop. (1986E) 882,814.

Maracaibo Zulia, Venezuela. 10 40N 71 37W. State cap. and port sit. on NW shore of L. Maracaibo. Exports inc. oil, sugar, coffee and cocoa. A channel for ocean-going vessels connects L. Maracaibo with the Caribbean Sea. Pop. (1981E) 890,553.

Maracaibo, Lake Zulia, Venezuela. 9 50N 71 30W. Sit. in NW and surrounded by rich oilfields. Area, 5,000 sq. m. (13,000 sq. km.).

Maracay Aragua, Venezuela. 10 15N 67 36W. State cap. in central N just NE of L. Valencia. Manu. textiles. Pop. (1981E) 388,000.

Maradi Niger. 13 29N 7 06E. Cap. of dept. in s, centre for groundnut and cotton producing district. Pop. (1983E) 65,100.

Maraisburg *see* **Roodepoort-Maraisburg**

Marajó Island Pará, Brazil. 1 00S 49 30W. Sit. in Amazon R. delta. Area, 18,519 sq. m. (47,964 sq. km.).

Marakei Kiribati, W Pacific. Atoll of the N Gilbert group, area 4 sq. m. (10 sq. km.). The main product is sponges. Pop. (1980E) 2,335.

Maramba Southern Province, Zambia. 17 50S 25 53E. Formerly Livingstone. Town near Victoria Falls and Zambezi R. Airport and tourist centre. Industries inc. motor vehicle assembly, clothing and blanket manu. Engineering and foundries. Pop. (1987E) 94,637

Maranhão Brazil. A maritime state in the NE bordered by the Atlantic Ocean. Area, 126,897 sq. m. (328,663 sq. km.). In the s and w it is hilly and the main occupation is stock rearing. The low coastal plain extending N is thickly forested but cotton, sugarcane, rice and maize are grown in the R. valleys. Cap. São Luís. Pop. (1980C) 3,996,404.

Marañón River Peru. 4 30S 73 35W. Rises in the Andes WNW of Cerro de Pasco and flows *c.* 1,000 m. (1,600 km.) NWW through the Andes towards the Ecuador frontier, then NE through the Pongo de Manseriche gap and E past Barranca and Nauta to join the Ucayali R. 55 m. (88 km.) SSW of Iquitos. Together they form Amazon R. Main tribs. Santiago, Morona, Pastaza, Tigre and Huallaga Rs. Navigable to Pongo de Manseriche.

Maraş Maraş, Turkey. 37 36N 36 55E. Cap. of province in s central Turkey sit. at the foot of Mount Taurus. Trades in wheat, cotton and carpets. Pop. (1980C) 178,557.

Marazion Cornwall, England. 50 08N 5 28W. Resort sit. E of Penzance on Mount's Bay. At low tide a causeway joins it to St. Michael's Mount. Pop. (1988E) 1,400.

Marburg Hessen, Federal Republic of Germany. 50 49N 8 46E. Town sit. N of Frankfurt on Lahn R. Manu. inc.

precision instruments, photographic equipment, pottery and soap. Pop. (1989E) 71,358.

March Cambridgeshire, England. 52 33N 0 05E. Town sit. N of Cambridge on Nene R. Pop. (1988E) 15,930.

Marches, The Italy. Regions of central Italy bounded E by the Adriatic Sea. Area, 3,744 sq. m. (9,697 sq. km.). Chief cities Ancona (cap.), Ascoli Piceno, Pésaro. Crossed NW to SE by the Apennines, with mountain valleys descending to a narrow coastal plain. Mainly an agric. and stock farming area with sulphur mining in the N. Main products wheat, maize, grapes, olives, fruit, vegetables, raw silk, livestock, fish, timber. Industries inc. paper making, shipbuilding, textiles, ceramics, engineering.

Mar del Plata Buenos Aires, Argentina. 38 01S 57 35W. Resort sit. between Buenos Aires and Bahía Blanca on the Atlantic Ocean. Industries inc. fish canning, meat packing, flour milling and tourism. Pop. (1984E) 350,000.

Mardin Mardin, Turkey. 37 18N 40 44E. Cap. of province in the SE, near the border with Syria. Manu. textiles. Trades in cereals and wool.

Maré New Caledonia, South West Pacific. 21 30S 168 00E. Island sit. S of Lifu in the Loyalty Islands. Area, 248 sq. m. (642 sq. km.). Coral, rising to 300 ft. (91 metres) a.s.l. The main products are copra, oranges. Pop. (1983C) 4,610.

Maree, Loch Scotland. 57 40N 5 30W. Sit. in the NW, to the NE of Loch Torridon, it drains into Loch Ewe *via* R. Ewe.

Maremma Tuscany, Italy. 42 45N 11 15E. District mainly sit. in Grosseto province and bounded W by the Tyrrhenian Sea. Area, *c.* 1,930 sq. m. (4,999 sq. km.). Former malarial swamps reclaimed as agric. land producing cereals, grapes, olives, fruit, livestock. Mineral deposits inc. copper, lignite, iron and mercury. Important source of boric acid.

Mareotis, Lake Egypt. A salt L. sit. in Nile R. delta with a narrow spit separating it from the Mediterranean Sea.

Margate Kent, England. 51 24N 1 24E. Resort in the SE sit. in the Isle of Thanet. Pop. (1981C) 53,280.

Mariana Islands West Pacific. 16 00N 145 30E. Group sit. E of the Philippines. Area, 391 sq. m. (1,012 sq. km.), comprising a chain of islands extending 500 m. (800 km.) N to S. Main islands are Agrihan, Aguiguan, Alamagan, Anatahan, Asuncion, Guam, Guguan, Maug, Medinilla, Pagan, Pajaros, Rota, Saipan, Sariguan, Tinian. Volcanic and mountainous, rising to 3,166 ft. (1,038 metres) a.s.l. on Agrihan. Main products sugar-cane, coffee, coconuts. Guam is a U.S. depend. with important U.S. bases. Other islands form the Commonwealth of Northern Mariana Islands, with U.N. Trust Territory. Area, 182 sq. m. (471 sq. km.). Cap. Saipan. Pop. (1980C) 16,780, N Marianas only.

Mariánské Lázně Západočeský, Czechoslovakia. 49 59N 12 43E. Spa in the W, sit. WNW of Pilsen.

Mari Autonomous Soviet Socialist Republic Russian Soviet Federal Socialist Republic, U.S.S.R. 56 30N 48 00E. Area, 8,955 sq. m. (23,193 sq. km.). Constituted an Autonomous Rep. 1936. Cap. Yoshkar-Ola, sit. ENE of Moscow. Chief industries are metal working, timber, paper, woodworking and food processing. Grain is the main crop occupying 70% of the cultivated land. There are minerals inc. coal at Pechora. Pop. (1989E) 750,000.

Maribor Slovenia, Yugoslavia. 46 33N 15 39E. Town in NW sit. on Drava R. Manu. inc. textiles, chemicals, footwear and leather goods. Trades in timber, grain and wine. Pop. (1981C) 185,699.

Marie-Galante Guadeloupe, West Indies. 15 56N 61 16W. Island sit. E of Basse-Terre. Area, 61 sq. m. (158 sq. km.). Chief town Grand-Bourg. Low-lying limestone rock. The main product is sugar-cane. Pop. (1982c) 13,750.

Mariinsk Canal   Russian Soviet Federal Socialist Republic, U.S.S.R. Canal extending 5 m. (8 km.) between Vytegra (NW) and Kovzha (SE) Rs. as part of the Mariinsk system joining Volga R. and Rybinsk Reservoir with Neva R.

Marion   Indiana, U.S.A. 40 33N 85 40W. City sit. SW of Fort Wayne on Mississinewa R. Manu. inc. vehicle bodies and parts, paper, glass and electrical equipment. Pop. (1988E) 34,980.

Marion   Ohio, U.S.A. City sit. N of Columbus. Manu. inc. farm, excavating and road construction machinery. Pop. (1980c) 37,040.

Maritime Provinces   Canada. The three provinces on the Atlantic coast, New Brunswick, Nova Scotia and Prince Edward Island.

Maritsa River   Bulgaria/Turkey/Greece. 40 52N 26 12E. Rises on Stalin Peak in the Rila range and flows 300 m. (480 km.) first E and SE past Pazardzhik, Plovdiv and Svilengrad to the frontier with Greece and Turkey near Edirne, Turkey. It then forms the Greek/Turkish frontier and flows S and SW to enter the Aegean Sea below Enez. Used for power and irrigation of a fertile valley. Main tribs. Bacha, Asenovitsa, Arda, Topolnitsa, Strema, Tundzha and Ergene Rs.

Market Drayton   Shropshire, England. 52 55N 2 30W. Town sit. NE of Shrewsbury on R. Tern. Manu. agric. implements and meat products. Pop. (1985E) 8,500.

Market Harborough   Leicestershire, England. 52 29N 0 55W. Town sit. N of Northampton on R. Welland. Manu. inc. general light industry, textiles, electrical equipment, rubber goods and brushes. Pop. (1987E) 16,258.

Markham   Ontario, Canada. 43 52N 79 16W. Town sit. NE of Toronto. Industries inc. flour milling, dairy and timber. Pop. (1981c) 81,932.

Markinch   Fife Region, Scotland. 56 12N 3 09W. Town sit. W of Leven on Leven R. Industries inc. coalmining, textiles, paper making. Pop. (1980E) 2,500.

Marlborough   Wiltshire, England. 51 26N 1 43W. Market town sit. SSE of Swindon on R. Kennet. Industries inc. engineering and tanning. Tourist centre. Pop. (1985E) 6,900.

Marlborough   South Island, New Zealand. 41 45S 173 33E. A district in the NE. Area, 4,220 sq. m. (10,930 sq. km.). Mainly agric. producing vineyards, sheep, grain and fruit. The chief town is Blenheim; the chief port, Picton. Pop. (1989E) 4,300.

Marlow   Buckinghamshire, England. 51 35N 0 48W. Town sit. S of High Wycombe on Thames R. Residential with some industry. Pop. (1984E) 14,000.

Marmara, Sea of   Turkey. 40 45N 28 15E. Sit. between Europe and Asia and connected with the Aegean Sea by the Dardanelles, and with the Black Sea by the Bosphorus.

Marne   France. 49 00N 4 10E. Dept. in the Champagne-Ardenne region. Area, 3,152 sq. m. (8,163 sq. km.). It is chalky and dry in the N producing champagne wines. The principal R. is Marne. Industries inc. iron and copper founding, brewing and tanning. Cap. Châlons-sur-Marne. Pop. (1982c) 543,627.

Marne River   France. 48 49N 2 24E. Rises S of Langres and flows 325 m. (520 km.) NW past Chaumont, Châlons-sur-Marne and Meaux to enter Seine R. at Charenton-le-Pont above Paris. Much of its course is in canals. The middle course waters the Champagne region. Main tribs. Blaise, Petit-Morin, Grand-Morin, Saulx and Ourcq.

Maroni River   French Guiana/Suriname. 5 45N 53 58W. Rises in the Tumuc-Humac range of French Guiana, near the Brazilian frontier, flowing *c.* 450 m. (720 km.) N along the French Guiana/Suriname frontier through tropical forest to enter the

Atlantic at Galibi Point. The upper course is called Itany or Litany R., the middle course is called Aoua R. Main trib Tapanahoni R.

Maroua   Cameroon. 10 36N 14 20E. Cap. of far N province. Agric. centre for cotton, millet, maize and ground-nuts. Textile industry. Pop. (1981E) 81,861.

Marple   Greater Manchester, England. 53 24N 2 03W. Town sit. E of Stockport. Pop. (1981C) 24,010.

Marquezas Islands   French Polynesia. 9 00S 139 30W. A volcanic group to the NE of Tahiti. Area, 492 sq. m. (1,274 sq. km.). The largest islands are Nukuhiva, Hiva-Oa and Uapou. The chief town is Taiohae, on Nukuhiva. The chief product is copra. Pop. (1983C) 6,548.

Marrakesh   Morocco. 31 38N 8 00W. Town in the W centre sit. at the foot of the High Atlas Mountains. Tourist centre. Manu. carpets and leather goods. Pop. (1982C) 439,728.

Marsala   Sicilia, Italy. 37 48N 12 26E. Port in the W. Industries inc. fishing and the manu. of bottles. Trades in Marsala wine, olive oil and agric. produce. Pop. (1980E) 85,000.

Marseille   Bouches-du-Rhône, France. 43 18N 5 24E. City sit. on the Mediterranean. Major port exporting wine, liqueurs, olive oil, soap, cement products, ochre, sugar, metal goods, vegetables; imports inc. vines, olives, citrus fruit, spices, hides, oil seeds and nuts, mainly produce of North Africa. Industries inc. chemicals, smelting iron, bauxite and copper, shipbuilding, engineering, oil refining, sugar refining, soap, brick, tile, glass, cigarette and match manu. Pop. (1982C) 878,689 (agglomeration, 1,110,511).

Marshall Islands   Pacific. 9 00N 168 00E. Sovereign state controlling own foreign policy, with the U.S. controlling defence. Arch. sit. N of Auckland, New Zealand. Area, 70 sq. m. (181 sq. km.) consisting of 34 atolls and coral islands in parallel chains extending NW to SE and about 130 m. (208 km.) apart. Main islands are Kwajalein, Majuro (cap.), Jaluit. The chief product is copra, also breadfruit, papaya, arrowroot. Pop. (1988E) 40,069.

Marske-by-the-Sea   Cleveland, England. 54 35N 1 00W. Town sit. E of Middlesbrough on the North Sea coast. Dormitory town for Teesside area with some light engineering. Pop. (1990E) 9,050.

Martaban, Gulf of   Burma. 16 05N 96 30E. An inlet of the Indian Ocean to the E of the Irrawaddy R. delta. The Rs. Salween and Sittang drain into it.

Martha's Vineyard   Massachusetts, U.S.A. 41 25N 70 40W. An island and resort off the SE coast of Cape Cod. Pop. (1989E) 12,000, in the summer can reach 90,000.

Martigues   Bouches-du-Rhône, France. 43 24N 5 04E. Port at w end of Étang de Berre, site of oil refineries and indust. plants. Pop. (1982C) 42,039.

Martina Franca   Puglia, Italy. 40 42N 17 21E. Town sit. NNE of Taranto. Wine and olive oil are produced. Manu. inc. hosiery. Pop. (1980E) 41,000.

Martinique   West Indies. 14 40N 61 00W. Island between Dominica (N) and St. Lucia (S) in the Windward Islands, forming an overseas dept. of France. Martinique has been in French possession since 1635, except during the Seven Years' War (1762–3) and the French Revolution and Empire (1794–1802, 1809–1815) when it was under British occupation. Area, 417 sq. m. (1,079 sq. km.). Cap. Fort-de-France. Tropical, humid climate with high rainfall. Volcanic and mountainous, rising to 4,429 ft. (1,094 metres) a.s.l. in Mont Pelée, and thickly forested (N). Most cultivable land is under sugar-cane. Chief products sugar, rum, coffee, cacao, tobacco, cotton, fruit. Pop. (1988E) 336,000.

Martos   Jaén, Spain. 37 43N 3 58W. Town sit. WSW of Jaén. Manu. inc. textiles, pottery, soap and cement. Trades in wine and olive oil.

Maruy   Turkmenistan, U.S.S.R. Town in an oasis ESE of Ashkhabad sit. on Murghab R. Manu. inc. cotton, carpets, beer and food products.

Maryborough   Queensland, Australia. 25 35S 152 40E. Port in SE of state sit. on Mary R. Industries inc. tourism, railway, heavy engineering, sawmilling, sugar milling, timber and market garden crops. Pop. (1989E) 22,900.

Maryland   U.S.A. 39 10N 76 40W. State on the Atlantic coast bounded N by Pennsylvania, S and W by West Virginia, Virginia and the District of Columbia, E by the Atlantic and Delaware. Area, 10,577 sq. m. (27,394 sq. km.). Chief cities Annapolis (cap.), Baltimore, Dundalk. The N boundary (the Mason-Dixon line *see* below) extends *c.* 200 m. (320 km.) W to E and the state extends S from it for widely varying distances, 4 m. (7 km.) in the middle W, 125 m. (200 km.) in the E, to an irregular S boundary on the Potomac R. Mainly coastal plain divided N to S by Chesapeake Bay, with its many inlets, and rising W to the Appalachian Mountains. Main products livestock, dairy products, grain and soybeans, coal, tobacco, fish esp. shell-fish. The main industries are food processing and canning, with most manu. industry in Baltimore, inc. iron and steel, engineering, electrical and transport equipment, chemicals, textiles. Coalmining is in the W, esp. at Cumberland. Pop. (1980C) 4,216,975.

Maryport   Cumbria, England. 54 43N 3 30W. Port sit. WSW of Carlisle at the mouth of R. Ellen on Solway Firth. Industries inc. shoemaking and soft drinks, and the manu. of chemicals. Pop. (1973E) 11,560.

Masaka   Buganda, Uganda. 0 21N 31 45E. Town sit. SW of Kampala. Trading centre for an agric. area, handling cotton, coffee, bananas, millet, maize. The main industry is coffee processing. Pop. (1980E) 15,000.

Masaya   Masaya, Nicaragua. 11 59N 86 06W. Cap. of dept. sit. between L. Managua and L. Nicaragua. It is the centre of a rich agric. district producing cotton, coffee and tobacco. The main products are Indian handicrafts, footwear and cigars. Pop. (1984E) 38,000.

Masbate   Philippines. 12 10N 123 30E. An island sit. between Luzon and Negros. Area, 1,562 sq. m. (4,046 sq. km.). It was formerly a gold mining district. Coconuts, rice, corn and hemp are grown.

Maseru   Lesotho. 29 19S 27 29E. Cap. of kingdom sit. just inside the border and E of Bloemfontein, Rep. of S. Africa, and near Caledon R. There is light industry and manu. inc. carpets, pottery and sheepskin products, diamond processing and there is some tourism. Pop. (1986C) 109,382.

Mashonaland   Zimbabwe. A region in the NE inhabited by the Mashona. The chief town is Harare.

Mashreq   The states of Egypt, Jordan and Syria.

Mason City   Iowa, U.S.A. 43 09N 93 12W. Sit. NNE of Des Moines. Industries inc. meat processing, printing, engineering and the manu. of ice-making machines, puddings and soft drinks, cement, bricks and tiles. Pop. (1980C) 30,144.

Mason-Dixon Line   U.S.A. Boundary between Maryland (S) and Pennsylvania (N). Surveyed 1763–7 by Charles Mason and Jeremiah Dixon. Extended 1779 between Pennsylvania and West Virginia. The 'slave states' before the Civil War were S of the line. It is still regarded as the N to S boundary.

Massachusetts   U.S.A. 42 25N 72 00W. State of New England bounded E by Atlantic, S by Connecticut and Rhode Island, W by New York. Area, 8,257 sq. m. (21,386 sq. km.). Chief cities Boston (cap.), Worcester, Springfield, New Bedford, Cambridge, Fall River, Lowell, Newton, Lynn. First settled in 1620 and was the 6th State to ratify the U.S. Constitution in 1788. Uplands (W) descend to depressions on the E coast. Industrialized state with woodland and mixed farming in rural areas, producing dairy products, hay, corn, oats and potatoes. Market gardening is important. Industries inc. electrical engineering, textiles, chemicals, paper making, shipbuilding, printing and publishing, engineering, fishing and fish-processing. About half the pop. is concentrated in the Boston-centred conurbation. Pop. (1985E) 5,819,087.

Massawa Eritrea, Ethiopia. 15 38N 39 28E. Port sit. partly on Massawa Island and partly on the mainland on the Red Sea coast. Exports inc. hides, coffee, oil seeds and cattle. Tourist centre.

Massillon Ohio, U.S.A. 40 48N 81 32W. City sit. s of Cleveland, on the Tuscarawas R. Industrial city in an agric. area. Industries inc. metal working, engineering and food processing. Pop. (1980C) 30,557.

Masterton North Island, New Zealand. 40 57S 175 39E. Town 50 m. (80 km.) ENE of Wellington. Industries inc. meat packing, woollen milling and dairy farming. Pop. (1980E) 20,000.

Masvingo Zimbabwe. 20 10S 30 49E. Town sit. in s central Zimbabwe. Industries inc. gold and asbestos mining. A tourist centre for Kyle Dam and Game Reserve, the Mushandike National Park and the Zimbabwe ruins. Pop. (1982C) 31,000.

Mat River Albania. Rises E of Klos and flows 60 m. (96 km.) NW then W past Burrel to enter the Adriatic in Drin Gulk just SSW of Lesh. Main trib. Fan R.

Matabeleland Zimbabwe. A region in the SW inhabited by the Matabele. The chief town is Bulawayo.

Matadi Bas-Zaïre, Zaïre. 5 49S 13 27E. Port in SW, near the border with Angola, sit. just below Livingstone Falls on Zaïre R. Exports inc. coffee, cacao, cotton and palm oil. Pop. (1976C) 162,396.

Matagalpa Matagalpa, Nicaragua. 12 53N 85 57W. Cap. of dept. sit. NNE of Managua. Industries inc. flour milling and coffee processing.

Matamoros Tamaulipas, Mexico. 25 53N 97 30W. Town sit. on Rio Grande opposite Brownsville in Texas, U.S.A., and E of -Monterrey. Industries inc. cotton ginning, distilling, tanning and the processing of vegetable oil. Pop. (1980C) 238,840.

Matanzas Matanzas, Cuba. 23 03N 81 35W. Cap. of province and port in the NW. Industries inc. sugar refining, tanning and the manu. of footwear and textiles. Sugar is exported. Pop. (1980E) 87,000.

Matapan, Cape Peloponnesus, Greece. Sit. at the point of the Matapan peninsula between the Gulfs of Laconia and Messenia.

Mataró Barcelona, Spain. 41 32N 2 27E. Port sit. NE of Barcelona. Manu. inc. textiles, hats, paper and soap. Trades in wine. Pop. (1990E) 100,000.

Matera Basilicata, Italy. 40 40N 16 37E. Cap. of Matera province sit. WNW of Taranto. Manu. inc. pottery and macaroni. Trades in cereals and olive oil.

Mathura Uttar Pradesh, India. 27 30N 77 41E. Town NW of Agra sit. on the Jumna R. It is an important Hindu religious centre and birthplace of Krishna. Industries inc. chemicals, textiles, oil refining and paper. Trades in cotton, grain and oilseeds. Pop. (1981C) 159,458.

Matlock Derbyshire, England. 53 08N 1 32W. Principal town in the Derbyshire Dales sit. N of Derby, a spa and former hydropathic resort with mineral springs. Tourism is main industry but textiles and high technology important. Pop. (1990E) 15,000.

Mato Grosso Brazil. 14 00S 55 00W. State of central Brazil, divided into 2 states in 1979, bounded w by Bolivia, SW and S by Paraguay. Area, 340,155 sq. m. (881,001 sq. km.). Cap. Cuiabá. Plateau, descending w and s to flood plains of Paraguay R. Drained N by tribs. of Amazon R. and s by Paraná and Paraguay Rs. Mainly a stock raising area with some agric. land producing sugar, rice, beans, tobacco.

The forests of the N yield timber, rubber and medicinal plants. Minerals are considerable; inc. gold, diamonds, manganese. Pop. (1980C) 1,138,691.

Mato Grosso do Sul   Brazil. A state created in 1979 from part of Mato Grosso. Cap. Campo Grande. Area, 135,347 sq. m. (350,548 sq. km.). Pop. (1980C) 1,369,567.

Matopo Hills   Zimbabwe. 20 36S 28 28E. A range about 50 m. (80 km.) in length sit. to the S of Bulawayo. Over 5,000 ft. (1,524 metrès) a.s.l.

Matozinhos   Porto, Portugal. 41 11N 8 42W. Port and resort NW of Oporto sit. at mouth of Leça R. Industries inc. fishing and fish canning. Pop. (1981C) 26,404.

Matrah   Oman. 23 38N 58 34E. Town sit. just W of Muscat. It is a terminus of caravan routes and trades in pearls, dates and fruit. Pop. (1980E) 15,000.

Matsuyama   Shikoku, Japan. 33 50N 132 45E. Cap. of Ehime prefecture and port in the NW. Industries inc. oil refining, fruit canning, and the manu. of agric. machinery, chemicals and textiles. Pop. (1990E) 442,717.

Matterhorn   Valais, Switzerland. 45 59N 7 43E. A peak in the Pennine Alps, on the border with Italy and just SW of Zermatt, 14,690 ft. (4,478 metres) a.s.l.

Maubeuge   Nord, France. 50 17N 3 58E. Industrial centre with steelworks, breweries, chemical works. Pop. (1982C) 36,156.

Mauchline   Strathclyde Region, Scotland. 55 31N 4 24W. Town sit. ENE of Ayr. Industries inc. agric. implements, curling stones, glass engraving, food products. Pop. (1985E) 3,500.

Mauke   (Formerly Parry Islands) Cook Islands, S Pacific. 20 09S 157 23W. Volcanic island sit. NE of Rarotonga, most E of the Cook group. Area, 7 sq. m. (18.4 sq. km.). Fertile, producing fruit, copra. Pop. (1981E) 681.

Maule   Chile. Region of S central

Chile, bounded N by Maule R., W by Pacific. Area, 11,783 sq. m. (30,518 sq. km.). Cap. Talea. Lowland rising to Andean foothills. Wine growing area, also producing wheat, maize, potatoes, lentils, peas. Forests produce valuable timber for building. Industries inc. lumbering, sawmilling, wine making, flour milling, shipbuilding. Pop. (1982C) 723,224.

Maupiti   Society Islands, South Pacific. 16 27S 152 15W. Island sit. W of Bora-Bora. Area, 26 sq. m. (67 sq. km.). Volcanic, with deposits of black basalt.

Mauritania   West Africa. 20 50N 10 00W. The Islamic Republic of Mauritania became independent in 1960, after having been a French protectorate (1903) and colony (1920). Bounded N by Algeria, N and NW by Western Sahara, E and S by Mali, and SW by Senegal and the Atlantic Ocean. Area, 398,000 sq. m. (1,030,700 sq. km.). Cap. Nouakchott. Desert broken by mountain ridges running NE to SW with some large oases. Most cultivated land is on Senegal R. (S). Elsewhere the pop. is nomadic. Main products dates, salt, gum arabic, fish, salted and dried fish are exported. Pop. (1988E) 1,894,000

Mauritius   Indian Ocean. 20 17S 57 33E. Island sit. E of Madagascar forming, with Rodriguez and their dependencies, an independent state of the Commonwealth. Mauritius was known to Arab navigators probably not later than the 10th cent. It was probably visited by Malays in the 15th cent., and was discovered by the Portuguese between 1507 and 1512, but the Dutch were the first settlers (1598). In 1710 they abandoned the island, which was occupied by the French under the name of Ile de France (1715). The British occupied the island in 1810, and it was formally ceded to Great Britain by the Treaty of Paris, 1814. Mauritius attained independence in 1968. Area, 720 sq. m. (1,865 sq. km.). Cap. Port Louis. Plateau rising to mountains around its edges and then descending abruptly to coastal lowlands. There are numerous Rs. Tropical climate with rainfall up to 175 ins. and frequent cyclones. The main crop is sugar-cane, also tobacco, hemp, tea. Industries inc. distilling rum, manu. tobacco products. Pop. (1988E) 1,077,187.

May, Isle of   Scotland. 56 11N 2 34W. Sit. at entrance to Firth of Forth, near the SE coast.

Mayaguana   Bahamas, West Indies. Island and district between Acklins Island (W) and the Caicos Islands, SE of Nassau. Area, 96 sq. m. (249 sq. km.). Chief settlements Abraham's Bay, Pirates' Well. Low-lying, wooded, with salt pans on the coast. Pop. (1980C) 476.

Mayagüez   Puerto Rico, U.S.A. 18 12N 67 09W. Port sit. on the W coast. Industries inc. sugar refining, brewing, rum distilling and the manu. of tobacco products. Noted for embroidery. Pop. (1980C) 96,193.

Maybole   Strathclyde Region, Scotland. 55 21N 4 41W. Town sit. S of Ayr. Industries inc. engineering, packaging and printing. Pop. (1981C) 4,785.

Mayenne   France. 48 20N 0 38W. Dept. in Pays-de-la-Loire region, sit. mainly in Loire R. basin. Area, 1,997 sq. m. (5,171 sq. km.). It is generally low-lying with hills in the NE. The principal R. is Mayenne. Cattle and horse breeding are important, and cereals, sugar-beet and cider apples are grown. There are iron mines, and marble and slate quarries. Cap. Laval. Pop. (1982C) 271,784.

Mayenne River   France. 47 30N 0 33W. Rises just E of Pré-en-Pail and flows 125 m. (200 km.) S past Mayenne and Château-Gontier to join Sarthe R. above Angers; together they form Main R. Main tribs. Varenne and Oudon Rs.

Mayfair   Greater London, England. 51 30N 0 08W. Residential and commercial district of W central London and once the most fashionable district bounded W by Hyde Park.

Mayo   Connacht, Ireland. 53 47N 9 07W. A maritime county in NW bordered by the Atlantic Ocean. Area,

2,084 sq. m. (5,397 sq. km.). It is mainly mountainous in the N and W rising to over 2,000 ft. (610 metres) a.s.l. with a rugged coastline. In the lower central and E part cattle and pigs are reared, and oats and potatoes are grown. The principal Rs. are Moy, Robe and Owenmore. There are many loughs. County town Castlebar. Pop. (1988E) 115,184.

Mayotte   Indian Ocean. French dependency. In 1979 a *collectivé particulière*, an intermediate status prior to becoming an Overseas Department. An island with off-shore islets, easternmost of the Comoro group and NW of Madagascar. Chief towns, Dzaoudi (cap.) and Mamoundzou. Area, 144 sq. m. (374 sq. km.). Fertile, producing sugar, rum, sisal, spices and oils. Pop. (1984E) 56,000.

Mázándárán   Iran. Province of N Iran on S shore of Caspian Sea and bounded S by the Alburz mountains. Cap. Sarí. Area, 17,937 sq. m. (46,456 sq. km.). Humid, subtropical climate which is unhealthy but supports rich crops of rice, tea, citrus fruit, sugarcane, tobacco, silk, cotton and jute. Pop. (1986C) 3,419,346.

Mazár-i-Sharif   Mazár-i-Sharif, Afghánistán. 36 41N 67 00E. Cap. of province sit. NW of Kábul. Industries inc. flour milling, brick making and the manu. of textiles. Trades in carpets and skins. Pop. (1984E) 118,000.

Mazatenango   Suchitepéquez, Guatemala. 14 32N 91 30W. Cap. of dept. sit. WSW of Guatemala. It is the centre of an agric. district producing cotton, coffee and sugar-cane. Pop. (1984E) 21,000.

Mazatlan   Sinaloa, Mexico. 23 13N 106 25W. Port and resort sit. at SE end of the Gulf of California. Industries inc. sugar refining, distilling, flour milling and the manu. of textiles. Exports inc. tobacco and bananas. Pop. (1984E) 174,000.

Mbabane   Swaziland. 26 18S 31 06E. Cap. of kingdom sit. WSW of Maputo, Mozambique, at 3,750 ft. (1,143 metres) a.s.l. It became cap. of the British Protectorate of Swaziland in 1902 having been founded as a trading station. The principal industry is tin mining. Pop. (1986E) 32,290.

Mbale   Eastern Province, Uganda. 1 05N 34 10E. Town sit. NE of Jinja at the foot of Mount Elgon. Trading centre for an agric. area.

Mbandaka   Equateur, Zaïre. 0 04N 18 16E. Cap. of region on Congo R. with river and air ports. Commercial and trading centre.

Mbuji-Mayi   Kasai-Oriental, Zaïre. 6 09S 23 38E. Cap. of region on Bushimaie R. Centre for diamond mining. Airport.

Meath   Leinster, Ireland. 53 32N 6 40W. A maritime county bordering on the Irish Sea. Area, 903 sq. m. (2,340 sq. km.). It is mainly flat, rising in the W. The principal Rs. are Boyne and Blackwater. Agric. is important, esp. cattle and horse breeding, and the cultivation of oats and potatoes. County town Trim. Pop. (1988E) 103,881.

Meaux   Seine-et-Marne, France. 48 57N 2 52E. Town in NE, sit. on Marne R. and Ourcq Canal. Manu. inc. chemicals, cheese and mustard. Trades in agric. produce. Pop. (1982C) 45,873.

Mecca   Hejaz, Saudi Arabia. 21 27N 39 49E. City sit. E of Jidda in the Hejaz hills. Joint cap. (with Riyadh) and cap. of Hejaz. The most sacred city of Islam, birthplace of Mohammed and site of the Kaaba shrine. A pilgrimage city with few industries, which has always been a centre of caravan routes. Pop. (1986E) 618,006.

Mechelen   Antwerp, Belgium. 51 02N 4 28E. Town in central N sit. on Dyle R. Industries inc. railway engineering, and the manu. of furniture, transformers, textiles, paper and machinery. Pop. (1990E) 77,000.

Mecklenburg   German Democratic Republic. Formerly a *Land* but since 1942 it has been divided between the districts of Neubrandenburg, Rostok and Schwerin.

Medan   North Sumatra, Indonesia.

3 35N 98 40E. Cap. of province sit. on Deli R. Trades in timber, rubber and tobacco. Pop. (1980C) 1,378,955.

Medellín   Antioquia, Colombia. 6 15N 75 35W. Cap. of dept. sit. on a trib. of Cauca R. in the Central Cordillera at 5,000 ft. (1,524 metres) a.s.l. It is the main textile centre of Colombia. Other manu. inc. steel, cement, glass, pottery and leather. It is also the principal centre for gold and silver mining, and coffee growing and is known as the drugs capital of South America. Pop. (1980E) 1,664,000.

Medford   Massachusetts, U.S.A. 42 25N 71 07W. City sit. NNW of Boston on Mystic R. Manu. inc. machinery, packaging, adhesives, scientific instruments and food products. Pop. (1980C) 58,076.

Medford   Oregon, U.S.A. 42 19N 122 52W. City in SW of state sit. on Bear Creek. Industries inc. fruit growing and packaging, timber and tourism. Trades in agric. produce. Pop. (1987E) 45,290.

Medicine Hat   Alberta, Canada. 50 03N 110 40W. Town in SE of province sit. on S. Saskatchewan R. Industries inc. railway engineering, and the manu. of glass and pottery. Trades in agric. produce. Pop. (1989E) 42,290.

Medina   Hejaz, Saudi Arabia. 24 28N 39 36E. Town in central W sit. in an oasis. It is a sacred Moslem town with important mosques. Trades in dates, cereals and fruit. Pop. (1986E) 500,000.

Medinipur   West Bengal, India. Town sit. W of Calcutta. Commercial centre trading in grain and groundnuts. Industries inc. textiles; manu. copper and brass ware. Pop. (1981E) 86,118.

Mediterranean Sea   36 00N 15 00E. Inland sea extending E from the Straits of Gibraltar, which connect it to the Atlantic, to the coast of SW Asia. Area, 965,000 sq. m. (2,499,000 sq. km.). Bounded N by Spain, France, Monaco, Italy, Yugoslavia, Albania, Greece and Turkey; E by Syria, Lebanon, Israel; S by Egypt, Libya, Tunisia, Algeria, Morocco. Containing the islands of Sicily, Sardinia, Corsica, Crete, Cyprus and Malta and the Balearic, Ionian and Aegean islands, with many small off-shore islands on the NE coasts. It receives a strong, cold current from the Atlantic and loses warm and highly salt water by the same strait at a lower level. Tides vary *c.* 1 ft. (0.3 metres). The characteristic climate has mild winters, warm, dry, sunny summers, with strong winds, the Sirocco from Africa, the Mistral and Bora from continental Europe. Chief ports Barcelona, Marseilles, Genoa, Naples, Trieste, Piraeus, Smyrna, Beirut, Haifa, Port Said, Alexandria, Tunis, Algiers, Oran.

Medjerda River Tunisia/Algeria. 37 07N 10 13E. Rises in the Medjerda range (NE Algeria) near Souk-Ahras, and flows 230 m. (368 km.) ENE entering Tunisia at Ghardimaou and continuing past Medjez-el-Bab to enter the Mediterranean 20 m. (32 km.) N of Tunis. Waters an important wheat growing area. Main tribs. Mellegue and Siliana Rs.

Médoc Gironde, France. 45 10N 0 56W. A district sit. along Gironde R. in the SW, and noted for its wines.

Medway River England. 51 27N 0 44E. Rises in headstreams in E Sussex and SE Surrey and flows 70 m. (112 km.) NE through Kent, past Maidstone, then N to enter Thames R. by a long estuary with Rochester on its W bank, and Chatham and Gillingham on its E bank. Navigable up to Maidstone.

Meerut Uttar Pradesh, India. 29 01N 77 50E. Town sit. NE of Delhi. Noted as a former Brit. military cantonment where the Indian mutiny began in 1857. Industries inc. flour and oilseed milling; manu. tyres, textiles, sugar, electronics, cotton goods, chemicals, soap and pottery. Pop. (1981C) 417,395.

Megara Attica, Greece. 38 01N 23 21E. Town sit. W of Athens. Trading centre for wine and olive oil.

Meghalaya India. 25 50N 91 00E. State founded 1970 as a State with Assam and became an independent

State of the Union in 1972. Sit. in NE India and bounded S by Bangladesh. Area, 8,657 sq. m. (22,429 sq. km.). Chief towns Shillong (cap.), Tura. Hilly, with no Rs., well forested. Minerals inc. sillimanite, coal, dolomite, sandstone, limestone, clay, corundum. Chief products are timber, cement and resins, with some potatoes, fruit, rice, maize, tea and cotton. Pop. (1981C) 1,335,819.

Meissen Dresden, German Democratic Republic. 51 10N 13 28E. Town sit. in SE on Elbe R. The main industry is manu. porcelain from local kaolin, also sugar refining and brewing; manu. matches, glass and textiles. Pop. (1989E) 37,100.

Meknès Morocco. 33 54N 5 33W. City in NW sit. at 1,700 ft. (518 metres) a.s.l. to the N of the Middle Atlas range. Former cap. of Morocco. Industries inc. manu. pottery, leather and carpets. Pop. (1982C) 319,783.

Mekong River South East Asia. 10 33N 105 24E. Rises in the Tanghla Range of Qinghai, NW China, flows 2,500 m. (4,000 km.) SE across Chamdo province of E Tibet into W Yunnan province, then S to the S frontier of China; then SW along the Burma/Laos frontier and SE again on, or parallel to, the Laos/Thailand frontier and S into Cambodia. It turns SW to Phnom Penh, then SE into Vietnam, and enters the South China Sea by an extensive delta. Navigable for *c.* 300 m. (480 km.) of its lower course. The delta is a rice growing area.

Melaka (Malacca), Peninsular Malaysia, Malaysia. 2 11N 102 15E. (i) State in the SW. Area, 637 sq. m. (1,650 sq. km.). Mainly consists of a low-lying coastal plain producing rubber, rice and coconuts. Pop. (1980C) 453,153. (ii) Cap. of state of same name and port sit. on Strait of Malacca. Rubber and copra are exported. Pop. (1980C) 88,073.

Melanesia W Pacific Ocean. 13 00S 164 00E. One of the three main divisions of islands, inc. Fiji, New Caledonia, New Hebrides, Loyalty Islands, Solomon Islands, Santa Cruz Islands, Admiralty Islands and the Louisade and Bismarck Archs.

Melbourne Victoria, Australia. 37 49S 144 58E. State cap. sit. on Yarra R. near the mouth of Hobson's Bay in S centre of state. Founded in 1835 and named after Lord Melbourne, Prime Minister of England. Commercial centre. Industries inc. engineering, petroleum and chemical products, vehicles, aircraft, and other transport equipment, electrical products, scientific, medical technical and information technology, food processing, beverages, fertilizers, soap and woollen textiles. Port, inc. Port Melbourne and Williamstown, exports inc. wool, petroleum products, foodstuffs and non-ferrous metals. Pop. (1987E) 2,964,800.

Melilla North Africa. 35 19N 2 57W. Town sit. ESE of Ceuta on the Mediterranean, forming a Spanish enclave in Morocco. Area, 5 sq. m. (12 sq. km.). Garrison town. Industries inc. fishing. Sea port exporting iron ore from the Beni bu Ifrur mines. Pop. (1981E) 58,449.

Melitopol Ukraine, U.S.S.R. 46 50N 35 22E. Town sit. S of Zaporozhye on Molochnaya R. Industries inc. meat packing and flour milling; manu. agric. machinery, diesel engines and clothing. Pop. (1985E) 170,000.

Melksham Wiltshire, England. 51 23N 2 09W. Market town sit. E of Bath on Avon R. Industries inc. engineering, agric. products, rubber products. Pop. (1985E) 10,000.

Melrose Borders Region, Scotland. 55 36N 2 44W. Town in SE Scot. sit. on Tweed R. Market town and tourist centre. Pop. (1981C) 2,080.

Melrose Massachusetts, U.S.A. 42 27N 71 04W. Residential city sit. N of Boston. Pop. (1980C) 30,055.

Melton Mowbray Leicestershire, England. 52 46N 0 53W. Market town sit. NE of Leicester. Industries inc. manu. food products and pet foods. Pop. (1981C) 24,500.

Melun Seine-et-Marne, France. 48 32N 2 40E. Cap. of dept. on Seine R. Agric. centre. Pop. (1982C) 36,218.

Memel *see* **Klaipéda**

Memphis Tennessee, U.S.A. 35 08N 90 03W. City in extreme SW of state sit. on the Chickasaw Bluffs above Mississippi R. where it forms the Tennessee/Arkansas boundary. Market for cotton and timber. Industries inc. manu. agric. machinery, satellite communications, tourism, tyres and other rubber products, glass and textiles. R. port. Pop. (1988E) 645,190.

Menado North Sulawesi, Indonesia. 1 32N 124 55E. Cap. of province sit. at E end of the NE peninsula of Sulawesi. Sea port exporting copra, coffee and nutmegs. Pop. (1980E) 170,000.

Menai Strait Wales. 53 12N 4 12W. Channel extending 15 m. (24 km.) SW to NE between the NW Wales mainland to the Isle of Anglesey. Crossed by a road bridge and railway bridge near Menai Bridge.

Menam Chao Phraya River Thailand. Rises in the N highlands near the frontier of Laos, and flows 750 m. (1,200 km.) S on a winding course past Bangkok to the Gulf of Siam. Its lower course has a parallel branch, the Tachin R., from above Sing Buri. Navigable for small boats to the W, used for transporting rice and teak.

Mende Lozère, France. 44 30N 3 30E. Cap. of dept. on Lot R. Pop. (1982C) 12,113.

Menderes River Turkey. Name of several Turkish Rs., notably Büyük Menderes R. in the SW, rising near Uşak, Anatolia, and flowing 250 m. (400 km.) WSW on a winding course. Its Greek name *Maeander* was the origin of the verb 'to meander'.

Mendip Hills England. 51 15N 2 40W. Range extending 23 m. (37 km.) NW to SE midway between Bath and Bridgwater, and inc. Cheddar Gorge. The highest point is Blackdown, 1,068 ft. (326 metres) a.s.l. Mainly limestone.

Mendoza Mendoza, Argentina. 32 54S 68 50W. Cap. of province in W sit. in Mendoza R. valley at 2,500 ft. (762 metres) a.s.l. Commercial centre of an irrigated agric. and vine growing area. Pop. (1980C) 597,000.

Mendoza River　Argentina. Rises on Mount Aconcagua and flows 200 m. (320 km.) E and N across Mendoza province to enter L. Guanacashe. Used for hydroelectric power and irrigation.

Menton　Alpes-Maritimes, France. 43 47N 7 30E. Resort in extreme SE, sit. on the Mediterranean coast. Pop. (1982c) 25,449.

Mentor　Ohio, U.S.A. 41 40N 81 20W. City NE of Cleveland. Residential suburb. Industries inc. manu. of fork lifts, arc welding products, pressure gauges. Pop. (1990E) 47,000.

Menûfiya　Egypt. Governorate in the Nile Delta bounded N by Gharbiya, E and W by Damietta and Rosetta branches of Nile. Area, 613 sq. m. (1,588 sq. km.). Main towns Shibin el Kom (cap.), Ashmun, Tala, Minuf, Sirs el Laiyana. Agric. area producing cotton, flax, cereals. Industries inc. cotton ginning, textiles. Otherwise spelt Minûfiya. Pop. (1976c) 1,532,100.

Merano　Trentino-Alto Adige, Italy. 46 40N 11 09E. Town in NE. Industries inc. tourism, fruit canning, chemicals and pottery.

Mercedes　Buenos Aires, Argentina. 34 40S 59 30W. Town sit. W of Buenos Aires. Industries inc. manu. metal goods and footwear. Pop. (1984E) 40,000.

Mercedes　San Luis, Argentina. 33 40S 65 30W. Town in NE. Commercial centre trading in wheat, maize and alfalfa. Pop. (1984E) 20,000.

Mercedes　Soriano, Uruguay. 33 16S 58 01W. Town in SW sit. on Negro R. Industries inc. tourism. R. port trading in cereals, livestock and wool. Pop. (1985E) 37,000.

Mer de Glace　Haute-Savoie, France. 45 55N 6 55E. Glacier 4½ m. (7 km.) long on the N slope of Mont Blanc and formed by the union of the Talèfre, Leschaux and Tacul glaciers. Source of Arveyron R. which flows from it into Arve R.

Mergui Archipelago　Burma. 12

00N 98 00E. Group of islands in the Andaman Sea off the Tenasserim coast. Mountainous and forested.

Mérida　Yucatán, Mexico. 20 58N 89 37W. State cap. in NW of the Yucatán Peninsula. Commercial centre, trading, through its port Progreso, in indigo, sugar, hides, timber and henequen. Industries inc. manu. ropes, sacks and twine from local henequen sisal. Pop. (1984E) 285,000.

Mérida　Mérida, Venezuela. 8 36N 71 08W. State cap. in NW sit. at the foot of the Cordillera de Mérida at 5,385 ft. (1,641 metres) a.s.l. Industries inc. textiles; manu. footwear. Pop. (1981E) 143,000.

Meriden　Connecticut, U.S.A. 41 32N 72 48W. City sit. NNE of New Haven. Industries inc. manu. clothing, aircraft and vehicle parts, tools, pumps, gauges, confectionery, equipment for offshore gas wells, dies, presses and electrical equipment. Pop. (1980c) 57,118.

Meridian　Mississippi, U.S.A. 32 22N 88 42W. City in central E of state. Industries inc. electronics, automobile and aircraft parts, paper converters, metal fabrication. Pop. (1980c) 46,577.

Merksem　Antwerp, Belgium. 51 14N 4 29E. Town just NE of Antwerp. Industries inc. manu. glassware. Pop. (1982E) 42,000.

Merrimack River　U.S.A. 42 49N 70 49W. Formed by the union of Rs. Pemigewasset and Winnipesaukee and flowing 110 m. (176 km.) S through New Hampshire, then ENE across NE Massachusetts to enter the Atlantic below Newbury.

Merseburg　Halle, German Democratic Republic. 51 21N 11 59E. Town S of Halle sit. on Saale R. Industries inc. tanning and brewing; manu. paper and machinery. Pop. (1989E) 46,500.

Mersey River　England. 53 25N 3 00W. Formed by union of Rs. Goyt and Etherow, flowing 70 m. (112 km.) W past Warrington, Widnes and Runcorn to enter the Irish Sea by a broad estuary to the NE with Liverpool and Birkenhead to the SW on its banks. Main tribs. are R. Irwell and Weaver. Linked to the Manchester Ship Canal.

Merseyside   England. 53 25N 2 55W. Metropolitan county in NW Eng. created under 1974 re-organization with the districts of Sefton, Liverpool, St. Helens, Knowsley and Wirral. Inc. land on both sides of Mersey R. estuary. Area, 251 sq. m. (652 sq. km.). The Mersey estuary is an important industrial area particularly for the chemical industry. Light industry has also been attracted to the new towns of Skelmersdale and Runcorn (Cheshire). The vehicle and electrical industries are also important. Pop. (1988E) 1,448,100.

Mersin   İçel, Turkey. 36 48N 34 38E. Cap. of province in extreme central S, sit. on the S coast. The main industry is oil refining. Sea port exporting cotton, wool and chrome. Pop. (1980C) 216,308.

Merthyr Tydfil   Mid Glamorgan, Wales. 51 45N 3 23W. Town in SE Wales sit. N of Cardiff on Taff R. Industries inc. engineering and chemicals; manu. washing machines and toys. Pop. (1982E) 60,000.

Merton   Greater London, England. 5 24N 0 11W. District of SW London, mainly residential. Pop. (1988E) 164,000.

Mesa   Arizona, U.S.A. 33 25N 111 50W. City E of Phoenix sit. on Salt R. Commercial centre. Industries inc. cotton ginning, fruit packing and manu. helicopters. Pop. (1980C) 152,453.

Mesa Verde   Colorado, U.S.A. High plateau in SW of state with anc. Indian settlements and many early excavated sites, preserved in the Mesa Verde National Park.

Meshed   (Mashhad) Khurasan, Iran. 36 18N 59 36E. Cap. of province in NE sit. at 3,200 ft. (975 metres) a.s.l. Commercial centre trading in carpets and cotton goods. Industries inc. tanning, flour milling and rug making. A

sacred city and place of pilgrimage for Shia Moslems. Pop. (1976C) 670,180.

Messenia   Greece. Prefecture in SW Peloponnessos bounded W and S by the Ionian Sea. Area, 1,155 sq. m. (2,991 sq. km.). Cap. Kalamáta. Mainly agric. The main products are citrus fruits, vines, olives and cereals. Pop. (1981C) 159,818.

Messina   Sicilia, Italy. 38 11N 15 33E. Cap. of Messina province in NE sit. on the W shore of the Strait of Messina. Industries inc. chemicals; manu. pasta and soap. Port exporting olive oil, wine and fruit. Pop. (1983E) 263,924.

Messina, Strait of   Italy. 38 15N 15 35E. Strait extending 20 m. (32 km.) SW to NE between the Regions of Calabria and Sicilia with Messina on its W shore and Calabria on its E. Width 2 m. (3 km.) at its narrowest.

Meta   Colombia. Dept. of central Colombia, bounded N by Meta R. and S by Guaviare R. Area, 32,903 sq. m. (85,219 sq. km.). Cap. Villavicencio. Mainly grassland and thick forest rising to the Cordillera Oriental. Cattle raising area with coffee growing on higher ground. Forests produce vanilla, gums, timber, resins. Minerals inc. gold, salt. Pop. (1983E) 384,800.

Meta River   Colombia/Venezuela. 6 12N 67 28W. Rises S of Bogotá, Colombia, and flows 620 m. (992 km.) NE and E across the plains to the Venezuelan border, which it forms from below San Rafael to Puerto Paez, where it joins Orinoco R. Subject to severe flooding from May–Oct.

Metz   Moselle, France. 49 08N 6 10E. Cap. of dept. in NE sit. at confluence of R. Moselle and Seille. Industries inc. tanning, brewing and food preserving; manu. footwear, cement and metal goods. Pop. (1982C) 118,502 (agglomeration, 186,437).

Meudon   Hauts-de-Seine, France. 48 48N 2 14E. Suburb of SW Paris on Seine R. Industries inc. manu. munitions and electrical equipment. Pop. (1982C) 49,004.

Meurthe-et-Moselle   France. 48

52N 6 00E. Dept. in Lorraine region, bounded N by Belgium and Luxembourg. Area, 2,021 sq. m. (5,235 sq. km.). Main towns are Nancy (the cap.), Longwy, Briey, Lunéville and Toul. Mainly plateau drained by Rs. Moselle and Meurthe. Important for its iron ore deposits and metallurgical industries. Pop. (1982C) 716,846.

Meuse France. 49 08N 5 25E. Dept. in Lorraine region. Area, 2,402 sq. m. (6,220 sq. km.). Chief towns are Barle-Duc (the cap.), Commercy and Verdun. Drained by Meuse R., with two ridges, the Argonne to the W and Côtes de Meuse to the E running on either side of the R. from SSE to NNW. Mainly agric. Pop. (1982C) 200,101.

Meuse River France/Belgium/ Netherlands. 51 49N 5 01E. Rises on the Plateau de Langres in Haute-Marne dept. and flows 580 m. (928 km.) NNW past Sedan and Mézières-Charleville, then N through the Ardennes, NE into Belgium, past Namur and Liège and into the Netherlands (as Maas R.) above Maastricht. It forms the Belgian/Dutch frontier to above Roermond, continues past Venlo and turns NW then W to join Waal R. The lower course is extensively used for transport.

Mexborough South Yorkshire, England. 53 29N 1 18W. Town sit. NE of Sheffield on Don R. in a coalmining area. Industries inc. mining and manu. of furniture. Pop. (1982E) 15,825.

Mexicali Baja California, Mexico. 32 40N 115 29W. City in NW on the U.S. frontier adjoining Calexico, California, and E of San Diego, U.S.A. Commercial centre for an irrigated agric. area. Industries inc. cotton ginning; manu. cottonseed oil and soap. Pop. (1984E) 500,000.

Mexico 20 00N 100 00W. Rep. at the S extremity of N. America bounded N by U.S.A., W and SW by Pacific, S by Guatemala and Belize and E by Gulf of Mexico. The history of Mexico falls into 4 epochs: the era of the Indian empires (before 1521), the Spanish colonial phase (1521–1810), the period of national formation

(1810–1910), which includes the war of independence (1810–21) and the long presidency of Porfirio Díaz (1876–80, 1884–1911), and the present period which began with the social revolution of 1910–21 and is regarded by Mexicans as the period of social and national consolidation. Area, 756,198 sq. m. (1,958,201 sq. km.). Chief cities are Mexico City (the cap.), Guadalajara, Monterrey, Puebla, Ciudad Juárez and Léon. There are 29 states, 2 territories, and a Federal District around Mex. City. The Rio Grande forms the N boundary for over 1,100 m. (1,760 km.).The peninsula of Yucatán lies to E and to the W is the peninsula of Lower California. The coastal strip is bordered by mountain ranges including the Sierra Madre, 9,000 ft. (2,743 metres) a.s.l. to the W and a line of heights of 16,000 ft. (4,877 metres) a.s.l. to the E, sometimes known as the Sierra Madre Oriental. Low ground extends along the peninsula of Yucatán. Between the Sierra Madre ranges stretches the great central plateau with heights varying from 7,000–8,000 ft. (2,134–2,438 metres) a.s.l. There are many dormant and extinct volcanoes and in the S a few are still active. On the plateau the soil is fertile and where irrigated 2 crops a year can be grown. In the N herds of cattle, horses, sheep and pigs are reared and exported, mainly to U.S.A. Grains occupy 58% of cultivated land and of this 43% is maize and 5% wheat. The other important crops are, beans, cotton, coffee, sorghum, sugar and rice. Forestry is important and timber includes pine, spruce, cedar, mahogany, logwood and rosewood. Wood products include chicle, pitch, resins, turpentine, fibres, vegetable waxes and tan-barks. Mexico is one of the richest mineral countries in the world and deposits include gold, silver, coal, oil, copper, lead, zinc, antimony, graphite, quicksilver, arsenic, bismuth, cadmium, sulphur, cement, fluorite. Pop. (1989E) 84,278,992.

México Mexico. 19 20N 99 10W. State on the central plateau. Area, 8,284 sq. m. (21,455 sq. km.). Cap. Toluca. Encloses the Federal District which is administered separately. The main occupations are stock farming, agric. and mining. Pop. (1989E) 12,013,056.

Mexico, Gulf of   North America. 25 00N 90 00W. Sea almost enclosed by Florida peninsula, U.S.A., s coast of the U.S.A., E coast of Mexico and Yucatán Peninsula, Mexico. Its entrance is protected by the w end of Cuba, with access by the Florida Strait in the E, and the Yucatán Channel in the W. Area, 700,000 sq. m. (1,813,000 sq. km.).

Mexico City   Federal District, Mexico. 19 25N 99 10W. Cap. of rep. at s of the central plateau at 7,350 ft. (2,200 metres). The city is sit. on the site of the ruins of the Aztec Tenochtitlan. Commercial centre. Industries inc. textiles, gold and silver refining, brewing and assembling motor vehicles; manu. tobacco products, glass and tyres. Pop. (1986E) 18,748,000 (metropolitan area).

Mezötur   Szolnok, Hungary. 47 00N 20 41E. Town SE of Budapest, sit. in an agric. area. Trades in cereals and livestock. Industries inc. flour milling, pottery and brick making. Pop. (1984E) 22,000.

Miami   Florida, U.S.A. 25 46N 80 12W. City in SE of state sit. at mouth of Miami R. on Biscayne Bay. Resort, port and airport. Industries inc. tourism; manu. clothing. Pop. (1980C) 346,865.

Miami Beach   Florida, U.S.A. 25 47N 80 08W. City opposite Miami on Biscayne Bay and connected to it by causeway and bridge. Holiday resort. Pop. (1980C) 96,296.

Miaoli   Taiwan. 24 37N 120 49E. Town sit. SSW of Sinchu. Centre of the Miaoli oilfield, with production centre at Chukwangkeng, to the SE. Pop. (1985E) 84,326.

Miass River   U.S.S.R. 54 59N 60 06E. Rises in the s Urals and flows for 300 m. (480 km.) across the Chelyabinsk and Kuzan regions to join Iset R.

Michigan   U.S.A. 44 00N 85 40W. State on the Great Lakes, divided into 2 parts by L. Michigan, bounded N by L. Superior and E by L. Huron, and by Ontario, Canada, and L. Erie. Area, 58,216 sq. m. (150,780 sq. km.). Chief towns are Lansing (the cap.), Detroit, Flint, Grand Rapids and Warren. The Upper Peninsula, N of L. Michigan, is separated from the Lower Peninsula by the L. and the Straits of Mackinac. Low-lying, but rising to 2,023 ft. (622 metres) a.s.l. in the Upper Peninsula. The main industry is vehicle manu. The main natural products are livestock, cereals, sugar-beet, potatoes, fruit, timber, iron ore and petroleum. Pop. (1986E) 9,145,000.

Michigan, Lake   U.S.A. 44 00N 87 00W. The only one of the Great Lakes entirely in U.S.A., and the third largest. Area, 22,400 sq. m. (58,000 sq. km.). Used extensively by shipping, the main ports being Chicago and Milwaukee.

Michigan City   Indiana, U.S.A. 41 43N 86 54W. Town sit. ENE of Gary on L. Michigan. Industries inc. manu. machinery and furniture. Resort and port. Pop. (1988E) 35,330.

Michoacán de Ocampo   Mexico. State on the Pacific coast. Area, 23,138 sq. m. (59,928 sq. km.). Cap. Morelia, sit. WNW of Mexico City. Mountainous with a narrow coastal plain, and many Ls. The main products are cereals, sugar-cane, rice, tobacco, silver, lead and copper. Pop. (1989E) 3,424,235.

Michurinsk   Russian Soviet Federal Socialist Republic, U.S.S.R. 52 54N 40 30E. Town sit. WNW of Tambov on Voronezh R. Industries inc. engineering. Noted as a horticultural research centre.

Micronesia   W Pacific. 17 00N 160 00E. One of 3 main divisions of islands, inc. the Caroline, Marshall, Mariana and Kiribati, Tuvalu, and Nauru.

Micronesia, Federated States of   W Pacific. Former U.N. Trust Territory. It consists of 4 States: Kosrae, Pohnpei, Truk, and Yap. Area, 297 sq. m. (769 sq. km.). Cap. Kolonia. Pop. (1988E) 86,094.

Middelburg Zeeland, Netherlands. 51 30N 3 37E. Cap. of province sit. NNE of Flushing on Walcheren Island. Market town. Industries inc. electrical and electronic equipment, chemicals, iron works and construction. Pop. (1990E) 39,319.

Middelburg Transvaal, Republic of South Africa. 31 30S 28 00E. Town sit. E of Pretoria at 5,000 ft. (1,524 metres) a.s.l. in an extensive coalfield. Industries inc. stainless steel and ferro-alloy. Pop. (1989E) 33,070.

Middleback Range South Australia, Australia. Range in S of state extending 40 m. (64 km.) N to S parallel with the E coast of Eyre Peninsula. Rich in iron ore, with workings at Iron Knob, Iron Monarch and Iron Baron.

Middle East Area of the E Mediterranean especially Israel and the Arab countries from Turkey to N Africa and E to Iran.

Middlesbrough Cleveland, England. 54 35N 1 14W. Town in NE sit. on S bank of Tees R. estuary. The main industries are iron and steel, also manu. chemicals and fertilizers. Port handling iron, steel, machinery, chemicals, iron ore and timber. Pop. (1983E) 148,400.

Middleton Great Manchester, England. 53 33N 2 12W. Town sit. NNE of Manchester on Irk R. Industries inc. garment making, chemicals, rubber, warehousing and distribution. Pop. (1986E) 51,800.

Middletown Connecticut, U.S.A. 41 33N 72 39W. City sit. NNE of New Haven on Connecticut R. Industries inc. insurance, manu. aircraft engines and hardware. Pop. (1980C) 39,040.

Middletown New Jersey, U.S.A. Township in E New Jersey S of the Hudson estuary. Pop. (1980C) 62,574.

Middletown Ohio, U.S.A. 39 29N 84 25W. City sit. SW of Dayton on Great Miami R. Industries inc. steel and paper making. Pop. (1980C) 43,719.

Middle West, The U.S.A. Area comprising the states of Illinois, Indiana, Iowa, Michigan, Minnesota, Ohio and Wisconsin. Important wheat growing area.

Middlewich Cheshire, England. 53 12N 2 28W. Town sit. N of Crewe. Industries inc. salt refining; manu. plastics, clothing, chemicals. Pop. (1987E) 9,600.

Mid Glamorgan Wales. 51 20N 3 20W. County in S Wales created under 1974 re-organization with the districts of Ogwr, Rhondda, Cynon Valley, Taff-Ely, Merthyr Tydfil and Rhymney Valley. Area, 393 sq. m. (1,019 sq. km.). The S area is agric. and the N industrial. Iron and steel production has now finished and coalmining has been much reduced. Much light industry has been established in the county inc. furniture, cosmetics, engineering and electrical goods, and vehicle manu. Pop. (1989E) 535,500.

Midhurst West Sussex, England. 50 59N 0 44W. Town sit. ESE of Winchester on Rother R. in a farming area. Mainly residential. Pop. (1988E) 4,545.

Midland Michigan, U.S.A. 43 37N 84 17W. City sit. W of Bay City. Industries inc. chemical research and production. Pop. (1980C) 37,250.

Midland Texas, U.S.A. 32 00N 102 05W. City sit. WNW of San Angelo in an oil producing and stock farming area. Industries inc. oil refining; manu. oilfield equipment and chemicals. Pop. (1980C) 70,525.

Midlands, The England. General term applied to the central counties of England.

Midlothian Scotland. Former county now part of Lothian Region.

Mid-Pyrénées France. 42 45N 0 18E. Region in SW comprising depts. of Ariège, Aveyron, Gers, Haute-Garonne, Haute-Pyrénées, Lot, Tarn and Tarn-et-Garonne. Area, 17,523 sq. m. (45,382 sq. km.). Chief towns Toulouse, Tarbes, Albi and Montauban. Pop. (1982C) 2,325,319.

Midway Islands North Pacific

Ocean. 28 13N 177 22W. Two small
islands, Sand and Eastern 1,150 m.
(1,852 km.) NW of Honolulu. Area, 2
sq. m. (5.1 sq. km.). U.S. naval air
base. Pop. (1981E) 2,300.

Miercurea-Ciuc   Romania. 46 22N
25 42E. Town sit. NNW of Bucharest
on Olt R. Trading centre for a lumbering area; industries inc. sawmilling,
flour milling; manu. vinegar. Tourist
centre. Pop. (1982E) 40,674.

Migiurtinia   Somali Republic. Also
called Mijirtein. Region of N Somalia
bounded N by the Gulf of Aden, E by
the Indian Ocean. Narrow, coastal
plain rises to a hot, dry plateau at
1,500–3,000 ft. (457–914 metres)
a.s.l. Main occupations fishing, stock
farming. Main products salt, gum
arabic, tunny fish, mother of pearl.

Mihailovgrad   Vratsa, Bulgaria. 43
25N 23 13E. Town sit. NW of Vratsa
on Ogosta R. Centre of a farming
area, trading in cattle.

Mikkeli   Finland. 61 41N 27 15E.
(i) County of SE Finland. Area, 6,310
sq. m. (16,343 sq. km.), about 30% of
the area is water. Chief cities, Mikkeli, Savonlinna, Heinola. Low-lying,
marshy area with many Ls. and Rs.
Main occupations fishing, lumbering,
quarrying. Industries inc. wood and
metal working. Pop. (1983E) 209,062.
(ii) City sit. NE of Helsinki at the head
of W arm of L. Saimaa. County cap.
Resort with industries inc. manu.
plywood, machinery and granite quarrying. Pop. (1983E) 29,243.

Milan   Lombardia, Italy. 45 28N
9 12E. Cap. of Milano province sit. on
Olona R. on the Lombardy plain.
Commercial and route centre. Industries inc. printing, publishing,
engineering, chemicals and vehicle
manu. Pop. (1988E) 1,464,127.

Mildenhall   Suffolk, England. 52
21N 0 30E. Town NE of Cambridge
sit. on Lark R. Market town and important airfield. Pop. (1983E) 12,040.

Mildura   Victoria, Australia. 34
12S 142 09E. Town NW of Melbourne
sit. on Murray R. Commercial centre
of an irrigated farming area. Industries

inc. citrus dried fruit processing, wine
and tourism. Pop. (1990E) 19,350.

Milford   Connecticut, U.S.A. 41
13N 73 04W. Town and resort in S
Connecticut on Long Island Sound,
SW of New Haven. Industries inc.
oyster fisheries; manu. metal goods,
aircraft parts, chemicals and packaging. Pop. (1987E) 52,100.

Milford   Delaware, U.S.A. 38 55N
75 25W. City sit. S of Dover, near the
Atlantic coast. Industries inc. manu.
dental supplies, air conditioning units,
processing chickens and seafood,
farming and haulage. Pop. (1980C)
5,400.

Milford Haven   Dyfed, Wales. 51
42N 5 03W. Town in SW sit. on the N
coast of Milford Haven inlet. The
main industries are oil refining and
fishing. Pop. (1985E) 14,000.

Millwall   Greater London, England.
District of E London on N shore of
Thames R. *See* Tower Hamlets.

Milne Bay   Papua New Guinea.
Bay sit. SE of Port Moresby on E
coast. Length 30 m. (48 km.) and
width 15 m. (24 km.).

Milngavie   Strathclyde Region,
Scotland. 55 57N 4 20W. Town sit.
NNW of Glasgow. Industries inc. paper
making, bleaching and dyeing. Pop.
(1980E) 12,000.

Milton Keynes   Buckinghamshire,
England. 52 03N 0 47W. 'New city'
sit. NW of London. Industries inc.
electronics, clothing, gas appliances,
packaging, printing, pumps for chemicals. Headquarters of the Open University. Pop. (1989E) 141,800.

Milwaukee   Wisconsin, U.S.A. 43
02N 87 55W. City in SE of state sit. on
L. Michigan. Industries inc. brewing,
engineering (especially internal com-
bustion engines, vehicle parts and
other machinery), meat packing,
manu. electrical equipment and metal
goods. L. port. Pop. (1980C) 636,212.

Mina al Fahal   Oman. Oil pipeline
terminal and port immediately W of
Muscat sit. on the Gulf of Oman.

Mina Hassan Tani   Morocco. Town
NE of Rabat sit. on Sebou R. Industries inc. textiles, fish processing,
fertilizers, and tobacco products. Port
exporting grain and cork. Formerly
named Kénitra. Pop. (1971E) 139,206.

Minami Tori Shima   *see* **Marcus
Island**

Minas   Lavalleja, Uruguay. 34 23S
55 14W. Cap. of dept. ENE of Montevideo sit. in a quarrying area. Industries inc. brewing. Pop. (1980E)
25,000.

Minas Gerais   Brazil. 18 50S 46
00W. State in E Braz. Area, 224,701
sq. m. (581,975 sq. km.). Cap. Belo
Horizonte. Plateau, rich in minerals,
esp. iron ore, but mainly agric. The
main products are livestock, dairy
produce, maize, beans, coffee, cotton,
tobacco and fruit. Pop. (1980C)
13,378,553.

Minch, The   Scotland. 58 10N 5
50W. Arm of the Atlantic extending
NE to SW between Lewis, Outer
Hebrides, and the W coast of the mainland. Width 20–46 m. (32–74 km.).

Mindanao   Philippines. 7 30N 125
00E. Second largest island at the S end
of the group. Area, 36,537 sq. m.
(94,631 sq. km.). Chief towns Davao,
and Zamboanga. Mountainous and
forested, rising to Mount Apo, 9,500
ft. (3,115 metres) a.s.l. The main products are Manila hemp, coconuts, rice,
maize and gold. Pop. (1980E) 9m.

Mindelo   Cape Verde. 16 53N 25
00W. Town and port on São Vicente
island. Pop. (1980C) 36,746.

Minden   North Rhine-Westphalia,
Federal Republic of Germany. 52 17N
8 55E. Town in central N sit. on
Weser R. and Mittelland canal. Industries inc. chemicals and brewing;
manu. soap and glass. Pop. (1985E)
80,000.

Mindoro   Philippines. 13 00N 121
10E. Island S of Luzon. Area, 3,759
sq. m. (9,735 sq. km.). Chief town
Calapan. Mountainous, rising to 8,000
ft. (2,438 metres) a.s.l. The main products are rice, maize and coconuts.

Pop. (1980E) 500,000 inc. adjacent
islands.

Minehead   Somerset, England. 51
13N 3 29W. Town in SW Eng. sit. on
the Bristol Channel. Market town and
resort. Pop. (1985E) 8,500.

Minho   Portugal. Area S of R.
Douro. Chief towns, Oporto and
Braga. Forms the N part of the former
province of Entre Minho e Douro.
Produces cereals, fruit, olives, vines.
Wine is exported.

Minicoy Island   Lakhshadweep,
India. Southernmost island of the
Lakhshadweep group, separated from
the Laccadives by the Nine Degree
Channel. Area, 1.8 sq. m. (4.66 sq.
km.). Main product coconuts. Pop.
(1981C) 6,658.

Minneapolis   Minnesota, U.S.A. 44
59N 93 13W. City in SE Minnesota sit.
on Mississippi R. Industries inc.
manu. non-electric machinery, food
products and fabricated metal products. Pop. (1980E) 370,951.

Minnesota   U.S.A. 46 40N 94 00W.
State in N bounded N by Canada, E by
L. Superior and Wisconsin. Area,
84,068 sq. m. (217,736 sq. km.). Chief
cities are St. Paul (the cap.) Minneapolis, Duluth, and Bloomington.
First explored in the 17th cent. and
settled after 1819, it was admitted to
the Union, with its present boundaries, in 1858. Mostly prairie, with
many Ls. Mainly agric. The chief products are livestock, dairy foods,
cereals and iron ore. Pop. (1988E)
4,306,550.

Minorca   Spain. 40 00N 4 00E.
Island in the Balearic group, Mediterranean Sea, sit. NE of Majorca. Area,
266 sq. m. (689 sq. km.). Chief town
Mahon. Low-lying, rising to 1,107 ft.
(337 metres) a.s.l. in El Toro. The
main occupations are farming and
tourism. Pop. (1990E) 62,000.

Miño River   Spain/Portugal. 41 52N
8 51W. Rises in the Sierra de Meira in
Galicia, NE Spain, flows SSW to
Orense, then WSW to the Spain/Portu-
gal frontier which it forms until it
enters the Atlantic Ocean at Caminha,
Portugal.

Minot North Dakota, U.S.A. 48 14N 101 18W. City in NW centre of state. Route and commercial centre in a farming and lignite-mining area. Pop. (1980C) 32,843.

Minsk Belorussia, U.S.S.R. 53 54N 27 34E. Cap. of S.S.R. in central Belorussia sit. on Svisloch R. Industries inc. manu. motor vehicles, machinery, electrical equipment, textiles and furniture. Pop. (1989E) 1,589,000.

Minya, El Minya, Egypt. Cap. of governorate sit. on W bank of Nile R. Industries inc. sugar refining and cotton ginning. Port trading in cotton and cereals. Pop. (1986E) 203,300

Miquelon *see* **St. Pierre and Miquelon**

Miranda Venezuela. State of N Venezuela bounded N by the Federal District. Area, 3,070 sq. m. (7,951 sq. km.). Cap. Los Teques. Mountainous except for lower Tuy R. valley. Agric. area producing coffee on high ground, cacao, sugar, maize, rice, yucca, fruit. Industries inc. lumbering, sawmilling, sugar refining. Pop. (1980E) 1,110,215.

Mirdite Albania. Region of N Albania, trad. tribal territory in highlands drained by Fan R. Stock farming and dairying area.

Mirfield West Yorkshire, England. 53 40N 1 41W. Town NE of Huddersfield sit. on Calder R. Industries inc. textiles and engineering. Pop. (1981C) 18,686.

Mirzapur Uttar Pradesh, India. 25 10N 82 34E. Town sit. WSW of Varanasi on Ganges R. Industries inc. carpets and aluminium. Noted as a place of pilgrimage. Pop. (1981C) 127,787 with Vindhyachal.

Misima Papua New Guinea. 10 40S 152 45E. Island of the Louisiade Arch. sit. SE of New Guinea. Area, 100 sq. m. (259 sq. km.). Chief town Bwagaoia. The main product is gold.

Misiones Argentina. Province of NE Argentina bounded SE by Uruguay R.,

W by Paraná R. and N by Iguassú R. Area, 11,514 sq. m. (29,821 sq. km.). Cap. Posadas. Sub-tropical and thickly forested area. Agric. areas produce maté, tobacco, citrus fruit, cereals, cotton, sugar. Industries inc. lumbering, processing maté and tobacco. Pop. (1980C) 588,977.

Miskolc Borsod-Abaúj-Zemplén, Hungary. 48 06N 20 47E. Cap. of county sit. ENE of Budapest on Sajó R. Industries inc. iron and steel, textiles, flour milling and wine making; manu. railway rolling stock. Trades in cattle, wine and tobacco. Pop. (1984E) 212,000.

Mississauga Ontario, Canada. 43 35N 79 37W. City on W boundary of Metropolitan Toronto with Lester B. Pearson intern. airport. Industries inc. computers, electronics, pharmaceuticals, chemicals and transportation. Pop. (1985E) 355,000.

Mississippi U.S.A. 33 00N 90 00W. State in S bounded SE by the Gulf of Mexico. Area, 47,716 sq. m. (123,584 sq. km.). Settled in 1716 and acquired its present boundaries in 1817. Chief cities are Jackson (the cap.), Biloxi and Meridian. Mainly plain, drained by Yazoo R. and its tribs. and Big Black R. into Mississippi R. and by Pearl, Pascagoula and Tombigbee R. systems into the Gulf of Mexico. Heavily forested. The main occupations are farming, oil and gas extraction, and lumbering. Main agric. products are cotton, livestock inc. poultry. Pop. (1985E) 2,656,600.

Mississippi River U.S.A. 29 00N 89 15W. Rises in streams draining into L. Itasca in N Minnesota, flows 2,348 m. (3,757 km.) S past Minneapolis to the Minnesota/Wisconsin border, and continues as an inter-state border past Iowa, Missouri, Arkansas and Louisiana on the W bank and Illinois, Kentucky, Tennessee and Mississippi on the E. At St. Louis it receives the Missouri R. and at Cairo, Illinois, the Ohio R. On its lower course it meanders through a flood plain and is banked by artificial levées against flooding.

Missolonghi Aetolia and Acarnania,

Greece. 38 21N 21 17E. Cap. of prefecture sit. WNW of Patras on N side of the Gulf of Patras. Market for agric. produce. Noted for its part in the Turkish wars 1822–26 and as the scene of the death of Lord Byron. Pop. (1981C) 10,164.

Missoula   Montana, U.S.A. 46 52N 114 01W. City in W mountains near confluence of Clark, Fork, Blackfoot and Bitterroot Rs. Pop. (1980C) 33,388.

Missouri   U.S.A. 38 25N 93 30W. State in central U.S.A. bounded E by the Mississippi R. Area, 69,674 sq. m. (180,456 sq. km.). Chief cities are Jefferson City (the cap.), St. Louis and Kansas City. Mainly prairie in the N which is bounded in the W, and then crossed NW to SE, by the Missouri R. Hilly in the S with the Ozark Mountains at an average 1,100 ft. (335 metres) a.s.l. except in the extreme SE which is in the Mississippi R. flood plain. The main occupations are farming with some lead mining, and industries (mainly transport equipment and food products) in the cities. Main products are livestock, cereals, soybeans, zinc and lead. Pop. (1988E) 5,139,000.

Missouri River   U.S.A. 38 50N 90 08W. Rises near Three Forks in SW Montana and flows 2,714 m. (4,342 km.) N and E through Montana, S and SE through North Dakota and South Dakota to the Nebraska border of which it forms the E end, then turns SSE at Sioux City to form the Nebraska border with Iowa and NW Missouri; it continues SE as the Missouri/Kansas border as far as Kansas City, then E across Missouri to join the Mississippi R. above St. Louis. Important for irrigation and water supply with several reservoirs. High water is in April and June.

Misurata   Tripolitania, Libya. 32 23N 15 06E. Cap. of division in the NW. Oasis; trading centre for dates and cereals. Industries inc. carpet making.

Mitcham   Greater London, England. 51 24N 0 09W. District of SW London. Industries inc. manu. paints, varnishes and pharmaceutical goods.

Mitiaro   Cook Islands, S Pacific. Volcanic island sit. NE of Rarotonga. Area, 8.6 sq. m. (22.3 sq. km.). The main products are copra, sandalwood. Pop. (1981E) 256.

Mitchell   South Dakota, U.S.A. 43 43N 98 01W. City sit. N of Omaha. Industries inc. agric. and tourism. Pop. (1980C) 13,916.

Mittelland Canal   Federal Republic of Germany. 52 16N 11 41E. Canal system across F.R.G. from Ems R. to Havel R. extending 273 m. (437 km.) W to E and linking Rs. Ems, Rhine, Elbe and Havel.

Miyazaki   Kyushu, Japan. 31 54N 131 25E. Town sit. SE of Nagasaki. Market town. Industries inc. manu. porcelain, chemicals, wood pulp and charcoal. Pop. (1988E) 285,427.

Mizoram   India. State created from the Mizo Hills district of Assam. Area, 8,139 sq. m. (21,081 sq. km.). Chief towns, Aizawl (cap.), Lungtei, Kolasib, Saiha. Mountainous and 66% is forested. Timber products and handicrafts are manu. Pop. (1981C) 493,757.

Mjösa, Lake   Norway. 60 40N 11 00E. L. in SE Norway. Area, 141 sq. m. (365 sq. km.), the largest L. in the country. Fed by Lågen R. and drained by Vorma R. to Glomma R.

Mmabatho   Bophuthatswana. 25 49S 25 30E. Newly built cap. of Rep. of Bophuthatswana. Airport opened 1984.

Mobile   Alabama, U.S.A. 30 42N 88 05W. City in SW of state sit. on Mobile Bay on the Gulf of Mexico. Industries inc. shipbuilding, meat packing, textiles and paper making; manu. food products and clothing. The state's only seaport, exporting cotton, timber, coal and steel products. Pop. (1980C) 200,452.

Mobile River   U.S.A. 31 15N 87 58W. Formed by confluence of Tombigbee and Alabama Rs. above Mobile, and flows 45 m. (72 km.) S through a delta to enter Mobile Bay at its head.

Mobutu Sese Soko, Lake Africa. 1 30N 31 00E. Formerly named L. Albert. Sit. on the Uganda/Zaïre border within the W branch of the Great Rift Valley. Area, 2,064 sq. m. (5,346 sq. km.), *c*. 100 m. (160 km.) N to S.

Moçambique *see* **Mozambique**

Mocha Republic of Yemen. 13 19N 43 15E. Port SSW of San'a sit. on the Red Sea. Pop. (1985E) 8,000.

Modder River Cape Province, Republic of South Africa. Resort sit. SSW of Kimberley, near confluence of Rs. Riet and Modder.

Modder River Republic of South Africa. Rises NW of Caledon on the Lesotho border and flows 190 m. (304 km.) NW and W, through the Orange Free State into Cape Province, to join Riet R. near Modder River, Cape Province.

Modena Emilia-Romagna, Italy. 44 40N 10 55E. Cap. of Modena province in central Italy sit. on the Aemilian Way. Industries inc. agric. engineering; manu. motor vehicles, pasta and glass. Pop. (1983C) 178,985.

Modesto California, U.S.A. 37 39N 121 00W. City SE of San Francisco; trade and service centre, diverse manu. Pop. (1980C) 106,602.

Modica Sicilia, Italy. 36 51N 14 47E. Town sit. WSW of Syracuse. Market town trading in cattle, poultry, olive oil, wine and cheese. Pop. (1990E) 50,000.

Moers North Rhine-Westphalia, Federal Republic of Germany. 51 27N 6 37E. Town sit. WNW of Duisburg. Industries inc. coalmining; manu. machinery. Pop. (1985E) 100,574.

Moeskroen *see* **Mouscron**

Moffat Dumfries and Galloway Region, Scotland. 55 20N 3 27W. Town in central S sit. on Annan R. Resort and spa. Pop. (1985E) 2,000.

Mogadishu Somali Republic. 2 02N 45 21E. Cap. of rep. sit. on SE coast. Mogadishu was taken by the Sultan of Zanzibar in 1871 and sold to the Italians in 1905 who lost it to the British in 1941. It became the cap. on independence in 1960. Commercial centre and seaport. Pop. (1982E) 377,000.

Mogilev Belorussia, U.S.S.R. 53 54N 30 21E. Town in central W U.S.S.R. sit. on Dnieper R. Industries inc. manu. machinery, rayon, clothing and leather. Pop. (1987E) 359,000.

Mohales Hoek Mohales Hoek, Lesotho. 30 07S 27 26E. Village sit. S of Maseru on the main N to S road. District cap.

Mohéli *see* **Mwali**

Mojave Desert California, U.S.A. 38 00N 117 30W. Arid region in S California, S of the Sierra Nevada and part of the Great Basin. Area, 15,000 sq. m. (38,500 sq. km.). Mountainous with intervening wide basins. The Mojave R. is the only stream; it flows mainly underground for *c*. 100 m. (160 km.).

Moji *see* **Kitakyushu**

Mokpo Cholla South, South Korea. 34 48N 126 22E. Town sit. SSW of Seoul on SW coast. Industries inc. cotton ginning and rice milling. Seaport. Pop. (1980E) 200,000.

Mold Clwyd, Wales. 53 10N 3 10W. Town sit. WSW of Chester, Eng., on Alyn R. Market town. Industries inc. light industry and agric. Pop. (1989E) 8,935.

Moldavia Romania. District of NE Romania lying mainly between Rs. Prut and Siret.

Moldavian Soviet Socialist Republic U.S.S.R. 47 00N 28 00E. Constituent rep. bounded NE and S by the Ukraine and W by Romania, across Prut R. Area, 13,000 sq. m. (33,670 sq. km.). Chief towns are Kishinev (the cap.), Tivaspol and Beltsy. Mainly lowland with a fertile black soil. Drained by Dniester R. The main occupation is viticulture, with sturgeon fishing in the S, and agric. Pop. (1989E) 4·3m.

Molfetta Puglia, Italy. 41 12N 16

36E. Town sit. WNW of Bari on the Adriatic Sea. Industries inc. manu. soap, pottery and cement. Port exporting wine and olive oil.

Moline Illinois, U.S.A. 41 30N 90 31W. City sit. WSW of Chicago on Mississippi R. Industries inc. manu. agric. machinery, tools and furniture. Pop. (1980C) 46,407.

Mollendo Arequipa, Peru. 17 02S 72 01W. Town in SW sit. on the Pacific coast. Industries inc. fishing, fish canning and brick making. Port. Pop. (1980E) 14,650.

Mölndal Göteborg and Bohus, Sweden. 57 39N 12 01E. Town in SW. Industries inc. textiles and paper making. Pop. (1983E) 48,327.

Molokai Hawaii, U.S.A. 21 07N 157 00W. Island sit. E of Oahu. Area, 261 sq. m. (676 sq. km.). Mountainous, rising to Mount Kamakon, 4,970 ft. (1,515 metres) a.s.l. Main occupations stock farming, growing pineapple. Pop. (1980E) 6,049.

Molotov *see* **Perm**

Moluccas Indonesia. 2 00S 128 00E. Large island group between Sulawesi to the W and New Guinea to the E forming Maluku province. Area, 33,315 sq. m. (86,285 sq. km.). Cap. Amboina. The N group inc. Halmahera, Morotai, Obi and Sula Islands; the S inc. Ceram, Beru, Aru, Tanimbar Islands and Wetar. Most are mountainous and volcanic. The main products are spices and copra.

Mombasa Kenya. 4 03S 39 40E. City sit. on E coast, partly on an offshore island. Industries inc. oil refining, coffee curing and brewing; manu. cement, glass and soap. Exports are coffee, cotton, hides, pyrethrum, sisal, soda and tea, through Kilindini harbour. Pop. (1984E) 425,634.

Monaco 43 46N 7 23E. Principality forming an enclave in Alpes-Maritime dept., France, and bounded S by the Mediterranean. Area, 481 acres (195 hectares), comprising Monaco (the cap.), Monte Carlo, Fontvieille and La Condamine. The main industries are

tourism and legalized gambling; also fruit growing; manu. perfume. Pop. (1989E) 28,000.

Monadnock, Mount New Hampshire, U.S.A. 42 52N 72 07W. Mountain in the SW of state, 3,165 ft. (965 metres) a.s.l., in an isolated position. The name is now used for any hill of hard rock which resists erosion.

Monagas Venezuela. State of NE Venezuela bounded SE by Orinoco R., NE by San Juan R. and Gulf of Paria. Area, 11,160 sq. m. (28,904 sq. km.). Cap. Maturín. Coastal mountains descend to grasslands and marsh on the Orinoco delta. Tropical climate. Cattle farming area with agric. on the highlands, producing coffee, tobacco, sugar, cacao, cotton, yucca, maize. Forests yield hardwoods and palm products. Minerals inc. petroleum, coal, zinc, cadmium, asphalt, sulphur, salt. Pop. (1981C) 388,536.

Monaghan Ireland. 54 15N 6 58W. (i) County in NE bounded W, N and E by Northern Ireland. Area, 498 sq. m. (1,290 sq. km.). Chief towns are Monaghan (the county town), Clones, Carrickmacross and Castleblaney. Undulating, rising to 1,200 ft. (366 metres) a.s.l. in Slieve Beagh in the NW. Drained by Rs. Finn and Blackwater. The main occupation is farming, esp. dairy farming, poultry, mushrooms and soft fruit. Pop. (1988E) 52,379. (ii) County town in NE. Industries inc. food processing and manu. of footwear. Pop. (1981C) 6,177.

Monchegorsk Russian Soviet Federal Socialist Republic, U.S.S.R. 67 54N 32 58E. Town S of Murmansk sit. on the NW shore of L. Imandra. The main industry is smelting copper and nickel from local ores.

Mönchengladbach North Rhine-Westphalia, Federal Republic of Germany. 51 12N 6 28E. Town W of Düsseldorf sit. on Niers R. Industries inc. electrical and mechanical engineering, aircraft, steel and light metal construction, textiles, clothing, food and printing. Pop. (1990E) 252,000.

Monfalcone Friuili-Venezia Giulia, Italy. 45 49N 13 32E. Town sit. NW of

Trieste. The main industry is shipbuilding; also oil refining and chemicals.

Mongolian People's Republic Bounded N by Siberia, U.S.S.R., E, S and W by China. Area, 604,250 sq. m. (156,500 sq. km.). Cap. Ulan Bator. Severe climate. Mainly plateau at 3,000–4,000 ft. (923–1,231 metres) a.s.l. rising to the Altai and Khangai Mountains in the W with desert in the S. About 84% is pasture and the main occupation is rearing livestock. The main products are cattle, horses, wool, hides, meat and butter. Pop. (1989C) 2,001,000.

Mongu Barotse, Zambia. 15 15S 23 09E. Town sit. NW of Maramba near Zambesi R. Provincial cap. and centre of a farming area.

Monmouth Gwent, Wales. 51 50N 2 43W. Town sit. at confluence of Rs. Wye and Monnow. Market town, with some light industry inc. packaging. Pop. (1985E) 7,500.

Monongahela River U.S.A. 40 27N 80 00W. Formed by union of Rs. Tygart and West Fork in West Virginia, flows 128 m. (205 km.) N into Pennsylvania to join Allegheny R. at Pittsburgh where together they form Ohio R. Used extensively for transport.

Monopoli Puglia, Italy. 40 57N 17 19E. Town sit. ESE of Bari. Industries inc. food canning, oil mills, shipyards; manu. pasta and plastic goods. Port exporting olive oil, wine, vegetables, timber and fruit. Pop. (1990E) 47,000.

Monroe Louisiana, U.S.A. 32 33N 92 07W. City NE of Baton Rouge sit. on Ouachita R. Industries inc. manu. chemicals, carbon black, wood pulp, paper and furniture. Pop. (1980C) 57,597.

Monrovia Liberia. 6 20N 10 46W. Cap. of rep. sit. near mouth of St. Paul R. Founded for freed slaves in 1822 and named after the U.S. President, James Monroe (1758–1831). Port exporting rubber, iron ore, diamonds, palm oil and kernels. Industries inc. cement, hardware, rubber goods, paper, footwear. Pop. (1985E) 500,000.

Mons Hainaut, Belgium. 50 27N 3 56E. Cap. of province SW of Brussels. Sit. in a coalmining area. Industries inc. sugar refining, textiles, chemicals and cement. Pop. (1988E) 89,515.

Montana U.S.A. State in NW bounded N by Canada. Area, 147,138 sq. m. (381,087 sq. km.). First settled in 1809. It was made a Territory (from parts of Idaho and Dakota) in 1864 and a State in 1889. Chief towns are Helena (the cap.), Great Falls and Billings. Mountainous in the W with the Rocky Mountains rising to 12,850 ft. (3,917 metres) a.s.l. in Granite Peak in the S. Plains in the E at an average 2,000–4,000 ft. (610–1,220 metres) a.s.l. Severe winters, and an annual rainfall of 10–15 in. (25–38 cm.). The main occupations are farming, mining and oil extracting. Main products are wheat, barley, sugar-beet, petroleum, copper, silver and gold. Pop. (1986E) 819,000.

Montaña Clara Canary Islands, Atlantic Ocean. Islet of the N Canaries, sit. NE of Las Palmas. Area, 275 acres, rising in a peak 790 ft. (240 metres) a.s.l.

Montauban Tarn-et-Garonne, France. 44 01N 1 21E. Cap. of dept. in SW sit. on Tarn R. Market town trading in fruit and poultry. Industries inc. food processing and textiles; manu. furniture. Pop. (1982C) 53,147.

Mont Blanc Haute-Savoie, France. 45 55N 6 55E. Highest mountain in the Alps, 15,781 ft. (4,810 metres) a.s.l., in the Mont Blanc Massif, which lies on the French/Italian/Swiss borders. First climbed by Dr. Michel Paccard and Jacques Balmat, of Chamonix, in 1786. There is a French/Italian road tunnel beneath it, $7^{1}/_{2}$ m. (12 km.) long.

Mont Cenis France. 45 15N 6 54E. Alpine Pass in Savoie dept. SE France, between Lanslebourg, France and Susa, Italy at 6,831 ft. (2,082 metres) a.s.l.

Montclair New Jersey, U.S.A. 40 49N 74 13W. Town sit. NNW of Newark, mainly residential. Industries inc. healthcare, chemicals; manu.

paints and metal goods. Pop. (1988E) 38,721.

Monte Bello Islands South Pacific Ocean. 20 28S 115 32E. Group of uninhabited islands just off the NW coast of Western Australia. The main island is Barrow Island, 12 m. (19 km.) by 5 m. (8 km.).

Monte Carlo Monaco. 43 44N 7 25E. Town sit. ENE of Nice, France. Resort noted for its gambling casino, and for an annual motor car rally. Pop. (1988E) 28,000.

Montecristi Dominican Republic. (i) Province of NW bounded W by Haiti, N by the Caribbean. Area, 1,150 sq. m. (2,979 sq. km.). Sit. in Yaque del Norte R. valley bounded N and S by mountains. Fertile forested area. Farm products inc. cereals, fruit, cacao, dairy products, livestock, hides, skins. Lumbering is important. Pop. (1987E) 83,407. (ii) Town sit. WNW of Santiago near mouth of Yaque del Norte R. on the Caribbean coast. Provincial cap. Trading centre and port with a good harbour, handling hides, skins, rice, cotton, coffee, bananas, livestock.

Monte Gargano Puglia, Italy. Mountainous peninsula in the SE extending 35 m. (56 km.) E into the Adriatic Sea and rising to 3,461 ft. (1,055 metres) a.s.l. Chief town, Monte Sant' Angelo.

Montego Bay Jamaica. 18 30N 77 50W. Town on NW coast, sit. WNW of Kingston. The main industry is tourism. Pop. (1980E) 43,000.

Montélimar Drôme, France. 44 34N 4 45E. Town in SE sit. on Roubion R. Industries inc. manu. confectionery, esp. nougat. Pop. (1990E) 32,000.

Montenegro Yugoslavia. 42 30N 19 19E. Constituent rep. bounded SE by Albania and SW by the Adriatic. Area, 5,333 sq. m. (13,812 sq. km.). Chief towns are Titograd (the cap.), and Cetinje. Mountainous, rising to 8,000 ft. (2,438 metres) a.s.l., descending in the S to L. Shkodër. The main occupation is stock rearing. Pop. (1988E) 632,000.

Monterey California, U.S.A. 36 37N 121 55W. City on Pacific coast at N end of Santa Lucia range, c. 100 m. (160 km.) S of San Francisco. Pop. (1989E) 31,529.

Monterey Park California, U.S.A. 34 04N 118 07W. Residential city sit. E of Los Angeles. Pop. (1988E) 63,900.

Montería Bolivar, Colombia. 8 46N 75 53W. Town sit. SSW of Cartagena on Sinú R. R. port and trading centre for a lumbering and stock farming area. Airfield. Cattle research station. Pop. (1980E) 226,000.

Monte Rosa Italy/Switzerland. 45 57N 7 53E. The highest mountain group in the Pennine Alps, sit. SSW of Brig, Switzerland, and inc. 10 peaks of which the highest is Dufourspitze 15,217 ft. (4,638 metres) a.s.l.

Monterrey Nuevo León, Mexico. 25 40N 100 19W. State cap. in NE sit. in valley of Santa Catarina R. Industries inc. iron and steel, lead smelting, textiles and chemicals; manu. glass and tobacco products. Pop. (1980C) 1,916,472.

Montevideo Montevideo, Uruguay. 34 55S 56 10W. Cap. of rep. and of dept. sit. on N shore of Río de la Plata. Founded in 1726 by the Spanish its industries inc. meat packing, tanning, flour milling and textiles; manu. footwear, soap and matches. Seaport exporting wool, hides, skins and meat. Pop. (1985C) 1,246,500.

Montgomery Alabama, U.S.A. 32 23N 86 18W. State cap. in E centre of state sit. on Alabama R. Industries inc. manu. glass, wiring harnesses for cars, cotton goods and fertilizers. Pop. (1980C) 177,857.

Montgomery Powys, Wales. 52 33N 3 03W Town sit. SW of Shrewsbury, Eng. Market town in a farming area. Pop. (1985E) 1,000.

Montluçon Allier, France. 46 20N 2 36E. Town in central France sit. on Cher R. Industries inc. machinery and chemicals; manu. glass. Pop. (1982C) 51,765.

Montmartre Paris, France. District in N Paris, on the *Butte de Montmartre* with many artistic associations.

Montparnasse Paris, France. District in W Paris noted for the Pasteur Institute and cafés where artists and writers meet. The cemetery contains the tombs of many writers, musicians and artists.

Mont Pelée Martinique. Volcano 4,428 ft. (1,350 metres) a.s.l. sit. NW of Fort de France. The eruption of 1902 destroyed the then chief town of St. Pierre.

Montpelier Vermont, U.S.A. 44 16N 72 35W. City sit. NW of Barre on Winooski R. State cap. Industries inc. quarrying, textiles, printing, manu. wood and concrete products, machinery. Pop. (1980c) 8,241.

Montpellier Hérault, France. 43 36N 3 53E. Cap. of dept. in central S sit. on Lez R. Commercial centre trading in wine and brandy. Industries inc. manu. soap, electronics and advanced technology, chemicals, computer research, perfumes, pharmaceuticals and confectionery. Pop. (1989E) 220,000.

Montreal Quebec, Canada. 45 31N 73 34W. City on Montreal island at confluence of Rs. Ottawa and St. Lawrence. Industries inc. manu. aircraft, railway equipment, plastics, clothing, footwear and cement. Port exporting grain, timber and paper; ice-free from May–Nov. Pop. (1986c) 1,015,420.

Montreuil-sous-Bois Seine St. Denis, France. Suburb of E Paris noted for growing peaches; manu. metal goods, glass, paints and chemicals. Pop. (1982c) 93,394.

Montreux Vaud, Switzerland. 46 26N 6 55E. Town near the E end of L. Geneva. Tourist centre with a 4 m. (6.5 km.) long frontage on the L. Pop. (1989E) 20,278.

Montrose Tayside Region, Scotland. 56 43N 2 29W. Town in central E sit. on a peninsula between the North Sea and Montrose Basin. In-dustries inc. oil and allied services, fisheries, canning, chemicals, distillery and tourism. Pop. (1982E) 11,000.

Mont St. Michel Manche, France. Islet in the Bay of St. Michel, an inlet of the English Channel, in NW France. Rocky peak 260 ft. (79 metres) high and 3 acres in area. Accessible by a causeway. Noted for its Benedictine monastery. Pop. (1982c) 80.

Montserrat West Indies. 16 45N 62 12W. Island in the Leeward Islands sit. SW of Antigua. Area, 38 sq. m. (98 sq. km.). Cap. Plymouth. Volcanic, rising to 3,002 ft. (915 metres) a.s.l. in Soufrière. The main occupation is agric. esp. cotton growing and tourism. Pop. (1985E) 11,852.

Monza Lombardia, Italy. 45 35N 9 16E. Town in N sit. on Lambro R. Industries inc. manu. textiles, carpets and machinery. Important car race track. Pop. (1983E) 122,476.

Moorea Society Islands, South Pacific. 17 32S 149 50W. Island sit. NW of Tahiti in the Windward group. Area, 51 sq. m. (132 sq. km.). Chief town Afareaitu. Chief products vanilla, copra, coffee. Pop. (1983E) 7,000.

Moose Jaw Saskatchewan, Canada. 50 23N 105 32W. Town in central S of province. Industries inc. meat packing, flour milling and clothing. Pop. (1984E) 35,118.

Mopti Mali. 14 30N 4 12W. Cap. of region sit. on Niger R. Agric. centre. Pop. (1976c) 53,885.

Moquegua Peru. 17 20S 70 55W. (i) Dept. of S Peru, bounded W by the Pacific. Area, 6,245 sq. m. (16,174 sq. km.). Chief towns, Moquegua, Ilo. Crossed by mountain ridges in the E; fertile and irrigated land in the W. Mainly agric., producing vines, olives, cotton, sugar-cane, figs, corn, cereals, potatoes. Stock farming on higher ground. There are associated processing industries. Pop. (1981c) 101,610. (ii) City sit. SE of Lima on Moquegua R. Dept. cap. Trading and processing centre of an agric. area; industries inc.

cotton ginning, wine making, flour milling, pressing olive oil. Exports wine, olive oil, cotton, copper, lead, through the Pacific port of Ilo. Pop. (1984E) 10,460.

Moradabad   Uttar Pradesh, India. 28 50N 78 50E. Town sit. ENE of Delhi on Ramganga R. Industries inc. manu. brassware, cotton goods and carpets. Pop. (1981C) 330,051.

Morar, Loch   Highland Region, Scotland. 56 57N 5 40W. L. in w, its w end lying s of Mallaig, 11 m. (18 km.) by 1.5 m. (2.4 km.) and 987 ft. (301 metres) deep. The deepest loch in Scot.

Morava River   Czechoslovakia. 48 10N 16 59E. Rises in the Sudeten Mountains and flows 230 m. (368 km.) s through Moravia to the Austrian/Czech. border, continuing down the border to join Danube R. just above Bratislava.

Morava River   Yugoslavia. 44 43N 21 03E. Formed by union of S. and W. Morava R. near Stalac, and flowing 130 m. (208 km.) N to join Danube R. ESE of Belgrade.

Moravia   Czechoslovakia. Central region of Czech. bounded w by Bohemia, E by Slovakia. Mainly plateau drained by Rs. Morava and Oder. Rises to the Sudeten Mountains in the N and the Carpathian Mountains in the E. Chief towns are Brno, Ostrava, Olomouc and Gottwaldov. Mainly agric., with important mineral resources.

Moray Firth   Scotland. 57 50N 3 30W. Inlet of the North Sea in NE Scot. with Inverness sit. at its head.

Morayshire   Scotland. Former county now part of Grampian Region.

Morbihan   France. 47 55N 2 50W. Dept. in Brittany region on the Bay of Biscay. Area, 2,611 sq. m. (6,763 sq. km.). Chief towns are Vannes (the cap.) and Lorient. Much of it is barren. The main products are cereals, potatoes and cider apples. Noted for the megaliths of Carnac and Locmariaquer. Pop. (1982C) 590,889.

Mordovian Autonomous Soviet Socialist Republic   54 20N 44 30E. Russian Soviet Federal Socialist Republic, U.S.S.R. Autonomous rep. in the bend of Volga R. ESE of Moscow. Area, 10,110 sq. m. (26,185 sq. km.). Cap. Saransk, sit. ESE of Moscow. Mainly agric. The chief products are grain, sugar-beet, sheep and dairy products. Industries inc. manu. electrical goods, textiles, furniture and building materials. Pop. (1989E) 964,000.

Morecambe and Heysham   Lancashire, England. 54 04N 2 53W. Town and seaside resort in NW sit. on Morecambe Bay. Tourism is the chief industry. Pop. (1982E) 43,000.

Morecambe Bay   England. 54 07N 3 00W. Inlet of the Irish Sea on the NW coast between the Furness peninsula and Lancashire.

Morelia   Michoacán, Mexico. 19 42N 101 07W. State cap. sit. WNW of Mexico City. Industries inc. textiles, flour milling, sugar refining and tanning; manu. tobacco products. Pop. (1980C) 353,055.

Morelos   Mexico. State in central Mex. Area, 1,923 sq. m. (4,980 sq. km.). Cap. Cuernavaca, sit. s of Mexico City. Mainly agric. The main products are maize, rice, wheat, coffee and sugar-cane. Pop. (1980C) 947,089.

Morena, Sierra   Spain. 38 00N 5 00W. Mountain range in s Spain between Rs. Guadiana and Guadalquivir. Forms a barrier between Andalusia and N Spain, at an average height of 2,500 ft. (762 metres) a.s.l.

More og Romsdal   Norway. County of w Norway bounded w by North Sea, E by the Dovrefjell and Trollheimen mountains. Area, 5,810 sq. m. (15,049 sq. km.). Chief towns, Kristiansund (cap.), Alesund, Molde, Andalsnes, Orstavik. Mountainous with high valleys, fjords and offshore islands. Main occupations farming, fishing. Industries inc. tourism. Pop. (1984E) 237,548.

Moreton Bay   Queensland, Australia. 27 20S 153 15E. Bay in SE of state receiving Brisbane R. Site of the original penal settlement of 1824–39.

Morgantown West Virginia, U.S.A. 39 38N 79 57W. City sit. s of Pittsburgh, Pennsylvania, on the Monongahela R. Employment in U.S. Dept. of Energy, West Virginia hospitals, Monongahela General Hospital and West Virginia University; manu. taps and valves, shirts. Pop. (1985E) 26,500.

Morioka Honshu, Japan. 39 42N 141 09E. Cap. of Iwate prefecture sit. N of Sendai. Commercial centre. Industries inc. manu. iron goods and toys.

Morlaix Finistère, France. 48 35N 3 50W. Town in NW sit. on an inlet of the English Channel. Industries inc. tanning, brewing and flour milling; manu. tobacco products. Port exporting fruit, vegetables and dairy products. Pop. (1982C) 19,541.

Morley West Yorkshire, England. 53 46N 1 36W. Town sit. SSW of Leeds. Industries inc. distribution, light and heavy engineering, electrical and auto parts assembly. Pop. (1981C) 44,019.

Morocco 31 00N 6 00W. Kingdom bounded N by Mediterranean, E by Algeria, s by the disputed Western Sahara, w by Atlantic. Area, inc. Western Sahara, 274,461 sq. m. (710,850 sq. km.). Chief towns are Rabat (cap.), Casablanca, Meknès, Fez, Oujda and Marrakesh. The N Mediterranean region is separated from the s Sahara region by the Atlas Mountains, rising to 13,664 ft. (4,165 metres) a.s.l. in Djebel Toubkal. Rainfall varies from 30–35 in. (76–89 cm.) (annual rainy season) in the N to under 10 in. (25 cm.) annually in the s. Mainly agric. The chief products are cereals, vines, olives, almonds, fruit, livestock, cork and fish. The main exports are phosphates from Khouribga and Youssoufia, iron, manganese and lead ores. Pop. (1987E) 23,557,000.

Morogoro Tanzania. 4 49S 37 40E. Town sit. W of Dar es Salaam. Centre of an agric. area, with mining in the vicinity.

Morón de la Frontera Sevilla, Spain. 37 08N 5 27W. Town sit. SE of Seville. Trading centre for cereals, olive oil and wine; manu. cement, ceramics and soap.

Moroni Comoros, Indian Ocean. 11 40S 43 16E. Cap. on Njazídja. Trading centre for vanilla, cacao, coffee. Passenger port. Noted for numerous mosques. Pop. (1980C) 20,112.

Morpeth Northumberland, England. 55 10N 1 41W. Town in NE sit. on Wansbeck R. Market town. Industries inc. pharmaceuticals, light engineering and mineral water manu. Pop. (1983E) 16,000.

Mortlake Greater London, England. 51 27N 0 16W. Residential district of SW London on Thames R.

Mortlock Solomon Islands, South Pacific. Atoll sit. NE of Bougainville consisting of about 20 inlets on a reef. Area, 205 acres. Sometimes called Taku.

Moscow Russian Soviet Federal Socialist Republic, U.S.S.R. 55 45N 37 42E. Cap. of the U.S.S.R., the Russian Soviet Socialist Rep., and of Moscow Region, sit. on Moskva (Moscow) R. Moscow was first mentioned in 1147. It became the cap. of a separate principality in the 13th cent. and of the grand principality of Vladimir in the 14th cent. Peter the Great transferred the cap. to what is now Leningrad in 1712 but Moscow became cap. again in 1918. The Kremlin or Citadel is a triangular walled enclosure in the city centre. A commercial centre since the middle ages its industries inc. manu. textiles, steel, locomotives, vehicles, aircraft, machinery, machine tools, chemicals, paper and tobacco products. River port with 2 harbours on the Moskva R. and 1 on the Moscow–Volga canal. Pop. (1989E) 8,967,000.

Moselle France. 48 59N 6 33E. Dept. in Lorraine region, bounded N by Luxembourg and NE by Federal Republic of Germany. Area, 2,399 sq. m. (6,214 sq. km.). Chief towns are Metz (the cap.) and Thionville. Contains part of the Lorraine coal and iron fields and has important metallurgical industries. Pop. (1982c) 1,007,189.

Moselle River   France/Luxembourg/ Federal Republic of Germany. 50 22N 7 36E. Rises in the Vosges Mountains and flows 340 m. (544 km.) NW past Metz and Thionville to form part of the Luxembourg/F.R.G. frontier, then turns NE in F.R.G. past Trier to join Rhine R. at Coblenz. Moselle grapes are grown on the banks of its meandering lower course. Navigable for small boats below Frouard. Main tribs. are Rs. Meurthe and Saar.

Moskva (Moscow) River   Russian Soviet Federal Socialist Republic, U.S.S.R. 55 05N 38 50E. Rises on the w boundary of Moscow Region and flows 310 m. (496 km.) E past Moscow and Kolomna to join Oka R. Linked with Volga R. by Moscow Canal.

Mosquito Coast   Nicaragua. Coastal strip, *c.* 40 m. (64 km.) wide along the Caribbean coast, named after the original Meskito Indian inhabitants. The main occupations are lumbering and fruit growing, esp. bananas.

Mossel Bay   Cape Province, Republic of South Africa. 34 12S 22 08E. Town half-way between Cape Town and Port Elizabeth, on the s coast. Industries inc. oyster fishing and tourism.

Mossley   Greater Manchester, England. 53 30N 2 02W. Town sit. ENE of Manchester on Tame R. Industries inc. textiles, engineering and metal working. Pop. (1987E) 10,272.

Most   Bohemia, Czechoslovakia. 50 32N 13 39E. Town sit. NW of Prague in a lignite-mining area. Industries inc. chemicals and metal working; manu. ceramics and glass. Pop. (1983E) 61,000.

Mostaganem   Algeria. 35 56N 0 05E. Town ENE of Oran sit. on the Gulf of Arzew. Port exporting wine, fruit and wool. Pop. (1982E) 170,000.

Mostar   Bosnia-Herçegovina, Yugoslavia. 43 20N 17 49E. Town sw of Sarajevo sit. on Neretva R. Market

town for a fruit and wine producing area. Industries inc. textiles; manu. tobacco products. Pop. (1981c) 110,377.

Mosul   Mosul, Iraq. 36 20N 43 08E. Cap. of governorate in the N sit. on Tigris R. Commercial centre for a farming and oil producing area, trading in grain, fruit, livestock and wool. Industries inc. tanning and flour milling.

Motala   Östergötland, Sweden. 58 33N 15 03E. Town on NE shore of L. Vättern, where the Motala R. leaves the L. Industries inc. television sets, torpedoes, steel, refrigerators and kitchen stoves. Pop. (1990E) 41,715.

Motala River   Sweden. Leaves L. Vättern and flows 60 m. (96 km.) E through Ls. Bor, Rox and Gla, and past Norrköping to enter the Baltic Sea.

Motherwell   Strathclyde Region, Scotland. 55 48N 4 00W. Town in s, sit. near R. Clyde. Industries inc. steelmaking and milling, engineering, food processing and the manu. of diesel engines, trucks, electronics, earthmoving equipment and machinery. Pop. (1985E) 30,287.

Moulins   Allier, France. 46 34N 3 20E. Cap. of dept. in central France sit. on Allier R. Industries inc. brewing and tanning; manu. hosiery and furniture. Pop. (1982c) 25,548.

Moulmein   Tenasserim, Burma. 16 30N 97 38E. Town sit. ESE of Rangoon on the Salween R. estuary. Industries inc. sawmilling and rice milling. Port exporting rice and teak. Pop. (1983E) 202,967.

Moundou   Logone Occidental, Chad. 8 34N 16 05E. Cap. of prefecture and centre for cotton-growing district.

Mountain Ash   Mid Glamorgan, Wales. 51 42N 3 24W. Town sit. NNW of Cardiff on the Cynon R. Pop. (1983E) 14,000.

Mountain View   California, U.S.A. 37 23N 122 04W. City sit. NW of San

José. Industries inc. computer software, genetic engineering, biological and medical research; manu. electronic equipment. Pop. (1980c) 58,655.

Mount Gambier   South Australia, Australia. 37 50S 140 46E. Town sit. SSE of Adelaide. City trading in agric. and hort. produce, clothing, textiles and timber. Industries inc. tourism and transport. Pop. (1989E) 28,250.

Mount Isa   Queensland, Australia. 20 44S 139 30E. Town in W of state. Industries inc. mining zinc, lead, silver and copper, and smelting, cattle and tourism. Pop. (1988E) 24,104.

Mount Lyell   Tasmania, Australia. 42 03S 145 38E. Town sit. WSW of Launceston. The main industry is mining and processing copper.

Mount Vernon   New York, U.S.A. 40 54N 73 50W. Suburb of New York immediately N of the Bronx. Industries inc. electronics, chemicals, plastics, special machinery, printing presses and paint. Pop. (1980c) 68,500.

Mourne Mountains   Down, Northern Ireland. 54 10N 6 04W. Range extending 14 m. (22 km.) NE to SW from Dundrum Bay to Carlingford Lough, rising to Slieve Donard, 2,796 ft. (852 metres) a.s.l.

Mouscron   West Flanders, Belgium. 50 45N 3 12E. Town sit. SW of Ghent. Industries inc. manu. cotton goods and carpets. Pop. (1983E) 54,315.

Moyobamba   San Martín, Peru. 6 02S 76 58W. Town sit. N of Lima in the Amazon basin. Dept. cap. Centre of a fertile agric. area. Industries inc. making wine, spirits and alcohol; manu. straw hats. Hot springs, gold and oil deposits nearby. Pop. (1984E) 14,000.

Mozambique   19 00S 36 00E. Republic on SE coast of Africa opposite Madagascar. Area, 308,642 sq. m. (799,380 sq. km.). Cap. Maputo. Coastal plain rising to an inland plateau at an average 1,000–2,000 ft. (308–615 metres) a.s.l. and rising to 7,000 ft. (2,154 metres) a.s.l. near the Malawi frontier to the NW. Drained by Zambezi R. delta. The main occupation is farming; main products are sugar, nuts, copra, cotton, sisal and rice. Pop. (1988E) 14,907,000.

Mozambique Channel   Indian Ocean. 19 00S 41 00E. Strait extending over 1,000 m. (1,600 km.) SSW to NNE between Mozambique and Madagascar. Width 250–600 m. (400–960 km.).

Mubende   Buganda, Uganda. 0 35N 31 23E. Town sit. WNW of Kampala. Trading centre of an agric. area, with tungsten, beryl, tantalite deposits.

Mudugh   Somalia. Region of central Somalia bounded E by the Indian Ocean, W by Ethiopia. Arid plain watered (W) by Webi Shebeli R. Some agric. along R., elsewhere the main occupation is stock raising.

Mugla   Turkey. 37 12N 28 22E. (i) Province of SW Turkey bounded W by Aegean Sea, S by Mediterranean, E by Elmali range. Area, 4,917 sq. m. (12,735 sq. km.). Drained by Ak, Dalaman, and Koca Rs. Thickly forested with rich mineral deposits. Agric. areas produce sesame, millet, tobacco, olives, wheat. Pop. (1986c) 486,290. (ii) Town sit. SSE of Smyrna. Provincial cap. and trading centre for a farming area.

Muharraq   Bahrain. 26 15N 50 39E. Island near NE shore of Bahrain. Airport and seaplane base, connected to Bahrain island by a causeway. Pop. (1981c) 61,853.

Mühlhausen   Erfurt, German Democratic Republic. 51 12N 10 27E. Town in SW sit. on Unstrut R. Industries inc. manu. textiles, machinery and furniture. Pop. (1981E) 43,000.

Mukden   *see* **Shenyang**

Mülheim-an-der-Ruhr   North Rhine-Westphalia, Federal Republic of Germany. 51 24N 6 54E. Town WSW of Essen sit. on Ruhr R. Industries inc. power station construction, tube manu. leather and commerce. Pop. (1988E) 176,300.

Mulhouse Haut-Rhin, France. 47 45N 7 20E. Town in the extreme E sit. on Ill R. and Rhône-Rhine Canal. The main industry is textiles; also manu. chemicals from local potash, paper and machinery. Pop. (1982C) 113,794 (agglomeration, 220,613).

Mull Strathclyde Region, Scotland. 56 27N 6 00W. Island in the Inner Hebrides separated from the mainland by the Sound of Mull and the Firth of Lorne. Area, 351 sq. m. (909 sq. km.). The chief town is Tobermory. The island is mountainous and the main occupations are crofting and fishing.

Mullingar Westmeath, Ireland. 53 32N 7 20W. County town in centre of county sit. on Brosna R. and Royal Canal. Market and industrial centre. Manu. vinyl floor coverings, tobacco products, computer equipment. Pop. (1986E) 12,000.

Multan Multan, Pakistan. 30 10N 71 36E. Cap. of Multan division in central Pakistan. Industries, powered by Sui natural gas, inc. manu. of textiles, hosiery, fertilizers and carpets. Pop. (1981E) 732,000.

Munankuan *see* **Yuyijuan**

München *see* Munich

Muncie Indiana, U.S.A. 40 11N 85 23W. City sit. NE of Indianapolis, on White R. Industries inc. manu. glassware, furniture and electrical equipment. Pop. (1988E) 73,320.

Munger Bihar, India. Town sit. ESE of Patna on Ganges R. Commercial centre. Industries inc. manu. tobacco products. Pop. (1981C) 129,260.

Munich Bavaria, Federal Republic of Germany. 48 08N 11 35E. Cap. of province sit. on Isar R. Commercial and route centre. Industries inc. printing and publishing, brewing; manu. motor vehicles, electronics, precision instruments, machinery, clothing and food products. Pop. (1987E) 1,188,800.

Münster North Rhine-Westphalia, Federal Republic of Germany. 51 57N 7 37E. Town in NW sit. on Dortmund-Ems Canal. Industries inc. chemical, mechanical and medical engineering, biotechnology. Manu. light machinery, pharmaceuticals. Pop. (1989E) 252,000.

Munster Ireland. Province in SE, comprising the counties of Clare, Cork, Kerry, Limerick, Tipperary and Waterford. Pop. (1988E) 1,020,577.

Muntafiq *see* **Nasiriyah**

Murano Veneto, Italy. Town built on islets in the Venice Lagoon, N of the city. The main industry is glass making of which there are 284 producers, 50 of them large concerns. Pop. (1990E) 6,717.

Mureş River Hungary/Romania. 46 15N 20 13E. Rises in the Carpathian Mountains in NE Romania and flows 550 m. (880 km.) WSW past Targu-Mureş and Arad to join Tisza R. at Szeged, Hungary.

Mufulira Copperbelt, Zambia. 12 30S 28 12E. Town sit. just SW of Zaïre border in the main copper-mining area. Industries inc. smelting and refining copper. Pop. (1980E) 149,778.

Murghab River Afghánistán/ U.S.S.R. 38 10N 73 59E. Rises in the W Hindu Kush and flows 450 m. (720 km.) W and NW across NW Afghánistán to the SE Turkmen Soviet Socialist Rep. It turns N across the Kara Kum desert, supplies the Mary Oasis and then peters out in the desert.

Murmansk Russian Soviet Federal Socialist Republic, U.S.S.R. 68 58N 33 05E. City on E shore of Kola Bay in the extreme NW of the U.S.S.R. Main industries are shipbuilding, fishing, fish canning and sawmilling. Sheltered port, on an inlet of the Barents Sea and ice-free, exporting timber and apatite. Pop. (1987E) 432,000.

Muroran Hokkaido, Japan. 42 18N 140 59E. Town sit. NNE of Hakodate. The main industry is iron and steel, also oil refining; manu. cement. Port exporting iron and steel, coal and timber.

Murray River Australia. 35 22S 139 22E. Rises in the Australian Alps near Mount Kosciusko, New South Wales, and flows 1,600 m. (2,560 km.) w along the New South Wales/ Victoria border into South Australia, where it turns s at Morgan and enters Encounter Bay through L. Alexandrina. The whole of its w course is through an extensive level basin containing two-thirds of the irrigated land in Australia which produces grain, sheep and fruit. Used for water storage, hydroelectric power and irrigation. Main tribs. are R. Lachlan, Murrumbidgee, Goulburn and Darling.

Mürren Bern, Switzerland. 46 34N 7 54E. Resort SE of Bern sit. near the Jungfrau, at 5,415 ft. (1,650 metres) a.s.l. Pop. (1990E) 320.

Mur River Austria/Yugoslavia. 46 18N 16 53E. Rises at w end of the Niedere Tauern, Austria, and flows 300 m. (480 km.) E to Leoben, then SE past Graz to the Yugoslav border which it follows E for *c.* 25 m. (40 km.); enters Yugoslavia briefly and continues on the Yugoslav/Hungarian border to join Drava R. at Legrad. Navigable below Graz.

Murshidabad West Bengal, India. Former cap. of Bengal-Bihar-Orissa sit. N of Calcutta. Robert Clive fought a war with Nabob Siraj-ud-dowla at Plassey 27 m. (43 km.) from Murshidabad, which estab. British rule in 1757. The palace with 1,000 doors still stands on the bank of Bhagirathi R., known for silk, ivory and brassware industries, and also for varieties of mangoes. Pop. (1981C) 21,946.

Murrumbidgee River Australia. 39 43S 143 12E. Rises in the Great Dividing Range in New South Wales and flows 1,050 m. (1,680 km.) N and then w to join Murray R. on the border with Victoria NE of Yungera. The main trib. is Lachlan R. Used to irrigate an agric. basin producing sown pastures, cereals, fruits, vines and vegetables.

Muş Turkey. 38 44N 41 30E. (i) Province of E Turkey. Area, 2,946 sq. m. (7,630 sq. km.). Mountainous and barren, drained by Murat R. Pop.

(1986C) 339,492. (ii) Town sit. S of Erzurum. Provincial cap.

Muscat and Oman *see* **Oman**

Musgrave Ranges Australia. 26 10S 131 50E. Mountain ranges extending *c.* 200 m. (320 km.) w to E along the border between NW South Australia and the s of Northern Territory, rising to Mount Woodroffe, 4,970 ft. (1,529 metres) a.s.l. Heavily eroded.

Muskegon Michigan, U.S.A. 43 14N 86 16W. City sit. at mouth of Muskegon R. on L. Michigan, and NW of Grand Rapids. Industries inc. aerospace, automotive parts, furniture, paper, chemicals and tourism. Pop. (1980C) 40,823.

Muskegon River Michigan, U.S.A. Rises in L. Houghton in the centre of state and flows SW to L. Michigan.

Muskogee Oklahoma, U.S.A. 35 45N 95 22W. City sit. SE of Tulsa. Service and trading centre for an agric. area; processing and manu. industries. Pop. (1980C) 40,011.

Musoma Lake Province, Tanzania. Town sit. NE of Mwanza at mouth of Mara R. on SE shore of L. Victoria. Centre of a farming and gold mining area.

Mussau Islands Papua New Guinea. Island group in the Bismarck Arch. NW of New Zealand, consisting of Mussau and Emirau Islands, both volcanic, and coral islets. Area, of Mussau, 160 sq. m. (414 sq. km.), Emirau, 20 sq. m. (52 sq. km.). The main product is coconuts.

Musselburgh Lothian Region, Scotland. 55 57N 3 04W. Town E of Edinburgh sit. at mouth of Esk R. on Firth of Forth. Industries inc. metal manu., paper and sawmills, nets, printing, stone cleaning and carving and manu. of electrical products. Pop. (1981C) 18,861.

Mutare Zimbabwe. 19 00S 34 40E. City sit. near border with Mozambique, at 3,550 ft. (1,082 metres) a.s.l. Main industries inc. board and paper mills, vehicle assembly, textiles, milling and gold mining in vicinity. Pop. (1982E) 69,621.

Muzaffarpur   Bihar, India. 26 07N 85 32E. Town NNE of Patna sit. on Burhi Gandak R. Trading centre for grain, tobacco and sugar-cane. Industries inc. rice and sugar milling. Pop. (1981C) 190,416.

Mwali   (formerly Monéli) Comoros, Indian Ocean. 12 15S 43 45E. Island SE of Njazídja in the Mozambique Channel. Area, 112 sq. m. (290 sq. km.). Chief town Fomboni. Fertile, producing palms, cacao, vanilla, copra. Pop. (1980C) 17,194.

Mwanza   Tanzania. 2 30S 32 54E. Town sit. N of Tabora on S shore of L. Victoria. Provincial cap. and shipping centre for cotton, nuts and rice. Pop. (1978C) 110,611.

Mweru, Lake   Zaïre/Zambia. 9 00S 28 45E. L. on the frontier between Zaïre and Zambia extending 75 m. (120 km.) SW to NE and 30 m. (48 km.) wide, at a height of 3,050 ft. (930 metres) a.s.l. Fed by Luapula R. in S and drained by Luvua R. in N.

Myanmar (Burma)   South Asia. Rep. bounded E by China, Laos and Thailand, W by Indian Ocean, Bangladesh and India. Area, 261,789 sq. m. (678,000 sq. km.). Its divisions are Myanmar proper, the states of Kachin, Shan, Kayah and Karen and Chin division. Cap. Rangoon (Yangon). Agric. is the most important occupation and rice the most important crop. There are over 35,000 sq. m. (89,600 sq. km.) of reserve forests. Over 3,000 elephants help to extract the teak and hardwoods from the forests. Minerals include oil, silver, zinc, copper, lead, nickel, antimony, tin and tungsten. Pop. (1988E) 39,840,000.

Myitkyina   Kachin, Burma. 25 24N 97 26E. State cap. sit. NNE of Mandalay on Irrawaddy R. Market town. Pop. (1985E) 15,000.

Mymensingh   Dacca, Bangladesh. 24 25N 90 24E. Cap. of district sit. N of Dacca on an old channel of the Brahmaputra R. Industries inc. jute pressing and engineering; manu. electrical goods. Pop. (1984E) 175,000.

Mysore   *see* **Karnataka**

Mysore   Karnataka, India. 12 17N 76 41E. State cap. sit. SW of Bangalore. Industries inc. paper, electronics, wood products, ivory, textiles; manu. paints and fertilizers. Pop. (1981C) 479,081.

Mytishchi   Russian Soviet Federal Socialist Republic, U.S.S.R. 55 55N 37 46E. Town sit. NNE of Moscow. Industries inc. textiles; manu. railway rolling stock.

N

Naaf River Burma/Bangladesh. Tidal inlet of the Bay of Bengal extending 30 m. (48 km.) from Taungbro past Maungdaw.

Naaldwijk Zuid Holland, Netherlands. Town sit. SW of The Hague. Centre of an agric. area, trading in fruit and potatoes. Pop. (1989E) 27,910.

Naantali Turku-Pori, Finland. Town sit. W of Turku on the Gulf of Bothnia. Railway terminus and seaside resort with some industry inc. paper making.

Naas Kildare, Ireland. 53 12N 6 40W. Town sit. WSW of Dublin. Punchestown racecourse is sit. to the SE. Pop. (1981C) 8,345.

Naberejnye-Chelny Russian Soviet Federal Socialist Republic, U.S.S.R. 55 42N 52 19E. Town sit. ESE of Yelabuga in the Tatar Autonomous Soviet Socialist Rep. on Kama R. Grain trading and distribution centre. Industries inc. flour milling, sawmilling, metal-working; manu. bricks, railway sleepers, wine. Pop. (1987E) 480,000.

Nablus Israeli-occupied Jordan. 32 13N 35 15E. Town sit. N of Jerusalem between Mount Gerizim and Mount Ebal. Manu. olive oil, soap.

Nacka Stockholm, Sweden. 59 18N 18 10E. Suburb of SE Stockholm. Industries inc. manu. steam turbines, diesel motors, drills. Pop. (1989E) 63,145.

Nafud Saudi Arabia. A desert area consisting of the Great Nafud sit. S of the Syrian Desert, and the Little Nafud to the SE.

Naga Hills India. 26 00N 95 00E. An area of forested hills in the NE on the Burma border. A wild undeveloped region.

Nagaland India. 26 00N 95 00E. State of NE India bounded E by Burma, N and W by Assam, S by Manipur. Area, 6,401 sq. m. (16,579 sq. km.). Cap. Kohima. Wild, mountainous area, little developed, with a tribal pop. Minerals inc. coal, clay, limestone and sand. The main occupation is rice growing, much of it on hill terraces. Industries inc. sugar, paper, timber and distilling. Pop. (1981C) 774,930.

Nagano Honshu, Japan. 36 40N 138 10E. Cap. of Nagano prefecture sit. NW of Tokyo. Manu. machinery, textiles and food products. Trades in silk. Pop. (1983E) 327,000.

Nagapattinam Tamil Nadu, India. 10 46N 79 50E. Port and town sit. E of Tiruchirapalli on the Coromandel coast. Industries inc. railway engineering. Exports inc. groundnuts, tobacco and cotton goods. Pop. (1981C) 82,828.

Nagar Haveli India. Enclave forming part of the Territory of Dadra and Nagar Haveli on the W coast at the foot of the W Ghats. Area of the Territory, 189 sq. m. (490 sq. km.), bounded N by Gujarat, S by Maharashtra, and 25 m. (40 km.) inland from Daman. Mainly agric. with *c.* 17,000 hectares under crops, much of it terraced. Main crop rice. Thickly forested with teak and kahir. Pop. of the Territory (1981C) 103,676.

Nagasaki Kyushu, Japan. 32 48N 129 55E. Cap. of Nagasaki prefecture, tourist city and port, sit. on SW coast. Industries inc. shipbuilding, engineering, fishing and the manu. of machine tools and tortoiseshell goods. Exports inc. shipping machinery, and metal goods. The second atomic bomb was dropped on this city in 1945. Pop. (1988E) 446,000.

Nagercoil Tamil Nadu, India. 8 11N

77 26E. Town sit. SE of Trivandrum.
Manu. inc. mats and ropes. Pop.
(1981C) 171,648.

**Nagorno-Karabagh Autonomous
Region** U.S.S.R. Part of the Azerbaijan Soviet Socialist Republic.
Area, 1,700 sq. m. (4,403 sq. km.).
From 18th cent. formed a separate
khanate, but estab. as an autonomous
region 1923. Cap. Stepanakert, sit.
WSW of Baku. Chief industries silk,
wine, dairying and building materials.
Also agric. inc. cotton, grapes and
winter wheat, with 48,000 hectares
under crops. Pop. (1989E) 188,000.

Nagoya Honshu, Japan. 35 10N
136 55E. Cap. of Aichi prefecture and
port sit. E of Kyoto. Industries inc. engineering, metal working and the
manu. of machine tools, sewing
machines, bicycles, watches, textiles,
chemicals and porcelain. Pop. (1983E)
2,058,000.

Nagpur Maharashtra, India. 21
08N 79 10E. Town sit. ENE of Bombay. Orange-growing area. Manu. inc.
pharmaceuticals, cotton goods and
hosiery. Trades in oranges. Pop.
(1981C) 1,219,461.

Nagykanizsa Zala, Hungary. 46
27N 17 00E. Town sit. SW of Budapest. Industries inc. distilling, brewing, flour milling and the manu. of
footwear. Trades in grain and livestock.

Nagykörös Szolnok, Hungary. 47
05N 19 48E. Town sit. SE of Budapest.
Industries inc. distilling, canning and
flour milling. Trades in wine and fruit.
Pop. (1984E) 27,000.

Naha Okinawa, Japan. 26 13N 127
40E. Port and largest town sit. on SW
coast of island. Manu. inc. textiles and
pottery. Sugar and dried fish are exported. Pop. (1983E) 302,000.

Nahichevan Nahichevan Autonomous Soviet Socialist Republic,
U.S.S.R. 39 12N 45 24E. Town sit. SE
of Erevan near Aras R. Rep. cap. Industries inc. cotton ginning, wine
making, food processing, manu. metal
goods, furniture, clothing, building
materials. Anc. city ruled in turn by

Persians, Arabs, Mongols, Turks and
Armenians.

**Nahichevan Autonomous Soviet
Socialist Republic** U.S.S.R. Part of
the Azerbaijan Soviet Socialist Rep.
sit. on borders of Turkey and Iran near
the SW slopes of the Zangezur mountains. Area, 2,120 sq. m. (5,490 sq.
km.). Annexed by Russia 1828 and
constituted as an Autonomous Rep.
1924. Cap. Nahichevan, sit. SE of
Erevan. Chief industries are agric.
(mainly cotton and tobacco), silk,
clothing, cotton, canning and meat-
packing; also fruit and grape-growing.
Mulberry trees are grown for sericulture. Pop. (1989E) 295,000.

Nahuel Huapí, Lake Neuquén/Río
Negro, Argentina. 41 00S 71 32W.
Sit. near the Chilean border with San
Carlos de Bariloche on its S shore.
Area, 205 sq. m. (530 sq. km.). Now a
National Park and tourist area.

Nairn Highland Region, Scotland.
57 35N 3 53W. Town and port sit. on
Moray Firth at mouth of R. Nairn. The
main occupation is fishing. Pop.
(1985E) 10,000 (district).

Nairn River Scotland. Rises NE of
Fort Augustus and flows NE for 38 m.
(61 km.) to enter the Moray Firth at
Nairn.

Nairnshire Scotland. Now part of
Highland Region.

Nairobi Central Province, Kenya. 1
17S 36 49E. City sit. NW of Mombasa.
It replaced Mombasa as cap. in 1907.
Cap., commercial, industrial and
administrative centre, in a densely
populated upland farming region, with
a temperate climate. Industries inc.
vehicle assembly, publishing, brewing, meat packing, fruit preserving,
canning, flour milling, paper making;
manu. soap, chemicals, textiles, soft
drinks, foodstuffs, furniture, leather
and tyres. Nairobi National Park is a
game reserve. Pop. (1989E) 1,500,000.

Naivasha, Lake Kenya. 0 46S 36
21E. Sit. in the Great Rift Valley, at
6,187 ft. (1,886 metres) a.s.l. Area,
108 sq. m. (279 sq. km.). The resort of
Naivasha is sit. on its NE shore.

Najaf Iraq. Holy city sit. S of Baghdad near Euphrates R. A centre of pilgrimage.

Najin *see* **Rajin**

Nakhichevan *see* **Nahichevan**

Nakhodka Russian Soviet Federal Socialist Republic, U.S.S.R. 42 48N 132 52E. Port sit. ESE of Vladivostok on the Pacific coast. Industries inc. ship repairing, food processing, fishing, sawmilling and the manu. of plywood and matches.

Nakhon Pathom Nakhon Pathom, Thailand. Cap. of province, sit. W of Bangkok. Trades in sugar-cane and rice. Pop. (1970E) 34,000.

Nakhon Ratsima (Korat) Nakhon Ratchasima, Thailand. 15 00N 102 06E. Cap. of province sit. NE of Bangkok on Mun R. An important commercial centre trading in livestock and rice. Pop. (1982E) 89,000.

Nakhon Sawan Nakhon Sawan, Thailand. 15 42N 100 06E. Cap. of province, and port, sit. N of Bangkok below confluence of Rs. Nan and Ping on R. Menam Chao Phraya. Trades in teak. Pop. (1970E) 46,100.

Nakhon Si Thammarat Nakhon Si Thammarat, Thailand. 8 26N 99 58E. Cap. of province sit. S of Bangkok near the E coast. Trades in rice, fruit and coconuts. Pop. (1970E) 41,000.

Nakuru Rift Valley, Kenya. 0 17S 36 04E. Cap. of province in the SW, sit. on the N shore of L. Nakuru at 6,000 ft. (1,846 metres) a.s.l. The centre of an important agric. district producing wheat, maize, coffee, sisal, pyrethrum and dairy produce. Pop. (1984E) 130,000.

Nakuru, Lake Great Rift Valley, Kenya. 0 22S 36 05E. A salt L. 8 m. (13 km.) long. Noted for flamingoes. The surrounding area is studded with extinct volcanic craters.

Nalaikha Central Aimak, Mongolia. Town sit. SE of Ulan Bator. Major coalmining centre. Railway.

Nalchik Russian Soviet Federal Socialist Republic, U.S.S.R. 43 29N 43 37E. Cap. of the Kabardino-Balkar Autonomous Soviet Socialist Rep. sit. W of Grozny. Industries inc. meat packing, flour milling and the manu. of oilfield equipment, textiles, footwear and furniture. Pop. (1987E) 236,000.

Namangan Uzbekistan, U.S.S.R. 41 00N 71 40E. Town sit. E of Tashkent in the Fergana valley. Manu. cotton and food products. Trades in livestock and fruit. Pop. (1987E) 291,000.

Namaqualand South Africa/Namibia. A desert coastal region consisting of Great Namaqualand in the N and Little Namaqualand in the S, separated by the Orange R. Copper, diamonds and tungsten are found.

Namchi Sikkim, India. 27 09N 86 23E. Town sit. SW of Gangtok in the Himalaya foothills. Centre of an agric. area noted for Buddhist monasteries and Sangachelling, Pemiongchi and Tashiding. Pop. (1981C) 1,444.

Namen *see* **Namur**

Namib Desert Namibia. Desert coastal region 800 m. (1,280 km.) long and 30–100 m. (48–160 km.) wide.

Namibia Independent Republic, formerly UN trusteeship territory (South West Africa). Bounded on N by Angola, E by Botswana and Republic of South Africa. The territory (excluding Walvis Bay and certain islands) was proclaimed a German protectorate in 1884, but was surrendered to the Forces of the Union of South Africa in 1915 at Khorab. The administration was vested in the Government of the Union of South Africa by mandate of the League of Nations in 1920. In 1921 the Governor-General delegated certain of his functions to the Administrator of the Territory, who was assisted by an Advisory Council and, from 1925, by an Executive Committee and the Legislative Assembly. In 1971 the International Court of Justice ruled in an advisory opinion that the Republic of South Africa's presence in Namibia was illegal. Became independent in 1990.

Area, 318,827 sq. m. (825,762 sq. km.). Chief town, Windhoek. Namibia is essentially a stock-raising country, the scarcity of water and poor rainfall rendering agriculture, except in the northern and north-eastern portions, almost impossible. Generally speaking, the southern half is suited for the raising of small stock, while the central and northern portions are better fitted for cattle. The chief minerals found are diamonds. Pop. (1988E) 1,288,000.

Nampa　Idaho, U.S.A. 43 34N 116 34W. City sit. W of Boise. Processing and distribution centre for an agric. and dairying area. Industries inc. processing and packing food and seed. Manu. computer chips, mobile homes, campers, wood mouldings and trim. Pop. (1989E) 29,900.

Nampula　Mozambique. 15 09S 39 14E. (i) Province in NE. Area, 30,218 sq. m. (78,265 sq. km.). Pop. (1982E) 2,498,800. (ii) Cap. of (i) sit. W of Mozambique. Provincial cap. and agric. trading centre. Pop. (1980E) 146,000.

Namur　Belgium. 50 28N 4 52E. (i) A province in the SW bordering on France. Area, 1,413 sq. m. (3,660 sq. km.). The N is fertile producing fruit, and the S is wooded with important quarries of marble, slate, sandstone and limestone. The principal Rs. are Meuse, Sambre and the Lesse. Pop. (1987E) 415,326. (ii) Cap. of province of same name sit. at confluence of Rs. Meuse and Sambre. Industries inc. tanning, flour milling and the manu. of cutlery, porcelain, pottery and glass. Pop. (1988E) 103,104.

Nanaimo　British Columbia, Canada. 49 10N 123 56W. City sit. W of Vancouver on the Vancouver Island. Port exporting lumber and pulp. Industries inc. pulp, lumber, fishing and tourism. Pop. (1983E) 49,347.

Nanchang　Jiangsu, China. 28 41N 115 53E. Cap. of province in E sit. on Gan R. Manu. machinery, chemicals, porcelain, pottery, glass, paper and textiles. Trades in rice, timber, tea, tobacco, cotton and hemp. Pop. (1987E) 1,190,000.

Nancowrie　Nicobar Islands, India. Land-locked harbour between Nancowrie Island (S) and Camorta Island, NNW of Great Nicobar. Centre of inter-island trade. Pop. (1981C) 14,968.

Nancy　Meurthe-et-Moselle, France. 48 41N 6 12E. Cap. of dept. in NE, sit. on Meurthe R. and Marne-Rhine Canal. An important centre of the iron and steel industry and also manu. machinery, textiles, glass, pottery, and footwear. There are many fine 18th-cent. buildings. Pop. (1982C) 99,307 (agglomeration, 306,982).

Nanda Devi　Uttar Pradesh, India. 30 22N 79 59E. A Himalayan peak in the Garhwal district. 25,645 ft. (7,817 metres) a.s.l.

Nanga Parbat　Jammu and Kashmir. 35 14N 74 35E. A Himalayan peak in W Kashmir sit. W of Skardu. 26,660 ft. (8,126 metres) a.s.l.

Nanjing　*see* **Nanking**

Nanjing (Nanking)　Jiangsu, China. 32 03N 118 47E. Cap. of province and R. port in E sit. on Yangtse-kiang R. An industrial centre with manu. of textiles, paper, machine tools and fertilizers. The cloth 'nankeen' originated here. Pop. (1987E) 2,229,000.

Nanking　*see* **Nanning**

Nanning　Guangxi Zhuang Autonomous Region, China. 22 48N 108 20E. Cap. of Region and port on S coast sit. NNW of Pakhoi on Siang R. Industries inc. sugar refining, tanning and the manu. of textiles and chemicals. Trades in spices, rice, hides and tobacco. Pop. (1982C) 866,000.

Nanterre　Hauts-de-Seine, France. 48 53N 2 13E. Cap. of dept. and a suburb of Paris sit. near Seine R. at the foot of Mont Valérien. Manu. inc. metal and electrical goods and chemicals. Pop. (1982C) 90,371.

Nantes　Loire-Atlantique, France. 47 13N 1 33W. Cap. of dept. in NW and port sit. at confluence of Rs. Loire, Erdre and Sèvre Nantaise. Industries inc. oil refining, boatbuilding, sugar refining, tobacco processing, vegetable canning and the manu. of soap, fertilizers, textiles, vegetable oils and chocolate. Pop. (1990E) 264,857 (agglomeration, 473,841).

Nantucket   Massachusetts, U.S.A. 41 16N 70 03W. An island sit. s of Cape Cod, mainly a summer resort. Pop. (1988E) 7,000.

Nantwich   Cheshire, England. 53 04N 2 32W. Town sit. sw of Crewe on R. Weaver. Industries inc. light engineering and clothing manu. Noted for brine baths. Pop. (1978E) 11,500.

Nanumanga   Tuvalu, Pacific. Atoll in the N Tuvalu group. Area, 1 sq. m. (2.5 sq. km.). The main product is copra. Pop. (1983E) 760.

Nanumea   Tuvalu, Pacific. Northern-most atoll of the Tuvalu group. Area, 1.4 sq. m. (3.5 sq. km.). Formerly St. Augustine. Pop. (1983E) 910.

Nanyuki   Central Province, Kenya. 0 01N 37 00E. Town sit. NNE of Nairobi at 9,000 ft. (2,743 metres) at the foot of Mount Kenya. Rail terminus, resort and centre of a farming area. Pop. (1984E) 27,000.

Napier   North Island, New Zealand. 39 29S 176 54E. City and port sit. on sw shore of Hawke Bay. Centre of an important farming and industrial area. It exports wood pulp and lumber, wool, meat, fruit and food products. Industries inc. pulp mill, tobacco products, electronics and tourism. Pop. (1985E) 50,100.

Naples   Campania, Italy. 26 08N 81 48W. City sit. SE of Rome on the Bay of Naples. Sea port, the second largest after Genoa, with heavy passenger traffic. Exports food products, gloves, textiles. Tourist and commercial centre. Archbishopric. Industries inc. food processing, railway repair shops, ship-building, iron and steel, aircraft assembly, textiles, tanning, oil refining, manu. gloves and shoes. Pop. (1988E) 1,212,387.

Nara   Honshu, Japan. 34 40N 135 49E. Cap. of Nara prefecture sit. E of Osaka. Manu. inc. textiles, dolls and

fans. There are many temples and shrines and tourism is important. Pop. (1988E) 344,000.

Narbonne   Aude, France. 43 11N 3 00E. Town in s, near the Gulf of Lions. Industries inc. brandy distilling, sulphur refining and the manu. of barrels, pottery, bricks and tiles. Trades in wine, brandy and honey. Pop. (1982C) 42,657.

Narew River   Poland/U.S.S.R. 52 26N 20 42E. Rises NW of Pruzhany, U.S.S.R., flows 275 m. (440 km.) WNW into Poland past Lomza, then SSW past Erock and WSW to enter Vistula R. w of Nowy Dwor. Navigable in the lower course. Main tribs. Western Bug, Wkra, Orzyc, Omulew, Biebrza and Suprasl Rs.

Nariño   Colombia. Dept. of sw Colombia bounded w by Pacific, s by Ecuador. Area, 11,548 sq. m. (29,562 sq. km.). Chief towns Pasto (cap.), Ipiales, Tumaco. Lowland near the coast rises to E to the Andes, with two ranges divided by Patía R. Climate varies greatly according to altitude. Agric. lowlands produce cereals, maize, coffee, cacao, sugar, bananas, tobacco, cotton. Thick forests yield rubber, quinine, tagua nuts, resins, varnish, timber. Livestock are grazed on the uplands. The chief mineral is gold. Pop. (1985E) 848,618.

Narragansett Bay   Rhode Island, U.S.A. 41 20N 71 15W. An inlet of the Atlantic enclosing a number of islands inc. Rhode, Prudence and Conanicut.

Narsapur   Tamil Nadu, India. Town sit. ESE of Ellore in the Godavari R. delta. Rail terminus trading in tobacco, sugar, coconuts. Industries inc. rice milling.

Narva   Estonia, U.S.S.R. 59 23N 28 12E. Town and port sit. on Narva R. near the Gulf of Finland. It has an important textile industry. Pop. (1980E) 74,000.

Narvik   Nordland, Norway. 68 26N 17 25E. Port in N sit. on Ofot Fjord. It is a technological centre with advanced industry production. Main occupations are exporting iron ore. Pop. (1990E) 18,500.

Naryn River   U.S.S.R. 40 54N 71 45E. Rises in several headstreams in the Tien Shan mountains with the main stream rising in the Petrov Glacier, and flows 449 m. (718 km.) w through fertile wheatlands past Naryn, then N and w through the Ketmen-Tyube valley to Toktogul, then sw into Fergana Valley and enters the Uzbek Soviet Socialist Rep. to join Kara Darya R. near Balykchi. Together they form Syr-Darya R. Used for irrigation. Main tribs. Lesser Naryn, Son-Kul and Kokomeren Rs.

Nashik   Maharashtra, India. Town sit. NE of Bombay on upper Godavari R. A holy place for Hindus, with many temples and shrines. Agric. products inc. grapes and onions. Manu. inc. brass and copper goods and soap. There is an aeronautics industry in the vicinity. Pop. (1981C) 262,428.

Nashua   New Hampshire, U.S.A. 42 46N 71 27W. City s of Manchester sit. at confluence of Rs. Merrimack and Nashua. Industries inc. computers, electronics. Manu. paper, machinery, plastics, greeting cards and hardware. Pop. (1980C) 67,817.

Nashville   Tennessee, U.S.A. 36 09N 86 48W. City sit. NE of Memphis on Cumberland R. in the fertile bluegrass region. State cap. Commercial, financial, communications and industrial centre. Industries inc. food processing, printing and publishing, transport equipment, aerostructures, clothing, chemicals, air-conditioning equipment and insurance. Pop. (1988E) 481,380.

Nassau   New Providence Island, Bahamas. 25 00N 77 30W. Cap. of Bahamas and port sit. on NE coast. Exports inc. pulpwood, cucumbers, salt and crawfish. Tourism is the most important industry, with over 1m. visitors a year. Finance and banking is another important industry. Pop. (1980C) 135,437.

Nässjö   Jönköping, Sweden. 57 38N 14 45E. Town sit. ESE of Jönköping. Railway junction. Industries inc. metal working, wood working, textiles, paper making; manu. clothing. Pop. (1990E) 30,749.

Natal   Rio Grande do Norte, Brazil. 5 47S 35 13W. Cap. of state and port in NE sit. on Potengi R. near its mouth. Industries inc. salt refining and the manu. of textiles. Exports inc. sugar, cotton, hides, salt and carnauba wax. Pop. (1980C) 376,446.

Natal   Republic of South Africa. 28 30S 30 30E. Province bounded s by Cape Province, sw by Lesotho, NW by the Transvaal, N by Swaziland and Mozambique, E by the Indian Ocean. Area, 33,578 sq. m. (86,967 sq. km.) Chief cities Pietermaritzburg (cap.), Durban. Coastal plain rises (W) to the Drakensberg mountains. Drained by Tugela, Buffalo and Pongola Rs. Main occupations farming, esp. sugar, stock raising, and mining. Main products are sugar, cotton, tea, tobacco, coal, gold. Industries are concentrated on Durban and the cap. Tourism is important on the coast. Pop. (1985E) 2,145,018 (excluding Zwa Zulu).

Natanya   Plain of Sharon, Israel. 32 20N 34 51E. Town and resort sit. NNE of Jaffa-Tel Aviv. Industries inc. food processing, diamond cutting and the manu. of chemicals and textiles.

Natchez   Mississippi, U.S.A. 31 34N 91 23W. City sit. ssw of Vicksburg on bluffs above the Mississippi R. where it forms the Louisiana border. Trading and distribution centre for a cotton growing and stock farming area. Industries inc. meat packing, food canning; manu. tyres, pulp, timber and paper products. Noted for colonial and ante-bellum houses. Pop. (1984E) 22,438.

Natick   Massachusetts, U.S.A. 42 17N 71 21W. Town sit. wsw of Boston. Manu. inc. metal goods, boxes, corrugated paper and footwear. Pop. (1980C) 29,461.

Natitingou   Atacora, Benin. 10 19N 1 22E. Cap. of dept. in N of Benin. Pop. (1979E) 50,800.

Natron, Lake   Great Rift Valley, Tanzania. 2 25S 36 00E. Sit. on the Kenya frontier, it has salt and soda deposits.

Natural Bridge   Virginia, U.S.A.

37 40N 79 40W. Village sit. W of Richmond. It takes its name from an arch of limestone 215 ft. (66 metres) high with a span of 90 ft. (27 metres) which straddles Cedar Creek.

Naumburg   Halle, German Democratic Republic. 51 09N 11 48E. Town sit. SSW of Halle on Saale R. Industries include tourism, metal goods and the manu. leather goods, textiles and toys. Pop. (1990E) 30,000.

Nauplion   Peloponnessos, Greece. 37 34N 22 48E. Cap. of the Argolis prefecture and port sit. on the gulf of Argolis, and WSW of Athens. Trades in fruit, vegetables and tobacco. Pop. (1981C) 10,609.

Nauru   South West Pacific. 0 32S 166 56E. Atoll forming an independent Commonwealth Rep., sit. NE of Australia. Area, 8 sq. m. (21 sq. km.). Fertile land rising to central plateau. The main product is phosphates, mining and processing being the main industry. It is estimated that the phosphate deposits will be exhausted by the end of the century. Pop. (1983E) 8,100.

Navsari   Gujarat, India. 20 51N 72 55E. Town sit. SSE of Surat, in W central India. Manu. inc. copper and brass goods and textiles. Trades in millet and cotton. Pop. (1981C) 106,793.

Naxos   Aegean Islands, Greece. 37 02N 25 35E. The largest island in the Cyclades. Area, 169 sq. m. (438 sq. km.). Noted in anc. times for wine and the cult of Dionysus. Products inc. wine, olive oil, fruit and emery.

Nayarit   Mexico. 22 00N 105 00W. Maritime state on the Pacific. Area, 10,417 sq. m. (26,979 sq. km.). Cap. Tepic, sit. NW of Guadalajara. Mainly mountainous with a narrow coastal plain producing cattle, cereals, cotton, sugar-cane, coffee and tobacco. There are some lead and silver deposits. The principal R. is Santiago. Pop. (1989E) 857,359.

Nazareth   Lower Galilee, Israel. 32 42N 35 17E. Town sit. ESE of Haifa. Associated with the early life of Jesus Christ. A dormitory town for Haifa.

Ndola   Western Province, Zambia. 12 58S 28 38E. Cap. of province sit. in central Zambia, in the Copperbelt. An industrial and railway centre concerned with copper and cobalt refining. Other industries inc. sugar refining and brewing. Pop. (1980E) 282,439.

N'djaména   Chari-Baguirmi, Chad. 12 10N 14 59E. Formerly Fort Lamy. Cap. sit. near the Cameroon frontier at confluence of Shari and Logone Rs. A communication centre trading in dates, livestock, millet and salt. Pop. (1986E) 511,700.

Neagh, Lough   Northern Ireland. 54 38N 6 24W. Sit. between Londonderry, Antrim, Down, Armagh and Tyrone, it is the largest L. in the Brit. Isles. Area, 153 sq. m. (396 sq. km.). Rs. Blackwater, Upper Bann and Ballinderry run into it, and it is drained to the Atlantic by Lower Bann R.

Neanderthal   North Rhine-Westphalia, Federal Republic of Germany. A valley sit. E of Düsseldorf noted for the discovery of the skeleton of Neanderthal Man.

Neath   West Glamorgan, Wales. 51 40N 3 48W. Town sit. ENE of Swansea on R. Neath, providing sea wharf facilities. Major manu. centre for metal industries inc. aluminium products, cans, industrial and car components. Pop. (1984E) 26,100.

Nebit-Dag   Turkmen Soviet Socialist Republic, U.S.S.R. 39 30N 54 22E. Town sit. ESE of Krasnovodsk on the Trans-Caspian railway. Oil refining centre serving the Vyshka oilfield.

Nebraska   U.S.A. 41 30N 100 00W. State of central U.S.A. bounded N by South Dakota, W by Wyoming, E by Missouri and Iowa. Area, 77,237 sq. m. (20,044 sq. km.). Sold by France to the U.S. as part of the Louisiana Purchase of 1803 and settled in 1847. Chief cities Lincoln (cap.), Omaha. Sit. on the Great Plains, sloping gently W to SE, with table-land (W) reaching 5,340 ft. (1,628 metres) a.s.l. Drained by Niobrara, Platte and Nemaha Rs. into Missouri R. Continental climate. Mainly agric., chief crops wheat, maize, sorghum, rye, hay. Fruit, vegetables and livestock farming are also important. There are associated processing industries. Pop. (1987E) 1,594,000.

Nechako River   Canada. 53 56N 122 42W. Rises in the Ootsa and Tetachuk Ls., British Columbia, and flows 250 m. (400 km.) NE and E to join Fraser R. at Prince George.

Neckar River   Federal Republic of Germany. 49 31N 8 26E. Rises SW of Schwenningen in the Black Forest and flows 228 m. (365 km.) NE past Rothenburg to Plochingen, then NNW past Bad Canstatt to Eberbach, then W past Heidelberg to enter Rhine R. at Mannheim. Its valley is noted for scenic beauty. Navigable up to Bad Canstatt.

Necochea   Buenos Aires, Argentina. 38 30S 58 50W. Port E of Bahia Blanca sit. at mouth of Quequén Grande R. Grain is exported. Pop. (1984E) 50,000.

Needles, The   England. 50 39N 1 35W. Chalk cliffs off W tip of Isle of Wight in the English Channel, SSW of Southampton. There are 3 cliffs *c.* 100 ft. (30 metres) high.

Neembucu   Paraguay. Dept. of S Paraguay bounded W and S by Argentina, NE by L. Ypoá. Area, 5,354 sq. m. (13,706 sq. km.). Cap. Pilar. Marshy lowland area between Paraguay and Paraná Rs. and drained by Tebicuary R. Main occupations lumbering, farming, esp. cattle, sugar, cotton, maize, fruit. Pop. (1982E) 70,689.

Negeri Sembilan   Malaysia. State of Peninsular Malaysia on the Strait of Malacca. Area, 2,565 sq. m. (6,644 sq. km.). Cap. Seremban. Crossed NW to SE by foothills of the central Malayan range, drained by Linggi and Muar Rs. Main products rubber, rice, coconuts, tin. Pop. (1980C) 563,955.

Negev   Israel. 30 45N 34 50E. A triangular region in the S bordering on Egypt and Jordan. Mainly semi-desert but agric. has been made possible by irrigation and cereals, sugar-beet, fruit and vegetables are now grown. Some

chemicals are obtained from the Dead Sea, and oil is produced at Heletz.

Negombo   Western Province, Sri Lanka. 7 13N 79 51E. Port sit. N of Colombo on W coast. Manu. inc. brassware. Pop. (1981C) 61,376.

Negro, Río   Argentina. 41 02S 62 47W. R. formed by confluence of Rs. Neuquén and Limay near Nequén in N. Patagonia, flows 400 m. (640 km.) E and SE through a fruit-growing area and enters the Atlantic below Viedma. Used for irrigation.

Negro, Río   Brazil. 3 08S 59 55W. R. rising in E Colombia as Guainia R. and flowing 1,300 m. (2,080 km.) E then S as part of the Colombia/Venezuela border, it enters Brazil and turns E to Barcelos, then SE to join Amazon R. below Manaus. Flows through dense forests in the Amazon basin. Linked with Orinoco R. by Casiquiare R.

Negro, Río   Uruguay. 26 01S 50 30W. R. rising in Brazil near Bagé and flowing 500 m. (800 km.) WSW across Uruguay to enter Uruguay R. below Fray Bentos. Used for storage and hydroelectric power, with a dam creating an 87 m. (122 km.) L. in its middle course.

Negros   Philippines. 10 00N 123 00E. Sit. between Cebu and Panay, it is the fourth largest island. Area, 5,278 sq. m. (13,670 sq. km.). Mainly mountainous with a coastal plain producing sugar-cane, coffee, rice and tobacco. The chief town is Bacolod.

Neiba   Bahoruco, Dominican Republic. 18 28N 71 25W. (i) Town sit. W of Santo Domingo near L. Enriquillo. Provincial cap. and centre of an area producing sugar-cane and timber. (ii) Sierra de Neiba, range extending *c.* 60 m. (96 km.) E from the Haiti border to the Yaque del Sur R. along the N side of L. Enriquillo. Rises to 5,545 ft. (1,690 metres).

Nei Monggol   *see* **Inner Mongolia**

Neisse River   Czechoslovakia/Poland/German Democratic Republic. (i) 52 04N 14 46E. Lusatian Neisse R.

rises NE of Liberec in N Bohemia and flows 140 m. (224 km.) s then WNW into G.D.R. and continues N along the G.D.R./Polish frontier to join Oder R. just N of Guben. (ii) Glatzer Neisse 50 49N 17 50E. rises in the Sudeten highlands of SW Poland near the Czech. frontier and flows 120 m. (192 km.) N past Klodzko (Glatz) then E past Nysa and NNE to join Oder R. just ESE of Brzeg.

Neiva Huila, Colombia. 2 56N 75 18W. Cap. of dept. sit. SSW of Bogotá on the upper Magdalena R. Manu. panama hats. Trades in cattle, coffee and tobacco. Pop. (1984E) 180,000.

Nejd Saudi Arabia. Province of central and E Saudi Arabia. Area, 450,000 sq. m. (1,166,000 sq. km.). Chief cities Riyadh (cap. and national joint cap.), Buraida, Anaiza. Main settlements are in oases above the Jabal at Tuwaiq escarpment (E centre) and the Arma plateau. Main agric. crops, dates, cereals, fruit. Nomadic stock rearing is important. Pop. (1970E) 3m.–4m.

Nellore Andhra Pradesh, India. 14 27N 79 59E. Town sit. NNW of Madras on Penner R. The main occupations are rice and oilseed milling. Pop. (1981C) 237,065.

Nelson Lancashire, England. 53 51N 2 13W. Town sit. NNE of Burnley. Industries inc. textiles, furniture, wallcoverings, precision engineering, food processing, printing and hospital disposables. Pop. (1981C) 30,455.

Nelson South Island, New Zealand. 41 17S 173 17E. (i) District in NW. Area, 3,937 sq. m. (10,197 sq. km.). Mainly mountainous but the Waimea Plain in the N produces fruit. Timber and coal are produced. Pop. (1986C) 69,648. (ii) Cap. of district of same name and port sit. on the N coast on Tasman Bay. Industries inc. food processing and timber-related industries. Fruit, vegetables, timber and tobacco are exported. Pop. (1989E) 45,400.

Nelson River Canada. 57 04N 92 30W. Issues from L. Winnipeg, N. Manitoba, and flows 400 m. (640 km.) NE to enter the Hudson Bay at Port

Nelson. Trad. waterway of fur-traders to the Hudson Bay Company's York factory.

Neman River U.S.S.R. 55 18N 21 23E. Rises SSW of Minsk in the Belorussian Soviet Socialist Rep., flows 597 m. (955 km.) W to Grodno, then N into the Lithuanian Soviet Socialist Rep. past Neman and Sovetsk to enter Courland lagoon by a delta. Main tribs. Viliya, Nevezys, Dubysa, Shchara and Sheshupe Rs. Used for logging.

Nemi, Lake Lazio, Italy. 41 43N 12 42E. Lake sit. SE of Rome in the Alban Hills.

Nenagh Tipperary, Ireland. 52 52N 8 12W. Town in the S Midlands sit. on R. Nenagh. Industries inc. lead and zinc mining, glass cutting, and the manu. of aluminium, electronics, farm machinery and textiles. Trades in potatoes and dairy produce. Pop. (1981E) 5,871.

Nene River England. 52 38N 0 13E. Rises SSW of Daventry and flows 90 m. (144 km.) NE past Northampton and Peterborough to enter the Wash just N of Sutton Bridge. Navigable, by artificial channelling, to Peterborough.

Nenetz National Area Russian Soviet Federal Socialist Republic, U.S.S.R. Sit. in the Arkhangelsk Region in the N. The principal occupations are reindeer breeding and fishing. Cap. Naryan-Mar, sit. NE of Arkhangelsk.

Nepál South Asia. 28 00N 84 30E. Kingdom bounded S and W by India, N by Tibet, E by Sikkim and India. Area, c. 56,136 sq. m. (145,391 sq. km.). Chief cities Káthmándu (cap.), Patan, Bhadgaon. Himalayan ranges traverse the country across the N, with spurs running out SSW and gradually descending to the Terai swamps and jungles, and some cultivated land. About one-third is under forest; 9.6m. acres are under paddy, 3.7m. under maize, wheat and millet. Towards the Tibetan border the mountains rise to high peaks and include Everest, Kanchenjunga, Dhaulagiri. Minerals inc. bismuth, antimony, cobalt, sapphire, rubies, corundum, iron ore, bauxite, sulphur. Industries, recently developed, inc. jute and sugar milling, lumbering and associated manu., chemicals, leather working. Pop. (1985E) 16.63m.

Neskaupstadur Sudur-Mula, Iceland. 65 10N 13 43W. Town on Nord Fjord; fishing and fish processing centre. Pop. (1988E) 1,714.

Ness, Loch Scotland. 57 15N 4 30W. A long, narrow and deep Loch extending NE from Fort Augustus, and lying SW of Inverness. It is drained to the Moray Firth by R. Ness.

Netherlands North Europe. 52 00N 5 30E. Kingdom bounded W and N by North Sea, S by Belgium, E by Federal Republic of Germany. Land area, 13,102 sq. m. (33,935 sq. km.). William of Orange (1533-84), as the German count of Nassau, inherited vast possessions in the Netherlands and the Princedom of Orange in France. He was the initiator of the struggle for independence from Spain (1568–1648). The Congress of Vienna joined the Belgian provinces, the 'Austrian Netherlands' before the French Revolution, to the Northern Netherlands. The union was dissolved by the Belgian revolution of 1830, and the Treaty of London, in 1839, constituted Belgium an independent kingdom. Chief cities Amsterdam (cap.), Rotterdam, 's-Gravenhage (The Hague), Utrecht, Eindhoven. (The Hague is the administrative cap.) Low-lying, with *c*. 25% of its area b.s.l. and protected by dykes and drainage systems. Land reclamation, esp. in the Zuider Zee, has added *c*. 700 sq. m. (*c*. 1,813 sq. km.) since 1920. Drained by Rhine R. delta and an extensive network of canals. There are three main divisions: sand-dunes (W) rising to 100 ft. (30 metres) and occupied by a chain of towns; agric. land producing cereals, vegetables; heathland with peat bogs and some fertile areas (E) rising to the hills of Limburg province (SE). A densely populated and industrial country with major commercial centres at Amsterdam and Rotterdam, the latter being the most important European port. Minerals inc. coal,

coke and some crude oil, none in any quantity. Industries inc. metalworking and diamond cutting, food processing, chemicals, printing, textiles, manu. pottery, glass and wood products, clothing, paper, leather and rubber products. Pop. (1989E) 14,805,240.

Netherlands Antilles West Indies. 12 10N 69 00W. Two groups of islands, the Leeward and Windward Islands; the Leewards are sit. N of Venezuela; the Windwards are E of Puerto Rico. Area, 308 sq. m. (800 sq. km.). Chief town, Willemstad. The Leeward Islands are Curaçao, Bonaire. The Windwards Islands are St. Maarten (N half owned by France), St. Eustatius, Saba. All the islands are rocky and infertile. Fishing and tourism are important. Curaçao has an oil refinery which handles Venezuelan oil and employs 20% of the work force; the Leewards also have textile, petrochemical, electronics and tobacco factories. Rum is distilled on St. Maarten. Pop. (1988E) 192,866.

Nettilling, Lake Northwest Territories, Canada. 66 30N 71 00W. L. in SW Baffin Island, NE Canada, within the Arctic circle. Area, 70 m. (112 km.) by 65 m. (104 km.). Drains (W) into Foxe Basin.

Neubrandenburg German Democratic Republic. 53 33N 13 15E. (i) County bounded on E by Poland. Area, 4,227 sq. m. (10,948 sq. km.). Pop. (1988E) 620,500. (ii) Cap. of county of same name sit. in NE on Tollense R. Manu. inc. machinery, chemicals and paper. Pop. (1988E) 90,000.

Neuchâtel Switzerland. 47 00N 6 55E. (i) Canton in the Jura Mountains bordering on France. It joined the confederation in 1815. Area, 308 sq. m. (797 sq. km.). Vine growing and dairy farming are carried on in the valleys. Pop. (1988E) 157,434. (ii) Cap. of canton of same name sit. on N shore of the L. of Neuchâtel. Manu. inc. watchmaking, precision machinery, electronics, cigarettes and chocolate. Pop. (1988E) 31,423.

Neuchâtel, Lake of Switzerland.

46 52N 6 50E. Sit. at the S foot of the
Jura Mountains, it is the largest L. in
the country. Area, 83 sq. m. (214 sq.
km.).

Neuilly-sur-Seine   Hauts-de-Seine,
France. A NW suburb of Paris sit. on
Seine R. Manu. inc. motor-cars, machine tools and perfumes. Pop.
(1982c) 64,450.

Neumünster   Schleswig-Holstein,
Federal Republic of Germany. 54 04N
9 59E. Town in N. Industries inc. railway engineering, and the manu. of
machinery, paper and textiles. Pop.
(1984E) 79,200.

Neunkirchen   Saarland, Federal
Republic of Germany. 49 20N 7 10E.
Town sit. NE of Saarbrücken. Commercial and shopping centre. Pop.
(1989E) 51,500.

Neuquén River   Argentina. 38 55S
68 55W. Rises in the Andes on the
Chilean frontier and flows 320 m.
(512 km.) S and SE past Añelo to Neuquén, where it joins Limay R. to form
Río Negro. Used for irrigation in its
lower course. Main trib. Agrio R.

Neuruppin   Potsdam, German
Democratic Republic. Town and resort in central N sit. on L. Ruppin. Industries inc. printing and the manu. of
chemicals. Pop. (1988E) 27,000.

Neuss   North Rhine-Westphalia,
Federal Republic of Germany. 51 12N
6 41E. Town sit. on Rhine R. opposite
Düsseldorf. Industries inc. paper milling, food processing and the manu. of
agric. machinery. Pop. (1984E)
144,800.

Neustadt-an-der-Weinstrasse Rhineland-Palatinate, Federal Republic of
Germany. 49 21N 8 08E. Town SW of
Ludwigshafen sit. at the foot of the
Haardt Mountains. It is an important
centre of the wine trade and manu.
agric. and viticultural implements as
well as textiles. Pop. (1990E) 54,000.

Neustrelitz   Neubrandenburg, German Democratic Republic. 53 21N 13
04E. Town and resort NNW of Berlin
sit. between Zierker See and Glam-
becker See. Industries inc. food

processing, woodwork, metal engineering. Important railway junction.
Pop. (1990E) 27,000.

Neuwerk   Frisian Islands, Federal
Republic of Germany. Island just NW
of Cuxhaven. Area, 1.3 sq. m. (3.4 sq.
km.). Wreck-salvage station and
lighthouse.

Nevada, Sierra   Spain. 37 05N
3 10W. Mountain range in S Spain extending *c*. 60 m. (96 km.) W to E and
rising to Mulhacén, 11,421 ft. (3,481
metres) a.s.l. Many summits have
permanent snow.

Nevada, Sierra   California, U.S.A.
39 00N 120 30W. Mountain range in
E California extending 400 m. (640
km.) NW to SE and rising to Mount
Whitney, 14,495 ft. (4,418 metres)
a.s.l. Inc. three National Parks:
Yosemite, Sequoia and King's Canyon.

Nevada   U.S.A. 39 20N 117 00W.
State of W U.S.A. bounded W and S by
California, SE by Arizona, E by Utah,
N by Idaho and Oregon. Area,
110,540 sq. m. (286,289 sq. km.).
First settled (as part of Utah) in 1851
and became a State in 1864. It was enlarged with land from Utah and Arizona 1866–67. Chief cities Carson
City (cap.), Reno, Sparks, Las Vegas.
Mainly sit. in the arid Great Basin,
with eroded mountain ranges running
N to S and intervening alluvial flat-
lands. Peaks rise to Boundary Peak,
13,145 ft. (4,007 metres) (SW) and descend (SE) to lowlands near Colorado
R. Most of the Rs. are intermittent and
flow into interior basins, sinks and Ls.
The Federal Govt. own *c*. 86% of the
land area. About 9m. acres are farmland, mainly supporting cattle with
some sheep and hay crops, cereals and
root crops. Irrigation is essential.
Mineral deposits are valuable, and inc.
copper, gold, sand and gravel, gypsum,
iron ore, mercury. Service industries,
mining and smelting, chemicals and
lumbering are the main industries;
legalized gambling is the main industry in Reno and Las Vegas and a major
state revenue. Pop. (1989E) 1,195,700.

Neva River   U.S.S.R. Issues from
L. Ladoga and flows 46 m. (74 km.) W

to enter the Gulf of Finland by a delta at Leningrad. Main tribs. Mga, Tosna and Izhora Rs. Connected by canals with Volga R. and White Sea.

Nevers Nièvre, France. 47 00N 3 09E. Cap. of dept. sit. at confluence of Rs. Loire and. Nièvre, in central France. Manu. inc. metal goods, pharmaceutical products, pottery and wine. Pop. (1989E) 47,611.

Nevis Leeward Islands, West Indies. 17 00N 62 30W. Island just SE of St. Kitts with which it forms a political unit. Area, 50 sq. m. (130 sq. km.). Cap. Charlestown. Rises in the centre to Nevis Peak, 3,596 ft. (1,096 metres). The main crop is sea-island cotton, also sugar, maize, fruit. Pop. (1980C) 9,300.

New Albany Indiana, U.S.A. 38 18N 85 49W. City sit. on Ohio R. opposite Louisville, Kentucky. Manu. inc. plywood, veneers and refrigerated bakery products. Pop. (1988E) 37,540.

New Amsterdam Guyana. 6 17N 57 36W. Port sit. at mouth of Berbice R. The centre of an agric. district growing rice and sugar-cane. Pop. 18,000.

Newark Delaware, U.S.A. 39 41N 74 45W. City sit. SW of Wilmington on the White Clay and Christina Creeks. Industries inc. vehicle assembly, chemicals, paper. Pop. (1980C) 25,247.

Newark New Jersey, U.S.A. 40 44N 74 10W. City and port sit. W of lower Manhattan on Passaic R. and Newark Bay. An important transportation centre and industries inc. insurance and banking. Pop. (1980C) 329,248.

Newark Ohio, U.S.A. 40 04N 82 24W. City sit. ENE of Columbus on Licking R. Industries inc. railway engineering, and the manu. of lawn mowers, fibre glass and plastics. Pop. (1980C) 41,200.

Newark-on-Trent Nottinghamshire, England. 53 05N 0 49W. Town in central Eng. sit. at confluence of R. Devon and a branch of R. Trent. Industries inc. engineering, textiles,

food processing and vehicles and the manu. of building materials from local gypsum and limestone. Pop. (1988E) 24,365.

New Bedford Massachusetts, U.S.A. 41 38N 70 56W. Port and resort sit. S of Boston on Buzzards Bay. Manu. textile machinery, electrical equipment, footwear, soap and cotton goods. Pop. (1980C) 98,478.

New Brighton England *see* **Wallasey**

New Britain Australian Territories. 6 00S 150 00E. The largest island in the Bismarck Arch., New Guinea Islands. Area, 14,100 sq. m. (36,520 sq. km.). Mountainous with several active volcanoes, the highest being The Father, 7,500 ft. (2,286 metres) a.s.l. Copra, cocoa and timber are exported. The principal town is Rabaul. Pop. (1980E) 227,700.

New Britain Connecticut, U.S.A. 41 40N 72 47W. City, sit. NNE of New Haven. 50% of the country's hardware is manu. here. Other manu. inc. ball bearings, electrical equipment and clothing. Pop. (1980C) 73,840.

New Brunswick Canada. 46 50N 66 30W. Province of E Canada bounded W by Maine, U.S.A., S by the Bay of Fundy, SE by Nova Scotia, E by the Gulf of St. Lawrence and Northumberland Strait, N by Chaleur Bay. Area, 27,633 sq. m. (71,569 sq. km.). Visited by Jacques Cartier in 1534, New Brunswick was first explored by Samuel de Champlain in 1604. It was ceded by the French in the Treaty of Utrecht in 1713 and became a permanent British possession in 1763. It was separated from Nova Scotia and became a province in June 1784, as a result of the great influx of United Empire Loyalists. Chief cities Fredericton (cap.), St. John, Moncton. Low-lying, rising to highland extension of the Appalachian range (NW) and with an indented coastline and numerous Ls. Severe winter climate and heavy rainfall. Forests are extensive and varied. Fertile (S and centre) land supports dairy farming and fodder crops, with some grain and vegetables. Minerals inc. coal, oil shale, natural gas, gypsum, zinc, lead, copper. Industries are mainly based on hydroelectric power and inc. lumbering, footwear and clothing, pulp milling, fishing, fish and food processing, shipbuilding, woodworking, chemicals, textiles. St. John is an ice-free port. Pop. (1989E) 718,500.

New Brunswick New Jersey, U.S.A. 40 29N 74 27W. City sit. SW of Newark on Raritan R. Manu. inc. machinery, motor vehicles, hospital supplies, chemicals, pharmaceutical goods, clothing and dairy products. Pop. (1980C) 41,442.

Newburgh Fife Region, Scotland. 56 20N 3 15W. Port sit. ESE of Perth on the Firth of Tay. Industries inc. boatbuilding, quarrying, clothing and manu. of coated fabrics. Pop. (1983E) 2,180.

Newburgh New York, U.S.A. 41 30N 74 01W. Port sit. N of New York on Hudson R. Manu. inc. machinery, textiles, leather and aluminium goods. Pop. (1980C) 10,183.

Newbury Berkshire, England. 51 25N 1 20W. Town sit. W of Reading on R. Kennet. Industries inc. engineering and electronics. Noted for its racecourse. Pop. (1989E) 28,113.

New Caledonia South West Pacific. 21 30S 165 30E. Overseas Territory of France sit. E of Australia. Area, inc. dependencies, 7,172 sq. m. (18,576 sq. km.). Dependencies of New Caledonia include the Isles of Pines, the Loyalty Islands, Huon Islands, Bélep Archipelago, Chesterfield Islands and Walpole. Cap. Nouméa. Mountainous, rising to Mont Panié, 5,412 ft. (1,665 metres) and subject to hurricanes. Minerals are many and valuable, inc. nickel, chrome, iron, manganese, silver, gold, cobalt, lead and copper. The main industry is smelting. The main agric. crop is coffee. Pop. (1989C) 164,173.

Newcastle New South Wales, Australia. 32 55S 151 45E. Town and port sit. at mouth of Hunter R. It is the centre of the country's principal coalfield with important iron and steel works. Manu. inc. railway rolling stock, locomotives, ships, girders,

chemicals, fertilizers and textiles. Coal, wheat, wool and dairy produce are exported. Pop. (1990E) 500,000.

Newcastle County Down, Northern Ireland. 54 13N 5 54W. Resort sit. S of Belfast on Dundrum Bay. Pop. (1985E) 6,035.

Newcastle Natal, Republic of South Africa. 27 45S 29 58E. Town NNW of Durban sit. at the foot of the Drakensberg at 3,900 ft. (1,189 metres) a.s.l. Industries inc. rubber, iron and steel works and the manu. of stoves, steel windows, clothing, kitchen units, bricks and tiles. Pop. (1985E) 38,000.

New Castle Pennsylvania, U.S.A. 41 00N 80 20W. City sit. NNW of Pittsburgh at the confluence of Neshannock and Shenango Rs. Manu. china, fireworks, steel dioxidizing agents, clay, glass and stone products, metal goods, leatherwork. Pop. (1980C) 33,621.

Newcastle-under-Lyme Staffordshire, England. 53 00N 2 14W. Town sit. W of Stoke-on-Trent. Industries inc. computers, electric motors, cable systems, industrial control machinery, and coalmining. Pop. (1988E) 117,500.

Newcastle upon Tyne Tyne and Wear, England. 54 59N 1 35W. City and port sit. on R. Tyne opposite Gateshead. Industrial centre for engineering, defence equipment and pharmaceuticals. Pop. (1981C) 277,829.

New England U.S.A. Region of NE U.S.A. comprising 6 states: Maine, New Hampshire, Vermont, Massachusetts, Rhode Island, Connecticut. Heavily industrialized area with rocky land rising (W) to the Appalachian mountains and supporting mainly stock farming, dairying and some vegetables. The most important non-manu. industries are fishing and tourism.

New Forest Hampshire, England. 50 53N 14 0W. Anc. royal hunting ground, still partly wooded, in SW Hampshire. Area, 145 sq. m. (376 sq. km.), *c.* 75% woodland and heathland. Noted for breeds of ponies and cattle. Chief towns are Lyndhurst and Ringwood.

Newfoundland   Canada. 48 28N 56
00W. Province of E Canada consisting
of an island in the Atlantic Ocean and
the adjoining coast of Labrador,
across the Strait of Belle Isle. Area,
156,649 sq. m. (405,720 sq. km.).
Newfoundland was discovered by
John Cabot in 1497, and was soon fre-
quented in the summer months by the
Portuguese, Spanish and French for its
fisheries. It was formally occupied in
1583 by Sir Humphrey Gilbert on behalf of the English Crown, but various
attempts to colonize the island remained unsuccessful. Although British
sovereignty was recognized in 1713
by the Treaty of Utrecht, disputes
over fishing rights with the French
were not finally settled until 1904,
when France retained sovereignty of
the off-shore islands of St. Pierre and
Miquelon. In a National Convention
in 1948, confederation with Canada
was decided by a small majority. Chief
cities St. John's (cap.), Corner Brook.
Rocky and deeply indented coastline
with numerous small islands; inland
on Newfoundland Island is plateau
with hills rising to the Long Range
Mountains (NW) with Gros Morne,
2,673 ft. (815 metres). Numerous Ls.
and Rs., with fertile land beside them.
Both the island and Labrador are
thickly forested with valuable timber.
Main minerals are lead, copper, zinc,
gold and silver, and important iron
deposits in SW Labrador, esp. near
Wabush and Labrador City. Industries
other than mining inc. lumbering, pulp
milling, paper-making, fishing and
fish processing, boat-building. Pop.
(1986C) 568,349.

New Georgia   Solomon Islands,
West Pacific. 8 30S 157 20E. Island
group sit. NW of Guadalcanal. Area,
2,170 sq. m. (5,621 sq. km.). Main
island New Georgia; chief town Hobu
Hobu. Main product copra.

New Guinea *see* **Papua New Guinea**

Newham   Greater London, England.
51 32N 0 03E. Boro. of E London on
N bank of Thames R. formed 1965 by
merger of West Ham and East Ham.
Mainly residential, inc. former Royal
Docks (now closed). Industries inc.
chemicals,   sugar   refining,   light
engineering. Pop. (1988E) 206,900.

New Hampshire   U.S.A. 43 40N 71
40W. State in New England bounded
N by Quebec, Canada, E by Maine, SE
by the Atlantic, S by Massachusetts
and W by Vermont across the Connecticut R. Area, 9,304 sq. m. (24,097 sq.
km.). Chief cities Concord (cap.),
Manchester, Nashua. Hilly and well
wooded with numerous Ls., and rising
(N) to the White Mountains of the Appalachian system. Main farm products
are hay, vegetables, dairy products,
apples from *c.* 426,000 acres. Industries in the S inc. manu. leather products, electrical goods and machinery.
Pop. (1988E) 1,085,000.

Newhaven   East Sussex, England.
50 47N 0 03E. Port sit. at mouth of R.
Ouse on the English Channel. There is
some light industry. It has a roll
on/roll off ferry to Dieppe. Pop.
(1989E) 11,200.

New Haven   Connecticut, U.S.A. 41
18N 72 56W. Port sit. on Long Island
Sound. Manu. inc. electrical equipment,   clocks,   cutlery,   hardware,
firearms and sewing machines. Yale
University is sit. here. Pop. (1980C)
126,109.

New Hebrides *see* **Vanuatu**

New Jersey   U.S.A. 40 30N 74 10W.
State of E U.S.A. bounded E by the
Atlantic, N and NE by New York, W by
Delaware and Pennsylvania, S by Delaware Bay. Area, 7,532 sq. m. (19,508
sq. km.). Chief cities Trenton (cap.),
Newark, Jersey City, Paterson, Elizabeth, Camden. Sandy coastline with
marshes and lagoons leads to central
plain and highlands in the N and NW.
An industrial state, but farmland and
market gardens, produce maize, potatoes, sweet potatoes, and fruit, with
mixed farming and dairying in the uplands. The most important mineral is
glass sand. Industries inc. food processing, electrical engineering, oil
refining, chemicals, meat packing,
shipbuilding, manu. vehicles, vehicle
parts, paints and varnishes. Pop.
(1985E) 7,562,000.

New London   Connecticut, U.S.A.
41 21N 72 07W. Port and resort sit. E
of New Haven on Thames R. estuary.
Industries inc. shipbuilding and repair.

Manu. collapsible tubes, dentifrice, floor coverings, doors and windows, clothing and turbines. Pop. (1980c) 28,842.

Newmarket Suffolk, England. 52 15N 0 25E. Town in East Anglia. It is the centre of Brit. horse-racing with a famous racecourse. Pop. (1988E) 16,830.

New Mexico U.S.A. 34 30N 74 10W. State of SW U.S.A., bounded W by Arizona, S by Mexico and Texas, E by Texas and Oklahoma, N by Colorado. Area, 121,412 sq. m. (314,457 sq. km.). The first settlement was established in 1598. Until 1771 New Mexico was the Spanish kings' 'Kingdom of New Mexico'. In 1771 it was annexed to the northern provinces of New Spain. When New Spain won its independence in 1821, it took the name of Republic of Mexico and established New Mexico as its northernmost department. When the war between the U.S. and Mexico was concluded in 1848 New Mexico was recognized as belonging to the U.S., and in 1850 it was made a Territory. Part of the Territory was assigned to Texas; later Utah was formed into a separate Territory; in 1861 another part was transferred to Colorado, and in 1863 Arizona was disjoined, leaving to New Mexico its present area. New Mexico became a state in 1912. Chief cities Santa Fé (cap.), Albuquerque, Las Cruces, Roswell. Crossed N to S by Rio Grande valley, with the Continental Divide (w) and the Southern Rocky Mountains (NE). E of the Rockies is extensive lowland drained by Pecos R. Highest point is Wheeler Peak in the Sangre de Cristo mountains, 13,150 ft. (4,046 metres). Average altitude, 5,700 ft. (1,737 metres) a.s.l. Farms and ranches cover 44.5m. acres, but soil erosion is widespread and much of it is only suited to grazing. Cattle and sheep farming are important, with some crops inc. cotton, hay and fodder. Mineral deposits are considerable and inc. uranium, also perlite, potassium salts, petroleum, natural gas, copper and zinc. Pop. (1988E) 1,507,000.

New Orleans Louisiana, U.S.A. 29 58N 90 07W. City sit. above mouth of

Mississippi. Commercial centre, major R. port and sea port. Important cotton market and communications centre (R. canal, sea, air and rail). Industries inc. oil and gas, ship repairing, processing and manu. Tourist centre noted for its French Quarter and the Mardi Gras Festival. Pop. (1990E) 600,000.

New Plymouth North Island, New Zealand. 39 04S 174 04E. Cap. of Taranaki district and port sit. on W coast. Trades in and exports dairy produce, esp. cheese. Industries inc. oil, gas and tourism. Pop. (1990E) 67,000.

Newport Isle of Wight, England. 50 42N 1 18W. Town and port on R. Medina just above its mouth, on the N shore. Parkhurst Prison is nearby. Pop. (1988E) 25,700.

Newport Shropshire, England. 52 47N 2 22W. Town serving an agric. area, sit. ENE of Shrewsbury. Manu. valves, garden furniture, construction equipment. Pop. (1989E) 9,607.

Newport Kentucky, U.S.A. 39 06N 84 29W. City sit. at confluence of Rs. Ohio and Licking opposite Cincinnati. Manu. inc. metal goods, food products and printing. Pop. (1980c) 21,587.

Newport Rhode Island, U.S.A. 41 13N 71 18W. Resort and naval base sit. on Narragansett Bay. Industries inc. shipbuilding, tourism, fishing, and the manu. of electrical equipment, precision instruments and jewellery. Pop. (1980c) 29,259.

Newport Gwent, Wales. 51 35N 3 00W. Town sit. on R. Usk. There are extensive docks and major steelworks. The area is developing as a major centre for the electronics industry. Pop. (1983E) 130,200.

Newport News Virginia, U.S.A. 37 04N 76 28W. Port sit. on estuary of James R. and Hampton Roads, SE of Richmond. Shipbuilding is a major industry. Manu. inc. machinery, woodpulp and paper. Coal, petroleum and tobacco are exported. Pop. (1980c) 144,903.

New Providence Bahamas, West

Indies. 25 25N 78 35W. Island between Andros (w) and Eleuthera (e) ese of Miami, Florida. Area, 80 sq. m. (207 sq. km.). Chief town Nassau, cap. of the Bahamas. The most important island of the group. The main industry is tourism, also fishing and vegetable growing. Pop. (1980c) 135,437.

Newquay   Cornwall, England. 50 25N 5 05W. Port and resort on w coast. Pop. (1980e) 14,000.

New River   U.S.A. 38 10N 81 21W. Rises as South Fork R. in Watauga county, North Carolina, flows 320 m. (572 km.) nne into sw Virginia past Radford, then nnw through the Alleghenies into w Virginia to enter Gauley R. at Gauley Bridge. Together they form Kanawha R. Used for water storage at Bluestone Dam and power at Claytor L.

New Rochelle   New York, U.S.A. 40 55N 73 47W. A suburb of New York sit. on Long Island Sound. Mainly residential. Manu. medical supplies and heating equipment. Pop. (1980c) 70,794.

Newry   Down, Northern Ireland. 54 11N 6 20W. Market town in se sit. on R. Newry and Newry Canal (closed to commercial traffic) at the head of Carlingford Lough. Manu. inc. clothing, rope, electrical equipment and food products. Pop. (1981e) 20,000.

New Siberian Islands   Russian Soviet Federal Socialist Republic, U.S.S.R. 75 00N 142 00E. An arch. in the Arctic Ocean sit. between the Laptev Sea and the E. Siberian Sea. Area, 11,000 sq. m. (28,500 sq. km.). The islands are uninhabited, the three largest being Kotelny, Faddeyev and New Siberia.

New South Wales   Australia. 33 00S 146 00E. State of se Australia bounded se by the Tasman Sea, e by the Pacific, n by Queensland, w by South Australia, sw by Victoria. Area, 309,432 sq. m. (801,428 sq. km.). New South Wales became a British possession in 1770; the first settlement was established at Port Jackson in 1788; a partially elective Council

was established in 1843, and responsible government in 1856. New South Wales federated with the other Australian states to form the Commonwealth of Australia in 1901, excluding Australian Capital Territory which is an enclave. Chief cities Sydney (cap.), Newcastle, Wollongong. Fertile coastal lowland rises to the Great Dividing Range running n to s. Beyond the mountains are extensive plains reaching to the Darling R. and rising again to hills nw. The main rural occupations are sheep farming and dairying, with agric. on *c.* 12m. acres of cropland producing wheat, barley, maize, cotton, rice, hay, tobacco, sugar-cane, grapes and fruit. Minerals inc. coal, silver, lead, zinc, sulphur, titanium, zircon, copper, gold. Industries inc. iron and steel and other metalworking industries based on the Newcastle and Port Kembla coalfields, food processing, chemicals, textiles, paper making, printing. Pop. (1986c) 5,401,881.

Newton   Massachusetts, U.S.A. 42 21N 71 11W. Town sit. w of Boston on Charles R. Mainly residential. Economy based on services and retailing. Pop. (1983c) 83,622.

Newton Abbot   Devon, England. 50 32N 3 36W. Town in se of county sit. at head of R. Teign estuary. Industries inc. precision engineering, clay extraction and processing, publishing, clothing manu. and light industry. Pop. (1988e) 22,678.

Newton Aycliffe   Durham, England. 54 36N 1 33W. District, designated a new town, sit. n of Darlington. Industries include the manu. of washing machines, lawnmowers, pvc resins, electrical and electronic equipment, telephone equipment and vehicle axles. Pop. (1985e) 25,500.

Newton-le-Willows   Merseyside, England. 53 28N 2 40W. Town sit. e of St. Helens. Industries inc. railway engineering, calico printing, sugar refining, and the manu. of machinery, paper, glass and biscuits. Pop. (1980e) 22,000.

Newton Stewart   Dumfries and Galloway Region, Scotland. 54 57N

4 30W. Town in sw sit. on R. Cree. Trades in cattle and sheep. Pop. (1987E) 3,212.

Newtownabbey   County Antrim, Northern Ireland. 54 40N 5 55W. District sit. N of Belfast and formed in 1958 by amalgamation of 7 villages. Light industry is expanding inc. food processing, textiles, tyres, telecommunications equipment. Pop. (1988E) 72,800.

Newtownards   County Down, Northern Ireland. 54 36N 5 41W. Town sit. E of Belfast near Strangford Lough. Industries inc. engineering, rose growing, dairy products, textiles, manu. hosiery. Noted for nearby 6th-cent. abbey founded by St. Finian. Pop. (1981C) 20,531.

New Westminster   British Columbia, Canada. 49 12N 122 55W. Port sit. E of Vancouver on Fraser R. Industries inc. wood products, manu. paper and allied products, metal fabricating, machinery and equipment manu. Timber, grain, ores and fruit are exported. Pop. (1981E) 38,550.

New York   U.S.A. 42 40N 76 00W. State of E U.S.A. bounded E by Vermont, Massachusetts and Connecticut, on the SE corner by the Atlantic, S by New Jersey and Pennsylvania, W, NW and N by Canada across L. Erie and L. Ontario and the St. Lawrence R. Area, 49,576 sq. m. (128,402 sq. km.). From 1609 to 1664 the region now called New York was claimed by the Dutch; then it came under the rule of the English, who governed the country till the outbreak of the War of Independence. In 1777 New York adopted a constitution which transformed the colony into an independent state; in 1788 it ratified the constitution of the U.S., becoming one of the 13 original states. New York dropped its claim to Vermont after the latter was admitted to the Union in 1791. With the annexation of a small area from Massachusetts in 1853, New York assumed its present boundaries. Chief cities Albany (cap.), New York, Buffalo, Rochester, Yonkers, Syracuse. Lowlands of the St. Lawrence valley (N) rise to the Adirondack Mountains; S of these is a lowland trough traversing

the state E to W and carrying the Barge Canal which links Buffalo and the Ls. to the Hudson R. (E). S again is the Allegheny plateau at 1,500–2,000 ft. (457–610 metres) a.s.l. Farmland covers 11.4m. acres and produces dairy products, maize, winter wheat, oats, hay, vegetables, fruit. Minerals inc. sand and gravel, salt, zinc, petroleum, lead, stone, titanium, talc, garnet, wollastonite, emery. The main industries, 60% in the New York City area, are manu. clothing, food processing, printing and publishing, metals, machinery, electrical goods. Pop. (1985E) 17,783,000.

New York   New York, U.S.A. 40 43N 74 01W. City on New York bay at mouth of Hudson R. Largest city and commercial centre of the U.S.A. Comprises 5 boros.: Manhattan, on Manhattan Island; the Bronx, NE of Manhattan across Harlem R.; Queens, E of Manhattan on W Long Island; Brooklyn adjoining Queens on SW Long Island, on East R. and New York Bay; State Island, SW of Manhattan on Staten Island across Upper New York Bay. Major sea port for freight and passenger traffic, with an excellent natural harbour. Cultural, educational and entertainment centre. Industries inc. manu. clothing, furs, leather goods, jewellery, tourism, banking, finance, advertising, media, printing and publishing—all these concentrated in Manhattan. Manu. industries in the other boros. inc. food processing, metal-working, textiles, shipbuilding, pharmaceuticals, manu. machinery, wood products, scientific equipment, vehicles. Pop. (1980C) 7,071,639.

New Zealand   Australasia. 40 00S 176 00E. State of the Commonwealth 1,200 m. (1,920 km.) SE of Australia. Area, 103,515 sq. m. (268,103 sq. km.). North Island 44,318 sq. m. (114,785 sq. km.); South Island 59,196 sq. m. (153,318 sq. km.). Comprises North and South Islands, Stewart Island, Chatham Island and smaller offshore islands. The first European to discover New Zealand was Tasman in 1642. The coast was explored by Capt. Cook in 1769. From about 1800 onwards, New Zealand became a resort for whalers and traders, chiefly from Australia. By the Treaty of Waitangi, in 1840, between Governor William Hobson and the representatives of the Maori race, the Maori chiefs ceded the sovereignty to the British Crown and the islands became a British colony. Then followed a steady stream of British settlers.

The Maoris are a branch of the Polynesian race, having emigrated from the eastern Pacific before and during the 14th cent. Between 1845 and 1848, and between 1860 and 1870, misunderstandings over land led to war, but peace was permanently established in 1871, and the development of New Zealand has been marked by racial harmony and integration. Chief cities Wellington (cap.), Auckland, Dunedin, Christchurch, Invercargill, Hamilton. Mountainous the length of the W coast of South Island, with ranges extending E between the smaller Waimea Plains (SE) and extensive Canterbury Plains (E), and continuing up the W coast as the Southern Alps to spread E again to the N of the Waimakariri R. Highest point Mount Cook, 12,349 ft. (3,764 metres) a.s.l., in the Southern Alps. Mountain ranges continue into the centre and NE of North Island, with volcanoes inc. Ruapehu, 9,175 ft. (2,797 metres) and large areas of hot springs. Two-thirds of the total land area can be farmed or grazed; there are c. 23m. acres under cultivation. Dairying, beef production and sheep farming are most important; main arable crops, wheat, oats and barley. Minerals inc. gold, clays and stone, esp. limestone, coal, petroleum, natural gas. The main industries are meat processing, manu. dairy products, paper making, sawmilling, engineering, manu. clothing, vehicles and metal products. The main exports are wool, meat and butter. Pop. (1986E) 3,301,852.

Ngaoundéré   Cameroon. 7 19N 13 35E. Cap. of Adamaoua province sit. NE of Yaoundé. Important agric. market and experimental farms. Pop. (1981E) 47,508.

Ngozi   Burundi. 2 54S 29 50E. Town sit. NNW of Kitega. Food market for a cattle farming area. Industries inc. manu. bricks. tiles.

Nguru   Borno, Nigeria. 12 52N 10 27E. Town sit. ENE of Kano. Railway terminus serving a farming area. Industries inc. salt manu.

Niagara Falls   Ontario, Canada. 43 06N 79 04W. Resort sit. on Niagara R. opposite Niagara Falls, U.S.A. Manu. inc. machinery, wood pulp, paper and fertilizers. Pop. (1981E) 71,000.

Niagara Falls   New York, U.S.A. 43 06N 79 02W. Town sit. NNW of Buffalo on Niagara R. opposite Niagara Falls, Canada. Industries include many electro-chemical plants located near electricity supply. Manu. inc. machinery, metal goods, electrical equipment, paper and chemicals. Pop. (1980C) 71,384.

Niagara Falls   Canada/U.S.A. 43 15N 79 04W. Cataract on the frontier between the cities of Niagara Falls, Ontario, and Niagara Falls, New York. The Canadian (Horseshoe) Falls are c. 160 ft. (49 metres) high and c. 2,500 ft. (762 metres) wide, and are separated by Goat Island from the American Falls, c. 167 ft. (51 metres) high and c. 1,000 ft. (305 metres) wide. Intern. agreements control the exploitation of the Falls for power, of which they are an important source.

Niagara River   U.S.A./Canada. 43 15N 79 04W. Issues from L. Erie between Buffalo, U.S.A. and Fort Erie, Canada, and flows c. 34 m. (55 km.) N along the Canadian/U.S.A. frontier, round Grand Island, U.S.A., and over Niagara Falls into L. Ontario. Navigable 7 m. (11 km.) upstream from Lewiston, and again 20 m. (32 km.) upstream from the Falls. Linked with the New York State Barge Canal at Tonawanda. The Falls are by-passed by the Welland Ship Canal, Ontario. Bridged at Buffalo, Grand Island and Rainbow Bridge.

Niamey   Niger. 13 27N 2 06E. Cap. of Niger sit. on E bank of Niger R. It became the cap. in 1926. A terminus of the trans-Saharan motor routes, trading in livestock, hides and skins. Pop. (1983E) 399,100.

Nicaragua   Central America. 11 40N

85 30W. Rep. bounded N by Honduras, E by the Caribbean, S by Costa Rica, w by the Pacific. Area, *c.* 57,143 sq. m. (148,000 sq. km.). Chief cities Managua (cap.), Léon, Matagalpa. (Managua was almost completely destroyed by an earthquake in 1972, and evacuated.) Swampy and low-lying (E), the land rises to mountains of which the main ranges run NW to SE with spurs running w to E. The lowlands (E) are drained by Río Grande, Escondido, Coco and San Juan Rs., the last forming the Caribbean outlet for a wide depression w of the mountains which encloses L. Nicaragua and L. Managua. The Ls. are separated from the Pacific by a 12 m. (19 km.) wide isthmus with a chain of volcanic peaks, some active. Tropical climate (E) with *c.* 200 ins. (508 cm.) of rain annually; healthier (w) with *c.* 60 ins. (152 cm.) May–December. The main occupations are mining and farming. Chief minerals are gold, silver, copper, tungsten. Chief crops are coffee, cotton, cacao, bananas, rice, wheat, tobacco, sesame seed, sugar-cane. There are some processing industries. Pop. (1987E) 3·5m.

Nicaragua, Lake Nicaragua. 11 35N 85 25W. Sit. near the Pacific Ocean, it is the largest L. in Central America, being 100 m. (160 km.) long and up to 45 m. (72 km.) wide. L. Managua discharges into it *via* Tipitapa R. and it is drained into the Caribbean Sea by San Juan R.

Nice Alpes-Maritimes, France. 43 42N 7 15E. Cap. of dept. and resort sit. at mouth of Paillon R. on the Baie des Anges of the Mediterranean Sea. Manu. inc. olive oil, perfume, soap, textiles, furniture and liqueurs. Trades in fruit, flowers and essences. Pop. (1982C) 338,486 (agglomeration, 449,496).

Nickerie River Suriname. Rises in the Guiana Highlands in S Suriname and flows 200 m. (320 km.) N then WNW to enter the Atlantic below Nieuw Nickerie. Navigable for 60 m. (96 km.). Linked to Coppename R. Fertile rice lands on its lower course.

Nicobar Islands *see* **Andaman Islands**

Nicosia Cyprus. 35 11N 33 21E. Cap. of Cyprus sit. on the Messaoria plain in central N. Manu. textiles, plastics, detergents, cosmetics, furniture, clothing, vegetable oils, paints, brewing, tiles, leather goods, footwear and cigarettes. There are also 2 tanneries and a car assembly plant. Trades in cereals, wine and olive oil. Pop. (1988E) 166,900.

Nicoya Guanacaste, Costa Rica. 9 45N 85 40W. (i) Town sit. S of Liberia, one of the oldest towns of Costa Rica. Centre of an agric., stock farming and lumbering area. (ii) Peninsula extending 75 m. (120 km.) NW to SE into the Pacific along the seaward side of the Gulf of Nicoya. Width 20–30 m. (32–48 km.). Chief town, Nicoya.

Nidd River England. 54 01N 1 12W. Rises on Great Whernside and flows 50 m. (80 km.) SE past Knaresborough to enter Ouse R. NW of York.

Nidwalden Switzerland. 46 50N 8 15E. Half-canton on the S shore of L. Lucerne. Area, 107 sq. m. (276 sq. km.). Cap. Stans. Meadowland (N) rising to mountains (S). Main occupations farming, fruit growing, tourism. Industries inc. wood working, glass making. Pop. (1988E) 32,763.

Niederwald Hessen, Federal Republic of Germany. A hill sit. in the SW Taunus Hills, w of Wiesbaden, 1,080 ft. (329 metres) a.s.l.

Nieuport West Flanders, Belgium. 51 08N 2 45E. Port sit. SW of Ostend on Yser R. Industries inc. fishing and the manu. of ropes. sails and nets.

Nièvre France. 47 10N 5 40E. Dept. in Burgundy region, consisting mainly of former province of Nivernais. Area, 2,640 sq. m. (6,837 sq. km.). In the E are the Morvan Mountains rising to 2,790 ft. (850 metres) a.s.l. The principal Rs. are Loire and Yonne. Large areas are forested. Crops inc. cereals, potatoes and vines and livestock is raised. Coal is mined in the Decize area. Cap. Nevers. Pop. (1982C) 239,635.

Niğde Turkey. (i) Province of S

central Turkey. Area, 5,891 sq. m.
(15,258 sq. km.). Drained by Kizil
Irmak R. Farming and mining are the
main occupations; main products lignite, lead, mohair, cereals, vegetables,
raisins. Pop. (1980C) 512,071. (ii)
Town 75 m. (120 km.) NNW of Adana.
Provincial cap. and trading centre for
cereals and vegetables. The main industry is tile making. Pop. (1970C)
26,936.

Niger   West Africa. 13 30N 10 00E.
Rep. bounded N by Algeria and Libya,
E by Chad, S by Nigeria, SW by Benin
and Burkina Faso and W by Mali.
Area, 458,075 sq. m. (1,186,408 sq.
km.). Chief cities Niamey (cap.),
Zinder. The S has wells and intermittent streams; the SW is watered by
Niger R. and its tribs. Both regions
support agric.; chief crops are millet,
groundnuts, beans, cotton, rice. The
centre receives enough rain to support
pasture for cattle, sheep and goats.
The N and by far the largest zone is
desert. Minerals inc. uranium which is
mined in the Air mountains N of
Agades. There are also deposits of
salt, natron and tin. Mining and processing ores are the only industries.
Pop. (1988E) 7.19m.

Niger   Nigeria. State chiefly formed
from the Niger province of the former
N region. Area, 25,111 sq. m. (65,037
sq. km.). The main ethnic group is the
Nupe people. Chief towns Minna
(cap.) and Bida. Pop. (1983E) 1.8m.

Nigeria   West Africa. 8 30N 8 00E.
Nigeria is an independent Rep. of the
Commonwealth bounded N by Niger,
NE by Chad, E and SE by Cameroon, S
by the Gulf of Guinea and W by Benin.
Area, c. 356,669 sq. m. (923,773 sq.
km.). Nigeria comprises a number of
areas formerly under separate admin-
istrations. Lagos, ceded in Aug. 1861
by King Docemo, was placed under
the Governor of Sierra Leone in 1866.
In 1874 it was detached, together with
the Gold Coast Colony, and formed
part of the latter until Jan. 1886, when
a separate 'colony and protectorate of
Lagos' was constituted. Meanwhile
the National African Company had
established British interests in the
Niger valley, and in July 1886 the
company obtained a charter under the

name of the Royal Niger Company.
This company surrendered its charter
to the Crown in 1899, and in 1900 the
greater part of its territories was
formed into the protectorate of Northern Nigeria. Along the coast the Oil
Rivers protectorate had been declared
in June 1885. This was enlarged and
renamed the Niger Coast protectorate
in 1893; and in 1900, on its absorbing
the remainder of the territories of the
Royal Niger Company, it became the
protectorate of Southern Nigeria. In
1906 Lagos and Southern Nigeria were
united into the 'colony and protectorate of Southern Nigeria', and in 1914
the latter was amalgamated with the
protectorate of Northern Nigeria to
form the 'colony and protectorate of
Nigeria', under a Governor. In 1954
Nigeria became a federation under a
Governor-General. In 1960 the Federation of Nigeria became sovereign
and independent and a member of the
Commonwealth of Nations and in
1963 Nigeria became a republic.
Nigeria is divided into 21 states and a
capital territory. (Akwa Ibom, Anambra, Bauchi, Bendel, Benue, Borno,
Cross River, Gongola, Imo, Kaduna,
Kano, Katsima, Kwara, Lagos, Niger,
Ogun, Ondo, Oyo, Plateau, Rivers,
Sokoto, Abuja). Chief cities Lagos
(cap.), Ibadan, Ugo, Ogbomosho,
Kano, Oshogbo. A new capital is to
be built near Abuja in Niger state.
Mangrove swamps on the coast give
way to tropical rain forest and oil-
palm bush extending c. 50–100 m.
(80–160 km.) inland. Beyond this the
land rises to savannah and woodland,
with desert to the N. The N plateau
rises to 5,000 ft. (1,524 metres) and
there are mountains on the E frontier.
Drained by Niger, Benne and Cross
Rs. Tropical climate with rains (N)
April–September and (S) March–
November and Harmattan desert wind
during dry seasons. The N produces
groundnuts, tobacco, cotton, soybeans, hides; the S palm products,
cocoa, timber, rubber. Minerals inc.
tin at Jos, coal at Enugu, colombite,
gold, tantalite, petroleum. Industries
other than mining are mainly manu.
for home consumption. Pop. (1988E)
118.7m.

Niger River   West Africa. 5 33N
6 33E. Rises in foothills of the Fouta

Djallon highlands near the Sierra Leone/Guinea frontier; flows *c.* 2,600 m. (4,160 km.), first NE through Guinea to enter Mali below Siguiri, past Bamako and Timbuktu to meet the Trans-Saharan highway at Bourem, where it turns SE and runs parallel to the highway into SW Niger, past Niamey and S to the Niger/Benin border which it forms up to the Nigerian border. It enters Nigeria and bends S through the Kainji reservoir, then SE to Lokoja where it receives Benue R. It continues S past Onitsha to enter the Gulf of Guinea by a wide marshy delta beginning at Abo, 80 m. (128 km.) upstream from the mouths W of Port Harcourt, which extend *c.* 200 m. (320 km.) along the coast and have broad mangrove swamps. The Bonny channel of the delta (E) has been dredged to provide deep-water access to Port Harcourt.

Nightingale Islands St. Helena, Atlantic Ocean. Island group sit. just SSW of Tristan da Cunha and a depend. of St. Helena.

Niigata Honshu, Japan. 37 58N 139 02E. Cap. of Niigata prefecture, and port sit. NNW of Tokyo on W coast. Industries inc. oil refining, and the manu. of machinery, machine tools, chemicals and textiles. Oil is exported. Pop. (1988E) 470,000.

Niihau Hawaii, U.S.A. 21 55N 160 10W. Island SW of Kauai across Kaulakahi Channel. Area, 72 sq. m. (186 sq. km.). Privately owned. Arid lowland with a rocky E coast. Main occupations cattle raising, making rush matting. Pop. (1980C) 226.

Nijmegen Gelderland, Netherlands. 51 50N 5 50E. Town in E sit. on Waal R. and Maas-Waal Canal. Industries inc. engineering, sugar refining and the manu. of bricks, electrical equipment, chemicals, leather goods and pottery. Pop. (1988E) 145,405.

Nikolayev Ukraine, U.S.S.R. 46 58N 32 00E. Port sit. N of Black Sea on estuary of Rs. Bug and Ingu. It is an important shipbuilding centre and a naval base. Flour milling is also important and some of the largest grain elevators in Europe are sit. here.

Manu. inc. machinery and footwear. Grain, timber, ores and sugar are exported. Pop. (1987E) 501,000.

Nikopol Ukraine, U.S.S.R. 47 35N 34 25E. Town SSW of Dnepropetrovsk on Dnieper R. Sit. in a rich manganese-mining area, it is a major centre of metallurgy and engineering, besides supplying the Donbas and Dnepropetrovsk steel plants.

Nikunau Kiribati, W Pacific. Atoll of the S Gilbert group. Area, 7 sq. m. (18 sq. km.). Pop. (1980E) 1,829.

Nile River North East Africa. 30 10N 31 06E. Issues from L. Victoria (White Nile) and L. Tana (Blue Nile) and flows 3,485 m. (5,576 km.) (from L. Victoria) N through Uganda, the Sudan and Egypt to enter the Mediterranean by a delta below Cairo. Headstreams above L. Victoria extend to the source of the Luvironza R. SE of Matana, Burundi. The length from these is over 4,150 m. (6,640 km.). The White Nile leaves L. Victoria as the Victoria Nile and descends from 3,720 ft. (1,134 metres) over Ripon and Owen Falls, continuing NW through L. Kyoga to Murchison Falls where it drops 400 ft. (122 metres), and enters L. Mobutu Sese Soko, at 2,030 ft. (618 metres) a.s.l. in the W Great Rift Valley. It then flows N into the Sudan at Nimule where it becomes known as the Bahr el Jebel, and continues N past Juba into the sudd swamps, losing *c.* half its water through dispersal and evaporation. It receives the Bahr el Ghazal and L. No, whence it is known again as the White Nile, continuing N to Khartoum where it joins the Blue Nile. The Blue Nile issues from L. Tana in the Ethiopian highlands S of Gondar, at 6,000 ft. (1,829 metres) a.s.l., flows SE then NW 1,000 m. (1,600 km.) into the Sudan, past Wad Medani to Khartoum. From Khartoum the Nile flows NNE to the 5th cataract below Berber, then NNW almost to El Kab where it turns SW through the 4th cataract to Merowe and Korti, then N past Dongola to the 3rd cataract, N into L. Nasser (Egypt) and the 2nd cataract, with the 1st cataract N of the Aswan High Dam: the R. drops 935 ft. (285 metres) between Khartoum and Luxor. From Luxor it flows NNW to Al Minya, then N to Cairo and the delta which has 2 main channels, the Rosetta (W) and Damietta (E) Rs. Navigable for most of its course except between Aswan and Khartoum and for other short stretches where its course is impeded by cataracts. The volume of water is increased 16 times in the flood season, August–September, from its lowest level in April. The flow of the White Nile is constant, the extra volume comes from the seasonal rains in Ethiopia which swell the Blue Nile. The deposits of alluvial soil washed down by the floods come also from the Blue Nile's course. Irrigation and flood control have been practised since 4,000 B.C. Control is now effected by numerous barrages diverting water for irrigation. Main storage schemes are the Aswan High Dam, Sennar and Jebel Aulia dams.

Nilgiri Hills   Tamil Nadu, India. A plateau sit. in SW at an average height of 6,500 ft. (2,000 metres) a.s.l. The highest point is Doda Betta, 8,640 ft. (2,658 metres) a.s.l. Tea, coffee and eucalyptus grown.

Nîmes   Gard, France. 43 50N 4 21E. Cap. of dept. in central S France. A tourist and commercial centre with many fine Roman remains. Manu. inc. agric. machinery, carpets, textiles, footwear and clothing. Trades in brandy, wine and grain. Pop. (1982C) 129,924.

Ninghsia Hui   *see* **Ningxia Hui**

Ningxia Hui (Ninghsia Hui) Autonomous Region   China. 38 00N 106 00E. Sit. in NE. Area, 25,640 sq. m. (66,400 sq. km.). Mainly mountainous and traversed by Hwang-ho R. in N. Wheat, beans and rice are grown in the R. valley. Cap. Yinchuan, sit. WSW of Beijing. Pop. (1987E) 4,240,000.

Niort   Deux-Sèvres, France. 46 19N 0 27W. Cap. of dept. in W central France sit. on Sèvre Niortaise R. Manu. footwear, gloves and brushes. Pop. (1982C) 60,230.

Nipe, Sierra de   Cuba. Range extending 25 m. (40 km.) S from Nipe Bay on the NE coast forming a divide

between Nipe (W) and Mayarí (E) Rs. Rises to 3,200 ft. (975 metres) and has iron deposits (N) with ironworks at Felton.

Nipigon, Lake   Ontario, Canada. 49 50N 89 30W. Sit. N of L. Superior and drains into it *via* Nipigon R. Area, 66 m. (106 km.) long and 46 m. (74 km.) wide.

Nipissing, Lake   Ontario, Canada. 46 17N 80 00W. Sit. between L. Huron and Ottawa R. It drains into L. Huron *via* French R.

Niš   Serbia, Yugoslavia. 43 19N 21 54E. Town sit. SSE of Belgrade on Nišava R. It is an important transportation centre with large railway workshops. Manu. inc. chemicals, textiles and tobacco products. Trades in grain and wine. Pop. (1981E) 230,711.

Nishapur   Iran. 36 12N 58 50E. Town in NE sit. in a fertile area producing cereals and cotton. Pottery is manu. Omar Khayyám was born and buried here. The famous turquoise mines are in the vicinity.

Nishinomiya   Honshu, Japan. 34 43N 135 20E. Town sit. just NW of Osaka in Hyogo prefecture. Noted for saké. Manu. machinery, chemicals and soap. Pop. (1983E) 401,000.

Nishino Shima   *see* **Rosario Islands**

Nissan   Solomon Islands, South Pacific. Atoll sit. E of New Ireland comprising Nissan, Barahun, Sirot and Han islands on a 10 m. (16 km.) reef. The main product is coconut.

Niterói   Rio de Janeiro, Brazil. 22 53S 43 07W. State cap. and resort sit. on E shore of Guanabara Bay opposite Rio de Janeiro city to which it is linked by ferry. Mainly residential. Industries inc. shipbuilding and the manu. of metal goods, textiles and matches. Pop. (1980C) 382,736.

Nith River   Scotland. 55 20N 3 05W. Rises just NE of Dalmellington, Strathclyde Region, and flows 80 m. (128 km.) E into Dumfries, past Sanquhar then SE to Dumfries town and into the Solway Firth by an estuary.

Nitra Slovakia, Czechoslovakia. 48 20N 18 05E. Town sit. ENE of Bratislava on Nitra R. Provincial cap. Centre of a farming area. Industries inc. processing agric. produce. Noted as a centre of Christianity, seat of a 9th cent. bishopric.

Niuafo'ou Tonga, South Pacific. 15 34S 175 40W. Northernmost island of Tonga. Area, 3.5 m. by 3 m. (6 km. by 5 km.). Volcanic with crater L. and hot springs.

Niue Island South Pacific. 19 02S 169 52W. A self-governing depend. of New Zealand. A coral island sit. WNW of Rarotonga. Area, 100 sq. m. (259 sq. km.). The principal town and port is Alofi. Copra and bananas are exported. Pop. (1986E) 2,531.

Niulakita Tuvalu, Pacific. Island of the S Tuvalu group. Area, 104 acres. Formerly called Sophia Island. Pop. (1983E) 90.

Niutao Tuvalu, Pacific. Atoll of the N Tuvalu group. Area, 625 acres (2.1 sq. km.). The main product is copra. Pop. (1983E) 920.

Nizhni Novgorod *see* **Gorky**

Nizhni Tagil Russian Soviet Federal Socialist Republic, U.S.S.R. 57 55N 59 57E. Town sit. NNW of Sverdlovsk near Tagil R. It is an important metallurgical and engineering centre with manu. of railway rolling stock, aircraft, agric. machinery, machine tools and chemicals. Pop. (1987E) 427,000.

Njazídja (formerly Grande Comore) Comoros, Indian Ocean. 12 10S 44 15E. Island sit. off the coast of Mozambique in the Mozambique channel. Area, 443 sq. m. (1,148 sq. km.). Chief town, Moroni. Volcanic. The main occupations are farming and lumbering; main products vanilla, cacao, cloves, copra, rice, vegetables, oilseeds, timber. Pop. (1980C) 192,177.

Nkongsamba Cameroon. 4 57N 9 56E. Town in Coast province sit. NW of Yaoundé. Railway terminus and airport; trading centre. Industries inc.

sawmilling, palm oil production, railway repair shops. Pop. (1981E) 86,870.

Nocera Inferiore Campania, Italy. 40 44N 14 38E. Town sit. ESE of Naples. Industries inc. tomato canning and the manu. of textiles.

Nógrad Hungary. 48 00N 19 30E. County of N Hungary bounded W and N by Ipoly R. Area, 982 sq. m. (2,544 sq. km.). Cap. Salgotarjan. Mountainous with thick forest (W). Agric. areas produce cereals, potatoes; sheep and pig farms on pasture land; minerals inc. coal and lignite. Industries are centred on the cap. Pop. (1984E) 237,000.

Nome Alaska, U.S.A. 64 30N 165 24W. Port sit. on the S coast of the Seward Peninsula. The principal occupations are mining, fishing, fur trapping and tourism. Pop. (1989E) 4,303.

Nomo Peninsula Kyushu, Japan. Peninsula extending 17 m. (27 km.) S from the Hizen Peninsula between the Amakusa Sea (E) and E China Sea. Chief town Nagasaki, at the base, on the W coast.

Nongkhai Nongkhai, Thailand. Town sit. N of Udon on Mekong R. where it forms the Laos frontier. Provincial cap. Trades in rice, beans, tobacco.

Nonouti Kiribati, W Pacific. Atoll of the S Gilbert group. Area, 9.8 sq. m. (25 sq. km.) on a 24 m. (38 km.) reef. The main product is copra. Pop. (1980E) 2,284.

Noord-Brabant Netherlands. 51 40N 5 00E. Province of S Netherlands bounded N by Maas R., S by Belgium. Area, 1,894 sq. m. (4,913 sq. km.). Chief towns 's Hertogenbosch (cap.), Eindhoven, Tilburg, Breda. Low-lying, fertile near the Maas, sandy soil elsewhere. Industrial, esp. textiles, electrical engineering, tanning and manu. leather products, pharmaceuticals. Pop. (1989E) 1,099,622.

Noord-Holland Netherlands. 52 30N 4 45E. Province of NW bounded

w by the North Sea, N by the Wadden-
zee, E by the Ijsselmeer, S by Zuid
Holland and Utrecht provinces. Area,
1,913 sq. m. (4,956 sq. km.). Chief
cities Haarlem (cap.), Amsterdam,
Ijmuiden, Zaandam, Alkmaar, Helder,
Hilversum. Fenland and reclaimed
land drained by numerous Rs. and
canals. Agric. areas produce cereals,
vegetables, fruit and flowers; dairy
farming on reclaimed land. Industries
are concentrated in Amsterdam and
Zaanstreek and inc. steel. shipbuilding, food processing, sugar refining,
manu. machinery, paper and wood
products, cattle oil and feed cake, railway equipment. Pop. (1989E)
2,365,160.

Noordoostpolder   Netherlands. 52
45N 5 45E. Reclaimed area of the
Ijsselmeer on w coast of Overijssel
province. Area. 185 sq. m. (479 sq.
km.).

Nordaustlandet   Svalbard, Norway.
79 55N 23 00E. Island of the Spitsbergen group NE of w Spitsbergen
across the Hinlopen Strait. Area,
5,710 sq. m. (14,789 sq. km.). Glacial
with inland ridges rising to 2,000 ft.
(610 metres).

Nord   France. 50 15N 3 30E. Dept.
in Nord-Pas-de-Calais region, bounded by the North Sea and Belgium.
Area, 2,215 sq. m. (5,738 sq. km.).
Mainly low-lying with numerous
canals and tribs. of the principal Rs.,
Scheldt and Sambre. Part of the
Franco-Belgian coalfield lies in this
dept. and there are important metallurgical and textile industries. Agric.
products inc. cereals, sugar-beet, hops
and flax. Cap. Lille. Pop. (1982C)
2,520,526.

Nordenham   Lower Saxony, Federal
Republic of Germany. 53 29N 8 28E.
Port sit. SW of Bremerhaven on Weser
R. Industries inc. tourism, lead and
zinc smelting. Pop. (1988E) 28,695.

Nordhausen   Erfurt, German Democratic Republic. 51 30N 10 47E.
Town sit. NNW of Erfurt at the foot of
the Harz Mountains. Centre of a
potash-mining district with printing,
textile and distilling industries. Pop.
(1989E) 48,300.

Nordland   Norway. 65 40N 13 00E.
County of N Norway bounded SE by
Sweden, NW by the North Sea and inc.
the Lofoten islands, some of the
Vesteralen islands and other off-shore
islands. Area, 14,798 sq. m. (38,327
sq. km.). Chief towns Bodø (cap.),
Narvik. About two-thirds is within the
Arctic Circle. Mountainous interior
with a rocky and deeply indented
coastline. Mainly forest and marsh.
Main occupations fishing, lumbering,
quarrying, mining, farming. Pop.
(1989E) 239,611.

Nord-Pas-de-Calais   France. 50
15N 3 30E. Region in N comprising
Nord and Pas-de-Calais. Area, 4,778
sq. m. (12,377 sq. km.). Chief towns
Lille, Valenciennes, Lens, Douai,
Béthune, Dunkerque, Maubeuge,
Calais, Boulogne-sur-Mer and Arras.
Pop. (1982C) 3,932,939.

Nord-Trøndelag   Norway. County
of central Norway, bounded E by
Sweden, w by the North Sea and Sør-
Trøndelag. Area, 8,673 sq. m. (22,463
sq. km.). Chief towns Steinkjer (cap.),
Namsos, Levanger. Mountainous (E)
with a coastline indented by fjords.
Main occupations fishing, farming,
lumbering, mining (pyrites). Pop.
(1989E) 126,750.

Nore, The   England. 51 29N 0 51E.
A sandbank sit. at mouth of R. Thames
and marked by numerous buoys.

Norfolk   England. The county is
bounded by the North Sea to N and E,
the Wash to NW, Lincolnshire and
Cambridgeshire to w and Suffolk to S.
It contains the districts of Breckland,
Broadland, Great Yarmouth, King's
Lynn and West Norfolk, North Norfolk, Norwich, South Norfolk; also
the Broads Authority with responsibility for the Broads. Mostly flat good
arable land; low coast line with sand
dunes. Drained by Rs Yare, Bure,
Waveney and Ouse. County town,
Norwich, other towns Great Yarmouth
and King's Lynn. Economy traditionally agricultural, now wide variety of
industries inc. international financial
services, high technology vehicle
manu., biogenetic engineering and
processed foods. Pop (1988E)
744,300.

Norfolk Virginia, U.S.A. 38 40N
76 14W. Port and naval base sit. on
Hampton Roads and Elizabeth R.
Headquarters of the U.S. Atlantic
Fleet. Financial and educational
centre. Port-related industries, manu.,
medical services, trade, tourism and
conventions. Pop. (1987E) 280,800.

Norfolk Island Australia. 29 02S
167 57E. A small volcanic island sit.
in S. Pacific Ocean. Area, 14 sq. m.
(36 sq. km.). It is very fertile and
grows and exports many fruits, esp.
citrus. Pop. (1980E) 1,700.

Norilsk Russian Soviet Federal
Socialist Republic, U.S.S.R. 69 20N
88 06E. Town sit. ESE of the port
Dudinka. It is an important mining
centre dealing with nickel, copper,
platinum and gold. Pop. (1987E)
181,000.

Normal Illinois, U.S.A. *see*
Bloomington

Norman Oklahoma, U.S.A. 35 13N
97 26W. City sit. SSE of Oklahoma
City. Service and trading centre for an
agric. area. Pop. (1980C) 68,020.

Normandy France. 48 45N 0 10E.
Ancient province of N France bounded N by the English Channel. Now
forms 2 regions of Basse-Normandie
and Haute-Normandie. Drained by
lower Seine, Béthune and Orne Rs.
Fertile agric. area. The chief industry
is textile milling. Chief cities Rouen,
Le Havre, Cherbourg, Caen, Dieppe.

Norrbotten Sweden. 66 30N 22
30E. County of Swedish Lapland
bounded N and NW by Norway, NE and
E by Finland, SE by the Gulf of Bothnia. Area, 38,193 sq. m. (98,919 sq.
km.). Chief towns Luleå (cap.),
Boden, Kiruna, Haparanda, Piteå.
About half is N of the Arctic Circle.
Fertile lowland (E) rises to mountains
(W) with Kebmiekaise at 6,965 ft.
(2,123 metres). Numerous Rs. and
large Ls. Main occupations are lumbering, sawmilling and iron mining.
Pop. (1988E) 261,536.

Norristown Pennsylvania, U.S.A.
40 07N 75 21W. Town sit. NW of
Philadelphia on Schuylkill R. Manu.

inc. textiles, metal goods, plastics,
drugs, furniture and food products.
Pop. (1980C) 34,684.

Norrköping Östergötland, Sweden.
58 36N 16 11E. Port in SE sit. on
Motala R. near its confluence with the
Bråvik, a branch of the Baltic Sea. Industries inc. electrotechnical, paper
milling and chemicals. Manu. inc.
microwaves, cold storage plant, plastics and confectionery. Wood pulp
and paper are exported. Pop. (1989E)
119,926.

Norrtälje Sweden. Town sit. NE of
Stockholm at head of Norrtälje Bay
on the Baltic Sea. Industries inc. fishing, farming and manu. printing paper,
giant television screens, rotosign systems, medals, precision engineering
tools, boats and electronics. Pop.
(1990E) 45,000.

Norte de Santander Colombia.
Dept. of N Colombia bounded E by
Venezuela across Táchira R. Area,
8,297 sq. m. (21,240 sq. km.). Chief
towns Cúcuta (cap.), Pamplona,
Ocaña. Maracaibo lowlands rise to the
Cordillera Central which occupies
most of the area; the climate varies
according to altitude. Agric. area
around Cúcuta produces coffee, cacao,
sugar-cane, tobacco, cotton, maize.
Minerals inc. petroleum, coal, kaolin,
sulphur, gypsum, iron, copper, lead,
tin. Forests are commercially valuable. Pop. (1983E) 878,300.

Northallerton North Yorkshire,
England. 54 20N 1 26W. Town in NE.
Industries inc. trailer manu., light
engineering and colour finishes for
leather. Pop. (1985E) 13,800.

North America 45 00N 100 00W.
Northern continent of the W Hemi-
sphere connected to South America
by the Isthmus of Panama. Bounded
SE by the Caribbean, E by the Atlantic
Ocean, N by the Arctic Ocean and its
numerous inlets, W by the Pacific
Ocean. Area of the mainland and off-
shore islands inc. Central America, *c.*
8.4m. sq. m. (21.5m. sq. km.). Divi-
ded politically between Canada,
U.S.A., Mexico, Guatemala, Belize,
Honduras, Nicaragua, El Salvador,
Costa Rica and Panama. Separated

(NW) from the NE tip of Asia (Cape Dezhnev, U.S.S.R.) by the 55 m. (88 km.) Bering Strait, and from Greenland (NE) by Baffin Bay. Mountains extend N to S down the W coast with the Rocky mountains running parallel inland, from the Arctic Ocean to N Mexico. The coastal and Rocky ranges are separated by the plateaux of the Yukon, the Columbia and Colorado plateaux and the Great Basin of W U.S.A. In Mexico the Sierra Madre ranges continue S to join the Central American highlands. The Rockies form the continental divide: to the E are the Great Plains of Canada and the U.S.A. extending from central Saskatoon to the Gulf of Mexico at 500–1,000 ft. (152–305 metres) a.s.l. Major Rs. draining the lowlands are Sasketchewan, Nelson and Red Rs. (N) into Hudson Bay, St. Lawrence to the Atlantic, and Missouri–Mississippi system into the Gulf of Mexico. Lowland continues down the E coast of Mexico and central America and is drained into the Caribbean. The highlands of the E continent are lower than the W ranges. They extend from the Laurentian mountains, Canada, through the Adirondack and Appalachian ranges of the E U.S.A., with sandy plain between the foothills and the Atlantic. Exceptionally long navigable Rs. provide access to the centre of the continent *via* St. Lawrence Seaway and the Great Lakes on the U.S./Canadian border, Mississippi and Ohio R. systems. Widely varying climate.

Northampton Northamptonshire, England. 52 14N 0 54W. Town in S Midlands sit. on R. Nene. Industries inc. a wide range of engineering, cosmetics, footwear, shoe machinery, leather goods and motor car accessories. Pop. (1989E) 183,167.

Northampton Massachusetts, U.S.A. 42 19N 72 38W. City sit. W of Boston on Connecticut R. Manu. electro-optical instruments, brush and wire products, plastics. Pop. (1980C) 29,286.

Northamptonshire England. 52 16N 0 55W. Under 1974 re-organisation the county consists of the districts of Corby, Kettering, East Northamptonshire, Daventry, Northampton, Wellingborough and South North-

amptonshire. Bounded in N by Leicestershire, E by Cambridgeshire and Bedfordshire, S by Buckinghamshire and Oxfordshire and W by Warwickshire. County town, Northampton. Drained by Rs. Welland and Nene (to the Wash). Traditional industries such as footwear and agric. are still important but commerce, distribution, brewing and food processing are now prominent. Pop. (1988E) 570,300.

North Bay Ontario, Canada. 46 19N 79 28W. City on L. Nipissing sit. N of Toronto. Summer resort, indust. inc. timber, high technology and mining equipment. Pop. (1988E) 51,598.

North Berwick Lothian Region, Scotland. 56 04N 2 44W. Port and resort sit. ENE of Edinburgh on S shore of Firth of Forth. Pop. (1980E) 4,500.

North Beveland Zeeland, Netherlands. An island sit. in Scheldt estuary. Area, 35 sq. m. (91 sq. km.). Crops inc. wheat and sugar-beet.

North Borneo *see* **Sabah**

North Cape Finnmark, Norway. 71 15N 25 40E. A headland sit. on the island of Mageröy in the extreme N.

North Carolina U.S.A. 35 30N 80 00W. State of E U.S.A. bounded E by the Atlantic, N by Virginia, W by Tennessee, S by Georgia and Carolina. Area, 52,712 sq. m. (135,524 sq. km.). Chief cities Raleigh (cap.), Charlotte, Greensboro, Winston-Salem, Durham. Coastal plain rises slowly through a hilly area to the Appalachian ridges (W). Drained to the Atlantic by Catawba, Yadkin, Tar, Roanoke, Neuse and Cape Fear Rs; into the Gulf of Mex. by Hiwassee, Little Tennessee, French Broad and Watauga Rs. Humid subtropical climate. Farmland covers 14m. acres; the main crop is tobacco, also maize, cotton, peanuts, soybeans, cereals, fruit. Dairy and poultry farming are important. The forests produce valuable timber and cover *c*. 66% of the land area. The main industry is textile milling. Other industries inc. manu. tobacco products, electrical machinery, food, chemicals, furniture and bricks. Pop. (1986E) 6,331,000.

North Dakota U.S.A. 47 30N 100
00W. State of NW U.S.A. bounded N
by Canada, W by Montana, S by South
Dakota, E by Minnesota. Area, 69,457
sq. m. (179,894 sq. km.). Chief cities
Bismarck (cap.), Fargo, Grand Forks,
Minot. Land rises gently from valley
of Red R. of the North (E) across
prairies at 1,300–1,600 ft. (396–488
metres) to the Missouri Plateau and
Rocky Mountains (W), highest point
Black Butte (SW), 3,468 ft. (1,057
metres) a.s.l. Drained by Missouri R.,
its trib. James R. and Sheyenne R.,
trib. of Red R. Farmland covers 42m.
acres, mainly arable. Chief cash crops
are wheat, barley and flaxseed. Beef
cattle are grazed (W). Farming is no
longer the main occupation; there are
more people employed in trade and
govt. Minerals inc. petroleum, natural
gas, lignite. The main industries are
processing food products and manu.
machinery, electrical goods, bricks
and tiles. Pop. (1984E) 686,000.

North Downs *see* **Downs**

**North East Frontier Agency
(India)** *see* **Arunachal Pradesh**

Northern Ireland United Kingdom.
54 40N 6 45W. The 6 counties of NE
Ireland forming part of the United
Kingdom of Great Britain and Northern Ireland. Area, 5,238 sq. m.
(13,409 sq. km.). Chief cities Belfast
(cap.), Londonderry. The 6 counties
are: Antrim, Armagh, Down, Fermanagh, Londonderry, Tyrone. Often
called Ulster, the N province of anc.
Ireland in which it lies. Mainly plateau with hills around it and an indented coastline. Agric., dairying and
stock farming area. Industries inc. textiles, esp. linen, shipbuilding and engineering in Belfast. Lignite is found
on the E shores of L. Neagh in County
Antrim. Chief exports dairy products.
Pop. (1988E) 1,578,100.

**Northern Marianas, Commonwealth
of** 17 00N 145 00E. A territory in
the W Pacific, close to and formerly
part of the U.S. Trust Territory of the
Marshall, Caroline (and N Mariana)
Islands. Separated from the Trust Territory in 1976; now a commonwealth
in union with the U.S.A. The islands
stretch over 500 m. N–S; land area

only 182 sq. m. The main island, and
cap. is Saipan. Pop. (1988E) 20,591.

Northern Rhodesia *see* **Zambia**

Northern Territory Australia. 16
00S 133 00E. Territory bounded N by
the Timor and Arafura Seas, S by
South Australia, NE by the Gulf of
Carpentaria, E by Queensland, W by
Western Australia. Area, 520,280 sq.
m. (1,346,000 sq. km.). Chief towns
Darwin (cap.), Alice Springs. Desert
(SE) giving way to grassy plains which
rise to the Macdonnel Range (SW) and
the Barkly Tableland (E). The aboriginal reserve of Arnhemland is NE, and
there are about 15 smaller reserves.
Tropical climate with rains November–April, monsoons December–January. The most important products are
minerals; bauxite in Arnhemland, iron
ore at Francis Creek, copper at Ten-
nant Creek, uranium in the Alligator
R. district, also manganese, gold, bismuth, petroleum. The main occupations apart from mining are beef cattle
farming, with some lumbering and
fishing. Pop. (1987E) 156,700.

North Esk River *see* **Esk River,
North**

Northfleet Kent, England. 51 26N
0 20E. Town sit. W of Gravesend on
R. Thames. Manu. paper products,
chemicals, cables, engineering, and
cement. Pop. (1981C) 26,300.

North Foreland Kent, England. 55
22N 1 28E. A chalk headland just N of
Ramsgate.

North Holland *see* **Noord-Holland**

North Island New Zealand. 39 00S
176 00E. Smaller of the 2 main
islands separated from South Island
by Cook Strait. Area, 44,319 sq. m.
(114,785 sq. km.). Chief cities Auckland, Wellington. Mountain chains
rise to volcanoes of up to 9,175 ft.
(2,797 metres) and descend again to
fertile coastal plains. Drained by Waikato R. Mainly a dairy farming area.
Minerals inc. coal, gold. Pop. (1986C)
2,438,249.

North Korea *see* **Korea, North**

North Las Vegas Nevada, U.S.A.

36 12N 115 07W. City immediately N of Las Vegas in S Nevada. Residential, with employment in Las Vegas gambling industry, local warehousing, distribution, transport, energy research. Nellis Air Force Base is here. Pop. (1984E) 47,370.

North Little Rock Arkansas, U.S.A. 34 46N 92 14W. City sit. on Arkansas R. opposite Little Rock. Industries inc. railway engineering, sawmilling and cottonseed-oil processing. Pop. (1980C) 64,288.

North Miami Florida, U.S.A. 25 53N 80 10W. City sit. N of Miami. Mainly residential with retailing and services. Pop. (1980C) 42,566.

North Olmsted Ohio, U.S.A. 41 25N 80 09W. City sit. SW of Cleveland. Residential, commercial and retail centre. Employment inc. graphic arts. Pop. (1980C) 36,486.

North Ossetian Autonomous Soviet Socialist Republic U.S.S.R. Part of Russian Soviet Federal Socialist Republic. Area, 3,088 sq. m. (7,980 sq. km.). In 1924 constituted as an Autonomous Region and 1936 as an Autonomous Rep. Cap. Ordzhonikidze, sit. SE of Rostov. Chief industries are nonferrous metals (mining and metallurgy), maize-processing, timber and woodworking, textiles, building materials, distilleries and food processing; also varied agric. Pop. (1989E) 634,000.

North Platte Nebraska, U.S.A. 41 08N 100 46W. City in W Nebraska sit. NE of Denver, Colorado, on N Platte R. Main employers Union Pacific railway, local services, packing, construction. Pop. (1980C) 24,479.

North Platte River U.S.A. 41 15N 100 45W. Rises in Park Range, N Colorado, and flows 680 m. (1,088 km.) N past Casper, Wyoming, then E and SE into Nebraska to North Platte City where it joins South Platte R. to form Platte R. Main tribs. Sweetwater, Medicine Bow and Laramie Rs. Used for power and irrigation in SE Wyoming and W Nebraska.

North Rhine-Westphalia Federal Republic of Germany. 51 45N 7 30E. A *Land* comprising the former province of Westphalia, the districts of Aachen, Cologne, Düsseldorf, and Lippe. Area, 13,148 sq. m. (34,054 sq. km.). A very industrialized region producing coal, iron and steel, and textiles. Cap. Düsseldorf. Pop. (1988E) 16,874,059.

North Sea North Europe. 56 00N 4 00E. Shallow arm of the Atlantic Ocean extending 600 m. (960 km.) N to S between Britain (W) and Norway, Denmark, Fed. Rep. of Germany and the Netherlands. Linked with the Atlantic by the English Channel, which it joins off Belgium, and the straits between the Orkney and Shetlands Islands, off N Scotland. Merges N with the Norwegian Sea. Linked with the Baltic Sea by the Skagerrak and Kattegat straits and the Kiel canal. Width 400 m. (640 km.), average depth 180 ft. (59 metres). Area, *c.* 200,000 sq. m. (518,000 sq. km.). Valuable for cod and herring fisheries and sub-marine oil and gas deposits. Chief ports Rotterdam, Hamburg, Bremen, Antwerp, London.

North Solomons Papua New Guinea. Province (formerly Bougainville) comprising Bougainville and Buka islands with several atolls. Area, 4,100 sq. m. (10,619 sq. km.). Mountainous, rising to 8,500 ft. (2,591 metres) in Mount Balbi (volcanic), and heavily forested. Main products copra, tagua nuts. Pop. (1984E) 137,600.

North Tonawanda New York, U.S.A. Port sit. N of Buffalo at confluence of Niagara R. and State Barge Canal. Its principal industry is timber. Other manu. inc. iron and metal goods, paper, plastics and chemicals. Pop. (1980C) 35,760.

North Uist *see* **Uist, North and South**

Northumberland England. 54 59N 1 35W. Under 1974 re-organization the county consists of the districts of Berwick-upon-Tweed, Alnwick, Tynedale, Castle Morpeth, Wansbeck and Blyth Valley. The most N county in England bounded in N by Scotland, W by Cumbria, s by Durham and Tyne and Wear and E by North Sea. Area, 1,943 sq. m. (5,032 sq. km.). Chief towns, Alnwick, Hexham, Blyth, Berwick-on-Tweed, Cramlington, Ashington and Morpeth. Shores are generally low. Off the coast lie Holy Isle, Farne Islands and Coquet Island. Inland it is mainly rugged and broken with undulating hills and moors rising to the Cheviot Hills (highest point, Cheviot at 2,476 ft. (755 metres) a.s.l). Drained by Rs. Tweed, Till, Aln, Coquet, Wansbeck, Blyth and Tyne. Important sheep-rearing county. Cattle also raised. Crops include barley and oats. Industries inc. electronics, textiles, agriculture, pharmaceuticals and engineering. Pop. (1989E) 303,500.

North Vietnam   *see* **Vietnam**

Northwest Passage   Canada. 74 40N 100 00W. Sea passage through the Arctic Arch. between the Atlantic and Pacific Oceans. Three possible routes connected by numerous N to S straits, all subject to ice and severe weather. The most N route is from N Baffin Bay through Jones Sound, Norwegian Bay, Belcher Channel, Hazen Strait and Ballantyne Strait to the Beaufort Sea. The most S route is through Davis Strait, Hudson Strait, Foxe Channel, Foxe Basin, Fury and Hecla Strait, the Gulf of Boothia, Prince Regent Inlet, Bellot Strait, Franklin Strait, Victoria Strait, Queen Maud Gulf, Dease Strait, Coronation Gulf, Dolphin and Union Strait, Amundsen Gulf, Beaufort Sea.

Northwest Territories   Canada. 67 00N 110 00W. Canadian territory bounded S by Lat. 60°N, W by Yukon, E by Hudson Bay and the Atlantic, N by the Arctic Ocean, and inc. the islands of the Arctic Arch. Area, 1,304,903 sq. m. (3,376,698 sq. km.). Chief towns Yellowknife (cap.), Inuvik, Hay River, Fort Smith, Frobisher Bay. Desolate and underdeveloped except for mining, lumbering, fishing and trapping. The Rockies extend N through the W beyond the Great Slave and the Great Bear Ls.; the centre is barren plain rising again to plateau NE of Hudson Bay. There are numerous Ls. and forests of conifers, birch and poplar. The Indian and Eskimo pop. live by trapping, fishing. Minerals inc.

gold, silver, lead, zinc, copper and oil. Mining, oil refining and fish processing are the main industries. Pop. (1986E) 51,384.

Northwich   Cheshire, England. 53 16N 2 32W. Town sit. ENE of Chester at confluence of Rs. Weaver and Dane. The principal industry is the manu. of chemicals, the works being one of the largest in the world. Salt is also produced. Other industries inc. printing, pet food manu., food processing and construction. Pop. (1988E) 54,000.

Northwood   *see* **Ruislip-Northwood**

North Yorkshire   England. 54 15N 1 25W. The largest county in England and Wales. Area, 3,209 sq. m. (8,309 sq. km.). Under 1974 re-organization the county consists of the districts of Richmondshire, Hambleton, Ryedale, Scarborough, Craven, Harrogate, York and Selby. There is a large farming sector and a wide variety of industry inc. manu. of mechanical and electrical equipment, vehicles, clothing, footwear, plastics and foodstuffs. A major tourist area containing National Parks and 50 m. (80 km.) of coastline. Pop. (1990E) 713,100.

Norwalk   Connecticut, U.S.A. 41 07N 73 27W. City sit. SW of New Haven on Long Island Sound. Manu. inc. electronics, batteries, toys. Commercial and service centre. Pop. (1980C) 77,767.

Norway   Scandinavia. 63 00N 11 00E. Kingdom bounded N by Arctic Ocean, NW and W by the Norwegian Sea, SW by the North Sea, S by the Skagerrak Strait and E by Sweden, Finland and the U.S.S.R. Area, 125,182 sq. m. (323,886 sq. km.). By the Treaty of 1814 Norway was ceded to the King of Sweden by the King of Denmark, but the Norwegian people declared themselves independent and elected Prince Christian Frederik of Denmark as their king. The foreign Powers refused to recognize this election, and a convention proclaimed the independence of Norway in a personal union with Sweden in 1814, which was repealed in 1905. After a plebiscite, Prince Carl of Denmark was formally elected King in 1905, and took the name of Haakon VII. Chief cities Oslo (cap.), Bergen, Trondheim, Stavanger. Mountainous and mainly unsuitable for crops, but with extensive pasture for sheep, dairy and beef cattle, and goats. Narrow strips of arable land in valleys or L. shores yield cereals, hay and potatoes. Minerals inc. iron ore, copper and pyrites. The most valuable natural resources are oil, timber and fish. Productive forest covers *c.* 25,714 sq. m. (66,600 sq. km.), 80% of it conifer trees providing the raw material for the paper industry. The fishing industry is based mainly on cod, mackerel, coalfish, herring, haddock, prawns. Industries are mainly powered by hydroelectricity. The chief industries are oil, pulp and paper making, fish processing, chemicals, manu. basic metals, machinery, transport equipment. Pop. (1989E) 4,220,686.

Norwich　Norfolk, England. 52 38N 1 18E. City sit. on Wensum R. Cathedral city with many medieval buildings. Agric. market town with industries inc. manu. shoes, electrical and electronic goods, food and drink, chemicals, printing, insurance and bookbinding. Pop. (1981C) 119,759.

Norwich　Connecticut, U.S.A. 41 32N 72 05W. City sit. NNE of New Haven on R. Thames. Manu. inc. textiles, paper and leather goods, clothing, chemicals, furniture, cable, printing, aerospace and marine equipment. Pop. (1980C) 38,074.

Norwood　Ohio, U.S.A. 39 10N 84 28W. City forming a suburb of Cincinnati, but administered separately. Industries inc. manu. vehicles, machinery, electrical goods, aircraft parts, metal products, furniture, railway equipment. Pop. (1980C) 26,342.

Nossi-Bé　Madagascar. 13 20S 48 15E. A volcanic island just off the NW coast. Area, 130 sq. m. (337 sq. km.). The chief town and port is Hellville. Crops inc. rice, maize, sugar, coffee and bananas.

Nottingham　Nottinghamshire, England. 52 58N 1 10W. City sit. NNW of London, on Trent R. Industries inc. textiles, manu. tobacco products, lace,

leather and leather goods, hosiery, pharmaceuticals, electrical equipment. Pop. (1981C) 271,080.

Nottinghamshire　England. 53 10N 1 00W. Under 1974 re-organization the county consists of the districts of Bassetlaw, Mansfield, Newark and Sherwood, Ashfield, Broxtowe, Nottingham, Gedling and Rushcliffe. Midland county sit. in R. Trent basin. Bounded in N by South Yorkshire and Humberside, in E by Lincolnshire, in S by Leicestershire and W by Derbyshire. Agric. is important and it is an arable area. Nottinghamshire contains some of the most important coalfields in Britain and other minerals include gypsum and limestone. Chemicals, bicycles, electronics, tobacco products and knitwear are manu. Trad. associated with Robin Hood. Pop. (1990E) 1,007,000.

Nouakchott　Mauritania. 18 09N 15 58W. Cap. of Mauritania, sit. NNE of Dakar, Senegal near the Atlantic coast. A commercial centre on caravan routes, trading in grain and gums. Pop. (1985E) 500,000.

Nouméa　New Caledonia. 22 16S 166 27E. Cap. and port sit. on SW coast. Nickel, chrome, iron and manganese ores are exported Pop. (1983C) 60,112.

Nouvelle Amsterdam　Southern and Antarctic Territories. 37 52S 77 32E. A volcanic island sit. in the Indian Ocean at 37° lat., 70°E long. Area, 21 sq. m. (54 sq. km.). Research station at Base Martin de Vivies. Pop. (1983C) 33.

Nova Goa　*see* **Panjim**

Nova Lisboa　*see* **Huambo**

Novara　Piemonte, Italy. 45 28N 8 38E. Cap. of province of same name in NW. Industries inc. rice milling, map making and printing and the manu. of chemicals and textiles. Pop. (1981E) 102,086

Nova Scotia　Canada. 45 10N 63 00W. Province of E Canada occupying a peninsula and island off the SE coast of New Brunswick. Bounded W by the

Bay of Fundy, s and e by the Atlantic, ne by Cabot Strait and n by the Gulf of St. Lawrence. Area, 21,425 sq. m. (55,000 sq. km.). Chief cities Halifax (cap.), Dartmouth, Sydney. Deeply indented coastline with numerous inlets, chiefly the Minas Basin, (w) and Bras d'Or L. which almost cuts Cape Breton Island (ne) in two. About 15,500 sq. m. (39,680 sq. km.) are forested with conifers, birch, oak, maple, poplar, ash. The main types of farming are dairying, poultry and stock farming, with some fruit growing. Minerals inc. coal, gypsum, salt, sand and gravel. The main industries are fish and other food processing, manu. transport equipment, lumbering sawmilling, pulp and paper making. Pop. (1989e) 886,800.

Novaya Zemlya Russian Soviet Federal Socialist Republic, U.S.S.R. 74 00N 57 00E. An arctic land mass sit. between the Barents Sea and the Kara Sea, consisting of two main islands separated by the narrow strait of Matochkin Shar. Fishing, sealing and hunting are the main occupations.

Novgorod Russian Soviet Federal Socialist Republic, U.S.S.R. 58 31N 31 17E. Cap. of Novgorod region sit. sse of Leningrad on Volkhou R. Industries inc. sawmilling, brewing, flourmilling, and the manu. of footwear and clothing. Pop. (1987e) 228,000.

Novi Sad Vojvodina, Yugoslavia. 45 15N 19 50E. Cap. of federal unit and port sit. nw of Belgrade on Danube R. Manu. inc. agric. machinery, electrical equipment, textiles and pottery. Pop. (1981e) 258,000.

Novocherkassk Russian Soviet Federal Socialist Republic, U.S.S.R. 47 25N 40 06E. Town sit. ene of Rostov. Manu. inc. locomotives, machinery, machine tools, mining equipment and textiles. Trades in timber, grain and wine. Pop. (1987e) 188,000.

Novokuznetsk Russian Soviet Federal Socialist Republic, U.S.S.R. 55 00N 85 05E. Town sit. sse of Kemerovo on Tom R. An important transportation and industrial centre with one of the largest iron and steel works

in the world. Manu. inc. locomotives, machinery, metal and aluminium products, chemicals and cement. Pop. (1987e) 589,000.

Novomoskovsk Russian Soviet Federal Socialist Republic, U.S.S.R. 54 05N 38 13E. Town sit. sse of Moscow. The centre of a lignite mining district with manu. of chemicals and machinery.

Novorossiisk Russian Soviet Federal Socialist Republic, U.S.S.R. 44 45N 37 45E. Port sit. on ne shore of Black Sea. An important centre of the cement industry, possibly the largest in the U.S.S.R. Other manu. inc. agric. machinery, bicycles and machine tools. Cement, petroleum and grain are exported. Pop. (1987e) 179,000.

Novoshakhtinsk Russian Soviet Federal Socialist Republic, U.S.S.R. 47 47N 39 56E. Town sit. nne of Rostov. The principal industry is anthracite mining.

Novosibirsk Russian Soviet Federal Socialist Republic, U.S.S.R. 55 02N 82 55E. Cap. of Novosibirsk region sit. on Ob R. An important industrial centre and the largest town in Siberia. Industries inc. sawmilling, flour milling, brewing, and the manu. of lorries, bicycles, agric. and mining machinery, machine tools, and textiles. Trades in grain, meat and dairy produce. Pop. (1987e) 1,423,000.

Novo Troitsk Russian Soviet Federal Socialist Republic, U.S.S.R. 51 12N 58 20E. Town sit. w of Orsk in Orsk-Khalilovo industrial centre on Ural R. The main industry is manu. steel, also mining nickel, cobalt, chromium.

Novy Bohumin Moravia, Czechoslovakia. Town sit. nne of Ostrava on Oder R. where it forms the Polish frontier. Industrial centre with iron and steel works, chemical plants and oil refinery. Main products pig-iron, steel, pipes, cables and wires, soap, candles, pharmaceutical goods and chemical products.

Nowa Huta Kraków, Poland. 50 03N 19 55E. Suburb sit. e of Kraków

and now part of that city on Vistula R.
It is an important steel centre with
associated industries.

Nowgong   Assam, India. 25 04N 79
27E. Town sit. SW of Sadiya. Manu.
chemicals, paper and pharmaceuticals.
There is also a distillery. Trades in
rice and tea.

Nowy Dwor Mazowiecki Warszawa,
Poland. 52 26N 20 44E. Town sit. NW
of Warsaw on Vistula R. Navy base
and R. port. Industries inc. brewing,
sawmilling, flour milling; manu.
soaps and earthenware.

Nowy Sącz   Kraków, Poland. 49
38N 20 42E. Town sit. SE of Kraków
in the Dunajec R. valley. Industries
inc. railway engineering, and the
manu. of agric. implements, chemicals
and textiles. Pop. (1983E) 68,000.

Nubian Desert   Sudan. 21 30N 33
30E. Sit. in NE between Red Sea and
Nile Valley, it consists mainly of a
sandstone plateau.

Nueva Esparta   Venezuela. Off-
shore state comprising Margarita
Coche and smaller islands off NE
Venezuela. Area, 444 sq. m. (1,150
sq. km.). Chief towns La Asunción,
Porlamar. Margarita is horse-shoe
shaped, mountainous with rough grazing (W), fertile agric. and resort area
(E). Agric. yields sugar, cotton, tobacco, maize, coconuts. Industries inc.
pearl fishing, deep-sea fishing, mining
(magnesite), fish canning. Pop.
(1980E) 145,923.

Nueva San Salvador   *see* **Santa
Tecla**

Nueva Segovia   Nicaragua. Dept. of
NW Nicaragua bounded SE by Coco
R., N by Honduras. Area, 1,595 sq. m.
(4,131 sq. km.). Chief towns Ocotal
(cap.), El Jicaro. Rises NW to the
Dipilto and Jalapa ranges. Farming
area, producing tobacco, sugar-cane,
coffee, livestock in higher ground,
cacao, maize and vegetables in the
lowlands. Minerals inc. gold and silver at El Jicaro. Pop. (1981C) 97,765.

Nuevo Laredo Tamaulipas, Mexico.
27 30N 99 31W. City sit. on Rio
Grande opposite Laredo, U.S.A. It is
an important entry point into Mexico.
Manu. inc. textiles. Trades in agric.
produce. Pop. (1980C) 203,286.

Nuevo León   Mexico. 25 00N 100
00W. State in NE bordering on the
U.S.A. Area, 25,067 sq. m. (64,924
sq. km.). Mountainous in W but
mainly flat in E and agric. is possible
with irrigation. Crops produced inc.
cereals, sugar-cane, fruit and cotton;
cattle raising is important. There are
considerable deposits of gold, silver,
lead, copper, phosphates and manganese. Cap. Monterrey. Pop. (1980C)
2,513,044.

Nui   Tuvalu, Pacific. Atoll of the N
Tuvalu group. Area, 1.3 sq. m. (3.3 sq.
km.). The main product is copra. Pop.
(1983E) 650.

Nuku'alofa   Tongatabu, Tonga. 21
08S 175 12W. Town on N Tongatabu.
Cap. and seat of royal palace. Port exporting bananas, coconut oil, desic-
cated coconut and vanilla beans. Pop.
(1986C) 94,649.

Nukufetau   Tuvalu, Pacific. Atoll
of the central Tuvalu group. Area,
1 sq. m. (1.6 sq. km.), consisting of
numerous islets on a 24 m. (38 km.)
reef. Pop. (1983E) 740.

Nukuhiva   Marquesas Islands, South
Pacific. 8 54S 140 06W. Island sit. N
of Hiva Oa. Area, 46 sq. m. (118 sq.
km.). Chief town Taiohae. Fertile,
with wooded hills rising to 4,000 ft.
(1,219 metres). Main products copra,
fruit.

Nukumanu   Solomon Islands, South
Pacific. Atoll sit. ENE of Bougainville,
consisting of *c.* 40 islets on 11 m.
(18 km.) reef.

Nukunonu   Tokelau, South Pacific.
Atoll sit. WNW of Fakaofo. Area,
1,370 acres, consisting of 24 islets.
The main product is copra.

Nukus   Kara-Kalpak Autonomous
Soviet Socialist Republic, U.S.S.R. 42
50N 59 23E. Cap. of Kara-Kalpak sit.
WNW of Tashkent at head of Amu
Darya R. delta. Manu. inc. cotton and
food products, footwear and furniture.

Nullarbor Plain   South Australia/ Western Australia, Australia. 31 20S 128 00E. A treeless and riverless area sit. between the Great Victoria Desert and the Great Australian Bight.

Nuneaton   Warwickshire, England. 52 32N 1 28W. Town sit. NNE of Coventry on R. Anker. Industries inc. light engineering, electronics, distribution services and the manu. of textiles. Pop. (1986E) 71,300.

Nuremberg   Bavaria, Federal Republic of Germany. 49 26N 11 05E. Town in SE sit. on Pegnitz R. Manu. inc. machinery, electrical and office equipment, precision instruments and esp. toys. Pop. (1990E) 473,910.

Nusa Tenggara   Indonesia. 9 30S 122 00E. Otherwise called the Lesser Sunda Islands; arch. extending E of Java and forming a province of Indonesia. Chief islands are Bali, Lombok, Sumbawa, Sumba, Flores, the Solor and Alor islands, Timor. Mountainous. The main occupation is farming with some fishing. Main products copra, hides, tobacco, rice, timber. Pop. (1980C) 5·45m.

Nutley   New Jersey, U.S.A. 40 49N 74 10W. Town sit. N of Newark. Mainly residential. Manu. inc. textiles, chemicals and paper. Pop. (1980C) 28,998.

Nyamlagiria   Zaïre. 1 25S 29 12E. Active volcano at 10,000 ft. (3,048 metres) NE of Sake and N of L. Kivu.

Nyanza   Kenya. Province of W Kenya bounded W by L. Victoria, S by Tanzania. Area, 6,240 sq. m. (16,162 sq. km.). Chief towns Kisumu, Londi-

ani, Kisii. Mainly plateau descending towards L. Victoria. Farming area with L. fisheries, and some gold mining. Pop. (1980C) 3,508,500.

Nyasa, Lake   *see* **Malawi, Lake**

Nyasaland   *see* **Malawi**

Nyíregyháza   Szabolcs-Szatmár, Hungary. 47 59N 21 43E. Cap. of county sit. WNW of Budapest. The centre of an agric. district noted for potatoes and tobacco. Manu. inc. soap, furniture and cement. Pop. (1984E) 114,000.

Nyköping   Södermanland, Sweden. 58 45N 17 00E. Cap. of county and port in SE sit. on the Baltic Sea. Manu. inc. machinery and textiles. Timber, wood-pulp, iron and zinc ores are exported. Pop. (1988E) 64,739.

Nyland   (Uusimaa), Finland. County of S Finland. Area, 3,822 sq. m. (9,898 sq. km.). Chief cities Helsinki, Borga, Lovisa, Ekenas, Hango. Lowlying, well wooded. Main occupations lumbering, quarrying, farming. Industries inc. sawmilling, paper making, wood working, textiles, metalworking; manu. glass, cement, machinery, fishing. Pop. (1983E) 1,162,871.

Nzani   (formerly Anjouan) Comoros, Indian Ocean. Island in the Mozambique Channel off NW Madagascar, between Mayotte and Njazídja. Area, 164 sq. m. (424 sq. km.). Chief town, Mutsamuda. The island is hilly, rising to Mount Tingue. The main products are coffee, vanilla, copra, essential oils, sweet potatoes and peanuts. Indust. inc. sisal processing and sugar milling. Pop. (1980C) 137,621.

O

Oahu   Hawaii. 21 30N 158 00W. Island of the Hawaiian group NW of Molokai across Kaiwi Channel and SE of Kauai across Kauai Channel. Area, 618 sq. m. (1,601 sq. km.). Chief towns Honolulu (Hawaiian cap.), Waipahu, Wahiawa, Kaneohe, Kailua, Lanikai. Mountains (not volcanoes) rise to Kaala, 4,046 ft. (1,233 metres) with fertile valleys. The most developed and commercially important island. Army and navy bases, esp. at Pearl Harbor. Industries inc. sugar planting and processing, pineapple growing, tourism. Pop. (1980C) 762,534.

Oakengates   *see* **Telford**

Oakham   Leicestershire, England. 52 40N 0 43W. Town sit. E of Leicester. Industries inc. engineering and manu. plastics and hosiery. Pop. (1987E) 9,782.

Oakland   California, U.S.A. 37 47N 122 13W. City sit. on mainland side of San Francisco Bay. Manu. cars and lorries, electronic and office equipment, metal products, chemicals, food products and paint. Pop. (1980C) 339,337.

Oak Park   Illinois, U.S.A. 41 53N 87 48W. Village sit. to W of Chicago. Mainly residential with some light industry. Pop. (1990E) 55,006.

Oak Ridge   Tennessee, U.S.A. 36 01N 84 16W. Town sit. WNW of Knoxville. The Oak Ridge National Laboratory is sit. here and two uranium-processing plants. Pop. (1988E) 27,710.

Oakville   Ontario, Canada. 43 27N 79 41W. Town sit. SW of Toronto. Manu. inc. automobiles. Pop. (1981C) 76,720.

Oamaru   South Island, New Zealand. 45 06S 170 58E. Port sit. on E coast. Industries inc. engineering, tex-tiles, confectionery, sawmilling and market gardening. Vegetables and limestone are exported. Pop. (1990E) 12,500.

Oaxaco   Mexico. 17 03N 96 43W. (i) State in S bordering the Pacific Ocean. Area, 36,275 sq. m. (93,952 sq. km.). Very mountainous with the Sierra Madre del Sur stretching W to E, the highest peak being Zempoalté-petl, 11,145 ft. (3,397 metres) a.s.l. Mainly agric. esp. in the fertile valleys. Crops inc. cereals, rice, coffee, sugar, cotton, tobacco, fruit and rubber. There are large mineral deposits. Pop. (1989E) 2,669,120. (ii) Cap. of state of same name sit. near Atoyac R. at the foot of San Felipe de Agua, and NW of Salina Cruz. Industries inc. food processing, textile manu. and Indian handicrafts esp. leather goods and pottery. Pop. (1980C) 157,284.

Oban   Strathclyde Region, Scotland. Resort sit. on Oban Bay, an inlet of the Firth of Lorne. Manu. glassware, pottery and whisky. Pop. (1981C) 8,134.

Obeid, El   Kordofan Sudan. 13 08N 31 10E. Cap. of province in central Sudan. Trades in cattle, cereals and gum arabic. Pop. (1982C) 140,024.

Oberammergau   Bavaria, Federal Republic of Germany. 47 35N 11 03E. Village sit. SSW of Munich on Ammer R. Noted for its Passion Play, performed every 10 years. It is also a centre of wood carving. Pop. (1980E) 4,900.

Oberhausen   North Rhine-Westphalia, Federal Republic of Germany. 51 28N 6 50E. Town sit. NE of Düsseldorf on Rhine-Herne Canal. Industries inc. plant and mechanical engineering construction, and the manu. of chemicals and steel. There is still one coal pit working. Pop. (1989E) 224,500.

427

Ob River   U.S.S.R. 66 45N 69 30E. Formed SW of Bisk, Altai Territory, by union of Biya and Katun Rs. Flows 2,113 m. (3,381 km.) through W Siberia, first NW past Kamen and Novosibirsk into swampy forests and on to Narym, then E past Surgut to receive Irtysh R. near Khanty-Mansisk, then N dividing into numerous arms and entering Ob Bay on the Kara Sea ENE of Salekhard. Frozen for 6 months of the year, but is otherwise an important trade route. Main tribs. Irtysh, Tom, Chulym, Ket, Vakh, Kazam, Vasyugan, Konda and N Sosva Rs. Navigable throughout. With the Irtysh it forms a 3,230 m. (5,168 km.) waterway.

Obuasi   Ghana. 6 12N 1 40W. Town sit. S of Kumasi. Important gold mining centre on the Kumasi railway.

Obwalden   Switzerland. Half-canton comprising the W part of Unterwalden canton. Area, 190 sq. m. (486 sq. km.). Cap. Sarnen. Inc. the commune of Engelberg which is separated from it by a strip of Nidwalden canton. Dairy farming area rising to the Alps (S). Industries inc. woodworking, tourism. Pop. (1978E) 25,200.

Oceania   Pacific. Collective name for the island groups of central, W and S Pacific, comprising 3 groups: Polynesia (French Polynesia, Tonga, Samoa, Hawaii, Tuvalu, Niue, Cook and Tokeau islands): Melanesia (New Caledonia, Solomon Islands, Fiji, New Hebrides): Micronesia (U.S. Trust Territories, Guam, Nauru, Gilbert Islands). The term also inc. Papua New Guinea and the Bismarck Arch.

Ocean Island   (Banaba) Kiribati, West Pacific. 0 52S 169 35E. Island sit. SW of Tarawa in the Gilbert group. Area, 2.2 sq. m. (5.5 sq. km.). Formerly worked for phosphates now exhausted. Pop. (1983E) 70.

Ochil Hills   Scotland. 56 14N 3 40W. A range of hills stretching from Bridge of Allan, N of Stirling, to the Firth of Tay. The highest peaks are Ben Cleuch, 2,363 ft. (720 metres) a.s.l., and King's Seat, 2,111 ft. (649 metres) a.s.l.

Ocotepeque   Honduras. Dept. of W Honduras bounded SW by Salvador, NW by Guatemala. Area, 649 sq. m. (1,680 sq. m.). Chief towns Nueva Ocotepeque (cap.), San Marcos. Mountainous, drained by Lempa and Jicatuyo Rs. Agric. area producing cereals, maize, coffee, tobacco, rice, fruit. Main industries flour milling; manu. straw and tobacco products. Pop. (1983E) 64,151.

Oda   Ghana. Town sit. NNW of Winneba. Diamond mining centre.

Odense   Fyns, Denmark. 55 24N 10 23E. Cap. of county and port sit. on Odense R. in N of Fyn Island. Industries inc. shipbuilding, iron founding, sugar refining, and the manu. of machinery, textiles, glass and electrical equipment. Pop. (1989E) 174,943.

Odenwald   Federal Republic of Germany. 49 40N 9 00E. Hilly district bounded S by R. Neckar, E by Tauber, N by Main and W by the Rhine plain. Wooded hills rise to 2,054 ft. (626 metres) at Katzenbuckel. The W slopes have orchards and vines. Chief town, Michelstadt.

Oder River   Poland/Czechoslovakia. 53 33N 14 38E. Rises ENE of Olomouc in the Oder mountains, Czechoslovakia. Flows 563 m. (901 km.) generally NE past Novy Bohumin into Poland, then NW past Raciborz, Kozle, Wrocław and Glogow to Nowa Sol, W to Krosno Odrzanskie and on to German Democratic Republic border which it forms (N) for the rest of its course, passing Frankfurt to the W and Szczecin to drain into the Baltic by Szczecin lagoon. Paralleled from Kostrzyn onward by other arms, some channelled, which rejoin SE of Police and enter the lagoon by a canal. Linked by canal with Havel R., Spree R., Upper Silesian industrial area around Katowice: also with Vistula R. by various waterways. Extensively controlled by locks and artificial channels with reservoirs fed by tribs. of which chief are Lusatian and Glatzer Neisse, Olza, Klodnica, Baryz, Warta, Ihna, Olawa, Katzbach and Bobrawa Rs. Used extensively for transport, esp. iron, coal and coke.

Odessa   Texas, U.S.A. 31 51N 102

22W. City sit. WSW of Fort Worth. An important centre of the oil industry with oil refining and the manu. of oil-field equipment. Also manu. synthetic rubber, chemicals and carbon black. Pop. (1980C) 90,027.

Odessa Ukraine, U.S.S.R. 46 28N 30 44E. Cap. of Odessa region and port sit. on the Black Sea. Industries inc. oil and sugar refining, engineering and the manu. of machine tools, agric. machinery, chemicals and bricks. Timber, grain, sugar and wool are exported. Pop. (1989E) 1,148,000.

Oeno Pitcairn, South Pacific. Uninhabited islet off Pitcairn Island, forming part of Pitcairn Island district.

Offaly Leinster, Ireland. 53 20N 7 30W. An inland county bounded on W by Shannon R. and on SE by the Slieve Bloom Mountains. Area, 771 sq. m. (1,997 sq. km.). Mainly flat and traversed by Rs. Barrow, Brosna and Nore, and the Grand Canal. The chief crops are oats, barley, sugar-beet, wheat and vegetables. Cattle, pigs and sheep are reared. County town, Tullamore. Pop. (1981E) 58,000.

Offenbach Hessen, Federal Republic of Germany. 50 08N 8 47E. Town in S central F.R.G. sit. on Main R. A centre of the leather industry with a leather museum. Also manu. machinery, chemicals and metal products. Pop. (1989E) 113,150.

Ofoten Norway. Region surrounding Ofot Fjord, an inlet of Vest Fjord opposite Lofoten Islands. Chief town Narvik.

Ofu American Samoa, South Pacific. Island of the Manua group. Rises to 1,585 ft. (483 metres).

Ogaden Harar, Ethiopia. Region of SE Ethiopia bounded E and S by Somalia. Arid plateau at 1,500–3,000 ft. (457–914 metres) watered by intermittent Rs. Nomadic farming is the main occupation.

Ogasawara Gunto *see* **Bonin Islands**

Ogbomosho Oyo, Nigeria. 8 08N 4 15E. Town sit. NE of Lagos. The main occupation is cotton weaving. Pop. (1988E) 527,000.

Ogden Utah, U.S.A. 41 14N 111 58W. City in N central Utah. An industrial centre of an agric. district esp. concerned with livestock and meat packing. Manu. iron castings, cement and clothing. Pop. (1980C) 64,407.

Ogowe River Gabon/Congo. Rises just S of Zanaga, Congo; flows 560 m. (896 km.) NW into Gabon, past Moanda and Lastoursville to join Ivindo R., then W past Booué into a region of Ls. around Lambaréné and enters the Gulf of Guinea by a delta E and SE of Port Gentil. Navigable all the year to Lambaréné, seasonally to N'Djole.

Ogun Nigeria. State formed from the Abeokuta and Ijebu provinces of the former W region. Area, 6,472 sq. m. (16,762 sq. km.). Inhabitants primarily of the Egba, Egbado and Ijebu sub-divisions of the Yoruba people. Main towns Abeokuta (cap.), Ijebu-Ode and Shagamu. Pop. (1988E) 3,397,900.

Ohio U.S.A. 38 00N 86 00W. State of E central U.S.A. bounded N by Michigan and L. Erie, W by Indiana, S by Kentucky, SE by West Virginia, E by Pennsylvania. Area, 41,222 sq. m. (106,765 sq. km.). Chief cities Columbus (cap.), Cleveland, Cincinnati, Toledo, Akron, Dayton, Youngstown, Canton, Parma. Plateau in the E, a section of the Allegheny plateau at 900–1,400 ft. (274–427 metres) a.s.l., descending W to lowland plain. Drained by Muskingum, Scioto and Great Miami Rs. to Ohio R. and to L. Erie by Maumel and Sandusky Rs. Continental climate. Farmland covers *c.* 17m. acres, producing cereals, soy beans, vegetables, beef and dairy cattle, sheep, pigs and poultry. The most important mineral is coal from the E and SE Appalachian coalfield; also sand and gravel, limestone, natural gas, petroleum. The main industry is iron and steel, based on Canadian ores shipped to the L. ports, also manu. metal products, vehicles, paper, clothing, chemicals, cement, electrical goods, foodstuffs, esp. meats. Pop. (1986E) 10,752,000.

Ohio River U.S.A. 36 59N 89 08W. Formed at Pittsburgh, Pennsylvania, by confluence of Allegheny and Monongahela Rs. Flows 980 m. (1,568 km.) NW to the Ohio boundary, where it turns generally SW forming the boundaries of several states, West Virginia, Kentucky to E and S; Ohio, Indiana, Illinois ·to W and N, before joining the Mississippi at Cairo, Illinois. Main tribs. Beaver, Muskingum, Hocking, Scioto, Miami, Wabash, Kentucky, Green, Cumberland and Tennessee Rs. Used to transport coal, coke, stone, cement, iron, steel, oil and timber. Liable to flooding.

Ohrid *see* **Okhrida**

Oise France. 49 28N 2 30E. Dept. in Picardy region. Area, 2,261 sq. m. (5,857 sq. km.). Mainly flat with large wooded areas esp. the Forest of Compiègne. The principal R. is Oise. Agric. products inc. wheat, sugar-beet and fruit and it is noted for cattle-rearing. Cap. Beauvais. Pop. (1982C) 661,781.

Oise River Belgium/France. 49 30N 2 56E. Rises in the Ardennes S of Chimay, Belgium and flows 186 m. (298 km.) SW, entering France N of Hirson and continuing past Compiègne to enter the Seine above Andrésy. Navigable up to Compiègne and from there on paralleled by canal. Linked by canal with Somme, Escaut and Sambre Rs. Main trib. Aisne R.

Oita Kyushu, Japan. 33 15N 131 36E. Port sit. ENE of Nagasaki on NE coast. Manu. textiles, paper and metal goods. Pop. (1988E) 393,000.

Ojos del Salado Chile/Argentina. 27 00S 68 40W. Mountain peak in the Andes sit. WSW of Cerro Incahuasi, at 22,550 ft. (6,873 metres) a.s.l. The second highest peak in the W hemisphere, after Aconcagua, sit. on the Chile/Argentina frontier.

Okanagan Lake British Columbia, Canada. A narrow L. 80 m. (129 km.) long drained by Okanagan R., and *c.* 160 m. (256 km.) ENE of Vancouver.

Oka River U.S.S.R. 56 20N 43 59E. (i) Rises W of Maloarkhangelsk

in the central Russian Upland; flows 918 m. (1,469 km.) N past Orel, Beler and Chekalin to Kaluga, then E to Ryazan and NE past Murom to enter Volga R. at Gorki. Navigable for large vessels below Kolomna 550 m. (880 km.) but subject to flooding in spring, low water in summer and ice 220–240 days of the year. Carries grain, timber. (ii) Rises in the E Sayan range in Buryat-Mongol Autonomous Soviet Socialist Rep.; flows 500 m. (800 km.) N to enter Angara R. at Bratsk.

Okavango River Angola/Namibia. 18 50S 22 25E. Rises on Bié plateau E of Nova Lisboa, flows *c.* 1,000 m. (1,600 km.) SE to the South West Africa border, then E and SE across the Caprivi Zipfel strip into Botswana where it disappears in the Okavango Marshes NW of Maun.

Okayama Honshu, Japan. 34 40N 133 64E. Port sit. W of Osaka on an inlet of the Inland Sea. Manu. cotton goods, native matting, agric. implements and porcelain. Pop. (1988E) 576,000.

Okazaki Honshu, Japan. 34 57N 137 10E. Town sit. E of Osaka. Manu. inc. machinery, fabricated metals, textiles and chemicals. Pop. (1988E) 292,000.

Okeechobee, Lake Florida, U.S.A. 26 55N 80 45W. Sit. in S it is the second largest freshwater L. in the U.S.A. Area, 730 sq. m. (1,891 sq. km.). The Kissimmee R. runs into it in N, and it drains to the Atlantic Ocean *via* the Everglades.

Okefenokee Swamp Georgia/Florida, U.S.A. 30 42N 82 20W. Mainly the Okefenokee National Wildlife Refuge, containing many native species.

Okhotsk, Sea of U.S.S.R. 55 00N 145 00E. Arm of the NW Pacific W of the Kamchatka Peninsula and the Kurile Islands. Area, 590,000 sq. m. (1,528,100 sq. km.). Connects SW with the Sea of Japan, by the Tatar and La Pérouse straits, SE with the Pacific between the Kurile Islands. Average depth 2,750 ft. (838 metres). Icebound November–June, and subject to thick fog.

Okhrida Macedonia, Yugoslavia. 41 06N 20 49E. Town and resort sit. on NE shore of L. Okhrida in SW, near the border with Albania. Railway terminus, tourist and trading centre. Industries inc. fishing. Anc. town once a centre of Slav Christian culture. Pop. (1981E) 64,000.

Okhrida, Lake Yugoslavia. 41 06N 20 49E. Deepest L. in the Balkans on the Yugoslav/Albanian border at 2,280 ft. (695 metres) a.s.l. Depth 938 ft. (286 metres). Connected to L. Prespa by underground channels. Area, 134 sq. m. (343 sq. km.). Important fishing industry, esp. carp. eel, trout. Tourist centre in a beautiful and historic area.

Okinawa Ryukyu Islands, Japan. The largest island sit. SSW of Kyushu. Area, 454 sq. m. (1,176 sq. km.). Cap. Naha City. Crops inc. rice, sugar-cane and sweet potatoes.

Oklahoma U.S.A. 35 30N 98 00W. State of SW U.S.A. bounded N by Kansas, E by Missouri and Arkansas, S and SW by Texas, W by New Mexico and NW by Colorado. Area, 69,919 sq. m. (181,091 sq. km.). Chief cities Oklahoma City (cap.), Tulsa, Lawton. Plains, sloping SE and drained to Arkansas and Red Rs. Some highland (NW) rises to Black Mesa, 4,978 ft. (1,517 metres). The Boston and Ouchita Mountains extend into the E from Arkansas and the Arbuckle and Wichita Mountains rise from the plain (S). Farming is the largest industry, covering 31m. acres. The main products are beef and dairy cattle, wheat, cotton, grain sorghums. Soil erosion is a serious problem to agric.; much former arable land is being turned into pasture. The chief mineral is petroleum, also helium, natural gas, coal, copper, silver. Pop. (1987E) 3,272,000.

Oklahoma City Oklahoma, U.S.A. 35 28N 97 32W. State cap. sit. on N. Canadian R. The commercial and industrial centre of an important oil-producing and agric. area. Industries inc. meat packing and the manu. of aircraft, automobiles, machinery, iron and steel, and paint. Pop. (1980C) 403,213.

Oktemberyan Armenian Soviet Socialist Republic, U.S.S.R. Town sit. W of Erivan in Aras R. valley. Centre of an irrigated agric. area. Industries inc. cotton ginning, fruit canning, tanning, oil pressing, metalworking.

Oktyabrski Russian Soviet Federal Socialist Republic, U.S.S.R. 54 28N 53 28E. Town sit. WSW of Ufa in the W Bashkir Autonomous Soviet Socialist Rep. on Ik R. Petroleum centre in the Tuimazy oilfield, linked by pipeline to Chernikovsk and Urussu.

Olafsjordur Eyjafjardar, Iceland. 66 06N 18 38W. Town sit. NE of Reykjavik on a SW arm of Eyja Fjord. Fishing port. Pop. (1988E) 1,179.

Olancho Honduras. Dept. of E central Honduras bounded SE by Oro R. and Nicaragua. Area, 9,402 sq. m. (24,350 sq. km.). Chief towns Juticalpa (cap.), Catacamas. Mountainous, with forest covering *c.* 50% of the area. There is stock and dairy farming in the Olancho Valley; agric. areas elsewhere produce coffee, tobacco, sugar-cane, rice. Lumbering is the main occupation in the E. Other industries inc. tanning, sugar refining, tobacco processing. Pop. (1983E) 228,122.

Öland Island Kalmar, Sweden. 56 45N 16 38E. A narrow island sit. in Baltic Sea and separated from the mainland by Kalmar Sound. Area, 520 sq. m. (1,344 sq. km.). Cap. Borgholm. The chief occupations are farming, fishing and limestone quarrying.

Oldbury West Midlands, England. 52 30N 2 00W. Town sit. WNW of Birmingham. Manu. iron, steel, aluminium, machinery, chemicals and glass. Pop. (1981C) 46,450.

Oldenburg Lower Saxony, Federal Republic of Germany. 53 08N 8 13E. City sit. WNW of Bremen on Hunte R. at its confluence with the Ems-Hunte canal. One of the largest garrison cities. R. port trading in building materials, timber, grain, cattle, horses. Industries inc. food processing, printing; manu. electrical appliances, dyes, varnishes and waxes. Pop. (1985E) 138,700.

O Oldham

Oldham   Greater Manchester, England. 53 33N 2 07W. Sit. NE of Manchester. Industries include electronics, aircraft, textiles, food processing and mail order distribution. Pop. (1985E) 220,000.

Olekma River   U.S.S.R. Rises in the Yablonovy mountains and flows 794 m. (1,270 km.) N to enter Lena R. below Olekminsk. Gold mining along its course. Main tribs. Nyukzha, Tungir and Chara Rs.

Oléron   Charente-Maritime, France. 45 56N 1 15W. An island sit. in Bay of Biscay opposite mouth of Charente R. Area, 68 sq. m. (175 sq. km.). Its main products are corn, vegetables, wine and oysters. Pop. (1982E) 12,598.

Olhäo   Faro, Portugal. 37 02N 8 50W. Town sit. E of Faro on the Atlantic. Important fishing and fish processing centre, esp. sardine, tuna.

Olives, Mount of   Jordan. 31 47N 35 15E. Sit. to E of Jerusalem and separated from it by the Kidron valley at 2,680 ft. (819 metres) a.s.l.

Olomouc   Moravia, Czechoslovakia. 49 36N 17 16E. Town in central Czech. sit. on Morava R. Manu. agric. machinery, textiles, cement and food products. There are some fine medieval buildings. Pop. (1983E) 103,000.

Olosega   American Samoa, South Pacific. Island in the Manua group. Fertile and mountainous, rising to 2,095 ft. (639 metres) a.s.l.

Ólsztyn   Olsztyn, Poland. 53 48N 20 29E. Cap. of province in N central Poland. Manu. inc. machinery and leather. Pop. (1983E) 144,700.

Olt River   Romania. 43 43N 24 51E. Rises E of Gheorgheni and flows 348 m. (557 km.) S past Sfantu-Gheorghe, then W past Fagaros and S through the Turnu-Rosu Pass in the Transylvanian Alps to enter Danube R. S of Turnu-Magurele and opposite Nikpol (Bulgaria). Used for logging.

Olympia   Washington, U.S.A. 47 03N 122 53W. State cap. and port sit. at S end of Puget Sound, and SE of Tacoma. Timber and fish are exported. Employment is mainly in government health services, timber and paper industries. Pop. (1980C) 27,447.

Olympus   Thessaly/Macedonia, Greece. 40 05N 22 21E. A snow-capped mountain range sit. near the Aegean Sea coast. 9,600 ft. (2,926 metres) a.s.l.

Omagh   Tyrone, Northern Ireland. 54 36N 7 18W. County town in W central N. Ireland sit. on R. Strule. Manu. shirts and dairy products. Trades in livestock, oats and potatoes. Pop. (1981E) 41,137.

Omaha   Nebraska, U.S.A. 41 16N 95 57W. City sit. on Missouri R. An important industrial and transport centre. Industries inc. food processing, insurance, telecommunications. Manu. telephone equipment, irrigation equipment, agric. machinery, office furniture and transport equipment. Pop. (1980C) 314,255 (metropolitan area, 567,614).

Oman   Arabia. 23 00N 58 00E. Formerly Muscat and Oman. Sultanate bounded NE by the Gulf of Oman, E and S by the Arabian Sea, SW by the Republic of Yemen, W by Saudi Arabia, NW by the United Arab Emirates. Area, *c.* 130,000 sq. m. (300,000 sq. km.). Chief towns Muscat (cap.), Mutrah, Salalah. Coastal plain up to 10 m. (16 km.) wide rises to hills, mainly barren except for the Jebel Akhdar highlands which are very fertile. Beyond the hills there is plateau extending to the edge of the Empty Quarter. The coastline is fertile S and NW. Main crops are dates, other fruit, and coconuts in Dhofar (S). The chief mineral is oil. Concessions are worked in Dhofar and the Gulf of Masirah. There is an oil terminal port at Mina al Fahal. Oil refining is the main industry, with the construction also developing rapidly. Pop. (1987E) 1·3m.

Oman, Gulf of   Oman/Iran. 24 30N 58 30E. A branch of the Arabian Sea, 300 m. (480 km.) long and 200 m. (320 km.) wide.

Omdurman   Khartoum, Sudan. 15 40N 32 28E. Town sit. on White Nile

R. Trades in livestock, hides and gum arabic. Pop. (1983E) 526,000.

Omiya   Honshu, Japan. 35 54N 139 38E. (i) Town sit. NNW of Mito; mining centre for gold, silver and copper. (ii) Town sit. NNW of Tokyo and immediately N of Urawa. Trading centre for an agric. area, handling rice, wheat and raw silk. Noted for an anc. Shinto shrine to Susano-wo, brother of the Sun goddess. Pop. (1988E) 384,000.

Omsk   Russian Soviet Federal Socialist Republic, U.S.S.R. 55 00N 73 24E. Cap. of Omsk region sit. at confluence of Rs. Irtysh and Om. A centre of industry, transport and oil-refining, as well as a terminus of the oil pipeline from Tuimazy. Manu. inc. locomotives, motorcars, agric. machinery, tyres, synthetic rubber and food products. Pop. (1989E) 1,148,000.

Ondo   Nigeria. 7 04N 4 47E. State, formerly the Ondo province of the W region. Area, 8,092 sq. m. (20,959 sq. km.). Main cities Akure (cap.), Ado-Ekiti, Ikerre-Ekiti, Owo, Ondo, Effon-Alaiye, Oka-Akoko and Ikare. Pop. (1983E) 3.6m.

Onega Bay   U.S.S.R. 64 30N 37 00E. Inlet of the White Sea in NW European U.S.S.R., W of the Onega peninsula. The Kem, Vyg and Onega Rs. flow into it. 100 m. (160 km.) long and 30–50 m. (48–80 km.) wide, with the Solovetskiye Islands at the entrance. Main ports Kem, Onega, Belomorsk.

Onega, Lake   Russian Soviet Federal Socialist Republic, U.S.S.R. 61 30N 35 45E. Sit. in Karelia, in NW U.S.S.R. it is the second largest L. in Europe. Area, 3,817 sq. m. (9,886 sq. km.). It is drained by Svir R. into L. Ladoga and is connected by canals with White Sea and Volga R.

Ongole   Andhra Pradesh, India. Town sit. SSW of Guntur. Industries inc. tobacco processing, rice and oilseed milling; manu. leather goods. Pop. (1981C) 85,302.

Onitsha   Anambra, Nigeria. 6 09N 6 47E. Town in SE sit. on Niger R. Trades in maize, cassava, yams, palm oil and kernels. Pop. (1983E) 268,700.

Onotoa   Kiribati, W Pacific. Atoll in S Gilbert Group. Area, 5.2 sq. m. (13 sq. km.). The chief product is copra. Formerly called Clerk Island. Pop. (1980E) 2,034.

Ontario   Canada. 52 00N 88 10W. Province of E central Canada bounded N by Hudson Bay, E by Quebec, S by the U.S.A. across the Great Lakes and Rainy R., W by Manitoba. Area, 412,582 sq. m. (1.6m. sq. km.). Chief cities Toronto (cap.), Hamilton, Ottawa (Canadian cap.), Windsor, London, Kitchener, Sudbury. Thickly wooded and sparsely populated (N) with most settlement on the Great Lakes (S). Farmland covers *c.* 7.5m. acres, producing grain, dairy and beef cattle, fruit, vegetables and tobacco. Mineral deposits are extensive, esp. nickel, copper, iron ore, gold, uranium. Forests supply important pulp and paper making industries. Other industries inc. metalworking; manu. machinery, vehicles, railway equipment, textiles, leather goods, food products. Pop. (1988E) 9·1m.

Ontario   California, U.S.A. 34 04N 117 39W. City sit. E of Los Angeles. Manu. inc. aircraft parts and electrical equipment. It is the commercial centre of a fruit growing area. Pop. (1980C) 88,820.

Ontario, Lake   Canada/U.S.A. 43 40N 78 00W. Smallest and furthest E of the Great Lakes. Area, 7,540 sq. m. (19,529 sq. km.). It receives the outflow of the Ls. system through the Niagara R. and Falls, and discharges NE through the St. Lawrence R. Canals by-pass the Falls (S) into L. Erie. Its shores are fertile with a modified climate. Main ports are Hamilton, Toronto and Kingston (Can.), and Rochester and Oswego (U.S.A.). Mainly ice-free.

Onverwacht   Suriname. Village sit. S of Paramaribo on the Paramaribo railway. Produces coffee, rice, fruit.

Oosterhout   Noord-Brabant, Netherlands. Town sit. NE of Breda. Manu. audio- and videotapes, confectionery, pipes, concrete and steel furniture. Pop. (1990E) 48,500.

Ootacamund   Madras, India. 11

433

25N 76 43E. Town and tourist resort sit. in the Nilgiri Hills, ENE of Kozhikode, at over 7,000 ft. (2,134 metres) a.s.l.

Opatija Croatia, Yugoslavia. 45 21N 14 19E. Resort sit. W of Rijeka on the Kvarner Gulf. Pop. (1981E) 30,000.

Opava Severomoravsky, Czechoslovakia. 49 56N 17 54E. Town in N central Czech. sit. on Opava R. Manu. textiles and woollen goods. Pop. (1983E) 61,000.

Opole Opole, Poland. 50 41N 17 55E. Cap. of province in SW, sit. on Oder R. Manu. inc. machinery, textiles, chemicals, cement and soap. Trades in livestock and grain. Pop. (1983E) 122,000.

Oporto *see* **Porto**

Oppland Norway. 61 15N 9 30E. County of S central Norway. Area, 9,753 sq. m. (25,259 sq. km.). Chief towns Lillehammer (cap.), Gjovik. Mountainous in the N with the Jotunheim, Dovrefjell and Rondane mountains rising to the Galdhopiggen, 8,068 ft. (2,468 metres). Traversed N to S by Gudbrandsdal valley, drained by Lagen R., where most of its settlement has developed. Main occupations dairy, stock and arable farming and quarrying (slate, talc, steatite). Tourist centre. Pop. (1988C) 182,510.

Orádea Bihor, Romania. 47 03N 21 57E. Cap. of District sit. NE of Szeged, Hungary on Crişul Repede R. Manu. machinery, pottery, glass and textiles. Trades in livestock, fruit and wine. Pop. (1983E) 206,000.

Oran Oran, Algeria. 35 43N 0 43W. Cap. of dept., port and the second largest town in Algeria, sit. on the Gulf of Oran. Industries inc. iron smelting, fruit and fish canning, and the manu. of textiles, glass, footwear and cigarettes. Exports inc. wheat, vegetables, wine, wool and esparto grass. Pop. (1984E) 450,000.

Orange New South Wales, Australia. 33 17S 149 06E. Town sit. WNW of Sydney. It is the centre of a sheep-

rearing and fruit growing district. Manu. woollen goods, domestic appliances, confectionery, wool scouring. Pop. (1990E) 33,000.

Orange Vaucluse, France. 44 08N 4 48E. Town in SE. Noted for its Roman remains. Manu. food products. Pop. (1982C) 27,502.

Orange California, U.S.A. 33 47N 117 51W. City sit. N of Santa Ana, and SE of Long Beach. The main occupation is orange growing and packing. Other industries inc. computers, oil well drilling equip., furnace manu. Pop. (1987E) 105,000.

Orange New Jersey, U.S.A. 40 46N 74 14W. City sit. NW of Newark. Manu. chemicals, clothing, cosmetics and electronic equipment. Pop. (1980C) 31,136.

Orange Texas, U.S.A. 30 01N 93 44W. City and port sit. E of Beaumont on Sabine R. Industries inc. the manu. of chemicals, boats, and paper. Pop. (1980C) 23,628.

Orange Free State Republic of South Africa. 28 30S 27 00E. Province bounded by Vaal R. and Transvaal (N), Natal (E), Lesotho (SE), Cape Province (SW and W). Area, 49,418 sq. m. (127,999 sq. km.). Chief towns Bloemfontein (cap.), Kroonstadt, Bethlehem, Harrismith, Ficksburg. Mainly plateau at 4,000–5,000 ft. (1,219–1,524 metres) a.s.l. drained by tribs. into Orange and Vaal Rs. Agric., producing grain, beef and dairy cattle, sheep, fruit, vegetables, tobacco. The chief mineral is gold, also diamonds (W) and coal near the Vaal R., the major gold fields are in the Odendaalsrust. Industries other than mining inc. processing coal, manu. fertilizers, agric. equipment, woollen goods, food products, cement, bricks and tiles. Pop. (1989E) 2,349,556.

Orange River Republic of South Africa/Lesotho. 28 41S 16 28E. Rises in the Drakensberg mountains of NE Lesotho; flows *c.* 1,300 m. (2,080 km.) SW to the border with the Orange Free State then W and NW along the Orange Free State/Cape Province border to the mouth of the Vaal R., then

SW to Prieska, NW to Upington and then generally W in a winding course along the South African/Namibian border into the Atlantic at Alexander Bay. Navigation blocked by sand bars and several cataracts on the lower course. Main tribs. Vaal, Caledon and Molopo Rs.

Ord River   Australia. 15 30S 128 21E. Rises SE of Hall's Creek, Western Australia, and flows 300 m. (480 km.) N through mountains to enter the Cambridge Gulf near Wyndham.

Ordu   Turkey. 41 00N 37 53E. (i) Province of N Turkey on Black Sea, bounded S by the Canik Mountains. Area, 2,076 sq. m. (5,377 sq. km.) Well forested area with deposits of zinc, copper and iron. Pop. (1985C) 763,857. (ii) Town sit. W of Trebizond on the Black Sea near mouth of Melet R. Provincial cap. Port and market town.

Ordzhonikidze   Russian Soviet Federal Socialist Republic, U.S.S.R. 43 03N 44 40E. Cap. of the North Ossetian Autonomous Soviet Socialist Rep. sit. on Terek R. in the Caucasus Mountains, W of Makhachkala. Industries inc. metallurgical plants, food processing, woodworking and glass manu. Zinc, lead and silver are obtained from the Sadon mines. Pop. (1985E) 303,000.

Örebro   Örebro, Sweden. 59 27N 15 00E. Cap. of county in S central Sweden sit. at W end of L. Hjälmar. It is the centre of a footwear industry. Also manu. machinery, chemicals, paper and soap. Pop. (1990E) 120,000.

Oregon   U.S.A. 44 00N 121 00W. State of NW U.S.A. bounded N by Washington, W by the Pacific, S by California and Nevada, E by Idaho. Area, 96,981 sq. m. (251,181 sq. km.). Chief cities Salem (cap.), Portland, Eugene. Mountainous W and NE with the Cascade Range extending parallel to the coast and the Blue Mountains and Wallowa Mountains rising NE beyond the Inland Empire plains of the Columbia Basin (N). The E is plateau, merging with the arid Great Basin (SE). Farming is divided in type by the Cascade Range; W of it is a fertile area

suitable for all crops; E of it the land is suitable for stock raising and wheat, but extensive irrigation is necessary for root crops or fruit. Farms cover *c*. 32% of the land area. Forests cover *c*. 50% of it, providing about half the U.S. production of plywood, a fifth of the softwood and a quarter of the hardboard. Sawmilling, pulp milling and associated manu. are the chief industries, also food processing, flour milling, fishing. Pop. (1987E) 2,690,000.

Orekhovo-Zuyevo   Russian Soviet Federal Socialist Republic, U.S.S.R. 55 49N 38 59E. Town sit. E of Moscow on Klyazma R. Industries inc. cotton milling, weaving and dyeing, flour milling, sawmilling, metalworking, manu. plastics.

Orel   Russian Soviet Federal Socialist Republic, U.S.S.R. 52 59N 36 05E. City sit. SSW of Moscow on Oka R. Oblast cap. Communications and industrial centre; manu. tractor parts, textile machinery, beer, spirits, footwear, iron products. Pop. (1987E) 335,000.

Orem   Utah, U.S.A. 40 19N 111 42W. Town immediately N of Provo. Centre of an irrigated market gardening area. The main industries are electronic components and computer software. The Geneva Steel works is sit. to the W on Utah L. Pop. (1990E) 68,000.

Orenburg   Russian Soviet Federal Socialist Republic, U.S.S.R. 51 45N 55 06E. Town sit. on Ural R., and NNE of Caspian Sea. An important industrial and transport centre trading in livestock, meat, hides, grain, wool and textiles. Industries inc. railway engineering, sawmilling, flour milling and the manu. of metal goods and clothing. Pop. (1989E) 547,000.

Orense   Spain. 42 20N 7 51W. (i) Province in Galicia bordering on Portugal. Area, 2,810 sq. m. (7,278 sq. km.). Mainly mountainous but agric. is possible in the valleys, and cereals, flax and potatoes are grown. The principal Rs. are Miño and Sil. Pop. (1986C) 399,378. (ii) Cap. of province of same name in NW, sit. on Miño R.

Manu. ironware, textiles, leather and chocolate. An interesting 7-arched 13th cent. bridge crosses the R. Pop. (1986c) 102,455.

Orgeev Moldavian Soviet Socialist Republic, U.S.S.R. Town sit. N of Kishinev on Reut R. Centre of a fertile agric. area. Industries inc. flour milling, brewing, fruit and tobacco processing.

Orinoco River South America. 8 37N 62 15W. Rises in the Parima Mountains on the NW Brazil border, flows 1,200–1,700 m. (1,920–2,735 km.) depending on the seasonal variations in the volume of its water. First NW across S Venezuela to San Fernando de Atabapo, whence it turns N up the Venezuela/Colombia frontier to Puerto Carreño, then NE through N Venezuela past Ciudad Bolivar and Ciudad Guayana to enter the Atlantic in a broad delta SW of Trinidad. Its course is through tropical rain forests and savannahs, and the volume of its water is immense. Navigable for *c.* 1,000 m. (1,600 km.), its course obstructed by the Atures and Maipures rapids on the Colombia border. The delta spread over 7,745 sq. m. (20,060 sq. km.), beginning 100 m. (160 km.) from the sea, and the surrounding lowlands become swamps in the rainy season. Ciudad Bolivar is the main R. port, 260 m. (416 km.) upstream and accessible to ocean-going vessels. Linked by the Casiquiare waterway with Negro R. and Amazon R. Main tribs. Guaviare, Vichada, Meta, Capanaparo, Arauca, Apure, Ventuari, Caura, Caroni and Paragua Rs.

Orissa India. 21 00N 85 00E. State of E India bounded N by Bihar, NE by West Bengal, E by the Bay of Bengal, S by Andhra Pradesh, W by Madhya Pradesh. Area, 60,119 sq. m. (155,707 sq. km.). Chief cities Bhubaneswar, Cuttack, Rourkela. Hilly region at 1,500–3,000 ft. (457–914 metres) a.s.l., with the E Ghats extending into the S and the Chota Nagpur plateau rising N. Fertile coastal strip irrigated by the flood waters of Mahanadi, Brahmani and Baitarani Rs. Forests cover *c.* 43% of the land area. Agric. land is *c.* 8.7m. hectares, and *c.* 80% of the pop. is engaged in growing rice,

with 4.5m. hectares under paddy. Other crops inc. pulses, jute, cotton, tobacco, sugar-cane, turmeric. Minerals inc. iron, manganese, limestone, dolomite, chromite, bauxite, graphite, china clay, vanadium. Industries inc. lumbering, fishing and fish processing, textiles, food processing; manu. iron and steel, refractories, pharmaceuticals, synthetic fibres, cables, televisions, paper, aluminium, fertilizers, cement, heavy water plant, ferrocrome and ferro-manganese. Pop. (1981c) 26,370,271.

Orizaba Veracruz, Mexico. 18 51N 97 06W. Town sit. WSW of Veracruz at 4,200 ft. (1,280 metres) a.s.l. It is the chief textile centre in Mexico. Other industries are railway engineering, brewing and the manu. of paper and tobacco products. Trades in coffee, sugar, tobacco and maize. Pop. (1980c) 114,848.

Orizaba, Pico de Veracruz, Mexico. An extinct volcano sit. N of Orizaba. It is the highest peak in the country, 18,700 ft. (5,700 metres) a.s.l.

Orkney Islands United Kingdom. 59 00N 3 00W. Arch. off N coast of Scotland across the Pentland Firth. Area, 376 sq. m. (974 sq. km.). Chief towns Kirkwall (cap.), Stromness. Main islands are Mainland, South Ronaldsay, Westray, Sanday, Stronsay, Hoy. There are 90 islands altogether, 10 of which are inhabited. The main occupations are fishing, fish curing, sheep, cattle and poultry farming, woollen weaving and knitting. The most important mineral is off-shore oil; considerable reserves are being exploited, with the growth of associated industries. Pop. (1985E) 19,351.

Orlando Florida, U.S.A. 28 32N 81 23W. City near Atlantic coast E of Kennedy Space Centre. Industries include film making, tourism (Walt Disney World and Epcot Center are attractions), citrus fruit, agric. and high technology. Pop. (1990E) 167,000.

Orléanais France. 48 00N 2 00E. Ancient province of central France on Loire R. Agric. area inc. the Beauce (N) and Gâtinais (E) fertile regions, the Forest of Orleans and the Sologne reclaimed marshland.

Orléans Loiret, France. 47 55N 1 54E. Cap. of dept. sit. SSW of Paris on Loire R. Prefecture; commercial and communications centre. Industries inc. food processing, textiles; manu. clothing, blankets, agric. equipment, wines. Historic city associated with St. Joan of Arc. Pop. (1982C) 105,589 (agglomeration, 220,478).

Orly Val-de-Marne, France. S suburb of Paris containing intern. airport. Pop. (1982C) 23,886.

Ormskirk Lancashire, England. 53 35N 2 53W. Town sit. NNE of Liverpool. Industries inc. engineering, services and electrical. Pop. (1984E) 16,975.

Orne France. 48 40N 0 00E. Dept. in the Basse-Normandie region. Area, 2,355 sq. m. (6,100 sq. km.). The W is hilly with the Perche and Normandy Hills, while the E has fertile valleys growing cider apples and cereals. It is noted for Percheron horses. The principal R. is Orne. Cap. Alençon. Pop. (1982C) 295,472.

Örnsköldsvik Västernorrland, Sweden. 63 17N 16 40E. Town sit. NE of Sundsvall at mouth of Ångerman R. on the Gulf of Bothnia. Sea port handling pulp and timber. Industries inc. woodworking, engineering, timber products. Pop. (1983E) 60,105.

Orontes River South West Asia. 36 02N 35 58E. Rises near Baalbek in the Bekaa valley, Lebanon, flows 240 m. (384 km.) NNE into W Syria past Homs and Hamah and on N to the Syrian/Turkish frontier below Jisr ash Shughur, turning W to Antioch then SW into the Mediterranean at Suveydiye. Unnavigable. Used for irrigation esp. at L. Homs, Syria.

Oroya Junín, Peru. Town in W central Peru sit. at confluence of Rs. Mantaro and Yauli at 12,200 ft. (3,719 metres) a.s.l. An important metallurgical centre working with silver, copper, zinc and lead. Pop. (1980E) 36,000.

Orpington Greater London, Eng-

land. 51 22N 0 05E. Town sit. ESE of Bromley on Cray R. Residential. Gave its name to a breed of hen.

Orsk Russian Soviet Federal Socialist Republic, U.S.S.R. 51 12N 58 34E. Town sit. on Ural R., and NE of Caspian Sea. Industries inc. oil refining, nickel-smelting, heavy engineering, meat packing and the manu. of locomotives and agric. machinery. Pop. (1985E) 266,000.

Orta, Lake Italy. 45 49N 8 24E. Lake sit. NNW of Novara in Piedmont. Area, 7 sq. m. (18 sq. km.). Maximum depth 469 ft. (143 metres). Drains through an outlet (N) to Toce R. into L. Maggiore.

Oruro Oruro, Bolivia. 17 59S 67 09W. Cap. of dept. sit. SE of La Paz in the Altiplano at 12,160 ft. (3,706 metres) a.s.l. Centre of a tin-mining area, working also with silver, copper and tungsten. Pop. (1982E) 132,213.

Orvieto Umbria, Italy. 42 43N 12 07E. Town sit. NNW of Rome on an isolated rock near Paglia R. Noted for wine and pottery.

Osage River Missouri, U.S.A. 38 35N 91 57W. Formed by union of Marais des Cygnes and Little Osage Rs., SE of Rich Hill; flows 360 m. (576 km.) SE and E past Osceola then NE into the L. of the Ozarks (for *c.* 130 m. (208 km.)) then NE to enter Missouri R. E of Jefferson City. Used for power at Bagnell Dam at the NE end of the L.

Osaka Honshu, Japan. 34 30N 135 30E. Port, cap. of Osaka prefecture and the third largest city in Japan. An important industrial and commercial centre. Main manu. inc. heavy and chemical industries, publishing and printing, machinery, steel and metal goods, electrical and electronics equip. Steel, machinery, chemicals, electrical and electronic goods are exported. Pop. (1989E) 2,633,008.

Osh Kirghizia, U.S.S.R. 40 33N 72 48E. Cap. of the Osh region sit. ESE of Tashkent at the E end of the fertile Fergana valley. Manu. silk and food products. Pop. (1987E) 209,000.

Oshawa   Ontario, Canada. 43 54N 78 51W. City and port sit. ENE of Toronto on L. Ontario. Industries inc. high technology, manu. automobiles and vehicle accessories, plastics, and general consumer goods. Pop. (1984E) 119,653.

Oshkosh   Wisconsin, U.S.A. 44 01N 88 33W. City sit. NNW of Milwaukee on the W shore of L. Winnebago. Manu. packaging, printing and graphics, transport equip., machinery, timber and wood products, paper and metal products. Pop. (1989E) 53,534.

Oshogbo   Oyo, Nigeria. 7 47N 4 34E. Town sit. NE of Lagos. Its main industry is cotton weaving. Trades in palm oil, kernels and cacao. Pop. (1983E) 344,500.

Osijek   Croatia, Yugoslavia. 45 33N 18 41E. Town and R. port sit. NW of Belgrade on Drava R. Manu. textiles and furniture. Trades in agric. produce. Pop. (1981C) 158,790.

Oslo   Norway. 59 55N 10 45E. City at head of Oslo Fjord. Cap., commercial, industrial and cultural centre, founded in 1048. Ice-free sea port handling timber, paper, fruit, cars. Industries inc. metal-working, chemicals, electronics, computers, engineering, brewing and distilling. Pop. (1989E) 456,124.

Osmanabad   Maharashtra, India. Town sit. N of Sholapur. District cap. and trading centre for agric. produce. Noted for nearby anc. Jain and Vishnuite caves. Pop. (1981C) 39,068.

Osnabrück   Lower Saxony, Federal Republic of Germany. 52 16N 8 02E. Town sit. NE of Dortmund on Haase R. Manu. iron, steel, machinery, paper, turbines and textiles. Pop. (1984E) 154,700.

Osorno   Osorno, Chile. 40 34S 73 09W. Town sit. SSE of Valdivia. Centre of an agric. and livestock rearing district with associated industries. Pop. (1987E) 122,462.

Osorno, Mount   Chile. A volcanic peak on E shore of L. Llanquihue in S Chile, at 8,790 ft. (2,679 metres) a.s.l.

Oss   Noord-Brabant, Netherlands. 51 46N 5 31E. Town sit. E of 's Hertogenbosch. Manu. inc. electrical equipment, pharmaceutical products, carpet, food, transport, electronics and cars. Pop. (1988E) 50,987.

Ossett   West Yorkshire, England. 53 41N 1 35W. Town sit. S of Leeds. Industries inc. woollen mills; manu. textile machinery, leather, leather products, paint, pharmaceutical goods. Pop. (1980E) 20,000.

Ostend   West Flanders, Belgium. 51 13N 2 55E. Port and resort sit. on North Sea. It is an important port for ferry and jet foil services to Britain. Industries inc. shipbuilding, fishing and allied industries. Pop. (1989E) 68,366.

Östergötland   Sweden. 58 24N 15 34E. County of SE Sweden bounded E by the Baltic, W by L. Vätter. Area, 4,079 sq. m. (10,566 sq. km.). Chief cities Linköping, Norrköping, Motala, Mjolby. Traversed E to W by Göta canal across a fertile plain with several Ls. Agric. areas produce cereals, sugar beet, dairy and beef cattle. Minerals inc. iron, zinc, lead and copper. Metallurgical and general manu. industries are concentrated on Norrköping, Linköping and Motala. Pop. (1988E) 396,919.

Östersund   Jämtland, Sweden. 63 11N 14 39E. Cap. of county in central Sweden sit. on E shore of L. Storsjö. Manu. machinery and furniture. Pop. (1988E) 57,281.

Östfold   Norway. 59 25N 11 25E. County of SE Norway bounded E and S by Sweden, W by Oslo Fjord. Area, 1,614 sq. m. (4,180 sq. km.). Chief cities Moss (cap.), Frederikstad, Sarpsborg, Halden. Level agric. country, drained by Glomma R. with main cities to the S. Main occupations are lumbering, farming, quarrying and fishing. Industries inc. paper making, textiles, shipbuilding, fish canning, chemicals. Pop. (1989E) 237,997.

Ostrava   Moravia, Czechoslovakia. 49 50N 18 17E. City sit. E of Prague above confluence of Oder and Ostravice Rs. Railway junction and centre of the metallurgical industry based on Moravian, Slovak and Swedish ores and local anthracite and other coal. Produces iron, steel, parts for ships, vehicles, railway equipment, bridge sections and cranes, tinplate, dies, light alloys, printing plates. Pop. (1989E) 331,000.

Ostrów Wielkopolski Poznań, Poland. 51 39N 17 49E. Town sit. SE of Poznań. Manu. inc. chemicals and agric. machinery. Trades in agric. produce.

Oswestry Shropshire, England. 52 52N 3 04W. Town sit. NW of Shrewsbury. Industries inc. light engineering, clothing, hot air balloons and tourism. Pop. (1981C) 13,114.

Oświęcim Kraków, Poland. 50 03N 19 12E. (Auschwitz). Town sit. W of Kraków. Manu. metal goods and chemicals. Pop. (1975E) 40,000.

Otago South Island, New Zealand. 44 44S 169 10E. District of S comprising a fertile coastal plain (SE) and Otago Peninsula (E), Fiordland National Park (SW). Area 25,220 sq. m. (64,230 sq. km.). Chief city Dunedin. Produces wool, fruit, stone, quartz and gold.

Otaru Hokkaido, Japan. 43 13N 141 00E. Port sit. N of Hakodate on W coast. Industries inc. engineering, fishing, fish processing and the manu. of rubber products. Coal, timber and fertilizers are exported. Pop. (1990E) 165,000.

Otley West Yorkshire, England. 53 54N 1 41W. Town sit. NW of Leeds on R. Wharfe. Manu. printing machinery, woollen goods and leather. Pop. (1981C) 13,700.

Otranto Puglia, Italy. 40 09N 18 30E. Port and tourist centre sit. SE of Brindisi on the Otranto Strait.

Otranto Strait Italy. 40 00N 19 00E. Sit. between Cape Otranto, the 'heel' of Italy, and Albania, and connecting the Ionian Sea with the Adriatic Sea. 45 m. (72 km.) in width.

Ottawa Ontario, Canada. 45 27N 75 42W. City sit. WSW of Montreal on Ottawa R. Federal cap. Founded 1826 and chosen as cap. of Canada 1858. Govt. absorbs much of the labour force; other industries inc. advanced technology, clothing, food processing, metals, plastics, paper making, woodworking; manu. clocks and watches. Pop. (1986E) 300,763.

Ottawa River Canada. 45 20N 73 58W. Issues from Grand Lake Victoria in SW Quebec and flows 696 m. (1,114 km.) W through L. Simard and L. des Quinze to L. Timiskaming, then S and SE down the Ontario/Quebec border to Mattawa, Pembroke, Ottawa and Hawkesbury to enter the St. Lawrence W of Montreal. There are numerous rapids and Ls. along its lower and middle courses. Linked with L. Ontario by the Rideau Canal. Used for power. Main tribs. Rouge, North Nation, Lièvre, Gatineau, Coulonge, Mattawa, South Nation, Mississippi, Madawaska, Petawawa and Rideau Rs.

Ottumwa Iowa, U.S.A. 41 01N 92 25W. City sit. SW of Cedar Rapids on Des Moines R. Manu. agric. and mining machinery, tools, furniture and dairy products. Pop. (1980C) 27,381.

Ötztal Alps Austria/Italy. 46 45N 10 55E. A range sit. mainly in the Austrian Tirol. The highest peak is Wildspitze, 12,382 ft. (3,774 metres) a.s.l.

Ouachita River U.S.A. 31 38N 91 49W. Rises in Ouachita Mountains of W Arkansas; flows *c.* 605 m. (968 km.) SE and S past Arkadelphia into Louisiana, past Monroe to enter Red R. *c.* 35 m. (56 km.) above its mouth. Called Black R. between Red R. and Jonesville, Louisiana.

Ouagadougou Burkina Faso. 12 25N 1 30W. Cap., since 1954, and commercial centre of an agric. district trading in livestock, millet and groundnuts. Manu. inc. textiles, soap and vegetable oil. Pop. (1985C) 442,223.

Ouarzazate Marrakesh, Morocco. Town sit. SE of Marrakesh on the S slope of the High Atlas. Oasis and former fortress town of French North Africa. Tourist centre.

Oubangui River *see* **Ubangi River**

Oudenaarde East Flanders, Belgium. 50 51N 30 36E. Town sit. SSW of Ghent on Scheldt R. Manu. textiles. Pop. (1990E) 27,067.

Oudh Uttar Pradesh, India. Area nearly co-extensive with the anc. Kosala Kingdom, central Uttar Pradesh. Area, 24,070 sq. m. (62,341 sq. km.). Chief town, Lucknow.

Oudtshoorn Cape Province, Republic of South Africa. 33 35S 22 14E. Town sit. ENE of Cape Town on Little Karoo R. Centre of an ostrich-farming district trading in ostrich feathers, cereals, fruit, wine and tobacco. Manu. inc. footwear, furniture and tobacco products. The limestone Cango Caves are in the vicinity.

Oued Zem Casablanca, Morocco. 32 55N 6 33W. Town sit. SE of Casablanca; railway terminus and agric. trading centre. Industries inc. flour milling, with phosphate and iron mining in the district. Pop. (1982E) 59,000.

Oujda Oujda, Morocco. 34 40N 1 54W. Cap. of Region sit. ESE of Tangier near the Algerian frontier. An important commercial centre trading in sheep, wool, cereals, fruit and wine. Pop. (1982E) 260,082.

Oulu Oulu, Finland. 65 01N 25 28E. Cap of province and port sit. at mouth of Oulu R. on the Gulf of Bothnia. Industries inc. products of sawn timber, pulp and other wood products, chemicals, cables and electronic equipment, sawmilling, tanning and fertilizer manu. Timber and wood products are exported. Pop. (1988E) 98,993.

Oulu River Finland. 64 25N 27 00E. Emerges from NW end of L. Oulu and flows NW for 65 m. (104 km.) to enter the Gulf of Bothnia at Oulu.

Oundle Northamptonshire, England. 52 29N 0 29W. Town sit. SW of Peterborough on R. Nene. Pop. (1980E) 4,000.

Ouse River Sussex, England. 50 47N 0 03E. Rises SSW of Crawley and flows E and S for 30 m. (48 km.) to enter the English Channel at Newhaven.

Ouse River Yorkshire, England. 54 03N 0 07E. Formed by confluence of R. Swale and R. Ure near Boroughbridge. The main stream flows for 60 m. (96 km.) generally SE to join R. Trent, and these two Rs. form the Humber estuary.

Ouse River, Great England. 52 47N 0 22E. Rises near Brackley in Northamptonshire and flows for 160 m. (256 km.) through the S Fenland to reach The Wash.

Ouse River, Little England. 52 25N 0 50E. A trib. of Great Ouse R. in S Norfolk and flows for 24 m. (38 km.) forming part of the Norfolk/Suffolk boundary.

Outer Mongolia *see* **Mongolian People's Republic**

Ovalau Fiji, West Pacific. 17 40S 178 48E. Island sit. E of Viti Levu. Area, 39 sq. m. (101 sq. km.). Chief town Levuka. Volcanic, rising to 3,000 ft. (914 metres). Main products bananas, pineapples.

Ovamboland Namibia. 17 20S 16 30E. A district in the N bordering on Angola. Area, *c.* 16,220 sq. m. (*c.* 42,010 sq. km.). The principal occupations are cattle rearing and maize growing. The Ovambos are a Bantu race and still possess tribal organization to its full extent. Pop. (1988C) 641,000.

Overijssel Netherlands. 52 25N 6 35E. Province in the E bordering on Federal Republic of Germany. Area, 1,469 sq. m. (3,805 sq. km.). Mainly flat and agric. Dairy farming is important. The principal Rs. are Ijssel and the Vecht. There is a textile industry in the Twente district. Cap. Zwolle. Pop. (1989E) 1,014,949.

Overland Park Kansas, U.S.A. 38 59N 94 40W. City forming part of the Kansas City metropolitan area. Employment largely office, retail or service-oriented. There is some light industry. Known for extensive park system. Pop. (1980C) 84,643.

Oviedo   Spain. 43 22N 5 50W. Town in NW. Manu. inc. iron, steel, metal goods, armaments, chemicals, glass and cement. Pop. (1986C) 190,651.

Owensboro   Kentucky, U.S.A. 37 46N 87 07W. City and R. port sit. WSW of Louisville on Ohio R. Manu. building materials, electrical equipment, machinery, furniture and food products. It is an important centre of oil and trade. Pop. (1980C) 54,450.

Oxelösund   Södermanland, Sweden. 58 43N 17 15E. Town sit. SE of Nyköping on the Baltic. Sea port, ice-free, shipping iron ore. Industries inc. iron and coke ovens. Pop. (1989E) 13,159.

Oxford   Oxfordshire, England. 51 46N 1 15W. City sit. on Thames R., where it is called the Isis R., at its confluence with the Cherwell R. Seat of Oxford University, estab. in the 12th cent. Famous buildings apart from colleges inc. the Radcliffe Camera, Bodleian Library, Sheldonian Theatre, Radcliffe Observatory. Cowley suburb (SE) is industrial; manu. vehicles, pressed steel products, paper, electrical goods. Pop. (1983E) 116,000, which inc. 18,000 full-time students.

Oxfordshire   England. 51 45N 1 15W. A South-Midland county bounded on the S by R. Thames and Berkshire, E by Buckinghamshire, NE by Northamptonshire, NW by Warwickshire and W by Gloucestershire and Wiltshire. The NW has the Cotswold Hills and the SW the Chiltern

Hills. The R. Thames flows through much of the county. Industries include the manu. of motor vehicles, blankets, agric. implements, paper, and modern technological industries. Agric. is very important. Under 1974 reorganization the county consists of the districts of West Oxfordshire, Cherwell, Oxford City, Vale of White Horse and South Oxfordshire. Pop. (1988E) 578,900.

Oxnard   California, U.S.A. 34 12N 119 11W. City sit. WNW of Los Angeles. Centre of an agric. area with associated industries. Pop. (1980C) 108,195.

Oyapoc   French Guiana. Port at mouth of R. of same name.

Oyapoc River   French Guiana/Brazil. 4 08N 51 40W. Rises in the Tumuc-Humoc mountains and flows for *c.* 300 m. (480 km.) to the Atlantic Ocean, forming the French Guiana/Brazil border throughout its course.

Oyo   Nigeria. 7 46N 3 56E. State comprising the Ibadan and Oyo provinces of the former W region. Area, 14,558 sq. m. (37,705 sq. km.). The pop. is almost entirely Yoruba. Chief cities Ibadan (cap.), Ogbomosho, Oshogbo, Ilesha, Iwo, Ede, Ife, Oyo, Ila, Iseyin, Ilobu and Shaki. Pop. (1988E) 11,412,300.

Oyo   Oyo, Nigeria. 7 50N 3 56E. Town sit. NNE of Lagos. Trades in palm oil, kernels and cacao. Pop. (1983E) 185,300.

Ozark Mountains   U.S.A. 37 00N 93 00W. Area of highlands in states of Arkansas, Illinois, Kansas, Missouri and Oklahoma, with several summits over 2,000 ft. (610 metres) a.s.l. Summer resort.

P

Paama Vanuatu, South West Pacific. Island sit. s of Ambrym. Area, 6 m. (10 km.) by 2 m. (3 km.). Volcanic.

Paarl Cape Province, Republic of South Africa. 33 45S 18 56E. Town in sw sit. on Great Berg R. in a vine and fruit-growing area. The main industry is wine-making, also fruit canning and jam-making; manu. tobacco products and textiles. Pop. (1980E) 71,370.

Pabna Rajshahi, Bangladesh. 24 00N 89 15E. Cap. of district sit. wnw of Dacca on n bank of Ganges R. Industries inc. rice milling and engineering; manu. hosiery. Pop. (1984E) 190,000.

Pacaraima, Sierra Brazil/Venezuela. 5 30N 60 40W. Mountain range forming the watershed between the basins of Rs. Orinoco and Amazon, extending c. 385 m. (616 km.) along the frontier between Brazil and Venezuela. Rises to Mount Roraima, 9,219 ft. (2,810 metres) a.s.l.

Pachuca de Soto Hidalgo, Mexico. 20 07N 98 44W. City in e central Mex. sit. at 8,000 ft. (2,438 metres) a.s.l. on the central plateau in a silver-mining area. The main industry is mining and refining silver; also manu. woollen goods and leather. Pop. (1980C) 135,248.

Pacific Ocean 10 00N 140 00W. The world's largest expanse of water, bounded e by Americas and w by Asia and Australia. Connected in n with the Arctic Ocean by the Bering Strait, and merging in the s with the Southern Ocean, between Polynesia and Antarctica. Area, with peripheral seas, c. 70m. sq. m. (c. 181m. sq. km.), being c. 11,000 m. (17,600 km.) at its widest w to e and c. 10,000 m. (16,000 km.) at its longest n to s. Average depth c. 14,000 ft. (4,267 metres) with deep trenches around its edges. Contains many coral and volcanic islands. Linked with Atlantic in e by the Panama Canal.

Padang Sumatra, Indonesia. 0 57S 100 21E. Town sit. on central w coast. Port exporting coffee, Copra and rubber. Pop. (1980C) 196,000.

Paddington Greater London, England. 51 31N 0 10W. District of w London containing an important railway terminus. Mainly residential.

Paderborn North Rhine-Westphalia, Federal Republic of Germany. 51 43N 8 45E. Town in n centre. Industries inc. agric. engineering and textiles; manu. cement. Pop. (1990E) 123,000.

Padiham Lancashire, England. 53 47N 2 18W. Town sit. wnw of Burnley. Industries inc. engineering and textiles. Pop. (1986E) 9,222.

Padstow Cornwall, England. 50 33N 4 56W. Resort in n centre of county sit. on Camel R. estuary. Pop. (1980E) 2,500.

Padua Veneto, Italy. 45 25N 11 53E. Cap. of Padova province in NE. Industries inc. manu. machinery, chemicals, plastics, electrical equipment and furniture. Town of special historical interest. Pop. (1981E) 234,678.

Paducah Kentucky, U.S.A. 37 05N 88 36W. City in sw of state sit. on Ohio R. at its confluence with Tennessee R. Trading centre for tobacco. Industries inc. manu. hosiery and textile machinery. Pop. (1990E) 31,019.

Pagalu *see* **Annobón**

Pago Pago American Samoa, South Pacific. 14 16S 170 42W. Town in se of Tutuila Island sit. on island's only good harbour for ocean-going ships. Naval base, radio station and airport. Pop. (1980E) 3,000.

Pahandut   *see* **Palangka Raja**

Pahang   Peninsular   Malaysia,
Malaysia. State on E coast consisting
mainly of the Pahang R. basin. Area,
13,873 sq. m. (35,931 sq. km.). Cap.
Kuala Lipis. The chief products are
rubber and tin. Pop. (1980C) 770,644.

Paignton   Devon, England. 50 26N
3 34W. Town sit. on Tor Bay SSW of
Torquay. Seaside resort.

Pailen   Battambang,   Cambodia.
Town sit. SW of Battambang on the
Thai frontier. Brahman sanctuary.
Deposits of topaz and sapphire in the
neighbourhood.

Paisley   Strathclyde Region, Scotland. 55 50N 4 26W. Town in SW sit.
on White Cart Water near its confluence with Clyde R. Industries inc. textiles, engineering, bleaching and dyeing; manu. cotton thread, starch and
cornflour. Pop. (1981C) 84,789.

Paita   Piura, Peru. 5 05S 81 00W.
Pacific port in NW forming the outlet
for a cotton-growing area, exporting
cotton, hides and wool.

Pajakumbuh   Sumatra, Indonesia.
Town sit. NNE of Padang in the
Padang Highlands. Railway terminus
and airfield; centre of an agric. area
trading in tea, coffee, tobacco, rubber.

Pakistan   30 00N 70 00E. Rep.
bounded W by Iran, NW by Afghánistán, N by U.S.S.R. and China, E by
India and S by the Arabian Sea. Area,
307,293 sq. m. (796,095 sq. km.).
Chief cities are Islamabad (the cap.),
Karachi, Hyderabad, Lahore and
Rawalpindi. Crossed NNE to SSW by
Indus R.; mountainous in the N with
the Hindu Kush rising to Tirich Mir,
25,260 ft. (7,699 metres) a.s.l., and in
the W. Fertile plains in the centre
watered by the Indus R. and its tribs.;
desert in the E. Pakistan was constituted as a Dominion in 1947, under the
provisions of the Indian Independence
Act, 1947. The Dominion consisted of
the following former territories of
British India: Balúchistán, East Bengal (including almost the whole of
Sylhet, a former district of Assam),
North-West Frontier, West Punjab

and Sind; and those States which had
acceded to Pakistan. In 1956 an Islamic republic was proclaimed. In 1970
the Awami League based on E Pakistan gained a majority in the Provincial
Assembly. Martial law was introduced
and in 1971 civil war broke out. The
war ended and the E province declared
itself an independent state, Bangladesh. In 1972 Pakistan withdrew from
the Commonwealth. The entire area in
the north and west is covered by great
mountain ranges. The rest of the province consists of a fertile plain watered
by 5 big rivers and their tributaries.
Agriculture is the occupation of a vast
majority of the pop., and is dependent
almost entirely on the irrigation system based on these rivers. The main
crops are wheat, cotton, barley, sugar-
cane, millet, rice, maize and fodder
crops, while the Quetta and Kalat
divisions are known for their fruits
and dates. Minerals inc., coal, chromite, gypsum, iron-ore. Oil has been
found at Kot Sarong and natural gas at
Sui. Industry employs 25% of the
pop. Pop. (1989E) 105·4m.

Paknam   Samutprakan, Thailand.
Town sit. SSE of Bangkok at mouth of
Chaophraya R. on the Gulf of Thailand. Port and provincial cap. Industries inc. fisheries. The official name
is Samutprakan.

Pakse   Champassac, Laos. 15 07N
105 47E. Town sit. SE of Savannakhet
on Mekong R. at mouth of Se Done R.
Provincial cap, communications and
trading centre for the Boloven Plateau, handling cardamom, cotton,
hemp, tobacco, cattle. Pop. (1984E)
44,860.

Pakwach   Northern   Province,
Uganda. Town sit. SE of Arua on
White Nile R. R. port trading in cof-
fee, cotton, tobacco.

Palangka Raja   Kalimantan, Indonesia. Town sit. NNW of Bandjarmasin. Cap. of Central Kalimantan
province.

Palau, Republic of   West Pacific.
7 30N 134 30E. Rep. within U.N.
Trust Territory inc. 9 volanic and
coral islands in the W Caroline Islands
sit. E of the Philippines. The main pro-

ducts are copra, tapioca and dried fish. Area, 178 sq. m. (461 sq. km.). Cap. Koror. Pop. (1980C) 12,177.

Palawan Philippines. 9 30N 118 30E. Island between the South China Sea to W and the Sulu Sea to E, the largest island of the Philippines group. Area, 5,747 sq. m. (14,885 sq. km.). Cap. Puerto Princesa. Mountainous, rising to Mount Mantalingajan, 6,839 ft. (2,084 metres) a.s.l., and forested. The main products are rice, rubber and coconuts.

Palembang South Sumatra, Indonesia. 3 00S 104 45E. Cap. of province in SE sit. on Musi R. Commercial centre and port handling petroleum products from nearby refineries, and rubber. Pop. (1975E) 475,000.

Palencia Spain. 42 01N 4 32W. (i) Province in N central Spain. Area, 3,100 sq. m. (8,029 sq. km.). Mountainous in N with the Cantabrian Mountains crossing W to E, but otherwise arid plateau. The valleys of Rs. Pisuerga and Carrión are fertile and cultivated. The main products are cereals, fruit and vegetables. Pop. (1986C) 188,472. (ii) Cap. of province of same name sit. on Carrión R. Industries inc. railway engineering, tanning, flour milling and textiles. Pop. (1986C) 76,707.

Palermo Sicilia, Italy. 38 08N 12 23E. Cap. of island and region sit. on NW coast. Industries inc. shipbuilding, steel and chemicals; manu. glass and furniture. Port exporting wine, olive oil and fruit. Tourist resort. Pop. (1981E) 702,000.

Palimé *see* **Kpalimé**

Palisadoes, The Port Royal, Jamaica. Peninsula bounding Kingston Harbour E to W. Length 7.5 m. (12 km.). Port Royal is sit. at the W end, Palisadoes airport, centre. Coextensive with Port Royal Parish.

Palk Strait South Asia. 9 30N 79 30E. Channel extending *c.* 60 m. (96 km.) W to E between SE India and N Sri Lanka; 33 m. (53 km.) at its widest point.

Palma Majorca, Spain. 39 34N 2 39E. Cap. of the Balearic Islands sit. at the head of the Bay of Palma in SW Majorca. Industries inc. tourism and textiles; manu. cement, paper, footwear and glass. Seaport handling almonds, fruit and wine. Pop. (1989E) 315,693.

Palmas, Las *see* **Las Palmas**

Palm Beach Florida, U.S.A. 26 42N 80 02W. Winter resort sit. N of Miami on E shore of L. Worth, a sea lagoon. Pop. (1980C) 9,729.

Palmer Alaska, U.S.A. 61 36N 149 07W. Town in the Matanuska Valley, sit. NE of Anchorage. Trade centre. Pop. (1989E) 3,116.

Palmerston Cook Islands, South Pacific. 18 04S 163 10W. Atoll in N, sit. W of Rarotonga. Area, 0.8 sq. m. (2 sq. km.). Produces copra. Pop. (1981C) 51.

Palmerston North North Island, New Zealand. 40 21S 175 37E. City sit. NNE of Wellington on Manawatu R. Railway junction in a farming area. Manu. plastics, trucks and coach bodies. Pop. (1986C) 66,821.

Palmira Valle del Cauca, Colombia. 3 32N 76 16W. Town in WSW sit. at 3,500 ft. (1,067 metres) a.s.l. Commercial centre trading in tobacco, coffee, sugar-cane, cereals and rice. Pop. (1983E) 160,000.

Palm Springs California, U.S.A. 33 50N 116 33W. City sit. E of Los Angeles. Resort and inland spa with sulphur springs. Pop. (1980C) 36,500.

Palo Alto California, U.S.A. 37 27N 122 09W. Residential city sit. SE of San Francisco. Extends from the S end of San Francisco Bay into Santa Cruz foothills. Stanford University is here. Part of 'Silicon Valley' high technology area. Pop. (1980C) 55,225.

Palomar, Mount California, U.S.A. Mountain sit. NNE of San Diego; 6,126 ft. (1,867 metres) a.s.l. Site of the Mount Palomar Observatory.

Pamir Central Asia. Plateau extending through the Gorno-Badakhshan Autonomous Region, U.S.S.R. into NE Afghánistán and Sinkiang-Uigur, China. A series of high mountain valleys at 12,000–14,000 ft. (3,657–4,267 metres) a.s.l. with ranges inc. the Akademia Nauk, rising to Kommunizma Peak at 24,590 ft. (7,495 metres) a.s.l. The high valleys have grass on which sheep and goats are reared.

Pamplona   Norte de Santander, Colombia. 7 23N 72 39W. Town S of Cúcuta sit. in the E. Cordillera. Commercial centre trading in coffee, cacao and cereals. Industries inc. textiles, brewing and distilling. Pop. (1983E) 23,200.

Pamplona   Spain. 42 49N 1 38W. Town sit. in central N on Arga R. Industries inc. textiles, flour milling, tanning and soap-making. Pop. (1989E) 183,423.

Panaji   Goa, India. 15 29N 73 50E. Cap. of state sit. on the W coast SE of Bombay. Port and commercial centre trading in rice, fish and salt. Industries inc. fishing. Pop. (1981E) 77,226.

Panama   8 48N 79 55W. Rep. bounded N by the Caribbean, E by Colombia, S by the Pacific and W by Costa Rica. A revolution, inspired by U.S.A., led to the separation of Panama from the United States of Colombia and the declaration of its independence in 1903. Area, 28,575 sq. m. (74,009 sq. km.). Cap. Panama City. Mountainous interior with narrow coastal lowlands. The active volcano Chiriqui rises to 11,410 ft. (3,477 metres) a.s.l. The climate is tropical on the coast, with *c.* 70 in. (178 cm.) rain annually on the Pacific coast and 160 in. (406 cm.) on the Caribbean; temperate in the mountains. The main products are timber, bananas, rice, maize, coffee and shrimps. Local industries inc. cigarettes, clothing, food processing, shoes, soap, cement factories; foreign firms are being encouraged to establish industries, and a petrol refinery is operating in Colón. Pop. (1990E) 2,417,955.

Panama, Gulf of   Panama. 8 00N

79 10W. Inlet of the Pacific Ocean in SE Panama, 120 m. (192 km.) wide.

Panama, Isthmus of Panama. 9 20N 79 30W. Narrow neck of land between the Pacific Ocean and the Caribbean Sea crossed by the Panama Canal.

Panama Canal   Central America. 9 00N 79 37W. Ship canal, linking the Pacific Ocean and the Caribbean Sea and extending 40 m. (64 km.) SE to NW across the Isthmus of Panama. The passage takes 7–8 hours, in either direction, through double locks. From the Caribbean the course is raised 85 ft. (26 metres) a.s.l. by the Gatun Locks, crosses L. Gatun and the Gaillard Cut, and descends through the Pedro Miguel Locks to the last sea-level canal.

Panama City   Panama. 9 00N 79 25W. Cap. of Panama founded in 1519, sit. at the Pacific Ocean end of the Panama Canal. Industries inc. food processing, plastics, clothing and oil. Pop. (1980C) 386,393.

Panama City   Florida, U.S.A. 30 10N 85 41W. City SW of Tallahassee sit. on Gulf of Mex. Industries inc. paper-making, metalworking, boat building and support for a military base. Port and resort. Pop. (1980C) 33,346.

Panay   Philippines. 11 10N 122 30E. Island between Mindoro to the NW and Negros to the SE. Area, 4,744 sq. m. (12,287 sq. km.). Chief town Iloilo. Mountainous in the W, rising to 6,724 ft. (2,049 metres) a.s.l. Fertile lowland in the E. The main products are copra and rice.

Pančevo   Vojvodina, Yugoslavia. 44 52N 20 40E. Town sit. ENE of Belgrade. Industries inc. flour milling. Pop. (1981C) 123,791.

Panch Mahals   Gujarat, India. District of SE Gujarat. Area, 1,608 sq. m. (4,165 sq. km.). Chief towns Godhra, Dohad. Agric. and forest area, producing rice, maize, grain, peanuts, cotton, teak. Chief minerals are manganese and bauxite.

Pando   Bolivia. 34 30S 56 00W.

Dept. of NW Bolivia bounded E by Brazil across Madeira R., w by Peru. Area, 23,876 sq. m. (61,839 sq. km.). Chief towns, Cobija (cap.), Puerto Rico. Tropical forest region with many Rs., important source of rubber, also producing timber, furs, bananas, rice. Transport is mainly by R. Pop. (1981c) 42,594.

Pandu   Assam, India. Town sit. w of Gauhati and opposite Amingaon; rail ferry terminal on the Brahmaputra R.

Panevėžys   Lithuanian Soviet Socialist Republic, U.S.S.R. 55 44N 24 21E. Town sit. NNE of Kaunas on Nevezys R. Railway centre with repair shops; industries inc. flour milling, textiles, meat packing, sugar refining, tobacco processing; manu. metal products, paints, turpentine, cement, soap.

Pantelleria   Italy. 36 47N 12 00E. Island in the Mediterranean sit. off SW coast of Sicily. Area, 32 sq. m. (83 sq. km.). Chief town Pantelleria. Volcanic with hot springs, and fertile, but with inadequate fresh water. The main occupations are farming and fishing The main products are cereals and wine. Pop. (1980E) 10,000.

Paotow   (*now* Baotau) Inner Mongolia, China. 40 40N 109 59E. Town sit. WNW of Beijing. Route and commercial centre, trading in wool, hides and cereals. Industries inc. steel; manu. rugs and soap. Pop. (1982E) 1,042,000.

Papar   Sabah, Malaysia. Town sit. SSW of Kota Kinabula. Railway centre for an agric. area, handling fruit, rice and sago. The main industry is rice milling.

Papeete   Society Islands, South Pacific. 17 32S 149 34W. Town on NW coast of Tahiti. Cap. of French Polynesia. Port exporting copra, phosphates, vanilla, mother of pearl. Industries inc. sugar refining, manu. soap. Pop. (1983c) 23,496.

Paphos   Paphos, Cyprus. 34 45N 32 23E. Town sit. WSW of Nicosia on SW coast. District cap. Trades in nuts, wine, olives, bananas, citrus and vege-

tables. Tourism is important. Pop. (1985E) 17,000.

Papua New Guinea   SW Pacific Ocean. 8 00S 145 00E. An independent state sit. off N coast of Queensland, the S coast of Papua being 60 m. (96 km.) N of Cape York Peninsula. It comprises the E half of the large island of New Guinea, of which the w half is part of Indonesia, and neighbouring smaller islands. Total area, 178,703 sq. m. (462,840 sq. km.). Chief town Port Moresby. The small islands consist of an arch. of 4 main islands and *c.* 100 islets extending E from the E coast of the mainland. The main islands are New Britain, New Ireland, Admiralty Islands and 5 of the Solomon Islands. Pop. (1989E) 3·59m.

Pará   Brazil. 3 20S 52 00W. State in N central Braz. sit. in Amazon R. basin. Area, 481,744 sq. m. (1,247,717 sq. km.). Cap. Belém. Mainly rain forest, with little communication other than by water. The main products are rubber, hardwoods, nuts, jute and skins. Pop. (1989E) 4,862,775.

Paracel Islands   Vietnam. Group in the South China Sea, 150 m. (240 km.) SE of Hainan Island (China). Inc. numerous coral islands and reefs, mainly the Amphitrite and Crescent groups, Lincoln and Triton Islands. The main product is phosphates.

Paraguari   Paraguay. 25 38S 57 09W. Town sit. SE of Asunción. Commercial centre of an agric. and lumbering area. Industries inc. rice and flour milling, tanning, oil extracting, manu. pottery. Pop. (1980E) 6,000.

Paraguay   27 18S 58 38W. Rep. bounded N and NW by Bolivia, E by Brazil and SE, S and W by Argentina. Paraguay gained its independence from Spain in 1811. Area, 157,047 sq. m. (406,752 sq. km.). Cap. Asunción. Divided N to S by Paraguay R. Scrub forest to the w, with plains rising to hills and forested plateau in the E. The climate is sub-tropical. Agric. products inc. cotton, potatoes, maize, tobacco, wheat and soybeans. *Yerba maté*, or strongly flavoured Paraguayan tea, continues to be produced but is declining in importance. Cattle rearing is extremely important. There are huge reserves of hardwoods and cedars that have scarcely been exploited. Palms, tung and other trees are exploited for their oils. The Japanese are experimenting with mulberries for silk growing. Pines and firs have been introduced under a U.N. project. In the Chaco the accessible Quebracho forests have nearly been worked out but plans are being made to open up new areas. Iron, manganese and other minerals have been reported but have not been shown to be commercially exploitable. Salt, limestone, kaolin and apatite are worked and prospecting for natural gas is taking place. There are meat-packing plants and other factories producing vegetable oils. A textile industry in Pilar and Asunción meets a large part of local needs. There is a cement works at Vallemi. The oil refinery at Villa Elisa has been in operation since 1966. There are some flour-mills and small match, pharmaceutical, soap, cigarette, footwear, furniture and building materials industries. Pop. (1988E) 4,039,161.

Paraguay River   South America. 27 18S 58 38W. Rises near Diamantina in Mato Grosso, Brazil, and flows 1,300 m. (2,080 km.) generally S, forming the Braz./Paraguay frontier between Bahía Negra and Porto Sastre, then through Paraguay to form the Paraguay/Argentina frontier below Asunción, joining with the Alto Parana R. as it enters Argentina and forming the Paraná R. Important for communication, esp. in Paraguay; navigable by large vessels to Concepción, Paraguay, and by small boats up to Cáceres, Brazil.

Paraíba   Brazil. 7 00S 36 00W. State in NE Braz. bounded E by the Atlantic Ocean. Area, 21,760 sq. m. (56,358 sq. km.). Cap. João Pessoa. Coastal plain rising to a hilly, wooded interior. The main product is cotton; sugar-cane, pineapples and tobacco are also produced. Pop. (1989E) 3,200,430.

Paraíba do Norte River   Brazil. 3 00S 41 50W. Rises near the Paraíba/Pernambuco state border in the NE,

and flows 200 m. (320 km.) ENE to the Atlantic below João Pessoa.

Paraíba do Sul River   Brazil. 21 37S 41 03W. Rises in the Serra do Mar near São Paulo in the SE, and flows 600 m. (960 km.) NE to the Atlantic below Campos, 140 m. (224 km.) NE of Rio de Janeiro.

Parakou   Borgou, Benin. 9 21N 2 37E. Cap. of dept. sit. N of Porto-Novo. Rail terminus serving an agric. area, handling cotton, nuts, beans, kapok, butter, rice and soya. Industries inc. cotton, and kapok ginning. Pop. (1982E) 65,945.

Paramaribo   Suriname. 5 50N 55 10W. Cap. of Suriname sit. on Suriname R. near its mouth. It was first settled in 1630 by the British. The port exports coffee, bauxite, timber and citrus fruits. Pop. (1988E) 192,109.

Paraná   Entre Rios, Argentina. 31 45S 60 30W. R. port in central E sit. on E bank of Paraná R. Commercial centre with important trade in grain and cattle. Pop. (1983E) 200,000.

Paraná   Brazil. 24 30S 51 00W. State in S Braz. bounded W by Paraguay across the Paraná R. and E by the Atlantic. Area, 77,027 sq. m. (199,500 sq. km.). Narrow lowland strip on the coast, rising to wooded plateau. Fertile in the N. The main products are coffee, cotton, citrus fruits, maté and timber. Pop. (1989E) 8,935,142.

Paranaguá   Paraná, Brazil. 25 31S 48 30W. Town sit. E of Curitiba on Paranaguá Bay, Atlantic Ocean, at the foot of the Serra do Mar. Railway terminus and port shipping maté, pine, coffee, bananas, sugar. Industries inc. coffee and maté processing; manu. matches, pencils, waxes, soap. Pop. (1980E) 68,366.

Paranam   Suriname. Village sit. SSE of Paramaribo on Suriname R. Bauxite mining and shipping centre, accessible to ocean-going vessels.

Paraná River   Brazil/Paraguay/Argentina. 33 43S 59 15W. Formed as Alto Paraná R. from Rio Grande and Paranaíba R. on the border between the states of Mato Grosso and São Paulo, Brazil. Flows 1,800 m. (2,880 km.) S to form the Braz./Paraguay frontier, then W to form the Paraguay/ Argentina frontier at the SW end of which it joins Paraguay R. Together they form Paraná R. which flows SSW then SE through Argentina past Santa Fé, Paraná and Rosario to enter the Río de la Plata above Buenos Aires.

Pará River Brazil. 1 30S 48 55W. An arm of the Amazon R. delta S and E of Marajó Island, extending *c*. 200 m. (320 km.), and *c*. 40 m. (64 km.) wide at its mouth.

Parbhani Maharashtra, India. 19 16N 76 47E. Town sit. NW of Hydera-bad (Andhra Pradesh). Railway centre for an agric. area. District cap. In-dustries inc. cotton ginning. Pop. (1981C) 109,364.

Pardubice Bohemia, Czechoslova-kia. 50 02N 15 47E. Town in N sit. on Elbe R. Industries inc. oil refining and brewing. Pop. (1989E) 96,000.

Parentis-en-Born Landes, France. Village sit. SSW of Bordeaux. In-dustries inc. lumbering, extracting timber resin, lignite mining. Pop. (1982C) 4,254.

Paricutín, Mount Michoacán, Mexico. 19 28N 102 15W. Volcano in SW Mex., formed 1943–4 in a culti-vated field at Paricutín village, sit. at 7,380 ft. (2,249 metres) a.s.l., and burying the village under lava. The peak rose 820 ft. (250 metres) before activity ceased in 1952.

Parika Demerara, Guyana. Village sit. W of Georgetown on Essequibo R. Railway terminus and R. landing-stage, shipping timber.

Paris Paris, France. 48 50N 2 20E. Cap. of rep. and dept. in Île-de-France region, on Seine R. below its conflu-ence with Marne R. Area, 41 sq. m. (105 sq. km.). A settlement existed in the 5th cent. B.C., but Paris today was planned by Baron Haussmann in the mid-19th cent. It is a cultural, com-munications and commercial centre and an inland port. Industries inc. manu. aircraft, cars, luxury goods,

metal goods and chemicals. Tourism is extremely important. Pop. (1990E) 2m. (agglomeration, 10·4m).

Paris Basin France. Depression in N France containing the Île de France and Paris, drained by Rs. Seine, Somme and Loire. Fertile area. The main products are wheat and dairy produce.

Parkersburg West Virginia, U.S.A. 39 17N 81 32W. City sit. SE of Col-umbus, Ohio, on Little Kanawha R. near its confluence with Ohio R. on the Ohio border. Industries inc. manu. chemicals, metals and metal products and glass. Pop. (1980C) 39,967.

Park Ridge Illinois, U.S.A. City and suburb of NW Chicago near the in-tern. airport at O'Hare Field. Residen-tial community. Pop. (1980C) 38,704.

Parma Emilia-Romagna, Italy. 44 48N 10 20E. Town in N central Italy sit. on Parma R. Market town trading in farm produce inc. Parmesan cheese. Industries inc. manu. food products, machinery and glass. Pop. (1989E) 172,313.

Parma Ohio, U.S.A. 41 23N 81 43W. City immediately S of Cleve-land. Industries inc. manu. motor car parts and machine tools. Pop. (1980C) 92,548.

Parnaíba Piauí, Brazil. 2 54S 41 47W. Town sit. near mouth of Parnaíba R. on NE coast. Commercial centre and port handling cotton, sugar, cattle and carnauba wax. Pop. (1980E) 79,000.

Parnaíba River Brazil. 3 00S 41 50W. Rises in the Serra das Manga-beiras in NE Braz. and flows 750 m. (1,200 km.) NNE along the boundary between the states of Piauí and Maranhao to enter the Atlantic below Parnaíba.

Parnassus, Mount Boeotia, Greece. 38 32N 22 35E. Mountain sit. NW of Korinthos, 8,061 ft. (2,457 metres) a.s.l. Sacred to the anc. Greeks.

Pärnu Estonia, U.S.S.R. 58 24N 24 32E. Port S of Tallinn sit. at mouth of

Pärnu R. on the Gulf of Riga. Industries inc. textiles and sawmilling; manu. leather goods. Exports timber and flax. Tourist resort.

Parsipanny   New Jersey, U.S.A. 40 52N 74 26W. Reservoir L. sit. NNE of Morristown, *c.* 2 m. (3 km.) long. Water supply for Jersey City formed by the Boonton Dam on the Rockaway R.

Pasadena   California, U.S.A. 39 09N 118 09W. City sit. N of Los Angeles. Mainly residential with light industry. Pop. (1980C) 118,550.

Pasadena   Texas, U.S.A. 29 43N 95 13W. City immediately SE of Houston. Manu. and trading centre. Pop. (1980C) 112,560.

Pascagoula   Mississippi, U.S.A. 30 21N 88 33W. City and port at mouth of Pascagoula R. on Gulf of Mexico. Industries inc. fishing, paper, chemicals, shipbuilding. Pop. (1989E) 30,265.

Pasco   Peru. Dept. of central Peru in the Andes. Area, 11,654 sq. m. (30,184 sq. km.). Chief towns Cerro de Pasco, Chacayán, Huachón, Oxapampa. Mountainous; Peru's principal mining area producing copper, silver, gold, lead and zinc (Cerro de Pasco, Huarón); coal (Goyllarisquizga, Quishuarcancha); vanadium (Mina Ragra). Pop. (1982C) 213,125.

Pascua, Isla de   *see* **Easter Island**

Pas-de-Calais   France. 50 30N 2 30E. Dept. in Nord-Pas-de-Calais region, bounded N by the Strait of Dover and NE by Belgium. Area, 2,563 sq. m. (6,639 sq. km.). Main towns Arras (the cap.), Calais, Boulogne, Béthune, St. Omer and Lens. Mainly industrial with coalfields supporting mining, metallurgical and textile industries. Pop. (1982C) 1,412,413.

Pasir Mas   Kelantan, Peninsular Malaysia. 6 02N 102 08E. Town sit. SW of Kota Bharu on Kelantan R. Railway junction on the E coast line, handling rice.

Passaic   New Jersey, U.S.A. 40 52N 74 08W. City N of Newark sit. on Passaic R. Industries inc. textiles; manu. radio and television equipment, rubber and leather products. Pop. (1980C) 52,463.

Passau   Bavaria, Federal Republic of Germany. 48 35N 13 28E. Town in extreme SE sit. at confluence of Rs. Danube, Inn and Ilz. Industries inc. tourism, agric. engineering, papermaking, brewing and tanning; manu. textiles and lenses. Pop. (1984E) 52,100.

Pastaza River   Ecuador/Peru. Formed by junction of Patate and Chambo Rs. W of Baños in the Ecuadorian Andes. Flows 400 m. (640 km.) SE and S through tropical forest into Peru, to enter Marañón R. below Bacranca.

Pasto   Nariño, Colombia. 1 13N 77 17W. Cap. of dept. in SW. Commercial centre for a farming area. Industries inc. food processing and hat making. Pop. (1980E) 219,000.

Patagonia   Argentina. 45 00S 69 00W. The most southerly part of the S. American continent extending *c.* 1,000 m. (1,600 km.) S from Rs. Limay and Negro to the Strait of Magellan. Area, *c.* 300,000 sq. m. (*c.* 777,000 sq. km.). Arid plateau rising to the base of the Andes. The main occupation is sheep farming, with cattle farming in the W where the climate is not so dry. Contains the major oilfield of Comodoro Rivadavia and coalfields at Río Turbio in the S.

Pátan   (Lalitpur) Nepál. 27 40N 85 20E. Town sit. S of Káthmándu. Trading centre for cereals and vegetables, noted for numerous temples esp. the Macheudranath and Mahabuddha temples. Anc. city which was cap. of the first king of Nepál. Pop. (1971E) 48,577.

Paternó   Sicilia, Italy. Resort immediately S of Mount Etna sit. in a fruit and vine growing area.

Paterson   New Jersey, U.S.A. 40 55N 74 10W. City sit. N of Newark on Passaic R. Industries inc. textiles; manu. textile machinery, plastics and clothing. Pop. (1980C) 137,970.

Pathankot Punjab, India. 32 18N 75 45E. Town sit. NE of Gurdaspur. Trading centre for agric. produce. Industries inc. fruit processing, handloom weaving. Pop. (1981c) 110,039.

Patiala Punjab, India. Town sit. NNW of Delhi. Industries inc. textiles; manu. footwear, railway carriages and metal goods. Pop. (1981c) 206,254.

Patna Bihar, India. 25 36N 85 06E. State cap. in N central India sit. on Ganges R. Industries inc. manu. brassware, carpets and furniture. Pop. (1981c) 813,963.

Patras Akhaïa, Greece. 38 15N 2 44E. Cap. of prefecture sit. on Gulf of Patras in NW Peloponnessos. Industries inc. textiles and flour milling. Sea port exporting currants, olive oil, wine and tobacco. Pop. (1981c) 141,529.

Patras, Gulf of Greece. Inlet of the Ionian Sea on W coast, continuing through a narrow strait as the Gulf of Korinthos.

Pau Pyrénées-Atlantiques, France. 43 18N 0 22W. Cap. of dept. in SW sit. on Gave de Pau R. Industries inc. tourism, engineering, tanning, brewing, flour milling and textiles; manu. footwear. Pop. (1982c) 85,766 (agglomeration, 131,265).

Paulo Alfonso Falls Alagoas/ Bahia, Brazil. Falls on lower São Francisco R. 195 m. (312 km.) from its mouth on the Atlantic coast of Alagoas. The R. descends 270 ft. (83 metres) by rapids and 3 cascades. Used for hydroelectric power.

Pavia Lombaria, Italy. 45 10N 9 10E. Town in N sit. on Ticino R. Market for the Po R. valley agric. area. Industries inc. textiles and agric. engineering; manu. furniture and sewing machines. Pop. (1981E) 85,000.

Pavlodar Kazakhstan, U.S.S.R. 52 18N 76 57E. Cap. of Pavlodar region sit. ENE of Tselinograd on Irtysh R. Industries inc. meat packing, milk canning and flour milling. Pop. (1987E) 331,000.

Pawtucket Rhode Island, U.S.A. 41 53N 71 23W. City N of Providence sit. on Blackstone R. Industries inc. the manu. of toys, wire and cables. Pop. (1980c) 71,204.

Paysandú Paysandú, Uruguay. 32 19S 58 05W. Cap. of dept. in central W sit. on Uruguay R. Industries inc. meat packing and textiles; manu. leather and footwear. R. port. Pop. (1980E) 80,000.

Pays-de-la-Loire France. Region in W comprising departments of Loire-Atlantique, Maine-et-Loire, Mayenne, Sarthe and Vendée. Area, 12,404 sq. m. (32,126 sq. km.). Chief towns Nantes, Angers and Le Mans. Pop. (1982c) 2,930,398.

Pazardzhik Plovdiv, Bulgaria. 42 12N 24 20E. Town sit. W of Plovdiv on Maritsa R. Railway junction and processing centre serving a fertile agric. and vine-growing area. Manu. textiles, rubber and leather goods. Pop. (1987E) 81,513.

Peace River Canada. 59 00N 111 25W. Formed by union of R. Finlay and Parsnip at Finlay Forks, Brit. Columbia, and flows 1,195 m. (1,912 km.) E into Alberta, then N and NE to join Slave R. near the W end of Athabasca L.

Peak District, The Derbyshire, England. Hilly district of the Pennine range in N Derbyshire, rising to Kinder Scout, 2,088 ft. (636 metres) a.s.l. Limestone caves inc. the Peak Cavern near Castleton.

Pearl Harbor Hawaii, U.S.A. 21 22N 157 58W. Harbour on S coast of Oahu Island with a major U.S. naval and air base.

Pearl Islands Panama. 8 20N 79 02W. Arch. of *c.* 180 islands extending *c.* 40 m. (64 km.) NNE to SSW through the Gulf of Panama; the chief islands are San Miguel, San José and Pedro González. The main occupations are pearl fishing and sea angling.

Peć Serbia, Yugoslavia. 42 40N 20 19E. Town sit. SE of Niš, noted for the Turkish influences in its streets, mosques and other buildings. Pop. (1981E) 111,071.

Pechora River U.S.S.R. 68 13N 54 10E. Rises in the N Ural Mountains and flows 1,100 m. (1,760 km.) N and W to enter the Gulf of Pechora on the Barents Sea by a delta. Used for transporting coal, timber and furs. Ice-free from June–Sept.

Peckham Greater London, England. 51 28N 0 04W. Mainly residential area of S London.

Pecos River U.S.A. 29 42N 101 22W. Rises in the Sangre de Cristo Range in N New Mexico and flows 740 m. (1,184 km.) SSE through New Mexico and Texas, to join Rio Grande on the Mexican border above Del Rio. Used to irrigate an extensive agric. area.

Pécs Baranya, Hungary. 46 05N 18 13E. Cap. of county sit. SW of Budapest near the Yugoslav frontier; sit. in a coalmining area. Trading centre for wine. Industries inc. manu. clothing, leather goods, porcelain, tobacco products and soap. Pop. (1989E) 183,000.

Pedernales Barahona, Dominican Republic. 18 02N 71 44W. Town sit. WSW of Barahona on the SW coast on the Haiti border. Centre of an irrigated agric. area.

Pedregal Chiriqui, Panama. Village sit. S of David on the Pacific. Road, rail terminus and port for David, exporting coffee, bananas, cacao, but much declined.

Pedro Juan Caballero Amambay, Paraguay. 22 34S 55 37W. Town sit. NE of Concepción in the Sierra de Amambay. Dept. cap. and trading centre for a cattle raising area. Pop. (1983C) 39,000.

Peebles Borders Region, Scotland. 55 39N 3 12W. County town in central S sit. on Tweed R. Market town. Industries inc. manu. woollen goods, photographic processing laboratory. Pop. (1989E) 7,000.

Peeblesshire Scotland. Former county and now part of Borders Region.

Peel Isle of Man, United Kingdom. 54 13N 4 40W. Town sit. on W coast. Industries inc. fishing and tourism. Pop. (1986C) 3,660.

Peenemünde Rostock, German Democratic Republic. Village in NE, on the NW of Usedom Island, and ENE of Greifswald, sit. at the entrance to the Peene R. estuary. Site of a research station for guided missiles in World War II.

Pegu Pegu, Burma. 17 20N 96 29E. Cap. of province sit. NE of Rangoon on Pegu R. Industries inc. pottery. Noted for the Shwe-mawdaw pagoda, 324 ft. (99 metres) high. Pop. (1983C) 254,761.

Pegu Yoma Burma. 19 00N 95 50E. Range of hills extending 250 m. (400 km.) N to S between Rs. Irrawaddy and Sittang. Average height below 1,000 ft. (305 metres) a.s.l. except for Mount Popa, 4,985 ft. (1,519 metres) a.s.l., and covered largely by teak forests.

Peipus, Lake U.S.S.R. 58 45N 27 30E. L. on the frontier between Estonia and the Russian Soviet Socialist Rep., comprising 2 Ls. joined by a 15 m. (24 km.) strait. The larger is L. Peipus (L. Chudskoye) to the N, with L. Pskov to the S. Total area, 1,356 sq. m. (3,512 sq. km.). Drained by Narova R. into the Gulf of Finland. Frozen Dec.–March.

Pekalongan Central Java, Indonesia. 6 53S 109 40E. Town W of Semarang sit. on the Java Sea. Port serving a hinterland of rice and sugar cane plantations. Pop. (1980E) 145,000.

Pekin Illinois, U.S.A. 40 35N 89 40W. City S of Peoria sit. on Illinois R. Industries inc. grain distillation, podiatry products, hybrid seed corn, chemical products and metal goods. R. port trading in grain and livestock. Pop. (1990E) 34,000.

Peking (Beijing) Hebei, China. 39 55N 116 25E. Cap. of the Rep. sit. in NE, and its commercial and cultural centre. Industries inc. engineering, food processing, printing and publishing, tanning and textiles. The Outer

City to the S and Inner City to the N are adjacent, and otherwise known as the Chinese and Tartar cities. The Inner City contains the Imperial City, which in turn contains the Forbidden City. Pop. (1987E) 5,970,000, metropolitan area, 9,750,000.

Pelagosa Islands Yugoslavia. Otherwise called Pelagruž. Group sit. WSW of Dubrovnik in the Adriatic Sea.

Pella Macedonia, Greece. *Nome* in N bounded N by Yugoslavia. Area, 968 sq. m. (2,506 sq. km.). Cap. Edessa, 50 m. (80 km.) WNW of Thessaloníki. Mountainous. The main products are cotton, wheat and tobacco. Pop. (1981C) 132,386.

Peloponnessos Greece. 37 10N 22 00E. The S peninsula of Greece joined to the rest by the Isthmus of Korinthos. Area, 8,354 sq. m. (21,637 sq. km.). Mountainous. The main occupations are growing vines, fruit and olives, and rearing sheep and goats.

Pelotas Rio Grande do Sul, Brazil. 31 46S 52 20W. Town in extreme SE, sit. on the São Gonçalo Canal near the entrance to the Laga dos Patos. Industries inc. meat packing, flour milling and tanning; manu. soap, furniture and footwear. Exports are meat products, hides and wool. Pop. (1980E) 197,000.

Pemba Tanzania. 5 10S 39 48E. Island in the Indian Ocean sit. NNE of Zanzibar. Area, 380 sq. m. (984 sq. km.). Cap. Chake Chake. Coral island producing cloves and copra. Pop. (1985E) 257,000.

Pembroke Dyfed, Wales. 51 41N 4 55W. Town in SW sit. on the shore of Milford Haven. Market town for a farming area. Industries inc. engineering. Pop. (1981C) 15,618 with Pembroke Dock.

Penarth South Glamorgan, Wales. 51 27N 3 11W. Town S of Cardiff sit. at mouth of Taff R. on the Bristol Channel. Seaside resort. Pop. (1981C) 22,983.

Pendembu South Eastern Province, Sierra Leone. Town sit. E of Freetown

near the Liberian border. Rail terminus trading in palm products, cacao, coffee.

Penge Greater London, England. 51 25N 0 04W. Area of SE London forming a dormitory suburb. Industries inc. the manu. of electrical equipment. Formerly the site of the Crystal Palace, 1854–1936.

Peninsular Malaysia Malaysia. 4 00N 102 00E. The S end of the Malay Peninsula formerly named Western Malaysia, and bounded N by Thailand. Area, 50,806 sq. m. (131,588 sq. km.).

Penistone South Yorkshire, England. 53 31N 1 38W. Town sit. NW of Sheffield on Don R. The main industry is steel. Pop. (1981C) 8,608.

Penmaenmawr Gwynedd, Wales. 53 16N 3 56W. Town SW of Llandudno sit. on Conwy Bay. Seaside resort. Pop. (1990E) 4,000.

Pennine Range England. 54 50N 2 20W. Range of hills extending N from Trent R. valley, Derbyshire, to Tyne R. valley near the Scottish border. Rises to Cross Fell, 2,930 ft. (893 metres) a.s.l., and forms the watershed of the main R. of N Eng. Mainly moorland and rough grazing, with tourist areas.

Pennsylvania U.S.A. 40 50N 78 00W. State in the NE bounded NW by L. Erie. Area, 45,333 sq. m. (117,412 sq. km.). Chief cities Philadelphia, Pittsburgh, Harrisburg (the cap.), Erie, Scranton and Allentown. Mostly within the Appalachian Mountains, with areas of the Allegheny Mountains inc. Mount Davis 3,213 ft. (979 metres) a.s.l. Rich in mineral deposits esp. coal (including anthracite), natural gas and petroleum. Farming covers about 9m. acres which inc. dairying, tobacco, and fruit and vegetables. The main industries are iron and steel, with Pittsburgh producing *c*. 30% of the national output, and textiles, centred on Philadelphia. Pop. (1987E) 11,936,396.

Penonomé Coclé, Panama. 8 31N 80 21W. Town sit. SW of Panama City. Provincial cap. Industries inc.

manu. Panama hats, soap. Trades in rubber, coffee, cacao. Pop. (1980E) 30,913.

Penrhyn Cook Islands, South Pacific. 9 00S 158 00W. Atoll sit. NE of Rarotonga, in the N group. Area, 4,000 acres (1,619 hectares). Main products pearls, copra. Pop. (1981C) 608.

Penrith Cumbria, England. 54 40N 2 44W. Town sit. SSE of Carlisle. Industries inc. tourism, agric., and light engineering. Pop. (1990E) 12,500.

Penryn Cornwall, England. 50 09N 5 06W. Market town NW of Falmouth sit. at head of Penryn R. estuary. Industries inc. quarrying granite. Pop. (1980E) 6,000.

Pensacola Florida, U.S.A. 30 25N 87 13W. City in extreme W of state sit. on Gulf of Mexico. Industries inc. fishing, fish canning, furniture and paper making. Naval air base and port. Pop. (1980C) 57,619.

Pentecost Vanuatu, South Pacific. 15 42S 168 10E. Island E of Espiritu Santo. Area, 190 sq. m. (492 sq. km.). Volcanic. Produces copra, coffee.

Penticton British Columbia, Canada. 49 30N 119 35W. Town sit. on Okanagan R. near the S end of Okanagan L., and E of Vancouver. Commercial centre of a fruit-growing area. Industries inc. food processing, trucks, software, tourism, forestry and mining. Pop. (1989E) 25,312.

Pentland Firth Scotland. 58 44N 3 13W. Channel extending 20 m. (32 km.) E to W between the NE coast of Caithness and the Orkney Islands. Noted for rough seas. Contains the islands of Stroma and Swoma and the Pentland Skerries.

Pentland Hills Scotland. 55 48N 3 23W. Range extending 16 m. (26 km.) NE to SW across W Lothian Region, rising to Scald Law, 1,898 ft. (578 metres) a.s.l.

Pentonville Greater London, England. 51 32N 0 06W. District of N London containing Pentonville Prison.

Penza Russian Soviet Federal Socialist Republic, U.S.S.R. 53 13N 45 00E. Cap. of Penza region NNW of Saratov sit. on Sura R. Industries inc. engineering, sawmilling and paper making; manu. watches, cement and matches. Pop. (1987E) 540,000.

Penzance Cornwall, England. 50 07N 5 33W. Town in extreme SW Cornwall sit. on Mount's Bay. Industries inc. tourism, fishing, ship repair yard, and the distribution of horticultural produce from the hinterland and from the Isles of Scilly. Pop. (1981C) 19,579.

People's Democratic Republic of Yemen *see* **Yemen**

Peoria Illinois, U.S.A. 40 42N 89 36W. City in N centre of state sit. on Illinois R. Industries inc. manu. agric. machinery, electrical goods and food products. Pop. (1980C) 124,160.

Perak Peninsular Malaysia, Malaysia. 5 00N 101 00E. State bounded N by Thailand and W by the Strait of Malacca. Area, 8,110 sq. m. (21,019 sq. km.). Chief towns Taiping (the cap.) and Ipoh. Lies mainly in the basin of the Perak R. rising to 7,000 ft. (2,133 metres) a.s.l. in the interior. The main product is tin. Pop. (1980C) 1,762,288.

Pereira Caldas, Colombia. 4 49N 75 43W. Town in central W. Commercial centre trading in coffee and cattle. Industries are coffee processing and brewing; manu. clothing. Pop. (1980E) 270,000.

Perekop Isthmus Ukraine, U.S.S.R. 46 00N 33 00E. Isthmus joining the Crimea peninsula to the Ukrainian mainland, 4 m. (7 km.) at its narrowest point.

Périgord France. 45 00N 0 40E. Area of SW France, mainly in the Dordogne dept., consisting of dry limestone plateau and the fertile Dordogne and Isle R. valleys. Noted for truffles.

Perigueux Dordogne, France. 45 11N 0 43E. Cap. of dept. in SW sit. on Isle R. Industries inc. chemicals; manu. hardware and cutlery, and *pâté* *de foie gras.* Pop. (1982C) 35,392 (agglomeration, 57,826).

Perim Republic of Yemen. 12 40N 43 25E. Island off the SW coast of the Arabian peninsula in the strait of Bab-el-Mandeb. Area, 5 sq. m. (13 sq. km.). Rocky and barren.

Perlis Peninsular Malaysia, Malaysia. State in NW bounded NE and NW by Thailand. Area, 310 sq. m. (803 sq. km.). Cap. Kangar. The main products are rice, tin and rubber. Pop. (1980C) 147,726.

Perm Russian Soviet Federal Socialist Republic, U.S.S.R. 58 00N 56 15E. Cap. of Perm region sit. on Kama R. Industries inc. engineering, tanning and sawmilling; manu. aircraft parts, tractor parts, agric. and construction equipment, fertilizers, paper and matches. R. port Pop. (1987E) 1,075,000.

Përmet Albania. 40 14N 20 20E. Town sit. S of Berat on Vijosë R. Commercial centre.

Pernambuco Brazil. 8 00S 37 00W. State in NE Braz. bounded E by the Atlantic. Area, 37,458 sq. m. (97,016 sq. km.). Cap. Recife. A mainly arid state descending to a more humid coastal zone. The main products are cotton, sugar and tropical fruit. Pop. (1989E) 7,238,280.

Pernik Sofia, Bulgaria. 42 36N 23 02W. Town WSW of Sofia sit. on Struma R. Industries inc. coalmining, iron and steel, cement, engineering and glass making. Pop. (1982E) 94,859.

Perpignan Pyrénées-Orientales, France. 42 41N 2 53E. Cap. of dept. sit. on Têt R. near its mouth on the Gulf of Lions, and near the Spanish border. Commercial centre trading in wine, olives and fruit. Industries inc. tourism. Pop. (1975C) 107,971.

Pershore Hereford and Worcester, England. 52 06N 2 05W. Town SE of Worcester sit. on the Avon R. Market town in a fruit-growing and market gardening area.

Persia *see* **Iran**

Persian Gulf Arabian Sea. 27 00N 51 00E. Generally called 'The Gulf'. Arm of the Arabian Sea extending mainly NW between Iran to the E and Saudi Arabia to the W, entered from the Sea by Gulf of Oman and the Strait of Hormuz. Area, 90,000 sq. m. (233,000 sq. km.). Receives Shatt-al-Arab R. at its N end into which flow Rs. Tigris, Euphrates and Karun.

Perth Western Australia, Australia. 31 56S 115 50E. State cap. on the W coast. Commercial centre serving an extensive hinterland of farming, cattle rearing and gold mining areas. Industries inc. textiles and flour milling; manu. clothes, furniture, cars and cement. Pop. (1981E) 809,035.

Perth Tayside Region, Scotland. 56 24N 3 28W. County town in Central Scot. sit. on Tay R. Industries inc. textiles, glassmaking, whisky and bottling. Pop. (1989E) 44,000.

Perth Amboy New Jersey, U.S.A. 43 31N 74 16W. City SSW of Newark sit. at mouth of Raritan R. Industries inc. steel wire bar, oil refining, chemicals and plastics; manu. clothing and cosmetics. Pop. (1990E) 45,000.

Perthshire Scotland. Former county now part of Tayside and Central Region.

Peru 8 00S 75 00W. Rep. bounded on W by the Pacific, N by Ecuador and Colombia, E by Brazil, SE by Bolivia and S by Chile, with a 1,400 m. (2,240 km.) Pacific coastline. Area, 496,525 sq. m. (1,286,000 sq. km.). Peru, formerly the most important of the Spanish vice-royalties in South America, declared its independence in 1821; but it was not till after the war, protracted till 1824, that the country gained its actual freedom. Chief towns Lima (cap.), Callao, Huancayo, Cerro de Pasco, Cuzco, Arequipa, Chiclayo and Piura. The dry coastal plain rises to the inland Sierra, at *c.* 13,000 ft. (4,000 metres) a.s.l., which covers about half of the country. The forested Andes beyond it extend to the Brazilian border. There are 4 zones: the coast strip, with an average width of

50 m. (80 km.); the Sierra or Uplands, formed by the coast range of mountains and the Andes proper; the Montaña or high wooded region which lies on the eastern slopes of the Andes, and the jungle in the Amazon Basin, known as the Selva. There are 4 fertilizer factories, near Callao and in Cuzco. Peru is a substantial importer of foodstuffs, chiefly wheat, but also fats and oil, meat and dairy products. Nearly half of the population is dependent on agriculture. Products include sugar, cotton, coffee and wool. Peru suffers periodically from severe droughts particularly in the N zone. Peru is the world's foremost fishing nation in terms of value of catch, most of which is anchoveta which is reduced into fishmeal for export as animal feed. Minerals inc. lead, copper, iron, silver, zinc and petroleum. Pop. (1988E) 21,255,900.

Perugia Umbria, Italy. 43 08N 12 22E. Cap. of Perugia province sit. in central Italy above Tiber R. valley. Industries inc. manu. woollen goods, furniture and chocolate. Pop. (1981E) 142,348.

Pervouralsk Russian Soviet Federal Socialist Republic, U.S.S.R. 56 54N 59 58E. Town WNW of Sverdlovsk in the central Urals on Chusovaya R. Railway junction. Industries inc. metal-working, brick making, sawmilling.

Pesaro Marche, Italy. 43 54N 12 55E. Town in E central Italy sit. on the Adriatic coast. Industries inc. tourism and pottery; manu. agric. machinery. Pop. (1981E) 90,000.

Pescadores Taiwan. 23 20N 119 30E. Islands off the W coast of Taiwan. Area, 49 sq. m. (127 sq. km.). There are 64 islands, the main island being Penghu. The main occupations are fishing and fish processing. Tourism is a growth industry. Pop. (1985E) 102,930.

Pescara Abruzzi e Molise, Italy. 42 28N 14 13E. Cap. of Pescara province sit. at mouth of Pescara R. on the Adriatic. Industries inc. tourism and textiles; manu. furniture, soap and glass. Pop. (1981E) 131,000.

Peshawar Pakistan. 34 01N 71 40E. Cap. of Peshawar division sit. NW of Lahore. Commercial centre trading with Afghánistán through the Khyber Pass. Industries inc. textiles, food processing, rice milling and pottery; manu. copperware and leather goods. Pop. (1981E) 506,000.

Pessac Gironde, France. 44 48N 0 38W. SW suburb of Bordeaux. Pop. (1982C) 50,543.

Pest Budapest, Hungary. 44 29N 19 05E. City on left bank of Danube R. opposite Buda. Together they form Budapest. Commercial centre; almost completely rebuilt after flooding in 1838. Trad. the business centre of the city, Buda being the residential sector. Pop. (1984E) 983,000.

Petah Tiqva Israel. 32 05N 34 53E. City sit. E of Tel Aviv on the Plain of Sharon. Industries inc. textiles, chemicals and furniture making. Pop. (1982E) 120,000.

Pétange Luxembourg. 49 33N 5 33E. Town sit. NW of Esch-sur-Alzette on Chiers R. Iron mining centre. Pop. (1984E) 6,416.

Petén Guatemala. Dept. of Guatemala bounded N and W by Mexico. Area, 13,843 sq. m. (35,853 sq. km.). Cap. Flores. Forested lowland with stretches of savannah and large Ls. Drained by San Pedro, Azul and Pasión Rs. Sparse settlement. The main occupation is lumbering. Pop. (1982E) 102,803.

Peterborough Ontario, Canada. 44 18N 78 19W. City sit. NE of Toronto on Otonabee R. Manu. food products, auto parts, paper, electrical machinery, hardware and dairy equipment. Pop. (1990E) 62,500.

Peterborough Cambridgeshire, England. 52 35N 0 15W. Town in E central Eng. sit. on Nene R. Industries inc. engineering, electronics, domestic appliances, printing, publishing, beet sugar refining and brickmaking. Pop. (1985E) 138,500.

Peterhead Grampian Region, Scotland. 57 30N 1 49W. Largest fishing

port in the European Community and
also white fish market sit. in NE on
North Sea. The main industry is food
processing, also woollen cloth manu.,
fish canning, light engineering and oil
rig repairs. Pop. (1986E) 18,150.

Peterlee   Durham, England. 54 46N
1 18W. Town sit. E of Durham. Founded as a 'new town' in 1948. There is
considerable light industry to replace
the declining coalmining industry.
Pop. (1990E) 23,500.

Peter I Island   Norway. Island in
the Bellingshausen Sea off Thurston
Peninsula, Antarctica, was sighted in
1821 by the Russian explorer, Admiral
von Bellingshausen. The first landing
was made in 1929 by a Norwegian expedition which hoisted the Norwegian
flag. Area, 69 sq. m. (178 sq. km.).
Depend. of Norway. Uninhabited.

Petersburg   Virginia, U.S.A. 37
13N 77 24W. City sit. on Appomattox
R. Manu. tobacco products, textiles,
pharmaceuticals, machine parts, metal
goods, computer software, luggage
and optical equipment. Pop. (1980C)
41,055.

Petersfield   Hampshire, England.
51 00N 0 56W. Town sit. NNE of
Portsmouth. Market town for a farming area. Pop. (1980E) 9,300.

Petropavlovsk   Kazakhstan,
U.S.S.R. 54 54N 69 06E. City in N
Kazakhstan on the Trans-Siberian
Railway. Commercial centre trading
in grain, furs and textiles. Industries
inc. meat packing, flour milling and
tanning. Pop. (1987E) 233,000.

Petropavlovsk-Kamchatski   Russian Soviet Federal Socialist Republic, U.S.S.R. 53 01N 158 39E. City in
NE sit. on the SE coast of the Kamchatka Peninsula on the Bering Sea.
Industries inc. shipbuilding, sawmilling and fish canning. Naval base and
sea port. Pop. (1987E) 252,000.

Petrópolis   Rio de Janeiro, Brazil.
22 31S 43 10W. Mountain resort in
SE. Industries inc. textiles and brewing. Pop. (1980E) 149,000.

Petrozavodsk   Russian Soviet Fed-

eral Socialist Republic, U.S.S.R. 61
47N 34 20E. Town sit. on W shore of
L. Onega. Cap. of the Karelian
Autonomous Soviet Socialist Rep. Industries inc. sawmilling and mica processing; manu. cement, furniture and
machinery. Pop. (1987E) 264,000.

Pforzheim   Baden-Württemberg,
Federal Republic of Germany. 48 52N
8 42E. Town in SW sit. on Enz R. The
main industry is jewellery, also manu.
clocks and watches, precision instruments, machinery and tools. Pop.
(1990E) 109,500.

Phenix City   Alabama, U.S.A. 32
29N 85 01W. City in E of state, sit.
opposite Columbus, Georgia, on Chattahoochee R., and forming a suburb of
it. Industries inc. manu. wood products. Pop. (1980C) 26,928.

Philadelphia   Pennsylvania, U.S.A.
39 57N 75 07W. City in extreme SE of
state sit. on Delaware R. at its confluence with Schuylkill R. Industries
inc. textiles, oil refining, shipbuilding,
metal working, printing and publishing, engineering and paper making;
manu. machinery and electrical goods.
Sea port, second in importance to
New York, exporting petroleum products, coal, grain, timber, flour and
manu. goods. Large U.S. Navy dockyard. Pop. (1980C) 1,688,210.

Philippines, Republic of the   12
00N 123 00E. Rep. consisting of over
7,000 islands of the Malay Arch. in
the SW Pacific. The main islands are
Luzon in the N and Mindanao in the S,
with Samar, Negros, Palawan, Panay,
Mindoro, Leyte, Cebu, Bohol and
Masbate. The Philippines were discovered by Magellan in 1521 and conquered by Spain in 1565. Following
the Spanish-American war, the islands
were ceded to the U.S.A. in 1898,
after the Filipinos had tried in vain to
establish an independent republic in
1896. The Republic of the Philippines
came into existence in 1946, by agreement with the U.S. Government.
Area, 115,707 sq. m. (299,681 sq.
km.). Main towns Quezon City (the
cap.), Manila, Cebu, Davao, Basilan
and Iloilo. Mountainous and volcanic,
rising to the active volcano Mount
Apo in SE Mindanao, 9,690 ft. (2,953

metres) a.s.l. The climate is tropical, with frequent typhoons and earthquakes. About 45% of the land is forest, the rest produces rice, maize, tobacco, coconuts, sugar-cane and abacá. Minerals inc., gold, silver, lead, zinc, copper, manganese, quicksilver, molybdenum and salt. Pop. (1989E) 60m.

Phnom Penh   Cambodia. 11 33N 104 55E. Cap. of rep. sit. at confluence of Rs. Mekong and Tonlé Sap at the head of the Mekong delta. Founded as the cap. in 1434 it is a commercial centre and R. port, accessible to smaller ocean-going ships through Vietnam. Industries inc. textiles and rice milling. Pop. (1988E) 500,000. Formerly it had a pop. of 2·5m.

Phoenix   Arizona, U.S.A. 33 27N 112 05W. State cap. in the s centre of state. Commercial centre for an extensive agric. area, trading in cotton and citrus fruits. Industries inc. tourism and food processing. Pop. (1980C) 789,704.

Phoenix Islands   Kiribati, Central Pacific Ocean. 3 30S 172 00W. Group of 8 islands *c.* 1,300 m. (2,080 km.) NE of Fiji. Area, 11 sq. m. (28 sq. km.). The group comprises Canton (Kanton), Enderbury, Birnie, McKean, Rawaki, Manra, Nikumaroro and Orona. Coconuts formerly grown by Micronesian colonists on last 3. Now all uninhabited.

Phokis   Greece. *Nome* in central Greece bounded s by Gulf of Korinthos. Area, 819 sq. m. (2,121 sq. km.). Cap. Amphissa, sit. WNW of Athens. Mountainous with ranges surrounding the plain of Crisa. The main occupation is stock farming; also some wheat, olives and wine. The chief mineral is bauxite. Pop. (1981C) 44,222.

Phthiotis   Greece. *Nome* in central Greece bounded E by the Gulf of Euboea. Area, 1,686 sq. m. (4,368 sq. km.). Cap. Lamia, sit. NW of Athens. Mountainous, but fertile in the Sperkhios R. valley. The main occupation is farming, esp. wheat and cotton. Pop. (1980C) 161,995.

Phuket   Phuket Island, Thailand. 8 00N 98 28E. Town SSW of Bangkok sit. on an island just off the w coast of the Malay peninsula. Sea port exporting tin ore from the island's resources, and rubber.

Piacenza   Emilia-Romagna, Italy. 45 01N 9 40E. Cap. of Piacenza province in central N sit. on Po R. Industries inc. food processing, manu. pasta, leather goods and agric. and petrochemical machinery. Pop. (1989E) 104,023.

Piandzh River   U.S.S.R./Afghánistán. Formed by junction of Pamir and Wakhan Rs. near Qala Panja, on the Afghan/U.S.S.R. frontier. Flows *c.* 400 m. (640 km.) SW then NW along the frontier and SW again as far as Nizhui Piandzh where it joins Vakhsh R. to form the Amu Darya R.

Piarco   Trinidad, West Indies. Village sit. ESE of Port of Spain, site of an intern. airport.

Piatra Neamt   Neamt, Romania. 46 56N 26 22E. Cap. of Neamt district N of Bucharest sit. on Bistrita R. Industries inc. textiles, woodworking and food processing; manu. pharmaceutical goods. Pop. (1983E) 100,549.

Piauí   Brazil. State in NE extending inland SSW from a 35 m. (56 km.) coastline on the NE coast, and bounded W by Parnaíba R. Area, 96,261 sq. m. (249,316 sq. km.). Cap. Teresina. Grassy plateau; the main occupation is stock rearing, also cultivating cotton and tobacco. Pop. (1980C) 2,139,021.

Piave River   Italy. 45 32N 12 44E. Rises in the Carnic Alps and flows 140 m. (224 km.) SW past Belluno, then SE to the Adriatic Sea ENE of Venice.

Picardy   France. 50 00N 2 15E. Region of N France comprising Aisne, Somme and Oise depts. Area, 7,494 sq. m. (19,410 sq. km.). Fertile area, mainly agric., with a textile industry in Amiens. Pop. (1982C) 1,740,321.

Pickering   North Yorkshire, England. 54 15N 0 47W. Town sit. NE of

York. Market town for a farming area and a tourist centre. Pop. (1981c) 6,019.

Pico Pico Island, Azores. 38 28N 28 20W. Active volcano sit. ESE of Horta, on Faial Island. Highest peak in the Azores, 7,611 ft. (2,319 metres) a.s.l.

Piedmont *see* **Piemonte**

Piemonte Italy. 45 00N 7 30E. Region in N Italy bounded NE by Switz. and W by France. Area, 9,817 sq. m. (25,426 sq. km.). Mainly in the upper Po R. basin. Agric., with industries in Turin. Pop. (1980E) 4,531,141.

Pieria Greece. 40 13N 22 25E. *Nome* in Macedonia bordering on the Aegean Sea. Area, 593 sq. m. (1,535 sq. km.). Cap. Katerini, sit. W of Thessaloniki. Narrow coastal plain rising to a mountainous hinterland with Mount Olympus, 9,550 ft. (2,910 metres) a.s.l. Pop. (1981c) 106,859.

Pierre South Dakota, U.S.A. 44 22N 100 21W. City and state cap. sit. WNW of Sioux Falls, on Missouri R. at 1,440 ft. (438 metres) a.s.l. Commercial centre for an agric. area, employment is in government, and trading in grain and livestock. Pop. (1980c) 11,973.

Pietermaritzburg Natal, Republic of South Africa. 29 37S 30 16E. Cap. of province in the central E. Important railway centre. Industries inc. manu. of furniture, footwear, metal goods, carpet yarn and textiles. Pop. (1989E) 185,000.

Pietersburg Transvaal, Republic of South Africa. 23 54S 29 25E. Town in the N. Industries inc. mining, esp. silica, gold, diamond cutting and asbestos, food processing, manu. radios and television. Pop. (1990E) 35,000.

Piet Retief Transvaal, Republic of South Africa. Town sit. SE of Ermelo near the Natal and Swaziland borders. Centre of an agric. area, with gold deposits nearby.

Piggs Peak Swaziland. Village sit. NNE of Mbabane. Mining centre for gold and asbestos.

Pikes Peak Colorado, U.S.A. 38 51N 105 03W. Mountain in centre of state and in the Front Range of the Rocky Mountains, 14,110 ft. (4,341 metres) a.s.l.

Pila Poznań, Poland. 53 10N 16 44E. Town in NW. Industries inc. textiles; manu. agric. machinery. Pop. (1983E) 65,000.

Pilar Ñeembucú, Paraguay. 26 52S 58 23W. Town sit. SSW of Asunción on Paraguay R. where it forms the Argentine frontier. Dept. cap., commercial and processing centre for a fertile area. R. port handling sugarcane, cotton, maize, oranges, onions, hides, timber. Industries inc. cotton ginning, sawmilling, textiles, distilling. Pop. (1983c) 25,600.

Pilatus, Mount Switzerland. Mountain at the boundaries of 4 cantons in central Switz. and 5 m. (8 km.) SSW of Luzern, 6,994 ft. (2,132 metres) a.s.l. with a rack-and-pinion railway.

Pilcomayo River South America. 25 21S 57 42W. Rises near L. Poopó in the Bolivian Andes and flows 1,000 m. (1,600 km.) to the Argentine border at Fort D'Orbigny, forms the Argentina/Paraguay frontier for *c.* 200 m. (320 km.), and enters Paraguay to join Paraguay R. at Asunción.

Pilsen *see* **Plzeň**

Pimlico Greater London, England. 51 29N 0 08W. District of central London on N bank of R. Thames. Residential, containing Victoria railway terminus.

Pinang Peninsular Malaysia, Malaysia. 5 24N 100 19E. State consisting of the island of Penang in the Strait of Malacca, and a mainland strip on the NW coast. Area, 398 sq. m. (1,031 sq. km.). Cap. Penang. Hilly, rising to 2,700 ft. (831 metres) a.s.l., with rubber and coconut plantations on low ground. Pop. (1980c) 911,586.

Pindus Mountains Greece. 39 49N 21 14E. Range extending *c.* 100 m. (160 km.) NNW to SSE on the Epirus/Thessaly border, rising to Smólikas, 8,652 ft. (2,637 metres) a.s.l. in N.

Pine Bluff Arkansas, U.S.A. 34 13N 92 01W. City in s centre of state sit. in a cotton growing area. Industries inc. railway engineering, paper, radiator caps, steel radial cord for tyres and chemicals. Pop. (1990E) 63,000.

Pingtung Taiwan. Town sit. ENE of Kaohsiung. Agric. inc. rice and vegetables. Pop. (1985E) 198,961.

Pinner England. 51 35N 0 24W. Mainly residential district of NW Greater London.

Pinsk Belorussia, U.S.S.R. 52 07N 26 04E. Town SSW of Minsk, sit. in the Pripet Marshes. Industries inc. manu. paper, furniture, soap, matches and leather.

Piotrków Trybunalski Lódź, Poland. 51 25N 19 42E. Town in central Poland. Industries inc. textiles, agric. engineering, flour milling, sawmilling, tanning and brewing. Pop. (1983E) 77,000.

Piparia Madhya Pradesh, India. Town sit. E of Hoshangabad. Industries inc. flour, oilseed and dal milling. Pop. (1981C) 25,319.

Piraeus Attica, Greece. 37 57N 23 38E. City SW of Athens sit. on the Saronic Gulf. Industries inc. shipbuilding, oil refining and textiles; manu. fertilizers. Major port, exporting wine and olive oil, and handling all the traffic for Athens. Pop. (1981C) 196,389.

Pirmasens Rhineland-Palatinate, Federal Republic of Germany. 49 12N 7 36E. Town sit. SW of Ludwigshafen. Manu. footwear and leather goods. Pop. (1984E) 47,200.

Pirna Dresden, German Democratic Republic. 50 58N 13 56E. Town SE of Dresden sit. on Elbe R. Industries inc. manu. rayon, paper, glass, pottery and electrical equipment. Pop. (1989E) 46,000.

Pisa Toscana, Italy. 43 43N 10 23E. Cap. of Pisa province in central W sit. on Arno R. Industries inc. glass, motorcycles, motor boats and yachts. Tourist centre noted for the buildings of the Piazza del Duomo, inc. the 'Leaning Tower'. Pop. (1984E) 104,213.

Pishpek *see* **Frunze**

Pistoia Toscana, Italy. 43 55N 10 54E. Cap. of Pistoia province in central Italy sit. at the foot of the Apennines. Manu. pasta, textiles, footwear, agric. machinery and musical instruments. Pop. (1981E) 92,000.

Pitcairn Island South Pacific Ocean. 25 04S 130 05W. Volcanic island *c*. half-way between New Zealand and Panama. Area, 2 sq. m. (5 sq. km.). The main occupation is growing fruit and vegetables. Inhabited by descendants of the crew of H.M.S. *Bounty* who settled there in 1790 with women from Tahiti. Administered from the Brit. High Commission in New Zealand. In 1902 the uninhabited islands of Henderson, 12 sq. m. (31 sq. km.), Ducie, 2.5 sq. m. (6.5 sq. km.) and Oeno, 2 sq. m. (5 sq. km.) were annexed as part of the Pitcairn group. Pop. (1989E) 54.

Piteä Norrbotten, Sweden. 65 20N 21 30E. Town sit. SW of Luleå on Pite R. near its mouth on the Gulf of Bothnia. Sea port, shipping timber and processed goods. Industries inc. sawmilling, plastics and metal working. Tourism. Pop. (1985E) 38,797.

Pitești Argeș, Romania. 44 52N 24 52E. Cap. of Argeș district sit. on Argeș R. Market town. Industries inc. textiles, pottery and flour milling. Pop. (1985E) 154,112.

Pitlochry Tayside Region, Scotland. 56 42N 3 43W. Town sit. NNW of Perth on Tummel R. Highland resort with trout and salmon fishing. Industries include tourism and a distillery. Pop. (1981C) 2,650.

Pittsburg Kansas, U.S.A. 37 25N 94 43W. City sit. NNW of Joplin (Missouri), in extreme SE of Kansas. Manu. inc. mining equipment, plastics, agric. feeds and fertilizers, clothing, food processing, electrical equipment and textile machinery. Pop. (1980C) 18,770.

Pittsburgh Pennsylvania, U.S.A. 40 26N 80 00W. City in SW of state at confluence of Rs. Allegheny and Monongahela which form Ohio R. Sit. in an extensive coalfield. The main industries inc. government, engineering, iron and steel, scientific and technological research and general manu. Pop. (1980C) 423,938.

Pittsfield Massachusetts, U.S.A. 42 27N 73 15W. City sit. W of Boston. Industries inc. tourism, textiles, chemicals and paper making; manu. electrical equipment. Pop. (1980C) 51,974.

Piura Piura, Peru. 5 12S 80 38W. Cap. of dept. in NW. Commercial centre of a large agric. area irrigated by Piura R.; trading in long-staple cotton. Industries inc. cotton ginning; manu. cottonseed oil. Pop. (1981C) 186,354.

Plainfield New Jersey, U.S.A. 40 37N 74 26W. City sit. SW of Newark. Industries inc. manu. machinery, esp. printing presses, tools, concrete products and clothing. Pop. (1980C) 45,555.

Plata, La Buenos Aires, Argentina. 34 52S 57 55W. Cap. of province sit. near Río de la Plata. Industries inc. oil refining, meat packing, and cement and textile works. Exports grain, meat and wool. Pop. (1980C) 455,000.

Plata, Río de la Argentina/Uruguay. 18 29N 66 15W. Estuary of R. Paraná and Uruguay extending *c.* 160 m. (256 km.) SE to the Atlantic between Argentina and Uruguay. Width 20 m. (32 km.), broadening to 60 m. (96 km.) in the centre, and 140 m. (224 km.) at its mouth. Dredging to remove silt is continuous.

Plate, River *see* **Plata, Río de la**

Plateau Nigeria. State, formerly Plateau province and the N part of Benue province of the N region. Area, 22,405 sq. m. (58,030 sq. km.). Important mining area with major deposits of tin and columbite. Main towns Jos (cap.), Lafia and Keffi. Pop. (1988E) 4,385,100.

Platte River U.S.A. 41 04N 95

53W. Formed by union of N. Platte and S. Platte R. at North Platte, Nebraska, and flows 310 m. (496 km.) E through an agric. area to join Missouri R. 15 m. (24 km.) S of Omaha. Un-navigable, but used for irrigation and power.

Plauen Karl-Marx-Stadt, German Democratic Republic. 50 30N 12 08E. Town in central S sit. on Elster R. Industries inc. textiles; manu. textile machinery, machine tools and paper. Pop. (1989E) 77,200.

Plenty, Bay of North Island, New Zealand. 37 45S 177 00E. A wide inlet on NE coast containing several small islands.

Pleven Pleven, Bulgaria. 43 26N 24 37E. Cap. of Pleven province sit. NE of Sofia. Trading centre for wine and cattle. Industries inc. heavy engineering, lorries, oils, food processing. Manu. clothing, cement, wire and tobacco products. Pop. (1987E) 133,747.

Plock Warsaw, Poland. 52 33N 19 43E. Town in N central Poland sit. on Vistula R. Industries inc. agric. engineering, flour milling and fruit canning.

Ploeşti Prahova, Romania. 44 56N 26 02E. Cap. of district sit. N of Bucharest. Centre of the Romanian oil industry with pipelines to Giurgia, Bucharest and Constanta. Other industries inc. textiles and chemicals; manu. hardware, cardboard and leather. Pop. (1987E) 234,021.

Plovdiv Plovdiv, Bulgaria. 42 09N 24 45E. Cap. of province in S centre sit. on Maritsa R. Commercial centre trading in wheat, tobacco and attar of roses. Industries inc. chemicals, textiles, flour milling and tanning; manu. tobacco products. Pop. (1987E) 356,596.

Plymouth Devon, England. 50 23N 4 10W. Town at head of Plymouth Sound in extreme SW of county. Industries inc. engineering, boatbuilding, brewing, chemicals and food processing. Important as a naval base. Seaport. Pop. (1981C) 243,895.

Plymouth Montserrat. 16 42N 62

13W. Cap. and chief port. Pop. (1990E) 3,500.

Plymouth Massachusetts, U.S.A. 41 58N 70 41W. Town sit. on Plymouth Bay. Industries inc. fishing, light manu. and tourism. The first permanent colony in New England. Pop. (1990E) 44,000.

Plymouth Sound England. 50 20N 4 10W. Inlet of the English Channel, 3 m. (5 km.) wide at the entrance, extending inland to Plymouth and Devonport to E and Saltash to W between Devon and Cornwall. Sheltered roadstead, receiving R. Tamar and Plym.

Plynlimmon Wales. 52 29N 3 47W. Mountain 2,468 ft. (752 metres) a.s.l. sit. ENE of Aberystwyth on the border between Dyfed and Powys counties. The source of Rs. Wye and Severn

Plzeň Západočeský, Czechoslovakia. 49 45N 13 23E. Cap. of region in W sit. at confluence of Rs. Radbuza and Mze. Commercial centre of an agric. area, trading in cereals and cattle. The main industries are brewing and manu. armaments, cars, locomotives and machinery. Pop. (1984E) 174,000.

Pobé Benin. Town sit. N of Porto-Novo. Rail terminus for an agric. area with agric. research institute and a meteorological station.

Pobeda Peak Kirghiztan, U.S.S.R. Highest peak of the Tien Shan Mountains in the E Kirghiz Soviet Socialist Rep., 24,406 ft. (7,438 metres) a.s.l.

Pocatello Idaho, U.S.A. 42 52N 112 27W. City sit. N of Salt Lake City, Utah. Industries inc. railway engineering, microchips, cheese processing and natural gas production. Pop. (1980C) 46,420.

Pocklington Humberside, England. 53 56N 0 48W. Town sit. ESE of York. Market town for an agric. area. Pop. (1987E) 6,040.

Podgorika   *see* **Titograd**

Podolia Ukraine, U.S.S.R. Region

between R. Dniester to W and S Bug to E. Chief towns Vinnitsa and Kamenets-Podolski.

Podolsk Russian Soviet Federal Socialist Republic, U.S.S.R. 55 26N 37 33E. Town S of Moscow sit. on Pakhra R. Industries inc. railway engineering; manu. oil refining plant, cables, sewing machines, lime and cement. Pop. (1987E) 209,000.

Podor Senegal. Town sit. ENE of Saint Louis on Senegal R. R. port serving an agric. area, with some fishing.

Podrinje Bosnia, Yugoslavia. Plain extending 30 m. (48 km.) N to S from Loznica and partly bounded E by Drina R. Cattle raising and fruit-growing area, with lead mining near Krupanj.

Pogradec Albania. 40 54N 20 40E. Town sit. NNW of Koritsa on SW shore of L. Ochrida. Trading centre in a fruit-growing area. Industries inc. fishing.

Point Barrow   *see* **Barrow, Point**

Pointe-à-Pitre Grande Terre, Guadeloupe. 16 14N 61 32W. Principal town and chief port of department on SW coast of Grande Terre Island. Industries inc. sugar milling and distilling rum. Sea port exporting bananas, sugar cane, rum and coffee. Pop. (1982E) 53,000.

Pointe-Noire Congo. 4 48S 11 51E. Town WSW of Brazzaville sit. on the Atlantic coast. Major airport. Manu. aluminium goods. Port exporting palm products, cotton, rubber and timber, and handling trade for Congo and Gabon. Pop. (1980E) 185,105.

Poitiers Vienne, France. 46 35N 0 20E. Cap. of dept. in central W sit. at confluence of Rs. Clain and Boivre. Market town trading in wine and wool. Industries inc. chemicals, wire, engineering and tyres. Home of the 'Futuroscope'. Pop. (1982C) 82,884 (agglomeration, 103,804).

Poitou-Charentes France. 46 25N 0 15W. Region of W France comprising the depts. of Deux-Sèvres and Vienne Charente, Charente-Maritime. Area, 9,957 sq. m. (25,790 sq. km.). Pop. (1982C) 1,568,230.

Poland   52 00N 20 00E. Rep. bounded N by the Baltic Sea, NE and E by the U.S.S.R., S by Czech. and W by the Federal Republic of Germany and the German Democratic Republic. Area, 120,727 sq. m. (312,683 sq. km.) In 1966 Poland celebrated its millennium, but modern Polish history begins with the partitions of the once-powerful kingdom between Russia, Austria and Prussia in 1772, 1793 and 1795. After the creation by Napoleon I of a semi-independent Grand Duchy of Warsaw, the country was again partitioned at the Congress of Vienna in 1815 between Russia (Congress Poland), Austria and Prussia (Grand Duchy of Posen), and the free city of Crakow. The Polish revolution of 1830–1 caused the suppression of the 1815 constitution and made 'Congress Poland' virtually a Russian province. The revolution of 1846–8 led to the incorporation of Crakow in Austria, the abolition of the Grand Duchy of Posen and further repression in 'Congress Poland', which was intensified after the revolution of 1863–4. During the First World War Russian Poland was occupied by the Austro-German forces. In 1939 Germany invaded Poland and after the German attack on Russia, the Germans occupied the whole of Poland. Chief cities Warsaw (cap.), Bydgoszcz, Wroclaw, Crakow, Lódz and Lublin. Mainly plain with glacial Ls. in N. Plateau in S rises to mountains along the Czech. border, with the Sudeten Mountains and Carpathian Mountains rising to 8,000 ft. (2,438 metres) a.s.l. Drained by Rs. Vistula, Oder and Neisse. Continental climate. Chief agric. crops include, wheat, rye, barley, oats, potatoes and sugar-beet. Poland is a leading copper producer. Other minerals inc. coal, iron ore, lead and zinc. Industrial production inc. coke, pig-iron, crude steel, rolled steel, cement, sulphuric acid, fertilizers, aluminium, ships and cars. Output of light industry is increasing. Pop. (1988E) 37,775,000.

Polotsk   Belorussia, U.S.S.R. 55 31N 28 46E. Town NNE of Minsk sit.

on W. Dvina R. Industries inc. sawmilling, oil refining and flour milling.

Polperro   Cornwall, England. 50 19N 4 31W. Fishing village and resort sit. WSW of Plymouth on S coast. Pop. (1990E) 1,675.

Poltava   Ukraine, U.S.S.R. 49 35N 34 34E. Town ESE of Kiev sit. on Vorskia R. Commercial centre of an agric. area. Industries inc. textiles, brewing, meat packing, tanning and flour milling. Pop. (1987E) 309,000.

Poltoratsk   *see* **Ashkhabad**

Polyghyros   Khalkidiki, Greece. Town sit. SE of Salonika. *Nome* cap. Trading centre for wheat, olive oil, wine and timber. Pop. (1981C) 4,075.

Polynesia   Pacific Ocean. 10 00S 162 00W. One of the three main divisions of islands in the central and SE Pacific, lying within the three points New Zealand/Hawaii/Easter Island and inc. Hawaii, Samoa, Tonga, Tokelau, Tubuai, Tuamotu, Society, Marquesas, Cook, Tuvalu, Kiribati and Easter Islands.

Pomerania   *see* **Pomorze**

Pomona   California, U.S.A. 39 04N 117 45W. City sit. E of Los Angeles. Commercial centre for a fruit-growing and agric. area, residential and industrial centre of Pomona Valley. Pop. (1980C) 102,257.

Pomorze   German Democratic Republic/Poland. Region extending W from lower Vistula R. across NW Poland into G.D.R., inc. Rügen Island and Stralsund. Bounded N by the Baltic. Low-lying agric. area, with main products of rye, potatoes and oats.

Ponape   Micronesia, West Pacific. 6 55N 158 15E. Volcanic island group with outlying atolls, forming state within the Federated States of Micronesia. Area, 163 sq. m. (422 sq. km.). Volcanic, rising to Mount Tolocolme, 2,579 ft. (786 metres) a.s.l. Cap. Colonia. Pop. (1980C) 22,319 (district).

Ponce   Puerto Rico. 18 01N 66 37W. City WSW of San Juan sit. on S coast. Industries inc. textiles, sugar refining, distilling rum, brewing and fruit canning. Port exporting sugar, tobacco, textiles and rum. Pop. (1980c) 189,046.

Pondicherry   India. 11 59N 79 50E. (i) Union Territory on the Coromandel coast of SE India. Area, 190 sq. m. (492 sq. km.). The main occupation is rice growing; other crops are sugar and groundnuts. Pop. (1981c) 604,471. (ii) Cap. of Territory sit. on the Coromandel coast. Commercial centre. Industries inc. sugar, paper, soda, ceramics, textiles, esp. cotton. Port exporting groundnuts. Pop. (1981c) 604,471.

Ponta Delgada   São Miguel, Azores. City on S coast of São Miguel Island. District cap. and largest city of the Azores. Port with good artificial harbour. Commercial centre trading in fruit, tea, wine, cereals, vegetables and dairy produce. Industries inc. sugar refining, distilling, processing foodstuffs. Transatlantic cable station.

Ponta Grossa   Paraná, Brazil. 25 05S 50 09W. Town in SE of state, in SE Braz. Commercial centre of a farming area, trading in timber and *maté*. Industries inc. meat packing and saw-milling. Pop. (1984E) 187,000.

Pontchartrain, Lake   Louisiana, U.S.A. 30 10N 90 10W. L. N of New Orleans, 40 m. (64 km.) long and 25 m. (40 km.) wide. Linked by canal with Mississippi R.

Pontefract   West Yorkshire, England. 53 42N 1 19W. Town sit. SE of Leeds. Market town in a coal mining area. Industries inc. brewing, tanning and making liquorice sweets (Pontefract cakes). Pop. (1981c) 31,971.

Pontevedra   Spain. 42 26N 8 38W. (i) Province in NW Spain on the Atlantic coast, bounded S by Portugal across Miño R. Area, 1,729 sq. m. (4,477 sq. km.). Mountainous. The chief occupations are agric. and fishing. Pop. (1986c) 884,408. (ii) Cap. of province of same name sit. at head of Pontevedra Bay. Commercial centre trading in wine, livestock, fruit and cereals. Industries inc. fishing and

boatbuilding; manu. leather and pottery.

Pontiac   Michigan, U.S.A. 42 37N 83 18W. City NW of Detroit sit. on Clinton R. Important centre of the car manu. industry, and also manu. heavy vehicles, rubber goods and paint. Pop. (1980c) 76,715.

Pontianak   West Kalimantan, Indonesia. 0 02S 109 20E. Cap. of province on W coast of Borneo sit. near mouth of Kapuas R. Sea port exporting timber and copra. The main industries are shipbuilding and the processing of palm oil, rubber and sugar. Pop. (1980E) 218,000.

Pontine Marshes   Lazio, Italy. Reclaimed marsh-land extending NW to SE along the Tyrrhenian coast from Cisterna di Latina to Terracina. Main towns Latina and Sabaudia. The main occupation is farming, esp. cereals, sugar-beet, vines, fruits and vegetables.

Pontoise   Val-d'Oise, France. 49 03N 2 06E. Cap. of dept. sit. NW of Paris. Pop. (1982c) 29,411.

Pontresina   Graubünden, Switzerland. 46 28N 9 53E. Town and resort E of St. Moritz sit. at 5,915 ft. (1,802 metres) a.s.l. in the Upper Engadine. Pop. (1980c) 1,700.

Pontypool   Gwent, Wales. 51 43N 3 02W. Town sit. NNW of Newport. Main industries are pharmaceuticals, motor brake and suspension components and nylon. Pop. (1981c) 36,761.

Pontypridd   Mid Glamorgan, Wales. 51 39N 3 22W. Town NW of Cardiff sit. on Taff R. at confluence with Rhondda R. Industries inc. engineering, light industry, electronics and chemicals. Pop. (1990E) 33,600.

Poole   Dorset, England. 50 43N 1 59W. Town on S coast sit. on Poole Harbour, an inlet of the English Channel. Industries inc. engineering, commerce, boatbuilding, ferry and cargo port and tourism. Pop. (1989E) 133,000.

Poona   *see* **Pene**

Poopó, Lake Oruro, Bolivia. 18 45S 67 07W. L. in W central Bolivia at 12,100 ft. (3,688 metres) a.s.l. on the Altiplano. Area, 980 sq. m. (2,538 sq. km.). Fed by Desaguadero R. from L. Titicaca.

Popayán Cauca, Colombia. 2 27N 76 36W. Town in SW sit. at the foot of the volcano Puracé. Commercial centre of a coffee growing area. Industries inc. flour milling and tanning. Pop. (1980E) 93,000.

Poplar Greater London, England. 51 30N 0 01W. District of E London on N bank of Thames R. *See* Tower Hamlets.

Popocatépetl, Mount Mexico. 19 02N 98 38W. Dormant volcano in central Mex., the second highest Mexican peak, 17,887 ft. (5,452 metres) a.s.l. The last eruption was in 1920.

Popondetta Papua New Guinea. 8 46S 148 14E. Town in Papua sit. N of Higaturu with an agric. station and airfield. Pop. (1975E) 5,000.

Porbandar Gujarat, India. 21 38N 69 36E. Town SW of Ahmedabad sit. on Kathiawar peninsula. Birthplace of Mahatma Gandhi. Industries inc. textiles; manu. cement. Port exporting cotton and salt. Pop. (1981E) 115,182.

Pori Turku-Pori, Finland. 61 29N 21 47E. Town in SW sit. on Kokemäki R. near its mouth on the Gulf of Bothnia. Industries inc. copper refining, oil rigs and offshore products; manu. wood pulp, paper, metal goods and titanic oxide. Port exporting timber and timber products. Pop. (1985C) 79,000.

Po River Italy. 44 57N 12 04E. Rises on Mount Viso in the Cottian Alps and flows 417 m. (667 km.) E and NE past Turin and Chivasso, then E past Piacenza and Cremona to enter the Adriatic Sea by a delta. It drains *c.* 17,000 sq. m. (*c.* 44,000 sq. km.) of the plain between the Alps and the Apennines. In its lower course its bed is higher than the surrounding country; it is artificially reinforced and the waters used for irrigation.

Porsgrunn Telemark, Norway. 59 09N 9 40E. Town sit. SW of Oslo. Industries inc. manu. fertilizers and electrical equipment. Pop. (1985E) 32,000.

Port Adelaide South Australia, Australia. 34 51S 138 30E. Town sit. NW of Adelaide on Gulf St. Vincent. Industries inc. chemicals; manu. cement. Chief sea port of S. Australia, exporting wheat, barley and wool. Pop. (1985E) 37,000.

Portadown Armagh, Northern Ireland. 54 26N 6 27W. City, now part of Craigavon, in S centre sit. on Bann R. Industries inc. manu. linen, clothing, electronics, furniture, light industry and carpets. Pop. (1990E) 62,000 (Craigavon City).

Portage La Prairie Manitoba, Canada. 49 57N 98 25W. City W of Winnipeg sit. on Assiniboine R. Industries inc. food processing and agric. Pop. (1983C) 13,086.

Port Alberni British Columbia, Canada. 49 14N 124 48W. Town sit. NW of Victoria on Vancouver Island. Industries inc. fishing and lumbering. Port exporting timber. Pop. (1990E) 18,500.

Portalegre Portalegre, Portugal. 39 15N 7 40W. Cap. of district sit. ENE of Lisbon on W slopes of the Serra de São Mamede, near the Spanish frontier. Trading centre for agric. produce. Industries inc. textiles, cork processing.

Portarlington Laoighis, Ireland. 53 10N 7 10W. Town on R. Barrow where it forms the border with Offaly. Market town in an agric. area.

Port Arthur *see* **Thunder Bay**

Port Arthur Texas, U.S.A. 29 55N 93 55W. City sit. on W shore of L. Sabine SE of Beaumont. Industries inc. oil refining; manu. chemicals and rubber products. Port exporting petroleum, grain and timber. Pop. (1980C) 61,251.

Port Augusta South Australia, Australia. 32 30S 137 46E. City at head of

Spencer Gulf. Industries inc. tourism and engineering. Port exporting wool and wheat. Pop. (1985E) 17,000.

Port-au-Prince   Haiti. 18 40N 72 20W. Cap. of rep. and main port founded by the French in 1749, sit. on the Gulf of Gonaïves on w coast. Industries inc. textiles, distilling, brewing and sugar refining. Pop. (1988E) 1,153,626.

Port Blair   Andaman Islands, India. Town sit. sw of Rangoon (Burma) on s Andaman Island. Cap. of the Andaman and Nicobar Islands Territory of India. Port exporting timber, coconuts, copra. Industries inc. sawmilling, boatbuilding, fishing, food processing. Pop. (1981C) 49,634.

Port Buet   Côte d'Ivoire. 5 15N 3 58W. Town opposite Abidjan on Atlantic coast across Ebrié Lagoon. Railway terminus and port. Site of Abidjan's international airport.

Port-de-Paix   Nord-Ouest, Haiti. 19 57N 72 50W. Town sit. wnw of Cap-Haïtien on Atlantic coast opposite Tortuga Island. Port shipping agric. produce, hides, timber. Industries inc. fishing. Pop. (1980E) 21,500.

Port Dickson   Negri Sembilan, Peninsular Malaysia. 2 32N 101 48E. Town sit. sw of Seremban on the Strait of Malacca. Port shipping rubber, copra, tin. Resort. Pop. (1980E) 24,000.

Port Elizabeth   Cape Province, Republic of South Africa. 33 58S 25 40E. City on s coast sit. on w shore of Algoa Bay. Industries inc. motor-car assembly and components, fruit and jam canning, and flour milling; manu. tyres, leather, diesel locomotives, confectionery, clothing and furniture. Railway junction and sea port exporting fruit, hides, skins, mohair, steel and ore, and wool. Pop. (1990E) 700,000.

Port Francqui   *see* **Ilebo**

Port-Gentil   Ogooué-Maritime, Gabon. 0 43S 8 47E. Town sit. ssw of Libreville on an island between mouths of Ogooué R. Port shipping

crude oil, timber, cacao. Industries inc. sawmilling, fishing, fish processing. Pop. (1983E) 123,300.

Port Glasgow   Strathclyde Region, Scotland. 55 56N 4 14W. Town ese of Greenock sit. on Clyde R. estuary. Industries inc. shipbuilding and engineering; manu. rope, twine and canvas. Pop. (1981C) 21,554.

Port Harcourt   Rivers, Nigeria. 4 46N 7 01E. Town in se sit. on Bonny R., an arm of the Niger delta. Industries inc. manu. cement, tyres and tobacco products. Port exporting palm products, groundnuts, cacao, tin and coal. Pop. (1983E) 296,200.

Porthcawl   Mid Glamorgan, Wales. 51 28N 3 42W. Town sit. se of Swansea on Bristol Channel. Port and seaside resort. Pop. (1989E) 17,000.

Porthmadog   Gwynedd, Wales. 52 55N 4 13W. Town sse of Caernarvon sit. on Tremadoc Bay. Market town for an agric. and quarrying area. Industries inc. tourism. Port exporting slate. Pop. (1990E) 4,000.

Port Huron   Michigan, U.S.A. 42 58N 82 27W. City and port ne of Detroit sit. on St. Clair R. between L. Huron and L. Erie. Industries inc. tourism; manu. paper, vehicle parts and metal goods. Pop. (1980C) 33,981.

Portici   Campania, Italy. 40 49N 14 20E. Town se of Naples sit. on Bay of Naples. Industries inc. fishing, tanning and silk weaving.

Portimão   Faro, Portugal. 37 08N 8 32W. Town sit. wnw of Faro at mouth of the Arade R. estuary on s coast. Important fishing and fish canning centre, esp. sardine, tuna. Pop. (1980C) 19,600.

Portishead   Avon, England. 51 30N 2 46W. Town sit. wnw of Bristol on Severn R. estuary. Resort and sea port with docks forming part of the Port of Bristol. Pop. (1990E) 12,000.

Port Jackson   New South Wales, Australia. 33 51S 151 15E. Deep inundated valley in Sydney forming the city's harbour. Crossed by Sydney Harbour bridge, 1,650 ft. (503 metres) long.

Port Kelang   Selangor, Malaysia. 3 02N 101 26E. Formerly Port Swettenham. Port sit. on Strait of Malacca sw of Kuala Lumpur mainly handling rubber. Pop. (1980c) 192,080.

Port Kembla   New South Wales, Australia. 34 29S 150 56E. Town sit. ssw of Sydney forming part of Greater Wollongong. Industries inc. iron and steel, and copper refining; manu. tin plate, wire and cables, and fertilizers.

Portland   Maine, U.S.A. 43 39N 70 17W. City in sw of state sit. on a peninsula on Casco Bay. Industries inc. manu. paper, furniture and footwear. Port exporting timber and grain. Pop. (1987E) 66,337.

Portland   Oregon, U.S.A. 45 33N 122 36W. City in NW of state sit. on Willamette R. near its confluence with Columbia R. Industries inc. manu. of instruments, machinery, electrical equipment, food, metals, heavy vehicles, lumber, wood products and clothing. Port exporting timber, barley and aluminium. Pop. (1989E) 432,175.

Portland, Isle of   Dorset, England. 50 33N 2 27W. Rocky peninsula on Dorset coast, extending s into the English Channel and connected to the mainland by a ridge of shingle known as Chesil Beach. Area, 4½ sq. m. (12 sq. km.). Mainly limestone which is quarried for building. Main town Portland Harbour, a naval base.

Portlaoighise   Laoighis, Ireland. 53 02N 7 17W. County town in w centre of the Rep. Industries inc. peat, tennis balls, rubber, paper, provender milling, concrete products and oil recycling. Pop. (1990E) 9,500.

Port Lincoln   South Australia, Australia. 34 44S 135 52E. City and port s of Port Augusta sit. at the s tip of the Eyre Peninsula on Boston Bay. Port exports cereal grain, inc. wheat, barley and oats. Industries inc. fishing and tourism. Pop. (1986E) 13,100.

Port Louis   Mauritius. 20 18S 57 31E. Cap. and chief port of Mauritius founded in 1736 by the French and named after King Louis XV, it is sit. on NW coast. Industries inc. tourism, sugar, railway engineering, diamond polishing, ship repair; manu. clothing, knitwear, plastics, footwear, jewellery, furniture and food processing. Exports sugar. Pop. (1987E) 139,038.

Portmadoc   *see* **Porthmadog**

Port Moresby   Papua New Guinea. 9 24S 147 08E. Cap. on s coast, sit. on the Coral Sea. Sea port exporting coffee, timber, copra and rubber. Pop. (1989c) 152,100.

Porto   Porto, Portugal. Port and second largest town sit. on Douro R. near the mouth. Noted for port wine which is exported to UK, France, Italy and the Netherlands. Other exports inc. fruit and cork. Manu. inc. textiles, pottery, glass, paper, food and tobacco products. There are many interesting buildings inc. the granite Torre dos Clerigos, 246 ft. (75 metres) high. Pop. (1990E) 500,000.

Pôrto Alegre   Rio Grande do Sul, Brazil. 30 04S 51 11W. State cap. in SE. Commercial centre. Industries inc. meat packing, tanning, brewing, textiles and chemicals. Sea port exporting meat products, hides and wool. Pop. (1980c) 1,114,867.

Port of Spain   Trinidad and Tobago. 10 40N 61 31W. Cap. of Trinidad and Tobago sit. on Gulf of Paria. Industries inc. sawmilling, food processing, printing, furniture, brewing, iron and steel and distilling; manu. cement and cigarettes. Assembling automobiles, furniture and household appliances. Sea port exporting asphalt, petroleum products, sugar, rum, molasses, animal feed, electrical fittings and cacao. Pop. (1989E) 59,200.

Porto Novo   Ouémé, Benin. 6 23N 2 42E. Cap. of rep. in E sit. on N shore of a coastal lagoon. The Portuguese used it as a base for the slave trade with Brazil, becoming a French Protectorate in 1863. Port exporting palm products, cotton and kapok. Pop. (1982E) 208,258.

Pôrto Velho Rondônia, Brazil. 8 46S 63 54W. Cap. of federal territory in W sit. at head of navigation on Madeira R. Trading centre for rubber, timber and medicinal plants. Pop. (1984E) 150,000.

Portoviejo Manabí, Ecuador. 1 02S 80 25W. Town sit. NNW of Guayaquil. Market town trading in coffee and cacao; manu. baskets and panama hats. Pop. (1982C) 167,070.

Port Phillip Bay Victoria, Australia. 38 07S 144 48E. Large inlet off Bass Strait in S Victoria, 35 m. (56 km.) N to S and 40 m. (64 km.) W to E, with Melbourne sit. at its head.

Port Pirie South Australia, Australia. 33 11S 138 01E. Town in SE of state sit. on Spencer Gulf, N of Adelaide. Industries inc. smelting metals from Broken Hill; manu. chemicals. Port exporting wheat and barley grain. Pop. (1985E) 16,030.

Portree Highland Region, Scotland. 57 25N 6 11W. Town on Island of Skye sit. NW of Kyle of Lochalsh, on the mainland. Fishing port and resort. Pop. (1981C) 1,505.

Portrush County Antrim, Northern Ireland. 55 13N 6 40W. Town sit. on central N coast. Tourist resort esp. for the Giant's Causeway basalt formation. Pop. (1981C) 5,114.

Port Said Egypt. 31 16N 32 18E. City at Mediterranean entrance to Suez Canal, sit. on a strip of land between L. Manzala and the sea. Industries inc. chemicals and salt panning; manu. tobacco products. Fuelling station and commercial centre. Port, exporting cotton. Pop. (1986E) 382,000.

Portslade-by-Sea East Sussex, England. 50 50N 0 13W. Town immediately W of Hove sit. on the English Channel. Industries inc. manu. electrical equipment. Pop. (1982E) 89,000 (borough of Hove).

Portsmouth Hampshire, England. 50 48N 1 05W. City in SE Hampshire at entrance to Portsmouth Harbour. Industries inc. computers and electronics. Major naval base. Pop. (1987E) 179,400.

Portsmouth New Hampshire, U.S.A. 43 04N 70 46W. City ENE of Manchester sit. at mouth of Piscataqua R. Industries inc. tourism; manu. clothing, machinery and instruments. The state's only sea port, and a naval base. Pop. (1980C) 26,214.

Portsmouth Ohio, U.S.A. 38 45N 82 59W. City SE of Cincinnati sit. on Ohio R. where it forms the border with Kentucky. Industries inc. railway engineering, and iron and steel; manu. footwear and furniture. Pop. (1980C) 25,943.

Portsmouth Virginia, U.S.A. 36 52N 76 24W. City in SE of state sit. on Elizabeth R. Industries inc. food processing, railway engineering, chemicals and fishing; manu. cottonseed oil, fertilizers and hosiery. Sea port exporting cotton and tobacco. Important naval dockyard. Pop. (1989E) 111,731.

Port Stewart County Londonderry, Northern Ireland. Seaside resort on N coast, NE of Londonderry. Pop. (1981C) 5,312.

Port Swettenham *see* **Port Klang**

Port Talbot West Glamorgan, Wales. 51 36N 3 47W. Town ESE of Swansea sit. on Swansea Bay. Industries inc. steel, electronics, engineering and chemicals. Deep-water harbour serving steel works. Pop. (1988E) 49,000.

Portugal 40 00N 7 00W. Rep. occupying W part of the Iberian peninsula, bounded N and E by Spain, W and S by the Atlantic. Area, 35,561 sq. m. (91,985 sq. km.). Portugal has been an independent state since the 12th cent.; until 1910 it was a monarchy; In 1910 the Republic was proclaimed. Chief cities, Lisbon (cap.), and Oporto. Chief agric. crops are wheat, maize, oats, barley, rye, rice, beans, potatoes. Wine production, particularly port, is important to the economy. The forests produce pine, cork oak, oak, chestnut, eucalyptus. Portugal surpasses the rest of the world in the production of cork. Other forestry products inc. resin and turpentine. Main indust. inc. food processing, textiles, clothing, pottery, metalworking, ship repairing. Pop. (1987E) 10·27m.

Portuguesa Venezuela. State of w Venezuela. Area, 5,870 sq. m. (15,203 sq. km.). Chief towns Guanare (cap.), Acarigua. Lowland grasslands rising w into foothills of the Andes. Drained into Orinoco R. Humid tropical climate; extensive forest. Main products cattle, coffee, cotton, rice, maize, timber. Subject to frequent flooding by tribs. of Orinoco R. Pop. (1980E) 375,854.

Portuguese Guinea *see* **Guinea-Bissau**

Posadas Misiones, Argentina. 27 25S 55 50W. Cap. of province in NE sit. on Alto Paraná R. Industries inc. meat packing and flour milling. R. port handling rice, tobacco and *maté*. Pop. (1984E) 130,000.

Posets, Mount France/Spain. Mountain peak sit. w of Pico de Aneto in the Pyrénées on the French/Spanish frontier. Height 11,046 ft. (3,366 metres) a.s.l.

Posillipo Campania, Italy. Ridge extending SW from Naples through Campania. In anc. times the road from Naples to Pozzuoli went through one of its many tunnels, the Grotto of Posillipo.

Potchefstroom Transvaal, Republic of South Africa. 26 46S 27 01E. Town WSW of Johannesburg sit. on Mooi R. Centre of an agric. area.

Potenza Basilicata, Italy. 40 38N 15 49E. Cap. of region sit. E of Salerno on a hill above Basento R. Agric. centre and market town. Pop. (1981E) 64,000.

Poti Georgia, U.S.S.R. 42 09N 41 40E. Town N of Batumi sit. on Black Sea at mouth of Rion R. Industries inc. fish canning and ship repairing. Port exporting manganese from the Chiatura mines.

Potomac River U.S.A. 38 00N 76 18W. Formed by union of N. and S. Potomac R. rising in the Allegheny Mountains, flowing 287 m. (459 km.)

NE and then SE along the boundary between Maryland and W. Virginia, and then between Maryland and the District of Columbia in the NE and Virginia in the SW. Enters a long estuary below Washington, D.C. and thence into Chesapeake Bay. Navigable up to Washington, D.C. Main trib. Shenandoah R. The Great Falls are 15 m. (24 km.) above Washington, consisting of cascades and rapids in a 200 ft. (61 metres) gorge.

Potosí Potosí, Bolivia. 19 35S 65 45W. Cap. of dept. in S central Bolivia. Industries inc. mining, esp. tin; manu. footwear and furniture. Pop. (1982E) 103,183.

Potsdam German Democratic Republic. 52 24N 13 04E. (i) District formed from part of the *Land* of Brandenburg in 1952. Area, 4,849 sq. m. (12,560 sq. km.). (ii) Cap. of district of same name sit. on Havel R. Industries inc. engineering and chemicals; manu. precision instruments, soap and furniture. Pop. (1989E) 142,000.

Potteries, The England. District in upper Trent valley in NW Midlands extending *c.* 9 m. (14 km.) NW to SE and *c.* 3 m. (5 km.) across, inc. Stoke-on-Trent, Hanley, Burslem, Tunstall, Longton and Fenton. The centre of English china and earthenware manu. using local coal and clay, with china clay imported from Cornwall and Dorset, and finished wares trad. (from 1769) transported by canal.

Potters Bar England. 51 41N 0 11W. Small residential town forming a suburb of N London.

Poughkeepsie New York, U.S.A. 41 42N 73 56W. City N of New York sit. on Hudson R. Banking, service and retailing centre for an area manu. electronic equipment, ball bearings and precision instruments. Pop. (1980C) 29,757.

Poverty Bay North Island, New Zealand. Inlet on E coast with Gisborne on its N coast. Scene of Capt. Cook's landing, 1769.

Powell Wyoming, U.S.A. 44 45N 108 46W. City sit. NE of Cody at

4,390 ft. (1,338 metres) a.s.l. Trading centre for an irrigated agric. area, handling sugar-beet, sweet clover seed, alfalfa, potatoes. Industries inc. seed processing. Pop. (1980C) 5,310.

Powys   Wales. 52 20N 3 30W. County of E Wales bounded E by Eng. and formed by 1974 re-organization with the districts of Montgomery, Radnor and Brecknock. Area, 1,960 sq. m. (5,077 sq. km.). A picturesque area with chief Rs. Usk, Wye and Severn. The Ls. provide water for the cities of Liverpool and Birmingham. Tourism is important and Llandridnod Wells is the county town. Pop. (1989E) 115,000.

Poyang, Lake   Jiangxi, China. 29 00N 116 25E. L. in N of province connected by canal to Yangtse R. whose flood waters feed it in summer. Area, summer, 80 m. (128 km.) by 40 m. (64 km.). In winter it is considerably smaller. Fed also by Gan Jiang R.

Poznań   Poznań, Poland. 52 25N 16 55E. Cap. of province in central W sit. on Warta R. Industries inc. brewing and distilling; manu. machinery, railway rolling stock, boilers, bicycles, tyres, glass and chemicals. Pop. (1985E) 553,000.

Pozzuoli   Campania, Italy. 40 49N 14 07E. Town WSW of Naples sit. on N coast of the Bay of Naples. Industries inc. iron and steel. Port.

Prague   Czechoslovakia. 50 05N 14 22E. Cap. of rep. in NE sit. on Vltava R. Commercial and route centre. It was founded in the 9th cent. and became the cap. of Czech. in 1918 when the country was created. Industries inc. brewing, printing and publishing, and food processing; manu. cars, aircraft, clothing, paper and chemicals. Pop. (1989E) 1,211,000.

Prahova River   Romania. Rises SW of the Predeal Pass in the Transylvanian Alps and flows 80 m. (128 km.) S past Sinaia then SE past Campina to enter Ialomita R. SE of Ploesti. Its valley is an important corridor between the Transylvanian Alps and Moldavian Carpathians.

Praia   Cape Verde, Atlantic. 14 53N 23 30W. Cap. of rep. sit. W of Dakar (Senegal) on São Tiago Island. Port shipping oranges, sugar-cane, coffee, castor oil. Industries inc. fishing and fish curing, distilling; manu. straw hats. Airfield. Pop. (1980E) 36,676.

Prato   Toscana, Italy. 43 53N 11 06E. Town WNW of Florence sit. on Bisenzio R. The main industry is woollen textiles; also manu. textile machinery and furniture. Pop. (1989E) 165,889.

Přerov   Moravia, Czechoslovakia. 49 27N 17 27E. Town in central Czech. sit. on Bečva R. Industries inc. manu. machinery and textiles. Pop. (1984E) 54,000.

Prescot   Merseyside, England. 52 26N 2 48W. Town sit. E of Liverpool. Industries inc. manu. electric cables and tools. Pop. (1985E) 11,438.

Presidente Hayes   Paraguay. Dept. of central and W Paraguay bounded S and W by Argentina across Pilcomayo R. Area, 22,579 sq. m. (58,480 sq. km.). Cap. Villa Hayes. Low-lying, marshy pasture land with slow-moving Rs. and swamps. Fertile SE near Paraguay R. Main product sugar-cane, also maize, tobacco, cotton, alfalfa, livestock. Main industries are distilling, sugar refining. Pop. (1982E) 43,787.

Prešov   Slovakia, Czechoslovakia. 49 00N 21 15E. Town NE of Bratislava sit. on Torysa R. Market town. Industries inc. distilling; manu. linen. Pop. (1976E) 76,000.

Prespa, Lake   Yugoslavia. 40 55N 21 00E. L. mainly in S Yugoslavia with its SW part in Albania and its SE part in Greece. Area, 110 sq. m. (285 sq. km.). Drained by underground streams to L. Okhrida, just to the W.

Prestatyn   Clwyd, Wales. 53 20N 3 24W. Town ENE of Rhyl sit. on Irish Sea. Market town and resort. Pop. (1982E) 16,000.

Prestea   Ghana. 5 27N 2 08W. Town sit. NW of Tarkwa on Ankobra R. Gold mining centre.

Preston Lancashire, England. 53 46N 2 42W. Town in NW part of Central Lancashire New Town sit. on Ribble R. Industries inc. engineering; manu. machinery, electrical equipment, plastics and chemicals. Pop. (1990E) 125,800.

Prestonpans Lothian Region, Scotland. 55 57N 3 00W. Town E of Edinburgh sit. on Firth of Forth. Industries inc. oilskin manu., precision engineers, specialist fabrication and engineering. Pop. (1981C) 7,620.

Prestwich Greater Manchester, England. 53 32N 2 17W. Town sit. NNW of Manchester of which it forms a suburb. Industries inc. textiles. Pop. (1981E) 32,000.

Prestwick Strathclyde Region, Scotland. 55 30N 4 37W. Town in SW sit. on Firth of Clyde. Intern. airport which is a transatlantic terminal. Manu. light aircraft. Also a seaside and golfing resort. Pop. (1989E) 14,052.

Pretoria Transvaal, Republic of South Africa. 25 44S 28 12E. City in NE sit. at 4,600 ft. (1,402 metres) a.s.l., founded in 1855. Administrative and provincial cap. and a railway centre. Industries inc. railway engineering, brewing and chemicals; manu. automobiles, iron and steel, cement and leather. Pop. (1985C) 822,925.

Préveza Greece. 38 57N 20 44E. (i) *Nome* in Epirus bordering on the Ionian Sea. Area, 425 sq. m. (1,101 sq. km.). Mainly mountainous. The main occupations are fishing and farming, esp. cereals, olives and citrus fruits. Pop. (1981E) 55,915. (ii) Cap. of *nome* of same name at N side of entrance to the Gulf of Arta, and sit. NE of Patrai. Port exporting olives, olive oil and citrus fruits. Pop. (1981E) 12,662.

Pribilof Islands U.S.A. 57 00N 170 00W. Group of 4 islands in Bering Strait off SW Alaska. Area, 65 sq. m. (168 sq. km.). The main islands are St. Paul and St. George, no others being inhabited. Centre of the trade in seal fur.

Prijedor Bosnia, Yugoslavia. Town

sit. NW of Banja Luka on Sana R. and railway. Distribution centre for iron ore from Ljubija. The main industry is manu. wood pulp. Pop. (1981E) 108,868.

Primorye Territory Russian Soviet Federal Socialist Republic, U.S.S.R. Territory in SE Siberia sit. on Sea of Japan. Area, 65,000 sq. m. (168,350 sq. km.). Cap. Vladivostok. Mountainous in E with the Sikhote Alin Range, lowland in W, drained by Ussuri R. The main products are coal and timber.

Prince Albert Saskatchewan, Canada. 53 12N 104 46W. Town in S centre of province sit. on N. Saskatchewan R. Industries inc. mining, agric., forestry and tourism. Pop. (1985E) 32,800.

Prince Edward Island Canada. 46 20N 63 20W. Island in the Gulf of St. Lawrence forming the smallest Canadian province. Area, 2,184 sq. m. (5,656 sq. km.). It was first settled by the French, but was taken by the British in 1758. It was annexed to Nova Scotia in 1763, and constituted a separate colony in 1769. Prince Edward Island entered the Confederation in 1873. Cap. Charlottetown. Low-lying with a deeply indented coastline. The main occupations are dairy farming, cattle rearing and fishing. Pop. (1986E) 126,646.

Prince George British Columbia, Canada. 54 20N 130 11W. City in central S of province sit. N of Vancouver on Fraser R. at its confluence with Nechako R. Major distribution point. Industries inc. forestry, tourism, agric., pulp and paper, mining, forest-related manu., petro-chemicals. Pop. (1989E) 65,451.

Prince Rupert British Columbia, Canada. 54 19N 130 19W. City and tourist centre, near mouth of Skeena R., just S of the border with Alaska, U.S.A., sit. on an island with a bridge to the mainland. Trading centre and port serving a mining and lumbering area. Industries inc. fishing for halibut and salmon, fish canning and pulp milling. Pop. (1989E) 15,595.

Princeton New Jersey, U.S.A. 40

21N 74 40W. Town sit. NE of Trenton on Millstone R. Seat of Princeton University, with associated service and research industry. Pop. (1980C) 12,035.

Prince William Sound　Alaska, U.S.A. Sound of the Gulf of Alaska, lying N of Middleton Island, extending *c.* 120 m. (*c.* 190 km.) NW from Kayak Island.

Principe Island　São Tomé e Príncipe, Gulf of Guinea. Island, area 42 sq. m. (110 sq. km.), W of Gabon forming (with São Tomé Island) the state of São Tomé e Príncipe. Chief town Santo António. Volcanic, mountainous (S). Chief products cacao, coffee, coconuts. Pop. (1981C) 5,255.

Pripet River　U.S.S.R. 51 21N 30 09E. Rises in NW Ukraine and flows 500 m. (800 km.) ENE into Belorussia, then turns E and SE to join Dnieper R. above Kiev. Most of it is navigable and linked by canals to Rs. Bug and Neman.

Privas　Ardèche, France. 44 45N 4 37E. Cap. of dept. sit. NNW of Marseille. Pop. (1982C) 10,638.

Prokopyevsk　Russian Soviet Federal Socialist Republic, U.S.S.R. 53 53N 86 45E. Town in S central Siberia sit. WNW of Novokuznetsk and in the Kuznetsk basin. The main industry is coalmining. Pop. (1987E) 278,000.

Prome　Burma. 18 45N 95 30E. R. port sit. NNW of Rangoon on Irrawaddy R. Pop. (1983E) 148,123.

Prostějov　Moravia, Czechoslovakia. 40 29N 17 07E. Town in central Czech. Industries inc. brewing and textiles; manu. clothing and footwear.

Provence Côte d'Azur　France. 43 40N 5 45E. Region in SE France bounded E by Italy, S by the Mediterranean and W by the Rhône R. Inc. the depts. of Bouches-du-Rhône, Var, Alpes-de-Haute-Provence, Hautes-Alpes, Alpes-Maritimes and Vaucluse. Area, 12,137 sq. m. (31,435 sq. km.). Mountainous; dry and sunny with fertile R. valleys. The main pro-

ducts are vines, olives and mulberries. Chief towns: Marseille, Nice and Toulon. Pop. (1982C) 3,965,209.

Providence　Rhode Island, U.S.A. 41 50N 72 25W. State cap. sit. on Providence R. at head of Narragansett Bay. Industries inc. services, government, distribution. Educational centre. Manu. jewellery. Pop. (1980C) 156,804.

Provo　Utah, U.S.A. 40 14N 111 39W. City SSE of Salt Lake City sit. at 4,549 ft. (1,386 metres) a.s.l. Trading centre for an agric. and tourist area. Industries inc. computer hardware, hi-tech, aerospace and software. Pop. (1989E) 82,000.

Prudhoe Bay　Alaska, U.S.A. 70 18N 148 19W. Bay of N coast in Beaufort Sea, with important arctic oil field; small town with oil pipeline terminal opposite Jones Island.

Prut River　Romania, U.S.S.R. 45 28N 28 12E. Rises in Carpathian Mountains in SW Ukraine and flows 530 m. (848 km.) N, then E past Kolomyya and Chernovtsy, then SSE forming the Romania/U.S.S.R. frontier, to join Danube R. 8 m. (13 km.) E of Galati, Romania.

Przemyśl　Rzeszow, Poland. 49 47N 22 47E. Town in SE Poland sit. on San R. near the Ukrainian frontier, and S of Lublin. Industries inc. flour milling, saw-milling and tanning; manu. machinery. Pop. (1983E) 64,100.

Przhevalsk　Kirghiz Soviet Socialist Republic, U.S.S.R. 42 29N 78 24E. Town sit. E of Frunze near SE shore of L. Issyk-kul. Food processing centre in a wheat growing area; manu. wines, beverages, machinery, furniture.

Pskov　Russian Soviet Federal Socialist Republic, U.S.S.R. 57 50N 28 20E. Town in W sit. on Velikaya R. Cap. of Pskov region. Industries inc. manu. linen from local flax, rope, leather and agric. machinery. Pop. (1987E) 202,000.

Pucallpa　Loreto, Peru. 8 20S 74

30W. Town NE of Lima sit. on
Ucayali R. Industries inc. sawmilling.
There is a pipeline (47 m., 76 km.)
from the Ganzo Azul oilfield. Pop.
(1980E) 90,000.

Pudsey   West Yorkshire, England.
53 48N 1 40W. Town sit. E of Bradford. Industries inc. woollen textiles,
engineering and tanning. Pop. (1981C)
39,952.

Puebla   Mexico. 19 03N 98 12W.
(i) State on central plateau. Area,
13,090 sq. m. (33,902 sq. km.).
Mountainous, rising to the three highest peaks in Mex., Pico de Orizaba,
18,700 ft. (5,700 metres) a.s.l., Popocatépetl, 17,887 ft. (5,452 metres)
a.s.l. and Ixtaccihuatl, 17,342 ft.
(5,286 metres) a.s.l. An agric. area.
The main products are maize, wheat,
sugar-cane, tobacco, gold, silver and
copper. Pop. (1989E) 4,139,609. (ii)
Cap. of state of same name sit. at
7,100 ft. (2,164 metres) a.s.l. Industries inc. textiles, cement, glazed tiles
and pottery. Pop. (1980C) 835,759.

Pueblo   Colorado, U.S.A. 38 16N
104 37W. City in S centre of state sit.
near Arkansas R. Industries inc. steel
manu. and high-tech. Pop. (1990E)
103,949.

Puerto Ayacucho   Amazonas, Venezuela. 5 40N 67 35W. Town and cap.
of Amazonas territory, sit. S of
Caracas on Orinoco R. where it forms
the Colombia border. Trading centre
for a forest region, handling balata,
rubber. Pop. (1981E) 15,000.

Puerto Barrios   Izabal, Guatemala.
15 43N 88 36W. Town NE of Guatemala City sit. on Caribbean Sea.
Industries inc. oil refining. Port exporting coffee and bananas. Pop.
(1989E) 338,000.

Puerto Cabello   Carabobo, Venezuela. 10 28N 68 01W. Town N of
Valencia sit. on Caribbean Sea. Industries inc. flour milling; manu. soap
and candles. Port exporting coffee,
cacao and hides. Pop. (1980E) 71,200.

Puerto de Santa Maria   Cádiz,
Spain. 36 36N 6 13W. Town on NW
coast sit. at mouth of Guadelete R. In-

dustries inc. tanning and fish canning;
manu. alcohol, liqueurs and soap. Port
with an important trade in sherry. Pop.
(1981E) 57,000.

Puerto Deseado   Patagonia, Argentina. 47 45S 66 00W. Town sit. SE of
Comodoro Rivadavia at mouth of
Deseado R. on the Atlantic. Railway
terminus and port serving a sheep
farming area. Exports wool and skin.
Industries inc. meat packing; manu.
furniture. Pop. (1981E) 3,750.

Puerto La Cruz   Anzoátegui, Venezuela. 10 13N 64 38W. Town NNE of
Barcelona, and sit. on the Caribbean.
The main industry is refining and exporting oil from the E Ilanos. Pop.
(1980E) 85,000.

Puerto Montt   Chile. 41 28S 72
57W. Town sit. in S central Chile.
Holiday resort for mountainous area
and sea port. Pop. (1982C) 81,353.

Puerto Plata   Dominican Republic.
19 40N 70 45W. (i) Province of N
bounded N by the Atlantic. Area, 809
sq. m. (2,095 sq. km.). Chief towns
Puerto   Plata,   Altamira,   Imbert,
Luperón. Fertile agric. area crossed by
the Cordillera Central. Main products
coffee, cacao, tobacco, sugar-cane,
rice, dairy cattle, maize, fruit. Pop.
(1980E) 186,112. (ii) Town sit. N of
Santiago. Officially called San Felipe
de Puerto Plata. Sea port serving an
agric. area. Processing and shipping
centre for tobacco, coffee, cacao, rice,
sugar-cane, bananas, hides, timber.
Manu. dairy products, matches, lard,
pasta, essential oils. Pop. (1980E)
45,348.

Puerto Rico   West Indies. 18 15N
66 30W. The most easterly of the
Greater Antilles, forming a territory of
the U.S.A. Area, 3,435 sq. m. (8,897
sq. km.). Puerto Rico, by the treaty of
1898, was ceded by Spain to the U.S.
The name was changed from Porto
Rico to Puerto Rico by an Act of Congress approved in 1932. Cap. San
Juan, other towns, Bayamón, Ponce,
Carolina, Caguas, Mayaguez. Mountainous, rising to over 4,000 ft. (1,219
metres) a.s.l. Main products, sugar-
cane, pineapples, citrus fruits, coffee
and tobacco. Pop. (1980C) 3,196,520.

Puerto Varas   Chile. 41 19S 72 59W. Town N of Puerto Montt in S central Chile, sit. on shore of L. Llanquihue. Tourist centre. Pop. (1980E) 35,000.

Puget Sound   Washington, U.S.A. 47 50N 122 30W. Inlet of the Pacific Ocean in W of state, extending 100 m. (160 km.) S from the Juan de Fuca Strait to Olympia, 28 m. (45 km.) WSW of Tacoma.

Puglia   Italy. 41 00N 16 30E. Region in S Italy on the Adriatic Sea. Area, 7,470 sq. m. (19,347 sq. km.). The N consists of a plain on which sheep, cattle and horses are raised. In S there is a limestone plateau where wheat, used in the manu. of macaroni, barley, maize and pulses are grown. Olive oil, wine and fruits are also produced. Marble is quarried in Monte Gargano. Chief ports; Bari (the cap.), Brindisi and Taranto. Pop. (1981C) 3,871,617.

Pukapuka   *see* **Danger Islands**

Pula   Croatia, Yugoslavia. 44 52N 13 50E. Town in NW sit. near S tip of the Istria peninsula. Port and tourist resort.

Pune   Maharashtra, India. 18 31N 73 54E. City in W central India sit. at 1,850 ft. (564 metres) a.s.l. in the Western Ghats. Commercial centre for grain and cotton. Industries inc. engineering, chemicals and textiles; manu. automobiles, soap and paper. Military centre. Pop. (1981C) 1,203,351.

Punjab   India. 31 00N 76 00E. State in NW bounded NE by Himachal Pradesh, SE by Haryana and S by Rajasthan. Area, 19,673 sq. m. (50,362 sq. km.). Cap. Chandigarh. Mainly agric. with textile industries concentrated mainly in Chandigarh, Amritsar (famous for its Golden Temple of the Sikh community), Jalandhar and Ludhiana. Other manu. inc. tractors, batteries, polyester fibres, scooters, sewing machines, sports goods, electronics, fertilizers, beer and engineering products. Pop. (1981C) 16,788,915.

Punjab   Pakistan. 30 00N 72 00E. Province in N bounded NE by Kashmir and E by Punjab, India. Area, 70,178 sq. m. (181,761 sq. km.). Cap. Lahore. The main products, from irrigated lands, are wheat, cotton, maize, rice and sugar-cane.

Puno   Peru. 15 50S 70 02W. (i) Dept. of SE Peru on the Bolivian border. Area, 27,947 sq. m. (72,382 sq. km.). Chief towns Huancané, Juliaca, Juli, Puno. Plateau between high mountain ranges, drained by Coata, Ramis and Inambari Rs. Main products grain, livestock, tobacco. Minerals inc. silver (at Lampa), gold (Poto and Santo Domingo), oil (Pirin). Pop. (1981C) 890,258. (ii) Town sit. ENE of Arequipa on NW shore of L. Titicaca. Dept. cap., communication and commercial centre for SE Peru. L. Port, railway terminus and customs point. Trade in wool and fur. Pop. (1981C) 45,348.

Punta Arenas   Chile. 53 09S 70 55W. Town sit. in extreme S on the Strait of Magellan, in a sheep farming area. Port exporting wool and meat. Pop. (1982C) 98,785.

Puntarenas   Puntarenas, Costa Rica. 9 58N 84 50W. Town W of San José sit. on Pacific coast. Industries inc. tuna fishing and soap making. Port exporting coffee and bananas. Pop. (1984E) 47,851.

Purbeck, Isle of   Dorset, England. 50 40N 2 05W. Peninsula in S Dorset extending 12 m. (19 km.) W to E into the English Channel along the S side of Poole Harbour. The main product is limestone. Chief towns are Swanage and Corfe Castle.

Puri   Orissa, India. 19 50N 85 58E. Town SW of Calcutta and sit. on Bay of Bengal. Noted for the temple of the God Jagannath (Juggernaut) whose image is dragged through the streets on a float during the annual festival. Pop. (1981C) 100,942.

Purmerend   Nord-Holland, Netherlands. Town sit. N of Amsterdam on the S edge of Beemster Polder. Market for a farming and agric. area. Industries inc. manu. refrigerators, wood products, food products, tiles. Pop. (1985E) 49,950.

Purús River   Brazil/Peru. 3 42S 61 28W. Rises in Peruvian Andes and flows 2,000 m. (3,200 km.) NE through NW Braz. to join Amazon R. WSW of Manáus, Brazil.

Pusan   South Korea. 35 06N 129 03E. City sit. on SE coast. Industries inc. railway engineering, shipbuilding, rice milling, salt refining and textiles. Port handling rice, fish and soybeans. Pop. (1983C) 3,395,171.

Putamayo River   Colombia/Peru/ Brazil. Rises in Colombian Andes and flows 980 m. (1,568 km.) ESE forming the boundary between Colombia and Peru, entering Braz. (as Içá R.) and joining Amazon R.

Puteaux   Hauts-de-Seine, France. Suburb of W Paris on Seine R. Industries inc. engineering and printing; manu. perfumes. Pop. (1982C) 36,143.

Putney   Greater London, England. 51 27N 0 13W. Area of SW London on S bank of Thames R. Putney Bridge is the starting point of the annual Oxford and Cambridge boat race.

Putumayo   Colombia. Intendency of SW Colombia bounded S by Peru and Ecuador. Area, 9,608 sq. m. (24,885 sq. km.). Cap. Mocoa. Thickly forested plains rising W to the Andes. Tropical climate. A mainly Indian area producing some rubber, resins, timber and medicinal plants, with some cattle farming on the Andean slopes.

Puy-de-Dôme   France. 45 46N 2 57E. Dept. in Auvergne region in northern part of the Massif Central. Area, 3,071 sq. m. (7,955 sq. km.). Chief towns, Clermont-Ferrand (the cap.) and Thiers. In the W the Auvergne Mountains rise to Puy de Sancy, 6,187 ft. (1,904 metres) a.s.l. In the E the Forez Mountains rise to 5,380 ft. (1,639 metres) a.s.l. The fertile plain of the Limagne in between is watered by Rs. Allier and Dore. The main occupations are viticulture and farming. Pop. (1982C) 594,365.

Puy, Le   Haute-Loire, France. City and cap. of dept. in S central France

sit. on Loire R. Industries inc. tourism, lace and liqueurs. Pop. (1982C) 25,968.

Puyo   Napo-Pastaza, Ecuador. Town sit. ESE of Ambato. Trading centre for the nearby Mera oilfield.

Pwllheli   Gwynedd, Wales. 52 53N 4 25W. Town in NW sit. on Cardigan Bay. Seaside resort. Pop. (1973E) 4,020.

Pyatigorsk   Russian Soviet Federal Socialist Republic, U.S.S.R. 44 03N 43 04E. Town and spa sit. ESE of Krasnodar in the North Caucasus. Industries inc. metal working; manu. clothing and furniture. Pop. 70,000.

Pyongan North   North Korea. Province of NW bounded N by China across the Yalu R. Area, 10,982 sq. m. (28,443 sq. km.). Cap. Sinuiju. Mountainous and sparsely settled except NW. The main occupation is lumbering, with some gold and coalmining. Pop. (1975E) 1·76m.

Pyongan South   North Korea. 39 05N 125 50E. Province of NW bounded N by N. Pyongan. Area, 5,764 sq. m. (14,929 sq. km.). Chief towns Pyongyang (cap.), Chinnampo. Lowlying on coast, rising to mountainous interior. Drained by Taedong R. Mainly agric. producing millet, soybeans, fruit, rice. Extensive coal fields and some gold mines. Mountainous area supports stock farming, with some lumbering. Pop. (1981E) 1·28m.

Pyongyang   South Pyongan, North Korea. 39 00N 125 30E. Cap. and city sit. NW of Seoul on Taedong R. Founded in 108 B.C. and then called Lolang, it is an industrial and market centre of an agric. and mining area. Industries inc. sugar refining, chemicals, textiles, brewing; manu. matches, machinery. Pop. (1984E) 2,639,448.

Pyrénées-Atlantiques   France. 43 15N 0 45W. Dept. of Aquitaine region bounded W by the Atlantic, S by the Spanish border. Area, 2,946 sq. m. (7,629 sq. km.). Chief towns Pau (prefecture), Bayonne. Drained by Adour R. and its tribs. Mountain streams are used for hydro-electric power. The N lowlands are mainly agric., producing maize, wheat, fruit and vines. The S highlands are stock farming areas. The coast has many tourist resorts inc. Biarritz, Hendaye, St.-Jean-de-Luz. Pop. (1982c) 555,696.

Pyrénées   France/Spain. 42 40N 1 00E. Mountain range extending *c.* 275 m. (440 km.) W to E from the Bay of Biscay to the Mediterranean Sea along the French/Spanish frontier. The central Pyrénées are the highest and largest, rising to 11,168 ft. (3,404 metres) a.s.l. The E Pyrénées fall to the Mediterranean from 9,000 ft. (2,743 metres) a.s.l., the W Pyrénées are in general 3,000–4,000 ft. (914–1,219 metres) a.s.l. Noted for mountain torrents and cascades, and natural amphitheatre structures. There are few passes.

Pyrénées-Orientales   France. 42 35N 2 25E. Dept. in Languedoc-Roussillon region, bounded E by the Gulf of Lions and S by Spain. Area, 1,578 sq. m. (4,086 sq. km.). Cap. Perpignan. The W part is mountainous, rising to Pic Carlitte, 9,583 ft. (2,921 metres) a.s.l. there is a coastal plain in the E. Mainly a wine-growing and farming area. Pop. (1982c) 334,557.

Pyrgos   Elis, Greece. 37 41N 21 28E. Town sit. SSW of Patras near mouth of Alpheus R. Commercial centre. Industries inc. manu. beverages, tobacco products. Exports fruit and wine through the port of Katokolon. Pop. (1981c) 21,958.

Q

Qachas Nek Qachas Nek, Lesotho. Area SE of Maseru in the Drakensberg Mountains, on the South African frontier. Area, 1,520 sq. m. (3,935 sq. km.).

Qatar 25 30N 51 15E. The State of Qatar, which includes the whole of the Qatar peninsula, extends on the landward side from Khor al Odeid to the boundaries of the Saudi Arabian province of Hasa. Area, approx. 4,416 sq. m. (11,437 sq. km.). Cap. Doha. Main product petroleum. Iron, steel, and fertilizers are also produced. Dukhan in W is the chief oilfield. Pop. (1987E) 371,863.

Qattara Depression Egypt. 29 30N 27 30E. Depression sit. S of El Alamein; approx. 200 m. (320 km.) long and 75 m. (120 km.) wide and its lowest point 440 ft. (134 metres) b.s.l.

Qena (Qinâ) Egypt. 26 10N 32 43E. Cap. and market town in governorate of same name. Sit. N of Luxor on E bank of R. Nile. Area of governorate 699 sq. m. (1,811 sq. km.). Routes through the E desert begin here, inc. the pilgrimages to Mecca. Pop. (1986E) 141,700.

Qingdao (Tsingtao) Shandong, China. 36 05N 120 20E. Port and resort SE of Beijing sit. at entrance to Kiaochow Bay. Manu. inc. machinery, soap, textiles and cement. Soya beans and groundnuts are exported. Pop. (1987E) 1,270,000.

Qinghai (Chinghai) China. 36 00N 98 00E. (i) L. in NE Tibet, known in Tibetan as Koko Nor, in S foothills of the Nan Shan near the Great Wall. Area, 1,630 sq. m. (4,222 sq. km.). (ii) Province on the NE border of Tibet. Area, 278,378 sq. m. (721,000 sq. km.). Cap. Sining. Mainly grazing land, swampy in the N, mountainous elsewhere, with the Kunlun Mountains crossing the centre E to W. Main occupation nomadic livestock farming. Pop. (1987E) 4,120,000.

Qiqihar (Tsitsihar) Heilongjiang, China. 47 26N 124 00E. Town in NE sit. on Nun-kiang R. Industries inc. soya bean processing, flour milling and the manu. of chemicals and matches. Pop. (1982C) 1,222,000.

Qishm Iran. 26 45N 56 45E. Island in (Persian) Gulf approx. 70 m. (112 km.) long and 2–15 m. (3–24 km.) wide. Chief town, Qishm. Main occupations fishing, salt, cultivating cereals and dates.

Quantock Hills Somerset, England. 51 08N 3 10W. Sit. NNE Taunton. Highest point 1,262 ft. (385 metres) at Will's Neck.

Quebec Canada. 50 00N 70 00W. Province, bounded on W and S by Ontario, the U.S.A. and New Brunswick, on NE by Labrador. Quebec was formerly known as New France or Canada from 1535–1763; as the province of Quebec from 1763–90; as Lower Canada from 1791–1846; as Canada East from 1846–67, and when, by the union of the four original provinces, the Confederation of the Dominion of Canada was formed, it again became known as the province of Quebec (Québec). The Quebec Act, passed by the British Parliament in 1774, guaranteed to the people of the newly conquered French territory in N America security in their religion and language, their customs and tenures, under their own civil laws. Area, 594,860 sq. m. (1,540,668 sq. km.), of which 523,000 sq. m. (1,354,570 sq. km.) is land and 71,000 sq. m. (183,890 sq. km.) is water. Principal cities, Quebec (cap.); Montreal; Laval; Sherbrooke; Verdun; Trois-Rivières and Hull. Pop. (1986C) 6,532,461.

Quebec Quebec, Canada. 46 49N

71 13W. Cap. of Province. Sit. on St. Lawrence R. at confluence of R. St. Charles. Manu. wood pulp and paper, newsprint and paper products, clothing, food products, electrical goods, footwear, tobacco. Pop. (1986C) 164,580.

Quedlinburg Halle, German Democratic Republic. 51 48N 11 09E. Market town on R. Bode sit. SW of Magdeburg. Main occupation horticulture. Pop. (1989E) 29,000.

Queenborough Kent, England. 51 26N 0 45E. Town on Isle of Sheppey, sit. S of Sheerness. Industries inc. glassmaking, plastics, pharmaceuticals, fertilizers and boatbuilding, car and vehicle importation and pre-sales preparation. Pop. (1981C) 3,700.

Queen Charlotte Islands British Columbia, Canada. 51 30N 129 00W. Approx. 150 islands off the coast of Canada. Area, 3,780 sq. m. (9,790 sq. km.). Main occupation fishing and lumbering.

Queen Elizabeth Islands Northwest Territories, Canada. 78 00N 95 00W. Nineteen islands in Canadian arctic. The largest are Axel Heiberg, Devon, Ellesmere and Melville.

Queen Maud Land Norwegian Antarctica. 72 30S 12 00E. Area of the Antarctic continent between 20° W and 45° E, forming a depend. of Norway.

Queens New York, U.S.A. 40 34N 73 52W. Boro. of New York City forming E district. Pop. (1980C) 1,891,325.

Queensferry Lothian Region, Scotland. 56 00N 3 25W. Town sit. WNW of Edinburgh on the Firth of Forth. South Queensferry on the S shore is a port and market town for agric. produce. It is linked by the Forth Railway Bridge to North Queensferry on the N shore (Fife Region). Pop. (1983E) 5,500.

Queensland Australia. 22 00S 142 00E. Second largest state of Australia, bounded N by Gulf of Carpentaria and Torres Strait, E by S. Pacific, S by New South Wales, and W by North West Territory. Divided into two areas by Great Dividing Range, which follows coastline at distance of from 100–300 m. (160–480 km.); country between range and coast consists of alluvial areas and rich river valleys. Area, 666,873 sq. m. (1,727,200 sq. km.), with seaboard of 3,236 m. (5,178 km.). Cap. Brisbane. Food processing is the main industry. Main exports are meat, sugar, wool, dairy produce and minerals. Principal minerals copper, coal, lead, zinc, silver, gold and bauxite. Oil was discovered at Moonie in S Queensland in 1961. A pipeline has been laid from Moonie to Brisbane, where refineries are operating. Large natural gas reserves have been proved in S and Central Queensland and a pipeline has been laid from Roma to Brisbane. Approx. one-third of the secondary production of the State is from works processing primary products, the most important being sugar-mills, meat works, butter factories and sawmills. Pop. (1989E) 2,830,198.

Queenstown Cape Province, Republic of South Africa. 31 52S 26 52E. Town sit. NW of E. London. Agric. centre for wheat, cattle and wool.

Quelimane Zambezia, Mozambique. 17 53S 36 51E. Cap. of province. Pop. (1970C) 71,786.

Quemoy Quemoy Island, Taiwan. 24 27N 118 23E. Town on Quemoy (or Kinmen) Island sit. E of Xiamen (on the Chinese coast) in the Formosa Strait. Trading centre for rice and wheat; kaolin is quarried nearby. Manu. inc. sorghum liquor (known as Kailiang).

Que Que *see* **Kwekwe**

Quercy France. Area of SW France sit. in Lot and Tarn-et-Garonne depts. Farming area producing vines, fruit and sheep. Drained by the Dordogne, Lot and Aveyron Rs. in fertile valleys. Chief town, Cahors.

Querétaro Mexico. 20 36N 100 23W. (i) State sit. mainly in central plateau. Area, 4,420 sq. m. (11,489 sq.

km.). Crops inc. cereals and fruit. Minerals inc. mercury and opals. Pop. (1980C) 739,605. (ii) Cap. of the State of the same name. Pop. (1980E) 203,586.

Quetta Balúchistan, Pakistan. 30 15N 66 55E. Provincial cap. Military station and commercial centre on the route through the Bolan Pass to Afghánistán. Trades in carpets and wool. Manu. include engineering and flour milling. Other occupations inc. fruit growing and coalmining. The Indian Army Staff College was estab. here in 1907. Quetta was nearly destroyed in an earthquake in 1935. Pop. (1981E) 281,000.

Quezaltenango Guatemala. 14 50N 91 31W. Cap. of Quezaltenango dept., sit. WNW of Guatemala City. Main occupations textile manu., flour milling and brewing. Pop. (1989E) 246,000.

Quezon City Philippines. 14 38N 121 00E. Cap. since 1948 and laid out in 1940. Sit. just NE of Manila, the old cap. Mainly residential and named after the first Pres. Manuel Luis Quezon (1878–1944). Pop. (1980C) 1,165,865.

Quibdó Chocó, Colombia. 5 42N 76 40W. Cap. of Chocó dept. Sit. on Atrato R. WNW of Bogotá. Mining is important occupation esp. for platinum and gold. Pop. (1980C) 34,000.

Quiché Guatemala. (i) Dept. of W central Guatemala bounded N by Mexico. Area, 3,235 sq. m. (8,378 sq. km.). Mainly mountainous, with the Sierra de Chaucús and the Cuchumatanes range sloping to lowland (N). Drained by Chixoy and Montagua Rs. Farming area with some lumbering; main products maize, beans, coffee, tobacco, sugar-cane, potatoes, livestock. Pop. (1982E) 430,003. (ii) Town otherwise known as Santa Cruz del Quiché, sit. NW of Guatemala. Dept. cap. Market town for an agric. area, noted for the nearby ruins of Utatlán, anc. Indian cap.

Quilmes Argentina. 34 43S 58 15W. Seaside resort sit. SE of Buenos Aires on R. Plate. Important industries

inc. oil refining, textiles, glass and metal goods manu. Pop. (1983E) 150,000.

Quilon Kerala, India. 8 50N 76 38E. Port sit. on Malabar coast NW of Trivandrum. Manu. textiles, rope and mats. Pop. (1981C) 167,598.

Quimper Finistère, France. 47 59N 4 04W. Important tourist resort in Brittany. Manu. pottery. Pop. (1990E) 62,000.

Quincy Illinois, U.S.A. 39 56N 91 23W. City sit. W of Springfield on R. Mississippi. Manu. agric. machinery, footwear, clothing and chemicals. Pop. (1980C) 42,554.

Quincy Massachusetts, U.S.A. 42 15N 71 01W. Industrial city sit. SSE of Boston. Shipbuilding, quarrying and foundry products are among important industries. Pop. (1980C) 84,743.

Quindío Nevada, Colombia. 5 00N 75 00W. Mountain in W central Colombia near Nevado del Tolima in the Cordillera Central. Height, 16,900 ft. (5,151 metres) a.s.l. At the S foot is Quindío Pass, the main trans-Cordillera route.

Quintana Roo Mexico. 19 00E 88 00W. Territory sit. on Yucatán Peninsula. Cap. Chetumal. Area, 19,387 sq. m. (50,212 sq. km.). Pop. (1980C) 225,985.

Quito Pichincha, Ecuador. 0 15S 78 35W. Cap. of rep. and of province of Pichincha, founded in 1534 it is sit. on plateau of the Andes at 9,350 ft. (2,850 metres). Manu. inc. textiles, clothing, leather goods and carpets. Pop. (1982C) 1,110,248.

Qum (Qom) Markazi, Iran. Place of pilgrimage and sacred to the Shia Moslems, the official religion of Iran. Sit. S of Tehrán. Manu. footwear and pottery. Pop. (1986C) 543,139.

Qytet Stalin Albania. 40 47N 19 57E. Town sit. NNW of Berat on Devoll R. Oil centre, linked by pipeline to Valona, with important refineries.

R

Raasay Highland Region, Scotland. 52 25N 6 04W. Island in Inner Hebrides separated from the mainland by the Inner Sound and from the E coast of Skye by the Sound of Raasay. Area, 28 sq. m. (72 sq. km.). Hilly, rising to 1,456 ft. (444 metres) a.s.l. in the S. Chief occupations crofting and fishing. Pop. (1985E) 170.

Rabat Morocco. 34 02N 6 48W. Cap. of Morocco sit. on Atlantic coast at mouth of Bou Regreg R. It was founded in the 12th century and it became cap. in 1913, replacing Fez. Industries inc. flour milling; manu. textiles, rugs, cement, potteries and bricks. Pop. (1982C) 518,616.

Rabaul New Britain, Papua New Guinea. 4 12S 152 15E. Town on NE coast of island. Sea port exporting copra and cacao, sit. in a volcanic area. Pop. (1980E) 14,954.

Rabigh Hejaz, Saudi Arabia. Town sit. N of Jidda on the Sherm Rabigh inlet of the Red Sea. Port with coastal trade. Date growing centre.

Racibórz Opole, Poland. 50 06N 18 13E. Town in SW, near border with Czech., sit. on Oder R. Industries inc. manu. electrical and other machinery and soap. R. port.

Racine Wisconsin, U.S.A. 42 43N 87 48W. City sit. SSE of Milwaukee on L. Michigan. Industries inc. manu. electrical equipment, agric. machinery, hardware, paints and varnishes. L. port. Pop. (1980C) 85,725.

Radcliffe Greater Manchester, England. 53 34N 2 20W. Town in Bury Metropolitan Borough. Industries inc. engineering, textiles, chemicals and paper manu. Pop. (1986E) 32,040.

Radom Kielce, Poland. 51 25N 21 10E. Town sit. S of Warsaw. Indus-

tries inc. manu. agric. machinery, leather, glass, wire and nails. Pop. (1982E) 198,000.

Ragged Islands Bahamas. Arch. sit. SE of Nassau and N of Cuba. A chain of cays extending c. 70 m. (112 km.) N, main islands Great Ragged Island, Little Ragged Island. Area, 5 sq. m. (13 sq. km.). Pop. (1980C) 146.

Ragusa Sicily, Italy. 36 55N 14 44E. Town WSW of Syracuse sit. on Irminio R. Industries inc. oil extracting and refining and asphalt mining; manu. textiles and cement. Pop. (1981E) 64,000.

Rahway New Jersey, U.S.A. 40 37N 74 17W. City sit. SSW of Newark. Industries inc. manu. pharmaceuticals, cleaning appliances, car engine filters. Pop. (1978C) 26,723.

Raiatea French Polynesia, South Pacific. 16 50S 151 25W. Island in Leeward group of the Society Islands. Area, 92 sq. m. (238 sq. km.). Cap. Uturoa. Volcanic and mountainous, rising to 3,388 ft. (1,033 metres) a.s.l. in Mount Temehani. Produces copra, tobacco, kapok, fruit. Industries at Uturoa inc. fruit canning.

Raichur Karnataka, India. Town sit. SW of Hyderabad. Commercial centre for a farming area, trading in cereals and cotton. Industries inc. cotton ginning. Pop. (1981C) 124,762.

Rainier, Mount Washington, U.S.A. 46 52N 121 46W. Mountain peak in the Cascade Range 14,408 ft. (4,392 metres) a.s.l., and centre of Mount Rainier National Park. It has permanent icefields and 26 extensive glaciers.

Rainy Lake Canada/U.S.A. 48 42N 93 10W. L. on the boundary between Minnesota, U.S.A. and Ontario, Canada, drained by Rainy R. which flows

85 m. (136 km.) w along the frontier into the L. of the Woods. Area, 345 sq. m. (894 sq. km.).

Raipur Madhya Pradesh, India. 28 50N 79 05E. Town sit. E of Nagpur. Market town for an agric. area, trading in rice and oilseeds. Industries inc. engineering, rice milling and oilseed milling. Pop. (1981C) 338,245.

Raivavae Austral Islands, French Polynesia. Island sit. SE of Tubuai Island. Area, 6 sq. m. (16 sq. km.). Chief town Amaru. Volcanic. Produces coffee, arrowroot.

Rajahmundry Andhra Pradesh, India. 17 01N 81 48E. Town sit. ESE of Hyderabad on Godavari R. Market centre trading in rice and salt. Industries inc. textiles, esp. cotton; manu. paper and tiles. Pop. (1981C) 216,851.

Rajasthan India. 27 00N 74 00E. State in NW India bounded W by Pakistan and S by Gujarat. Area, 133,677 sq. m. (342,214 sq. km.). Chief towns Jaipur (the cap.), Jodhpur, Ajmer, Bikaner, Beawar, Kota and Udaipur. Desert in W, mountainous in centre with the Aravalli Range rising in S to 5,650 ft. (1,722 metres) a.s.l., hilly and fertile in E. Chief occupation farming; chief products wheat, tobacco, jawar, bajra, maize, millet and cotton. Minerals inc. gypsum, silver, copper, lead, zinc, emeralds, garnets, salt rock, phosphate, marble, oil and gas, asbestos, felspar and mica. Industries inc. sugar, cement, glass, pesticides, railway carriages, sulphuric acid, nylon, caustic soda and tyrecord. Pop. (1981C) 34,261,862.

Rajin Hamgyong North, North Korea. Hamgyong, otherwise called Najin. Town sit. E of Chongjin on the Sea of Japan. Ice-free fishing port and naval base. Terminus of the coastal and N frontier railway from Wonsan.

Rajkot Gujarat, India. 22 15N 70 56E. Town sit. wsw of Ahmedabad in the Kathiawar peninsula. Market centre trading in cotton and grain. Industries inc. tanning, oilseed and flour milling; manu. chemicals. Pop. (1981C) 445,076.

Rajshahi Rajshahi, Bangladesh.

Cap. of division sit. WNW of Dacca. Market town trading in rice, jute, oilseeds, wheat and cane. Industries inc. milling, esp. rice, atta, flour and sugar, and soap manu. Pop. (1981E) 1,539,000.

Raleigh North Carolina, U.S.A. 35 47N 78 39W. State cap. sit. in N centre of state. Industries inc. printing, publishing and textiles; manu. cottonseed oil. Trading centre for tobacco. Pop. (1980C) 150,255.

Rama Zelaya, Nicaragua. Town sit. WNW of Bluefields on Escondido R. Trading centre for bananas, sugar, livestock. R. port and terminus of the road from the Pacific coast.

Ramat Gan Israel. 32 05N 34 48E. City sit. NE of Jaffa–Tel Aviv on the Plain of Sharon. Industries inc. textiles; manu. food products and furniture. Pop. (1982E) 118,000.

Rameswaram Tamil Nadu, India. Island in Palk Straits at the extreme SE limit of the Indian peninsula, an important centre of pilgrimage for the Hindus, associated with Lord Rama. Pop. (1981C) 27,928.

Rampur Uttar Pradesh, India. 28 50N 79 05E. Town sit. E of Delhi. Commercial centre trading in cotton, grain and sugar-cane. Industries inc. sugar refining and metal working; manu. chemicals and pottery. Pop. (1981C) 204,610.

Ramsbottom Greater Manchester, England. 53 36N 2 20W. Town sit. N of Bury on Irwell R. Industries inc. engineering and textiles; manu. paper and soap.

Ramsey Cambridgeshire, England. 52 27N 0 07W. Town sit. NNE of Huntingdon. Market town for an agric. area. Pop. (1977E) 5,870.

Ramsey Isle of Man. 54 20N 4 21W. Town sit. NNE of Douglas on NE coast. Industries inc. tourism, light engineering and fishing. Pop. (1986C) 5,778.

Ramsgate Kent, England. 51 20N 1 25E. Town in NE Kent and in the Isle of Thanet. Industries inc. tourism and fishing. Cross channel passenger and freight terminal. Pop. (1981c) 39,642.

Rancagua   Chile. 34 10S 70 45W. City sit. in central Chile. Market and manu. centre for an agric. area. Industries inc. agric. engineering, flour milling and food canning. It has a railway link with the El Teniente copper mine. Pop. (1982E) 137,773.

Rance River France. 48 31N 1 59W. Rises in Côte du Nord dept. and flows 62 m. (99 km.) E and N to enter the Gulf of St. Malo by an estuary 12 m. (19 km.) long between St. Malo and Dinard, on the N coast. There is a marine power station in the estuary.

Ranchi Bihar, India. 23 19N 85 27E. Town sit. WNW of Calcutta on Chota Nagpur plateau, at 2,000 ft. (610 metres) a.s.l. Commercial centre trading in rice, maize, cotton and oilseeds. Industries inc. steel and wire rope. Pop. (1981c) 489,626.

Randers Åarhus, Denmark. 56 28N 10 03E. Town and port in NE, sit. at mouth of Gudenaa R. and at the head of Randers Fjord. Industries inc. brewing and railway engineering; manu. combine harvesters, gloves. Pop. (1990E) 61,013.

Randfontein Transvaal, Republic of South Africa. 26 11S 27 42E. Town sit. W of Johannesburg on the Witwatersrand at 5,600 ft. (1,707 metres) a.s.l. Industries inc. engineering, food processing, gold-mining. Pop. (1985E) 28,000.

Rangiroa Tuamotu Archipelago, French Polynesia. 15 10S 147 35W. Atoll sit. N of Tahiti. Largest and most populated of arch.

Rangoon (Yangon) Burma. 16 45N 96 20E. Cap. of Burma sit. on Rangoon R. near its mouth. Industries inc. food processing, oil refining and engineering; manu. matches and soap. Sea port exporting rice, cotton and timber, esp. teak. Pop. (1983E) 2,458,712.

Rangpur Rajshahi, Bangladesh. 25 45N 89 21E. Cap. of district sit. NW of

Dacca. Trading centre for rice, jute, tobacco and oilseeds. Industries inc. manu. electrical machinery, bricks and cutlery. Pop. (1981E) 1,708,000.

Rannoch, Loch Scotland. 56 41N 4 20W. Loch in central Scot. extending 9 m. (14 km.) E to W. Width up to 1 m. (1.5 km.). Fed by R. Ericht and drained by R. Tummel.

Rapa French Polynesia, South Pacific. 27 36S 144 20W. Island sit. SE of Raivavae in the Austral group. Area, 16 sq. m. (41 sq. km.). Mountainous, rising to 2,077 ft. (633 metres) a.s.l. Chief product copra.

Rapallo Liguria, Italy. 44 21N 9 14E. Town sit. on Gulf of Genoa in a fertile area producing olives and flowers. Industries inc. tourism; manu. olive oil and cement. Pop. (1990E) 30,000.

Rapid City South Dakota, U.S.A. 44 05N 103 14W. Sit. in E centre of state, in the Black Hills. Commercial centre of an agric. and mining area. Industries inc. milling flour and timber; manu. cement, bricks and tiles. Pop. (1980c) 46,492.

Rappananock River Virginia, U.S.A. 37 34N 76 18W. Rises in the Blue Ridge Mountains and flows 210 m. (336 km.) SE to Chesapeake Bay. Main trib. Rapidan R.

Rarotonga Cook Islands, S Pacific. 21 14S 159 46W. Largest and furthest SW of the Cook Islands. Area, 26 sq. m. (67 sq. km.). Chief town Avarua. Volcanic, rising to 2,110 ft. (643 metres) a.s.l. Exports fruit and copra from 2 harbours. Pop. (1981c) 9,530.

Ras al Khaima United Arab Emirates. 25 47N 55 57E. Furthest NE of the sheikhdoms of the United Arab Emirates, on a peninsula extending NE from the S coast of the Gulf into the Strait of Hormuz. Area 400 sq. m. (1,036 sq. km.). Chief town Ras Al Khaima. Bounded NE by the Russ al Jibal district of N Oman, which the United Arab Emirates separates from the rest of Oman, E by the Gulf of Oman, W by the Gulf. Low-lying with desert coastline rising to hills inland.

Chief products petroleum, dates. Pop. (1985C) 116,470.

Rasht Gilan, Iran. 37 20N 49 40E. Cap. of province in NW sit. near the Caspian Sea and the Alburg Mountains. Trading centre for an agric. area, trading in farm produce esp. rice. Silk is also important. Manu. textiles, hosiery and carpets. Pop. (1983E) 260,000.

Ras Tanura Al Hasa, Saudi Arabia. 26 40N 50 10E. Town sit. N of Dhahran on the (Persian) Gulf. The main industry is refining and shipping oil from the fields at Dhahran, Abqaiq and Qatif.

Rathenow Potsdam, German Democratic Republic. 52 36N 12 20E. Town sit. in W centre, on Havel R. Industries inc. manu. scientific precision instruments, agric. machinery and chemicals.

Rathlin Island Antrim, Northern Ireland. 55 18N 6 14W. Island off the N coast of Antrim, 6 m. (10 km.) long.

Ratlam Madhya Pradesh, India. Town sit. NW of Indore. Commercial centre trading in cotton and grain. Industries inc. textiles. Pop. (1981C) 155,578.

Ratnagiri Maharashtra, India. Town sit. SSE of Bombay on the Arabian Sea. District cap. Port, exporting timber, fish, rice and bamboo. The main industries are fishing, fish processing, tanning. Pop. (1981C) 47,036.

Rauma Turku-Pori, Finland. 61 08N 21 30E. Town sit. NW of Turku on the Gulf of Bothnia. Port, handling timber and cellulose. Industries inc. shipbuilding, sawmilling, pulp and cellulose. Trad. noted for lace. Pop. (1990E) 30,000.

Ravenna Emilia-Romagna, Italy. 44 25N 12 12E. Cap. of province of same name connected with the Adriatic Sea by a 5 m. (8 km.) canal. Trading centre for agric. produce. Industries inc. chemicals. Noted for its examples of Byzantine art. Pop. (1989E) 136,166.

Ravensburg Baden-Württemberg,

Federal Republic of Germany. 47 47N 9 37E. Town sit. SSW of Ulm. Industries inc. engineering; manu. textiles. Pop. (1985E) 43,400.

Ravi River India/Pakistan. 30 35N 71 48E. Rises in the Pir Panjal Range in Himachal Pradesh, India, and flows 450 m. (720 km.) W and SW past Lahore, Pakistan, to join Chenab R. *c.* 34 m. (54 km.) N of Multan. Navigable up to Lahore: feeds the Upper Bari Doab canal.

Rawalpindi Rawalpindi, Pakistan. 33 38N 73 08E. City sit. NNW of Lahore. Cap. of Rawalpindi division. Military station and trading centre for grain, wool and timber. Industries inc. textiles, oil refining, chemicals and railway engineering; manu. furniture. Pop. (1981C) 928,000.

Rawmarsh South Yorkshire, England. 53 27N 1 20W. Town NNE of Rotherham sit. in a coalmining and steel working district. Industries inc. food processing. Pop. (1988E) 18,450.

Rawson Chubut, Argentina. 43 18N 65 06W. Cap. of province sit. SSW of Bahia Blanca near S Atlantic coast. Market town for an agric. area, trading in grain and sheep. Industries inc. milling. Pop. (1980C) 52,000.

Rawtenstall Lancashire, England. 53 42N 2 18W. Town sit. SSW of Burnley on Irwell R. Industries inc. engineering, textiles, esp. cotton and woollen goods; manu. furniture, carpets and footwear. Pop. (1981C) 22,231.

Raxaul Bihar, India. Town sit. N of Motihari. Border station and railway junction with the main customs point for Nepál trade. Pop. (1981C) 102,612.

Razgrad Ruse, Bulgaria. 43 32N 26 31E. Town sit. SE of Ruse on Beli Lom R. Trading centre for the W Deliorman agric. area; handles cattle, grain, legumes, sunflowers, timber. Manu. starch, leather.

Ré, Île de Charente-Maritime, France. Island off La Rochelle in the Bay of Biscay, separated from the mainland by the Pertuis Breton. Area

30 sq. m. (78 sq. km.). Chief occupations fishing and farming. Pop. (1982c) 9,660.

Reading   Berkshire, England. 51 28N 0 59W. County town in central S sit. at confluence of Rs. Thames and Kennet. Industries inc. engineering, brewing, printing and computing. Pop. (1985E) 138,000.

Reading   Pennsylvania, U.S.A. 40 20N 75 56W. City sit. NW of Philadelphia on Schuykill R. Industries inc. railway engineering; manu. metal goods, machinery, optical goods and hosiery. Pop. (1980c) 78,686.

Recife   Pernambuco, Brazil. 8 03S 34 48W. State cap. in NE, on the Atlantic coast. Sit. partly on the mainland, partly on a peninsula, and partly on an island. Industries inc. sugar refining, cotton milling, pineapple canning and textiles. Sea port, naval station and airport. Pop. (1980c) 1,183,391.

Recklinghausen   North Rhine-Westphalia, Federal Republic of Germany. 51 36N 7 13E. Town sit. NW of Dortmund in the Ruhr district. Industries inc. coalmining, iron founding and brewing; manu. machinery, coaltar products, soap and cement. Pop. (1988E) 119,590.

Redbridge   Greater London, England. 51 35N 0 07E. Boro. of E London N of Thames R. It inc. Ilford, Wanstead and Woodford. Pop. (1984E) 226,500.

Redcar   Cleveland, England. 54 37N 1 04W. Town sit. SE of Middlesbrough on North Sea. Resort. Industries inc. manu. steel and chemicals. Pop. (1990E) 40,000.

Red Deer   Alberta, Canada. 52 16N 113 48W. Town sit. N of Calgary on Red Deer R. Market town for a farming, petroleum and petrochemical area. Pop. (1989E) 55,947.

Redditch   Hereford and Worcester, England. 52 19N 1 56W. Town S of Birmingham. Industries inc. manu. fishing tackle, needles, small metal goods, batteries, compressors, control

systems for aerospace and computer software. Pop. (1989E) 78,023.

Rede River   England. 55 08N 2 12W. Rises on Carter Fell in the Cheviot Hills and flows 21 m. (34 km.) SE and S to join Tyne R. at Redesmouth, NW of Newcastle on Tyne.

Redhill   Surrey, England. 51 14N 0 10W. Town sit. S of London at the foot of the North Downs. Residential. Pop. (1971c) 56,088 with Reigate.

Redlands   California, U.S.A. 34 03N 117 11W. City sit. E of Los Angeles in a fruit growing area. Industries inc. fruit packing, medical and financial services. Manu. aircraft batteries and electric vehicles. Pop. (1980c) 43,619.

Redondo Beach   California, U.S.A. 33 51N 118 23W. City sit. SW of Los Angeles on the Pacific coast. Resort and residential city with some light industry. Pop. (1980c) 57,102.

Red River   U.S.A. 31 00N 91 40W. Rises on the Llano Estacado and flows 1,000 m. (1,600 km.) E across NW Texas to form the Texas/Oklahoma boundary, then turns briefly into SW Arkansas and S again into Louisiana, past Shreveport, and SE past Alexandria to join R. Mississippi. Used for water storage, particularly in L. Texoma on the Texas/Oklahoma border, area 223 sq. m. (578 sq. km.). Navigation impeded by silt.

Red River   U.S.A./Canada. 31 00N 91 40W. Formed by confluence of Rs. Otter Tail and Bois de Sioux in E North Dakota, and flows 350 m. (560 km.) N to form the North Dakota/Minnesota border and enters Canada at Pembina; continues N across Manitoba to receive Assiniboine R. at Winnipeg, and empty into L. Winnipeg. Its valley is fertile, esp. for wheat.

Redruth   Cornwall, England. 50 14N 5 14W. Town sit. ENE of Penzance. Market town. Industries inc. tin mining, electronics, clothing manu., pottery, sheepskin products, food processing and light engineering. Pop. (1985E) 2,000.

Red Sea   25 00N 36 00E. Extends 1,500 m. (2,400 km.) SSE from Suez to the Strait of Bab el Mandeb between NE Africa and SW Asia. The N end divides into the Gulf of Aqaba and the Gulf of Suez, separated by the Sinai peninsula. Linked to the Mediterranean Sea by the Suez Canal. Chief ports Suez, Egypt; Port Sudan, Sudan; Massawa, Ethiopia; Jidda, Saudi Arabia.

Redwood City   California, U.S.A. 37 29N 122 13W. Sit. SSE of San Francisco on the peninsula between the Pacific Ocean and San Francisco Bay. Industries inc. manu. cement, leather and electronic equipment. Port exporting oil and salt. Pop. (1980C) 54,951.

Ree, Lough   Ireland. 53 35N 8 00W. L. on Shannon R. between County Roscommon to W, and County Longford and County Westmeath to E. Length 17 m. (27 km.), width up to 7 m. (11 km.). Noted for trout fishing.

Regensburg   Bavaria, Federal Republic of Germany. 49 01N 12 06E. Town in SE sit. on Danube R. Industries inc. brewing and sugar refining; manu. machinery, electronic products, clothing, snuff and vehicles. R. port. Pop. (1983E) 132,000.

Reggio de Calabria   Calabria, Italy. 38 07N 15 39E. Town in the Strait of Messina, in extreme SW tip of Italy. Cap. of province of same name. Industries inc. fruit canning; manu. olive oil, pasta and furniture. Sea port exporting citrus fruits, olive oil and wine. Pop. (1983E) 132,000.

Reggio nell'Emilia   Emilia-Romagna, Italy. 44 43N 10 36E. Town in central N, and cap. of province of same name. Industries inc. railway, agric. and motor engineering; manu. wine and cement. Pop. (1980C) 173,486.

Regina   Saskatchewan, Canada. 50 25N 104 39W. Cap. of province sit. in central S, in an extensive wheat-growing area. Commercial centre with important distribution trades, esp. agric. implements, cars and hardware, and access to deposits of light oil and potash. Industries inc. oil refining; manu. cement, paints and varnishes. W headquarters of the Mounted Police. Pop. (1986C) 175,064.

Regnitz River   Federal Republic of Germany. 49 54N 10 49E. Formed by union of Rs. Pegnitz and Rednitz at Fürth, and flows 40 m. (64 km.) N to join Main R. below Bamberg.

Rehoboth Gebiet   Namibia. Area of central SW Africa immediately S of Windhoek district. Area, *c.* 12,420 sq. m. (*c.* 32,168 sq. km.). Chief town Rehoboth. Occupied by the Basters, a race of mixed European-Nama descent.

Rehovoth   Judaea, Israel. 31 53N 34 48E. Town sit. SSE of Tel Aviv on the Judaean plain. Centre of an area growing citrus fruit. Manu. dairy products, metal products, glass, pharmaceutical goods, fruit juice.

Reichenbach   Karl-Marx-Stadt, German Democratic Republic. 50 37N 12 18E. Town sit. SW of Zwickau. Industries inc. textiles, esp. cotton, woollen and rayon goods, dyeing and printing; manu. machinery. Pop. (1989E) 26,000.

Reigate   Surrey, England. 51 14N 0 13W. Town in SE sit. at the foot of the North Downs. Residential. Pop. (1988E) 50,550.

Reims   Marne, France. 49 15N 4 02E. Town in NW, sit. on Vesle R. Centre of the champagne industry, also manu. bottles and equipment for champagne production. Industries also inc. engineering and flour milling. Pop. (1982C) 181,985 (agglomeration, 199,388).

Reindeer Lake   Canada. 57 15N 102 40W. L. extending 145 m. (232 km.) NE to SW through NW Manitoba and NE Saskatchewan. Width up to 40 m. (64 km.). Drained, SW, by Reindeer R. into Churchill R.

Remscheid   North Rhine-Westphalia, Federal Republic of Germany. 51 11N 7 11E. Town sit. S of Wuppertal on Wupper R. Industries inc. steel; manu. machine tools, cutlery, agric. implements and hand tools. Pop. (1985E) 125,000.

Renala Khurd　　Punjab, Pakistan. Village sit. NE of Okara on the Lower Bari Doab canal. Market for wheat, cotton. Industries inc. dairy farming, cotton ginning, fruit canning.

Rendsburg　　Schleswig-Holstein, Federal Republic of Germany. 54 18N 9 40E. Town in NW, sit. on Kiel canal. Industries inc. shipbuilding and iron-founding; manu. machinery and ferti-lizers. Pop. (1984E) 31,200.

Renfrew　　Strathclyde Region, Scot-land. 55 52N 4 24W. County town sit. WNW of Glasgow on Clyde R. Indus-tries inc. shipbuilding and engineer-ing. Pop. (1984E) 22,927.

Renfrewshire　　Scotland. Former county now part of Strathclyde Region.

Renkum　　Gelderland, Netherlands. 51 58N 5 45E. Municipality of 5 vil-lages sit. W of Arnhem. Industries inc. manu. rubber and paper products. Pop. (1990E) 33,582.

Rennes　　Ille-et-Vilaine, France. 48 05N 1 41W. Cap. of dept. in Brittany, sit. at confluence of Rs. Ille and Vilaine. Commercial centre and rail-way junction. Industries inc. electron-ics, engineering, printing and tanning; manu. cars, chemicals, footwear and hosiery. Pop. (1982C) 200,390 (agglo-meration (1990E) 320,000).

Reno　　Nevada, U.S.A. 39 31N 119 48W. Town sit. on Truckee R. at 4,500 ft. (1,372 metres) a.s.l. In-dustries inc. legalized gaming, meat packing, warehousing. Manu. fork lifts, skis, cars, plastics and electro-nics. Pop. (1980C) 100,756.

Renton　　Washington, U.S.A. 47 30N 122 11W. City immediately SSE of Seattle on Cedar R. Industries inc. air-craft and heavy manu. Pop. (1980C) 30,612.

Republican River　　U.S.A. 39 03N 96 48W. Rises in NE Colorado and flows ESE for 450 m. (720 km.) through Nebraska and Kansas to join

Smoky Hill R. at Junction City. To-gether they become Kansas R.

Resina　　*see* **Ercolano.**

Resistencia　　Chaco, Argentina. 27 30S 58 59W. Cap. of province in NE. Trading centre, through the port of Barranqueras on the Paraná R., for cotton, timber, hides and livestock. In-dustries inc. meat packing and saw-milling. Pop. (1980C) 218,000.

Reşita　　Caraş-Severin, Romania. 45 17N 21 53E. Town sit. WNW of Bucharest. Industries inc. iron and steel; manu. machinery and electrical equipment. Pop. (1983E) 105,000.

Restigouche River　　Canada. Rises in NW New Brunswick and flows 130 m. (208 km.) NE to the New Brunswick/ Quebec border, of which it forms the E end, before emptying into Chaleur Bay. Noted for salmon fishing.

Retalhuleu　　Retalhuleu, Guatemala. 14 32N 91 41W. Cap. of dept. sit. W of Guatemala. Commercial centre of a coffee and cane growing area.

Retford　　Nottinghamshire, England. 53 19N 0 56W. Town sw of Gains-borough, sit. on R. Idle. Market town. Industries inc. textiles, rubber pro-ducts, paper and other light industry. Pop. (1989E) 20,145.

Rethymnon　　Crete, Greece. 35 23N 24 28E. (i) *Nome* in central Crete. Area, 578 sq. m. (1,496 sq. km.). Chief occupation farming. Pop. (1981C) 62,634. (ii) Cap. of (i) sit. W of Irak-lion. Port trading in cereals, wine and olive oil. Pop. (1981C) 17,736.

Réunion　　Indian Ocean. 21 06S 55 36E. Island sit. sw of Mauritius, form-ing an overseas dept. of France. Area, 969 sq. m. (2,512 sq. km.). Cap. St. Denis. Volcanic, rising to 10,069 ft. (3,069 metres) a.s.l., and subject to violent storms. Chief occupation farming; chief products sugar and rum. Pop. (1988E) 574,800.

Reus　　Tarragona, Spain. 41 09N 1 07E. Town sit. wsw of Barcelona. Trading centre for wine and fruit. In-dustries inc. agric. engineering and textiles; manu. leather goods. Pop. (1986C) 85,251.

Reutlingen   Baden-Württemberg, Federal Republic of Germany. 48 29N 9 11E. Town in SW sit. on N edge of Jura Mountains. Industries inc. manu. textiles and textile machinery, paper and leather. Pop. (1984E) 96,300.

Revelstoke   British Columbia, Canada. 50 59N 118 12W. Town sit. E of Kamloops on Columbia R. and Trans-Canada Highway. Supply centre for a lumbering area, and tourist centre for the Mount Revelstoke National Park. Pop. (1990E) 8,500.

Revere   Massachusetts, U.S.A. 42 24N 71 01W. City sit. NE of Boston on the Atlantic coast. Industries inc. tourism and chemicals; manu. electrical equipment. Pop. (1980C) 42,423.

Rewa   Madhya Pradesh, India. 29 33N 81 25E. Town sit. SSW of Allahabad. District cap. and trading centre for grain, timber, building stone. Pop. (1981C) 100,641.

Reykjavik   Iceland. 64 10N 21 57W. Cap. of Iceland, founded in 874 A.D., and sit. on Faxa Fjord on SW coast. Chief commercial centre. Industries inc. fishing and fish processing, cement, fertilizers. Sea port exporting fish. Pop. (1988E) 95,811.

Rhayader   Powys, Wales. 52 19N 3 30W. Town in central Wales sit. on Wye R. Market town for a farming area, trekking and tourist centre. Pop. (1978E) 1,450.

Rheims   *see* **Reims**

Rheinfelden   Aargau, Switzerland. Town sit. E of Basel on Rhine R. Tourist centre with some industry, inc. manu. clothing, barrels, beer, tobacco products. Saline baths.

Rheingau   Hesse, Federal Republic of Germany. Wine growing district extending *c.* 15 m. (24 km.) along E bank of Rhine R. from Biebrich to Assmannshausen, WSW of Frankfurt.

Rheinhausen   North Rhine-Westphalia, Federal Republic of Germany.

51 24N 6 44E. Town opposite Duisburg on Rhine R. sit. in a coalmining area. Industries inc. mining, iron and steel; manu. mining machinery and concrete products.

Rhine-Herne Canal   Federal Republic of Germany. Canal 24 m. (38 km.) long from Duisburg to Herne in the Ruhr district.

Rhineland   Federal Republic of Germany. Region on banks of Rhine R. comprising parts of North Rhine-Westphalia, Rhineland-Palatinate, Hesse and Baden-Württemberg.

Rhineland-Palatinate   Federal Republic of Germany. 50 00N 7 00E. Province in extreme W bounded E by Rhine R. and W by Belgium, Luxembourg and France. Area, 7,662 sq. m. (19,845 sq. km.). Cap. Mainz. Drained by Rs. Moselle and Nahe into Rhine R. Chief occupations farming and viticulture. Pop. (1988E) 3,653,155.

Rhine River   West Europe. 51 52N 6 02E. Rises as the Hinter Rhein and Vorder Rhein, the former on the Rheinwaldhorn in the Adula Mountains in SE Switz., the latter in L. Toma below the St. Gotthard Pass. Flows 820 m. (1,312 km.) NE to the borders of Liechtenstein and Austria and into L. Constance; W along the Swiss/Federal Republic of Germany border to Basel, and then generally N along the Franco/FRG border and into FRG near Pforzheim, passing Mannheim, Mainz and Wiesbaden and turning NW past Bonn, Cologne, Düsseldorf and Duisburg. It enters the Netherlands below Emmerich and divides into 2 main branches; Lek R. to the N flows W to Rotterdam where it becomes New Maas R. and enters the North Sea at the Hook of Holland. The Waal R. to the S flows WSW to empty into the Hollandsch Diep, 6 m. (10 km.) SSE of Dordrecht. Main tribs. Rs. Ill, Neckar, Main and Moselle. It is the most important waterway in Europe for commercial traffic; chief cargoes coal, ore and grain. Navigable up to Basel.

Rhode Island   U.S.A. 41 38N 71 37W. State in New England bounded S by the Atlantic Ocean and the smallest State in U.S.A. The first of the settlements was made in 1636, settlers of every creed being welcomed. In 1647 a patent was granted for the government of the settlements, and in 1663 a charter was executed recognizing the settlers as forming a body corporate and politic by the name of the 'English Colony of Rhode Island and Providence Plantations, in New England, in America'. Area, 1,214 sq. m. (3,144 sq. km.). Chief towns Providence (the cap.), Pawtucket, Warwick and Cranston. Low-lying with a continental climate. Drained by Blackstone R. into Narrangansett Bay. Mainly urban. Chief industries metal working, jewellery, engineering and textiles; small-scale (100 acres) farming is important in rural areas. Pop. (1987E) 986,000.

Rhodes Greece. 36 10N 28 00E. (i) Island in the Dodecanese group, in the Aegean Sea, lying 10 m. (16 km.) from the Turkish coast. Area, 542 sq. m. (1,404 sq. km.). Mountainous, well-watered and fertile. Chief products cereals, wine, fruit. (ii) Cap. of (i) and of Dodecanese *nome*, sit. at NE tip of island. Trade in agric. produce. Industries inc. manu. wine and tobacco products and tourism. Pop. (1981C) 40,392.

Rhodesia *see* **Zimbabwe**

Rhodope *see* **Rodopi**

Rhodope Mountains Greece/Bulgaria. 41 30N 24 30E. Mountain system extending *c.* 120 m. (192 km.) ESE from SW Bulgaria along the Greek frontier, rising to Mount Musala, 9,596 ft. (2,925 metres) a.s.l.

Rhondda Wales. 51 40N 3 27W. District of Mid-Glamorgan in SE Wales extending along the Rhondda Fawr and Rhondda Fach R. valleys in the S Wales coalfield. There is still some coalmining although most pits have closed since 1948.

Rhône France. 45 54N 4 35E. Dept. in Rhône-Alpes region, bounded E by Rs. Rhône and Saône. Area, 1,241 sq. m. (3,215 sq. km.). Chief towns Lyons (the cap.) and Villefranche-sur-Saône. Mountainous, rising to 3,000 ft.

(914 metres) a.s.l. in the Monts du Beaujolais. Chief occupations farming and viticulture. Pop. (1982C) 1,445,208.

Rhône-Alpes France. Region in SW France comprising depts. of Ain, Ardèche, Drôme, Haute-Savoie, Isère, Loire, Rhône and Savoie. Area, 16,868 sq. m. (43,694 sq. km.). Chief cities Lyon, Grenoble, Saint-Etienne and Valence. Pop. (1982C) 5,015,947.

Rhône River France/Switzerland. 43 28N 4 50E. Rises in Rhône Glacier in E Valais, Switz., and flows 504 m. (806 km.) WSW and then NW to L. Geneva, leaving it at Geneva to enter France. It passes by gorges between the Jura Mountains and the Alps to Lyons, and turns S along the W side of the French Alps past Valence and Avignon to Arles, where it divides; the two branches, Grand and Petit Rhône, enter the Gulf of Lions by the Camargue delta. Used for hydroelectric power, esp. at Génissiat. Navigation is limited.

Rhyl Clwyd, Wales. 53 19N 3 29W. Town on N coast sit. at mouth of Clwyd R. Industries inc. tourism. Pop. (1980E) 23,000.

Rhymney Mid Glamorgan, Wales. 51 32N 3 17W. Town ENE of Merthyr Tydfil sit. on Rhymney R. in a coalmining area. Industries inc. engineering. Pop. (1980E) 8,000.

Riau Archipelago *see* **Riouw Archipelago**

Ribble River England. 53 44N 2 50W. Rises in the Pennine Hills below Whernside, and flows 75 m. (120 km.) S past Settle, then SW into Lancashire below Clitheroe, past Preston and into the Irish Sea by a broad estuary.

Ribe Ribe, Denmark. 55 21N 8 46E. Town sit. SE of Esbjerg in SW Jutland. Flourishing medieval port now 4 m. (6 km.) from the sea. Industries inc. plastics and iron. Pop. (1988E) 17,964.

Ribeirão Prêto São Paulo, Brazil. 21 10S 47 48W. Town in SE. Commercial centre of a coffee-growing area, trading in coffee, cotton, sugar and grain. Industries inc. cotton milling, distilling, brewing, steel-making and agric. engineering. Pop. (1980C) 300,828.

Riberalta Vaca Diez, Bolivia. 10 59S 66 06W. Town sit. NNW of Trinidad at confluence of Madre de Dios and Beni Rs. R. port, shipping company base and airport. Important centre for distributing rubber, nuts, fruits and general Brazilian imports. Pop. (1984E) 20,000.

Richland Washington, U.S.A. 46 17N 119 18W. City sit. WNW of Pasco on a govt. reservation on Colombia R. next to the U.S. Centre for Hanford atomic energy plant. Pop. (1980C) 33,578.

Richmond Surrey, England. 51 28N 0 18W. District WSW of city centre, sit. on Thames R. Residential resort with Richmond Park. Pop. (1989E) 162,824.

Richmond North Yorkshire, England. 54 24N 1 44W. Town in central N, sit. on Swale R. Market town for a stock-farming area. Pop. (1981C) 7,596.

Richmond California, U.S.A. 37 57N 122 22W. City sit. NNW of Oakland on San Francisco Bay. Industries inc. oil-refining, petrochemical research and chemicals; manu. electronic equipment and metal goods. Pop. (1980C) 74,676.

Richmond Indiana, U.S.A. 39 50N 84 54W. City sit. E of Indianapolis. Industries inc. manu. machine tools, motor-car parts and agric. implements. Pop. (1988E) 39,200.

Richmond Virginia, U.S.A. 37 30N 77 28W. State cap. sit. SSW of Washington D.C. on James R., at the head of navigation. Commercial centre and port handling tobacco, coal and grain. Industries inc. tobacco processing, chemicals and textiles; manu. food products and paper. Pop. (1980C) 219,214.

Rickmansworth   Hertfordshire,

England. 51 39N 0 29W. Town sit. WSW of Watford at confluence of Rs. Chess and Gade with Colne R. Residential. Pop. (1976E) 29,574.

Ridderkerk Zuid-Holland, Netherlands. Town sit. SE of Rotterdam on Ijsselmonde island. Industries inc. shipbuilding. Fruit and vegetable market for Rotterdam. Pop. (1984E) 47,124.

Riesa Dresden, German Democratic Republic. 51 18N 13 17E. Town sit. NW of Dresden on Elbe R. Industries inc. sawmilling; manu. iron and steel, glass, soap, furniture and beer. R. port trading in grain and petroleum. Pop. (1989E) 47,300.

Rieti Lazio, Italy. 42 24N 12 51E. Town in central Italy sit. on Velino R. Cap. of Rieti province. Manu. textiles, pasta and olive oil.

Rif, Er Morocco. Mountain range extending 200 m. (320 km.) S and E from Ceuta (a Spanish enclave) on the Strait of Gibraltar to the Moulouya valley near the Algerian border. Inhabited by the Riff Berbers.

Riga Latvia, U.S.S.R. 56 57N 24 06E. City on Dvina R. Cap. of Latvia. Industries inc. shipbuilding and textiles; manu. cement, footwear, rubber products, paper and telephone equipment. Sea port trading in flax, timber, paper, butter and eggs; the harbour is open *c.* 8 months of the year. Pop. (1989E) 915,000.

Riga, Gulf of U.S.S.R. 57 30N 23 35E. Inlet of the Baltic Sea off the coasts of Latvia and Estonia. Length *c.* 100 m. (160 km.); width up to 60 m. (96 km.). Ice-free May to Dec.

Rigi Switzerland. Mountain ridge in the Alps to the N of L. Luzern rising to Kulm, 5,908 ft. (1,801 metres) a.s.l.

Riihimäki Häme, Finland. 60 45N 24 46E. Town sit. N of Helsinki. Railway junction. The main industry is glass making, also manu. clothing. Pop. (1983E) 24,196.

Rijeka (Fiume) Croatia, Yugoslavia.

45 20N 14 27E. City in NW, sit. on Adriatic coast. Industries inc. shipbuilding, oil refining and chemicals; manu. tobacco products. Chief sea port of Yugoslavia. Pop. (1983E) 193,044.

Rijswijk Zuid-Holland, Netherlands. 52 04N 4 22E. Town immediately SE of The Hague. Residential with some manu. inc. film, furniture. Pop. (1990E) 48,500.

Rimac River Peru. Rises in the W Cordillera of the Andes and flows 80 m. (128 km.) to the Pacific near Lima. Used for hydroelectric power and irrigation. Main trib. Santa Eulalia R.

Rimatara Austral Islands, French Polynesia. Island sit. W of Rurutu. Chief town Amaru. Volcanic; produces taro.

Rimini Emilia-Romagna, Italy. 44 04N 12 34E. Town sit. on N Adriatic coast. Industries inc. tourism and textiles; manu. footwear and pasta. Pop. (1988E) 130,644.

Ringerike Buskerud, Norway. 60 09N 10 16E. Region of SE Norway centred on Honefoss extending between Rands Fjord (N) and Tyri Fjord (S). Lumbering and pulp milling district, industries powered by Begua and Rand Rs. Pop. (1983E) 27,000.

Riobamba Chimborazo, Ecuador. 1 40S 78 38W. Town in central Ecuador sit. at 9,000 ft. (2,743 metres) a.s.l. in the Andes, SE of Mount Chimborazo. Industries inc. textiles; manu. carpets and footwear. Pop. (1982C) 149,757.

Río Cuarto Córdoba, Argentina. 33 08S 64 20W. Town in central Argentina, sit. on Río Cuarto. Commercial centre trading in agric. produce. Industries inc. textiles and manu. cement. Garrison town. Pop. (1980C) 75,000.

Rio de Janeiro Brazil. 22 00S 42 30W. State of E Braz. on Atlantic coast. Area, 17,092 sq. m. (44,268 sq. km.). Cap. Niterói. Mountainous, crossed by the Serra do Mar and drained by Paraíba R. Industrial, with agric. areas producing coffee, fruit and sugar-cane. At Volta Redonda

there are steel works. Pop. (1989E) 13,845,243.

Rio de Janeiro Guanabara, Brazil. 22 54S 43 15W. State cap. sit. on the Atlantic coast. Cap. of Brazil until inauguration of Brasília as cap. in 1960. Industries inc. flour milling, sugar refining and railway engineering; manu. chemicals, clothing, furniture and tobacco products. Sea port exporting coffee, sugar and iron ore. Noted for its spectacular mountain setting and land-locked harbour on the SW shore of Guanabara Bay. Pop. (1980C) 5,090,700.

Río Gallegos Santa Cruz, Argentina. 51 37S 69 10W. Town sit. SSW of Buenos Aires on Gallegos R. estuary. Trading centre for wool and sheepskins. Distribution centre linked by rail to the El Turbio coal mines. Pop. (1980C) 42,000.

Rio Grande Rio Grande do Sul, Brazil. 32 02S 52 05W. Town in extreme SE sit. at entrance to the Lagôa dos Patos. Industries inc. meat packing, fish and vegetable canning, oil refining and textiles; manu. footwear. Sea port exporting hides and meat products. Pop. (1980E) 125,000.

Rio Grande Mexico/U.S.A. *see* **Grande, Rio**

Rio Grande do Norte Brazil. 5 40S 36 00W. State in NE Braz. bounded N and E by Atlantic Ocean. Area, 20,469 sq. m. (53,015 sq. km.) Cap. Natal. Semi-arid plateau dropping to sandy coastal plain. Salt deposits are important. Other products are cotton, sugarcane and hides. Pop. (1989E) 2,277,672.

Rio Grande do Sul Brazil. 30 00S 54 00W. Most S state, bounded S by Uruguay and E by Atlantic Ocean. Area, 110,895 sq. m. (282,184 sq. km.). Cap. Pôrto Alegre. Main occupations stock farming and viticulture, with commercial forestry in the N. Pop. (1989C) 9,026,725.

Ríohacha Magdalena, Colombia. 11 33N 72 55W. Town sit. ENE of Santa Marta on Caribbean near mouth of Ranchería R. Commercial centre for an agric. area with saltworks and fishing industry. Pop. (1984E) 100,000.

Rio Muni   Equatorial Guinea. 1 30N 10 00E. The mainland area bounded N by Cameroon, E and S by Gabon, W by the Gulf of Guinea. It inc. the off-shore islands of Corisco, Elobey Grande, Elobey Chico. Area, 10,045 sq. m. (26,017 sq. km.). Chief town Bata. Densely forested region with a humid tropical climate. The only industry is lumbering. Pop. (1984E) 245,000.

Rion River   Georgia, U.S.S.R. Rises in the Caucasus Mountains and flows 180 m. (288 km.) WSW past Kutaisi to enter the Black Sea at Poti. Used for hydro-electric power.

Riouw Archipelago   Indonesia. 1 00N 104 30E. Arch. off E coast of Sumatra at S end of the Strait of Malacca. Area, 2,280 sq. m. (5,905 sq. km.). Chief town, Tandjungpinang. Chief products tin and bauxite.

Ripley   Derbyshire, England. 54 03N 1 34W. Town sit. NNE of Derby. Industries inc. iron-working, engineering and knitwear manu. Pop. (1990E) 19,000.

Ripon   North Yorkshire, England. 54 08N 1 31W. Town in central N Eng. sit. on Ure R. Industries inc. paints and varnishes, agric. engineering, tourism and food processing. Pop. (1987E) 13,090.

Risalpur   North-West Frontier Province, Pakistan. Town sit. ENE of Peshawar. Military station and site of airforce training school.

Risaralda   Caldas, Colombia. Town sit. WNW of Manizales in Cauca R. valley. Centre of an agric. area, producing rice, cacao, bananas.

Risdon   Tasmania, Australia. Town sit. opposite Hobart on Derwent R. Main industry is electro-metallurgical work. The oldest settlement in Tasmania (1803).

Rivas   Nicaragua. 11 26N 85 51W. (i) Dept. of SW Nicaragua bounded S

by Costa Rica, W by the Pacific and E by L. Nicaragua. Area, 850 sq. m. (2,202 sq. km.). Stock raising area with dairy farming and agric., esp. coffee, cacao, tobacco, sesame, sugar-cane, cotton, fruit. Associated processing industries. Pop. (1981E) 108,913. (ii) Town sit. SW of Managua on the Inter-American Highway. Dept. cap. and processing centre for an agric. area. Industries also inc. tanning, manu. rubber. Pop. (1980E) 21,000.

Rivera   Rivera, Uruguay. 30 54S 55 13W. Cap. of dept. sit. N of Montevideo on the Brazilian frontier. Market town trading in grain, cattle and fruit. Industries inc. textiles. Pop. (1980E) 25,000.

Riverina   New South Wales, Australia. 35 30S 145 30E. Region bounded N by Rs. Lachlan and Murrumbidgee and S by Murray R. Grassy plains and fertile arable land important for sheep farming and wheat.

River Cess   Grand Bassa, Liberia. Town sit. SE of Buchanan at mouth of Cess R. on the Atlantic coast. Port exporting cacao, copra, cassava, rice.

Rivers   Nigeria. State formed from the Degema, Port Harcourt and Yenagoa provinces of the former E region. Area, 8,436 sq. m. (21,850 sq. km.). The inhabitants are mainly Ijaw. Major oilfields providing 95% of Nigeria's exports. The cap. and major port is Port Harcourt. Pop. (1983E) 2·1m.

Riverside   California, U.S.A. 33 59N 117 22W. City sit. E of Los Angeles on Santa Ana R. Residential and industrial city and commercial and service centre for agric. and recreational area. Industries inc. electronics, aeronautics, food processing. Manu. inc. mobile homes and irrigation equipment. Pop. (1989E) 211,000.

Riverton   Wyoming, U.S.A. 43 02N 108 23W. City sit. NE of Lander where Wind and Popo Agie Rs. meet to form Bighorn R. Mining centre (oil, gas, iron ore, uranium); service centre for a pastoral and irrigated agric. area. Pop. (1980C) 9,980.

Riviera   France/Italy. 46 15N 8 58E. Coastal strip between the Mediterranean Sea and the mountains, extending E from Hyères, France, to La Spezia, Italy. Noted as a tourist area because of its climate, flowers and scenery, and inc. the resorts of Nice, Cannes, Menton, Antibes, Juan-les-Pins, Monte Carlo, San Remo, Bordighera, Imperia, Ventimiglia, Rapallo, Portofino and Sestri Levante.

Riyadh   Nejd, Saudi Arabia. 24 41N 46 42E. Joint cap. (the religious cap. is Mecca) of Saudi Arabia, and cap. of Nejd, sit. in central S, in an oasis. Commercial centre. Industries inc. oil processing. Pop. (1988E) 2m.

Riyak   Bekaa, Lebanon. Town sit. E of Beirut near the Syrian frontier. Railway centre for Aleppo, Damascus and Beirut.

Rize   Turkey. 41 02N 40 31E. (i) Province of NE Turkey on the Black Sea, bounded S by the Rize Mountains. Mining and farming area; chief products zinc, iron, manganese, tea, olives, maize. Pop. (1985C) 374,206. (ii) Town sit. NW of Erzurum on the Black Sea. Provincial cap. Port handling citrus fruit, maize, tea, manganese.

Road Town   British Virgin Islands. 18 26N 64 32W. Cap. and port on the island of Tortola. Tourism is important with cruise ships calling. Pop. (1983E) 3,000.

Roanne   Loire, France. 46 02N 4 04E. Town in E central France sit. on Loire R. Industries inc. textiles; manu. paper, leather goods and tiles. Pop. (1982C) 49,638 (agglomeration, 81,786).

Roanoke   Virginia, U.S.A. 37 16N 79 57W. City sit. WSW of Washington D.C., at the end of the Shenandoah Valley W of the Blue Ridge Mountains. Banking, insurance, medical and distribution centre for W Virginia. Pop. (1980C) 100,200.

Roanoke Island   North Carolina, U.S.A. 35 53N 75 39W. Island off the coast of N Carolina at the N end of Pamlico Sound. Scene of an attempt at colonization by Sir Walter Raleigh, 1585 and 1587. About 12 m. (19 km.) long and 3 m. (5 km.) wide.

Roanoke River   U.S.A. 35 56N 76 43W. Rises in SW Virginia and flows 410 m. (656 km.) ESE into N Carolina to enter Albemarle Sound.

Roatán   Bay Islands, Honduras. 16 18N 86 35W. Town of S coast of Roatán Island, sit. NE of La Ceiba. Dept. cap. Industries inc. fishing, shipbuilding, fish canning, fruit and meat packing. Port exports coconuts.

Robin Hood's Bay   North Yorkshire, England. 54 26N 0 31W. Village sit. SE of Whitby on the North Sea coast. Industries inc. tourism and fishing. Pop. (1981C) 1,387.

Roboré   Santa Cruz, Bolivia. Village sit. ESE of San José on the Corumba–Santa Cruz railway. Military headquarters for E Bolivia.

Robson, Mount   Canada. 53 07N 119 09W. Peak in E British Columbia, 12,972 ft. (3,954 metres) a.s.l., the highest peak in the Canadian Rockies. Centre of the Mount Robson Provincial Park, extending 65 m. (104 km.) by 20 m. (32 km.).

Roçadas   Huíla, Angola. Town sit. SE of Sá da Bandeira. District cap.

Rocha   Uruguay. 34 29S 54 20W. (i) Dept. of SE Uruguay on the Atlantic coast. Area, 4,281 sq. m. (11,089 sq. km.). Low-lying area with marshes and fresh water lagoons. The main occupation is stock farming; esp. cattle, sheep, horses. Pop. (1985C) 68,500. (ii) Town sit. ENE of Montevideo. Dept. cap. and trading centre for a stock farming area. Pop. (1985C) 23,910.

Rochdale   Greater Manchester, England. 53 37N 2 09W. Metropolitan district of Greater Manchester sit. on Roch R. The main industries are engineering and distribution services. Pop. (1988E) 206,000.

Rochefort-sur-Mer   Charente-Maritime, France. 45 57N 0 58W. Town and port in central W sit. above mouth of Charente R. on the Bay of Biscay. Industries inc. fish processing and sawmilling. Pop. (1982C) 27,716.

Rochelle, La   Charente-Maritime, France. 46 10N 1 09W. Cap. of Dept. Fishing and international leisure port sit. on the Bay of Biscay. Pop. (1982C) 78,231 (agglomeration, 102,143).

Rochester   Minnesota, U.S.A. 44 02N 92 29W. City in SE of state. Commercial centre for an agric. area. Industries inc. high technology, manu. foodstuffs, esp. dairy produce. Noted for the Mayo clinic. Pop. (1989E) 64,797.

Rochester   New York, U.S.A. 43 10N 77 36W. Town and port sit. on L. Ontario at mouth of Genesee R. in a fruit growing and market-gardening area. Industries inc. horticulture, manu. cameras, photographic equipment, optical appliances and office equipment. Pop. (1980C) 241,741.

Rochester upon Medway   Kent, England. Includes the City of Rochester, and towns of Chatham and Strood, sit. ESE of London on Medway R. estuary. Industries inc. navigational equipment, electronics engineering, cement and insurance. Pop. (1983E) 146,400.

Roche-sur-Yon, La   Vendée, France. Cap. of dept. in central W sit. on Yon R. Market town. Industries inc. flour milling and tanning. Pop. (1982C) 48,156.

Rockall   Scotland. 57 35N 13 48W. The uninhabited island of Rockall lies W of St. Kilda, the most W of the Hebrides, and about 290 m. (464 km.) from the nearest point on the mainland of Scotland. Rising almost sheer from the ocean to a height of 70 ft. (21 metres) a.s.l., it is only about 100 ft. (30 metres) by 80 ft. (24 metres) in area.

Rockford   Illinois, U.S.A. 42 17N 89 06W. City sit. WNW of Chicago on Rock R. Industries inc. agric. engineering, manu. machine tools, hardware, furniture, hosiery and paints. Pop. (1980C) 139,712.

Rockhampton   Queensland, Australia. 23 23S 150 31E. City in E sit. up Fitzroy R. from Keppel Bay. Commercial centre for an extensive stock-raising hinterland. Industries inc. meat processing. Pop. (1986E) 55,700.

Rock Hill   South Carolina, U.S.A. 34 56N 81 01W. Town sit. N of Columbia. Industries inc. textiles and paper-making. Pop. (1980C) 35,344.

Rock Island   Illinois, U.S.A. 41 30N 90 34W. Town sit. opposite Davenport, Iowa, on Mississippi R. Industries inc. railway engineering; manu. agric. and electrical equipment, hardware and clothing. On the island offshore is the main U.S. arsenal. Pop. (1980C) 47,036.

Rock Springs   Wyoming, U.S.A. 41 35N 109 13W. City in SW of state at 6,271 ft. (1,913 metres) a.s.l. on trib. of Green R. Trad. mining town, now producing trona, oil, gas and coal. Service and railway centre for a live-stock-farming area. Pop. (1980C) 19,458.

Rockville   Maryland, U.S.A. 39 05N 77 09W. City NW of Washington D.C. Pop. (1980C) 43,811.

Rocky Mount   North Carolina, U.S.A. 35 57N 77 48W. City sit. ENE of Raleigh on Tar R. Industries inc. railway engineering, tobacco processing, textiles, pharmaceuticals, electronics, diesel engines, clothing, furniture, food processing and cotton milling. Pop. (1980C) 41,283.

Rocky Mountains   North America. 48 00N 116 00W. Extensive mountain system stretching from Yukon Territory, Canada, to New Mexico, U.S.A. The Canadian Rockies extend *c.* 900 m. (1,440 km.), partly along the border between British Columbia and Alberta. The U.S. Rockies consist of over 20 ranges stretching through Montana, Wyoming, Colorado, Utah and New Mex. The N section (Montana) has 26 peaks over 9,000 ft. (2,743 metres) a.s.l. The central section (Wyoming) has the Teton Range with Grand Teton, 13,747 ft. (4,190 metres) a.s.l. The S section rises to Mount Elbert, 14,431 ft. (4,339 metres) a.s.l., in the Sawatch Range. Some geographers consider the Yukon and Alaska ranges as part of the system, rising to Mount Logan, 19,850 ft. (6,050 metres) a.s.l.

Rodez   Aveyron, France. 44 22N 2 34E. Cap. of dept. Chief indust. woollens and tourism. Pop. (1982C) 26,346.

Rodopi   Greece. 41 30N 24 30E. *Nome* in Thrace bounded N by Bulgaria and S by the Aegean Sea. Area, 984 sq. m. (2,549 sq. km.). Cap. Komotini. Mountainous in the N, coastal plain in the S. Chief occupation farming, esp. tobacco. Pop. (1981C) 107,957.

Rodriguez   Mauritius. 19 42S 63 25E. Island in the Indian Ocean sit. ENE of Mauritius. Area, 40 sq. m. (104 sq. km.). Chief town, Port Mathurin. Volcanic, rising to Mount Limon, 1,300 ft. (396 metres) a.s.l. Chief products maize, fish, fruit and tobacco. Pop. (1988E) 36,465.

Roermond   Limburg, Netherlands. 51 12N 6 00E. Town sit. NNE of Maastricht on Maas R. at its confluence with Roer R. Trading centre for agric. produce. Industries inc. textiles, paper-making, flour milling, chemicals, electronics, plastics, metals. Pop. (1985E) 38,260.

Roeselare   West Flanders, Belgium. 50 57N 3 08E. Town sit. NW of Kortrijk. The most important industries are textiles, coachwork, electrotechnics. Pop. (1985E) 51,884.

Rogaland   Norway. 59 12N 6 20E. County of SW Norway bounded W by North Sea, E by Ruven and Bykle ranges. Area, 3,529 sq. m. (9,140 sq. km.). Chief towns Stavanger (cap.), Egersund, Sandnes, Haugesund. Varied region of mountains, fjords and lowlands. Main occupations farming, fishing and mining including copper on Karmoy Island. Bokna Fjord and its many inlets occupy *c*. 25% of the county (NW). Pop. (1989E) 333,351.

Rohtak   Haryana, India. 28 54N 76 35E. Town sit. NW of Delhi. Commercial centre trading in grain, cotton and oilseeds. Industries inc. engineering goods and cotton ginning. Pop. (1981C) 166,767.

Roissy-en-France   Val-d'Oise, France. Site of Paris's Charles-de-Gaulle intern. airport. Pop. (1982C) 1,411.

Rolla   Missouri, U.S.A. 37 57N 91 46W. City sit. SE of Jefferson City in the Ozark Mountains. Medical and retail trade centre for a farming and mining area. Site of the University of Missouri Schools of mines, metallurgy and engineering. Pop. (1980C) 13,303.

Roma, Campagna di   Lazio, Italy. Plain in Rome province bounded SW by the Tyrrhenian Sea. Drained by Tiber R. Neglected land much restored by drainage.

Roma   Queensland, Australia. 26 35S 148 47E. Town sit. WNW of Brisbane serving an agric. area. The largest industries are connected with food processing. Natural gas was found in the 1960s and is used locally and is also piped to Brisbane. Oil wells are found in the vicinity. Pop. (1988E) 7,000.

Roman   Neamt, Romania. 46 55N 26 56E. Town sit. WSW of Iasi on Moldava R. near its confluence with Siret R. Railway junction. Industries inc. flour milling and sugar refining.

Romana, La   La Altagracia, Dominican Republic. 18 27N 68 57W. Port sit. E of Santo Domingo. Industries sugar refining, and coffee and tobacco processing. Pop. (1982E) 99,000.

Romania   46 00N 25 00E. Rep. bounded N and E by U.S.S.R. and the Black Sea, S by Bulgaria, W by Yugoslavia and Hungary. The Carpathian mountains have 3 main branches; the E Carpathians, the Transylvanian Alps and the Bihor Mountains. To the S and E of the Carpathians is the plain of the R. Danube. Area, 91,671 sq. m. (237,428 sq. km.). Chief cities Bucharest (cap.), Constanta, Cluj, Timisoara, Galati and Ploesti. Romania is primarily an agric. country and crops include wheat, barley, maize, potatoes, sunflower seed and sugar-beet. The principal minerals are oil and natural gas, salt, brown coal, lignite, iron and copper ores, bauxite, chromium, manganese and uranium. The oilfields are in the Prahova, Bǎcau, Gorj, Crişana and Agreş districts. Refining capacity exceeds production of crude and some crude is imported. Salt is mined in the lower Carpathians and in Transylvania. Pop. (1987E) 22,940,430.

Rome Lazio, Italy. 41 54N 12 30E. Cap. of Italy sit. on Tiber R. Also cap. of Lazio region and Roma province. Following the unification of Italy Turin (1861) and Florence (1865) were capitals before Rome (1870). Vatican City, the seat of the Papacy, is on the N bank of the R., the main part of the city on the S. Cultural, commercial and communications centre. Industries inc. administration, printing, publishing, tourism and film production; manu. machinery, furniture, glass and cement. Pop. (1988E) 2,816,474.

Rome Georgia, U.S.A. 34 16N 85 11W. City sit. NW of Atlanta at confluence of Rs. Etowah and Oastanaula. Industries inc. manu. cotton and rayon goods, hosiery, paper and agric. machinery. Pop. (1980C) 29,654.

Rome New York, U.S.A. 43 13N 75 27W. City NE of Syracuse on Mohawk R. The main industry is copper and brass working and tourism; also manu. cables and wire. Pop. (1980C) 50,418.

Romford Greater London, England. 51 35N 0 12E. District in E London N of Thames R. Industries inc. engineering and brewing; manu. pharmaceutical and plastic products. *Now* Havering.

Romney Marsh Kent, England. 51 03N 0 55E. Stretch of reclaimed marshland extending along the English Channel coast from Winchelsea in the W to Hythe in the E. Used as pasture for sheep.

Romsdal Norway. Valley in SW Norway made by Rauma R. extending *c.* 60 m. (96 km.) between mountains

to the head of Romsdal Fjord. The surrounding peaks rise to 5,102 ft. (1,555 metres) a.s.l. in Romsdalshorn.

Romsey Hampshire, England. 50 59N 1 30W. Town sit. NW of Southampton on Test R. Tourist centre and town serving an agric. area. Pop. (1990E) 14,733.

Ronda Málaga, Spain. 36 44N 5 10W. Town sit. W of Málaga on Guadiaro R. Trading centre for olives, wine and leather. Industries inc. flour milling and tanning. Pop. (1981E) 31,000.

Rondônia Brazil. 11 00S 63 00W. State in W Braz. formerly called Guaporé, bounded W and S by Bolivia. Area, 93,839 sq. m. (243,044 sq. km.). Cap. Pôrto Velho. Mainly rain forest, crossed by the Madeira–Mamora railway. Chief products rubber and Brazil nuts. Pop. (1989E) 1,057,237.

Rongai Rift Valley, Kenya. Town sit. NW of Nakuru at 6,137 ft. (1,871 metres) a.s.l. Railway junction serving L. Solai and a farming area producing wheat, maize, coffee and dairy products.

Ronse East Flanders, Belgium. 50 45N 3 35E. Town sit. WSW of Brussels. The main industry is textiles. Pop. (1989E) 23,997.

Roodepoort-Maraisburg Transvaal, Republic of South Africa. 26 11S 27 54E. Town sit. W of Johannesburg on the Witwatersrand. A goldmining centre containing the towns of Roodepoort, Florida and Maraisburg. Pop. (1980E) 165,315.

Roosendaal North Brabant, Netherlands. 51 32N 4 28E. Town sit. W of Tilburg. Railway junction. Industries inc. railway engineering, chemicals, food processing and sugar refining. Pop. (1989E) 60,000.

Roosevelt, Teodoro, River Brazil. Rises near Velhena in SE Rondônia and flows 500 m. (800 km.) N to join Aripuana R., a trib. of Madeira R.

Roquefort-sur-Soulzon Aveyron, France. 43 59N 2 59W. Village sit.

ssw of Millau, noted for its cheese. Pop. (1982E) 880.

Roraima   Brazil. 5 12N 60 44W. Federal territory in N Braz., formerly called Rio Branco, bounded N and W by Venezuela and E by Guyana. Area, 86,879 sq. m. (225,017 sq. km.). Cap. Boa Vista. Mainly rain forest, drained by Rio Branco and its tribs.; some stock is reared on the savannah around the cap. Pop. (1989E) 116,765.

Rosario   Santa Fé, Argentina. 35 57S 60 40W. Town in central E sit. on Paraná R. Railway, commercial and industrial centre, esp. for sugar refining, meat packing, brewing and flour milling; manu. bricks and furniture. Port exporting wheat, meat and hides from the Pampas. Pop. (1980E) 1m.

Rosario   West Pacific. Island W of the Bonin Islands, U.S. depend.

Roscoff   Finistère, France. 48 44N 4 00W. Town and car ferry port sit. NW of Morlaix on N coast of Brittany. Industries inc. fishing and tourism. Trades in horticultural produce, esp. onions, for the Eng. market. Pop. (1982C) 3,787.

Roscommon   Ireland. 53 38N 8 11W. (i) County in central Ireland bounded E by Shannon R. Area, 984 sq. m. (549 sq. km.). Hilly in N, rising to the Bralieve Mountains, 1,082 ft. (330 metres) a.s.l., with many Ls. Main occupations mixed farming and light industry. Pop. (1988E) 54,592. (ii) County town of (i). Market for livestock and farm produce. Pop. (1981C) 3,533.

Roseau   Dominica, Windward Islands, West Indies. 15 20N 61 24W. Cap. and port sit. on SW coast of Dominica NNW of Fort-de-France, Martinique. Exports fruit, oils, vegetables, spices. Picturesque resort beneath high mountains. Pop. (1981E) 20,000.

Roseirès   Blue Nile, Sudan. 11 52N 34 23E. Town sit. SSE of Singa on E bank of Blue Nile near the Ethiopian border. R. port at the head of navigation, handling cotton, sesame, maize, durra.

Rosenheim   Bavaria, Federal Republic of Germany. 47 51N 12 07E. Town sit. in SW on Inn R. Industries inc. chemicals, textiles, electrical engineering, paper, food processing. Pop. (1988E) 54,000.

Rosetta   Beheira, Egypt. 31 24N 30 25E. Town sit. NE of Alexandria on Rosetta branch of Nile Delta. Trading centre for rice. Industries inc. rice milling, fishing.

Roseville   Michigan, U.S.A. 42 30N 82 56W. City sit. NE of Detroit and forming part of Macomb county industrial area. Pop. (1980C) 54,311.

Roskilde   Zealand, Denmark. 55 39N 12 05E. Town sit. at S end of Roskilde Fjord. Industries inc. agric. engineering, distillery, meat canning, concrete, kitchen units. Pop. (1988E) 48,968.

Ross and Cromarty   Scotland. Former county now part of Highland Region except the burgh of Stornoway and the district of Lewis both in the Hebrides.

Ross Dependency   Antarctica. 80 00S 180 00E. New Zealand territory in the Antarctic between long. 160°E and 150°W, and S of lat. 60°S, inc. the mainland area, the coastal regions of Victoria Land and King Edward VII Land, Ross Sea and its islands. Area, c. 160,000 sq. m. (c. 414,000 sq. km.). The Ross Sea inlet provides a near approach to the Pole for survey and other expeditions.

Ross Island   Ross Dependency, Antarctica. 77 30S 168 00E. Island in the Ross Sea off Victoria Land. Volcanic, rising to Mount Erebus which is active, 13,202 ft. (4,024 metres) a.s.l., and Mount Terror, 10,750 ft. (3,277 metres) a.s.l.

Rosslare   Wexford, Ireland. 52 17N 6 23W. Town sit. on St. George's Channel. Rosslare Harbour to the SE is a ferry-service port for Wales and Europe. Pop. (1980E) 600.

Rosso   Mauritania. Town sit. NE of Saint Louis, Senegal, on Senegal R. Market for gum arabic, maize, millet, beans, melons. Pop. (1984E) 16,466.

Ross-on-Wye   Hereford and Worcester, England. 51 55N 2 35W. Town sit. SSE of Hereford. Market town for a farming and cider-making area, Industries inc. tourism. Pop. (1990E) 9,000.

Ross Sea   Ross Dependency, Antarctica. 74 00S 178 00E. Large inlet between Cape Adare and Cape Colbeck in Ross Depend. The N half is occupied by the Ross Ice Shelf.

Rostock   German Democratic Republic. 54 05N 12 07E. (i) District in N bounded N by the Baltic Sea. Area, 2,729 sq. m. (7,068 sq. km.). Pop. (1988E) 468,800. (ii) Cap. of (i) and port sit. near the Baltic coast at the mouth of the Warnow R. estuary. Commercial centre. Industries inc. shipbuilding, fishing and fish processing, brewing, diesel engines and chemicals. Pop. (1988E) 253,990.

Rostov-on-Don   Russian Soviet Federal Socialist Republic, U.S.S.R. 47 17N 39 39E. Cap. of Rostov region sit. NE of the Sea of Azov on Don R. Communications and commercial centre. Industries inc. shipbuilding, railway and agric. engineering, textiles and chemicals; manu. tobacco and leather products. Port, handling overseas trade through Taganrog. Sit. on the Sea of Azov. Pop. (1989E) 1,020,000.

Roswell   New Mexico, U.S.A. 33 23N 104 32W. City sit. in SE of state on Rio Hondo. Commercial centre for an irrigated farming area, trading in wool. Industries inc. oil refining and food processing. Pop. (1988E) 59,300.

Rosyth   Fife Region, Scotland. 56 03N 3 26W. Naval base and dockyard on the N shore of the Firth of Forth. Pop. (1988E) 11,540.

Rota   Cádiz, Spain. 36 37N 6 20W. Town sit. NNW of Cádiz on the Atlantic. Port exporting red wine, fish and vegetables. Industries inc. fishing, fish processing, wine making, distilling.

Rotherham   South Yorkshire, England. 53 26N 1 20W. Metropolitan district of South Yorkshire sit. on Don R. Industries inc. coal, iron and steel;

manu. brassware, machinery and glass. Pop. (1988E) 251,800.

Rotherhithe   Greater London, England. 51 30N 0 03W. District of SE London on S bank of Thames R.

Rother River   England. (i) Rises in NE Derbyshire and flows 21 m. (34 km.) N into South Yorkshire to join Don R. at Rotherham. (ii) Rises 5 m. (8 km.) NNE of Petersfield in Hampshire and flows 24 m. (38 km.) S and E past Midhurst, W. Sussex, to join Arun R. 7 m. (11 km.) N of Arundel. (iii) Rises W of Mayfield, E. Sussex, and flows 31 m. (50 km.) E to the Kent/E. Sussex boundary then S past Rye to enter the English Channel. It once flowed across Romney Marsh to enter the sea at New Romney but its course was diverted by a severe storm in 1287.

Rothesay   Strathclyde Region, Scotland. 55 51N 5 03W. Town on E coast of island of Bute on Clyde estuary. Industries inc. tourism, electronics and fabric manu. Pop. (1980E) 6,400.

Rothwell   Northamptonshire, England. 52 25N 0 48W. Town sit. WNW of Kettering. Manu. clothing, welding equipment, plastics, steel fabricators and footwear. Pop. (1987E) 6,700.

Rotorua   North Island, New Zealand. 38 08S 176 15E. Town sit. near N coast and at S end of L. Rotorua, in a district of hot springs. Health resort; the main industries are tourism, farming and forestry. Pop. (1989E) 63,000.

Rotterdam   Zuid-Holland, Netherlands. 51 55N 4 28E. City in SW, sit. on New Maas R. Industries inc. shipbuilding, oil refining, engineering, brewing and distilling; manu. food products, soap and chemicals. It is the world's largest port handling 292m. tonnes of goods in 1989. Pop. (1989E) 576,218.

Roubaix   Nord, France. 50 42N 3 10E. Town in NE. Together with the town of Tourcoing, immediately NNW, it forms the centre of the woollen industry; also manu. textile machinery, other textiles, rubber and plastic products, clothing and carpets. Pop. (1982C) 101,886.

Rouen   Seine-Maritime, France. 49
26N 1 05E. Cap. of dept. in N central
France, sit. on Seine R. Industries inc.
textiles, esp. cotton, paper making and
chemicals. Port handling manu. goods,
wines   and   spirits.   Pop.   (1988E)
101,945 (agglomeration, 354,704).

Roulers   West Flanders, Belgium.
50 57N 3 08E. Town sit. WSW of
Ghent. Industries inc. manu. textiles,
esp. linen, and carpets, electronics,
cattle feed, plastics, coachwork. Pop.
(1989E) 52,310.

Rourkela   Orissa, India. Town sit.
WSW of Jamshedpur. The main industry is iron and steel. Pop. (1981C)
214,521.

Rouyn   Quebec, Canada. 48 14N 79
01W. Town NW of Cobalt, Ontario,
sit. in a mining area. The main industry is copper mining, also gold, zinc
and silver. Pop. (1985E) 20,000.

Rovaniemi   Lappi, Finland. 66 34N
25 48E. Town in NW sit. on Kemi R.,
just S of the Arctic Circle. Commercial and tourist centre. Pop. (1988E)
32,915.

Rovereto   Trentino-Alto   Adige,
Italy. 45 53N 11 02E. Town sit. SSW
of Trento. Industries inc. tourism,
chemicals, textiles and paper making.
Pop. (1990E) 33,000.

Rovigo   Veneto, Italy. 45 04N 11
47E. Cap. of province of same name
in NE. Industries inc. sugar refining;
manu. rope and furniture.

Rovno   Ukraine, U.S.S.R. 50 37N
26 15E. Cap. of region of same name
sit. ENE of Lvov. Railway junction. Industries inc. textiles; manu. machinery
and   food   products.   Pop.   (1987E)
233,000.

Rowley Regis   West Midlands, England. 52 29N 2 03W. Town sit. WNW
of Birmingham on Stour R. in a quarrying and mining area. Industries inc.
mining, metal working and engineering.

Roxburghshire   Scotland.   Former
county now part of Borders Region.

Royal Leamington Spa   *see* **Leamington Spa**

Royal Oak   Michigan, U.S.A. 42
30N 83 08W. City sit. NNW of Detroit,
mainly residential. Manu. tools and
paint. Pop. (1980C) 70,893.

Royston   Hertfordshire, England. 52
03N 0 01W. Market town for an agric.
area. Industries inc. chemicals. Pop.
(1989E) 14,000.

Royston   South Yorkshire, England.
53 36N 1 28W. Town NNE of Barnsley sit. in a coalmining area. Manu.
clothing. Pop. (1981C) 10,512.

Ruapehu   North Island, New Zea-
land. 39 18S 175 35E. Mountain peak,
volcanic and sometimes active, in the
Tongariro National Park SSW of L.
Taupo. The highest peak on the N.
Island, 9,175 ft. (2,797 metres) a.s.l.

Rub'al Khali   Southern Arabia. 18
00N 48 00E. The 'Empty Quarter';
desert mainly in S Saudi Arabia. Area,
300,000 sq. m. (800,000 sq. km.).
Contains both stony and sandy desert,
with large dunes in the E.

Rubicon River   Emilia-Romagna,
Italy. R. rising NW of San Marino Rep.
and flowing 25 m. (40 km.) NE to
enter the Adriatic NW of Rimini.
Crossed by Julius Caesar when it
formed the S boundary of Cisalpine
Gaul.

Rubtsovsk   Russian Soviet Federal
Socialist Republic, U.S.S.R. 51 33N
81 10E. Town sit. NNE of Semipalatinsk near the border with Kazakhstan.
Industries inc. flour milling and agric.
engineering. Pop. (1987E) 168,000.

Ruda Slaska   Silesia, Poland. 50
18N 18 51E. Town formed, in 1959,
by the amalgamation of the towns of
Ruda and Nowy Bytom. Pop. (1983E)
163,000.

Rudolf, Lake   *see* **Turkana, Lake**

Rudolstadt   Gera, German Democratic Republic. 50 43N 11 20E.
Town sit. WSW of Gera on Saale R. Industries inc. tourism, porcelain and
chemicals. Pop. (1989E) 32,300.

Rueil-Malmaison Hauts-de-Seine, France. 48 53N 2 11E. Suburb of NW Paris. Industries inc. manu. photographic equipment and pharmaceutical goods. Pop. (1982C) 64,545.

Rufiji River Tanzania. 8 00S 39 20E. Rises in S central Tanzania and flows 375 m. (600 km.) NE and then E to enter the Indian Ocean by a delta 90 m. (144 km.) S of Dar-es-Salaam. Navigable by small craft. Main trib. Great Ruaha R.

Rufisque Senegal. 14 43N 17 17W. Town sit. ENE of Dakar. Industries inc: tanning and engineering; manu. cement, cotton goods, groundnut oil, footwear and pharmaceutical products.

Rugby Warwickshire, England. 52 23N 1 15W. Town in central Eng. Industries inc. engineering. Noted for a public school where the game of Rugby football originated. Pop. (1989E) 60,000.

Rugeley Staffordshire, England. 52 47N 1 56W. Town sit. ESE of Stafford on Trent R. Market town. Industries inc. electronics, engineering, coalmining and power station. Pop. (1981E) 24,366.

Rügen Rostock, German Democratic Republic. 54 25N 13 24E. Island in Baltic Sea off Stralsund and separated from it by the Strelasund. Area, 358 sq. m. (927 sq. km.). Chief town, Bergen. Main occupations farming, fishing and tourism. There is a causeway to the mainland.

Ruhr North Rhine-Westphalia, Federal Republic of Germany. Industrial area bounded N by Lippe R. and S by Wupper R. and sit. mainly in Ruhr R. valley. Extends W to E from Duisburg to Dortmund in a continuous industrial conurbation.

Ruhr River Federal Republic of Germany. 51 27N 6 44E. Rising in the Saarland and flowing 145 m. (232 km.) N and W past Mülheim to enter Rhine R. at Duisburg.

Ruislip-Northwood Greater London, England. 51 35N 0 26W. Town

sit. WNW of central London. Residential with some industry in Ruislip.

Rukwa, Lake Tanzania. 8 00S 32 25E. L. in SW Tanzania, mainly in Mbeya, extending 80 m. (128 km.) NW to SE, width up to 20 m. (32 km.). Fed by Rs. Songwe and Momba; it has no outlet.

Rum Highland Region, Scotland. 57 00N 6 20W. Island in the Inner Hebrides S of Skye. Area, 42 sq. m. (109 sq. km.). Mountainous, rising to 2,659 ft. (810 metres) a.s.l.

Rum Jungle Northern Territory, Australia. 13 01S 131 00E. Mining centre in NW of state. The industry is uranium mining.

Runcorn Cheshire, England. 53 20N 2 44W. A new town estab. 1964 grafted on to an old town, sit. ESE of Liverpool on Mersey R. and Manchester Ship Canal. Industries inc. chemicals, brewing and light engineering. R. port. Pop. (1985E) 64,600.

Rungwe, Mount Tanganyika, Tanzania. Volcanic peak 20 m. (32 km.) SE of Mbeya in S Highlands province. Height 9,713 ft. (2,961 metres) a.s.l.

Rurutu Austral Islands, French Polynesia. 22 26S 151 20W. Volcanic island 310 m. (500 km.) S of Tahiti. Chief town Moerai. Produces taro, arrowroot, copra and vanilla.

Ruse Ruse, Bulgaria. 43 50N 25 57E. Cap. of province sit. NE of Sofia on Danube R. Commercial centre trading in cereals. Industries inc. flour milling, tanning, sugar refining and textiles; manu. tobacco products. Port. Pop. (1987E) 190,450.

Rushden Northamptonshire, England. 52 17N 0 37N. Town sit. ENE of Northampton. Industries inc. footwear and light engineering. Pop. (1988E) 23,700.

Russian Soviet Federal Socialist Republic U.S.S.R. 62 00N 105 00E. Largest of the constituent reps. containing over 76% of the total area and *c*. 55% of the total pop. The Russian Soviet Federal Socialist Republic consists of: (i) Territories: Altai, Khabarovsk, Krasnodar, Krasnoyarsk, Primorye, Stavropol. (ii) Regions: Amur, Archangel, Astrakhan, Belgorod, Briansk, Chelyabinsk, Chita, Gorki, Irkutsk, Ivanovo, Kaluga, Kalinin, Kaliningrad, Kamchatka, Kemerovo, Kirov, Kostroma, Kuibyshev, Kurgan, Kursk, Leningrad, Lipetsk, Magadan, Moscow, Murmansk, Novgorod, Novosibirsk, Omsk, Orel, Orenburg, Penza, Perm, Pskov, Rostov, Ryazan, Sakhalin, Saratov, Smolensk, Sverdlovsk, Tambov, Tomsk, Tula, Tyumen, Ulyanovsk, Vladimir, Volgograd, Vologda, Voronezh, Yaroslavl. (iii) Autonomous Soviet Socialist Republics: Bashkir, Buryat, Checheno-Ingush, Chuvash, Daghestan, Kabardino-Balkar, Kalmyk, Karelian, Komi, Mari, Mordovian, North Ossetia, Tatar, Tuva, Udmurt, Yakut. (iv) Autonomous Regions: Adygei, Karachayevo-Cherkess, Gorno-Altai, Jewish, Khakass. (v) National Areas: Aginsky, Buryat, Chukot, Evenki, Khanty-Mansi, Komi-Permyak, Koryak, Nenetz, Taimyr (Dolgano-Nenetz), Ust-Ordynsky Buryat, Yamalo-Nenetz. Area, 6,501,500 sq. m. (16,838,900 sq. km.). Cap. Moscow; Leningrad is the second cap. The rep. produces *c.* 70% of the total industrial and agric. output of the U.S.S.R. The republic has a variety of climates (ranging from arctic to sub-tropical) and geographical conditions (tundra, forest lands, steppes and rich agricultural soil). It also contains great mineral resources: iron ore in the Urals, the Kerch Peninsula and Siberia; coal in the Kuznetz Basin, Eastern Siberia, Urals and the sub-Moscow Basin; oil in the Urals, Azov–Black Sea area and Bashkiria. It also has abundant deposits of gold, platinum, copper, zinc, lead, tin and rare metals. Pop. (1989E) 147·4m.

Rustavi Georgian Soviet Socialist Republic, U.S.S.R. 41 33N 45 02E. Town sit. SSE of Tiflis on Kura R. The main industries are iron and steel milling and manu. metal products. Industry is based on coal from Tkibuli and Tkvarcheli, ore from Dashkesan.

Rustenburg Transvaal, Republic of South Africa. 25 37S 27 08E. Town sit. W of Pretoria at the foot of the

Magaliesberg mountains. Commercial centre of a farming and mining area, handling cotton, fruit, tobacco, chrome, platinum and nickel. Industries inc. tobacco processing. Pop. (1990E) 46,000 (excluding Blacks who are citizens of the Rep. of Bophuthatswana).

Rutba Iraq. Staging post on the desert routes from Baghdad to Damascus, in W central Iraq, in the Syrian desert. Service industries on a small scale.

Rutherglen Strathclyde Region, Scotland. 55 50N 4 12W. Former burgh of Lanarkshire now part of Glasgow.

Ruthin Clwyd, Wales. 53 07N 3 18W. Town WNW of Wrexham sit. on Clwyd R. Market town for a farming area. Pop. (1981C) 4,480.

Rutland England. Former English county now part of Leicestershire.

Rutland Vermont, U.S.A. 43 36N 72 59W. City, sit. N of Bennington on Otter Creek. Railway centre and winter sports resort sit. between the Green and Taconic Mountains. Industries inc. marble cutting, food processing, manu. machinery, aircraft parts, tools, scales, cement products, timber products. Pop. (1990E) 20,000.

Ruvu Eastern Province, Tanzania. Town sit. W of Dar-es-Salaam on Ruvu R., where it is called Kingani R. Trading centre for sisal, cotton, rice.

Ruvuma River Mozambique/Tanzania. 10 29S 40 28E. Rises in S Tanzania, E of L. Nyasa and flows 500 m. (800 km.) E to form the Mozambique/Tanzania border, entering the Indian Ocean just S of Mtwara, Tanzania.

Ruwenzori Zaïre/Uganda. 0 23N 29 54E. Mountain massif extending 70 m. (112 km.) on the borders of Zaïre and Uganda, between L. Albert and L. Idi Amin Dada, rising to Mount Stanley, 16,794 ft. (5,119 metres) a.s.l. with equatorial forests at the foot, and snowfields and glaciers among the peaks.

Rwanda 2 00S 30 00E. Rep. bounded N by Uganda, E by Tanzania, S by Burundi and W by Zaïre. From the 16th century to 1959 the Tutsi kingdom of Rwanda shared the history of Burundi. In 1959 an uprising of the Hutu destroyed the Tutsi feudal hierarchy and led to the departure of the Mwami Kigeri V. A republic was proclaimed by the Parmehutu in 1961. Area, 10,169 sq. m. (26,338 sq. km.). Cap. Kigali. Plateau at 4,000–6,000 ft. (1,219–1,829 metres) a.s.l. Subsistence agriculture dominates. Staple food crops are beans, cassava, maize, sweet potatoes, peas, groundnuts and sorghum. The main cash crop is *aravica* coffee as in Burundi. Tea and pyrethrum are also produced on a limited scale. There is a pilot rice-growing project. There is no general industrial development apart from mining of cassiterite and wolframite. There are 4 hydro-electric installations, a large modern brewery and over 100 small-sized manu. enterprises. Methane gas is abundant under Lake Kivu. Pop. (1988E) 6·71m.

Ryazan Russian Soviet Federal Socialist Republic, U.S.S.R. 54 38N 39 44E. Town sit. SE of Moscow on S bank of Oka R. Industries mainly serve the surrounding agric. area. Pop. (1989E) 515,000.

Rybachi Russian Soviet Federal Socialist Republic, U.S.S.R. Town sit. NE of Zelenogradsk on the lagoon inland of Courland Spit on the Baltic coast near the Lithuanian Soviet Socialist Rep. border. Resort and fishing port.

Rybinsk (Andropov) Russian

Soviet Federal Socialist Republic, U.S.S.R. 53 03N 38 50E. Town sit. NE of Moscow, at SE end of Rybinsk Reservoir on Volga R. Linked to Leningrad by the Moriinsk Canal. Industries inc. shipbuilding, wire, matches etc. Important trade in timber, grain and petroleum. Pop. (1987E) 254,000.

Rybinsk Reservoir Russian Soviet Federal Socialist Republic, U.S.S.R. 58 30N 38 00E. L. sit. NE of Moscow fed by Rs. Maloga, Chagoda, Suda, Sheksna and Volga.

Ryde Isle of Wight, England. 50 44N 1 09W. Town SW of Portsmouth sit. on NE coast of the Isle of Wight. Seaside resort. Pop. (1988E) 27,100.

Rye East Sussex, England. 50 57N 0 44E. Town in SE sit. near the English Channel coast and on Rother R. Pop. (1985E) 4,490.

Ryukyu Islands Japan. 26 30N 128 00E. Islands in the East China Sea extending SW to NE off the E coast of Taiwan. Area, 848 sq. m. (2,196 sq. km.). The main island is Okinawa. Mainly agric. and fishing is of increasing importance. Pop. (1980E) 1·2m.

Rzeszów Poland. 50 03N 22 00E. Town sit. E of Kraków. Industries inc. aeronautical and agric. engineering.

Rzhev Russian Soviet Federal Socialist Republic, U.S.S.R. 56 16N 34 20E. Town sit. SW of Kalinin on S bank of Volga R. Industries inc. agric. engineering, paper, distilling.

S

Saale River Germany. 51 57N 11 55E. Rises in the Fichtelgebirge in Bavaria and flows *c*. 265 m. (424 km.) N past Jena and Halle to join R. Elbe 25 m. (40 km.) SE of Magdeburg. Navigable confluence with R. Elbe.

Saalfeld Gera, German Democratic Republic. 50 39N 11 22E. Town sit. SW of Jena on Saale R., in iron-mining area. Manu. textiles, machine tools, electrical equipment. Pop. (1989E) 32,300.

Saarbrücken Saarland, Federal Republic of Germany. 49 14N 6 59E. Cap. of Saarland since 1919, sit. on Saar R. where it forms frontier between France and F.R.G. Industrial centre of coalmining area. Industries inc. iron and steel, cement, machinery, manu. of clothing, paper and printing. Ceded to Prussia from France 1815. Pop. (1984E) 189,600.

Saaremaa Estonia, U.S.S.R. 58 25N 22 30E. Baltic island at mouth of Gulf of Riga. Area, 1,050 sq. m. (2,720 sq. km.). Chief town Kuresaare. Low-lying. Main occupations farming and fishing.

Saarland Federal Republic of Germany. 49 20N 0 75E. Province of Federal Rep. bordered W by Luxembourg, SW and S by France. Area, 993 sq. m. (2,571 sq. km.). Cap. Saarbrücken. Important coalfield with iron and steel industries. Possession disputed between France and Germany 1919 until integration in Federal Rep. 1959. Pop. (1988E) 1,054,142.

Saar River France/Germany. 49 42N 6 34E. Rises in the Vosges and flows 149 m. (238 km.) N through Moselle and Saarland to join Moselle R. SW of Trier. Navigable from confluence with Moselle R. to Sarreguemines in Moselle dept.

Saba Netherlands Antilles, West Indies. 17 38N 63 10W. Island sit. NW of St. Eustatius in the NW Leeward Islands. Area, 5 sq. m. (13 sq. km.). Chief town Bottom. A single extinct volcano, with settlements in the crater. Main occupations fishing, lace making, small farming. Pop. (1984E) 1,000.

Sabadell Barcelona, Spain. 41 33N 2 06E. Town sit. N of Barcelona. Manu. centre with important textile industry. Pop. (1981C) 184,943.

Sabah Malaysia. 6 00N 117 00E. State of the Federation of Malaysia, formerly N. Borneo. Bounded W by Sarawak, S by Kalimantan, Indonesia. Area, 28,460 sq. m. (73,711 sq. km.). Cap. Kota Kinabalu. Mountainous with highest peak in Borneo, Mount Kinabalu, 13,455 ft. (4,101 metres) a.s.l., in N. Hot and very humid with heavy rain. Chief products timber, rubber, copra. Pop. (1988E) 1,371,000.

Sabaragamuwa Sri Lanka. Province of central Sri Lanka bounded E by Walawe Ganga R. Area, 1,893 sq. m. (4,903 sq. km.). Cap. Ratnapura. Extends E and S into the hills. Agric. area with important mines for precious and semi-precious stones, at Ratnapura and Pelmadulla, iron ore, at Balangoda, and graphite. Pop. (1981C) 1,482,031.

Sabinas Coahuila, Mexico. 27 51N 101 07W. Town sit. NW of Monterrey on the Inter American Highway. Railway junction and distribution centre for a silver, lead and gold mining area. Industries inc. coke ovens. Pop. (1980E) 22,000.

Sabine Hills Italy. Range in the Apennines NE of Rome extending SE from lower Salto R. The highest point is Monte Pellecchia, 4,488 ft. (1,368 metres) a.s.l.

Sabine River   Texas/Louisiana, U.S.A. 30 00N 93 45W. Rises in NE Texas and flows 500 m. (800 km.) SE and then S, from Loganport onwards forming the boundary of Texas and Louisiana. Discharges through the Sabine L. into the Gulf of Mexico. A canal system links it with Mississippi R. at New Orleans.

Sable Island   Nova Scotia, Canada. 43 55N 59 50W. Island in the Atlantic sit. ESE of Halifax. Area, 20 m. (32 km.) long by 1 m. (1.5 km.) wide. It is exposed and is a gradually shrinking part of a large sandbank and the cause of many shipwrecks. 2 lighthouses.

Sabunchi   Azerbaijan Soviet Socialist Republic, U.S.S.R. Town sit. NE of Baku on the central Apsheron Peninsula of the W Caspian coast. The main industry is oil.

Sabya   Asir, Saudi Arabia. Town sit. S of Abha, centre of the Tihama district with a port (W) at Qizan on the Red Sea.

Sacatepéquez   Guatemala. Dept. of S central Guatemala. Area, 180 sq. m. (466 sq. km.). Chief towns Antigua (cap.), Sumpango, Ciudad Vieja, Santa Maria. Highland area drained by Guacalate R., rising to Agua and Fuego volcanoes. Farming area producing maize, coffee, beans, fodder, livestock. Pop. (1982E) 137,815.

Sackville   New Brunswick, Canada. 45 54N 64 22W. Town sit. SE of Moncton in an area of reclaimed coastal marshes. Industries inc. woodworking, printing; manu. leather, paper and plastic products, stoves and machinery. Pop. (1990E) 6,000.

Sacramento   California, U.S.A. 38 35N 121 30W. State cap. sit. on Sacramento R. and American R. Industries are mainly food processing, high technology, manu., service and government. Pop. (1980C) 274,105.

Sacramento River   California, U.S.A. 38 03N 121 56W. Rises in the Klamath Mountains and runs 390 m. (624 km.) S through the N half of the Great Central Valley to enter San Francisco Bay. The Shosta and Kes-

wick dams are important for irrigation and hydroelectric power. Chief tribs. are Rs. McCloud, Pit, Feather and American.

Sá de Bandeira   *see* **Lubango**

Safaniyah   Hasa, Saudi Arabia. Oilfield in the (Persian) Gulf off Cape Ras Safaniyah, sit. SE of Ras Misha'ab.

Saffron Walden   Essex, England. 52 01N 0 15E. Anc. market town in a rural area. Pop. (1990E) 14,500.

Safi   Morocco. 32 18N 9 14W. Town on the Atlantic coast. Important port for fishing and exporting phosphates which come by rail from Youssoufia mines. Industries inc. sardine canning, boatbuilding, manu. chemical fertilizers, sulphuric acid. Pop. (1982E) 197,616.

Sagaing   Sagaing, Burma. 21 55N 95 56E. Town sit. SW of Mandalay on Irrawaddy R. Divisional and district cap. and trading centre for cotton, sesame, salt, fruit. Pop. (1980E) 16,000.

Sagamihara   Honshu, Japan. 35 32N 139 23E. Town sit. NW of Yokohama. Industries inc. machinery, agric. and electronics. Pop. (1985E) 477,000.

Sagar   Karnataka, India. Town sit. WNW of Shimoga. Trading centre for rice, timber, betel nuts. The main industry is tile making. Pop. (1981C) 35,648.

Sagar   Madhya Pradesh, India. Town SSE of Agra and sit. in the Vindhya Hills at 1,700 ft. (518 metres) a.s.l. Trading centre for an agric. area, with trade in wheat, cotton and oilseeds. Pop. (1981C) 174,770.

Saginaw   Michigan, U.S.A. 43 25N 83 58W. City on Saginaw R. above its mouth on L. Huron, and sit. SSW of Bay City. Commercial centre for agric. area. Manu. inc. machinery, tools, iron castings, food processing, furniture, paper. Pop. (1985E) 73,081.

Saginaw River   Michigan, U.S.A. Rises near Pontiac in E Michigan, and flows 22 m. (35 km.) NW and then NNE past Saginaw into Saginaw Bay, an arm of L. Huron.

Saguenay River　Quebec, Canada. 48 10N 69 45W. Leaves L. St. John, whence its furthest headstream is the Peribonca R., through the Grande Décharge and Petite Décharge which then join. Flows 475 m. (760 km.) ESE to enter St. Lawrence R. near Tadoussac. There are important hydroelectric projects on the Grande Décharge. Widens below Chicoutimi to 2 m. (3 km.).

Sagunto　Valencia, Spain. 39 41N 0 16W. Town sit. NNE of Valencia. Manu. fertilizers, cement, iron and steel, hardware and tiles. Exports iron ore. Pop. (1989E) 57,300.

Sahara　North Africa. 23 00N 5 00E. The largest desert in the world, extending from the Atlantic coast of Western Sahara to the Red Sea coast of Egypt and Sudan. Area, *c.* 3m. sq. m. (*c.* 7.7m. sq. km.). Bounded N by Atlas Mountains in Algeria, Morocco and Tunisia, and by Mediterranean Sea in Libya and Egypt, and taking in the Libyan and Nubian Deserts. Mainly plateau at 1,000 ft. (305 metres) a.s.l. with highest points in the Ahaggar, Aïr and Tibesti highlands, and depressions b.s.l. in NE Algeria and Tunisia. Annual rainfall irregular, under 10 in. (25 cm.). Temp. ranges from 100°F+ (38°C) by day to near-freezing at night. Types of ground: shifting dunes, hard rock, loose boulders. Cultivation is only possible in oases. Motor routes run from Béchar in Algeria to Ségou in Mali and into Benin, and from Laghouat and Touggourt in Algeria to Kano in Nigeria. Sparse, nomadic population.

Saharanpur　Uttar Pradesh, India. 29 58N 77 33E. Town sit. NNE of Delhi on the railway line between Amritsar, Meerut and Delhi. Industries inc. railway engineering, paper-milling, cigarette and hosiery manu. Trading centre for grain and sugar-cane. Pop. (1981C) 295,355.

Sahel　Africa. Six sub-Sahara countries of Chad, Mali, Mauritania, Niger, Senegal and Burkina Faso.

Saida　Lebanon. (Sidon.) 33 33N 35 22E. Town on Mediterranean coast sit. SSW of Beirut. Oil pipeline terminal. Main industry refining oil from Saudi Arabia. Exports citrus fruit. Pop. (1984E) 24,740.

Saigon　Vietnam (Ho Chi Minh City). 10 45N 106 40E. City sit. on Saigon R., on the S. China Sea. Since 1932, joint administration with Cholon to the SW. Two cities together make the country's main industrial centre. Industries inc. rice milling, brewing, distilling, manu. of rubber products, textiles and soap. Pop. (1989E) 4m.

St. Albans　Hertfordshire, England. 51 46N 0 21W. City sit. on Ver R., with important abbey and Roman site. Market town with some industries inc. printing, computing and light engineering. Pop. (1981C) 55,719.

St. Andrews　Fife Region, Scotland. 56 20N 2 48W. City on North Sea coast. Market town and resort, with the Royal and Ancient Golf Club, headquarters of Golf. University founded 1411. Pop. (1987E) 13,830.

St. Asaph　Clwyd, Wales. 53 16N 3 26W. Village sit. SSE of Rhyl. Episcopal seat in rural area. Industries inc. plastics and glass. Pop. (1988E) 3,400.

St. Augustine　Florida, U.S.A. 29 53N 81 18W. City on Atlantic coast. Oldest city in the U.S.A. founded by Pedro Menéndez de Avilés in 1565. Main industry, tourism. Pop. (1985E) 11,985.

St. Austell　Cornwall, England. 50 20N 4 48W. Town sit. ENE of Truro. Main industry china clay extraction. Pop. (1985E) 39,240, with Fowey.

Saint Barthélemy　Guadeloupe, West Indies. 17 55N 62 50W. Island sit. NW of Guadeloupe of which it is a depend. Area, 8 sq. m. (21 sq. km.). Chief town Gustavia. Hilly, rising to 990 ft. (302 metres) a.s.l. Produces cotton, fruit, livestock, fish and salt, all in small quantities. Pop. (1982C) 3,059.

St. Bernard Passes　France/Switzerland/Italy. (i) Great St. Bernard, alpine pass at 8,100 ft. (2,469 metres) a.s.l. on the Swiss–Italian border between Martigny and Aosta. Famous hospice founded 1050 by St. Bernard of Menthon. A road tunnel was opened in 1964. (ii) Little St. Bernard, alpine pass at 7,178 ft. (2,188 metres) a.s.l. on the Franco-Italian border between Bourg St. Maurice and Aosta. Hospice also founded by St. Bernard of Menthon.

St. Boniface Manitoba, Canada. 49 55N 97 06W. Town opposite Winnipeg on Red R. Main industries meat packing, flour milling, oil refining, manu. paint and soap. Centre of French-Canadian culture. Pop. (1982E) 49,000.

St. Brieuc Côtes-du-Nord, France. 48 31N 2 47W. Cap. of dept. sit. on R. Gouët near its mouth on English Channel. Manu. hosiery and furniture. Pop. (1982C) 51,399.

St. Catharine's Ontario, Canada. 43 10N 79 15W. City sit. ESE of Hamilton. Centre of a fruit-growing region. Industries inc. engineering, machinery, transport equipment, paper and printing, wood products, fruit and vegetable canning, manu. electrical equipment, auto parts. Pop. (1986E) 123,455.

St. Charles Missouri, U.S.A. 38 47N 90 29W. Town sit. NW of St. Louis on N bank of Missouri R. Centre of an agric. area. Industries inc. sawmilling; manu. steel dies, railway stock, diesel engines, metal products, footwear. Deposits of sand, coal and gravel nearby. Noted as the first permanent white settlement on the Missouri R., made by French traders, 1769 A.D. Pop. (1980C) 37,379.

St. Christopher *see* **St. Kitts**

St. Clair, Lake Canada/U.S.A. 42 04N 146 10E. L. on the border between SE Ontario, Can., and NE Michigan, U.S.A. Area, 432 sq. m. (1,119 sq. km.). Fed from L. Huron by St. Clair R., which has a deepened shipping channel, and drained by Detroit R. into L. Erie.

St. Clair Shores Michigan, U.S.A.

42 30N 82 53W. City sit. NE of Detroit, on W shore of L. St. Clair. L. marinas. Pop. (1980C) 76,210.

St. Cloud Hauts-de-Seine, France. Suburb of Paris sit. W of the city centre. Industry, Sèvres porcelain factory. Pop. (1982C) 28,760.

St. Cloud Minnesota, U.S.A. 45 33N 94 10W. City in centre of state on Mississippi R. Commercial centre of an agric. region. Industries inc. vegetable canning, railway engineering, granite quarrying. Pop. (1980C) 42,566.

St. Croix West Indies. 17 45N 64 45W. Island in the U.S. Virgin Islands. Area, 82 sq. m. (212 sq. km.). Cap. Christiansted. Tourist centre. Main occupations cattle rearing, growing vegetables, distilling rum; small manu. Pop. (1980C) 49,013.

St. David's Dyfed, Wales. 51 54N 5 16W. Village sit. NW of Milford Haven on St. Bride's Bay. Episcopal seat with 12th cent. cathedral. Pop. (1981C) 1,800.

Saint-Denis Seine-et-Denis, France. 48 56N 2 22E. Town forming N suburb of Paris on Seine R. Industries inc. railway engineering, manu. barges, boilers, diesel engines, chemicals, glass, pottery. Has an anc. abbey with tombs of French Kings. Pop. (1982C) 91,275.

Saint-Denis Réunion. 20 52S 55 28E. Town sit. ENE of Pointe-des-Galets on N coast. Cap. and port. Commercial and processing centre for an agric. area. Industries inc. food canning, manu. cigarettes. Pop. (1982C) 126,323.

St. Dizier Haute-Marne, France. 48 38N 4 57E. Town in NE France on Marne R. Manu. nails, wire, tools, hardware. Pop. (1982C) 39,815.

St. Elias Mountains Canada/U.S.A. 60 18N 140 55W. Range extending S *c.* 200 m. (320 km.) from Mount Logan (19,850 ft./6,050 metres a.s.l., the highest peak in Can.) and Mount Elias, 18,008 ft. (5,489 metres) a.s.l., along the borders between Alaska and

Yukon/Brit. Columbia to Mount Fair-weather, 15,320 ft. (4,670 metres) a.s.l. The flow of ice from the range forms the largest non-polar icefield which inc. the Malaspine Glacier, 1,500 sq. m. (3,885 sq. km.).

Saintes, Les West Indies. Arch. of small islands between Dominica and Guadeloupe. Area, 5¹/₂ sq. m. (14 sq. km.). Principal islands Terre-de-Bas and Terre-de-Haut. Main occupations stock-rearing, fishing. Pop. (1982C) 2,901.

St. Étienne Loire, France. 45 26N 4 24E. Cap. of dept. sit. on Furens R. in SE France. Industrial centre of coal-mining area, manu. iron and steel, machinery, armaments, silk and rayon goods, glass and pottery. The School of Mines is famous. Pop. (1982C) 206,087.

St. Francis River Missouri/ Arkansas, U.S.A. 34 38N 90 35W. Rises in SE Missouri and flows S for 470 m. (752 km.) through NE Arkansas to join Mississippi R. 50 m. (80 km.) SW of Memphis.

St. Gallen Switzerland. 47 25N 9 23E. (i) Canton of NE Switz., bordered by canton of Thurgau and L. Constance to N and Austria and Liechtenstein to E. Area, 777 sq. m. (2,012 sq. km.). Cap. St. Gallen. Mountainous in S, industrial in N and E. Manu. muslin, lace, confectionery. Pop. (1980C) 391,995. (ii) Town of same name and cap. of (i). Manu. cotton, chemicals, confectionery. Pop. (1983E) 74,000.

Saint George's Grenada, West Indies. 12 00N 61 43W. Town on SW coast and sit. N of Port of Spain, Trinidad. Cap. and port with a land-locked harbour. Industries inc. light manu., sugar refining, rum. Exports cacao, nutmeg, mace and other spices. Pop. (1981E) 29,369.

St. George's Channel British Isles. 52 00N 6 00W. Channel linking the Atlantic Ocean and the Irish Sea between SW Wales and SE Ireland. Narrowest point 46 m. (74 km.) wide.

St. Germain-en-Laye Yvelines, France. 48 54N 2 05E. Town forming

outer suburb of Paris sit. WNW of the city centre, on Seine R. Residential town and resort for the forest of St. Germain. Manu. hosiery, confectionery, musical instruments. Pop. (1982C) 40,829.

St. Gotthard Pass Switzerland. 46 34N 8 31E. Alpine pass between cantons of Uri and Ticino in the Lepontine Alps. 6,916 ft. (2,108 metres) a.s.l. The St. Gotthard Railway Tunnel beneath it extends 9¹/₄ m. (15 km.).

St. Helena South Atlantic. 15 57S 5 42W. Island sit. W of Angola. Area, 47 sq. m. (122 sq. km.). Cap. Jamestown. Volcanic and mountainous. Chief occupation flax growing. Manu. rope and twine. Famous for the exile of Napoleon I, 1815–21. Pop. (1988E) 5,564.

St. Helier Jersey, Channel Islands. 49 12N 2 06W. Town on St. Aubin's Bay on S coast. Chief town of Jersey. Holiday resort, conference centre and market for horticultural area. Offshore finance centre. Pop. (1988E) 82,536.

St. Hyacinthe Quebec, Canada. 45 38N 72 57W. Town sit. ENE of Montreal on Yamaska R. Industrial centre, manu. paper products, food processing, wire. Pop. (1985C) 38,500.

St. Ives Cambridgeshire, England. 52 20N 0 05W. Town sit. E of Huntingdon on Great Ouse R. Market town for agric. area. Pop. (1985E) 14,000.

St. Ives Cornwall, England. 50 12N 5 29W. Town sit. on Atlantic coast. Industries fishing and tourism. Pop. (1981E) 11,065.

St. Jean Quebec, Canada. 45 18N 73 16W. Town sit. SE of Montreal on Richelieu R. Manu. textiles, paper, sewing machines. Pop. (1981E) 34,000.

St. Jean de Luz Pyrénées-Atlantiques, France. 42 23N 1 40W. Town sit. SW of Bayonne on the Bay of Biscay. Industry fishing. Popular resort. Pop. (1982C) 12,921.

Saint John New Brunswick, Canada. 45 16N 66 03W. City at mouth of

St. John R. on the Bay of Fundy. Industries oil and sugar refining, dock yards; manu. cotton goods, bricks and tiles, wood pulp and paper. Exports timber, grain, minerals and is an important sea-port, ice-free in winter. Pop. (1986E) 121,580.

Saint John United States Virgin Islands, West Indies. 18 20N 64 45W. Island sit. E of San Juan, Puerto Rico. Area, 19 sq. m. (49 sq. km.). Cap. Cruz Bay. Hilly, with an indented coastline and good harbours. Mainly a stock farming area, also producing bay leaves. Pop. (1980C) 2,360.

St. John, Lake Quebec, Canada. 48 35N 72 00W. L. in SE Quebec. Area, 375 sq. m. (971 sq. km.). Fed by R. Peribonca, Mistassibi, Mistassini and Ashuapmuchuan and drained by Saguenay R. into St. Lawrence R.

St. John River Canada/U.S.A. 45 16N 66 03W. Rises with headstreams in NE Maine, U.S.A. and S Quebec, Can. Flows 418 m. (669 km.) NE and then E to form frontier between U.S.A. and Canada, then SE through New Brunswick to enter the Bay of Fundy at St. John (*see* above). Navigable from its mouth to Fredericton. Strong tides in the Bay of Fundy cause the Reversing Falls near St. John.

Saint John's Antigua, West Indies. 17 06N 61 51W. Town on NW coast. Cap. and port in a sheltered bay, which processes and exports rum, cotton and sugar. Resort. Pop. (1982E) 30,000.

St. John's Newfoundland, Canada. 47 34N 52 43W. City and cap. of province, sit. on E coast of the Avalon peninsula. Industries inc. retail services, manufacturing construction, transport and government services. Pop. (1986E) 161,900.

St. John's Wood Greater London, England. 51 32N 0 11W. Residential district and NW suburb of London 3½ m. (6 km.) from city centre (Hyde Park Corner). Famous for Lord's Cricket Ground, headquarters of the M.C.C.

St. Joseph Michigan, U.S.A. 42 06N 86 29W. City sit. E of Chicago and on SE shore of L. Michigan. Commercial, professional and government centre. Manu. electric washers and dryers, vehicle parts, electronics and instruments. Pop. (1980C) 9,622.

St. Joseph Missouri, U.S.A. 39 46N 94 51W. City sit. NNW of Kansas City on Missouri R. Commercial centre for livestock; industries meat-packing, flour milling. R. port and important railway centre. Pop. (1980C) 76,691.

St. Joseph River Michigan/Indiana, U.S.A. 42 07N 86 29W. Rises in S Michigan, flows 210 m. (336 km.) w, with a curve S, into N Indiana, and enters L. Michigan between St. Joseph and Benton Harbor.

St. Just (St. Just-in-Penwith) Cornwall, England. 50 07N 5 41W. Town sit. w of Penzance. Market town for rural area. Pop. (1989E) 4,072.

St. Kilda Highland Region, Scotland. 57 49N 8 36W. Island in the Outer Hebrides, sit. WNW of N. Uist. Area, 3 m. (5 km.) by 2 m. (3 km.). Bird sanctuary. The pop. were evacuated at their own request in 1930.

St. Kitts (St. Christopher-Nevis) West Indies. 17 20N 62 45W. State within the Commonwealth in the Leeward Islands group sit. NW of Antigua. Area, 101 sq. m. (261 sq. km.). Cap. Basseterre. Volcanic and mountainous, rising to Mount Misery, 3,792 ft. (1,156 metres) a.s.l. Chief occupation sugar-cane cultivation. Pop. (1987E) 43,410.

St. Laurent Quebec, Canada. 45 31N 73 41W. City sit. w of Montreal on Montreal Island. Industries inc. engineering and high technology. Manu. inc. chemicals and textiles. Pop. (1984E) 65,190.

St. Lawrence River and Seaway Canada/U.S.A. 49 15N 67 00W. Greatest commercial waterway in Can. and one of the greatest in the world. Leaves the NE end of L. Ontario and flows 750 m. (1,200 km.) NE to the Gulf of St. Lawrence, forming the E end of the Seaway by which ocean-going vessels can reach the w end of

L. Superior. The first 114 m. (182 km.) form the border between Canada and U.S.A., after which it flows through Quebec. Main tribs. are Rs. Ottawa, St. Maurice, Saguenay, Chaudière and Richelieu. Below Quebec it forms a tidal estuary from 3 m. (5 km.) to 70 m. (112 km.) wide. The Seaway was opened in 1959 after narrow channels had been widened and others newly cut to avoid the 8 rapids between L. Ontario and Montreal. The whole seaway is 2,480 m. (3,968 km.) long.

St. Lô Manche, France. 49 08N 1 07W. Cap. of dept. on Vire R. Centre for agric. district. Pop. (1982c) 24,792.

Saint-Louis Senegal. 16 00N 16 30W. Town sit. on Senegal R. estuary. Manu. textiles; exports, hides, skins, groundnuts. Cap. of Senegal until replaced by Dakar in 1958. Pop. (1985E) 91,500.

St. Louis Missouri, U.S.A. 38 38N 90 11W. City sit. downstream from confluence of Missouri R. and Mississippi R. Major centre of commerce and communications. Trades in fur pelts, grain, timber, livestock, wool. Industries oil refining, brewing, meat packing; manu. aircraft, motor vehicles, machinery, electrical equipment, chemicals. Has 19 m. (30 km.) R. frontage. Pop. (1980c) 453,085.

St. Louis Park Minnesota, U.S.A. 44 56N 93 22W. Town immediately SW of Minneapolis. Residential with some light industry. Pop. (1980c) 42,931.

St. Lucia West Indies. 13 53N 60 58W. Independent country (from 1979) sit. S of Martinique. Area, 238 sq. m. (616 sq. km.). Cap. Castries. Volcanic and mountainous rising to Mount Gimie, 3,145 ft. (958 metres) a.s.l. Chief occupation, cultivating bananas, sugar, coconuts, tourism. Pop. (1988E) 146,600.

St. Malo Ille-et-Vilaine, France. 48 39N 2 01W. Town at mouth of Rance R. on the Eng. Channel. Industries fishing, tourism. Trade, mainly with Eng., in fruit, vegetables, dairy produce. Pop. (1982c) 47,324.

St. Martin (Sint Maarten) West Indies. 18 04N 63 04W. Island in Leeward Islands group sit. NW of St. Kitts. Area, 34 sq. m. (88 sq. km.). Part of the Neth. Antilles. Little cultivable land. Chief occupation extracting sea-salt. Pop. N part (1982c) 8,072; S part (1983E) 15,926.

St. Marylebone Greater London, England. 51 31N 0 09W. District of central London to N of Oxford Street. Bounded to N by Regent's Park.

St. Maur-des-Fossés Val-de-Marne, France. Suburb of Paris sit. SE of the city centre. Manu. hosiery, furniture, toys, electrical equipment. Pop. (1982c) 80,954.

St. Maurice River Quebec, Canada. 46 22N 72 32W. Rises in S Quebec, flows 325 m. (520 km.) through the Gouin Reservoir SE and then S to join St. Lawrence R. at Trois Rivières. Important for floating timber and hydroelectric power. Flows through many industrial developments inc. pulp and paper mills, plastics, chemical and aluminium plants. Main tribs R. Trenche and Vermilion.

St. Mawes Cornwall, England. 50 10N 5 01W. Town sit. E of Falmouth at the entrance to Carrick Roads. Industries fishing and tourism. Pop. (1983E) 1,500.

St. Michael's Mount Cornwall, England. 48 21N 3 57W. Island in Mount's Bay ½ m. (1 km.) from Marazion, with which it is connected by a natural causeway at low tide, and 2½ m. (4 km.) from Penzance by sea. Surmounted by a castle and chapel.

St. Moritz Graubünden, Switzerland. 46 30N 9 50E. Resort consisting of village and neighbouring spa in SE Switz. Industry tourism, esp. winter sports. Pop. (1989E) 6,000.

St. Nazaire Loire-Atlantique, France. 47 14N 2 12W. Town sit. at mouth of Loire R. Industries shipyards; manu. steel, fertilizers, marine engineering, brewing, vegetable canning. Sea port for Nantes. Pop. (1982c) 68,947.

St. Neots Cambridgeshire, England.

52 14N 0 16W. Town sit. SSW of Huntingdon on Ouse R. Market town for an agric. area. Industries brewing, plastics, packaging, clothing, electronics, and paper-making. Pop. (1981C) 22,500.

St. Nicolas *see* **Sint Niklaas**

St. Omer Pas-de-Calais, France. 50 45N 2 15E. Town sit. SE of Calais, on Aa R. Market centre for horticultural area. Industries inc. distilling, brewing; manu. hosiery. Pop. (1982C) 15,497.

Saintonge France. Area on W coast N of Gironde estuary. Cap. Saintes. Industries farming and viticulture, esp. for cognac.

St. Ouen Val-d'Oise, France. Suburb of Paris sit. NNW of the city centre. Industry railway engineering; manu. chemicals, machine tools. Pop. (1982C) 43,743.

St. Pancras Greater London, England. 51 32N 0 07W. District of London sit. N of city centre. Industries centred on 3 important railway termini at Euston, St. Pancras, King's Cross.

St. Paul Minnesota, U.S.A. 44 58N 93 07W. State cap. adjoining Minneapolis at confluence of Minnesota R. and Mississippi R. With Minneapolis, commercial and industrial centre for large agric. area. R. port with 29 m. (47 km.) shoreline and foreign trade zone; transport centre. Industries inc. engineering, processing raw materials, brewing, printing and publishing; manu. inc. electrical equipment, chemicals. Pop. (1980C) 270,230.

St. Paul Island Indian Ocean. 38 43S 77 29E. Island at lat. 39°S, long. 78°E, forming part of the French Southern and Antarctic Territories. Area, 3 sq. m. (7 sq. km.). Volcanic. Uninhabited.

St. Peter Port Guernsey, Channel Islands. 49 27N 2 32W. Chief town of Guernsey sit. on E coast. Market for horticultural area. Industry tourism. A growing financial services centre. Manu. electrical and electronic goods. Exports flowers and vegetables. Pop. (1986E) 16,085.

St. Petersburg Florida, U.S.A. 27 46N 82 38W. City sit. on Tampa Bay, on W coast of Florida. Main industries tourism and high technology. Pop. (1980C) 238,647.

St. Pierre Martinique, West Indies. 14 45N 61 11W. Town sit. NW of Fort de France on W coast at the foot of Mount Pelée. Industry distilling rum. The chief town until its destruction by an eruption of Mount Pelée in 1902.

St. Pierre Réunion. Town on S coast of island. Pop. (1982C) 90,627.

St. Pierre and Miquelon N. Atlantic Ocean. 46 55N 56 10W. Island group off S coast of Newfoundland, Canada, forming a French overseas department. Area, 93.5 sq. m. (242 sq. km.): area of St. Pierre group 10 sq. m. (26 sq. km.); area of Miquelon group 83 sq. m. (215 sq. km.). Cap. St. Pierre. Main occupation cod fishing. Pop. (1988E) 6,400.

St. Pölten Lower Austria, Austria. Town sit. W of Vienna. Industrial centre and railway junction. Manu. textiles, machinery. Pop. (1985E) 50,419.

St. Quentin Aisne, France. 49 51N 3 17E. Town in NE sit. on Somme R. Industry mainly textiles, esp. curtains, muslin. Also manu. chemicals, machinery. Pop. (1982C) 65,067.

St. Thomas United States Virgin Islands, West Indies. 18 21N 64 55W. Island of 28 sq. m. (73 sq. km.) with the group's only city and cap., Charlotte-Amalie. Main occupations govt. and tourism. Pop. (1980C) 44,218.

St. Tropez Var, France. 43 17N 6 38E. Town sit. SW of Cannes on the Riviera. Industries tourism and fishing. Pop. (1982C) 6,248.

St. Vincent and the Grenadines West Indies. 13 00N 61 10W. State within the Commonwealth in the Windward Islands group, sit. NE of Grenada. Area, 150 sq. m. (388 sq. km.) Cap. Kingstown. Volcanic and mountainous (inc. Soufrière). Chief occupations growing bananas, arrowroot, starch, tourism, sea-island cotton. Pop. (1987E) 112,614.

St. Vincent, Cape　Faro, Portugal. 37 01N 9 00W. Headland at SW tip of Portugal, with cliffs 175 ft. (53 metres) a.s.l. There is a lighthouse on the cliff-top.

St. Vincent, Gulf　South Australia, Australia. 35 00S 138 05E. Inlet of the Southern Ocean E of the Yorke Peninsula. Length *c.* 95 m. (152 km.) and greatest width 40 m. (64 km.).

Saipan　Mariana Islands, West Pacific. 15 10N 145 45E. Island sit. NNE of Guam. Area, 47 sq. m. (122 sq. km.). Cap. of Commonwealth of N Mariana Islands. Volcanic limestone, rising to extinct Mount Tapotchan, 1,554 ft. (474 metres) a.s.l. Main products sugar, coffee, copra, manganese and phosphates. Pop. (1983E) 16,532.

Sakai　Honshu, Japan. 34 35N 135 28E. Town immediately S of Osaka sit. on Osaka Bay. Manu. chemicals, fertilizers, machinery, aluminium products and hosiery. The harbour is silted up and no longer important. Pop. (1988E) 808,000.

Sakarya River　Turkey. Rises NNE of Afyonkarahisar in W Turkey and flows 490 m. (784 km.) in great curves E, NW and N past Osmaneli Geyve and Adapazari to enter the Black Sea at Karasu.

Sakhalin　Russian Soviet Federal Socialist Republic, U.S.S.R. 51 00N 143 00E. Island off E Siberia in the sea of Okhotsk, forming part of Khabarovsk Territory. Area, 29,700 sq. m. (76,923 sq. km.). Chief town Aleksandrovsk. Central valley runs N to S and is flanked by parallel mountain ranges. Mainly tundra and forest. Chief occupations fishing, growing rye, oats, potatoes and other vegetables.

Sakkara　Egypt. Village sit. SW of Cairo. Principal necropolis of anc. Memphis, together with Giza, and famous for step pyramids.

Sal　Cape Verde. Island in E of arch. containing Amilcar Cabral Intern.

Airport. Area, 83 sq. m. (216 sq. km.). Cap. Santa Maria. Pop. (1980C) 6,006.

Salado River　Argentina. 35 44S 57 22W. Rises near General La Madrid and flows 450 m. (720 km.) NE and then E through Buenos Aires province on a shallow course, to enter the Río de la Plata estuary.

Salado (Chadileufú, Desaguedero) River　Argentina. Rises in the Andes Mountains as Desaguedero R. and flows 850 m. (1,360 km.) SSE to form the boundary between Mendoza and San Luis provinces. It disappears in the marshes of La Pampa and re-emerges as Chadileufú R. which joins Colorado R. *c.* 50 m. (80 km.) upstream of Río Colorado.

Salado (Juramento) River　Argentina. Rises in the Andes Mountains as Juramento R. and flows 1,250 m. (2,000 km.) SE through Salta, Santiago del Estero and Santa Fé provinces to join Paraná R. at Santa Fé.

Salalah　Dhofar, Oman. 17 00N 54 06E. Cap. of Dhofar, port and trading centre, with additional new port at Raysut.

Salamanca　Spain. 40 58N 5 39W. (i) Province bordered N by the province of Zamora and W by Portugal. Area, 4,754 sq. m. (12,313 sq. km.). Mountain plateau. Main occupation farming. Pop. (1986C) 366,668. (ii) City of same name and cap. of (i) sit. on Tormes R. Important rail and road centre. Industries tanning, brewing, flour milling, manu. chemicals. Outstanding in historical and architectural interest. Pop. (1986C) 166,615.

Salamis　Greece. 37 59N 23 28E. Island in the Saronik Gulf, an arm of the Aegean Sea, off the coast of Attica and W of Athens. Area, 36 sq. m. (93 sq. km.). Main occupations tourism, fishing. Pop. (1981E) 19,000.

Salcombe　Devon, England. 50 14N 3 47W. Town sit. SW of Dartmouth on an inlet. Industries tourism and fishing. Pop. (1984E) 2,800.

Sale　Victoria, Australia. 38 06S 147 04E. Town sit. E of Melbourne.

Market centre for dairy-farming and beef cattle area, trade in grain and livestock. Industries oil related companies, flour and woollen mills, dairy produce and bacon factories. Linked by canal to the Gippsland L. and through it to Bass Strait. Pop. (1985E) 15,000.

Sale Greater Manchester, England. 53 26N 2 19W. Town forming suburb of Manchester sit. SW of the city centre. Residential. Pop. (1981E) 58,000.

Salé Morocco. 34 04N 6 48W. Town and port immediately NE of Rabat on the Atlantic coast. Industries fish canning, manu. carpets, pottery. Pop. (1982C) 289,391.

Salem Tamil Nadu, India. 12 00N 78 20E. Town sit. SW of Madras. Railway junction. Industries textiles, esp. cotton and stainless steel. Pop. (1981C) 361,394.

Salem Massachusetts, U.S.A. 42 31N 70 55W. City sit. NE of Boston on Massachusetts Bay. Industries textiles, manu. footwear, electric lamps. One of the oldest towns in New England. Pop. (1980C) 58,220.

Salem Oregon, U.S.A. 44 57N 123 01W. State cap. sit. in NW Oregon on Willamette R. Industries inc. food processing. Manu. metal goods, high technology equipment and silicon wafers. Pop. (1980C) 89,223.

Salerno Campania, Italy. 40 41N 14 47E. Town and port sit. on Gulf of Salerno. Cap. of Salerno province. Industries electrical engineering, tomato canning, flour milling, tanning, textiles. Pop. (1981C) 157,385.

Salford Greater Manchester, England. 53 28N 2 18W. Metropolitan district of Greater Manchester sit. on Irwell R. Industries electrical engineering, textiles; manu. textile machinery, chemicals, clothing, tyres. Important docks on the Manchester Ship Canal. Pop. (1980E) 130,000.

Salgótarján Pest, Hungary. 48 07N 19 48E. Town sit. NE of Budapest. Chief town of Nógrad county. Centre of a coalmining area. Industries iron and steel, manu. machinery. Pop. (1984C) 50,000.

Salina Kansas, U.S.A. 38 50N 97 37W. City sit. NNW of Wichita on Smoky Hill R. Commercial centre for agric. region. Industries flour milling, manu. agric. machinery, cement, bricks, tiles. Pop. (1988E) 45,011.

Salinas California, U.S.A. 36 40N 121 38W. City sit. S of San Francisco. Market centre for agric. area. Industries inc. food processing and shipping, retailing and electronics. Pop. (1980C) 80,479.

Salisbury Wiltshire, England. 51 05N 1 48W. Town sit. at confluence of Avon R. and Wylye R. Market centre for agric. area. Industries include light engineering and tourism. Pop. (1983E) 40,000.

Salisbury Zimbabwe *see* **Harare.**

Salisbury Plain Wiltshire, England. 51 13N 1 50W. Undulating chalk downs. Area, 200 sq. m. (518 sq. km.). Used as a military training ground. Important prehistoric sites, inc. Stonehenge.

Salop *see* **Shropshire.**

Salta Salta, Argentina. 24 47S 65 24W. Cap. of province of same name, sit. in NW. Commercial centre of an agric. region. Trades in tobacco, livestock, agric. produce. Industries meat packing, flour milling, tanning. Sit. on the narrow gauge Transandine Railway to Antofagasta in Chile. Pop. (1990E) 357,796.

Saltash Cornwall, England. 50 24N 4 12W. Town sit. NW of Plymouth on estuary of Tamar R. Industry tourism. Has Brunel's Royal Albert Railway Bridge and the Tamar Suspension Bridge over the estuary. Pop. (1989E) 17,000.

Saltburn Cleveland, England. 54 35N 0 58W. Dormitory town for Teesside sit. E of Middlesbrough on North Sea coast. Industry tourism. Pop. (1989E) 6,100.

Saltcoats Strathclyde Region, Scotland. 55 38N 4 47W. Town sit. NNW of Ayr on Firth of Clyde. Industry tourism. Pop. (1988E) 13,000.

Saltillo Coahuila, Mexico. 25 25N 101 00W. Cap. of state of same name sit. in NE Mexico. Commercial centre of a mining area. Industries inc. textiles, clothing, ceramics. Pop. (1980C) 321,758.

Salt Lake City Utah, U.S.A. 40 46N 111 53W. State cap. sit. on Jordan R. Communications centre. Industries oil refining, copper smelting and refining; manu. textiles, metal goods, food products. Centre for the Mormon religion. Pop. (1980C) 163,033.

Salto Salto, Uruguay. 31 23S 57 58W. Cap. of dept. of same name sit. on Uruguay R. Commercial centre for a stock-rearing and fruit growing area. Industries flour milling, meat packing. R. port at head of navigation on Uruguay R. Pop. (1985C) 80,787.

Salvador, El *see* **El Salvador**

Salvador Bahia, Brazil. 12 59S 38 31W. State cap. sit. on a peninsula between Todos Santos Bay and the Atlantic Ocean. Industries sugar refining, flour milling; manu. cigars, cigarettes, textiles, cement. Sea port exporting cocoa, sugar, tobacco. Pop. (1980C) 1,491,642.

Salween River South East Asia. 16 31N 97 37E. Rises in the Tanglha Range in E Tibet, China, and flows 1,750 m. (2,800 km.) first SE through a deep gorge, then S through Yunnan province, China, into Burma. It cuts a gorge through the Shan plateau and enters the Gulf of Martaban near Moulmein. Course frequently impeded by rapids. Navigable for 70 m. (112 km.) from the mouth.

Salzach River Austria. 48 12N 12 56E. Rises in the Hohe Tauern, and flows 130 m. (208 km.) E and N through Federal state of Salzburg and past Salzburg city to join Inn R.

Salzburg Austria. 47 48N 13 02E. (i) Federal state bordered by Germany to NW. Area, 2,762 sq. m. (7,154 sq.

km.). Mountainous, containing the Hohe Tauern and part of the Niedere Tauern in the S. Industries cattle farming, commercial forestry, tourism, salt mining. Pop. (1988E) 464,600. (ii) Cap. of state of same name, sit. on Salzach R. Industries tourism, brewing, manu. metal goods, textiles. Outstanding architectural interest. Pop. (1981C) 139,426.

Salzgitter Lower Saxony, Germany. 52 03N 10 22E. District forming part of Watenstedt-Salzgitter city, sit. SW of Brunswick which consists of 7 industrial districts incorporated in 1942. The main industry is mining iron and potash. Pop. (1984E) 108,400.

Salzkammergut Upper Austria, Austria. Region in the E Alps of N Austria. Area, *c.* 900 sq. m. (*c.* 2,331 sq. km.). Mountainous, rising to Dachstein, 9,830 ft. (2,996 metres) a.s.l. Inc. many Ls. Chief occupations tourism, stock rearing, forestry.

Samaná Samaná, Dominican Republic. Town sit. NE of Santo Domingo on the S coast of Samaná peninsula. Port and provincial cap. Trade in timber, cacao, coconuts, rice. Resort. Pop. (1984E) 54,420.

Samar Philippines. 12 00N 125 00E. Island separated from Luzon to N by the San Bernardino Strait. Area, 5,181 sq. m. (13,419 sq. km.). Cap. Catbalogan. Mountainous and forested, with frequent typhoons. Chief occupation growing rice, coconuts, abacá. Pop. (1980E) 1m.

Samara *see* **Kuibyshev.**

Samaria Israel. Region bounded W by the Plain of Sharon, E by the Jordan valley. Hilly area rising to Mount Ebal, 3,084 ft. (940 metres) a.s.l. Chief town Nablus.

Samarinda Kalimantan, Indonesia. 0 30S 117 09E. Town sit. NE of Banjermasin on Mahakam R. delta. Port handling coal, timber, rattan, skins, rubber, guttapercha. Pop. (1980E) 138,000.

Samarkand Uzbek Soviet Socialist Republic, U.S.S.R. 39 40N 66 58E.

City in the Zeravshan valley and sit. on the Trans-Caspian Railway. Cap. of the Samarkand region. Trading centre of a fertile area. Industries brewing, distilling, flour milling, tobacco processing; manu. textiles, chemicals, clothing, footwear. Outstanding historical interest. Pop. (1987E) 388,000.

Samarska Luka Russian Soviet Federal Socialist Republic, U.S.S.R. Region within the ox-bow of the Volga R. w of Kuibyshev. Area, 390 sq. m. (1,010 sq. km.). Agric. lowland on the R. rises to forested mountains in the centre. Chief product petroleum, with extracting centre at Zhigulevsk; also asphalt, dolomite, limestone.

Sambre River France/Belgium. 50 28N 4 53E. Rises in the Aisne Dept. of NE France, and flows 118 m. (189 km.) ENE past Maubeuge into Belgium, and then past Charleroi to join R. Meuse at Namur. Navigable from Namur to Landrecies. Linked by canal with Oise R. and Charleroi-Brussels canal.

Samoa South Pacific Ocean. 14 00S 171 00W. Group of islands sit. ENE of Fiji, forming Western Samoa and American Samoa. (i) Western Samoa. Independent state within the Commonwealth consisting of Savai'i and Upolu islands, two smaller islands, Manono and Apolima, and uninhabited islets. Area, 1,093 sq. m. (2,831 sq. km.). Cap. Apia, sit. on Upolu. Volcanic and mountainous. Chief occupations farming and fishing. Chief exports copra, bananas, cocoa. Pop. (1986E) 163,000. (ii) American Samoa, consisting of Tutuila and some smaller islands. Area, 76 sq. m. (197 sq. km.). Cap. Fagototo. Volcanic and mountainous. Forest covers *c.* 50% of land area. Chief occupations farming and fishing. Industry fish canning, esp. tuna. Pop. (1980C) 32,297.

Samos Greece. 37 48N 26 44E. Island in the Aegean Sea just off the w coast of Turkey. Area, 190 sq. m. (492 sq. km.). Mountainous but fertile. Industry farming, esp. vines, olives, tobacco, citrus fruits, and fishing. Pop. (1981E) 40,519.

Samothrace Greece. 40 30N 25 32E. Island in NE Aegean Sea sit. SSW of Alexandroúpolis. Area, 70 sq. m. (181 sq. km.). Mountainous, rising to Fengári, 5,249 ft. (1,601 metres) a.s.l., the highest peak in the Aegean islands. Chief occupations goat farming, sponge fishing.

Samsö Island Denmark. Island in the Kattegat between Jutland and Zealand. Area, 43 sq. m. (111 sq. km.). Industry mainly farming.

Samsun Samsun, Turkey. 41 17N 36 20E. Town sit. on the Black Sea. Cap. of province of same name. Port and trading centre for tobacco growing region. Industry tobacco processing. Pop. (1985C) 280,068.

San'a Republic of Yemen. 15 27N 44 12E. Cap. sit. in mountainous area at 7,500 ft. (2,286 metres) a.s.l. Handicraft industries inc. weaving and jewellery-making. Has important royal and religious buildings. Pop. (1986E) 427,150.

San Andrés y Providencia Islands Colombia. 13 00N 81 30W. Two Caribbean islands sit. off the coast of Nicaragua. Area, 21 sq. m. (54 sq. km.). Chief occupation cultivating coconuts, oranges. Pop. (1980E) 8,000.

San Angelo Texas, U.S.A. 31 28N 100 26W. City in central Texas sit. on Concho R. Market centre for a ranching and farming area. Important trade in wool and mohair. Industries cotton ginning, oil related, meat packing. Manu. surgical supplies, clothing. Pop. (1980C) 73,240.

San Antonio Chile. 33 40S 71 40W. Town and port sit. s of Valparaiso on Pacific coast, serving an agric. hinterland. Industry tourism. Pop. (1982E) 65,000.

San Antonio Texas, U.S.A. 29 28N 98 31W. City in s central Texas. Tourist centre. Commercial and communications centre for ranching area. Industries inc. electronics, general manu. and biotechnology. Pop. (1980C) 785,880.

San Bernardino California, U.S.A. 34 06N 117 17W. City sit. E of Los Angeles in the San Bernardino valley. Rail and highway distribution centre, manu. city and seat of local government. Pop. (1989E) 150,000.

San Bruno California, U.S.A. 37 37N 122 25W. City sit. S of San Francisco. Industry printing. San Francisco International Airport sit. here. Pop. (1980C) 35,417.

San Buenaventura (also called Ventura), California, U.S.A. 34 17N 119 18W. City sit. WNW of Los Angeles on Pacific coast. Market centre for an agric. and oil-producing region. Industries inc. tourism, oil and food processing. Pop. (1980C) 74,393.

San Carlos Costa Rica. 11 07N 84 47W. R. rising NW of San Ramón in the Aguacate mountains and flowing 70 m. (112 km.) NNE past Muelle de San Carlos to enter San Juan R. ENE of San Carlos del Norte. Navigable in the lower and middle course.

San Carlos Negros Occidental, Philippines. 10 29N 123 25E. Town sit. ESE of Bacolod on Tañon Strait opposite Refugio Island. Ferry port for Toledo, Cebu Island. Industries inc. sugar milling; sawmilling. Exports sugar, tobacco. Pop. (1980E) 91,627.

San Carlos Cojedes, Venezuela. 9 40N 68 36W. Town sit. WSW of Caracas on Tirgua R., sometimes called San Carlos R. State cap. and trading and processing centre for an agric. area. Industries inc. manu. dairy products, rice milling, sawmilling. Pop. (1971C) 21,029.

San Cristóbal San Cristóbal, Dominican Republic. 18 27N 70 07W. Town sit. WSW of Santo Domingo. Provincial cap. and trading centre for an agric. area, handling rice, coffee, vegetables, fruit, sugar. Arsenal manu. small arms and ammunition. Pop. (1982E) 128,000.

San Cristóbal Táchira, Venezuela. 7 46N 72 14W. State cap. sit. S of Maracaibo. Commercial centre of a coffee growing region. Industries tan-

ning, distilling. Manu. cement. Pop. (1980E) 164,000.

Sancti Spíritus Cuba. 21 56N 79 27W. Town sit. ESE of Matanzas. Commercial centre for an agric. region. Trade in sugar-cane, tobacco, cattle. Industry tanning, manu. cigars, dairy products. Pop. (1980E) 65,000.

Sandakan Sabah, Malaysia. 50 53N 118 05E. Town sit. on E coast. Industries sawmilling, fishing. Sea port exporting rubber, timber. Pop. (1980C) 113,496.

Sandbach Cheshire, England. 53 09N 2 23W. Market town sit. NE of Crewe. Manu. heavy commercial vehicles, light engineering, clothing and chemicals. Pop. (1989E) 16,500.

Sandefjord Vestfold, Norway. 59 08N 10 14E. Town sit. SSW of Oslo at the head of a small fjord near the mouth of Oslo Fjord. The main industries are shipping, and paint and varnish manu. Pop. (1989E) 35,800.

San Diego California, U.S.A. 32 43N 117 09W. City on San Diego Bay, N of border with Mexico. Industries electronics, biomedical, aerospace, tourism. Port, exporting agric. produce from a natural harbour. Important naval and marine base. Pop. (1989E) 1,086,000.

Sand Island *see* **Midway Islands**

Sandnes Rogaland, Norway. 58 51N 5 44E. Town sit. S of Stavanger at the head of Gands Fjord. Industrial town manu. metal products, bicycles, marine equipment, bricks, furniture, clothing and ceramics. Pop. (1984E) 38,859.

Sandown-Shanklin Isle of Wight, England. 50 39N 1 09W. Towns on the E coast S of Portsmouth. Industry tourism. Pop. (1980E) 16,000.

Sandusky Ohio, U.S.A. 41 27N 82 42W. City sit. W of Cleveland on L Erie. Industry tourism, manu. paper products, toys, crayons. Pop. (1980C) 31,360.

Sandviken Gävleborg, Sweden, 60

37N 16 46E. Town sit. near coast of Gulf of Bothnia. Manu. inc. speciality steel and cemented-carbide products. Pop. (1989E) 39,940.

Sandwich   Kent, England. 51 17N 1 20E. Town sit. E of Canterbury on R. Stour. Industries inc. electrical engineering, pharmaceuticals and tourism. Pop. (1981C) 4,184.

San Felipe   Aconcagua, Chile. 32 45S 70 44W. Town sit. N of Santiago. Cap. of province in central Chile. Commercial centre of agric. and mining area. Industry mining. Pop. (1980E) 42,000.

San Felipe   Yaracuy, Venezuela. 10 20N 68 44W. Town sit. W of Caracas on the Barquisimeto–Puerto Cabello road. State cap., railway terminus and market for timber, cacao, coffee, cotton, sugar-cane, rice, fruit. Industries inc. sugar refining. Pop. (1981C) 57,526

San Felipe de Puerto Plata   *see* **Puerto Plata**

San Fernando   Argentina. 34 26S 58 34W. Town forming NW suburb of Buenos Aires on Río de la Plata. Industry fish. canning; manu. footwear and furniture.

San Fernando   Colchagua, Chile. 34 35S 71 00W. Town sit. S of Santiago. Cap. of province. Market town for agric. area. Trade in cereals, wine, fruit. Pop. (1980E) 44,500.

San Fernando   Cádiz, Spain. 36 28N 6 12W. Town and naval dockyard sit. SE of Cádiz on the Isla de León. Industries distilling, tourism, processing salt. Pop. (1989E) 82,862.

San Fernando   Trinidad. 10 17N 61 28W. Town sit. S of Port of Spain on the Gulf of Paria. Industry sugar refining. Sea port exporting petroleum products and sugar. Pop. (1980C) 33,395.

San Fernando   Apure, Venezuela. 7 54N 67 28W. Town sit. S of Caracas on Apure R. State cap. Port for R. steamers, trading in cattle. Industries inc. meat canning. Airport. Pop. (1981C) 57,308.

San Francisco   California, U.S.A. 37 48N 122 24W. City sit. on a peninsula on the Pacific coast. Commercial, financial and industrial centre. Industries inc. oil refining, food processing, sugar refining, printing and publishing, tourism; manu. clothing and furniture. Sea port exporting iron and steel products, oil and canned foods, from fine land-locked harbour. Pop. (1980C) 678,974.

Sangihe Islands   Indonesia. 3 00N 125 30E. Group of islands NNE of Celebes. Area, 314 sq. m. (813 sq. km.). Cap. Tahuna. Chief products sago, nutmeg, copra.

Sangli   Maharashtra, India. 16 55N 74 37E. Town sit. SSE of Bombay on Kistna R. Railway terminus and market for agric. produce. Industries inc. cotton and oilseed milling, processing coffee, peanuts, dairy produce; manu. agric. implements. Pop. (1981C) 152,389.

Sangre de Cristo Mountains   U.S.A. 37 30N 105 15W. Range of Rocky Mountains extending 210 m. (336 km.) N and S through S Colorado and N New Mexico and rising to Blanca Peak, 14,317 ft. (4,364 metres) a.s.l.

San Ignacio de Velasco   Velasco, Bolivia. Town sit. NE of Santa Cruz. Provincial cap. Centre of the rubber trade.

San Joaquin River   U.S.A. 36 43N 120 10W. Rises in the Sierra Nevada, flows 320 m. (512 km.) SW and then NNW through the S half of the California central valley, to join Sacramento R. E of Suisun Bay. Its course lies through one of the most fertile regions of the U.S.A.

San José   San José, Costa Rica. 10 00N 84 02W. Cap. of Costa Rica and of San José province, founded in 1738, sit. on central plateau. Commercial centre. Industries sugar, coffee and cacao processing, flour milling, fruit and vegetable canning. Pop. (1984E) 245,550.

San José   Escuintla, Guatemala. Town sit. S of Escuintla on the Pacific. Rail terminus and port shipping coffee, timber, sugar. Also called Puerto de San José. Pop. (1980E) 8,000.

San José Uruguay. 34 20S 56 42W. (i) Dept. of S Uruguay on Río de la Plata, bounded E by Santa Lucia R. Area, 2,688 sq. m. (6,962 sq. km.). Stock farming and agric. area producing cattle, sheep, wheat, maize, flax, fruit. Pop. (1985C) 91,900. (ii) Town sit. NW of Montevideo on San José R. Dept. cap. and commercial centre. Industries inc. flour milling. Pop. (1985C) 31,732.

San José California, U.S.A. 37 20N 121 53W. City on the S tip of San Francisco Bay. Centre of the Silicon Valley high technology complex. Manu. semiconductors, electronic equipment, computers and guided missiles. Pop. (1980C) 629,442.

San Juan San Juan, Argentina. 31 30S 68 30W. Town in W central Argentina. Cap. of province. Commercial centre of a wine growing region irrigated by San Juan R. Trades in wine, meat, dried fruit and dairy produce. Pop. (1980C) 400,000.

San Juan Puerto Rico, U.S.A. 18 28N 66 07W. Cap. of Puerto Rico, sit. on NE coast. Industries sugar refining, rum distilling, brewing; manu. cigars, cigarettes. Sea port exporting coffee, sugar and tobacco to U.S.A. Pop. (1980C) 434,849.

San Juan del Sur Rivas, Nicaragua. Town sit. S of Rivas on the Pacific. Railway terminus and port handling coffee, livestock and sugar. Resort. Cable station. Pop. (1984E) 4,750.

San Juan River Nicaragua. Leaves L. Nicaragua and flows 120 m. (192 km.) ESE to form E half of the Nicaragua/Costa Rica border. It enters the Caribbean Sea at S end of San Juan del Norte Bay.

San Juan River Colorado/Utah, U.S.A. 37 18N 110 28W. Rises in San Juan Mountains of SW Colorado and flows 400 m. (640 km.) W to join Colorado R. at the entrance to L. Powell, Utah. Its meandering course through the Colorado plateau has incised deep, twisted canyons called the Goosenecks. Unnavigable, but used for irrigation.

San Leandoro California, U.S.A. 37 43N 122 09W. City immediately SE of Oakland, on San Francisco Bay. Industries food processing; manu. tractors, electrical equipment. Pop. (1980C) 63,952.

Sanlúca de Barrameda Cádiz, Spain. 36 47N 6 21W. Town sit. N of Cádiz on estuary of Guadalquivir R. Industries flour milling, distilling, tourism. Sea port exporting manzanilla wine.

San Luis San Luis, Argentina. 33 20S 66 20W. Town immediately S of the Sierra de San Luis. Centre of a quarrying and farming area. Trade in grain, wine, livestock. Pop. (1980C) 71,000.

San Luis Potosi Mexico. 22 09N 100 59W. (i) State in central Mex. Area, 24,417 sq. m. (63,240 sq. km.). Mountainous region crossed by the Sierra Madre Oriental. A silver-mining region, arid in N and fertile in SE where coffee and tobacco are grown on irrigated land. Industries smelting and refining metal. Pop. (1989E) 2,055,364. (ii) Cap. of (i) sit. at 6,158 ft. (1,877 metres) a.s.l. Centre of commerce and communications. Industries smelting, refining silver and arsenic; manu. textiles, clothing, footwear, brushes, ropes. Special architectural interest. Pop. (1980C) 406,630.

San Marcos Guatemala. 14 58N 91 48W. (i) Dept. of SW Guatemala on Mexican border. Area, 1,464 sq. m. (3,792 sq. km.). Chief towns San Marcos, San Pedro. Coastal plain on a short Pacific coastline rises to highlands inc. Tacaná and Tajumulco volcanoes. Drained by Suchiate and Naranjo Rs. Agric. area with stock farming in the highlands. Main industries textiles, salt working. Pop. (1985E) 590,152. (ii) Town sit. WNW of Quezaltenango at 8,136 ft. (2,408 metres) a.s.l. Dept. cap. and market centre for a coffee growing area. With the adjoining town of San Pedro, on the Inter-American Highway, sometimes called La Unión.

San Marino Independent rep. surrounded by Italy and near coast of Adriatic Sea. Area, 24 sq. m. (62 sq. km.). Main occupations agric., stock rearing. Industries textiles, quarrying, wine-making, pottery-making. Pop. (1988E) 22,746.

San Marino San Marino. 43 56N 12 25E. Cap. sit. on Mount Titano. Market centre for an agric. area. Pop. (1986E) 4,363.

San Martín Peru. Dept. of N Peru. Area, 17,452 sq. m. (45,201 sq. km.). Chief towns Moyobamba (cap.), Tarapoto. Traversed by ridges of the E Andes N to S it slopes E, to the R. Amazon basin. Humid, tropical climate with fertile agric. valleys and extensive forest. Chief products sugar-cane, rice, maize, cacao, coffee, tobacco, yucca, fruit, timber, balata, rubber, vanilla. Industries inc. distilling alcohol, manu. straw hats. Pop. (1980E) 319,751.

San Mateo California, U.S.A. 37 35N 122 19W. City sit. S of San Francisco on San Francisco Bay. Centre of a horticultural district. Residential. Pop. (1980C) 77,561.

San Miguel San Miguel, El Salvador. 13 29N 88 11W. Cap. of San Miguel dept. sit. ESE of San Salvador at the foot of the volcano San Miguel. Trades in cereals, cotton, sisal etc. Industries flour milling, manu. cotton goods, rope. Pop. (1981E) 161,156.

San Nicolás Buenos Aires, Argentina. 33 20S 60 13W. Town on Paraná R. Industries steel-making, generating hydroelectric power. R. port trading in wool, hides and grain. Pop. (1980E) 96,000.

San Pablo Luzon, Philippines. 14 04N 121 19E. Town sit. SSE of Manila. Commercial centre trading in rice and copra.

San Pedro Paraguay. 24 07S 56 59W. (i) Dept. of central Paraguay bounded W by Paraguay R. Area, 7,723 sq. m. (20,003 sq. km.). Chief towns San Pedro, Rosario, Hacurubí del Rosario, Antequera. Low-lying forest and marsh rising E towards the Brazilian plateau. Humid sub-tropical climate. Main products maté, oranges, timber. Pop. (1981E) 189,751. (ii) Town sit. NNE of Asunción. Dept. cap. and trading centre. Industries inc. sawmilling. Pop. (1982E) 4,500.

San Pedro de Macoris Dominican Republic. 18 27N 69 18W. Town sit. E of Santo Domingo on the Caribbean Sea. Cap. of province of same name. Industries sugar and flour milling, tanning; manu. clothing and soap. Port exporting sugar and molasses. Pop. (1982E) 115,000.

San Pedro Sula Cortés, Honduras. 15 27N 88 02W. Cap. of dept. sit. NW of Tegucigalpa. Commercial centre of an agric. region. Industries flour milling, tanning, brewing; manu. soap, furniture, cigarettes. Pop. (1986E) 399,700.

Sanquhar Dumfries and Galloway region, Scotland. 55 22N 3 56W. Town sit. NW of Dumfries on Nith R. Industries inc. carpets, nylon yarn and knitwear manu., aluminium extrusions and pyrotechnics. It is a special development area. Pop. (1981C) 2,170.

San Rafael Mendoza, Argentina. 34 40S 68 21W. Town and commercial centre of an irrigated area growing vines, cereals and fruit. Industries meat packing, wine-making, fruit processing. Pop. (1980C) 46,000.

San Remo Liguria, Italy. 43 49N 7 46E. Town sit. on the Riviera di Ponente, and near the border with France. Tourist centre. Trades in fruit, flowers, olives. Pop. (1989E) 60,000.

San Salvador (Watling or Watling's Island) Bahamas. Island in the central Bahamas. Area, 60 sq. m. (155 sq. km.). Chief occupation farming. Pop. (1980C) 804 (with Rum Cay).

San Salvador El Salvador. 13 40N 89 10W. Cap. of El Salvador, founded 1525, sit. below San Salvador dormant volcano which last erupted in 1917. Chief commercial and industrial centre. Industry milling flour and packing meat; manu. textiles, clothing, cigars and cigarettes, soap. Has suffered frequent earthquakes. Pop. (1985E) 972,810.

San Sebastián　Guipúzcoa, Spain. 43 19N 1 59W. Town and cap. of province, sit. at mouth of Urumea R. on the Bay of Biscay. Coastal resort. Industries fishing, manu. transformers, soap, cement, glass, paper. Pop. (1986E) 175,138.

San Severo　Puglia, Italy. 41 41N 15 23E. Town in SE Italy. Centre of a farming and wine growing area. Industries brick making, making wine, olive oil and pasta. Pop. (1981E) 60,000.

Santa Ana　Santa Ana, El Salvador. 13 59N 89 34W. Cap. of dept. sit. NW of San Salvador. Industries coffee processing, sugar milling, brewing; manu. textiles, leather goods and cigars. Pop. (1983E) 208,322.

Santa Ana　California, U.S.A. 33 43N 117 54W. City sit. SE of Los Angeles at foot of Santa Ana Mountains. Industries inc. telecommunications. Pop. (1980C) 203,713.

Santa Bárbara　Honduras. 14 53N 88 14W. (i) Dept. of W Honduras on Guatemala border. Area, 2,864 sq. m. (7,418 sq. km.). Chief towns Santa Bárbara, Naranjito, Colinas, Trinidad. Mountainous agric. area, producing coffee, sugar-cane, tobacco, maize, beans, livestock, with some lumbering. Pop. (1983E) 286,854. (ii) Town sit. SSW of San Pedro Sula. Dept. cap. and trading centre. Industries inc. manu. hats. Pop. (1980E) 6,000.

Santa Barbara　California, U.S.A. 34 25N 119 42W. City sit. WNW of Los Angeles on Pacific coast. Seaside resort with Spanish mission heritage. Pop. (1980C) 74,414.

Santa Barbara Islands　California, U.S.A. 33 23N 119 01W. Group of 8 islands, and other uninhabited islets, off Pacific coast extending *c*. 150 m. (240 km.) between Santa Barbara and San Diego. Main islands, San Miguel, Santa Rosa, Santa Cruz, Anacapa, Santa Catalina, San Clemente, San Nicolas, Santa Barbara. Santa Catalina is a tourist resort.

Santa Catarina　Brazil. 27 30S 48 30W. State in S Braz. on Atlantic coast. Area, 36,802 sq. m. (95,318 sq. km.). Cap. Florianópolis. Narrow coastal lowlands with the Serra do Mar to the W and forest to the N. Chief occupation maize growing and pig rearing; some coalmining in the SE. Pop. (1989E) 4,386,697.

Santa Clara　Las Villas, Cuba. 22 24N 79 58W. Cap. of province sit. ESE of Havana. Centre of commerce and communications in an agric. region, trading in coffee, sugar-cane and tobacco. Manu. cigars and leather goods. Pop. (1986E) 178,300.

Santa Clara　California, U.S.A. 37 21N 121 57W. City immediately NW of San José. Industry fruit packing and canning; manu. machinery, chemicals. Pop. (1980C) 87,746.

Santa Cruz　Argentina. 50 00S 68 50W. National territory in Patagonia bounded N by Deseado R., S by Strait of Magellan, E by Atlantic, W by Andes. Area, 94,186 sq. m. (243,943 sq. km.). Cap. Río Gallegos. Drained by Río Chico, Santa Cruz, Coyle and Gallegos Rs. Sheep-farming area with a dry, cold climate. Minerals inc. kaolin, platinum, coal, iron, salt, manganese. Main occupations farming, mining, wool processing, lumbering. Industries inc. flour milling, meat packing, fisheries. Pop. (1980C) 114,941.

Santa Cruz　Santa Cruz, Bolivia. 17 48S 63 10W. Cap. of dept. of same name in central Bolivia. Commercial centre trading in rice and sugar-cane. Industries sugar milling, tanning, distilling. Expanded greatly since being linked in 1953 with Cochabamba by road, and with Corumbá, Brazil, and Aguaray, Argentina, by rail. Pop. (1982E) 376,912.

Santa Cruz　California, U.S.A. 36 58N 122 01W. City sit. SSE of San Francisco on Monterey Bay. Holiday resort. Industries food processing and electronics. Pop. (1989E) 50,050.

Santa Cruz de Tenerife　Canary Islands. 28 27N 16 14W. (i) Spanish province of the Canary Islands in the N Atlantic Ocean, comprising four W islands of Tenerife, Gomera, Palma and Hierro. Area, 1,238 sq. m. (3,206 sq. km.). Chief occupation agric. and tourism. Pop. (1981c) 688,273. (ii) Cap. of province of same name, sit. on E coast of Tenerife. Important harbour and fuelling station. Industries tourism, oil refining. Pop. (1981c) 190,784.

Santa Cruz River Argentina. 50 10S 68 20W. Rises in L. Argentino and flows 200 m. (320 km.) E through Patagonia to Atlantic Ocean.

Santa Fé Santa Fé, Argentina. 31 40S 60 40W. Town sit. on Salado (Juramento) R. Market centre of a grain and stock farming area. Industries tanning, flour milling, manu. dairy produce. Pop. (1980c) 287,000.

Santa Fé New Mexico, U.S.A. 35 42N 105 57W. State cap. sit. in N central New Mex. at 7,000 ft. (2,134 metres) a.s.l. Commercial and communications centre. Exceptional historic interest. Pop. (1988E) 59,300.

Santa Isabel *see* **Malabo**

Santa Maria Rio Grande do Sul, Brazil. 29 41S 53 48W. Town sit. in S Brazil. Railway and market centre of agric. region. Industries railway engineering, brewing, tanning. Pop. (1980E) 150,000.

Santa Marta Magdalena, Colombia. 11 15N 74 13W. Cap. of dept. sit. on Caribbean coast at Cape de la Aguia. Port exporting bananas, coffee, hides. Pop. (1980E) 234,000.

Santa Monica California, U.S.A. 34 01N 118 30W. City sit. W of Los Angeles on Santa Monica Bay. Residential town with tourist and recreational facilities. Light manu. Pop. (1980c) 88,314.

Santander Colombia. Dept. of N central Colombia bounded W by Magdalena R. Area, 12,382 sq. m. (32,069 sq. km.). Chief towns Bucaramanga (cap.), Socovro, Piedecuesta, San Gil, Barrancabermeja. R-side lowlands rise to foothills of the Cordillera Oriental; climate varies with altitude. Agric. area with forests on the mountain slopes. Main products,

coffee, cacao, tobacco, sugar, rubber, cinchona, balsam, timber. The chief mineral is petroleum, at the Barrancabermeja oilfield. Pop. (1983E) 1,367,600.

Santander Spain. 43 28N 3 48W. City sit. on an inlet on the Bay of Biscay. Resort. Industries shipbuilding, oil refining, tanning. Manu. chemicals, cables, machinery. Port exporting minerals, wine and wheat. Pop. (1981c) 180,328.

Santarém Ribatejo, Portugal. 39 14N 8 41W. Town sit. NE of Lisbon on lower Tagus R. Commercial centre of an agric. area, noted for raising horses and fighting bulls, corn, viticulture, rice and tomato production, trading in olive oil, wine, fruit, grain, cork. Industries inc. manu. alcohol and fertilizers. Pop. (1985E) 24,733.

Santa Rosa La Pampa, Argentina. 14 10N 90 18W. Town sit. NW of Bahía Blanca. Cap. of La Pampa and of Santa Rosa dept. Agric. trading and meat packing centre. Pop. (1980c) 52,000.

Santa Rosa California, U.S.A. 38 26N 122 43W. City sit. NNW of San Francisco. Industries inc. electronics and wine. Pop. (1989E) 109,900.

Santa Tecla (Nueva San Salvador) La Libertad, El Salvador. Cap. of dept. sit. W of San Salvador at 3,000 ft. (914 metres) a.s.l. Commercial centre of a coffee growing and stock rearing district. Founded to replace San Salvador as cap. when the latter was destroyed by earthquake in 1854, but San Salvador was rebuilt and reinstated. Pop. (1980E) 52,563.

Santee River South Carolina, U.S.A. 33 14N 79 28W. Formed by confluence of Rs. Congarce and Wateree in SE Carolina and flowing 143 m. (229 km.) SE to the Atlantic. Forms part of a big hydroelectric and waterway development; the Santee Dam forms L. Marion, 40 m. (64 km.) long, up to 12 m. (19 km.) wide.

Santiago Chile. 33 24S 70 40W. Cap. of Chile sit. on Mapocho R., founded in 1541. Manu. textiles, clothing, footwear, chemicals, metal goods, pharmaceuticals and food products. Pop. (1987E) 4,858,342.

Santiago de Compostela La Coruña, Spain. 42 53N 8 33W. City sit. in NW Spain. Agric. trading centre, dealing in plywood, cereals, fruit. Industries brewing, distilling; manu. linen, soap, paper, TV parabolic aerials. The most famous place of pilgrimage in Spain; claims the tomb of St. James, Spanish patron saint. Pop. (1989E) 105,000.

Santiago de Cuba Oriente, Cuba. 20 01N 75 49W. Town sit. ESE of Havana on S coast. Industries sugar milling, tanning, rum-distilling, brewing; manu. cigars, soap, perfumes. Sea port on a narrow 5 m. (8 km.) channel, exporting tobacco products, sugar, rum and mineral ores. Pop. (1987E) 358,800.

Santiago del Estero Santiago del Estero, Argentina. 27 50S 64 15W. Cap. of province of same name sit. SSE of Tucumán on Dulce R. Commercial centre of agric. area, trading in cotton, cereals and livestock. Industries flour milling, tanning; manu. textiles. Pop. (1980C) 148,000.

Santiago de los Caballeros Dominican Republic. 19 27N 70 42W. Town sit. NW of Santo Domingo. Commercial centre of an agric. area trading in coffee, rice, tobacco. Industries coffee and rice milling; manu. tobacco products, furniture, pottery. Pop. (1980E) 278,638.

Santiniketan (Bolpur) West Bengal, India. Viva-Bharati University founded by Nobel Laureate Rabindranath Tagore, is located here, known for handicrafts and leather-work. Pop. (1981C) 38,386.

Santo Antao Cape Verde. 17 05N 25 10W. Most N island of group. Area, 301 sq. m. (779 sq. km.). Chief town Ribeira Grande. Pop. (1980C) 43,198.

Santo Domingo Dominican Republic. 18 30N 64 54W. Cap. of Dominican Republic, founded in 1496 by the brother of Christopher Columbus, sit. on coast. Industries distilling, brew-

ing, tanning, manu. soap. Sea port exporting sugar, cacao, coffee. Largely rebuilt after a hurricane in 1930. Re-named Ciudad Trujillo 1936–61; reverted to its former name on Pres. Trujillo's assassination. Tourism is an important industry. Pop. (1981C) 1,313,172.

Santos São Paulo, Brazil. 23 57S 46 20W. Town sit. on Atlantic coast. Sea port for São Paulo city with which it is linked by rail and road over the Serra do Mar. The world's leading coffee port, also exporting sugar, fruit, cotton. Pop. (1980C) 410,933.

San Vicente El Salvador. 13 41N 88 43W. Town sit. ESE of San Salvador at foot of volcano San Vicente. Market centre of an agric. area. Trade in coffee, sugar-cane, tobacco. Industries textiles, manu. leather goods. Pop. (1980E) 48,000.

São Francisco River Brazil. 10 30S 36 24W. Rises in the Serra da Canastra in SW Minas Gerais state and flows 2,000 m. (3,200 km.) NNE and then E, forming the boundaries between Bahia and Pernambuco and between Alagôas and Sergipe states. Enters Atlantic NE of Aracajú. Navigable only by small vessels in the middle course.

São Luís Maranhão, Brazil. 2 31S 43 16W. State cap. sit. ESE of Belém on São Luís island. Industries cotton, sugar refining distilling. Sea port exporting cotton, hides, skins, babassu oil. Pop. (1980E) 350,000.

São Miguel Azores, Atlantic. 37 47N 25 30W. Largest of the Azores islands sit. N of Santa Maria in E group. Area, 288 sq. m. (746 sq. km.). Chief towns Ponta Delgada (cap.), Ribeira Grande, Vila Franca do Campo. Mountainous volcanic ridges separated by central lowland, with Pico da Vara (E) rising to 3,625 ft. (1,105 metres) a.s.l. Mild climate and fertile soil, producing fruit, tea, vines, tobacco, cereals, vegetables, livestock. Tourism is important.

Saône-et-Loire France. 46 25N 4 50E. Dept. in Burgundy region. Bordered on W by R. Loire and extending across R. Saône E into Bresse. Area, 3,307 sq. m. (8,565 sq. km.). Cap. Mâcon. Agric., esp. viticulture, with coalmining and iron and steel industries in the SE. Pop. (1982C) 571,852.

Saône River France. 45 44N 4 50E. Rises in Monts Faucilles, in the Vosges, flows 280 m. (448 km.) SSW past Chalon-sur-Saône, Mâcon and Villefranche to enter Rhône R. at Lyons. Main tribs. Rs. Doubs and Ognon. Connected by canals with Rs. Moselle, Marne, Loire, Seine, Meuse and Rhine.

Sao Nicolau Cape Verde. 16 35N 24 15W. Island in N of arch. Area, 150 sq. m. (388 sq. km.). Chief town Ribeira Brava. Pop. (1980C) 13,575.

São Paulo Brazil. 23 32S 46 37W. (i) State in SE Braz. on Atlantic coast. Area, 95,852 sq. m. (248,256 sq. km.). Table-land with a small coastal strip, adequate rainfall and equable climate; the most highly developed state in Braz., and main coffee producing area. Beef-cattle ranching also of great importance. Industries mainly centre on the cap. Pop. (1989E) 33,361,701. (ii) Cap. of (i) sit. NW of its port, Santos, and at 2,700 ft. (823 metres) a.s.l. Industries textiles, clothing, manu. paper, chemicals, metal goods, cars, machinery. Pop. (1980C) 7,032,547.

São Tiago Cape Verde. 15 05N 23 40W. Largest island of group containing Praia, cap. of rep. Produces coffee, sugar-cane and bananas. Area, 383 sq. m. (991 sq. km.). Pop. (1980C) 145,923.

São Tomé São Tomé e Príncipe, Gulf of Guinea. 0 12N 6 39E. Cap. and port on the NE coast of São Tomé island. It exports cocoa and coffee. Pop. (1984E) 34,997.

São Tomé São Tomé e Príncipe, Gulf of Guinea. Island sit. W of Gabon forming (with Príncipe island) the state of São Tomé e Príncipe. Area, 330 sq. m. (854 sq. km.). Chief town, São Tomé. Pop. (1987E) 106,900.

São Tomé e Príncipe Gulf of

Guinea. Independent republic formerly Portuguese province comprising 2 islands, sit. off W African coast. Area, 372 sq. m. (964 sq. km.). Cap. São Tomé. Chief occupations growing coconuts and coffee. Pop. (1988E) 115,600.

São Vicente Cape Verde. 23 58S 46 23W. Main island in Barlovento group. Area, 88 sq. m. (227 sq. km.) (with uninhabited Santa Luzia). Cap. Mindelo. Pop. (1980C) 41,792.

Sapele Bendel, Nigeria. 5 54N 5 41E. Town sit. N of Warri on Benin R. at the head of navigation. Main industries inc. sawmilling; manu. plywood.

Sapporo Hokkaido, Japan. 43 03N 141 21E. Town sit. NNE of Hakodate. Cap. of Hokkaido prefecture. Industries inc. brewing, sawmilling; manu. agric. machinery. Pop. (1989E) 1,642,011.

Sarajevo Bosnia and Herzegovina, Yugoslavia. 43 52N 18 25E. Cap. of Federal unit sit. on Miljacka R. Commercial centre. Industries flour milling, brewing, engineering; manu. chemicals, carpets, pottery, tobacco products. Centre of the Moslem faith in Yugoslavia. Pop. (1982E) 448,500.

Sarandë Albania. 39 52N 20 00E. Town sit. SE of Valona on Ionian Sea. Port and commercial centre. Naval base.

Saransk Russian Soviet Federal Socialist Republic, U.S.S.R. 54 11N 45 11E. Cap. of the Mordovian Autonomous Soviet Socialist Republic, sit. SSE of Gorky. Centre of an agric. area. Industries processing grain, hemp, sugar-beet, dairy produce; manu. agric. machinery, electrical equipment. Pop. (1987E) 323,000.

Sarapul Udmurt Autonomous Soviet Socialist Republic, U.S.S.R. 56 28N 53 48E. Town sit. SW of Perm on Kama R. Trading centre for grain and timber. Manu. leather, footwear, rope, machine tools.

Sarasota Florida, U.S.A. 27 20N 82 34W. City sit. S of Tampa on Sarasota

Bay. Industries inc. construction, light manu., tourism, agric. and catering for the retired. Pop. (1988E) 53,259.

Saratoga Springs New York, U.S.A. 43 05N 74 47W. City sit. N of Albany. Resort and horse racing centre, with mineral springs. Pop. (1980C) 23,906.

Saratov Russian Soviet Federal Socialist Republic, U.S.S.R. 51 34N 46 02E. Cap. of Saratov Region sit. on Volga R. Industries oil refining, natural gas, flour milling, sawmilling; manu. agric. machinery, diesel engines, railway rolling stock. Pop. (1989E) 905,000.

Sarawak Malaysia. 2 00N 113 00E. Part of E Malaysia bounded NE by Brunei and Sabah, E and S by Kalimantan, Indonesia. Area, 48,050 sq. m. (124,449 sq. km.). Cap. Kuching. Low-lying coastlands with mountainous and forested hinterland. Mainly agric., chief products rubber, timber, pepper, sago. Chief towns Kuching and Sibu. Pop. (1988E) 1·59m.

Sarcelles Val-d'Oise, France. 49 00N 2 23E. N suburb of Paris. Manu. plastics and paints. Pop. (1982C) 53,732.

Sardegna (Sardinia) Italy. 40 00N 9 00E. Island and administrative Region in Mediterranean, separated from Corsica by the strait of Bonifacio. Area, 9,301 sq. m. (24,090 sq. km.). Cap. Cagliari. Mountainous, rising to the Monti del Gennargentu, 6,017 ft. (1,834 metres) a.s.l. Hot dry summers, annual rainfall 15–25 ins. (38–63 cm.) in lowlands, higher in the mountains. Fertile Campidano plain is agric. centre for cereals, grapes, olives, sheep and goats. Mining area, esp. for zinc and lead. Chief towns Cagliari, Sassari, Igleias. Pop. (1988E) 1,655,859.

Sardinia *see* **Sardegna**

Sarh Moyen-Chari, Chad. 9 08N 18 22E. Cap. of prefecture in S, formerly Fort-Archambault. Centre for fishing along Chari R., and textiles. Pop. (1985E) 124,000.

Sari Iran. 36 33N 53 06E. Town sit. E of Babul on Tajan R. near Caspian Sea. Centre of an agric. area growing rice, sugar-cane, oranges. Pop. (1983E) 125,000.

Sariwon Hwanghae, North Korea. 38 31N 125 44E. Town sit. S of Pyongyang. Railway junction and commercial centre for an agric. area. Industries inc. silk reeling, flour milling, cotton ginning.

Sark Channel Islands, United Kingdom. 49 26N 2 21W. Island sit. E of Guernsey in the English Channel. Area, 2 sq. m. (5 sq. km.). Consists of Great Sark and Little Sark, joined by an isthmus. Main occupations farming, fishing, tourism. Chief harbour is Maseline Jetty. Motor traffic, except tractors, forbidden. Pop. (1986E) 550.

Sarnia Ontario, Canada. 42 58N 82 23W. City sit. SW of Toronto on St. Clair R. immediately opposite Port Huron, Michigan, U.S.A. Industries oil refining and associated petrochemical industries, supplied by the Alberta oilfields. Natural gas pipeline to Toronto and Montreal. Pop. (1981C) 83,951.

Sarthe France. 47 58N 0 10E. Dept. in Pays-de-la-Loire region. Area, 2,398 sq. m. (6,210 sq. km.). Cap., Le Mans. Drained by Rs. Sarthe, Huisne and Loir, low-lying and agric. Produces cereals, apples, pears, cider, hemp. Chief towns Le Mans and La Flèche. Pop. (1982C) 504,768.

Sarthe River France. 47 44N 0 32W. Rises in Perche hills and flows 177 m. (283 km.) SW and S past Alençon and Le Mans to join Mayenne R. above Angers. Together they form Maine R.

Sasebo Kyushu, Japan. 33 10N 129 43E. Town sit. NNW of Nagasaki on NW coast. Industries shipbuilding, engineering. Naval station and sea port exporting coal. Pop. (1983E) 252,000.

Saskatchewan Canada. Central province bounded on S by Montana and North Dakota, U.S.A. Area, 220,120 sq. m. (570,113 sq. km.).

Cap. Regina. Forest, lake and swamp in N, open prairie in S, gradually rising to 4,546 ft. (1,386 metres) a.s.l. in SW. Winters long and cold, summers short and hot. Annual rainfall 10–15 ins. (25–38 cm.). Large wheat producer. Chief minerals petroleum in SW, uranium in NW and copper. Important towns Regina and Saskatoon. Pop. (1986C) 1,010,198.

Saskatchewan River Canada. 53 12N 99 16W. Formed by union of N. and S. Saskatchewan R. N. Saskatchewan rises near Mount Saskatchewan in the Rocky Mountains and flows 760 m. (1,216 km.) E through SW Alberta, past Edmonton and Prince Albert. The S. Saskatchewan is formed from Bow and Oldman Rs. in S Alberta, and flows 865 m. (1,384 km.) E and NE past Saskatoon to join N. Saskatchewan 30 m. (48 km.) E of Prince Albert. The combined R. flows 370 m. (592 km.) E into the NW end of L. Winnipeg.

Saskatoon Saskatchewan, Canada. 52 07N 106 38W. City sit. on S Saskatchewan R. Distribution centre for agric. area. Industries inc. electronics, biotechnology, fibre optics, telecommunications, agric. equipment, meat packing, printing and publishing, foundries, and provincial mining centre. Pop. (1989E) 185,000.

Sassandra Côte d'Ivoire. 4 57N 6 05W. Town sit. W of Abidjan at mouth of Sassandra R. on Gulf of Guinea. Port handling cacao, coffee, rubber, fruit, timber, palm oil and kernels. Industries inc. fishing, sawmilling.

Sassari Sardegna, Italy. 40 44N 8 33E. Town near N coast. Cap. of Sassari province. Manu. cheese, olive oil and pasta. Pop. (1981C) 119,596.

Satara Maharashtra, India. Town sit. SE of Bombay. District cap., route and trading centre for grain and millet. Industries inc. manu. plastics, metal products, with engineering nearby. Pop. (1981C) 83,336.

Satpura Range Maharashtra/Madhya Pradesh, India. Range of hills at 2,000–4,000 ft. (610–1,219 metres)

a.s.l. between Narmada R. and Tapti R. extending *c.* 600 m. (960 km.) partly along the Maharashtra/Madhya Pradesh border, and then through S Madhya Pradesh.

Satu Mare Satu Mare, Romania. 47 48N 22 53E. Cap. of district of same name sit. NW of Bucharest, on Somes R. Trading centre of agric. area. Manu. textiles, machinery, toys, furniture. Pop. (1985E) 128,115.

Saudi Arabia Kingdom occupying most of the peninsula of Arabia. 26 00N 44 00E. Area, 927,000 sq. m. (2,401,000 sq. km.). The kingdom has been welded together from Hejaz, Nejd, Asir and Al-Hassa. Caps. Riyadh (political) and Mecca (religious). The country is mainly desert. Some agric. is carried on in SW and in oases where dates, fruit, wheat and barley are grown. Oil is the main source of wealth. Pop. (1988E) 12m.

Sault Ste. Marie Ontario, Canada. 46 31N 84 20W. City sit. on St. Mary's R. between L. Superior and L. Huron, and directly opposite Sault Ste. Marie, Michigan, U.S.A. Industries railway engineering; manu. steel, wood pulp, paper. Holiday resort. Pop. (1981C) 82,902.

Sault Ste. Marie Michigan, U.S.A. 46 30N 84 21W. City sit. opposite Sault Ste. Marie, Ontario, Canada, on St. Mary's R. Site of 5 locks, connecting L. Superior to L. Huron. Industries inc. tourism, plastics, veneers, automobile parts, shipbuilding. Pop. (1989E) 17,842.

Saumur Maine-et-Loire, France. 47 16N 0 05W. Town sit. in NW France at confluence of Loire and Thouet Rs. and on an island of the Loire R. Market town in a wine producing district. Noted for mushrooms and the National Riding School. Manu. brandy, liqueurs, leather goods, rosaries. Pop. (1989E) 35,000.

Sauternes Gironde, France. Village sit. SSE of Bordeaux. Centre of a wine producing district.

Savannah Georgia, U.S.A. 32 04N 81 05W. City sit. near mouth of

Savannah R. Industries shipbuilding, sugar refining, pulp and paper milling; manu. cottonseed oil, aircraft, fertilizers, chemicals. Resort and port exporting naval stores and raw cotton. Pop. (1988E) 146,000.

Savannah River Georgia/South Carolina, U.S.A. 32 02N 80 53W. Formed by union of Tugaloo R. and Seneca R. on the Georgia/S. Carolina border and flowing 314 m. (502 km.) SE down the border past Augusta and Savannah to the Atlantic. Used for hydroelectric power. Navigable for barges below Augusta.

Savannakhet Savannakhet, Laos. 16 30N 104 49E. Town on Mekong R. where it forms the Thai frontier. Provincial cap. The main industry is sericulture. Pop. (1984E) 50,690.

Savé Benin. Town sit. N of Porto-Novo. Trading centre on the Niger road.

Save River Yugoslavia. 44 50N 20 26E. Rises in the Karawanken Alps and flows 580 m. (928 km.) ESE through Slovenia and Croatia to form part of the Croatia/Bosnia border, and joins Danube R. at Belgrade. Main tribs. Rs. ·Una, Vrbas, Bosna and Drina. Navigable below Sisak.

Savoie France. 45 26N 6 35E. Dept. in Rhône-Alpes region, forming S part of former Duchy of Savoy. Bounded E and SE by Italy. Area, 2,330 sq. m. (6,036 sq. km.). Cap. Chambéry. Mountainous, with the Savoy Alps rising to Mont Pourri, 12,428 ft. (3,788 metres) a.s.l. Chief products dairy produce and wine. Metallurgical and chemical industries. Chief towns Chambéry and Aix-les-Bains. Pop. (1982C) 323,675.

Savona Liguria, Italy. 44 17N 8 30E. Town and port on Gulf of Genoa. Cap. of Savona province. Industries iron and steel, manu. pottery, glass, bricks. Exports glass and pottery. Pop. (1981E) 75,000.

Savu Islands Indonesia. 9 40S 122 00E. Group of islands in the Lesser Sundas between Sumba and Timor. Area, 231 sq. m. (598 sq. km.). Main island, Savu. Chief products copra, rice, tobacco.

Sayan Mountains Russian Soviet Federal Socialist Republic, U.S.S.R. 52 45N 96 00E. Two mountain ranges in extreme S of U.S.S.R. The Eastern Sayan Mountains extend SE from Yenisei R. to the Mongolian border, rising to Munku Sardyk, 11,457 ft. (3,492 metres) a.s.l.; the Western Sayan Mountains extend ENE from the Altai Mountains, rising to 9,000 ft. (2,743 metres) a.s.l. and join the E range. Gold, silver, lead and coal are found as well as timber.

Scafell Cumbria, England. 54 27N 3 12W. Mountain mass in the Lake District of NW, consisting of Scafell Pike, 3,210 ft. (978 metres) a.s.l., the highest English peak, and three other peaks.

Scandinavia 64 00N 12 00E. Geographically, area in NW Europe comprising Norway, Sweden and Denmark; bounded N by Arctic Ocean, W by Norwegian Sea, E by Baltic Sea and Gulf of Bothnia. Culturally, Scandinavia inc. Iceland and the Faroes. Finland is often included.

Scapa Flow Orkney Islands, Scotland. 58 52N 3 06W. Area of sea almost enclosed by islands, bounded N by Mainland, E by Burray and Ronaldsay, W and SW by Hoy, S by Flotta and Swona. Area, 50 sq. m. (129 sq. km.). Historically important as a naval base.

Scarborough North Yorkshire, England. 54 17N 0 24W. Town in NE England sit. on North Sea coast, with 2 bays and an intervening headland. Seaside resort, conference centre and residential town. Industry coach building, food processing, fishing, tourism. Pop. (1989E) 50,908.

Schaffhausen Switzerland. 47 42N 8 38E. (i) Canton in N Switzerland to N of R. Rhine and bounded, except in SE, by Federal Republic of Germany. Area, 115 sq. m. (298 sq. km.). Commercial forest on a large scale; hydroelectric power from the Rhine falls. Industry mainly in the cap. Pop. (1988E) 70,313. (ii) Cap. of (i) sit.

NNE of Zürich on N bank of R. Rhine. Industries powered by the Rhine falls, textiles. manu. metal goods, armaments. Pop. (1980E) 34,250.

Schaumburg-Lippe Federal Republic of Germany. District of Lower Saxony between Hanover and North Rhine-Westphalia. Area, 131 sq. m. (339 sq. km.). Chief town Bückeburg, sit. ESE of Minden. Chief occupation agric. with coalmining in the Bückeburg region. Formerly a principality.

Schefferville Quebec, Canada. 54 47N 64 49W. Town in centre of Labrador Peninsula in NE Quebec. Centre of an iron ore region, founded recently to develop iron industry and transport ore to Sept-Îsles by rail.

Scheldt River France/Belgium/Netherlands. 51 22N 4 15E. Rises in Aisne dept., NE France. Flows 270 m. (432 km.) NNE past Valenciennes into Belgium, and then past Ghent and Antwerp into the Netherlands. Its estuary once had two branches, E and W, dividing Beveland and Walcheren islands from the mainland, to which they are now joined, and the E and W Scheldt are linked by canal. A 3 m. (5 km.) bridge crosses E Scheldt linking N. Beveland and Schouwen. Main trib. Lys R.

Schenectady New York, U.S.A. 42 47N 73 53W. City, sit. NW of Albany on Mohawk R. Manu. electrical equipment. Pop. (1980C) 67,972.

Scheveningen Zuid-Holland, Netherlands. 52 06N 4 18E. Town forming suburb of The Hague, sit. NW of the city centre. Industries fishing and tourism. Pop. (1980E) 22,000.

Schiedam Zuid-Holland, Netherlands. 51 55N 4 24E. Town sit. W of Rotterdam on Rhine R. Industries shipyards, chemical works; manu. gin, glass, bottles, crates, corks. R. port. Pop. (1989E) 69,496.

Schleswig Schleswig-Holstein, Federal Republic of Germany. 54 31N 9 33E. Town sit. NW of Kiel at W end of Schlei inlet on the Baltic Sea. Pop. (1989E) 26,817.

Schleswig-Holstein Federal Republic of Germany. 54 10N 9 40E. Province in the N bounded N by Denmark and sit. mainly on the Jutland peninsula. Area, 6,070 sq. m. (15,721 sq. km.). Cap. Kiel. Low-lying, drained by Rs. Elbe and Eide and crossed by Kiel canal, linking Elbe estuary with the Baltic Sea. Mainly agric. esp. rye, wheat, potatoes, cattle, pigs. Chief towns and sea ports Kiel, Lübeck and Flensburg, where industry is concentrated. Industries shipbuilding, manu. machinery, food products, electrical equipment. Pop. (1988E) 2,564,565.

Schönebeck Magdeburg, Germany. 52 01N 11 44E. Town sit. SSE of Magdeburg on Elbe R. Industries salt processing, manu. chemicals, explosives, machinery. Pop. (1989E) 45,000.

Schuylkill River Pennsylvania, U.S.A. 39 53N 75 12W. Rises in SE Pennsylvania and flows 130 m. (208 km.) SE through anthracite mining area, past Reading to join Delaware R. at Philadelphia. Navigable from the confluence for 100 m. (160 km.) by barges.

Schweinfurt Bavaria, Federal Republic of Germany. 50 03N 10 14E. Town in NW Bavaria. Manu. paints and dyes, ball bearings, machinery. Pop. (1984E) 51,500.

Schwerin Schwerin, German Democratic Republic. 53 38N 11 25E. Cap. of district of same name in NW, sit. on SW shore of L. Schwerin. Industries inc. engineering and pharmaceuticals, manu. furniture, cigarettes, soap. Pop. (1988E) 130,121.

Schwyz Switzerland. 47 02N 8 40E. (i) Canton in central Switzerland bordering Ls. Zürich, Luzern and Zug. Area, 351 sq. m. (909 sq. km.). Mountainous with extensive cattle-pasture and forest. Pop. (1988E) 106,409. (ii) Cap. of (i) sit. E of Lüzern at the foot of Gross Mythen, 6,240 ft. (1,902 metres) a.s.l. Industry tourism. Pop. (1980E) 12,100.

Sciacca Sicilia, Italy. 37 30N 13 06E. Town sit. SSW of Palermo. Tourist centre and port on SW coast, with sulphur springs.

Scilly Isles *see* **Isles of Scilly**

Scoresby Sound Greenland. 70 15N 23 15W. Deep inlet on the E coast *c*. 250 m. (400 km.) NNW of Iceland. Length *c*. 200 m. (320 km.). Many fjords receiving glaciers from the interior ice-cap.

Scotland United Kingdom. 57 00N 4 00W. The N part of the United Kingdom of Great Britain and N Ireland, bounded W and N by Atlantic Ocean, E by North Sea, S by England (across the Cheviot Hills from Solway (W) to Berwick (E). Area, 30,410 sq. m. (78,762 sq. km.). Chief cities Edinburgh (cap.), Glasgow, Aberdeen, Dundee. The Regions and Districts of Scotland from 1975 are Borders (Tweedale, Ettrick and Lauderdale, Berwickshire, Roxburgh); Central (Stirling, Clackmannan, Falkirk); Dumfries and Galloway (Merrick, Stewartry, Nithsdale, Annandale and Eskdale); Fife (North-East Fife, Kirkcaldy, Dunfermline); Grampian (Moray, Banff and Buchan, Gordon, Aberdeen City, Kincardine and Deeside); Highland (Caithness, Sutherland, Ross and Cromarty, Nairn, Inverness, Skye and Lochalsh, Lochaber, Badenoch and Strathspey); Lothian (West Lothian, Edinburgh City, Midlothian, East Lothian); Strathclyde (Argyll, Dumbarton, Clydebank, Bearsden and Milngavie, Bishopbriggs and Kirkintilloch, Cumbernauld, Monklands, Glasgow City, Renfrew, Inver-Clyde, Cunninghame, Kilmarnock and Loudon, Eastwood, East Kilbride, Hamilton, Motherwell, Lanark, Cumnock and Doon Valley, Kyle and Carrick); Tayside (Angus, Perth and Kinross, Dundee City). Highland N and W, with a deeply indented coastline and numerous lochs and off-shore islands; main island groups are the Inner and Outer Hebrides, W; the Orkneys and the Shetlands, N. The highland area rises to the Cairngorm mountains and Ben Nevis (W), 4,406 ft. (1,343 metres) a.s.l. Mountain ranges rise from high moorland, used for rough grazing, and watered by many short, fast flowing Rs. which provide hydroelectric power. The area is bisected by a chain of Ls. along the Great Glen fault which extends SW to NE across the country from Loch Linnhe to Moray Firth. Central Scotland is agric. and industrial lowland sit. mainly in Forth and Clyde Rs. basins, with extensive coalfields and supporting *c*. 80% of the pop. S of it is a hilly region with stock farming predominant and agric. in the R. valleys, esp. Tweed R. The agric. areas cover *c*. 24% of the land, with *c*. 60% as rough grazing. The chief crops are oats, barley, potatoes, vegetables, with fruit growing (E). Sheep and cattle breeding are important, esp. Ayrshire, Shorthorn and Aberdeen Angus cattle. Forestry and lumbering are important in the highlands. The most important minerals are coal and petroleum. Industries are concentrated in the lowland cities and on the coast, and inc. mining, engineering, shipbuilding, esp. on R. Clyde, fishing esp. herring, textiles, chemicals, brewing and distilling. The offshore oil industry has produced much employment. Trad. industries inc. spinning and weaving wool, metal working, quarrying and distilling. Pop. (1988E) 5,094,000.

Scottsdale Arizona, U.S.A. 33 30N 111 56W. City immediately NE of Phoenix. Employment is in services, wholesale and retail trade, some manu. Pop. (1980C) 88,412.

Scranton Pennsylvania, U.S.A. 41 24N 75 40W. City sit. NNW of Philadelphia on Lackawanna R. Industries inc. tourism. Manu. textiles, clothing, metal goods, plastics. Pop. (1980C) 88,117.

Scunthorpe Humberside, England. 53 36N 0 38W. Town sit. S of Humber estuary. Centre of iron-mining district. Industries iron and steel, engineering, manu. electronics, food processing, furniture and clothing. Pop. (1987E) 61,500.

Seaford East Sussex, England. 50 46N 0 06E. Town sit. W of Eastbourne on English Channel. Industry tourism. Pop. (1986E) 21,000.

Seaforth, Loch Outer Hebrides, Scotland. Inlet into SE coast of Lewis with Harris. Length 15 m. (24 km.). Forms part of the Lewis/Harris boundary.

Seaham County Durham, England. 54 52N 1 21W. Town and port sit. s of Sunderland. Centre of a coalmining area with increasing importance of light industry. Pop. (1987E) 22,560.

Sea Islands U.S.A. 31 20N 81 20W. Chain of islands in the Atlantic Ocean off S. Carolina/Georgia/Florida coasts, extending *c.* 160 m. (256 km.) NE to SW between the mouths of Rs. Santee and Saint John. Low-lying and sandy. Main occupation mixed farming. Formerly known for long-stapled 'Sea Island' cotton.

Seattle Washington, U.S.A. 47 36N 122 20W. City sit. on an isthmus between Puget Sound and L. Washington. Commercial, service, financial and industrial centre of the Pacific NW. Industries manu. aircraft, shipbuilding, food canning, tourism. Sea port trading with Alaska and the Far East, exporting timber, grain, fish, fruit. Pop. (1980C) 493,846.

Sebha Libya. 27 03N 14 26E. Town sit. SW of Hun at an oasis. Provincial cap. and communications centre. Airport. Industries inc. servicing motor caravans, small-engineering, flour milling, manu. ice. Pop. (1980E) 113,000.

Secunderabad Andhra Pradesh, India. 17 18N 78 30E. Suburb of Hyderabad, immediately N of the city centre. Commercial centre and railway junction. Pop. (1981C) 135,994.

Sedan Ardennes, France. 49 42N 4 57E. Town near border with Luxembourg, sit. on Meuse R. Industries inc. transport, tourism, woollen goods, machinery, mirrors, chemicals. It has the largest fortified castle in Europe. Pop. (1982C) 24,535.

Sedbergh Cumbria, England. 54 20N 2 31W. Town sit. E of Kendal. Market town for agric. and tourist area. Pop. (1980E) 2,300.

Ségou Mali. 13 27N 6 16W. Cap. of region sit. ENE of Bamako on Niger R. Trading centre for agric. produce. Industries inc. cotton ginning, rug making. Pop. (1976C) 64,890.

Segovia Spain. 40 57N 4 07W. (i) Province of central Spain, bounded SE by the Sierra de Guadarrama, rising to 8,100 ft. (2,469 metres) a.s.l. Area, 2,683 sq. m. (6,950 sq. km.). Mainly agric., esp. arable and sheep farming. Pop. (1986C) 151,520. (ii) Cap. of (i) sit. NW of Madrid on a ridge above Eresma R. Industries flour milling, tanning, manu. pottery. Pop. (1981E) 53,000.

Seinäjoki Vaasa, Finland. 62 47N 22 50E. Town sit. SE of Vaasa. Railway junction serving a farming and lumbering area. Pop. (1983E) 25,827.

Seine-et-Marne France. 48 45N 3 00E. Dept. in Île-de-France region. Area, 2,285 sq. m. (5,917 sq. km.). Cap. Melun. Drained by Rs. Seine and Marne and their tribs. Fertile agric. area. Chief products wheat and cheeses. Chief towns Melun and Fontainebleau. Pop. (1982C) 887,112.

Seine-Maritime France. 49 40N 1 00E. Dept. in Haute-Normandie region, bounded N and NW by English Channel, S by Seine R. Area, 2,415 sq. m. (6,254 sq. km.). Cap. Rouen. Mainly agric., esp. for flax and dairy cattle. Industry concentrated in chief towns Rouen, Le Havre and Dieppe. Pop. (1982C) 1,193,039.

Seine River France. 49 26N 0 26E. Rises on Plateau de Langres, in E central France, and flows 479 m. (766 km.) NW across Champagne, turning WSW below Troyes and again NW through Paris into Normandy, past Rouen and into the English Channel in an estuary between Le Havre and Honfleur. Navigable for ocean-going vessels to Rouen and for barges to Bar-sur-Seine. Linked by canal with Rs. Escaut (Scheldt), Meuse, Rhine, Rhône and Loire. Main tribs. Rs. Aube, Marne, Oise and Yonne.

Seine-St. Denis France. Dept. in Île-de-France region, NE of Paris. Area, 91 sq. m. (236 sq. km.). Cap. Bobigny. Pop. (1982C) 1,324,301.

Sekondi Western Region, Ghana. 4 55N 1 43W. Cap. of Region and port, sit. on the Gulf of Guinea. Since 1946 a single municipality with Takoradi. Pop. (1980E) 254,543.

Selangor Peninsular Malaysia, Malaysia. 3 20N 101 30E. State bordering on Strait of Malacca. Area, 3,074 sq. m. (7,962 sq. km.). Cap. Kuala Lumpur. Chief products rubber and tin. Chief towns Kuala Lumpur and Port Swettenham. Pop. (1980C) 1,467,441.

Selby North Yorkshire, England. 53 47N 1 05W. Town sit. on R. Ouse. Market town for a farming area and also centre for new coalfield development. Industries flour milling, beetsugar refining, shipbuilding. Manu. oilcake, chemicals and paper. Pop. (1981C) 10,726.

Selenga River Mongolia/U.S.S.R. 52 16N 106 16E. Rises in the Khangai Mountains of NW Mongolia, and flows 750 m. (1,200 km.) ENE to the Russian border near Altan Bulak, then N through Buriat Autonomous Soviet Socialist Republic, past Ulan Ude into L. Baikal. Navigable along the Russian section in summer.

Selenicë Albania. Town sit. NE of Valona on lower Vijose R. Mining centre for bitumen, ozocerite.

Selkirk Manitoba, Canada. 50 09N 96 52W. Town sit. on the W bank of the Red R. S of the lower tip of L. Winnipeg 18 m.(29 km.) N of Winnipeg. Indust. inc. steel, barge and boat building, light manu. Tourism is expanding. Pop. (1989E) 11,000.

Selkirk Borders Region, Scotland. 55 33N 2 50W. County town sit. SE of Edinburgh on Ettrick Water. Market town for agric. area. Manu. tweeds, woollen goods and electronics. Pop. (1981C) 5,829.

Selkirk Mountains British Columbia, Canada. 51 00N 117 40W. Range in SE British Columbia, parallel to, and W of, Rocky Mountains. Extends 200 m. (320 km.) NNW to Columbia R., rising to Mount Sir Sandford, 11,590 ft. (3,533 metres) a.s.l.

Selkirkshire Scotland. Former county now part of Borders Region.

Selma Alabama, U.S.A. 32 25N 87 01W. City sit. W of Montgomery on

Alabama R. Market centre of an agric. area. Industries inc. clothing, locks, switches, aircraft, wood pulp, cigars and foodstuffs. Pop. (1980C) 26,243.

Selsey West Sussex, England. 50 44N 0 47W. Town sit. S of Chichester on the English Channel. Industries tourism and fishing. Pop. (1988E) 8,555.

Semarang Java, Indonesia. 6 58S 110 25E. City and port sit. on Java Sea. Cap. of central Java province. Industries shipbuilding, railway engineering; manu. textiles, electrical equipment. Exporting sugar, copra, kapok, tobacco. Pop. (1980E) 647,000.

Semipalatinsk Kazakhstan, U.S.S.R. 50 28N 80 13E. Town sit. on Irtysh R. Cap. of Semipalatinsk region. Industries meat packing, flour milling, tanning. Pop. (1987E) 330,000.

Semmering Pass Austria. 47 38N 15 49E. Alpine pass sit. between Lower Austria and Styria and 23 m. (37 km.) SW of Weiner Neustadt at 3,215 ft. (980 metres) a.s.l.

Semnan Iran. 35 33N 53 24E. Town sit. E of Tehran at foot of Elburz Mountains. Commercial centre of tobacco growing area. Industry rug making.

Sendai Honshu, Japan. 31 49N 130 18E. City sit. in NE Honshu. Cap. of Miyagi prefecture. Industry food processing; manu. metal goods, textiles and pottery. Pop. (1988E) 866,000.

Senegal Rep. bounded N by Mauritania across Senegal R., E by Mali, S by Guinea and Guinea-Bissau, and W by Atlantic. Surrounds Gambia on three sides. Area, 75,750 sq. m. (196,192 sq. km.). Cap. Dakar. Mainly savannah; chief crops maize, millet, groundnuts. Industries processing foodstuffs and groundnut oil. Pop. (1988E) 6,982,000.

Senegal River West Africa. 15 48N 16 32E. Rises in the Fouta Djallon highlands in N Guinea (as Bafing R.), flows 1,050 m. (1,680 km.) NW into Mali, is joined by Falémé R., and forms the Mauritania–Senegal frontier, flowing W to enter the Atlantic below St. Louis, Senegal. Navigable July–Oct. up to Kayes in Mali.

Senegambia A Confederation of the States of Senegal and Gambia estab. 1982. The confederal state maintains its independence and sovereignty and aims at the integration of the armed security forces, economic and monetary union, co-operation in the fields of communications and external relations, and the establishment of joint institutions.

Senigallia Marche, Italy. 43 43N 13 13E. Town sit. WNW of Ancona on Adriatic coast. Industry manu. pasta. Port and resort.

Senlis Oise, France. 49 12N 2 35E. Town sit. NNE of Paris. Market town and tourist resort. Manu. furniture, rubber products. Pop. (1982c) 15,280.

Sennar Sudan. 13 30N 33 35E. Town sit. on W bank of Blue Nile R. Market centre for agric. area. Railway junction. The Sennar dam is important for irrigation.

Sens Yonne, France. 48 12N 3 17E. Town sit. in N central France on Yonne R. Market town for agric. area, trading grain and timber. Industries inc. electrical and mechanical engineering, flour milling, manu. brushes. Pop. (1982c) 26,961.

Senta Vojvodina, Yugoslavia. 45 56N 20 04E. Town sit. NE of Sombar on Tisza R. Railway junction. Industry flour milling.

Seoul South Korea. 37 31N 126 58E. Cap. of S. Korea sit. in Han R. valley and founded in 1394. Industries tanning, flour milling, railway engineering; manu. textiles. Pop. (1985E) 9,645,824.

Sept-Îles Quebec, Canada. 50 12N 66 23W. Town sit. on N shore of St. Lawrence estuary. Industry shipping. Pop. (1984E) 26,800.

Sequoia National Park California, U.S.A. 36 30N 118 30W. Park on W slopes of the Sierra Nevada extending

N to S from Upper Kings R. to Upper Tule R. Area, 602 sq. m. (1,559 sq. km.). Estab. to preserve groves of sequoia trees.

Seraing Liège, Belgium. 50 36N 5 29E. Town sit. SW of Liège on Meuse R. Industries coalmining, steel; manu. locomotives, machinery, glass. Pop. (1980E) 62,832.

Serampore West Bengal, India. 22 45N 88 21E. Town sit. NNW of Calcutta on Hooghly R. Industries silk printing, fertilizers, glass, jute and cotton milling. Pop. (1981c) 127,304.

Serbia Yugoslavia. 43 30N 21 00E. Constituent rep. of Yugoslavia, bounded N by Hungary, E by Romania and Bulgaria, and SW by Albania. Inc. the autonomous province of Vojvodina and the autonomous region of Kosovo-Metohija. Area, 34,107 sq. m. (88,337 sq. km.). Cap. Belgrade. Chief towns Belgrade, Niš, Kragujevac, Leskovac. Mountainous and mainly agric. Chief products wheat, maize, wine. Chief minerals copper and antimony. Pop. (1988E) 5·83m. (without Kosovo and Vojvodina).

Seremban Peninsular Malaysia, Malaysia. 2 44N 101 56E. Town sit. SE of Kuala Lumpur. Cap. of Negri Sembilan state. Centre of a tin and rubber producing area. Pop. (1980c) 136,252.

Sergipe Brazil. 10 30S 37 30W. State in NE Brazil on Atlantic coast. Area, 8,492 sq. m. (21,994 sq. km.). Cap. Aracajú. Low coastlands rising to a plateau. Mainly agric., esp. cattle-rearing inland. Chief products sugarcane, coconuts, beef, rice. Industry processing foodstuffs. Pop. (1989E) 1,392,934.

Seria Brunei. 4 39N 114 23E. Town sit. SW of Brunei city on N coast. Centre of an important oilfield. Industries oil processing and shipping. Pop. (1981E) 24,000.

Serov Russian Soviet Federal Socialist Republic, U.S.S.R. 59 29N 60 31E. Town sit. N of Sverdlovsk in the Sverdlovsk Region. Manu. metal goods, esp. special steels. Important metallurgical centre. Pop. (1974E) 100,000.

Serowe   Botswana. 22 25S 26 44E. Town sit. in E central Botswana. Cap. of the Bamangwato Tribe. Market town for an agric. area. Pop. (1980E) 16,000.

Serpukhov   Russian Soviet Federal Socialist Republic, U.S.S.R. 54 55N 37 25E. Town sit. S of Moscow at confluence of Rs. Oka and Nara. Commercial centre of agric. area trading in grain and timber. Industries textiles, sawmilling, metal-working. Pop. (1974E) 130,000.

Serres   Greece. (i) *Nome* in Macedonia, bounded N by Bulgaria. Area, 1,539 sq. m. (3,987 sq. km.). Mainly a plain drained by Struma R. Chief products cotton, tobacco, cereals. Industry textiles; manu. tobacco products. Pop. (1981C) 196,247. (ii) Cap. of (i) sit. NE of Thessaloníki. Commercial centre of fertile agric. plain. Manu. cotton goods, tobacco products. Pop. (1981C) 45,213.

Sète   Hérault, France. 43 24N 3 41E. Town and port sit. .on Gulf of Lions in the Mediterranean. Industries fishing, oil refining, distilling, chemicals; manu. cement, wine casks. Exporting wine, petroleum products. Terminus of the Canal du Midi. Pop. (1982C) 40,466.

Sétif   Algeria. 36 12N 5 24E. Town sit. WSW of Constantine at 3,500 ft. (1,067 metres) a.s.l. Market town in agric. area, esp. for grain and livestock. Industry flour milling. Pop. (1983E) 186,978.

Settat   Casablanca, Morocco. 33 00N 7 40W. Town sit. S of Casablanca. Commercial centre trading in wheat, skins, leather, wool, fruit and vegetables. Industries inc. flour milling, manu. soap. fruit preserves. Pop. (1982E) 65,000.

Settle   North Yorkshire, England. 54 05N 2 18W. Town sit. NW of Skipton on Ribble R. Market town and tourist centre in a limestone-hill district. Industries inc. paper making, creamery and quarrying. Pop. (1987E) 2,305.

Setúbal   Setúbal, Portugal. 38 30N 8 58W. (i) District S of the estuary of the R. Tagus and one of the main manu. areas of Portugal. Area 1,955 sq. m. (5,064 sq. km.). Pop. (1987E) 779,600. (ii) Town and port sit. SE of Lisbon on N shore of the Bay of Setúbal. Industries fishing, sardine canning, boatbuilding; manu. fertilizers, cement. Exporting oranges, grapes, wine, salt. Pop. (1987E) 77,885.

Sevastopol   Ukraine, U.S.S.R. 44 36N 33 32E. Town on SW point of the Crimea peninsula. Industries shipbuilding, fish processing, tanning, flour milling, tourism. Naval base and sea port. Pop. (1987E) 350,000.

Sevenoaks   Kent, England. 51 16N 0 12E. Town sit. SE of London. Residential and market town. Pop. (1980E) 19,000.

Severn River   Wales/England. 51 35N 2 40W. Rises on Plynlimmon in central Wales, flows 210 m. (336 km.) at first NE and then on a semi-circular course across the English border into Shropshire, past Shrewsbury, Worcester and Gloucester to enter the Bristol Channel. Main tribs. Rs. Vyrnwy, Stour, Teme and Warwickshire Avon. R. Wye and Bristol Avon join it at its estuary mouth. Subject to a high bore which may travel as far as Tewkesbury. Navigable for small vessels up to Gloucester with the aid of the Gloucester and Berkeley canal from Sharpness. Bridged between Aust and Beachley.

Sevilla   Spain. 37 25N 6 00W. (i) Province in SW, separated from the Atlantic by Huelva and Cádiz provinces. Area, 5,408 sq. m. (14,007 sq. km.). Drained by Guadalquivir R. Mainly agric. Chief products olive oil, grapes, cereals. Industry textiles, manu. pottery, pharmaceuticals. Pop. (1986C) 1,550,492. (ii) Cap. of (i) 54 m. (86 km.) from mouth of Guadalquivir R. Industry textiles, manu. pottery, soap, pharmaceuticals. R. port exporting wine, olives, olive oil, citrus fruits, cork. Special historic and architectural interest. Pop. (1986C) 668,356.

Sèvres   Hauts-de-Seine, France. 48 49N 2 12E. Suburb of Paris sit. WSW of city centre, on Seine R. Manu. porcelain. Pop. (1982C) 20,255.

Seward Alaska, U.S.A. 60 06N 149 26W. Town and port sit. on the Gulf of Alaska. Communications and supply centre for the interior. Port (exporting coal) with ship maintenance and repair facility, marine industry. Pop. (1989E) 3,149.

Seychelles Indian Ocean. 5 00S 56 00E. Republic, within the Commonwealth, consisting of a group of 85 islands and islets 600 m. (960 km.) NE of Madagascar. Area, 156 sq. m. (404 sq. km.). Cap. Victoria, on Mahé, the largest island, area, 56 sq. m. (145 sq. km.). The other islands are Praslin, La Digue and Silhouette. Main occupations farming and fishing. Chief products frozen fish, copra and cinnamon. Main industry, tourism. Pop. (1988E) 67,305.

Seyne-sur-Mer Var, France. Resort sit. SW of Toulon. Pop. (1988E) 63,960.

Sfax Tunisia. 34 39N 10 48E. City and port sit. on N shore of the Gulf of Gabès. Industries fishing, esp. for sponge and octopus. Manu. olive oil, soap. Exporting olive oil, phosphates, sponges. Pop. (1984C) 231,911.

Shaanxi (Shensi) China. 35 00N 109 00E. Province in E central China bounded N by Inner Mongolia, and E by Hwang-ho R. Area, 75,598 sq. m. (195,800 sq. km.). Cap. Xian. Mountainous, rising to the Tsin Ling Shan range in the S. Chief products wheat, millet, cotton, fruits. Large coal deposits. Pop. (1987E) 30,430,000.

Shaba Zaïre. Region of SE Zaïre. Area, 191,878 sq. m. (496,965 sq. km.). Chief cities, Lubumbashi (cap.), Likasi, Kalemie and Kolwezi. Important mining region, esp. for copper, manganese, zinc, cobalt and industrial diamonds. Pop. (1981E) 3,823,172.

Shabani *see* **Zvishavane**

Shadwell Greater London, England. 51 30N 0 03W. District in boro. of Tower Hamlets, on N bank of R. Thames.

Shaftesbury Dorset, England. 51 01N 2 12W. Town sit. WSW of Salisbury. Market town in an agric. area. There are light industries. Pop. (1989E) 7,221.

Shah Alam Selangor, Malaysia. Cap. of Selangor.

Shahjahanpur Uttar Pradesh, India. 27 53N 79 55E. Town sit. NE of Agra. Market trading in grain, sugar-cane. Industry sugar milling. Pop. (1981C) 187,934.

Shaker Heights Ohio, U.S.A. 41 29N 81 32W. City immediately SE of Cleveland and forming a residential suburb. Pop. (1980C) 32,487.

Shakhty Russian Soviet Federal Socialist Republic, U.S.S.R. 47 42N 40 13E. Town sit. NE of Rostov in Donets Basin, Rostov Region. Important coalmining centre. Manu. clothing, furniture, machinery. Pop. (1987E) 225,000.

Shandong (Shantung) China. 36 00N 118 00E. Province of E China on the coast. Area, 59,174 sq. m. (153,260 sq. km.). Cap. Jinan. Mountainous in the centre and on the Shandong peninsula, with serious soil erosion following deforestation. Fertile plain in N, watered by lower Hwang-ho R. Chief crops millet, wheat, kaoliang and groundnuts. Large deposits of iron ore and coal. Pop. (1987E) 77,760,000.

Shanghai China. 31 14N 121 28E. City sit. on Huangpu R., 18 m. (28 km.) from its confluence with Yangtse R. estuary. Industrial centre for the populous Chang Jiang valley hinterland. A comprehensive industrial base with 434 industrial sectors. Industries inc. metallurgy, machinery, electronics, petro-chemicals, chemicals, automobiles, textiles, shipbuilding and instruments. The most important sea port and largest city in China. Pop. (1987E) 7,330,000 (metropolitan area, 12,620,000).

Shanklin *see* **Sandown-Shanklin**

Shannon River Ireland. 52 36N 9 41W. Rises on Cuilcagh Mountain in

County Cavan, and flows 240 m. (384 km.) s through Loughs Allen, Boderg, Forbes, Ree and Derg to Limerick, then w into an estuary 60 m. (96 km.) long, to enter the Atlantic between Loop Head, to N, and Kerry Head, to s. Serves a hydroelectric power station N of Limerick. Chief tribs. R. Suck, Brosna, Little Brosna and Deel. Navigable.

Shansi *see* **Shanxi**

Shan State Burma. 21 30N 98 30E. State of E Burma bounded E by China, SE by Laos and Thailand. Area, 58,000 sq. m. (150,220 sq. km.). Mainly plateau, at an average height of 2,000–4,000 ft. (610–1,219 metres) a.s.l., cut N to S by the Gorge of the Salween R. Pop. (1983C) 3,718,706.

Shantung *see* **Shandong**

Shanxi (Shansi) China. 37 00N 112 00E. Inland province of N China, bounded N by Inner Mongolia, across the Great Wall. Area, 60,395 sq. m. (156,420 sq. km.). Cap. Taiyuan. Mountainous. Chief occupation farming, esp. wheat, millet. Coal deposits are extensive, with important mines at Tatung in N. Pop. (1987E) 26,550,000.

Shari River Central African Republic/Chad. Rises in N Central Afr. Emp. and flows 1,400 m. (2,240 km.) NW into Chad, past Fort Archambault and Bousso to cross into w Cameroon below N'djamena, and enters L. Chad by a delta.

Sharjah Sharjah, United Arab Emirates. 25 20N 55 26E. Town and port sit. NE of Dubai on the Gulf. Chief town of the Sheikhdom of Sharja and Kalba. Industries include fishing. Pop. (1980C) 125,149.

Sharon Israel. Plain in w Israel bounded w by the Mediterranean and E by the Samaria hills, extending N to s between Haifa and Tel Aviv. Fertile agric. area. Chief occupations mixed farming, poultry farming, fruit and vine cultivation.

Sharqiya Egypt. Province of Lower Egypt bounded E by the Suez Canal. Area, 1,908 sq. m. (4,942 sq. km.).

Chief towns Zagazig (cap.), Bilbeis. Agric. areas produce cotton, cereals, dates. The main industry is cotton ginning.

Shatt-al-Arab Iraq. Tidal river formed by union of Tigris and Euphrates Rs. below Al Qurnah; it flows 120 m. (192 km.) SE past Basra and Abu al Khasib to form the boundary between Iraq and Iran, and enters N end of the (Persian) Gulf.

Shawinigan Quebec, Canada. 46 33N 72 45W. Town sit. NW of Trois Rivières on St. Maurice R. Industries—powered by hydroelectric plant—textiles, chemicals, aluminium; manu. wood pulp and paper.

Sheboygan Wisconsin, U.S.A. 43 46N 87 44W. City sit. N of Milwaukee at mouth of Sheboygan R. on L. Michigan. Centre of a dairy farming region, trading in cheese. Manu. footwear, furniture, knitwear. Pop. (1980C) 48,085.

Sheerness Kent, England. 51 27N 0 45E. Town and port sit. on NW coast of the Isle of Sheppey. Roll-on, roll-off ferry link to the Netherlands. Industries tourism and steel manu. Pop. (1989E) 12,000.

Sheffield South Yorkshire, England. 53 23N 1 30W. Metropolitan district of South Yorkshire sit. on R. Don. Industries iron and steel, esp. special steels for cutlery, armour plate, rails, etc. Manu. cutlery, machinery, tools, silverware, glassware, optical instruments, food products. Pop. (1986E) 534,300.

Sheksna River U.S.S.R. 59 09N 37 50E. Rises in L. Beloye in w Russian Soviet Federal Socialist Republic, and flows 100 m. (160 km.) s to the Rybinsk Reservoir, forming part of the Mariinsk canal system.

Shellharbour New South Wales, Australia. 34 35S 150 52E. Town sit. s of Wollongong. Dormitory suburb for steel and coalmining industries. Tourism is being promoted. Pop. (1990E) 50,000.

Shenandoah National Park Virginia, U.S.A. 38 30N 78 30W. Park in N Virginia, in the Blue Ridge Mountains. Area, 302 sq. m. (782 sq. km.). Heavily forested, rising to Hawksbill Mountain, 4,049 ft. (1,234 metres) a.s.l.

Shenandoah River Virginia/West Virginia, U.S.A. 39 19N 77 44W. Formed by union of N. Fork and S. Fork R. and flows 55 m. (88 km.) NE through Virginia, enters W. Virginia briefly SSE of Charles Town and joins Potomac R. near Harper's Ferry.

Shendi Northern Province, Sudan. 16 46N 33 33E. Town sit. NNE of Ed Damer on Nile R. Industries inc. manu. cotton goods, dyeing, metal working.

Shensi *see* **Shaanxi**

Shenyang Liaoning, China. 41 50N 123 25E. Cap. of province sit. on Hun-ho R. in Manchuria. Commercial and communications centre for the NE, trading in grain, soy beans. Industries textiles, engineering, chemicals; manu. paper, matches. Pop. (1987E) 4,290,000.

Shepparton Victoria, Australia. 36 23S 145 25E. Town sit. NNE of Melbourne on Goulburn R. Centre of a rich irrigated agric. area trading in grain, wool, dairy produce, fruit, meat and wine. Industries inc. food processing. Pop. (1989E) 27,000.

Sheppey, Isle of England. 51 26N 0 45E. Island in Thames estuary off N coast of Kent and separated from it by the Swale channel. Area, 10 m. (16 km.) long and up to 4 m. (6 km.) wide. Low-lying and fertile; chief products sheep, cereals and vegetables. Chief towns Sheerness and Queenborough.

Shepshed Leicestershire, England. 52 47N 1 18W. Town sit. NNE of Leicester. Industries inc. quarrying and light engineering. Manu. hosiery, tiles, animal foodstuffs. Pop. (1981C) 11,070.

Shepton Mallet Somerset, England. 51 11N 2 31W. Town sit. SSW of Bath serving an agric. area. Industries inc. perry and cider-making, manu. foot-

wear and agric. machinery. Pop. (1983E) 6,538.

Sherborne Dorset, England. 50 57N 2 31W. Town sit. E of Yeovil on Yeo R. Market town for an agric. area. Industries glove-making, glass fibres, engineering. Pop. (1987E) 8,830.

Sherbrooke Quebec, Canada. 45 24N 71 54W. City sit. on St. François R. 30 m. (48 km.) from the border with U.S.A. Commercial and industrial centre. Industries inc. fabricated metal, machinery, electronic equipment, rubber, plastics, textiles, clothing, footwear and food. Pop. (1981C) 74,075.

Sheringham Norfolk, England. 52 56N 1 11E. Town sit. WNW of Cromer on North Sea coast. Industries fishing, esp. lobsters, and tourism. Pop. (1986E) 5,892.

's Hertogenbosch Noord-Brabant, Netherlands. 51 41N 5 19E. Town sit. SSE of Tilburg at confluence of Rs. Dommel and Aa. Industries inc. brewing; manu. tyres, compressors, cableshoes, tools, refrigerating equipment. Pop. (1989E) 89,991.

Sherwood Forest Nottinghamshire, England. 53 08N 1 08W. Anc. hunting ground between Nottingham and Worksop, area, *c.* 1,000 acres (405 hectares) originally, now only small tracts remain.

Shetland Islands Scotland. 60 30N 1 30W. Arch. of over 100 islands and islets NE of Orkney islands. Area, 567 sq. m. (1,468 sq. km.). The largest islands are Mainland, Yell and Unst. County town Lerwick. Bleak and infertile. Chief occupations fishing, fish farming, agric., knitting, tourism and servicing North Sea oil developments. Sullom Voe largest oil terminal in Europe. Pop. (1988E) 22,913.

Shiberghan Mazar-i-Sharif, Afghánistán. Town sit. W of Mazar-i-Sharif on Andkhui road and Sar-i-Pul R. Centre of an irrigated oasis.

Shibin el Kôm Menúfiya, Egypt. 30 33N 31 01E. Town sit. NW of Cairo on Nile delta. Cap. of governorate.

Centre of an agric. area trading in cereals and cotton. Industries textiles, manu. tobacco products. Pop. (1986E) 135,900.

Shikarpur Sind, Pakistan. 27 57N 68 38E. Town sit. SW of Quetta. Trading centre for grain and precious stones. Industries engineering, flour and rice milling; manu. cotton goods and carpets. Pop. (1981E) 88,000.

Shikoku Japan. 33 45N 133 30E. Smallest of the four main islands, lying S of Honshu and E of Kyushu. Area, 7,248 sq. m. (18,772 sq. km.). Chief towns Matsuyama, Takamatsu. Interior mountainous and forested. Chief lowland crops rice, tobacco, soya beans.

Shildon County Durham, England. 54 37N 1 39W. Town sit. SSE of Bishop Auckland. Industries inc. paint and clothing. Pop. (1987E) 11,379.

Shillelagh Wicklow, Ireland. 52 46N 6 32W. Village sit. SSW of Dublin on Shillelagh R., in an anc. oak forest. Origin of the trad. cudgel which was made first of oak, later of blackthorn.

Shillong Meghalaya, India. 25 35N 91 53E. State cap. sit. NE of Calcutta at 4,978 ft. (1,517 metres) a.s.l. in the Khasi Hills. Commercial centre trading in rice, cotton and fruit. Industry tourism. Pop. (1981C) 109,244.

Shimizu Honshu, Japan. 35 01N 138 29E. Town sit. NE of Shizuoka on NW Suruga Bay in a rice growing area. Port handling tea, oranges.

Shimla Himachal Pradesh, India. 31 02N 77 15E. State cap. in N India, and a hill-station sit. at 7,000 ft. (2,134 metres) a.s.l. Trading centre for grain and timber. Handicraft industries.

Shimoga Karnataka, India. 13 56N 75 35E. Town on Tunga R. Industries cotton ginning, rice milling. Pop. (1981C) 151,783.

Shimonoseki Honshu, Japan. 33 57N 130 57E. Town sit. in extreme SW of Honshu, connected with Kitakyushu by tunnels under the Shimonoseki Strait. Industries shipbuilding, engineering, metal working, fishing and fish processing; manu. textiles and chemicals. Pop. (1988E) 260,000.

Shipka Pass Bulgaria. 42 46N 25 19E. Pass through the Balkan Mountains between Kazanluk and Gabrovo, at 4,166 ft. (1,270 metres) a.s.l., and *c.* 95 m. (152 km.) W of Sofia.

Shipley West Yorkshire, England. 53 50N 1 47W. Town sit. NNW of Bradford on R. Aire. Industry engineering; manu. woollen and worsted goods. Pop. (1989E) 32,000.

Shiraz Iran. 29 36N 52 32E. Town in SW Iran sit. at 4,875 ft. (1,486 metres) a.s.l. Commercial centre of an agric. area, trading in cereals, sugarbeet and grapes. Industries textiles, sugar, fertilizers, cement; manu. carpets. Pop. (1986C) 848,289.

Shire Highlands Malawi. Uplands in S Malawi extending E from Shire R. and rising to 5,800 ft. (1,768 metres) a.s.l. Important for tobacco and tea cultivation.

Shire River Malawi/Mozambique. 17 42S 35 19E. Flows from L. Malawi 370 m. (592 km.) S with rapids and cataracts, inc. Murchison Falls, on its middle course. It enters Mozambique near Port Herald and joins Zambesi R. near Vila Fontes.

Shizuoka Honshu, Japan. 35 00N 138 30E. Town sit. SW of Tokyo. Cap. of prefecture of same name. Commercial centre of a tea-growing area. Industries tea processing and packing; manu. chemicals, machinery. Pop. (1988E) 470,000.

Shkodër (Scutari) Albania. 42 05N 19 30E. Town in NW Albania sit. at SE end of L. Shkodër on Bojana R. Trading centre for wool, grain and tobacco. Industries textiles and cement. Pop. (1983E) 71,000.

Shkodër, Lake Albania/Yugoslavia. L. on the Yugoslavia/Albania border. Area, 135 sq. m. (350 sq. km.). Fed by Morača R. and formerly an inlet of the Adriatic, now separated from the sea by an alluvial isthmus. Drained by Bojana R.

Shoa (Shewa) Ethiopia. Region of central Ethiopia bounded NW by Blue Nile R. and SW by Omu R. Area, 32,896 sq. m. (85,200 sq. km.). Chief towns Addis Ababa (cap.), Addis Alam, Ankober, Awash, Debra Birchan, Fiche. Mountainous, drained by Awash R. Agric. and stock rearing area. Chief products coffee, hides, beeswax. Pop. (1984C) 8,090,565.

Shoeburyness Essex, England. 51 31N 0 49E. Suburb of Southend-on-Sea on N shore of Thames estuary. Residential. Pop. (1980E) 9,000.

Shoreditch Greater London, England. 5 31N 0 05W. District in NE of City. Industries printing and furniture making.

Shoreham-by-Sea West Sussex, England. 50 49N 0 16W. Town sit. W of Brighton near mouth of Adur R. Residential town. Pop. (1981C) 20,705.

Shreveport Louisiana, U.S.A. 32 30N 93 45W. City sit. in NW of state on Red R. Centre of an oil and natural gas region. Industries oil refining, diverse manu. inc. vehicles, telephones and glassware. Pop. (1980C) 205,820.

Shrewsbury Shropshire, England. 52 42N 2 45W. Town sit. on Severn R. Market town for an agric. area. Industries inc. engineering, service industries, horticulture, agric. and tourism. Pop. (1987E) 91,900.

Shrirangapattana Karnataka, India. Town sit. SW of Bangalore on an island in Cauvery R. Market town in an agric. area. Historic importance. Pop. (1981C) 127,304.

Shropshire England. 52 36N 2 45W. County consisting of the districts of Oswestry, North Shropshire, Shrewsbury and Atcham, Wrekin, South Shropshire and Bridgnorth. Area, 1,347 sq. m. (3,490 sq. km.). Bounded N by Cheshire, E by Staffordshire, S by Hereford and Worcester and W by Powys and Clwyd. County town Shrewsbury, other towns inc. Telford, Newport, Ellesmere, Oswestry, Bridgnorth and Ludlow. The county is chiefly agric. with some engineering industry in Shrewsbury and engineering and electronics in Telford. Pop. (1988E) 400,800.

Shumen Bulgaria. Town sit. near Varna in NE Bulgaria. Provincial cap. Centre of grain and wine trade. Industries inc. metal, leather and woodworking. Pop. (1982E) 99,642.

Sialkot Pakistan. 32 32N 74 30E. Town sit. NNE of Lahore. Trading centre for grain and sugar-cane. Industry textiles; manu. sports goods, surgical instruments, cutlery, leather goods and carpets. Pop. (1981E) 302,000.

Sian *see* **Xian**

Šibenik Croatia, Yugoslavia. 43 44N 15 54E. Town sit. on Adriatic coast. Industries textiles and chemicals. Sea port, exporting timber and bauxite.

Siberia Russian Soviet Federal Socialist Republic, U.S.S.R. 60 00N 100 00E. Region extending W and E from the Ural Mountains to the Pacific Ocean, and N and S from the Arctic Ocean to the Central Asian mountain ranges. Area, *c.* 5.2m. sq. m. (*c.* 13.5m. sq. km.). Plains in W drained by R. Ob and Yenisei, and bounded S by the Altai and Sanai mountain ranges. The central area is a plateau, bounded E by Lena R. The E is mountainous. All Rs. except Amur flow into the Arctic Ocean and are frozen for most of the year. The climate is continental. Verkhoyansk in E Siberia has a mean Jan. temp. of −59°F. (−50°C.) Chief occupations of the N are timber felling, fur-trapping and fishing. In S and SW, where agric. is concentrated, cereal growing, cattle and sheep farming occur. Coal, oil, gold and iron ore exist in large quantities. Important industries are concentrated in the Kuznetsk coal basin and adjoining industrial region centred on Sverdlovsk and Chelyabinsk in the Ural Mountains. Other chief towns Novosibirsk, Omsk and Vladivostock on the Trans-Siberian railway.

Sibi Baluchistan, Pakistan. Town sit. SE of Quetta on the N Kachhi plain. District cap. and terminus of the road from Quetta through the Bolan Pass. Pop. (1981E) 32,000.

Sibiu　Sibiu, Romania. 45 48N 24 09E. Cap. of district of same name sit. NW of Bucharest. Industries textiles, tanning, brewing and distilling; manu. electrical equipment, machinery and paper. Pop. (1983E) 159,599.

Sibu　Sarawak, Malaysia. 2 19N 111 51E. Town sit. NE of Kuching on Rajang R. Commercial centre trading in rubber, sago and rice. R. port accessible to large steamers. Pop. (1984E) 140,000.

Sichuan　(Szechwan) China. 31 00N 104 00E. Province of S central China. Area, 222,000 sq. m. (574,980 sq. km.). Cap. Chengdu. The most populous province with density reaching 1,000 persons per sq. m. (485 per sq. km.) in the fertile Red Basin. Mountainous, rising to the Tibetan plateau in W. Drained by Chang Jiang R. Chief products are rice, maize, sugar-cane, beans and tobacco. The chief export is tung oil. Chief towns Chengdu and Chongqing. Pop. (1982C) 99,713,310.

Sicily　Italy. 37 30N 14 30E. Island forming an autonomous region of Italy in the Mediterranean, SW of Calabria on the mainland. The region inc. the small islands of Pantelleria, Ostica and the Lipari and Egadi islands. Area, 9,926 sq. m. (25,708 sq. km.). Cap. Palermo. Mountainous in N with ranges extending W to E across the main island, rising to the volcano of Mount Etna, 10,741 ft. (3,274 metres) a.s.l. Mainly agric. Chief products fruit, vines, olives, cereals and vegetables on the coastal plains. Sulphur and oil deposits. Industries centred on chief towns of Palermo, Catania and Messina. Pop. (1988E) 5,164,266.

Sidamo　Ethiopia. Region in S. Area, 45,290 sq. m. (117,300 sq. km.). Cap. Awasa. Pop. (1984C) 3,790,579.

Sidcup　Greater London, England. 51 25N 0 06E. District S of city. Residential area.

Sidi-bel-Abbès　Algeria. 35 12N 0 38W. Town sit. S of Oran. Commercial centre trading in cereals, wine, olives and livestock. Industry flour milling; manu. cement and furniture. Former Headquarters of the French Foreign Legion. Pop. (1983E) 186,978.

Sidlaw Hills　Scotland. 56 32N 3 10W. Range rising to 1,492 ft. (455 metres) a.s.l.

Sidmouth　Devon, England. 50 41N 3 15W. Town sit. ESE of Exeter at mouth of Sid R. in Lyme Bay. Industries tourism, farming and fishing. Pop. (1981C) 12,484.

Sidon　*see* **Saida**

Sidra, Gulf of　Libya. 31 30N 18 00E. Inlet of the Mediterranean extending 275 m. (440 km.) from Benghazi to Misurata. Important for tunny and sponge fisheries. Sirte is chief port.

Siebengebirge　North Rhine-Westphalia, Federal Republic of Germany. Range of volcanic hills extending along E bank of R. Rhine S of Bonn, rising to Drachenfels, 1,067 ft. (325 metres) a.s.l. and Ölberg, 1,522 ft. (464 metres) a.s.l.

Siedlce　Warsaw, Poland. 52 11N 22 16E. Town sit. E of Warsaw. Industry cement; manu. glass, soap and leather.

Siegburg　North Rhine-Westphalia, Federal Republic of Germany. 50 47N 7 12E. Town sit. ENE of Bonn on Sieg R. Manu. chemicals, machinery and furniture. Pop. (1985E) 35,600.

Siegen　North Rhine-Westphalia, Federal Republic of Germany. 50 52N 8 02E. Town sit. ENE of Bonn on Sieg R., in an iron-mining area. Industries iron and steel. Pop. (1987E) 115,000.

Siena　Toscana, Italy. 43 19N 11 21E. Town in central Italy. Cap. of Siena province. Commercial centre of an agric. region. Industry tourism. Of exceptional historic, artistic and architectural interest. Pop. (1989E) 58,384.

Sierra de Guadarrama　Spain. Range in central Spain extending *c.* 110 m. (176 km.) SW to NE along the border between Madrid and Segovia provinces, rising to 8,100 ft. (2,469 metres) a.s.l. Developed for timber and as a winter resort.

Sierra de Guadeloupe Spain. Range in W Spain extending 30 m. (48 km.) SW to NE through Cáceres province between Rs. Tagus and Guadiana. Rises to 4,734 ft. (1,454 metres) a.s.l.

Sierra Leone 9 00N 12 00W. Independent state in the Commonwealth bounded N and E by Guinea, SE by Liberia and SW by the Atlantic. Area, 27,925 sq. m. (72,326 sq. m.). Cap. Freetown. Flat coastlands rising to plateau inland with peaks over 6,000 ft. (1,829 metres) a.s.l. Hot and humid, rainfall (Freetown) 150 ins (375 centimetres) annually, but with dustladen Sahara winds in the dry season. Chief crops rice, palm kernels and timber. Important resources of diamonds and iron ore. Chief occupations fishing, farming and mining. Industries mining and sawmilling. Pop. (1982C) 3,354,000.

Sierra Madre Mexico. 17 00N 100 00W. Mountain system extending 1,500 m. (2,400 km.) SE from N frontier with U.S.A. in three main ranges enclosing the central Mexican plateau. The Sierra Madre Occidental runs parallel to the Gulf of California and Pacific coast; the Sierra Madre Oriental runs parallel to the Gulf of Mex.; the Sierra Madre del Sur in S Mex. follows the Pacific coast. The highest peak is Pico de Orizaba, 18,700 ft. (5,700 metres) a.s.l.

Sierra Morena Spain. Mountain range in S Spain between Rs. Guadiana and Guadalquivir. Forms a barrier between Andalusia and N Spain, at an average height of 2,500 ft. (762 metres) a.s.l. Has important minerals, esp. copper, lead, silver and mercury.

Sierra Nevada Spain. Mountain range in S Spain extending c. 60 m. (96 km.) W to E and rising to Mulhacén, 11,421 ft. (3,481 metres) a.s.l. Many summits have permanent snow.

Sierra Nevada California, U.S.A. 40 00N 120 00W. Mountain range in E California extending 400 m. (640

km.) NW to SE and rising to Mount Whitney, 14,495 ft. (4,418 metres) a.s.l. Inc. three national parks. Yosemite, Sequoia and King's Canyon.

Sierra Pacaraima Brazil/Venezuela/Guyana. Mountain range forming watershed between R. Orinoco and Amazon basins, extending c. 385 m. (616 km.) along the frontiers between Brazil and Venezuela, and Brazil and Guyana. Rises to Mount Roraima, 9,219 ft. (2,810 metres) a.s.l.

Siirt Turkey. 37 57N 41 55E. (i) Province of SE Turkey bounded S by Tigris R. Area, 4,383 sq. m. (11,352 sq. km.). Mountainous, infertile area. Pop. (1985C) 524,741. (ii) Town sit. E of Diyarbakir on Buhtan R. Provincial cap. Founded by the Armenian king, Tigranes.

Sikasso Mali. 11 19N 5 40W. Cap. of region in S, centre for agric. district. Pop. (1976C) 47,030.

Sikhote Alin Range Russian Soviet Federal Socialist Republic, U.S.S.R. 48 00N 138 00E. Mountain range in the Primorye and Khabarovsk territories, extending 750 m. (1,200 km.) along the Pacific coast from Vladivostock to Nikolayevsk. Rises to 6,000 ft. (1,829 metres) a.s.l. Forested, with mineral resources esp. coal, iron, lead and zinc.

Sikkim 27 50N 88 30E. State of India on S slopes of the E Himalayas, bounded N and E by Tibet, China, SE by Bhután, S by India and W by Nepál. Area, 2,740 sq. m. (7,096 sq. km.). Cap. Gangtok. Mountainous, rising to Kanchenjunga, 18,168 ft. (5,530 metres) a.s.l. on the W border. Minerals inc. copper, lead, zinc, gold and silver. Chief occupation farming; main crops ginger, oranges, rice, maize and fruits. Industries inc. electronics, brewing, matches, leather goods, cables and watches. Pop. (1981C) 316,385.

Silesia Region mainly in Poland, but partly in Czechoslovakia, in the Upper Oder basin, bounded S by the Sudeten Highlands. Important for coalfield in SE Polish Silesia. Metallurgical industries centred on Wroclaw, Katowice, Zabrze, Bytum, Chorzów, Gliwice, Sosnowice.

Silistra Ruse, Bulgaria. 44 06N 27 19E. Town sit. ENE of Ruse on Danube R. opposite Calarasi, Romania. R. port and rail centre handling grain, wool, wine, fish. Industries inc. weaving, tanning, market gardening.

Silsden West Yorkshire, England. 53 55N 1 55W. Town sit. NNW of Keighley. Industry textiles. Pop. (1980E) 6,000.

Silvassa Dadra and Nagar Haveli, India. Market town sit. SE of Damâo. Seat of administration for the territory. Pop. (1981C) 6,914.

Silver Spring Maryland, U.S.A. Town at N boundary of Washington D.C. Pop. (1980C) 72,893.

Simferopol Ukraine, U.S.S.R. 44 57N 34 06E. City sit. NE of Sevastopol on Salgir R. Cap. of Crimea region. Centre of a farming and horticultural area. Industries fruit and vegetable canning, flour milling, and tanning. Pop. (1987E) 338,000.

Simon's Town Cape Province, Republic of South Africa. 34 14S 18 26E. Formerly Simonstown. Town and port in extreme SW, on False Bay. Industries fishing, processing fish oil, animal feeds, margarine, soap and gemstone polishing. Naval base. Pop. (1985E) 6,500.

Simplon Pass Valais, Switzerland. 46 15N 8 02E. Alpine pass between the Pennine and Lepontine Alps, at 6,565 ft. (2,001 metres) a.s.l. on the road between Brig and Domodossola, Italy.

Sinai Egypt. 29 30N 34 00E. Peninsula in NE Egypt bounded E by Israel and Gulf of Aqaba, S and W by Gulf of Suez and Suez Canal, N by Mediterranean. Area, 23,000 sq. m. (59,570 sq. km.). Chief town El Arish. Sand dunes in N, with central plateau rising to mountains in S. Highest point is Jebel Katrin, 8,652 ft. (2,637 metres) a.s.l. Barren, with some minerals in W, esp. manganese, iron and oil.

Sinaia Prahova, Romania. 45 21N 25 33E. Town sit. NW of Ploesti at SE foot of the Bucegi Mountains. Resort at 2,953 ft. (900 metres) a.s.l. esp. for winter sports. Industries inc. manu. hardware, mosaics, plaster, cement. Former summer residence of the kings of Romania.

Sinaloa Mexico. 25 50N 108 20W. State in NW Mexico between the Sierra Madre Occidental and the Pacific Ocean. Area, 22,582 sq. m. (58,487 sq. km.). Cap. Culiacán. Chief occupations mining, esp. for silver and gold, and agric. Chief towns Culiacán, Mazatlán. Pop. (1989E) 2,425,006.

Sind Pakistan. Province of SE Pakistan mainly comprising lower Indus R. valley. Area, 50,000 sq. m. (129,500 sq. km.). Cap. Karachi. Hilly in S and bounded by Thar Desert in E. Arid except for irrigated areas. Chief crops wheat, rice, cotton. Industry textiles. Chief towns Karachi, Hyderabad, Sukkur and Shirkapur. Pop. (1980E) 19m.

Singapore 1 17N 103 51E. Rep. at S end of the Malay peninsula forming an independent state within the Commonwealth. Area, 225 sq. m. (583 sq. km.). Cap. Singapore city. An island separated from the peninsula by the Johore Strait, crossed by a causeway. Hot and humid; many market gardens and plantations. Industries tin smelting, rubber processing, fruit canning and sawmilling. The city is a sea port exporting rubber, tin and copra. Pop. (1988E) 2,647,100.

Sinkiang-Uighur Autonomous Region *see* **Xinjiang Uygur Autonomous Region**

Sinop Sinop, Turkey. 42 01N 35 09E. Town sit. NE of Ankara on the Black Sea. Port exporting tobacco, fruit, timber. Flourished under the Cimmerians, 7th cent. B.C., and under the Romans and until the 13th cent. A.D. Anc. Sinope.

Sint Maarten *see* **St. Martin**

Sint Niklaas East Flanders, Belgium. 51 10N 4 08E. Town sit. WSW of Antwerp. Market town for agric.

district of Waasland. Industry textiles; manu. carpets, bricks and pottery. Pop. (1989E) 69,923.

Sintra Lisboa, Portugal. 38 48N 9 23W. Town sit. WNW of Lisbon. Tourist centre in the Serra da Sintra. Pop. (1989E) 20,000.

Sion Valais, Switzerland. 46 14N 7 21E. Cap. of canton sit. SE of Lausanne on Rhône R. Trading centre for wine, fruit and vegetables. Industries printing, high technology and woodworking. Pop. (1989E) 25,000.

Sioux City Iowa, U.S.A. 42 30N 96 23W. City sit. WNW of Des Moines on Missouri R. Trading centre for livestock and grain. Industries meat packing and flour milling; manu. dairy products and clothing. Pop. (1980C) 82,003.

Sioux Falls South Dakota, U.S.A. 43 32N 96 44W. City sit. in SE of state, near Minnesota and Iowa borders on Big Sioux R. Hydroelectric power station. Industry meat packing; manu. biscuits and soap. Pop. (1980C) 81,343.

Siret River Romania/U.S.S.R. 45 24N 28 01E. Rises in E slopes of the Carpathian Mountains in Ukraine, and flows 280 m. (448 km.) SSE, entering Romania SW of Chernovtsy to join R. Danube above Galati.

Sistan Iran/Afghánistán. Region and inland L. depression of E Iran and SW Afghánistán containing the Hamun-i-Helmand reed lagoon extending 70 m. (112 km.) at flood water. Helmand R., Khash Rud, Farah Rud and Harut R. drain into it; Shelagh R. drains it. Supports cereals and cotton between floods.

Sitapur Uttar Pradesh, India. 27 38N 80 45E. Town sit. NNW of Lucknow. Commercial centre trading in grain and oilseeds. Pop. (1981C) 101,210.

Sitka Alaska, U.S.A. 57 03N 135 14W. Town in SE of state, sit. on W coast of Baranot Island in the Alexander Arch. Industries fishing, timber and tourism; manu. wood pulp. Sea port. Pop. (1980C) 7,803.

Sittard Limburg, Netherlands. 51 00N 5 53E. Town sit. NE of Maastricht. Market town. Industry tanning. Pop. (1989E) 45,000.

Sittingbourne Kent, England. 51 21N 0 44E. Town sit. ENE of Maidstone. Market town in a fruit growing district. Industries inc. paper and packaging. Pop. (1981C) 33,650.

Sittwe *see* **Akyab**

Sivas Sivas, Turkey. 39 43N 36 58E. Cap. of province sit. E of Ankara. Trading centre for agric. produce. Industries are textiles, carpet making. Pop. (1985C) 197,266.

Siwa Egypt. 29 11N 25 31E. Oasis sit. WSW of Alexandria in the Libyan Desert, *c.* 100 ft. (30 metres) b.s.l. Cultivation of dates and olives.

Siwalik Range India/Nepal. 31 00N 76 30E. Range of hills extending 1,000 m. (1,600 km.) WNW to ESE parallel to, and S of, the Himalayas. Stretches from Kashmir through Punjab and Uttar Pradesh into Nepal. Average height 2,000–3,500 ft. (610–1,067 metres) a.s.l.

Sjaelland *see* **Zealand**

Skagerrak Denmark/Norway. 57 45N 9 00E. Strait between Norway and Den. *c.* 80 m. (128 km.) wide, extending *c.* 170 m. (272 km.) SW to NE, and continuing S and SSE as the Kattegat, between Denmark and Sweden.

Skagway Alaska, U.S.A. 59 28N 135 19W. City in SE of state, near border with Canada, sit. at head of Chilkoot Inlet on Lynn Canal. Railway terminus and port. Important during the Klondike gold rush. Pop. (1990E) 712.

Skaraborg Sweden. 58 20N 13 30E. County of S Sweden between L. Väner and L. Vätter. Area, 3,065 sq. m. (7,938 sq. km.). Chief towns Mariestad (cap.), Lidköping, Falköping, Skara, Hjo, Tidaholm, Tkövde. Hilly area with many Ls. Agric. areas support cereals, beets and livestock; industries inc. textiles, paper making, lumbering, quarrying, metalworking. Pop. (1988E) 272,126.

Skeena River Canada. 54 09N 130 02W. Rises in the Stikine Mountains, and flows 360 m. (576 km.) S and SW through British Columbia, and enters the Hecate Strait of the Pacific Ocean SE of Prince Rupert.

Skegness Lincolnshire, England. 53 10N 0 21E. Town sit. just N of the Wash on the North Sea coast. Industries tourism and light engineering. Pop. (1985E) 14,553.

Skellefte River Sweden. Rises in N Sweden near the Norwegian frontier. Flows 250 m. (400 km.) SE through L. Hornavan, Uddjaur and Storavan to the Gulf of Bothnia. Important for logging.

Skellefteä Västerbotten, Sweden. 64 46N 20 57E. Port sit. near mouth of Skellefte R. Exports tar, timber and metallic ores. Pop. (1988E) 74,127.

Skelmersdale Lancashire, England. 53 33N 2 48W. New town sit. W of Wigan. Town created in 1961 as an overspill for Liverpool. Industries inc. engineering and electronics. Pop. (1985E) 41,800.

Skiddaw Cumbria, England. 54 39N 3 09W. Mountain in the Lake District sit. N of Keswick. Height 3,054 ft. (928 metres) a.s.l.

Skien Telemark, Norway. 59 12N 9 36E. Cap. of county in SE, sit. on Skien R. Industries tanning and sawmilling; manu. wood pulp and paper. Pop. (1984E) 46,693.

Skikda Algeria. 36 50N 6 58E. Town sit. on the Mediterranean coast. Sea port exporting citrus fruits, vegetables and wine.

Skipton North Yorkshire, England. 53 58N 2 01W. Town sit. NW of Bradford. Market town for a farming area. Industry limestone quarrying, general manu. and tourism. Pop. (1981C) 13,246.

Skopje Macedonia, Yugoslavia. 41 59N 21 26E. Cap. of Macedonia, sit. SSE of Belgrade on Vardar R. Commercial and communications centre for an agric. area. Industries brewing,

flour milling and wine making; manu. cement, carpets, tobacco products, iron and steel. Pop. (1981C) 506,547.

Skövde Skaraborg, Sweden. 58 24N 13 50E. Town sit. NNE of Gothenburg. Industry chemicals, manu. automobiles and cement. Pop. (1988E) 47,000.

Skye Highland Region, Scotland. 57 15N 6 10W. Largest island of the Inner Hebrides. Area, 643 sq. m. (1,655 sq. km.) Chief town Portree. Hilly, deeply indented by sea lochs. The Cuillin Hills in S rise to 3,309 ft. (1,009 metres) a.s.l. Sheep and cattle farming; industry tourism. Pop. (1985E) 7,500.

Skyros Greece. 38 53N 24 32E. Island in the North Sporades in the Aegean Sea. Area, 79 sq. m. (205 sq. km.). Mainly agric. Chief products wheat and olive oil.

Slave River Canada. 61 18N 113 39W. Leaves NW end of L. Athabasca, and flows 260 m. (416 km.) through the NW Territories NNW to enter Great Slave L. Main trib. Peace R.

Slavonia Yugoslavia. Region of Croatia mainly sit. between Rs. Drava and Sava. Chief town Osijek. Low-lying and fertile. Chief occupation farming.

Sleaford Lincolnshire, England. 53 00N 0 22W. Town sit. W of the Wash on Slea R. Market town for an agric. area. Industry agric. engineering. Pop. (1985E) 8,824.

Sligo Ireland. 54 17N 8 28W. (i) County in Connacht bounded N by Atlantic Ocean. Area, 694 sq. m. (1,797 sq. km.). High ground in W, with the Ox Mountains rising to 1,778 ft. (542 metres) a.s.l. Chief occupation farming, toolmaking and medical supplies. Pop. (1981C) 55,474. (ii) County town of (i) sit. at mouth of Garavogue R. Industries inc. toolmaking, general light manu., tourism and medical supplies. Pop. (1981C) 17,232.

Sliven Sliven, Bulgaria. 42 42N 26 19E. Cap. of province of same name, sit. SW of Varna. Industries wine making, manu. carpets and woollen goods. Pop. (1987E) 106,610.

Slobozia Ialomita, Romania. 44 34N 27 23E. Town sit. N of Calarasi on Ialomita R. Railway junction and cattle market. Industries inc. flour milling, manu. candles. Pop. (1983E) 41,175.

Slough Berkshire, England. 51 31N 0 36W. Town sit. W of central London. Industries inc. paints, plastics, food products, electronics, pharmaceuticals, aircraft and automobile parts, radio and television sets. Pop. (1981E) 97,000.

Slovakia Czechoslovakia. 48 30N 20 00E. Republic of Federal Czechoslovakia, bounded N by Poland, E by U.S.S.R., S by Hungary. Cap. Bratislava. Mountainous, rising to 8,737 ft. (2,663 metres) a.s.l. in the High Tatra, in N, and sloping to Danube valley in S. Principal occupation farming. Pop. (1989E) 5,263,541.

Slovenia Yugoslavia. Constituent rep. in NW bounded N by Austria, NE by Hungary, W by Italy. Area, 7,819 sq. m. (20,251 sq. km.). Chief towns, Ljubljana (cap.), Maribor. Mountainous, with the Karawanken Mountains in N and the Julian Alps in NW rising to Triglav, 9,396 ft. (2,864 metres) a.s.l. Drained by Rs. Sava and Drava. Chief crops, cereals, potatoes, vegetables. Important for commercial forestry and cattle raising. Industries mining, esp. coal, lead and mercury. Pop. (1988E) 1·94m.

Slupsk Koszalin, Poland. 54 28N 17 01E. Town sit. near Baltic Sea coast on Slupia R. Industries chemicals, agric. engineering. Pop. (1983E) 91,000.

Slyudyanka Russian Soviet Federal Socialist Republic, U.S.S.R. 51 40N 103 30E. Town sit. SSW of Irkutsk at SW end of L. Baikal, on the Trans-Siberian railway. Mining centre for mica, marble, quartz, with some metal-working.

Smethwick West Midlands, England. 52 30N 1 58W. Sit. W of Birmingham. Manu. glass and metal goods. Pop. (1981C) 55,950.

Smolensk Russian Soviet Federal Socialist Republic, U.S.S.R. 54 47N 32 03E. City in W central U.S.S.R. sit. on Dnieper R. Cap. of Smolensk Region. Communications centre. Industries brewing, flour milling, sawmilling; manu. textiles, textile machinery, clothing, footwear and furniture. Pop. (1987E) 338,000.

Snake River U.S.A. 46 12N 119 02W. Rises in Yellowstone National Park in NW Wyoming, and flows 1,038 m. (1,661 km.) S then W into Idaho. Flows E to W through Idaho past Idaho Falls and Twin Falls, turns N beyond Homedale, to form the N half of the Oregon/Idaho border, and the S end of the Washington/Idaho border. Turns NW at Clarkston and flows through Washington to join Columbia R. below Kennewick. Used for irrigation and hydroelectric power. Forms many canyons inc. Hells Canyon 100 m. (160 km.) long, greatest depth over 1 m. (1.6 km.).

Sneek Friesland, Netherlands. 53 02N 5 40E. Town sit. SSW of Leeuwarden on the Prinses Margriet canal. Market for dairy produce. Industries inc. paint, metal, cables, confectionery, ship repairs, printing, tinplate, machinery. Yachting centre. Pop. (1989E) 30,000.

Snowdon Gwynedd, Wales. 53 04N 4 05W. Mountain consisting of 5 peaks, rising to 3,560 ft. (1,085 metres) a.s.l., the highest in England and Wales.

Snowy Mountains Australia. 36 30S 148 20E. Range of Australian Alps mainly in New South Wales near the border with Victoria. Tableland at 3,000–6,000 ft. (914–1,829 metres) a.s.l. rising to Mount Kosciusko 7,328 ft. (2,234 metres) a.s.l.

Snowy River New South Wales/ Victoria, Australia. 37 48S 148 32E. Rises in Snowy Mountains and flows 260 m. (416 km.) S into Victoria and past Orbost into Bass Strait at Mario. Upper waters supply the Snowy Mountains hydroelectric scheme, feeding reservoirs from which they are diverted N through Tumur R. to Murrumbidgee R., and w to Murray R.

Sobat River   Ethiopia/Sudan. 9 22N 31 33E. Formed by union of Pibor R. and Baro R., flows 460 m. (736 km.) (from head of Baro R.) NW into Sudan to join White Nile R. *c.* 65 m. (104 km.) above Kodok. Navigable June–Dec. up to Gambela on Baro R. Flow retarded by swamps.

Sochi   Russian Soviet Federal Socialist Republic, U.S.S.R. 43 35N 39 45E. Town sit. SSE of Krasnodar on Black Sea coast. Manu. food and tobacco products. Leading seaside resort and spa. Pop. (1987E) 317,000.

Society Islands   French Polynesia. 17 00S 150 00W. Two groups of islands in S. Pacific Ocean, consisting of the Windward Islands (Îles du Vent) and Leeward Islands (Îles sous le Vent). Area, 650 sq. m. (1,685 sq. km.). Chief occupations farming and fishing. Chief products phosphates, copra and vanilla. Pop. (1983C) 142,129.

Socna   Tripolitania, Libya. Oasis sit. WSW of Hun on N slopes of the Geb-es-Soda at *c.* 820 ft. (250 metres) a.s.l.

Socorro   New Mexico, U.S.A. 34 04N 106 54W. Town S of Albuquerque on the Rio Grande. Pop. (1980C) 7,173.

Socotra   Indian Ocean. 12 30N 54 00E. Island lying ENE of Cape Guardafui, Somalia. Part of the People's Democratic Rep. of Yemen. Area, 1,400 sq. m. (3,600 sq. km.). Chief town Tamrida. Barren plateau rising to 4,700 ft. (1,433 metres) a.s.l. with agric. coastlands and valleys. Chief occupation farming, esp. livestock, dates and gums.

Södermanland   Sweden. County of E Sweden on the Baltic. Area, 2,340 sq. m. (6,060 sq. km.). Chief cities Nyköping (cap.), Eskilstuna, Katrineholm, Strängnäs, Torshälla, Oxelösund. Lowland fertile area with numerous Ls. Agric. areas produce grain, fruit, beef and dairy cattle.

Industries inc. iron mining, paper milling, wood-working. Pop. (1980C) 252,515.

Södertälje   Stockholm, Sweden. 59 12N 17 37E. Town and port sit. on Södertälje canal between L. Mälar and the Baltic Sea S of Stockholm. Industries textiles, chemicals; manu. pharmaceuticals, trucks, coaches, buses and automobile engines. Pop. (1989E) 80,660.

Soest   Utrecht, Netherlands. 52 09N 5 18E. Town sit. NE of Utrecht city. Centre of an agric. district. Manu. dairy products. Pop. (1985E) 40,298.

Soest   North Rhine-Westphalia, Federal Republic of Germany. 51 34N 8 07E. Town sit. E of Dortmund. Manu. electrical equipment, machinery, accumulators, wire, paper, wood products and soap. Pop. (1985E) 43,000.

Sofia   Bulgaria. 42 45N 23 20E. Cap. of Bulgaria, sit. on the Belgrade–Istanbul railway near the Yugoslav border. It was founded by the Romans. The modern city has been rebuilt since the Second World War. Industries textiles and chemicals; manu. machinery and electrical equipment. Pop. (1987E) 1,128,859.

Sofala   Mozambique. Province in E. Area, 26,262 sq. m. (68,018 sq. km.). Cap. Beira. Pop. (1987E) 1,257,710.

Sogne Fjord   Sogn og Fjordane, Norway. 61 06N 5 10E. Fjord penetrating almost to the Jotunheim Mountains, its mouth N of Bergen. Length 110 m. (176 km.), longest and deepest fjord in Norway.

Sohag   Sohag, Egypt. 26 27N 31 43E. Cap. of governorate of same name, sit. on w bank of Nile R. Centre of an agric. area. Industries cotton ginning and pottery. Pop. (1986E) 142,000.

Sohar   Batina, Oman. Town sit. NW of Muscat on the Gulf of Oman. Port shipping dates, hides, skins. Airfield.

Soho   Greater London, England. 5 13N 0 08W. District bounded w by

Regent Street, E by Charing Cross Road. Area of foreign settlement, mainly restaurateurs.

Soissonais France. Area NE of Paris in the Paris Basin, traversed by Aisne R. Agric. area. Chief products grain, sugar-beet and dairy produce.

Soissons Aisne, France. 49 22N 3 20E. Town sit. in NE on Aisne R. Market town for an agric. area. Industry agric. engineering; manu. boilers and rubber products. Pop. (1982C) 32,236.

Sokodé Centrale, Togo. 8 59N 1 08E. Cap. of region and chief town of N. Pop. (1980E) 33,500.

Sokolov Bohemia, Czechoslovakia. 50 10N 12 30E. Town sit. WSW of Carlsbad on Ohre R. Lignite mining centre, also manu. leather, chemical products. Pop. (1984E) 29,000.

Sokoto Nigeria. 50 12N 22 07E. State primarily formed from the Sokoto province of the former N region, comprising the emirates of Sokoto, Argungu, Gwandu and Yauri. Area, 39,589 sq. m. (102,535 sq. km.). Main cities Sokoto (cap.), Gusau, Kaura Namoda, Birnin Kebbi and Argungu. Pop. (1988E) 9·9m.

Sokoto Sokoto, Nigeria. 13 04N 05 16E. Town in NW, sit. on Kebbi R. Commercial centre trading in rice and cotton. Pop. (1983E) 143,000.

Solapur Maharashtra, India. 17 43N 75 56E. Town sit. SE of Bombay. Railway centre. Industry cotton; manu. carpets, glass, leather goods. Pop. (1981C) 514,860.

Soldeu Andorra. Village at 5,987 ft. (1,825 metres) a.s.l. sit. NE of Andorra la Vella. Lumbering and iron mining are the main occupations.

Solent, The England. 50 46N 1 20W. Channel forming an arm of the English Channel between the mainland and the Isle of Wight, *c.* 15 m. (24 km.) long. Main shipping route to Southampton and Portsmouth.

Solihull West Midlands, England.

52 25N 1 45W. Town forming a suburb of SE Birmingham, residential with some industries. Birmingham Airport and the National Exhibition Centre within the borough's boundries. Manu. cars, packaging, barges and machinery. Pop. (1981C) 198,287.

Solikamsk Russian Soviet Federal Socialist Republic, U.S.S.R. 59 39N 54 47E. Town sit. N of Perm. Centre of a mining area. Manu. chemicals.

Solingen North Rhine-Westphalia, Federal Republic of Germany. 51 10N 7 05E. Town sit. ESE of Düsseldorf. Industry inc. mechanical and electrical engineering; manu. mainly cutlery, flatware and surgical instruments. Pop. (1989E) 163,157.

Solna Stockholm, Sweden. 59 22N 18 01E. Town sit. NW of Stockholm. Industries inc. film-making, paper milling; manu. machinery, electrical and electronic goods, graphics, chocolate. Pop. (1983E) 49,080.

Sologne France. 47 50N 2 00E. Area in SSW Paris Basin between Rs. Loire and Cher. Plain with marshes and forest. Chief products cereals and vegetables. Industries minor.

Sololá Guatemala. 14 46N 91 11W. (i) Dept. of SW central Guatemala. Area, 410 sq. m. (1,062 sq. km.). Chief towns Sololá Panajachel, Atitlán, San Lucas. Highland area inc. Ls. Atitlán, and Atitlán, Tolimán and San Pedro volcanoes. Agric. area with livestock farming, L. fisheries. Pop. (1980E) 173,401. (ii) Town sit. WNW of Guatemala city on the Inter-American Highway. Dept. cap. and trading centre. Industries inc. flour milling. Pop. (1980E) 4,000.

Solomon Islands South West Pacific. 6 00S 155 00E. An independent country consisting of archipelago of volcanic islands SE of the Bismarck Arch. The main islands are Guadalcanal and Malaita. Area, 11,500 sq. m. (29,800 sq. km.). Cap. Honiara. Mountainous. Climate hot and humid. Chief occupations farming and fishing; chief products coconuts, bananas and sweet potatoes. Pop. (1986E) 285,796.

Solothurn Switzerland. 47 13N 7 32E. (i) Canton in NW Switzerland. Area, 306 sq. m. (792 sq. km). Crossed by the Jura Mountains in N and by fertile Aar R. valley in S. Agric. with industrial centre. Chief towns Solothurn, Olten, Grenchen. Pop. (1988E) 221,464. (ii) Cap. of (i) sit. S of Basel on Aar R. Industry textiles, manu. watches and precision instruments. Pop. (1983E) 15,400.

Solway Firth Scotland/England. 54 45N 3 38W. Inlet of the Irish Sea separating Cumbria, England, from Dumfries and Galloway Region, Scotland. Greatest width, 22 m. (35 km.). Rs. Esk and Eden drain into it at its head.

Somali 7 00N 47 00E. Bounded N by Gulf of Aden, E and S by Indian Ocean, SW by Kenya, W by Ethiopia. Area, 246,200 sq. m. (637,700 sq. km.). The Somali Republic came into being in 1960 as a result of the merger of the British Somaliland Protectorate and the Italian Trusteeship Territory of Somalia. Cap. Mogadishu. Coastal plains rise to internal plateau which reaches 7,900 ft. (2,408 metres) a.s.l. Hot and arid. Chief occupations nomadic stock farming, fishing and farming in irrigated plantations. Chief crops sugar-cane, maize, bananas and durra. There are a number of small meat and fish canneries, a small leather tanning and footwear industry Pop. (1988E) 6·22m.

Somerset England. 51 09N 3 00W. County in SW bounded N by Bristol Channel and Avon, E by Wiltshire, S by Dorset and W by Devon. County town Taunton. Uplands in W and SW—Exmoor, the Quantock Hills and Blackdown Hills—separated by a plain from the Mendip Hills in the NE. Farming important esp. dairy farming, cattle and sheep, cider orchards. Industry inc. agric. engineering, tourism, and printing; manu. leather products, footwear, aircraft, paper and beverages. Chief towns Taunton, Bridgwater, Yeovil. Pop. (1988E) 452,300.

Somersw·th New Hampshire, U.S.A. 43 N 70 52W. Town immediately S of Rochester in SE New Hampshire. Pop. (1980C) 10,350.

Somerville Massachusetts, U.S.A. 42 23N 71 06W. City sit. immediately NW of Boston on Mystic R. Industries meat packing and textiles, manu. furniture and paper products. Pop. (1980C) 77,372.

Somme France. 40 00N 2 15E. Dept. in Picardy region, on the English Channel. Area, 2,384 sq. m. (6,175 sq. km.). Chief towns Amiens (cap.), Abbeville. Crossed by Somme R. Low-lying, fertile farmland. Chief products wheat, vegetables, sugarbeet and cider apples. Industries include fertilizers and textiles. Pop. (1982C) 544,570.

Somme River France. 50 02N 2 04E. Rises in the Aisne dept. in NE and flows 152 m. (243 km.) W past St. Quentin, Amiens and Abbeville to enter the English Channel by an estuary. Linked by canals with Rs. Oise and Scheldt.

Songkhla Songkhla, Thailand. 7 12N 100 35E. Town sit. S of Bangkok on E coast of the Malay peninsula. Provincial cap. and port trading in rubber and copra. Industries inc. fishing, tin mining. Pop. (1982E) 78,000.

Sonora Mexico. 30 33N 100 37W. State of NW Mexico bounded N by Arizona, U.S.A., and W by Gulf of California. Area, 70,290 sq. m. (182,052 sq. km.). Cap. Hermosillo. Mountainous except for a narrow coastal plain; traversed by the Sierra Madre Occidental. Mainly arid, drained and irrigated by Rs. Magdalena, Sonora, Yagui and Mayo. Main occupations farming and mining. Chief crops sugar-cane, wheat, alfalfa and rice. Chief mineral deposits copper, esp. N at Cananea, silver, gold, lead and zinc. Chief towns Hermosillo, Nogales, Cananea, Agua Prieta, Ciudad Obregón Navojoa. Pop. (1989E) 1,828,390.

Son River Madhya Pradesh/Bihar, India. 25 40N 84 51E. Rises in NE Madhya Pradesh and flows 470 m. (752 km.) ENE across SE tip of Uttar Pradesh into Bihar, to join Ganges R. above Patna. Its course in Bihar is used for irrigation.

Sonsonate Sonsonate, El Salvador.

13 43N 89 44W. Cap. of dept. of same name sit. w of San Salvador. Commercial centre for an agric. area, trading in hides, livestock and other agric. produce. Manu. cotton goods and tobacco products. Pop. (1984E) 47,489.

Sopot Gdańsk, Poland. 54 28N 18 34E. Town sit. between Gdańsk and Gdynia on the Gulf of Gdańsk, forming part of a commercial and industrial conurbation.

Sopron Györ-Sopron, Hungary. 47 41N 16 36E. Town in NW near the Austrian frontier. Industries chemicals, textiles and sugar-refining.

Soria Spain. 41 46N 2 28W. (i) Province of NE Spain in the N Meseta plateau. Area, 3,977 sq. m. (10,300 sq. km.). Arid region crossed by Duero R. Chief occupation sheep farming. Chief towns Soria, Almazán. Pop. (1986C) 97,565. (ii) Cap. of (i) sit. on Duero R. at 3,450 ft. (1,052 metres) a.s.l. and 66 m. (106 km.) w of Zaragoza. Industries flour milling and tanning, manu. tiles. Pop. (1981E) 32,000.

Soriano Uruguay. Dept. of SW Uruguay bounded w by Uruguay R. Area, 3,561 sq. m. (9,223 sq. km.). Chief cities Mercedes (cap.), Soriano, Santa Catalina. Agric., sheep, and cattle farming area. Main products wool, grain. Industries inc. flour milling. Pop. (1980E) 80,114.

Sorocaba São Paulo, Brazil. 23 29S 47 27W. Town sit. w of São Paulo city. Trading centre for an agric. area, dealing in coffee, cotton and oranges. Industries railway engineering, textiles, chemicals, fruit packing. Pop. (1980E) 255,000.

Soroki Moldavian Soviet Socialist Republic, U.S.S.R. Town sit. NNW of Kishinev on Dniester R. Industries inc. flour and oilseed milling, brewing, soap manu.

Sorrento Campania, Italy. 40 37N 14 22E. Town sit. SE of Naples on S shore of the Bay of Naples. Tourism is the main occupation.

Sortavala Karelo-Finnish Soviet Socialist Republic, U.S.S.R. Town sit.

w of Petrozavodsk on N shore of L. Ladoga. Resort and commercial centre. Industries inc. sawmilling, manu. woollens, furniture, felt, leather, beer, dairy produce.

Sør-Trøndelag Norway. County of central Norway bounded E by Sweden. Area, 7,268 sq. m. (18,824 sq. km.). Chief cities Trondheim (cap.), Orkanger, Lokken. Varied area of mountain, valleys and fjords, drained by Ganla and Orkla Rs., with numerous Ls. Main occupations farming, fishing, lumbering, mining. Main products barley, potatoes, cod, copper, chromite, sulphur. Pop. (1983E) 246,206.

Sosnowiec Katowice, Poland. 50 20N 19 10E. Town in S central Poland, sit. in a coalmining area. Industries iron founding, chemicals, textiles and brick-making; manu. machinery. Pop. (1985E) 255,000.

Sound, The Denmark/Sweden. 55 50N 12 40E. The most easterly strait linking the Kattegat with the Baltic, sit. between Zealand, Denmark and Sweden. Length *c*. 70 m. (112 km.), 3 m. (5 km.) wide at its narrowest.

Souris River Canada/U.S.A. 49 39N 99 34W. Rises N of Weyburn, Saskatchewan, flows 450 m. (720 km.) SE into N. Dakota, U.S.A., and turns N into Manitoba, Canada, and then NE to join Assiniboine R. near Brandon.

Sousse Tunisia. 35 49N 10 38E. Town sit. on E coast on the Gulf of Hammamet. Industries olive processing and tourism. Sea port exporting olive oil and phosphates. Pop. (1984C) 83,509.

South Africa, Republic of 30 00S 25 00E. Bounded E by Indian Ocean, w by Atlantic. Area, without the 'independent' homelands, 347,860 sq. m. (900,957 sq. km.). Caps. Pretoria and Cape Town. Provinces Cape Province, Natal, Orange Free State, Transvaal. Chief cities Durban, Johannesburg, Bloemfontein, Port Elizabeth. Narrow plains on the coast rise to mountain ranges which surround the extensive high plateaux covering most of the area. Plateau areas are separated by ranges running roughly parallel to the coast, so that the land appears to rise in steps; the coastal range is at 500–600 ft. (152–183 metres) a.s.l., the plateau beyond at 1,500 ft. (457 metres) a.s.l.; the inner range is at 6,000–7,000 ft. (1,829–2,134 metres) a.s.l. and the plateau beyond at 2,000–3,000 ft. (610–914 metres) a.s.l. Beyond that is the range inc. the Nieuwveld, Sneeubergen and Drakensberg mountains, and the N is sit. on the High Veld plateau at 3,000–6,000 ft. (914–1,829 metres) a.s.l. The climate varies with altitude; in general the E coast is well watered and the climate becomes progressively drier to the W. Agric. areas produce tropical plants (E), cereals, sheep, dairy cattle, maize, citrus fruit, some cotton and tobacco, vines. The most important minerals are gold, silver, iron ore, iron pyrites, manganese, chrome, coal, copper, diamonds, phosphates, lime and limestone. The main mining centres are the Witwatersrand (gold), Kimberley (diamonds). Industries inc. food processing, canning and packing; manu. machinery, vehicles, paper, basic metals, metal products, chemicals, textiles. The fishing industry is important off Natal. Pop. (1986c) 27,607,000 (this figure excludes Transkei, Bophuthatswana, Venda and Ciskei).

Southall   Greater London, England. District sit. W of central London. Industries engineering, pharmaceuticals; manu. food products.

South America   12 25N to 56 00S and on longitude 70W. Continent in S of the W Hemisphere. Area, 6,872,000 sq. m. (17,798,500 sq. km.). Comprising the reps. of Argentina, Brazil, Bolivia, Chile, Colombia, Ecuador, French Guiana, Guyana, Paraguay, Peru, Suriname, Uruguay and Venezuela. Chief neighbouring islands: Falkland Islands (Brit.), off Argentina, and Galápagos Islands, off Ecuador. The Andean mountains system extends N to S along the whole W coast, varying in depth from 25 m. (40 km.) in Chile to 400 m. (640 km.) in Bolivia. The W range is unbroken; the parallel inland ranges are broken by plateaux at 10,000–14,000 ft. (3,048–4,267 metres) a.s.l. The Andean peaks

rise to Aconcagua, 23,081 ft. (7,135 metres) a.s.l., Chimborazo, 20,577 ft. (6,272 metres) a.s.l., Huascarán, 22,205 ft. (6,768 metres) a.s.l., and many others over 19,000 ft. (5,791 metres) a.s.l., numbers of them volcanic. To the E of the Andes is lowland extending N to S through the Orinoco basin, the Amazon basin and the Pampa-Chaco plain of Argentina. The E is highland, with the Guiana Highland (NE) and the Brazilian Highland which ends in steep escarpments dropping to the coastal plains. The principal Rs. are Amazon, Orinoco and La Plata systems.

Southampton   Hampshire, England. 50 55N 1 25W. City sit. on a peninsula between estuaries of R. Test and Itchen and at head of Southampton Water. Industry marine engineering; manu. petrochemicals, electrical equipment and cables. Sea port, and chief Brit. passenger port. Pop. (1981c) 204,604.

Southampton Island   Northwest Territories, Canada. 64 20N 84 40W. Island at entrance to Hudson Bay. Area, 16,936 sq. m. (43,864 sq. km.). Chief occupations hunting and fishing.

South Australia   32 00S 139 00E. State in central Southern Australia bounded N by Northern Territory, E by Queensland, New South Wales and Victoria, S by Indian Ocean, W by Western Australia. Area, 380,070 sq. m. (984,380 sq. km.). Chief towns Adelaide (the cap.), Elizabeth, Port Pirie, Port Augusta, Whyalla and Port Lincoln. Plateau in W, swamps and salt flats inland, with an inland drainage area feeding L. Eyre, 39 ft. (12 metres) b.s.l. Rainfall in N and W light and erratic, about 10 ins. (25 cm.) annually; the centre is arid; there is grassland and woodland in the moister SE. South of 32° lat. is an agric. area for sheep, wheat, barley, fruit, vines and vegetables. Iron ore is found in the Middleback Range and low-grade coal at Leigh Creek. Industries iron smelting and shipbuilding at Whyalla, metal and chemical industries at Port Pirie, general industry centred on Adelaide. Pop. (1986c) 1,345,945.

South Bend Indiana, U.S.A. 41 41N 86 15W. City sit. ESE of Chicago on St. Joseph R. Industry agric. engineering; manu. cars, aircraft equipment, clothing and toys. Pop. (1988E) 106,190.

South Beveland Netherlands. Peninsula in Scheldt estuary. Area, 135 sq.m. (350 sq. km.). Chief occupation horticulture, esp. vegetables, red currants, raspberries. Crossed by S. Beveland ship canal. Chief town Goes.

South Carolina U.S.A. 33 45N 81 00W. State in SW bounded by N. Carolina, E and SE by Atlantic Ocean, W by Georgia. Area, 31,055 sq. m. (80,432 sq. km.). Cap. Columbia. One of the original 13 states of the Union. Low-lying coastal plain occupies two-thirds of the land, rising to the Blue Ridge Mountains in NE with the highest point, Sassafras Mountain, 3,560 ft. (1,085 metres) a.s.l. Drained by R. Savannah, Santee, Pee Dee and Edisto into the Atlantic. Sub-tropical climate, annual rainfall 45–50 ins. (112–125 cm.). Main occupation agric., esp. tobacco growing, soya beans, wheat, cotton and stock farming. Industries inc. tourism and the manu. of textiles, clothing and chemicals. Chief towns Columbia, Charleston, Greenville. Pop. (1988E) 3,470,000.

South Dakota U.S.A. 45 00N 100 00W. State in N central U.S.A. in the Great Plains, bounded N by N. Dakota, S by Nebraska. Area, 77,047 sq. m. (199,552 sq. km.). Cap. Pierre. Settled as the Dakota Territory from 1857. Mainly prairie, rising to the Black Hills in SW and Harney Peak, 7,242 ft. (2,207 metres) a.s.l. Drained by Missouri R. and its tribs. Rs. Cheyenne, Grand, Morean, White and James. Continental climate; annual rainfall 15–25 ins. (38–63 cm.). Chief occupation farming, esp. stock-raising, maize, oats and wheat growing. Gold mining in the Black Hills. Other industries inc. engineering and food processing. Timber-based industries and printing also important. Chief towns Sioux Falls and Rapid City. Pop. (1986E) 708,000.

South Downs *see* **Downs**

Southend-on-Sea Essex, England. 51 33N 0 43E. Town sit. on N shore of Thames R. estuary. Industry tourism, light engineering. Industries inc. electronic and electrical goods, printing, finance and insurance. Pop. (1988E) 165,400.

South Esk River *see* **Esk River, South**

Southern Alps New Zealand. 43 30S 170 20E. Mountain range in W of South Island, extending 200 m. (320 km.) NE to SW. Average crest height over 8,000 ft. (2,438 metres) a.s.l. some peaks exceeding 10,000 ft. (3,048 metres) a.s.l. Central snowfields of Mount Cook, 12,349 ft. (3,764 metres) a.s.l., feed glaciers.

South Foreland England. 51 07N 1 23E. Chalk headland sit. ENE of Dover on E coast of Kent, with a lighthouse serving the Strait of Dover.

South Gate California, U.S.A. 33 57N 118 12W. City sit. S of Los Angeles. Manu. chemicals, tyres, building materials and furniture. Pop. (1980C) 66,784.

South Georgia South Atlantic Ocean. 54 15S 36 45W. Island *c.* 800 m. (1,280 km.) E of the Falkland Islands. Area, 1,600 sq. m. (4,144 sq. km.). Mountainous, with snow-covered peaks and glaciers. In use as a base, at King Edward Point, for the British Antarctic Survey.

South Glamorgan Wales. 51 30N 3 20W. County under 1974 re-organization sit. on coastal plain of S Wales and bounded on N by Mid Glamorgan, on W by Gwent and on S by Bristol Channel. County town and cap. of Wales, Cardiff. Other towns are Barry, Penarth and Cowbridge. Industries inc. steel manu., plastics, food processing, vehicle component manu. and light engineering. Agric. is carried on in the Vale of Glamorgan. Pop. (1988E) 403,400.

South Hadley Massachusetts, U.S.A. 42 16N 72 35W. Town sit. N of Springfield on Connecticut R. Industries inc. manu. bricks, paper, dairy products. Pop. (1980C) 4,137.

South Holland Netherlands *see* **Zuid-Holland**

South Ijssellakepolders Netherlands. Polders reclaimed from the Zuider Zee (now Ijsselmeer) forming good agric. land. Area, 207 sq. m. (536 sq. km.).

South Island New Zealand. 43 50S 171 00E. Larger of the 2 main islands separated from North Island by Cook Strait. Area, 59,196 sq. m. (153,318 sq. km.). Mountainous W with ranges running SW to NE parallel to the coast, and spurs extending E between the E plains. Highest point Mount Cook, 12,349 ft. (3,764 metres) a.s.l. Chief cities Christchurch, Dunedin. Main products grain and livestock (E), coal and gold (W). Pop. (1986C) 863,603.

South Korea *see* **Korea, South**

South Molton Devon, England. 51 01N 3 50W. Town sit. NW of Exeter. Sheep, cattle and pannier markets serving an agric. area S of Exmoor. Industries inc. light engineering, chipboard manu., wool-grading, agric. engineering, tourism, roadstone. Pop. (1987E) 3,804.

South Orkney Islands South Atlantic Ocean. 60 35S 45 30W. Island group SW of S. Georgia and part of British Antarctic Territory. Area, 240 sq. m. (622 sq. km.). Whaling base.

South Ossetian Autonomous Region Georgia, U.S.S.R. Autonomous region on S slopes of the Great Caucasus Mountains. Area, 1,505 sq. m. (3,898 sq. km.). Cap. Tskhinvali. Mountainous. Main occupations stock rearing and lumbering. Pop. (1989E) 99,000.

Southport Merseyside, England. 53 39N 3 01W. Town in NW, sit. on the Irish Sea coast. Industries tourism and engineering; manu. hosiery, knitwear and confectionery. Pop. (1981E) 89,756.

South Portland Maine, U.S.A. 43 38N 70 15W. Town immediately S of Portland on Atlantic coast. Pop. (1980C) 22,712.

South Sandwich Islands South At-

lantic Ocean. 57 45S 26 30W. Volcanic group of islands ESE of S. Georgia. Area, 130 sq. m. (337 sq. km.). Uninhabited.

South San Francisco California, U.S.A. 37 39N 122 24W. City sit. S of San Francisco on San Francisco Bay. Industries inc. meat packing; manu. steel, paint and chemicals. Pop. (1980C) 49,393.

South Shetland Islands South Atlantic Ocean. 62 00S 60 00W. Group of islands off NW coast of Graham Land forming part of British Antarctic Territory. Area, 1,800 sq. m. (4,662 sq. km.). Deception Island is volcanic. Uninhabited.

South Shields Tyne and Wear, England. 55 00N 1 25W. Town sit. opposite North Shields on R. Tyne estuary. Industries shipbuilding and marine engineering; manu. petro-chemicals, paint and biscuits. Sea port. Pop. (1989E) 155,700 (with Hebburn).

South Uist *see* **Uist, North and South**

Southwark Greater London, England. 51 30N 0 06W. District on S bank of Thames R.

Southwell Nottinghamshire, England. Town sit. NE of Nottingham. Industries inc. farming, general light engineering, garden centres. Pop. (1986E) 6,560.

South West Africa *see* **Namibia**

Southwold Suffolk, England. 52 20N 1 40E. Town sit. SSW of Lowestoft on North Sea coast. Industries brewing and tourism. Pop. (1988E) 1,690.

South Yemen *see* **Yemen, Republic of**

South Yorkshire England. 53 30N 1 20W. A metropolitan county under the 1974 re-organization. Bounded on N by West Yorkshire and North Yorkshire, E by Humberside, and S by Nottinghamshire and Derbyshire. Chief towns, Sheffield, Rotherham, Doncaster, Barnsley. Pop. (1988E) 1,292,700.

Sovetsk Russian Soviet Federal Socialist Republic, U.S.S.R. 55 05N 21 53E. Town sit. ENE of Kaliningrad on Neman R. in the Kaliningrad region. Trading centre for dairy produce. Manu. cheese, wood pulp, leather and soap.

Sowerby Bridge Calderdale, West Yorkshire, England. Town sit. SW of Halifax on Calder R. Industries include chemicals and engineering. Pop. (1981C) 11,280.

Soweto Transvaal, Republic of South Africa. 26 15S 27 52E. Black African township, its name being an acronym of South-West Townships. A residential area for blacks who travel to white areas for work. Pop. (1989E) 850,000.

Spa Liège, Belgium. 50 30N 5 52E. Town sit. S of Liège city. Health resort with mineral springs, giving its name to all others. Pop. (1989E) 10,000.

Spain 40 00N 5 00W. Rep. occupying most of the Iberian peninsula, bounded N by Bay of Biscay and Pyrénées, S and E by Mediterranean, W by Portugal and Atlantic Ocean. Area, inc. the Balearic and Canary Islands, 194,845 sq. m. (504,750 sq. km.). Cap. Madrid. Mainly plateau crossed by mountain ranges. The Andalusian mountains occupy the S, the Cantabrian mountains the N, the Pyrénées the NE, surrounding a central plateau at an average height of 2,600 ft. (792 metres) a.s.l. which is ridged and furrowed by smaller ranges and depressions, and drained by Ebro R. SE to the Mediterranean, Guadalquivir and Guadiana Rs. S to the Mediterranean, Tagus and Duero Rs. W to the Atlantic. Agric. areas cover *c.* 1,600,000 hectares, mostly under vines; other important crops are wheat, barley, potatoes, beet, tomatoes, olives, onions, oranges, mandarins. There are associated processing industries in all provinces. The chief minerals are coal, lignite, anthracite, iron ore, potash, zinc and lead; there is also some copper, sulphur, manganese, gold, silver, uranium. Industries other than mining and metallurgical industries inc. textiles, mainly NE, fishing,

fish processing, wine making, paper making, tourism. Chief cities are Barcelona, Valencia, Sevilla, Zaragoza, Bilbao, Málaga. Pop. (1986C) inc. the Balearic and Canary Islands, 38,891,313.

Spalding Lincolnshire, England. 52 47N 0 09W. Town sit. SSW of The Wash on Welland R. Market town for an agric. area. Industries inc. fruit and vegetable canning and engineering. Centre of a flower-growing region. Pop. (1985E) 19,000.

Spandau West Berlin, Federal Republic of Germany. Suburb of W. Berlin sit. at confluence of Rs. Havel and Spree. Industry engineering.

Spanish Sahara *see* **Western Sahara**

Sparks Nevada, U.S.A. 39 32N 119 45W. Town immediately E of Reno. Railway centre with repair shops. There is a gold and silver mining area nearby. Pop. (1980C) 40,780.

Spartanburg South Carolina, U.S.A. 34 57N 81 55W. City sit. NW of Columbia. Railway centre. Industries textiles, esp. cotton; manu. clothing, electrical equipment and food products. Pop. (1988E) 45,550.

Sparte Lakonia, Greece. Cap. of prefecture sit. SW of Athens on Eurotas R. Commercial centre of a farming area, trading in citrus fruits and olive oil. Pop. (1981C) 11,911.

Spenborough West Yorkshire, England. 55 44N 1 37W. District SW of Leeds in the Spen valley, inc. towns of Cleckheaton, Gomersal, Liversedge, Birkenshaw and Hunsworth. Industries manu. woollen goods and machinery. Pop. (1981C) 47,866.

Spencer Gulf South Australia, Australia. 34 00S 137 00E. Inlet of the Southern Ocean between Eyre Peninsula and Yorke Peninsula; 200 m. (320 km.) long and 75 m. (120 km.) at its widest point.

Spennymoor County Durham, England. 54 43N 1 35W. District sit. S of Durham. Coalmining area. Pop. (1980E) 19,000.

Speyer Rhineland-Palatinate, Federal Republic of Germany. 49 19N 8 27E. Town sit. s of Ludwigshafen on w bank of Rhine R. Industries sugar-refining and brewing, manu. paper products, footwear, bricks and tobacco products. Pop. (1989E) 45,089.

Spey River Scotland. 57 40N 3 06W. Rises 10 m. (16 km.) SSE of Fort Augustus, Highland Region, flows 107 m. (171 km.) NE through Strathspey, past Kingussie, Aviemore and Fochabers to enter the Moray Firth. Swift-flowing salmon river.

Spitalfields Greater London, England. Area in boro. of Tower Hamlets. Industries manu. furniture and footwear.

Spithead Hampshire/Isle of Wight, England. 50 45N 1 05W. Anchorage off Portsmouth; important for naval displays.

Spitsbergen Norway. 78 00N 17 00E. Arch. in the Arctic Ocean sit. N of the Norwegian coast. Area, 23,979 sq. m. (62,106 sq. km.). Consists of West Spitsbergen, North-East Land, Edge Island, Barents Island, Bear Island and smaller islands. West Spitsbergen, deeply indented with fjords, rises to Mount Newton, 5,633 ft. (1,717 metres) a.s.l. w coast ice-free April–Sept. Industry coalmining.

Split Croatia, Yugoslavia. 43 31N 16 27E. Town on the Dalmatian coast, sit. on a peninsula. Industries shipbuilding, fish canning and tourism; manu. cement and carpets. Important sea port. Noted for the palace of Diocletian. Pop. (1981C) 235,922.

Splügen Pass Graubünden, Switzerland. 46 30N 9 20E. Alpine pass between Splügen, Switz. and Chiavenna, Italy, at 6,944 ft. (2,117 metres) a.s.l. and 23 m. (37 km.) N of L. Como.

Spokane Washington, U.S.A. 47 40N 117 23W. City in NE Washington on Spokane R. Commercial, transport and service centre for an agric., lumbering and mining area. Pop. (1980C) 171,300.

Spoleto Umbria, Italy. 42 44N 12

44E. Town in central Italy. Industry textiles; manu. leather. Pop. (1981E) 20,000.

Sporades Greece. 37 14N 26 11E. Islands in the Aegean Sea in two groups, N. and S. Sporades, lying off the coasts of Euboia and w Turkey respectively. N. Sporades inc. Skiathos, Skópelos and Skíros; S. Sporades, commonly called the Dodecanese, inc. Ikaria, Samos, Gaidaro, Pátmos, Leros, Kálimnos, Kos, Nisiros and Tílos.

Spree River German Democratic Republic. 52 32N 13 13E. Rises in the Dresden district and flows 247 m. (395 km.) N and NW past Bautzen and Cottbus, and through Berlin to join Havel R. at Spandau, W. Berlin.

Springdale Arkansas, U.S.A. 36 11N 94 08W. Town in extreme NW Arkansas at w end of Ozark plateau. Pop. (1980C) 23,458.

Springfield Illinois, U.S.A. 39 47N 89 40W. State cap. sit. on Sangamon R. Industry agric. engineering; manu. electrical equipment, food products and footwear. Pop. (1980C) 99,637.

Springfield Massachusetts, U.S.A. 42 07N 72 36W. City sit. wsw of Boston on Connecticut R. Industries printing and publishing; manu. electrical machinery, machine tools and firearms. Pop. (1980C) 152,319.

Springfield Missouri, U.S.A. 37 14N 93 17W. City sit. in sw Missouri in the Ozark Mountains. Commercial centre for an agric. area. Manu. electronic components and television sets. Pop. (1980C) 133,116, metropolitan area (1980C) 207,894.

Springfield Ohio, U.S.A. 39 56N 83 49W. City sit. w of Columbus on Mad R. Manu. motor vehicles, agric. machinery and electrical appliances. Pop. (1980C) 72,563.

Springlands Berbice, Guyana. 5 53N 57 10W. Town sit. SE of New Amsterdam on Caribbean at mouth of Courantyne R. where it forms the border with Suriname. Port trading in cattle, rice, edible oil, canned fish, salt.

Springs Transvaal, Republic of South Africa. 26 13S 28 25E. Town sit. ESE of Johannesburg at 5,300 ft. (1,615 metres) a.s.l. Centre of a gold and coalmining area. Industry mining, inc. some uranium; manu. mining machinery, electrical equipment, glass and paper. Pop. (1980E) 153,974.

Spurn Head Humberside, England. 53 34N 0 08E. Headland at S end of Holderness; sand and shingle bar extending *c.* 4 m. (6 km.) NE to SW into Humber R. estuary.

Sri Lanka 7 30N 80 50E. South Asia. Formerly called Ceylon. Rep. within the Commonwealth, sit. immediately SE of India and separated from SE of Tamil Nadu, India, by the Gulf of Mannar and the Palk Strait, width 32 m.–120 m. (51–192 km.). Area, 24,959 sq. m. (64,644 sq. km.). The monarchical form of government continued until the beginning of the 19th cent. when the British subjugated the Kandyan Kingdom in the central highlands. In 1802 Ceylon was constituted a separate colony. Passing through various stages of increasing self-government, Ceylon reached fully responsible status within the British Commonwealth in 1948. Chief cities Colombo (cap.), Jaffna, Kandy, Galle. Undulating plain rises to mountains S 6,000–8,000 ft. (1,829–2,438 metres) a.s.l. The highest point is Pidurutalaga, 8,281 ft. (2,524 metres) a.s.l. Rainfall is heavy throughout the year except on the N plains which are subject to drought. Thick forests support a variety of animals and reptiles, inc. elephants, leopards, bears, monkeys and crocodiles. Agric. areas cover *c.* 2m. hectares. Chief products are rice, tea, coconuts. The most important minerals are graphite, ilmenite, iron ore, kaolin, glass-sand and salt. Precious stones and pearls are important, esp. sapphire, ruby, aquamarine, tourmaline, garnet, topaz, chrysoberyl. There are processing industries associated with agric. and mining, and some heavy industry inc. textiles, pharmaceuticals, oil refining. Pop. (1988E) 16·6m.

Srinagar Kashmir, India. 34 05N 74 49E. City sit. on Jhelum R. at 5,250 ft. (1,600 metres) a.s.l. Summer cap. of Jammu and Kashmir state. Industry, tourism; manu. carpets, leather goods, copperware. Pop. (1981C) 594,775.

Stade Schleswig-Holstein, Federal Republic of Germany. 53 36N 9 28E. Town sit. WNW of Hamburg on Schwinge R. Manu. chemicals and leather goods. R. Port. Pop. (1984E) 43,000.

Staffa Strathclyde Region, Scotland. 56 25N 6 20W. Island sit. W of Mull. Uninhabited, noted for basalt caves, inc. Fingal's Cave 227 ft. (69 metres) long with an entrance 66 ft. (20 metres) high.

Stafford Staffordshire, England. 52 48N 2 07W. Midlands town sit. on Sow R. Market town. Manu. footwear, chemicals and electrical goods. Pop. (1984E) 62,978.

Staffordshire England. County in central Eng. County town Stafford. Mainly undulating plain drained by Trent R. and its tribs., hilly in N rising to Axe Edge, 1,756 ft. (535 metres) a.s.l. Extensive coalfields. Chief industries, metal-working, pottery, brewing, general and electrical engineering, agric. Chief towns Stoke-on-Trent, Stafford, Newcastle-under-Lyme, Lichfield, Tamworth and Burton-on-Trent. Pop. (1988E) 1,032,000.

Staines Surrey, England. 51 26N 0 31W. Town sit. W of central London at confluence of Rs. Thames and Colne. Industry engineering; manu. paint and glass. Pop. (1981E) 51,949.

Stalybridge Greater Manchester, England. 53 29N 2 04W. Town sit. E of Manchester on Tame R. Industry engineering; manu. cotton goods, metal goods. Pop. (1987E) 21,932.

Stamford Lincolnshire, England. 52 39N 0 29W. Historic market town serving an agric. area sit. WNW of Peterborough on Welland R. Industry agric. and marine and electrical engineering. Pop. (1985E) 16,200.

Stamford Connecticut, U.S.A. 41 03N 73 32W. City sit. SW of New Haven on Long Island Sound. Com-

mercial centre, corporate headquarters, regional shopping centre.
Pop. (1980C) 102,453.

Standerton Transvaal, Republic of
South Africa. Town sit. SE of Johannesburg on Vaal R. at 5,000 ft. (1,524
metres) a.s.l. Market town for an
agric. area. Industries associated with
neigbouring coalmines, textiles, clothing and food processing. Pop. (1985E)
40,200.

Stanley East Falkland, Falkland
Islands. 51 40S 59 51W. Town on Port
William inlet of NE coast. Cap. of the
Falkland Islands, main port and commercial centre. Exports wool, skins;
imports foods, coal, oil, timber. Also
called Port Stanley. A new airport at
nearby Mount Pleasant was completed
in 1986. Pop. (1986E) 1,200.

Stanley Falls Zaïre. 0 12N 25 25E.
Series of 7 cataracts on Lualaba R.
above Kisangani; the fall is 200 ft. (61
metres) in 56 m. (90 km.).

Stanley Pool Congo/Zaïre. 4 15S
15 25E. L. formed by widening of
Zaïre R. 350 m. (560 km.) above the
mouth. Length E to W, 18 m. (29 km.),
width N to S 15 m. (24 km.). Kinshasa,
Zaïre, is on SW shore and Brazzaville,
Congo, on W. Contains the swamp
island of Bamu, area 70 sq. m. (181
sq. km.), dividing Zaïre R. into N and
S branches, the S branch always
navigable.

Stanleyville *see* **Kisangani**

Stann Creek Belize. 16 58N 88
13W. District with principal town
Dangriga SE of Belmopan at mouth of
Stann Creek on Caribbean. Port trading in bananas, coconuts, timber. Industries inc. fishing. Pop. (1988E)
17,700.

Stanmore Greater London, England. 51 37N 0 19W. Town within NW
Greater London. Residential.

Stanovoi Range Russian Soviet
Federal Socialist Republic, U.S.S.R.
56 20N 126 00E. Mountain range in
SE U.S.S.R. extending 500 m. (800
km.) E from the Olekma R. and rising
to Skalisty Mountain, 8,143 ft. (2,682

metres) a.s.l. Forms part of the
watershed between rivers flowing to
the Arctic and those flowing to the
Pacific.

Stara Zagora Stara Zagora, Bulgaria. 42 25N 25 38E. Cap. of
province of same name, sit. ENE of
Plovdiv. Trading centre for an agric.
area, trading in wheat, wine and attar
of roses. Industries brewing, tanning,
distilling and flour milling; manu. textiles and fertilizers. Pop. (1987E)
156,441.

Stassfürt Magdeburg, Germany. 51
51N 11 34E. Town sit. SSW of Magdeburg city. Centre of a mining area
for potash and rock salt. Industry
chemicals. Pop. (1989E) 27,500.

Staten Island New York, U.S.A. 40
35N 74 09W. Island in SE of New
York state. Area, 57 sq. m. (148 sq.
km.). Separated from New Jersey by
the Kill van Kull channel and Arthur
Kill channel and from Long Island,
New York, by the Narrows. Industries
shipbuilding, oil refining, printing and
tourism; manu. paper. Pop (1980C)
360,000.

Stavanger Rogaland, Norway. 58
58N 5 45E. Town in SW sit. on Bökn
Fjord. Industries oil, fish canning,
shipbuilding and woodworking. Sea
port. Pop. (1989E) 96,948.

Staveley Derbyshire, England. 53
16N 1 20W. Town sit. NE of Chesterfield. Industries scientific instruments
and chemicals. Pop. (1981C) 17,750.

Stavropol Russian Soviet Federal
Socialist Republic, U.S.S.R. 45 05N
42 00E. (i) Territory N of the
Caucasus Mountains and separated
from the Caspian Sea by Dagestan
Autonomous Soviet Socialist Republic. Area, 29,500 sq. m. (76,400 sq.
km.). Dry; irrigated and drained by
Rs. Kuma, Kuban and Yegorlyk.
Main occupations growing wheat,
maize; rearing sheep. Some natural
gas deposits. Industries centred on
chief towns Stavropol, Pyatigorsk. (ii)
Cap. of (i) sit. SE of Rostov. Market
town for an agric. area, trading in
grain and livestock. Industries tanning, flour milling, agric. engineering
and textiles. Pop. (1987E) 306,000.

Steinkjer Nord-Trøndelag, Norway. 64 01N 11 30E. Town sit. NE of Trondheim at mouth of Steinkjer R. on Trondheim Fjord. Industries inc. sawmilling, tanning, dyeing, wool spinning. Pop. (1984E) 20,654.

Stellenbosch Cape Province, Republic of South Africa. 33 58S 18 50E. Town sit. E of Cape Town. Centre of a fruit growing and vineyard area. Industries wine making, sawmilling; manu. footwear, bricks and tiles. Pop. (1989E) 55,914.

Stelvio Pass Lombardia, Italy. Alpine pass just S of border with Switzerland and 34 m. (54 km.) SW of Merano; N of the Ortler mountains, at 9,052 ft. (2,759 metres) a.s.l.

Stendal Magdeburg, Germany. 52 36N 11 51E. Town and railway centre. Industries engineering, metal working and sugar refining. Pop. (1989E) 48,700.

Stepanakert Azerbaijan Soviet Socialist Republic, U.S.S.R. 39 49N 46 44E. Town sit. SSW of Yevlakh. Oblast cap. Industries inc. food processing, silk mills, wine making.

Stepney Greater London, England. 51 31N 0 04W. District N of Thames. Contains the Tower of London. *See* Tower Hamlets.

Sterling Heights Michigan, U.S.A. 42 34N 83 01W. City sit. W of L. St. Clair on the Clinton R. Industrial and commercial, inc. vehicle assembly. Pop. (1980C) 108,999.

Sterlitamak Russian Soviet Federal Socialist Republic, U.S.S.R. 53 37N 55 58E. Town sit. S of Ufa on Belaya R. in Bashkir Autonomous Soviet Socialist Republic. Industry heavy engineering; manu. synthetic rubber, chemicals, cement, clothing and food products. R. port. Pop. (1987E) 251,000.

Stettin *see* **Szczecin**

Stettiner Haff German Democratic Republic/Poland. 53 50N 14 25E. Lagoon 34 m. (54 km) W to E separated from the Baltic Sea by Usedom

and Wolin islands. Receives Oder R. below Szczecin, Poland.

Steubenville Ohio, U.S.A. 40 22N 80 37W. City sit. W of Pittsburgh on Ohio R. Manu. steel, tinplate, electrical equipment and pottery. Pop. (1980C) 26,400.

Stevenage Hertfordshire, England. 51 55N 0 14W. Town sit. N of London. Market town developed post-war as U.K.'s first 'New Town'. High technology industrial base inc. space and communications. Pop. (1985E) 75,700.

Stewart Island New Zealand. 47 00S 167 52E. Island off South Island coast across the Foveaux Strait. Area, 670 sq. m. (1,735 sq. km.). Volcanic and mountainous. Chief settlement Oban. Chief occupations salmon farming and fishing. Pop. (1989E) 450.

Stewarton Strathclyde Region, Scotland. Town sit. N of Kilmarnock. Industries inc. knitwear manu. and light engineering. Pop. (1989E) 6,500.

Steyning West Sussex, England. 50 53N 0 20W. Town sit. WNW of Brighton. Town for an agric. area. Pop. (1985E) 4,500.

Steyr Upper Austria, Austria. 48 03N 14 25E. Town in N central Austria, sit. at confluence of Rs. Enns and Steyr. Industrial centre for iron and steel; manu. machinery, motor vehicles, bicycles and ball-bearings. Pop. (1988E) 38,942.

Stillwater Oklahoma, U.S.A. 36 07N 97 04W. Town sit. NNE of Oklahoma City. Trading and service centre for a farming area. Industries inc. marine and automobile engineering, publishing, manu. wire products, electronics, business stationery and business forms. Pop. (1989E) 39,000.

Stilton Cambridgeshire, England. 52 29N 0 17W. Village sit. SSW of Peterborough. Gave its name to Stilton cheese, now manu. mainly in Leicestershire.

Stirling Central Region, Scotland

56 07N 3 57W. County town sit. on Forth R. Industries inc. agric. engineering, high technology. Pop. (1985E) 29,238.

Stirlingshire   Scotland. Former county now part of Central Region.

Stockholm   Stockholm, Sweden. 59 20N 18 03E. Cap. of Sweden sit. on E coast between E end of L. Mälar and the Saltsjön, its outlet to the Baltic Sea, and extending from the mainland to cover a group of 20 islands. Commercial and cultural centre. Industries inc. brewing, engineering, building, graphics, electronics, chemicals. Sea port, with ferries to Finland, exporting timber and paper. Pop. (1988E) 669,485.

Stockport   Greater Manchester, England. 53 25N 2 10W. Metropolitan district sit. at confluence of Rs. Tame and Goyt which together form Mersey R. Industries inc. aerospace, computers, electronics, engineering, printing and food processing. Pop. (1981C) 288,980.

Stocksbridge   South Yorkshire, England. Town sit. NW of Sheffield. Steelmaking centre mainly special and high-alloy steels. Pop. (1978E) 17,000.

Stockton   California, U.S.A. 37 57N 121 17W. City sit. SSE of Sacramento on San Joaquin R., at head of navigation. Centre of an agric. area with food processing industries, agric. and marine engineering. R. port exporting agric. produce. Pop. (1980C) 149,779.

Stockton-on-Tees   Cleveland, England. 54 34N 1 19W. Town sit. on Tees R., which developed with the opening of the Stockton and Darlington railway (the first passenger line) in 1825. Industries inc. chemicals and engineering. Pop. (1989E) 175,300.

Stoke-on-Trent   Staffordshire, England. 53 01N 2 10W. Town sit. on Trent R. Centre of the pottery industry, also manu. tyres and cables; some engineering. Pop. (1981C) 252,509.

Stolberg   North Rhine-Westphalia, Federal Republic of Germany. 50 46N 6 13E. Town sit. E of Aachen. Industries inc. metal- and glass-working, electronics and chemicals. Pop. (1989E) 58,900.

Stone   Staffordshire, England. 52 55N 2 10W. Town sit. S of Stoke-on-Trent, on Trent R. Market town, manu. pottery, computer systems and measuring equipment. Pop. (1989E) 12,600.

Stonecutters Island   Hong Kong. 22 19N 114 08E. Small island off W coast of Kowloon peninsula, and N of Victoria. Area, 1 m. (1.6 km.) by 0.25 m. (0.4 km.).

Stonehaven   Grampian Region, Scotland. 56 58N 2 13W. Town sit. on North Sea coast. Industries inc. tourism and joinery. Pop. (1988E) 9,040.

Stornoway   Lewis, Scotland. 58 12N 6 23W. Town on E coast of Lewis in the Outer Hebrides. Administrative centre for the Western isles, chief town and industrial centre for the islands. Industries inc. fishing, fish farming, light engineering and tourism; manu. tweeds. Pop. (1989E) 8,400.

Storstrøms   Vordingborg Bay, Denmark. Strait between Masnedo and Falster Islands, crossed by a rail and road bridge from Vordingborg.

Stourbridge   West Midlands, England. 52 27N 2 09W. Town sit. WSW of Birmingham on Stour R. Manu. inc. glass, engineering and steel-based industries. Pop. (1981E) 54,659.

Stourport-on-Severn   Hereford and Worcester, England. 52 21N 2 18W. Town sit. N of Worcester at confluence of R. Severn and Worcestershire Stour. Industries inc. manu. electrical equipment, plastics, vinegar, carpets and iron goods. R. and canal port. Pop. (1981C) 19,054.

Stour River   Essex/Suffolk, England. 51 57N 1 18E. Rises as a number of headstreams, and flows 47 m. (75 km.) E along the Essex/Suffolk border to enter the North Sea at Harwich.

Stour River Oxfordshire/Warwickshire, England. Rises in NW Oxfordshire sit. SW of Banbury, and flows 20 m. (32 km.) NW to join Avon R. below Stratford-upon-Avon.

Stour River West Midlands/Hereford and Worcester, England. 52 25N 2 13W. Rises near West Bromwich and flows 20 m. (32 km.) W and SW past Stourbridge to join Severn R. at Stourport-on-Severn.

Stour River Wiltshire/Dorset, England. 50 48N 2 07W. Rises in SW Wiltshire 4 m. (6 km.) NW of Mere, enters Dorset and flows SE for 55 m. (88 km.) past Sturminster Newton and Blandford Forum to enter Avon R. at Christchurch.

Stour River, Great Kent, England. 51 18N 1 22E. Rises in two headstreams which join at Ashford, flows 40 m. (64 km.) NE past Canterbury to Sarre and divides into N branch, which enters the North Sea at Reculver, and S branch which enters the Strait of Dover at the head of Pegwell Bay.

Stowmarket Suffolk, England. 52 11N 1 00E. Town sit. on Gipping R. Market town for an agric. area, manu. chemicals and agric. implements. Pop. (1989E) 14,250.

Strabane Tyrone, Northern Ireland. 54 49N 7 27W. Town in NW sit. at confluence of Rs. Mourne and Finn, where they form Foyle R. Market town for an agric. area, manu. shirts and collars, and a salmon-fishing centre. Pop. (1986E) 12,100.

Stralsund Rostock, German Democratic Republic. 54 19N 13 05E. Town sit. opposite Rügen Island on an arm of the Baltic Sea. Industries inc. shipbuilding, fish curing, metal working and sugar refining; manu. machinery. Pop. (1989E) 75,600.

Stranraer Dumfries and Galloway Region, Scotland. 54 55N 5 02W. Town in SW sit. at head of Loch Ryan. Main vehicle ferry port for Northern Ireland, with daily crossings to Larne. Industries inc. textile manu. Pop. (1981E) 10,170.

Strasbourg Bas-Rhin, France. 48

35N 7 45E. Cap. of dept. in extreme E, sit. on Ill R. and near W bank of Rhine R. Industries inc. brewing, tanning, milling, fruit and vegetable canning, paper-making and printing; manu. electrical goods, electrical equipment, chemicals, soap and foodstuffs, esp. *pâté de foie gras*. R. port trading in iron ore, potash, wines and vegetables, through its two Rs. and the Marne-Rhine and Rhône-Rhine canals. Pop. (1982E) 252,264 (agglomeration, 373,470).

Stratford Ontario, Canada. 43 22N 80 57W. Town sit. WNW of Hamilton on Avon R. Industries inc. railway engineering, flour milling, textiles, agric. engineering, scientific instruments and furniture-making. Pop. (1985E) 27,000.

Stratford Connecticut, U.S.A. 41 14N 73 07W. Town sit. at mouth of Housatonic R. on Long Island Sound. Industries inc. tourism, boatbuilding and chemicals; manu. hardware. Pop. (1989E) 52,625.

Stratford-upon-Avon Warwickshire, England. 52 12N 1 41W. Market town sit. on Avon R. Industries inc. tourism, mainly visitors to the birthplace of William Shakespeare and the Royal Shakespeare Theatre, and light industry. Pop. (1986E) 20,860.

Strathclyde Scotland. 55 30N 5 00W. Region of SW Scot. bounded on W by North Channel, N by Highland region, E by Tayside, Central, Lothian and Borders regions and S by Dumfries and Galloway region. The most populous of the administrative regions with 45%. Industry is concentrated on the Clydeside conurbation. Coal is still important in Ayr, but mining has declined in Lanark. Light industries, particularly electronics and computing, are growing. Tourism is important and agric. in the upper Clyde valley. Pop. (1989E) 2,311,110.

Strathmore Tayside Region, Scotland. 58 23N 4 40W. Fertile vale extending *c*. 40 m. (64 km.) SW to NE along N side of the Sidlaw Hills from Tay R. to N. Esk R.

Straubing Bavaria, Federal Republic of Germany. 48 53N 12 34E. Town sit. E of Regensburg on Danube R. Industries inc. electrical and electronic engineering, steel, transport equipment, sports equipment and food processing. Pop. (1989E) 41,000.

Streatham   Greater London, England. 51 26N 0 07W. Residential district of S London.

Street   Somerset, England. 51 07N 2 43W. Town sit. E of Bridgwater. Company town serving an agric. area, manu. footwear and leather goods. Pop. (1980E) 9,000.

Stresa   Piemonte, Italy. 45 53N 8 32E. Village on W shore of L. Maggiore, WNW of Como. The main industry is tourism.

Stretford   Greater Manchester, England. 53 27N 2 19W. Town sit. SW of Manchester on Bridgewater Canal. Industries inc. engineering, electrical products, chemicals, tractor manu. and food processing. Pop. (1983E) 49,900.

Stromboli   Sicilia, Italy. 38 48N 15 13E. Island in the Lipari group sit. NW of Messina, in the Tyrrhenian Sea. Noted for its active volcano.

Stromness   Orkney Islands, Scotland. 58 57N 3 18W. Town on SW coast of Mainland. Principal ferry terminal for the islands. Supply base for lighthouses. The main industries inc. crafts, knitting, bakery, fishing, fish and seafood processing. Pop. (1985E) 2,160.

Stroud   Gloucestershire, England. 51 44N 2 12W. Town sit. S of Gloucester on R. Frome. Market town for an agric. area. Industries inc. manu. service uniform cloth, food products, pianos, plastics, and oil drilling equipment. Pop. (1981C) 20,930.

Sturgeon Bay   Wisconsin, U.S.A. 44 50N 87 23W. City at head of Sturgeon Bay on the Door Peninsula between Green Bay and L. Michigan; sit. on a ship canal which cuts across the peninsula from the Bay to the L. Industries inc. tourism, fisheries, light manu., shipyards, dairies and fruit canning factories. Pop. (1980C) 8,847.

Stuttgart   Baden-Württemberg, Federal Republic of Germany. 48 46N 9 11E. Cap. of province in SW sit. on Neckar R. Railway junction and commercial centre. Industries inc. publishing, textiles and engineering; manu. scientific and optical instruments, clocks and watches, musical instruments, leather goods and chemicals. Also important wine and fruit-growing area in vicinity. Pop. (1987E) 552,300.

Styria   Austria. 47 56N 15 00E. Federal state in SE, bounded S by Yugoslavia. Area, 6,326 sq. m. (16,384 sq. km.). Cap. Graz. Mountainous, esp. in N, sloping to SE and drained by Rs. Mur, Enns, Mürz and Raab. Forestry is important, also mining, esp. for iron and lignite, with industry concentrated around Graz. Pop. (1988E) 1,181,000.

Suakin   Kassala, Sudan. 19 07N 37 20E. Town sit. on Red Sea coast. The main occupation is fishing.

Subansiri River   India. Rises SW of Konam Dzong, Tibet, and flows 226 m. (361 km.) W, S, then SE into NE Assam, India, where it turns S to enter Brahmaputra R. NNW of Golaghat.

Subotica   Yugoslavia. 46 06N 19 39E. Town sit. NW of Belgrade near border with Hungary. Market town for an agric. area. Industries inc. flour milling, meat packing, furniture making and chemicals; manu. footwear. Pop (1981C) 154,611.

Suceava   Suceava, Romania. 47 39N 26 19E. District cap. sit. WNW of Iaşi on Suceava R. Market town for an agric. area. Industries inc. flour milling and tanning. Pop. (1985E) 92,690.

Suchitépequez   Guatemala. Dept. of SW Guatemala on the Pacific Ocean. Area, 960 sq. m. (2,486 sq. km.) Chief towns Mazatenango (cap.), Cuyotenango, San Antonio. Coastal plain rising to mountains inland, drained by Nahualate and Madre Vieja Rs. Agric. area producing coffee, sugar-cane and tobacco on high ground; maize, beans and bananas on the coast. Industries inc. sugar refining, lumbering, textiles, cotton ginning. Pop. (1985E) 327,763.

Sucre Chuquisaca, Bolivia. 19 00S 65 15W. Legal cap., founded 1538, of Bolivia and seat of the judiciary, also regional cap., sit. at 9,000 ft. (2,743 metres) a.s.l. Commercial centre for an agric. area. Pop. (1985E) 86,609.

Sucré Venezuela. State of NE Venezuela on the Caribbean. Area, 4,560 sq. m. (11,810 sq. km.). Traversed by spurs on the coastal range W to E; its coastline extends (W) as the Peninsula de Araya and (E) as the Peninsula de Paria. Agric. area, producing cacao, sugar-cane, coffee, tobacco, maize, cotton, rice, yucca and bananas. Fishing is important. Minerals inc. asphalt, at Guanoco, salt, on Araya, gypsum, Macuro, sulphur, Casanay, some iron, coal, petroleum and copper. Processing industries are centred on Cumaná. Pop. (1981C) 585,698.

Sudan 15 00N 30 00E. Rep. in NE Africa bounded N by Egypt, E by Red Sea and Ethiopia, S by Kenya, Uganda and Zaïre, W by Central Afr., Chad and Libya. Area, 967,500 sq. m. (2,505,800 sq. km.). Chief towns, Khartoum (cap.), Wad Medani, Omdurman, El Obeid, Port Sudan. Mainly plateau at under 600 ft. (183 metres) a.s.l. in N, rising to 1,500 ft. (457 metres) a.s.l. near the Ugandan border in S, with the Lolebai Mountains rising to 10,456 ft. (3,187 metres) a.s.l. in extreme S, and the Etbai Mountains on the Red Sea coast rising to Jebel Erba, 7,273 ft. (2,217 metres) a.s.l. Drained mainly by Rs. White Nile and Blue Nile with their tribs. The climate is tropical; annual rainfall is very low in N, 5 ins. (12.5 cm.) per annum at Khartoum, and 40–50 ins. (100–125 cm.) per annum in S, with desert in N and wooded savannah in S. Irrigated land supports agric. esp. cotton, millet and gum arabic. Pop. (1987E) 25·56m.

Sudbury Ontario, Canada. 46 30N 81 00W. City in N central Ontario, sit. N of L. Huron. Centre of a rich mining area, producing most of the world's nickel, and also lead, zinc, copper, gold, silver and platinum. Industries are mainly associated with mining, esp. nickel smelting and refining, and also some engineering and sawmilling. Pop. (1981C) 91,388.

Sudbury Suffolk, England. 52 02N 0 44E. Town sit. S of Bury St. Edmunds on Stour R. Market town for an agric. area, manu. diesel injectors, textiles and silk weaving.

Sudeten Mountains Poland/Czechoslovakia. 50 30N 16 00E. Ranges extending along the NE frontier of Czech. with Poland *c*. 130 m. (208 km.) SE from the Riesengebirge and rising to Sněžka (Schneekopf) 5,259 ft. (1,603 metres) a.s.l., rich in minerals, esp. coal.

Suez Egypt. 29 58N 32 33E. Town sit. at S end of the Suez Canal, at head of the Gulf of Suez. An important oil-fuelling station with 2 oil refineries. Manu. fertilizers.

Suez Canal Egypt. 29 55N 32 33E. Ship canal in NE Egypt between the Mediterranean Sea at Port Said and the Gulf of Suez at Suez opened to shipping in 1869. In 1967 the canal was blocked and became unusable but in 1974–75 it was cleared of obstructions. 103 m. (165 km.) long, minimum width 197 ft. (60 metres), passing through L. Timsah and the Great and Little Bitter L. The canal takes vessels of up to 37 ft. (11 metres) draught and has no locks.

Suez, Gulf of Egypt. NW arm of Red Sea between the Arabian Desert and the Sinai Peninsula, 170 m. (272 km.) long and up to 25 m. (40 km.) wide, and linked with the Mediterranean Sea by the Suez Canal.

Suffolk England. 52 16N 1 00E. County in East Anglia bounded N by Norfolk and S by Essex. Consists (1974 re-organization) of the districts of Waveney, Forest Heath, St. Edmundsbury, Mid Suffolk, Suffolk Coastal, Babergh and Ipswich. Generally low-lying rising to 420 ft. (128 metres) a.s.l. in SW chalk hills, and drained by Rs. Stour, Little Ouse, Waveney, Deben, Orwell and Alde. Mainly agric., esp. wheat, barley and sugar-beet. Industries inc. electronics, heavy engineering, telecommunication research and development, insurance and banking, food processing, chemicals, textiles, printing, plastics and shipbuilding. The chief towns are

Ipswich, Lowestoft, Felixstowe and Bury St. Edmunds. Pop. (1988E) 636,580.

Suhl Suhl, German Democratic Republic. 50 37N 10 41E. District cap. sit. SW of Erfurt in the Thuringian Forest. Manu. cycles, motor cycles, hunting and sports equip., and pottery. Pop. (1990E) 57,400.

Sui Baluchistan, Pakistan. 28 37N 69 19E. Natural gas field 345 m. (552 km.) from Karachi, supplies Karachi and Multan.

Suir River Ireland. 52 15N 7 00W. Rises in N Tipperary and flows 85 m. (136 km.) S and then E past Clonmel to Waterford, forming the border between Tipperary and Waterford, and then Kilkenny and Waterford. Below Waterford town it joins Barrow R. to enter Waterford Harbour, an inlet of St. George's Channel.

Suita Honshu, Japan. 34 45N 135 32E. Town sit. N of Osaka. The main industry is brewing. Pop. (1988E) 342,000.

Sukh Bator Selenga, Mongolia. Town sit. NNW of Ulan Bator on Selenga R. Trading centre on an important route between Mongolia and the U.S.S.R., head of navigation on Selenga R., road and rail centre.

Sukhumi Abhazia, U.S.S.R. 43 00N 41 02E. Cap. of Abhazian Autonomous Soviet Socialist Rep. sit. NNW of Batumi on the Black Sea. Industries inc. metal working, fruit, tobacco and fish processing, and tourism. Sea port and health resort. Pop. (1981E) 114,000.

Sukkur Sind, Pakistan. 27 42N 68 54E. Town sit. NE of Karachi on Indus R. Market centre for an agric. area, trading in grain and oilseeds. Industries inc. textiles; manu. leather goods, fertilizers, and cement. Pop. (1981E) 191,000.

Sulaimaniya Sulaimaniya, Iraq. Cap. of governorate sit. ESE of Mosul. Market town sit. on routes of trade with Iran. Pop. (1985E) 279,424.

Sulaiman Range Pakistan. 30 30N

70 10E. Mountain range extending 180 m. (283 km.) N to S along E border of Quetta province, rising to Takht-i-Sulaiman, 11,085 ft. (3,379 metres) a.s.l. at N end.

Sulawesi Indonesia. 2 00S 120 00E. Island E of Borneo. Area, 88,459 sq. m. (229,108 sq. km.). Divided in 2 provinces of N and S Sulawesi. Cap. and chief port Vjung Padang. Products include copra, coffee and spices. Pop. (1980C) 10,409,533.

Sullom Voe Shetland Islands, Scotland. 60 27N 1 20W. Sit. N of Lerwick on the N part of Mainland. It is a deep coastal inlet and the site of Europe's largest oil and liquified gas terminal. Pop. (summer 1985E) 1,000.

Sulu Archipelago Philippines. 5 30N 121 30E. Group of islands off SW Mindanao extending *c.* 270 m. (432 km.) SW to within 20 m. (32 km.) of E coast of Sabah, Malaysia. Area, 146 sq. m. (379 sq. km.). Chief town Jolo. Most of the islands are volcanic, others are coral. Chief occupations pearl fishing, growing rice and coconuts.

Sumatra Indonesia. 0 05S 102 00E. Island separated from Malay Peninsula by Strait of Malacca, and from Java to SE by Sunda Strait. Area, 202,354 sq. m. (524,097 sq. km.) divided into the provinces of Atjeh, N. Sumatra, W. Sumatra, Djambi and S. Sumatra. The island lies across the equator and has a hot climate with heavy rain in the SW and NE monsoons. Mainly forest with mountains along W coast, chiefly volcanic and rising to Kerintji, 12,484 ft. (3,805 metres) a.s.l. Chief occupation growing rice, rubber, tobacco, tea, coffee and coconuts; some deposits of coal and petroleum. Chief towns Palemban, Telanaipura, Bukit Tinggi, Medan and Bedan Atjeh. Pop. (1980C) 28,016,160.

Sumba Indonesia. 10 00S 120 00E. Island in Lesser Sunda group separated from Flores and Sumbawa by Sumba Strait. Area, 4,306 sq. m. (11,152 sq. km.). Chief town Waingapu. Chief occupation farming.

Sumbawa Indonesia. 8 40S 118

00E. Island in Lesser Sunda group between Lombok and Flores. Area, 5,695 sq. m. (14,750 sq. km.). Cap. Raba. Mountainous rising to Mount Tambora, 9,353 ft. (2,851 metres) a.s.l. Chief occupation farming. Chief towns are Raba, Bima and Sumbawa.

Sumgait   Azerbaijan, U.S.S.R. 40 36N 49 38E. Town sit. NW of Baku on Caspian Sea. Industries are dependent on the Baku oilfields and inc. manu. chemicals and synthetic rubber. Pop. (1987E) 234,000.

Summerside   Prince Edward Island, Canada. 46 24N 63 47W. Town sit. WNW of Charlottetown on Bedeque Bay. Port shipping dairy products, seed potatoes. Main industries farming, fishing and tourism. Pop. (1989E) 12,000.

Sumy   Ukraine, U.S.S.R. 50 55N 34 45E. Cap. of Sumy region, sit. ENE of Kiev. Centre of an agric. area, esp. wheat and sugar-beet. Industries sugar refining, tanning, sawmilling, agric. engineering and textiles; manu. food products and fertilizers. Pop. (1987E) 268,000.

Sunbury-on-Thames   Surrey, England. 51 25N 0 26W. Town sit. WSW of central London on Thames R. Mainly residential with some industrial estates. Pop. (1981C) 28,240.

Sunda Islands   Indonesia. 2 00S 110 00E. Island group consisting of W part of the Malay Arch. between South China Sea and Indian Ocean. The Greater Sunda Islands are Borneo, Java, Sumatra and Celebes. The Lesser Sunda Islands are those E of Java, esp. Bali, Lombok, Sumbawa, Sumba, Flores and Timor.

Sundarbans   Bangladesh. 22 00N 89 00E. Swampy region in S Ganges delta next to Indian border, extending 150 m. (240 km.) along the coast and up to 50 m. (80 km.) inland, with creeks, tidal R. and mangrove swamps.

Sunday Island   Kermadec Islands, New Zealand. Otherwise Raoul Island. Largest and only inhabited island of the Kermadec group, S Pacific. Area, 11 sq. m. (28 sq. km.). Volcanic and mountainous with thick forest. Rises to 1,723 ft. (525 metres) a.s.l.

Sunderland   Tyne and Wear, England. 54 55N 1 23W. Town sit. SE of Newcastle on Tyne at mouth of Wear R. on the North Sea. Coalmining area. Manu. chemicals, glass, machinery, shipbuilding and repairing. Site of first Nissan manu. plant in U.K. Pop. (1983E) 299,400.

Sundsvall   Västernorrland, Sweden. 62 23N 17 18E. Town in E central Sweden sit. on Gulf of Bothnia. Industries inc. aluminium, sawmilling; manu. fibreboard, wood pulp and paper. Sea port exporting timber and wood pulp, but icebound in winter. Pop. (1989E) 93,233.

Sundyberg   Stockholm, Sweden. Town sit. NW of Stockholm. Industries inc. banking and high technology. Pop. (1989E) 31,600.

Sungari River   China. 47 42N 132 30E. Rises in the Changpai Shan in SE Kirin and flows 1,150 m. (1,840 km.) NW and then ENE. It passes the Fengman Dam and hydroelectric station and joins Nun R. on the Kirin/Heilungkiang border, flowing ENE through Heilungkiang, past Harbin and Kiamusze, to join Amur R. above Tungkiang. It is ice free May–Oct.

Sunnyvale   California, U.S.A. 37 23N 122 01W. City sit. WNW of San José in a fruit growing area. Industries inc. fruit canning; manu. electronic equipment. Pop. (1980C) 106,618.

Sunyani   Ashanti, Ghana. 7 20N 2 20W. Town sit. NW of Kumasi. Industries inc. brick and tile making. Pop. (1980E) 62,000.

Suoyarvi   Karelo-Finnish Soviet Socialist Republic, U.S.S.R. Town sit. WNW of Petrosavodsk on N shore of L. Suoyarvi. Manu. paper, cardboard.

Superior   Wisconsin, U.S.A. 46 44N 92 05W. City sit. at W end of L. Superior forming, with Duluth, Minnesota, the W end of the St. Lawrence Seaway. Industries inc. oil refining, railway engineering, flour milling, shipping iron ore, coal, grain and general cargo. Pop. (1980C) 29,571.

Superior, Lake Canada/U.S.A. 48 00N 88 00W. Bounded N and E by Ontario, Canada, S by Michigan and Wisconsin, U.S.A., W by Minnesota, U.S.A. Area, 32,483 sq. m. (84,131 sq. km.), the largest fresh-water L. in the world. Navigable April–Dec., but never completely frozen. Ships pass by canals at Sault Ste. Marie at E end, where the L. drains into L. Huron by St. Mary's R. It is fed by *c.* 200 Rs. inc. R. St. Louis and Nipigon.

Sur Oman. 46 44N 59 32E. Port sit. SE of Muscat on the Gulf of Oman. Industries inc. shipbuilding and the export of dates.

Surabaya East Java, Indonesia. 7 15S 112 45E. Cap. of province on the Madura Strait sit. near mouth of Kali Mas R. Industries inc. engineering, oil refining, shipbuilding and ship-repairing, textiles and chemicals; manu. glass and tobacco products. Sea port exporting *c.* half Java's produce, esp. sugar, tobacco and coffee. Naval base. Pop. (1981C) 2,027,913.

Surakarta Java, Indonesia. 7 35S 110 50E. Town sit. ENE of Djarkarta on Solo R. Commercial centre trading in rice, sugar and tobacco. Industries inc. tanning and textiles. Pop. (1970C) 367,626.

Surakhany Azerbaijan Soviet Socialist Republic, U.S.S.R. Town sit. ENE of Baku on the Apsheron peninsula of the W Caspian coast. There are oilfields. Other industries inc. metalworking.

Surat Gujarat, India. 21 12N 72 55E. City sit. N of Bombay on Gulf of Cambay, near Tapti R. Industries inc. engineering; manu. cotton goods, paper, silk and soap. Port trading in cotton and grain. Pop. (1981C) 776,876.

Surbiton Greater London, England. 51 24N 0 19W. District in SW, on Thames R. Mainly residential with some industry inc. light engineering.

Surendranagar Gujarat, India.

Town sit. ENE of Rajkot on the Kathiawar peninsula. District cap. and trading centre for cotton, grain, ghee, cloth and hides. Industries inc. cotton ginning, manu. soap. Pop. (1981C) 91,770.

Suresnes Hauts-de-Seine, France. Suburb of W Paris on Seine R. Industries inc. chemicals; manu. cars, bicycles and perfumes.

Suriname 4 00N 56 00W. An independent Rep. on N coast of S. America, bounded E by French Guiana, S by Brazil and W by Guyana. Area, 70,087 sq. m. (181,525 sq. km.). Chief towns Paramaribo (the cap.), Coronie, Nieuw Nickerie, Paranam and Kwakoegron. The land rises from coastal lowlands through savannah to forested highlands in the S. Climate tropical and humid. The chief occupations are farming, extracting bauxite, together with shrimps, fish and bananas which are the chief exports; other products from the coastal lowlands are sugarcane, rice and citrus fruits. Pop. (1990E) 416,839.

Surkhan Darya River U.S.S.R. 37 12N 67 20E. Rises in 2 branches in the Gissar mountains, Uzbek Soviet Socialist Rep., and flows *c.* 150 m. (240 km.) SSW through a cotton growing area, past Dzhar-Kurgan to enter Amu Darya R. near Termez. Linked by canal to Kafirnigan R.

Surrey England. 51 16N 0 30W. County in SE consisting of the boroughs of Elmbridge, Epsom and Ewell, Guildford, Reigate and Banstead, Runnymede, Spelthorne, Surrey Heath, Waverley and Woking and the districts of Mole Valley and Tandridge. Urban in NE, crossed E to W in centre by North Downs rising to Leith Hill 965 ft. (294 metres) a.s.l. Drained by Rs. Wey and Mole flowing into Thames R. Horticulture and farming are important in S, industry mainly in urban N. Pop. (1988E) 999,752.

Susquehanna River New York/ Pennsylvania, U.S.A. 39 33N 76 05W. Rises in Otsego L., New York, and flows 444 m. (710 km.) S entering Pennsylvania briefly above Binghamton and crossing into it again at Sayre, flowing past Wilkes-Barre and Harrisburg to enter the head of Chesapeake Bay. It is unnavigable, but used for hydroelectric power.

Sutherland   Scotland.   Former county now part of Highland Region.

Sutlej River   China/India/Pakistan. 29 23N 71 02E. Rises in L. Manasarowar in SW Tibet, China, and flows 900 m. (1,440 km.) WNW and then SW, turning into Himachal Pradesh, India, at E end of the Zaskar Mountains, entering Punjab at SW end of the Bhakra reservoir and going on to form S half of the Punjab, India and Pakistan border. It flows into Chenab R. near Alipur and both join R. Indus *c.* 53 m. (*c.* 85 km.) further on. Important for irrigation and hydroelectric power.

Sutton   Greater London, England. 51 22N 0 12W. District SSW of the city centre; a mainly residential suburb.

Sutton Coldfield   West Midlands, England. 52 34N 1 48W. Town sit. NE of Birmingham. Industries inc. manu. machinery and pharmaceutical products. Pop. (1981C) 86,494.

Sutton-in-Ashfield   Nottinghamshire, England. 53 08N 1 15W. Town sit. NNW of Nottingham. Industries inc. engineering and textiles. Pop. (1990E) 40,235.

Suva   Viti Levu, Fiji. 18 06S 178 30E. Cap. of Fiji sit. on the S coast. Industries inc. manu. coconut oil and soap. Sea port exporting sugar, gold and coconut oil. Pop. (1986C) 71,608.

Suwaiq   Batina, Oman. Town sit. WNW of Muscat on the Gulf of Oman. Port serving a fertile date growing area. Industries inc. fishing.

Suwannee River   Georgia/Florida, U.S.A. 29 18N 83 09W. Rises in the Okefenokee swamp in SE Georgia and flows 250 m. (400 km.) S across N Florida to enter the Gulf of Mexico.

Suzdal   Russian Soviet Federal Socialist Republic, U.S.S.R. 56 29N 40 26E. Town sit. NE of Moscow in the Vladimir Region. Important as a religious and monastic centre.

Svay Rieng   Svay Rieng, Cambodia. 11 05N 105 48E. Town sit. SE of Pnom Penh on West Vaico R. near the Vietnam frontier. Centre of a rice growing area.

Sveagruva   Svalbard, Norway. Settlement sit. SE of Longyear City on NE shore of Van Mijen Fjord, West Spitsbergen. Coalmining village with an ice-free port July–September.

Svendborg   Fyn, Denmark. 55 05N 10 35E. Cap. of county sit. on S coast of Fyn Island. Industries inc. shipbuilding, tanning, tobacco, furniture, machinery, electronics, textiles and engineering. It is also a sea port, tourist centre and yachting centre. Pop. (1989E) 40,871.

Sverdlovsk   Russian Soviet Federal Socialist Republic, U.S.S.R. 56 51N 60 36E. City in E foothills of the central Ural Mountains. Cap. of Sverdlovsk region. Industries are mainly metallurgical, esp. manu. steel; mining and heavy engineering equipment, railway rolling stock, ball-bearings, aircraft; smelting copper, and cutting and polishing gems. Important junction at terminus of the Trans-Siberian Railway. Formerly called Ekaterinburg, it was the place of execution of Tsar Nicolas II, 1918. Pop. (1989E) 1,367,000.

Svir River   U.S.S.R. Rises in L. Onega and flows 140 m. (224 km.) WSW past the Svirstroi hydroelectric plant to L. Ladoga, forming part of the Mariinsk Canal system.

Svolvaer   Lofoten Islands, Norway. 68 14N 14 34E. Town on SE coast of Austvågöy island, sit. N of Bodø, on the mainland. Industries are mainly food processing, salmon fishing, fishing for cod and manu. codliver oil and fertilizers. There are also ship repair yards. Pop. (1989E) 4,000.

Swadlincote   Derbyshire, England. 52 47N 1 33W. Town sit. SE of Burton-upon-Trent. Principal industries inc. coalmining, gravel extraction, agric. Manu. of earthenware, and engineering. Pop. (1988E) 25,430

Swakopmund   Namibia. 22 41S 14

34E. Town sit. on Atlantic coast. Tourist centre. Uranium mine in vicinity. Industries inc. salt, brewing and tanning. Pop. (1990E) 18,000.

Swale River North Yorkshire, England. 54 06N 1 20W. Rises in the Pennines near border with Cumbria, and flows 75 m. (120 km.) E past Richmond, then SSE through the Vale of York to join Ure R., together forming Ouse R.

Swanage Dorset, England. 50 37N 1 58W. Town sit. S of Poole on Swanage Bay: seaside resort. Pop. (1987E) 9,484.

Swan Hill Victoria, Australia. 35 21S 143 34E. Town sit. NNW of Melbourne on Murray R. Processing centre for a sheep farming, agric. and fruit-growing area. Pop. (1989E) 9,600.

Swanland Western Australia, Australia. SW region of W. Australia, together with Perth and Fremantle on Swan R. the most fertile region of the state with a Mediterranean climate producing wheat, wool, timber, fruit, vines, tobacco and dairy produce.

Swansea West Glamorgan, Wales. 51 38N 3 57W. District of West Glamorgan at mouth of Tawe R. Industries inc. metal foundries and oil refining; manu. tin-plate and chemicals. Pop. (1981C) 167,796.

Swatow (Shantou) Guangdong, China. 23 23N 116 41E. Port in SE at mouth of Han R. on the Nan Hai. Manu. pottery, pharmaceutical products, matches and cigarettes. Exporting sugar, tobacco and fruit. Pop. (1984E) 746,000.

Swaziland 26 30S 31 30E. Kingdom in SE Africa bounded NE by Mozambique and on all other sides by Rep. of S. Afr. Area, 6,705 sq. m. (17,366 sq. km.). Cap. Mbabane. The country divides into the high veld, rising to 4,000 ft. (1,219 metres) a.s.l. in W, and the middle veld and low veld at 1,000 ft. (305 metres) a.s.l. in E. Rainfall is high on the high veld, light on the low. Chief occupation, farming, esp. maize, cattle, sugar, rice and cotton. Pop. (1986E) 681,059.

Sweden 57 00N 15 00E. Kingdom occupying E part of the Scandinavian peninsula, bounded E by Finland, the Gulf of Bothnia and the Baltic Sea, SW by the Sound, the Skagerrak and the Kattegat, and W by Norway. Area, 173,731 sq. m. (449,964 sq. km.). Chief cities Stockholm (cap.), Göteborg, Malmö, Uppsala, Västerås, Norrköping, Örebro, Jönköping, Linköping, Hälsingborg. Mountains extend along the Norwegian frontier. Most of the country is table-land at 300–800 ft. (91–244 metres) a.s.l., with an indented coastline and 96,000 Ls. Thickly forested N covering c. half the area. Supports c. one-seventh of the pop. on a stony, infertile soil. The central Svealand region has many Ls. inc. the extensive Ls. Väner, Vätter, Mälar and Hjälmar, and connecting Rs. and canals. The soil is more fertile and mineral deposits are considerable. The S is fertile plain, mainly agric. Chief crops are hay, barley, oats, potatoes, sugar-beet, wheat. Dairy farming is important. Forests (N) produce coniferous timber, pulpwood and fuel from c. 57% of the land area. Minerals inc. iron, mainly in Svealand and the far N, copper, lead, zinc, pyrites and alum shale containing oil and uranium. The chief industries are iron and steel milling, other metal working, sawmilling, pulp milling, manu. machinery and vehicles, chemicals. Pop. (1989E) 8,526,000.

Swedru Western Province, Ghana. Town sit. NNW of Winneba. Route centre and market for cacao.

Świdnica Wrocław, Poland. 50 51N 16 29E. Town in SW, near the border with Czechoslovakia. Market town for an agric. area. Industries inc. tanning, brewing, textiles, chemicals and agric. engineering. Pop. (1983E) 59,000.

Swift Current Saskatchewan, Canada. 50 17N 107 50W. Town sit. W of Regina on Swift Current Creek. Major distribution centre. Manu. plastics, tempered glass, metal fasteners, hooks, agric. equip. Pop. (1989E) 16,000.

Swilly, Lough Donegal, Ireland. Sea inlet in N coast extending c. 30 m. (48 km.) inland from its mouth between Fanad Head and Dunaff Head.

Swindon Wiltshire, England. 51
34N 1 47W. Town in sw. Manu. inc.
clothing, car bodies, financial services, pharmaceuticals and electronic
equipment. Pop. (1985E) 129,300.

Świnoujście Szczecin, Poland. 53
53N 14 14E. Bamo Sea port on E
coast of Usedltic Island; a fishing port
and holiday resort.

Swinton South Yorkshire, England.
52 28N 1 20W. Town sit. NE of Sheffield. Industries inc. coalmining,
manu. glass and pottery.

Swinton and Pendlebury Greater
Manchester, England. 53 31N 2 20W.
Town sit. NW of Manchester. Industries inc. engineering and textiles, esp.
cotton; manu. electrical equipment
and chemicals. Pop. (1984E) 45,000.

Switzerland 46 30N 8 00E. Rep.
bounded N by Federal Republic of
Germany, E by Austria and Liechtenstein, S by Italy and W by France.
Area, 15,943 sq. m. (41,293 sq. km.).
Chief towns Bern (cap.), Zürich,
Basel, Geneva, Lausanne. The Jura
mountains extend SW to NE through
NW Switzerland, and the central
uplands are divided by numerous Ls.
and stretch SW to NE from L. Geneva
to L. Constance as a broad plain,
bounded SE by the Alps, considerably
higher than the Jura and rising to
Monte Rosa, 15,216 ft. (4,638 metres)
a.s.l. The main Rs. are Rhône, Rhine,
Aare and Ticino. Agric. and pasture
land covers *c*. 2.1m. hectares, of
which *c*. 287,000 hectares are under
cultivation. Chief crops are cereals,
vines, potatoes, sugar-beet, fruit.
Dairy and pig farming are important.
Forests produce timber from 999,795
hectares. The chief minerals are salt,
iron ore, manganese. The main industries are food processing, tourism, textiles, chemicals, manu. basic metals,
metal products, precision instruments,
clocks and watches, chemical and
pharmaceutical goods. Pop. (1988E)
6,619,973.

Sydney New South Wales, Australia. 33 52S 151 13E. State cap. on S
shore of Port Jackson. Commercial
centre for a vast agric. hinterland, and
sit. on a large coalfield. Industries inc.

electronics, metal working, oil refining, food processing and brewing;
manu. machinery, scientific equipment, textiles, chemicals, paper, furniture and clothing. Trading centre for
wool and wheat, important sea port.
Pop. (1987E) 3,525,850.

Sydney Nova Scotia, Canada. 46
09N 60 11W. City in SE, sit. on NE
coast of Cape Breton Island, in a coalmining area. Industries inc. mining;
manu. steel, chemicals and bricks. Sea
port handling coal, iron ore and limestone. Pop. (1986C) 27,754.

Sydney Mines Nova Scotia, Canada. 46 14N 60 14W. Town sit. NNW
of Sydney on E coast of Cape Breton
Island. Industries inc. coal reclamation and processing of waste. Pop.
(1988E) 8,001.

Syktyvkar Russian Soviet Federal
Socialist Republic, U.S.S.R. 61 40N
50 46E. Cap. of Komi Autonomous
Soviet Socialist Rep. sit. on Vychegda
R. Important trading centre for timber;
industries inc. sawmilling, boatbuilding, fur processing, wood pulp and
paper making. Pop. (1987E) 224,000.

Sylhet Sylhet, Bangladesh. 25 43N
91 51E. District cap. sit. NE of Dacca.
Airport. Trades in rice, tea, oilseeds,
jute and cotton. Tea processing is the
main industry, also manu. cotton cloth
and electrical goods and sawmilling
Pop. (1984E) 150,000.

Symond's Yat Hereford and
Worcester, England. 51 50N 2 37W.
Point on Wye R. sit. SW of Ross-on-
Wye, where R. flows through a narrow gorge in a 5 m. (8 km.) loop
round Huntsham Hill, returning to
within 600 yds. (0.6 km.) of its former
course.

Syracuse Siracuse, Italy. 37 04N 15
17E. Cap. of Siracuse province, sit. on
SE coast of Sicily. Industries inc.
chemicals, fisheries and saltworks.
Port exporting wine and olive oil. Historical interest is strong. Pop. (1981C)
117,615.

Syracuse New York, U.S.A. 43
03N 76 09W. City in centre of state,
sit. at S end of L. Onondaga. Manu.

communications equip., pharmaceuticals, medical products, air conditioning and refrigeration parts. Pop. (1980c) 170,105.

Syr Darya River   U.S.S.R. 46 03N 61 00E. Rises as Naryn R. in the E Kirghiz Soviet Socialist Rep. and flows 1,400 m. (2,240 km.) generally W and SW to irrigate the Fergana valley in the Uzbek Soviet Socialist Rep. It turns NW bounding Kyzyl Kum desert to W, and enters Aral Sea by a delta.

Syria   35 00N 38 00E. Rep. bounded N by Turkey, E by Iraq, S by Jordan and W by Israel, Lebanon and the Mediterranean. Area, 71,081 sq. m. (185,100 sq. km.). Chief cities Damascus (cap.), Aleppo, Homs, Hama, Lattakia. In W there are mountain ranges parallel to the coast. The highest point is Jebel ed Druz, 5,564 ft. (1,696 metres) a.s.l. W of them is a fertile coastal strip with moderate rainfall; E of them is a fertile depression watered by Orontes R., and further E the ground rises again to the plateau of the Syrian Desert which extends NE to Euphrates R. Agric. land covers *c*. 3m. hectares, of which *c*. 540,000 hectares are irrigated. Chief crops are wheat, barley, olives, sugarbeet, citrus fruit, cotton. Minerals inc. some phosphate, gypsum, bitumen, oil and natural gas. Industries inc. manu. food products, textiles, sugar, leather, tobacco, soap and cement. Pop. (1988E) 11,338,000.

Syros   Greece. 37 26N 24 54E. Island in the Cyclades group in the Aegean Sea. Area, 33 sq. m. (85 sq. km.). Chief town Syros. Mainly mountainous, producing grain, fruit and wine. Pop. (1980E) 19,000.

Syzran   Russian Soviet Federal Socialist Republic, U.S.S.R. 53 09N 48 27E. Cap. of Kuibyshev region, sit. SE of Moscow on Volga R. Industries inc. oil refining; manu. machinery,

building materials and clothing. Important as a R. port and railway centre, handling grain and oil. Pop. (1987E) 174,000.

Szczecin   Szczecin, Poland. 53 24N 14 32E. Cap. of province in extreme NW, sit. near the mouth of Oder R. Industries inc. shipbuilding; manu. synthetic fibres, paper, fertilizers and cement. Port with access to the Baltic Sea. Pop. (1982E) 390,000.

Szechwan   *see* **Sichuan**

Szeged   Csongrád, Hungary. 46 15N 20 09E. Town sit. SSE of Budapest near confluence of Rs. Tisza and Maros. Commercial centre for an agric. area. Industries inc. flour milling, brewing and sawmilling; manu. textiles, footwear and tobacco products. Pop. (1989E) 189,000.

Székesfehérvár   Fejér, Hungary. 47 12N 18 25E. Cap. of county sit. SW of Budapest. Market town trading in wine and tobacco. Industries inc. aluminium processing and footwear manu. Pop. (1989E) 114,000.

Szekszard   Tolna, Hungary. 46 21N 18 42E. Market town sit. SSW of Budapest. County cap. Industries inc. making red wine, distilling alcohol, making bricks. Pop. (1989E) 39,000.

Szolnok   Szolnok, Hungary. 47 10N 20 12E. Cap. of county sit. ESE of Budapest, at confluence of Rs. Tisza and Zagyra. Industries inc. flour milling, sawmilling and brickmaking. Railway junction and R. port. Pop. (1989E) 82,000.

Szombathely   Vas, Hungary. 47 14N 16 38E. Cap. of county in W Hungary near Austrian frontier sit. in a fruit-growing region. Industries inc. sawmilling, flour milling, textiles and agric. engineering. Pop. (1989E) 88,000.

T

Tabar Islands New Ireland, Bismarck Archipelago, South West Pacific. A single volcanic island and scattered coral islets, rising to 1,500 ft. (457 metres) a.s.l.

Tabasco Mexico. 17 45N 93 30W. State in SE sit. on the Gulf of Campeche and bordered by Guatemala. Area, 9,522 sq. m. (24,661 sq. km.). Mainly flat with semi-tropical forest and many swamps and lagoons. Livestock raising is a major occupation and crops inc. cacao, rice, sugar, coffee and tobacco. Oil and natural gas are found. Cap. Villa Hermosa, sit. ESE of Mexico City. Pop. (1989E) 1,322,613.

Tabiteuea Kiribati, W Pacific Ocean. Atoll in S Gilberts group guarded by a 50 m. (80 km.) long reef. Area, 15 sq. m. (38 sq. km.). Exports inc. copra. Pop. (1985C) 4,493.

Table Mountain Cape Province, Republic of South Africa. 33 57S 18 25E. A flat-topped mountain overlooking Cape Town and Table Bay, 3,567 ft. (1,087 metres) a.s.l. It is frequently covered with a white cloud called 'The Table Cloth'. A cable railway runs from Cape Town to the summit, Maclear's Beacon.

Tabora Tabora, Tanzania. 5 01S 32 48E. Cap. of region in W centre sit. at 3,900 ft. (1,189 metres) a.s.l. A railway junction and agric. market. Trades in groundnuts, cotton and millet. Pop. (1978C) 67,382.

Tabriz E Azerbáiján, Iran. 38 05N 46 18E. Cap. of province and fourth largest town of Iran, sit. in NW and to N of the volcanic Kuh-e-Sahand, at 4,400 ft. (1,341 metres) a.s.l. Industrial centre of an agric. region and noted for carpets. Other manu. are textiles, leather and soap. Agric. products inc. almonds and dried fruit. Pop. (1986C) 971,482.

Tabuaeran *see* **Fanning Island**

Táchira Venezuela. 8 07N 72 21W. An inland state bordering on Colombia. Area, 4,285 sq. m. (11,098 sq. km.). Mainly mountainous, traversed by the Sierra Nevada de Mérida; with the Maracaibo lowlands in the NW. It is the principal coffee growing area of the country. Other crops inc. cereals, sugar, fruit, vegetables and tobacco. Cattle raising and forestry are also important. There are deposits of gold, coal, petroleum, copper and sulphur. Cap. San Cristóbal. Pop. (1981C) 629,499.

Tacna Tacna, Peru. 18 01S 70 15W. Cap. of dept. in extreme S sit. on Tacna R. Products inc. tobacco, cotton, sugar-cane and sulphur. Pop. (1983E) 46,250.

Tacoma Washington, U.S.A. 47 15N 122 27W. City and port in W centre of state sit. on Commencement Bay in Puget Sound. One of the major container ports on the Pacific coast with exports inc. timber, aluminium, scrap metal, agric. machinery and fruit. Industries inc. boatbuilding and the manu. of chemicals, wood and paper products, food and metals. Pop. (1989E) 162,100.

Tacuarembó Uruguay. 31 44S 55 59W Dept. sit. in N and bordered by Río Negro in S. Area, 8,114 sq. m. (21,015 sq. km.). Cap. Tacuarembó. Mainly agric. producing cattle, sheep, cereals and vegetables. An important hydroelectric plant is sit. at Rincón del Bonete on Río Negro. Pop. (1985C) 86,200.

Tadoussac Quebec, Canada. Village sit. near confluence of Rs. Saguenay and St. Lawrence, and NNW of Rivière du Loup. The earliest European settlement in Canada was probably sit. here.

Tadzhikistan U.S.S.R. 35 30N 70

00E. Tadzhik Soviet Socialist Republic was formed from those regions of Bokhara and Turkestan where the pop. consisted mainly of Tadzhiks. It was admitted as a constituent rep. of the Soviet Union in 1929. Area, 55,240 sq. m. (143,100 sq. km.). Cap. Dushanbe. Horticulture and cattle breeding are important as well as other farming. There are rich deposits of brown coal, lead and zinc. Oil is found in N of the Rep. Other minerals found are asbestos, mica, corundum and emery, lapis lazuli, potassium salts and sulphur. Pop. (1989E) 5·1m.

Taegu   South Korea. 35 52N 128 36E. Town sit. NNW of Pusan. Products inc. grain, fruit, tobacco and textiles. Pop. (1985C) 2,030,649.

Taganrog   Russian Soviet Federal Socialist Republic, U.S.S.R. 47 12N 38 56E. Port sit. on Gulf of Taganrog in the Sea of Azov. An important industrial centre manu. iron and steel goods inc. agric. machinery, aircraft, boilers, machine tools and hydraulic presses. There are three harbours exporting coal and grain. Pop. (1987E) 295,000.

Tagus River   Spain/Portugal. 38 40N 9 24W. Rises in Teruel, E Spain WSW of Albarracín and flows *c.* 600 m. (960 km.) generally SW through mountainous country to the central plateau, then W past Aranjuez, Toledo and Talavera de la Reina into Estremadura, where its course widens into long Ls., and on to the Portuguese frontier which it forms (E to W) for *c.* 30 m. (48 km.). It turns SW past Abrantes and Santarém to Vila Franca, and below forms an estuary which narrows again to a channel, with Lisbon on the N shore before entering the Atlantic. Its course flows through many deep mountain gorges. The R. is used for power. Chief tribs. Jarama, Tajuña, Alberche, Tiétar, Alagón, Guadiela and Almonte Rs.

Tahaa   Leeward Islands, French Polynesia. 16 38S 151 30W. A mountainous volcanic island sit. N of Raiatea. Area, 34 sq. m. (88 sq. km.). The highest peak is Mount Ohiri, 1,936 ft. (590 metres) a.s.l.

Tahiti   French Polynesia, South Pacific Ocean. 17 37S 149 27W. The largest and most important of the Windward Islands. Area, 402 sq. m. (1,042 sq. km.). Cap. Papeete. Generally mountainous in the interior with extinct volcanoes, and having a fertile coastal belt. Exports inc. copra, phosphates, pearl shell and vanilla. Pop. (1988E) 115,820.

Tahoua   Niger. 14 54N 5 16E. Cap. of dept. in SW, sit. NE of Niamey. Centre of agric. and grazing area. Pop. (1981E) 38,000.

Taichung   Taiwan. 24 09N 120 41E. Town sit. SW of Taipei. Centre of a cultural and light industry district. Industries inc. footwear and watches. Pop. (1985E) 660,001.

Taimyr   U.S.S.R. 75 00N 100 00E. (i) Peninsula extending NE into Arctic Ocean from the central Siberian plateau, bounded W by Yenisei and E by Khatanga Rs., and inc. L. Taimyr on Taimyr R. Extends N to Cape Chelyuskin. Tundra area. (ii) Taimyr National Area, Russian Soviet Federal Socialist Rep., N section of Krasnoyarsk Territory extending to Arctic coast across the Taimyr Peninsula. Area, 316,700 sq. m. (820,253 sq. km.). Capital Dudinka. Tundra and forest area within the Arctic Circle. Main occupations nomadic reindeer herding, hunting, fishing and some mining (at Norilsk).

Tainan   Taiwan. 23 00N 120 11E. Town sit. in SW. Industries inc. iron and aluminium works and rice and sugar mills, food processing. Manu. inc. plastics, electrical appliances, chemicals, clothing, footwear, metal products, construction materials. Pop. (1985E) 633,607.

Taipa Islands   Macao. Two islands sit. SSE of Macao peninsula. Taipa port is sit. on SW coast of the larger island. The main occupation is fishing.

Taipei   Taiwan. 25 02N 121 30E. Cap. of Taiwan sit. in extreme N on Tanshi R. Industries inc. electronics, textiles, machinery and plastics. Pop. (1988E) 2·68m.

Taitung Taiwan. 22 45N 121 09E.
Port sit. E of Kaohiung on E coast.
Industries inc. agric. (rice, maize, tea,
pineapple, and Shi-chia), sugar milling, paper. The coastal sea abounds
in lobster and abalone. Pop. (1985E)
110,469.

Taiwan 24 00N 121 00E. East
Asia. Island off SE coast of China
across Formosa Strait. Area, 13,899
sq. m. (36,000 sq. km.). Together with
the Pescadores Island and Quemoy
Island it forms the Rep. of China
under the Nationalist Party, but considered to be a province of the People's
Republic of China by the Beijing
authorities. Chief cities Taipei (cap.),
Keelung, Kaohsiung, Hsinchu,
Chiayi, Taichung, Tainan. Mountains
rise steeply along E coast and slope
gradually to W alluvial plain which
supports most of the pop. Highest
points Mount Morrison (Yu Shan), *c.*
12,965 ft. (3,952 metres) a.s.l., Mount
Sylvia and many others over 10,000
ft. (3,048 metres) a.s.l. Tropical
climate with typhoons July–September. Drained by Choshui, Tanshui and
Lower Tanshui Rs. Thickly forested
(E). Agric. land covers 894,974 hectares, of which 483,514 hectares are
under paddy. Main crops are rice,
bananas, pineapple, sugar-cane, sweet
potatoes. Forests cover 1.87m. hectares. Chief products oak, cypress,
cedar, camphor and cork. The chief
minerals are coal, gold, copper, sulphur, oil and natural gas. Industries
are concentrated on W coast and inc.
iron and steel, aluminium, shipbuilding, sugar refining, paper making, textiles, cement, electrical machinery and
appliances, and fertilizers. Pop.
(1988E) 19·9m.

Taiyuan Shanxi, China. 37 55N
112 30E. Cap. of province in central E
sit. on Fen-ho R. An important centre
of heavy industry inc. coalmining,
iron and steel works and paper mills.
Manu. agric. and textile machinery,
cement and textiles. Pop. (1987E)
1,930,000.

Ta'iz Ta'iz, Republic of Yemen.13
38N 44 04E. Town sit. NW of Aden.
Centre of a coffee growing area. Industries inc. cotton weaving, tanning
and the manu. of jewellery.

Takamatsu Shikoku, Japan. 34 20N
134 03E. Port and cap. of Kagawa
prefecture, sit. WSW of Osaka. Manu.
pulp and paper, fans, parasols and cotton goods. Exports inc. rice and tobacco. Pop. (1988E) 328,000.

Takaoka Honshu, Japan. 36 45N
137 01E. Town sit. WNW of Tokyo.
Manu. inc. cotton goods and lacquer-
ware.

Takatsuki Honshu, Japan. 34 51N
135 37E. Town sit. NW of Tokyo.
Manu. machinery and textiles. Pop.
(1988E) 354,000.

Takeo Takeo, Cambodia. Town sit.
S of Pnom Penh. Agric. products inc.
rice, lac, sugar-cane, pepper and
hardwoods. Sericulture is practised.

Takla Makan Xinjiang Uygur,
China. 39 00N 83 00E. A sandy desert
sit. in Tarim Basin and forming part
of the Gobi Desert. Tarim R. lies
along its N border and Khotan R.
traverses it. Uninhabited owing to
shifting sand.

Takoradi *see* **Sekondi**

Taku Solomon Islands. An atoll
120 m. (192 km.) NE of Bougainville.
Area, 205 acres.

Talara Piura, Peru. 4 35S 81 25W.
Port in NW of Piura and sit. in NW
desert. An important centre for oil
with pipelines to the oilfields. Pop.
(1980E) 44,500.

Talca Talca, Chile. 35 26S 71 40W
Cap. of province in S central Chile. Industrial centre of an agric. district
producing wine. Manu. matches,
paper, footwear, tobacco products and
flour. Pop. (1987E) 164,482.

Talcahuano Concepción, Chile. 36
43S 73 07W. Port in S central Chile
sit. on Concepción Bay. A naval station with an excellent harbour, it
exports timber, hides and wool, and
imports iron ore for its steel industry.
Other industries are fishing and fish-
canning, flour milling and oil refining.
Pop. (1987E) 231,356.

Tallahassee Florida, U.S.A. 30 25N

84 16W. State cap. in NW of state. City area doubled after 1980. Industries inc. government, education, high technology, and the manu. of wood and food products. Pop. (1988E) 182,531.

Tallinn   Estonia, U.S.S.R. 59 25N 24 45E. Cap. of Estonia and port sit. on S coast of the Gulf of Finland. Icebreakers keep the port open during most of the winter, and timber, textiles and paper are exported. Industries inc. woodworking, engineering and the manu. of textiles and food products. Pop. (1987E) 478,000.

Tamana   Kiribati, W Pacific. Atoll in S Gilberts group. Area, 2 sq. m. (5 sq. km.). Pop. (1985C) 1,348.

Tamar River   Tasmania, Australia. 41 04S 146 47E. Formed by union of N. and S. Esk Rs. at Launceston, and flows NNW for 40 m. (64 km.) to enter Bass Strait.

Tamar River   England. 50 22N 4 10W. Rises near Morwenstowe in N Cornwall and flows 60 m. (96 km.) SE along county border between Cornwall and Devon, to enter the English Channel by Plymouth Sound. Navigable up to Launceston. The estuary is called the Hamoaze.

Tamatave   *see* **Toamasina**

Tamaulipas   Mexico. 24 00N 99 00W. Maritime state in NE sit. on Gulf of Mexico. Area, 30,650 sq. m. (79,384 sq. km.). In the interior there are several small mountain ranges, while the coastal plain has many lagoons. Many Rs. run W to E into the Gulf of Mex. Crops inc. cotton, maize, beans, sugar-cane and tobacco, and cattle are raised. Fishing and lumbering are important and oil is found in S. Cap. Ciudad Victoria, sit. NW of Tampico. Pop. (1989E) 2,294,680.

Tambacounda   Senegal. Cap. of region in E of country. Pop. (1979E) 29,054.

Tambov   Russian Soviet Federal Socialist Republic, U.S.S.R. 52 43N 41 25E. Cap. of Tambov region, sit. SE of Moscow on Tsna R. Industrial

centre of an agric. district producing grain, potatoes, sugar-beet and sunflowers. Industries inc. sugar refining, distilling and flour milling and the manu. of machinery, textiles, chemicals and synthetic rubber. Pop. (1987E) 305,000.

Tamil Nadu   India. 11 00N 77 00E. State of SE India formerly called Madras. Bounded E by Indian Ocean along the Coromandel Coast, N by Andhra Pradesh and Kharnataka, E by Kerala, S by Indian Ocean W of Sri Lanka. Area, 50,215 sq. m. (130,058 sq. km.). Chief cities Madras (cap.), Madurai, Coimbatore, Tiruchchirappalli, Salem. Coastal plains S and E are watered by Cauvery and Vaigai Rs. Mountainous N and W rising to the S apex of the Deccan plateau. Agric. land covers 7·1m. hectares, *c.* 3.5m. of which are irrigated. Main crops are rice, maize, jawar, bajra, chillies, bananas, pulses, millet, sugar-cane, oilseeds, cotton, tobacco, coffee, tea, rubber and pepper. Forests cover *c.* 8,494 sq. m. (22,000 sq. km.) producing mainly sandalwood. The main industry is cotton textiles, also manu. railway coaches, high pressure boiler plant, films, oil refining, teleprinters, fertilizers, heavy vehicles, textile machinery, tanning, engineering, manu. cement and sugar. The main mineral is lignite, mined at Neyveli, also manganese, mica, quartz, feldspar, salt, bauxite and gypsum. Pop. (1981C) 48,408,077.

Tampa   Florida, U.S.A. 27 57N 82 27W. City and port on W coast of state sit. on Tampa Bay. Pop. (1988E) 285,225.

Tampere   Häme, Finland. 61 30N 23 45E. Town in SW. The principal industrial trade centre of Finland with manu. of engineering, textiles, clothing, packaging, footwear and leather, and plastics. Pop. (1985E) 167,500.

Tampico   Tamaulipas, Mexico. 22 13N 97 51W. Port sit. on Pánuco R. in NE, and 7 m. (11 km.) inland from the Gulf of Mexico. Industries oil refining, fishing and fish-processing, sawmilling and boat-building. Petroleum and petroleum products are exported. Pop. (1980C) 267,957.

Tamworth New South Wales, Australia. 31 05S 150 55E. City in NE of state sit. on Peel R. Serving rich pastoral and agric. area with associated industries. Pop. (1984E) 34,000.

Tamworth Staffordshire, England. 52 39N 1 40W. Town sit. NE of Birmingham on R. Tame. Manu. cars and sanitary products, paper, clothing, bricks and tiles. Pop. (1989E) 65,000.

Tana, Lake Ethiopia. 12 00N 37 20E. Sit. S of Gondar, in NW, at 6,004 ft. (1,830 metres) a.s.l. it is the largest L. in Ethiopia. Area, *c.* 1,200 sq. m. (3,108 sq. km.). Little Abbai R. runs into it in SW and it is drained by Blue Nile R. in SE.

Tananarive *see* **Antananarivo**

Tana River Kenya. 2 32S 40 31E. Rises in Aberdare Mountains, and flows E and S for *c.* 500 m. (800 km.) to enter Indian Ocean at Kipini.

Tandil Buenos Aires, Argentina. 37 20S 59 05W. Resort in central E sit. in a dairy-farming area. There are granite quarries in the vicinity. Some light industry exists. Pop. (1980E) 70,000.

Tanga Tanzania. 5 04S 39 06E. Port in NE sit. on Indian Ocean. Railway terminus. Exports inc. coffee, sisal and copra. Pop. (1978E) 103,409.

Tanga Islands New Ireland, Bismarck Archipelago, South West Pacific. A group of volcanic, coral islands sit. NE of New Ireland. Area, 54 sq. m. (140 sq. km.). The four inhabited islands are Malendok, Boang, Lif and Tefa. The chief product is coconuts.

Tanganyika, Lake East Africa. 6 00S 29 30E. L. in Great Rift Valley, SW of L. Victoria on the Zaïre–Tanzania frontier, extending over 400 m. (640 km.) NNW to SSE and 15–50 m. (24–80 km.) wide. Fed by Ruzizi and Malagarasi Rs., drained irregularly by Lukuga R., which is often silted up, into Zaïre R. basin. The deepest L. in Africa with steep shores rising to 9,000 ft. (2,743 metres) a.s.l. in N. Chief ports Kigoma (Tanzania), Kalemie (Zaïre), Bujumbura (Burundi).

Tangier Morocco. 35 48N 5 45W. City sit. SW of Gibraltar on N coast. Sea port; commercial, industrial and tourist centre and major port at the entrance to the Mediterranean. Exports citrus fruit. Formerly the centre of the Intern. Zone of Tangier, and a free port. It was incorporated in Morocco in 1957 and its free status abolished, to be restored in 1962. Pop. (1982C) 266,342.

Tantâ Gharbiya, Egypt. 30 47N 31 00E. Town sit. NNW of Cairo between Rosetta and Damietta branches of Nile. Provincial cap. and rail centre. Industries inc. cotton ginning, cotton seed processing, wool spinning. Burial place of the 13th cent. Moslem Ahmed al-Badawi. Pop. (1986E) 373,500.

Tanzania East Africa. 6 40S 34 00E. Rep. consisting of Tanganyika, Zanzibar and Pemba, on E coast of Africa, bounded NE by Kenya, N by L. Victoria and Uganda, NW by Rwanda and Burundi, W by L. Tanganyika, SW by Zambia and Malawi, S by Mozambique. Area, 364,884 sq. m. (945,050 sq. km.). Chief cities Dar es Salaam, Dodoma (cap.), Tabora, Kigoma, Mwanza, Lindi, Tanga. The mainland is plateau bounded N and W by highland and E by a coastal plain. The mountains rise N to Kilimanjaro, 19,455 ft. (5,930 metres) a.s.l. and Meru. The central plateau is drained by Ugalla R. into L. Tanganyika, and by Great Ruaha and Rufiji Rs., into the Indian Ocean. The highlands are more fertile than the plateau, which is arid with bush-veld vegetation. The agric. uplands produce coffee, cotton, sisal, tobacco and oil seeds. Stock-farming is important. The chief minerals are gold, diamonds, tin and salt. Zanzibar and Pemba are both coral islands with very fertile soil. On Pemba the main crop is cloves; on Zanzibar, coconuts and some cloves, fruit and tobacco. Chief subsistence crops throughout the rep. are millet (Tanganyika), rice (Zanzibar), pulses, maize and sorghum. The chief industries are lumbering, fishing, processing food products. Pop. (1987E) 23·2m.

Taormina Sicilia, Italy. 37 52N 15

17E. Resort sit. SSW of Messina on E coast.

Taoyuan   Taiwan. Town sit. WSW of Taipei. Agric. products inc. rice, tea, sweet potato, vegetables, and livestock. Industries inc. electronics, machinery, vehicles, metals, plastics and chemicals. Pop. (1985E) 202,377.

Tapajoz River   Brazil. 2 24S 54 41W. Formed by union of Arinos and Juruena Rs. E of the Serra do Norte in NW Mato Grosso, flows *c.* 900 m. (1,440 km.) N through the Capoeiras and Chacorão Falls, then NE past Bacabal, Aveiro and Belterra to enter Amazon R. at Santarém. Main tribs. Jamanxim and Teles Pires (or São Manuel) Rs.

Tapti River   India. 21 05N 72 40E. Rises near Betul in S Madhya Pradesh and flows 440 m. (704 km.) generally W into Maharashtra and Gujarat, to enter the Gulf of Cambay below Surat.

Taranaki   North Island, New Zealand. 39 05S 174 51E. Province in SW devoted to agric. and pastoral farming, particularly dairy. There are petrochemical plants including synthetic fuels at Motunui resulting from on-and offshore oil and gas fields. Area, 3,760 sq. m. (9,739 sq. km.). Cap. and port New Plymouth. Pop. (1981E) 103,747.

Taranto   Puglia, Italy. 40 28N 17 15E. Cap. of Ionio province and port sit. on Gulf of Taranto. A naval base with shipyards. Industries oil refining, oyster and mussel fishing and the manu. of furniture and glass. Pop. (1988C) 244,694.

Tarapacá   Chile. Region in N bordering on Peru and Bolivia. Area, 22,422 sq. m. (58,073 sq. km.). The Andes in E give way to the Atacama Desert in W coastal plain. Nitrates are found in the Atacama and other mineral deposits inc. iodine, copper, sulphur, borax and salt. There are guano deposits on the coast. The chief Rs. are Lluta and Azapa and limited agric. is possible in their valleys. Crops inc. cotton, corn and fruit. Cap. Iquique. Pop. (1982C) 273,427.

Tarascon   Bouches-du-Rhône,

France. 43 48N 4 40E. Town sit. SW of Avignon on Rhône R. opposite Beaucaire. Manu. furniture and textiles. Trades in olive oil, fruit and vegetables. Pop. (1982C) 11,024.

Tarawa   Kiribati, W Pacific. 1 30N 173 00E. Atoll of the N Gilberts group inc. cap. on Bairiki islet. Area, 12 sq. m. (31 sq. km.). Exports inc. mother-of-pearl and copra. Pop. (1985C) 25,598.

Tarbes   Hautes-Pyrénées, France. 43 14N 0 05E. Cap. of dept. in SW sit. on Adour R. Noted for horses, and has a school of artillery. Manu. armaments, furniture, footwear and harness. Pop. (1982C) 54,055.

Taree   New South Wales, Australia. 31 54S 152 28E. Town sit. NE of Newcastle on Manning R. Industries inc. dairy products, timber, tourism. Pop. (1989E) 17,000.

Târgu-Mureş   Mureş, Romania. 46 33N 24 33E. Town sit. ESE of Cluj on Mureş R. District cap. and trading centre for timber, sugar and oil. Industries inc. engineering. Pop. (1985E) 157,411.

Tarifa   Cádiz, Spain. Port sit. on Strait of Gibraltar. Industries fishing and fish canning esp. anchovies. Trades in cereals and oranges.

Tarija   Tarija, Bolivia. 21 31S 64 45W. Cap. of dept. in SE sit. in basin of Guadalquivir R. A market town trading in agric. produce. Pop. (1982E) 54,001.

Tarim Basin   Xinjiang Uygur, China. 39 40N 83 55W. Depression in centre of the province surrounded by the Tien Shan mountains to NW and N, the Qurug Tagh and Altyn Tagh to E, the Kunlun Shan to S. Area, 700 m. (1,120 km.) E to W, 200–300 m. (320–480 km.) N to S. Mainly a sandy waste which receives enough water from surrounding hills to feed Tarim R. and a number of oases, but the streams drain into the sand towards the E.

Tarn   France. 43 50N 2 08E. Dept. in Midi-Pyrénées region, sit. with the Massif Central on E and the Aquitaine

Basin on w. Area, 2,220 sq. m. (5,751 sq. km.). Hilly in E and SE rising to 4,134 ft. (1,260 metres) a.s.l. in the Monts de Lacaune. The principal Rs. are Tarn and Agout. Agric. products inc. cereals, vines and vegetables. Cattle and sheep are raised. Coal, zinc and iron are mined. Cap. Albi. Pop. (1982C) 339,345.

Tarn-et-Garonne   France. 44 08N 1 20E. Dept. in Midi-Pyrénées region. Area, 1,435 sq. m. (3,716 sq. km.). Mainly an alluvial plain formed by Rs. Garonne, Tarn and Aveyron. Principally agric. producing cereals, wine, fruit and truffles. Cap. Montauban. Pop. (1982C) 190,485.

Tarnów   Kraków, Poland. 50 01N 21 00E. Town sit. E of Kraków on Biala R. Manu. agric. machinery and glass.

Tarn River   France. 44 05N 1 06E. Rises on Mont Lozère in the Cévennes and flows W for 234 m. (374 km.) to join Garonne R. below Moissac.

Tarquinia   Lazio, Italy. 42 15N 11 45E. Town sit. NW of Rome near the mouth of Marta R. Manu. cement, pottery, paper and matches. Etruscan ruins are sit. to SE.

Tarragona   Spain. 41 07N 1 15E. (i) Province in NE bordering on the Mediterranean Sea. Area, 2,426 sq. m. (6,283 sq. km.). Mainly mountainous with a coastal plain containing Ebro R. delta. Agric. products inc. wine, olive oil, fruit, cereals, almonds and hemp. Silver, copper, lead and marble are found. Pop. (1986C) 531,281. (ii) Cap. of province of same name, and port, sit. on NE coast. Manu. pharmaceutical goods, electrical equipment, wine and Chartreuse liqueur. Pop. (1986C) 109,557.

Tarrasa   Barcelona, Spain. 41 34N 2 01E. Town sit. NNW of Barcelona. Manu. textile machinery, textiles, fertilizers and dyes. Trades in agric produce, wine and oil. Pop. (1986C) 159,530.

Tarsus   Içel, Turkey. 36 55N 34 53E. Town sit. W of Adana. Trades in cereals and fruit.

Tartous   Lattakia, Syria. Town sit. S of Lattakia on the Mediterranean. The main occupations are fishing and agric. Crops inc. cotton, cereals, olives and fruit.

Tartu   Estonia, U.S.S.R. 58 23N 26 43E. Town sit. to W of L. Chudskoye and on Ema R. Manu. agric. machinery, metal and food products, textiles, cigars and cigarettes. Pop. (1980E) 106,000.

Tashkent   Uzbek Soviet Socialist Republic, U.S.S.R. 41 20N 69 18E. City in E Uzbekistan on Kazakhstan border, in an oasis on Chirchik R. Regional cap. and trading centre on important routes. Centre of an irrigated area producing fruit, cotton, rice and tobacco. Industries inc. manu. textiles, flour, oils, agric. and textile machinery, fertilizers. Pop. (1989E) 2,073,000.

Tasmania   Australia. 43 00S 147 00E. State of SE Australia consisting of an island off the S coast of Victoria. Area, 26,383 sq. m. (68,332 sq. km.). Chief cities, Hobart (cap.), Launceston. The State of Tasmania inc. the Furneaux Islands (NE), King Island and Hunter Island (NW) and Bruny Island (SE). Rugged highland rising to Mount Ossa (W), 5,305 ft. (1,617 metres) a.s.l., and Legges Tor (E), 5,160 ft. (1,573 metres) a.s.l., and divided by ravines, precip. and mountain Ls. The N coast has undulating hill country and there are some fertile valleys and coastal strips. Agric. land occupies 1·87m. hectares, of which only 916,800 hectares are under crops. Sheep and dairy farming are the main occupations, with some arable farms producing barley, apples, hops, oats, hay, wheat and peas. Forests are extensive, producing timber for building and pulp. The chief minerals are iron, coal, zinc, copper, lead, gold and silver. The main industries are mining and refining metals, lumbering, sawmilling and paper making, food processing; most industries are powered by hydroelectricity. Pop. (1986C) 436,353.

Tasman Sea   Australia. 42 30S 168 00E. A branch of the Pacific Ocean sit. between Australia and Tasmania in W and New Zealand in E.

Tatabánya Komárom, Hungary. 47 34N 18 26E. Cap. of county sit. w of Budapest. Centre of a lignite-mining district. Manu. bricks. Pop. (1984E) 77,000.

Tatar Autonomous Soviet Socialist Republic Russian Soviet Federal Socialist Republic, U.S.S.R. 55 30N 51 30E. Area, 26,250 sq. m. (67,987 sq. km.). Constituted as an Autonomous Rep. 1920. Cap. Kazan, 450 m. (720 km.) E of Moscow. Chief industries are engineering, oil and chemicals, while timber, building materials, textiles, clothing and food are also expanding. 3.7m. hectares are under crops. Pop. (1989E) 3,640,000.

Tatra Mountains Poland/Czechoslovakia. 49 14N 20 00E. The High Tatra are part of the Carpathians, extending in an arc w to E along the Czech./Polish frontier NE of Ružomberok, rising to Gerlachovka, 8,737 ft. (2,663 metres) a.s.l. The Low Tatra lie s of them, across upper Vah R. and rise to Djumbir, 6,709 ft. (2,045 metres) a.s.l.

Ta'u Island American Samoa. Sit. E of Tutuila. Area, 17 sq. m. (44 sq. km.). Cone-shaped with steep forested slopes rising to 3,056 ft. (931 metres) a.s.l.

Taunton Somerset, England. 51 01N 3 06W. Town in SW sit. on R. Tone. County town and admin. and shopping centre. Manu. bedding, aerosols, computer software, precast concrete, textile goods, aeronautical and optical instruments, hydrographic charts, electronic and engineering products and cider. Pop. (1987E) 53,285.

Taunton Massachusetts, U.S.A. 41 54N 71 06W. City sit. s of Boston on Taunton R. Manu. machinery, textiles, plastics, pottery and jewellery. Pop. (1980C) 45,001.

Taunus Federal Republic of Germany. 50 10N 8 15E. Hills extending E from Rhine R. between Main and Lahn Rs., rising to Great Feldberg, 2,886 ft. (880 metres) a.s.l. Wellwooded, with vineyards and orchards on the s slopes. Tourist area with spas inc. Wiesbaden, Nauheim, Homburg.

Taupo, Lake North Island, New Zealand. 38 48S 175 55E. L. sit. in centre in a volcanic area of hot springs. Area, 234 sq. m. (606 sq. km.). It is the source of the Waikato R.

Tauranga North Island, New Zealand. 37 42S 176 10E. Port sit. SE of Auckland on the Bay of Plenty. Centre of an agric. area noted for subtropical and kiwi fruits and exporting dairy produce, meat and timber. Pop. (1989E) 62,000.

Taurus Mountains Turkey. 37 00N 33 00E. Mountain system forming s and SE edge of the Anatolian plateau and extending parallel to the Mediterranean coast W to E as far as Seyhan R. basin. Divided into the Lycian Taurus, from w coast to the Gulf of Antalya, and the Cilician Taurus from the Gulf to the Seyhan basin.

Taveuni Fiji, South West Pacific. 16 51S 179 58E. A volcanic island sit. opposite Vanua Levu across the Somosomo Strait. Area, 168 sq. m. (435 sq. km.). Mountainous, rising to 4,072 ft. (1,241 metres) a.s.l. The main products are copra, cattle and coffee. The chief town is Waiyevo.

Tavistock Devon, England. 50 33N 4 08W. Town N of Plymouth sit. on the R. Tavy. An agric. centre with manu. of agric. implements and chemicals. Pop. (1981C) 9,271.

Tavoy Tenasserim, Burma. 14 07N 98 18E. Port s of Moulmein sit. on Tavoy R. estuary, in a rice-producing area with tin, wolfram and iron mines.

Tavoy Island Mergui Archipelago, Burma. Sit. in the Andaman Sea s of Moulmein. Mountainous, rising to 2,254 ft. (687 metres) a.s.l.

Taw River England. 50 58N 3 58W. Rises near Yes Tor on Dartmoor, Devon, and flows 50 m. (80 km.) NE then NW to enter the Bristol Channel by an estuary below Barnstaple.

Tayport Fife Region, Scotland. 56 27N 2 52W. Town sit. SE of Dundee across Firth of Tay, on s bank of mouth. Small port and resort. Industries inc. textiles and metal manu. Pop. (1983E) 3,190.

Tay River   Scotland. 56 35N 3 50W. Rises in headstreams on Ben Lui on the Argyll/Perth border and flows 118 m. (189 km.) NE through L. Tay, past Aberfeldy to Ballinluig, where it turns SE to the Strathmore valley then SW to Perth and SE again to form Firth of Tay estuary into the North Sea. Main tribs. Tummel and Earn Rs. Noted for salmon.

Tayside   Scotland. Region of E Scot. bounded on W by North Sea, N by Grampian and Highland regions, E by Strathclyde region and S by Central and Fife regions. Area, 3,000 sq. m. (7,770 sq. km.) comprising the districts of Angus, Dundee and Perth and Kinross. Main industries inc. service, textiles and electronic, medical and computing technologies. Agriculture (soft fruit, grains, vegetables, potatoes and beef) and sea fishing are important. Pop. (1990E) 394,000.

Taza   Fez, Morocco. 34 16N 4 01W. Town ENE of Fez sit. on a pass between the Rif Mountains in N and the Middle Atlas in S. 1,800 ft. (549 metres) a.s.l. Manu. inc. carpets, footwear and building materials. Pop. (1982E) 77,216.

Tbilisi   Georgia, U.S.S.R. 41 43N 44 49E. Cap. of Georgian Soviet Socialist Rep. sit. on Kura R. A commercial and transportation centre with engineering and wood-working industries. Also manu. textile machinery, machine tools, electrical equipment, textiles and food products. Many anc. buildings still exist and the hot sulphur springs in the vicinity make it a spa. Pop. (1987E) 1,194,000.

Tczew   Gdańsk, Poland. 54 06N 18 47E. Town SSE of Gdańsk sit. on Vistula R. Manu. agric. machinery, bricks and beer. Industries inc. railway engineering and sugar refining. Pop. (1983E) 56,700.

Te-au-o-tu   Cook Islands, New Zealand. An atoll sit. NE of Rarotonga which comprises two islets. The one to E is Te-au-o-tu, W one is Manuae. Copra is exported.

Tebessa   Algeria. 35 28N 0 09E. Town in NE sit. in the E Saharan Atlas

Mountains. Manu. carpets. Phosphates are mined in the area.

Teddington   Greater London, England. 51 25N 0 20W. District of Twickenham on Thames R. at limit of tidal flow upstream. Mainly residential.

Tees River   England. 54 34N 1 16W. Rises on Cross Fell, Cumbria, and flows 80 m. (128 km.) SE down the Cumbria/Durham border and across Durham to the N Yorkshire border which it forms (E) as far as Yarm, turning NE past Thornaby and Stockton to enter the North Sea by an estuary below Middlesbrough. Its lower course lies through a heavily industrialized area.

Teesside   Cleveland, England. 54 37N 1 13W. Area surrounding estuary of Tees R. Formed in 1967 and became part of Cleveland in 1974. Industries inc. chemicals, iron and steel, engineering, shipbuilding. Chief towns, Stockton, Thornaby, Middlesbrough, Billingham.

Tegucigalpa   Francisco Morazan, Honduras. 14 05N 87 14W. Cap. of Rep. since 1880 and dept. sit. on Choluteca R. in SW. Manu. textiles, clothing, food products, soap and building materials. Gold and silver have been mined since the 16th cent. Pop. (1986E) 604,600.

Tehrán   Iran. 35 40N 51 26E. Province of N Iran bounded N by the Alburz Mountains. Area, 7,381 sq. m. (19,118 sq. km.). Fertile in N, watered from the slopes of the Alburz, producing cereals, fruit, vines and livestock. Pop. (1986C) 8,712,087.

Tehrán   Iran. 35 44N 51 30E. City at foot of the Alburz range. Cap. and provincial cap., communications centre. Rebuilt by Riza Shah from 1925. Industries inc. chemicals, textiles, carpet making, tanning, glass making. The Pahlevi palace is on the outskirts. Pop. (1986C) 6,042,584.

Tehuantepec, Isthmus of   Mexico. 17 00N 94 30W. Sit. in S between the Gulf of Campeche, an inlet of the Gulf of Mexico, in N and the Gulf of

Tehuantepec on the Pacific Ocean in s. There are important oilfields near the port of Coatzacoalcos in N.

Teignmouth   Devon, England. 50 33N 3 30W. Resort sit. at mouth of R. Teign, s of Exeter. Pop. (1985E) 13,500.

Teign River   England. 50 41N 3 42W. Rises near Chagford on Dartmoor and flows SE for 30 m. (48 km.) to enter English Channel at Teignmouth.

Tekirdağ   Tekirdağ, Turkey. Port sit. w of Istanbul. The commercial centre of an area producing grain, flax and linseed.

Tel Aviv-Jaffa   Israel. 32 03N 34 46E. Town in central w sit. on the Mediterranean coast. An important administrative, industrial and cultural centre of Israel with very varied industries. Pop. (1988E) 320,000.

Telford   Shropshire, England. 52 35N 2 35W. New town designed to house the overspill pop. of the West Midlands, sit. NW of Birmingham. Originally designated as Dawley New Town in 1963, expanded in area and renamed Telford in 1968. Covers an area of 19,240 acres and inc. the estab. communities of Dawley, Madeley, Ironbridge, Wellington and Oakengates. Industries inc. metal and vehicle manu., electronics, plastics and miscellaneous light indust. Pop. (1984E) 107,700.

Tema   Ghana. 5 38N 0 01E. Port sit. ENE of Accra. Industries inc. oil refining and fishing. Pop. (1982E) 324,000.

Témbi Gorge   Thessaly, Greece. Otherwise spelt Tempe; deep valley of Piniós R. flowing E to Aegean Sea between Mount Olympus (N) and Mount Ossa (S).

Tempe   Arizona, U.S.A. 33 25N 111 56W. City E of Phoenix sit. on Salt R. A commercial centre. Pop. (1980C) 106,743.

Temple   Texas, U.S.A. 31 06N 97 21W. City sit. N of Austin. Centre of an agric. area with manu. of furniture,

metal products, machinery, plastics, clothing and electronics. Pop. (1980C) 42,483.

Temuco   Cautín, Chile. 38 44S 72 36W. Cap. of province in s central Chile sit. on Cautín R. Centre of an agric. region producing timber, cereals and apples. Pop. (1987E) 217,789.

Tenasserim   Lower Burma, Burma. 12 06N 99 03E. Division in s sit. E of the Bay of Bengal and extending from Moulmein, the cap., in N to Victoria Point in s. Area, 31,588 sq. m. (81,813 sq. km.). The chief products are rice, tin, tungsten and teak. Pop. (1983C) 917,628.

Tenasserim River   Burma. Rises in the Tenasserim range in s Burma, on the narrow coastal strip at the head of the Malay Arch. Flows 250 m. (400 km.) s to Tenasserim, then NW to enter the Andaman Sea by a long estuary.

Tenby   Dyfed, Wales. 51 41N 4 43W. Resort E of Pembroke sit. on a promontory in Carmarthen Bay. Pop. (1973E) 4,930.

Tenerife, Santa Cruz de   Canary Islands. 28 19N 16 34W. The largest island sit. between Gomera in w and Gran Canaria in E. Area, 1,239 sq. m. (3,208 sq. km.). The volcanic Pico de Teide rises to 12,172 ft. (3,710 metres) a.s.l. Very fertile, producing fruit and vegetables, esp. tomatoes. Cap. Santa Cruz. Pop. (1986C) 759,388.

Tennessee   U.S.A. 36 00N 86 30W. State of SE, bounded E by North Carolina, s by Georgia, Alabama and Mississippi, w by Arkansas and Missouri, N by Kentucky, NE by Virginia. Area, 42,244 sq. m. (109,412 sq. km.). Chief cities Nashville (cap.), Memphis, Knoxville, Chattanooga. Mountainous (E), levelling to the Cumberland plateau and a central basin with surrounding uplands; in w is plateau sloping gradually to Mississippi R., the w boundary. Drained N into Ohio R. by Tennessee and Cumberland Rs. Farmland covers *c.* 12.6m. acres, mainly producing beef cattle, dairy cattle, cotton, tobacco, cereals. Forests cover *c.* 50% of the state. Chief minerals are coal, zinc, ball clay, pyrites, phosphate rock, marble, copper. The most important industries are chemicals, textiles (synthetic fabrics), electrical goods, iron and steel, manu. food products, sawmilling. Pop. (1988E) 4,895,000.

Tennessee River U.S.A. 37 04N 88 33W. Rises in headstreams from the Appalachians which join E of Knoxville, Tennessee, and flows 652 m. (1,043 km.) in a deep southward bend through Tennessee, past Chattanooga into N Alabama, past Decatur and Sheffield and back to the Tennessee border where it joins NW Alabama and NE Mississippi. Then N through Kentucky L. into W Kentucky and on to enter Ohio R. above Paducah on the Kentucky/Illinois border.

Tepelenë Albania. 40 18N 13 50E. Town sit. ESE of Valona on Vijosë R. Coal and iron are found in the vicinity.

Tepic Nayarit, Mexico. 21 30N 104 54W. State cap, NW of Guadalajara sit. beneath the extinct volcano Sangangüey, 7,716 ft. (2,352 metres) a.s.l. Centre of an agric. area with cotton spinning, sugar refining and rice milling. Pop. (1980C) 177,007.

Teplice Czechoslovakia. 50 39N 13 48E. Town and spa in NW sit. E of Usti-nad-Labem. Manu. paper, glass, pottery and textiles. Pop. (1984E) 54,000.

Teraina *see* **Washington Island**

Teramo Abruzzi e Molise, Italy. 42 39N 13 42E. Cap. of Teramo province in S central Italy sit. on Tordino R. Manu. inc. textiles and food products. Pop. (1981E) 51,000.

Terceira Azores, Atlantic Ocean. 38 43N 24 13W. A mountainous volcanic island sit. ENE of São Jorge. Area, 153 sq. m. (396 sq. km.). Mainly agric. producing cattle, cereals, fruit and wine. The chief town and port is Angra do Heroísmo.

Terek River U.S.S.R. 43 44N 46 33E. Rises in glaciers near Mount

Kazbek, Georgia, and flows 382 m. (611 km.) through the Daryal Pass to Nalchik, then E to enter the Caspian Sea by a delta 70 m. (112 km.) wide. Used for irrigation, esp. for rice on its lower course.

Terengganu Peninsular Malaysia, Malaysia. State sit. on E coast. Area, 5,002 sq. m. (12,955 sq. km.). A narrow coastal plain rises to densely forested mountains over 5,000 ft. (1,524 metres) a.s.l. Produces rubber, copra and rice. Cap. Kuala Terengganu, sit. NE of Kuala Lumpur. Pop. (1980C) 542,280.

Teresina Piauí, Brazil. 5 05S 42 49W. State cap. in NE sit. on Parnaíba R. Centre of an agric. area producing rice, cotton and cattle. Manu. textiles and soap. Pop. (1980C) 339,264.

Terni Umbria, Italy. 42 34N 12 37E. Cap. of Terni province in central Italy sit. on Nera R. An industrial centre with important steelworks and iron foundries. Manu. inc. locomotives, armaments, chemicals and textiles. Pop. (1981C) 111,564.

Ternopol Ukraine, U.S.S.R. 49 34N 25 36E. Cap. of the Ternopol region sit. ESE of Lvov. Manu. agric. machinery, cement and food products, esp. sugar refining.

Terracina Lazio, Italy. 41 17N 13 15E. Port and resort SE of Rome sit. on the Gulf of Gaeta.

Terre Adélie *see* **Adélie Land**

Terre Haute Indiana, U.S.A. 39 28N 87 24W. City sit. WSW of Indianapolis on Wabash R. Manu. digital audio and video discs, electronic components, plastics and chemicals. Pop. (1988E) 56,330.

Tete Mozambique. 16 13S 33 33E. Province in NW, area, 38,886 sq. m. (100,714 sq. km.). Cap. Tete. Pop. (1982E) 864,240.

Teton Range Wyoming, U.S.A. 43 50N 110 55W. Range of the Rocky Mountains sit. to S of Yellowstone National Park. Forms part of the Grand Teton National Park, the highest peak being Grand Teton, 13,766 ft. (4,196 metres) a.s.l.

Tetuán Tetuán, Morocco. 35 34N 5 23W. Cap. of province sit. SE of Tangier near the Mediterranean coast. Manu. textiles, tiles, leather goods and soap. Pop. (1982C) 199,615.

Tewkesbury Gloucestershire, England. 51 59N 2 09W. Town NE of Gloucester sit. at confluence of Rs. Severn and Warwickshire Avon. There is some light industry and a retail market. Tourism is important. Pop. (1981C) 9,554.

Texarkana Texas/Arkansas, U.S.A. 33 26N 94 03W. Town sit. astride the border near Red R. Manu. textiles, wood and food products, cotton-seed oil and fertilizers. Pop. (1980C) 52,730 (Texas, 31,271; Arkansas, 21,459).

Texas U.S.A. 31 40N 93 30W. State of S U.S.A. bounded SE by Gulf of Mexico, E by Louisiana, NE by Arkansas, N by Oklahoma, NW by New Mexico, SW and S by Rio Grande forming the Mexican frontier. Area, 267,339 sq. m. (692,408 sq. km.). Chief cities Austin (cap.), Houston, Dallas, San Antonio, Fort Worth, El Paso. The State declared itself independent of Mexico in 1836; it was the independent Republic of Texas until joining the Union in 1845. The coast is lowland with numerous islands, sand bars and lagoons; inland is prairie which rises to 2,500 ft. (762 metres) a.s.l. in the SW, and beyond that plateau rising (W) to 8,751 ft. (2,667 metres) a.s.l. at S end of the Rocky Mountains. The climate is temperate on the coast, continental on the prairies and cool in the mountains. Drained by Pecos R. into the Rio Grande, Nueces, San Antonio, Colorado, Brazos, Trinity and Sabine Rs. into the Gulf. The N is drained into Mississippi R. by Can. and Red Rs. Mainly a farming state; farms cover *c.* 136m. acres, with soil erosion a serious problem. Chief products are beef cattle, sheep, pigs, cotton, cereals, peanuts and fruit. The most important mineral is oil, also natural gas, lime, salt, sulphur, mercury, helium. The main industries are oil,

petrochemicals, food-processing and distribution, engineering and electronics. Pop. (1988E) 16,841,000.

Texas City Texas, U.S.A. 29 23N 94 54W. City and port sit. NNW of Galveston. Industries inc. oil refining, tin smelting and chemicals. Pop. (1980C) 41,403.

Thailand East Asia. 16 00N 102 00E. Kingdom on mainland of SE Asia extending S along the Malay Peninsula to the N frontier of Peninsular Malaysia. Bounded W by Burma and Indian Ocean, E by Gulf of Thailand, Cambodia, E and N by Laos. Area, 198,250 sq. m. (513,468 sq. km.). Chief cities Bangkok (cap.), Thonburi, Chiengmai. Mountainous and forested (N) with parallel ridges extending N to S and Rs. flowing S to form the Menam Chao Phraya R., which waters the extensive alluvial plain of central Thailand. The E is scrubby plateau, the S is hilly, hot and humid, extending along the peninsula with fertile valleys and hills rising to 5,000 ft. (1,524 metres) a.s.l. The chief agric. crop is rice, grown mainly on the central plain over *c.* 18m. acres. Other crops inc. maize, sugar-cane, coconuts. About 28% of the land area is forest, yielding valuable hardwoods, esp. teak, yang. Mineral deposits are varied, the most important being tin and wolfram. Rubber is grown on the peninsula. Industries are not highly developed, but there is some textile manu. esp. silk, sugar refining, tobacco processing and paper milling. Pop. (1988E) 54,465,056.

Thame Oxfordshire, England. 51 44N 0 58W. Town sit. SW of Aylesbury on R. Thame. There is a variety of light industry. Pop. (1989E) 10,441.

Thame River England. 51 35N 1 08W. Rises in the Chiltern Hills and flows 30 m. (48 km.) W to enter R. Thames above Wallingford.

Thames Ditton Surrey, England. 51 24N 0 20W. Town sit. SW of London on R. Thames. Mainly residential. Manu. vehicles. Pop. (1981C) 8,731.

Thames River England. 51 28N 0 43E. Rises in the Cotswold Hills in SE

Gloucestershire and flows *c*. 220 m. (352 km.) first ENE to Oxford, where it is called the Isis R., then SE past Reading and Windsor to London and on E to enter the North Sea by a long estuary between Essex (N) and Kent (S). Navigable by small boats, mainly for pleasure, and as far as London for ocean-going ships. Main tribs. Cherwell, Thame, Colne, Lea, Roding, Kennet and Medway Rs.

Thane Maharashtra, India. Town sit. NNE of Bombay. Manu. textiles, engineering goods and matches. Trades in sugar-cane and rice. Pop. (1981C) 309,897.

Thanet, Isle of England. 51 22N 1 20E. Area of NE Kent separated from the rest by Stour R. dividing into 2 channels, and bounded N and E by the North Sea. Area, 42 sq. m. (109 sq. km.). Chief towns Westgate, Margate, Ramsgate, Broadstairs, all coastal resorts.

Thanjavur Tamil Nadu, India. 10 45N 79 15E. Town sit. SSW of Madras. Manu. textiles, copperware and jewellery. Pop. (1981C) 184,015.

Thar Desert India/Pakistan. 28 00N 72 00E. A sandy desert sit. in NW Rajasthan and the adjoining part of Pakistan. Chief town Bikaner, sit. NW of Jaipur.

Thasos Kavala, Greece. 40 41N 24 47E. An island sit. in the N Aegean Sea WSW of Alexandroúpolis. Area, 154 sq. m. (399 sq. km.). Products inc. olive oil and wine.

The Hague *see* **Hague, The**

Thesprotia Greece. A *nome* in N bounded by Albania and the Strait of Corfu. Area, 588 sq. m. (1,515 sq. km.). The principal Rs. are Acheron and Thyamis. Mainly agric. producing barley, corn, olive oil and almonds Lumbering. Cap. Egoumenitsa. Pop. (1981C) 41,278.

Thessaloníki Macedonia, Greece. 40 38N 22 56E. (i) A prefecture sit. to N of the Khalkidiki peninsula. Area, 1,375 sq. m. (3,560 sq. km.). Products inc. cereals and vines. Pop. (1972C)

871,580. (ii) Cap. of prefecture of same name and port sit. at head of Gulf of Thessaloníki, NNW of Athens. An industrial and transportation centre and the second town of Greece, it also has important exports of metallic ores, hides and tobacco. Manu. inc. textiles, chemicals, metal products and cigarettes. Pop. (1981C) 406,413.

Thessaly Greece. 39 30N 22 00E. Region of E Greece comprising the 4 *nomoi* of Karditsa, Larissa, Magnessia and Trikkala. Area, 14,000 sq. m. (36,260 sq. km.). Chief towns Lárisa, Vólos. Mainly agric. area producing cotton, tobacco, wheat, maize. Pop. (1981E) 511,401.

Thetford Norfolk, England. 52 25N 0 45E. Town in East Anglia sit. at confluence of Rs. Thet and Little Ouse. There are 5 industrial estates occupied by over 100 companies engaged in light industry. Pop. (1987E) 19,883.

Thetford Mines Quebec, Canada. 46 05N 71 18W. Town sit. S of Quebec. The largest asbestos deposits in the world are in the vicinity giving rise to an important asbestos industry.

Thiès Senegal. 14 48N 16 56W. Town sit. ENE of Dakar. Manu. fertilizers from local phosphates. Trades in groundnuts, fruit and rice. Pop. (1985E) 156,200.

Thimphu Bhután. 27 31N 89 45E. Cap. of Bhután, sit. in the Himalayas. Since 1962 the official seat of the royal govt. Pop. (1987E) 15,000.

Thionville Moselle, France. 49 22N 6 10E. Town in NE sit. on Moselle R. Manu. iron and steel, chemicals and cement. Pop. (1982C) 41,448 (agglomeration, 138,034).

Thirlmere Cumbria, England. A lake sit. SSE of Keswick. It is used as a reservoir for Manchester.

Thirsk North Yorkshire, England. 54 15N 1 20W. Town sit. NNW of York. It has a noted cattle market and racecourse. Pop. (1980E) 3,000.

Thompson Manitoba, Canada. 55

45N 97 45W. City sit. N of Winnipeg, a northern regional service centre for other communities in N Manitoba and the Central Arctic of Canada. Indust. inc. mining, milling, smelting and refining for nickel, cobalt, copper and some precious metals. Pop. (1984E) 14,500.

Thonburi   Thailand. 13 50N 100 36W. Town sit. on Chao Phraya R. opposite Bangkok to which it is joined by 3 bridges. The second town of Thailand with industries inc. rice milling and sawmilling.

Thornaby-on-Tees   Cleveland, England. 54 34N 1 18W. Town WSW of Middlesbrough sit. on R. Tees. Industries inc. heavy engineering, iron founding, bridge building and the manu. of wire rope. Pop. (1981E) 26,000.

Thornton Cleveleys   Lancashire, England. Resort comprising the towns of Thornton and Cleveleys sit. N of Blackpool on the Fylde coast. Pop. (1985E) 26,700.

Thousand Islands   Canada/U.S.A. 44 22N 75 55W. A group of about 1,700 small islands sit. in St. Lawrence R. near E end of L. Ontario.

Thrace   Greece. 41 09N 25 30E. Region of NE Greece comprising the 3 *nomoi* of Evros, Xanthi and Rodopi. Area, 3,520 sq. m. (8,800 sq. km.), bounded N by Bulgaria and E by Turkey. Chief towns Alexandroúpolis, Kamotini, Xanthi. Mainly agric. area, producing cotton, tobacco.

Three Kings Islands   New Zealand. 34 09S 172 09E. An uninhabited group of 3 islands sit. WNW of North Island.

Thun   Bern, Switzerland. 46 45N 7 37E. Town sit. on Aar R. 1 m. (1.5 km.) from its exit from L. Thun, SSE of Bern. Manu. metal goods, watches, slates, bricks and clothing. Pop. (1989E) 38,000.

Thun, Lake   Bern, Switzerland. A L. in central Switz. 11 m. (18 km.) long, traversed by Aar R:, sit. at 1,837 ft. (560 metres) a.s.l.

Thunder Bay   Ontario, Canada. 48 25N 89 14W. Industrial city and lake port 862 m. (1,387 km.) from Toronto, 435 m. (700 km.) from Winnipeg sit. on the NW shore of L. Superior, the largest fresh water lake in the world. Tourist area. Industries inc. grain export, pulp, paper, and wood products, major bulk handling facilities for coal, iron ore, potash, etc., manu. of light and heavy rail products. Pop. (1984E) 115,000.

Thurgau   Switzerland. 47 34N 9 10E. Canton in NE bordering L. Constance and Rhine R. Area, 388 sq. m. (1,006 sq. km.). A rich agric. area producing cereals and vines. Industries inc. textiles. Cap. Frauenfeld, sit. NE of Zürich. Pop. (1980C) 183,795.

Thuringian Forest   German Democratic Republic. A wooded range sit. between Rs. Werra and Thuringian Saale. The highest point is Beerberg, 3,222 ft. (982 metres) a.s.l.

Thurles   Tipperary, Ireland. 52 41N 7 49W. Town in central S NNW of Clonmel sit. on Suir R. Industries inc. a sugar-beet refinery, food processing and mineral water factories, and packaging and wool spinning Pop. (1981E) 7,600.

Thursday Island   Queensland, Australia. 10 35S 142 13E. A small island NW of Cape York sit. in Torres Strait. The main industry is pearl fishing. Chief town Port Kennedy.

Thurso   Caithness, Highland Region, Scotland. 58 35N 3 32W. Town and port in NE sit. at mouth of Thurso R. on Thurso Bay. W is Dounreay, an experimental atomic power station. Pop. (1989E) 9,000.

Tianjin (Tientsin)   Hebei, China. 39 08N 117 12W. City and port in NE sit. at junction of Grand Canal and Paiho R. The chief industrial centre in the N and the third largest town in China, it is also an important railway junction. Manu. inc. iron and steel, chemicals, textiles, leather and food products. Pop. (1987E) 5·46m.

Tiberias   Israel. 32 47N 35 33E. Town and health resort sit. E of Haifa on W shore of the Sea of Galilee at 680 ft. (207 metres) b.s.l.

Tiber River Italy. 41 44N 12 14E. Rises in the Tuscan Apennines above Sansepulero and flows 244 m. (390 km.) S past Perugia to Rome, and on to enter Tyrrhenian Sea by 2 mouths near Ostia; the Fiumara mouth to the S is now silted up and the Fiumicino is made navigable by an artificial channel. Navigable for small ships to Rome, but is prone to rapid flooding.

Tibesti Highlands Chad/Libya. 21 30N 17 30E. A mountain region within the Sahara. Of volcanic origin with fertile oases growing cereals and dates. The highest point is Emi Koussi, Chad, 11,200 ft. (3,414 metres) a.s.l.

Tibet (Xizang), Central Asia. 32 00N 88 00E. Autonomous Region of China bounded S by India, Nepál and Bhután, W by India, N by Sinkiang-Uighur Autonomous Region, E by China. Area, 471,660 sq. m. (1,221,600 sq. km.). Chief town, Lhasa (cap.). Mountainous area rising S to the Himalayas, N to the Kunkin range, and crossed E to W by the Tanglha, Aling Kangri and Nyerichen Tanglha ranges. There are numerous mountain Ls., mostly salt, which have no outlet. The climate is very hot for a short summer, very cold for the rest of the year. The economy is based on nomadic stock farming; there is agric. in the upper valleys of the Brahmaputra and Indus Rs. producing cereals, fruit and potatoes. Minerals inc. gold and coal. Industries are chiefly small manu. plants making agric. implements, textiles, cement and paper. Pop. (1987E) 2,030,000.

Ticino Switzerland. 46 20N 8 45E. Canton in SE Switz. sit. on S slopes of the Alps. Area, 1,085 sq. m. (2,811 sq. km.). Mainly mountainous but the valleys produce cereals, fruit, vines and tobacco. The principal Rs. are Ticino and the Maggia. Cap. Bellinzona, sit. near from the border with Italy. Pop. (1980C) 265,899.

Ticino River Switzerland. 45 09N 9 14E. Rises in the Lepontine Alps and flows S for 161 m. (258 km.) to enter Po R., Italy, just below Pavia.

Tien Shan Central Asia. 42 00N 80 00E. Mountain system extending E from the Pamirs N of the Tarim Basin across Xijiang Uygur. The highest point is Tengri Khan, 23,620 ft. (7,199 metres) a.s.l.

Tientsin *see* **Tianjin**

Tierra del Fuego Argentina/Chile. 54 00S 69 00W. An arch. sit. just S of S. America and separated from the mainland by the Strait of Magellan. The W side belongs to Chile and the E to Argentina. The main island, Isla Grande, area, 18,000 sq. m. (46,620 sq. km.), is flat in N with the Andean mountains in S. The principal industries are oil production, sheep farming and lumbering. The chief towns are Punta Arenas, Chile, and Ushuaia, Argentina.

Tigré Ethiopia. Region in N bordering on Eritrea. Area, 25,400 sq. m. (65,900 sq. km.). Mainly mountainous rising to Mount Alaji, 11,279 ft. (3,438 metres) a.s.l. The principal Rs. are Mareb, Takkaze, Tsellari, Ererti, Gheva and Weri. The main occupations are stock rearing, agric. crops inc. cereals, cotton, coffee and vegetables, potash mining and salt extracting. There are gold deposits S of Aksum. Cap. Makale. Pop. (1984C) 2,409,700.

Tigris River South West Asia. 34 30N 44 00E. Turkey /Syria/Iraq. Rises in Kurdistan, E Turkey, NW of Ergani and flows 1,150 m. (1,840 km.) SE past Diyarbakir to form, briefly, the Syrian border before entering Iraq, where it continues SSE past Mosul and Baghdad to join Euphrates R. at Al Qurnah. It is used extensively for irrigation.

Tijuana Baja California, Mexico. 32 32N 117 01W. Town and resort SE of San Diego, U.S.A., sit. on Tijuana R. and the U.S.A./Mex. border. There are tourist attractions inc. casinos and racecourses. Pop. (1980C) 461,257.

Tilburg Noord-Brabant, Netherlands. 51 34N 5 05E. Town in central E. An important textile centre and railway junction. Manu. soap. Pop. (1988E) 155,110.

Tillicoultry Central Region, Scotland. 56 09N 3 44W. Town sit. NE of Alloa at the foot of the Ochil Hills. The main industry is woollen milling. Pop. (1984E) 4,000.

Timaru South Island, New Zealand. 44 24S 171 15E. Port and resort on central E coast. Centre of an agric. district with allied trades. Exports inc. grain, wool and meat. Pop. (1983E) 28,900.

Timbuktu *see* **Tombouctou**

Timișoara Timiș, Romania. 45 45N 21 13E. Cap. of district sit. WNW of Bucharest. Manu. textiles, chemicals and footwear. Pop. (1985E) 318,955.

Timmins Ontario, Canada. 48 28N 81 20W. Town N of Sudbury sit. on Mattagami R. Centre of the Porcupine gold-mining district where silver, zinc and tin have been discovered. Pop. (1981E) 46,000.

Timor *see* **Indonesia**

Tinto River Huelva, Spain. 37 12N 6 55W. Rises near the famous copper-pyrite mines at Ríotinto and flows 60 m. (96 km.) S and SW to join Odiel R., 3 m. (5 km.) below Huelva, which then flows into the Gulf of Cádiz.

Tipperary Munster, Ireland. 52 58N 8 10W. An inland county, mainly mountainous, containing the Knockmealdown Mountains in S, the Galty Mountains in SW and the Slieveardagh Hills in SE. Area, 1,643 sq. m. (4,255 sq. km.) divided into N and S Ridings. The principal Rs. are Shannon and Suir. Lough Derg lies on the NW boundary. The fertile Golden Vale is sit. in SW producing barley, oats, potatoes and sugar-beet. Dairy farming and cattle raising are important occupations. Coal, copper, zinc, lead, slate and limestone are found. The county town is Clonmel. Pop. (1988E) 136,619; N Riding, 59,522, S Riding, 77,097.

Tipperary Tipperary, Ireland. 52 29N 8 10W. Town in W of county sit. in the Golden Vale. Agric. products inc. cheese and condensed milk, mineral waters. Manu. inc. computers. Pop. (1981C) 4,982.

Tipton West Midlands, England. 52 32N 2 05W. Town sit. SSE of Wolverhampton. Industries inc. heavy and light engineering, iron founding, and the manu. of motorcar parts, metal goods, machine tools, electrical equipment and glass. Pop. (1981C) 34,500.

Tirana Tirana, Albania. 41 18N 19 49E. Cap. of rep. in central W sit. on the edge of a fertile plain. It was founded by the Turks in the 17th cent. Manu. textiles, cigarettes and soap. Pop. (1983E) 206,000.

Tîrgu Jiu Gorj, Romania. 45 02N 23 17E. Town sit. NW of Bucharest on Jiu R. Manu. inc. furniture, textiles, wood products and bricks. Trades in lumber, livestock and cheese. Pop. (1985E) 85,058.

Tîrgu Mureș Mureș, Romania. 46 33N 24 33E. Town sit. NW of Bucharest on Mureș R. Industries inc. oil refining, brewing, sugar refining and the manu. of furniture, clay products, soap, confectionery and food products. Trades in petroleum, timber, grain, fruit, wine and tobacco. The huge 'Cultural Palace' is sit. here. Pop. (1985E) 157,411.

Tirol Austria. 47 03N 10 43E. Province of W Austria bounded S by Italy and N by Germany. Area, 4,884 sq. m. (12,650 sq. km.). Chief towns Innsbruck (cap.), Landeck. Mountain ranges extend E to W on either side of Inn R. valley. Mainly a forest and pastureland area, with dairy farming important and some agric. in Inn R. valley, producing cereals and maize. The chief mineral is salt. Tourism, particularly winter sports, is the most important industry. Pop. (1988E) 613,700.

Tiruchirapalli Tamil Nadu, India. 10 45N 78 45E. Town in SE sit. on Cauvery R. Important railway junction with railway workshops. Manu. textiles, gold and silver ware and cigars. Pop. (1981C) 362,045.

Tirunelveli Tamil Nadu, India. 8 45N 77 43E. Town sit. SSW of Madurai. It is an important missionary centre. Industries inc. sugar refining. Pop. (1981C) 128,850.

Tirupati Andhra Pradesh, India. Important centre of pilgrimage for the Hindus, known for its temples. Pop. (1981C) 115,244.

Tisza River U.S.S.R./Hungary/ Yugoslavia. 45 15N 20 17E. Rises below the Pass of the Tartars in the E Carpathians, Ukrainian Soviet Socialist Republic, and flows 840 m. (1,344 km.) W into Hungary, turning SW across the Hungarian plain to Szeged and on into Yugoslavia where it joins Danube R. above Belgrade. Used for power and irrigation, esp. in NE Hungary.

Titicaca, Lake Bolivia/Peru. 15 50S 69 20W. L. on W Bolivian frontier and SE Peruvian frontier, the largest in S America. Area, 3,200 sq. m. (8,288 sq. km.), at over 12,500 ft. (3,810 metres) a.s.l. It consists of L. Chocuito (N) and L. Uinamarca (S) joined together by a strait. Chief ports Puno, Huancane (Peru), Guaqui (Bolivia).

Titograd Montenegro, Yugoslavia. 42 26N 19 14E. Cap. of federal unit SSW of Belgrade sit. at confluence of Rs. Morača and Ribnica. Manu. inc. tobacco products. Pop. (1981C) 132,290.

Tiverton Devon, England. 50 55N 3 29W. Town in E centre of county sit. on Rs. Exe and Lowman. Manu. inc. beer bottling, agric. machinery, sawmill machinery, woven and knitted fabrics and other light industry. Pop. (1988E) 17,002.

Tivoli Lazio, Italy. 41 58N 12 48E. Town ENE of Rome sit. on Aniene R. Noted for its anc. ruined buildings and beautiful gardens. Manu. paper, footwear and wine.

Tlaxcala Mexico. 19 19N 98 14W. (i) The smallest state, sit. in centre, and extremely mountainous. Area, 1,551 sq. m. (4,016 sq. km.). The altitude varies from 7,000 ft. (2,134 metres) to over 14,000 ft. (4,268 metres) a.s.l. Mainly agric. but depen-

dent on artificial irrigation. Cereals, vegetables, alfalfa and maguey are produced, the latter being used to make the liquor *pulque*. Industries inc. textiles, flour milling and food processing. Pop. (1989E) 676,446. (ii) Cap. of state of same name, officially known as Tlaxcala de Xicohténcatl, E of Mexico City sit. at 7,350 ft. (2,240 metres) a.s.l. Commercial centre manu. textiles. Pop. (1980E) 15,000.

Tlemçen Tlemçen, Algeria. 34 52N 1 19W. Cap. of dept. in NW sit. near Morocco frontier. Manu. carpets, leather goods, brassware, hosiery, footwear and furniture. Exports, through its port Rashgun, inc. blankets, olive oil and alfalfa. It has many fine synagogues and mosques dating from the Middle Ages. Pop. (1983E) 146,089.

Toamasina Madagascar. 18 10S 49 23E. Formerly Tamatave. Cap. of province of same name, and port, sit. ENE of Antananarivo. Exports inc. sugar, rice and coffee. Pop. (1982E) 82,907.

Tobago Trinidad and Tobago, West Indies. 11 10N 60 30W. Island sit. N of Trinidad and administratively united with it. Area, 116 sq. m. (300 sq. km.). Chief town, Scarborough. Chief products, copra, rubber. Pop. (1980C) 39,529.

Tobata *see* **Kitakyushu**

Tobermory Mull, Strathclyde Region, Scotland. 56 37N 6 05W. Port and resort sit. on N coast of the island of Mull. Main occupations are tourism and fishing. Pop. (1989E) 800.

Tobol River U.S.S.R. 58 10E 68 12E. Rises in S Ural mountains in N Kazakh Soviet Socialist Republic and flows 800 m. (1,280 km.) NE past Kustanai and Kurgan to enter Irtysh R. at Tobolsk. Navigable up to Kurgan.

Tobolsk Russian Soviet Federal Socialist Republic, U.S.S.R. 58 12N 68 16E. Town in W Siberia sit. at confluence of Rs. Irtysh and Tobol. Centre of a lumbering region and trading in fish and furs. An anc. craft of bone carving is still carried on.

Tobruk East Cyrenaica, Libya. 32 05N 23 59E. Port in NE linked by pipeline to the Sarir oilfield.

Tocantins River Brazil. 1 45S 49 10W. Rises in 2 headstreams, one in the Federal District N of Planaltina, one SW of it near Anapolis, and flows 1,500 m. (2,400 km.) N through Goias state to Carolina, whence it forms the state boundary NW to São João and enters Pará state, turning N at Marabā to enter Pará estuary. Main trib. Araguaia R. Impeded by several cataracts and rapids.

Tocopilla Antofagasta, Chile. 22 05S 70 12W. Port in N. Industries inc. copper smelting and fishing. Exports inc. nitrates, iodine and copper ore. Pop. (1980E) 32,000.

Todmorden West Yorkshire, England. 53 43N 2 05W. Town sit. NE of Manchester. Industries inc. textiles, furniture and engineering. Pop. (1981C) 14,665.

Togliatti Russian Soviet Federal Socialist Republic, U.S.S.R. 53 32N 49 24E. R. port E of Moscow sit. near S end of the Kuibyshev Reservoir. An industrial centre with ship repairing, engineering and food processing. Manu. vehicles, synthetic rubber, chemicals and furniture. Pop. (1989E) 630,000.

Togo West Africa. 6 15N 1 35E. Rep. bounded E by Benin, N by Burkina and W by Ghana on the Gulf of Guinea. The Republic of Togo became independent in 1960, after having been a German protectorate (1894–1914, subsequently divided between the French and the British), a mandate of the League of Nations (1922) and a trusteeship territory of the United Nations (1946). Area, 21,925 sq. m. (56,785 sq. km.). Cap. Lomé. Hilly and forested area rising to 3,600 ft. (1,097 metres) a.s.l. with some dry plains and arable land. Chief crops are coffee, cocoa, palms, copra, cotton, manioc. Stock raising is important, esp. cattle, pigs, sheep and goats. The most important minerals are phosphate and bauxite; there are deposits of iron and limestone. Pop. (1988E) 3,246,000.

Tokaj Borsod-Abaúj-Zemplén, Hungary. 48 07N 21 25E. Town ENE of Budapest sit. at confluence of Rs. Tisza and Bodrog. Tokay wine is produced in the neighbourhood.

Tokelau New Zealand. 9 00S 171 45W. A group of three atolls in S. Pacific Ocean: Atafu, Fakaofo and Nukunono, sit. N of Western Samoa. Area, 4 sq. m. (10 sq. km.). Copra is exported. Pop. (1980E) 1,620.

Tokushima Shikoku, Japan. 34 03N 134 34E. Port and cap. of Tokushima prefecture sit. SW of Osaka. Manu. inc. *saké* and cotton goods. Pop. (1988E) 258,000.

Tokyo Honshu, Japan. 35 45N 139 45E. Metropolis inc. city centre area (which was originally founded in 1457 as Edo) and its suburbs. Centre city area at head of Bay of Tokyo on R. Sumida. Cap., sea port and commercial centre built on both banks of R. and intersected by canals. It was renamed Tokyo in 1868 and today inc. 26 cities. The port handles domestic and foreign trade, as an amalgamation of the two ports of Tokyo and Yokohama. Industries inc. shipbuilding, engineering, manu. chemicals, textiles, electrical goods, vehicles. The city is subject to earth tremors. The Imperial Palace is sit. in the central city area. Pop. (1988E) 8,156,000.

Tolbukhin Shumen, Bulgaria. 43 34N 27 51E. Town NNW of Varna on railway into Romania. Market town for an agric. area. Pop. (1987E) 111,037.

Toledo Spain. 39 52N 4 01W. (i) Province sit. in R. Tagus basin S of Madrid. Area, 5,934 sq. m. (15,368 sq. km.). Mainly mountainous with the Montes de Toledo rising to over 4,000 ft. (1,219 metres) a.s.l. in S. The principal R. is Tagus. Chiefly agric. producing cereals, wine and olive oil and esp. noted for rearing of fighting bulls. Pop. (1986C) 487,844. (ii) Cap. of province of same name in central Spain sit. on Tagus R. Noted for the manu. of swords. Other manu. inc. firearms and textiles. Contains many fine buildings and art treasures. Pop. (1981E) 58,000.

Toledo Ohio, U.S.A. 41 39N 83 32W. Port w of Cleveland sit. on Maumee Bay of L. Erie. An important railway and industrial centre trading in coal, iron ore, oil and agric. produce. Industries inc. shipbuilding, oil refining and the manu. of cars, machinery, machine tools, glass and electrical equipment. Pop. (1980c) 354,635.

Toliary Madagascar. (Formerly Tuléar.) 23 20S 43 41E. Cap. of province of same name and port sit. sw of Antananarivo on the Mozambique Channel. Exports inc. agric. products, mother-of-pearl, tortoise shell and edible molluscs. Pop. (1982E) 48,929.

Tolima Colombia. 4 40N 75 19W. A dept. in central w bounded by the Cordillera Oriental and the Cordillera Central. Area, 9,006 sq. m. (23,325 sq. km.). Cap. Ibagué. The principal R. is Magdalena. Mainly agric., the chief products being coffee and livestock. Other crops inc. cereals, rice, sugar-cane, tobacco and vegetables. There are deposits of petroleum, coal, silver, copper, sulphur and mercury. Pop. (1983E) 1,128,900.

Tolna Hungary. 46 25N 18 48E. (i) County in central w. Area, 1,429 sq. m. (3,702 sq. km.). Mainly hilly. The principal Rs are Danube, Sarviz and Sio. Crops inc. cereals, potatoes, hemp and beans. Chief town Szekszard. Pop. (1984E) 268,000. (ii) Town in county of same name sit. NE of Szekszard. Industries inc. textiles, distilling and brick making. Pop. (1989E) 262,000.

Toluca Mexico, Mexico. 19 17N 99 40W. Cap. of federal district wsw of Mexico City, sit. at 8,660 ft. (2,640 metres) a.s.l. Manu. textiles, pottery and food products. Noted for bulls. Pop. (1980E) 357,071.

Tomaszów Mazowiecki Lódź, Poland. 51 32N 20 01E. Town SE of Lódź sit. on Pilica R. Manu. textiles, bricks and iron.

Tombouctou Mali, West Africa. 16 49N 3 01W. Town in central Mali near N bank of Niger R. Important trading centre on trad. route from N Africa to the Atlantic coast states. Pop. (1976E) 20,483.

Tomsk Russian Soviet Federal Socialist Republic, U.S.S.R. 56 30N 84 58E. Cap. of Tomsk region in w central Siberia sit. on Tom R. An important transportation and industrial centre. Manu. machinery, chemicals, electrical equipment, ball-bearings, wood products and matches. It is also a major cultural centre of Siberia. Pop. (1989E) 502,000.

Tonbridge Kent, England. 51 12N 0 16E. Town sw of Maidstone sit. on R. Medway. Industries inc. printing, tanning, tar-distilling and the manu. of cricket balls, bricks and plastic goods. Pop. (1981c) 30,457.

Tonga South Pacific. 20 00S 173 00W. (Friendly Islands). Kingdom of islands E of Fiji, consisting of 3 main groups, the Tongatapu, Ha'apai and Vava'u islands, and 4 smaller groups. The kingdom of Tonga attained unity under Taufa'ahau Tupou (George I). By the Anglo-German Agreement of 1899, subsequently accepted by the U.S.A., the Tonga Islands were left under the Protectorate of Great Britain. Area, 290 sq. m. (748 sq. km.). Cap. Nuku'alofa on Tongatapu. The E groups are low lying limestone formations; the w groups are volcanic and hilly rising to 3,433 ft. (1,046 metres) a.s.l. The volcanoes are active. The climate is temperate. Chief products copra, bananas. Pop. (1988E) 95,200.

Tongareva *see* **Penrhyn**

Tonlé Sap, Lake Cambodia. 12 50N 104 00E. Sit. in NW plains this large tidal L. is linked with Kekong R. *via* Tonlé Sap R. During the rainy season (July–Oct.) the Mekong R. discharges into the L., and this process is reversed during the rest of the year. Area, c. 1,000 sq. m. (2,590 sq. km.) at low water. It is the centre of the Rep.'s fishing industry.

Toowoomba Queensland, Australia. 27 33S 151 58E. Town in SE of state sit. in Darling Downs. Commercial centre of a rich agric. area trading in meat, wool, wheat and dairy produce. Industries inc. iron foundries, engineering, pecan nut processing, and clothing factories, with agric. equipment manu., saddlery and coalmining in the vicinity. Pop. (1985E) 78,500.

Topeka Kansas, U.S.A. 39 03N 95 41W. State cap. in NE of state sit. on Kansas R. Industrial centre of an agric. area with associated industries. Other industries inc. railway engineering, distribution, health care, printing and publishing and the manu. of tyres. Pop. (1987E) 122,000.

Torbay Devon, England. 50 28N 3 30W. An inlet of the English Channel sit. between Hope's Nose and Berry Head. Also an administrative area. The towns of Torquay, Paignton and Brixham are sit. on its shores.

Torne River Finland/Sweden. 65 48N 24 08E. Rises in N Sweden in L. Torneträsk and flows 320 m. (512 km.) SE to the Finnish border which it forms (S) for the rest of its course, entering head of Gulf of Bothnia at Torneå. Main trib. Muonio R.

Toronto Ontario, Canada. 43 39N 79 23W. Provincial cap. and largest Canadian city sit. on N shore on Don and Humber Rs. Provincial cap. Commercial and industrial centre, manu. machinery, clothing, petroleum products, metal products. Meat packing centre. Other industries inc. automobile manu., communications, publishing, food processing. L. port with a natural harbour ice-free April-November, trading in wheat and coal. Pop. (1986E) 700,000 (metropolitan area (1986E) 3,427,168).

Torquay Devon, England. 50 28N 3 30W. Resort in SW sit. on Torbay, an inlet of the English Channel. A noted holiday and yachting centre.

Torrance California, U.S.A. 33 50N 118 19W. City sit. SSW of Los Angeles. Industries inc. computer components, aerospace and electronics research, and oil refining. Pop. (1980C) 131,000.

Torre Annunziata Campania, Italy. 40 45N 14 27E. Port sit. SE of Naples on Bay of Naples. Manu. firearms, iron and macaroni.

Torre del Greco Campania, Italy.

40 47N 14 22E. Port and resort SE of Naples sit. on Bay of Naples. Industries inc. boat-building, coral fishing and the manu. of cameos and macaroni. Pop. (1981C) 103,605.

Torreón Coahuila, Mexico. 25 33N 103 26W. Town in central N. An important railway junction and centre of the Laguna wheat and cotton growing district. Manu. textiles and chemicals. There are silver, zinc and copper mines in the vicinity. Pop. (1980C) 363,886.

Torres Islands Vanuatu, South West Pacific. 13 15S 166 37E. A group of islands sit. N of Espiritu Santo, comprising Hiu, the largest island, Tegua, Toga and Lo.

Torres Strait Australasia. 10 25S 142 10E. A channel, about 90 m. (144 km.) wide, sit. between New Guinea and Australia. It connects the Arafura Sea in W with the Coral Sea in E.

Torrington England *see* **Great Torrington**

Torrington Connecticut, U.S.A. 41 48N 73 08W. Town NNE of New Haven sit. on Naugatuck R. Manu. hardware, machinery, tools, electrical equipment, anti-friction bearings, bronze and gray iron castings, corrugated containers and brushes. Pop. (1989E) 35,000.

Tortola British Virgin Islands. 18 26N 64 37W. The main island of the group, sit. between St. John and Virgin Gorda. Area, 21 sq. m. (54 sq. km.). Rises to 1,781 ft. (543 metres) a.s.l. in Mount Sage. The main occupations are stock rearing and the growing of sugar-cane, fruit and vegetables. Charcoal is also produced and exported. Pop. (1980C) 9,322.

Tortosa Tarragona, Spain. 40 48N 0 31E. Port in NE sit. above mouth of Ebro R. Manu. porcelain, glass, soap and leather. Trades in wine, olive oil and rice.

Torún Bydgoszcz, Poland. 53 02N 18 35E. Town in N central Poland sit. on Vistula R. Manu. machinery, textiles and chemicals. Trades in timber and grain. Pop. (1982E) 183,000.

Toscana Italy. 43 30N 11 05E. Central region bordering on the Ligurian and Tyrrhenian Seas, and comprising the provinces of Apuania, Arezzo, Florence, Grosseto, Leghorn, Lucca, Pisa, Pistoia and Siena. Area, 8,877 sq. m. (22,991 sq. km.). Mainly mountainous with a narrow coastal plain. The principal Rs. are Arno, Sieve, Pesa, Elsa, Era, Ombrone, Cecina and upper Tiber. It is chiefly mountainous but cereals, vines, olives and tobacco are grown. There are deposits of iron, copper, mercury, lignite and manganese. Cap. Florence. Pop. (1981C) 3,581,051.

Totnes Devon, England. 50 25N 3 41W. Town wsw of Torquay sit. on R. Dart. A market town with industries inc. boat-building, sawmilling, brewing, country crafts, and the production of bacon and cream. Pop. (1981C) 5,627.

Totonicapán Guatemala. 14 55N 91 22W. (i) A dept. in w highlands. Area, 410 sq. m. (1,062 sq. km.). Mainly agric. growing cereals, fodder grasses and beans, and rearing sheep. Industries inc. flour milling, and the manu. of textiles, pottery and furniture. Pop. (1985E) 249,067. (ii) Cap. of dept. of same name sit. ENE of Quezaltenango. Manu. inc. textiles, and local handicrafts. Pop. (1980E) 55,000.

Tottenham Greater London, England. 55 36N 0 04W. District of N London. Residential and industrial.

Tottori Honshu, Japan. Port and cap. of Tottori prefecture WNW of Kyoto sit. on Sea of Japan. Manu. inc. textiles and paper.

Touggourt Algeria. 33 06N 6 04E. Town and oasis in NE. A commercial centre trading in dates, which grow in profusion. Sit. on the oil pipeline from Hassi Messaoud to Bejaia.

Toulon Var, France. 43 07N 5 56E. Cap. of dept., port and naval base in SE sit. on Mediterranean Sea. Industries inc. shipbuilding and repairing, oil refining, textiles and the manu. of armaments and chemicals. Pop. (1989E) 185,000 (agglomeration, 450,000).

Toulouse Haute Garonne, France. 43 36N 1 26E. City sit. SE of Bordeaux, on Garonne R. Cap. of dept. and market trading in wine and wheat. Commercial city with important banks. Industries inc. publishing, chemicals, technological research, aeronautics and aerospace, electrical engineering, manu. clothing, footwear, armaments. Pop. (1982C) 354,289 (agglomeration, 541,271).

Toungoo Burma. Town N of Rangoon sit. on Sittang R. Industries are lumbering and the production of rice and rubber.

Touquet, Le Pas-de-Calais, France. 50 30N 1 36E. Resort sit. s of Boulogne, on English Channel coast. With Paris-Plage, it is known as Le Touquet-Paris-Plage. Pop. (1982C) 5,425.

Tourane Vietnam *see* **Da Nang**

Tourcoing Nord, France. 50 43N 3 09E. Town sit. NNE of Lille. Its textile industry dates from the 12th cent. and together with Toubaix, just SSE, it produces most of the country's woollen products. Pop. (1982C) 97,121.

Tournai, (Doornik) Hainaut, Belgium. 50 36N 3 23E. Town in w sit. on Scheldt R. Manu. carpets, pottery, hosiery, leather and cement. Pop. (1988E) 66,749.

Tours Indre-et-Loire, France. 47 23N 0 41E. Town sit. E of Angers on Loire R. Cap. of dept. and archbishopric. Tourist centre and market town for wine and agric. produce. Industries inc. textiles, railway repair shops, manu. cement and food products. Pop. (1982C) 136,483 (agglomeration, 262,786).

Tower Hamlets Greater London, England. 51 32N 0 03W. Boro. of E London on N shore of Thames R. Inc. districts of Bethnal Green, Bow, Limehouse, Millwall, Poplar, Shadwell, Stepney, Wapping and Whitechapel. The heart of London's 'East End'. Pop. (1988E) 161,800.

Townsville Queensland, Australia. 19 16S 146 48E. Port in NE of state sit. on Cleveland Bay. An important commercial, defence and · tourist centre, exporting beef, wool, sugar, minerals and metals, inc. copper, lead, nickel, cobalt and silver. Also an important centre for international coral reef research. Pop. (1985E) 104,000.

Towson   Maryland, U.S.A. 39 26N 76 34W. Town immediately N of Baltimore. Pop. (1980C) 51,083.

Towyn   *see* **Tywyn**

Toyama   Honshu, Japan. 30 40N 137 15E. Cap. of Toyama prefecture NW of Tokyo sit. on Toyama Bay. An important centre for the manu. of pharmaceutical products. Also manu. machinery, textiles and chemicals. Pop. (1988E) 316,000.

Toyohashi   Honshu, Japan. 34 46N 137 23E. Town sit. E of Osaka in the Aichi prefecture. Industries inc. metal working, food processing, and the manu. of cotton goods. Pop. (1988E) 326,000.

Toyonaka   Honshu, Japan. 34 47N 135 28E. Town sit. N of Osaka. Wheat, rice, and flowers are grown. Pop. (1988E) 406,000.

Trabzon   Trabzon, Turkey. 41 00N 39 45E. Cap. of province and port in NE sit. on Black Sea coast. Exports inc. beans, nuts, tobacco, opium and hides. Pop. (1980C) 108,403.

Trail   British Columbia, Canada. 49 06N 117 42W. Town in extreme S of province sit. on Columbia R. Industries inc. a large metallurgical plant and the manu. of chemicals and fertilizers. Pop. (1981C) 9,600.

Tralee   Kerry, Ireland. County town in SW sit. on R. Lee near Tralee Bay, and joined to the sea by canal. Industries inc. textiles, light engineering, meat products and milling. Pop. (1987E) 19,500.

Trani   Puglia, Italy. 41 17N 16 26E. Port ESE of Barletta sit. on Adriatic Sea. Agric. products inc. wine, olive oil and grain. Manu. furniture.

Transcaucasia   U.S.S.R. 42 00N 44 00E. Territory S of the Caucasus, separated from rest of European Russia; comprising reps. of Georgia, Azerbáiján and Armenia.

Transkei   South Africa. 32 15S 28 15E. Homeland of the Xhosa nation. It was granted the status of an independent state by the Republic of South Africa in 1978. Sit. N of the Great Kei R. Area, 16,910 sq. m. (43,798 sq. km.). Cap. Umtata. Subsistence agric. is being diversified by such cash crops as tea, coffee and flax. There are large mineral deposits and afforestation is being undertaken. 80% of the population works outside Transkei. Pop. (1985E) 2,876,122.

Transvaal   Republic of South Africa. 25 00S 29 00E. Province of NE South Africa, bounded N by Zimbabwe, E by Mozambique and Swaziland, NW by Botswana, S by Vaal R. Area, 101,351 sq. m. (262,499 sq. km.) inc. Gazan kulu, Lebowa, Ka Ngwane and Kwa Ndebele. Chief cities Pretoria (cap.), Johannesburg, Germiston, Benoni, Springs, Krugersdorp. Mainly plateau at *c*. 3,000 ft. (914 metres) a.s.l. with the Rand (S) rising to 5,000 ft. (1,524 metres) a.s.l. and the land sloping slowly from there to the Limpopo R. (N). Some grassland, but mainly arid ground with bush. The High Veld supports agric. in spite of drought, but most crop growing depends on irrigation. Stock farming is important. The chief minerals are gold, in the Rand to the S, and coal. The Rand area is highly industrialized, manu. iron, brass, brick and tile products, pottery, soap, candles, clothing, vehicles, tobacco products, machinery, foods. Pop. (1985C) 7,532,179.

Trapani   Sicilia, Italy. 38 01N 12 31E. Port and cap. of Trapani province in extreme W. Products inc. wine, salt, macaroni and tunny fish. Pop. (1981E) 72,000.

Trasimeno, Lake   Umbria, Italy. 43 08N 12 06E. The largest lake in central Italy, lying W of Perugia. Area, 50 sq. m. (129 sq. km.). Drains into a trib. of Tiber R. *via* an underground tunnel.

Trawden   Lancashire, England. 53 50N 2 07W. Town sit. NE of Burnley. Industries inc. textiles and light engineering. Pop. (1981C) 1,912.

T Treasury Islands   *World Gazetteer*

Treasury Islands Solomon Islands, South West Pacific. A group of islands SE of Bougainville, comprising Mono, Stirling and some smaller islets.

Tredegar Gwent, Wales. 51 47N 3 16W. Town sit. NW of Newport. Centre of a former coalmining district with light engineering industries. Pop. (1985E) 17,000.

Treinta y Tres Uruguay. 33 14S 54 23W. (i) Dept. in central E bounded by Brazil. Area, 3,683 sq. m. (9,539 sq. km.). The principal R. is Olimar. Mainly agric. producing cattle, sheep, cereals and linseed. Pop. (1985C) 45,500. (ii) Cap. of dept. of same name sit. NE of Montevideo on Olimar Grande R. Trades in agric. products. Pop. (1985C) 30,956.

Trelew Chubut, Argentina. Town in SE. The centre of a sheep-farming area, it was founded by Welsh settlers. Pop. (1980E) 60,000.

Trengganu *see* **Terengganu**

Trentino-Alto Adige Italy. 46 30N 11 00E. Region comprising provinces of Trento and Bolzano-Bozen, sit. on borders of Switzerland and Austria, and once part of the latter. Area, 5,256 sq. m. (13,613 sq. km.). Pop. (1981C) 873,413.

Trento Trentino-Alto Adige, Italy. 46 04N 11 08E. Cap. of Trento province in NE, sit. on Adige R. and on Brenner Pass route. Manu. chemicals, electrical products, silk, pottery, wine and cement. Pop. (1980E) 100,000.

Trenton New Jersey, U.S.A. 40 13N 74 45W. State cap. and port sit. on Delaware R. at head of navigation. An important industrial centre concerned mainly with steel and steel products. Other manu. inc. textiles, rubber, pottery and paper. Pop. (1980C) 92,124.

Trent River England. 53 42N 0 41W. Rises S of Biddulph in N Staffordshire and flows 170 m. (274 km.) SE to Ashby-de-la-Zouch, then NE into Derbyshire and along N border of Leicestershire into Nottinghamshire, past Nottingham and Newark and N along Lincolnshire border into Humberside to enter Humber near Whitton.

Its upper and middle course is through an industrial area; many industrial towns are linked to it by canal. Main tribs. Blythe, Dove, Derwent, Tame and Soar Rs.

Tres Arroyos Buenos Aires, Argentina. 38 20S 60 20W. Town in central E. Centre of an agric. and livestock rearing area, and manu. furniture. Pop. (1980C) 40,000.

Treviso Veneto, Italy. 45 40N 12 15E. Cap. of Treviso province in NE. Manu. inc. agric. machinery, brushes and paper. Pop. (1980E) 100,000.

Trier Rhineland-Palatinate, Federal Republic of Germany. 49 45N 6 38E. Town, the oldest in Germany, in extreme central W sit. on Moselle R. Manu. inc. textiles, precision instruments and leather goods, and esp. Moselle wine. Pop. (1989E) 94,119.

Trieste Friuli-Venezia-Giulia, Italy. 45 40N 13 46E. City at NE corner of the Adriatic Sea opposite Venice at NW corner. Regional cap. and sea port with important passenger traffic. Pop. (1981C) 252,369.

Triglav, Mount Slovenia, Yugoslavia. 46 23N 13 50E. The highest peak in Yugoslavia, 9,396 ft. (2,864 metres) a.s.l. in the Julian Alps and NW of Ljubljana.

Trikkala Greece. 39 34N 21 47E. (i) A prefecture in Thessaly with the Pindus Mountains in W. Area, 1,289 sq. m. (3,338 sq. km.). Cereals and olives are produced. Pop. (1981C) 134,207. (ii) Cap. of prefecture of same name sit. SSE of Thessaloníki. Trades in cereals, tobacco and cotton. Pop. (1981C) 40,857.

Trincomalee Eastern Province, Sri Lanka. 8 38N 81 15E. Port sit. on NE coast with one of the finest natural harbours in the world. Pop. (1981E) 44,913.

Tring Hertfordshire, England. 51 48N 0 40W. Town sit. WNW of St. Albans. Trades in agric. produce and has light engineering industries. Pop. (1978E) 12,000.

Trinidad Beni, Bolivia. 14 47S 64

47W. Town ENE of La Paz sit. near Mamoré R. A commercial and transportation centre. Industries inc. sugar refining and alcohol distilling. Trades in rice, cotton, sugar-cane, cattle, furs and valuable feathers. Pop. (1980E) 36,200.

Trinidad and Tobago West Indies. 10 30N 61 20W. Island off N coast of Venezuela opposite mouth of Orinoco R. Forms an independent Commonwealth State. Trinidad was discovered by Columbus in 1498 and ceded to Great Britain by the Treaty of Amiens in 1802. Trinidad and Tobago were joined in 1889. Area, 1,864 sq. m. (4,828 sq. km.). Cap. Port of Spain. Hilly and fertile with a hot, humid climate. Agric. areas produce sugar, cacao, coconuts, citrus fruits; some rice is grown in irrigated areas. The most important minerals are oil and asphalt. Industries inc. oil refining, local and Venezuelan oil, asphalt production, food processing. Pop. (1988E) 1,243,000.

Tripoli Lebanon. 34 26N 35 51E. Port sit. NNE of Beirut on Mediterranean coast. It is a terminus of the oil pipeline from Iraq. Manu. inc. textiles, cement and soap. Exports inc. citrus fruits and tobacco. Pop. (1980E) 175,000.

Tripoli Tripolitania, Libya. 32 49N 13 07E. City on Mediterranean coast of NW Libya. Cap. and sea port founded by the Phoenicians. Industries inc. fishing, fish processing, tobacco processing, textiles, manu. bricks, salt and leather goods. Pop. (1984E) 990,697 (municipality).

Tripolis Arcadia, Greece. 37 31N 22 21E. Cap. SE of Patras sit. on E slopes of the Maenalon Mountains. Industries inc. woodworking, tanning and the manu. of textiles and cheese. Trades in agric. produce. Pop. (1981C) 141,529.

Tripura India. 24 00N 92 00E. State in NE, mainly hilly and covered in jungle. Area, 4,049 sq. m. (10,486 sq. km.). Cap. Agartala, 56 m. (90 km.) ENE of Dhaka, Bangladesh. Products inc. tea, timber, jute, cotton and rice. Industries inc. tea, handicrafts,

natural gas, handlooms. Pop. (1981C) 2,053,058.

Tristan da Cunha Atlantic Ocean. 37 04S 12 19W. Islands midway between Argentina and the Republic of South Africa, forming a depend. of Saint Helena. Area, 45 sq. m. (117 sq. km.), consisting of Tristan da Cunha, Gough, Inaccessible and Nightingale Islands. Tristan da Cunha island is a volcano, 6,760 ft. (2,060 metres) a.s.l. high with a plateau of 12 sq. m. (31 sq. km.) on the NW side which is the site of settlement and cultivation (mainly potatoes). The pop. were evacuated to U.K. in 1961 when the volcano, believed extinct, erupted. Many returned in 1963. Main occupations are farming and fishing. There is a meteorological and radio station. Pop. (1988E) 313.

Trivandrum Kerala, India. 8 31N 77 00E. State cap. in SW. Manu. inc. textiles, copra, soap, coir ropes, electronics and mats. Pop. (1981C) 499,531.

Trnava Slovakia, Czechoslovakia. 48 23N 17 35E. Town in central S. Industries inc. sugar refining, brewing and the manu. of textiles and fertilizers. Pop. (1989E) 72,000.

Trois-Rivières Quebec, Canada. 46 21N 72 33W. City and port in central S of province sit. at confluence of St. Lawrence R. and St. Maurice R. An important industrial centre with textile and metallurgical plants. It is also a leading producer of newsprint with vast paper and pulp mills whose hydroelectric power is derived from the St. Maurice R. Pop. (1981E) 50,466 (metropolitan area, 111,453).

Trollhättan Älvsborg, Sweden. 58 16N 12 18E. Town in SW sit. on Göta R. near SW end of L. Väner. The Trollhätte falls, descending 108 ft. (33 metres) supply power both to the town and to much of S Sweden. One of the largest industrial towns in S Sweden. Manu. inc. lorries, transport equipment. Pop. (1988E) 50,296.

Tromsø Troms, Norway. 69 40N 18 58E. Cap. of county and port sit. on an island of the same name just off NW coast. It is the largest town N of the Arctic Circle and a base for Arctic expeditions. Centre for space and atmospheric research. Meteorological station for N Norway with 4 Arctic outstations. Administration, communications and educational centre for N Norway. Main industries are fishing, shipyards, brewing, prefabricated houses, high technology and tourism. Pop. (1989E) 50,027.

Trondheim Sør-Trøndelag, Norway. 63 25N 10 25E. Cap. of county and port in W centre sit. at mouth of Nid R. on Trondheim Fiord. An important centre of fishing and fishcanning. Manu. metal products, soap and margarine. Exports wood pulp, paper, fish and metal goods. It has an 11th cent. cathedral. Pop. (1989E) 136,601.

Troon Strathclyde Region, Scotland. 55 32N 4 40W. Resort and port sit. N of Ayr on the Firth of Clyde. Industries inc. fishing, shipbuilding and breaking. Pop. (1989E) 14,641.

Trouville-sur-Mer Calvados, France. 49 22N 0 05E. Resort and port on central N coast sit. at mouth of Touques R. opposite Deauville, to which it is joined by a bridge. Pop. (1982C) 6,012.

Trowbridge Wiltshire, England. 51 20N 2 13W. Town sit. ESE of Bath. Industries inc. brewing, printing, seed processing and the manu. of woollen goods, gloves, mattresses, heating and ventilating appliances, and food processing. Pop. (1981C) 22,984.

Troy Michigan, U.S.A. 42 37N 83 09W. City N of Detroit in industrial Oakland county. Pop. (1980C) 67,102.

Troy New York, U.S.A. 42 43N 73 40W. City sit. NNE of Albany on Hudson R. Manu. agric. machinery, aircraft and car parts, fire hydrants, high technology components, and clothing, esp. shirts. Pop. (1980C) 56,638.

Troyes Aube, France. 48 18N 4 05E. Cap. of dept. in NW sit. on Seine R. An industrial and transportation centre with manu. of machinery, textiles, paper, hosiery and food products. Noted for its medieval fairs and gave its name to 'troy weight'. Pop.

(1982C) 64,759 (agglomeration, 125,240).

Trucial States *see* **United Arab Emirates**

Trujillo Colón, Honduras. 15 55N 86 00W. Cap. of dept. in NE and first cap. of Spanish colony of Honduras (1525–61). Commercial centre and port for vessels up to 12,000 GRT at Puerto Castillo (12 miles, 19 km. N). Pop. (1985E) 6,971.

Trujillo La Libertad, Peru. 8 10S 79 02W. Cap. of dept. in NW. Commercial and transportation centre of an irrigated area producing sugar-cane and rice. Manu. soap, candles, food products and cocaine. Pop. (1981C) 354,557.

Trujillo Venezuela. 9 20N 70 38W. (i) State in W on L. Maracaibo. Area, 2,860 sq. m. (7,407 sq. km.). Mainly mountainous. The principal R. is Motatán. Mainly agric. with large forests. Crops inc. cereals, coffee, sugar-cane and tobacco. There are deposits of petroleum. Pop. (1981E) 461,416. (ii) Cap. of state of same name sit. NE of Mérida. Sit. at 2,600 ft. (792 metres) a.s.l. Trades in agric. produce. Pop. (1971C) 25,921.

Truk Islands Micronesia, W Pacific. 7 23N 151 46E. Volcanic island group with outlying atolls, forming state within Federated States of Micronesia. Exports copra and dried fish. Area, 45 sq. m. (116 sq. km.). Cap. Moen. Pop. (1980C) 37,742.

Truro Cornwall, England. 50 16N 5 03W. County town in SW sit. on Truro R. Mainly administrative, shopping and service centre. Manu. inc. fertilizers from calcified seaweed and light engineering. Pop. (1982E) 16,040.

Tselinograd Kazakhstan, U.S.S.R. 51 10N 71 30E. Cap. of Tselinograd region in central N of Kazakhstan, sit on Ishim R. An important industrial and railway centre with asphalt and reinforced concrete works, meat packing plants and engineering works. Pop. (1987E) 276,000.

Tsévié Maritime, Togo. Town sit. N

of Lomé. Trades in palm oil, kernels, cacao and cotton. Pop. (1983E) 17,000.

Tsinan   *see* **Jinan**

Tsinghai   *see* **Qinghai**

Tsingtao   *see* **Qingdao**

Tsitsihar   *see* **Qiqihar**

Tskhinvali   South Ossetian Autonomous Region, U.S.S.R. 42 14N 43 58E. Cap. of region sit. NW of Tbilisi. Industries inc. fruit canning.

Tsushima   Japan. 34 20N 129 20E. A group of five islands sit. between Kyushu and S. Korea, and WNW of Kitakyushu. Area, 271 sq. m. (702 sq. km.). Fishing is the chief occupation. Chief town Izuhara.

Tuamotu Islands   French Polynesia. 19 00S 142 00W. An arch. comprising 80 small flat atolls. Area, 299 sq. m. (774 sq. km.). The largest is Rangiroa, and the most important Hao. Copra and pearl shell are produced. Pop. (1983C) 11,793 inc. Gambier Island.

Tuapse   Russian Soviet Federal Socialist Republic, U.S.S.R. 44 07N 39 05E. Port S of Krasnodar sit. on the Black Sea. Industries inc. ship-repairing, engineering, food processing, and oil refining and exporting.

Tübingen   Baden-Württemberg, Federal Republic of Germany. 48 31N 9 02E. Town SSW of Stuttgart sit. on Neckar R. Industries inc. the manu. of precision instruments, electrical household goods, clothing and textiles. Pop. (1985E) 75,000 (inc. 24,000 university students).

Tubuai   Austral Islands, French Polynesia. 23 00S 150 00W. Volcanic island sit. S of Tahiti. Chief town, Mataura. Area, 18 sq. m. (47 sq. km.). Produces coffee, oranges, coconuts and arrowroot.

Tucson   Arizona, U.S.A. 32 13N 110 58W. City and resort in S centre of state. sit. at 2,400 ft. (732 metres) a.s.l. Industries inc. services and government, and the manu. of missile, computer, aeronautical electrical and

electronic equipment. Pop. (1988E) 412,590.

Tucumán   Tucumán, Argentina. Cap. of province in central N sit. at foot of the Sierra de Aconquija in the E Andes. An industrial centre of a district producing sugar-cane, maize and rice, with associated processes. Pop. (1980C) 497,000.

Tucipita   Delta Amacuro, Venezuela. Town sit. NE of Ciudad Bolívar. Centre of an agric. area growing corn, cacao, sugar-cane, tobacco and bananas.

Tugela River   Natal, Republic of South Africa. 29 14S 31 30E. Rises in the Drakensberg and flows *c*. 300 m. (480 km.) E past Colenso to enter the Indian Ocean NE of Stanger. Noted for spectacular waterfalls. Main tribs. Blood and Buffalo Rs.

Tula   Russian Soviet Federal Socialist Republic, U.S.S.R. 54 12N 37 37E. Cap. of Tula region in central European Russia. The chief centre of the Moscow Coal Basin with metallurgical and engineering industries. Firearms are manu., the first gun factory being estab. in 1595. Pop. (1989E) 540,000.

Tulcán   Carchi, Ecuador. 0 49N 77 43W. Town NE of Quito sit. in the Andes. Industries inc. tanning, cattle rearing and the manu. of woollen goods and carpets. Trades in agric. produce. Pop. (1980E) 33,000.

Tulcea   Galati, Romania. 45 11N 28 48E. Port sit. ESE of Galati on Sfantu-Gheorghe R. Industries inc. flour milling, woodworking, tobacco processing and the manu. of sulphuric acid and cordage. Trades in agric. produce. Pop. (1983E) 75,127.

Tuléar   *see* **Toliary**

Tulle   Corrèze, France. 45 16N 1 46E. Cap. of dept. sit. SSE of Limoges. Manu. textiles and armaments. The fabric *tulle* was first produced here. Pop. (1982C) 20,642.

Tulsa   Oklahoma, U.S.A. 36 09N 95 58W. City in NE of state sit. on Arkansas R. A principal centre of the oil industry, sometimes known as the 'oil capital of the world'. Manu. are very varied ranging from oilfield equipment and aircraft to plastics, paint and food products. Pop. (1980C) 360,919.

Tumaco Nariño, Colombia. 1 49N 78 46W. Port in extreme SW sit. on a small offshore island. Exports inc. coffee, tobacco, nuts and gold. Pop. (1980E) 100,000.

Tumbes Peru. 3 30S 80 25W. (i) A dept. in NW bordered by Ecuador and the Pacific Ocean. Area, 1,827 sq. m. (4,731 sq. km.). It is mainly flat and is the chief tobacco growing district of Peru. The principal Rs. are Zarumilla and Tumbes. Crops inc. cotton and rice, and cattle are reared. Industries inc. oil refining, charcoal burning and salt extraction. Pop. (1988E) 135,900. (ii) Cap. of dept. of same name sit. NNW of Lima on Tumbes R. Industries inc. rice milling, charcoal burning and tourism. Trades in tobacco, cotton and rice. Pop. (1988E) 61,100.

Tummel River Perthshire, Scotland. 56 43N 3 55W. Rises in L. Rannoch and flows E to enter Tay R. at Ballinluig. Main trib. Garry R. Used for power.

Tunbridge Wells Kent, England. 51 08N 0 16E. Town and spa in SE. There is an expanding light industrial estate. Pop. (1981C) 96,051.

Tunceli Turkey. 39 05N 39 30E. Province in central E. Area, 3,229 sq. m. (8,363 sq. km.). The principal Rs. are Peri, Murat, Euphrates and Munzur. There are deposits of tin and copper. Cap. Cemiskezek. Pop. (1985C) 151,906.

Tung Ting, Lake *see* **Dongting Hu**

Tungurahua Ecuador. Province in central Ecuador in the Andes. Area, 1,237 sq. m. (3,204 sq. km.). Cap. Ambato. Very mountainous and volcanic. The principal Rs. are Patate and Chambo. Noted for fruit; other crops inc. cereals, sugar-cane, vegetables and cinchona. Pop. (1982C) 324,286.

Tunis Tunisia, North Africa. 36 50N 10 11E. Cap. and sea port with a

modern harbour at the head of Gulf of Tunis with an outport at La Goulette. Just N of La Goulette are the ruins of Carthage. Commercial, tourist and fishing centre. It trades in olives and agric. produce, iron ore and phosphates. Pop. (1984C) 556,654.

Tunisia North Africa. 39 30N 9 00E. Rep. on Mediterranean coast bounded N and E by Mediterranean, SE and S by Libya, W by Algeria. Area, 63,362 sq. m. (164,150 sq. km.). Chief cities Tunis (cap.), Sfax, Bizerta. Mountainous with broad, fertile valleys in N, sloping down to a central grassy plateau which gives way (S) to desert. About 9m. hectares are productive. Chief crops cereals, citrus fruit, dates, olives, vines, nuts. Sheep and cattle are reared on the plateau. The most important minerals are phosphates, with some iron and lead. Industries inc. manu. iron and steel at Menzel Bourguiba, refining oil in Bizerta, manu. cellulose at Kassénrie. Pop. (1988E) 7,745,500.

Tunja Boyaca, Colombia. 5 31N 73 22W. Cap. of dept. in N centre, and in the E. Cordillera sit. at 9,250 ft. (2,819 metres) a.s.l. Occupations are mainly agric. and mining. Pop. (1984E) 75,000.

Tupelo Mississippi, U.S.A. 34 16N 88 43W. Town in NE Mississippi; Convention and route centre on upper Tombigbee R. Pop. (1989E) 63,262.

Turino Piedmont, Italy. 45 03N 7 40E. City at confluence of Po and Dora Riparia Rs. sit. WSW of Milan. Communications, commercial and industrial centre; regional and provincial cap. Industries inc. manu. vehicles, textiles, paper, metal products, chemicals, leather goods. Centre of engineering and publishing. Pop. (1981C) 1,117,154.

Turkana, Lake Ethiopia/Kenya. 4 10N 36 10E. L. in the Great Rift Valley at 1,230 ft. (375 metres) a.s.l., 155 m. (248 km.) long and up to 35 m. (56 km.) wide; sit. in NW Kenya with the N tip across the Ethiopian border. Saline, and shrinking by evaporation, with no known outlet.

Turkestan Central Asia. 43 18N 68

15E. Term describing an extensive area comprising the Chinese province of Sinkiang-Uighur, the Russian reps. of Kazakhstan, Uzbekistan, Kirgizia, Turkmenistan, Tadzhikistan.

Turkey 39 00N 36 00E. Rep. occupying a small area NW of the Bosphorus in Europe, and otherwise extending E to W between the Black Sea and the Mediterranean, with the S frontier continuing E across the N frontiers of Syria and Iraq. Bounded by Iran, NE by the U.S.S.R. Area, 300,947 sq. m. (779,452 sq. km.). Chief cities Ankara (cap.), Istanbul, Izmir, Adana. The Anatolian plain occupies W, bounded SE by the Taurus Mountains and N by the Pontine Mountains, and rises (E) to mountains in Kurdistan, inc. Mount Ararat, 16,916 ft. (5,156 metres) a.s.l. Drained by Sakarya and Kizli Irmak Rs. into the Black Sea, and Tigris and Euphrates Rs. into the (Persian) Gulf. Main crops are cotton, tobacco, wheat, olives, fruit. Stock farming is important on the plain, esp. goats, sheep and cattle. The chief minerals are chrome, coal, lignite, iron ore, petroleum. The main industries are iron and steel, processing agric. produce, paper making, silk weaving. Pop. (1985C) 50,664,458.

Turkmen Soviet Socialist Republic U.S.S.R. 39 00N 59 00E. Constituent rep. of central Asian U.S.S.R. bounded W by Caspian Sea, N by Kazakhstan and Uzbekistan, S by Iran. Area, 186,400 sq. m. (482,776 sq. km.). Cap. Ashkhabad. About 80% desert, rising to the Kopet Dagh foothills (S) and settled in a chain of oases. Irrigation is highly developed, and agric. depends on it. Main crops are cotton, maize, fruit and vegetables, vines and olives. Livestock raised inc. Karakul sheep, horses and cattle. Mineral deposits are extensive, inc. oil, coal, ozocerite, sulphur, salt and magnesium. Industries inc. mining, oil refining, engineering, textiles, chemicals. Pop. (1989E) 3·5m.

Turks and Caicos Islands West Indies. 21 20N 71 20W. Island group forming a British crown colony lying 400 m. (640 km.) NE of Jamaica. Area, 166 sq. m. (430 sq. km.), consisting of Grand Turk, Salt Cay, West Caicos, Providenciales, North, Grand, East and South Caicos. Chief products sponges, shell, salt, sisal. Pop. (1989E) 13,000.

Turku Turku-Pori, Finland. 60 27N 22 17E. Cap. of province and port in SW sit. on the Gulf of Bothnia. Industries inc. metal, engineering, chemical, food processing and textiles. Pop. (1989E) 160,000 (metropolitan area, 250,000).

Turnhout Antwerp, Belgium. 51 19N 4 57E. Town sit. ENE of Antwerp. Noted for the manu. of paper and playing cards. Other manu. inc. textiles, agric. machinery, bricks, pottery, lace and confectionery. Pop. (1989E) 37,000.

Turnu Severin Romania. 44 18N 22 39E. Town on Danube R. near the Iron Gate; site of Trajan's bridge. Industries inc. engineering, food processing and tourism.

Turriff Grampian Region, Scotland. 57 32N 2 28W. Town in NE. Manu. agric. machinery and implements.

Tuscaloosa Alabama, U.S.A. 33 12N 87 34W. Town in NW centre of state sit. on Black Warrior R. Manu. metal and cotton goods, paper and tyres. Pop. (1980C) 75,000.

Tuscany *see* **Toscana**

Tuticorin Tamil Nadu, India. 8 50N 78 09E. Port S of Madurai sit. on Gulf of Mannar. Industries inc. salt works and cotton factories. Coffee and cotton goods are exported. Pop. (1981C) 192,949.

Tuttlingen Baden-Württemberg, Federal Republic of Germany. Town SSW of Stuttgart sit. on Danube R. Manu. surgical instruments and footwear. Pop. (1984E) 30,684.

Tuva Autonomous Soviet Socialist Republic Russian Soviet Federal Socialist Republic, U.S.S.R. 52 00S 176 00E. Sit. to NW of Mongolia, bounded to E, W and N by Siberia and to S by the Republic of Mongolia. Incorporated in the U.S.S.R. as an Autonomous Region in 1944 and later to an Autonomous Republic in 1961. Area, 65,810 sq. m. (170,448 sq. km.). Cap. Kizyl. Chief industries pastoral farming. Also mining of gold, cobalt and asbestos. Pop. (1989E) 309,000.

Tuvalu Pacific Ocean. 8 00S 178 00E. Independent state formerly Ellice Islands comprising 8 atolls and a volcanic island. Area, 9.3 sq. m. (24 sq. km.). Only export copra. Pop. (1985E) 8,229.

Tuxtla Gutiérrez Chiapas, Mexico. 16 45N 93 07W. State cap. sit ESE of Mexico City. Commercial centre of an agric. area producing coffee, cacao, sugar-cane, tobacco and bananas. Pop. (1985E) 200,000.

Tuz, Lake Turkey. 38 45N 33 25E. A salt L. 65 m. (104 km.) SSE of Ankara, lying at 2,900 ft. (884 metres) a.s.l.

Tuzla Bosnia and Hercegovina, Yugoslavia. 44 32N 18 41E. Town in NW centre sit. on Jala R. There are coal, lignite, salt and timber industries Pop. (1981C) 121,717.

Tver *see* **Kalinin**

Tweed River Scotland. 55 46N 2 00W. Rises near Tweedswell and flows 97 m. (155 km.) E to the English border which it forms (NE) till its confluence with the Whiteadder R., when it turns E to enter the North Sea at Berwick-upon-Tweed. It is tidal for 10 m. (16 km.). Main tribs, Yarrow, Ettrick, Teviot and Till Rs.

Twickenham Greater London, England. 51 27N 0 20W. District of W London on Thames R. Mainly residential.

Tyler Texas, U.S.A. 32 21N 95 18W. City sit. ESE of Dallas. Industries inc. oil refining, food processing, and the growing of roses and ornamental shrubs. Pop. (1980C) 70,508.

Tyne and Wear England. 54 55N 1 36W. Metropolitan county; under 1974 re-organization consists of Newcastle upon Tyne, North Tyneside, Gateshead, South Tyneside and Sunderland. There are 2 New towns, Killingworth and Washington. Impor-

tant industries are shipbuilding and engineering. Other industries inc. light engineering. Pop. (1988E) 1,130,600.

Tynemouth Tyne and Wear, England. 55 01N 1 24W. Town and port in NE, sit. at mouth of R. Tyne. Industries are ship-repairing, engineering and the manu. of plastics, clothing, furniture and confectionery. Pop. (1985E) 9,442.

Tyne River England. 55 01N 0 26W. Formed NW of Hexham by confluence of S and N Tyne Rs. and flows 30 m. (48 km.) E past Corbridge, Blaydon and Newcastle upon Tyne to enter the North Sea by an estuary between Tynemouth (N) and South Shields (S). The N Tyne rises in the Cheviot Hills on the Scottish border SE of Carter Fell and Peel Fell. The S Tyne rises on Cross Fell in E Cumbria. The course of the N and S Tyne Rs. is rural and picturesque. The Tyne proper flows through a coalfield and serves industrial towns inc. Newcastle, Jarrow, Wallsend, South Shields and Gateshead. The main tribs. are Derwent and Team Rs. Navigable up to Blaydon, 8 m. (13 km.) above Newcastle.

Tyrone Northern Ireland. County sit. to W of Lough Neagh. Area, 1,260 sq. m. (3,263 sq. km.). Mainly hilly with the Sperrin Mountains, 2,240 ft. (683 metres) a.s.l., in N, and the Slievebeagh, 1,255 ft. (383 metres) a.s.l. in S. The principal Rs. are Blackwater, Foyle and Strule. Oats, potatoes, flax and turnips are grown, and cattle and sheep are reared. County town Omagh. Pop. (1971C) 138,975.

Tyumen Russian Soviet Federal Socialist Republic, U.S.S.R. 57 09N 65 32E. Cap. of Tyumen region in W Siberia sit. on Tura R. Industries inc. boat building, sawmilling, tanning, and the manu. of chemicals and carpets. Pop. (1987E) 456,000.

Tywyn Gwynedd, Wales. 52 36N 4 05W. Town sit. on Cardigan Bay N of Aberystwyth. Holiday resort. Small industrial estate producing: honey, puppets, printing, electronics and model engineering. Pop. (1985E) 2,800.

U

Uapou Marquezas Islands, French Polynesia. Volcanic island rising to 4,043 ft. (1,232 metres). Area, 25 sq. m. (66 sq. km.). Most populous of group. Chief town, Hakahau.

Ubangi River Zaïre. 0 30S 17 42E R. formed by union of R. Mbomu and R. Uele. Chief N trib. of R. Congo.

Ube Yamaguchi, Japan. 33 56N 131 15E. Town sit. WSW of Hiroshima, on the Inland Sea with good harbour. Important centre for chemicals and heavy industry.

Uberaba Minas Gerais, Brazil. 19 45S 47 55W. Town sit. in the centre of an agric. region. Manu. lime. Pop. (1984E) 140,000.

Ucayali River Peru. 4 30S 73 30W. Chief headstream of R. Amazon formed by R. Apurímac and R. Urubamba. Its length is 1,200 m. (1,920 km.) and half is navigable.

Udaipur Rajasthan, India. 24 26N 73 44E. Town sit. SW of Jaipur at S end of the Aravalli Hills. L. Pichola is adjacent with fine anc. buildings on its many islands. Trade in cotton and grain. Pop. (1981C) 232,588.

Uddevalla Göteborg and Bohus, Sweden. 58 21N 11 55E. Industrial town and port on the Byfjord, and W of the S tip of L. Väner. Manu. textiles, wood-pulp, paper and matches. Also shipbuilding and sugar refining. Pop. (1988E) 46,489.

Udine Italy. 46 03N 13 14E. Town in N of state. Manu. textiles, chemicals and bells. Pop. (1983E) 102,021.

Udmurt Autonomous Soviet Socialist Republic Russian Soviet Federal Socialist Republic, U.S.S.R. 57 30N 52 30E. Area, 16,250 sq. m. (42,087 sq. km.). In 1920 constituted an Autonomous Region (then named

Votyak) and then an Autonomous Republic 1934. Cap. Izhevsk, sit. ENE of Moscow. Chief industries are the manu. of locomotives, machine tools and other engineering products; also varied light industries. Pop. (1989E) 1,609,000.

Uèle River Zaïre. Rises as R. Kibali NW of L. Mobutu Sese Seko, formerly L. Albert, near the border of Zaïre and Uganda and flows for 750 m. (1,200 km.) to join Mbomu R. to form Ubangi R. at Yakoma, NNE of Lisala.

Ufa Bashkiria, U.S.S.R. 54 44N 55 56E. Cap. of the Bashkirian Autonomous Soviet Socialist Republic sit. at confluence of Ufa R. and Belaya R. and E of Moscow. One of the largest industrial centres of the Urals, with important industries of heavy and light engineering; oil refining with pipe-lines to the Volga-Ural coalfield; the largest petrochemical plant in the U.S.S.R.; sawmilling, food processing and many others. Pop. (1989E) 1,083,000.

Ufa River U.S.S.R. Rises in S Urals, flows NW and SSW for 450 m. (720 km.) to join Belaya R. at Ufa.

Uganda 2 00N 32 00E. Bounded on N by Sudan, on E by Kenya, on S by Tanzania and W by Zaïre. Area, 91,343 sq. m. (236,860 sq. km.). The surface is a plateau over 3,000 ft. (914 metres) a.s.l. with mountains in S and W rising to 20,000 ft. (6,096 metres). The chief R. is Nile. Cap. Kampala. For administrative purposes Uganda is divided into 4 regions: (i) the Eastern Region, comprising the districts of Bugisu, Bukedi, Busoga, Mbale Township, Sebei and Teso; (ii) the Western Region, comprising the districts of Bunyoro, Toro, Ankole and Kigezi; (iii) Buganda Region, with islands in L. Victoria, comprising the districts of Mengo, Masaka and

Mubende; and (iv) the Northern Region, comprising the districts of Karamoja, Lango, Acholi and West Nile. Cotton and coffee are principal exports. Other cash crops, tea, tobacco, groundnuts, maize, castor seed, sisal and sugar. Exploitable forests consist almost entirely of hardwoods. With its 13,600 sq. m. (35,224 sq. km.) of L. and many Rs., Uganda possesses one of the largest fresh-water fisheries in the world. Fish farming (esp. carp and tilapia) is a growing industry. Minerals produced inc. copper, cement and tin. About 3m. Africans speak Bantu languages; there are a few Congo pygmies living near the Semliki R.; the rest of the Africans belong to the Hamitic, Nilotic and Sudanese groups. Ki- Swahili is generally understood in trading centres. Pop. (1989E) 17m.

Uist, North and **South**   Western Isles, Scotland. 57 22N 7 28W. Two islands in the Outer Hebrides. (i) N. Uist lies SW of Harris, across the Little Minch. Area, 118 sq. m. (305 sq. km.). Mainly hilly, Ben Eaval in the E rising to 1,138 ft. (347 metres) a.s.l. In the E there are the two sea lochs, Maddy and Eport. Chief village Lochmaddy. Pop. (1973E) 1,840. (ii) S. Uist is sit. S of N. Uist, with the island of Benbecula lying between. Area, 141 sq. m. (365 sq. km.). Ben More rises to 2,034 ft. (619 metres) a.s.l. There are three sea lochs, Boisdale, Eynort and Skiport, on the E coast. Chief village Loch-boisdale.

Uitenhage   Cape Province, Republic of South Africa. 33 40S 25 28E. Town with railway engineering and car assembly plant. Manu. textiles, car components and tyres. Pop. (1985E) 200,500.

Ujiji   Tanzania. Port on the E shore of L. Tanganyika, and sit. SE of Kigoma. Formerly the centre of the Arab slave trade. Dr. Livingstone was found here by Stanley in 1871.

Ujjain   Madhya Pradesh, India. 23 11N 75 50E. Town sit. SSE of Ajmer on Sipra R. Manu. textiles. A sacred Hindu city celebrating a bathing festival, the Kumbh Mela, every twelfth year. Pop. (1981C) 282,203.

Ujung Padang   Sulawesi, Indonesia. 5 09S 119 28E. Sea port and cap. of Sulawesi province sit. on SW coast. Formerly Makassar. Exports inc. coffee, copra, vegetable oils, spices, gums and teak. Pop. (1980E) 435,000.

Ukraine   U.S.S.R. 49 00N 32 00E. The Ukrainian Soviet Socialist Republic was proclaimed in 1917 and was finally estab. two years later. In 1920 it concluded a military and economic alliance with the Russian Soviet Federal Socialist Republic and in 1923 formed, together with the other Soviet Socialist Reps., the U.S.S.R. Area is 231,990 sq. m. (603,700 sq. km.). Cap. Kiev. Contains rich land and is therefore important agric. Coal and iron ore mined extensively so industrial production is high. Other important minerals are manganese, oil, gypsum and alabaster. Pop. (1989E) 51·7m.

Ulan Bator   Mongolian People's Republic. 47 55N 106 53E. Cap. and the main industrial and commercial centre. Industries inc. food processing and building materials, manu. textiles, prefabricated houses and leather goods. Connected by railway with Ulan-Ude in the U.S.S.R. and Beijing, China. Pop. (1988E) 500,000.

Ulan-Ude   Siberia, U.S.S.R. 51 50N 107 37E. Cap. of the Buryat Autonomous Soviet Socialist Republic. Sit. at confluence of Selenga R. and Uda R. Important railway centre. Industries are railway engineering, ship-repairing, and sawmilling. Manu. glass and food products. Pop. (1985E) 335,000.

Ullswater   Cumbria, England. 54 34N 2 52W. The second largest L. in Eng. sit. E of Keswick. Length 7$\frac{1}{2}$m. (12 km.), width $\frac{1}{2}$m. (1 km.) and depth 210 ft. (64 metres).

Ulm   Baden-Württemberg, Federal Republic of Germany. 48 24N 10 00E. Industrial town sit. on left bank of Danube R. Manu. machinery, textiles, leather goods and cement. Pop. (1989E) 105,000.

Ulster   Ireland. Anc. province of Ireland consisting of the counties of Cavan, Donegal, Monaghan, and the six counties now forming Northern Ireland. The name is mainly used as a synonym for Northern Ireland, a political unit of the United Kingdom. The province of Ulster in the Republic of Ireland consists of counties Cavan, Donegal and Monaghan.

Ulveston Cumbria, England. 54 12N 3 06W. Market town sit. NE of Barrow-in-Furness. Industries engineering and brewing. Manu. pharmaceuticals, lead crystal, underwater electronic control systems and electrical goods. Pop. (1981C) 11,970.

Ulyanovsk U.S.S.R. 54 20N 48 24E. (i) Region in the Russian Soviet Federal Socialist Republic. Main R. is Middle Volga. Chiefly agric. (ii) Cap. of region of same name sit. NW of Kuibyshev between Volga R. and Sviyaga R. Important transportation centre. Industries engineering, sawmilling, tanning and food production. Lenin's birthplace. Pop. (1989E) 625,000.

Umbria Italy. 42 53N 12 30E. Region of the Apennines in central Italy, composed of the provinces of Perugia and Terni. Area, 3,265 sq. m. (8,456 sq. km.). Mainly agric. producing cereals, olive oil and wine. Largest R. is upper Tiber. Chief town and cap. Perugia. Pop. (1981C) 807,552.

Umeå Västerbotten, Sweden. 63 50N 20 15E. Cap. of Västerbotten county sit. at mouth of Ume R. Manu. and exports associated with wood. Pop. (1989E) 89,589.

Umm al Qaiwain United Arab Emirates. 25 35N 55 34E. A former Trucial State stretching for 15 m. (24 km.) along the Coast of the Gulf. Area, 200 sq. m. (518 sq. km.). Chief town, Umm al Qaiwain. Pop. (1980C) 12,300.

Umm-An-Nassan Island Bahrain. Sit. in the Gulf off the NW coast. Area, 3.5 m. (6 km.) by 2.5 m. (4 km.).

Umtali *see* **Mutare**

Umtata Transkei, Republic of South Africa. 31 36S 28 49E. Cap. sit.

NNE of East London on Umtata R. Railway terminus. Pop. (1988E) 45,000.

Ungava Quebec, Canada. 60 00N 74 00W. District in N. inc. the Ungava Peninsula, to the E of Hudson Bay. Area, 351,780 sq. m. (911,110 sq. km.). Large iron-ore deposits.

Union of Soviet Socialist Republics 60 00N 100 00E. Until 1917 the territory now forming the U.S.S.R., together with that of Finland, Poland and certain tracts ceded in 1918 to Turkey, but less the territories then forming part of the German, Austro-Hungarian and Japanese empires—E Prussia, E Galicia, Transcarpathia, Bukovina, E Sakhalin and Kurile Islands—which were acquired during and after the Second World War, was constituted as the Russian Empire.

In 1917 a revolution broke out, a Provisional Govt. was appointed, and in a few months a rep. was proclaimed. Late in 1917 power was transferred to the second All-Russian Congress of Soviets. This elected a new govt., the Council of People's Commissars, headed by Lenin.

Early in 1918 the third All-Russian Congress of Soviets issued a Declaration of Rights of the Toiling and Exploited Masses, which proclaimed Russia a Rep. of Soviets of Workers', Soldiers' and Peasants' Deputies; and in the middle of 1918 the fifth Congress adopted a Constitution for the Russian Socialist Federal Soviet Republic. In the course of the civil war other Soviet Reps. were set up in the Ukraine, Belorussia and Transcaucasia. These first entered into treaty relations with the Russian Soviet Federal Socialist Republic and then, in 1922, joined with it in a closely integrated Union. Area, 8.65m. sq. m. (22.4m. sq. km.). Cap. Moscow.

The U.S.S.R., until 1928 predominantly agric. in character, has become an industrial-agric. country. The total area under cultivation was (1990) 210·3m. hectares. Of the 814m. hectares of forest land of the U.S.S.R., a large portion is administered and worked by the state, and the other, about 19m. hectares in extent, is granted for use to the peasantry free of charge. The largest forested areas are

515m. hectares in the Asiatic part of the U.S.S.R., 51.4m. along the N seaboard, 25.4m. in the Urals and 17.95m. in the NW.

The U.S.S.R. is rich in minerals and claims that it contains 58% of the world's coal deposits, 58.7% of its oil, 41% of its iron ore, 76.7% of its apatite, 25% of all timber land, 88% of its manganese, 54% of its potassium salts and nearly one-third of its phosphates.

In the 1930s practically all Soviet oil came from the Caucasian fields, of which the Baku fields yielded 75–80% and the Grozny and Maikop fields between them 15%. Since then, the distribution has considerably changed. The Ural-Volga area, the 'Second Baku', has 4 large centres in operation, at Samarska Luka (Kuibyshev), Tuimazy (Bashkiria), Ishimbaev (Bashkiria) and Perm. A large new oilfield has been developed in the Trans-Volga area of the Saratov region. The Tyumen (West Siberian) complex accounted for over 50% of the U.S.S.R.'s oil output in 1990. Pop. (1989E) 286·7m.

Union Township New Jersey, U.S.A. Town sit. SW of New York City. Kean College is here. Residential, business and manu. town. Pop. (1988E) 51,342.

United Arab Emirates 25 50N 54 00E. The United Arab Emirates consists of the former Trucial States; Abu Dhabi, Dubai, Sharjah, Ajman, Umm al Qaiwain, Ras al Khaimah and Fujairah. The small state of Kalba was merged with Sharjah in 1952. The area is approx. 32,300 sq. m. (83,660 sq. km.).

The largest and most important town of the United Arab Emirates is Dubai (about 60,000 inhabitants). Abu Dhabi has a pop. in excess of 60,000 and has expanded rapidly since it became an important oil producer. Trad. the inhabitants of the United Arab Emirates have depended for their livelihood on trading, fishing and pearling. In recent years there has been a drift towards the towns, and the oil industry, particularly in Abu Dhabi, has become the major employer.

From Sha'am, 35 m. (56 km.) SW of Ras Musam dam, for nearly 400 m.

(640 km.) to Khor al Odeid at the SE end of the peninsula of Qatar, the coast, formerly known as the Pirate Coast, of the (Persian) Gulf (together with 50 m. (80 km.) of the coast of the Gulf of Oman) belongs to the rulers of the 7 Trucial States. In 1820 these rulers, after committing acts of hostility against the East India Co., signed a treaty prescribing peace with the Brit. Govt. and perpetual abstention from plunder and piracy (specifically inc. the slave trade) by land and sea. This treaty was followed by further agreements providing for the suppression of the slave trade and by a series of other engagements, of which the most important are the Perpetual Maritime Truce (May 1853) and the Exclusive Agreement (March 1892). Under the latter, the shaikhs, on behalf of themselves, their heirs and successors, undertook that they would on no account enter into any agreement or correspondence with any power other than the Brit. Govt., receive foreign agents, or cede, sell or give for occupation any part of their territory save to the Brit. Govt. Brit. forces withdrew from the Gulf in Dec. 1971 and the treaties whereby Brit. had been responsible for the defence and foreign relations of the Trucial States were terminated, being replaced on 2 Dec. 1971 by a treaty of friendship between Brit. and the United Arab Emirates. The United Arab Emirates, comprising 6 of the 7 Trucial States, came into existence on the same day. The remaining state, Ras al Khaimah, joined in Feb. 1972. Pop. (1988E) 1·6m.

United Arab Republic *see* **Egypt**

United Kingdom 55 00N 3 00W. Kingdom consisting of England, Wales, Scotland, Northern Ireland. Bounded S by the English Channel, E by the North Sea, N and W by the Atlantic Ocean and the Irish Republic. Area, 94,500 sq. m. (244,755 sq. km.). Chief cities London (cap.), Cardiff (cap. of Wales), Edinburgh (cap. of Scotland), Belfast (cap. of Northern Ireland), Birmingham, Glasgow, Liverpool, Manchester, Sheffield, Leeds, Bristol, Coventry, Nottingham, Bradford, Wolverhampton, Derby, Hull, Leicester, Newcastle upon

Tyne, Southampton, Stoke-on-Trent, Sunderland, Dundee, Aberdeen, Swansea, Portsmouth. The mainland is low-lying in the SE rising to highland in the Pennine ridge which extends N to S up central England. Beyond it is broad lowland watered by the Dee and Severn Rs. and rising again to mountains in N and central Wales and in NW England. At the N end of the Pennines the hills of the Scottish border slope down to a central Scottish lowland which rises (N) to mountains with Ben Nevis, 4,406 ft. (1,343 metres) a.s.l., the highest peak. The NW coast is deeply indented with many offshore islands. Northern Ireland is fertile plain bounded NW, NE and SE by hills. NW England and NW Scotland have numerous Ls. and sea-inlets. The main Rs. are Severn, Thames, Great Ouse, Nene, Welland, Trent, Dee, Mersey, Aire, Swale, Tees, Tyne, Tweed, Clyde, Forth, Tay, Wye, Usk, Teifi and Bann. The industrial midlands of England have a canal network linking navigable rivers, but these are not used for transport as much as in the past. Farm land covers *c.* 18·8m. hectares. Chief crops are cereals, esp. barley, wheat and oats, potatoes, sugar-beet and fodder crops. Dairying, beef-cattle and sheep-farming are important. Vegetable crops are grown extensively in E England and also in E Scotland. Manu. exists throughout the U.K. but there is a great concentration in the midlands, particularly vehicle production, and the N of England. The chief minerals are coal (central Scotland, NE England, central England and S Wales, with smaller deposits in Kent, Cumbria and NE Wales), iron, mainly in E central England; limestone, chalk, clays, salt, fluorspar, gypsum, barytes, lead, and slate. There are submarine deposits of oil and natural gas in the North Sea. Coastal fisheries produce cod, haddock, whiting, coal-fish, hake, mackerel, sole, herring, plaice, halibut, turbot and shell-fish. Pop. (1981C) 55·78m.

United States of America   37 00N 96 00W. Federal Republic; bounded N by Canada, E by Atlantic, S by Gulf of Mexico and Mexico, and W by Pacific. Alaska to NW of Can. is a detached territory. Area, 3,536,855 sq. m.

(9,160,454 sq. km.). Cap. Washington, District of Columbia. Largest cities, New York, Chicago, Los Angeles, Philadelphia, Detroit and Houston. The surface is divided into three natural regions; (i) Appalachian Mountains along E coast, with coastal plain (200 m. (320 km.) wide); (ii) Central Plains which are divided into Gulf Plain extending inland from Gulf of Mex., Prairies N of Gulf Plain stretching to Great Ls., and Great Plains at a higher elevation, W of Gulf Plains and Prairies. Central Plains occupy largest area of country, and are broken by Ozark Mountains between Gulf Plain and Great Plains; (iii) Rocky Mountain system in W, of great elevation, extreme height, 14,898 ft. (4,541 metres). The NE is drained by streams flowing to Great Ls. and St. Lawrence R.; coastal strip by Penobscot, Connecticut, Hudson, Delaware, Susquehanna, Potomac, James, Roanoke and Savannah Rs.; Central Plain by Mississippi–Missouri R. system, Alabama, Sabine and Trinity Rs., flowing direct to Gulf of Mex.; SW by Rio Grande del Norte (flowing into Gulf of Mex.) and Colorado R. into Gulf of California; NW by Columbia R., which breaks across coast mountains.

The great Rs. of Central Plains are navigable for long distances; Hudson R. is navigable, but other E rivers are chiefly important for power and harbour purposes; W rivers mostly flow through deep gorges and, except Columbia R., are of no commercial importance. E coast as far S as Long Island is rocky, with many good harbours; further S it is low and sandy, and broken only by great openings of Delaware and Chesapeake Rs.; along Gulf of Mex. it is low and swampy, while near the middle the Mississippi R. delta projects some 50 m. (80 km.) out to sea. The long line of W cliffs is broken only in extreme N, and near centre, at Golden Gate, San Francisco.

Climate varies from winter snows and cold of central N, to mildness of Pacific slopes and sub-tropical conditions of S and SE. Greatest range of temp. is in interior, and is much greater on NE coast than on Pacific Coast. Rainfall in E half varies from over 60 in. (152 cm.) in S to 20 in. (51 cm.) at base of Rocky Mountains;

along Pacific coast-strip it is abundant in N and scanty in S; some parts of W hills and W plateaux are desert.

Agric. is chief industry, but others are increasing rapidly in importance. Of country E of 100° long., nearly two-thirds is under crops—N districts producing crops of ordinary grains and green crops, centre maize and tobacco, S cotton, with rice and sugarcane in much smaller proportion on shores of Gulf of Mex. Chief grain crops are maize, wheat and oats; barley, flax, rye, buckwheat and rice are also important crops. Chief wheat-growing states lie in a central strip and in NW. Chief maize-producing states lie S and W of Great Ls. Great cotton states are in SE; area under cotton is over 45m. acres. Chief tobacco states lie roughly N of cotton belt. Sugarcane is grown in Louisiana, Georgia, etc.; over 1m. acres (chiefly in Louisiana and Texas) are under rice. Vines succeed best in California. Among temperate fruits are apples, peaches, plums and pears; many tropical and sub-tropical fruits are cultivated in California and several S states. Stock rearing is important in W half of country (chiefly cattle, pigs, sheep, horses), the crops being produced chiefly for feeding. There are great tracts of forest land on the Atlantic side, the Pacific coast, the Rocky Mountains and W central States, occupying about one-fifth of whole country. Salmon fishing of W, fisheries off Grand Bank of Newfoundland and Alaska.

Minerals are also of very great importance. A large proportion of world's petroleum is produced in U.S.A.; other important minerals inc. gold, silver, copper, lead, zinc, aluminium, iron, coal and natural gas. Manu. are very important in NE esp. in states of New York and Pennsylvania. Pop. (1980C) 226,545,805.

University City Missouri, U.S.A. 38 40N 90 20W. Residential suburb immediately NW of St. Louis, Missouri. Pop. (1980C) 42,690.

Unna North Rhine-Westphalia, Federal Republic of Germany. 51 32N 7 41E. Town sit. E of Dortmund. Industrial and coalmining centre. Manu. metal goods. Pop. (1985E) 58,000.

Unterwalden Switzerland. 46 50N 8 15E. Forest canton in central Switz. sit. S of L. Lucerne and comprising Obwalden, 190 sq. m. (492 sq. km.), and Nidwalden, 106 sq. m. (274 sq. km.). Mainly agric. Pop. (1980C) 54,482.

Upington Cape Province, Republic of South Africa. 28 25S 21 15E. Town sit. on Orange R. Railway junction, commercial, educational and tourist centre. Pop. (1989E) 56,100.

Upolu Island Western Samoa, South Pacific. 13 55S 171 45W. A volcanic island sit. SE of Savaii, it is the second in size but most important island of Samoa. Area, 433 sq. m. (1,126 sq. km.). Mountainous, rising to 3,608 ft. (1,100 metres) a.s.l. in Vaaifetu. The lowlands are fertile producing rubber, cacao, coconuts and bananas. The chief town is Apia. Pop. (1981C) 114,980.

Upper Arlington Ohio, U.S.A. Suburb of Columbus in central Ohio on the Scioto R. Residential community; a computer information company is the largest employer. Pop. (1980C) 35,648.

Upper Austria Austria. 48 10N 14 00E. Federal State in N bordering Bavaria on W and Czech. on N. Area, 4,625 sq. m. (11,978 sq. km.). Cap. Linz. Chief Rs. are Danube, Inn and Enns. Mainly agric. inc. forestry. Pop. (1988E) 1,300,300.

Upper Nile Sudan. Province in central S bordered by Ethiopia. Area, 35,625 sq. m. (92,269 sq. km.). Mainly grassland with forests in the S. The principal Rs. are Bahr el Jebel and Sobat. The most important occupation is stock rearing, mainly cattle, sheep and goats. Peanuts, corns and durra are grown. Cap. Malakal. Pop. (1983C) 1,599,605.

Upper Volta see **Burkina Faso**

Uppingham Leicestershire, England. 52 35N 0 43W. Market town sit. W of Peterborough. Light industry inc. leather goods. Pop. (1985E) 3,500.

Uppsala Uppsala, Sweden. 59 53N

17 38E. Cap. of county of same name. Cultural centre and university town. Many fine buildings inc. Gothic cathedral, dating from 13th cent., and famous library. Industries inc. pharmaceuticals, building and the biggest employer is the Academic hospital. Pop. (1988E) 161,828.

Ural Mountains   U.S.S.R. 60 00N 60 00E. Mountain range running N to s from the Arctic Ocean, and forming part of the boundary between Europe and Asia. It extends for over 1,400 m. (2,240 km.) of mainly low elevation, the highest peak being Narodnaya, 6,214 ft. (1,894 metres) a.s.l. There are rich mineral resources inc. iron, copper, nickel, manganese, gold and platinum. Extensive mixed and coniferous forests.

Ural River   U.S.S.R. 47 00N 51 48E. Rises in s of Ural Mountains and flows for 1,400 m. (2,240 km.) s to enter the Caspian Sea through a delta at Chapaev sw of Guryev.

Uralsk   Kazakhstan, U.S.S.R. 51 14N 51 22E. Cap. of Uralsk region, sit. on Ural R. Centre of an agric. area with chief occupations meat packing and wool processing. Pop. (1985E) 192,000.

Uranium City   Saskatchewan, Canada. 59 34N 108 36W. Town sit. N of L. Athabasca. Important for uranium mining. Industries include fishing and tourism. Pop. (1988E) 209.

Urawa   Saitama, Japan. 35 51N 139 39E. Cap. of Saitama prefecture sit. NNW of Tokyo. Pop. (1988E) 392,000.

Urbana   Illinois, U.S.A. 40 07N 88 12W. City sit. ssw of Chicago. Manu. scientific instruments. Pop. (1980C) 35,978.

Urbino   Marche, Italy. 43 43N 12 38E. Town sit. in E central Italy. Industries inc. tourism, textiles, lace and majolica.

Ure River   England. 54 20N 1 25W. Rises in the Pennines and flows E and SE for 50 m. (80 km.) to join R. Swale and form R. Ouse.

Urfa   Urfa, Turkey. 37 12N 38 50E. Cap. of province of same name sit. N of border with Syria. Anc. city of Edessa. Now centre of wheat industry. Pop. (1988E) 147,488.

Uri   Switzerland. 46 43N 8 35E. Forest canton in central Switz. Area, 415 sq. m. (1,075 sq. km.). Cap. Altdorf. Bounded in N by L. Lucerne and in s by the St. Gotthard Pass. Principal R. is Reuss. Mainly glacial rocks but some agric. carried on. Pop. (1988E) 33,544.

Urmia, Lake   Iran. 37 40N 45 30E. Saline L. sit. in the NW with no outlet. Area, 1,500 sq. m. (3,885 sq. km.), increasing occasionally to over 2,300 sq. m. (5,960 sq. km.).

Urmston   Greater Manchester, England. 53 27N 2 21W. District sit. wsw of Manchester. Industries inc. food processing. Pop. (1981E) 44,000.

Uruapán   Michoacán, Mexico. 19 25N 102 04W. Town sit. w of Mexico City, at 5,300 ft. (1,615 metres) a.s.l. Tourist centre. Pop. (1980C) 146,998.

Urubamba River   Peru. 10 43S 73 48W. Rises in the SE Andes and flows NNW through gorges for 450 m. (720 km.) to join Apurímac R. and form Ucayali R.

Uruguaiana   Rio Grande do Sul, Brazil. 29 45S 57 05W. Town sit. on Uruguay R. and on the Argentina border. Centre of cattle industry. Manu. leather goods and soap. Pop. (1980E) 75,000.

Uruguay   32 30S 55 30W. Bounded in NE by Brazil, on SE by Atlantic, on s by Rio de la Plata and w by Argentina. Area, 72,172 sq. m. (186,926 sq. km.). Cap. Montevideo. Surface is flat grassland with low ridges of hills in N. It is drained by Rio Negro and other tribs. of Uruguay R. Uruguay is primarily a pastoral country. Of the total land area of 46m. acres some 41m. are devoted to farming, of which 90% to livestock and 10% to crops. Wool, sheep skins and hides are important exports. Crops grown are; wheat, linseed, oats, barley, rye, maize, sunflower, ground nuts, cotton, rice and sugar (cane and beet). Industries inc. meat packing, lumbering, oil refining, cement manu. and general light industry. Pop. (1988E) 3,080,000.

Uruguay River   South America. 34 12S 58 18W. Rises in the Serra do Mar, SE Brazil and flows for *c.* 1,000 m. (1,600 km.) W and S. It divides Argentina from Brazil and then from Uruguay and flows into the Rio de la Plata. It is navigable for *c.* 200 m. (320 km.) to the cataract at Salto, Uruguay.

Urumchi   *see* **Urumqi**

Urumqi (Urumchi)   Xinjiang Uygur, China. 43 48N 87 35E. Sit. N of T'ienshan on E bank of Urumqi R. at 9,000 ft. (2,743 metres) a.s.l. Transportation centre. Manu. textiles and chemicals. Pop. (1982C) 947,000.

Usak   Kutahya, Turkey. 38 43N 29 28E. Town sit. E of Smyrna. Manu. carpets. Sugar refining.

Usedom   German Democratic Republic/Poland. 53 50N 13 55E. Island across the mouth of the Szczecin harbour on the Baltic coast, bisected by the G.D.R./Polish frontier. The Polish name is Uznam. Area, 170 sq. m. (440 sq. km.). Chief town Swinoujście (Polish). Low-lying, with important fisheries. The main indust. is shipbuilding. It has some tourist resorts.

Ushant   Finistère, France. Island sit. WNW of Brest off the W coast of Brittany. Sole port is Ouessant (Ushant) on the SW. Chief occupation is fishing.

Ushuaia   Tierra del Fuego, Argentina. 54 47S 68 20W. Cap. of Tierra del Fuego province sit. on the Beagle Channel, SE of Punta Arenas, Chile. Occupations are lumbering, sheep-rearing and fishing. The most S town in the world. Pop. (1980E) 11,000.

Usk   Gwent, Wales. 51 43N 2 54W. Town sit. NNE of Newport on R. Usk serving agric. area. Pop. (1976E) 2,033.

Usk River   Wales. 51 36N 2 58W. Rises SE of Llandovery and flows E,

SE and S for 60 m. (96 km.) to enter the Bristol Channel at Newport. The Usk valley is noted for its scenery and fishing.

Uspallata Pass   South America. Pass at 12,600 ft. (3,840 metres) a.s.l. in the Andes Mountains linking Argentina and Chile between Mount Aconcagua in N and Mount Tupungato in S.

Ússuriisk   Russian Soviet Federal Socialist Republic, U.S.S.R. 43 48N 131 59E. Industrial town sit. N of Vladivostock in S Primorye Territory. It stands at the junction of the Trans-Siberian and Chinese Eastern Railways. Industries railway engineering, sawmillling, soy bean processing.

Ússuri River   U.S.S.R. 48 27N 135 04E. Rises 50 m. (80 km.) from Ussuri Bay, an inlet of Sea of Japan, and flows for 560 m. (896 km.) N to Amur R. SW of Khabarovsk.

Ústi-nad-Labem   Bohemia, Czechoslovakia. 50 40N 14 02E. Cap. of Severočeský region on Labe (Elbe) R. Transportation centre. Manu. chemicals, machinery and textiles. Trades in coal, timber and sugar-beet. Pop. (1983E) 90,000.

Ustinov (Izhevsk)   Russian Soviet Federal Socialist Republic, U.S.S.R. 56 49N 53 11E. Cap. of the Udmurt Autonomous Soviet Socialist Rep. sit. ENE of Gorki on the Izh R. An important metallurgical centre with manu. of firearms, motorcycles, lathes and machinery. Pop. (1987E) 631,000.

Ust-Kamenogorsk   Kazakhstan, U.S.S.R. 49 58N 82 38E. Town sit. ESE of Semipalatinsk on Irtysh R. Industrial centre of a district mining zinc, copper and lead, it has the largest non-ferrous metallurgical plant in the U.S.S.R. At Ablaketka, 10 m. (16 km.) to the S, there is a large hydro-electric power station. Pop. (1987E) 321,000.

Ust-Ordynsky Buryat National Area   Russian Soviet Federal Socialist Republic, U.S.S.R. A National Area in the Irkutsk Region, S Siberia. Cap. Ust-Ordynsky sit. NNE of Irkutsk. Mainly wooded, with cattle rearing and some coalmining. Pop. (1989E) 136,000.

Ust Urt Kazakhstan/Kara-Kalpak, U.S.S.R. A desert plateau in the SW between the Aral Sea and the Caspian Sea. Area, 90,000 sq. m. (233,100 sq. km.).

Usulután Usulután, El Salvador. 13 21N 88 27W. Cap. of province of same name, sit. at the foot of the volcano Usulután in the SE. Mainly agric. Pop. (1981E) 69,355.

Utah U.S.A. 39 30N 111 30W. Inland mountain State lying among W ridges of the Rocky Mountains and bounded by Idaho, Wyoming, Colorado, Arizona and Nevada R. Area, 84,916 sq. m. (219,932 sq. km.). Acquired by the U.S. during the Mexican war and settled by Mormons in 1847. Cap. Salt Lake City. Other towns inc. Ogden and Provo. The Wasatch Mountains which run roughly N and S reach *c.* 13,000 ft. (3,962 metres), and the Uinta range in NE rises to *c.* 14,000 ft. (4,267 metres); in NW is Great Salt L.; watered by Colorado R. and tribs., and by R. Sevier and other streams. Rainfall is very small and irregular. Agric. carried on in fertile regions where water is obtainable and erosion is serious; chief crops, wheat, oats, potatoes, hay; sugar-beet is an important product. Fruits and vegetables are grown; horses, cattle, sheep and pigs reared. Minerals inc. copper, silver, lead, gold, petroleum and zinc. Industries of importance are copper refining, other primary metals, weapons, vehicles, foods, machinery, metal goods and petroleum products. Pop. (1985E) 1,645,000.

Utica New York, U.S.A. 43 05N 75 14W. City sit. E of Syracuse on Mohawk R. Manu. textiles and clothing. Pop. (1980C) 75,632.

Utrecht Netherlands. 52 05N 5 08E. (i) Smallest province sit. S of the Ijsselmeer. Mainly agric. in the fertile W. Area, 526 sq. m. (1,362 sq. km.). Pop. (1989E) 1,004,632. (ii) Cap. of province of same name, sit. on Old Rhine R. Transportation centre. Manu.

food and allied products, printing, textiles, metal, chemicals, tobacco and organs. Pop. (1988E) 230,634.

Utsunomiya Honshu, Japan. 36 33N 139 52E. Cap. of Tochigi prefecture sit. N of Tokyo. Tourist centre. Pop. (1988E) 417,000.

Uttar Pradesh India. 27 00N 80 00E. State lying in the upper Ganges basin along the N boundary of India. In 1833 the then Bengal Presidency was divided into two parts, one of which became the Presidency of Agra. In 1836 the Agra area was styled the North-West Province and placed under a Lieutenant-Governor. The two provinces of Agra and Oudh were placed, in 1877, under one administrator, styled Lieutenant-Governor of the North-West Province and Chief Commissioner of Oudh. In 1902 the name was changed to 'United Provinces of Agra and Oudh', under a Lieutenant-Governor, and the Lieutenant-Governorship was altered to a Governorship in 1921. In 1935 the name was shortened to 'United Provinces'. On Independence, the states of Rampur, Banaras and Tehri-Garhwal were merged with United Provinces. In 1950 the name of the United Provinces was changed to Uttar Pradesh. Area, 113,672 sq. m. (294,411 sq. km.). Principal cities, Agra, Aligarh, Allahabad, Kanpur, Lucknow (cap.), Varanasi, Rampur. Agric. occupies 75% of the pop., crops inc. rice, wheat, pulses, sugar-cane and oilseeds. Industries inc. cotton and woollen textiles, synthetic fibres, petrochemicals, leather and footwear, distilleries and breweries, paper, chemicals, glass, fertilizers, cement, oil refining, aluminium, electronics, aeronautics and diesel locomotive manu. Pop. (1981C) 110,862,013.

Uttoxeter Staffordshire, England. 52 53N 1 50W. Town sit. NW of Burton-on-Trent. Manu. agric. machinery and biscuits. Pop. (1989E) 10,000.

Uturoa Raiatéa, Leeward Islands, French Polynesia. 16 44S 151 26W. Cap. and chief port of Leeward Islands. Fruit canning.

Uusimaa Finland. 60 25N 23 00E.

Province on the Gulf of Finland. Area, 3,807 sq. m. (9,859 sq. km.). Mainly low, level country drained by short streams. Industries inc. lumbering, sawmilling, dairying, stock rearing and agric., fishing, textile milling and metalworking. Cap. Helsinki. Pop. (1988E) 1,226,344.

Uvéa Loyalty Islands, New Caledonia. 20 30S 166 35E. Sit. NW of Lifu. Area, 51 sq. m. (132 sq. km.). A fertile coral island producing copra and coconuts. The chief town is Fayahoué. Pop. (1983C) 2,772.

Uvéa Wallis and Futuna Islands, South Pacific. 13 22S 176 12W. Volcanic island also known as Wallis. Area, 62 sq. m. (159 sq. km.). Chief products copra, yams, taro roots and bananas. Pop. (1983E) 8,072.

Uxbridge Greater London, England. 51 32N 0 29W. District sit. WNW of central London on R. Colne. Industries engineering, iron-founding, brick-making, sawmilling and brewing.

Uzbekistan U.S.S.R. 41 30N 65 00E. In 1917 the Tashkent Soviet assumed authority, and in the following years estab. its power throughout Turkestan. The semi-independent Khanates of Khiva and Bokhara were first (1920) transformed into 'People's Republics', then (1923–4) into Soviet Socialist Republics, and finally merged in the Uzbek Soviet Socialist Republic and other reps. The Uzbek Soviet Socialist Republic was formed in 1924 from lands formerly inc. in Turkestan. Area is 172,819 sq. m. (447,600 sq. km.). Cap. Tashkent. Intensive farming and a high cotton producer. Agric. very important inc. many orchards and vineyards. Oil, coal and copper are among mineral resources. Pop. (1989E) 19·9m.

V

Vaal River Republic of South Africa. 129 04S 23 38E. Rises in the Drakensberg and flows 780 m. (1,248 km.) first N through the Orange Free State through Vaal Dam to S border of the Transvaal which it follows SW to the end before joining Harts R. in Cape Province. It is called the Madikwe when it flows through Bophuthatswana. There are irrigation works, esp. at Vaal Barrage and Vaal Dam near Vereeniging. The main tribs. are Klip and Wilge Rs.

Vaasa Vaasa, Finland. 63 06N 21 36E. Cap. of province and port sit. on the Gulf of Bothnia. Manu. inc. machinery, diesel engines, textiles, electrical appliances, plastics and footwear. Pop. (1989E) 54,000.

Vadodara Gujarat, India. Town in central W. An industrial and railway centre trading in cotton. Manu. inc. metal goods, petrochemicals, chemicals, textiles and matches. Pop. (1981C) 734,473.

Vaduz Liechtenstein. 47 08N 9 32E. Cap. of principality sit. near Rhine R. and Swiss-Austrian frontier. Industries inc. tourism and metalworking. Trades in agric. produce and is an important financial centre. Pop. (1988E) 4,919.

Valais Switzerland. 46 12N 7 45E. Canton in S bordering Italy and France. Area, 2,020 sq. m. (5,231 sq. km.). Very mountainous but cereals and vines are cultivated in Rhône valley. It joined the Confederation in 1815. Cap. Sion. Pop. (1980C) 218,707.

Valdai Hills Russian Soviet Federal Socialist Republic, U.S.S.R. 57 00N 33 30E. Sit. to NW of Moscow and rising to 1,050 ft. (320 metres) a.s.l.

Val-de-Marne France. Dept. sit. in Île-de-France region. Area, 94 sq. m.

(244 sq. km.). Cap. Créteil. Pop. (1982C) 1,193,655.

Valdepeñas Ciudad Real, Spain. 38 46N 3 23W. Town sit. NE of Córdoba. Noted for its red wine. Other products are olive oil, vinegar and leather.

Valdez Alaska, U.S.A. 61 07N 146 16W. Oil pipeline terminal at head of Prince William Sound, Gulf of Alaska, receiving oil from Beaufort Sea.

Valdivia Valdivia, Chile. 39 48S 73 04W. City in S centre, on Valdivia R. near the mouth where its port Corral is sit. Industries inc. shipbuilding, tanning, flour milling, brewing, sugar refining and the manu. of metal goods, paper and food products. Pop. (1982E) 113,565.

Val d'Oise France. 49 05N 2 00E. Dept. sit. in Île-de-France region. Area, 482 sq. m. (1,249 sq. km.). Cap. Pontoise. Pop. (1982C) 920,598.

Valdosta Georgia, U.S.A. 30 50N 83 17W. City sit. SW of Savannah. Centre of an agric. area producing timber, cotton and tobacco. Manu. inc. wood and metal products, textiles and cigars. Pop. (1988E) 48,000.

Valence Drôme, France. 44 56N 4 54E. Cap. of dept. in SE sit. on Rhône R. Manu. inc. textiles, leather and food products, and furniture. Trades in cereals, fruit, olives and wines. Pop. (1982C) 68,157 (agglomeration, 106,041).

Valencia Spain. 39 28N 0 22W. (i) Province of SE Spain bounded E by Mediterranean. Area, 4,239 sq. m. (10,763 sq. km.). Chief cities Valencia, Sagunto, Algemesi, Jativa. Traversed SSE by Guadalaviar R. where it is called the Turia R. and by Júca R. Densely populated coastal plain rises to mountains inland. The chief products are oranges, vines and rice. Pop. (1986C) 2,079,762. (ii) City near mouth of Turia R. Provincial cap., commercial centre, port and resort. An anc. city noted for baroque architecture. The port exports oranges, raisins and rice. Industries inc. textiles, shipbuilding, oil-refining and the manu. of metals and metal products. Pop. (1986C) 738,575.

Valencia Carabobo, Venezuela. 10 11N 68 00W. State cap. sit. WSW of Caracas on Cabriales R. Centre of an important agric. area producing esp. cotton and sugar-cane. Industries inc. meat packing, sugar refining, tanning and the manu. of textiles and leather goods. Pop. (1981E) 752,000.

Valenciennes Nord, France. 50 21N 3 32E. Town in extreme NE sit. on Escaut R. Industries inc. coalmining, engineering, sugar refining, and the manu. of chemicals and textiles. Formerly noted for lace. Pop. (1982C) 40,881 (agglomeration, 349,505).

Valentia Kerry, Ireland. 51 52N 10 20W. An island just off SE coast, with a meteorological observatory.

Valladolid Spain. 41 35N 4 40W. (i) A province in N. Area, 3,167 sq. m. (8,202 sq. km.). It is mainly open fertile country producing cereals, fruit, wine, olive oil and honey. The principal R. is Duero. Pop. (1986C) 503,306. (ii) Cap. of province of same name sit. at confluence of Rs. Pisuerga and Esgueva. Industries inc. railway engineering, flour milling, tanning, brewing, and the manu. of textiles, chemicals and paper. Pop. (1986C) 341,194.

Valle Honduras. Dept. on the Gulf of Fonseca. Area, 604 sq. m. (1,565 sq. km.). Hilly in S with a coastal plain producing rice, cotton, corn and beans. Industries inc. gold and silver mining, salt-works and fishing. The principal Rs. are Goascarán and Nacaome. Cap. Nacaome. Pop. (1981E) 125,640.

Valle d'Aosta Italy. 45 45N 7 22E. Region of NW Italy bounded NW by Mont Blanc. Area, 1,259 sq. m. (3,262 sq. km.). Cap. Aosta. Mountain region

with valleys enclosed by the Pennine and Graian Alps. Agric. and resort area. Pop. (1988E) 114,760.

Valle de Cauca Colombia. Dept. on the Pacific Ocean. Area, 8,548 sq. m. (22,140 sq. km.). Mainly mountainous, crossed by the Cordillera Occidental and the Cordillera Central. Very fertile producing cattle, sugar-cane, rice, cotton, coffee, cacao, tobacco, corn and vegetables. There are rich deposits of gold, silver and platinum. Cap. Cali. Pop. (1982E) 2,848,000.

Vallejo California, U.S.A. 38 07N 122 14W. Port sit. N of Oakland on San Pablo Bay. Industries inc. shipbuilding, meat packing, sawmilling and flour milling. Pop. (1980C) 80,303.

Vallenar Atacama, Chile. 28 34S 70 45W. Town in N central Chile and in the Huasco valley. Trades in agric. produce and wine. Pop. (1984E) 42,000.

Valletta Malta. 35 54N 14 30E. Cap. sit. on the E coast at the end of a ridge dividing the bay into Grand Harbour and Marsamxett. Sea port, trading centre and tourist resort. It is noted for historical associations with the Knights of Malta and named after their 16th cent. Grand Master Jean de la Vallette. Pop. (1985C) 9,210 (urban harbour area) 101,210.

Valleyfield (Salaberry-de-Valleyfield), Quebec, Canada. 45 15N 74 08W. City and port sit. SW of Montreal on St. Lawrence R. and at E end of L. St. Francis. Industries inc. zinc refining, distilling, flour milling, and the manu. of textiles. Pop. (1981E) 29,574.

Valley Stream New York, U.S.A. 40 39N 73 45W. A residential suburb of New York sit. on SW coast of Long Island. Pop. (1980C) 35,769.

Valois France. Area in Paris basin bounded by the forests of Compiègne, Chantilly and Villers-Cotterets. Chief towns Crépy-en-Valois, and Senlis. A rich agric. area.

Valparaíso Valparaíso, Chile. 33

02S 71 38W. Port in central Chile sit. on the Pacific coast. Industries inc. sugar refining and the manu. of textiles, chemicals, leather goods and clothing. Copper, silver, nitrates and wheat are exported. Pop. (1982E) 266,577.

Valsad Gujarat, India. Town sit. S of Surat on Gulf of Cambay. Port exporting timber, cotton, silk fabrics, millet, wheat, molasses and tiles. Industries inc. coach building, engineering; manu. bricks and pottery. Pop. (1981C) 82,697.

Valtellina Lombardia, Italy. 46 11N 9 55E. The valley of the Upper Adda R. stretching from the Ortler mountains to L. Como. It produces wine, figs, mulberries and wheat. The chief town is Sondrio.

Van Turkey. 38 28N 43 20E. (i) Province in SE. Area, 6,857 sq. m. (17,760 sq. km.). The principal R. is Buhtan. Coal, rye and apples are produced. Pop. (1980C) 468,646. (ii) Cap. of province of same name sit. SE of Erzurum. The commercial centre of a wheat-growing area. Pop. (1985E) 50,000.

Van, Lake Turkey. 38 33N 42 46E. L. in Turkish Armenia. It has no outlet but is a salt L. Area, 75 m. (120 km.) by 50 m. (81 km.). Produces salt, soda and fish.

Vancouver British Columbia, Canada. 49 16N 123 07W. City on Burrard Inlet on W coast. Third city of Canada; seaport with an ice-free natural harbour, commercial centre and fishing port, sit. on both S and N shores of the inlet. Industries inc. fish canning, oil refining, sawmilling, shipbuilding, brewing, distilling, sugar refining, and the manu. of pulp and paper machinery, mining machinery, sawmill equipment, and trucks and trailers. Pop. (1989E) Greater Vancouver, 1,498,980.

Vancouver Washington, U.S.A. 45 39N 122 40W. Port sit. on Columbia R. opposite Portland in Oregon. Manu. aluminium, electronics, pulp and paper. Timber, steel, aluminium, and grain are exported. Pop. (1980C) 42,434.

Vancouver Island British Columbia, Canada. 49 45N 126 00W. Island just off SW coast across narrow channels. Area, 12,408 sq. m. (31,204 sq. km.). Main towns Victoria (provincial cap.), Nanaimo, Port Alberni, Esquimalt. Mountainous and thickly forested, rising to Mount Victoria, 7,484 ft. (2,281 metres) a.s.l. The chief minerals are iron and coal. The main indust. are lumbering, the manu. of wood pulp, dairying, fishing and fruit farming.

Väner, Lake Sweden. 58 55N 13 30E. Sit. in SW, it is the largest L. in Sweden. Area, 2,156 sq. m. (5,585 sq. km.). The Göta Canal joins it to L. Vätter and the Baltic Sea, and the Göta R. drains it to the Kattegat.

Vannes Morbihan, France. 47 39N 2 46W. Cap. of dept. and port sit. SW of Rennes on the Gulf of Morbihan, an inlet of the Bay of Biscay. Industries inc. boat-building, tanning, ropemaking and textile manu. Pop. (1982C) 45,397.

Vanua Levu Fiji. 16 33S 179 15E. Volcanic island sit. NE of Viti Levu. Area, 2,137 sq. m. (5,535 sq. km.). Rises to 3,139 ft. (957 metres) a.s.l. in Mount Thurston. The principal R. is Ndreketi. The main products are sugar, rice and copra. There are deposits of gold.

Vanuatu South West Pacific. 15 00S 168 00E. Island group forming the Republic of Vanuatu, independence achieved in 1980, formerly an Anglo-French condominium 1,100 m. (1,760 km.) E of Australia and 500 m. (800 km.) W of Fiji. Area, 5,700 sq. m. (14,760 sq. km.). Main islands Espiritu Santo, Efate, Maewo, Epi, Malo, Malekula, Pentecost, Tanna, Aoba, Ambrym, Erromanga, Aneityum, Banks Islands, Torres Islands. Cap. Vila, on Efate. Volcanic and mountainous, rising to 6,195 ft. (1,888 metres) on Espiritu Santo. Main products copra, coffee, cocoa, mother of pearl. Pop. (1989E) 142,630.

Var France. 43 27N 6 18E. Dept. in Provence–Côte d'Azur region. Area, 2,316 sq. m. (5,999 sq. km.). Mainly hilly and wooded with the Alpes de

Provence in N and the Monts des Maures in S. The principal R. is Argens. Crops inc. vines, olives and fruit. There are deposits of bauxite, lead, zinc and salt. Cap. Toulon. Pop. (1982C) 708,331.

Varanasi Uttar Pradesh, India. 25 20N 83 00E. Town in central N sit. on Ganges R. it is an important religious centre with many temples and bathing ghats. Manu. inc. brassware, jewellery and handmade textiles. Pop. (1981C) 720,755.

Varberg Halland, Sweden. 57 17N 12 20E. Port and resort sit. S of Göteborg on the Kattegat. Industries inc. a nuclear power station, pulp mill and sawmill and manu. bicycles, metal goods, shoes and linen. Pop. (1983E) 45,367.

Vardar River Yugoslavia/Greece. 40 35N 22 50E. Rises in SW Serbia and flows 20 m. (32 km.) NNE, then SSE, into Greece and across Macedonia to enter the Aegean Sea in the Gulf of Thessaloníki.

Varese Lombardia, Italy. 45 48N 8 48E. Cap of Varese province sit. NW of Milan. Manu. inc. machinery, motorcars, textiles, footwear and furniture.

Varkaus Kuopio, Finland. 62 19N 27 55E. Town sit. S of Kuopio on one of the Saimaa Ls. Industries inc. paper and pulp, lumbering, engineering, water and air anti-pollution systems. Pop. (1984E) 24,724.

Värmland Sweden. 59 45N 13 00E. County sit. in W. Area, 6,788 sq. m. (17,582 sq. km.). Forested with numerous Ls., the principal R. being Klar. Industries inc. iron and steel works, pulp and paper mills, and lumbering. Cap. Karlstad. Pop. (1988E) 280,694.

Varna Varna, Bulgaria. 43 13N 27 55E. Cap. of province and port in NE sit. on the Black Sea. Industries inc. shipbuilding, engineering, tanning and food processing. Trades in livestock, grain and canned fish. Pop. (1987E) 305,891.

Var River Alpes-Maritimes, France.

43 39N 7 12E. Rises in the Maritime Alps, and flows SSE for 84 m. (134 km.) to enter the Mediterranean Sea SW of Nice.

Vas Hungary. 47 10N 16 55E. County in W. Area, 1,290 sq. m. (3,340 sq. km.). Mainly hilly and heavily forested in the SW. The principal Rs. are Raba and Gyöngyös. Agric. products inc. pigs, wheat, clover, potatoes and fruit. Cap. Szombathely. Pop. (1989E) 276,000.

Västerås Västmanland, Sweden. 59 37N 16 33E. Cap. of county sit. on N shore of L. Mälar. Manu. inc. electrical motors, nuclear power equipment, generators, transformers, turbines and other electrical equipment. Pop. (1988E) 117,717.

Västerbotten Sweden. 64 58N 18 00E. County in N sit. between the Gulf of Bothnia and the Norwegian frontier. Area, 21,390 sq. m. (55,401 sq. km.). Hilly and forested in W. The principal Rs. are Ångerman, Vindel, Ume and Skellefte. Industries inc. lumbering, pulp milling, metal smelting, stock rearing and agric. The main crops are rye and oats. Cap. Umeå Pop. (1988E) 247,521.

Västernorrland Sweden. 63 30N 17 40E. County in central NE on the Gulf of Bothnia. Area, 8,383 sq. m. (21,711 sq. km.). Mainly hilly with a low coastal strip, it is the most important lumbering region of Sweden. The principal Rs. are Ljunga, Indal, and Ångerman. Industries inc. sawmilling, aluminium milling and the manu. of cellulose. Cap. Härnösand. Pop. (1988E) 259,964.

Västervik Kalmar, Sweden. 57 43N 16 43E. Port 69 m. (110 km.) SSE of Norrköping on Gamleby Bay. Industries inc. agric. machinery, fishing, tourism. Manu. household appliances, steel nails and chains, and grinding equipment. Pop. (1989E) 22,000.

Västmanlands Sweden. 59 45N 16 20E. County in central Sweden bounded by L. Mälar in S. Area, 2,433 sq. m. (6,302 sq. km.). Hilly in N sloping to fertile lowlands in the SE. The principal Rs. are Kolbäck and Arboga.

The main industry is iron mining; copper, lead and feldspar are also mined. Other industries inc. lumbering and steel milling. Cap. Västerås. Pop. (1988E) 254,847.

Vatican City Italy. 41 54N 12 27E. Independent papal state within Rome. For many centuries the Popes bore temporal sway over a territory stretching across mid-Italy from sea to sea and comprising some 17,000 sq. m. (44,030 sq. km.), with a population finally of over 3m. In 1859–60 and 1870 the Papal States were incorporated with the Italian Kingdom. The consequent dispute between Italy and successive Popes was only settled in 1929 by three treaties between the Italian Government and the Vatican. The Treaty has been embodied in the Constitution of the Italian Republic of 1947. A revised Concordat between the Italian Republic and the Holy See came into force in 1985. The Vatican City State is governed by a Commission appointed by the Pope. The reason for its existence is to provide an extra-territorial, independent base for the Holy See, the government of the Roman Catholic Church. The area of the Vatican City is 109 acres. Sit w of Castel Sant'Angelo on w bank of R. Tiber. It includes the Piazza di San Pietro (St. Peter's Square), which is to remain normally open to the public and subject to the powers of the Italian police. It has its own railway station postal facilities, coins and radio. Twelve buildings in and outside Rome enjoy extra-territorial rights including the Basilicas of St. John Lateran, St. Mary Major, St. Paul without the Walls and the Pope's summer villa at Castel Gandolfo. In 1951 extra-territorial rights were also granted to a new Vatican radio station on Italian soil. Pop. (1990E) 1,000.

Vatnajökull Iceland. 64 30N 16 48W. A large icefield sit. in SE containing some active volcanoes. Area, 3,140 sq. m. (8,133 sq. km.).

Vätter, Lake Sweden. 58 24N 14 36E. Sit to SE of L. Väner, to which it is joined by the Göta Canal. It is the second largest L. in Sweden. Area, 738 sq. m. (1,912 sq. km.). It is drained to the Baltic Sea by Motala R.

Vaucluse France. 44 03N 5 10E. Dept. in Provence–Côte d'Azur region bounded by Rs. Rhône and Durance. Area, 1,377 sq. m. (3,566 sq. km.). In E are the Alpes de Provence. The fertile Rhône valley is in the w producing cereals, olives, mulberries and fruit. It is esp. noted for wine. Cap. Avignon. Pop. (1982C) 427,343.

Vaud Switzerland. 46 35N 6 30E. Canton in w bordering on France and sit. between L. Neuchâtel and L. Geneva. Area, 1,240 sq. m. (3,211 sq. km.). Mainly mountainous in SE but fertile and noted for wine. Salt is mined near Bex. It joined the Confederation in 1803. Cap. Lausanne. Pop. (1988E) 565,181.

Vaupés Colombia. A commissary in SE bordering on Venezuela and Braz. Area, 34,990 sq. m. (90,625 sq. km.). Mainly forested lowland, largely undeveloped, producing rubber and balata gum. The principal Rs. are Vaupés and Inírida. Cap. Mitú.

Vauxhall Greater London, England. 51 28N 0 09W. District of central London on s bank of Thames R. formerly noted for its pleasure gardens.

Vava'u Tonga, South Pacific. 18 40S 174 00W. A coral island group. The largest and most important island is Vava'u. Cap. Neiafu.

Växjö Kronoberg, Sweden. 56 52N 14 49E. Cap. of county in central S. University and administrative centre and some manu. industry. Pop. (1989E) 68,700.

Veenendaal Utrecht, Netherlands. Town sit. ESE of Utrecht. Manu. textiles. Trades in agric. produce. Pop. (1989E) 47,000.

Vejle Vejle, Denmark. 55 42N 9 32E. Cap. of county and port sit. on Vejle Fjord, on E coast of Jutland. Manu. inc. machinery, hardware, textiles, leather and soap. Trades in dairy produce. Pop. (1981E) 43,300.

Velbert North Rhine-Westphalia, Federal Republic of Germany. 51 20N 7 02E. Town sit. s of Essen. Manu. inc. motor-car accessories, locks and windows. Pop. (1984E) 90,000.

Veldhoven Noord-Brabant, Netherlands. Town sit. SW of Eindhoven. Industries include high-tech. Pop. (1990E) 39,000.

Velebit Mountains Croatia, Yugoslavia. A range in the Dinaric Alps in NW stretching for about 100 m. (162 km.) along the Adriatic Sea. The highest peak is Vaganjski Vrh, 5,768 ft. (1,758 metres) a.s.l.

Velebit Strait Croatia, Yugoslavia. A channel in NW, running parallel to the Velebit Mountains, between the mainland and the islands of Pag and Rab.

Vélez Málaga Málaga, Spain. 36 48N 4 05W. Town sit. ENE of Málaga on Vélez R. Industries inc. sugar refining, tanning and the manu. of soap. Trades in wine, fruit and sugarcane.

Vella Lavella Solomon Islands. Protectorate. A volcanic island 225 m. (360 km.) NW of Guadalcanal. Area, *c.* 250 sq. m. (648 sq. km.). Copra is produced.

Vellore Tamil Nadu, India. 12 56N 79 09E. Town sit. WSW of Madras on Palar R. Trades in agric. produce and sandalwood. Pop. (1981C) 174,247.

Velsen Noord-Holland, Netherlands. 52 27N 4 39E. Town and port WNW of Amsterdam sit. on the North Sea Canal. Industries inc. shipbuilding, iron and steel works, and the manu. of chemicals, fertilizers and cement. Pop. (1985E) 58,000.

Venda An independent homeland granted independence by the Rep. of South Africa in 1979. Area, 2,510 sq. m. (6,500 sq. km.). About 85% of the country is only suitable for raising livestock. Industry is in early stages. Pop. (1985C) *de jure* 651,393, *de facto* 459,986.

Vendée France. 46 40N 1 20W. Dept. in Pays-de-la-Loire region. Area, 2,595 sq. m. (6,721 sq. km.). Mainly low-lying and agric. The principal Rs. are Sèvre Nantaise, Sèvre Niortaise and Vendée. Crops inc. wheat, sugar-beet, vines and fruit.

Cattle and horses are reared. Cap. La Roche-sur-Yon. Pop. (1982C) 483,027.

Veneto Italy. 45 30N 11 45E. Region of NE Italy bounded by Austria, comprising 7 provinces: Belluno, Padua, Rovigo, Treviso, Venezia, Verona and Vicenza. Area, 7,090 sq. m. (18,364 sq. km.). Chief cities Venice (cap.), Vicenza, Treviso and Padua. Mountainous in N with the Dolomites and the Carnic Alps in S, watered by Po, Adige, Brenta and Piave Rivers. Farming area. Tourism is important in N and in Venice. Indust. are concentrated in Mestre, Vicenza, Padua and Verona. Pop. (1988E) 4,380,587.

Venezuela South America. 8 00S 65 00W. Republic of N South America bounded N by the Caribbean, E by Guyana, S by Brazil and W by Colombia. Area, *c.* 325,143 sq. m. (912,050 sq. km.). There is a Federal District, 20 States and 2 territories. Chief cities Caracas (cap.), Maracaibo, Valencia, Barquisimeto, Ciudad Bolivar, La Guaira and Puerto Cabello. The Andes occupy the NW with coastal spurs extending E in parallel ranges; highest peaks are in the Mérida range, 15,000–16,000 ft. (4,572–4,877 metres) a.s.l. The Maracaibo lowlands is a depression surrounded by the mountains and inc. L. Maracaibo which occupies *c.* one-third of it. Beyond this the Orinoco R. plain extends E to the foothills of the Guiana Highland, densely forested, in the SE. Drained by the Orinoco R. system; the Angel Falls, highest in the world, 3,212 ft. (979 metres) a.s.l. are on a trib. of the Caroni R. Climate varies with altitude from tropical to cold. The central prairie plains are hot and wet April–Oct., cooler and drier Nov.–Mar. Earthquakes are fairly frequent. The main occupations are farming, mining, extracting oil. The cultivated area is 5.5m. acres, producing beans, cocoa, yucca, maize, coffee, sugar and bananas. Beef-cattle ranching and pig-farming are important. Forests cover *c.* one-third of the country and produce rubber, balatá, vanilla and resins. Minerals, apart from oil, inc. gold (SE), diamonds (Amazonas), manganese, phosphates, sulphur, coal, salt, nickel and iron. Indust. inc. oil processing, food processing, chemicals, textiles, iron and steel, and engineering. Pop. (1988E) 18·77m.

Venice　Veneto, Italy. 45 27N 12 21E. City E of Milan on the Adriatic coast, built on a number of islands divided into 2 groups by the Grand Canal which runs NW to SE as the main route, with a network of smaller canals branching off it. Commercial centre, seaport and tourist resort noted for its unique character and the numerous historic buildings, mostly built along the canals. Indust. are concentrated in the suburb of Marghera and inc. metallurgical indust., oil refining, and the manu. of metal products. Trad. indust. are the manu. of glassware, jewellery, silk and cotton textiles and lace. Pop. (1989E) 328,249.

Venlo　Limburg, Netherlands. 51 24N 6 10E. Town and port sit. on Maas R. near frontier of Federal Republic of Germany. Manu. inc. chemicals, paper, electric lamps, photocopiers and office machinery. Trades in agric. produce. Pop. (1988E) 63,820.

Venray　Limburg, Netherlands. Town sit. NW of Venlo. Industries inc. photocopiers, metal goods, leather goods, cardboard and cigars. Pop. (1988E) 34,172.

Ventimiglia　Liguria, Italy. 43 49N 7 36E. Port and resort sit. W of San Remo on the Gulf of Genoa, and near the border with France. Trades in flowers and nearby are the Hanbury Gardens. Pop. (1989E) 27,003.

Ventnor　Isle of Wight, England. 50 36N 1 11W. Resort sit. on S coast. Pop. (1985E) 7,000.

Ventura　*see* **San Buenaventura**

Veracruz　Mexico. 19 00N 96 15W. State in E on the Gulf of Mex. Area, 27,683 sq. m. (71,699 sq. km.). It is very mountainous in W where the Sierra Madre Oriental are sit., and descends to the coastal plain in E. Mainly agric. producing cotton, rice, sugar-cane, coffee, maize, fruit and tobacco. Rubber and timber are obtained from the forests. There are many important oilfields in the coastal strip. Cap. Jalapa, sit. E of Mexico City. Pop. (1989E) 6,798,109.

Veracruz　Veracruz, Mexico. 19 12N 96 08W. Port sit. on the Gulf of Mex. Manu. inc. chemicals, textiles, sisal, soap and food products. Exports inc. coffee, tobacco and vanilla. Pop. (1980C) 305,456.

Veraguas　Panama. A province in central W, extending from the Caribbean Sea to the Pacific Ocean. Area, 4,635 sq. m. (12,005 sq. km.). It is traversed by the Veraguas Mountains which rise to over 6,000 ft. (1,829 metres) a.s.l. The main occupations are stock-rearing, lumbering and agric., the chief crops being coffee, sugar-cane, rice and corn. There are deposits of gold, mercury, zinc, magnesium, lead and iron. Cap. Santiago. Pop. (1989E) 218,867.

Vercelli　Piemonte, Italy. 45 19N 8 25E. Cap. of Vercelli province sit. on Sesia R. Industries inc. rice and flour milling, chemicals, sugar refining, engineering and the manu. of textiles. Pop. (1988E) 51,103.

Verdun　Meuse, France. 49 10N 5 23E. Town in NE sit. on Meuse R. Manu. inc. metal products, textiles, alcohol and furniture. Pop. (1982C) 24,120.

Vereeniging　Transvaal, Republic of South Africa. 26 38S 27 57E. Town sit. S of Johannesburg on the Vaal R. It is an important coalmining centre with manu. of iron and steel, cables, wire, nuts, bolts, bricks and tiles. Pop. (1985E) 61,000.

Verkhoyansk　Russian Soviet Federal Socialist Republic, U.S.S.R. 67 35N 133 27E. Town sit. on Yana R. Formerly used as a place of exile, it is one of the coldest towns on earth, a temp. of nearly −100°F having been recorded. The main occupations are tin-mining and fur trapping. Pop. (1985E) 1,500.

Verkhoyansk Range　Russian Soviet Federal Socialist Republic, U.S.S.R. 67 00N 129 00E. Sit. in the Yakut Autonomous Soviet Socialist Rep. to

E of Lena R. It stretches W and N for about 1,000 m. (1,600 km.).

Vermont U.S.A. 43 40N 72 50W. State of New England bounded N by Canada, E by New Hampshire, W by New York, S by Massachusetts. Area, 9,267 sq. m. (24,002 sq. km.). Chief cities Montpelier (cap.), and Burlington. Settled in 1724 onwards and became 14th State of the Union in 1791. Crossed N to S by the Green Mountains rising to 4,000 ft. (1,219 metres) a.s.l. and sloping (W) to Hudson R. and (NW) to L. Champlain. Mainly agric. with maple syrup indust. based on the N woodland, stone quarrying (inc. marble) and some mining (asbestos). The chief products are hay, apples and maple. Indust. inc. stone-working, syrup, paper-making, dairying and tourism. Pop. (1986E) 541,000.

Vernon British Columbia, Canada. 50 16N 119 16W. City sit. NE of Vancouver in the Okanagan Valley, between 3 lakes, Swan L. to the N, Okanagan L. (80 m., 129 km. long) to the W and Kalamalka, famous for its changing colours; to the S. Indust. inc. forestry, agric. (orchards, beef and dairy cattle), tourism and manu. Pop. (1989E) 20,678.

Verona Veneto, Italy. 45 27N 11 00E. City sit. W of Venice on Adige R. Provincial cap.; important grain market and tourist centre noted for anc. buildings. Route centre at junction of the Brenner Pass road with the Milan–Venice road. Pop. (1988E) 258,724.

Verria Imathia, Greece. Cap. of prefecture sit. WSW of Thessaloníki. Manu. textiles. Trades in agric. produce. Pop. (1981C) 37,087.

Versailles Seine-et-Oise, France. 48 48N 2 08E. Town sit. SW of Paris. Cap. of Dept. and bishopric, noted for the Palace of Versailles designed by Le Vau and Mansart for Louis XIV, with gardens designed by Le Notre. Pop. (1982C) 95,240.

Vesoul Haute Saône, France. 47 38N 6 10E. Cap. of dept. Market centre for agric. produce. Pop. (1982C) 20,269.

Verviers Liège, Belgium. 50 35N 5 52E. Town ESE of Liège sit. on Vesdre R. It is an important centre of the woollen industry. Other manu. are textile machinery, chemicals, leather goods and chocolate. Pop. (1988E) 53,355.

Vest-Agder Norway. County in S. Area, 2,811 sq. m. (7,280 sq. km.). Mainly wooded. The principal Rs. are Audna, Lygna, Sira and Kvina. The main occupations are engineering, oil-based industries, refineries, civil engineering, financial services, and agric. in the lower R. valleys. Cap. Kristiansand. Pop. (1989E) 143,350.

Vestfold Norway. 59 15N 10 00E. County in SE bordering on Oslo Fjord. Area, 855 sq. m. (2,215 sq. km.). The main occupations are lumbering, fishing and dairy farming. Cap. Tonsberg. Pop. (1989E) 196,099.

Veszprém Veszprém, Hungary. 47 06N 17 55E. Cap. of county SW of Budapest sit. in the Bakony Forest. Manu. inc. textiles and wine. Trades in agric. produce. Pop. (1989E) 66,000.

Vevey Vaud, Switzerland. 46 28N 6 51E. Resort sit. ESE of Lausanne on NE shore of L. Geneva. Tourist centre and manu. inc. machinery, watches, cigars, leather and chocolate. Pop. (1985E) 15,000.

Viana do Castelo Portugal. 41 42N 8 50W. (i) District in N bordered by Spain and the Atlantic Ocean. Area, 859 sq. m. (2,225 sq. km.). Mountainous in E. The principal R. is Lima. Pop. (1987E) 266,400. (ii) Cap. of district of same name and port sit. N of Oporto. Manu. inc. carpets, flour products and chocolate. Has an important trade in cod.

Viareggio Toscana, Italy. 43 52N 10 14E. Resort sit. WNW of Lucca on the Ligurian Sea. Manu. hosiery. Pop. (1985E) 60,000.

Viborg Viborg, Denmark. 56 26N 9 24E. Cap. of county sit. W of Randers. Industries inc. windmills, heating equipment, distilling, brewing and the manu. of textiles. Pop. (1984E) 39,300.

Vicenza Veneto, Italy. 45 33N 11 33E. Cap. of Vicenza province in NE. Manu. inc. gold, iron and steel, ceramics, machinery, textiles, glass and furniture. It has many fine medieval, gothic and renaissance buildings. Pop. (1989E) 109,932.

Vichada Colombia. 5 00N 69 30W. A commissary in E bordering on Venezuela. Area, 38,212 sq. m. (98,970 sq. km.). Mainly undeveloped. The principal Rs. are Tomo and Vichada. Forest products inc. gums, resins and vanilla. Crops inc. corn and yucca and some cattle are reared. Cap. Puerto Carreño.

Vichy Allier, France. 46 08N 3 26E. Spa in central France sit. on Allier R. Vichy waters are bottled and exported. Manu. cosmetics, electronics and mineral waters. French 'Vichy' government estab. here, 1940–4. Pop (1982C) 30,554.

Vicksburg Mississippi, U.S.A. 32 14N 90 56W. Port in W centre of state sit. at confluence of Rs. Mississippi and Yazoo. Industries inc. engineering, steel slitting, sawmilling, food canning and the manu. of cotton, cottonseed oil, chemicals and clothing. Pop. (1980C) 25,434.

Victoria Australia. 21 16S 149 03E. State of SE Australia bounded N by New South Wales, W by South Australia. Area, 87,884 sq. m. (227,600 sq. km.). Chief cities Melbourne (cap.), Ballarat, Bendigo, Geelong, Hamilton, Sale and Echuca. Highlands in the centre form a W extension of the Australian Alps and descend N to the Murray R. Plain. A S valley is separated from the coast at Port Phillip Bay by hills which extend E to W as a plateau from Mount Ararat (W), 3,827 ft. (1,166 metres) a.s.l. Drained by Snowy, Glenelg and Yarra Rs. Farming is the main occupation; chief crops, wheat, oats, barley, fodder, potatoes, vines, fruit. Stock-farming is important. Minerals inc. gold, coal, lignite. Manu. indust. are concentrated in the Melbourne area and inc. textiles, soap, furniture, machinery, electrical equipment and vehicles. Pop. (1986E) 4,183,500.

Victoria British Columbia, Canada.

48 25N 123 22W. Cap. of province and port sit. at SE end of Vancouver Island overlooking the Strait of Juan de Fuca. Industries inc. tourism, shipbuilding, sawmilling, computer software, fish canning and high technology. Wood products and fish are exported. Pop. (1989E), Greater Victoria, 273,242.

Victoria Hong Kong. 22 17N 114 09E. Cap. and port sit. on N shore of the island. Industries inc. cotton milling and engineering. Exports inc. iron, steel, coal, hides, food products and tungsten.

Victoria Mahé, Seychelles. 04 37S 55 28E. Cap., tourist centre and port sit. on NE coast. Exports inc. copra, cinnamon, vanilla, tortoiseshell and guano. Pop. (1985E) 23,000.

Victoria Texas, U.S.A. 28 48N 97 00W. City sit. SW of Houston on Guadalupe R. Industries are oil refining and the manu. of chemicals and cottonseed oil. Pop. (1980C) 50,595.

Victoria Falls Central Africa. 17 58S 25 52E. Falls on Zambezi R. on the frontier between Zambia (N) and Zimbabwe (S), caused by a 400 ft. (122 metres) rift in the river bed into which the water pours in 3 cataracts, separated by islands, and from which it escapes by a narrow channel. The Main, Rainbow and East cataract falls vary from 350–420 ft. (107–128 metres) high.

Victoria Island Northwest Territories, Canada. 71 00N 114 00W. The second largest island in the Arctic Arch. it is separated from the mainland by the Dolphin and Union Strait, Coronation Gulf and Queen Maud Gulf. Area, 80,340 sq. m. (208,080 sq. km.).

Victoria, Lake East Africa. 10 0S 33 00E. L. bounded SE by Tanzania, N and W by Uganda, NE by Kenya. Area, 26,828 sq. m. (69,485 sq. km.), the largest African L., sit. at 3,700 ft. (1,128 metres) a.s.l. in the Great Rift Valley. Fed by Kagera, Mara and Nzoia Rivers and drained by Nile R. (the White Nile). Chief ports, Jinja, Port Bell, Entebbe, Bukoba, Mwanza and Kisumu.

Viedma Río Negro, Argentina. 40 50S 63 00W. Cap. and port sit. SW of Buenos Aires on Río Negro. Trades in agric. produce. Pop. (1980C) 24,000.

Vienna Vienna, Austria. 48 12N 16 22E. City near the Czechoslovak border on Danube R. at NE edge of the Wienerwald wooded hills. Cap., commercial and cultural centre, former seat of the Hapsburg dynasty. It is the headquarters of the Organization of Petroleum Exporting Countries and the UN International Atomic Agency. Tourist centre noted for fine streets and buildings. Manu. inc. ceramics, clothing, electronics, machinery, pharmaceuticals, textiles, furniture and chemicals. Pop. (1988E) 1,482,800.

Vienne France. 49 31N 4 53E. Dept. in Poitou-Charentes region. Area, 2,697 sq. m. (6,985 sq. km.). Mainly flat and low-lying producing cereals, potatoes, fruit and wine. The principal Rs. are Vienne and the Clain. Cap. Poitiers. Pop. (1982C) 371,428.

Vienne Isère, France. 45 31N 4 52E. Town in SE central France sit. on Rhône R. Industries inc. iron founding, distilling and the manu. of woollen and leather goods. Pop. (1982C) 29,050.

Vienne River France. 47 13N 0 05E. Rises in N Corrèze and flows 230 m. (368 km.) N and NW past Limoges, Châtellerault and Chinon to enter Loire R. near Fontevrault.

Vientiane Laos. 17 58N 102 36E. Cap. of Laos and port sit. on the N bank of the Mekong R. at the border of Thailand. Trades in timber, textiles, resins and gums. Pop. (1985C) 377,409.

Vieques Island Puerto Rico. 18 08N 65 25W. Sit. 9 m. (14 km.) E of the main island. Area, 51 sq. m. (132 sq. km.). Industries inc. dairying, sugar milling and the manu. of charcoal. The chief town is Isabela Segunda or Vieques. Pop. (1982E) 7,662.

Viersen North Rhine-Westphalia, Federal Republic of Germany. 51 15N 6 23E. Town NNW of München-Glad-

bach sit. on Niers R. Manu. textiles and paper. Pop. (1984E) 79,200.

Vierzon Cher, France. 47 13N 2 05E. Town in central France sit. at confluence of Rs. Cher and Yèvre. Manu. agric. machinery, bricks, tiles and porcelain. Pop. (1982C) 34,886.

Vietnam South East Asia. 16 00N 108 00E. Independent republic forming a narrow strip along the W shore of the South China Sea, bounded N by China, W by Laos and Cambodia. Area, 127,246 sq. m. (329,566 sq. km.) with a coastline of 1,600 m. (2,560 km.). The cap. is Hanoi, other cities are Ho Chi Minh City (Saigon), Da Nang and Haiphong. Agric. is the most important industry, mainly rice production although coconuts, tea, maize, coffee, oil palms are also cultivated. Some minerals such as coal, tin, zinc are exploited. Pop. (1989E) 64m.

Vigevano Lombardia, Italy. 45 19N 8 51E. Town SW of Milan sit. near Ticino R. Manu. metal and mechanical goods, plastics and footwear. Tourist centre. Pop. (1989E) 64,000.

Vigo Pontevedra, Spain. 42 14N 8 43W. Port and resort sit. on S shore of Vigo Bay, and near N border of Portugal. Industries inc. boatbuilding, oil refining, distilling, vehicles, fish processing and sugar refining. Manu. chemicals, paper, leather, soap, cement, clothing and brandy. It is the chief port for transatlantic shipping. Pop. (1988E) 275,580.

Vijayavada Andhra Pradesh, India. 16 31N 80 37E. Town sit. ESE of Hyderabad on Krishna R. Industries inc. engineering, rice and oilseed milling. Pop. (1981C) 461,772.

Vila Vanuatu. 17 45S 168 18E. Cap. and port on SW coast of the island of Éfaté. Pop. (1989E) 19,400.

Vila Real Portugal. 41 18N 7 45W. (i) A district in N bordered by Spain. Area, 1,671 sq. m. (4,328 sq. km.). Mountainous in N. The principal Rs. are Douro and Corgo. The main industry is viticulture. Pop. (1987E) 262,900. (ii) Cap. of district of same name sit. ENE of Oporto on Congo R. Manu. inc. ceramics, buttons and flour products. It is the centre of the wine industry.

Villach   Carinthia, Austria. 46 36N 13 50E. Town in central S sit. on Drau R. Manu. iron, lead and cellulose products. It is the centre of the timber trade with Italy. Pop. (1981C) 52,692.

Villahermosa   Tabasco, Mexico. 17 59N 92 55W. State cap. sit. ESE of Veracruz on Grijalva R. Industries inc. distilling and sugar refining. Trades in agric. produce. Pop. (1980E) 110,000.

Villarrica   Guairá, Paraguay. 39 16S 72 13W. Cap. of province in central S. Industries inc. sawmilling, sugar refining and flour milling. Trades in cotton, sugar-cane, tobacco and other agric. products. Pop. (1980E) 21,203.

Villavicencio   Meta, Colombia. 4 09N 73 37W. Cap. of dept. in central Colombia sit. at foot of the E Cordillera. Industries inc. rice milling. Trades in cattle and hides. Pop. (1980E) 160,000.

Villejuif   Val-de-Marne, France. S suburb of Paris. Manu. glass and aircraft parts. Pop. (1982C) 52,488.

Villeneuve d'Ascq   Nord, France. E suburb of Lille. Pop. (1982C) 59,868.

Villeurbanne   Rhône, France. 45 46N 4 53E. A suburb of Lyon. Industries inc. food processing, tanning and the manu. of textiles, metal goods and chemicals. Pop. (1982C) 118,330.

Vilnius   Lithuania, U.S.S.R. 54 41N 25 19E. City sit. NW of Minsk, Russian Soviet Federal Socialist Republic. Cap. of the Lithuanian Soviet Socialist Republic, education centre, railway centre trading in grain. Indust. inc. chemicals, sawmilling, food processing, paper-making, and the manu. of leather and matches. Pop. (1989E) 582,000.

Vilvoorde   Brabant, Belgium. 50 56N 4 26E. Town sit. NNE of Brussels on Willebroek Canal. Manu. inc. oil refining, chemicals, textiles, electrical equipment, food products, hardware, varnish and glue. Pop. (1988E) 32,893.

Viña del Mar   Valparaíso, Chile. 33 02S 71 34W. Resort and suburb of Valparaíso, central Chile. Industries inc. sugar refining and the manu. of textiles and paint. Trades in fruit and vegetables. Pop. (1982E) 290,014.

Vincennes   Val-de-Marne, France. A suburb of E Paris. Manu. inc. chemicals, rubber and electrical products, and perfume. Pop. (1982C) 43,086.

Vindhya Range   India. 23 00N 77 00E. Mountains extending ENE from N shore of Narbada R., Maharashtra, and rising to 4,000–5,000 ft. (1,219–1,524 metres) a.s.l.

Vineland   New Jersey, U.S.A. 39 29N 75 02W. City in S New Jersey. Pop. (1980C) 53,753.

Vinnitsa   Ukraine, U.S.S.R. 49 14N 28 29E. Cap. of the Vinnitsa region SW of Kiev sit. on S. Bug R. Industries inc. meat packing, flour milling, engineering and the manu. of machinery, electrical equipment and fertilizers. Pop. (1987E) 383,000.

Virgin Gorda   British Virgin Islands. 18 29N 64 24W. An island sit. E of Tortola. Virgin Peak, in the centre, rises to 1,371 ft. (418 metres) a.s.l. Products inc. charcoal, livestock and vegetables. Pop. (1984E) 1,000.

Virginia   U.S.A. 37 45N 78 00W. State of E U.S.A. on the Atlantic coast, bounded NE by Potomac R., E by the Atlantic, NW by West Virginia, SW by Kentucky, S by Tennessee and North Carolina. Area, 39,780 sq. m. (103,030 sq. km.). Chief cities Richmond (cap.), Norfolk, Roanoke, Hampton and Newport News. A former British colony and one of the 13 original States of the Union. Low-lying coastal plain with a deeply-indented coastline inc. SE the Great Dismal Swamp, area 700 sq. m. (1,813 sq. km.). Inland of this is rolling country which rises W to the Appalachian Mountains. drained by Potomac, James, York and Rappahannock Rs. Farmland covers 10.65m.

acres. Chief products are beef cattle, dairy cattle, pigs, sheep, tobacco, maize, cereals, potatoes, sweet potatoes, peanuts, apples. The most important mineral is coal, with some lead, zinc, lime, titanium, stone, sand and gravel. The main indust. are tobacco processing, synthetic fibre and shipbuilding. Pop. (1988E) 5,996,000.

Virginia Beach   Virginia, U.S.A. 36 51N 75 58W. City sit. SE of Richmond on Atlantic coast. Resort and military base with convention trade, electronics and data processing industry, some farming. Pop. (1985E) 321,000.

Virgin Islands   U.S.A. 18 20N 64 50W. Group of islands between Puerto Rico (W) and the British Virgin Islands (ENE). The islands were formerly known as the Danish West Indies, were purchased by the U.S.A. from Denmark for $25m. in a Treaty ratified by both nations and proclaimed in 1917. Their value was wholly strategic, inasmuch as they commanded the Anegada Passage from the Atlantic Ocean to the Caribbean Sea and the approach to the Panama Canal. Although the inhabitants were made U.S. citizens in 1927, the islands are, constitutionally, an 'unincorporated territory'. The main islands are St. Thomas, St. John, St. Croix. Area, 134 sq. m. (347 sq. km.). Cap. Charlotte Amalie (St. Thomas). The main occupations are farming and tourism. The main product is bay rum. Pop. (1987E) 106,000.

Virgin Islands, British   *see* **British Virgin Islands**

Vis   Yugoslavia. 43 03N 16 12E. An island in the Adriatic Sea. Area, 39 sq. m. (101 sq. km.). The main occupations are fishing and vine growing. Pop. (1985E) 8,000.

Visakhapatnam   Andhra Pradesh, India. 17 45N 83 20E. Port ENE of Hyderabad sit. on the Bay of Bengal. Industries are shipbuilding and oil refining. Manganese and oilseeds are exported. Pop. (1981C) 584,166.

Visby   Gotland, Sweden. 57 38N 18 18E. Cap. of county and port sit. on W coast of Gotland island, in the Baltic

Sea. Industries inc. metal working, sugar refining and cement making. Pop. (1984E) 20,000.

Viseu   Portugal. 40 39N 7 55W. (i) A district in central N. Area, 1,938 sq. m. (5,019 sq. km.). The principal R. is Vouga. The main occupations are viticulture, agric. and lumbering. Pop. (1987E) 423,300. (ii) Cap. of district of same name sit. SE of Oporto. Industries inc. textile milling, tanning and the manu. of explosives. Trades in agric. produce.

Vistula River   Poland. 54 22N 18 55E. Rises in the Carpathian mountains on the Czecholovak border and flows 670 m. (1,072 km.) NE past Kraków to Zawichost then N to Pulawy and NW past Otwock, Warsaw, Wloclawek and Torun before turning NNE past Grudziadz to enter the Gulf of Danzig by several branches. Linked by canal with Oder, Neman and Bug Rs. used for transporting freight.

Vitebsk   Belorussia, U.S.S.R. 55 12N 30 11E. Cap. of Vitebsk region sit. in W on W Dvina R. Manu. inc. agric. machinery, machine tools, textiles, glass, footwear and furniture. Pop. (1987E) 347,000.

Viterbo   Lazio, Italy. 42 25N 12 06E. Cap. of Viterbo province in central Italy. Manu. inc. textiles, pottery, furniture and food products. Pop. (1981E) 58,000.

Viti Levu   Fiji. 18 00S 178 00E. Sit. SW of Vanua Levu, it is the largest island. Area, 4,010 sq. m. (10,386 sq. km.). Mountainous, rising to 4,341 ft. (1,323 metres) a.s.l. in Mount Victoria. The principal R. is Rewa whose fertile deltas produce rice, sugar, cotton and pineapples. Sugar is exported. There are deposits of gold.

Vitim River   Russian Soviet Federal Socialist Republic, U.S.S.R. 59 26N 113 18E. Rises in a small L. E of L. Baikal in the Buriat Autonomous Soviet Socialist Republic, and flows 1,100 m. (1,760 km.) S then NE to Ust Karengi, then NW to enter Lena R. at Vitim. The plateau between the Vitim and Olekma Rivers is an important source of gold.

Vitória   Espírito Santo, Brazil. 20
19S 40 21W. State cap. and port sit.
on an island in the Atlantic and joined
to the mainland by a bridge. Industries
inc. sugar refining and the manu. of
textiles and cement. Iron ore, timber
and coffee are exported. Pop. (1980E)
144,000.

Vitoria   Alava, Spain. 42 51N
2 40W. Cap. of province in N sit. at
1,750 ft. (533 metres) a.s.l. Industries
inc. tanning, flour milling and the
manu. of paper, furniture and leather
goods. Trades in agric. produce. Pop.
(1986C) 207,501.

Vitry-sur-Seine   Val-de-Marne,
France. SE suburb of Paris. Pop.
(1982C) 85,820.

Vittoria   Sicilia, Italy. 36 57N 14
32E. Town sit. WSW of Syracuse.
Manu. alcohol and macaroni. Trades
in wine and gypsum.

Vittorio Veneto   Veneto, Italy. 45
59N 12 18E. Town in NE sit. N of Treviso. Manu. inc. textiles, esp. silk, and
cement. Pop. (1974E) 25,500.

Vizcaya   Spain. 43 15N 2 55W. Province on the Bay of Biscay. Area, 856
sq. m. (2,217 sq. km.). It has important deposits of iron, lead, copper and
zinc. Agric., stock-rearing and fishing
are also carried on. Cap. Bilbao. Pop.
(1986C) 1,168,405.

Vlaardingen   Zuid-Holland, Netherlands. 51 54N 4 21E. Port sit. w of
Rotterdam on New Maas R. Industries
inc. fishing, fish processing, and the
manu. of rope, sails, soap and fertilizers. Pop. (1984E) 76,466.

Vladimir   Russian Soviet Federal
Socialist Republic, U.S.S.R. 56 10N
40 25E. Cap. of the Vladimir region
ENE of Moscow sit. on Klyazma R.
Manu. inc. tractors, chemicals, textiles, machine tools and precision
instruments. Pop. (1987E) 343,000.

Vladivostock   Russian Soviet Federal Socialist Republic, U.S.S.R. 43
10N 131 56E. Cap. of the Primorye
Territory and port in extreme SE sit.
on the Pacific coast between Amur
Bay and the Golden Horn. It is an im-

portant transport centre and naval
base. Industries inc. shipbuilding,
sawmilling, food processing, fish canning, fishing and whaling. Exports
inc. coal, timber, fish, oilcake and
soybean oil. Pop. (1987E) 615,000.

Vlissingen (Flushing)   Zeeland,
Netherlands. 51 26N 3 35E. Port and
container terminal 87 m. (139 km.) sw
of The Hague sit. at mouth of w
Scheldt R. Industries inc. oil-refining,
ship repair yards, iron and steelworks,
chemicals, fishing, and engineering.
Pop. (1989E) 43,947.

Vlorë   Vlorë, Albania. Cap. of district and port sit. SSW of Tirana. Industries inc. oil-refining, there being a
pipeline to the Kuçovë oilfield, fishing and fish canning. Manu. inc. olive
oil and cement. Petroleum is exported.
Pop. (1989E) 61,000.

Vltava River   Czechoslovakia. 50
21N 14 30E. Rises in the Bohemian
Forest and flows 265 m. (424 km.) SE
forming a L., then N past Ceske
Budějovice and Prague to join the
Elbe R. at Meluik. Main tribs. Sazava
and Beraun Rs. Used for power.

Vojvodina   Serbia, Yugoslavia. 45
18N 19 52E. Province of NE Yugoslavia bounded N by Hungary, E by
Romania. Area, 8,303 sq. m. (21,506
sq. km.). Chief cities Novi Sad (cap.),
Subotica, Zrenjanin, Pančevo. Fertile
plain area watered by Danube R. Important farming area producing wheat,
fruit and vegetables. Pop. (1981C)
2,034,772.

Volcano Islands   Japan. 25 00N 141
00E. A group of volcanic islands 750
m. (1,200 km.) s of Tokyo in the w
Pacific Ocean. Area, 11 sq. m. (28 sq.
km.). There are sulphur mines and
sugar plantations.

Volga Heights   Russian Soviet Federal Socialist Republic, U.S.S.R. 51
00N 46 00E. A range of hills stretching along Volga R. between Gorky
and Volgograd.

Volga River   U.S.S.R. 45 55N 47
52E. Rises in the Valdai Hills *c.*
200 m. (320 km.) NW of Moscow and
flows 2,325 m. (3,720 km.) first SE to

Rzhev then NE to Kalinin, through the Volga Reservoir and on into L. Rybinsk Reservoir, whence it turns SE past Yaroslavl, Kostroma, Kineshma, Gorky and Cheboksary to Kazan. Below Kazan it receives Vyatka and Kama Rs and flows S past Ulyanovsk to Kuibyshev, SSW past Saratov to Volgograd and finally SE to enter the Caspian Sea by a delta below Astrakhan. Linked by canal with Don R. at Volgograd, with Moscow and with Neva R. It is the most important Russian R., used for transport, water storage and power. The Greater Volga development scheme provided channelling, hydroelectric plants. reservoirs and dams. Ice-bound for 3–5 months of the year, prone to spring flooding and shoals in summer. Its middle course is important for irrigation in wheat-growing areas.

Volgograd (Formerly Stalingrad) Russian Soviet Federal Socialist Republic, U.S.S.R. 48 44N 44 25E. Cap. of the Volgograd region and port sit. on Volga R. near its junction with Volga–Don Canal. Industries inc. oil refining, sawmilling, iron and steelworks, and the manu. of tractors, oilfield machinery, railway equipment, machine tools, footwear, clothing and cement. Pop. (1989E) 999,000.

Volkhov River Russian Soviet Federal Socialist Republic, U.S.S.R. 60 08N 32 20E. Rises in L. Ilmen and flows 140 m. (224 km.) NE to enter L. Ladoga. There is an important power station at Volkhov town.

Völklingen Saarland, Federal Republic of Germany. 49 15N 6 50E. Town sit. WNW of Saarbrücken on Saar R. Industries coalmining, engineering and the manu. of steel, electrical equipment and cement. Pop. (1984E) 43,800.

Vologda Russian Soviet Federal Socialist Republic, U.S.S.R. 59 12N 39 55E. Cap. of the Vologda region sit. on Vologda R. and NE of the Rybinsk Reservoir. It is an industrial and railway centre. Industries inc. railway engineering and the manu. of agric. machinery, textiles, glass and cement. Trades in dairy produce. Pop. (1987E) 278,000.

Vólos Magnessia, Greece. 39 21N 22 56E. Cap. of prefecture and port sit. NNW of Athens on the Gulf of Vólos. Industries inc. chemicals, textiles, confectionery, wines, vegetables, salted fish and cement. Olive oil and tobacco are exported. Pop. (1981C) 182,222.

Volta Redonda Rio de Janeiro, Brazil. 22 32S 44 07W. Town sit. WNW of Rio de Janeiro on Paraíba R. The largest steel plant in S. America is sit. here. Pop. (1984E) 178,000.

Volta River *see* **Burkina River**

Voorburg Zuid-Holland, Netherlands. Town sit. E of The Hague. Manu. inc. machinery and ball-bearings. Pop. (1989E) 40,500.

Vorarlberg Austria. 47 20N 10 00E. Federal state in the W bordering on Federal Republic of Germany, Switzerland and Liechtenstein. Area, 1,004 sq. m. (2,601 sq. km.). Very mountainous and wooded. The principal occupations are forestry, dairy farming, fruit growing and tourism. Cap. Bregenz, sit. WNW of Innsbruck. Pop. (1988E) 316,300.

Voronezh Russian Soviet Federal Socialist Republic, U.S.S.R. 51 40N 39 10E. Cap. of the Voronezh region sit. on Voronezh R. near its confluence with Don R. It is the principal industrial centre of the black-earth area. Manu. inc. machinery, chemicals, excavators, electrical equipment, machine tools, tyres, synthetic rubber and food products. Pop. (1989E) 887,000.

Vosges France. 48 30N 7 10E. Dept. in Lorraine region sit. to W of the Vosges Mountains. Area, 2,267 sq. m. (5,871 sq. km.). Mainly hilly. The principal Rs. are Moselle, Meuse and Meurthe. There are deposits of coal, lead, iron and copper. Agric. products inc. cereals, vines, hops and potatoes. Cap. Épinal. Pop. (1982C) 395,769.

Vosges Mountains France. 48 20N 7 10E. Range in E France near the Federal German border, running parallel to the Black Forest ranges across R. Rhine. Thickly wooded and well watered.

Votkinsk Russian Soviet Federal Socialist Republic, U.S.S.R. 57 03N 53 59E. Town sit. SW of Perm. Industries inc. railway engineering, and the manu. of boilers and agric. machinery. Tchaikovsky was born here.

Vršac Vojvodina, Yugoslavia. 45 07N 21 18E. Town sit. ENE of Belgrade. Industries inc. meat packing, distilling and flour milling. Trades in brandy and wine.

Vyborg Russian Soviet Federal Socialist Republic, U.S.S.R. 60 42N 28 45E. Port sit. NW of Leningrad on Vyborg Bay. Manu. inc. agric. machinery and electrical equipment. Timber and wood products are exported. Pop. (1970E) 65,000.

Vyrnwy, Lake Powys, Wales. 52 48N 3 30W. Artificial L. which serves as a reservoir for Liverpool, and lies ENE of Barmouth.

Vyrnwy River Wales. Rises S of Bala and flows ESE for 35 m. (56 km.) to join R. Severn WNW of Shrewsbury.

Vyshni Volochek Russian Soviet Federal Socialist Republic, U.S.S.R. 57 34N 34 23E. Town sit. NW of Kalinin. Industries inc. sawmilling and the manu. of textiles and glass.

W

Waltham Cross *see* **Vaud**

Wabash River U.S.A. 37 46N 88 02W. Rises in Grand Lake, W Ohio and flows W and S for 475 m. (760 km.) to form the boundary between Indiana and Illinois before joining Ohio R.

Waco Texas, U.S.A. 31 55N 97 08W. City sit. on Brazos R. An aero modification centre, with light manu. Pop. (1989E) 106,000.

Wadi Halfa Sudan. 21 56N 31 20E. Town sit. NNW of Atbara on R. Nile near the Egyptian frontier. Important railway and steamer terminus.

Wadi Medani Blue Nile, Sudan. 14 26N 33 28E. Cap. of the Blue Nile province and sit. on Blue Nile R. Centre of important cotton-growing area made possible by Gezira irrigation scheme. Pop. (1984E) 141,065.

Wageningen Gelderland, Netherlands. 51 58N 5 40E. Port sit. W of Arnhem on Lower Rhine R. Manu. inc. wood and leather goods, bricks, tiles, millstones and cigars. There is a shipbuilding research station. Pop. (1988E) 32,370.

Wagga Wagga New South Wales, Australia. 35 07S 147 22E. Town sit. on Murrumbidgee R. Centre of an important agric. area. Pop. (1988E) 51,480.

Waikato River New Zealand. 37 23S 174 43E. Rises on E slopes of Mount Ruapehu in North Island and flows for 220 m. (352 km.) N to L. Taupo and then NNW to Port Waikato on the Tasman Sea.

Wainganga River Madhya Pradesh, India. 18 50N 79 55E. Rises in central India in the Satpura Range, and flows S for 400 m. (640 km.) before joining Wardha R., and becoming Pranhita R.

Wakamatsu *see* **Kitakyushu**

Wakatipu, Lake South Island, New Zealand. 45 04S 168 35E. L. 52 m. (83 km.) long in W Otago, in a glacial valley 90 m. (144 km.) NW of Dunedin, at 1,016 ft. (310 metres) a.s.l. Drained by Kawarau R.

Wakayama Honshu, Japan. 34 00N 135 20E. Town and cap. of prefecture of same name, sit. SSW of Osaka on the Kii peninsula. Manu. iron and steel, chemicals and textiles. Pop. (1988E) 401,000.

Wakefield West Yorkshire, England. 53 42N 1 29W. Sit. on R. Calder. A centre for the woollen industry since the 16th cent. Manu. mining machinery, machine tools, soft drinks and chemicals. Pop. (1988E) 75,750.

Wake Island North Pacific Ocean. 19 17N 166 36E. Atoll sit. 2,300 m. (3,680 km.) W of Hawaii. Pop. (1980C) 302.

Wakenaam Island Guyana. Sit. WNW of Georgetown in estuary of R. Essequibo. Crops inc. rice and coconuts.

Walbrzych Wrocław, Poland. 50 46N 16 17E. Town in SW Poland in the foothills of the Sudeten Mountains near border with Czechoslovakia. Centre of a coalmining region. Manu. glass, china and linen goods. Pop. (1983E) 137,000.

Walchensee Bavaria, Federal Republic of Germany. 47 36N 11 23E. L. in the Bavarian Alps sit. SSW of Munich at 2,630 ft. (802 metres) a.s.l. Has an important hydroelectric plant.

Walcheren Zeeland, Netherlands. 51 30N 3 35E. Island sit. in Scheldt estuary and joined to the mainland and S. Beveland island by causeways. Area 80 sq. m. (207 sq. km.). Chief towns are Middelburg and Vlissingen (Flushing). Mainly agric. producing sugar-beet and root vegetables.

Wales   United Kingdom. 52 30N 3 30W. Principality bounded E by England, S by Bristol Channel, W by St. George's Channel, N by Irish Sea, and inc. the island of Anglesey off NW coast, across the Menai Strait. Area, 8,016 sq. m. (20,761 sq. km.). Since 1974 Wales has been divided into 8 counties: Gwynedd, Clwyd, Powys, Dyfed, West Glamorgan, Mid Glamorgan, South Glamorgan and Gwent. Chief cities Cardiff (cap.), Swansea, Newport, Merthyr Tydfil, Rhondda (all in the indust. S). The N coast has an almost unbroken line of built-up areas between the resorts of Prestatyn, Rhyl, Colwyn Bay, Llandudno, and Aberconwy. N and central Wales are mountainous rising to Snowdon, 3,560 ft. (1,085 metres) a.s.l. The mountains extend E into England and end SE in the deep parallel valleys of the formerly important S Wales coalfield. The SW coast has lowland across the head of the peninsula, and there is a coastal strip of low ground up the W coast to the Mawddach estuary and on the Lleyn peninsula (NW). Anglesey is low-lying. The climate is cool with heavier rainfall than England. Chief rivers are Severn, Dee, Clwyd, Conway, Dovey, Teifi, Tawe, Towy, Taff, Usk and Wye. Farmland covers *c*. 4m. acres, inc. rough grazing land (1.5m.) and permanent pasture (1.8m.). Sheep farming is important, also dairy farming. Chief crops are hay, oats, barley, turnips and potatoes. The most important minerals are coal, inc. anthracite although declining, slate, lead, limestone and chalk, with some gold. Industries are concentrated in the S and inc. iron and steel, tin-plate, engineering, oil refining, textiles. Elsewhere the main occupations are farming, forestry, tourism, quarrying and fishing. Pop. (1988E) 2,857,000.

Wallasey   Merseyside, England. 53 26N 3 03W. Town opposite Liverpool on Mersey R. estuary, forming a residential suburb and including New Brighton and Egremont. Pop. (1981C) 92,987.

Wallingford   Oxfordshire, England.

51 37N 1 08W. Market town sit. SSE of Oxford on R. Thames. Mainly residential and a boating resort. Pop. (1981C) 6,230.

Wallingford   Connecticut, U.S.A. 41 27N 72 50W. Town sit. NNE of New Haven on Quinnipiac R. Industries inc. the manu. of silverware, metal products, electrical goods, plastics, chemicals and clothing. Pop. (1988E) 41,710.

Wallis   *see* **Valais**

Wallis and Futuna Islands   South West Pacific Ocean. 13 18S 176 10W. French dependency of 2 groups of islands sit. NE of Fiji comprising Uvéa in the N and Futuna and Alofi in the S. Area, 106 sq. m. (275 sq. km.). Cap. Matautu, on Uvéa. Pop. (1983E) 12,391.

Wallsend   Tyne and Wear, England. 55 00N 1 31W. Town sit. ENE of Newcastle upon Tyne, on R. Tyne. Sit. at E end of Hadrian's Wall, from which it derives its name. Industries shipbuilding, rope-making, engineering, coalmining and allied trades. Pop. (1981C) 51,207.

Walpole Island   New Caledonia. 22 37S 168 57E. Sit. E of the Isle of Pines. Area, $\frac{1}{2}$ sq. m. (1 sq. km.). Uninhabited.

Walsall   West Midlands, England. 52 35N 1 58W. Town sit. NW of Birmingham. Coalmining and engineering centre, industries inc. machine tool manu., leather, aircraft parts, locks and builders' hardware, plastics, electronics and chemicals. Pop. (1981C) 265,922.

Waltham   Massachusetts, U.S.A. 42 23N 71 14W. City sit. W of Boston on Charles R. One of the world's largest watch-factories is sit. here. Other manu. are precision instruments, tools, electronic equipment and textiles. Pop. (1981C) 58,200.

Waltham Abbey   Essex, England. 51 40N 0 01E. Town sit. SSE of Hertford on R. Lea. Manu. plastics, chemicals, general light industry and tiles. Noted for anc. abbey, part of which is still used as a church. Pop. (1986E) 18,811.

Waltham Cross Hertfordshire, England. 51 42N 0 01E. Town sit. SE of Hertford on R. Lea opposite Waltham Abbey. Manu. insecticides. Pop. (1981E) with Cheshunt, 50,000.

Walthamstow Greater London, England. 51 35N 0 01W. Mainly residential but industries of engineering and brewing.

Walton-le-Dale Lancashire, England. Town sit. S of Preston on R. Ribble. Textile manu. and engineering main occupations. Pop. (1973E) 27,660.

Walton-on-Thames Surrey, England. 51 23N 0 25W. Residential town sit. SW of London. Pop. (1988E) 20,011.

Walton-on-the-Naze Essex, England. 51 52N 1 17E. Resort sit. S of Harwich on the N. Sea. Pop. (1980E) 5,000.

Walvis Bay Namibia/Republic of South Africa. 22 59S 14 31E. Sea port in W central Namibia on the Atlantic coast, WSW of Windhoek. Namibia is strategically and economically reliant on the deep-water port, but it is administered by the Rep. of S. Afr., as part of Cape Province, and regards it as part of the Rep.'s sovereign territory. Industries inc. fishing and allied trades. Pop. (1980E) 25,000.

Wandsworth Greater London, England. 51 27N 0 11W. District of SW London on S bank of Thames R. opposite Fulham. Residential and indust. Pop. (1981C) 255,723.

Wanganui North Island, New Zealand. 39 56S 175 02E. Port on SW coast, sit. near mouth of Wanganui R. Centre of an agric. and pastoral region. Industries tourism, engineering, meat freezing works, woollen mills. Manu. steel pipes, footwear, biscuits, furniture and clothing. Pop. (1989E) 42,000.

Wanganui River North Island, New Zealand. 39 57S 174 59E. Rises in the

Rangitoto Range and flows S for 140 m. (226 km.) to the South Taranaki Bight.

Wangaratta Victoria, Australia. 36 22S 146 20E. Town sit. NE of Melbourne on Ovens R. Centre of an agric. district producing tobacco and hops. Manu. textiles. Pop. (1981E) 16,202.

Wankie   *see* **Hwange**

Wanne-Eickel North Rhine-Westphalia, Federal Republic of Germany. Town sit. NW of Bochum on Rhine-Herne Canal. Centre of a coalmining area. Manu. chemicals. Pop. (1971E) 97,100.

Wanstead and Woodford Greater London, England. 51 35N 0 02E. Mainly residential. *See* Redbridge.

Wantage Oxfordshire, England. 51 36N 1 25W. Town sit. SW of Oxford sit. on the downs. Trades in agric produce. Pop. (1985E) 8,753.

Wapping Greater London, England. 51 30N 0 53W. Riverside district sit. between R. Thames and the former London Docks. *See* Tower Hamlets.

Warangal Andhra Pradesh, India. 18 00N 79 35E. Town sit. NE of Hyderabad. Industries carpet-making, printing and cotton milling. Pop. (1981C) 335,150.

Wardha Maharashtra, India. Town sit. SW of Nagpur. District cap. Rail junction and trading centre for agric. produce, esp. cotton. Indust. inc. cotton-milling. Pop. (1981E) 88,495.

Ware Hertfordshire, England. 51 49N 0 02W. District sit. ENE of Hertford on R. Lea. Manu. plastics, chemicals, stationery, gloves and coach bodies. Pop. (1980E) 15,000.

Wareham Dorset, England. 50 41N 2 07W. Town sit. WSW of Poole on R. Frome. Pop. (1987E) 6,229.

Warminster Wiltshire, England. 51 13N 2 12W. Town sit. WNW of Salisbury on the edge of Salisbury Plain. Manu. agric. machinery, gloves and food products. Pop. (1984E) 15,000.

Warragul Victoria, Australia. Town sit. SE of Melbourne. Market and processing centre of a dairy-farming area. Pop. (1981E) 7,712.

Warrego River Australia. 30 24S 145 21E. Rises in the Carnarvon Range, Queensland, and flows 495 m. (792 km.) SSW past Charleville and Cunnamulla to enter Darling R. SW of Bourke, New South Wales.

Warren Michigan, U.S.A. 42 31N 83 02W. City NE of Detroit in industrial area. Manu. vehicles and parts. Pop. (1980C) 161,134.

Warren Ohio, U.S.A. 41 14N 80 52W. City sit. NW of Youngstown on Mahoning R. Manu. steel, car parts and assembly, machinery and electrical appliances. Pop. (1980C) 56,629.

Warrenpoint Down, Northern Ireland. 54 06N 6 15W. Resort sit. SE of Newry near head of Carlingford Lough. A tourist centre for the Mourne Mountains. Pop. (1981C) 4,798.

Warrington Cheshire, England. 53 24N 2 37W. Town sit. on R. Mersey. An important trading centre since Roman times. High-tech industries in science parks inc. computer, electronics and nuclear related businesses. Industries inc. engineering, brewing and chemicals. Pop. (1989E) 188,000.

Warrnambool Victoria, Australia. 38 23S 143 03E. Port in SW Victoria, sit. W of Melbourne in a rich agric. and pastoral area with exports of agric. produce inc. potato and onion growing. A major tourist centre and manu. woollens, textiles and dairy products. Pop. (1985E) 22,500.

Warsak North West Frontier Province, Pakistan. Village sit. NW of Peshawar on Kabul R. Centre of an irrigation and hydroelectric power scheme known as the Warsak Project.

Warsaw Poland. Province and city on Vistula R. in E central Poland. 52 15N 21 00W. Cap. of rep. and a commercial and industrial centre. The city founded in the 13th cent. was largely destroyed in the Second World War but much reconstruction has taken place on the medieval street patterns. Communication centre with rail, road and water traffic for Russia, Poland, Berlin and W Europe. Industries inc. the manu. of locomotives, tractors, machine tools, metals, textiles and clothing. Pop. (1985E) 1,649,000.

Warta River Poland. 52 35N 14 39E. Rises NW of Cracow and flows 445 m. (712 km.) NNW past Poznan to join Oder R. at Küstrin. Main tribs Prosna and Noteć Rivers. Linked by canal with Vistula R.

Warwick Queensland, Australia. 28 12S 152 00E. Town and tourist resort sit. SW of Brisbane on Condamine R. Railway centre serving an agric. area producing wheat, sorghum, maize and fruit. Industries inc. food processing, furniture manu. Pop. (1985E) 10,000.

Warwick Warwickshire, England. 52 17N 1 34W. Town sit. on R. Avon. Contains a 14th cent. castle and many fine medieval buildings. Chief occupations are agric. and tourism. Industries inc. light engineering and carpet manu. Pop. (1981C) 21,936.

Warwick Rhode Island, U.S.A. 41 42N 71 28W. City sit. S of Providence on Narragansett Bay. Manu. textiles, metal goods and machinery. Pop. (1980C) 87,123.

Warwickshire England. 52 17N 1 34W. County of central England bounded N by West Midlands, E by Leicestershire and Northamptonshire, S by Oxfordshire and Gloucestershire and W by Hereford and Worcester. Under 1974 reorganization the county consists of the districts of North Warwickshire, Nuneaton and Bedworthy, Rugby, Stratford-upon-Avon and Warwick. Mainly agric. but with industry located in towns such as Nuneaton, Rugby, Leamington Spa and Warwick. Pop. (1988E) 484,600.

Wasatch Range Idaho/Utah, U.S.A. 40 30N 111 15W. A range of the Rocky Mountains stretching N to S for 200 m. (320 km.). Contains at least four peaks over 11,000 ft. (3,353 metres) a.s.l., the highest being Mount

Timpanogos, 12,008 ft. (3,660 metres) a.s.l.

Wash, The England. 52 58N 0 20W. An inlet of the N. Sea between the counties of Lincolnshire and Norfolk. Generally low marshy coasts and very liable to silting. The principal Rs. draining into it are Welland, Witham, Ouse and Nene.

Washington Tyne and Wear, England. 54 54N 1 31W. New town designated 1964, now within Sunderland. Home of 400 companies manu. electronics, electrical products, motor vehicles and components. Pop. (1987E) 60,000.

Washington U.S.A. 47 45N 120 30W. State in NW, bounded N by Canada, E by Idaho, S by Oregon and W by Pacific Ocean. Washington was formerly part of Oregon and was created a Territory in 1853 and admitted into the Union as a State in 1889. The surface is crossed from N to S by the Cascade Mountains rising to 14,400 ft. (4,389 metres) a.s.l. to E of which is a high plateau. To W the state is heavily wooded with fertile prairies and valleys watered by R. Columbia and tribs. Area, 68,192 sq. m. (176,617 sq. km.). Chief towns Olympia (cap.), Seattle, Spokane and Tacoma. Trad. the state's economy has been based on agric., forestry, fishing and mining. Manu. led by the aircraft and aerospace industry, has steadily increased and has become the State's leading primary industry. Principal minerals, sand and gravel, stone, zinc, coal, lead, clay, peat. Uranium ore is also mined. Pop. (1987E) 4,481,000.

Washington District of Columbia, U.S.A. 38 54N 77 02W. City co-terminous with the District of Columbia with an area of 69 sq. m. (179 sq. km.) sit. on the N bank of the Potomac R. It is the federal cap. and is the administrative, judicial and legislative centre for the U.S. Most employment is provided by govt. It contains the White House, the Capitol (the 2 houses of Congress), the Library of Congress and the Smithsonian Institute. Seat of several universities, over 700 parks and national monuments and the U.S. military base, the Pentagon, sit. on the

w bank of the Potomac R. Pop. (1983E) 623,000.

Washington Pennsylvania, U.S.A. 40 10N 80 15W. City sit. SW of Pittsburgh. Manu. metal goods, wool, glass and chemicals. Pop. (1980C) 18,500.

Washington Island (Teraina) Kiribati. 4 43S 160 24W. An atoll NW of Tabuaeran. Copra is produced. Pop. (1980E) 416.

Washington, Mount New Hampshire, U.S.A. 44 15N 71 15W. The highest peak of the White Mountains in the Presidential Range, 6,288 ft. (1,917 metres) a.s.l. Also the highest point in the NE U.S.A.

Washoe Lake Nevada, U.S.A. L. 16 m. (26 km.) S of Reno and E of L. Tahoe at S end of the Virginia Mountains. Area, 3½ m. (5 km.) by 2 m. (3 km.).

Wassenaar Zuid-Holland, Netherlands. Town sit. NE of The Hague. Mainly residential with bulb growing. Pop. (1984E) 26,950.

Wast Water Cumbria, England. 54 26N 3 18W. L. sit. SW of Keswick. Depth 260 ft. (79 metres). Overlooked by Scafell, sit. at N end of Wasdale valley.

Watchet Somerset, England. 51 10N 3 20W. Port and resort sit. NW of Taunton on the Bristol Channel. Pop. (1980E) 3,000.

Watenstedt-Salzgitter Lower Saxony, Federal Republic of Germany. 52 02N 10 22E. Town sit. SSW of Brunswick. Industries iron and potash mining, iron and steel, manu. of machinery, chemicals and textiles.

Waterbury Connecticut, U.S.A. 41 33N 73 03W. City sit. NNW of New Haven on Naugatuck R. Centre of the brass industry since the early 19th cent. Manu. machinery, clocks and watches, plastics and clothing. Pop. (1980C) 103,266.

Waterford Munster, Ireland. 52 15N 7 06W. (i) County bordered on S by Atlantic Ocean. Area, 710 sq. m. (1,839 sq. km.). Main occupations tourism, dairying and cattle breeding. Pop. (1988E) 50,622. (ii) City and port of same name. Industries inc. glass, pharmaceuticals, precision and light engineering. Pop. (1986E) 41,054.

Waterloo Ontario, Canada. 43 28N 80 32W. City sit. W of Toronto. Industries inc. high technology and insurance. Pop. (1985E) 62,000.

Waterloo Iowa, U.S.A. 42 30N 92 21W. City sit. NW of Cedar Rapids on Cedar R. Site of the annual National Dairy Cattle Congress. Industries railway engineering, meat packing and soy-bean processing. Manu. farm machinery, cement-mixers, leather, concrete and wood products. Pop. (1980C) 75,985.

Waterlooville *see* **Havant and Waterlooville**

Waterton-Glacier International Peace Park Canada/U.S.A. A park comprising the adjacent Lakes Waterton Alberta National Park, and the Glacier National Park, Montana. Contains many glaciers and Ls. The highest peak is Mount Cleveland, 10,448 ft. (3,185 metres) a.s.l.

Watertown Massachusetts, U.S.A. 42 22N 71 11W. City sit. W of Boston on Charles R. Residential but there is light industry. Pop. (1989E) 32,189.

Watertown New York, U.S.A. 43 59N 75 55W. Town sit. NNE of Syracuse on Black R. Manu. paper, clothing, instruments and plumbing goods. Pop. (1980C) 27,261.

Watertown South Dakota, U.S.A. 44 54N 97 07W. City sit. NNW of Sioux Falls on Big Sioux R. Railway junction. Distribution and processing centre for a farming area and river resort. Indust. inc. the manu. of cement and granite products, metal fabrication, electronics, food products and beverages. Pop. (1980C) 15,649.

Waterville Maine, U.S.A. 44 34N 69 38W. City sit. N of Augusta on Kennebec R. at Ticonic Falls. Market for agric. produce. Indust. inc. timber

products, metals, machinery, stone, clay, glass and concrete products, transport equipment, instruments. Pop. (1985E) 16,750.

Watford Hertfordshire, England. 51 40N 0 25W. Town and regional shopping centre sit. NW of central London on R. Colne. Industries inc. printing, engineering and electronics. Pop. (1981C) 74,500.

Watling's Island *See* **San Salvador**

Watson Lake Yukon, Canada. 60 12N 129 00W. Village on Watson L. near the British Columbia border on a branch of the Alaska Highway. Trading post, police post, radio and meteorological station.

Wattenscheid North Rhine-Westphalia, Federal Republic of Germany. 51 29N 7 08E. Town sit. E of Essen. Centre of a coalmining area. Manu. metal goods, electrical equipment and footwear. Pop. (1971E) 81,300.

Waukegan Illinois, U.S.A. 42 22N 87 50W. City and port sit. NNW of Chicago on L. Michigan. Manu. outboard motors, electrical components, lasers, chemical and pharmaceutical products. Pop. (1989E) 71,293.

Waukesha Wisconsin, U.S.A. 43 01N 88 14W. City sit. W of Milwaukee on Fox R. Industries iron and steel, food processing and bottling of mineral spring water. Pop. (1980C) 50,319.

Wausau Wisconsin, U.S.A. 44 58N 89 38W. City in N central Wisconsin, on Wisconsin R. Industries agric. and cheese-making, granite quarrying and brewing. Manu. paper, wood products, plastics and chemicals. Pop. (1980C) 32,426.

Wauwatosa Wisconsin, U.S.A. 43 03N 88 00W. City adjacent to, and W of, Milwaukee. Manu. concrete block, motors, chemicals and timber products. Pop. (1988E) 49,673.

Waveney River Suffolk/Norfolk, England. 52 24N 0 24E. Rises near Diss and flows for 50 m. (80 km.) NW to join R. Yare at SW end of Breydon

Water. It is connected to the Norfolk Broads and is used by small craft.

Waverly Iowa, U.S.A. 42 44N 92 29W. City sit. NNW of Waterloo on Cedar R. Railway junction; distribution and processing centre of an area producing poultry and dairy products. Pop. (1980c) 8,444.

Waziristan Pakistan. A frontier region bounded on w by Afghánistán.

Weald, The England. 51 05N 0 25E. SE area comprising parts of Kent, Surrey and Sussex, and sit. between the N. and S. Downs. Previously forested, the wood being used for the iron industry in the Middle Ages. Now agric. and pastoral, producing hops, fruit and vegetables.

Wear River Durham, England. 54 55N 1 22W. Rises in the Pennines and flows for 67 m. (107 km.) to join the North Sea at Sunderland. The lower reaches are navigable and are used by the coal trade.

Weaver River Cheshire, England. 53 19N 2 45W. Rises near Crewe and flows for 50 m. (80 km.) to join R. Mersey at Runcorn. The last 20 m. (32 km.) below Runcorn, carries considerable quantities of salt and chemicals.

Webster Groves Missouri, U.S.A. 38 35N 90 21W. City immediately w of St. Louis. The main indust. is the manu. of petroleum products. Pop. (1980c) 23,097.

Wednesbury West Midlands, England. 52 34N 2 00W. Town sit. SE of Wolverhampton. Industries heavy and light engineering.

Wednesfield West Midlands, England. 52 36N 2 03W. Town sit. ENE of Wolverhampton. Manu. nuts and bolts.

Weert Limburg, Netherlands. Town sit. SE of Eindhoven. Manu. textiles and metal goods. Pop. (1984E) 39,402.

Weiden Bavaria, Federal Republic of Germany. 49 41N 12 10E. Town sit. SE of Bayreuth. Manu. textiles, china and glass. Pop. (1988E) 41,750.

Weimar Erfurt, German Democratic Republic. 50 59N 11 19E. Town sit. E of Erfurt, former seat of the Republican Government in 1919. Indust. inc. the manu. of machinery, textiles, food products. Noted for 17th- and 18th-cent. architecture. Pop. (1989E) 63,700.

Weinheim Baden-Württemberg, Federal Republic of Germany. 49 33N 8 39E. Town sit. NE of Mannheim. Industries tanning and soap making. Pop. (1984E) 41,200.

Weirton West Virginia, U.S.A. 40 25N 80 35W. City sit. w of Pittsburgh on Ohio R. Industries steel, cement and chemicals. Pop. (1980c) 24,736.

Weissenfels Halle, German Democratic Republic. 51 12N 11 58E. Town sit. WSW of Leipzig on Saale R. Industries lignite-mining, engineering and paper manu. Pop. (1989E) 39,000.

Weisshorn Pennine Alps, Switzerland. 46 06N 7 42E. Peak 6 m. (10 km.) NNW of Zermatt. 14,792 ft. (4,509 metres) a.s.l.

Welkom Orange Free State, Republic of South Africa. 27 59S 26 45E. Town sit. NNE of Bloemfontein, at 4,500 ft. (1,372 metres) a.s.l. Centre of a gold-mining area. Pop. (1984E) 228,000.

Welland Ontario, Canada. 42 59N 79 15W. Town and canal port sit. on Welland R. and Welland Ship Canal. Manu. iron and steel and agric. machinery. Trades in fruit. Pop. (1981E) 45,000.

Welland River England. 52 53N 0 02E. Rises near Market Harborough and flows for 70 m. (112 km.) NE to join The Wash.

Welland Ship Canal Ontario, Canada. 43 14N 79 13W. Canal 28 m. (61 km.) long joining L. Erie and L. Ontario. Capable of being used by ships up to 730 ft. (223 metres) in length, it is now one of the busiest inland waterways in the world.

Wellesley Massachusetts, U.S.A. 42 18N 71 17W. Town sit. WSW of

Boston on Waban L. Residential with some manu. inc. building supplies, electrical goods, hosiery and paper goods. Pop. (1980c) 27,209.

Wellingborough Northamptonshire, England. 52 19N 0 42W. Town sit. at confluence of R. Nene and R. Ise. Industries include footwear manu. and light engineering. Pop. (1983E) 47,600.

Wellington *see* **Telford**

Wellington Somerset, England. 50 59N 3 14W. Town sit. wsw of Taunton. Manu. woollen goods, furniture and bedding. Pop. (1985E) 13,437.

Wellington Tamil Nadu, India. Town sit. NW of Coonoor, forming a suburb. Military station with ordnance depot nearby. Indust. inc. the manu. of chemical products and cordite. Pop. (1981c) 19,638.

Wellington North Island, New Zealand. 41 19S 174 46E. Cap., city and port in sw of North Island founded in 1840. Industries in the Wellington-Hutt area inc. the manu. of vehicles, footwear, machinery, metal products, chemicals, soap and food processing. Pop. (1984E) 135,000.

Wells Somerset, England. 51 13N 2 39W. City sit. s of Bristol and just s of the Mendip Hills. The 12th-cent. cathedral is noted for its w front. There are many other fine medieval buildings associated with the cathedral. Industries inc. high technology, animal foodstuffs, engineering, printing, electronics, baby foods. Pop. (1989E) 10,000.

Wells-next-the-Sea Norfolk, England. 55 57N 0 51E. Port and resort sit. NW of Norwich. Industries include tourism and fishing, especially shellfish. Pop. (1989E) 2,226.

Wels Upper Austria, Austria. 48 10N 14 02E. Town sit. in N central Austria on Traun R. Manu. machinery and food products. Pop. (1981c) 51,060.

Welshpool Powys, Wales. 52 40N 3 09W. Town sit. near R. Severn and border with England. There is a variety of light industry. Pop. (1990E) 8,000.

Welwyn Garden City Hertfordshire, England. 51 50N 0 13W. Town sit. NE of St Albans. About 100 various industries inc. pharmaceuticals, chemicals, plastics and food products. Pop. (1981c) 40,496.

Wembley Greater London, England. 51 33N 0 18W. District of NW London, residential and indust., containing Wembley Stadium and the Empire Pool, built for the Wembley Exhibition of 1924–1925. Indust. inc. chemicals, and the manu. of glass and electric cables.

Wendover Buckinghamshire, England. 51 46N 0 45W. Town sit. SE of Aylesbury, and in the Chiltern Hills. Pop. (1989E) 7,500.

Wensleydale North Yorkshire, England. 54 18N 2 00W. The upper valley of the R. Ure. Noted for long-woolled sheep and cheese.

Wernigerode Magdeburg, German Democratic Republic. 51 50N 10 47E. Resort sit. sw of Magdeburg at the foot of the Harz Mountains. Industries engineering, paper and chemical works. Pop. (1989E) 36,500.

Werribee Victoria, Australia. 37 54S 144 40E. Town sit. wsw of Melbourne near NW shore of Port Phillip Bay. Processing centre of a dairy farming area, with dairy farming research and training institute.

Wesel North Rhine-Westphalia, Federal Republic of Germany. 51 40N 6 38E. Town sit. NNW of Duisburg near union of Rs. Rhine and Lippe. Manu. machinery. Pop. (1984E) 55,300.

Weser River Federal Republic of Germany. 53 32N 8 34E. Formed by confluence of R. Werra and Fulda and flows for 310 m. (496 km.) mainly through Lower Saxony, and enters the North Sea near Bremerhaven. Through a canal system it is connected to Rs. Rhine, Eins and Elbe.

West Allis Wisconsin, U.S.A. 43 01N 88 00W. City adjacent to, and w of, Milwaukee. Manu. heavy machinery, tanks, lorries, and castings. Pop. (1980C) 63,982.

West Bengal India. 23 00N 88 00E. State in NE India, area, 34,267 sq. m. (88,752 sq. km.). Chief cities, Calcutta (cap.), Howrah, Barddhaman, Asansol, Baharampur, Siliguri, Medinipur, Durgapur. Agric. inc. rice, jute, tea, betel leaf, tobacco and sugarcane, with allied industries. Important minerals found are coal, china clay, dolomite, rock phosphate, fire-clay, limestone, felspar, copper and iron. Industries and manu. inc. tyres, electronics, batteries, cigarettes, glass, explosives, aluminium foil, textiles, engineering goods, leather, oil refining, fertilizers, paper, railway electric locomotives, cables, iron and steel, alloy steel, graphite, carbon black, chemicals, pharmaceuticals and automobiles. Pop. (1981C) 54,580,647.

West Bromwich West Midlands, England. 52 31N 1 56W. Town in the 'Black Country'. Coal-mining centre and specializes in manu. of metal goods of all kinds. Pop. (1981C) 92,000.

Westbury Wiltshire, England. 51 16N 2 11W. Town sit. SE of Bath. Industries inc. cement and glove making. Pop. (1989E) 10,000.

West Covina California, U.S.A. 34 05N 117 58W. City sit. E of Los Angeles. Mainly residential. Pop. (1980C) 80,291.

West End Bahama Islands, West Indies. Town at w tip of Grand Bahama Island, sit. E of the Florida coast. Fishing port with fish processing plant.

Westerham Kent, England. 51 16N 0 05E. Town sit. w of Sevenoaks on R. Darent. Tourism and some light industry. Pop. (1981C) 4,255.

Western Australia Australia. 25 00S 118 00E. Most w and largest state of Australia. In 1791 Vancouver, in the *Discovery*, took formal possession of the country about King George

Sound. In 1901 Western Australia became one of the six Federated States within the Commonwealth of Australia. The coastline is deeply indented in NW. The surface is generally tableland with mountain ranges on w coast reaching 4,024 ft. (1,227 metres) a.s.l. at Mount Bruce. It is drained by Rs. Ord, Fitzroy, Ashburton, Gascoyne, De Grey, Fortescu, Murchison, Greenough and Swan. Area, 975,920 sq. m. (2,526,933 sq. km.). Chief towns, Perth (cap.) and Fremantle. Important crops are wheat, oats, barley, hay, potatoes, apples, pears, oranges, currants and raisins. Beef cattle rearing is very important. The mining industry has been for many years of considerable significance in the Western Australian economy. Until the mid-1960s the major mineral produced was gold. However, in recent years gold has been displaced by iron ore, crude oil and nickel in terms of value. Pop. (1986C) 1,459,019.

Western Desert Egypt. An ill-defined area of desert, part of the Libyan Desert, on the borders of Cyrenaica.

Western Isles (Outer Hebrides) United Kingdom. 57 30N 7 10W. Scottish Islands Area administrative unit. The islands inc. Lewis, Harris, N and S Uist, Benbecula, Barra, as well as other smaller uninhabited islands. Industries inc. tourism, fishing, weaving and crofting. Pop. (1989E) 31,834.

Western Sahara West Africa. Moroccan-occupied area, formerly a Spanish province (Spanish Sahara). Area, 102,680 sq. m. (266,769 sq. km.). Cap. El Aaiún. Pop. (1986E) 180,000.

Western Samoa   *see* **Samoa**

Westfield Massachusetts, U.S.A. 42 08N 72 45W. City sit. wsw of Boston on Westfield R. Manu. machinery, bicycles, prefabricated houses, paper and tobacco. Pop. (1980C) 36,465.

Westfield New Jersey, U.S.A. 40 39N 74 21W. Town sit. sw of Newark. Mainly residential. Industries inc. computer systems, foods. Pop. (1980C) 30,447.

West Germany   *see* **Germany, Federal Republic of**

West Glamorgan   Wales. 51 45N 3 58W. County of S Wales, mainly urban. Bounded by Dyfed NW, Powys N, Mid-Glamorgan E, and the Bristol Channel and Irish Sea S. Area, 316 sq. m. (818 sq. km.). Under 1974 reorganization the county consists of the districts of Swansea, Lliw Valley, Neath, Afan. Chief towns, Swansea, Port Talbot and Neath. Industries include coal-mining, steel, tinplating, chemicals and aluminium. Much light industry has been attracted to the county. Pop. (1988E) 363,600.

West Ham   Greater London, England. 51 32N 0 02E. District of E London on N bank of R. Thames. *See* Newham.

West Hartford   Connecticut, U.S.A. 41 46N 72 45W. Town sit. N of New Haven. Mainly residential and educational centre. Pop. (1985E) 61,850.

West Haven   Connecticut, U.S.A. 41 16N 72 57W. City sit. just SW of New Haven. Industries inc. aeroplane parts, machinery, drugs, x-ray equipment, food processing, door and printing services. Pop. (1986E) 54,050.

Westhoughton   Greater Manchester, England. Town sit. SW of Bolton. Industries inc. engineering, pharmaceuticals, paper tube and clothing manu. Pop. (1986E) 23,000.

West Indies   22 30N 65 00W. Islands to N and E of the Caribbean Sea. They extend from Yucatán, Mexico and Florida, U.S.A. to Gulf of Paria between Trinidad and Venezuela, and along N coast of Venezuela, and comprise the Bahamas; the four large islands of Cuba, Jamaica, Haiti and Puerto Rico known as the Greater Antilles; the Leeward and Windward Islands, known as Lesser Antilles, and Trinidad off Venezuelan coast. Many of islands mountainous in character. Others are of coral formation. Climate generally tropical and soil fertile. The islands produce sugar, bananas, and other fruits, cotton, coffee, cocoa, ginger, sponges, arrowroot, tobacco and asphalt from Trinidad. Tourism

has become a dominating occupation for many islands. Area, 91,819 sq. m. (237,811 sq. km.).

West Irian   Indonesia. 4 00S 137 00E. The W half of the islands containing Papua New Guinea, almost divided in two by McCluer Gulf (W). Area, 162,927 sq. m. (421,981 sq. km.). Cap. Jajapura. Mountainous in centre, rising to 16,400 ft. (4,990 metres) a.s.l. and descending to swampy lowland. Agric. is little developed. The main product is oil. Pop. (1980C) 1,173,875.

Westland   Zuid-Holland, Netherlands. Region extending *c.* 10 m. (16 km.) SW from The Hague to Hook of Holland. Fruit and vegetable growing area. The main towns are Naaldwijk, Poeldijk and Monster.

Westland   Michigan, U.S.A. 42 18N 83 23W. City SW of Detroit in industrial Wayne county. Pop. (1980C) 84,603.

West Lothian   Scotland. 55 55N 3 35W. Former county now part of Lothian Region.

West Malaysia   *see* **Peninsular Malaysia**

Westmeath   Leinster, Ireland. 53 30N 7 30W. Inland county bounded by Cavan, Meath, Offaly and Roscommon. Area, 681 sq. m. (1,764 sq. km.). Mainly low-lying with many loughs which are noted for trout. Chief towns, Mullingar (county town) and Athlone. Principal Rs. Shannon, Inny and Boyne. Main industries agric. and dairy farming. Pop. (1988E) 63,379.

West Memphis   Arkansas, U.S.A. 35 08N 90 11W. City sit. W of Memphis, Tennessee, on Mississippi R. Indust. inc. sawmilling, distilling, cotton-milling, cottonseed processing. Pop. (1980C) 28,138.

West Midlands   England. 52 30N 1 58W. Under 1974 re-organization the metropolitan county consists of the former county boroughs of Wolverhampton, Walsall, Dudley, Sandwell, Birmingham, Solihull and

Coventry. Area, 347 sq. m. (899 sq. km.) Wide variety of industry but dominated by the car and aircraft manu. and their related trades. The last steel making in the county was in 1990. Pop. (1988E) 2,617,300.

Westminster Greater London, England. 51 30N 0 07W. Metropolitan boro, which extends from banks of R. Thames to Hyde Park. Westminster contains Hyde Park, St. James's Park and Green Park. Important buildings inc. Houses of Parliament, Westminster Hall, Westminster Abbey, Westminster Cathedral (R.C.), Cenotaph, Buckingham Palace, and St. James's Palace. Westminster Abbey is one of finest examples of Early Eng. Architecture and is the burial-place of eighteen sovereigns of Eng. Pop. (1988E) 169,700.

Westminster Colorado, U.S.A. 39 50N 105 02W. City forming a NW suburb of Denver metropolitan area. Employment is in electrical indust. and medical services. Pop. (1985E) 61,000.

Westmoreland Jamaica. A parish in SW. Area, 320 sq. m. (829 sq. km.). Mainly fertile plains producing livestock, rice, sugar-cane, coffee and fruit. The principal R. is Cabaritta. Cap. Savanna-la-Mar. Pop. (1977E) 121,600.

Westmorland England. 54 30N 2 40W. Now part of Cumbria under 1974 re-organization.

West New York New Jersey, U.S.A. 40 47N 74 04W. Town sit. on Hudson R. opposite Manhattan. Manu. textiles, leather goods, clothing and toys. Pop. (1980C) 39,194.

Weston-super-Mare Avon, England. 51 21N 2 59W. Town and resort sit. on the Bristol Channel. Pop. (1984E) 170,479.

West Orange New Jersey, U.S.A. 40 47N 74 14W. Town sit. NW of Newark. Site of the Edison electrical equipment plant. Other manu. metal goods, tiles and clothing. Pop. (1980C) 39,510.

West Palm Beach Florida, U.S.A.

26 43N 80 04W. Town sit. N of Miami on the Atlantic coast. Resort. Pop. (1980C) 63,305.

West Point New York, U.S.A. 41 23N 73 57W. Site of the U.S. Military Academy sit. N of New York city, inc. Constitution Island in Hudson R.

Westport Mayo, Ireland. 53 44N 9 31W. Town and port sit. on an inlet of Clew Bay. Steamer and railway centre. Manu. clothing, hosiery and leather goods.

West Sussex England. 50 55N 0 30W. Under 1974 re-organization the county consists of the districts of Crawley, Chichester, Horsham, Mid-Sussex, Arun, Worthing and Adur. Bordered by Surrey to N, East Sussex to E, the English Channel to S, and Hampshire to W. Area, 778 sq. m. (2,015 sq. km.). Chief towns, Chichester, Worthing, Crawley, and Horsham. Agric. is a major industry and tourism important. Pop. (1989E) 714,100.

West Virginia U.S.A. 39 00N 81 00W. State of E U.S.A. bounded E and S by Virginia, SW by Kentucky, NW by Ohio, N by Pennsylvania and Maryland. Area 24,282 sq. m. (62,890 sq. km.). Chief cities Charleston (cap.), Huntington. It split from Virginia after the latter's secession from the Union in 1862. Mountainous E with the Allegheny Mountains extending SW to NE and descending W to plateau which occupies two-thirds of the state and drains W into Ohio R. along the W boundary. Farmland covers *c.* 5m. acres. Livestock farming is the more important, esp. beef cattle, sheep and poultry. Main crops are hay and fruit. The chief mineral is coal; *c.* half the state is on a coalfield. There is also oil, natural gas, lime, salt, sand and gravel. The main indust. are the manu. of metals and metal products, machinery, glass, timber products, textiles, clothing and chemicals. Pop. (1986E) 1,919,000.

Westward Ho! Devon, England. 51 02N 4 16W. Resort sit. NW of Bideford on Bideford Bay. Pop. (1985E) 1,500.

West Yorkshire England. 54 00N

2 00W. Metropolitan county of NE England part of the former West Riding. Under 1974 reorganization the metropolitan county consists of Bradford, Leeds, Calderdale, Kirklees and Wakefield. Area, 787 sq. m. (2,038 sq. km.). Textile manu. and coalmining are two dominant industries but efforts to attract light industry have been successful. On the edge of two national parks. Pop. (1988E) 2,056,500.

Wetaskiwin Alberta, Canada. 52 58N 113 22W. Town sit. s of Edmonton. Railway junction for freight only serving a farming area. Industries inc. food processing, cranes, plastics, furniture, transport and agric. equipment. Pop. (1989E) 10,101.

Wete Pemba, Tanzania. Town on NW coast, chief town of the island and trading centre for cloves.

Wetzlar Hessen, Federal Republic of Germany. 50 33N 8 30E. Town sit. ESE of Bonn on Lahn R. Manu. optical instruments, cameras, machine tools and textiles. Pop. (1984E) 49,900.

Wewak Papua New Guinea. 3 35S 143 40E. Town sit. NNW of Port Moresby. District cap. Main product, coconuts.

Wexford Leinster, Ireland. 52 20N 6 27W. (i) Maritime county in SE bordered by St. George's Channel. Area, 908 sq. m. (2,352 sq. km.). Hilly in NW rising to Mount Leinster, 2,610 ft. (796 metres) a.s.l. Chief towns Wexford, New Ross and Enniscorthy. Principal Rs. Barrow and Slaney. Chief industries agric., dairy farming and fishing. Pop. (1988E) 102,552. (ii) County town of same name and fishing port sit. on Wexford Harbour. Industries iron founding, manu. of agric. machinery, bacon-curing and tanning.

Weybridge Surrey, England. 51 22N 0 28W. Town on Wey R. near its confluence with Thames R. Residential, educational centre, noted for the Brooklands motor-racing circuit now used by aircraft industry.

Weyburn Saskatchewan, Canada. 49 41N 103 52W. City sit. SE of Regina.

Industries inc. manu. of wire and cable, plastic pipes, steel buildings, agric. and tractor manu. Pop. (1986E) 10,153.

Weymouth Dorset, England. 50 36N 2 28W. Port with cross channel ferries to the Channel Islands and France and resort sit. at mouth of R. Wey. There is some light engineering. On the Isle of Portland to s there is quarrying, a naval base and a prison. Pop. including Portland (1989E) 62,500.

Weymouth Massachusetts, U.S.A. 42 13N 70 58W. Town sit. SSE of Boston on Massachusetts Bay. Dormitory town to Boston. Pop. (1989E) 58,226.

Wey River England. 51 19N 0 29W. Rises near Alton, Hampshire, and flows for 35 m. (56 km.) through Farnham and Guildford to join R. Thames at Weybridge.

Whangarei North Island, New Zealand. 35 43S 174 20E. Town and port sit. on E coast of Auckland Peninsula. Export centre of an agric. and fruit growing district. Manu. glass and cement. Pop. (1980E) 39,700.

Wharfe River England. 53 51N 1 07W. Rises in s of Yorkshire Dales and flows 80 m. (128 km.) E, SE and then E through Ilkley, Otley and Tadcaster to join R. Ouse near Cawood. Wharfedale, the upper valley, is renowned for its beauty.

Wheat Ridge Colorado, U.S.A. 39 46N 105 05W. City immediately w of Denver. Residential. Pop. (1980C) 30,293.

Wheeling West Virginia, U.S.A. 40 05N 80 42W. City sit. SW of Pittsburgh on Ohio R. Industries chemicals, coal, iron and steel inc. tin-plate and sheet-metal production. Manu. textiles, plastics, tobacco and food products. Pop. (1980C) 43,070.

Whipsnade Bedfordshire, England. 51 51N 0 32W. Village sit. SSW of Luton famous for its Zoological Park.

Whitby North Yorkshire, England. 54 29N 0 37W. Port and resort sit. on

N. Sea coast, at mouth of R. Esk. Industries boatbuilding, fishing and fish-curing, plastics and food processing. Pop. (1981C) 13,403.

Whitchurch Shropshire, England. 52 58N 2 42W. Town sit. N of Shrewsbury. Manu. milk products. Pop. (1989E) 7,500.

Whitechapel Greater London, England. 51 31N 0 05W. District in E London. Manu. clothing. *See* Tower Hamlets.

Whitehaven Cumbria, England. 54 33N 3 35W. Town and port in NW Eng. on the Irish Sea. Industries inc. chemical, detergents and printing. Manu. textiles. Pop. (1981C) 26,714.

Whitehorse Yukon Territory, Canada. 60 43N 135 03W. Cap. of Territory sit. 90 m. (144 km.) from border with Alaska, U.S.A., on Yukon R. just below Mites Canyon. Industries inc. mining, tourism, fur-trapping and sport fishing. Pop. (1989E) 20,438.

White Horse, Vale of England. 51 34N 1 34W. Valley extending W to E through W Oxfordshire from White Horse Hill, Uffington, which has an earthwork and the figure of a horse cut out of the turf, exposing white chalk beneath.

White Mountains U.S.A. 44 10N 71 35W. (i) A range in New Hampshire. It is part of the Appalachian Mountains, the highest peak being Mount Washington, 6,288 ft. (1,917 metres) a.s.l. (ii) A range extending through E California and SW Nevada, the highest peak being White Mountain, 14,242 ft. (4,341 metres) a.s.l.

White Plains New York, U.S.A. 41 02N 73 46W. City sit. NNE of New York City on Bronx R. Corporate office and shopping centre. Pop. (1980C) 46,999.

White Russia *see* **Belorussia**

White Sea Russian Soviet Federal Socialist Republic, U.S.S.R. 66 00N 40 00E. A gulf of the Barents Sea, the entrance to which lies between the peninsulas of Kola and Kanin. Area,

36,680 sq. m. (95,001 sq. km.). Principal port Arkhangelsk. Rs. Onega, N Dvina and Mezen discharge into it. Herring and cod fisheries and seal to N.

Whitley Bay Tyne and Wear, England. 55 03N 1 25W. Resort sit. NE of Newcastle upon Tyne. Industries inc. tourism. Pop. (1985E) 38,133.

Whitney, Mount California, U.S.A. 36 35N 118 18W. Peak sit. in the Sierra Nevada. At 14,495 ft. (4,418 metres) a.s.l. it is the highest peak in the U.S.A., excluding Alaska.

Whitstable Kent, England. 51 22N 1 02E. Resort sit. NNW of Canterbury on Thames Estuary. Noted for oysters. Pop. (1985E) 27,284.

Whittier California, U.S.A. 33 59N 118 02W. City sit. SE of Los Angeles. Founded as a Quaker colony in 1887. Manu. gas and oil heaters, photographic equipment and bed frames. Pop. (1980C) 69,717.

Whittlesey Cambridgeshire, England. 52 34N 0 08W. Town sit. ESE of Peterborough on Nene R. Manu. bricks and tiles. Pop. (1988E) 13,240.

Whyalla South Australia, Australia. 33 02S 137 35E. Town and port sit. NNW of Port Pirie on NW shore of Spencer Gulf. Important steelworks, other indust. inc. gas liquids, salt and clothing. Water for town is piped from Murray R. Pop. (1984E) 31,500.

Wichita Kansas, U.S.A. 37 42N 97 20W. City sit. at confluence of Arkansas and Little Arkansas R. Centre of a wheat-growing and oil producing district. Industries oil refining, flour milling, meat packing and the manu. of chemicals. Pop. (1980C) 279,272.

Wichita Falls Texas, U.S.A. 33 54N 98 30W. City sit. near border with Oklahoma on Wichita R. Industries inc. oil and gas-related manu. Pop. (1980C) 94,201.

Wick Highland Region, Scotland. 58 26N 3 06W. Town and port NE of Inverness sit. at mouth of Wick R. Industries fishing, distilling and manu. of glass and electronic equip. Pop. (1989E) 8,500.

Wicklow Leinster, Ireland. 53 00N 6 30W. (i) County in SE bordering on the Irish Sea. Area, 782 sq. m. (2,025 sq. km.). Mainly mountainous, with many fine valleys, rising to Lugnaquillia in the Wicklow Mountains, 3,039 ft. (926 metres) a.s.l. Chief towns Wicklow, Arklow and Bray. Principal R., Liffey. The main occupation is agric. Pop. (1988E) 94,542. (ii) County town and port of same name sit. at mouth of R. Vartry.

Widnes Cheshire, England. 53 22N 2 44W. Town sit. E of Liverpool on R. Mersey. Manu. chemicals, timber, fertilizers and metal castings. Pop. (1983E) 54,900.

Wien *see* **Vienna**

Wiener Neustadt Lower Austria, Austria. 47 49N 16 15E. Town sit. S of Vienna. Manu. locomotives, rolling stock, cars, machinery, textiles and leather goods. Pop. (1981C) 35,006.

Wiesbaden Hessen, Federal Republic of Germany. 50 05N 8 14E. Cap. of province sit. at foot of the Taunus Hills. A spa since Roman times. Manu. chemicals, textiles, plastics, pottery and sparkling wine. Pop. (1986E) 267,000.

Wigan Greater Manchester, England. 53 33N 2 38W. Metropolitan borough sit. between Manchester and Liverpool on R. Douglas. Industries inc. cloth manu., glass, fibre products, mail order, engineering, food products. Pop. (1988E) 307,600.

Wight, Isle of England. 50 40N 1 20W. County and island off coast of Hampshire and Dorset. Area, 147 sq. m. (376 sq. km.). Sit. in the mouth of Southampton Water, crossed E to W by a range of chalk hills ending in the Needles cliffs, *c.* 100 ft. (30 metres) a.s.l. Yachting and health resort. Home of the Hovercraft and boat designers and builders. Manu. electronics. Main towns Newport, Ryde, Cowes, Shanklin and Ventnor. Drained by Medina R. Pop. (1989E) 132,000.

Wigston Leicestershire, England. 52 36N 1 05W. Town sit. SE of Leicester on Soar R. Indust. inc. the manu. of textiles, footwear, food processing, plastics, foundry working, precision engineering, pattern-making, knitwear, heavy metal fabrication, electrical engineering, and hosiery. Pop. (1987E) 31,210.

Wigtown Dumfries and Galloway Region, Scotland. 54 53N 4 45W. Town and port sit. on W shore of Wigtown Bay in SW Scot.

Wigtownshire Scotland. Former county now part of Dumfries and Galloway Region.

Wilhelmshaven Lower Saxony, Federal Republic of Germany. 53 31N 8 08E. City and port sit. on an inlet of N. Sea at E end of Ems-Jade Canal. Industrial centre and naval base. Manu. agric. machinery, refrigerators and chemicals. Pop. (1984E) 98,200.

Wilkes-Barre Pennsylvania, U.S.A. 41 14N 75 53W. City sit. NNW of Philadelphia on Susquehanna R. Industries inc. tourism and convention trade. Pop. (1984E) 50,677.

Willamette River Oregon, U.S.A. 43 39N 122 46W. Formed by union of Coast Fork R. and Middle Fork R., flows N for 183 m. (293 km.) to join Columbia R. below Portland.

Willemstad Curaçao, Netherlands Antilles. 12 06N 68 56W. Cap. and port sit. on SW coast, N of Venezuela. Possesses one of the best harbours in the Caribbean, and one of the largest oil refineries in the world. Pop. (1983E) 50,000.

Willenhall West Midlands, England. 52 36N 2 03W. Town sit. E of Wolverhampton. Now part of Walsall. Centre of a coal-mining area. Manu. locks, bolts, tools and car radiators.

Willesden Greater London, England. 51 33N 0 14W. Mainly residential district. Industries food processing and manu. of car parts.

Williamsburg Virginia, U.S.A. 37 16N 79 43W. City sit. ESE of Richmond now inc. in the Colonial National Historical Park. Many buildings now restored to early colonial appearance. Pop. (1980c) 9,870.

Williamsport Pennsylvania, U.S.A. 41 14N 77 00W. City sit. NW of Philadelphia on W. Susquehanna R. Manu. aeroplane parts, steel and iron goods, textiles, furniture, leather and electrical products. Pop. (1980c) 33,401.

Williamstown Victoria, Australia. 37 52S 144 54E. Town sit. SW of Melbourne on SW shore of Hobson's Bay and mouth of the Yarra R. Indust. inc. railway workshops, tourism, oil storage, shipbuilding and repair. Pop. (1986E) 24,000.

Williamstown Massachusetts, U.S.A. Town sit. W of North Adams on Hoosic R. Indust. inc. the manu. of wire and photographic equipment. Summer, and winter ski, resort. Pop. (1980c) 8,741.

Wilmington Delaware, U.S.A. 39 44N 75 33W. City and port sit. SW of Philadelphia on Delaware R. Industries inc. banking, shipbuilding, car assembly, oil refining, meat packing and tanning. Manu. inc. chemicals, textiles, iron and steel products, and paper. Pop. (1980c) 70,195.

Wilmington North Carolina, U.S.A. 34 13N 77 55W. City and port sit. SSE of Raleigh, on Cape Fear R. 10 m. (16 km.) from the Atlantic Ocean. Industries fibre optics, aircraft engine parts, dacron, cranes. clothing, paper, nuclear fuel. Pop. (1989E) 60,000.

Wilmslow Cheshire, England. 53 19N 2 14W. Town sit. S of Manchester on R. Bollin. Manu. textiles and clothing. Pop. (1981c) 30,055.

Wilson North Carolina, U.S.A. 35 44N 77 55W. Town sit. E of Raleigh. Important tobacco trade. Manu. inc. chemicals, tyres, vehicle parts, electronics, textiles and clothing. Pop. (1980c) 34,424.

Wilson, Mount California, U.S.A. 37 55N 105 03W. Peak NE of Los Angeles in the San Gabriel Mountains. 5,700 ft. (1,737 metres) a.s.l.

The Mount Wilson Observatory is sit. here.

Wilton Cleveland, England. Village sit. WSW of Redcar, near mouth of R. Tees. Chemical manu. centre.

Wilton Wiltshire, England. 51 05N 1 52W. Town WNW of Salisbury sit. at confluence of R. Wylye and Nadder. Noted since the 16th cent. for the manu. of carpets. Pop. (1980E) 4,100.

Wiltshire England. 15 20N 2 00W. Under 1974 reorganization the county consists of the districts of North Wiltshire, Thamesdown, West Wiltshire, Kennet and Salisbury. Bounded by Gloucestershire to N, Oxfordshire and Berkshire to E, Hampshire to E and S, Dorset to S and Somerset and Avon to W. Predominantly agric., there are vast chalk downs within the county and from Salisbury Plain and Marlborough downs radiate all the chalk ridges of S and E England. Apart from Swindon and Salisbury towns are small in size. Pop. (1989E) 564,000.

Wimbledon Greater London, England. 51 26N 0 13W. Mainly residential district.

Winchester Hampshire, England. 51 04N 1 19W. Cathedral city and anc. cap. of Eng. sit. on R. Itchen. Occupation of the city area can be traced back to 1800 B.C. but organized settlements appeared later. Alfred the Great made Winchester a centre of education. William the Conqueror estab. Winchester as his cap., and also compiled the Domesday Book here. Winchester remained the cap. for many years, but its decline as such began with the civil war between Stephen and Matilda; and by 1338 it had lost its favourable position. Tourism is important and there is some light industry. Pop. (1985E) 35,500.

Windermere Cumbria, England. 54 24N 2 56W. Resort in Lake District, sit. on shore of L. Windermere. Pop. (1981c) 8,636.

Windermere, Lake Cumbria, England. 54 23N 2 54W. The largest L. in Eng., 11 m. (18 km.) long. It has wooded shores and several small islands. Drains into Morecambe Bay through R. Leven.

Windhoek Namibia. 22 35S 17 04E. Cap. of Namibia, sit. in centre of country at 5,400 ft. (1,650 metres) a.s.l. Founded by Germans in 1890 its industries inc. food processing, clothing, textiles, timber products, printing, diamonds, cutting and polishing semi-precious stones. Trades in karakul (Persian lamb) skins and semi-precious stones. Pop. (1988E) 114,500.

Windsor Berkshire, England. 51 28N 0 36W. Town on R. Thames. Contains Windsor Castle, a royal residence, with St. George's Chapel, containing many royal tombs. In the park is mausoleum of Frogmore. Windsor is connected with Eton by a bridge across river. There is some light industry. Pop. (1986E) 31,225.

Windsor Ontario, Canada. 42 18N 83 01W. City and port sit. on Detroit R. opposite Detroit, Michigan, U.S.A., to which it is linked by the Ambassador Bridge, the Detroit-Canada Tunnel, and barge services. Centre of Canada's car and pharmaceutical industries. Also manu. machinery, tools, steel products, paint, food and beverage products and clothing. Pop. (1984E) 192,338.

Windward Islands French Polynesia, South Pacific. Group inc. Tahiti, Moorea, Maiao, Mehetia, Tetiaroa. Area, 445 sq. m. (1,178 sq. km.). The main island is Tahiti, which rises to 7,618 ft. (2,322 metres) a.s.l. Cap. (of the Society Islands) Papeete, on Tahiti. Chief products are copra, sugar, rum, mother-of-pearl and vanilla. Pop. (1983C) 123,000.

Windward Islands West Indies. 13 00N 61 00W. A group of volcanic islands stretching s from the Leeward Islands and comprising Dominica, Martinique, St. Lucia, St. Vincent, the Grenadines and Grenada, all of which are in association with Britain except Martinique which is a French possession.

Winneba Ghana. 5 20N 0 37W. Town sit. WSW of Accra on the Gulf of Guinea. Port trading in agric. produce. The main indust. is fishing. Exports inc. cacao.

Winnipeg Manitoba, Canada. 49 53N 97 09W. Cap. of Province sit. at confluence of Red R. and Assiniboine R. The principal commercial centre of the Prairie Provinces and one of the world's largest grain markets. Industries railway works, stock-yards, flour mills, meat packing and food processing plants. Manu. agric. machinery, cotton, clothing, furniture, jute and fur products. Pop. (1986C) 625,304.

Winnipeg, Lake Manitoba, Canada. 52 00N 97 00W. Sit. in S. Area, 9,094 sq. m. (23,533 sq. km.). The surplus water from L. Manitoba, L. Winnipegosis and the Lake of the Woods drains into it and thence *via* Nelson R. to Hudson Bay.

Winnipegosis, Lake Manitoba, Canada. 52 30N 100 00W. Sit in SW. Area, 2,086 sq. m. (5,403 sq. km.). Drained by Waterhen R. to L. Manitoba.

Winnipeg River Canada. 50 38N 96 19W. Flows from the Lake of the Woods for 475 m. (760 km.) to NW to L. Winnipeg.

Winona Minnesota, U.S.A. 44 03N 91 39W. City sit. SE of St. Paul on bluffs overlooking Mississippi R. Industries inc. manu. of vehicle controls, plastics, knitwear, electronics, food and construction equipment. Pop. (1980C) 25,075.

Winooski Vermont, U.S.A. 44 29N 73 11W. City immediately E of Burlington on Winooski R. Indust. inc. the manu. of textiles, timber and metal products. Pop. (1980C) 6,318.

Winsford Cheshire, England. 53 12N 2 31W. Town sit. SW of Manchester on the R. Weaver. Industries salt, chemicals, textiles, light engineering and animal feeds. Pop. (1980E) 28,500.

Winston-Salem North Carolina, U.S.A. 36 06N 80 15W. Industrial city formed in 1913 by the union of the towns Winston and Salem, sit. on

Yakdin R. Manu. tobacco products, clothing, beer, telephone equipment, air conditioning, metal products. Pop. (1980C) 131,885.

Winterswijk Gelderland, Netherlands. Town sit. E of Arnhem. Manu. furniture and cotton goods. Pop. (1989E) 28,018.

Wintherthur Zürich, Switzerland. 47 30N 8 43E. Town in NE Switz. Manu. diesel engines, electric locomotives, textiles and soap. Pop. (1980E) 107,732.

Wirksworth Derbyshire, England. 53 05N 1 34W. Town sit. NNW of Derby. Industries limestone quarrying, hosiery and textiles. Pop. (1989E) 5,900.

Wirral Merseyside, England. 53 25N 3 00W. Metropolitan district and peninsula of Merseyside.

Wisbech Cambridgeshire, England. 52 40N 0 10E. Town and port sit. s of The Wash on R. Nene. Centre of an agric. and horticultural area with associated industries. Also printing, brewing, general engineering, vehicle body building. Pop. (1988E) 18,320.

Wisconsin U.S.A. 44 30N 90 00W. State in N central U.S.A. Wisconsin was settled in 1670 by French traders and missionaries. In 1836 it became part of the Territory of Wisconsin, which also inc. the present states of Iowa, Minnesota and parts of N. and S. Dakota. It was admitted into the Union with its present boundaries in 1848. Bounded on N by L. Superior and Michigan, on E by L. Michigan, on s by Illinois, on w by Iowa and Minnesota. Surface an undulating plain with elevation of from 600–1,000 ft. (183–305 metres) a.s.l. In s are prairie lands. It is drained mainly by R. Mississippi on w boundary, and its tribs., R. St. Croix, Chippewa, Black and Wisconsin. There are numerous Ls., largest being L. Winnebago, in E. Area, 56,154 sq. m. (145,439 sq. km.) inc. 1,439 sq. m. (3,727 sq. km.) of inland water, but excluding any part of the Great Lakes. Chief towns, Madison (cap.), and Milwaukee. Agric. state inc. dairy farm-

ing, livestock, feeds and vegetables. Forest industries are important. Sand and gravel, stone, lime and iron ore are the chief mineral products. There is much heavy industry particularly in the Milwaukee area; machinery is the main industrial product, followed by foods. Tourism is important. Pop. (1989E) 4,862,554.

Wismar Rostock, German Democratic Republic. 53 53N 11 28E. Port and town on Wismar Bay, an inlet of Baltic Sea. Industries shipbuilding, railway engineering and sugar refining. Pop. (1989E) 57,900.

Witbank Transvaal, Republic of South Africa. 25 56S 29 07E. Town sit. E of Pretoria at 5,300 ft. (1,615 metres) a.s.l. Important coalmining centre. Manu. chemicals.

Witham Essex, England. 51 48N 0 38E. Town sit. NE of Chelmsford. Manu. metal windows, light engineering, chemicals and fertilizers and fruit juices. Seed producing is also important. Pop. (1989E) 26,624.

Witham River England. 51 48N 0 38E. Rises in Lincolnshire and flows for 90 m. (144 km.) by Grantham and Lincoln to The Wash.

Witney Oxfordshire, England. 51 48N 1 29W. Town sit. WNW of Oxford on R. Windrush. Has some fine medieval buildings. Noted for the manu. of blankets. There is some light industry. Pop. (1989E) 18,500.

Witten North Rhine-Westphalia, Federal Republic of Germany. 51 26N 7 20E. Town sit. SW of Dortmund on Ruhr R. Manu. machinery, steel, glass and coaltar products. Pop. (1984E) 102,900.

Wittenberg Halle, Democratic German Republic. 51 52N 12 39E. Town and port sit. SW of Berlin on Elbe R. Manu. paper, chemicals, machinery, soap and foodstuffs. Pop. (1989E) 53,600.

Wittenberge Schwerin, Democratic German Republic. 53 00N 11 44E. Town sit. on Elbe R. Industries railway engineering, textile mills and metal working. Pop. (1989E) 30,200.

Witwatersrand Transvaal, Republic of South Africa. Watershed of the Transvaal plateau draining N to Limpopo and S to Vaal Rivers. Usually known as the Rand. The economic centre of the Transvaal, with a gold-bearing reef producing nearly half the world's supply and extending *c.* 60 m. (96 km.). Chief city, Johannesburg.

Włocławek Bydgoszcz, Poland. 52 39N 19 02E. Town sit. on Vistula R. Manu. paper, cellulose, machinery and fertilizers. There are lignite deposits in the vicinity.

Woburn Massachusetts, U.S.A. 42 31N 71 12W. City sit. NNW of Boston. Manu. chemicals, machinery, leather goods and food products. Pop. (1980C) 36,626.

Woking Surrey, England. 51 20N 0 34W. Town sit. SW of London on Wey R. Mainly residential but important business/commercial centre with light indust. inc printing. Pop. (1989E) 82,679.

Wokingham Berkshire, England. 51 25N 0 51W. Town sit. ESE of Reading. Industries inc. electronics. Pop. (1989E) 29,694.

Wolfenbüttel Lower Saxony, Federal Republic of Germany. 52 10N 10 32E. Town sit. S of Brunswick on Oker R. Its library contains many anc. manuscripts. Manu. agric. machinery and soap. Pop. (1984E) 49,200.

Wolfsburg Lower Saxony, Federal Republic of Germany. 52 25N 10 47E. Town sit. NE of Brunswick on the Weser-Elbe Canal. Centre of the Volkswagen motor-car industry. Pop. (1989E) 131,000.

Wolin Poland. 53 55N 14 31E. Island sit. N of Szczecin on the Baltic coast. Area, 95 sq. m. (246 sq. km.). Chief town Wolin. Main occupations, tourism and fishing.

Wollo Ethiopia. Region in NE. Area, 30,656 sq. m. (71,200 sq. km.). It consists of highlands and deserts bounded by Rs. Takkaze, Blue Nile, Tsellari and Awash. It is drained by Golima R. The main occupations are stock rear-

ing, salt extracting, and agric. The main crops are cereals and cotton. Cap. Dessye. Pop. (1984E) 3,609,918.

Wollongong New South Wales, Australia. 34 25S 150 54E. Town sit. on coast. Industries coalmining, shipbuilding, heavy and light engineering and chemicals manu. Pop. (1981E) 208,651.

Wolverhampton West Midlands, England. 52 36N 2 08W. Town in 'Black Country' sit. NW of Birmingham. Centre for wide range of metal and engineering manu. Pop. (1981C) 255,000.

Wombwell South Yorkshire, England. 53 31N 1 24W. Town sit. ESE of Barnsley. Industries coalmining, engineering and glass manu.

Wonsan South Pyongan, North Korea. 39 09N 127 25E. Cap. of province and port sit. on E coast. Industries railway engineering, oil refining and fishing. Pop. (1984E) 350,000.

Woodbridge Suffolk, England. 52 06N 1 19E. Town sit. ENE of Ipswich on estuary of R. Deben. There are many fine medieval buildings. Industries boatbuilding, horticulture, fruit and vegetable canning. Pop. (1988E) 7,770.

Woodbridge New Jersey, U.S.A. Township SE of Elizabeth forming part of urban area W of Hudson R. Pop. (1980C) 90,074.

Wood Buffalo National Park Alberta/North West Territories, Canada. Large park, area, 17,300 sq. m. (44,807 sq. km.), estab. in 1922. Contains variety of natural fauna.

Wood Green Greater London, England. 51 36N 0 07W. Town sit. N of central London forming a residential suburb. Contains Alexandra Palace in Alexandra Park, a general entertainment and exhibition centre.

Woodhall Spa Lincolnshire, England. 53 10N 0 12W. Town sit. SW of Horncastle. Pop. (1989E) 2,900.

Woodlark Island Papua New

Guinea. 9 06S 152 50E. A volcanic island 175 m. (280 km.) SE of New Guinea. Gold is mined.

Woods, Lake of the   Canada/U.S.A. 49 30N 94 30W. Lake sit. in Minnesota, U.S.A. and in Manitoba, Canada. Area, 1,485 sq. m. (3,802 sq. km.). The main R. entering the L. is Rainy, flowing from L. Rainy, and the L. is drained by Winnipeg R. flowing into L. Winnipeg.

Wookey Hole   Somerset, England. 51 13N 2 41W. Sit. near village of Wookey, SE of Weston-super-Mare, containing large natural caves where R. Axe leaves Mendip Hills. Prehistoric relics have been found. Manu. paper. Pop. (1980E) 1,000.

Woolwich   Greater London, England. 51 30N 0 04E. District in SE on R. Thames. Noted for the Royal Arsenal founded in 1805.

Woomera   South Australia, Australia. 31 31S 137 10E. Town sit. N of Torrens. Site of the Long Range Weapons Establishment. Pop. (1984E) 1,800.

Woonsocket   Rhode Island, U.S.A. 41 60N 71 31W. City sit. NNW of Providence on Blackstone R. Manu. textiles, woollens, rubber, paper and metal products. Pop. (1980C) 45,914.

Wootton Bassett   Wiltshire, England. 51 32N 1 55W. Town sit. WSW of Swindon. Centre of a dairy-farming district. Pop. (1981C) 9,134.

Worcester   Hereford and Worcester, England. 52 11N 2 13W. Town sit. SW of Birmingham on Severn R. Cathedral city and market town with historical associations. Noted for the triannual Three Choirs Festival. Indust. inc. engineering, mail order, the manu. of china, machinery, machine tools, vehicle parts, Worcestershire sauce and cricket bats. Pop. (1987E) 79,900.

Worcester   Cape Province, Republic of South Africa. 33 39S 19 27E. Town sit. ENE of Cape Town near Hex River Mountains. Centre of a fruit-growing and wine-making district

with associated industries. Industries inc. tourism, textiles, food processing, wine. Pop. (1989E) 52,000.

Worcester   Massachusetts, U.S.A. 42 16N 71 48W. City sit. WSW of Boston on Blackstone R. Manu. machine tools, electrical goods, textiles, leather goods and paper. Pop. (1980C) 161,799.

Worcestershire   *see* **Hereford and Worcester**

Workington   Cumbria, England. 54 39N 3 35W. Port and town sit. SW of Carlisle at mouth of R. Derwent. Industries inc. steel rail making, bus manu., packaging, heavy and light engineering Pop. (1983E) 27,537.

Worksop   Nottinghamshire, England. 53 18N 1 07W. Town sit. NNE of Mansfield on R. Ryton. Industries inc. glassware, refractors, light engineering, coalmining, sawmilling, food processing and flour milling. Manu. chemicals, glass and hosiery. Tourism is important. Pop. (1988E) 36,195.

Worms   Rhineland-Palatinate, Federal Republic of Germany. 49 38N 8 22E. Town and port sit. on left bank of Rhine R. Centre of an important wine-producing district. Manu. machinery, chemicals, plastics, detergents, furniture, cereal products. Pop. (1989E) 73,000.

Worsley   Greater Manchester, England. 53 30N 2 23W. Town sit. N of Manchester. Indust. inc. coalmining, cotton-milling.

Worthing   West Sussex, England. 50 48N 00 23W. Town sit. on S coast. Mainly residential and holiday resort. Industries inc. engineering, pharmaceuticals, electronics, plastics, furniture and horticulture, and is the headquarters of several financial companies. Pop. (1989E) 97,680.

Wrangel Island   Russian Soviet Federal Socialist Republic, U.S.S.R. 71 00N 179 30W. Island sit. about 80 m. (128 km.) from coast of NE Siberia. Area, 1,800 sq. m. (4,662 sq. km.). Mainly tundra.

Wrath, Cape Highland Region, Scotland. 58 38N 5 00W. The NW extremity of the Scottish mainland extending in a 523 ft. (159 metres) cliff from N coast of Sutherland.

Wrekin, The England. 52 41N 2 35W. An isolated volcanic hill sit. ESE of Shrewsbury, Shropshire.

Wrexham Clwyd, Wales. 53 03N 3 00W. Borough sit. SW of Chester, England. Indust. inc. engineering, electronics, food processing, the manu. of bricks, pharmaceutical goods and metal and paper products. Pop. (1989E) 115,800.

Wrocław Wrocław, Poland. 51 06N 17 00E. Cap. of province sit. on Oder R. R. port and transportation centre. Industry railway engineering. Manu. textiles, machinery, chemicals, food products and pottery. Pop. (1985E) 636,000.

Wuhan Hebei, China. 30 36N 114 17E. City sit. at confluence of Han and Yangtze Rivers. Administrative centre for central and S China. Communications centre on the Beijing–Guangzhou railway. Pop. (1987E) 3·49m.

Wuppertal North Rhine-Westphalia, Federal Republic of Germany. 51 16N 7 11E. Town sit. on Wupper R., in the Ruhr region. Formed in 1929 by an amalgamation of smaller towns, it is the centre of a large and important textile industry, which has been operating since the 16th cent. Manu. machinery, tools, chemicals, paper, rubber and metal goods. Pop. (1989E) 380,000.

Württemberg Federal Republic of Germany. Former kingdom and republic. Now part of Baden-Württemberg.

Würzburg Bavaria, Federal Republic of Germany. 49 48N 9 56E. Town sit. on Main R. An important wine-producing centre with many fine medieval buildings. Manu. machine tools, chemicals, metal and paper goods, and furniture. Pop. (1984E) 129,700.

Wyandotte Michigan, U.S.A. 42 12N 83 10W. Town sit. SSW of Detroit on Detroit R. An important chemicals industry is based on local deposits of salt. Manu. gaskets, paint, automobile parts, paper products and barrels. Pop. (1980C) 34,006.

Wye River England. (i) Rises in Buckinghamshire and is a trib. of R. Thames which it joins at Bourne End. (ii) Rises near Buxton and flows for 20 m. (32 km.) SE through Miller's Dale to join R. Derwent.

Wye River Wales/England. 51 37N 2 39W. Rises on Plynlimmon in central Wales, and flows for 130 m. (208 km.) passing Rhayader, Builth Wells, Hereford, Ross-on-Wye and Monmouth to join R. Severn near Chepstow.

Wyoming U.S.A. 42 48N 109 00W. Mountain state of U.S.A. Bounded by Montana, S. Dakota, Nebraska, Colorado, Utah, and Idaho; traversed N to S by Rocky Mountains (Fremont Peak, 13,790 ft. (4,203 metres) a.s.l.), flanked on either side by plateaux from 7,000–8,000 ft. (2,134–2,484 metres) a.s.l.; drained by Rs. Green, Snake, Yellowstone, Big Horn, Powder and Platte. Pine forests among the hills, and in NW is Yellowstone National Park. Wyoming, first settled in 1834, was admitted into the Union in 1890. Area, 97,914 sq. m. (253,597 sq. km.) of which 711 sq. m. (1,841 sq. km.) are water. Chief towns, Cheyenne (cap.), Casper and Laramie. Wyoming is semi-arid, and agric. is carried on by irrigation and by dry farming. Crops inc. hay, wheat, sugar-beet, barley and beans. Cattle and sheep farming is important. Main mineral, oil, but other minerals inc. coal, trona, uranium, iron ore, feldspar, gypsum, limestone and phosphate. Tourism (hunters and fishermen) is important. Manu. inc. food, timber products and machinery. Pop. (1989E) 480,012.

Wyoming Michigan, U.S.A. 42 54N 85 42W. City just S of Grand Rapids. Pop. (1980C) 59,616.

X

Xanthi Greece. 41 08N 24 53E. (i) *Nome* of Thrace. Tobacco important. Pop. (1981C) 88,777; (ii) cap. of (i) sit. ENE of Thessaloníki. Pop. (1981C) 31,541.

Xauen Morocco. City sit. SE of Tangier, founded in 15th cent. after the expulsion of Moors from Granada, Spain. Non-moslems were not freely admitted until 1922.

Xenia Ohio, U.S.A. 39 41N 83 56W. City sit. ESE of Dayton near Little Miomi R. Manu. inc. rope, twine, electrical equipment, furniture and shoes. Pop. (1980C) 24,653.

Xiamen (Amoy) Fujian, China. 24 25N 118 04E. Port on Xiamen Island in the Formosa Strait between China and Taiwan. A former treaty port and first traded with the west in 1842. Pop. (1980E) 500,000.

Xian (Sian) Shaanxi, China. 34 15N 108 52E. Cap. of province in E central China, sit. in Wei-ho valley. Communications and commercial centre. Industries flour milling and tanning; manu. iron and steel, textiles, chemicals and cement. Pop. (1987E) 2,390,000.

Xingu River Brazil. 1 30S 51 53W.

Rises in Mato Grosso flowing 1,200 m. (1,920 km.) to delta of R. Amazon.

Xinjiang Uygur (Sinkiang-Uighur) Autonomous Region China. 42 00N 86 00E. Autonomous region in NW China bounded NE by Mongolia, NW and W by U.S.S.R., S by Tibet and Kashmir. Area, 636,000 sq. m. (1,647,000 sq. km.). Cap. Urumqi. Crossed W to E by the Tien Shan, N of which is Dzungaria, and S the Tarim Basin which contains the Takla Makan desert. Scanty rainfall and extreme temps. make cultivation difficult and limited to oases and mountain valleys. Chief crops wheat, maize, cotton and fruit. Livestock reared by nomads in Dzungaria. Pop. (1987E) 13,840,000.

Xizang *see* **Tibet**

Xochimilco District Federal, Mexico. 19 16N 99 06W. Town sit. S of Mexico City on L. Xochimilco. Noted for its 'floating gardens'.

Xuzhou Jiangsu, China. 34 16N 117 11E. Town in E central China sit. ESE of Kaifeng. An important railway junction on the lines from Tianjin to Shanghai. Industries inc. flour milling and textiles. Pop. (1982C) 773,000.

Y

Yablonovy Mountain Range Eastern Siberia, U.S.S.R. 53 00N 114 00E. Range running NE and SW for about 700 m. (1,120 km.) at an average height of 5,000 ft. (1,524 metres) a.s.l. from E of L. Baikal, dividing basins of R. Lena and the R. Amur and forming part of the watershed of rivers flowing to the Arctic and Pacific Oceans. The highest peak is Sokhondo at 8,200 ft. (2,499 metres) a.s.l. The watershed is continued in the NE by the Stanovoi Mountains.

Yacuiba Gran Chaco, Bolivia. 22 00S 63 25W. Town sit. ESE of Tarija on the Argentine border. Railway terminus and customs station. Provincial cap. and trading centre for farm produce. Pop. (1984E) 11,000.

Yakima Washington, U.S.A. 46 36N 120 31W. City sit. SE of Seattle on Yakima R. Centre of an irrigated agric. region producing fruit (especially apples), hops and mint. Indust. inc. fruit canning. Pop. (1980C) 49,826.

Yakut Autonomous Soviet Socialist Republic Part of Russian Soviet Federal Socialist Republic, U.S.S.R. 66 00N 125 00E. Area, 1,197,760 sq. m. (3,102,198 sq. km.). Constituted an Autonomous Republic 1922. Cap. Yakutsk. Chief industries are mining inc. gold, tin, mica, and coal, and livestock breeding; also trapping and breeding of fur-bearing animals. Pop. (1989E) 1,081,000.

Yakutsk Yakut, U.S.S.R. 62 13N 129 49E. Cap. of Yakut Rep. sit. near Lena R. Mean temp. (Jan.) –46°F. (–43°C.). Commercial centre trading in furs, ivory and hides. Industries, sawmilling, tanning, brick-making and food products. Pop. (1984E) 149,000.

Yallourn Victoria, Australia. 38 11S 146 21E. Town built in 1921 and since the early 1980s no longer exists. Coal seams are now mined where the town stood.

Yalta Ukraine, U.S.S.R. 44 30N 34 10E. Town sit. ESE of Sevastopol on the coast. Centre of the Crimean health resorts. Industries, fish canning and wine making. The historic conference between Churchill, Roosevelt and Stalin took place at Yalta in 1945.

Yalu River China/North Korea. 48 34N 122 09E. Rises in Paiktu-San, Manchuria, and flows 490 m. (784 km.) S and W, forming the boundary between China and North Korea. It enters Korea Bay at Tatungkow 125 m. (200 km.) SSE of Shenyang (Mukden), and is only navigable near the mouth and is frozen from Nov. to March.

Yamagata Yamagata, Japan. 37 55N 140 20E. Cap. of region of same name. Sit. NNE of Tokyo. Centre of a large rice producing area and also manu. machines and metal goods. Pop. (1985E) 245,158.

Yamaguchi Honshu, Japan. 34 10N 131 32E. Cap. of region of same name, sit. WSW of Hiroshima. St. Francis Xavier estab. a mission in 1550.

Yamalo-Nenetz National Area Tyumen, U.S.S.R. 66 00N 76 00E. Sit. between the Gulf of Ob and the Kara Sea, and inc. the Yamal Peninsula. Area, 259,000 sq. m. (670,810 sq. km.). Cap. Salekhard. Chief occupations are fishing, fur trapping and reindeer breeding.

Yambol Yambol, Bulgaria. 42 28N 26 30E. Cap. of province of same name. Sit. SW of Varna on Tundzha R. Manu. metal goods, textiles and tanning. Pop. (1987E) 94,951.

Yamoussoukro Côte d'Ivoire.

Eventual new cap. Pop. (1984E) 120,000.

Yampi Sound Western Australia, Australia. 16 08S 123 38E. Coastal inlet N of Derby. Islands inc. Cockatoo with haematite deposits.

Yana River Yakut, U.S.S.R. 71 30N 136 00E. Rises in the Verkhoyansk Range and flows 750 m. (1,200 km.) N to the Laptev Sea.

Yangon *see* **Rangoon**

Yangtse-kiang *see* **Chang Jiang**

Yannina Greece. (i) A *nome* in Epirus. Area, 1,949 sq. m. (4,990 sq. km.). The principal Rs. are Aoos and Arachthus. Noted for its dairy products Pop. (1981C) 147,304. (ii) Cap. of *nome* of same name sit. NW of Athens. Manu. inc. silks and brocades. Trades in agric. produce. Pop. (1981C) 44,829.

Yao Osaka, Japan. 34 37N 135 36E. A residential suburb of Osaka.

Yaoundé Cameroon. 3 50N 11 35E. Cap. of rep. and Centre province, sit. on railway to Douala. Commercial centre. Manu. inc. cigarettes and soap. Pop. (1986E) 653,670.

Yap Micronesia, W Pacific. 9 30N 138 10E. Group of atolls forming state within the Federated States of Micronesia, inc. Ngulu, Yap, Ulithi, Fais, Sorol in W, and Eauripik, Wolsai, Ifalik, Faraulep, Gaferut, Olimarao, Elato, Lomo trek, Pigailoe, Satawal and Pikelot in E. Area, 47 sq. m. (122 sq. km.). Cap. Yap. Pop. (1980C) 8,172.

Yaracuy Venezuela. A state in the N. Area, 2,740 sq. m. (7,097 sq. km.). Cap. San Felipe. Crops inc. sugarcane, rice, cacao, cotton and tobacco, all grown in the fertile Yaracuy valley. Coffee is grown on the hill slopes. There are deposits of copper, lead, platinum, coal and marble. Pop. (1980E) 276,153.

Yare River Norfolk, England. 52 35N 1 44E. Rises near East Dereham, flows 50 m. (81 km.) E to enter the N.

Sea at Great Yarmouth. Chief tribs. are Wensum, Waveney and Bure which connect it with the Norfolk Broads.

Yarmouth Nova Scotia, Canada. 43 50N 66 07W. Town sit. WSW of Halifax at the entrance to the Bay of Fundy. Seaport, resort and fishing port. Indust. inc. fishing and fish processing, boat building, textiles, knitwear, tin-mining, wood products and transportation equipment. Pop. (1985E) 8,000.

Yarmouth Isle of Wight, England. 50 42N 1 29W. Port sit. WSW of Cowes on the Solent. Connected by ferry with Lymington in Hampshire. Pop. (1988E) 878.

Yarmouth Norfolk, England *see* **Great Yarmouth**

Yaroslavl Russian Soviet Federal Socialist Republic, U.S.S.R. 57 37N 39 52E. (i) Region N of Moscow, adjacent to Rybinsk Reservoir. Area, 14,250 sq. m. (36,908 sq. km.). Partly forested with peat deposits. Agric. products inc. flax, potatoes and dairy goods. (ii) Cap. of region of same name at confluence of Volga R. and Kotorosl R. Industrial centre with large engineering, textile, chemical and rubber plants. Pop. (1989E) 633,000.

Yarrow Water Scotland. 53 40N 2 49W. R. which rises in the Moffat Hills, flows 24 m. (38 km.) ENE through St. Mary's Loch and the Loch of the Lowes to enter the Ettrick Water SW of Selkirk.

Yasnaya Polyana Russian Soviet Federal Socialist Republic, U.S.S.R. Village sit. S of Tula. Birthplace of Tolstoy.

Yavatmal Maharashtra, India. Town sit. SW of Nagpur. District cap. and trading centre for cotton. Indust. inc. cotton milling and sawmilling. Pop. (1981E) 89,071.

Yawata *see* **Kitakyushu**

Yellowhead Pass Canada. 53 00N 118 30W. Sit. between Alberta and

British Columbia in the Rocky Mountains at 3,700 ft. (1,128 metres) a.s.l.

Yellowknife   Northwest Territories, Canada. 62 27N 114 21W. Town and territorial cap. sit. on N shore of Great Slave L. in Yellowknife Bay. Centre of a gold mining region. Pop. (1985E) 11,077.

Yellow Sea   *see* **Hwang-Hai**

Yellowstone National Park   U.S.A. 44 35N 110 00W. A Reservation mainly in NW Wyoming, and projecting about 2 m. (3 km.) into Idaho and Montana. Area, 3,458 sq. m. (8,956 sq. km.). The whole region is volcanic, at an average height of 8,000 ft. (2,438 metres) a.s.l. with many hundreds of hot springs and geysers. The Yellowstone R. with its famous Grand Canyon feeds Yellowstone L. There are many large herds of wild animals, and a great variety of protected species, inc. bears.

Yemen, Republic of   Arabian Peninsula. 15 00N 48 00E. Formed in 1990 with the merging of the Yemen Arab Republic (North) and the People's Democratic Republic of Yemen (South). Bounded on W by the Red Sea, N by Saudi Arabia, E by Oman and S by the Gulf of Aden. Area, 183,602 sq. m. (475,529 sq. km.), comprising the mainland and the islands of Kamaram and Perim. Chief towns San'a (cap.), Aden (economic and commercial cap.), Hodeida, Othman, Mukalla and Ta'iz. Arid with some fertile valleys. Agric., the most important occupation, is by terrace cultivation and by irrigation. Chief occupations farming and fishing. Main crops, millet, cotton, fruit, sesame, wheat and barley. Beyond the mountains is desert. There are no important minerals or developed industries. Pop: (1990E) 13m.

Yenakiero   Ukraine, U.S.S.R. 48 14N 38 13E. Town immediately SE of Gorlovka in the Donbas indust. area, on the oil pipeline from the Caucasus to Gorlovka.

Yenbo   Hejaz, Saudi Arabia. 24 05N 38 03E. Town sit. NNW of Jidda on the Red Sea. Seaport for Medina, carrying pilgrim traffic, exporting dates.

Yendi   Northern Territories, Ghana. 9 30N 0 01W. Town sit. E of Tamale. Road junction and trading centre for nuts, millet, durra, yams and livestock.

Yenisei River   Siberia, U.S.S.R. 71 50N 82 40E. Rises in the Sayan Mountains, in two headstreams which unite at Kyzyl. Flows for 2,364 m. (3,782 km.) N to enter the Kara Sea at Yenisei Bay. Main tribs. are Angara, Stony Tunguska and the Lower Tunguska. Navigable throughout most of its length, despite being frozen for a considerable part of the year. One of the longest Rs. of the world.

Yeovil   Somerset, England. 50 51N 2 39W. Town on R. Yeo. Manu. helicopters, aircraft control systems, light engineering and electronics and leather products. There is a large livestock market. Pop. (1989E) 37,980.

Yerevan   Armenia, U.S.S.R. 40 11N 44 30E. Cap. of Armenia sit. on Zonga R. Important industrial centre producing chemicals, synthetic rubber, electrical equipment, machinery, textiles and plastics. Famous for brandy. Many old Turkish and Persian buildings. Pop. (1989E) 1,199,000.

Yerim   Ibb, Republic of Yemen. Town sit. S of San'a on the central plateau. Centre of a stock-farming area.

Yezd   Yezd, Iran. (i) Province in centre. Area, 24,673 sq. m. (63,905 sq. km.). Pop. (1986C) 574,028. (ii) Cap. of province of same name sit. SE of Esfahán on an oasis in a hot, arid basin subject to sandstorms. Indust. inc. the manu. and distribution of silk. Pop. (1986C) 230,483.

Ynys Môn (Anglesey)   Gwynedd, Wales. 53 17N 4 20W. Island off NW coast separated from Gwynedd by the Menai Strait. Linked to mainland by road and rail bridges. Area, 276 sq. m. (715 sq. km.). Chief town Holyhead. Sheep rearing and agric. main occupations. From Holyhead ferry service to Dun Laoghaire (Rep. of Ireland). Pop. (1984E) 68,500.

Yoho National Park   British Col-

umbia, Canada. 51 26N 116 30W. Sit.
in the SE of the province in the Rocky
Mountains. Area, 507 sq. m. (1,313
sq. km.).

Yokkaichi Honshu, Japan. 34 58N
136 37E. Port sit. SW of Osaka. Manu.
textiles, porcelain and chemicals. Oil
refining. Exports inc. cotton goods.
Pop. (1988E) 268,000.

Yokohama Honshu, Japan. 35 45N
139 35E. Cap. and chief port of
Kanagawa prefecture, on the W shore
of Tokyo Bay. Almost destroyed by
an earthquake in 1923, it was quickly
rebuilt, and is now Japan's fourth largest city. The main exports are silk,
rayon and canned fish, and about 30%
of the total foreign trade is handled
here. Industries inc. shipbuilding, oil
refining, heavy and light engineering
and the manu. of chemicals, clothing,
glass, furniture and firearms. Pop.
(1988E) 3,122,000.

Yokosuka Kanagawa, Japan. 35
18N 139 40E. Port and naval base sit.
SW of Yokohama. Shipbuilding important. Pop. (1988E) 431,000.

Yonghung Hamgyong, North
Korea. Town sit. SW of Hungnam.
Centre of a farming, lumbering and
graphite mining area.

Yonkers New York, U.S.A. 40 60N
73 52W. City and N residential suburb
of New York, sit. on E bank of Hudson R. and stretching to the W bank of
Bronx R. Employment is from retailing, commerce and waterfront industry. Pop. (1980C) 195,351.

Yonne France. 47 50N 3 40E. Dept.
sit. in Burgundy region. Area, 2,867
sq. m. (7,425 sq. km.). Principal R. is
Yonne. Mainly agric. producing
cereals, sugar-beet and wine. The
chief towns are Auxerre (cap.), Aval-
lon, Chablis and Sens. Pop. (1982C)
311,019.

Yonne River France. 48 23N 2 58E.
Rises in the Monts du Morvan, flows
182 m. (291 km.) NNW to join Seine
R. near Montereau.

York North Yorkshire, England. 53
58N 1 05W. City and archiepiscopal

seat. The National Railway Museum
was opened in 1975. Industries inc.
manu. of chocolate, railway coaches,
scientific instruments and sugar.
Tourism is important. Pop. (1989E)
99,910.

York Pennsylvania, U.S.A. 39 58N
76 44W. City sit. ESE of Pittsburgh.
Centre of a rich agric. area. Manu. inc.
agric. machinery, refrigerators, turbines and building materials. Pop.
(1980C) 44,619.

Yorke Peninsula South Australia,
Australia. 35 00S 137 30E. Peninsula
sit. between the Gulfs of Spencer and
St. Vincent. Occupations mainly
sheep-rearing and wheat growing. The
chief ports are Port Pirie and
Wallaroo.

Yorkshire, reorganized in 1974 *see*
**Humberside, North Yorkshire,
South Yorkshire, West Yorkshire**

Yorkton Saskatchewan, Canada. 51
13N 102 28W. Town sit. ENE of
Regina on Yorkton R. Distribution
and processing centre for E Saskatchewan. Manu. inc. agric. equipment,
food-processing and concrete products. Pop. (1986E) 15,574.

Yoro Honduras. 15 09N 87 07W.
(i) Dept. in the N bounded by R.
Sulaco, Comayagua, Ulúa and Sierra
de Nombre de Dios. Mainly mountainous and drained by Aguán R.
Livestock, coffee, grain and sugar-
cane are produced in the highlands;
bananas are grown in the valleys. Pop.
(1983E) 304,310. (ii) Cap. of dept. of
same name sit. N of Tegucigalpa.
Trades in agric. produce. Manu.
ceramics.

Yosemite National Park California, U.S.A. 37 51N 119 33W. A
granite region, inc. the Yosemite Valley, in central California. Area, 1,183
sq. m. (3,064 sq. km.). Famous for
superb waterfalls, inc. Ribbon Fall,
1,612 ft. (491 metres) the second
highest in the world, and the Nevada,
Yosemite, Silver Strand and Bridal-
veil Falls. There are many high peaks
and tall trees inc. the well-known
Wawona tree, through which an 11 ft.
(3.4 metres) wide road was tunnelled.

Yoshkar-Ola Mari Autonomous Soviet Socialist Republic, U.S.S.R. 56 38N 47 52E. Cap. of the rep. sit. NW of Kazan. Main industries are wood and food processing. Pop. (1987E) 243,000.

Youghal Cork, Ireland. 51 58N 79 52W. Port and resort on the Blackwater estuary. The main industry is the manu. of rayons and the local lace is well known. Other occupations are carpet making, fishing and the manu. of bricks and earthenware.

Youngstown Ohio, U.S.A. 41 06N 80 39W. Industrial city and an important centre for iron and steel, and allied products. There are limestone quarries in the vicinity. Manu. steel and aluminium products, rolling mill equipment, vehicles and parts, paints, plastics, hydraulic machinery. Pop. (1980C) 115,436.

Yozgat Turkey. (i) A province in the centre to the NW of the Ak Mountains. Area, 5,291 sq. m. (13,704 sq. km.). The principal Rs. are Delice and Cekerek. Products inc. lead, grain, wool, mohair and gum tragacanth. Pop. (1985C) 545,301. (ii) Cap. of province of same name sit. E of Ankara. Trades in agric. produce.

Ypres West Flanders, Belgium. 50 51N 2 53E. Town sit. WSW of Ghent on Yperlée R. Important in the Middle Ages as a centre of the Flanders cloth trade, its chief manu. are now linen and biscuits. Pop. (1985E) 35,000.

Yucatán Mexico. 21 30N 86 30W. SE state forming the N part of the Yucatán Peninsula. Area, 14,868 sq. m. (38,508 sq. km.). Cap. Mérida. Mainly flat with poor soil and tropical climate. The chief products are sisal fibre, hardwoods, sugar, tobacco and maize. Pop. (1989E) 1,327,298.

Yucatán Peninsula Mexico. 19 30N 89 00W. Peninsula in SE Mexico and inc. parts of Belize (formerly British Honduras) and N Guatemala. It separates the Gulf of Mexico from the Caribbean Sea. Area, *c.* 70,000 sq. m. (*c.* 181,300 sq. km.). An important centre of Mayan civilization with many relics, esp. at Uxmal and Chichén Itzá. Generally low-lying with underground rivers running through the limestone. There are tropical forests in the S.

Yugoslavia 44 00N 20 00E. Federal Republic sit. in SE Europe and bordered on N by Austria and Hungary, NE by Romania, E by Bulgaria, S by Greece and W by Albania, the Adriatic Sea and Italy. The Federation consists of Bosnia and Hercegovina, Montenegro, Croatia, Macedonia, Slovenia, and Serbia with Vojvodina and Kosovo. Cap. Belgrade. Area, 98,766 sq. m. (255,804 sq. km.). Chief crops are maize, wheat, barley, rye, tobacco, hemp, sunflower and potatoes. There are considerable mining resources inc. coal (chiefly brown coal), iron, copper ore, gold, lead, chrome, antimony and cement. The most important iron mines are at Vares and Ljubija in Bosnia, and there are also considerable siderite and limonite iron ores between Prijedor, Sanski Most and Topusko. Copper ore is exploited chiefly at Bor (Serbia). The principal lead mines are at Trepca and Mezice. Chrome mines are in S Serbia (Kosovo) and Macedonia (Skopje Kumanovo). There are 2 antimony mines in W Serbia (Podrinje). The major industries are sit. in the NW of the country and manu. inc. pig-iron, steel, cement, sulphuric acid, nitric acid, fertilizers, iron castings, steel castings, cotton, woollens, rayon and hemp. Pop. (1990E) 24,107,000.

Yukon Canada. 63 00N 135 00W. Territory of NW Canada, bordered by Alaska on the W. Area, 207,076 sq. m. (536,000 sq. km.). Cap. Whitehorse. Mountainous in the N and W with the highest peak in Can., Mount Logan, 19,850 ft. (6,050 metres) a.s.l. in the SW. Mining is the principal occupation, the chief minerals being gold (Klondike gold rush 1898), silver, lead and zinc. Other industries are fishing, lumbering and fur trapping. Pop. (1989E) 29,845.

Yukon River Canada/U.S.A. 62 33N 163 59W. The R. rises in Tagish L. It flows for *c.* 2,300 m. (3,680 km.) NW into Alaska then SW. It empties as a great delta into the Bering Sea. Chief tribs. are R. Stewart and R. Klondike.

Yuma Arizona, U.S.A. 32 43N 114 37W. City sit. wsw of Phoenix on Colorado R. where it shares the border with Mexico. An irrigated agric. area. Indust. inc. winter tourism. Pop. (1985E) 70,000 metropolitan area.

Yunnan China. 25 04N 102 41E. s central province bounded on s by Laos and N. Vietnam and on w by Burma. Area, 168,417 sq. m. (436,200 sq. km.). Cap. Kunming Mainly a mountainous plateau ranging from 7,000 ft. (2,134 metres) a.s.l. in the s to 17,000 ft. (5,182 metres) in the N. There are fertile valleys and plains growing tea, rice, cereals and tobacco. Silk is also produced. Mineral deposits are large, inc. tin and copper ore, gold, silver, lead, jade and anthracite. The chief Rs. are Mekong, Yangtse and Salween. Pop. (1987E) 34,560,000.

Yuscarán El Paraíso, Honduras. 13 55N 86 51W. Town sit. ESE of Tegucigalpa at the E foot of Mount Mon-serrat. Dept. cap. and trading centre for grain and fruit. Silver mining nearby. Pop. (1980E) 1,250.

Yuzhno-Sakhalinsk Russian Soviet Federal Socialist Republic, U.S.S.R. 46 58N 142 42E. Town sit. s of Aleksandrovsk on the E coast railway, Siberia. Railway junction in an agric. area. Indust. inc. pulp and paper milling, sugar refining, brewing and the manu. of plastic goods. Airport.

Yvelines France. 48 40N 1 45E. Dept. w of Paris in Île-de-France region. Area, 877 sq. m. (2,271 sq. km.). Pop. (1982E) 1,196,111.

Yverdon-les-Bains Vaud, Switzerland. 46 47N 6 39E. Town and spa sit. at sw end of L. Neuchâtel. Industries inc. the manu. of office machines, electrical appliances and mechanical equipment. Pestalozzi estab. his famous school at Yverdon in 1806 and there is a Pestalozzi Centre of Research. Pop. (1989E) 21,600.

Z

Zaanstad Noord-Holland, Netherlands. Town sit. NW of Amsterdam on R. Zaan. Formed in 1974 by joining Zaandam, Koog a/d Zaan, Zaandijk, Wormerveer, Krommenie, Westzaan and Assendelft. Main occupations are food processing, metal forming and machinery manu. Pop. (1989E) 129,650.

Zabrze Katowice, Poland. 50 18N 18 46E. Town sit. in S central Poland. Main occupations are coalmining, iron and steel, chemicals and glass manu. Pop. (1985E) 198,000.

Zacapa Zacapa, Guatemala. 14 58N 89 32W. Market town and cap. of dept. sit. ENE of Guatemala City. Pop. (1989E) 35,769.

Zacatecas Mexico. 22 47N 102 35W. (i) State on the central plateau. Area, 28,973 sq. m. (75,040 sq. km.). Mining, particularly silver. Pop. (1989E) 1,259,407. (ii) Cap. of state of same name. Pop. (1980E) 120,000.

Zacatecoluca El Paz, El Salvador. 13 30N 88 52W. Market town and cap. of El Paz dept. sit. SE of San Salvador. Manu. cotton goods and cigars. Pop. (1981E) 58,000.

Zadar Croatia, Yugoslavia. 44 07N 15 14E. Port and seaside resort. Maraschino liqueur manu. Pop. (1981E) 116,174.

Zagazig Sharqîya, Egypt. 30 35N 31 31E. Cap. of Sharqîya governorate on R. Nile delta, and sit. NNE of Cairo. Centre for cotton and grain trade at a railway and canal junction. Also some cotton manu. Pop. (1986E) 274,400.

Zagreb Croatia, Yugoslavia. 45 48N 15 58E. Cap. of Croatia and second city of Yugoslavia. Sit. on R. Sava. Chief occupations manu. machinery, paper, asbestos, textiles, chemicals and carpets. Pop. (1981C) 1,174,512.

Zagros Mountains Iran. 33 40N 47 00E. System forming the SW frontier of Iran with many parallel chains, cut by rivers in deep gorges, and rising to 17,000 ft. (5,182 metres) a.s.l. There are oilfields in the vicinity.

Záhedán Báluchestán, Iran. 29 30N 60 50E. Cap. sit. ESE of Kerman. It is an important transport centre. Pop. (1983E) 165,000.

Zahlé Bekaa, Lebanon. 33 50N 35 53E. Town sit. ESE of Beirut on the Beirut–Damascus railway at 3,100 ft. (945 metres) a.s.l. Summer resort in a vine-growing area.

Zaïre Central Africa. 3 00S 23 00E. Formerly Congo (Kinshasa) and until 1960 Belgian Congo. Bounded on N by the Central African Empire and Sudan, E by Uganda and Tanzania, S by Angola and Zambia and W by Congo. Area, 905,365 sq. m. (2,344,885 sq. km.). Until the middle of the 19th cent. the territory drained by the Congo River was practically unknown. When Stanley reached the mouth of the Congo in 1877, King Leopold II of the Belgians recognized the immense possibilities of the Congo Basin and took the lead in exploring and exploiting it. The Berlin Conference of 1884–5 recognized King Leopold II as the sovereign head of the Congo Free State. The country became independent in 1960. The departure of the Belgian administrators on the day of independence left a vacuum which speedily resulted in complete chaos. Civil war started and it was not until 1965 that complete peace was restored with the help of United Nations forces. The surface is generally a depression and probably was formerly occupied by an inland sea. Great parts are covered by trackless primeval forests surrounded by highlands and hills. Chief towns, Kinshasa (cap.), Kananga, Lubumbashi, Mbuji-Mayi, Kisangani, Bukavu, Likasi, Kikwit. Agric. products include palm-oil, coffee, rubber, cacao, tea, cotton fibre, sugar-cane and ivory. The most important industry is mining particularly for copper, zinc, manganese, cobalt, cassiterite and industrial diamonds. The most important mining area is in the region of Shaba (formerly Katanga). Pop. (1988E) 32,564,000.

Zaïre River   Central Africa. 6 04S 12 24E. Formerly Congo River. Rises in S Zaïre near the Zambian border and flows *c.* 2,300 m. (3,680 km.), first N along the W side of the Mitumba range, then NNW to Kisangani and W to the Congo border which it forms (SW) to below Kinshasa, and enters the Atlantic Ocean at S. Antonia de Zaïre. Its course is much impeded by falls, cataracts and swamp.

Zakynthos   Greece. 37 47N 20 57E. Southernmost island in Ionian group sit. S of Kephalonia. Area, 157 sq. m. (406 sq. km.). Fertile plain with hills in N. Cultivation mainly consists of currants, citrus fruits and olives. Subject to earthquakes. Pop. (1981C) 30,014.

Zalaegerszeg   Hungary. 46 51N 16 51E. Cap. of county of Zala. Sit. sw of Budapest on R. Zala. Pop. (1984E) 60,000.

Zambézia   Mozambique. A central province bounded on SE by the Mozambique Channel. Area, 39,722 sq. m. (102,880 sq. km.). Cap. Quelimane. A coastal plain rises to highlands in the NW. The principal Rs. are Zambezi and lower Shire. Sugar, sisal and cotton are grown. Pop. (1982E) 2,600,208.

Zambezi River   Africa. 18 55S 36 04E. The R. was discovered by Livingstone and rises in the NW of Zambia near the frontiers of Zaïre and Angola, and flows SE to enter the Mozambique Channel NE of Beira. The length is *c.* 2,000 m. (3,200 km.) and is the fourth longest R. in Africa, crossing E Angola and Barotseland, forming the frontiers between Zambia and the Caprivi Strip in SW Africa, and between Zambia and Zimbabwe, before finally entering Mozambique at Zumbo. Only short stretches are navi-

gable owing to the frequent rapids and falls. The Victoria Falls and Kariba Dam form part of this R. The Cabora Bassa dam near Tete was completed in 1974 to provide power and irrigation from a lake *c.* 2,000 sq. m. (5,180 sq. km.). Principal tribs. are R. Lungwebungu, Luanginga, Chobe, Shangani, Sanyati, Kafre, Luangwa and, Shire. The last trib. drains L. Malawi.

Zambia   15 00S 28 00W. The independent Rep. of Zambia (formerly Northern Rhodesia) came into being on 24 Oct. 1964 after 10 months of internal self-govt. following the dissolution of the Federation of Rhodesia and Nyasaland in 1963. It is bordered by Tanzania in N, Malawi in E, Mozambique in SE, and Zimbabwe and Namibia in S. Area, 290,586 sq. m. (752,262 sq. km.). Cap. Lusaka. Principal agric. production consists of maize, tobacco (Chipata centre for tobacco growing), groundnuts, cotton and sugar. Zambia is an important copper mining country and the chief mines are Kabwe, Roan Antelope, Nkana, Mufulira and Nchanga. Other important minerals, zinc, lead, manganese and cobalt. Pop. (1987E) 7·12m.

Zamboanga   Mindanao, Philippines. 6 54N 122 05E. Cap. of province of same name, Zamboanga is also a port, exporting copra, rubber, timber and coconuts. Pop. (1980E) 344,000.

Zamora   Zamora-Chinchipe, Ecuador. 4 04S 78 58W. Provincial cap. sit. ESE of Loja on Zamora R. Pop. (1984E) 88,000.

Zamora   Spain. 41 30N 5 45W. (i) Province in NW bordering on Portugal. Area, 4,082 sq. m. (10,572 sq. km.). The main R. is Duero. Sheep-rearing is important, much of Spain's merino wool being produced here. Pop. (1986C) 221,560. (ii) Cap. of province of same name. Sit. WSW of Valladolid. Manu. textiles, cement, pottery and soap. Also wine making and flour milling.

Zanesville   Ohio, U.S.A. 39 56N 82 01W. Industrial city sit. E of Columbus on Muskingum R. Manu. tiles, pottery, iron and steel, cement and glass. Pop. (1980C) 28,655.

Zanján Gilán, Iran. 36 40N 48 29E. Town sit. WNW of Tehrán. Manu. rugs, cotton goods and matches. Pop. (1986C) 215,261.

Zante Zakynthos, Greece. 37 52N 20 44E. Cap. SW of Patras sit. on the SE coast. Trades in agric. produce. Industries inc. flour milling and the manu. of soap. Pop. (1981C) 9,764.

Zanzibar Tanzania. 6 10S 39 11E. Island forming part (with Pemba and Tanganyika) of the United Rep. of Tanzania. Sit. off the mainland across the Zanzibar Channel, 22 m. (35 km.) wide. Area, 640 sq. m. (1,658 sq. km.). Chief town, Zanzibar. Coral with a fertile soil. The main products are cloves, fruit and copra. Pop. (1985E) Island, 571,000; town (1978C) 110,669.

Zaporozhye Ukraine, U.S.S.R. 47 50N 35 10E. (i) Region in the SE, mainly lowland steppe with mineral deposits of lignite, iron ore, manganese and kaolin. Largely industrial area though some wheat and cotton are grown. Area, 10,400 sq. m. (26,936 sq. km.). (ii) Cap. of region of same name, sit. on Dnieper R. Important industrial and transport centre with a large hydroelectric power station. Mainly known for engineering and metallurgical industries with allied products. Pop. (1989E) 884,000.

Zaragoza Spain. 41 38N 0 53W. (i) Province in the NW, mainly arid plain with extremes of temp. and low rainfall, therefore only restricted agric. possible. The main R. is Ebro. Area, 6,615 sq. m. (17,132 sq. km.). Pop. (1986C) 845,832. (ii) Cap. of province of same name, sit. at confluence of Ebro R., Huerva R. and Gállego R. An important railway junction and industrial centre manu. machinery, chemicals, glass, porcelain, soap and cement. Sugar refining, flour milling and wine making are also carried on. Pop. (1986C) 596,080.

Zárate Buenos Aires, Argentina. 34 05S 59 02W. Town sit. NW of Buenos Aires on Paraná R. A busy R. port,

linked by ferry with Ibicuy, with meat refrigeration plants and paper mills. Pop. (1984C) 54,000.

Zaria Kaduna, Nigeria. 11 07N 7 44E. Market town sit. SW of Kano. Railway junction and important centre for groundnuts and cotton. Other indust. railway engineering, tanning and printing. Pop. (1983E) 274,000.

Zarka Jordan. Town sit. NE of Amman on the Hejaz railway. Site of a power plant and phosphate mines, with yellow ochre nearby. Pop. (1984E) 265,700.

Zealand Denmark. Largest island of Den. bounded by the Kattegat, the Great Belt, the Sound and the Baltic. Area, 2,709 sq. m. (7,016 sq. km.). Chief towns are Copenhagen and Roskilde. The main occupations are farming and fishing.

Zeebrugge West Flanders, Belgium. 51 19N 3 12E. Port sit. N of Bruges, to which it is connected by canal. Exports inc. coke and chemicals.

Zeeland Netherlands. 51 30N 3 50E. Province in the SW consisting of five islands in the Scheldt estuary, one of them Walcheren, and the mainland region adjoining E. Flanders, Belgium. Area, 689 sq. m. (1,785 sq. km.). A fertile area though mainly b.s.l. and protected by dykes. Chief towns are Middelburg (cap.) and Flushing. Pop. (1989E) 355,585.

Zeist Utrecht, Netherlands. 52 05N 5 15E. Town sit. E of Utrecht. Manu. porcelain, candles and soap. Pop. (1988E) 59,431.

Zeitz Halle, German Democratic Republic. 51 03N 12 08E. Town sit. SSW of Leipzig on White Elster R. Manu. textiles, pianos, chemicals, also sugar refining. Pop. (1989E) 42,700.

Zelaya Nicaragua. A maritime dept. stretching along the Caribbean coast. Area, 27,145 sq. m. (70,306 sq. km.). Mainly tropical forest with a low swampy coastal belt. The principal Rs. are Coco, Huahua, Prinzapolka, Escondido, San Juan and Rio Grande. The main occupations are lumbering and gold mining. There is limited agric., the main crops being rice, sugar-cane, corn and beans. Coconuts and bananas are exported. Cap. Bluefields. Pop. (1981E) 202,462.

Zemun Serbia, Yugoslavia. Suburb of Belgrade sit. NW of the city centre on Danube R. across mouth of Sava R. Port and international airport. Industries inc. the manu. of aircraft, textiles, penicillin, glue, leather goods and asbestos goods.

Zeravshan River U.S.S.R. 39 22N 63 45E. Rises in the Zeravshan Mountains, flows 450 m. (720 km.) W through Tadzhikistan and Uzbekistan before disappearing in the desert near the Amu-Darya R.

Zermatt Valais, Switzerland. 46 02N 7 45E. Village sit. at the head of the Zermatter Valley, just below the Matterhorn at 5,315 ft. (1,620 metres) a.s.l. One of the main summer and winter sports resorts in the Swiss Alps. Pop. (1989E) 4,200.

Zhdanov Ukraine, U.S.S.R. 47 06N 37 33E. Port sit. on the Sea of Azov. Important industrial centre manu. iron and steel, machinery and chemicals. Fishing and fish processing are carried on, and coal, grain and salt are the chief exports. Pop. (1985E) 522,000.

Zhejiang (Chekiang) China. 21 15N 110 20E. Province on coast of E China Sea. Area, 39,300 sq. m. (101,800 sq. km.). Chief town Hangchow (the cap.). A mountainous region, the chief crops being tea, rice and cotton. Pop. (1987C) 40,700,000.

Zheleznovdsk Russian Soviet Federal Socialist Republic, U.S.S.R. Town sit. NNW of Pyatigorsk at the S foot of Zheleznaya mountain, in the N Caucasus. Health resort with mineral springs and sanatoria.

Zhengzhou (Chenghow) Henan, China. 34 45N 113 34E. Cap. of province in central E, sit. W of Kaifeng. Route and commercial centre trading in grain, hides and skins. Industries inc. flour milling, textiles and food processing; manu. machinery. Pop. (1987C) 1,610,000.

Zhitomir Ukraine, U.S.S.R. 50 16N 28 40E. (i) Region in the W. Agric. products inc. grain, sugar-beet and potatoes. (ii) Cap. of region of same name. Sit. WSW of Kiev. Industrial and transportation centre, manu. metal goods, furniture and clothing. Pop. (1987E) 287,000.

Zhob River Pakistan. Rises near Kand Peak in the Toba-Kakar range, Baluchistan, and flows *c.* 230 m. (368 km.) E past Hindubagh then NE to enter the Gumal R. NNE of Fort Sandeman. Its flow is subject to wide seasonal variation.

Zhu Jiang River formed by the confluence of the Xi Jiang, Bei Jiang and the Dong Jiang Rs. It flows 110 m. (176 km.) into a broad delta to enter the South China sea near Macao. Also known as the Pearl R.

Zielona Góra Poland. 51 56N 15 31E. (i) Province in the W bordering German Democratic Republic. The main Rs. are Oder, Lusatian Neisse and the Bobrawa. The chief occupation is agric. Area, 3,424 sq. m. (8,868 sq. km.). Pop. (1989E) 651,000. (ii) Cap. of province of same name, sit. WSW of Poznán. Centre for lignite mining and railway engineering. Manu. textiles and wine. Pop. (1983E) 108,000.

Zifta Gharbiya, Egypt. Town sit. ESE of Tanta on Damietta branch of Nile. Centre of an irrigated cotton-growing area.

Ziguinchor Senegal. 12 35N 16 16W. Port and cap. of region sit. SSE of Dakar, on estuary of Casamance R. Exports inc. groundnuts. Pop. (1979E) 79,464.

Žilina West Slovakia, Czechoslovakia. 49 14N 18 46E. Town sit. NE of Bratislava on Vah R. Industrial and railway centre, manu. textiles, paper, matches and fertilizers. Pop. (1983E) 87,000.

Zilupe Latvia, U.S.S.R. Town sit. ESE of Rezekne on the border with the Russian Soviet Federal Socialist Republic.

Zimbabwe National Park Zimbabwe. Site of a ruined 'fortress' sit. SE of Masvingo. Formerly believed to date from the Middle Ages the ruins are now regarded as 19th cent. Ming porcelain and gold have been discovered in the many stone houses, and there are many relics in museums in various African towns.

Zimbabwe   20 00S 30 00E. Bounded N by Zambia across Zambezi R., W by Botswana, S by the Republic of South Africa and E by Mozambique. Area, 150,872 sq. m. (390,759 sq. km.). Before 1923 Southern Rhodesia, like Northern Rhodesia, was under the administration of the British South Africa Company. In 1922 Southern Rhodesia voted in favour of responsible government and in 1923 the country was formally annexed to His Majesty's Dominions, government was established under a governor, assisted by an executive council, and a legislature, with the status of a self-governing colony. After the dissolution of the Federation of Rhodesia and Nyasaland in 1963 Southern Rhodesia reverted to the status of a self-governing colony within the Commonwealth, but, at the same time, became responsible for those powers which had been surrendered to the federal government on its formation and which, once again, became its responsibility. In 1965 the Prime Minister of Rhodesia issued a unilateral declaration of independence. Chief towns Harare (cap.) and Bulawayo. The most important single food crop in Zimbabwe is maize, the staple food of a large proportion of the population. The livestock industry is second to tobacco as regards its export potential. Dairying forms the foundation of many mixed farms. Fish farming is being developed and large catches are taken from Lake Kariba, where a fish freezing plant was completed in 1964. Sugar is being produced in the Triangle and Hippo Valley estates.

The citrus estates of the British South Africa Company, the state-owned deciduous orchards at Nyanga and a scheme for large-scale citrus growing at Hippo Valley form the basis of the citrus fruit industry. However, many parts of the .country between 2,500 ft. (769 metres) and 4,000 ft. (1,231 metres) a.s.l. are suitable for citrus culture, and large numbers of deciduous fruit trees planted in the Chimanimani and Nyanga areas are coming into production. There are large tea plantations. Other crops grown in substantial quantities include small grains (sorghums and millet), rice, groundnuts, cassava. These crops form the basis of much subsistence farming undertaken by the African population. Tobacco is the most important single product. Cotton and wheat are also important. Minerals include asbestos, gold, chrome ore, coal and copper. Pop. (1989E) 9,122,000.

Zinder   Niger. 13 48N 8 59E. Cap. of region sit. in S central Niger. A walled city, terminus of a motor-route across the Sahara, and trading in salt, spices, groundnuts, hides and skins. Pop. (1983E) 82,800.

Zion National Park   Utah, U.S.A. 37 20N 113 05W. Region in SW, estab. 1919, containing Zion Canyon. Area, 230 sq. m. (596 sq. km.).

Zipaquirá   Cundinamarca, Colombia. 5 00N 74 00W. Town sit. N of Bogotá in the S Cordillera. Large quantities of rock salt are mined and chemicals manu. Pop. (1984E) 40,850.

Zittau   Dresden, German Democratic Republic. 50 54N 14 47E. Town sit. SSW of Görlitz, near the frontiers of Poland and Czechoslovakia. Railway junction. Manu. machinery, textiles and commercial vehicles. Pop. (1989E) 38,700.

Zlatoust   Chelyabinsk, U.S.S.R. 55 10N 59 40E. Industrial town of Chelyabinsk sit. in the S Urals. Notable metallurgical centre producing special steels and manu. tools, precision instruments and cutlery. Pop. (1987E) 206,000.

Zlin   *see* **Gottwaldov**

Zomba   Malawi. 15 23S 35 18E. City sit. S of L. Malawi on the lower slopes of Zomba Mountain 3,000 ft. (914 metres) a.s.l., 41 m. (66 km.) from Blantyre. There are many tobacco estates in the area, and some dairy farming. Industries inc. clothing manu. Pop. (1985E) 53,000.

Zorzor   Western Province, Liberia. 7 46N 9 28W. Town sit. SSE of Vonjama. Centre of a farming area producing cattle, palm products, cotton and pineapples.

Zoutpansberg   Republic of South Africa. Mountain range in the Transvaal. Sit. S of Limpopo R. it extends 100 m. (162 km.) E to W and rises to 5,700 ft. (1,737 metres) a.s.l. Chief town Louis Trichardt.

Zrenjanin   Vojvodina, Yugoslavia. 45 23N 20 24E. Town sit. N of Belgrade on Begej R. R. port and railway junction. Manu. agric. machinery, chemicals, soap, leather and food products. Pop. (1981E) 139,300.

Zuara   Tripolitania, Libya. 32 56N 12 06E. Town sit. W of Tripoli on the coastal road in an oasis. Port serving an irrigated agric. area. Railway terminus. Indust. inc. sponge fishing, olive-oil pressing and flour milling.

Zueitina   Cyrenaica, Libya. Village sit. S of Benghazi on the Gulf of Sidra. Port serving a farming area which produces vegetables, fruit, sheep and goats. The main indust. is fishing.

Zug   Switzerland. 47 10N 8 31E. (i) Canton in N central Switz. Area, 93 sq. m. (241 sq. km.). Mountainous in the S and SE but agric. carried on in the Reuss basin. Pop. (1988E) 83,419. (ii) Town in canton of same name, sit. S of Zürich at the foot of the Zugerberg on the NE shore of L. Zug. Industries are printing, woodworking and the manu. of electrical equipment and metal goods. Pop. (1980E) 22,000.

Zug, Lake   Switzerland. 47 07N 8 35E. Lake N of L. Lucerne, lying partly in Zug Canton and partly in Schwyz Canton. Area, 21 sq. m. (54 sq. km.). The Rigi, 5,908 ft. (1,801 metres) a.s.l. stands to the S.

Zugspitze   Bavaria, Federal Republic of Germany. 47 25N 10 59E. Highest peak in the Bavarian Alps at 9,721 ft. (2,963 metres) a.s.l. Sit. on the frontier with Austria, it can be ascended from both sides by cable railway.

Zuider Zee   Netherlands. 52 45N 5 25E. Gulf touching on 5 provinces of the Netherlands: Friesland, Gelderland, Overijssel, Utrecht and Noord Holland. Area, *c.* 2,000 sq. m. (5,180 sq. km.). Reclamation began in 1923 behind two barrages which cut off an interior L. for reclamation in 4 'polders', surrounding a water area now called the Ijsselmeer. The exterior area beyond the barrages is called the Wadden Zee.

Zuid-Holland   Netherlands. 52 00N 4 35E. Province bounded W by the North Sea, S by Zeeland and N. Brabant. Area, 1,122 sq. m. (2,907 sq. km.). Cap. The Hague. Mainly b.s.l. and protected by dunes, drained by R. Rhine delta. Dairy farming and horticulture are important in the rural districts, but the province is the most densely populated with industries centred on numerous manu. towns, chiefly Rotterdam, Schiedam, Ulaardingen, Delft, Leiden and Dordrecht. Pop. (1989E) 3,200,408.

Zulia   Venezuela. A maritime state on the Caribbean Sea. Area, 24,360 sq. m. (63,092 sq. km.). Mainly lowland encircling L. Maracaibo, around which lies one of the world's richest oil producing areas. Agric. is carried on in the S, the main crops being sugar-cane, cotton, cacao, coconuts and fruit. Other industries inc. lumbering and fishing. Cap. Maracaibo. Pop. (1981C) 1,674,252.

Zululand   Natal, Republic of South Africa. Area, 10,375 sq. m. (26,871 sq. km.), bounded by Mozambique in N, Swaziland in NW and Tugela R. in S. Cap. Eshowe, sit. NNE of Durban. A fertile coastal plain growing cotton and sugar-cane gives way to a higher plateau where cattle are reared. There are several game reserves.

Zürich   Switzerland. 47 23N 8 32E. (i) Canton in NE, bounded in N by R. Rhine. Area, 668 sq. m. (1,730 sq. km.). A large part of Switzerland's heavy textile and electrical industry is sit. in Zürich. Pop. (1988E) 1,141,494. (ii) Cap. of canton of same name. Sit. on Limmat R. where it leaves L. Zürich. The largest city of Switz. and centre for commerce, industrial banking and culture. Manu. inc. textiles, machinery, chemicals, paper and printing. Tourism is a growing industry. Pop. (1988E) 346,800.

Zürich, Lake Switzerland. 47 13N 8 45E. Sit. mainly in Zürich canton, 1,332 ft. (406 metres) a.s.l. Length 25 m. (40 km.) and with a maximum width 2.5 m. (4 km.) it is fed by Linth R. and drained by Limmat R.

Zutphen Gelderland, Netherlands. 52 08N 6 12E. Industrial town sit. NE of Arnhem. Manu. silk products, metalwork, paper, printing and bricks. Pop. (1989E) 31,143.

Zvishavane Zimbabwe. Town and rail terminus sit. E of Bulawayo. Asbestos mining centre. Pop. (1982C) 27,000.

Zweibrücken Rhineland-Palatinate, Federal Republic of Germany. 49 15N 7 21E. Town sit. W of Saarbrücken.

Manu. machinery, footwear, textilesand electrical equipment. In the 18th cent. a printing press was founded, which is now famous. Pop. (1984E) 33,700.

Zwickau Karl-Marx-Stadt, German Democratic Republic. 50 44N 12 29E. Town on Zwickauer R. Centre of a coalmining area, manu. inc. textiles, machinery, tractors, paper and chemicals. Pop. (1988E) 121,749.

Zwijndrecht Zuid-Holland, Netherlands. Town sit. NW of Dordrecht on Old Maas R. Industries inc. shipbuilding, salt and rice processing, jute spinning and the manu. of chemicals and edible fats. Pop. (1988E) 41,357.

Zwolle Overijssel, Netherlands. 52 30N 6 05E. Cap. of Overijssel province, on Zwartewater R. Important transport centre with shipyards. Manu. vehicles, chemicals, graphics and dairy products. Pop. (1989E) 91,000.

Geographical Terms

A

abrasion Wearing of rock surfaces by friction, where abrasive material is transported by running water, ice, wind, waves, etc.

abrasion platform Coastal rock platform worn nearly smooth by abrasion.

absolute humidity Amount of water vapour per unit volume of air.

abyssal Ocean deeps between 1,200 and 3,000 fathoms, where sunlight does not penetrate and there is no plant life.

accessibility Nearness or centrality of one function or place to other functions or places measured in terms of distance, time, cost, etc.

acre-foot Amount of water required to cover 1 acre of land to depth of 1 ft. (43,560 cu. ft.).

adiabatic Relating to change occurring in temp. of a mass of gas, in ascending or descending air masses, without actual gain or loss of heat from outside.

afforestation Deliberate planting of trees where none ever grew or where none have grown recently.

aftershock Vibration of earth's crust caused by minor adjustments of rocks after main earthquake waves have passed.

age-sex pyramid Graphical representation of population structure showing number of males and females by age groups.

agglomerate Angular fragmented volcanic material cemented together by heat.

agglomeration Large tract of essentially urban and industrial land resulting from growth and physical expansion of formerly separate neighbouring settlements.

agronomy Agric. economy, including theory and practice of animal husbandry, crop production and soil management.

aiguille Prominent needle-shaped rock-peak, usually above snow-line and formed by frost action.

ait Small island in river or lake.

alfalfa Deep-rooted perennial plant, largely used as fodder-crop since its deep roots withstand drought. Produced mainly in U.S.A. and Argentina.

allocation-location problem Problem of locating facilities, services, factories etc. in any area so that transport costs are minimized, thresholds are met and total population is served.

alluvial cone Form of alluvial fan, consisting of mass of thick coarse material.

alluvial fan Mass of sand or gravel deposited by stream where it leaves constricted course for main valley.

alluvium Sand, silt and gravel laid down by rivers, especially when in flood or when their velocity is checked.

almwind Warm Föhn type wind, blowing s over Tatra Mountains into s Poland.

alp High summer pasture on shoulder of mountain, especially in Switzerland, where marked change of slope occurs above glaciated valley.

altocumulus Layer of fleecy cloud, at altitude of 8,000–20,000 ft. (2,380–6,096 metres) in groups or lines, often sign of good weather.

altostratus Mid-altitude cloud in form of extensive grey sheet, usually heralding rain associated with warm front.

aluminium   Extracted from bauxite; metal of remarkable lightness, high strength and resistant to corrosion.

aluminum   *see* **aluminium**

anabatic wind   Local wind blowing up-valley in afternoon, when convection currents on mountain slopes draw air upwards.

anemometer   Instrument for measuring or recording force of wind.

aneroid barometer   Instrument for measuring atmospheric pressure, comprising metallic box, almost exhausted of air, whose flexible sides expand and contract with changing air pressure. Such movements are magnified and recorded on calibrated dial.

Antarctic Circle   Parallel of lat. 66°32' s, enclosing s Polar Regions.

anthracite   Hard, shiny smokeless coal of high carbon content, producing great heat.

anticline   Arch or crest of fold in rock strata.

anticyclone   Area of high pressure with winds blowing clockwise round it in N hemisphere and anti-clockwise in s hemisphere. Moves slowly in temperate lat., with winds light or variable near centre. Produces fine settled weather in summer, but cold frosty weather or fog in winter.

antipodes   Any two places on opposite sides of earth, so that line joining them passes through centre of earth.

apogee   Part of orbit of moon or planet which is farthest from earth.

Appleton Layer   Part of ionosphere, about 150 m. (240 km.) above earth, reflecting short radio-waves back to earth.

aquifer   Rock holding or allowing the passage of water.

archipelago   (*i*) sea studded with islands; (*ii*) group of many islands, continental or oceanic.

Arctic Circle   Parallel of lat. 66°32' N, enclosing N Polar Regions.

areal differentiation   Differences between areas in terms of relief, climate, soils, vegetation, population, economy etc. Geography is often defined as the study of areal differentiation.

arête   Narrow ridge of spur or mountain, usually above snow-line and formed at crest between two cirques.

arroyo   Small stream channel made by occasional rainfall in arid districts (Latin America, sw U.S.A.).

artesian basin   Synclinal structure containing aquifer between two impermeable beds.

artesian well   Boring put down into aquifer in artesian basin; named from wells of this type in Artois, N France.

asbestos   Fibrous mineral which can be spun and woven into material which is fire-proof, heat-resistant and of low conductivity.

asteroids   Belts of small heavenly bodies revolving in Solar System between Mars and Jupiter.

Atlantic coast   *see* **discordant coast**

atmosphere   Envelope of air surrounding earth, consisting of 78% nitrogen, 21% oxygen, very small quantities of carbon dioxide and other gases and varying quantities of water vapour.

atoll   Circular or irregularly-shaped coral reef, partly submerged to enclose lagoon.

attrition   Constant wearing-down of load of rock-material by friction when transported by running water, wind and waves.

aurora   Luminous phenomena caused by electrical discharge and visible in sky in high lat.

avalanche   Fall of mass of snow, ice or rock down mountainside and capable of great destruction.

axis   (*i*) diameter between N and S Poles, about which earth rotates; (*ii*) central line of fold from which strata dip or rise in opposing directions.

Ayala   Strong, warm wind in Massif Central, France.

Azores high   Sub-tropical anti-cyclone situated over E side of N Atlantic, often extending to affect Britain and W Europe.

B

bad   German, bath. Common prefix attached to German towns which are spas.

badlands   High barren country in semi-arid regions with dry loose soils. Rain, when it falls, is heavy and washes out deep gullies so that little vegetation or animal life remains.

Baguio   Tropical storm of Philippine Islands, mainly occurring between July and Nov.

bahada   Sloping expanse of scree, gravel and sand round margins of inland basin or at foot of mountains in semi-arid regions, formed by coalescence of series of adjacent alluvial cones.

bajada   *see* bahada

Bantu   Native of tribe speaking one of 200 languages or dialects used in Africa S of line Cameroon–L. Victoria.

bar   (*i*) unit of atmospheric pressure equal to 1 m. dynes per sq. cm. At 45° N., in temp. of 0°C. at sea level, 1 bar = 29.53 ins. or 750.1 mm. of mercury. It is divided into 1,000 millibars (mb.) for weather map purposes; (*ii*) ridge of sand and rock fragments across mouth of river or entrance to bay, *see* **off-shore bar, tombolo**; (*iii*) navigational obstruction where alluvium, sand and gravel deposits occur in stream channels.

barchan   Crescent-shaped sand dune

transverse to wind direction, with horns extending down-wind.

barkhan   *see* **barchan**

barograph   Self-recording aneroid barometer tracing ink line on moving drum.

barometer   Instrument for measuring atmospheric pressure, also heights, and indicating weather changes. Commonest type employs column of liquid, usually mercury, in graduated tube. *See also* **aneroid barometer**

barrage   Natural or artificial obstacle to flow of stream.

barrier lake   L. produced by formation of natural dam across valley, caused by landslides, avalanches, deltas, terminal moraines, or dams of vegetation, ice, lava or calcium carbonate.

barrier reef   (*i*) coral reef parallel to coast but separated from it by lagoon; (*ii*) coral reef wholly or partly enclosing coral island.

barrow   Mound of earth or stones, covering anc. burial chamber, particularly common on chalk land of SE England.

barysphere   Heavy central core of earth below lithosphere.

basalt   Fine-grained, dark-coloured

igneous rock, formed of solidified lava extruded by volcanic and fissure eruptions: flows readily to form extensive sheets and may solidify to form hexagonal columns.

base-level Lowest level to which stream can erode its bed. Permanent base-level is the sea, but lake may provide temporary base-level.

basic slag Blast-furnace by-product in iron smelting, contains phosphates and is crushed for fertilizers.

basin (*i*) large-scale depression occupied by ocean; (*ii*) area of land drained by single river system; (*iii*) area enclosed by higher land, with or without outlet to sea; (*iv*) part of dock system or canal or navigable river.

batholith Very large intrusion of igneous rock, usually granite, originating at great depth and extending over many miles. Upper surface may ultimately be exposed by erosion.

bathylith *see* **batholith**

bathyorography Map detail depicting altitude of land and depth of sea by layer-colouring.

bathysphere (*i*) central core of earth; (*ii*) diving apparatus used for submarine operations.

bauxite Impure aluminium hydroxide occurring widely in clay deposits, chief ore of aluminium.

bayou Marshy creek or sluggish back-water in lower Mississippi and along Gulf Coast of U.S.A.

beach Accumulation of sand and shingle along coast, occupying space between low-water mark and highest point reached by storm waves.

bearing Horizontal angular direction of any place or object, taken from a fixed point.

Beaufort scale Scale of wind forces ranging from 0 (calm) to 12 (hurricane). Introduced by Admiral Sir Francis Beaufort, 1805.

beck Small rapid stream.

bed-rock Solid rock of earth's surface, usually covered by soil or other superficial deposits.

bedding Arrangement of rock strata in bands of varying thickness and character.

bedding-plane Structural feature of sedimentary rocks parallel to the original surface of deposition.

beet sugar Sugar from variety of beet grown in temperate zone; after sugar extraction, refuse, leaves and tops are pulped for cattle food, molasses and alcohol.

bench Narrow terrace, step or ledge produced by denudation, wave-cut bench, step-fault, or by mining and quarrying.

bench-mark Defined and located point of reference used in triangulation survey.

berg German hill or mountain, African range. Also used in combinations for physical features, *see* **bergschrund, Berg wind**.

bergschrund Gap left round upper rim of snowfield or glacier as snow or ice moves downstream.

Berg wind Warm, dry Föhn wind, blowing mainly in winter from s African Plateau to coast.

bight (*i*) wide curve or recess of coast, often between headlands; (*ii*) large bend of river.

bill Long narrow promontory.

billabong Backwater in temporary flowing stream in Australia.

Bise Cold, dry N or NE wind blowing in winter in Switzerland, N Italy and s France. Similar to Mistral or Tramontana.

bitumen Pitch or tar or asphalt.

bituminous coal Free-burning coal with high hydro-carbon content, includes most household and gas coals.

black earth   *see* **chernozem**

block mountain   *see* **horst**

blow-hole   Near-vertical cleft leading from sea-cave to cliff-top, formed by hydraulic action of wave-compressed air; spray is sometimes blown from it into air.

blow-out   Hollow in sandy terrain formed by deflation (wind eddying), particularly where vegetation has been destroyed. In rock deserts, may develop from break in resistant surface layer.

bluff   Headland formed by steep slopes bordering river or lake.

bocage   Landscape of NW France, with small fields enclosed by dry-stone walls or low banks with hedges.

Bohorok   Hot dry wind on N coast of Sumatra during NW Monsoon (June–Sept.).

bolson   Basin of interior drainage in arid or semi-arid regions. Floor may contain salt lake or alluvial fans round flanks of surrounding mountains.

Bora   Violent, cold, dry wind blowing from N or NE into Adriatic, mostly in winter; can produce rain or snow when depression lies to the S.

Boraccia   Particularly violent form of Bora.

bore   Steep, wall-like wave in river with funnel-shaped estuaries, flowing rapidly upstream on rising tide. Highest at spring tides.

boreal   (*i*) N; (*ii*) general term for climatic zone with short summers and snowy winters; (*iii*) climatic period from 7,500–5,500 B.C.

Borino   Weaker summer form of Bora.

boss   Small batholith, more or less circular in plan.

Boswash   Name given to the urban area extending from Boston (Massachusetts) to Washington (D.C.).

boulder clay   Unstratified glacial deposits of varying composition but usually consisting of clays, sand and stones of all shapes and sizes, probably representing ground moraine of ice-sheet.

bourne   Temporary or intermittent stream flowing in dry valley of chalk country after period of heavy rain.

brachycephalic   Broad headed. Cephalic Index > 83.1.

brae   Sloping bank or hillside (Scot.).

braided river channel   Course of river split into complicated channels because of inability to transport its load.

brash ice   Accumulation of crushed ice near shore, not thick enough to hinder navigation.

Brave West Winds   *see* **Roaring Forties**

break in bulk   Location at which materials or products are transferred from one mode of transport to another.

breaker   Mass of broken water rushing up beach or where mass of rock lies near surface of sea.

breccia   Angular fragments of rock of any type, cemented together in matrix.

breckland   Area on borders of Norfolk and Suffolk comprising heathland, thicket and bracken.

Brickfielder   Hot, dry, dusty, squally wind in SE Australia, caused by S movement of mass of tropical air, especially in summer.

British Summer Time   Period in year during which clocks are advanced one hour ahead of Greenwich Mean Time.

broad   Sheet of reed-fringed shallow water beside or forming part of sluggish river near estuary, particularly in E Anglia.

buffer state   Small, independent autonomous country between two or more powerful ones.

bund   (*i*) artificial embankment; (*ii*) main quay in some Far Eastern ports.

Buran Cold, strong NE wind in Central Asia, producing blizzards in winter, but can also blow in summer.

burn Small stream (Scot.).

bush Large uncultivated tract, overgrown with bushes and small trees, usually in semi-arid regions.

bushveld Type of savanna in sub-

tropical and tropical Africa, with trees scattered, forming open parkland, but sometimes dense so as to resemble forest.

butte Isolated flat-topped hill, usually capped with resistant rock and rising abruptly from plain.

Buys Ballot's Law *see* **Ferrel's Law**

C

caatinga Thorn forest or open woodland with scrub in arid parts of NE Brazil.

cacao Evergreen tree of equatorial or tropical regions, with large pods containing beans, the source of cocoa and chocolate.

caldera Large, shallow, circular cavity remaining after removal of former volcanic peak by immense eruption; sometimes contains lake.

caliche Impure deposits of nitrogen salts found in deserts of Chile or Peru; valuable fertilizer in 19th cent.

campos (Plural) open plains of Central Brazil with savanna climate and vegetation.

cane sugar Tall tropical or subtropical plant whose stem yields sugar.

canyon Spanish, cañon. (*i*) deep, steep-sided gorge with river at bottom, mainly found in arid or semi-arid areas; (*ii*) submarine trough on, and beyond, continental shelf.

'Cape Doctor' Strong wind blowing from plateau to coast in S Africa.

carr Fen containing reeds, shrubs and willows.

carse Level tract of fertile alluvial lands bordering estuaries in Scotland.

cartouche Panel on map, usually decorative, containing title, scale, etc.

cash crop Crop cultivated for market.

cassava Plant with tuberous roots, used as food in Central or S America.

caste Hindu hereditary social group.

cataract (*i*) waterfall running over sheer precipice; (*ii*) series of rapids, as on R. Nile.

catch crop Quick-growing crop (*i*) between two main crops in a rotation; (*ii*) between rows of main crop; (*iii*) in place of failed crop.

catchment Area served by a function, firm, establishment or central place.

catchment area Self-contained drainage basin collecting all rainfall in area.

causses (*i*) French term used for limestone country in general; (*ii*) limestone district in S of Massif Central, France, resembling Yugoslavian karst.

cavern Large cave or subterranean chamber, naturally produced, with entrance, from surface; common feature of karst landscape.

cay Low island or reef of sand and coral in Caribbean Sea.

Celsius scale Internationally accepted name for Centigrade scale of temp. where 0° is melting point of ice and 100° boiling point of water, after inventor Anders Celsius (1701–44).

Celtic field Roughly sq. field, varying in size from half to one and a half acres, used before Saxons introduced strip-field system.

centrality Relative importance of a settlement with regard to region it serves.

central place A settlement providing goods, services, etc. for a surrounding area (catchment).

central place system Spatial distribution of central places.

central place theory Theory of location of settlements and functions devised by W. Christaller (1933).

Cephalic Index Mathematical index to describe shape of skull = maximum breadth of skull ÷ maximum length × 100. *See also* **brachycephalic, dolichocephalic**.

chaco Forest land in parts of s America, *see* **Gran Chaco**.

chaparral Spanish, evergreen shrub vegetation of NW America, resembling garigue and maquis of Mediterranean Europe.

chernozem Russian, 'Black Earth' soil of loose, crumbly texture, rich in humus and lime, covering large areas of temperate lands where natural vegetation was grass, now forms great wheat growing areas of world.

chicle Gum obtained from latex of several trees in s and Central America, chief ingredient of chewing gum.

chili Hot, dry s wind over N Africa.

china clay *see* **kaolin**

chine Narrow cleft in cliffs of soft material, containing stream; mainly found in Hampshire and Isle of Wight.

chinook American Indian 'snow-eater', dry, warm SW wind descending E slopes of Can. or N U.S.A. Rocky Mountains and warmed adiabatically; in spring, causes swift rise in temp.

chorochromatic map Map showing non-quantitative phenomena by means of colouring or shading.

choropleth map Map showing quantities per unit of area by means of scaled shading or colouring (e.g. map of population density by parishes).

cinchona Tropical evergreen tree of s America whose bark yields quinine.

cirque Steep-sided, flat-floored, horseshoe shaped rock-basin of glacial origin, marking site of cirque glacier, but in post-glacial conditions often contains small, deep lake.

cirro-cumulus 'Mackerel sky'; type of cloud above 20,000 ft. (6,096 metres), in lines of ripples, containing ice-crystals.

cirro-stratus Layer or veil of sheet cloud, above 20,000 ft. (6,096 metres), of milky appearance, causing sun or moon to appear with halo; marks approach of warm front.

cirrus Delicate, wispy clouds at 20,000–40,000 ft. (6,096–12,192 metres), containing ice-particles. When drawn out into 'mare's tail' indicates strong upper winds.

citrus fruit Obtained from evergreen trees like grape-fruit, lemon, lime and orange; widely grown in sub-tropical or warm temperate regions, especially Mediterranean climates.

cliff High, steep face of rock mass, especially along sea coast, where marine denudation is active.

climate Total complex of weather conditions, average characteristics and range of variation over large areas of earth's surface, considered over period of 30–35 years.

clinometer Instrument for measuring vertical angle between two points or for measuring dip of strata.

clint Low, flat-topped ridge between fissures or grikes on surface of carboniferous limestone plateau, forming limestone pavement as result of solution.

cloud Mass of condensed water vapour or ice-particles formed by condensation on atmospheric nuclei, such as dust, smoke particles, pollen, etc. They float in masses from near ground level, as mist or fog, to over 40,000 ft. (12,192 metres).

cluse French for steep-sided valley cutting transversely across limestone ridges in Savoy Alps and Jura.

coal measures Geological term covering series of coal seams and intervening strata.

coast General name for zone of contact between land and sea.

coastline (*i*) general term for edge of land viewed from sea; (*ii*) high-water mark of medium tides; (*iii*) base of cliffs.

coca Bolivian shrub whose dried leaves provide stimulant when chewed.

cocoa *see* **cacao**

coconut Nut of palm from tropical regions of E Indies and Pacific Islands whose dried flesh produces oil for margarine and soap; *see* **copra**

coffee Beverage produced from roasting beans of berries produced by trees in tropical uplands.

coir Outer covering of coconut used to make matting and cordage.

col (*i*) high pass between two higher summits; (*ii*) area of lower pressure between two anticyclones.

cold front Rear of warm sector of depression, marking boundary between warm air and advancing wedge of cold air; undercutting causes appreciable temp. drop and heavy showers.

cold occlusion Where overtaking air in occlusion is colder than air mass in front.

'Cold Pole' Term applied to point with lowest temp. usually referring to Verkhoyansk, NE Asia, where mean Jan. temp. is −58°F. (−50°C.).

collective farming Type of agric. organization started in U.S.S.R., but now practised in E Europe and SE Asia. Land compulsorily amalgamated into large holdings, run by manager and worked by directed labour; proportion of crops go to Govt., workers are paid a share of proceeds of marketing remainder.

combe (*i*) Small, generally narrow valley; (*ii*) glacial cirque, cwm.

common Area of land, often unenclosed, over which members of community have certain rights of access, pasture, turf-cutting, etc.

compass Instrument used to determine direction, either by free-swinging magnetized needle or by rotating gyro-wheel.

concordant coast Running parallel to general grain of relief, giving straight and regular coastline.

condensation Process whereby substance changes from vapour to liquid, as with cloud formation, mist, etc.

condominium Territory governed jointly by two or more countries.

cone Volcanic peak with broad base tapering to summit.

confluence Point where trib. joins main stream.

conglomerate Rock composed of rounded, water-worn pebbles cemented together in matrix.

connectivity A measure of the number of routes connecting one place with other places.

continent (*i*) large, continuous mass

of land; (*ii*) mainland, as applied to Europe.

continental climate Climate of continental interiors in temperate zone, with seasonal extremes of temp. and low rainfall, mainly in summer.

continental drift Theory that continental masses have changed relative positions. First put forward in 19th cent. and now further substantiated through palaeomagnetism.

continental island One rising from continental shelf and structurally related to that continent.

continental shelf Gently sloping margins of continent, submerged beneath sea and extending to point where continental slope begins; in general less than 100 fathoms deep.

continental slope Steep slope descending from continental shelf to abyssal plain.

contour lines Lines on map connecting all points at same relative elevation above or below specific datum, usually sea-level.

contour ploughing Measure to combat soil erosion by ploughing along slope to reduce run-off.

conurbation Area of extensive urban development where existing separate towns have expanded so as to coalesce.

convection Transmission of heat through liquids and gases by movements of their particles.

convectional rain Caused by moisture-laden air being warmed, rising, expanding and cooling adiabatically to its dew-point.

coombe *see* **combe**

copra Dried flesh of coconut, source of coconut oil, shredded coconut and cattle food.

coral Hard, calcareous substance formed from skeletons of some marine polyps living in warm intertropical seas.

coral reef Reef composed of coral limestone, *see* **atoll, barrier reef**.

cordillera Mountain system comprising several ranges, broadly parallel in trend.

core Central mass of earth, radius about 2,160 m. (3,456 km.), probably consisting of metallic mass of nickel-iron or nife.

Coriolis Force *see* **Ferrel's Law**

corona (*i*) series of coloured rings surrounding sun or moon resulting from diffraction of light by water droplets; (*ii*) fringe of radiant light visible round circumference of sun at total eclipse.

corrasion Frictional wearing down of rock surface by material moved by gravity, running water, ice, wind and waves.

corrie *see* **cirque**

corrosion Chemical solution of rock by water.

côte (*i*) escarpment in France; (*ii*) section of coast (France).

coteau Name given by French explorers in N America to sharp ridge of hills or prominent escarpment.

cotton Annual, sub-tropical shrub whose bolls contain fibrous material used to make cotton cloth; seeds are crushed for oil, cooking fat, margarine and cattle food.

cove (*i*) small creek or inlet or rounded bay; (*ii*) steep-walled, semi-circular opening at head of valley.

cover crop Quick-maturing crop ploughed in as fertilizer or grown as protective mat between main crops to reduce soil erosion.

crag and tail Characteristic of glaciated area, where mass of resistant rock in path of advancing ice has been attacked on upstream side, to produce crag, but has protected land on downstream side, leaving a gentler slope or tail, to leeward.

crater (*i*) rounded, funnel-shaped hollow at summit of volcano; (*ii*) depression caused by impact of meteorite.

creep Slow, downward movement of soil under influence of gravity, lubricated by running water.

Creole Person born in W Indies but not of aboriginal descent; initially implying no question of colour, now mainly describes people of mixed blood.

crevasse Deep fissure in glacier, with transverse trend where slope increases or longitudinal where glacier spreads out as valley widens.

croft Small farm-holding in Scotland, reckoned to supply all family's needs in food, fuel and clothing.

cromlech Celtic form of megalithic burial chamber, with large flat stone resting on stone uprights.

crust Upper (sial) and lower (sima) layers of earth, forming outermost shell or lithosphere, about 10–30 m. (16–48 km.) thick.

cuesta Ridge with steep scarp slope and gentle dip slope, *see* **escarpment**.

cumec Unit for measuring river discharge; number of cu. m. per second passing particular point, approx. 19m. gallons per day.

cumulo-nimbus Cumulus cloud developing to immense vertical height with towering summits, sometimes spreading into anvil shape, and associated with thunderstorms, hail and heavy rain.

cumulus Convectional cloud, growing vertically from flat base to produce white domed summits, often reaching 25,000 ft. (7,620 metres).

cutoff Channel cut by river across neck of land at acute meander, leaving oxbow.

cwm *see* **combe**

cyclone Tropical low pressure system, similar to depression of temperate lat. but smaller in extent and more violent in effect, marked by very high winds, torrential rain and thunder.

D

dale Wide, open valley, particularly in N England.

Dalmatian coast *see* **concordant coast**

Date Line *see* **International Date Line**

datum Zero from which altitudes and depths are determined; *see* **ordnance datum**.

deflation Removal of fine deposits of dust and sand from earth's surface by wind.

delta Alluvial tract forming at mouth

of river where deposition of part of river load exceeds rate of removal by tidal or other currents; covered by network of branching channels or distributaries.

demersal fish Species living near bottom of shallow seas, caught by trawling.

demography Branch of anthropology, dealing with statistics of births, deaths, etc.

denudation Destructive process modifying surface of earth by all natural agencies.

deposition The laying down of material transported by all agencies; complementary to denudation.

depression (*i*) low-lying portion of earth's surface; (*ii*) low-pressure system in mid or high lat., formerly called cyclone, now known as cyclonic depression or 'low'; has diameter of 300–2,000 m. (480–3,200 km.) with winds blowing round it anti-clockwise in N hemisphere and clockwise in S. *See* **fronts, occlusion**.

desert Area of land with scanty vegetation and low rainfall or low temp. Types of desert include (*i*) hot, Trade Wind; (*ii*) coastal, in lat. 15°–30°, with cool offshore currents; (*iii*) mid-lat. of continental interiors; (*iv*) polar.

determinism Philosophy that man's actions are essentially determined by physical condition of his environment.

detritus Material shed from rock surfaces by disintegration.

dew Condensed droplets from atmospheric vapour on surfaces cooled by nocturnal radiation.

dew-point Lowest temp, to which air can be cooled without causing condensation.

dew-pond Small hollow of special construction to collect water from atmospheric condensation.

dike *see* **dyke**

dip-slope Gentle slope of cuesta.

discordant Cutting across general lines of structure of land.

discordant coast Running transverse to general grain of relief, giving dissected coastline.

dispersed city Group of large towns of similar size found in close proximity to each other, and tending to function in a complementary manner. Although the total population contained in its area is sufficient to support one large centre offering a full range of highest order goods and services, such a centre does not exist and these functions are shared between the constituent centres.

dispersed settlement Scattered farms and hamlets, with no concentration of population in village or larger settlement.

distributary Individual channel into which river may divide in delta, leading to sea without rejoining main stream.

diurnal range Difference in 24 hour period between maximum and minimum value of element such as temp.

Doctor Popular name for Harmattan wind, West Africa.

doldrums Equatorial belt of low pressure between two Trade Wind belts, marked by light winds or calms, high temp. and humidity.

dolichocephalic Long headed. Cephalic Index < 75.

doline Shallow, saucer-shaped depression in karst landscape.

drainage (*i*) act of removing water from poorly-drained area; (*ii*) discharge of water from any area through natural stream systems.

drift (*i*) superficial deposits, usually glacial; (*ii*) horizontal passage in mine following vein of ore; (*iii*) slow movement of surface ocean water under influence of prevailing winds; (*iv*) surface movement of loose material by wind, such as snow.

drumlin Low, oval-shaped hill of glacial origin, formed of boulder clay and elongated in direction of ice flow.

dry farming Agric. in semi-arid parts, without irrigation, where moisture is conserved by mulching, maintenance of fine tilth and cropping in alternate years so as to increase total water.

dry valley Valley, usually in chalk or limestone, with no permanent flowing stream.

dunes Low ridges of wind-blown sand, found in deserts and along low-lying coasts above high water mark.

dust bowl Originally applied to parts of SW and W U.S.A., but now refers to any area where soil erosion occurs as result of over-grazing or over-cropping.

dust devil Short-lived swirling wind resulting from convection over sandy area, lifting dust and sand but not to any great height.

dust storm Caused by winds in

semi-arid areas causing dense clouds of dust to rise to considerable heights.

dyke (*i*) mass of intrusive rock running discordantly across landscape; (*ii*) drainage ditch or watercourse; (*iii*) artificial embankment preventing flooding of low ground; (*iv*) embankment in flood plain of river parallel to its course; (*v*) man-made defensive earthwork.

E

eagre *see* **bore**

earth pillar Column of soft earth, often 20–30 ft. (6–9 metres) high, protected by capping of rock; product of sub-aerial erosion.

earthquake Movement or tremor of earth's crust causing series of shock waves to move outwards in all directions.

eclipse Partial or total obscuring of light of (*i*) moon, when earth comes between it and sun; (*ii*) sun, when moon comes between it and earth.

ecliptic Apparent movement of sun through sky during course of year; plane of ecliptic is tilted at angle of about $66\frac{1}{2}°$ to earth's axis.

entrepôt Centre or port to which goods in transit are sent for temporary storage before re-export.

environmentalism Philosophy that environmental factors determine man's actions. *See* **determinism**.

environmental perception The environment as perceived by individuals or groups rather than as it is in reality.

epicentre Point on earth's surface vertically above origin of earthquake shock.

equator Great circle, lat. 0°, midway between Poles, in plane perpendicular to earth's axis.

equinox Period of year when day and night are of equal length all over world, occurring about March 21 and Sept. 23, when sun crosses equator.

erg Arabic for sandy deserts of Sahara with dunes and sand-sheets.

erosion Wearing away of land surfaces by various natural agents, chiefly running water, ice, wind and waves but, unlike denudation, excludes weathering.

erratic block Piece of rock carried by glacier or ice-sheet and deposited some distance from parent outcrop when ice melted.

escarpment (*i*) *see* **cuesta**; (*ii*) term also applied to any steep slope breaking general uniformity of landscape.

esker Elongated ridge or mound formed from glacial and sub-glacial sands and gravels, deposited by sub-glacial streams.

esparto Grasses grown in N Africa and Iberian Peninsula, used in making paper, cordage, matting and sacks.

estancia Spanish, large farm in

Argentina used for extensive cattle rearing.

estuary Mouth of river where channel broadens into V-shape and tidal movements occur.

étang French, shallow lake in sanddune areas, formed by beach material thrown up by sea.

Etesian wind Strong N to NW wind at intervals in summer in E Mediterranean.

eucalyptus Gum tree of Australia.

eustasy World-wide change of sea level, caused by Ice Age, movement of ocean floors or sedimentation of ocean basins.

evaporation Physical process by which substance changes from liquid to vapour state, caused mainly by heat.

everglade Marshy area with grasses

and trees in Florida and along Gulf Coast of U.S.A.

exfoliation Alternate heating and cooling of rock surfaces causing peeling off in scales.

expanded town Town outside the metropolitan area whose expansion has been purposefully accelerated by Government agencies (e.g. by financial assistance towards development of industries, housing) to curtail further expansion of metropolitan regions and to alleviate over-crowding elsewhere (e.g. Basingstoke).

extensive agriculture Method of farming based on large farm units with little labour but with a high degree of mechanization (e.g. Canadian Wheatlands).

extrusive rocks Igneous rocks reaching the surface as lava, flowing from fissures and then solidifying.

eyot *see* **ait**

F

Fahrenheit Graduated scale of temp. where 32° is melting point of ice and 212° boiling point of water, after A. D. Fahrenheit (1686–1736).

fall-line Narrow belt, marked by line of waterfalls, where rivers descend from plateau edge to lowland, as in Fall Line in SE U.S.A.

fan Fine wind- or water-borne material, deposited in shape of fan with low angle of slope.

fathom Nautical measurement of depth, equal to 6 ft. (1.9 metres); 1,000 fathoms equal 1 nautical m.

fault Fracture or rupture in earth's crust, where movement has displaced strata on each side.

fauna Animal life of region or geological period.

fell Norwegian, bare, uncultivated hill or mountain in N England.

fen Water-logged land of reed and peat, providing excellent agric. land when drained, as in Fens of E Anglia.

Ferrel's Law Postulated by American scientist W. Ferrel in 1856, stating that bodies in free motion are deflected to right in N hemisphere and to left in S, (Coriolis Force) as result of earth's rotation.

fetch Extent of open water over which wind blows, instrumental in determining height and force of waves.

fiard   Swedish fjard; large open area of water, surrounded by islands, occurring in rocky, glaciated lowlands.

fiord   Norwegian fjord; drowned glacial valley, being long and narrow, steep-sided and deep, apart from threshold near mouth.

firn   German, snow accumulating in layers above glacier, partially compacted by freeze and thaw action.

firth   Estuary or strait or arm of sea (Scotland).

fjard   *see* **fiard**

fjord   *see* **fiord**

flash-flood   Sudden, short-lived torrent in semi-arid areas after intensive rainfall, carrying large load, product of desert weathering.

flax   Annual plant, yielding fibre for linen and seed for linseed oil, paint and cattle-cake.

fleet   (*i*) small inlet or lagoon of brackish water behind coastline, separated from it by sand or shingle bank; (*ii*) small creek or inlet.

flood-plain   Floor of river valley over which alluvium is spread in flood time.

flora   Plant life of region or geological period.

flume   (*i*) narrow ravine or gorge; (*ii*) artificial channel to provide water for power or to float logs.

fluvio-glacial   Effects of melt-water stream issuing from glacier-front or ice-sheet margin.

fog   Thick mist or haze in surface layer of atmosphere resulting from condensation of moisture with smoke and dust particles in suspension; visibility less than 0.6 m. (1 km.).

Föhn   German, warm, dry, adiabatic wind blowing down lee slopes of mountains, especially in Alps, when depression to N draws in air from Mediterranean.

fold   Bending of strata by compressive forces in earth's crust, usually along lines of weakness.

force   Waterfall in N England.

foreshore   Part of shore lying between high and low-water line of mean spring tides.

fosse   (*i*) ditch or trench round anc. earthwork; (*ii*) waterfall in N England, *see* **force**; (*iii*) in U.S.A., depression between valley-glacier side and valley wall.

free port   Port where goods can be unloaded, held in bond and re-exported without customs payments.

Friagem   Strong, cold winter wind on campos of Brazil and E Bolivia.

fringing reef   Uneven coral platform attached to coast, with or without lagoon intervening, sloping steeply seawards.

front   Plane of separation between warm and cold air masses.

frost   (*i*) condition of air whose temp. is at or below freezing point of water; (*ii*) minute crystals of ice formed from water vapour when dew-point is below freezing.

frost-pocket   Low-lying area into which cold air drains by gravity to produce freezing conditions when air on higher slopes is still above freezing; fruit-growers avoid such areas.

fumarole   Italian, small vent in surface, pouring forth steam, hydrochloric acid, sulphur dioxide and other gases.

G

gap *see* **water gap, wind gap**

garrigue French, stunted evergreen scrub separated by bare limestone rock, found in S France, Corsica, Sardinia and Malta.

geest German, area of coarse sand, gravel and heath vegetation in N Germany, Netherlands, Denmark and Poland.

geo Norse gya, a creek; long, narrow, steep-sided inlet in cliff, derives originally from cave enlarged by marine erosion.

geosyncline Linear depression in earth's crust, of considerable width and several m. deep, product of slow, continuous down-warping of ocean floor, accompanied by sedimentation of material worn away from neighbouring land masses.

geyser Icelandic geysir, 'gusher' or 'roarer'; hot spring ejecting superheated steam and hot water at intervals from volcanic sources.

ghat Indian, (*i*) river landing place; (*ii*) mountain pass; (*iii*) range of mountains.

ghaut *see* **ghat**

ghee Hindu, butter made from buffalo milk.

ghetto Part of urban area inhabited almost exclusively by one race or type of people.

Ghibli Arabic, hot, dry, S sirocco wind of Libya or Tunisia.

Gibli *see* **Ghibli**

glacier French, mass of compacted snow and ice, accumulated on ground above snow-line, moving downstream under gravity and exerting great erosive force.

glacio-fluvial *see* **Fluvio-glacial**

glen Long valley, steep-sided and flat-bottomed (Scotland).

gneiss Coarse-grained crystalline rock.

Gondwanaland Hypothetic single land-mass in anc. geological times, from which present S continents were formed as outcome of continental drift.

gorge Steep-sided valley, deep in relation to width.

graben German, narrow fault-trough.

Gran Chaco 'Great hunting ground', forest or savanna region on borders of Bolivia and Paraguay.

grasslands Region where rainfall is insufficient for tree growth but is less scanty and irregular than in desert regions; major types include tropical or savanna, temperate or steppe or prairie or pampas or veld, and mountain.

graticule Network of parallels and meridians drawn on map or globe.

great circle Circle on earth's surface whose plane passes through centre of earth; arc of great circle provides shortest distance between any two points on earth's surface.

Green Belt Planned and maintained zone of open country round town or separating adjacent towns, free from building or industrial development.

Greenwich Mean Time Standard time for British Isles and parts of W Europe, *see* **British Summer Time**

Greenwich Meridian Prime meridian, passing through old Royal Observatory at Greenwich; long. of a place is expressed E or W of this, up to 180°.

Gregale N cool-season winds of Central Mediterranean.

greywether Scattered blocks of sandstone found over parts of chalk-lands of S England, so-called because of their resemblance to sheep.

grid Network of sq. covering map series, based on lines drawn parallel and at right angles to central axis, from which position of any point can be stated; *see* **National Grid**

grike Deep groove, bounded by ridges or clints in area of karst landscape, where limestone pavement results from solution.

ground moraine Debris carried at base of glacier or ice-sheet and deposited as ice melts.

groundnut Tropical or sub-tropical plant whose pods ripen below ground; when crushed, these produce oil for margarine, cooking fat, other food-stuffs and animal fodder.

ground water Body of water contained in soil, subsoil and underlying rocks above impermeable layer.

gryke *see* **grike**

guano Accumulation of bird drop-pings forming valuable phosphatic fertilizer.

Gulf Stream Warm ocean current flowing NE from Gulf of Mexico towards Europe; *see* **North Atlantic Drift**

gulley erosion Accelerated erosion of land by concentrated surface run-off after sudden rainstorm; major cause of soil erosion.

H

haar Summer sea-fog blowing onto E coast of N England and Scotland.

hachure Lines of shading on relief map to show differences in slope.

hacienda Spanish, large agric. estate, ranch or plantation, chiefly in S America.

haff German, shallow, fresh-water lagoon, formed by growth of spit or nehrung across river mouth, notably in Baltic.

hamada Arabic, rock desert of N Africa, Australia, Gobi.

hanging valley Trib. entering main valley from considerable height, often with waterfalls; common in glaciated regions, where main valley has been over-deepened by glacier.

Hanse Towns Political and commercial league of Germanic towns.

Harmattan Hot, dry NE wind, blowing off Sahara to W African coast in Dec.–Feb. providing relief from humidity on coastal plain, but dry and dusty inland.

haze Obscurity in atmosphere caused by condensation of moisture on dust or salt or smoke particles or by heat refraction; visibility from 1 to 2 m. (1.5–3 km.).

headland Steep crag or cliff projecting into sea.

head of navigation Farthest point up-river reached by trading vessels.

Heaviside Layer Section of iono-sphere, 60–75 m. (96–120 km.) high, highly ionized and able to reflect long radio-waves back to earth.

Helm   Strong, cold E to NE wind blowing off Pennines in N England.

hemp   Fibre-producing plant, for twine, string, rope, matting etc. whose flowers and leaves produce drugs and whose seeds yield oil.

Hevea Brasiliensis   Tree, native to equatorial forests of Brazil, now grown in plantations in Malaysia or E Indies; source of commercial rubber; *see* **latex**

hinterland   (*i*) 'back country' of early coastal settlement; (*ii*) area served by port, taking its imports and supplying it with its exports.

hoar-frost   Thin deposit of ice crystals on surface cooled below freezing point by radiation.

hook   Sand bar ending in recurved spit.

horizon   (*i*) visible horizon, boundary of earth's surface, as viewed from point where sky appears to meet earth or sea; (*ii*) true horizon, great circle whose plane passes through centre of earth and is parallel to visible horizon; (*iii*) geological horizon, plane of stratified surface or bed; (*iv*) soil horizon, main layer or zone within soil profile.

horn   German, pyramid peak formed by development of back-to-back cirques on mountainside.

horse latitudes   Sub-tropical high pressure belts of atmosphere, over oceans in lat. 30–35°, giving calm conditions and stable weather.

horst   German, block of earth's crust (*i*) left standing after adjacent areas have sunk by faulting; (*ii*) uplifted between faults.

hot springs   *see* **thermal springs**

Humboldt Current   Cool ocean current flowing N along W coast of S America.

humidity   State of atmosphere with regard to water vapour it contains; *see* **absolute humidity, relative humidity**

humus   Vegetable mould derived from decomposing organic matter in soil.

hurricane   (*i*) tropical revolving storm with violent cyclonic winds in Caribbean Sea and off Queensland coast; (*ii*) wind of Force 12 on Beaufort Scale.

hydro-electric power   Electrical energy obtained by harnessing waterfalls or by damming river.

hydrological cycle   Endless interchange of water between sea, air and land, by evaporation, condensation, precipitation, etc.

hygrometer   Instrument for measuring relative humidity of air.

I

iceberg   Large mass of floating ice, broken off from edge of ice-sheet or tongue of tidal glacier and floating under influence of currents and winds; those in N hemisphere mainly irregular in shape, those in S are tabular.

ice-cap   Permanent mass of ice,

smaller than ice-sheet, covering plateaus and high-lat. islands.

icefall Heavily crevassed part of glacier where it meets change of slope of valley floor.

ice-field Large, continuous area of pack-ice.

ice-sheet Mass of permanent ice and snow of great thickness, covering large areas, as in Antarctica and Greenland; also refers to much greater areas in past glacial eras.

ideographic Studying individual cases and situations; *opposite* **nomothetic**.

igneous rock One formed by solidification of molten rock or magma; *see* **extrusive** and **intrusive rocks**

incised meander Pattern of meander maintained at progressively lower levels by downcutting of river resulting from rejuvenation.

industrial inertia Condition whereby industries, once established at a central location, remain there due to existing investment in capital, services, etc., even though some other location may ultimately appear more attractive in terms of certain other criteria.

infield Farmland immediately surrounding a farm or hamlet farmed intensively. *See also* **outfield**.

input-output analysis Analysis of the flows of goods or services, usually in terms of their cash value, which take place between all sectors of a regional economy. Often used to predict the overall effect of changes in the regional economy.

inselberg German, mountain rising abruptly from uniformly level surface, esp. in semi-arid and arid climates.

insolation Amount of radiant solar energy reaching earth's surface.

interfluve Area of land between two rivers.

interlocking spurs Alternate projecting spurs in upland valley where river follows winding course.

International Date Line Imaginary line closely following 180° meridian, where date changes by one day as it is crossed, to make up for cumulative effect of time zone changes; westbound travellers lose a day, eastbound ones gain a day.

intrusive rocks (*i*) those resulting from injection of magma into existing rocks; (*ii*) particular type of igneous rock thus formed; *see* **batholith, dyke, laccolith, sill.**

ionosphere Part of atmosphere above stratosphere, marked by distinctive layers reflecting electromagnetic waves; *see* **Appleton Layer, Heaviside Layer**

irrigation Artificial application of water to land in arid regions to promote plant growth.

island Piece of land, smaller than continent, surrounded by water.

isobar Line on map joining points of equal atmospheric pressure; figures reduced to sea-level for comparison.

isobath Line on map or chart joining submarine points of equal depth.

isochrone Line on a map joining all places of equal travel time from a given point.

isohyet Line on map joining points having same rainfall over given period.

isoneph Line on map joining points having same amount of cloudiness.

isopleth Line on map joining points having same value for particular phenomena, *see* **isobar**, etc.

isostasy State of equilibrium in surface crust of earth, whereby equal mass underlies equal surface area.

isotherm Line on map or chart joining points having same air or sea temp.

isotropic surface A theoretical area of broad, featureless plain on which transport is equally available in all directions, population is uniformly distributed and has uniform tastes and preferences. Used as a basic assumption in many locational theories (such as central place theory).

isthmus Narrow neck of land between two seas or joining peninsula to mainland.

J

jarrah Eucalyptus tree of SW Australia, resembling mahogany and very durable in water.

jet stream High altitude air movement of strong winds in narrow belt.

joint Plane of division in rock mass, usually transverse to bedding-plane, caused by tension or compression.

joran Cold, dry wind blowing at night from Jura Mountains to L. Geneva.

jungle Wild, uncultivated land; now widely used to refer to tropical forest with dense undergrowth.

jute Fibre from bark of plant grown in Bengal and used in making sacks, carpets, etc.

K

kame Hummocky deposits of sand and gravel laid down by melt-water along edge of ice-sheet, usually parallel to ice-front.

kaolin Hydrated silicate of alumina derived mainly from decomposed granites; known as china clay, used in manu. of pottery, china and paper.

Karaburan Hot NE wind in Central Asia, sweeping up clouds of sand, the lighter particles being carried great distances and ultimately deposited as loess.

karri W Australian eucalyptus, providing very hard timber.

karst Italian, carso, Serbo-Croat,

kars; proper name for rugged limestone plateau of NW Yugoslavia, but used generally to describe areas of carboniferous limestone with typical phenomena of limestone pavements, sink holes, underground drainage, caverns, gorges, etc.

katabatic wind Cold wind, blowing downhill, usually at night, caused by gravity flow of dense air from upper slopes, chilled by radiation.

kay *see* **cay**

kettle hole Circular hollow in glacial drift, caused by melting of large block of ice separated from glacier.

key *see* **cay**

Khamsin Arabic, hot, dry, sirocco wind of Egypt and SE Mediterranean.

kloof Afrikaans, deep, narrow ravine or mountain pass in S Africa.

knot (*i*) nautical measurement of speed, 1 nautical m. per hour; (*ii*) junction of two or more ridges in fold mountain system.

kopje Afrikaans, prominent, small, isolated hill in S Africa.

kraal Afrikaans, village, cattle-pen or enclosure in Africa, surrounded by thorn fence.

kyle Gaelic, channel or sound or strait between mainland and island or between two islands.

L

labour coefficient Ratio of labour cost to all other inputs or outputs of productive process.

laccolith Intrusive mass of magma forced up a pipe and spreading laterally in lenticular form, causing arching of the overlying strata.

lacustrine Pertaining to a lake.

ladang Slash and burn shifting agric. practised in SE Asia.

lagoon (*i*) sheet of salt water separated from open sea by sand or single bank; (*ii*) sheet of water between offshore reef and mainland; (*iii*) sheet of water within an atoll.

lake Extensive sheet of water enclosed by land, occupying hollow in earth's surface.

land and sea breezes Resulting from unequal heating of land and water masses. By day, with land heated more rapidly, sea-breeze blows on-shore; by night, land-breeze blows off-shore as land loses heat by radiation.

landes French, wastelands; proper name of lowlands of sand-dunes and lagoons in SW France.

lapiés French, bare limestone surface with joints widened by solution to form clints and grikes.

lapse-rate Rate of temp. decrease in atmosphere with increase in height. Environmental lapse-rate approx. $-21°F.$ ($6°C.$) per 3,250 ft. (1,000 metres).

laterite Red soil consisting of decomposed and weathered igneous rocks, found as soil horizon in tropical climates.

latex Milky fluid of rubber tree (*Hevea Brasiliensis*), extracted by tapping bark to provide raw material for manu. natural rubber.

latitude Angular distance N or S of Equator of any point along a meridian.

lava Molten rock or magma extruded onto surface of earth and subsequently solidifying.

leaching Removal of soluble salts from upper layers of soil by water percolation in humid climates.

least cost location Location at which costs of production of any economic activity are at a minimum compared with costs at all other possible locations.

Levanter Spanish, mild and humid E wind affecting SE Spain, Balearics and Straits of Gibraltar when depressions form in W Mediterranean.

Leveche Spanish, hot, dry S wind blowing from Morocco to Spain when E-moving depressions enter Mediterranean.

leveé (*i*) broad low ridge of alluvium built up by river in flood to form bank; (*ii*) artificial embankment along river to check flooding.

ley Area of cultivated grass or clover in an arable rotation.

liana Climbing plant, rooted in ground, usually in tropical forests.

Libeccio Italian, strong W wind of Corsica, most frequent in summer.

light-year Distance light travels in 1 year at rate of 186,326 m. (298,123 km.) per second, i.e. approximately 5,878,310,400,000 m. (9,405,296,000,000 km.).

lignite Low-grade type of coal, midway between peat and sub-bituminous coal in carbon content, mainly used in thermal-electric generators.

limestone pavement *see* **clint**

limon French, superficial deposit of fine material, often wind-blown and resembling loess, found in Belgium and the Paris Basin.

lithosphere Solid crust of earth.

littoral (*i*) relating to sea-shore; (*ii*) zone between high and low water marks.

llanos Spanish, tropical grasslands or savanna of Guiana Plateau, S America.

load Material transported by natural agent as part of denudation process, but applied mostly to river.

loam Soil consisting of a friable, permeable mixture of clay, sand and humus.

locational analysis Analysis of the location of human and economic activity.

location theory Attempt to isolate and analyse theoretically those factors which influence the location of economic activity.

loch Scottish, lake or long narrow arm of sea.

loess German, fine-grained, friable, porous dust, usually removed from arid regions by wind and deposited finally in broad sheets, as in Asia, Central Europe and N America.

longitude Angular distance of a place E or W of prime meridian.

longitudinal coast *see* **concordant coast**

longshore drift Drift of material along a beach as result of waves breaking at angle to shore.

lough (Irish), *see* **loch**

lucerne *see* **alfalfa**

lynchet Man-made terrace on hillside, parallel to contours, part of anc. cultivation practice.

M

maelstrøm Dutch (*i*) whirlpool; (*ii*) powerful eddy in tidal current in restricted channel, as Maelstrøm in Lofoten Islands, Norway.

Maestrale (Italian) *see* **Mistral**

magma Molten material under earth's crust, from which igneous rocks are formed.

magnetic pole Two extremities of earth's magnetic field, lying in polar regions, indicated by free-swinging magnetic needle in compass.

maize Cereal crop, originating in New World, grown in sub-tropical and warm temperate climates, mainly used to fatten livestock; known as corn in U.S.A.

mallee Dense scrubby growth of dwarf eucalyptus in SE and SW Australia.

mangrove Tree with short trunk and maze of aerial roots, forming almost impenetrable obstacle. Mangrove swamps are found mainly in tropical and sub-tropical regions, close to river mouths.

manila hemp Plant grown largely in Philippines for its fibre, used in making rope and fabrics.

manioc *see* **cassava**

mantle (*i*) layer of igneous rock, some 1,800 m. (2,880 km.) thick, lying between crust and core of earth; (*ii*) surface accumulation of soil and weathered rock, or regolith.

maquis French, low scrub of Mediterranean area, characterized by dense xerophytic vegetation, found mainly in S France, Italy (macchia) and Corsica.

marin Moist, warm SE wind over coast of S France in spring and autumn when depressions enter Gulf of Lyons.

marine terrace Shoreline above present sea-level.

marketing principle Theoretical spatial arrangement of settlements, suggested by W. Christaller (1933), whereby the supply of goods from central places to other settlements involves the least transport cost.

market location Location chosen by firm, industry etc. which is nearer to its major market(s) than to its raw materials, suppliers, etc.

marl (*i*) mixture of clay and calcium-carbonate; (*ii*) geological name of particular types of rock.

massif French, (*i*) compact mountainous area with relatively uniform characteristics and clearly defined margins; (*ii*) mountain group, as Mont Blanc massif.

maté tea Made from dried leaves of S American evergreen shrub.

meander Curve in course of river moving slowly across flat country.

Mediterranean climate Warm temperate climate of W margins of continents in lat. 30–40°, with hot, dry, sunny summers and mild, moist winters.

megalith Large stone or stones, erected singly or in circles as monuments, mainly Neolithic.

megalopolis Very large urbanized metropolitan area. Originally applied to the Boston–Washington region U.S.A. *see also* **Boswash.**

Melanesia Group of islands in W Pacific.

melt-water Water formed from melting snow or ice, as in melt-water stream of glacier.

menhir Single upright stone of Neolithic origin, marking burial place.

Mercator projection Cylindrical projection used by German Mercator for world map of 1569. Meridians of long. and parallels of lat. are all shown as straight lines, causing distortion of shape in high lat.

mere Small lake usually circular, common in Cheshire where removal of underground salt deposits causes subsidence.

meridian Line of long. part of Great Circle. Prime Meridian is Greenwich Meridian (0°).

merino Sheep with long-stapled, high-quality fleece, native to Spain and N Africa, but now abundant in S Africa and Australia.

mesa Spanish, small tableland capped with resistant rock, denuded remnant of former plateau.

meseta Extensive inland plateau forming about three-quarters of Spain.

mesosphere (*i*) layer of atmosphere between stratosphere and ionosphere; (*ii*) obsolete name for mantle.

metamorphism Alteration in rock structure resulting from increase in heat or pressure or both combined.

meteor Body of matter travelling through space, being heated by friction with atmosphere, hence 'shooting stars'.

meteorite Body of stone and iron which has survived passage through atmosphere to land on earth.

meteorology Scientific study of atmosphere.

Micronesia Groups of Pacific Islands, mainly volcanic, between Equator and 40°N and 130° and 180°E.

midnight sun Phenomenon of polar lat. observed about mid-summer, when sun does not set in 24-hour period.

millet Species of grain grown in places where soil is too poor for wheat and rainfall insufficient for rice, as in Africa and India.

millibar One-thousandth of a bar.

mirage Optical illusion caused by refraction of light, particularly in hot deserts.

misfit river One too small for present valley, either as result of river capture, change of climate, or glaciation.

mist Obscuring of ground layers of atmosphere by condensation of water particles, reducing visibility to between 0.6 and 1.2 m. (1 and 2 km.).

Mistral French, strong, cold, dry N wind blowing from Central France to Mediterranean, especially down lower Rhône Valley.

Mohorovičč Discontinuity Discontinuity between earth's crust and mantle, greatly affecting speed of earthquake waves. Lies approx. 20 m. (32 km.) deep under continents, but 4–6 m. (6–10 km.) under oceans.

Mollweide projection Equal-area projection with central meridian as a straight line half scale-length of Equator.

monadnock Residual mountain rising above peneplain; named after Mount Monadnock, New Hampshire, U.S.A.

monsoon From Arabic word for season, referring particularly to reversal of pressure systems and winds over Arabian Sea and neighbouring land masses. Now applied generally over Asia, and also used to describe 'the rains' of the monsoon season.

montana Spanish, forested slopes of E Andes in equatorial lat.

moon Sole natural satellite of earth, revolving round it in one sidereal month of approx. $27\frac{1}{4}$ days. Contributes to gravitational forces responsible for earth's tides.

moraine French, (*i*) masses of boulder-clay and stones carried and deposited by glacier; (*ii*) arrangement of this material to form particular landform.

mortlake *see* **ox-bow lake**

mountain Landform of marked elevation, bounded by steep slopes and rising to prominent ridges or peaks.

mud-flat Area of fine silt, usually covered and uncovered by tides, found in sheltered estuaries or behind bars and spits.

mud-volcano Ejection of hot water and mud from volcanic vent, producing short-lived cone, as in Sicily, Iceland and New Zealand.

mulga Dense thicket of acacia-scrub on margins of central desert of Australia.

muskeg Water-logged land of N and NW Canada, with characteristic vegetation cover of sphagnum moss.

N

nappe French, over-thrust mass of rock in near-horizontal fold.

National Grid Rectangular system of lines printed on Ordnance Survey maps, providing method of precise reference to points on it by means of numbered co-ordinates.

national park Area set aside for preservation of scenery, vegetation, wild life and historic objects, for scientific purposes and for general public.

natural gas Free hydrocarbons in gaseous state, usually associated with crude mineral oils.

natural region Region defined in terms of physical characteristics (relief, climate, vegetation, etc.).

nautical mile Unit of distance used in navigation, equal to 6,080 ft. (1,853 metres) or length of 1 minute of arc on Great Circle.

naze Promontory or headland.

neap tides Tides of low amplitude, about time of first and last quarters of moon.

nehrung German, spit enclosing or almost enclosing a haff.

Nevados Cold wind blowing down valley from Andes to high valleys of Ecuador.

névé French, accumulation area of firn.

Newlyn *see* **Ordnance Datum**

new town Town created and developed as result of deliberate government planning to relieve overcrowding and population pressure in large cities (e.g. Crawley, Skelmersdale).

nife Mass of nickel-iron believed to comprise core of earth.

nimbo-stratus Low thick cloud, dark grey, from which continuous rain falls.

nimbus General term for clouds from which rain is falling.

nivation Rotting or disintegration of rocks beneath and round margins of patch of snow by chemical weathering and alternate freeze-thaw action.

nodal point (settlement) *see* **node**

node Settlement upon which surrounding settlements are dependent for goods, services, etc. or upon which lines of communication converge.

nomad Member of social group continually changing habitation to find food for human and animal needs.

nome Province of modern Greece; originally a territorial division of ancient Egypt.

nomothetic Studying generalities rather than individual cases, so as to develop laws, principles, etc. *opposite* **ideographic.**

Norte (*i*) cold dry wind of Central America; (*ii*) cold winter wind of E Spain.

North Atlantic Drift Movement of surface water of N Atlantic E from Grand Banks, Newfoundland.

North East Trades Trade winds of N hemisphere.

Norther Cold dry wind over N American continent.

Nor'-Wester Föhn wind in s Island of New Zealand.

nucleated settlement Locational pattern of rural settlement where houses and farmsteads are clustered together in hamlets or villages; *opposite* **dispersed settlement.**

nuée ardente French, mass of hot gas, superheated steam and incandescent volcanic dust, producing glowing avalanche down flank of volcanic mountain.

nullah Indian, normally dry watercourse becoming temporary stream after heavy rains.

nunatak Eskimo, prominent rockpeak projecting above surface of icesheet in Greenland and Antarctica.

O

oasis Area in desert made fertile by presence of water sufficient for permanent plant growth and human settlement.

oblast Administrative and territorial division in some republics of U.S.S.R.

occlusion Overtaking of one front by another in atmospheric depression. *See* **cold front, warm front.**

ocean Great body of sea-water surrounding land-masses and covering over 70% of earth's surface.

ocean currents Movements of surface water of the oceans, mainly clockwise in N hemisphere and anticlockwise in s. They are labelled warm or cool according to their temp. relative to the waters to which they flow.

Oceania Islands of Pacific and seas surrounding them.

oceanic islands Those of volcanic or coral formation unconnected with any continent.

oceanography Study of oceans, including nature of water, its movements, temp. and depth, as well as its flora and fauna.

odometer Wheel used by surveyors to record distance traversed.

off-shore bar Bar formed off coast of low-lying mainland.

onion-weathering *see* **exfoliation**

opencast mining Form of excavation where mineral deposits lie near surface.

Ordnance Datum   Mean sea-level calculated from hourly tidal observations at Newlyn, Cornwall, from which all heights on official British maps are derived.

ore   Mineral containing one or more metals in sufficient quantity to warrant mining.

orogenesis   Phase of mountain building, when rock material is compressed, forming folds.

orography   Description or depiction of relief.

orrery   Working model of sun, moon, earth and planets to show relative movements of these bodies.

outcrop   Portion of rock stratum projecting above earth's surface.

outfield   Land beyond infield, nor-

mally extensive in use esp. in Saxon times.

outlier   (*i*) outcrop of rock whose surface is completely surrounded by older rocks; (*ii*) small isolated hill lying beyond main scarp and formed of same material.

outport   Port near mouth of estuary, affording deeper water and closer proximity to open sea than previous river port.

outwash plain   Alluvial plain formed by melt-water stream from glacier.

overspill   Population from urban area moved to new settlement area because of overcrowding, slum clearance and population pressure.

ox-bow lake   Surviving portion of former meander; also called cut-off or mortlake.

P

Pacific coast   *see* **concordant coast**

packet port   One from which regular specified daily sailings occur, carrying cargo, mail and passengers.

pack-ice   Masses of ice floating on sea to form almost continuous cover with little open water.

paddy   Rice, either whole plant or grain in outer husk.

paddy field   Flooded rice field.

palaeomagnetism   Study of past magnetism of rocks, thus of earth's magnetic field, enabling geologists to gauge amount of movement of continents in past.

pamir   (*i*) poor grassland of high plateau in Central Asia; (*ii*) proper

name of mountain complex in Central Asia.

pampa   Extensive monotonous plain in Argentina–Uruguay; now used to describe temperate grasslands of s America.

Pampero   Dry, cold s to sw wind over Argentina–Uruguay.

Papagayo   Spanish, dry, cold NE wind affecting Mexican coasts.

'parallel roads'   Series of wave-cut terraces on valley sides thought to mark former glacial lake levels.

paramo   Spanish, high, bleak plateau of Andes between puna and snowline.

parish   (*i*) ecclesiastical unit in care

of single priest; (*ii*) smallest local govt. sub-division in Britain.

pass Col or gap through mountain range.

paternoster lakes Series of rock-basin lakes in former glacial valley.

peat Partially decomposed vegetation accumulated under waterlogged conditions and used as fuel or for horticultural purposes.

pedalfer Soil in humid climates where leaching has removed calcium compounds, leaving aluminium and iron.

pediment Large, gently sloping rock-platform at base of mountain slope in arid or semi-arid region.

pedocal Soil in dry climate where leaching has not removed calcium compounds.

pelagic Belonging to open sea, between littoral and abyssal regions.

peneplain Landscape evolved in last stage of cycle of erosion, presenting an extensive, almost level plain.

peninsula Elongated projection of land into sea or lake.

perched block Boulder carried by glacier and left balanced on rock surface after ice has melted.

perigee Part of orbit of moon or planet which is closest to earth.

permafrost Permanently frozen soil, sub-soil or bed-rock.

petroleum Mixture of hydrocarbons found in earth in liquid, gaseous or solid form.

phosphates One of three main elements needed by plants and used as fertilizer.

phylloxera Aphid, highly destructive of grape-vine.

physical planning Planning of land use (not to be confused with economic planning).

piedmont glacier Mass of ice where several glaciers converge at base of mountain.

pike Mountain peak.

pitch (*i*) inclination of axis of fold; (*ii*) resinous substance formed in distillation of coal tar.

plain Extensive area of relatively level country at low altitude.

planetary winds General atmospheric circulation producing series of lat. wind belts.

plankton Minute floating or drifting organisms of plant and animal life found in oceans, lakes and ponds.

plantation (*i*) estate in tropical regions, usually employing scientific methods and providing paid labour to produce cash crops for temperate regions; (*ii*) areas of trees planted for commercial purposes.

plateau French, large, elevated, relatively level area of land, often bounded by steep slopes.

playa Spanish, basin of inland drainage in arid regions, containing shallow, fluctuating lake, often saline.

plucking Tearing action of glacier ice on rock surface.

plug Mass of acid lava occupying vent of dormant or extinct volcano, usually exposed by denudation.

plutonic rocks Igneous rocks cooled slowly deep in earth's crust, thus highly crystallized.

podsol Russian, leached soil of cool, humid climatic regions, where natural vegetation is coniferous or sandy heath.

polar front Line of discontinuity, marking meeting of polar and tropical air masses.

polders Dutch, land near or below sea level and reclaimed by draining and construction of dykes.

681

poles (*i*) geographic; N and S extremities of earth's axis; (*ii*) magnetic; poles of earth's magnetic field.

Polynesia Islands of Pacific, approx. lying between long. 120°w and 170°w.

Ponente Cool w wind on coast of Corsica and Mediterranean.

population lapse rate Rate at which population density decreases within increased distance from city, town, etc.

population potential (*i*) Maximum population that can be maintained at a reasonable living standard in relation to resources available in given region. (*ii*) Mathematical measure of total size of population found within given distance of a given location.

portage Carrying a boat, with its contents, overland where a natural interruption to navigation occurs.

possibilism Philosophy that environment offers a number of possible courses of development which man can take without serious consequences ensuing.

pot-hole Circular hole in rocks of river bed, caused by scouring action of water laden with pebbles; incorrectly used for swallow-hole in limestone districts.

prairie Extensive tract of temperate grassland in N America.

precipitation Deposition of atmospheric moisture on earth's surface, as dew, hail, sleet, snow.

pressure *see* **atmosphere**

prevailing wind One which, in any given area, blows most frequently from a specific direction.

primary industry Activities involved in collecting or winning natural resources (e.g. mining, forestry, fishing, agriculture).

Prime Meridian Generally accepted as Greenwich Meridian, from which all longitude is measured E and W.

prismatic compass Magnetic compass with prism, allowing observer to read bearing of distant object.

Probabilism Philosophy that there are several possible courses of action open to man within his environment of which some are more probable than others.

projection Geometrical method used by cartographers to represent the curved earth's surface on a plane.

promontory Headland or cliff or crag projecting into sea.

Puna S American, high, bleak plateau of S America with cover of coarse grass, at altitude of 12,000–16,000 ft. (3,658–4,877 metres).

pumice Surface scum of volcanic lava, solidified to cellular rock.

Purga *see* **Buran**

puy French, small volcanic cone, frequent in Auvergne, France.

pyramid peak *see* **horn**

Q

qanat Persian, underground channel for irrigation water.

quagmire Bog which shakes under weight of man or beast.

quartz Crystalline silica or silicon oxide, essential element of granite.

quebracho Very hard wood found in Gran Chaco of Argentina or Paraguay.

R

radial drainage Pattern of streams flowing outwards down sides of uplifted dome or volcanic cone.

radiation Process by which body emits radiant energy.

radiation fog Layers of fog forming over low-lying areas during settled weather.

radiosonde French, self-recording instrument carried to high altitudes by balloon, for transmitting meteorological data back to earth.

rain Condensed atmospheric moisture falling in separate drops.

rain gauge Instrument consisting of funnel leading to collecting vessel, carefully sited clear of obstructions and with rim of funnel 1 ft. (30 cm.) above ground.

rain shadow Area of comparatively low rainfall on lee side of mountain area.

raised beach Beach lifted by earth movements to form narrow coastal plain, often bounded by inland cliffs.

range (*i*) line of mountains or hills;
(*ii*) open, unfenced area of grazing in High Plains of U.S.A.; (*iii*) difference between highest and lowest values of temp. or pressures, at given place, over given period of time; (*iv*) limit of habitat of plant or animal.

rapid Area of broken, fast-flowing water in river, where slope of bed increases or where gently dipping band of hard rock crosses bed.

rattan Stem of long, flexible climbing palm used in Indonesia for ropes, nets, baskets, mats, etc.

raw material Substance forming basis of manu. product.

reef (*i*) rock or coral formation of considerable extent, at or just above low-water mark; (*ii*) vein of metal or ore.

reg Arabic, extensive area of stony desert in Sahara, from which fine sand has been blown.

régime French, (*i*) seasonal fluctuation in volume of river or glacier; (*ii*) seasonal pattern of climatic changes.

regionalism Feeling of regional

identity or consciousness held by the region's inhabitants.

regolith Greek, *regos*, blanket; *lithos*, stone; layer of disintegrated rock fragments below top-soil but above bed-rock.

rejuvenation Revival of erosive activity, usually by river, resulting from fall in sea-level or local uplift of land.

relative humidity Ratio of amount of water vapour present in air to amount needed for saturation at same temp.

relief Configuration of earth's surface.

remote sensing Gathering of information by aerial survey (aerial photographs, satellite recording, etc.).

resurgence Emergence of underground stream from cave.

rhyne or **rhine** Artificial drainage channel of Somerset Levels leading to larger channels or Drains.

ria Spanish, funnel-shaped indentation of coast caused by submerging of estuary or lower part of river valley, generally where rock-structure is at right angles to coast.

rice Chief cereal crop of Monsoon and SE Asia, usually grown in flat, swampy fields.

Richter scale Used for measuring earthquake magnitude from instrumental records, running from 0 to over 8·0.

Riding One of three former administrative districts of Yorkshire.

rift valley Narrow trough between parallel faults or between series of step-faults.

rime Accumulation of ice-crystals extending to windward of exposed objects when frost, fog and wind occur simultaneously.

rip Turbulence in sea or river where tidal streams meet or enter shallows.

river Large body of fresh water flowing with perceptible current in definite channel or course towards sea, lake, marsh, etc.

river capture Cutting-off of upper waters of one stream by neighbouring river system with greater erosional power.

Roaring Forties Uninterrupted stretch of ocean S of lat. 40°S where N to W winds blow with great strength and constancy.

roche moutonnée French, rock-mass projecting from floor of glacial valley, smooth and rounded on up-stream side, but rough and angular on downstream side.

rubber Substance, derived from latex, obtained by tapping bark of tree found originally in Amazon basin.

run-off Surface discharge of water from rainfall or snowmelt.

rural-urban fringe Transitional zone of edge of urban area where urban land uses (mainly housing, cemeteries and industrial estates) mix with mainly intensive rural land uses (e.g. market gardening).

S

saddle Broad, flat ridge between two summits.

saeter Norwegian, (*i*) farm on Norwegian alp, used only in summer; (*ii*) upland pasture in Norway.

sagebrush Scrub vegetation in semi-arid areas of W of N America.

salina Spanish, salt-encrusted surface of playa.

salinity Percentage of dissolved salts contained in sea water.

saltation Transport of solid material in series of hops (*i*) along bed of stream; (*ii*) by wind action in deserts.

salt-flat Horizontal stretch of salt-crust, formerly bed of salt-lake, temporarily or permanently dried out.

salting Marsh rich in salt, often suitable for grazing.

Samoon or **Samum** or **Simoon** Persian, warm, dry, descending wind in Persia; *see* **Föhn.**

sand Gritty particles, mainly of quartz, finer than gravel but coarser than silt.

sand-dunes *see* **dunes**

Santa Ana Spanish, hot, dry, dusty wind from N or NE, descending from Sierra Nevada in California, doing much damage to orchards by desiccation.

sarsen Large sandstone block, found in S Britain, left after erosion of continuous bed of which it formed part.

satellite town One associated with a larger town but separated from it by stretch of open land and dependent on it for goods and services (e.g. Slough, England, is a satellite town of London).

savanna or **savannah** or **savana** Tropical grasslands, with scattered trees and bushes and tall grasses, resulting from marked precipitation régime.

scar Rock-face of cliff, common in carboniferous limestone of N England.

scarp Steep face or slope of cuesta.

schist Fine-grained metamorphic rock.

scree Débris of rock, broken off by frost-shattering and forming steep slope at base of hill or mountain.

scrub Vegetation of stunted trees, bushes and brushwood on poor soils in semi-arid regions.

sea-breeze Diurnal movement of air, from sea to land, caused by differential heating.

sea-level (*i*) level which surface of sea would assume if uninfluenced by tides or waves or swell; (*ii*) mean level between high and low tide at any place.

seasons Divisions of year, determined by earth's position relative to sun.

secondary industry Activities processing the products of primary industries to manufactured goods.

sedentary agriculture Cultivation practised by a settled farmer in one place; *opposite* **shifting agriculture**.

sedimentary rocks Those formed of materials deposited in layers by water or wind, made up of remains of pre-existing rocks, and of vegetation and animal deposits.

seif dunes Arabic, long dunes aligned across desert in direction of prevailing wind, forming steep-sided ridges extending many m.

seismology   Study and interpretation of earthquakes.

selvas   Portuguese, rain-forest found in areas of heavy rainfall with high temp. all year.

sérac   French, pillars of ice in ice-fall where glacier reaches steep slope.

sericulture   Culture of silk-worms.

serir   Arabic, stony deserts of Egypt or Libya, covered by sheets of angular gravel.

shadoof or **shadouf** or **shaduf** Arabic, device for raising water from well.

Shamal   Arabic, dry, dusty NW wind over plains of Iraq in summer.

sheet flood   Flood resulting from violent rainfall, where water pours down slopes without forming individual channels.

shield   Large continental block remaining relatively stable since early geological times.

shieling   Summer pasture, often with rough dwelling, in Scottish highlands.

shifting cultivation   Primitive method used by nomadic peoples in tropics, whereby forest areas are cleared by burning, cultivated for a few years, then abandoned.

shire   Large territorial division of Anglo-Saxon Britain, largely corresponding to county divisions introduced by Normans.

shoreline   *see* **coastline**

shott   Arabic, shallow, brackish salt-lake in N Africa.

sial   Outer and lighter shell of earth's crust, composed largely of silica and alumina.

sidereal time   Time measured by apparent diurnal movement of stars.

sierra   Spanish, high range of mountains with serrated peaks, found

mainly in Spain, N, Central and S America.

silage   Layers of grass, clover or alfalfa, compressed in pit or silo, usually with addition of molasses and used as fodder.

sill   (*i*) near-horizontal sheet of igneous rock, formed from magma intruded between bedding-planes; (*ii*) submarine ridge separating adjacent ocean basins; (*iii*) rock ridge at entrance to fiord.

silt   Loose sedimentary material, coarser than clay and finer than sand, laid down in water.

sima   Thick, heavier layers of earth's crust, below sial, composed largely of silica and magnesia and forming large part of ocean floors.

sink hole   (*i*) solution hole or hollow or shaft in limestone or chalk country; (*ii*) hollow down which surface water disappears.

Sirocco   Italian, S to SE wind blowing from Sahara as hot, dry wind over N Africa coast, but reaching Malta, Sicily and S Italy as warm humid wind.

sisal   Tropical plant with long fleshy leaves, whose fibres are used in making twine.

site   Actual area occupied by town.

situation   Location of town in relation to its surroundings.

skerry   Low island, often forming series off-shore and parallel to coast, as in Scandinavia.

slag   *see* **basic slag**

slump   Mass-movement involving shearing of rocks, usually causing land-slips where massive rocks overlie clays or shales.

slurry   Flow of wet mud.

smallholding   Small unit of agric. land, intensively farmed by smallholder without paid labour, common in NW Europe.

smog   Radiation fog over a town,

where sulphur dioxide is present and soot acts as nuclei for condensation.

snout   Lower end of valley glacier, with melt-water stream issuing from cave.

snow   Precipitation as ice-crystals.

snow-line   Lower limit of permanent snow-cover.

soil   Topmost layer of earth, comprising disintegrated rock, decayed organic matter, living organisms, water and air.

soil erosion   Removal of soil by action of wind or water more rapidly than soil-forming processes can replace it; usually the result of over-cropping, over-grazing, dry farming or ploughing up and down steep slopes.

soke   Former small administrative region in Britain (Soke of Peterborough).

Solano   Spanish, hot and oppressive SE wind bringing occasional rain in summer to SE Spain.

solar system   Group of celestial bodies revolving round sun, including planets, asteroids and satellites.

solar time   Time reckoned by apparent motion of sun, recorded by sundial.

solfatara   Italian, vent giving off sulphurous gases from volcano approaching extinction.

solifluction   Downhill flow of surface deposits saturated with water, especially if thaw sets in at surface when ground beneath is still frozen.

solstice   Time during summer or winter when sun is vertically above point marking its farthest distance N or S of equator, i.e. the two tropics.

solution   Form of weathering where rock is dissolved by chemical action.

sound   (*i*) passage between two land areas; (*ii*) large inlet.

sounding   Recorded depth of water.

South East Trades   Trade winds of S hemisphere.

Southerly Burster   Strong, dry wind bringing unusually low temp. from polar regions to S Australia.

soya bean   Seeds rich in protein, from pods of leguminous plant, once indigenous to China, now common in U.S.A. and Europe.

spa   (*i*) mineral spring; (*ii*) town where mineral springs used for medicinal purposes.

spate   Sudden flood, usually in mountainous areas, resulting from heavy rainfall or sudden snow-melt.

speleology   Science of cave exploration and study of caves.

sphagnum   Bog-moss of tundra regions.

sphere of influence   *see* **catchment**

spillway   Outflow channel draining lake dammed up by ice-sheet.

spinifex   Australian desert grass with spiny sharp-pointed leaves.

spit   Long, narrow tongue of sand or gravel projecting from land into sea or across estuary mouth.

spot-height   Precise point where height above given datum is marked on map.

spring-line   Line of springs where water-table intersects surface.

spring tides   Those of considerable amplitude, about times of new and full moon.

stack   Tall pillar of rock, detached from headland by wave action.

staithe   Elevated stage from which wagons may discharge coal into ships.

stalactite   Icicle-shaped mass of carbonate of lime, deposited by percolating water and hanging from roof of cave.

stalagmite Stump-shaped mass of carbonate of lime, found on cave floor by deposition of water dripping from roof.

steppe Temperate grasslands of mid-lat. of Eurasia.

storm-beach Accumulation of coarse material well above high-water mark, produced by very powerful storm-waves.

storm-surge Rapid rise in level of sea above predicted tidal heights, caused by strong on-shore winds.

strait Narrow stretch of water linking two adjacent sea areas.

strandflat Wave-cut platform off coast of N Norway.

strata Layers or beds of sedimentary rock.

strath Broad, flat-floored valley in Scotland.

stratocumulus Continuous sheet of uniform dark grey cloud below 8,000 ft. (2,438 metres).

stratosphere Layer of atmosphere above troposphere, where temp. ceases to fall with increasing height.

stratus Thin, uniformly grey cloud sheet, found up to 8,000 ft. (2,438 metres).

striation Scratching and grooving of rocks by glacial action.

subsoil Partially decomposed rock between top-soil and bed-rock.

sudd Arabic, thick mass of floating vegetation on upper Nile.

Sumatra Squally wind in Malacca Straits, occurring suddenly, often at night, during SW monsoon.

Surazo Cold wind of S Brazil in anti-cyclonic conditions of winter.

surf Mass of breaking water when large wave approaches reef or steeply-shelving shore.

swallow-hole *see* **sink hole**

swash Mass of foaming water rushing up beach as wave breaks.

swell (*i*) regular undulating movement of waves in open ocean when they do not break; (*ii*) gently sloping elevation rising from sea-floor, but still well below ocean surface.

syncline Strata downfolded into form of trough by compressive forces.

T

tableland Broad, level-topped upland rising abruptly from lowland.

taiga Russian evergreen softwood forests of N continents where tree growth is slow, tree shapes give stability in wind and snow and needle leaves reduce transpiration.

talus *see* **scree**

tank Artificial pool or lake used in India or Sri Lanka to store irrigation water.

tarn Small pool or lake in mountainous areas, usually located in cirque.

tea Small evergreen shrub of tropical monsoon lands, whose leaves are picked and processed to form beverage.

tectonic Related to internal forces building up features of the earth's crust.

temperature Degree of heat or cold within atmosphere, measured on thermometric scale.

terminal moraine Crescent-shaped ridge of drift material and broken rock deposited by valley glacier at successive stages of retreat.

terrace (*i*) flat surface along valley side, marking former river level; (*ii*) embanked area along contour on steep sloping land, to prevent soil erosion.

terrace cultivation System where soil on steep slopes is retained by walls; often employs irrigation.

terra rossa Italian, reddish clay soils developed by chemical weathering in karst country under semi-arid conditions.

territorial waters Coastal waters over which bordering state retains jurisdiction; originally 3 m. (5 km.) wide, now often extended, to claimed, up to 12, 50 and even 200 m. (19, 80, 320 km.).

tertiary industry Activities providing services for primary and secondary industries (transport, finance, insurance, etc.).

thermal spring One with water temp. over 68°F. (20°C.) normally found in areas of active or recent vulcanism.

thermometer Instrument for measuring temp.

threshold population Minimum size of population necessary to make a single establishment of any function economically viable.

thrust-fault Reverse fault at very low angle, where upper beds are pushed forward over lower ones.

tide Periodic vertical movement of sea-level in response to gravitational forces of sun or moon.

till *see* **boulder clay**

time zone Long division within which mean time of meridian near centre of zone is used as standard for whole zone.

tombolo Italian, sand or shingle bar linking island with mainland.

tor Isolated mass of rock, usually granite, weathered into prominent shapes; common in Devon or Cornwall, England.

tornado (*i*) extremely violent whirlwind, of short duration but very destructive, affecting Mississippi basin; (*ii*) thundery squalls producing torrential rains over W Africa coasts.

trade winds Those blowing from sub-tropical high pressure cells to low pressure area of equatorial lat., noted for constancy of force or direction.

Tramontana Italian or Spanish, cold, dry N wind blowing down from mountains in W Mediterranean basin.

transhumance Seasonal movement of animals and man to new pastures, usually at different altitude.

transpiration Loss of water-vapour by plants through leaves.

transport principle Theoretical arrangement of settlements, where transport costs are significant so that as many important places as possible lie on one route between larger towns.

transverse coast *see* **discordant coast**

tropic One of parallels of lat. ($23\frac{1}{2}°$ N and S of Equator) marking limits of area where sun may be directly overhead.

troposphere Lower layers of atmosphere, where temp. fall with increasing altitude.

trough (*i*) elongated trench or ocean deep; (*ii*) glacial valley; (*iii*) depression between crests of two successive waves; (*iv*) narrow belt of low pressure between two areas of high pressure.

trough-end Abrupt slope at head of glacial valley.

truncated spur Former interlocking

spur of river valley, now cut away by glacier, causing straightening of valley.

tse-tse fly Small blood-sucking fly of Africa whose bite causes sleeping-sickness in man and is often fatal to cattle.

tsunami Japanese, large-scale waves caused by submarine earthquakes, capable of travelling vast distances very fast and causing considerable damage when surging ashore.

tumulus *see* **barrow**

tundra Lapp, zone of stunted vegetation, often marshy, in N hemisphere, between polar snow and ice and N limits of tree growth.

twilight zone Area of town in decay because of lack of amenities, poor condition of buildings, etc.

typhoon Small, intense tropical storm of China Sea and W Pacific margins.

U

underground drainage Feature of areas of well-jointed rocks, as in carboniferous limestone, where water flow may be directed below ground in distinct channels, from sink-hole to resurgence.

urban field Sphere of influence of town.

urban mesh Locational arrangement of central places.

urban renewal Planned renovation of buildings, services and facilities in an urban area.

urban sprawl Unplanned spread of

building by outward expansion of town, often leading to development of conurbation.

urstromtäler German, broad, shallow troughs eroded by melt-water from front of continental ice-sheet, notably in NW Europe, where they often contain parts of present post glacial river courses.

U-shaped valley Glaciated valley with flat floor, steep sides and straight course; may contain ribbon lake or mis-fit river.

uvala Depression in karst landscape, large than doline.

V

valley Elongated depression between hills or mountains, usually containing river.

variation Magnetic angle by which magnetic compass needle deviates from true N for given point on earth's surface.

veld Afrikaans, open grassland of S Africa, ranging from high veld above 5,000 ft. (1,524 metres) through middle veld to bushveld in low country.

ventifact Pebble polished and faceted by wind-blown sand.

vicuña Native animal of S America providing limited supplies of very fine, soft wool.

viticulture Cultivation of the vine.

volcanic eruption Ejection of solid or liquid or gaseous material from interior of earth on to its surface.

volcano Hill or mountain, with vent through which lava, gas, steam and cinders are expelled.

V-shaped valley Term for valley eroded by river.

vulcanicity All processes in which molten material is forced into earth's crust or onto the surface.

W

wadi Arabic, steep-sided rocky ravine in arid regions, usually dry except when heavy rains produce short-lived torrent.

warm front Front of warm sector of depression, marking boundary where advancing warm air rises over cold air to produce broad belt of rain.

warm occlusion Where overtaking cold air is not as cold as air-mass in front.

washland Embanked lowland bordering river, where excess water is collected in flood-time.

waterfall Steep fall of river water marking sudden change in gradient of river.

water-gap Narrow passage through ridge, cut by river.

water-power Originally referred to power generated by river turning mill-wheel, now synonymous with hydro-electric power.

watershed (*i*) line of separation between two adjacent drainage basins; (*ii*) in U.S.A., term to describe catchment area of single river system.

waterspout Intense, rapid moving low pressure system over seas, producing pillar of water, common in tropics.

water table Upper surface of zone of saturation in permeable rocks. The level varies seasonally with the amount of precipitation and percolation.

wattle Type of Australian acacia.

wave-cut bench Rock-shelf produced at base of cliff by wave action.

weather Condition of atmosphere at given place and time.

weathering Disintegration and decay of rocks in situ by climatic elements.

westerlies Prevailing winds of temperate lat., blowing from sub-tropical high pressure cells to low pressure zones in lat. 35–65°.

wheat Widely cultivated cereal, most commonly grown extensively in temperate grasslands, but intensively in NW Europe.

whirlwind Rapidly revolving storm round centre of low pressure, product of local heating.

Williwaw Violent squall in Straits of Magellan.

Willy-willy Intense cyclonic storm encountered off NW Australian coast.

wind Horizontal current of moving air.

wind-gap Gap in ridge of hills, originally caused by river erosion but now dry as result of river capture.

wind-rose Diagram of radiating lines, indicating by their length the proportion of winds blowing from each cardinal point.

wine Product of fermented grape juice, though can be made from variety of other fruits.

winterbourne Intermittent stream in chalk areas, generally flowing only in winter when water table is highest. At other times it represents a dry valley.

wold Open hilly country in chalk or limestone regions of Britain, such as Cotswolds and Yorkshire Wolds.

wood pulp Major softwood product for making paper, artificial silk and cellulose paints.

X

xerophyte Type of plant adapted to arid conditions.

Y

yam Plant whose thick tubers form staple food of W Africa and S Sea Islands.

yardangs Wind-scoured ridges of rock resulting from erosion in desert lands where rocks of differing resistance run parallel to prevailing winds.

yazoo Deferred junction of trib. stream, resulting from migration of meanders in main river.

Z

zenith   Point in heavens vertically above observer.

zeugen   German, tabular mass of resistant rock, up to 100 ft. (30 metres) high, whose protective capping makes it stand out from softer rocks removed by differential erosion, mainly in arid areas.

zonda   Warm, humid wind associated with depression over Argentina or Uruguay.

zone of assimilation   Area which is being developed for business and retail activity on edge of present central business district.

zone of discard   Area which is being vacated by business and retail activities on edge of central business district.